ENVIRONMENTAL
PERMITTING
HANDBOOK

ENVIRONMENTAL PERMITTING HANDBOOK

A. Roger Greenway, QEP, CCM Editor in Chief

RTP Environmental Associates, Inc.

McGRAW-HILL

New York   San Francisco   Washington, D.C.   Auckland   Bogotá
Caracas   Lisbon   London   Madrid   Mexico City   Milan
Montreal   New Delhi   San Juan   Singapore
Sydney   Tokyo   Toronto

Library of Congress Cataloging-in-Publication Data

Environmental permitting handbook / A. Roger Greenway, editor-in-chief.
     p.   cm.
   Includes index.
   ISBN 0-07-024824-9
   1. Environmental permits—United States. I. Greenway, A. Roger.

KF3775.E545   2000
344.73'046—dc21   00-032899

McGraw-Hill

A Division of The McGraw·Hill Companies

1 2 3 4 5 6 7 8 9 0 DOC/DOC 0 6 5 4 3 2 1 0

ISBN 0-07-024824-9

The sponsoring editor for this book was Scott Grillo, the editing supervisors were Margaret Lamb and David E. Fogarty, and the production supervisor was Maureen Harper. It was set in the HB1 design in Times Roman by Deirdre Sheean of McGraw-Hill's Professional Book Group composition unit, Hightstown, New Jersey.

Printed and bound by R. R. Donnelley & Sons Company.

McGraw-Hill books are available at special quantity discounts to use as premiums and sales promotions, or for use in corporate training programs. For more information, please write to the Director of Special Sales, Professional Publishing, McGraw-Hill, Two Penn Plaza, New York, NY 10121-2298. Or contact your local bookstore.

CONTENTS

Index follows Appendix Q

CONTRIBUTORS

Gail H. Allyn   *Pitney, Hardin, Kipp & Szuch, Florham Park, N.J.* (CHAP. 17)

Nicole E. Ambrose   *RTP Environmental Associates, Inc., Green Brook, N.J.* (CHAP. 19)

Richard B. Booth   *RTP Environmental Associates, Inc., San Diego, Calif.* (CHAPS. 11, 17)

Dean W. Broga   *Office of Environmental Health & Safety, Medical College of Virginia, Richmond* (CHAP. 14)

James N. Christman   *Hunton & Williams, Richmond, Va.* (CHAP. 14)

Donald F. Elias   *RTP Environmental Associates, Inc., Green Brook, N.J.* (CHAPS. 2, 17)

John A. Green   *Tate & Lyle North American Sugars Inc., New York, N.Y.* (CHAPS. 1, 3, 5, 7, 9, 10, 11, 16)

Amy S. Greene   *Amy S. Greene Environmental Consultants Inc., Flemington, N.J.* (CHAP. 12)

A. Roger Greenway   *RTP Environmental Associates, Inc, Green Brook, N.J..* (EDITOR IN CHIEF, CHAPS. 1, 2, 6, 15, 18)

Brian L. Lubbert   *RTP Environmental Associates, Inc., Green Brook, N.J.* (CHAPS. 4, 8, 17)

Paul E. Neil   *RTP Environmental Associates, Inc., San Diego, Calif.* (CHAP. 11)

Robert W. Pender   *RTP Environmental Associates, Inc., Green Brook, N.J.* (CHAP. 10)

Kenneth J. Skipka   *RTP Environmental Associates, Inc., Westbury, N.Y.* (CHAP. 2)

John Thonet   *Thonet Associates, Inc., South Orange, N.J.* (CHAP. 13)

PREFACE

The very concept of environmental permits strikes fear to the heart of many managers of industrial and commercial facilities. This is a common reaction because, while many managers at industrial facilities are trained in engineering and business disciplines, they are often not thoroughly trained in environmental disciplines. Other reasons that environmental permits are so troubling for many managers are that the regulations seem unfathomable, and strategies for successfully obtaining permits and complying with them are not within the realm of their general experience. Obtaining environmental permits is further complicated by the fact that many of the environmental permit requirements are state implementations of federal requirements. Although many states follow federal permitting requirements, environmental permitting programs in states also differ from federal requirements in significant ways.

The purpose of this book is to provide a resource to environmental managers, to serve as a reference to the federal legislative and regulatory programs which govern all environmental permitting activities in the United States, and to also provide useful tips based on real world experience in industrial environmental compliance or in environmental consulting.

A book of this scope has not been done before partly because it is difficult to cover all anticipated situations under which environmental permits might be required and to provide up-to-date information since regulatory programs are constantly changing and evolving. Additional difficulties involve the dichotomy between federal and state programs and the differences that pertain to state permitting requirements.

This book provides valuable assistance based on the premise that federal environmental laws and regulations form the overarching umbrella under which most, if not all, state environmental programs are based. This is because, for a state to obtain delegation under federal programs for a state program in an environmental media, for example, air quality, the state's program must be at least as stringent as the federal program. State programs may however be more stringent, so it is of course important to check with state regulatory agencies to determine where their program may differ from the federal requirements.

This book provides useful references to federal and state environmental programs and includes a detailed list of environmental agencies and key contacts for each environmental media throughout the United States. This contact information includes Internet addresses so that the reader may visit the Website for a state agency to view online or download the most recent up-to-date regulations and to obtain forms for filing permit applications.

This book also introduces the reader to the concept of electronic submission of permit applications and other environmental reports. This is an important trend in environmental regulations that has been developing over the past few years with leadership from the U.S. Environmental Protection Agency and the State of New Jersey. Both U.S. EPA and New Jersey have been using electronically based environmental permit applications, Emission Statement filings, Risk Management Plan filing sand R Form filings. This book discusses how these systems are used and provides some useful information on how to navigate and use them for permit applications and filing.

This book would not have been possible without the participation of key contributing authors including John A. Green, Director of Safety and Health at Tate & Lyle North American Sugars; Gail H. Allyn, Esq., a Partner with the law firm of Pitney, Hardin, Kipp & Szuch; James N. Christman, Esq., a partner with the law firm of Hunton & Williams; Dr. Dean W. Broga, Director of the Office of Environmental Health & Safety at the Medical College of Virginia; Amy S. Greene, President of Amy S. Greene Environmental Consultants Inc.; and John Thonet, President of Thonet Associates, Inc.

This book also would not have been possible without the dedication of a number of key RTP Environmental Associates Inc. Associates, Principals, and senior personnel.

A. Roger Greenway, QEP, CCM

CHAPTER 1
OVERVIEW: ENVIRONMENTAL PERMITTING IN AMERICA

A. Roger Greenway
Principal, RTP Environmental Associates Inc.

John A. Green
Director of Safety and Health
Tate & Lyle North American Sugars Inc.

1.1 BACKGROUND

Over the last 30 years, industrial operations and construction activities in the United States have been increasingly impacted by environmental regulations. Perhaps we should look at this from the point of view of the environment, which for more than a century, from the beginning of the industrial revolution to the late

1960s, appeared to exist solely for the benefit of man. In the 1960s there was a growing awareness that the environmental resources of the earth were impacted by man's activities. The seemingly infinite environmental resources that previous generations assumed would always exist were found not to be infinite after all, and, in fact, there was a growing awareness that, if some cautions were not taken to protect and preserve our environmental resources, future generations might not inherit the earth that we know.

There are really two aspects to our environmental concern: (1) if earth's resources are being damaged by man's activities, future generations might not have a healthy environment to live in and (2) our environment may not be as healthy as we assumed it to be all along (which in fact appears to be the case). This latter view was reinforced by the fact that many rivers cannot support salmon, striped bass, and other ocean-dwelling fish that use rivers for spawning. Additionally there's a worldwide decline in fish yields and loss of species diversity.

In response to these concerns, a number of states and the federal government began looking at legislation and regulations to bring some control, management, and balance to environmental impacts. This took two main directions:

- Requiring that the environment be taken into account in decisions on development of industrial, commercial, and large-scale residential projects.
- Controlling environmental releases from existing facilities, including air emissions, water discharges, and solid waste generation.

Growing concern over the cleanliness of drinking water supplies, especially groundwater, led to a realization that contaminated sites created by past industrial abuse should be studied and cleaned up.

In the 1990s, environmental concern took a new turn when government agencies began to adopt a concept called *sustainable development.* According to this concept, there is a maximum level of sustainable development, in countries and on continents. If development exceeds the ability of the environment to absorb the environmental impacts, the area becomes "poisoned" and, as a result, will no longer support the level of development it has been built to. This, in effect, is an argument meant to find favor even with those who couldn't care less about the environment. If there are, in fact, limits on sustainable development, we all must be concerned not to overdevelop.

Geographic borders of countries are arbitrary with respect to the environment. One only has to look at photographs of the earth from space and to wonder where national borders exist. Where they are not formed by oceans, rivers, or lakes, it is usually not possible to discern national borders from space.

The concept of sustainable development has as its root the premise that there is a level of development beyond which the environmental resources become so stressed and damaged that further development becomes impossible (i.e., unsustainable). The maximum level of development that is sustainable is presumably the level that maximizes man's use of a region of the earth without totally destroying the environment. The concept of sustainable development does not, however, necessarily mean that there is no environmental damage or that there is no loss of species diversity. It also does not mean that the "maximum" level of development can be defined, or that we can really be sure it is, in fact, sustainable. This concept is made to order for alarmist policies.

I hope the reader is beginning to understand that the questions posed when one looks at protecting the environment while taking into account man's objectives on earth and creating "sustainable development for the future" raises very difficult questions. No one thinks we have all of the answers; therefore, it is an evolving area in which environmental laws, regulations, and permitting requirements are constantly changing. For this reason, it is imperative that anyone concerned with the subject of environmental permitting be prepared to obtain the most recent permitting requirements from the state, county, and sometimes municipality in which the concern exists. It is important to remember that environmental programs are continually changing. Therefore, while this book is up to date at the time of its publication, the reader is cautioned to check with the federal government and state and local authorities as to the current details of environmental legislation and regulations.

Obtaining compliance with environmental regulations is further complicated by the fact that there are overlapping requirements at the federal and state levels of government. The federal government passes legislation and federal regulations concerning various environmental media. States, since we are a country that believes in states' rights, are free to pass their own environmental legislation and regulations. When there are federal requirements (which is the case for most environmental regulations) states must, if they are given delegation from the federal government to implement the federal requirements, develop regulations that are at least as stringent as the federal requirements. They are free, however, to impose regulations that are *more* stringent.

A new area of environmental legislation and regulation implementation that is evolving as this book is being prepared is the concept of *environmental justice* or *environmental equity.* This concept, which has been evolving over the last few years at the federal level, has as its objective the redress of a perceived environmental inequity involving minorities. The federal government believes that minority populations, particularly in inner cities, are unduly impacted by environmental assaults. The government's response has been to develop the environmental justice program, which is described in Chap. 19. The regulations being developed under the environmental justice legislation have, at their heart, the creation of special rights for minority populations, where minority populations exceed 50 percent of the total population in an area. These special rights include the ability to bring citizen suits to block development, particularly industrial development, that the community perceives could create undue environmental impact on its population. While the objective of this program is laudable, the implications of the program create significant problems. Under this program, industrial development can be ordered to stop even after all permits are approved by state and local authorities. This is because the environmental justice regulations give minority populations the right to bring citizen suits to block a project even after permits have been granted. All it takes to kill a project is a citizen suit and a judge deciding that a project cannot go forward. This creates immense uncertainty for industrial companies wishing to build or expand in minority areas. It also seems to work against the best interests of the minority population in that it will limit the generation of viable jobs in the area. It appears to the author that, as long as proposed industrial development meets the same environmental standards it would have to meet in upscale suburban or rural locations, it should be permitted in urban minority locations as well. It remains to be seen, however, how the final regulations will be developed and implemented by the states.

Because of the variation in permitting programs from state to state, it is important that the reader understand local permitting requirements. Over the last few years there has been a growing tendency for states to impose permitting requirements that closely parallel the federal regulations. This is for two reasons. First, the federal government will not delegate permitting authority to the states if the state program does not meet the federal requirements. Second, 10 to 15 years ago, when state programs were more different from each other than they are today, there was a concern in industry that the highly regulated states were not economically viable for new or expanded facilities because their environmental permitting requirements were more difficult to comply with. As a result of the mid-1980s recession, there has been a growing tendency in state government to be more industry-friendly in ways that are consistent with protecting the environment. States that tended to be more difficult to obtain permits have, for the most part, modified their permitting requirements to be consistent with federal programs. To illustrate specific state permitting requirements, this book includes chapters on environmental permitting requirements in New Jersey and California, two states with very active permitting programs. The reason these states are included is that many of the federal requirements for environmental approvals do not, in and of themselves, require permit forms or applications or licenses. In many environmental areas, these requirements are imposed by states, and the actual permit applications and permit application guidance are based on state laws and regulations.

This book will serve as a guide to environmental permitting. This chapter provides an overview of federal permitting requirements.

Table 1.1 is a list of state agencies. Appendix A lists agency name, telephone number, and addresses for each medium of environmental regulation. Note that these addresses may change from time to time but the agency name should not, and the telephone numbers should still work. You should plan on contacting your local agency for the most recent environmental regulations.

This introduction would not be complete without mentioning that the Internet has become an invaluable resource for obtaining regulatory and environmental information. The U.S. Environmental Protection Agency (EPA) has an excellent series of Web sites, which the reader can access by visiting the *www.epa.gov* home

TABLE 1.1   United States State and Local Environmental Agencies

State		Name
Alabama	DEM	Alabama Department of Environmental Management
Alaska	DEC	Alaska Department of Environmental Conservation
American Samoa	EQC	Environmental Quality Commission
Arizona	DEQ	Arizona Department of Environmental Quality
Arkansas	DPCE	Arkansas Department of Pollution Control and Ecology
California	ARG	California Air Resources Board
Colorado	DPHE	Colorado Department of Public Health & Environment
Connecticut	DEP	Connecticut Department of Environmental Protection
Delaware	DNREC	Department of Natural Resources and Environmental Control
District of Columbia		Department of Consumer & Regulatory Affairs, Environmental Regulation Administration
Florida	DEP	Florida Department of Environmental Protection
Georgia	DNR	Georgia Department of Natural Resources
Guam	EPA	Guam Environmental Protection Agency
Hawaii	DOH	Hawaii Department of Health, Environmental Management Division
Idaho	DEQ	Division of Environmental Quality
Illinois	EPA	Illinois Environmental Protection Agency
Indiana	DEM	Indiana Department of Environmental Management
Iowa	DNR	Iowa Department of Natural Resources
Kansas	DHE	Kansas Department of Health & Environment
Kentucky	DEP	Kentucky Department of Environmental Protection
Louisiana	DEQ	Department of Environmental Quality
Maine	DEP	Department of Environmental Protection
Maryland	DOE	Maryland Department of the Environment
Massachusetts	DEP	Department of Environmental Protection
Michigan	DEQ	Department of Environmental Quality
Minnesota	PCA	Minnesota Pollution Control Agency
Mississippi	DEQ	Mississippi Department of Environmental Quality
Missouri	DNR	Missouri Department of Natural Resources
Montana	DEQ	Department of Environmental Quality
Nebraska	DEQ	Nebraska Department of Environmental Quality

page. Links from there to various other pages will provide a wealth of information on the current status of federal regulations. At the same time, many states have developed, or are in the process of developing, enhanced Internet Web page resources for providing information to the regulated community on the status of environmental regulatory programs.

The Internet is also an important resource for obtaining information about the environmental quality of a region and the activities of the regulated community. More and more, the Internet is becoming indispensable in environmental compliance matters.

The Internet is also a valuable source of mapped geographic and environmental data. A number of federal and state agencies are putting environmental mapping information on line for access by the regulated communities. This includes census data, which is important in determining the nature of the impacted community for a planned project. This data, available from the Bureau of the Census, also includes information on environmental receptors such as parks, wildlife refuges, reservoirs and schools, hospitals, and other sensitive receptors. Environmental monitoring information from air and water monitoring programs is also being made available on the Internet. This information is invaluable in permit applications where the background levels of environmental parameters need to be disclosed.

Finally, the federal government under the Clinton administration has discovered that the court of public opinion can be a very effective tool in improving compliance. Therefore, under the Clinton administration and administrator Carol Browner, the EPA has pushed a number of initiatives to put environmental compliance information on the Internet where the regulated community, the public, and environmental action groups have easy access to it. Examples of this include the Superfund Amendment and the Reauthorization Act (SARA) Section 313 "R form" information. This information includes data on environmental discharges of hazardous and toxic substances to the air, water, solid waste, streams, and to off-site recycling. This information is available through the EPA Envirofacts database. Environmental action groups, including the Environmental Defense Fund (EDF) have exploited easy access to this information and have generated summarized information that is easily accessible through the EDF Web page (*edf.org*) and their Scorecard site. The Environmental Scorecard sites provides an easy way for the public to search for "polluters" in a county or community. With all companies meeting certain criteria identified, it is possible to see what their reported discharges to the environment are.

The court of public opinion helps improve environmental compliance because company reports of high levels of discharges to the environment are now available to the public through the Internet in an easy-to-find way within months or weeks, while in the past this information would sit in a file cabinet at a state or federal agency. Moreover, having this information available to environmental groups and the public makes the company more susceptible to citizen suits. If the information is easy to find, it is easier for independent parties to compare reported discharges and emissions to permit limits. The EDF, Natural Resources Defense Fund, and other environmental groups frequently bring citizen suit actions against companies for violations of their permits. The easier the access to the data, the easier it is to do this.

For these reasons, companies are looking for ways to minimize releases to the environment so the reported discharges and emissions on the R forms are lower. Since Risk Management Plans under the Clean Air Act Sec. 112(r) will also be filed on the Internet, companies are looking, where possible, to replace chemicals with unregulated chemicals, to reduce storage of chemicals, and to take other actions to opt out of the program. While filing risk management plans to the state or EPA is not a permit requirement, it is another way in which companies are required to disclose environmental releases. Penalties for noncompliance can range from fines of up to $25,000 per day to being ordered to shut down the facility until compliance is proved.

Exposing companies' activities to public scrutiny has proved to be a very effective way of encouraging them to reduce emissions and other discharges to the environment. It remains to be seen how much can be gained from this practice in the future, as there are limits on how much information can be released—limits, not on the physical capacity to make the information available, but on the amount of information available. Much information is already being released on most of the environmental regulatory areas. How much more can we disclose? How much deeper will government dig for information on our environmental practices? What is the tradeoff between environmental protection and freedom from government intervention? Too often in the past, industry has shown that without government regulations, it will not do the right thing. Too much regulation, on the other hand, leads to diminishing returns. There is an optimum level of regulation and permitting.

The recent emphasis on public release of information as a means to achieve environmental compliance reflects, to an extent, the growing sense that the previous 3 decades of environmental regulations, which rely heavily on what has come to be called "command and control," has produced about as much as most people expected. That approach has now reached a point where it is no longer as effective as it once was. The reason for this is easy to see. In the 1960s, when permitting programs required emission reductions of all new sources, the emphasis was on point sources of emissions. Chapter 2, on air permitting, discusses this in detail. As regulation of major point sources became tighter and tighter, improvements in monitored air quality were reflected as a result. Over the decade, however, about all that could be squeezed from the major source lemon was squeezed and agencies looked to regulate smaller and smaller sources. There was (and is) a tendency to exclude from regulation private residences and very small businesses. As a result, there are obviously limits to how much emission reductions can be achieved by regulating industrial facilities and power plants. Therefore, in the

1980s and 1990s, agencies looked more at other ways to improve air quality. These ways included transportation control measures on motor vehicles, regulation of energy efficiency, and other measures that could be productive in reducing emissions from the community. These types of sources are difficult to regulate under the command and control approach.

There was also a growing interest in looking at alternatives to command and control. One that evolved was the court of public opinion discussed above. Another approach, which several states are actively looking into, is the "Dutch model." The Dutch model involves a cooperative working relationship between industry and government in which environmental objectives are articulated by the government agency, and industries and the agency work cooperatively to achieve the objectives. This approach reflects the reality that government can't always know what's the best way to achieve the environmental objective and, at the same time, even if government does know, the best way to obtain an environmental objective for one industry, a one-size-fits-all approach, almost never works—the best way to achieve an environmental objective in one industry may not be the same as in another. For this reason, the Dutch model is being looked at very seriously by New Jersey and other states. It is a move away from the command and control philosophy, and its overall objective is to achieve environmental improvement while maintaining a cooperative working relationship between industry and government.

Environmental permitting is an evolving field. This book will provide an overview of federal permitting, licensing, and other approval programs, as well as detail on several state programs. Future editions will include additional states and will update these programs. The following chapters of this book will cover each environmental area of regulation in detail.

1.2 CLEAN AIR ACT

Unlike other environmental media, air quality is regulated in the United States primarily through the setting of air quality goals. These are the National Ambient Air Quality Standards. The difference in air quality, as compared to the other environmental media, is that one can choose not to drink the water by either using bottled water or filtering the water that comes into the house, however you cannot choose not to breathe the air, and you cannot filter the air you breathe. Therefore, the EPA concentrated on setting acceptable levels of air quality in the ambient air which is defined as the air in the outdoor environment that we breathe. An important distinction to keep in mind is that the indoor air quality environment in the commercial and industrial workplace is not regulated by the EPA but rather by the Occupational Safety and Health Administration (OSHA). EPA has recently begun regulating indoor air quality in the residential home environmental, however.

The regulation of ambient air quality, as opposed to other approaches, is significant in a number of ways. First, in other media, water for example, EPA sets performance standards that industries must achieve in order to obtain permission, i.e., permits to discharge pollutants to the waters of the United States. Regulating the concentration of pollutants in the air however, requires EPA to develop more comprehensive approaches. Performance standards for industries might not be adequate, for example, if an area has so many industries that even their meeting the most stringent performance standards would not lead to acceptable air quality. EPA's solution to this problem is to leave the regulation of air emissions primarily to the states. This is the way this works. The EPA sets ambient air quality standards and requires each state to develop a state implementation plan (SIP) which is designed to assure that the air quality in the state is as good or better than the air quality standards set by the EPA. States are also free to set air quality standards which are more stringent than the national standards. The national air quality standards, in effect, are minimum standards, which must be met by federal law in all states and territories of the United States.

The methods to be used by the states to meet the ambient air quality standards can be decided at the state level. As in the case of ambient air quality standards, however, there are minimum elements of state implementation plans that states must put in place. These are determined by federal regulations developed under federal legislation, the Clean Air Act and its amendments. The Clean Air Act and its amendments and federal regulations contain the federal requirements for the regulation of ambient air quality, most notably the New Source Review Program, and National Emission Standards for Hazardous Air Pollutants (NESHAP), and New Source Performance Standards for major new sources of air pollutants. Some states obtain delegated authority to implement these programs, while others rely on implementation and enforcement by the USEPA.

The evolution of air quality regulations, and permitting under those regulations over the last thirty years has produced a very complex set of laws, regulations, and permitting requirements. These are described in detail in Chapter 2. This introductory section provides an overview of the basis for these regulations and permitting requirements.

The minimum levels of air quality which were set under federal regulation for non-toxic air pollutants, commonly referred to as the Criteria Pollutants, have resulted in the National Ambient Air Quality Standards (NAAQS). There are actually two levels of ambient air quality standards. The first, the primary standards are set to be protective of human health. These standards are set by the USEPA and are revised on a regular basis. The standards are set based on medical and epidemiological information to be protective of the entire population. This means, not that they are protective of healthy young adults only, but that they are also protective of young children that are susceptible to the impacts of pollution and are also protective of older individuals who may have health conditions which make them also more susceptible to degraded air quality.

The second type of standard is the secondary ambient air quality standards. These are designed to be protective of other air quality related values such as agricultural crop yields, soiling of buildings, structures and visibility, for example in national parks such as the Grand Canyon.

Table 1.1A lists the primary and secondary ambient air quality standards for the criteria pollutants.

Another concept which must be understood is the distinction between nonreactive and reactive air pollutants. Nonreactive pollutants are pollutants such as sulfur dioxide, particulate matter, carbon monoxide, and lead, which when emitted into the atmosphere, get diffused and diluted transported and deluded by the atmosphere and are subsequently washed out by rain, settle out, or remain suspended in the atmosphere for long periods of time, but throughout this period do not react with other pollutants to form additional compounds. Reactive pollutants on the other hand, are pollutants such as nitrogen oxides and volatile organic substances which, in the atmosphere, combine with each other and in the presence of sunlight enter into chemical reactions forming ozone, regional haze, and smog.

An important concept relevant to understanding air quality regulations is that between emissions and ambient concentrations. Air pollutant emissions are the pollutants that are discharged from the stacks, vents, storage piles, and fixed and mobile sources throughout the country. This is the input to the atmosphere. Ambient concentrations, on the other hand, are the levels of the air pollutants in the atmosphere which individuals breathe and which impact agricultural yields, visibility, and contribute to the global loading of air pollutants.

The next concept which is important is that of attainment versus nonattainment. Ambient air quality standards, the NAAQS, were required to be met by certain key dates discussed later in this book. These dates were set in federal laws and regulations and states were required to achieve the ambient air quality standards by the dates that were set. Areas that are achieving the ambient air quality standards are considered to be attainment areas, that is, they are attaining the air quality standards. Areas that are not achieving the ambient air quality standards, however, are designated as nonattainment. This sounds logical, areas that are not meeting the standard are not attaining them, therefore they should be designated as nonattainment. The actual procedure for designating nonattainment areas are somewhat more complex however, in that whether an area is meeting an ambient air quality standard or not, must be determined based on some rational and defensible basis. Usually this is done through examining ambient air quality monitoring data. Problems arise however, in areas where there is not a large number of ambient air quality monitors. What does one do, for example, in a major metropolitan area when two out of eight monitors show violations of standards, but the other six show attainment of the standards. Should the area be designated nonattainment because the emissions are a regional problem and the fact that most of the monitors show attainment does not override the fact that two of them show nonattainment. How can the area be brought into attainment at these two monitoring locations if the entire area does not have strict emission controls? Further, the fact that even though there are eight monitors in the metropolitan area they may not adequately describe

TABLE 1.1A   National Ambient Air Quality Standards

Pollutant	Standard Value		Standard Type
Carbon Monoxide (CO)			
8-hour Average	9 ppm	(10 mg/m^3)**	Primary
1-hour Average	35 ppm	(40 mg/m^3)**	Primary
Nitrogen Dioxide (NO_2)			
Annual Arithmetic Mean	0.053 ppm	$(100 \text{ }\mu\text{g/m}^3)$**	Primary & Secondary
Ozone (O_3)			
1-hour Average*	0.12 ppm	$(235 \text{ }\mu\text{g/m}^3)$**	Primary & Secondary
8-hour Average	0.08 ppm	$(157 \text{ }\mu\text{g/m}^3)$**	Primary & Secondary
Lead (Pb)			
Quarterly Average		$1.5 \text{ }\mu\text{g/m}^3$	Primary & Secondary
Particulate , 10 micrometers (PM-10)			
Annual Arithmetic Mean		$50 \text{ }\mu\text{g/m}^3$	Primary & Secondary
24-hour Average		$150 \text{ }\mu\text{g/m}^3$	Primary & Secondary
Particulate , 2.5 micrometers (PM-2.5)			
Annual Arithmetic Mean		$15 \text{ }\mu\text{g/m}^3$	Primary & Secondary
24-hour Average		$65 \text{ }\mu\text{g/m}^3$	Primary & Secondary
Sulfur Dioxide (SO_2)			
Annual Arithmetic Mean	0.03 ppm	$(80 \text{ }\mu\text{g/m}^3)$**	Primary
24-hour Average	0.14 ppm	$(365 \text{ }\mu\text{g/m}^3)$**	Primary
3-hour Average	0.50 ppm	$(1300 \text{ }\mu\text{g/m}^3)$**	Secondary

*The ozone 1-hour standard applies only to areas that were designated nonattainment when the ozone 8-hour standard was adopted in July 1997. This provision allows a smooth, legal, and practical transition to the 8-hour standard. Visit our *AIRLinks* page for more information about the July 1997 revisions to the ozone and particulate matter standards.

**Parenthetical value is an approximately equivalent concentration.

air quality conditions throughout the entire metropolitan area. For this reason, the process of designating an area as nonattainment is one that involves a review of monitoring data for a number of years by the state environmental agency and a recommendation to the EPA administrator as to designation of attainment or nonattainment for the area. This has significant implications to permitting, since permitting new sources in nonattainment areas is significantly more difficult, expensive, and problematical as discussed in Chapter 2.

The final concept to be understood is that of sanctions. Should a state not provide for timely attainment of ambient air quality standards, the EPA's weapon against the states is the impositions of sanctions. These sanctions typically include loss of highway funds for new transportation projects and also more stringent permitting requirements including the requirement for increased offset ratios when permitting major sources in any nonattainment areas.

1.2.1 Clean Air Act Amendments

The Clean Air Act is amended on a regular basis, typically about once every 10 years. The Clean Air Act was last amended in 1990, so further changes or a complete set of amendments are due in the near future. In 1990, the driving factors behind the amendments to the act included chemical emergencies in Bhopal, India, and lesser emergencies in the United States, and a feeling in Congress that the country wanted more stringent regulation of air toxics. One of the keystones of the 1990 amendment of the Act was significant new regulations for the development of performance standards for sources that emit toxic air pollutants and the development of Risk Management Planning procedures to require facilities to identify toxic substances on site and to actively work to minimize a possibility of accidental releases. The 1990 Amendments also added significant regulations in the area of operating permits which required major sources throughout the country to obtain permits under this new federal permits program.

The objective of The Clean Air Act is to attain the national ambient air quality standards everywhere throughout the country through:

- State implementation plan regulations and permitting requirements developed at the state level
- New Source performance standards applicable to all major new sources of emissions
- Operating permit requirements for all major sources
- Increasingly stringent mobile sources controls

1.2.2 History

A number of states, including New Jersey, had air quality or air pollution control acts prior to the first federal clean air legislation. The first air quality act at the federal level was the 1967 Air Quality Act. This was a very basic act whose objective was improving ambient air quality conditions throughout the country. This was superseded in 1970 by the first Clean Air Act which for the first time added the concept of national ambient air quality standards, state implementation plans for states to show ways to meet these standards, new source performance standards, and national emission standards for hazardous air pollutants. The last is the federal requirement for performance-based standards for facilities that emit air toxics.

In 1977, a major series of amendments added the nonattainment area concept and its requirement for offsets for new sources. The offset provision is very significant because it says that new sources of air toxics cannot obtain permits to construct and operate unless they obtain offsets from other sources or from an emission bank in the area to more than compensate for the new emissions to be emitted by the facility.

The 1977 amendments also added the prevention of significant deterioration (PSD) program. The PSD program is a cornerstone of the EPA's current New Source Review program. It requires that all major new sources in areas that are not nonattainment (i.e., that are attainment areas), obtain permits to construct and operate under the PSD program. This program, described in detail in Chapter 2, is a significant program and requires facilities show that their emissions, in conjunction with the emissions of other new sources of emissions in the area, do not cause the PSD increments to be exceeded. This regulation came into being as a result of litigation between environmental groups and the EPA, the basis of which was that there are areas that are currently cleaner than the ambient air quality standard and should not be allowed to be degraded all the way until they meet the limit set by the ambient air quality standards. This program, in effect, limits the deterioration of air quality to increments which are set in the regulations to assure areas that are significantly cleaner than the ambient air quality standards are not polluted all the way up until the levels allowed by the standards. The problem with this program is that it is very complex, requiring significant demonstrations on the part of applicants for PSD permits, which can include obtaining 1 year of meteorological and air quality monitoring data at multiple locations in the vicinity of the proposed source before the permit application can be filed. This means that applicants for PSD permits may be facing an 18-month to 3-year time period for collecting data, completing permit applications, and allowing for agency review.

The 1990 Clean Air Act amendments added a number of key new provisions. These include:

- The Clean Fuels Program requiring specially formulated gasoline for motor vehicles
- Extended nonattainment area deadlines once again, however added strict new requirements for further progress and for increased levels of sanctions for states not meeting the new deadlines
- Significant new air toxics requirements, including a timetable with hammer provisions for EPA to develop air toxic regulations
- Significant regulation on the precursors of acid rain even though scientific evidence that showed direct relationship between emissions and acid rain is still lacking
- Stratospheric ozone requirements for reduction of emissions of pollutants, such as CFC refrigerants, which contribute to stratospheric ozone depletion
- Federal permitting and enforcement
- Increased the capability of federal enforcement particularly in the area of major sources through the federal operating permits program. This program is described in detail in Chapter 2.

The possibility of federal enforcement of industrial facilities was significantly increased by the 1990 amendments. This occurred for two reasons. First, in order not to be subject to the requirements of the operating permits program, facilities were encouraged to modify their existing state permits to limit their emissions below the major source threshold which would require the operating permit. This would, in effect, allow a facility to become a "synthetic minor" that is, a source that would be major if it operated at full capacity, but because its permit limited it to below major source threshold, was in effect a minor source and not required to file an operating permit application. The problem with this approach is that this worked only if the permits were deemed to be "federally enforceable." What this meant was that the state had to acknowledge that EPA had enforcement authority over state permits. The way this worked in many states is that the state took the position and made the public statements that since their permitting regulations were developed pursuant to the state implementation plan which is a federally approved document, their state permits were federally enforceable. While this is a good thing for sources wishing to opt out of the operating permits program, it gives EPA permission to enforce state permits. This opens the door to all facilities to federal enforcement personnel and to direct federal field enforcement of state air permits whether they are operating permit or nonoperating permits.

The 1990 amendments also increased the level of fines, increased the potential for criminal penalties, and increased the opportunity for public involvement in federal enforcement of permits.

The levels of nonattainment and target attainment dates for these areas are also shown earlier in Table 1.1A. This shows for ozone, marginal nonattainment areas have a design value of 0.121 ppm to 0.138 ppm in a targeted attainment date of November 15, 1993. The required offset-ratio for these areas is 1.1 to 1, meaning any source wishing to expand in the nonattainment area would need to obtain 110% of its new emissions from other sources to compensate for the pollutants to be emitted by the new action. There is a continuum of increasingly stringent requirements and increasingly distant dates ranging to extreme areas which have an attainment date of November 15, 2010, and a major source threshold of only 10 tons per year and an offset ratio of 1.5 to 1, meaning that sources in the extreme nonattainment area, which is limited in the United States to the Los Angeles basin area of 150% of the emissions.

1.2.3 Hazardous Air Pollutants (HAPs)

The Clean Air Acts Amendments of 1990 contain Title III which requires regulations of air toxic emissions. Title III is very significant since earlier versions of the act required EPA to develop regulations limiting toxic air pollutants. The 1977 amendments however, were never adequately implemented by the EPA by the late 1980s, which led to the 1990 amendments. The Congress had determined the EPA needed more legislative pressure to adequately regulate toxic air emissions. Therefore the 1990 amendments list 189 toxic substances to be regulated by the EPA, a detailed timetable as to when these regulations are required, and significant provisions to force EPA to develop these regulations in accordance with the schedule promulgated in the Act.

The most significant requirements of the Act include:

- The development of maximum achievable control technology (MACT) standards, which are effective for new sources and retroactively to existing sources of toxic air pollutants
- The requirement for EPA to look at the residual risk of toxic air emissions emitted by sources after imposition of the MACT standards
- Development of significant permitting requirements for facilities to show compliance with air toxic regulations before being granted permits from the state or EPA
- Development of new accidental release prevention regulations which became the Risk Management Planning regulations

Major sources of toxic hazardous air pollutants are defined as a source, or group of sources, that emits 10 tons per year of any of the listed hazardous air pollutants or 25 tons per year of any combination of hazardous air pollutants. A modification includes any physical change or change in method of operation of a major source that increases the emissions by any hazardous air pollutant by more than a diminimous amount.

Under the 1990 Amendments the existing NESHAPs stay in effect and, in addition, the EPA is required to develop MACT standards for new and existing sources in identified industrial categories and subcategories in accordance with a schedule contained in the Act. There are also requirements beyond the MACT standards. The EPA must identify the residual risk from the sources emitting toxic air pollutants, hazardous air pollutants and to look at whether the MACT standards need to be made more stringent to achieve the legislatively required improvement or reduction in risk posed by pollutants.

The Act also set stringent requirements under which EPA is to develop the MACT standards. For new sources, the MACT standards may be no less stringent than that achieved in practice by the best-controlled similar source. For existing sources where the standards are to be applied retroactively, the standards may be less stringent than for new sources, but must be no less stringent than:

- The average emission limitation achieved by the best performing 12% existing sources in the category, which means that EPA is required to look at the emission controls achieved by all other similar sources in the country, and to look at the twelfth percentile from the best to the worst so that the standards to be set will assure that 88% of existing sources will not be in compliance with proposed standard and therefore will require upgrades to achieve the new standard. For sources for which there are only a small number of similar sources in the country, less than 30, the legislation allowed EPA to set the standard at the fifth best source.

Finally, sources that have met lowest achievable emission rate control technology, which is the level of control required in nonattainment areas, also sets a floor for the MACT standard.

What types of approaches can be put in place for meeting the MACT reduction requirements?

- material changes
- process changes
- enclosure
- collect and treat
- operating practices
- combination of above

It is important to know that the MACT standards are performance standards as opposed to ambient air quality standards. This is because it is impossible to monitor the ambient air quality levels of 189 toxic pollutants at any meaningful number of locations. Therefore, the only approach available is to set performance standards that all individual sources must meet. It is important to note as well that, this does not necessarily mean collecting and treating the emissions that are generated. Changing the materials to eliminate the use of toxic substances has been a widespread reaction to industry to these standards. In other words, let's get out of the standard by not using the regulated substance anymore. An additional approach is to change the process to significantly reduce the pollutants emitted. This includes many industries looking at pollution prevention (P2) as an approach to reducing emissions. This has added benefits in the area of worker protection and in reducing costs.

The schedule for implementation required EPA to develop MACT standards for all of the categories within 10 years, or by November 15, 2000. The EPA is making significant progress on its schedule, but some slippage may occur. The EPA is also required to review and revise these standards at least every 8 years.

The MACT standards will also be required for existing sources. Therefore, if a source finds itself subject to a MACT standard, it must expeditiously install any required process modifications or control equipment that may be required. It is possible to achieve only a one year extension if installation of controls is required. This brings up an important point. Facilities must track the regulatory agenda in Washington and in their state environmental state agencies. One year to study, design, obtain bids and install controls that may be required under the MACT standard is not enough time for most companies. Typically, companies would want to study their operations, and to see if there are ways to get out of the standard by substituting materials. If installation of controls is required, companies would then want to study their options for controls, obtain quotes for engineering services to design the controls, obtain quotes from suppliers and installation companies to install them prior to budgeting the capital expenditure that might be required for the installation of the controls. This would normally take much more than one year. For these reason, it is critical that companies track the regulatory agenda in Washington and know what states have planned so that they can anticipate controls, influence the regulations if possible, and be in a position to have the capital funds available for any controls which are required.

There is one important provision for facilities which, prior to the promulgation of a MACT standard, have installed best achievable control technology or lowest achievable emission rate technology by virtue of a PSD permit application or New Source Review or a nonattainment area application. There is potentially available a 5-year delay for installation of a newly promulgated MACT standard for that source.

1.2.4   Area Source Regulation

The air toxic provisions do not automatically require regulations for area sources. Area sources are sources that emit their pollutants not through a defined stack, vent or discharge point, but emit emissions from material storage and handling, from leaking valves, fittings, and from other areas. The legislation requires EPA to study emissions from area source and to set emission control standards to assure compliance by November 15, 2000. These standards need not be MACT standards, but should be designed to achieve a 75% cancer risk reduction from area source emissions.

Finally with respect to residual risk, EPA is required, under the CAA legislation, to report to Congress within 6 years on methods to reduce to a residual risk remaining after application of each MACT standard, the significance of the residual risk and, if necessary, to revise the MACT standard to achieve the required emission reduction.

1.2.5   Hazardous Air Pollutants—Risk Management Planning

The 1990 CAA amendments required EPA to develop regulations to minimize the chances of accidental releases to the environment of hazardous and toxic substances. The regulations parallel those developed under the Clean Air Act by OSHA, the OSHA Process Safety Management Regulations, however they go beyond OSHA in regulating flammable and explosive substances as well as toxic substances and require facilities to look at the offsite consequences of potential releases of hazardous or toxic substances. The final RMP rule was published on June 20, 1996.

Under EPA's regulations, facilities are required to:

• Install and operate detection and control equipment
• Develop an Emergency Response Plan

- Develop a Risk Management Plan
- Prepare a Hazard Assessment of Regulated Substances
- Model the release of substances
- Disclose the 5-year accidental release history
- Describe the worst-case release scenario including potentially effected offsite parties.

Risk Management Plans were due in 1999, and are required to be amended any time a facility changes a regulated process or brings onsite a threshold quantity of a regulated substance in a process. Facilities are also required to audit their compliance with the RMP regulations every 3 years.

1.2.6   Operating Permits

The operating permits requirements under the Clean Air Act are a significant new regulatory program. The operating permit program described in detail in Chapter 2 applies to existing sources of criteria pollutants and hazardous air pollutants. It applies to all nonattainment area sources that exceed the major source threshold which ranges from 10 to 100 tons per year, all hazardous air pollutants sources that exceed 10 tons of any individual hazardous air pollutant per year, and 25 tons of any combination of hazardous air pollutants, to all new source performance standard sources, to all covered sources under the PSD program and to all sources covered by section 302 of the Clean Air Act Amendments.

A significant point of information is that sources of any regulated pollutant that do not have permits (grandfathered sources) are also covered under the operating permits program if they exceed the major source threshold. For permitted sources, the major source threshold test is to look at the permitted level of pollutants emitted (as long as permits are considered federally enforceable). For grandfathered sources the test is to look at the actual level of emissions that would result from operating the source 24 hours a day, 7 days a week, 52 weeks per year. This is considered to be the absence of a permit limiting the source to lesser hours of operation or lesser utilization to be the test for determining the emission level on an annual basis from unpermitted or grandfathered sources. This is the requirement that trips up many facilities that do not operate at that level of operation. Take for example a blending operation that operates 8 hours a day, 5 days a week. In the absence of permits, the annual emissions from such an operation would be considered at the maximum physical capacity of the equipment to process material, 24 hours a day, 7 days a week, 365 days per year. This could be 10 times the actual utilization. The solution to minimize the potential for Title V Operating Permit to be required is to obtain permits that limit the potential emissions to that which would be more in line with the actual utilization of the facility. Generally speaking, it is recommended that permitsnot be put in place for the actual current utilization, but that a buffer be provided to allow for a facility to be utilized at a higher level of production, for example, two shifts at 16 hours per day rather than 1 shift at 8 hours per day, therefore allowing the facility to double its current emissions.

The operating permits program is also a revenue-producing activity for state agencies. This is because the operating permits regulations developed under the Clean Air Act allow the agency to collect an application fee, as well as an annual fee, for all major sources. The suggested rate for this is $25.00 per ton per year for each pollutant emitted. This is also adjusted automatically with the consumer price index so that in 1999 this fee ranged up to $40 per ton. It is capped at 4000 tons annually; however, this can equate to $160,000.00 per year for a large source. Even for small to mid-size industrial facilities, this can range from $1,000.00 to $5,000.00 per year in additional state fees which, prior to the operating permits program, did not exist.

1.2.7   Enforcement

Enforcement under the Clean Air Act was also significantly increased under the 1990 Amendments. Since all criminal penalties of the Act are now considered felonies, it is much more attractive for EPA to pursue violations. Penalties range up to $25,000.00 per day per violation and EPA has the ability through its field enforcement staff to issue so-called traffic tickets for minor violations up to $5,000.00 per day. There is also under EPA's Credible Evidence Program the presumptions that violations continue until the facility demonstrates compliance. This credible evidence program, or "incredible evidence" (as it is sometimes referred to) is a significant factor in air compliance planning.

Another significant compliance problem for facilities is the potential for citizens suits. These are authorized under the 1990 Clean Air Act Amendments and, significantly, the 1990 Amendments overturned an earlier legal president the Gwaltney Clean Water Act case, which barred wholly past violations. The overturning of this precedent means that wholly past violations are not barred from enforcement. Consider a facility that had an air violation which has been corrected and is now discovered by a citizen. This could happen, for example, by an environmental group obtaining, through publicly available information, a facility's annual emission statement. Comparing this to publicly available permit information, the environmental group discovers that the facility in 1994 violated a permit. In 1995, 1996, 1997, and 1998, the Emission Statements show compliance with permits. This would be a "Wholly Past Violation" in that it did not continue past 1994. Before the 1990 Amendments, environmental groups would be barred from bringing civil suit to require agency enforcement for this violation. After the 1990 Amendments, however, there is no bar to enforcement against such wholly past violations.

With respect to criminal enforcement, the fact that after the 1990 Amendments all criminal violations of the Clean Air Act are considered felonies makes it much more attractive for EPA to prosecute violations. The standard is that a facility, or an individual, was in "knowing" violation of a Clean Air Act requirement. The standard is not that the person or company intended to violate the Act; the standard is that the person knew or should have known of an act or omission that violated the Clean Air Act. A further complication is the responsible corporate officer doctrine, which holds potentially liable persons in a position to know of, and direct environmental compliance. This means that the plant environmental manager, if he or she is directed by a higher corporate official, may not be the "designated felon" on the hook for violations, but rather the senior corporate official who knew or should have known, and was in a position to direct environmental compliance, may be held responsible.

1.2.8 Auditing

EPA has not adopted satisfactory guidance to encourage self-auditing of air compliance. EPA has waffled pro and con on this point and the latest policy is its penalty mitigation policy (60 CFR 66706-12/22/95). This program does not cover noncompliance discovered during due diligence for operating permit applications, but it does apply to violations discovered during a regular program by companies to identify and correct violations. The EPA policy, however, is very difficult to comply with and requires, among other things, that you have an active incentive program to reward employees for identifying and correcting noncompliance.

1.2.9 Credible Evidence

EPA's Credible Evidence Program, finalized on February 24, 1997, represents a fundamental change to enforcement. It authorizes use of any credible evidence (other than compliance test methods) to prove noncompliance. EPA's statements notwithstanding, it also increases the stringency of the underlining standards, in that prior to credible evidence, only reference test methods could be used to demonstrate noncompliance. Under Credible Evidence, any credible evidence such as emission statements, observations by the public, statements by the company, purchasing records or other material can be used to infer noncompliance. Whether the evidence is in fact credible, is determined not by the agency or by a state agency but by a judge in a court of law. Notwithstanding, a judge's legal credentials, very few of them have training and experience in environmental compliance matters. Therefore, making a judge an arbiter of what is credible is difficult.

Another problem with credible evidence is that it has a very onerous provision which says that if a violation is discovered through whatever credible evidence is available to the judge, that this violation will be assumed to have existed every day, back in time until the time when the source was last tested and found to be in compliance with its emission limitation. Absent a successful stack test showing compliance, this violation is assumed to have existed every day back until the first day the source began operation. In this way, EPA is able to calculate astronomical fines. Take for example a source which is found through some credible evidence to be in violation that has operated for 8 years without ever being tested. That 2,920 days at a $25,000.00 per day assessment level works out to a $73 million fine. Assuming that these fines are impossible because the Clean Air Act limits fines to $250,000, be aware that all the EPA administrator has to do is notify the Attorney General that she intends to levy a fine exceeding $250,000.00 and the administrator is free to do this. Also, within the last year, EPA levied a fine in the tens of millions of dollars against Toyota for violations of the Clean Air Act.

In practice, fines in the 10s of millions of dollars are unusual; however, fines in the hundreds of thousands of dollars to multiple million dollar fines are not that unusual and the EPA routinely imposes these on violators of the act. The violators, if they qualify for use of the penalty mitigation policy, will negotiate the EPA and ask for a reduction in the assessment. Even in situations where full compliance with the penalty mitigation policy cannot be demonstrated, it is usually possible to negotiate significant reductions in proposed fines. It is of small comfort to a small company, however, to reduce a $200,000.00 fine to $50,000.00 when this represents a significant percentage of its annual profits.

EPA indicates that frivolous lawsuits are not expected under credible evidence, because EPA will focus on violations that:

- Threaten harm to the public health or environment
- Are of significant duration or magnitude
- Present a pattern of noncompliance
- Involve a refusal to provide information
- Involve criminal conduct
- Allow a source to reap a windfall

Notwithstanding the above statements by EPA, these do not necessarily apply to environmental action groups who are not barred from bringing lawsuits that do not limit themselves to these types of violations.

The question of how credible evidence is possible under the Clean Air Act has been raised by many regulated companies. Since before the 1990 Amendments only reference methods were allowed to demonstrate noncompliance, how can credible evidence be legal under the Clean Air Act? The answer is that the 1990 Amendments give the EPA the ability to bring administrative, civil, or criminal actions "on the basis of information available to the administrator." It does not specify, on the basis of performance tests alone. This is the legislative language that EPA relies on as the foundation for its credible evidence program.

1.2.10   Other Enforcement Tools

Compliance certifications are the other key enforcement means available to EPA under the Clean Air Act and its amendments. Compliance certifications are required under the operating permits program in which facilities certify compliance with all applicable requirements and specifically list those which are not applicable. Other certification programs also exist, for example, in the nonattainment New Source Review program facilities are required to certify that they are in compliance with all other applicable requirements.

1.2.11   Types of Violations

All record keeping violations are considered criminal violations of the Clean Air Act since the 1990 Amendments. Record keeping violations include:

- False statements (including omissions)
- Failure to report
- Falsifications
- Penalties range up to $250,000.00 and/or 2 years in jail

Knowing endangerment release:

- Penalties range up to $250,000.00 fine and/or 15 years in jail.

Negligent release:

- Penalties range up to $250,000.00 fine and/or 1 year in jail.

Note, companies do not serve jail time, individuals representing the company do. So being the "designated felon" or the "responsible person" carries a huge responsibility for the individual responsible for air compliance. Failure to report, as noted above, can carry a 2-year jail sentence. The difference between knowing endangerment and a negligent release is that a release of a toxic substance that causes harm or potential harm to the public if deemed to be negligently released can carry a 1-year jail sentence and up to a $250,000.00 fine. A knowing endangerment release, on the other hand, involves a conscious decision to release a substance knowing that may cause harm to the public. This can carry a penalty of up to 15 years in jail. Remember that the monetary limit of $250,000.00 can be waived in an instant through a telephone call from the EPA Administrator to the Attorney General.

1.2.12   Compliance Management Tools

In order to maximize compliance with Clean Air Act requirements, companies should know what their facility emits, know how it is operated, and they should conduct strategic planning to know what needs to be done to remain at a high level of compliance. Companies should also track the regulatory agenda at EPA and at their state agencies through Internet resources, through trade publications, and through newsletters. They should build coalitions with regulatory agencies they deal with and with their publics. And they should interact with the agencies to know what is coming before it comes, and to be in a position to influence regulations as they develop.

There are a host of tools that are publicly available including Internet resources at the USEPA including the EPA www.epa.gov web page and its linked resources. There are also publications available through the Government Printing Office and available online through the EPA web site and through many state resources. See Appendix A of this book to see a list of state contacts and web site locations for guidance materials.

1.3 CLEAN WATER ACT

The basis of federal water pollution control laws and regulations in the United States is the Clean Water Act (CWA, 33 USC § 1251 et seq.), which establishes a comprehensive program for protecting our nation's waters. The United States Environmental Protection Agency (U.S. EPA), along with other federal, state, and local agencies, administers the numerous programs established under this Act, many of which are based on a permitting system. If a water-based waste is discharged into a storm ditch, culvert, sewer, or any waters of the United States, the need for applying in advance for the appropriate environmental permit has been established. The purpose of such an environmental permitting system is to assure the general public that the water it drinks, cooks with, and on which it enjoys recreational activities is clean and safe. These permits establish wastewater effluent limitations by which the discharge of pollutants in the waters of the United States can be controlled and/or minimized. These permitting systems include the National Pollutant Discharge Elimination System (NPDES) permit program, the dredge and fill permit program, and wastewater pretreatment programs.

The EPA administrator, in cooperation with other federal agencies, state water pollution control agencies, interstate agencies, and the municipalities and industries involved, is charged with developing comprehensive programs for preventing, reducing, or eliminating the pollution of the navigable waters and groundwaters and improving the sanitary condition of surface and underground waters. These permit programs give consideration to the improvements which are necessary to conserve such waters for the protection and propagation of fish and aquatic life and wildlife, recreational purposes, and the withdrawal of such waters for public water supply, agricultural, industrial, and other purposes.

This chapter provides an overview of the CWA, including its goals and objectives, while the focus of Chap. 3 is on the various permit programs mandated by the CWA that are most likely to affect industrial and municipal dischargers of pollutants into surface waters.

1.3.1   Clean Water Act Goals, Objectives, and Policies

Like many other federal environmental statutes, the CWA contains a statement of goals, objectives, and policies. The objective of the CWA is to restore and maintain the chemical, physical, and biological integrity of the nation's waters. In order to achieve this objective, the following two national goals are established by section 101 of the CWA:

* That the discharge of pollutants into navigable waters be eliminated by 1985.
* That, wherever attainable, an interim goal of water quality, which provides for the protection and propagation of fish, shellfish, and wildlife and provides for recreation in and on the water, be achieved by July 1, 1983.

Coincidentally, the following five national policies were also established by the CWA.

* That the discharge of toxic pollutants in toxic amounts be prohibited.
* That federal financial assistance be provided to construct publicly owned waste treatment works.
* That area-wide waste treatment management planning processes be developed and implemented to assure adequate control of sources of pollutants in each state.
* That a major research and demonstration effort be made to develop technology necessary to eliminate the discharge of pollutants into the navigable waters, waters of the contiguous zone, and the oceans.
* That programs for the control of nonpoint sources of pollution be developed and implemented in an expeditious manner so as to enable the goals of this Act to be met through the control of both point and nonpoint sources of pollution.

1.3.2   Elements of the Clean Water Act

In accordance with its statutory objectives and goals, the CWA consists of the following major elements that function as regulatory tools and mechanisms:

* A prohibition of discharges, except as in compliance with the Act (Sec. 301)
* A system for determining the limitations to be imposed on regulated discharges (Secs. 301, 306, 307)
* A system for preventing, reporting, and responding to spills (Sec. 311)
* A process for cooperative federal/state implementation (Secs. 401, 402)

- A permit program to authorize and regulate certain discharges (Sec. 402)
- Strong enforcement mechanisms (Secs. 309, 505)

1.3.3 Water Quality Effluent Limitations and Standards

Section 302 of the CWA (33 USC 1312), Water Quality Related Effluent Limitations, establishes effluent limitations whenever discharges of pollutants from a point source or group of point sources would interfere with the attainment or maintenance of that water quality in a specific portion of the navigable waters which shall assure protection of public health, public water supplies, and agricultural and industrial uses; assure the protection and propagation of a balanced population of shellfish, fish, and wildlife; and allow recreational activities in and on the water. Under such conditions, the effluent limitations (including alternative effluent control strategies) for such point source or sources shall be established that can reasonably be expected to contribute to the attainment or maintenance of such water quality.

Water quality standards are provisions of state or federal law that consist of a designated use or uses for the waters of the United States and water quality criteria for such waters based on such uses. Water quality standards are to protect the public health or welfare, enhance the quality of water, and serve the purposes of the Act. A water quality standard, as established by a state, defines the water quality goals of a water body, or portion thereof, by designating the use or uses to be made of the water and by setting criteria necessary to protect the uses. States adopt water quality standards to protect public health or welfare, enhance the quality of water, and serve the purposes of the Clean Water Act (the Act). "Serve the purposes of the Act" [as defined in Secs. 101(a)(2) and 303(c) of the Act] means that water quality standards should, wherever attainable, provide water quality for the protection and propagation of fish, shellfish, and wildlife and for recreation in and on the water and take into consideration their use and value of public water supplies; propagation of fish, shellfish, and wildlife; recreation in and on the water; and agricultural, industrial, and other purposes including navigation.

Such standards serve the dual purposes of establishing the water quality goals for a specific water body and serve as the regulatory basis for the establishment of pollutant-specific permitting programs, water-quality-based treatment controls, and strategies beyond the technology-based levels of treatment required by Secs. 301(b) and 306 of the Act.

1.3.4 Pollutants

In the context of water pollution control, the term *pollutant* can include such materials and/or properties as dredged spoils; solid waste; incinerator residue; filter backwash; sewage; garbage; sewage sludge; munitions; chemical wastes;

biological materials; radioactive materials [except those regulated under the Atomic Energy Act of 1954, as amended (42 USC 2011 et seq.)]; heat; wrecked or discarded equipment; rock; sand; cellar dirt; and industrial, municipal, and agricultural waste discharged into water. The term *pollutant* does *not* mean sewage from vessels; water, gas, or other material that is injected into a well to facilitate production of oil or gas, or water derived in association with oil and gas production and disposed of in a well (provided that the well used either to facilitate production or for disposal purposes is approved by authority of the state in which the well is located, and that the state determines that the injection or disposal will not result in the degradation of groundwater or surface water resources). For instance, examples of pollutants commonly introduced to our waters are pH, temperature, color, etc. Table 1.2 contains the list of the 65 toxic pollutants as found in 40 CFR Part 401.15 (*Note:* §401.15 added at 44 FR 44502, July 30, 1979; amended at 46 FR 2266, January 8, 1981; 46 FR 10724, February 4, 1981).

Table 1.3 lists the five conventional pollutants identified in 40 CFR Part 410.16. These are biochemical oxygen demand (BOD), total suspended solids (TSS), pH, fecal coliform, and oil and grease.

1.3.5 Statutory History

Although the origins of federal efforts to control the discharge of pollutants into surface waters can be traced all the way back to the Refuse Act of 1899, serious efforts at enforcing water pollution regulations did not come about until the mid-1950s, when the Federal Water Pollution Control Act was first passed in 1956. Approximately a decade later, in the mid-1960s, the Water Quality Act of 1965 was enacted. Subsequent to this legislation, the 1899 Refuse Act was modified to include a permit program that addressed the discharge of pollutants into surface waters. This concept of actually "permitting" the discharge of pollutants into a specific medium (i.e., surface waters) served as one of the basic premises for the subsequent enactment of the Federal Water Pollution Control Act (Public Law 92-500), one of the first truly comprehensive environmental laws. In 1972, Congress amended the Federal Water Pollution Control Act [referred to as the Clean Water Act (CWA)] to prohibit the discharge of any pollutant to waters of the United States from a point source unless the discharge is authorized by a National Pollutant Discharge Elimination System (NPDES) permit. During the years that followed the enactment of the CWA, the scope of this "permitting" concept grew to encompass such pollutants as materials with high biochemical oxygen demand (BOD) and hazardous and toxic materials. The Federal Water Pollution Control Act Amendments of 1972 replaced the previous language of the Act entirely, including the Water Quality Act of 1965, the Clean Water Restoration Act of 1966, and the Water Quality Improvement Act of 1970, all of which had been amendments of the Federal Water Pollution Control Act first passed in 1956. In 1977 these additional provisions and other changes were incorporated into the Clean

TABLE 1.2 Sixty-Five Toxic Pollutants

1.   Acenaphthene
2.   Acrolein
3.   Acrylonitrile
4.   Aldrin/dieldrin[1]
5.   Antimony and compounds[2]
6.   Arsenic and compounds
7.   Asbestos
8.   Benzene
9.   Benzidine[1]
10. Beryllium and compounds
11. Cadmium and compounds
12. Carbon tetrachloride
13. Chlordane (technical mixture and metabolites)
14. Chlorinated benzenes (other than dichlorobenzenes)
15. Chlorinated ethanes (including 1,2-dichloroethane, 1,1,1-trichloroethane, and hexachloroethane)
16. Chloroalkyl ethers (chloroethyl and mixed ethers)
17. Chlorinated naphthalene
18. Chlorinated phenols (other than those listed elsewhere; includes trichlorophenols and chlorinated cresols)
19. Chloroform
20. 2-chlorophenol
21. Chromium and compounds
22. Copper and compounds
23. Cyanides
24. DDT and metabolites[1]
25. Dichlorobenzenes (1,2-, 1,3-, and 1,4-dichlorobenzenes)
26. Dichlorobenzidine
27. Dichloroethylenes (1,1-, and 1,2-dichloroethylene)
28. 2,4-dichlorophenol
29. Dichloropropane and dichloropropene
30. 2,4-dimethylphenol
31. Dinitrotoluene
32. Diphenylhydrazine
33. Endosulfan and metabolites
34. Endrin and metabolites[1]
35. Ethylbenzene
36. Fluoranthene
37. Haloethers (other than those listed elsewhere; includes chlorophenylphenyl ethers, bromophenylphenyl ether, bis-(dichloroisopropyl) ether, bis-(chloroethoxy) methane and polychlorinated diphenyl ethers)
38. Halomethanes (other than those listed elsewhere; includes methylene chloride, methylchloride, methylbromide, bromoform, dichlorobromomethane)
39. Heptachlor and metabolites
40. Hexachlorobutadiene
41. Hexachlorocyclohexane

TABLE 1.2 Sixty-Five Toxic Pollutants (*Continued*)

42.	Hexachlorocyclopentadiene
43.	Isophorone
44.	Lead and compounds
45.	Mercury and compounds
46.	Naphthalene
47.	Nickel and compounds
48.	Nitrobenzene
49.	Nitrophenols (including 2,4-dinitrophenol, dinitrocresol)
50.	Nitrosamines
51.	Pentachlorophenol
52.	Phenol
53.	Phthalate esters
54.	Polychlorinated biphenyls (PCBs)[1]
55.	Polynuclear aromatic hydrocarbons (including benzanthracenes, benzopy- renes, benzofluoranthene, chrysenes, dibenzanthracenes, and indenopyrenes)
56.	Selenium and compounds
57.	Silver and compounds
58.	2,3,7,8-tetrachlorodibenzo-p-dioxin (TCDD)
59.	Tetrachloroethylene
60.	Thallium and compounds
61.	Toluene
62.	Toxaphene[1]
63.	Trichloroethylene
64.	Vinyl chloride
65.	Zinc and compounds

[1]Effluent Std. Promulgated (40 CFR Part 129).

[2]Includes organic & nonorganic compounds.

Water Act amendments. Permits can specify quantitative and qualitative limitations, guidelines, pretreatment standards, and new source performance standards for industrial dischargers.

In 1987, Congress amended the Clean Water Act with the Water Quality Act. The new law expanded and strengthened the Clean Water Act through a number of changes designed to improve water quality and environmental programs. Existing programs were expanded and new approaches were created to address emerging water pollution problems. The new law offered opportunities for regulatory agencies, the regulated community, and the public to expand ongoing programs with new initiatives, including programs to control toxic materials in surface waters, pollution from nonpoint sources, and pollution from storm water runoff, and to protect and restore lakes and estuaries.

In 1990 Congress enacted the Oil Pollution Act, which further focused on issues associated with maintaining and improving water quality standards.

TABLE 1.3   Five Conventional Pollutants

1. Biochemical oxygen demand (BOD)
2. Total suspended solids (nonfilterable) (TSS)
3. pH
4. Fecal coliform
5. Oil and grease

In order to be effective, today's laws and regulations that control water pollution must now comprehensively address permitting the discharge of pollutants from many different types of privately and publicly owned/operated facilities. The control of these pollutants entering our surface waters is accomplished by technology-based effluent limitations, wastewater treatment systems, field inspections by appropriate regulatory agencies, individual facility self-inspecting and oversight, the containment/cleanup of hazardous substance releases, and last, enforcement.

1.3.6   What Is a Wastewater Discharge Permit?

A permit means an authorization, license, or equivalent control document issued by EPA or an "approved state" to implement the requirements of 40 CFR Parts 122, 123, and 124. "Permit" includes an NPDES "general permit" (40 CFR §122.28). The term *permit* does not include any permit that has not yet been the subject of final agency action, such as a "draft permit" or a "proposed permit."

1.3.7   When Are Wastewater Discharge Permits Required?

As noted above, whenever a water-based waste is discharged into a storm ditch, culvert, sewer, or any waters of the United States, the need for applying in advance for the appropriate environmental wastewater discharge permit has been established. To assist in achieving the objective of the CWA, effluent permits (essentially licenses) that authorize the discharge of pollutants, as long as the discharge complies with established effluent limitations and other conditions, must be obtained from either the EPA or the appropriate state agency (that has been delegated the NPDES permitting program) for all discharges from point sources into one of the following:

- All navigable waters of the United States
- Tributaries of navigable waters of the United States
- Interstate waters
- Intrastate lakes, rivers, and streams that are used by interstate travelers for recreational or other purposes

- Intrastate lakes, rivers, and streams that are utilized for industrial purposes by industries in interstate commerce.

Permits must be obtained for all discharges of wastewaters to the waters of the United States or state waters, for all wastewater discharges from an industrial facility to a publicly owned treatment works (POTW) and selected wastewater treatment facilities.

Waters of the United States include the following:

- All waters that are used, were used, or may be used in interstate or foreign commerce, including all waters that are subject to the ebb and flow of the tide
- Interstate waters, including interstate wetlands
- All other waters, such as intrastate lakes, rivers, streams, intermittent streams, mudflats, sandflats, sloughs, prairie potholes, wet meadows, playa lakes, natural ponds, and wetlands that could affect interstate or foreign commerce when used, degraded, or destroyed
- Seas under the jurisdiction of the United States
- Tributaries of regulated waters
- Wetlands adjacent to waters (other than waters that are themselves wetlands)

1.3.8 Point Sources

All wastewater discharges of pollutants from a point source must be permitted. A *point source* is defined to include any discernible, confined, and discrete conveyance, including, but not limited to, any pipe, ditch, channel, tunnel, conduit, well, discrete fissure, container, rolling stock, concentrated animal feeding operation, or vessel or other floating craft from which pollutants are or may be discharged. Conveyance systems specifically designed for special and unique functions, such as a landfill leachate collection system, would also be defined as point sources.

1.3.9 Pollution from Nonpoint Sources

Although the common perception is that the majority of water pollution originates from point sources, nontraditional sources of pollution—especially stormwater runoff from urban/suburban areas and "nonpoint sources" of pollution originating from diffuse areas and land-use activities such as farming, logging, and construction—constitute some of the major causes of water quality problems. An example of this would be the nonpermitted runoff of nutrients, such as the nitrates found in fertilizers, from farmlands surrounding the Chesapeake Bay. The discharge of these nutrients has resulted in the overstimulation of aquatic plant and algae growth that chokes the water, uses up dissolved oxygen, and cuts off light in deeper waters.

As pollution control measures for industrial process wastewater and municipal sewage discharges from point sources were further developed, refined, and implemented, it became increasingly evident these efforts (i.e., permitting) alone would not be sufficient to halt the deterioration of our nation's waters. These nonpermitted discharges of pollutants from diffuse nonpoint sources of water pollution were very significant causes of deteriorating water quality and were impairing the healthy growth of invertebrates, leading to a decrease in animal and plant diversity, and affecting recreational uses of a body of water, including fishing, swimming, and boating. In 1987, Congress amended the CWA to require implementation of a comprehensive approach for addressing stormwater discharges under the NPDES program. Stormwater discharges have a number of environmental effects that can occur from land development, illicit discharges, construction site runoff, and improper disposal of materials.

1.3.10 Major Contributors of Water Pollution

The following 10 categories have been defined by the U.S. EPA as major contributors of water pollution:

- *Industrial,* including pulp and paper mills, chemical manufacturers, and food processing plants
- *Municipal,* including publicly owned sewage treatment works that may receive indirect discharges from small factories or businesses
- *Combined sewers,* including storm and sanitary sewers that, when combined, may discharge untreated wastes during storms
- *Storm sewers/runoff,* including runoff from streets, paved areas, and lawns that enters a sewer, pipe, or ditch
- *Agricultural,* including crop production, pastures, rangeland, and feedlots
- *Silvicultural,* including forest management, harvesting, and road construction
- *Construction,* including highway building and land development
- *Resource extraction,* including mining, petroleum drilling, and runoff from mine tailing sites
- *Land disposal,* including leachate or discharge from septic tanks, landfills, hazardous waste disposal sites
- *Hydrologic,* including channelization, dredging, dam construction, and stream bank modification

1.3.11 Wastewater Discharge Permit Process

Generally speaking, an application for a wastewater discharge permit must be completed on the forms issued by the appropriate agency (federal or state) and submitted at least 180 days prior to the commencement of the discharge. After receiving an application, the appropriate federal or state agency reviews it for completeness, and if information is found to be lacking, the applicant will be requested to provide the missing data. Before a final permit is issued, a draft permit is forwarded to the applicant by the agency office writing the permit. Usually the applicant has a period of 30 days to review and comment on any proposed changes to the draft permit. After the applicant and the agency office writing the permit agree on the conditions and/or limitations specified in the draft permit, a period of public notice begins and public comments are invited. If the permitting agency deems the degree of public interest to be high, a hearing can be scheduled. After the period of public notice ends and the public input is considered, the agency office writing the draft permit can recommend either approval or disapproval to the agency issuing the permit.

1.3.12   The National Pollutant Discharge Elimination System (NPDES) Permit Program

The NPDES Program is designed to regulate point source discharges. As the primary regulatory vehicle by which the discharge of pollutants into the waters of the United States is permitted, this program is the most significant enforcement mechanism contained in the entire Clean Water Act, regulating discharges into U.S. waters from point sources, including municipal, industrial, commercial, and certain agricultural sources. No permit may be issued unless it complies with all applicable sections of the Clean Water Act. Permit conditions may include specific monitoring and reporting requirements, including how sampling of the effluent should be done to check whether the effluent limitations are being met, the type of monitoring, sampling frequencies (e.g., on an hourly, daily, weekly, or monthly basis), and the way in which the results of the sampling are to be regularly reported to the EPA and/or appropriate state authorities. Usually these permits must be renewed at least once every 5 years. If a discharger fails to comply with the effluent limitations or monitoring and reporting requirements, EPA or the state may take enforcement action and actually revoke the permittee's right to discharge wastewaters.

1.3.13   Types of NPDES Permits

In addition to the NPDES permits that industrial and/or municipal dischargers of pollutants via point sources (including stormwater) must obtain, there are six other types of permits relating to the NPDES Program that are available under the Clean Water Act to control discharges of pollutants into surface waters or discharges of dredged or fill material into wetlands and navigable waters.

1.3.13.1   State NPDES Permit Programs.   NPDES permits can be obtained either from the Environmental Protection Agency or from a state authorized to issue NPDES permits. States, as defined in 40 CFR § 131.3, include the 50 states, the District of Columbia, Guam, the Commonwealth of Puerto Rico, Virgin Islands, American Samoa, the Trust Territory of the Pacific Islands, the Commonwealth of the Northern Mariana Islands, and Indian tribes that EPA determines to be eligible for purposes of a water quality standards program. Upon authorization of a state NPDES program, the state is primarily responsible for issuing permits and administering the NPDES permit program and is also responsible for reviewing, establishing, and revising water quality standards. Section 510 of the Clean Water Act recognizes the fact that authorized states may develop water quality standards more stringent than required by this regulation. State NPDES authority consists of the following four parts: municipal and industrial permitting, federal facilities permitting, pretreatment requirements, and general permitting. Currently, 41 states or territories have received NPDES enforcement authorization from the EPA and are authorized to, at a minimum, issue NPDES permits for municipal and industrial sources. EPA issues all NPDES permits in the 10 states and 6 territories that have not received NPDES enforcement authorization. In addition, EPA issues permits for discharges from federal facilities in 6 of the 41 states that are authorized to issue NPDES permits for municipal and industrial sources.

1.3.13.2   POTW NPDES Permits.   Federal- and/or state-issued NPDES permits are also issued to local sewage treatment plants, better known as *publicly owned treatment works* (POTWs). This type of permit can contain pretreatment requirements for industries that are covered by the 40 CFR Subpart N effluent discharge limitations to the POTW that are contained in 40 CFR Part 403. The purpose of these pretreatment criteria is to control concentrations of certain pollutants found in industrial wastewaters that cannot be adequately treated by the POTW receiving the wastewater discharge. If the discharge of these pollutants is not controlled, the POTW's treatment processes can be upset. Pollutants such as heavy metals, trace elements, and other materials are difficult to treat and control by the treatment processes found in many POTWs.

1.3.13.3   NPDES Permits for Sludge Use and Disposal.   In order to protect human health and the environment when sewage sludge is beneficially applied to land, placed in a surface disposal site, or incinerated, the sewage sludge permitting regulations found in 40 CFR Part 503 have been promulgated to:

• Establish standards, which consist of general requirements, pollutant limits, management practices, and operational standards, for the final use or disposal of sewage sludge generated during the treatment of domestic sewage in a treatment works. Standards are included in this part for sewage sludge applied to the land, placed on a surface disposal site, or fired in a sewage sludge incinerator. Also included in this part are pathogen and alternative vector attraction reduction requirements for sewage sludge applied to the land or placed on a surface disposal site.

• Establish standards for the frequency of monitoring and record keeping requirements when sewage sludge is applied to the land, placed on a surface disposal
site, or fired in a sewage sludge incinerator. Also included in this part are reporting requirements for Class I sludge management facilities, POTWs with a design
flow rate equal to or greater than 1 million gallons per day, and POTWs that
serve 10,000 people or more.

Generators, processors, users, and disposers of sewage sludge are all covered
by the regulations. Federal EPA regions will continue to issue all NPDES permits
for sludge use and disposal until states have received authorization by the U.S.
EPA to issue permits for the sludge use and disposal program. The sludge permitting program applies to any person who prepares sewage sludge, applies sewage
sludge to the land, or fires sewage sludge in a sewage sludge incinerator and to the
owner/operator of a surface disposal site.

1.3.13.4 Dredge and Fill Permits.   Section 404 of the CWA (33 USC 1344),
Permits for Dredged or Fill Material, provides for another type of permit program,
the Dredge and Fill Permits Program, that is jointly administered by EPA and the
U.S. Army Corps of Engineers under the guidelines found in 40 CFR Part 230.
Fundamental to these guidelines is the precept that dredged or fill material should
not be discharged into the aquatic ecosystem, unless it can be demonstrated that
such a discharge will not have an unacceptable adverse impact either individually
or in combination with known and/or probable impacts of other activities affecting the ecosystems of concern. From a national perspective, the degradation or
destruction of special aquatic sites—by conducting filling operations in wetlands,
for example—is considered to be among the most severe environmental impacts
covered by these guidelines. The guiding principle should be that degradation or
destruction of special sites may represent an irreversible loss of valuable aquatic
resources.

Under the Act, a permit must be obtained before discharging any dredge or fill
material into wetlands or navigable waters. Under Sec. 404(a), dredge and fill permits may be issued, after notice and opportunity for public hearings for the discharge of dredged or fill material into the navigable waters at specified disposal
sites. Not later than the fifteenth day after the date an applicant submits all the
information required to complete an application for a permit under this subsection,
the Secretary of the Army, acting through the Chief of Engineers, shall publish the
notice required by this subsection.

Under the Act, the discharge of dredged or fill material means the addition of
fill material into U.S. waters, and includes placement of fill that is necessary for
the construction of any structure; the building of any structure or impoundment
requiring rock, sand, dirt, or other materials for its construction; site-development
fills for recreational, industrial, commercial, residential, and other uses; cause-
way or road fills; dams and dikes; artificial islands; property protection and/or
reclamation devices such as riprap, groins, seawalls, breakwaters, and retaining

walls; beach protection; levees; fill for structures such as sewage treatment plants, intake and outfall pipes associated with power plants, and subaqueous utility lines; and artificial reefs.

1.3.13.5 General Permits. General permits are available under the NPDES program (40 CFR § 122.28) as well as the dredge or fill permit program. A general permit may be issued by EPA, a state, or the Corps for dischargers whose activities are similar in nature, will cause only minimal adverse environmental effects when performed separately, and will result in minimal cumulative adverse effects on the environment. The two main criteria considered in the issuance of a general permit are *area* and *source.*

The general permit can be written to cover a category of discharges or sludge use or disposal practices or facilities described below, except those covered by individual permits, within a geographic area. The area generally corresponds to existing geographic or political boundaries, such as:

- Designated planning areas under Secs. 208 and 303 of CWA
- Sewer districts or sewer authorities
- City, county, or state political boundaries
- State highway systems
- Standard metropolitan statistical areas as defined by the Office of Management and Budget
- Urbanized areas as designated by the Bureau of the Census according to criteria in 30 FR 15202 (May 1, 1974)
- Any other appropriate division or combination of boundaries.

In addition to stormwater point sources, a general permit may be written to regulate, within one of the areas described above, a category of point sources, or a category of "treatment works treating domestic sewage," if the sources or "treatment works treating domestic sewage" all:

- Involve the same or substantially similar types of operations
- Discharge the same types of wastes or engage in the same types of sludge use or disposal practices
- Require the same effluent limitations, operating conditions, or standards for sewage sludge use or disposal
- Require the same or similar monitoring
- Are more appropriately controlled under a general permit than under individual permits

1.3.13.6 Well Injection Permits. As further noted in Chap. 3 of this book, the regulations found in 40 CFR § 122.50 require that when part of a discharger's process wastewater is not being discharged into waters of the United States or contiguous zone because it is disposed into a well, into a POTW, or by land application, thereby reducing the flow or level of pollutants being discharged into waters of the United States, applicable effluent standards and limitations for the discharge in an NPDES permit will be adjusted to reflect the reduced raw waste resulting from such disposal.

In Chapter 3 we will review the requirements for CWA-mandated NPDES permits in more detail along with other permit programs that maintain the quality of our nation's waters. (Refer to App. B for a glossary of commonly used terms relating to the enforcement provisions of the Clean Water Act.)

1.4 NATIONAL ENVIRONMENTAL POLICY ACT

Chapter 6 provides an overview of the National Environmental Policy Act (NEPA). This act was passed on January 1, 1970. The purpose of NEPA is to set a national priority to encourage productive and enjoyable environmental use and harmony between man and his environment, to promote efforts to prevent or eliminate damage to the environment and the biosphere, to stimulate the health and welfare of man, to enrich the understanding of the ecological system and natural resources important to the nation, and to establish the Council on Environmental Quality.

NEPA has had far-reaching impacts, way beyond the scope of the original legislation, which is quite narrow. NEPA has stimulated environmental impact and environmental assessment requirements at state and local levels. Now, virtually any application for a major subdivision, major commercial or industrial development, or even large residential development carries with it a requirement to prepare an environmental impact statement or an environmental assessment. This requirement even has dropped down to the municipal level, where many townships have environmental assessment or impact statement ordinances that parallel the requirements of the National Environmental Policy Act.

1.5 SUPERFUND (CERCLA)

The Comprehensive Environmental Response, Compensation, and Liability Act of 1980 (CERCLA), commonly known as the *Superfund Act,* was the first federal legislation aimed at a number of issues related to the use and disposal of hazardous and toxic substances. Key provisions of the Act involve states and EPA designating sites for cleanup, potentially using federal funds, if the potentially responsible parties (PRPs) that have been shown to have created these sites are not in business or are not financially capable of providing the funds necessary to clean them up. This portion of the Act, although it is the most widely discussed portion, does not, in and of itself, require permitting, at least not for individual companies.

There is another aspect of the Act that does impact permitting and approvals for individual companies, and that is the portion known as the Emergency Planning and Community Right-to-Know Act (EPCRA). This portion, added to CERCLA in the Superfund Amendment and Reauthorization Act (SARA), requires companies to submit a number of filings with EPA or their states disclosing their locations and quantities of listed hazardous or toxic substances that they may have on site.

EPCRA also requires facilities to participate in local emergency planning committees and with state emergency planning commissions. The objective of EPCRA is to provide first responders with sufficient information so that, when they are called on to respond to a fire, explosion, or other emergency situation at an industrial facility, enough information about the hazardous or toxic substances that may be present on site is available.

Chapter 9 provides a detailed look at the requirements of the EPCRA regulations and discusses their impact to permitting for industrial facilities.

1.6 RESOURCE CONSERVATION AND RECOVERY ACT

The Resource Conservation and Recovery Act was passed in 1976. The Act was seen as a reaction to the ever-mounting increase in hazardous materials and waste generated and the potential and environmental health impacts of unsound storage and disposal practices.

This Act is truly one of the landmark environmental legislations of the last 30 years. EPA took almost 6 years after the Act had become law to finally develop and implement the first complete set of regulations covering this area. The result was a very complex system of regulations covering the initial generation of hazardous waste to its final disposal. The regulations are divided in the Code of Federal Regulations, Title 40, Protection of the Environment, into 19 parts, ranging from Part 60 to Part 80. These parts are listed in Table 1.4.

Basically, the requirements of RCRA can be crystallized into requirements governing several components of the cradle-to-grave hazardous waste generation, transport, storage, and disposal spectrum. These include:

- Generators of solid waste
- Transporters of hazardous waste
- Treatment, storage, and disposal facilities (TSDF)

RCRA also contains significant new regulations governing underground storage tanks. This is significant because leaking underground storage tanks have been found to be one of the principal contributors to soil and groundwater contamination in the United States.

The myriad regulations defining what constitutes hazardous waste (as opposed to intermediates or industrial raw materials); what quantities govern applicability of the regulations; what generators, transporters, and TSDF facilities need to do to obtain permits to operate; and what monitoring requirements and what potential cleanup requirements pertain to problems identified at these facilities are described in Chap. 11.

TABLE 1.4   Nineteen Parts of Title 40—Protection of the Environment

Part	Section name
Part 60	Standards of performance for new stationary sources
Part 61	National emission standards for hazardous air pollutants
Part 62	Approval and promulgation of State plans for designated facilities and pollutants
Part 63	National emission standards for hazardous air pollutants for source categories
Part 66	Assessment and collection of noncompliance penalties by EPA
Part 67	EPA approval of State noncompliance penalty program
Part 68	Chemical accident prevention provisions
Part 69	Special exemptions from requirements of the Clean Air Act
Part 70	State operating permit programs
Part 71	Federal operating permit programs
Part 72	Permits regulation
Part 73	Sulfur dioxide allowance system
Part 74	Sulfur dioxide opt-ins
Part 75	Continuous emission monitoring
Part 76	Acid rain nitrogen oxide emission reduction program
Part 77	Excess emissions
Part 78	Appeal procedures for Acid Rain Program
Part 79	Registration of fuels and fuel additives
Part 80	Regulation of fuels and fuel additives

CHAPTER 2
AIR QUALITY

Donald F. Elias

A. Roger Greenway

Kenneth J. Skipka

Principals, RTP Environmental Associates Inc.

2.1 AIR TOXICS AND RISK MANAGEMENT PLANNING

2.1.1   Air Toxics

The 1990 Clean Air Act Amendments regulated air toxics in Title III. This Title required EPA to develop a long list of emission standards for emitters' hazardous and toxic substances. The Act included a list of 189 chemicals because Congress had been disappointed that EPA, after the 1977 Amendments, had developed only a small handful of emission standards for hazardous air pollutants under its NESHAPs (National Emissions Standard for Hazardous Air Pollutants) Program. As a result, the Clean Air Act Amendments of 1990 required EPA to develop emission standards for emitters of hazardous pollutants.

2.1.2   Risk Management Planning Requirements

Title III of the 1990 Amendments also required EPA to develop its Risk Management Planning (RMP) regulations. These regulations parallel requirements developed by the Occupational Safety and Health Administration (OSHA) to regulate process safety management (PSM) of highly hazardous chemicals. Indeed, the OSHA program developed under Sec. 112(r) of the Clean Air Act Amendments of 1990 is the first example of occupational safety and health regulations being required by a piece of environmental legislation, that is, the Clean Air Acts (CAAs) of 1990. For a while, the proposed RMP regulations differed significantly from the existing OSHA PSM regulations. In the end, however, the final RMP regulations closely paralleled the PSM regulations. There are a number of areas of overlap, nevertheless, including the requirements to conduct off-site consequence analysis and to file RMP plans with the agency. As a note, OSHA PSM regulations require the development of plans but do not require these plans to be filed with the agency.

With RMP regulations coming as late as they did after the promulgation of the 1990 Amendments, EPA found itself in a situation where a number of states, including New Jersey and California, already had state programs of a similar nature. These include the Toxic Catastrophe Prevention Act Program (TCPA) in New Jersey and the Risk Management Prevention Program (RMPP) in California. Both states, having had programs in effect well before the final RMP regulations were promulgated in 1996, resisted changing their programs to be fully consistent with the RMP regulations. In the end, however, the states modified them significantly, so the differences between the programs are minimal.

The intent of Sec. 112(r) of the Clean Air Acts of 1990 is to prevent accidental releases to the air and to mitigate the consequences of such releases by focusing prevention measures on chemicals that pose the greatest risk to the public and the environment. These regulations came with the 1990 Amendments, just after a time when there had been a number of significant chemical accidents, including the Union Carbide accident in Bhopal, India, in which a large number of people were killed and an even larger number were injured. It came after similar but less extreme releases in the United States and elsewhere. As a result of this background, in 1986 EPA established its Chemical Accident Prevention Program to collect information on accidents and to work with other groups to increase the knowledge of prevention practices and to encourage prevention practices. EPA developed its Accident Release Information Program (ARIP) and the Voluntary Chemical Safety Audit Program.

However, minimizing the potential for releases and encouraging companies to study their processes to reduce the potential and severity of releases requires regulation of the processes themselves, including design, construction, maintenance, and training. This overlaps significantly with OSHA's regulatory domain. This is why the 1990 Clean Air Act Amendments required both agencies to develop programs aimed at improving process safety.

Additionally, a very important but sometimes overlooked requirement of Sec. 112(r) is the General Duty clause. The General Duty clause requires facilities to investigate their processes and plan for ways to minimize the consequences of any releases that may occur. It also requires facilities to coordinate with off-site responders. Importantly, the General Duty clause does not have a threshold for applicability, nor does it necessarily apply only to the list of regulated substances in the Clean Air Act Amendments and in the final RMP regulations.

Section 112(r) required EPA to develop regulations that cover the use, operation, repair, replacement, and maintenance of equipment to monitor, detect, inspect, and control accidental releases. Training is also required so that staff may be educated in the use and maintenance of equipment and in the conduct of periodic inspections.

Under the final regulations, facilities were required to register their RMP plans with EPA within 3 years of June 21, 1996. These plans were due then on June 21, 1999. The relevance to environmental permitting is that if such plans have not been filed, facilities will not be able to obtain permits to operate processes that involve regulated substances in greater than threshold quantities. In addition, RMP plans are required to be submitted or modified for processes that will begin operation using threshold quantities of regulated substances before they will be allowed to obtain permits for their operation.

Tables 2.1 and 2.2 list the regulated substances covered by the RMP regulations and the corresponding threshold quantities. A process with more than a listed threshold quantity was required to have filed its RMP plan by June 21, 1999. For new processes or existing processes that will utilize or manufacture regulated substances in greater than threshold quantities for the first time after that date, RMP plans must be filed with the state agency and/or U.S. EPA prior to use. Note that facilities are free to use regulated substances in underthreshold quantities as long as state permits for the use of the regulated substance are obtained prior to its actual use.

The U.S. EPA estimated that approximately 66,000 facilities nationwide are covered by the RMP regulations. This compares to over 87,000 facilities covered by the OSHA PSM regulations. The largest populations of regulated facilities include:

- Cold storage facilities (ammonia refrigerant)
- Public drinking water systems (chlorine)
- POTWs (chlorine)
- Manufacturers (for a list of substances)
- Propane retailers (at an undetermined threshold subject to a court stay presently in effect)
- Wholesalers and services industries (for a list of substances)
- Farm contractors (ammonia fertilizer)

The area or portion of a facility that is covered by the RMP regulations, and for which RMP compliance will affect the ability to obtain permits, is the only area of the facility or the process where the list of substances may be used, processed, stored, or otherwise handled in quantities exceeding the stated threshold quantities in the Title III List of Lists found in App. C. This list details the RMP threshold quantities and also lists regulated quantities for other hazardous and toxic chemical regulators.

For processes in which multiple substances are used in greater than threshold quantities, a single hazard analysis will be adequate for purposes of RMP compliance.

The RMP regulations require the calculation of the potential effects of the catastrophic and worst-case releases. A catastrophic release is defined in the regulations as "a major uncontrolled emission, fire or explosion, involving one or more regulated substances that presents immediate and substantial endangerment to public health and the environment." The worst-case release is defined in the regulations as "the release of the largest quantity of a regulated substance from a vessel or process line failure that results in a greater distance to an endpoint."

TABLE 2.1 Threshold Quantities of Regulated Toxic Substances for Accidental Release Prevention

CAS no.	Toxic	Threshold, lb
107-02-8	Acrolein [2-propenal]	5,000
107-13-1	Acrylonitrile [2-propenenitrile]	20,000
814-68-6	Acrylyl chloride [2-propenoyl chloride]	5,000
107-18-6	Allyl alcohol [2-propen-1-ol]	15,000
107-11-9	Allylamine [2-propen-1-amine]	10,000
7664-41-7	Ammonia (anhydrous)	10,000
7664-41-7	Ammonia (conc. 20% or greater)	20,000
7784-34-1	Arsenous trichloride	15,000
7784-42-1	Arsine	1,000
10294-34-5	Boron trichloride [borane, trichloro-]	5,000
7637-07-2	Boron trifluoride [borane, trifluoro-]	5,000
353-42-4	Boron trifluoride compound with methyl ether (1:1) [boron, trifluoro[oxybis[methane]]-, T-4	15,000
7726-95-6	Bromine	10,000
75-15-0	Carbon disulfide	20,000
7782-50-5	Chlorine	2,500
10049-04-4	Chlorine dioxide [chlorine oxide (ClO_2)]	1,000
67-66-3	Chloroform [methane, trichloro-]	20,000
542-88-1	Chloromethyl ether [methane, oxybis(chloro-)]	1,000
107-30-2	Chloromethyl methyl ether [methane, chloromethoxy-]	5,000
4170-30-3	Crotonaldehyde [2-butenal]	20,000
123-73-9	Crotonaldehyde, (E)-, [2-butenal, (E)-]	20,000
506-77-4	Cyanogen chloride	10,000
108-91-8	Cyclohexylamine [cyclohexanamine]	15,000
19287-45-7	Diborane	2,500
75-78-5	Dimethyldichlorosilane [silane, dichlorodimethyl-]	5,000
57-14-7	1,1-Dimethylhydrazine [hydrazine, 1,1-dimethyl-]	15,000
106-89-8	Epichlorohydrin [oxirane, (chloromethyl)-]	20,000
107-15-3	Ethylenediamine [1,2-ethanediamine]	20,000
151-56-4	Ethyleneimine [aziridine]	10,000
75-21-8	Ethylene oxide [oxirane]	10,000
7782-41-4	Fluorine	1,000
50-00-0	Formaldehyde (solution)	15,000
110-00-9	Furan	5,000
302-01-2	Hydrazine	15,000
7664-39-3	Hydrochloric acid (conc. 50% or greater)	1,000
7647-01-0	Hydrochloric acid (conc. 30% or greater)	15,000
74-90-8	Hydrocyanic acid	2,500
7647-01-0	Hydrogen chloride (anhydrous) [hydrochloric acid]	5,000
7664-39-3	Hydrogen fluoride/hydrofluoric acid (conc. 50% or greater) [hydrofluoric acid]	1,000
7783-07-5	Hydrogen selenide	500
7783-06-4	Hydrogen sulfide	10,000
13463-40-6	Iron, pentacarbonyl-[iron carbonyl (Fe(CO)$_5$), (TB-5-11)-]	2,500

TABLE 2.1 Threshold Quantities of Regulated Toxic Substances for Accidental Release Prevention (*Continued*)

CAS no.	Toxic	Threshold, lb
78-82-0	Isobutyronitrile [propanenitrile, 2-methyl-]	20,000
108-23-6	Isopropyl chloroformate [carbonochloride acid, 1-methylethyl ester]	15,000
126-98-7	Methacrylonitrile [2-propenenitrile, 2-methyl-]	10,000
74-87-3	Methyl chloride [methane, chloro-]	10,000
79-22-1	Methyl chloroformate [carbonochloridic acid, methylester]	5,000
60-34-4	Methyl hydrazine [hydrazine, methyl-]	15,000
624-83-9	Methyl isocyanate [methane, isocyanato-]	10,000
74-93-1	Methyl mercaptan [methanethiol]	10,000
556-64-9	Methyl thiocyanate [thiocyanic acid, methyl ester]	20,000
75-79-6	Methyltrichlorosilane [silane, trichloromethyl-]	5,000
13463-39-3	Nickel carbonyl	1,000
7697-37-2	Nitric acid (conc. 80% or greater)	15,000
10102-43-9	Nitric oxide [nitrogen oxide (NO)]	10,000
8014-95-7	Oleum (fuming sulfuric acid) [sulfuric acid, mixture with sulfur trioxide]	10,000
79-21-0	Peracetic acid [ethaneperoxoic acid]	10,000
594-42-3	Perchloromethylmercaptan [methanesulfenyl chloride, trichloro-]	10,000
75-44-5	Phosgene [carbonic dichloride]	500
7803-51-2	Phosphine	5,000
10025-87-3	Phosphorus oxychloride [phosphoryl chloride]	5,000
7719-12-2	Phosphorus trichloride [phosphorous trichloride]	15,000
110-89-4	Piperidine	15,000
107-12-0	Propionitrile [propanenitrile]	10,000
109-61-5	Propyl chloroformate [carbonochloridic acid, propylester]	15,000
75-55-8	Propyleneimine [aziridine, 2-methyl-]	10,000
75-56-9	Propylene oxide [oxirane, methyl-]	10,000
7446-09-5	Sulfur dioxide (anhydrous)	5,000
7783-60-0	Sulfur tetrafluoride [sulfur fluoride (SF_4), (T-4)-]	2,500
7446-11-9	Sulfur trioxide	10,000
75-74-1	Tetramethyllead [plumbane, tetramethyl-]	10,000
509-14-8	Tetranitromethane [methane, tetranitro-]	10,000
7750-45-0	Titanium tetrachloride [titanium chloride ($TiCl_4$)(T-4)-]	2,500
584-84-9	Toluene 2,4-diisocyanate [benzene, 2,4-diisocyanato-1-methyl-]	10,000
91-08-7	Toluene 2,6-diisocyanate [benzene, 1,3-diisocyanato-2-methyl-]	10,000
26471-62-5	Toluene diisocyanate (unspecified isomer) [benzene, 1,3-diisocyanatomethyl-]	10,000
75-77-4	Trimethylchlorosilane [silane, chlorotrimethyl-]	10,000
108-05-4	Vinyl acetate monomer [acetic acid ethenyl ester]	15,000

Note that these two definitions are not necessarily the same. A catastrophic release is not necessarily the worst-case release. Nor is the worst-case release always the release of the largest quantity. For example, a release near the property line—say, from a truck filling depot—of a quantity less than would be released by the largest process container at the main processing area at a facility could have a more significant off-site impact, including a greater distance to an off-site endpoint. An *endpoint* is defined in the regulations as a concentration of concern by EPA, the distance to the endpoint being the calculated or modeled distance from the property line to a point at which concentrations are projected to drop below a stated value. Table 2.3 lists the endpoint concentrations in the current RMP regulations. Note that the endpoint number is not a number that is necessarily potentially lethal or immediately dangerous to life and health (IDLH). The endpoints are, rather, fractions of the IDLH numbers. Typically (for chlorine, for example), the endpoint number is approximately one-third of the IDLH number. For explosives and flammable substances, EPA has defined the endpoint as follows: For explosive listed substances the endpoint is an overpressure from an explosion of 1 psi. For flammable substances the endpoint for heat from a fire is $5 kW/m^2$ for 40 seconds or the lower flammability limit.

2.1.2.1 The Prevention Program.
The final RMP regulations require facilities with more than a threshold quantity of a regulated substance to prepare a prevention program. The key elements of the prevention program include:

- Hazard review
- Process safety information
- Standard operating procedures (SOPs)
- Training
- Maintenance procedures
- Incident investigation

These elements of the prevention program will be discussed in the following sections.

2.1.2.2 Process Hazard Review.
The process hazard review includes a review of all processes with more than a threshold quantity of a regulated substance to determine the associated risks of operating the process, to identify potential areas for equipment malfunction, and to indicate where human error opportunities exist. It also encompasses identifying safeguards that may be instituted to reduce the potential for an accidental release, and detection and monitoring opportunities for releases that may occur. The process hazard review must be updated every 5 years.

2.1.2.3 Process Safety Information.
Process Safety Information must be based on up-to-date information. In many instances, facilities rely on material

TABLE 2.2   Threshold Quantities of Regulated Flammable Substances
for Accidental Release Prevention

CAS no.	Flammable substance	Threshold, lb
75-07-0	Acetaldehyde	10,000
74-86-2	Acetylene	10,000
598-73-2	Bromotrifluoroethylene	10,000
106-99-0	1,3-Butadiene	10,000
106-97-8	Butane	10,000
25167-67-3	Butene	10,000
590-18-1	2-Butene-cis	10,000
624-64-6	2-Butene-trans	10,000
106-98-9	1-Butene	10,000
107-01-7	2-Butene	10,000
463-58-1	Carbon oxysulfide	10,000
7791-21-1	Chlorine monoxide	10,000
590-21-6	1-Chloropropylene	10,000
557-98-2	2-Chloropylene	10,000
460-19-5	Cyanogen	10,000
75-19-4	Cyclopropane	10,000
4109-96-0	Dichlorosilane	10,000
75-37-6	Difluoroethane	10,000
124-40-3	Dimethylamine	10,000
463-82-1	2,2-Dimethylpropane	10,000
74-84-0	Ethane	10,000
107-00-6	Ethyl acetylene	10,000
75-04-7	Ethylamine	10,000
75-00-3	Ethyl chloride	10,000
74-85-1	Ethylene	10,000
60-29-7	Ethyl ether	10,000
109-95-5	Ethyl nitrite	10,000
75-08-1	Ethyl mercaptan	10,000
1333-74-0	Hydrogen	10,000
75-28-5	Isobutane	10,000
78-78-4	Isopentane	10,000
78-79-5	Isoprene	10,000
75-31-0	Isopropylamine	10,000
75-29-6	Isopropyl chloride	10,000
74-82-8	Methane	10,000
74-89-5	Methylamine	10,000
563-46-2	2-Methyl-1-butene	10,000
563-45-1	3-Methyl-1-butene	10,000
115-10-6	Methyl ether	10,000
107-31-3	Methyl formate	10,000
115-11-7	2-Methylpropene	10,000
504-60-9	1,3-Pentadiene	10,000
109-66-0	Pentane	10,000
109-67-1	1-Pentene	10,000
646-04-8	2-Pentene, (E)-	10,000

TABLE 2.2 Threshold Quantities of Regulated Flammable Substances for Accidental Release Prevention (*Continued*)

CAS no.	Flammable substance	Threshold, lb
627-20-3	2-Pentene, (Z)-	10,000
463-49-0	Propadiene	10,000
74-98-6	Propane	10,000
115-07-1	Propylene	10,000
74-99-7	Propyne	10,000
7803-62-5	Silane	10,000
116-14-3	Tetrafluoroethylene	10,000
75-76-3	Tetramethylsilane	10,000
10025-78-2	Trichlorosilane	10,000
79-38-9	Trifluorochloroethylene	10,000
75-50-3	Trimethylamine	10,000
689-97-4	Vinyl acetylene	10,000
75-01-4	Vinyl chloride	10,000
109-92-2	Vinyl ethyl ether	10,000
75-02-5	Vinyl fluoride	10,000
75-35-4	Vinylidene chloride	10,000
75-38-7	Vinylidene fluoride	10,000
107-25-5	Vinyl methyl ether	10,000

safety data sheets (MSDSs) provided by vendors of the hazardous material. Please note that these sheets are often not up to date. Therefore, it is important, even when these are used as a starting point, to contact the supplier of the material and to obtain the most up-to-date MSDS available. Manufacturers are starting to post MSDSs on the Internet's World Wide Web (WWW), thereby making obtaining the most current versions easier.

Information required about chemical safety includes the physical and chemical hazards of the materials, the process technology (for example, the process chemistry and normal range of parameter such as temperatures and pressures), and the equipment used in the process, including its specification and design aspects.

Drawings showing the layout of the process are also required. For significant processes, these should include piping and instrumentation diagram (P&ID) drawings; however, for simple processes, line drawings showing just the basic components of the process and their interrelationship may be sufficient. For processes that are not currently covered by the RMP regulations (either new processes or those that will use regulated substances in greater than threshold quantities for the first time), material and energy balances are required. These are not required for existing processes that may be overthreshold at the time the initial RMP is prepared.

2.1.2.4 Standard Operating Procedures. Most facilities have a standard operating procedure, or the somewhat similar batch sheet, for each regulated process. The standard operating procedures that facilities often have, however, are not fully

TABLE 2.3 Table of Toxic Endpoints

CAS no.	Toxic	Endpoint, mg/L
107-02-8	Acrolein [2-propenal]	0.0011
107-13-1	Acrylonitrile [2-propenenitrile]	0.076
814-68-6	Acrylyl chloride [2-propenoyl chloride]	0.00090
107-18-6	Allyl alcohol [2-propen-1-ol]	0.036
107-11-9	Allylamine [2-propen-1-amine]	0.0032
7664-41-7	Ammonia (anhydrous)	0.14
7664-41-7	Ammonia (conc. 20% or greater)	0.14
7784-34-1	Arsenous trichloride	0.010
7784-42-1	Arsine	0.0019
10294-34-5	Boron trichloride [borane, trichloro-]	0.010
7637-07-2	Boron trifluoride [borane, trifluoro-]	0.028
353-42-4	Boron trifluoride compound with methyl ether (1:1) [boron, trifluoro[oxybis[methane]]-, T-4	0.023
7726-95-6	Bromine	0.0065
75-15-0	Carbon disulfide	0.16
7782-50-5	Chlorine	0.0087
10049-04-4	Chlorine dioxide [chlorine oxide (ClO$_2$)]	0.0028
67-66-3	Chloroform [methane, trichloro-]	0.49
542-88-1	Chloromethyl ether [methane, oxybis[chloro-]	0.00025
107-30-2	Chloromethyl methyl ether [methane, chloromethoxy-]	0.0018
4170-30-3	Crotonaldehyde [2-butenal]	0.029
123-73-9	Crotonaldehyde, (E)-, [2-butenal, (E)-]	0.029
506-77-4	Cyanogen chloride	0.030
108-91-8	Cyclohexylamine [cyclohexanamine]	0.16
19287-45-7	Diborane	0.0011
75-78-5	Dimethyldichlorosilane [silane, dichlorodimethyl-]	0.026
57-14-7	1,1-Dimethylhydrazine [hydrazine, 1,1-dimethyl-]	0.012
106-89-8	Epichlorohydrin [oxirane, (chloromethyl)-]	0.076
107-15-3	Ethylenediamine [1,2-ethanediamine]	0.49
151-56-4	Ethyleneimine [aziridine]	0.018
75-21-8	Ethylene oxide [oxirane]	0.090
7782-41-4	Fluorine	0.0039
50-00-0	Formaldehyde (solution)	0.012
110-00-9	Furan	0.0012
302-01-2	Hydrazine	0.011
7647-01-0	Hydrochloric acid (conc. 30% or greater)	0.030
74-90-8	Hydrocyanic acid	0.011
7647-01-0	Hydrogen chloride (anhydrous) [hydrochloric acid]	0.030
7664-39-3	Hydrogen fluoride/Hydrofluoric acid (conc. 50% or greater) [hydrofluoric acid]	0.016
7783-07-5	Hydrogen selenide	0.00066
7783-06-4	Hydrogen sulfide	0.042
13463-40-6	Iron, pentacarbonyl-[iron carbonyl (Fe(CO)$_5$), (TB-5-11)-]	0.00044
78-82-0	Isobutyronitrile [propanenitrile, 2-methyl-]	0.14

TABLE 2.3 Table of Toxic Endpoints (*Continued*)

CAS no.	Toxic	Endpoint, mg/L
108-23-6	Isopropyl chloroformate [carbonochloride acid, 1-methylethyl ester]	0.10
126-98-7	Methacrylonitrile [2-propenenitrile, 2-methyl-]	0.0027
74-87-3	Methyl chloride [methane, chloro-]	0.82
79-22-1	Methyl chloroformate [carbonochloridic acid, methylester]	0.0019
60-34-4	Methyl hydrazine [hydrazine, methyl-]	0.0094
624-83-9	Methyl isocyanate [methane, isocyanato-]	0.0012
74-93-1	Methyl mercaptan [methanethiol]	0.049
556-64-9	Methyl thiocyanate [thiocyanic acid, methyl ester]	0.085
75-79-6	Methyltrichlorosilane [silane, trichloromethyl-]	0.018
13463-39-3	Nickel carbonyl	0.00067
7697-37-2	Nitric acid (conc. 80% or greater)	0.026
10102-43-9	Nitric oxide [nitrogen oxide (NO)]	0.031
8014-95-7	Oleum (fuming sulfuric acid) [sulfuric acid, mixture with sulfur trioxide]	0.010
79-21-0	Peracetic acid [ethaneperoxoic acid]	0.0045
594-42-3	Perchloromethylmercaptan [methanesulfenyl chloride, trichloro-]	0.0076
75-44-5	Phosgene [carbonic dichloride]	0.00081
7803-51-2	Phosphine	0.0035
10025-87-3	Phosphorus oxychloride [phosphoryl chloride]	0.0030
7719-12-2	Phosphorus trichloride [phosphorous trichloride]	0.028
110-89-4	Piperidine	0.022
107-12-0	Propionitrile [propanenitrile]	0.0037
109-61-5	Propyl chloroformate [carbonochloridic acid, propylester]	0.010
75-55-8	Propyleneimine [aziridine, 2-methyl-]	0.12
75-56-9	Propylene oxide [oxirane, methyl-]	0.59
7446-09-5	Sulfur dioxide (anhydrous)	0.0078
7783-60-0	Sulfur tetrafluoride [sulfur fluoride (SF_4), (T-4)-]	0.0092
7446-11-9	Sulfur trioxide	0.010
75-74-1	Tetramethyllead [plumbane, tetramethyl-]	0.0040
509-14-8	Tetranitromethane [methane, tetranitro-]	0.0040
7750-45-0	Titanium tetrachloride [titanium chloride ($TiCl_4$) (T-4)-]	0.020
584-84-9	Toluene 2,4-diisocyanate [benzene, 2,4-diisocyanato-1-methyl-]	0.0070
91-08-7	Toluene 2,6-diisocyanate [benzene, 1,3-diisocyanato-2-methyl-]	0.0070
26471-62-5	Toluene diisocyanate (unspecified isomer) [benzene, 1,3-diisocyanatomethyl-]	0.0070
75-77-4	Trimethylchlorosilane [silane, chlorotrimethyl-]	0.050
108-05-4	Vinyl acetate monomer [acetic acid ethenyl ester]	0.26

adequate to comply with RMP regulations. For example, the standard operating procedures at facilities normally do not identify the ranges of parameters, which could lead to accidental releases. The objective of the RMP program is that the standard operating procedures be enhanced to provide normal ranges of operating parameters such as temperature, pressure, and flow and to include the recommended responses to variations to these parameter values. For example, the standard range of temperatures for a process might be that from an initial ambient temperature, say 70 to 80°F, to a temperature produced by steady-state heating or an exothermic reaction, say 170°F. If a continued temperature rise to 200°F or more could indicate a runaway process or an impending emergency release, the standard operating procedure should identify this, should give acceptable deviations from the 170°F norm, and should identify what appropriate responses should be undertaken, depending on the extent of the deviation. For example, at 190°F the operator might be requested to take some corrective action to reduce the temperature. At 200°F the operator may be requested to take additional action, such as to notify an RMP manager or a first responder that there is a problem at the facility that could lead to an accidental release.

The advantage of having standard operating procedures is that they provide standard ranges of parameters to give operators advance warning that a dangerous condition may be developing so that they can notify first responders and, most important, take any required corrective action to reduce the possibility of a release.

Standard operating procedures are required to cover each operating phase of the process including start-up, normal operation, and shutdown. The level of detail of standard operating procedures needs to be appropriate for the process and for the phase of the process.

2.1.2.5 Training. Training required under the RMP regulations exceeds the training required under the OSHA PSM regulations. This is because, in addition to the training required under the OSHA PSM (which for areas that overlap is adequate for RMP purposes), operators must be trained in the standard operating procedures and in coordination with the off-site community in ways to minimize potential impact of an emergency release. They must be trained, in effect, to assure the effectiveness of the SOPs. In addition, personnel must be trained in the RMP process itself, including the management of change procedures, maintenance procedures, and other aspects. To the extent that these parallel PSM, and in most instances they do, PSM training would be adequate. However, in addition to the PSM requirements, it is required under the RMP program that management personnel who have responsibility for overseeing the RMP program also be trained in their responsibilities under the RMP requirements.

Refresher training is required at least every 3 years, and new employees must be trained before being assigned to processes with more than a threshold quantity of a regulated substance. Any modifications to the process would trigger new training requirements so that all personnel responsible for operation and management of the cover process understand the potential safety implications of the newer updated equipment. Finally, all training must be documented to show that the personnel trained have understood the training. There are two ways that understanding can be demonstrated. The first, through a written test, is the preferred method; the second, through observation of the employees in the operation of their duties as covered by the RMP program, is a fall-back approach in which emergency conditions are simulated and the employees' responses to the simulated conditions are observed.

2.1.2.6 Compliance Audits. The RMP regulations require that a compliance safety audit be conducted every 3 years. This is a check by management that the facility is operating within compliance of the RMP regulations and it is operating safely. It is also intended to be a confirmation that the SOPs are being followed and that they are appropriate. Finally, the compliance audit must document management's response to the findings. This potentially puts the environmental manager—if that is who conducts the audit and who is responsible for implementation of the program— in a difficult position: Management may not want to take the appropriate actions that the auditor identifies as being necessary. This can occur even if the auditor is a corporate environmental person or an environmental or RMP person from another facility, chosen to avoid potential conflict of interest and to ensure objectivity in assessing inadequacies in the program. The auditor is put in the difficult position of needing to document that management has refused to carry out audit-suggested activities to assure continued compliance with the RMP regulations. In these instances, the auditor has no choice but to document management's response and to accept the consequences that may affect the auditor's employment.

2.1.2.7 Accident Investigations. Any significant release must be investigated, and the investigation must start within 48 hours of the release. A *significant release* is defined as one that affected off-site receptors or that "could have" affected off-site receptors. The investigation must document the details of the event, indicate what was learned from the investigation, and indicate what was done to resolve problems that led to the release. Records must be retained for 5 years. Accident investigations are required to start quickly and be documented thoroughly, even though EPA has funded an accident investigation board, which it jointly administers with OSHA. The reason for this is that an accidental release may likely involve hazardous or toxic chemicals, since we are assuming it is a regulated process under the RMP program. Therefore, there is a potential that these potentially toxic substances could contaminate soil or groundwater, and there is an incentive to clean up the aftermath of an accident quickly. After an area is decontaminated and cleaned up, evidence related to the accident may be lost forever.

2.1.2.8 Emergency Response Plan. The RMP regulations require the facility to put in place an emergency response plan. While OSHA PSM requires a facility to have an emergency response plan for its employees in order to respond to accidental conditions to contain a release on site, the RMP regulations also require the employer to have an emergency response plan that includes interfacing with the community and potential off-site first responders. Note that in many instances, protective procedures, such as recommending that off-site residents stay in their houses with the windows closed, are often preferable to evacuation. Sheltering in a closed-up house can reduce exposure by as much as 80 percent, and the time required to notify the community of a hazardous release and to conduct an orderly, safe evacuation in the event may not be adequate anyway. Once a facility is aware of an accident and off-site responders have been notified, an initial exposure of the community may already be in progress. For this reason, a quick notification can be made and, assuming the community has been involved in an educational program by the facility, its members will know what protective measures may be required of them in the event of an accident. This is the best insurance that off-site impacts will be minimized.

2.1.2.9 Program Levels of RMP. There are three levels of compliance and plan complexity under the RMP regulations. Program 1 is reserved for facilities that have had no accidental releases within the last 5 years affecting any off-site receptor. Further, Program 1 facilities are facilities that do not have the potential to have a worst-case release that would affect an off-site receptor. Program 1 requires the least level of detail in its RMP plan. Program levels 2 and 3 must comply with all hazard assessment requirements including:

- Off-site consequence analysis
- Worst-case scenario analysis
- Alternative release scenario analysis
- Off-site impacts to population and environmental receptors
- Review and update of documentation
- Submission of a 5-year accident history

Program 3 facilities are those that fall in certain North American Industrial Classification Scheme (NAICS) codes as noted in Table 2.4. Program 2 facilities are all those that do not qualify for Program 1 and whose NAICS codes do not fall within the list of Table 2.4.

An important aspect of the RMP regulations is the estimation of off-site impacts from an accidental release. For Program 2 and 3 facilities, the final regulations require the calculation of off-site impacts. These can be done by using the EPA lookup tables, which are provided in EPA's *Offsite Consequence Analysis Guidance.* These lookup tables are based on very conservative assumptions, including:

- Realistic but fixed meteorological conditions (warmest day)
- Continuous release
- Ten-minute release time
- Passive mitigation systems credit is given if it can be shown to survive the event.

TABLE 2.4 Program Level 3 (NAICS Codes)

Code	Facility
3221	Pulp mills only
325181	Alkalies and chlorine
325188	Basic inorganic chemical manufacturing
325211	Plastics and resins
325192	Other; covers cyclic crudes and intermediates
32511	Petrochemical manufacturing
325311	Nitrogen fertilizer
32532	Pesticide and agricultural chemicals
32411	Petroleum refineries
325119	All other basic organic chemical manufacturing

Defining the relevant quantity of regulated substances for which to calculate potential off-site impacts requires that you look at the maximum quantity in an individual vessel or pipe inventory, considering administrative controls. A tank capable of holding more than a threshold quantity can be limited by administrative control to less than a regulated quantity. Therefore an RMP plan does not have to be filed for the tank. In the event that the facility needs to hold more than a regulated quantity in the tank, the RMP would need to be prepared and filed before the tank is filled above the threshold quantity.

Air permits for tanks holding regulated substances must be modified to reflect the administrative controls to keep a tank out of the RMP regulations. If the inventory then has to be increased above the threshold quantity, the air permit would need to be revised to reflect the added quantity in the tank.

2.2 NEW SOURCE REVIEW (NSR) PERMITTING

2.2.1 NSR Permitting

The New Source Review (NSR) permitting program applies to nearly all new air pollution sources and modifications. Generally, there are a few sources whose size (in tons per year of emissions) is not considered large enough by the U.S. EPA to warrant any type of approval to construct or modify. These are usually termed "exempt" sources. A few states have an intermediate category, which requires the source or modification to be "registered," but not to obtain a permit. Most sources and modifications, however, are required to obtain at least a *minor* source/modification permit ("minor permit"). New or modified sources that exceed thresholds specified in the New Source Review regulations must obtain *major* source/modification permits ("major permits").

The entire United States is classified in a number of different ways under the NSR program. One of the key classifications is based on the six criteria pollutants: PM-10 (particulate matter less than 10 μm), SO_2, NO_2, lead, CO, and ozone [volatile organic compounds (VOCs) and, in certain areas, NO_x are precursors for ozone and are the pollutants actually regulated].[1] For each pollutant (and each averaging time for each pollutant), an area is classified as attainment, nonattainment, or unclassifiable. *Attainment* means that the area is meeting that particular National Ambient Air Quality Standard (NAAQS); *nonattainment* means that it is not and that the area has been designated as a nonattainment area. Areas where there is a lack of information are presumed to be in attainment and are treated as if they are for NSR purposes. The attainment status for all areas is listed in 40 CFR 81.

Major NSR permits in areas that are in attainment with a National Ambient Air Quality Standard (NAAQS) or in areas that are "unclassifiable" are issued under the Prevention of Significant Deterioration (PSD) program. Major NSR permits in areas that are exceeding a NAAQS ("nonattainment" areas) are issued under the area's major Nonattainment NSR (NANSR) program.

Major permits require the application of the best available/lowest achievable emission rate, good control technology, and assurance that the air pollution impacts of the proposed project are acceptable. Minor permits require very little other than enforceable conditions that ensure the project's minor permit status.[2] As a result of these requirements, it is much more difficult to obtain a major permit than it is to obtain a minor one. Therefore, applicability (whether a source or modification is major or minor) is a key determination in the permitting process. Agencies will closely review minor NSR[3] permit applicability determinations to ensure that the analysis is correct.

2.2.1.1 Prevention of Significant Deterioration (PSD) Permitting.

New major sources or modifications required to obtain a PSD permit must apply Best Available Control Technology (BACT) to applicable emissions units and conduct one or more impact analyses to ensure that the impact of the proposed new source or modification is acceptable. Sources must conduct an additional-impacts analysis to determine the impact on soils, vegetation, and visibility. They may also be required to conduct analyses to determine whether the proposed project would cause or contribute to exceedances of a NAAQS or PSD increment. Finally, the source may have to conduct a Class I impact analysis to determine whether there will be an adverse impact on any Air Quality Related Values (AQRV). An important provision: These impact assessments may require the source to conduct 1 year of air quality and meteorological monitoring before it can submit its permit application.

---

[1]See 40 CFR Part 50 for a listing of the NAAQS and test methods.

[2]However, some states (e.g., Texas and, in some Air Pollution Control Districts, California) impose more stringent requirements on minor permits, such as best available technology or impact analyses.

[3]Minor NSR sources and modifications are described as either "synthetic" or "natural." As the term implies, a synthetic minor source or modification has an unrestricted potential to emit (PTE) high enough to be classified as major. However, the source has agreed to accept restrictions on emission rates, capacity, material or fuel composition, operating hours, etc. that synthetically reduce its PTE to minor status. A natural minor source or modification, on the other hand, needs no restrictions to be minor; its design, capacity, or some other immutable characteristic renders its PTE below the applicable threshold.

2.2.1.2 *Nonattainment Area New Source Review (NANSR) Permitting.* New major sources or modifications must apply lowest achievable emission rate (LAER), obtain offsets to balance out the proposed emissions increase, certify that all other sources in that state under common control are in compliance with the CAA (or under a compliance schedule), and conduct alternative site, size, process, and control system analyses. The offsets required increase with the severity of nonattainment.

2.2.1.2.1 Determining Applicability. Each of the regulatory requirements has different applicability criteria. There are also certain similarities, since many of the programs have borrowed concepts from each other. Often, however, the differences are important enough to affect applicability. Therefore, stationary sources must assess applicability separately for each regulatory program.

An applicability determination, as discussed in this section, is the process of determining whether a preconstruction review should be conducted of, and a permit issued to, a proposed new source or a modification of an existing source by the reviewing authority, pursuant to the appropriate NSR requirements.

Most new sources and modifications to existing sources must obtain the preconstruction permits mandated under the NSR program. The only exceptions are the exemptions allowed in the permitting agency's regulations. New sources and modifications that are not subject to major NSR (neither PSD nor NANSR) for a particular pollutant are still required to obtain a minor NSR permit unless specifically exempted by the permitting agency regulations. Note that a source or modification may be PSD for one pollutant, NANSR for another, minor NSR for a third, and exempt altogether from NSR permitting for a fourth. Each pollutant is assessed separately.

2.2.1.3 *Prevention of Significant Deterioration (PSD).* PSD applies to certain new sources and modifications at stationary sources. The specific actions subject to PSD are:

New sources with a potential to emit (PTE) of

- 100 tons per year (tpy) or more of any PSD-regulated pollutant if the source is one of the source categories listed in the PSD regulations
- 250 tpy or more of any PSD-regulated pollutant if the source is not one of the listed source categories

Any modification to a minor existing stationary source if the modification itself will have an emissions increase in a PSD-regulated pollutant that is equal to or greater than 100 or 250 tpy, for listed and nonlisted source categories, respectively

Any modification to a major existing stationary source if the net emissions increase from the modification equals or exceeds the significance level for that pollutant

2.2.1.4 *Definition of New Stationary Source ("Source").* A *stationary source* is defined as a building, structure, facility, or installation, that emits or may emit "any pollutant subject to regulation under the act." The source includes all emission units of the same industrial grouping, on contiguous or adjacent lands, and under common control.

The first step in defining the source is to determine what constitutes the site. This will consist of the land on which the proposed source or modification is located and all contiguous and adjacent land owned by the applicant. *Contiguous* means touching, so any other parcels of property whose boundaries touch the core parcel, even if only at a single point, are contiguous. *Adjacent* land is more difficult to identify. EPA considers any land that is "nearby" to be adjacent. Nearby is not defined, but is instead a case-by-case determination. Land separated by just a river, highway, railroad, pipeline, or easement is nearly *always* considered nearby. The factors used for less clear-cut cases are:

- Distance
- Interdependence
- Air impacts

The more distant two parcels of land, the less likely they are to be considered nearby. The 1980 PSD preamble stated directly that parcels 20 mi apart are too far to be considered adjacent.[4] EPA NSR *Guidance* provides an example of a "source" with two buildings 4300 ft apart, agreed to consider them one source, but almost entirely because they, in effect, acted as one source (auto chassis were partially assembled in one building and completed in the other). The distance appeared to be a factor against combining the buildings. The tenor of the Guidance is that at any distance over a mile, there should be very strong interdependence between the two facilities, in addition to an air impact overlap.[5]

The second step is to determine which emission units at the site are under common control. An agency will usually presume that all of the units at the site are under common control. Facilities wishing to be considered separate because of lack of common control will have to demonstrate this to the reviewing agency.

Except for the military, EPA considers a parent company to be a single owner; no matter how disparate or independent the divisions of a conglomerate or international company, it is considered a single owner. Department of Defense facilities, on the other hand, are considered separate sources if they are operated by different branches of the military (Army, Air Force, etc.).

---

[4]There are a few aberrant cases where an agency has determined two parcels to be adjacent that are more than 20 mi apart, but these determinations are inappropriate because they directly contradict the preamble.

[5]Without an air impact overlap, it does not appear to make sense to combine two facilities into one "source." At least one state makes the air impact test a subjective one by modeling the emissions from each source and defining a concentration "contour" line that represents half of the significant impact level of that pollutant/averaging time. If the two contour lines overlap, the two facilities are considered to have overlapping air impacts and, according to the state rule, would be aggregated as one source. The manner in which VOC and other pollutants without a significant impact level would be evaluated is left open. It would make more sense to have the overlapping air impacts a requirement for combining two facilities, but have the final decision rest on an overlapping air impact *plus* interdependence or very close proximity.

If there is more than one owner but the agency feels that at least some of the units with different owners are under common control, or if the owner disputes the presumption that all of the units it owns are under common control, the determination becomes more complicated. The two factors used to make these determinations are whether there is a common operator and whether a leasing or contractual agreement in effect establishes common control. These are generally case-by-case determinations, but there is some guidance.

The first guidance deals with how to interpret "control" when a company has a board of directors or voting stock. Generally, the majority of directors is considered in control, but if a director has the authority to veto purchase of control equipment, that person may be considered in control. In addition, if there is a 50-50 partnership for a facility and the facility is located at the same site as one of the partners (partner A), then partner A is considered to be in control of the new facility. The other guidance is for leasing or contractual agreements, where the key factor is who obtains the air permit and is responsible for violations. This determination often involves detailed scrutiny of the agreement by attorneys before a determination is reached.

The third step is to determine whether all of the units at the site under common control belong to the same "major group" as defined in the *Standard Industrial Classification (SIC) Manual*. This means having the same two-digit SIC code. Emissions units with a different two-digit SIC code are considered separate sources, unless they are considered support facilities. This means that there could be two or more sources at a site, even though the emissions units are on contiguous or adjacent land and under common control. Determining the number of separate sources from SIC codes is based on the number of primary activities at the site. Each primary activity that has a different two-digit SIC code is a separate source.

A primary activity produces or distributes a principal product or group of products, or renders a service. This means that a product is created for sale outside the source, or that the source provides a service to outside sources, such as storing or transporting a product, providing maintenance, etc. The difference between a primary activity and a support facility is that a support facility conveys, stores, or otherwise assists in the production of the principal product. The distinction is relatively straightforward unless a "product" is marketed both on and off site. In such cases, the language regarding primary activities in the 1980 preamble indicates that as long as more than a nominal portion of a product is marketed off site, that product is a separate primary activity.[6]

---

[6]It should be noted that some agencies have adopted (by policy or, in one case, by regulation) a much more stringent (and much less logical) concept of support facility. The policy makes any activity a support facility if 50 percent or more of its product is used on site by another primary activity. This interpretation appears to originate from the manner in which true support facility emissions are assigned if the support facility supports more than one primary activity.

EXAMPLE 1. *A boiler at a chemical plant provides the steam used in the production of the chemical. None of the steam is sold off site. The boiler is a support facility for the primary activity, which is chemical production.*

EXAMPLE 2. *A company operates an adhesives manufacturing plant and a plywood manufacturing plant at the same site. A significant portion of the adhesives (in this case, 30 percent) is sold off site and the remainder is used (through intracompany sales) in the manufacture of plywood at the site. In this case, there are two primary activities: adhesives manufacture and plywood manufacture. They have two different SIC codes, so they are separate stationary sources.*

When a support facility supports more than one primary activity, all of the emissions from the support activity are assigned to the primary activity that receives the majority of the support. The emissions are *not* proportioned. Generally, this involves estimating the output of the support facility to each activity.

EXAMPLE 3. *A boiler provides steam to both the adhesives plant and the plywood plant in Example 2. Approximately 70 percent of the steam is used in manufacturing plywood. Therefore, all of the boiler emissions are assigned to the plywood plant for applicability purposes.*

Step 4 is to determine which "PSD-regulated" pollutants are (or could be) emitted by the source. The term "pollutant subject to regulation under the act" was literally true prior to enactment of the 1990 Clean Air Act Amendments (CAAA). All of the pollutants regulated under the CAA from stationary sources[7] were also regulated under the PSD program, including the criteria pollutants and the pollutants regulated under New Source Performance Standards (NSPSs) and NESHAPs. The CAAA, however, prohibited EPA from regulating hazardous air pollutants (HAPs) under the PSD program.[8] This became effective November 15, 1990, the date of enactment of the CAAA. The reference to HAP is to the list of 189 (now 188)[9] hazardous air pollutants in §112(b) of the Act. EPA is considering extending this exclusion to all the pollutants listed under §112(r)[10] and, in effect, is reviewing PSD applicability as though such an exclusion is already in place.[11] The result is that the pollutants regulated under the PSD program include the criteria pollutants plus all other non-HAP pollutants regulated by non–Sec. 112 stationary source programs. The pollutants currently regulated are listed in Table 2.5.

---

[7] Hydrocarbons (HCs) were regulated as a criteria pollutant from 1971 to 1983 as a means of regulating ozone. VOC replaced hydrocarbons as an ozone precursor. Title II (mobile source) regulations still treat HC as a pollutant, but it is no longer treated as "regulated under the Act" by the NSR program, despite the fact that it is regulated (if only for mobile sources).

[8] NANSR regulates only criteria pollutants, so was unaffected by the 1990 CAAA.

[9] Caprolactan was removed from the list in response to a petition.

[10] NSR Reform proposal (June 23, 1996).

[11] To date, there has been no action by EPA to require PSD review for sources or modifications for any pollutants that are regulated only through §112(r).

Step 5 is to determine the potential to emit (PTE) of the new source. PTE, in tons per year, is calculated by taking the maximum emissions rate (for the worst-case fuel or material or product) times the maximum capacity of the unit, assuming operation 8760 hours per year. PTE is intended to reflect worst-case emissions, but based on the unit operating normally and in the manner in which it was intended to be operated. Emissions that could occur during startup, shutdown, malfunctions, and upsets are not included in this determination.[12] Limitations on emissions rate, production or use rates, hours of operation, work practices, and any other parameter that affects emissions can be taken into account in determining PTE as long as the limitations are enforceable as a practical matter.[13]

Fugitive emissions are counted in determining PTE if the source is classified as one of the 28 listed source categories or belongs to one of the source categories for which an NSPS or NESHAP was established prior to August 7, 1980. The source itself does not have to actually be regulated by an NSPS or NESHAP; it only has to belong to that category. For example, an automobile coating line constructed

TABLE 2.5 Pollutants Regulated

Federally regulated PSD pollutant	Significant emission rates, tons per year*
Particulate matter (PM)	25
SO_2	40
NO_2	40
VOC (ozone)	40
CO	100
Lead (Pb)(elemental)	0.6
Vinyl chloride	1
Mercury	0.1
Fluorides	3
Sulfuric acid mist	7
Total reduced sulfur (TRS) compounds including H_2S	10
Hydrogen sulfide (H_2S)	10
Municipal waste combustor (MWC) acid gases	36 mg/year
PM–10	15
MWC metals	14 mg/year
MWC organics	$3.5 \ 3 \ 10^{26}$ mg/year
Ozone-depleting substances (ODS)	Any increase
Municipal solid waste landfill emissions plus, for sources within 10 km of a class I area, any impact on that area ≥ 1 μmg/m^3, 24 average	50

*Unless otherwise noted.

---

[12] The permit will indicate what limits, if any, apply during these conditions.

[13] The regulations state that the limitations must be *federally* enforceable, but recent court decisions have vacated that requirement for PTE and netting purposes (but not for offsets).

prior to the effective date of the NSPS would be required to count fugitive emissions because it belongs to a source category regulated by the NSPS for automobile and light-duty truck coating lines, even though it is not subject to the NSPS.

There is some confusion about certain NSPSs that logically are regulating emissions units rather than "source categories." An example is boilers with a heat input capacity between 10 and 100 million Btu/h. Many sources have such units, but their "category" is not "boilers." Although some agencies require that fugitive emissions be counted for the *portion* of the source that the NSPS covers, it is more logical to use the primary activity of the source to determine whether the source falls within one of the listed categories. A can manufacturing plant, for example, may include a boiler, but the primary activity is the manufacture of cans. Can manufacturing is not one of the listed source categories, so fugitive emissions would not be counted in determining PTE.

A source that does not have to count fugitive emissions must still persuade the reviewing agency that certain emissions are, in fact, fugitive. The regulatory definition is that fugitive emissions could not "reasonably" pass through a stack, duct, vent, or equivalent opening. These determinations are inconsistent among agencies and tend to change over time.

Step 6 is to determine whether the new source is major. This is determined by comparing the source's PTE to the appropriate major source threshold level. A source belonging to one of the source categories (based on its primary activity) listed in 40 CFR 52.21(b)(1)(i)(a) is major if its PTE is 100 tpy or more for any PSD-regulated pollutant, regardless of the attainment status of the area (see Table 2.6). All other (nonlisted) sources are major if any PSD-regulated pollutant (regardless of the attainment status of the area) has a PTE of 250 tpy or more.

A major new source must obtain a PSD permit prior to construction. The pollutants subject to PSD review consist of all the PSD-regulated pollutants that the source has PTE in significant quantities *and* for which the area is *not* classified as nonattainment. This means that the source could be major only for a pollutant for which the area is nonattainment, but still be subject to PSD review for other pollutants that are emitted in significant quantities.

Significance levels for each pollutant are provided in Table 2.7. Pollutants with a PTE below the significance level may require a minor NSR permit, but are not subject to PSD review.

EXAMPLE 4. *A new electronics plant is proposed for an area that is attainment for all pollutants except ozone. PTE is estimated at 150 tpy VOC, 50 tpy SO$_2$, 260 tpy NO$_x$, 20 tpy PM, and 18 tpy PM-10. The source is major for PSD applicability purposes for NO$_x$, since it is above the 250-tpy threshold for nonlisted source categories. Because it is major, it is subject to PSD review for SO$_2$, NO$_x$, and PM-10. It is not subject to PSD review for PM because the PM PTE is below the 25-tpy significance threshold. VOC is not subject to PSD review, even though its PTE is significant, because it is a precursor for ozone and the area is nonattainment for ozone. Although NO$_x$ is also a precursor of ozone in nonattainment areas, the area*

TABLE 2.6   Source Categories Based on Primary Activity

Coal cleaning plants (with thermal dryers)
Kraft pulp mills
Portland cement plants
Primary zinc smelters
Iron and steel mills
Primary aluminum ore reduction plants
Primary copper smelters
Municipal incinerators capable of charging more than 250 tons of refuse per day
Hydrofluoric, sulfuric, or nitric acid plants
Petroleum refineries
Lime plants
Phosphate rock processing plants
Coke oven batteries
Sulfur recovery plants
Carbon black plants (furnace process)
Primary lead smelters
Fuel conservation plants
Sintering plants
Secondary metal production plants
Chemical process plants
Fossil-fuel boilers (or combination thereof) totaling more than 250 million Btu/h
  heat input
Petroleum storage and transfer units with a total storage capacity exceeding
  300,000 barrels
Taconite ore processing plants
Glass fiber processing plants
Charcoal production plants
Fossil fuel–fired steam electric plants of more than 250 million Btu/h heat input
Any other stationary source category that, as of August 7, 1980, is being
  regulated under Sec. 111 or 112 of the Act.

is attainment for NO_2, which makes the NO_x emissions both a PSD and a nonattainment pollutant. Note that the source is not PSD-major for VOC; nonlisted source categories have a PSD major source threshold of 250 tpy. (The source is nonattainment-major for VOC, since the nonattainment source threshold is no higher than 100 tpy for any ozone nonattainment area.)

2.2.1.4.1 "Source within a Source" Policy.   Because of the two different thresholds for a source to be PSD-major, certain source configurations could result in a "listed" source with a 100-tpy threshold escaping PSD review by locating within a source with a 250-tpy threshold. This occurs only when both the listed and the nonlisted source have a PTE between 100 and 250 tpy.

2.2.1.4.2 Secondary Emissions.   Secondary emissions are emissions that may be associated with a facility's operations, but that are not emitted from the source itself. For example, they could be emissions from transportation sources that bring

TABLE 2.7 Significant Emission Rates

Pollutant	Emision rate, tpy
Carbon monoxide	100
Nitrogen oxides	40
Sulfur dioxide	40
Ozone	40 (volatile organic compounds)
Lead	0.6

raw material or finished product to or from the operation, or they could be emissions from a related facility—for example, a company that makes automobile seats for installation at a car assembly facility. New Source Review defines "associated with" emissions as those that the facility would not emit except as a result of the construction or operation of the source being reviewed. For example, in the case of the automobile seat manufacturing company, this facility would not have been built and would not be manufacturing seats had it not been that a nearby auto assembly plant had been constructed and required the seats for installation in vehicles it produced.

It is important to note that secondary emissions are *not* included in the New Source Review applicability determination, but they are included in the impact analysis if the source is subject to major New Source Review. This means that the emissions related to the secondary source will need to be tabulated and taken into account in the impact assessment modeling. Therefore they do count toward PDS increment consumption compliance and National Ambient Air Quality Standards compliance determinations.

2.2.1.5 Modifications. New Source Review regulations define a *modification* as a physical or operational change that results in an emissions increase at the source being reviewed.

A *major* modification is a physical or operational change that results in a *significant* net emissions increase. Significant emission increases as defined under New Source Review are listed in Table 2.7.

Examples of modifications include:

A new emissions unit

Physical changes to existing units

Changes in the method of operating an emissions unit, process, or source, such as:

* Increase in capacity
* Increase in utilization
* Change in fuel or raw material

The New Source Review with respect to a modification is conducted first for the project, which may include new emission units or changes to existing units. It is done on a pollutant-by-pollutant basis for each pollutant regulated under New Source Review. Modifications generally mean going through the determination of applicability all over again for each pollutant. Each pollutant that has a significant emissions increase is subject to PSD review (unless netted out) and unless the source is in an area of nonattainment for that pollutant, in which case it undergoes Nonattainment Area New Source Review.

The treatment of currently minor or major sources differs with respect to modifications. For minor source modifications, these are subject to PSD only if the emissions increase from the modification itself is major—for example, greater than 100 or 250 tpy. No interfacility "netting" to net out of this requirement is allowed. (See Sec. 2.2.1.7 for a discussion of netting.) As an example, consider a 90-tpy modification and a 100-tpy modification. The 90-tpy modification would not be subject to PSD, assuming the source is one of the listed sources with a 100-tpy threshold; however, the 120-tpy modification would be subject to New Source Review. In both instances the source would be a major source after making the modification, assuming that the 90-tpy modification brought total facility emissions to more than 100 tpy.

Contrast the situation for an existing major source at which a major modification is defined as any physical or operational change that results in a net significant emissions increase (see Table 2.7) of a PSD-regulated pollutant.

2.2.1.6 Overview: Major NSR Applicability—Existing Major Sources.

Major sources are not subject to NSR unless modified. A modification is major if a proposed physical or operational change results in a significant net emissions increase. All other modifications are minor. Steps in making the determination include:

Step 1. *Is there a physical or operational change?* Only a few exceptions to the rule:

- Routine maintenance, repair, and replacement
- Use of alternative fuel or raw material if it could be accommodated prior to January 6, 1975 or is allowed by PSD permit
- Increasing operating hours or production rate (unless restricted by permit)
- Pollution control projects
- Can't increase capacity
- Can't improve efficiency

All other changes, including improved software, changing catalyst, decreasing downtime, and reducing manufacturing costs may be considered a physical or operational change. If there is such a change, it is necessary to determine whether it is major or minor. If there is no such change, then there is no modification.

Step 2. *Are the emissions from the proposed project significant?* Emissions are considered significant, as specified in the rule (Table 2.1), if their range is from

3×10^{-6} to 100 tpy, depending on the pollutant. The most common significance level is 40 tpy (VOC, NO_x, SO_2). The project emissions are also considered significant if they consist of the PTE of any new emissions units or if the anticipated emissions increase is from any modified existing emissions units. The default approach is to calculate the difference between actual emissions before the change and PTE after the change. Some "friendly" calculations are possible in certain cases, for example, if the emissions increase anticipated from debottlenecking any units up- or downstream, or if the anticipated emissions increase is from increased utilization of supporting units, such as steam from a boiler.

NOTE: When an emissions unit is being replaced, EPA considers the project to consist of the new unit's PTE only. The actual emissions decrease from the existing unit is not used unless the source wants to try to net out of review (see step 3).

If project emissions are not significant, the project is not major. If they are significant, the project can either go through major NSR or attempt to net out of major NSR.

Step 3. *Can the project net out of major NSR?* Netting is the algebraic summing of all creditable, contemporaneous increases and decreases at the source. The source *must* include *all* contemporaneous creditable increases and decreases, not just the project increase and a proposed decrease. *Contemporaneous* usually means 5 years back from the commence-construction date and forward to the commence-operation date. Reductions are actual emissions; increases can be PTE or actual.

If the net emissions increase is not significant, the project is minor. If it is significant, the project is major.

Step 4. *Determine whether change is physical or operational.* There is no definition of *physical* and *operational* in the rule. However, a change is not a physical or operational change *if* it is routine maintenance, repair, and replacement. "Routine" is based on the nature, extent, purpose, frequency, and cost of work (and other "relevant" factors).

A change is probably *not* routine if it:

- Involves replacement of *several major* components
- Significantly enhances the present efficiency and capacity of the plant
- Substantially extends the plant's useful economic life
- Is rarely (if ever) performed
- Is costly in both relative and absolute terms

"Routine" replacement is the replacement of parts of an emissions unit, within the limitations of reconstruction (PSD/70)—such parts as pumps, piping, and valves. Makes definition of the emissions unit important. Adding new equipment to what already exists at a source is never routine.

Use of alternative materials or fuel sources is considered a physical change unless it could be accommodated prior to January 6, 1975, or is authorized for use in an NSR permit.

Increases in production rate or operating hours (unless these are limited in permit) are considered operational changes.

Step 5. *Determine whether project is a Pollution Control Project (PCP).* PCPs are not subject to PSD. To qualify as a PCP, a project must meet EPA criteria. It must be instituted primarily for emission reductions, and it must be environmentally beneficial. It can increase some pollutant emissions so long as it is beneficial overall (e.g., incineration of VOC, producing NO_x). Most control equipment installations and switches to lower-polluting fuels qualify. Examples are low-NO_x burners and FCCU regenerator scrubber for Maximum Achievable Control Technology (MACT). Projects that increase capacity to improve economics are disqualified.

Step 6. *Determine whether change "results in" an emissions increase.* Many changes at a facility do not result in an emissions increase. For example, a new administration building or paving the parking lot would not logically "result in" an emissions increase. To determine whether a change results in an emissions increase, determine the emissions increase of the project. This may require the application of as many as four different components:

- New units or new activities associated with the project
- Modifications to existing units
- Emissions increases from debottlenecking units upstream and downstream from the modified unit
- Emissions increases from increased utilization of nonmodified units

Emission calculations for these four situations are straightforward, and it is relatively easy to determine the potential to emit of the new unit. If the PTE is below the significance level, the action is not subject to PSD. If the emissions increase is above significance level, the source has the option of "netting" by looking at emission reductions at other emission units at the facility.

Note that debottlenecking implies that a process is physically limited by at least one unit in the process to a level below that which at least one other unit can achieve. When the unit creating the bottleneck is modified to eliminate or partially alleviate the bottleneck, the process is considered debottlenecked. Recent EPA policy on calculating emissions increases for debottlenecking projects now requires applicants to use the actual-to-future potential to emit determination for each unit debottlenecked. The EPA policy and calculation in this regard is confusing. This policy has existed as a verbal guidance for years but has only recently been documented.

Step 7. *Calculating the Emissions Increase from a Modified Unit.* The determination of an emissions increase from a modified unit depends on the situation surrounding the nature of the modifications. EPA may insist that there are only two authorized calculations: (1) actual-to-potential comparison, which is the default recommendation, and (2) actual to future actual (for utilities only). However, other approaches are allowed in EPA policy memos.

EPA's default approach, actual to PTE, assumes that the change will increase utilization of a unit. Actual emissions equal the average of the most recent 2 years' emissions. Guidance allows applicants to use other years if the applicant can demonstrate they are more representative of normal emission levels.

Under this approach, the applicant must convince the state or EPA that the PTE equals the PTE of the unit after the change. You subtract the actual emissions from the PTE, the difference being the emissions increase. If the emissions increase is not significant, the action is not subject to PSD. If the increase is significant, the source has the option of electing to net the emissions increase. This calculation is very controversial. Most sources used below capacity would be subject to PSD when this calculation is made. The reason is because most sources are not operating at permit-allowable limits. Take, for example, a facility that has a prechange PTE of 400 tpy and actual emissions of 300 tpy (2-year average). The unit is modified to improve the economics of the process with a postchange PTE remaining at 400 tpy. The emissions increase in this example is the PTE of 400 (which also was the permit limit) minus the old actual of 300 resulting in an increase of 100 tpy, i.e., subject to PSD review.

If the unit has not yet begun normal operation, EPA allows a PTE-to-PTE comparison of emissions, but only in the situation where the unit has not begun normal operations.

Step 8. *Like-kind replacement.*   This approach can be used if the source is replacing components of an emissions unit with an equivalent component. The equipment replaced does not have to be identical, and the new emissions unit does not qualify for New Source Review. Examples include replacing pumps, duct work, electric motors, tubing, or steam turbines at a utility boiler. Note that the boiler is considered an emissions unit. Replacing a boiler cannot be considered a like-kind replacement.

Step 9. *Utilization unaffected.*   Utilization is the extent to which a unit is used. For example, 100 percent utilization means operation at full capacity for 8760 hours per year; 50 percent utilization could mean either half capacity for 8760 hours per year or full capacity for 4380 hours per year, or some other combination of capacity and hours. If a modification would not affect normal operation (so utilization is not affected), you can use the actual-to-actual comparison test to determine NSR applicability. Under this approach, the applicant calculates the emissions with the same utilization before and after the change, and the emission factor before and after may be different. If the difference in the emissions is significant, then the modification is considered major.

As an example, an industrial facility converted from oil to natural gas at several plants to save money. The company convinced EPA that the change would not result in increased utilization, as the cost savings per unit of production was very small, thereby not being enough to warrant switching production just for the cost savings.

Step 10. *Fuel/material substitution.*   This approach is used when one fuel or raw material is substituted for another and utilization may increase. The approach requires the applicant to calculate the PTE of the new fuel/material, assuming maximum substitution and utilization rate. The PTE for the old fuel is also calculated at maximum utilization rate, and the difference is the increase due to substitution.

Step 11. *Capacity increase.* This approach allows an alternative calculation for modifications that would increase the capacity of existing units, but that would *not* result in increased utilization of the unit. Under this approach, the applicant calculates the PTE of additional capacity, and if the PTE is significant, the modification is considered major. For example, a unit with a design capacity equivalent to 400 tpy PTE will be modified so that its capacity will be equivalent to 450 tpy PTE. The actual emissions are 300 tpy. The emission factors stay the same, and the economics are unaffected. The emissions increase based on this approach would be 50 tpy. Comparing an actual to future PTE, the increase would be 150 tpy.

Step 12. *Netting.* If the emissions increase as a result of a proposed modification at a facility is significant, the facility can consider one of two options:

- Go through PSD New Source Review.
- Net out of PSD (if reduction credits are available).

However, netting rules are complex, since they are intended to avoid penalizing a source if the net emissions increase is not significant.

2.2.1.7 Calculating Net Emissions Increases.
The calculation of net emissions increase must include all creditable emissions increases and decreases. Creditable emissions under the netting policy are those that are contemporaneous, that is, within 5 years prior to the date construction is expected to commence. Therefore the applicant must estimate the date of construction and the agency can reject the netting analysis if unreasonable processing time is assumed.

The timing on calculating the net emissions increase runs right up to the time the increased emissions from the proposed project are expected to occur. The definition of pollutant emissions relies on the same pollutant at a source and requires that the calculation be based on actual emissions as opposed to permit limit emissions for the existing source being netted against the potential emissions of the new source.

As in the case of offset, it is required that the emissions being netted be shown to be representative of the long-term emissions limits. For example, if, just before operation of a new unit, a facility "cranks up" emissions from an existing unit to increase the netting potential, this would probably not be acceptable to an agency. Increases in emissions that might be creditable against a future offset are those associated with permit/registration actions resulting from a physical or operational change. Increasing production rates or operating hours (unless a permit revision is required) is not a creditable increase.

Further restrictions govern whether emissions are creditable for purposes of netting. Emissions must be:

- Real (actual emissions decreases)
- Quantifiable
- Enforceable (federally)
- Permanent

If emissions decreases are relied on issuance of a major NSR permit, they can be "wiped out," that is, made unavailable for any further netting. Note also that emissions decreases from netting are included in assessing the impacts; that is, the net air quality benefit of the trade or offset must be demonstrated. Creditable decreases can be preserved if the applicant does not rely on them in netting out of NSR. In this instance, the applicant does not include the reduction in emissions in the impact analysis and preserves them for future possible netting. Note that in many states it is required that these emission reductions be banked, otherwise the applicant runs the risk of not having them available after a period of time, which could be as short as 18 months.

There are several common errors applicants make when they use netting to avoid New Source Review. The most common are using prechange allowable emissions instead of actual emissions (potential reductions instead of actual emission reductions), using a decrease at another unit and then netting against a planned increase without taking any contemporaneous increases into account, and not ensuring that the decreases are enforceable.

2.2.2 PSD Requirements

2.2.2.1 Prevention of Significant Deterioration. Regulations require PSD review only for pollutant areas designated attainment or unclassifiable and which will have a significant emissions increase from the proposed action.

The PSD permit review process requires application of Best Available Control Technology (BACT) and a demonstration that the project will have acceptable ambient air quality impacts in four areas:

1. *Ambient air quality standards compliance (the NAAQS).* This demonstration requires that the applicant show that the project will not create new violations of ambient air quality standards or contribute to existing violations.

2. *Compliance.* This demonstration requires computer dispersion modeling in which all other new sources, after the baseline date is triggered in the area, will not collectively cause the PSD increment to be exceeded within the impact area of the new source.

3. *Additional impacts.* Air quality–related values such as visibility in the national parks must be shown to be unaffected.

4. *The potential impacts of the operation in Class I areas.* Class I areas are pristine environmental areas including national parks of over 6000 acres that are designated Class I acres by the EPA. The allowable impact in these areas is much smaller than in other parts of the country.

2.2.2.2 Determining PSD Applicability. There are three basic criteria in determining PSD applicability. The first and primary criterion is whether the proposed project is sufficiently large (in terms of its emissions) to be a "major" stationary source or "major" modification. Source size is defined in terms of potential to emit, which is the source's capability at maximum design capacity to emit a pollutant, except as constrained by federally enforceable conditions (which include the effect of installed air pollution control equipment and restrictions on the hours of operation, or the type or amount of material combusted, stored, or processed).

A new source is major if it has the potential to emit any pollutant regulated under the Act in amounts equal to or exceeding specified major source thresholds (100 or 250 tpy), which are predicated on the source's industrial category. A major modification is a physical change or change in the method of operation at an existing major source that causes a significant "net emissions increase" at that source of any pollutant regulated under the Act.

The second criterion for PSD applicability is that a new major source would locate, or the modified source already is located, in a PSD area. A PSD area is one formally designated, pursuant to Sec. 107 of the Act and 40 CFR 81, by a state as "attainment" or "unclassifiable" for any criteria pollutant, i.e., an air pollutant for which a national ambient air quality standard exists.

The third criterion is that the pollutants emitted in, or increased by, "significant" amounts by the project are subject to PSD. A source's location can be attainment or unclassifiable for some pollutants and simultaneously nonattainment for others. If the project would emit only pollutants for which the area has been designated nonattainment, PSD would not apply.

The purposes of a PSD applicability determination are therefore:

1. To determine whether a proposed new source is a "major stationary source," or if a proposed modification to an existing source is a "major modification."

2. To determine if proposed conditions and restrictions that will limit emissions from a new source or an existing source that is proposing modification to a level that avoids preconstruction review requirements are legitimate and federally enforceable.

3. To determine for a major new source or a major modification to an existing source which pollutants are subject to preconstruction review.

In order to perform a satisfactory applicability determination, numerous pieces of information must be compiled and evaluated. Certain information and analyses are common to applicability determinations for both new sources and modified sources; however, there are several major differences. Consequently, two detailed discussions follow in this section: PSD applicability determinations for major new sources and PSD applicability determinations for modifications of existing sources. The common elements will be covered in the discussion of new source applicability. They are the following:

- Defining the source.
- Determining the source's potential to emit.
- Determining which major source threshold the source is subject to.

- Assessing the impact on applicability of the local air quality, i.e., the attainment designation, in conjunction with the pollutants emitted by the source.

2.2.2.3 Stationary Source. For the purposes of PSD, a *stationary source* is any building, structure, facility, or installation that emits or may emit any air pollutant subject to regulation under the Clean Air Act (the Act). "Building, structure, facility, or installation" means all the pollutant-emitting activities, belonging to the same industrial grouping, located on one or more contiguous or adjacent properties, and under common ownership or control. An *emissions unit* is any part of a stationary source that emits or has the potential to emit any pollutant subject to regulation under the Act.

The term "same industrial grouping" refers to the "major groups" identified by two-digit codes in the *Standard Industrial Classification (SIC) Manual,* which is published by the Office of Management and Budget. The 1972 edition of the *SIC Manual,* as amended in 1977, is cited in the current PSD regulations as the basis for classifying sources. Sources not found in that edition or the 1977 supplement may be classified according to the most current edition.

EXAMPLE 5. *A chemical complex under common ownership manufactures polyethylene, ethylene dichloride, vinyl chloride, and numerous other chlorinated organic compounds. Each product is made in separate processing equipment with each piece of equipment containing several emission units. All of the operations fall under SIC Major Group 28, "Chemicals and Allied Products"; therefore, the complex and all its associated emissions units constitute one source.*

In most cases, the property boundary and ownership are easily determined. A frequent question, however, particularly at large industrial complexes, is how to deal with multiple emissions units at a single location that do not fall under the same two-digit SIC code. In this situation the source is classified according to the primary activity at the site, which is determined by its principal product (or group of products) produced or distributed, or by the services it renders. Facilities that convey, store, or otherwise assist in the production of the principal product are called *support facilities.*

EXAMPLE 6. *A coal mining operation may include a coal cleaning plant that is located at the mine. If the sole purpose of the cleaning plant is to process the coal produced by the mine, then it is considered to be a support facility for the mining operation. If, however, the cleaning plant is collocated with a mine, but accepts more than half of its feedstock from other mines (indicating that the activities of the collocated mine are incidental) then coal cleaning would be the primary activity and the basis for the classification.*

EXAMPLE 7. *Another common situation is the collocation of power plants with manufacturing operations. An example would be a silicon wafer and semiconductor manufacturing plant that generates its own steam and electricity with fossil* *fuel–fired boilers. The boilers would be considered part of the source because the power plant supports the primary activity of the facility.*

An emissions unit serving as a support facility for two or more primary activities (sources) is to be considered part of the primary activity that relies most heavily on its support.

EXAMPLE 8. *A steam boiler jointly owned and operated by two sources would be included with the source that consumes the most steam.*

EXAMPLE 9. *As a corollary to the examples immediately above, suppose a power plant is co-owned by the semiconductor plant and a chemical manufacturing plant. The power plant provides 70 percent of its total output (in Btu per hour) as steam and electricity to the semiconductor plant. It sells only steam to the chemical plant. In the case of cogeneration, the support facility should be assigned to a primary activity based on pro rata fuel consumption that is required to produce the energy bought by each of the support facility's customers, since the emission rates in pounds per Btu are different for steam and electricity. In this example, then, the power plant would be considered part of the semiconductor plant.*

It is important to note that if a new support facility would by itself be a major source based on its source category classification and potential to emit, it would be subject to PSD review even though the primary source, of which it is a part, is not major and therefore exempt from review. The conditions surrounding such a determination are discussed further in the section on major source thresholds.

2.2.2.4 Pollutants Covered by NSR. The potential to emit must be determined separately for each pollutant regulated by the Act and emitted by the new or modified source. Twenty-six compounds, six of them "criteria" and twenty "noncriteria," are regulated as air pollutants by the Act as of December 31, 1989. Note that EPA has designated PM-10 (particulate matter with an aerodynamic diameter less than 10 μm) as a criteria pollutant by promulgating an NAAQS for this pollutant as a replacement for total PM. Thus, the determination of potential to emit PM-10 as well as total PM emissions (which are still regulated by many NSPSs) is required in applicability determinations. Several halons and chlorofluorocarbon (CFC) compounds have been added to the list of regulated pollutants as a result of the ratification of the Montreal Protocol by the United States in January 1989.

2.2.2.5 Role of Attainment Status. The air quality, i.e., attainment status, of the area of a proposed new source or modified existing source will impact the applicability determination in regard to the pollutants that are subject to PSD review. As previously stated, if a new source locates in an area designated attainment or unclassifiable for any criteria pollutant, PSD review will apply to any pollutant for which the potential to emit is major (or significant, if the source is major) so long as the area is not nonattainment for that pollutant.

EXAMPLE 10. *A kraft pulp mill is proposed for an attainment area for SO_2, and its potential to emit SO_2 equals 55 tpy. Its potential to emit total reduced sulfur (TRS), a noncriteria pollutant, equals 295 tpy. Its potential to emit VOC will be 45 tpy and PM/PM-10 will be 30/5 tpy; however, the area is designated nonattainment for ozone and PM. Applicability would be assessed as follows: The source would be major and subject to PSD review because of the noncriteria TRS emissions. The SO_2 emissions would therefore be subject to PSD because they are significant and the area is attainment for SO_2. The VOC and PM emissions would not be subject to PSD, even though they are significant, because the area is designated nonattainment for those pollutants. The PM-10 emissions are neither major nor significant and would therefore not be subject to review.*

2.2.2.6 Modifications.   Similarly, if the modification of an existing major source, located in an attainment area for any criteria pollutant, results in a significant increase in potential to emit and a significant net emissions increase, the modification is subject to PSD, unless the location is designated as nonattainment for that pollutant.

Note that if the source is major for a pollutant for which an area is designated nonattainment, all significant emissions or significant emissions increases of pollutants for which the area is attainment or unclassifiable are still subject to PSD review.

A modification is subject to PSD review only if:

- The existing source that is modified is "major."

- The net emissions increase of any pollutant emitted by the source as a result of the modification is "significant." This means that the emissions equal or exceed the emission rates given in Table 2.7 (unless the source is located in a nonattainment area for that pollutant). Note also that any net emissions increase in a regulated pollutant at a major stationary source that is located within 10 km of a PSD Class I area, and which will cause an increase of 1 $\mu g/m^3$ (24-hour average) or more in the ambient concentration of that pollutant within the Class I area, is considered significant.

EXAMPLE 11. *Examples include modifications such as replacing a boiler at a chemical plant, construction of a new surface coating line at an assembly plant, and a switch from coal to gas requiring a physical change in the plant, such as new piping.*

As discussed earlier, when a minor source—that is, one that does not meet the definition of major—makes a physical change or change in a method of operation that is by itself a major source, that physical or operational change constitutes a major stationary source that is subject to PSD review. Also, if an existing minor source becomes a major source as a result of a state implementation plan relaxation, then it becomes subject to PSD requirements just as if construction had not commenced on the source or the modification.

2.2.2.6.1 Activities That Are Not Modifications. The NSR regulations do not define "physical change" or "change in the method of operation" precisely. They do, however, exclude from those activities certain types of events such as:

- Routine maintenance, repair, and replacement.
- A fuel switch due to an order under the Energy Supply and Environmental Coordination Act of 1974 (or any superseding legislation) or due to a natural gas curtailment plan under the Federal Power Act.
- A fuel switch due to an order or rule under Sec. 125 of the Clean Air Act Amendments.
- A switch in a steam generating unit to a fuel derived at whole or in part from municipal solid waste.
- A switch to a fuel or raw material that:

  The source was capable of accommodating before January 6, 1975, as long as the switch would not be prohibited under any federally enforceable permit condition established after that date, under a federally approved SIP, or federal PSD permit.

  The source is approved to make under PSD permit.
- Any increase in the hours or rate of operation of a source, so long as the increase would not be prohibited by any federally enforceable permit condition established after January 6, 1975, under a federally approved SIP.
- A change in the ownership of a stationary source.

Notwithstanding these exemptions, if a significant increase in actual emissions of a regulated pollutant occurs at an existing major source as a result of a physical change or change in the method of operation of that source, the net emissions increase has been previously described.

2.3 MACT: MAXIMUM ACHIEVABLE CONTROL TECHNOLOGY

2.3.1 Overview

The Clean Air Act Amendments of 1990 required the U.S. EPA to regulate the emissions of toxic air pollutants. Congress directed the EPA to set national emission standards for hazardous air pollutants (NESHAPs) within listed source categories and to sharply reduce the emission of the 189 hazardous air pollutants listed in the 1990 Amendment. EPA was required to establish and phase in specific emission control standards that became known as Maximum Achievable Control Technology (MACT) Standards.

EPA began, and is continuing, to develop a source category list and is issuing NESHAPs on an industry-by-industry basis. The list of source categories is reviewed and updated at least every 8 years. There is a schedule for promulgation of MACT Standards. There are issues related to the protection of public and environmental health. There are requirements regarding modification, construction, and reconstruction of major HAP sources and the application of MACT. There are work practice standards, in certain cases, and compliance schedules. There are state and local agency requirements that apply in certain instances. In short, directives regarding air toxics regulation are one of the most extensive regulatory programs ever proposed because of the number of compounds, industries, and agencies involved.

The term *Maximum Achievable Control Technology* is defined in Sec. 112 (g) of the Clean Air Act. The EPA is directed by the Act to set MACT Standards on the following basis: For new sources, the MACT Standard is set at "the emission limitation achieved in practice by the best controlled similar source, and which reflects the maximum degree of reduction in emissions that the permitting authority, taking into consideration the cost of achieving such emission reduction and any non–air quality health and environmental impacts and energy requirements, determines is achievable by the constructed or reconstructed major source." The rules governing the construction or reconstruction of major sources became effective on January 27, 1996. After this date all owners and operators are required to install MACT, unless specifically exempted and provided they are located in a state with an approved Title V permit program. The rule also established the procedures to be followed to comply with Sec. 112 (g), provided guidance for permitting authorities in implementing Sec. 112 (g), and assured effective pollution control technology would be applied to new major sources of air toxics before EPA established national MACT Standards for each industrial category under 112 (d).

The MACT requirements will be reviewed in this section. Specific questions on how the rule applies to specific industrial categories or individual sources can be answered only by an in-depth analysis of the industrial category, the source, construction details, permitting authority actions, EPA MACT Standards, and timing.

2.3.2 Intent and Purpose

EPA was required by the Clean Air Act Amendments of 1990 to issue emission standards for all major sources of the 189 listed hazardous air pollutant (HAP) compounds. The HAPs (air toxics) were identified as pollutants known or suspected of causing cancer, nervous system damage, birth defects, or other serious health effects. Congress also recognized that it would be difficult for EPA to immediately develop and issue MACT standards for all industries. Therefore, MACT requirements were phased in over time. EPA has published an initial list of industries and has issued MACT guideline documents for several industries. That process is continuing, however. In the interim MACT is to be determined on a case-by-case basis for those industries without a relevant MACT Standard. The rule also requires MACT Standards to apply when major sources are modified. The specifics of which standard and when it applies will be discussed later.

2.3.3 Which Sources Must Comply?

There are two primary areas in Sec. 112 that require compliance with MACT Standards, Subparts (d) and (g). Subpart (h) of Sec. 112 provides for exceptions. Under 112 (d), the EPA must promulgate regulations that establish a NESHAP for each category or subcategory of major sources and area sources. The standards must require the maximum degree of emission reduction that EPA defines to be achievable by each particular category. Different criteria apply to new and existing sources and less stringent standards are allowed at the discretion of the administrator of EPA for area sources. The EPA has published tables containing the listed source categories and the details on the respective NESHAP for each source category. Once the NESHAP is published, there are specific timetables that become effective and dictate the schedule for compliance.

Under 112 (g), the source owner or operator of newly constructed, reconstructed, or modified sources that have HAP emissions above threshold limits are required to apply MACT if emission increases are above specified levels. The applicant must submit a preconstruction permit application proposing a source-specific MACT.

Under 112 (j), the control of HAP emissions is required even if EPA should miss a promulgation deadline. In essence, 18 months after the scheduled promulgation deadline, a source must propose a source-specific MACT, if EPA missed promulgating the source category MACT.

Under 112 (g), often referred to as *case-by-case MACT* or *interim guidance,* an affected source must first meet the definition of "construct a major source" or "reconstruct a major source." If these apply, then the owner/operator must demonstrate to the permitting authority that emissions will be controlled to a level consistent with the new source MACT definition in Sec. 112 (d) (3) of the Act. The specific requirements and procedures to demonstrate consistency are provided in Sec. 63.43 of the rule. Timing, however, is critical and is dependent on the individual state as to when the rule will be applicable. Each applicant should review the status of EPA Sec. 112 (d) determinations and the state's implementations program for 112 (g) to determine individual source requirements.

Interestingly, once a Sec. 112 (d) MACT Standard is issued for a source category, the source must comply with the specific MACT industry standard by the EPA-designated deadline. However, a major source, which is regulated under a 112 (g) determination, may be granted up to an 8-year extension to achieve compliance. EPA can also specify the length of this potential extension in the MACT Standard.

Another term, *presumptive MACT,* has been defined to address cases where EPA and the state agree on a preliminary MACT determination for a specific source category prior to the issuance of a formal MACT Standard. This provision is intended to give sources and state authorities the best information available on what MACT may eventually be for a specific category. Presumptive MACT determinations are provided by EPA and updated regularly on the EPA Web site.

2.3.4 MACT Applicability

MACT determinations for constructed or reconstructed major sources are required when the source is subject to a case-by-case MACT determination or when the reviewing authority uses one of the review options listed. The review options include: (1) requiring a Title V permit, (2) obtaining a Notice of MACT Determination, (3) applying for a MACT determination according to the requirements of the permitting authority, and (4) determining a case-by-case MACT for alternative operating scenarios after the source has complied with the requirements of Subparts (k) and (l) and all applicable requirements of Subpart A of Sec. 112. Modifications have yet to be regulated as of this writing.

2.3.5 MACT Determinations

There are two issues that constitute the process of a MACT determination for any major source of HAPs. The first is the specific control technology review that corresponds to selection of available MACT and the technical review procedures that establish the MACT emissions limitation for the source. The second is the administrative process for submitting, reviewing, and making the final MACT determination.

The process begins with a MACT analysis, which is outlined in Sec. 63.43 (d) of Chap. I, Title 40 of the Code of Federal Regulation (CFR). In general, these state that:

1. Control cannot be less stringent than what has been achieved in practice.
2. Controls must achieve maximum control, determined on the basis of available information, taking into account costs, non–air quality health and environmental impacts, and energy impacts.
3. The control technology is determined to be acceptable when the level of control required under Sec. 112 (h) (2) is not acceptable.
4. All presumptive or other relevant MACT requirements have been considered.

In a case where the MACT determination would require additional control technology or a change in control technology, the determination must contain sufficient technical data on an emission limitation consistent with the principles set forth in Sec. 112 (d). In addition, there are several other possible circumstances under which determinations are made, and these are also specifically addressed in the regulations.

The permitting authority has specific administrative procedures that must be followed during the permit review process. The authority is required to notify the applicant within 30 days whether the application is complete. Within an additional 45 days, the authority must initially approve the application or provide a written notice of intent to disapprove. The applicant can then provide additional information and the authority then must issue an initial approval or a final disapproval.

The Notice of MACT Approval has several requirements. The authority must specify any notification, operation, maintenance, performance testing, monitoring, reporting, and record keeping requirements. The requirements are federally enforceable. The Notice of MACT Approval also expires if construction or reconstruction has not commenced within 18 months, unless a 12-month extension is granted. The public has an opportunity to comment on the Notice of MACT Approval. EPA must be notified. Finally, the source must be in compliance with all requirements on or after the start-up date. There are also several instances in which the administrator can promulgate an emission standard under Sec. 112 (d) or 112 (h) or the permitting authority under Sec. 112 (j), which regulate sources deemed to be constructed or reconstructed.

2.3.6 Recommendations on Tracking MACT

The MACT requirements as discussed above are fairly complex and continue to be developed as all listed source categories are prescribed final MACT requirements. The permitting authority will be able to provide the specific status of each industrial source category in its region. Once the requirements are defined, facility owners and operators can begin the process of determining specific source or facility requirements under MACT. This is also the point at which the source can begin to determine the best course to meet the individual facility and work area requirements stipulated under the MACT/NESHAP rules. You can find lists of MACT Standards promulgated to date, and proposed as of the writing of this book, in App. D.

2.4 BACT: BEST AVAILABLE CONTROL TECHNOLOGY

A BACT determination is required for a major stationary source for each pollutant regulated under the Act that evinces a significant net emissions increase. The requirement to conduct a BACT analysis is set forth in Sec. 165(a)(4) of the Clean Air Act and federal regulations at 40 CFR 52.21(j), and in regulations setting forth the requirements for State Implementation Plan (SIP) approval at 40 CFR 51.166 (j), and in the State Implementation Plans of the various states at 40 CFR Part 52, Subparts A through FFF.

2.4.1 History and Definition

Best Available Control Technology determinations are one of the most contentious parts of PSD permit applications. They often involve subjective determinations by the permitting agency, and therefore offer a means for intervenors to contest permits.

The original PSD rules were published on December 5, 1974, in the *Federal Register.* The Clean Air Act, as amended in August 1977, substantially modified the PSD rules and defined Best Available Control Technology as follows:

> The term "best available control technology" means an emission limitation based on the maximum degree of reduction of each pollutant subject to regulation under this Act emitted from or which results from any major emitting facility, which the permitting authority, on a case by case basis, taking into account energy, environmental, and economic impacts and other costs, determines is achievable for such facility through application of production processes and available methods, systems, and techniques, including fuel cleaning or treatment or innovative fuel combustion techniques for control of each such pollutant. In no event should application of "best available control technology" result in emissions of any pollutants which will exceed the emissions allowed by any applicable standard established pursuant to Section 111 or 112 of this Act. [Section 169(3), Clean Air Act as amended August 1977.]

Obviously, the requirement to do case-by-case reviews, taking into account energy, environmental, and economic impacts, as well as other costs, affords considerable latitude for discretionary authority by the agency. By 1978, the definition of BACT had expanded and the revisions to 40 CFR Sec. 51.24 (b)(10) defined BACT as:

> "Best available control technology" means an emission limitation (including a visible emission standard) based on the maximum degree of reduction for each pollutant subject to regulation under the act, which would be emitted from any proposed major stationary source or major modification which the permitting authority, on a case by case basis, taking into account energy, environmental, and economic impacts and other costs, determines is achievable for such source or modification through application of production processes or available methods, systems, and techniques, including fuel cleaning or treatment or innovative fuel combustion techniques for control of such pollutant. In no event shall application of the best available control technology result in emissions of any pollutant which would exceed the emissions allowed by any applicable standard under 40 CFR Part 60 and Part 61. If the reviewing agency determines that technological or economic limitations on the application of measurement methodology to a particular class of Sources would make the imposition of an emissions standard infeasible, it may instead prescribe a design, equipment, work practice, or operational standard, or combination thereof, to require the application of best available control technology. Such standard shall, to the degree possible, set forth the emission reduction achievable by the implementation of such

design, equipment, work practice, or operation, and shall provide for compliance by means which achieve equivalent results.

This definition remains unchanged to this day.

2.4.2 Applicability

On August 7, 1980, U.S. EPA promulgated significant changes to the PSD regulations in response to a group of lawsuits combined under the title *Alabama Power, et al. vs. Douglas M. Costle as Administrator, Environmental Protection Agency, et al.* (No. 78-1006 U.S. Court of Appeals, District of Columbia Circuit). The major change to BACT requirements at this time was a change in applicability determination, not in definition. The BACT definition moved to Sec. 51.24(b)(12), with identical wording. However, prior to this time, BACT was applied to those pollutants, before controls, that were major under the Act themselves (i.e., greater than 100 tpy until 1978, and greater than 250 tpy after June 1978). The new 1980 rules and regulations required application of BACT to all pollutants emitted by a major source, which exceeded a significant emission increase after controls. *Significant* was defined as:

(i) "Significant" means in reference to a net emissions increase or the potential of a source to emit any of the following pollutants, a rate of emissions that would equal or exceed any of the following emission rates:

Pollutant	Emission Rate
Carbon Monoxide	100 tons per year (tpy)
Nitrogen Oxides	40 tpy
Sulfur Dioxide	40 tpy
Particulate Matter	25 tpy
Ozone	40 tpy of volatile organic compounds
Lead	0.6 tpy
Asbestos	0.007 tpy
Beryllium	0.0004 tpy
Mercury	0.1 tpy
Vinyl Chloride	1 tpy
Fluorides	3 tpy
Sulfuric Acid Mist	7 tpy
Hydrogen Sulfide (H_2S)	10 tpy
Total Reduced Sulfur (including H_2S)	10 tpy
Reduced Sulfur Compounds (including H_2S)	10 tpy

(ii) "Significant" means, in reference to a net emissions increase or the potential of a source to emit a pollutant subject to regulation under the Act that Paragraph (b)(23)(i) does not list, any emissions rate.

(iii) Notwithstanding Paragraph (b)(23)(i), "Significant" means any emissions rate or any net emissions increase associated with a major stationary source or major modification which would construct within 10 kilometers of a Class I Area, and have an impact on such area equal to greater than 1 microgram per cubic meter (24-hour average). [40 CFR 51.24(b)(23)]

Other than the addition of particulate matter less than 10 μm (PM-10), significance level of 15 tpy, which was added on July 1, 1987, and municipal waste combustor (MWC) organics at 3.5×10^{-6} tpy, MWC metals at 15 tpy, and MWC acid gases at 40 tpy added on August 12, 1991, the applicability levels have remained the same. In 1986, Sec. 51.24 was redesignated as Sec. 51.166 of Title 40 of the *Code of Federal Regulations* (CFR).

2.4.3   The BACT Process

Until September 1, 1987, a BACT analysis was conducted in what was termed a "bottom-up" fashion. Under this approach, the applicant identified the control technology or process that just demonstrated compliance with applicable emission standards (e.g., NSPS, NESHAPs) at a minimum cost. The applicant was then supposed to continue to evaluate other technologically feasible control measures and their environmental, economic, and energy costs for each significant pollutant until that control technology could be rejected on the basis of one of the three criteria (e.g., energy, environmental, or economic). An example of an unacceptable energy impact might be the requirement to use a fuel that was either unavailable or subject to curtailment due to short supply. This could be, for example, the use of natural gas for a thermal oxidizer in an area where natural gas supplies are either insufficient or unavailable.

Unacceptable environmental impacts could include the reduction in dispersion caused by a wet scrubber or the addition of significant levels of nonattainment pollutants such as may be the case for a thermal oxidizer in a severe nonattainment area. Cost or economic evaluations remain one of the more difficult determinations for the permitting agencies and applicants. Historically, costs in the range of $1000 to $3000 per ton of pollutant removed had been considered as the limit of acceptability by most agencies. However, no written guidance on the level of acceptability exists, and some permitting agencies have used values as high as $25,000 per ton of pollutant removed.

By the mid-1980s there was some concern that applicants were doing less than complete BACT analyses. Because of restrictions on agency resources, including staff experience levels, time, and budget, it was felt that inadequate BACT applications were being submitted and approved by permitting agencies. It was not possible for the agencies to become expert in control technology for every possible industry, including recent advances and costs. Therefore, the agencies switched to a "top-down" BACT approach. Ideally, either a bottom-up or top-down approach would produce the same results, however, a top-down approach puts the burden on the applicant rather than the agency to defend why a control technology is not applicable to a particular source.

The following sections describing the top-down process are from the U.S. EPA *New Source Review Workshop Manual, Prevention of Significant Deterioration and Nonattainment Area Permitting,* draft, October 1990. Although this document remains in draft form, it is still used in combination with interpretive memos as the most current, common source of BACT guidance. The interpretive memos are available from the Air and Waste Management Association (A&WMA) as part of a five-volume set of agency memos regarding the PSD program.

2.4.4 A Step-by-Step Summary of the Top-Down Process

Table 2.8 shows the five basic steps of the top-down procedure, including some of the key elements associated with each of the individual steps. A brief description of each step follows.

2.4.4.1 Step 1—Identify All Control Technologies. The first step in a top-down analysis is to identify, for the emissions unit in question (the term *emissions unit* also means *process* or *activity*), all "available" control options. Available control options are those air pollution control technologies or techniques with a practical potential for application to the emission unit and the regulated pollutant under evaluation. Air pollution control technologies and techniques include the production processes, methods, systems, and techniques, including fuel cleaning or treatment or innovative fuel combustion techniques for control of the affected pollutant. They include technologies employed outside the United States. As discussed later, in some circumstances, inherently lower-polluting processes are appropriate for consideration as available control alternatives. The control alternatives should include not only existing controls for the source category in question, but also (through technology transfer) controls applied to similar source categories and gas streams, and innovative control technologies. Technologies required under lowest achievable emission rate (LAER) determinations are available for BACT purposes and must also be included as control alternatives and usually represent the top alternative.

In the course of the BACT analysis, one or more of the options may be eliminated from consideration because they are demonstrated to be technically infeasible or have unacceptable energy, economic, or environmental impacts on a case-by-case (or site-specific) basis. However, at the outset, applicants should initially identify all control options with potential application to the emissions unit under review.

TABLE 2.8 Key Steps in the Top-down BACT Process

Step 1: Identify all control technologies
  • List is comprehensive (LAER included).
Step 2: Eliminate technically infeasible options
  • A demonstration of technical infeasibility should be clearly documented and should show, on the basis of physical, chemical, and engineering principles, that technical difficulties would preclude the successful use of the control option on the emission unit under review.
Step 3: Rank remaining control technologies by control effectiveness
Should include:
  • Control effectiveness (percent pollutant removed)
  • Expected emission rate (tons per year)
  • Expected emission reduction (tons per year)
  • Energy impacts (Btu, kWh)
  • Environmental impacts (other media and the emissions of toxic and hazardous air emissions) and economic impacts (total cost-effectiveness, incremental cost-effectiveness).
Step 4: Evaluate most effective controls and document results
  • Case-by-case consideration of energy, environmental, and economic impacts.
  • If top option is not selected as BACT, evaluate next most effective control option.
Step 5: Select BACT
  • Most effective option not rejected is BACT.

2.4.4.2 Step 2—Eliminate Technically Infeasible Options. In the second step, the technical feasibility of the control options identified in step 1 is evaluated with respect to the source-specific (or emissions unit–specific) factors. A demonstration of technical infeasibility should be clearly documented and should show, on the basis of physical, chemical, and engineering principles, that technical difficulties would preclude the successful use of the control option on the emissions unit under review. Technically infeasible control options are then eliminated from further consideration in the BACT analysis.

For example, in cases where the level of control in a permit is not expected to be achieved in practice (i.e., a source has received a permit but the project was canceled, or every operating source at that permitted level has been physically unable to achieve compliance with the limit), and supporting documentation why such limits are not technically feasible is provided, the level of control (but not necessarily the technology) may be eliminated from further consideration. However, a permit requiring the application of a certain technology or emission limit to be achieved for such technology usually is sufficient justification to assume the technical feasibility of that technology or emission limit.

2.4.4.3 Step 3—Rank Remaining Control Technologies by Control Effectiveness. In step 3, all remaining control alternatives not eliminated in step 2 are ranked and then listed in order of overall control effectiveness for the pollutant under review, with the most effective control alternative at the top. A list should be prepared for each pollutant and for each emissions unit (or grouping of similar units) subject to a BACT analysis. The list should present the array of control technology alternatives and should include the following types of information:

- Control efficiencies (percent pollutant removed)
- Expected emission rate (tons per year, pounds per hour)
- Expected emissions reduction (tons per year)
- Economic impacts (cost-effectiveness)
- Environmental impacts [includes any significant or unusual impacts on other media (e.g., water or solid waste) and, at a minimum, the impact of each control alternative on emissions of toxic or hazardous air contaminants]
- Energy impacts

However, an applicant proposing the top control alternative need not provide cost and other detailed information in regard to other control options. In such cases, the applicant should document, to the satisfaction of the review agency and for the public record, that the control option chosen is, indeed, the top, and review for collateral environmental impacts.

2.4.4.4 Step 4—Evaluate Most Effective Controls and Document Results. After the identification of available and technically feasible control technology options, the energy, environmental, and economic impacts are considered to arrive at the final level of control. At this point the analysis presents the associated impacts of the control option in the listing. For each option, the applicant is responsible for presenting an objective evaluation of each impact. Both beneficial and adverse impacts should be discussed and, where possible, quantified. In general, the BACT analysis should focus on the direct impact of the control alternative.

If the applicant accepts the top alternative in the listing as BACT, the applicant proceeds to consider whether impacts of unregulated air pollutants or impacts in other media would justify selection of an alternative control option. If there are no outstanding issues regarding collateral environmental impacts, the analysis is ended and the results proposed as BACT. In the event that the top candidate is shown to be inappropriate, because of energy, environmental, or economic impacts, the rationale for this finding should be documented for the public record. Then the next most stringent alternative in the listing becomes the new control candidate and is similarly evaluated. This process continues until the technology under consideration cannot be eliminated by any source-specific environmental, energy, or economic impacts that demonstrate that alternative to be inappropriate as BACT.

2.4.4.5 Step 5—Select BACT. The most effective control option not eliminated in step 4 is proposed as BACT for the pollutant and the emission unit under review.

2.4.5 Top-down Analysis Detailed Procedure: Identify All Control Technologies (Step 1)

2.4.5.1 Identify Alternative Emission Control Techniques. The objective in step 1 is to identify all control options with potential application to the source and pollutant under evaluation. Later, one or more of these options may be eliminated from consideration because they are determined to be technically infeasible or to have unacceptable energy, environmental, or economic impacts.

Each new or modified emission unit (or logical grouping of new or modified emission units) subject to PSD is required to undergo BACT review. BACT decisions should be made on the information presented in the BACT analysis, including the degree to which effective control alternatives were identified and evaluated. Potentially applicable control alternatives can be categorized in three ways:

- *Inherently lower-emitting processes/practices,* including the use of materials and production processes and work practices that *prevent* emissions and result in lower "production-specific" emissions.
- *Add-on controls,* such as scrubbers, fabric filters, thermal oxidizers, and other devices that *control* and *reduce* emissions after they are produced.
- *Combination of inherently lower emitting processes and add-on controls.* For example, the application of combustion and postcombustion controls to reduce NO_x emissions at a gas-fired turbine.

The top-down BACT analysis should consider potentially applicable control techniques from all three categories. Considerations of lower-polluting processes should be based on demonstrations that they manufacture identical or similar products from identical or similar raw materials or fuels. Consideration of add-on controls, on the other hand, should be based on the physical and chemical characteristics of the pollutant-bearing emission stream. Thus, candidate add-on controls may have been applied to a broad range of emission unit types that are similar, insofar as emissions characteristics, to the emissions unit undergoing BACT review.

2.4.5.2 Demonstrated and Transferable Technologies. Applicants are expected to identify all demonstrated and potentially applicable control technology alternatives. Information sources to consider include:

- EPA's BACT/LAER Clearinghouse and Control Technology Center
- Best Available Control Technology Guideline—South Coast Air Quality Management District
- Control technology vendors
- Federal/state/local new source review permits and associated inspection/performance test reports

- Environmental consultants
- Technical journals, reports and newsletters (e.g., *Journal of Air and Waste Management Association* and the McIvaine reports), and air pollution control seminars
- EPA's New Source Review (NSR) bulletin board

The applicant is responsible to compile appropriate information from available information sources, including any sources specified as necessary by the permit agency. The permit agency should review the background search and resulting list of control alternatives presented by the applicant to check that it is complete and comprehensive.

In identifying control technologies, the applicant needs to survey the range of potentially available control options. Opportunities for technology transfer lie where a control technology has been applied at source categories other than the source under consideration. Such opportunities should be identified. Also to be identified are technologies in application outside the United States to the extent that the technologies have been successfully demonstrated in practice on full-scale operations. Technologies that have not yet been applied to (or permitted for) full-scale operations need not be considered available; an applicant should be able to purchase or construct a process or control device that has already been demonstrated in practice.

To satisfy the legislative requirements of BACT, EPA believes that the applicant must focus on technologies with a demonstrated potential to achieve the highest levels of control. For example, control options incapable of meeting an applicable New Source Performance Standard (NSPS) or State Implementation Plan (SIP) limit would not meet the definition of BACT under any circumstances. The applicant does not need to consider them in the BACT analysis.

The fact that a NSPS for a source category does not require a certain level of control or particular control technology, does not preclude its consideration for control in the top-down BACT analysis. For example, postcombustion NO_x controls are not required under the Subpart GG of the NSPS for stationary gas turbines. However, such controls must still be considered available technologies for the BACT selection process and be considered in the BACT analysis. An NSPS simply defines the minimal level of control to be considered in the BACT analysis. The fact that a more stringent technology was not selected for NSPS (or that a pollutant is not regulated by an NSPS) does not exclude that control alternative or technology as a BACT candidate. When developing a list of possible BACT alternatives, the only reason for comparing control options to an NSPS is to determine whether the control option would result in an emissions level less stringent than the NSPS. If so, the option is unacceptable.

2.4.5.3 Innovative Technologies.
Although *not required* in step 1, the applicant *may* also evaluate and propose innovative technologies as BACT. To be considered

innovative, a control technique must meet the provisions of 40 CFR 52.21(b)(19) or, where appropriate, the applicable SIP definition. In essence, if a developing technology has the potential to achieve a more stringent emissions level than otherwise would constitute BACT or the same level at a lower cost, it may be proposed as an innovative control technology. Innovative technologies are distinguished from technology transfer BACT candidates in that an innovative technology is still under development and has not been demonstrated in a commercial application on identical or similar emission units. In certain instances, the distinction between innovative and transferable technology may not be straightforward. In these cases, it is recommended that the permit agency consult with EPA prior to proceeding with the issuance of an innovative control technology waiver.

In the past, only a limited number of innovative control technology waivers for a specific control technology have been approved. As a practical matter, if a waiver has been granted to a similar source for the same technology, granting of additional waivers to similar sources is highly unlikely since the subsequent applicants are no longer "innovative."

2.4.5.4 *Consideration of Inherently Lower Polluting Processes/Practices.*
Historically, EPA has not regarded the BACT requirement as a means to redefine the design of the source when available control alternatives are being considered. For example, applicants proposing to construct a coal-fired electric generator have not been required by EPA as part of a BACT analysis to consider building a natural gas–fired electric turbine, although the turbine may be inherently less polluting per unit product (in this case electricity). However, this is an aspect of the PSD permitting process in which states have the discretion to engage in a broader analysis if they so desire. Thus, a gas turbine normally would not be included in the list of control alternatives for a coal-fired boiler. However, there may be instances where, in the permit authority's judgment, the consideration of alternative production processes is warranted and appropriate for consideration in the BACT analysis. A production process is defined in terms of its physical and chemical unit operations used to produce the desired product from a specified set of raw materials. In such cases, the permit agency may require the applicant to include the inherently lower-polluting process in the list of BACT candidates.

In some cases, a given production process or emissions unit can be made to be inherently less polluting (e.g., the use of water-based versus solvent-based paints in a coating operation or a coal-fired boiler designed to have a low emission factor for NO_x). In such cases the ability of design to make the process inherently less polluting must be considered as a control alternative for the source. Inherently lower-polluting processes/practices are usually more environmentally effective because lower amounts of solid wastes and wastewater are generated compared to add-on controls. These factors are considered in the cost, energy, and environmental impacts analyses in step 4 to determine the appropriateness of the additional add-on option.

Combinations of inherently lower-polluting processes/practices (or a process made to be inherently less polluting) and add-on controls are likely to yield more effective means of emissions control than either approach alone. Therefore, the option to utilize an inherently lower-polluting process does not, in and of itself, mean that no additional add-on controls need be included in the BACT analysis. These combinations should be identified in step 1 of the top-down process for evaluation in subsequent steps.

2.4.5.5 Example. The process of identifying control technology alternatives (Step 1 in the top-down BACT process) is illustrated in the following hypothetical example.

2.4.5.5.1 Description of Source. A PSD applicant proposes to install automated surface coating process equipment consisting of a dip-tank priming stage followed by a two-step spray application and bake-on enamel finish coat. The product is a specialized electronic component (resistor) with strict resistance property specifications that restrict the types of coatings that may be employed.

2.4.5.5.2 List of Control Options. The source is not covered by an applicable NSPS. A review of the BACT/LAER Clearinghouse and other appropriate references indicates the following control options may be applicable:

Option 1: Water-based primer and finish coat. The water-based coatings have never been used in applications similar to this.

Option 2: Low-VOC solvent/high solids coating for primer and finish coat. The high solids/low-VOC solvent coatings have recently been applied with success to similar products (e.g., other types of electrical components).

Option 3: Electrostatic spray application to enhance coating transfer efficiency. Electrostatically enhanced coating application has been applied elsewhere on a clearly similar operation.

Option 4: Emissions capture with add-on control via incineration or carbon adsorber equipment. The VOC capture and control option (incineration or carbon adsorber) has been used in many cases involving the coating of different products, and the emission stream characteristics are similar to those of the proposed resistor coating process. This option is identified as an option available through technology transfer.

Since the low-solvent-coating, electrostatically enhanced application, and ventilation with add-on control options may be considered for use in combination to achieve greater emissions reduction efficiency, a total of eight control options are eligible for further consideration. The options include each of the four options listed above and the following four combinations of techniques:

Option 5: Low-solvent coating with electrostatic applications without ventilation and add-on controls

Option 6: Low-solvent coating without electrostatic applications with ventilation and add-on controls

Option 7: Electrostatic application with add-on control

Option 8: A combination of all three technologies

A no-control option also was identified but eliminated because the applicant's state regulations require at least a 75 percent reduction in VOC emissions for a source of this size. Because no control would not meet the state regulations, it could not be BACT and, therefore, was not listed for consideration in the BACT analysis.

2.4.5.5.3 Summary of Key Points. The example illustrates several key guidelines for identifying control options:

• All available control techniques must be considered in the BACT analysis.
• Technology transfer must be considered in identifying control options. The fact that a control option has never been applied to process emission units similar or identical to that proposed does not mean it can be ignored in the BACT analysis, if the potential for its application exists.
• Combinations of techniques should be considered to the extent they result in more effective means of achieving stringent emissions levels represented by the top alternative, particularly if the top alternative is eliminated.

2.4.6 Technical Feasibility Analysis (Step 2)

In step 2, the technical feasibility of the control options identified in step 1 is evaluated. This step should be straightforward for control technologies that are demonstrated—if the control technology has been installed and operated successfully on the type of source under review, it is demonstrated and it is technically feasible. For control technologies that are not demonstrated in the sense indicated above, the analysis is somewhat more involved.

Two key concepts are important in determining whether an undemonstrated technology is feasible: *availability* and *applicability.* As explained in more detail below, a technology is considered available if it can be obtained by the applicant through commercial channels or is otherwise available within the commonsense meaning of the term. An available technology is applicable if it can reasonably be installed and operated on the source type under consideration. A technology that is available and applicable is technically feasible.

Availability in this context is further revealed by tracking the technology through the following process, commonly used for bringing a control technology concept to reality as a commercial product:

• Concept stage
• Research and patenting
• Bench scale or laboratory testing

- Pilot scale testing
- Licensing and commercial demonstration
- Commercial sales

A control technique is considered available, within the context presented above, if it has reached the licensing and commercial sales stage of development. A source would not be required to experience extended time delays or resource penalties to allow research to be conducted on a new technique. Neither is it expected that an applicant would be required to experience extended trials to learn how to apply a technology on a totally new and dissimilar source type.

Consequently, technologies in the pilot scale testing stages of development would not be considered available for BACT review. An exception would be if the technology were proposed and permitted under the qualifications of an innovative control device consistent with the provisions of 40 CFR 52.21(v) or, where appropriate, the applicable SIP. In general, if a control option is commercially available, it falls within the options to be identified in step 1.

Commercial availability by itself, however, is not necessarily a sufficient basis for concluding that a technology is applicable and therefore technically feasible. Technical feasibility, as determined in step 2, also means a control option may reasonably be deployed on, or be "applicable" to, the source type under consideration.

Technical judgment on the part of the applicant and the review authority is to be exercised in determining whether a control alternative is applicable to the source type under consideration. In general, a commercially available control option will be presumed applicable if it has been or is soon to be deployed (e.g., is specified in a permit) on the same or a similar source type. Without a showing of this type, technical feasibility would be based on examination of the physical and chemical characteristics of the pollutant-bearing gas stream and comparison to the gas stream characteristics of the source types to which the technology has been applied previously.

Deployment of the control technology on an existing source with similar gas stream characteristics is generally a sufficient basis for concluding technical feasibility, barring a demonstration to the contrary.

For process-type control alternatives, the decision of whether or not it is applicable to the source in question would have to be based on an assessment of the similarities and differences between the proposed source and other sources to which the process technique has been applied previously. In the absence of an explanation of unusual circumstances by the applicant showing why a particular process cannot be used on the proposed source, the review authority may presume it is technically feasible.

In practice, decisions about technical feasibility are within the purview of the review authority. Further, a presumption of technical feasibility may be made by the review authority solely on the basis of technology transfer. For example, in the case of add-on controls, decisions of this type would be made by comparing the physical and chemical characteristics of the exhaust gas stream from the unit under review to those of the unit from which the technology is to be transferred. Unless significant differences between source types exist that are pertinent to the successful operation of the control device, the control option is presumed to be technically feasible unless the source can present information to the contrary.

Within the context of the top-down procedure, an applicant addresses the issue of technical feasibility in asserting that a control option identified in step 1 is technically infeasible. In this instance, the applicant should make a factual demonstration of infeasibility based on commercial unavailability and/or unusual circumstances that exist with application of the control to the applicant's emission units. Generally, such a demonstration would involve an evaluation of the pollutant-bearing gas stream characteristics and the capabilities of the technology. Also a showing of unresolvable technical difficulty in applying the control would constitute a showing of technical infeasibility (e.g., size of the unit, location of the proposed site, and operating problems related to specific circumstances of the source). Where the resolution of the technical difficulties is a matter of cost, the applicant should consider the technology as technically feasible. The economic feasibility of a control alternative is reviewed in the economic impacts portion of the BACT selection process.

A demonstration of technical infeasibility is based on a technical assessment considering physical, chemical, and engineering principles and/or empirical data showing that the technology would not work on the emission unit under review, or that unresolvable technical difficulties would preclude the successful deployment of the technique. Physical modifications needed to resolve technical obstacles do not, in and of themselves, provide a justification for eliminating the control technique on the basis of technical infeasibility. However, the cost of such modifications can be considered in estimating cost and economic impacts which, in turn, may form the basis for eliminating a control technology (see later discussions).

Vendor guarantees may provide an indication of commercial availability and the technical feasibility of a control technique and could contribute to a determination of technical feasibility or technical infeasibility, depending on circumstances. However, EPA does not consider a vendor guarantee alone to be sufficient justification that a control option will work. Conversely, lack of a vendor guarantee by itself does not present sufficient justification that a control option or an emission limit is technically infeasible. Generally, decisions about technical feasibility will be based on chemical and engineering analyses (as discussed above) in conjunction with information about vendor guarantees.

A possible outcome of the top-down BACT procedures discussed in this document is the evaluation of multiple control technology alternatives that result in essentially equivalent emissions. It is not EPA's intent to encourage evaluation of unnecessarily large numbers of control alternatives for every emissions unit.

Consequently, judgment should be used in deciding which alternatives will be evaluated in detail in the impact analysis (step 4) of the top-down procedure discussed in a later section. For example, if two or more control techniques result in control levels that are essentially identical, considering the uncertainties of emissions factors and other parameters pertinent to estimating performance, the source may wish to point this out and make a case for evaluation of only the less costly of these options. The scope of the BACT analysis should be narrowed in this way only if there is a negligible difference in emissions and the collateral environmental impacts between control alternatives. Such cases should be discussed with the reviewing agency before a control alternative is dismissed at this point in the BACT analysis because of such considerations. In this way, the applicant can be better assured that the analysis to be conducted will meet BACT requirements. The appropriate time to hold such a meeting during the analysis is after the completion of the control hierarchy discussed in the next section.

2.4.6.1 Summary of Key Points. In summary, important points to remember in assessing technical feasibility of control alternatives are:

- A control technology "demonstrated" for a given type or class of sources is assumed to be technically feasible unless source-specific factors exist and are documented to justify technical infeasibility.
- Technical feasibility of technology transfer control candidates generally is assessed on the basis of an evaluation of pollutant-bearing gas stream characteristics for the proposed source and other source types to which the control had been applied previously.
- Innovative controls that have not been demonstrated on any source types similar to the proposed source need not be considered in the BACT analysis.
- The applicant is responsible for providing a basis for assessing technical feasibility or infeasibility and the review authority is responsible for the decision for what is and is not technically feasible.

2.4.7 Ranking the Technically Feasible Alternatives to Establish a Control Hierarchy (Step 3)

Step 3 involves ranking all the technically feasible control alternatives that have been previously identified in step 2. For the regulated pollutant and emissions unit under review, the control alternatives are rank-ordered from the most to the least effective in terms of emission reduction potential. Later, once the control technology is determined, the focus shifts to the specific limits to be met by the source.

Two key issues that must be addressed in this process are:

- What common units should be used to compare emissions performance levels among options?

- How should control techniques that can operate over a wide range of emission performance levels (scrubbers, etc.) be considered in the analysis?

2.4.7.1 Choice of Units of Emission Performance to Compare Levels among Control Options. In general, this issue arises in comparing inherently lower-polluting processes to one another or to add-on control. For example, direct comparison of powder (and low-VOC) coatings and vapor recovery and control systems at a metal furniture finishing operation is difficult because of the different units of measure for their effectiveness. In such cases, it is generally most effective to express emission performance as an average steady-state emissions level per unit of product produced or processed. For example:

- Pounds VOC emissions per gallons of solids applied
- Pounds PM emissions per ton of cement produced
- Pounds SO_2 emissions per million Btu heat input
- Pounds SO_2 emissions per kilowatt of electric power produced

Calculating annual emissions levels (tons per year) with these units becomes straightforward once the projected annual production or processing rates are known. The result is an estimate of the annual pollutant emissions that the source or emissions unit will emit. Annual "potential" emission projections are calculated by using the source's maximum design capacity and full year-round operation (8760 hours), unless the final permit is to include federally enforceable conditions restricting the source's capacity or hours of operation. However, emissions estimates used for the purpose of calculating and comparing the cost-effectiveness of a control option are based on a different approach.

2.4.7.2 Control Techniques with a Wide Range of Emission Performance Levels. The objective of the top-down BACT analysis is to identify not only the best control technology, but also a corresponding performance level (or in some cases performance range) for that technology, considering source-specific factors. Many control techniques, including both add-on controls and inherently lower-polluting processes, can perform at a wide range of levels.

Scrubbers, high- and low-efficiency electrostatic precipitators (ESPs), and low-VOC coatings are just a few. It is not the EPA's intention to require analysis of each possible level of efficiency for a control technique, since such an analysis would result in a large number of options. Rather, the applicant should use the most recent regulatory decisions and performance data for identifying the emissions performance levels to be evaluated in all cases.

The EPA does not expect an applicant to necessarily accept an emission limit as BACT solely because it was required previously of a similar source type. While the most effective level of control must be considered in the BACT analysis, different levels of control for a given control alternative can be considered.[14] For example, the consideration of a lower level of control for a given technology may be warranted in cases where past decisions involved different source types. The evaluation of an alternative control level can also be considered where the applicant can demonstrate to the satisfaction of the permit agency that other considerations show the need to evaluate the control alternative at a lower level of effectiveness.

Manufacturer's data, engineering estimates, and the experience of other sources provide the basis for determining achievable limits. Consequently, in assessing the capability of the control alternative, latitude exists to consider any special circumstances pertinent to the specific source under review or to the prior application of the control alternative. However, the basis for choosing the alternative level (or range) of control in the BACT analysis must be documented in the application. In the absence of a showing of differences between the proposed source and previously permitted sources achieving lower emissions limits, the permit agency should conclude that the lower emissions limit is representative for that control alternative.

In summary, in reviewing a control technology with a wide range of emission performance levels, it is presumed that the source can achieve the same emission reduction level as another source unless the applicant demonstrates that there are source-specific factors or other relevant information that provides a technical, economic, energy, or environmental justification to do otherwise. Also, a control technology that has been eliminated as having an adverse economic impact at its highest level of performance may be acceptable at a lesser level of performance. For example, this can occur when the cost-effectiveness of a control technology at its highest level of performance greatly exceeds the cost of that control technology at a somewhat lower level (or range) of performance.

2.4.7.3 *Establishment of the Control Options Hierarchy.*

After determining the emissions performance levels (in common units) of each control technology option identified in step 2, a hierarchy is established that places at the top the control technology option that achieves the lowest emissions level. Each other control option is then placed after the top in the hierarchy by its respective emission performance level, ranked from the lowest emissions to the highest emissions (most effective to least effective emissions control alternative).

From the hierarchy of control alternatives, the applicant should develop a chart (or charts) displaying the control hierarchy and, where applicable:

---

[14]In reviewing the BACT submittal by a source, the permit agency may determine that an applicant should consider a control technology alternative otherwise eliminated by the applicant, if the operation of that control technology at a lower level of control (but still higher than the next control technology alternative) would no longer warrant the elimination of the alternative. For example, while a scrubber operating at 98 percent efficiency may be eliminated as BACT by the applicant because of source-specific economic considerations, the scrubber operating in the 90 to 95 percent efficiency range may not have an adverse economic impact.

- Expected emission rate (tons per year, pounds per hour)
- Emissions performance level (e.g., percent pollutant removed, emissions per unit product, pounds per million Btu, parts per million)
- Expected emissions reduction (tons per year)

The charts should also contain columns for the following information:

- Economic impacts (total annualized costs, cost-effectiveness, incremental cost-effectiveness)
- Environmental impacts (includes any significant or unusual other media impacts, e.g., water or solid waste, and the relative ability of each control alternative to control emissions of toxic or hazardous air contaminants)
- Energy impacts (indicate any significant energy benefits or disadvantages)

This should be done for each pollutant and for each emission unit (or grouping of similar units) subject to a BACT analysis. The chart is used in comparing the control alternatives during step 4 of the BACT selection process. Some sample charts are displayed in Tables 2.9 and 2.10.

2.4.8   The BACT Selection Process (Step 4)

After identifying and listing the available control options, the next step is the determination of the energy, environmental, and economic impacts of each option and the selection of the final level of control. The applicant is responsible for presenting an evaluation of each impact along with appropriate supporting information. Consequently, both beneficial and adverse impacts should be discussed and, where possible, quantified. In general, step 4 validates the suitability of the top control option in the listing for selection as BACT, or provides clear justification why the top candidate is inappropriate as BACT. If the applicant accepts the top alternative in the listing as BACT from an economic and energy standpoint, the applicant proceeds to consider whether collateral environmental impacts (e.g., emissions of unregulated air pollutants or impacts in other media) would justify selection of an alternative control option. If there are no outstanding issues regarding collateral environmental impacts, the analysis is ended and the results proposed to the permit agency as BACT. In the event that the top candidate is shown to be inappropriate, because of energy, environmental, or economic impacts, the rationale for this finding needs to be fully documented for the public record.

The determination that a control alternative is inappropriate involves a demonstration that circumstances exist at the source that distinguish it from other sources where the control alternative may have been required previously, or that argue against the transfer of technology or applications of new technology. Alternatively, where a control technique has been applied to only one or a very limited number of sources, the applicant can identify those characteristics unique

TABLE 2.9 Sample BACT Control Hierarchy

Pollutant	Technology	Range of control, %	Control level for BACT analysis, %	Emissions limit, ppm
SO_2	First alternative	80–95	95	15
	Second alternative	80–95	90	30
	Third alternative	70–85	85	45
	Fourth alternative	40–80	75	75
	Fifth alternative	50–85	70	90
	Baseline alternative	—	—	—

to those sources that may have made the application of the control appropriate in those cases, but not for the source under consideration. In showing unusual circumstances, objective factors dealing with the control technology and its application should be the focus of the consideration.

The specifics of the situation will determine to what extent an appropriate demonstration has been made regarding the elimination of the more effective alternatives as BACT. In the absence of unusual circumstance, the presumption is that sources within the same category are similar in nature, and that cost and other impacts that have been borne by one source of a given source category may be borne by another source of the same source category.

2.4.8.1 Energy and Impacts Analysis. Applicants should examine the energy requirements of the control technology and determine whether the use of the technology results in any significant or unusual energy penalties or benefits. A source may, for example, benefit from the combustion of a concentrated gas stream rich in volatile organic compounds; on the other hand, more often extra fuel or electricity is required to power a control device or incinerate a dilute gas stream. If such benefits or penalties exist, they should be quantified. Because energy penalties or benefits can usually be quantified in terms of additional cost or income to the source, the energy impacts analysis can, in most cases, simply be factored into the economic impacts analysis. However, certain types of control technologies have inherent energy penalties associated with their use. While these penalties should be quantified, so long as they are within the normal range for the technology in question, such penalties should not, in general, be considered adequate justification for nonuse of that technology.

Energy impacts should consider only *direct* energy consumption and not *indirect* energy impacts. For example, the applicant could estimate the direct energy impacts of the control alternative in units of energy consumption at the source (e.g., Btu, kWh, barrels of oil, tons of coal). The energy requirements of the control options should be shown in terms of total (and in certain cases also incremental) energy costs per ton of pollutant removed. These units can then be converted into dollar costs and, where appropriate, factored into the economic analysis.

TABLE 2.10 Sample Summary of Top-down BACT Impact Analysis Results

Pollutant/ emission unit	Control alternative	Emissions, lb/h, tpy	Emissions reduction,[a] tpy	Total annualized cost,[b] $/year	Average cost effectiveness,[c] $/ton	Incremental cost effectiveness,[d] $/ton	Toxics impact[e] (Yes/No)	Adverse environmental impacts[f]	Incremental increase over baseline[g] (million Btu/year)
NO_x unit A	Top alternative Other alternatives Baseline								
NO_x unit B	Top alternative Other alternatives Baseline								
SO_2 unit A	Top alternative Other alternatives Baseline								
SO_2 unit B	Top alternative Other alternatives Baseline								

[a]Emissions reduction over baseline level.

[b]Total annualized cost (capital, direct, and indirect) of purchasing, installing, and operating the proposed control alternative. A capital recovery factor approach using a real interest rate (i.e., without inflation) is used to express capital costs in present-day annual costs.

[c]Average cost effectiveness is total annualized cost for the control option divided by the emissions reductions resulting from the option.

[d]The incremental cost-effectiveness is the difference in annualized cost for the control option and the next most effective control option divided by the difference in emissions reduction resulting from the respective alternative.

[e]Toxics impact means there is a toxics impact consideration for the control alternative.

[f]Adverse environmental impact means there is an adverse environmental impact consideration with the control alternative.

[g]Energy impacts are the difference in total project energy requirements with the control alternative and the baseline expressed in equivalent million Btu per year.

As noted earlier, indirect energy impacts (such as energy to produce raw materials for construction of control equipment) generally are not considered. However, if the permit authority determines, either independently or on the basis of a showing by the applicant, that the indirect energy impact is unusual or significant and that the impact can be well quantified, the indirect impact may be considered. The energy impact should still focus on the application of the control alternative and *not* a concern over general energy impacts associated with the project under review as compared to alternative projects for which a permit is not being sought, or as compared to a pollution source that the project under review would replace (e.g., it would be inappropriate to argue that a cogeneration project is more efficient in the production of electricity than the power plant production capacity it would displace and, therefore, should not be required to spend equivalent costs for the control of the same pollutant).

The energy impact analysis may also address concerns over the use of locally scarce fuels. The designation of a scarce fuel may vary from region to region, but in general a scarce fuel is one that is in short supply locally and can be better used for alternative purposes, or one which may not be reasonably available to the source either at the present time or in the near future.

2.4.8.2 Cost/Economic Impact Analysis. Average and incremental cost-effectiveness are the two economic criteria that are considered in the BACT analysis. *Cost-effectiveness* is the dollars per ton of pollutant emissions reduced. *Incremental cost* is the cost per ton reduced and should be considered in conjunction with total average effectiveness.

In the economic impacts analysis, primary consideration should be given to quantifying the cost of control and not the economic situation of the individual source. Consequently, applicants generally should not propose elimination of control alternatives on the basis of economic parameters that provide an indication of the affordability of a control alternative relative to the source. BACT is required by law. Its costs are integral to the overall cost of doing business and are not to be considered an afterthought. Consequently, for control alternatives that have been effectively employed in the same source category, the economic impact of such alternatives on the particular source under review should be not nearly as pertinent to the BACT decision-making process as the average and, where appropriate, incremental cost-effectiveness of the control alternative. Thus, where a control technology has been successfully applied to similar sources in a source category, an applicant should concentrate on documenting significant cost differences, if *any,* between the application of the control technology on those other sources and the particular source under review.

Cost-effectiveness (dollars per ton of pollutant reduced) values above the levels experienced by other sources of the same type and pollutant are taken as an indication that unusual and persuasive differences exist with respect to the source under review. In addition, where the cost of a control alternative for the specific source reviewed is within the range of normal costs for that control alternative, the alternative, in certain limited circumstances, may still be eligible for elimination. To justify elimination of an alternative on these grounds, the applicant should demonstrate to the satisfaction of the permitting agency that costs of pollutant removal for the control alternative are disproportionately high when compared to the cost of control for that particular pollutant and source in recent BACT determinations. If the circumstances of the differences are adequately documented and explained in the application and are acceptable to the reviewing agency, they may provide a basis for eliminating the control alternative.

In all cases, economic impacts need to be considered in conjunction with energy and environmental impacts (e.g., toxics and hazardous pollutant considerations) in selecting BACT. It is possible that the environmental impact analysis or other considerations (as described elsewhere) would override the economic elimination criteria as described in this section. However, in the absence of concern over an overriding environmental impact or other considerations, an acceptable demonstration of an adverse economic impact can be an adequate basis for eliminating the control alternative.

2.4.8.3   Estimating the Costs of Control.   Before costs can be estimated, the control system design parameters must be specified. The most important item here is to ensure that the design parameters used in costing are consistent with emissions estimates used in other portions of the PSD application (e.g., dispersion modeling inputs and permit emission limits). In general, the BACT analysis should present vendor-supplied design parameters. Potential sources of other data on design parameters are bid documents used to support NSPS development, control technique guidelines documents, cost manuals developed by EPA, or control data in trade publications. Table 2.11 presents some example design parameters that are important in determining system costs.

To begin, the limits of the area or process segment to be costed are specified. This well-defined area or process segment is referred to as the *control system battery limits.* The second step is to list and cost each major piece of equipment within the battery limits. The top-down BACT analysis should provide this list of costed equipment. The basis for equipment cost estimates also should be documented, either with data supplied by an equipment vendor (i.e., budget estimates or bids) or by a referenced source (such as the *OAQPS Control Cost Manual,* fourth edition, EPA 450/3-90-006, January 1990, Table B-4). Inadequate documentation of battery limits is one of the most common reasons for confusion in comparison of costs of the same control applied to similar sources. For control options that are defined as inherently lower-polluting processes (and not add-on controls), the battery limits may be the entire process or project.

Design parameters should correspond to the specified emission level. The equipment vendors will usually supply the design parameters to the applicant, who in turn should provide them to the reviewing agency. In order to determine if the design is reasonable, the design parameters can be compared with those shown in documents such as the *OAQPS Control Cost Manual, Control Technology for*

TABLE 2.11 Example Control System Design Parameters

Control	Example design parameters
Wet scrubbers	Scrubber liquor (water, chemicals, etc.)
	Gas pressure drop
	Liquid/gas ratio
Carbon adsorbers	Specific chemical species
	Gas pressure drop
	Pounds carbon/pounds pollutant
Condensers	Condenser type
	Outlet temperature
Incineration	Residence time
	Temperature
Electrostatic precipitator	Specific collection area (ft^2/acfm)
Fabric filter	Air-to-cloth ratio
	Pressure drop
Selective catalytic reduction	Space velocity
	Ammonia to NO_x molar ratio
	Pressure drop
	Catalyst life

Hazardous Air Pollutants (HAPS) Manual (EPA 625/6-86-014, September 1986), and background information documents for NSPS and NESHAP regulations. If the design specified does not appear reasonable, then the applicant should be requested to supply performance test data for the control technology in question applied to the same source, or a similar source.

Once the control technology alternatives and achievable emission performance levels have been identified, capital and annual costs are developed. These costs form the basis of the cost and economic impacts (discussed later) used to determine and document whether a control alternative should be eliminated on grounds of its economic impact.

Consistency in the approach to decision making is a primary objective of the top-down BACT approach. In order to maintain and improve the consistency of BACT decisions made on the basis of cost and economic considerations, procedures for estimating control equipment costs are based on EPA's *OAQPS Cost Control Manual.* Applicants should closely follow these procedures.

Normally the submittal of very detailed and comprehensive project cost data is not necessary. However, where initial control cost projections on the part of the applicant appear excessive or unreasonable (in light of recent cost data), more detailed and comprehensive cost data may be necessary to document the applicant's projections. An applicant proposing the top alternative usually does not need to provide cost data on the other possible control alternatives.

Total cost estimates of options developed for BACT analyses should be on the order of ±30 percent accuracy. If more accurate cost data are available (such as specific bid estimates), these should be used. However, these types of costs may not be available at the time permit applications are being prepared. Costs should also be site-specific. Some site-specific factors are costs of raw materials (fuel, water, chemicals) and labor. For example, in some remote areas costs can be unusually high; remote locations in Alaska may experience a 40 to 50 percent premium on installation costs. The applicant should document any unusual costing assumptions used in the analysis.

2.4.8.4 *Cost-Effectiveness.*

Cost-effectiveness is the economic criterion used to assess the potential for achieving an objective at least cost. Effectiveness is measured in terms of tons of pollutant emissions removed. Cost is measured in terms of annualized control costs.

The cost-effectiveness calculations can be conducted on an average or incremental basis; both are discussed in this section. The resultant dollar figures are sensitive to the number of alternatives costed as well as the underlying engineering and cost parameters. There are limits to the use of cost-effectiveness analysis. For example, cost-effectiveness analysis should not be used to set the environmental objective. Second, cost-effectiveness should, in and of itself, not be construed as a measure of adverse economic impacts.

2.4.8.4.1 Average Cost-Effectiveness. Average cost-effectiveness (total annualized costs of control divided by annual emission reductions, or the difference between the baseline emission rate and the controlled emission rate) is a way to present the costs of control. Average cost-effectiveness is calculated by the following formula:

$$\text{Average cost effectiveness (dollars per ton removed)} = \frac{\text{control option annualized cost}}{\text{baseline emissions rate} - \text{control option emissions rate}}$$

Costs are calculated in (annualized) dollars per year ($/year) and emissions rates are calculated in tons per year (tpy). The result is a cost-effectiveness number in (annualized) dollars per ton ($/ton) of pollutant removed.

2.4.8.4.2 Calculating Baseline Emissions. The baseline emission rate represents a realistic scenario of upper-bound uncontrolled emissions for the source. The NSPS/NESHAP requirements or the application of controls, including other controls necessary to comply with state or local air pollution regulations, are not considered in calculating the baseline emissions. In other words, baseline emissions are essentially uncontrolled emissions, calculated by using realistic upper-boundary operating assumptions. In calculating the cost-effectiveness of adding postprocess emissions controls to certain inherently lower-polluting processes, baseline emissions may be assumed to be the emissions from the lower-polluting process itself. In other words, emission reduction credit can be taken for use of inherently lower-polluting processes.

Estimating realistic upper-bound emissions does not mean one should assume the emissions represent the potential emissions. For example, in developing a realistic upper-bound case, baseline emissions calculations can also consider inherent physical or operational constraints on the source. Such constraints should reflect the upper boundary of the source's ability to physically operate, and the applicant should verify these constraints. If the applicant does not adequately verify these constraints, then the reviewing agency should not be compelled to consider these constraints in calculating baseline emissions. In addition, the reviewing agency may require the applicant to calculate cost-effectiveness based on values exceeding the upper-boundary assumptions to determine whether or not the assumptions have a deciding role in the BACT determination. If the assumptions have a deciding role in the BACT determination, the reviewing agency should include enforceable conditions in the permit to assure that the upper-bound assumptions are not exceeded.

For example, VOC emissions from a storage tank might vary significantly with temperature, volatility of the liquid stored, and throughput. In this case, potential emissions would be overestimated if annual VOC emissions were estimated by extrapolating over the course of a year VOC emissions based solely on the hottest summer day. Instead, the range of expected temperatures should be considered in determining annual baseline emissions. Likewise, potential emissions would be overestimated if one assumed that gasoline would be stored in a storage tank being built to feed an oil-fired power boiler or that such a tank will be continually filled and emptied. On the other hand, an upper-bound case for a storage tank being constructed to store and transfer liquid fuels at a marine terminal should consider emissions based on the most volatile liquids at a high annual throughput level since it would not be unrealistic for the tank to operate in such a manner.

In addition, historic upper-bound operating data, typical for the source or industry, may be used in defining baseline emissions in evaluating the cost-effectiveness of a control option for a specific source. For example, if, for a source or industry, historical upper-bound operations call for two shifts a day, it is not necessary to assume full-time (8760 hours) operation on an annual basis in calculating baseline emissions. For comparing cost-effectiveness, the same upper-bound assumptions must, however, be used for both the source in question and other sources (or source categories) that will later be compared during the BACT analysis.

For example, suppose (on the basis of verified historic data regarding the industry in question) a given source can be expected to utilize numerous colored inks over the course of a year. Each color ink has a different VOC content ranging from high to relatively low. The source verifies that its operation will indeed call for the application of numerous color inks. In this case, it is more realistic for the baseline emission calculation for the source (and other similar sources) to be based on the expected mix of inks rather than an assumption that only one color (i.e., the ink with the highest VOC content) will be applied exclusively during the whole year.

In another example, suppose sources in a particular industry historically operate at most at 85 percent capacity. For BACT cost-effectiveness purposes (but *not* for applicability), an applicant may calculate cost-effectiveness using 85 percent capacity. However, in comparing costs with similar sources, the applicant must consistently use an 85 percent capacity factor for the cost-effectiveness of controls on those other sources.

Although permit conditions are normally used to make operating assumptions enforceable, the use of "standard industry practice" parameters for cost-effectiveness calculations (but *not* applicability determinations) can be acceptable without permit conditions. However, when a source project's operating parameters (e.g., limited hours of operation or capacity utilization, type of fuel, raw materials, product mix or type) are lower than standard industry practice or have a deciding role in the BACT determination, then these parameters or assumptions must be made enforceable with permit conditions. If the applicant will not accept enforceable permit conditions, then the reviewing agency should use the worst-case uncontrolled emissions in calculating baseline emissions. This is necessary to ensure that the permit reflects the conditions under which the source intends to operate.

For example, the baseline emissions calculations for an emergency/standby generator may consider the fact that the source does not intend to operate more than 2 weeks a year. On the other hand, baseline emissions associated with a base-loaded turbine would not consider limited hours of operation. This produces a significantly higher level of baseline emissions than in the case of the emergency/standby unit and results in more cost-effective controls. As a consequence of the dissimilar baseline emissions, BACT for the two cases could be very different. Therefore it is important that the applicant confirm that the operational assumptions used to define the source's baseline emissions (and BACT) are genuine. As previously mentioned, this is usually done through enforceable permit conditions, which reflect limits on the source's operation that were used to calculate baseline emissions.

In certain cases, such explicit permit conditions may not be necessary. For example, a source for which continuous operation would be a physical impossibility (by virtue of its design) may consider this limitation in estimating baseline emissions, without a direct permit limit on operations. However, the permit agency has the responsibility to verify that the source is constructed and operated in a manner consistent with the information and design specifications contained in the permit application.

For some sources, it may be more difficult to define what emissions level actually represents uncontrolled emissions in calculating baseline emissions. For example, uncontrolled emissions would theoretically be defined for a spray coating operation as the maximum-VOC-content coating at the highest possible rate of application that the spray equipment could physically process (even though use of such a coating or application rate would be unrealistic for the source). Assuming use of a coating with VOC content and application rate greater than expected is unrealistic and would result in an overestimate in the amount of emissions reductions to be achieved by the installation of various control option. Likewise, the cost-effectiveness of the options could consequently be greatly underestimated. To avoid these problems, uncontrolled emission factors should be represented by the highest realistic VOC content of the types of coatings and highest realistic application rates that would be used by the source, rather than by highest theoretical VOC content of coating materials or rate of application in general.

Conversely, if uncontrolled emissions are underestimated, emissions reductions to be achieved by the various control options would also be underestimated and their cost-effectiveness overestimated. For example, this type of situation would occur in the spray-coating example if the baseline for the coating operation were based on VOC content coating or application rate that is too low [when the source had the ability and intent to utilize (even infrequently) a higher-VOC-content coating or application rate].

2.4.8.4.3 Incremental Cost-Effectiveness. In addition to the average cost-effectiveness of a control option, incremental cost-effectiveness between dominant control options should also be calculated. The incremental cost-effectiveness should be examined in combination with the average cost-effectiveness in order to justify elimination of a control option. The incremental cost-effectiveness calculation compares the costs and emissions performance level of a control option to those of the next most stringent option, as shown in the following formula:

$$
\begin{array}{c}
\text{Incremental cost (dollars} \\
\text{per incremental ton} \\
\text{removed)}
\end{array}
=
\dfrac{
\begin{array}{c}
\text{total costs (annualized) of control option} \\
- \text{ total costs (annualized) of next control option}
\end{array}
}{
\begin{array}{c}
\text{next control option emission rate} \\
- \text{ control option emission rate}
\end{array}
}
$$

Care should be exercised in deriving incremental costs of candidate control options. Incremental cost-effectiveness comparisons should focus on annualized cost and emission reduction differences between *dominant* alternatives. Dominant sets of control alternatives are determined by generating what is called the *envelope of least-cost alternatives.* This is a graphical plot of total annualized costs for a total emissions reduction for all control alternatives identified in the BACT analysis (see Fig. 2.1).

For example, assume that eight technically available control options for analysis are listed in the BACT hierarchy. These are represented as *A* through *H* in Fig. 2.1. In calculating incremental costs, the analysis should be conducted only for control options that are dominant among all possible options. In Fig. 2.1, the dominant set of control options, *B, D, F, G,* and *H,* represents the least-cost envelope depicted by the curvilinear line connecting them. Points *A, C,* and *E* are inferior options and should not be considered in the derivation of incremental cost-effectiveness. Points *A, C,* and *E* represent inferior controls because *B* will buy more emissions reduction for less money than *A;* and similarly, *D* and *F* will buy smore reductions for less money than *C* and *E,* respectively.

Consequently, care should be taken in selecting the dominant set of controls for calculating incremental costs. First, the control options need to be rank-ordered in ascending order of annualized total costs. Then, as Fig. 2.1 illustrates, the most reasonable smooth curve of the control options is plotted. The incremental cost-effectiveness is then determined by the difference in total annual costs between two contiguous options divided by the difference in emissions reduction. An example is illustrated in Fig. 2.1 for the incremental cost-effectiveness of control option *F.* The vertical distance, Δ total costs annualized, divided by the horizontal

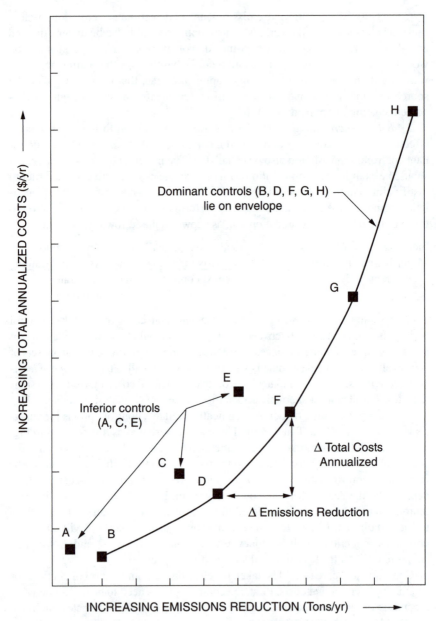

FIGURE 2.1 Least-cost envelope.

distance, Δ emissions reduced (tpy), would be the measure of the incremental cost-effectiveness for option *F.*

A comparison of incremental costs can also be useful in evaluating a specific control option over a range of efficiencies. For example, depending on the capital and operational cost of a control device, total and incremental cost may vary significantly (either increasing or decreasing) over the operation range of a control device.

As a precaution, differences in incremental costs among dominant alternatives cannot be used by themselves to argue that one dominant alternative is preferred to another. For example, suppose dominant alternatives *B, D,* and *F* on the least-cost envelope (see Fig. 2.1) are identified as alternatives for a BACT analysis. We may observe the incremental cost-effectiveness between dominant alternatives *B* and *D* is $500 per ton, whereas between dominant alternatives *D* and *F* it is $1000 per ton. Alternative *D* does not dominate alternative *F.* Both alternatives are dominant and hence on the least-cost envelope. Alternative *D* cannot legitimately be preferred to *F* on grounds of incremental cost-effectiveness.

In addition, in evaluating the average or incremental cost-effectiveness of a control alternative, reasonable and supportable assumptions regarding control efficiencies should be made. As mentioned above, unrealistically low estimates of the emission reduction potential of a certain technology could result in inflated cost-effectiveness figures.

In making a final decision regarding the reasonableness of calculated cost-effectiveness values, the review authority will consider previous regulatory decisions. Study cost estimates in BACT are typically accurate to ±20 to 30 percent. Therefore, control cost options that are within ±20 to 30 percent of each other should generally be considered to be indistinguishable in comparing options.

2.4.8.5 Determining Adverse Economic Impact. It is important to keep in mind that BACT is primarily a technology-based standard. In essence, if the cost of reducing emissions with the top control alternative, expressed in dollars per ton, is on the same order as the cost previously borne by other sources of the same type in applying that control alternative, the alternative should initially be considered economically achievable, and therefore acceptable as BACT. However, unusual circumstances may greatly affect the cost of controls in a specific application. If so, they should be documented. An example of an unusual circumstance may be the unavailability in an arid region of the large amounts of water needed for a scrubbing system. Acquiring water from a distant location might add unreasonable costs to the alternative, thereby justifying its elimination on economic grounds. Consequently, where unusual factors exist that result in cost-economic impacts beyond the range normally incurred by other sources in that category, the technology can be eliminated, provided the applicant has adequately identified the circumstances, including the cost or other analyses, that show what is significantly different about the proposed source.

Where the cost-effectiveness of a control alternative for the specific source being reviewed is within the range of normal costs for that control alternative, the alternative may also be eligible for elimination in limited circumstances. This may occur, for example, where a control alternative has not been required as BACT (or its application as BACT has been extremely limited) and there is a clear demarcation between recent BACT control costs in that source category and the control costs for sources in that category that have been driven by other constraining factors (e.g., need to meet a PSD increment or a NAAQS).

To justify an elimination of an alternative on these grounds, the applicant should demonstrate to the satisfaction of the permitting agency that costs of pollutant removal (e.g., dollars per total ton removed) for the control alternative are disproportionately high when compared to the cost of control for the pollutant in recent BACT determinations. Specifically, the applicant should document that the cost to the applicant of the control alternative is significantly beyond the range of recent costs normally associated with BACT for the type of facility (or BACT control costs in general) for the pollutant. This type of analysis should demonstrate that a technically and economically feasible control option is nevertheless, by virtue of the magnitude of its associated costs and limited application, unreasonable or otherwise not "achievable" as BACT in the particular case. Average and incremental cost-effectiveness numbers are factored into this type of analysis. However, such economic information should be coupled with a comprehensive demonstration, based on objective factors, that the technology is inappropriate in the specific circumstance.

The economic impact portion of the BACT analysis should not focus on inappropriate factors or exclude pertinent factors, as the results may be misleading. For example, the capital cost of a control option may appear excessive when presented by itself or as a percentage of the total project cost. However, this type of information can be misleading. If a large emissions reduction is projected, low or reasonable cost-effectiveness numbers may validate the option as an appropriate BACT alternative irrespective of the apparent high capital costs. In another example, undue focus on incremental cost-effectiveness can give an impression that the cost of a control alternative is unreasonably high, when in fact, the cost-effectiveness, in terms of dollars per ton removed, is well within the normal range of acceptable BACT costs.

2.4.8.6 Environmental Impacts Analysis.

The environmental impacts analysis is not to be confused with the air quality impact analysis (i.e., ambient concentrations), which is an independent statutory and regulatory requirement and is conducted separately from the BACT analysis. The purpose of the air quality analysis is to demonstrate that the source (using the level of control ultimately determined to be BACT) will not cause or contribute to a violation of any applicable national ambient air quality standard or PSD increment. Thus, regardless of the level of control proposed as BACT, a permit cannot be issued to a source that would cause or contribute to such a violation. In contrast, the environmental impacts portion of the BACT analysis concentrates on impacts other than impacts on air quality standards due to emissions of the regulated pollutant in question, such as solid or hazardous waste generation, discharges of polluted water from a control device, visibility effects, or emissions of unregulated pollutants.

Thus, the fact that a given control alternative would result in only a slight decrease in ambient concentrations of the pollutant in question compared to a less stringent control alternative should not be viewed as an adverse *environmental* impact justifying rejection of the more stringent control alternative. However, if the cost-effectiveness of the more stringent alternative is exceptionally high, it may be considered in determining the existence of an adverse *economic* impact that would justify rejection of the more stringent alternative.

The applicant should identify any significant or unusual environmental impacts associated with a control alternative that have the potential to affect the selection or elimination of a control alternative. Some control technologies may have potentially significant secondary (i.e., collateral) environmental impacts. Scrubber effluent, for example, may affect water quality and land use. Similarly, emissions of water vapor from technologies using cooling towers may affect local visibility. Other examples of secondary environmental impacts could include hazardous waste discharges, such as spent catalysts or contaminated carbon. Generally, these types of environmental concerns become important when sensitive site-specific receptors exist or when the incremental emissions reduction potential of the top control is only marginally greater than the next most effective option. However, the fact that a control device creates liquid and solid waste that must be disposed of does not necessarily argue against selection of that technology as BACT, particularly if the control device has been applied to similar facilities elsewhere and the solid or liquid waste problem under review is similar to those other applications. On the other hand, where the applicant can show that unusual circumstances at the proposed facility create greater problems than experienced elsewhere, this may provide a basis for the elimination of that control alternative as BACT.

The procedure for conducting an analysis of environmental impacts should be based on a consideration of site-specific circumstances. In general, however, the analysis of environmental impacts starts with the identification and quantification of the solid, liquid, and gaseous discharges from the control device or devices under review. This analysis of environmental impacts should be performed for the entire hierarchy of technologies (even if the applicant proposes to adopt the top, or most stringent, alternative). However, the analysis need only address those control alternatives with any significant or unusual environmental impacts that have the potential to affect the selection or elimination of a control alternative. Thus, the relative environmental impacts (both positive and negative) of the various alternatives can be compared with each other and the top alternative.

Initially, a qualitative or semiquantitative screening is performed to narrow the analysis to discharges with potential for causing adverse environmental effects.

Next, the mass and composition of any such discharges should be assessed and quantified to the extent possible, on the basis of readily available information. Pertinent information about the public or environmental consequences of releasing these materials should also be assembled.

2.4.8.7 Examples (Environmental Impacts).   The following paragraphs discuss some possible factors for consideration in evaluating the potential for an adverse impact on other media.

2.4.8.7.1 Water Impact.   Relative quantities of water used and water pollutants produced and discharged as a result of use of each alternative emission control system relative to the top alternative would be identified. Where possible, the analysis would assess the effect on groundwater and such local surface water quality parameters as pH, turbidity, dissolved oxygen, salinity, toxic chemical levels, temperature, and other important considerations. The analysis should consider whether applicable water quality standards will be met and the availability and effectiveness of various techniques to reduce potential adverse effects.

2.4.8.7.2 Solid Waste Disposal Impact.   The quality and quantity of solid waste (e.g., sludges, solids) that must be stored, disposed of, or recycled as a result of the application of each alternative emission control system would be compared with the quality and quantity of wastes created with the top emission control system. The composition and various other characteristics of the solid waste (such as permeability, water retention, rewatering of dried material, compression strength, leachability of dissolved ions, bulk density, ability to support vegetation growth, and hazardous characteristics) that are significant with regard to potential surface water pollution or transport into and contamination of subsurface waters or aquifers would be appropriate for consideration.

2.4.8.7.3 Irreversible or Irretrievable Commitment of Resources.   The BACT decision may consider the extent to which the alternative emission control systems may involve a tradeoff between short-term environmental gains at the expense of long-term environmental losses and the extent to which the alternative systems may result in irreversible or irretrievable commitment of resources (for example, use of scarce water resources).

2.4.8.7.4 Other Environmental Impacts.   Significant differences in noise levels, radiant heat, dissipated static electrical energy, or greenhouse gas emissions may be considered.

Another environmental impact that could be examined is the tradeoff between emissions of the various pollutants resulting from the application of a specific control technology. The use of certain control technologies may lead to increases in emissions of pollutants other than those the technology was designed to control. For example, the use of certain volatile organic compound (VOC) control technologies can increase nitrogen oxide (NO_x) emissions. In this instance, the reviewing authority may want to give consideration to any relevant local air quality concern relative to the secondary pollutant (in this case NO_x) in the region of the proposed source. For example, if the region in the example were nonattainment for NO_x, a premium could be placed on the potential NO_x impact. This could lead to elimination of the most stringent VOC technology (assuming it generated high quantities of NO_x) in favor of one having less of an impact on ambient NO_x concentrations. Another example is the potential for higher emissions of toxic and hazardous pollutants from a municipal waste combustor operating at low flame temperature to reduce the formation of NO_x. In this case, the real concern to mitigate the emissions of toxic and hazardous emissions (via high combustion temperatures) may well take precedent over mitigating NO_x emissions through the use of a low flame temperature. However, in most cases (unless an overriding concern over the formation and impact of the secondary pollutant is clearly present as in the example given), it is not expected that this type impact would affect the outcome of the decision.

Other examples of collateral environmental impacts would include hazardous waste discharges such as spent catalysts or contaminated carbon. Generally, these types of environmental concerns become important when site-specific sensitive receptors exist or when the incremental emissions reduction potential of the top control option is only marginally greater than the next most effective option.

2.4.8.8 *Consideration of Emissions of Toxic and Hazardous Air Pollutants.*
The generation or reduction of toxic and hazardous emissions, including compounds not regulated under the Clean Air Act, are considered as part of the environmental impacts analysis. Pursuant to the EPA administrator's decision in *North County Resource Recovery Associates,* PSD Appeal No. 85-2 (Remand Order, June 3, 1986), a PSD permitting authority should consider the effects of a given control alternative on emissions of toxics or hazardous pollutants not regulated under the Clean Air Act. The ability of a given control alternative to control release of unregulated toxic or hazardous emissions must be evaluated and may affect the BACT decision. Conversely, hazardous or toxic emissions resulting from a given control technology should also be considered and may, as appropriate, also affect the BACT decision.

Because of the variety of sources and pollutants that may be considered in this assessment, it is not feasible for the EPA to provide highly detailed national guidance on performing an evaluation of the toxic impacts as part of the BACT determination. Also, detailed information with respect to the type and magnitude of emissions of unregulated pollutants for many source categories is currently limited. For example, a combustion source emits hundreds of substances, but knowledge of the magnitude of some of these emissions or the hazards they produce is sparse. The EPA believes it is appropriate for agencies to proceed on a case-by-case basis using the best information available. Thus, the determination of whether the pollutants would be emitted in amounts sufficient to be of concern is one that the permitting authority has considerable discretion in making. However, reasonable efforts should be made to address these issues. For example, such efforts might include consultating the:

- EPA Regional Office
- Control Technology Center (CTC)
- National Air Toxics Information Clearinghouse
- Air Risk Information Support Center in the Office of Air Quality Planning and Standards (OAQPS)
- Current literature, such as EPA-prepared compilations of emission factors

Source-specific information supplied by the permit applicant is often the best source of information, and it is important that the applicant be made aware of its responsibility to provide a reasonable accounting of air toxics emissions.

Similarly, once the pollutants of concern are identified, the permitting authority has flexibility in determining the methods by which it factors air toxics considerations into the BACT determination, subject to the obligation to make reasonable efforts to consider air toxics. Consultation by the review authority with EPA's implementation centers, particularly the CTC, is again advised.

It is important to note that several acceptable methods, including risk assessment, exist to incorporate air toxics concerns into the BACT decision. The depth of the toxics assessment will vary with the circumstances of the particular source under review, the nature and magnitude of the toxic pollutants, and the locality. Emissions of toxic or hazardous pollutants of concern to the permit agency should be identified and, to the extent possible, quantified. In addition, the effectiveness of the various control alternatives in the hierarchy of controlling the toxic pollutants should be estimated and summarized to assist in making judgments about how potential emissions of toxic or hazardous pollutants may be mitigated through the selection of one control option over another. For example, the response to the administrator made by EPA Region IX in its analysis of the North County permitting decision illustrates one of several approaches [for further information see the September 22, 1987, EPA memorandum from Gerald Emison titled "Implementation of North County Resource Recovery PSD Remand" and the July 18, 1988, EPA memorandum from John Calcagni titled "Supplement Guidance on Implementing the North County Prevention of Significant Deterioration (PSD) Remand"].

Under a top-down BACT analysis, the control alternative selected as BACT will most likely reduce toxic emissions as well as the regulated pollutant. An example is the emissions of heavy metals typically associated with coal combustion. The metals generally are a portion of, or adsorbed on, the fine particulate in the exhaust gas stream. Collection of the particulate in a high-efficiency fabric filter, rather than a low-efficiency electrostatic precipitator, reduces criteria pollutant particulate matter emissions and toxic heavy metals emissions. Because in most instances the interests of reducing toxics coincide with the interests of reducing the pollutants subject to BACT, consideration of toxics in the BACT analysis generally amounts to quantifying toxic emission levels for the various control options.

In limited other instances, though, control of regulated pollutant emissions may compete with control of toxic compounds, as in the case of certain selective catalytic reduction (SCR) NO_x control technologies. The SCR technology itself results in emissions of ammonia, which increase, generally speaking, with increasing levels of NO_x control. It is the intent of the toxics screening in the BACT procedure to identify and quantify this type of toxic effect. Generally, toxic effects of this type will not necessarily be overriding concerns and will not likely affect BACT decisions. Rather, the intent is to require a screening of toxic emissions effects to ensure that a possible overriding toxics issue does not escape notice.

On occasion, consideration of toxic emissions may support the selection of a control technology that yields less than the maximum degree of reduction in emissions of the regulated pollutant in question. An example is the municipal solid waste combustor and resource recovery facility that was subject to the *North County* remand. Briefly, BACT for SO_2 and PM was selected to be a lime slurry spray drier followed by a fabric filter. The combination yields good SO_2 control (approximately 83 percent), good PM control (approximately 99.5 percent), and also removes acid gases (approximately 95 percent), metals dioxins, and other regulated pollutants. In this instance, the permitting authority determined that good balanced control of regulated and unregulated pollutants took priority over achieving the maximum degree of emissions reduction for one or more regulated pollutants. Specifically, higher levels (up to 95 percent) of SO_2 control could have been obtained by a wet scrubber.

2.4.9   Selecting BACT (Step 5)

The most effective control alternative not eliminated in step 4 is selected as BACT. It is important to note that, regardless of the control level proposed by the applicant as BACT, the ultimate BACT decision is made by the permit issuing agency after public review. The applicant's role is primarily to provide information on the various control options and, when it proposes a less stringent control option, provide a detailed rationale and supporting documentation for eliminating the more stringent options. It is the responsibility of the permit agency to review the documentation and rationale presented and (1) assure that the applicant has addressed all of the most effective control options that could be applied and (2) determine that the applicant has adequately demonstrated that energy, environmental, or economic impacts justify any proposal to eliminate the more effective control options. Where the permit agency does not accept the basis for the proposed elimination of a control option, the agency may inform the applicant of the need for more information regarding the control option. However, the BACT selection essentially should default to the highest level of control for which the applicant could not adequately justify its elimination based on energy, environmental, or economic impacts. The permit agency should proceed to establish BACT and prepare a draft permit based on the most effective control option for which justification for rejection was not provided.

2.4.10 Other Considerations

Once energy, environmental, and economic impacts have been considered, BACT can be made more stringent only by other considerations outside the normal scope of BACT analysis, as discussed under the above steps. Examples include cases where BACT does not produce a degree of control stringent enough to prevent exceedances of a national ambient air quality standard or PSD increment, or where the state or local agency will not accept the level of control selected as BACT and requires more stringent controls to preserve a greater amount of the available increment. A permit cannot be issued to a source that would cause or contribute to such a violation, regardless of the outcome of the BACT analysis. Also, states that have set ambient air quality standards at levels tighter than the federal standards may demand a more stringent level of control at a source to demonstrate compliance with the state standards.

Another consideration that could override the selected BACT are legal constraints outside the Clean Air Act requiring the application of more stringent technology (e.g., a consent decree requiring a greater degree of control). In all cases, regardless of the rationale for the permit requiring a more stringent emissions limit than would have otherwise been chosen as a result of the BACT selection process, the emissions limit in the final permit (and corresponding control alternative) represents BACT for the permitted source on a case-by-case basis.

The BACT emission limit in a new source permit is not set until the final permit is issued. The final permit is not issued until a draft permit has gone through public comment and the permitting agency has had an opportunity to consider any new information that may have come to light during the comment period. Consequently, in setting a proposed or final BACT limit, the permit agency can consider new information it learns, including recent permit decisions, subsequent to the submittal of a complete application. This emphasizes the importance of ensuring that, prior to the selection of a proposed BACT, all potential sources of information have been reviewed by the source to ensure that the list of potentially applicable control alternatives is complete (most importantly as it relates to any more effective control options than the one chosen) and that all considerations relating to economic, energy, and environmental impacts have been addressed. As noted in *Vermont Yankee Nuclear Power Corp. v. NRDC 435 U.S. 519* (1978) and the *Pennsauken County, New Jersey Resource Recovery Facility PSD Appeal No. 8-88 Remand,* dated November 10, 1988, the close of the public comment period typically represents the date when new BACT information (i.e., availability, costs, etc.) is no longer considered in setting BACT levels.

2.4.11 Enforceability of BACT

To complete the BACT process, the reviewing agency must establish an enforceable emission limit for each subject emission unit at the source and for each pollutant subject to review that is emitted from the source. If technological or economic limitations in the application of a measurement methodology to a particular emission unit would make an emission limit infeasible, a design, equipment, work practice, operation standards, or combination thereof may be prescribed. Also, the technology on which the BACT emission limit is based should be specified in the permit. These requirements should be written in the permit so that they are specific to the individual emission units subject to PSD review.

The emissions limits must be included in the proposed permit submitted for public comment, as well as in the final permit. BACT emission limits or conditions must:

- Be met on a continual basis at all levels of operation (e.g., limits written in pounds per million Btu or percent reduction achieved).
- Demonstrate protection of short-term ambient standards (limits written in pounds per hour).
- Be enforceable as a practical matter (contain appropriate averaging times, compliance verification procedures, and record keeping requirements).
- Show compliance or noncompliance (i.e., through monitoring times of operation, fuel input, or other indices of operating conditions and practices).
- Specify a reasonable compliance averaging time consistent with established reference methods, contain reference methods for determining compliance, and provide for adequate reporting and record keeping so that the permitting agency can determine the compliance status of the source.

2.4.12 Conclusion

As described above, the BACT process is a dynamic one that involves considerable subjective determinations by the permitting agency in the determination of what constitutes the Best Available Control Technology, as defined by the regulations. Individual sources have varied in the amount of effort expended for BACT determinations, depending on the sensitivity of the permitting agency and public interest in the project. Entire volumes have been submitted for single pollutants for some projects, while the BACT determination has consisted of less than a page or two for others. As in most regulatory processes, the typical way to expedite a determination is to select the most stringent control available in a top-down fashion. This completes the process and can typically be done in coordination with the permitting agency. ·

However, as long as the subjective nature of the process exists, intervenors can issue challenges about whether the proposed technology is appropriate. A number of PSD appeals have focused on BACT determinations involving both the proposed technology and the timing of the application process (i.e., does consideration of new processes terminate with the issuance of the completeness determination or at the completion of the public hearing, or at another time?). Therefore, BACT must be carefully considered and the scope of work determined both through agency consultations and public participation as necessary.

2.5 ASSESSING COMPLIANCE WITH CLEAN AIR REQUIREMENTS

Having a thorough knowledge of the Clean Air Act requirements is the necessary starting point for determining compliance for a facility. It is not, however, all that is needed. Assuming that a thorough knowledge of the Clean Air Act and state or local air quality regulations is available at a facility. It is often important to conduct a compliance audit to assess compliance. This is a requirement for a number of reasons. First, personnel at a facility often have grown up with the facility and are used to things the way they are. Old permits that have been in place for a number of years may not have been updated. The facility may have changed incrementally in small ways, so that sources are configured differently than shown in permits—and old sources lose their "grandfathering" status because they have been changed. This is often a critical issue for a facility. What *grandfathered* means basically is that sources in existence before certain permitting requirements went into effect are exempt from the new requirements. However, the catch 22 for a grandfathered source is that if it is modified in certain ways—even if replacements in kind occur— if the new equipment has a larger capacity or higher capacity for production and emissions (even if it is not operated at a higher capacity), the source may lose its grandfathered status. This means that permits are required for that source, and in many cases the source is required to meet the same state-of-the-art emissions control requirements that completely new sources would have to meet.

A problem in reviewing a facility's operations to determine its compliance status is that the people operating the facility, or the people responsible for environmental compliance at a facility, are often too close to these issues to be able to identify problems. Therefore, many companies conduct audits of facilities using corporate staff, consultants, or personnel from other facilities.

It is important in conducting an audit at a facility to define the relationship of the audit team to the facility staff and to define their role with corporate environmental, and potentially with environmental counsel. This is important because, invariably, such an audit will find compliance issues. In some instances, these issues require prompt reporting to regulatory agencies, while in others they may not. In some instances reporting compliance issues to counsel may provide a shield in terms of the client-attorney privilege which can enable more thought to be put into how violations are corrected, how they are reported, and how they are resolved.

Another key consideration in conducting an audit is the status of documents. Typically the audit team will be interested in reviewing not only facility operations, but also documents at the facility. Key documents to review include permit files and, importantly, permit applications. Permit applications are important to review because, in many instances, permits themselves are merely operating certificates identifying the source at the facility and validating its operation. Typically, one has to review the permit application itself to identify what emission limits were permitted and are now covered by the current operating certificate.

Another problem that arises in compliance audits is that permit applications may not be available. Permit files get corrupted over time, pages get lost, pages get misfiled, and, in some instances, operating certificates cannot be found. When current operating certificates cannot be found and permit applications are not in the files, what can be done? After a thorough review of a facility's permit files, if permits or applications are clearly unavailable, the only recourse is to conduct a file review at the state agency. States will permit this. This is often a good thing to do even if documentation at the facility appears to be complete. Why? Well, because even though the permits and applications appear complete, one can never tell whether modifications to sources have been filed if the paperwork corresponding to these changes is not present on site. A file review at the state agency will show everything that the state has.

Another problem that can develop is that permit modifications and applications may be found at the facility, but they may never have been filed with the state, or, if they were filed with the state, the state may have lost them and they may never have been processed. For all of these reasons, a file review at the state agency is something that should always be done when the compliance status of a facility is assessed.

Additional information that should be reviewed includes the compliance files. These include such items as inspection reports by state or local inspectors and any citations or notices of violation identifying any problems with existing permits or nonpermitted sources, which have been identified by state or local inspectors or enforcement officers. In case such Notices of Violation are discovered, it is important to look at documentation as to how they were resolved. Merely receiving a Notice of Violation (NOV) from an agency does not necessarily mean that a fine was imposed or that corrective action was required. In some instances the NOV is improperly written and an explanation to the agency as to why the facility really is in compliance may have resolved the issue.

Additional pieces of information that are required and may not be present in the permit files are dates that equipment was installed or modified. This gets to the grandfathering issue. If equipment is modified, it may have lost its grandfathered status. As an example, consider the case of a facility that operated oil-fired steam boilers since 1946. These boilers were clearly grandfathered, since requirements for permits for boilers came into existence in the early 1970s. However, on a compliance inspection, it was noted that the boilers are now dual-fueled boilers capable of burning both oil and natural gas. When queried as to when the gas conversion took place, the auditors are informed that it took place in 1984. To the follow-up question of "Well then, who applied for the air permits for the gas conversion?" the auditors were informed that it must have been the gas company because the facility didn't. Well, the gas company didn't apply for permits for the gas conversion and the boilers had lost their grandfathering status by virtue of a change to the method of operation, which included the ability to combust new fuel. Fortunately for the facility this audit was undertaken during one of those rare amnesty programs by a state agency. As a result, permit applications were completed and permits were received for the boiler with absolutely no question of retroactive enforcement. Interestingly, even if this audit had been undertaken not within an amnesty program, it is unlikely that the facility would have been cited or fined for failure to obtain these permits. This brings up an important point, namely that if problems are found during audit, as they often are, filing timely permit applications where required will typically not result in enforcement action being taken against a facility for failure to have permits that were discovered to be needed.

A more thorough discussion of this issue can be found in Sec. 2.9, "Liabilities and Enforcement." This also is relevant to EPA's audit policy, or, as it is now known, its *penalty mitigation policy.*

A final type of documentation that should be reviewed at the facility includes emission statements and other statements of environmental discharges. Many facilities are required to submit annual emission statements and others are required to submit the R form under the Superfund Amendment and Reauthorization Act of 1986, which includes a tabulation of air emissions of toxic substances. Why is it important to review these? The basic reason is that they may show emissions that violate permits. Comparing permit emission limits to the emission statements or R form disclosures is critical. If either the emission statements or R form statements show emission levels that violate permits, a compliance issue has been identified. In such instances, it is important to consider permit modifications for the subject sources. However, it is also important to verify that the emissions estimates that resulted in the disclosure in the reports are indeed accurate. Another issue that arises from reviewing emission statements and R forms is that pollutants may be disclosed on these forms that are not even listed on permit applications. This can happen because, when permit applications were prepared perhaps 5 to 10 years ago or more, these pollutants were not regulated and did not need to be disclosed on permit applications. Now, however, if these pollutants are regulated under state permitting regulations, permits need to be modified to reflect their emission.

It is important in conducting a document review to be clear with the facility on how documents are handled. In most instances, permit files, including the applications, are not taken from the facility. The facility needs to have these on site, and the auditor should not be responsible for potentially losing paperwork. Therefore, normally a physical review is done on site and Post-it notes or markers are used to identify documents that the auditors require copies of. The facility then usually arranges for these copies to be made and to be sent to the audit team after the on-site work is completed. In other instances, the auditors themselves will make copies at the plant and remove only the copies from the facility.

2.5.1 Plant Walk-through

After the file review, the next step is to conduct the plant walk-through. On the plant walk-through, several items of relevance are identified. Usually the audit team is accompanied by the plant environmental manager. The first task is to inspect the sources that have permits. Usually copies of the permit and application are taken along on the inspection because the objective is to review each source to determine its configuration and how closely it matches the permit application. This includes checking the capacity of the equipment, the nameplate rating for sources such as solvent coating, the materials used, the paints supplied, the solvents used, and so forth. Checking the ducts, fans, and roof vents, inspecting control devices, and looking for modifications for sources are also part of the inspection to see whether potential grandfathering implications have been missed by the facility.

Next, a second walk-through is conducted to look for unpermitted sources. This always includes, where possible, a rooftop inspection. Basically, the auditors will look for any pipes venting to the atmosphere. After making a tabulation of the pipes venting to the atmosphere, the auditors will go inside the facility and will look to see whether these are connected to sources. In some instances you will find that vents to the atmosphere are old and are disconnected inside the facility. However, in other cases, they will be connected to sources and the question is then whether these sources need permits. Don't forget, when conducting the inspections for unpermitted sources, to look for sources venting into buildings. Why would this be important? It is important because in many states sources venting into buildings, for example, dust collectors in woodworking areas, do require permits. The fact that they vent into the work area does not, in most cases, exclude sources from permitting requirements. Likewise, fugitive emissions often require permitting.

What are fugitive emissions? Fugitive emissions are emissions that are not vented through a stack or vent. In fact, the EPA defines fugitive emissions as emissions that could not reasonably be vented through a stack or vent. An example is emissions from woodworking equipment, which is often served by dust collectors.

For instance, a recent audit of a woodworking operation that had 34 sources venting to several dust collectors was undertaken by an audit team. The dust collection system was permitted, as were the 34 sources connected in various ways to the dust collector systems. After the audit review of permits, which had not been updated in almost 10 years, a review was prepared. How many sources were permitted correctly? Answer: three. This is not an atypical result. The reasons why permitted sources were not accurately permitted in this instance included things such as saws being replaced with sanders, sanders being replaced with drilling machines, saw blades being replaced with larger-diameter blades, and other issues. The replacement equipment still generated wood dust just like the old equipment; however, because it was equipment of different type and size, and often had been physically moved to new locations within the production area, the permit validity of each of those listed sources was called into question. Therefore, all of the sources except for three needed to have permits revised. Revised permits were submitted for this facility and it was subjected to no enforcement action.

2.5.2 Operating Characteristics

With the permit status for all permitted and unpermitted sources identified, the next step in determining compliance is to determine the operating characteristics of the equipment. Knowing its operating capacity is not enough. What's important in determining the validity of permits is to identify how the equipment is used and what the production rates are, and to calculate the actual and potential emissions for each piece of equipment. This requires holding discussions with production staff at the facility, not just environmental staff.

The next step is to identify the applicable emission factors and ways to calculate the emission rates for each piece of equipment identified. There are three principal ways to do this. The first involves using regulatory agency emission factors. Emission factors are available from the U.S. EPA, and also from trade associations and from other sources. Emission factors give you a potential emission rate as a function of the production level.

The second way to calculate emissions to the atmosphere is by using mass balance calculations. In its simplest terms, the mass balance approach takes the inputs to a process; calculates the outputs in terms of potential emissions that are present in the finished product, those which go out as waste in water, and solid waste; and assumes that the balance is an air emission.

The third way in which air emissions are estimated is through actual performance test or stack test data. This approach involves physically measuring the emissions from a stack or vent. All of these approaches have their advantages, and the extent to which they are acceptable to an agency also varies depending on the type of source and the reliability of the data, especially in the case of emission factors. Generally, stack test data are preferable to emission factors and mass balance calculations; however, if stack test data is used, it is important that it be based on EPA-approved test methods and that the facility be tested at its full operating capacity. If this is not possible, stack test data may be of limited value.

2.6 EMISSION ESTIMATING TECHNIQUES

2.6.1 Emission Factors

As an example of how emission factors are applied, consider estimating the emissions for a kiln that fires bricks. The kiln has a process input of 10 tons of clay per hour. It has a process output of 10 tons of bricks per hour. An EPA-published emission factor for kilns shows an emission factor for hydrogen fluoride (HF) of 0.2 per ton of production. This means for each ton of bricks produced by the kiln,0.2 ton of hydrogen fluoride emissions will be released. A second emission factor for kilns is for particulate matter (PM), and this number is 0.9 lb per ton of production. Figure 2.2 shows that the emission factor takes the process rate of 10 tons/h times the emission factor—in the case of HF, 0.2 lb/ton—and the resultant emission rate is 2 lb/h of HF. For particulate matter, it would be 10 tons of bricks times 0.9 lb of particulate matter per ton, yielding 9 lb/h of particulate matter.

A critical step in applying emissions factors is to do a dimensional analysis of the equation. Note that, in the above example, we have 10 tons/h of bricks times 0.2 lb/ton of HF. Our math training in high school algebra tells us the tons in the numerator and denominator can be crossed out, leaving pounds over hours on both sides of the equal sign, therefore the answer, being in pounds per hour, is dimensionally correct. In some instances, emission factors may be stated in pounds per 100 pounds or in pounds per 1000 pounds. Whatever the dimensions are, it is important that an appropriate correction factor be inserted in the equation if it is not inherently dimensionally correct, so that it ends up being dimensionally correct with the answer in pounds per hour.

2.6.2 Mass Balance Example

Consider a surface coating (painting) operation. In this instance, we calculate the volatile organic compound (VOC) emissions. First, we need to determine the paint

$$\begin{bmatrix} Process \\ Rate \end{bmatrix} \times \begin{bmatrix} Emission \\ Factor \end{bmatrix} = Emissions$$

$$\begin{bmatrix} 10\ \dfrac{tons}{hr}\ Bricks \end{bmatrix} \times \begin{bmatrix} 0.2\ \dfrac{lbs}{ton}\ HF \end{bmatrix} = 2\dfrac{lbs}{hr}\ HF$$

$$\begin{bmatrix} 10\ \dfrac{tons}{hr}\ Bricks \end{bmatrix} \times \begin{bmatrix} 0.9\ \dfrac{lbs}{ton}\ PM \end{bmatrix} = 9\dfrac{lbs}{hr}\ PM$$

FIGURE 2.2 Emission factor equation.

formulation in terms of its percent VOC composition. This is normally determined by reading the label on the paint can. The paint can will identify the percent VOC, percent solids, and the additional constituents. It is also important, however, to check the material safety data sheet (MSDS) for the coating to check the VOC content and to determine whether there are any hazardous substances present in the coating. The next point of information required is the density, or pounds per gallon, of the coating. Next, the application rate in terms of gallons per hour is needed.

Finally, it is important to estimate the evaporation percentage for the coating. In surface coating operations, this is normally taken to be 100 percent. Why would 100 percent be assumed if in fact only 70 to 80 percent of the volatiles evaporates within an hour or two of spraying. The reason is that the assumption is made that within a reasonable period of time, all of the volatiles in the coating will evaporate. If they don't evaporate in the spray booth or just outside the spray booth, they will evaporate in materials storage, processing, or warehousing. Therefore, it is conservative to assume that 100 percent of the volatiles evaporates at the facility.

The final point in mass balance is to consider whether any controls in place would reduce emissions. There are two issues in considering control effectiveness. It is obvious that the destruction or collection efficiency of the control device is important. For example, most would consider that a 98 percent control system would remove 98 percent of the emissions. This is true; however, another factor must be considered: the collection efficiency. Consider a spray booth in which metal parts are sprayed. The paint has a certain volatile component and may have certain hazardous materials in it. A carbon absorption unit is placed after the spray booth to control these emissions. The carbon unit will adsorb 98 percent of the volatile emissions. However, not all of the emissions from the paint in the spray booth are picked up by the ventilation system and taken to the carbon unit. The spray booth is typically only 85 percent efficient at collecting emissions. The 15 percent of the emissions that does not go to the control device is considered fugitive emissions. As an example, consider a paint formulation that is 50 percent VOC, as shown in Fig. 2.3. The density of the paint is 6 lb/gal and the application rate is 5 gal/h. We assume 100 percent evaporation and, running the math on this, find the paint yields 15 lb of VOC per hour. Again checking the dimensional correctness, we see the gallons can be crossed out of the denominator and numerator, leaving pounds per hour, the correct dimensions.

We see that if we take the 15 lb/h times the 85 percent capture efficiency of the control device and a 98 percent destruction efficiency, the control device yields 0.255 lb/h of VOC. And again, looking at the total emissions from this source, don't forget the fugitive emissions which, in this case, would be 15 percent of 15 lb/h, or 2.25 lb/h, which exceeds the emissions from the control device. Permits are required to reflect this total emission level.

$$\begin{bmatrix} Paint \\ Formulation \\ i.e.\ VOC\% \end{bmatrix} \times \begin{bmatrix} Density \\ of \\ Paint \end{bmatrix} \times \begin{bmatrix} Paint \\ Spray \\ Rate \end{bmatrix} \times \begin{bmatrix} \% \\ Evaporation \end{bmatrix}$$

$$= VOC\ Emissions$$

$$\begin{bmatrix} 50\% \\ VOC \end{bmatrix} \times \begin{bmatrix} 6.0\,\dfrac{lbs}{gal} \end{bmatrix} \times \begin{bmatrix} 5\,\dfrac{gal}{hr} \end{bmatrix} \times \begin{bmatrix} 100\ \% \\ evaporation \end{bmatrix}$$

$$= 15lbs\,\dfrac{VOC}{hr}$$

FIGURE 2.3 Mass balance equation for VOC emissions from a paint formulation.

2.6.3 Mass Balance Calculation

Consider a chemical reaction process in which methyl ethyl ketone (MEK) is added to a reaction vessel. We know that 5 gal/h of MEK is added to the reactor. We know that the product produced contains MEK. Measurements of the product show that an equivalent of 3 gal/h of MEK leave the reactor in product. We also know that the wastewater used to rinse out the reactor vessel contains MEK and that a calculation based on the measurement of the wastewater constituents shows that an equivalent of 1 gal/h of MEK leaves the reactor in the wastewater. What are the air emissions then? Figure 2.4 shows a schematic of the mass balance calculations. Figure 2.5 shows this in equation form. We have 5 gal/h of process input of MEK minus 3 gal/h of process output, minus 1 gal/h of wastewater, which equals 1 gal/h unaccounted for or lost. Extending this, we have 1 gal MEK per hour lost times 6.7 lb/gal, which is the density of MEK, which equals 6.7 lb/h of MEK. This is dimensionally consistent and therefore the 6.7 lb/h equals the air emissions of MEK. However, don't forget that of the 3 gal/h going out in the product, some may become an emission. If, for example, we are talking about a solid product, say pellets, which are stored in bags that are not airtight, some of this MEK will evaporate during product storage. If it goes out in a liquid product sealed in drums or cans, then it would not be a fugitive emission.

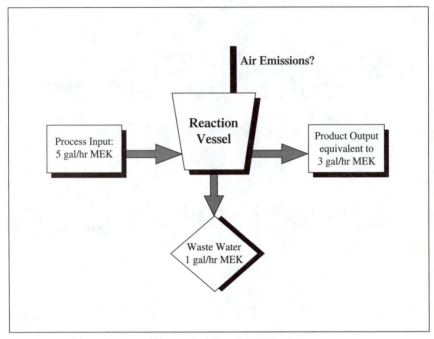

FIGURE 2.4   Schematic of mass balance calculations for MEK emissions.

2.6.4 Potential to Emit

The next step in determining compliance, after computing the emission rate, is to calculate the annual potential to emit. The potential to emit for permitted sources is based on the permit conditions, which if not specified explicitly in the permit itself, can be determined by reviewing the permit application. For unpermitted sources, the potential to emit for purposes of determining applicability of regulations such as Title V operating permits is based on 24-hour-day operations, 365 days per year (8760 hours), the maximum physical operating capacity of the equipment. This is something many people don't take into account when considering the applicability of these regulations. They know that the source does not operate that much and assume that its potential to emit is close to its actual emissions as recorded in an emission statement based on actual operating conditions and production levels. However, it is a requirement of the Title V program to base emissions for unpermitted sources on their maximum capacity.

The allowable potential to emit on permits can be less than 365 days per year, 24 hours per day, and less than maximum capacity, only if the permit limits the hours of operation or limits the maximum performance and production rates of the equipment. This can be done in a number of ways. First, permits may indicate a maximum percent utilization of the equipment. Second, they may indicate maximum throughput. This is commonly used for storage tanks, for example. And third, permits may

$$\left[\begin{array}{c} Process \\ Input \end{array}\right] - \left[\begin{array}{c} Product \\ Output \end{array}\right] - \left[\begin{array}{c} Waste \\ Water \end{array}\right] = Air\ Emissions$$

$$\left[5\frac{gal}{hr}Input \right] - \left[3\frac{gal}{hr}Output \right] - \left[1\frac{gal}{hr}Waste\ Water \right]$$

$$= 1\frac{gal}{hr}lost$$

$$\left[1\frac{gal}{hr}lost \right] \times \left[6.7\frac{lbs}{gal}MEK \right]$$

$$= 6.7\frac{lbs}{hr}MEK$$

FIGURE 2.5   Mass balance eqaution for MEK air emissions.

limit the hours of operation to, for example, 8 hours per day, 16 hours per day, 5 days per week, or a certain number of hours per quarter or per year.

2.6.5 Review Potentially Applicable Requirements

Having calculated the hourly and annual actual and potential emissions, and having identified the dates of construction and any modifications, the audit team is in a position to determine the compliance status of the facility. This requires reviewing the applicable state and any federal regulations regarding the type of sources operated at the facility. Next, it requires looking at the dates of construction and modifications to see if grandfathered sources need to be permitted because they have lost their grandfathered status.

The next step is to compare the permit limits and the emissions reported on annual emission statements or R forms to the applicable levels in state regulations for permitting. It also requires a comparison to emission limits specified for applicability of Reasonably Achievable Control Technology (RACT) for volatile organic substances and for nitrogen oxides. These regulations require existing major sources of VOC and NO_x to demonstrate compliance with RACT requirements. In many instances these demonstrations require simply submitting a form to the state showing that the facility is too small to be required to show RACT compliance or that it meets RACT requirements by virtue of the VOC content of coatings applied or, in the case of boilers (for NO_x), through annual tune-ups. In other instances, however, facilities are required to demonstrate facility-specific RACT compliance through an analysis. Where these are required, agencies required the facility to use a cost threshold that is typically in the range of $1500 per ton of emission removed to demonstrate that additional controls are not required because their cost would exceed the $1500 per ton removed threshold.

The next step in determining compliance is to identify any apparently noncompliant sources. This again is based on reviewing the emissions estimates and comparing them to regulatory requirements. Where noncompliant sources are identified, there are three potential requirements that ensue: (1) Permitted sources that are not in compliance will require modifications of existing permits. (2) For permits to be modified or for sources that need permits but are presently unpermitted, it is important to check whether there are any state-of-the-art emission control requirements that must be complied with. (3) Apply for new or modified permits as needed.

What should sources do when they identify permits that require modifications? The question here is whether an immediate notification to the agency is required, because now we know that the facility is not in compliance with its permits. The answer to this question is not straightforward and requires knowledge of the regulators, the regulations, penalties that may pertain for noncompliance, and other issues. This is where a consultant or attorney knowledgeable with the environmental compliance practices in the state can be helpful. It is the authors' experience that permit modifications or new permit applications for unpermitted sources typically will not result in enforcement action; however, it is always desirable to file these as soon as possible, because once a source knows it is operating without proper permits, it is in the status of a "knowing" violator. This is a criterion of violation that exposes a facility to potential criminal noncompliance under the Clean Air Act Amendments of 1990.

The strategy of how to file the permits and how to avoid liability for operating without permits until the permits are approved is also one that may require legal advice. In some instances, states will require the facility to enter into a consent agreement, which will basically allow it to operate for a specified period of time until the permits can be obtained. Typically this period of time would be 12 to 18 months. This is a topic to discuss with environmental counsel to determine whether a meeting with the agency is appropriate to review the compliance issues that have been found and the facility's plan to remedy these noncompliance issues. In exchange for coming forth voluntarily with these compliance issues and applying for permits, the facility would seek a specified period of time to be allowed to operate in violation of the requirements until permits can be put in place.

2.6.6 Emission Measurements

Conducting stack tests to measure emissions of substances is an additional method that can be used to determine the emission rate. In conducting stack tests, it is very important to be sure that you can operate the facility at its maximum permitted permit capacity. This is often not as simple as it seems because equipment is often operated way below its maximum capacity, and in some cases it is impossible to run the equipment at its full capacity. Consider the case of an incinerator that, to operate it at its maximum capacity, requires enough fuel (garbage) to combust at its maximum rate. Typically, equipment is oversized and permitted for its maximum capacity.

The next issue that must be addressed in conducting a stack test is where to withdraw the sample. EPA has published many protocols regarding stack testing that should be considered. They must be followed for compliance demonstrations. However, for estimating emissions, more flexibility may be allowed. It is, nonetheless, important that samples be extracted from a point in the gas stream that would meet certain criteria. Samples should be taken from a straight section of the ductwork that is not near a bend. It should not be taken too near the entrance or too near the discharge of the pipe so that it will be in an area where there is relatively undisturbed flow. It should be taken in a way that is consistent with stack testing recommendations. This generally requires that samples be extracted *isokinetically*, that is, at a velocity that matches the gas flow. If this is not done, the sample will not be considered representative. It is beyond the scope of this book to describe stack testing methods in detail, and the reader is advised to obtain copies of the U.S. EPA *Guidance* for the appropriate stack test being undertaken.

How are samples analyzed? Once a sample of the gas stream is obtained, it must be analyzed to determine the emission rate of a toxic or nontoxic substance. This can be done in a number of ways, but usually, the sample must be transferred to a laboratory in a physical state in which it can be analyzed. This involves, in the case of particulates, collecting them on filters, or, in the case of gases, collecting them in liquid or solid medium. There are a number of ways this can be done, listed in Table 2.12. Note that the sensitivity of these techniques ranges from less than 100 ppm to more than 1 ppb. The greater sensitivity correlates to greater costs.

The gravimetric technique uses filters to measure the emissions of solid particulate matter. This involves preweighing the filter before sampling to determine its weight, withdrawing a known volume of air through the filter and collecting particles on it, and then postweighing the filter in the laboratory. Knowing the mass

TABLE 2.12 Analytical Techniques for Air Toxics

Technique	Sensitivity
Gravimetric	$100 ppm
Titrimetric$100 ppm	
Potentiometric	$6 ppm
Combustion	$2 ppm
Conductivity	$1 ppm
Colorimetric	$1 ppm
Gas chromatography	$1 ppb
Spectrophotometry	$1 ppb

of particles collected on the filter by weighing it before and after exposure and knowing the volume of air that was sampled allows a simple calculation of the micrograms per cubic meter in the air sampled. It is important, since moisture can affect the weight of the filter, to weigh it before and after sampling at a uniform humidity and temperature. This is done by equilibrating the filter before and after use in the field.

Titrimetric analysis involves titrating a liquid sample in the laboratory with reagents to determine the concentration of a known contaminant. How are gases converted to liquid media? There are a number of ways to do this. The most common way is to use a bubbler chain, in which the extracted gas is bubbled through a series of jars containing a reagent. The bubbles are created by air stones much like those used in fish aquariums to aerate water. A chain of bottles is used instead of only one bottle to ensure that all the pollutant has been captured. Three or four bottles are used in series, and the objective is that the final bottle (or two) has no concentration of the pollutant, therefore the contents of the preceding bottles that do have concentration can be added together to determine the total concentration of the gas.

The potentiometric technique also involves bubblers and measures the electrical potential in the liquid after sampling. The electrical potential has a direct relationship to the concentration to the pollutant of interest.

The conductivity method works much the same; however, in this case the resistivity of the liquid varies in proportion to the concentration to the pollutant. The colorimetric technique relates the color change of the liquid, either as sampled or after adding reagents, to the concentration of the pollutant. The combustion technique involves using the gas (without transferring it to a liquid medium) and burning it. This works well for a large number of volatiles where the flame characteristics resulting from combusting the gas, with or without an additional heat source, can be related to the concentration of the pollutant of interest. Finally, gas chromatography and spectrophotometry involve sophisticated laboratory instruments in which the spectrum resulting from the breakdown of the gas can be directly correlated to the concentration of interest. These last techniques are very sensitive, very expensive, and the only techniques that allow you to *speciate,* that is, to identify the individual species of volatile organic, for example, in the laboratory.

Table 2.13 compares analytical techniques for applicability, sensitivity, accuracy, selectivity, and cost. Table 2.14 shows the cost of testing; the cost of stack testing ranges from about $3000 to almost $10,000 for a single sample, and for triplicate samples (which are often required) costs can range up to $20,000 or more. These costs assume that stack testing is already taking place at a facility, that audit personnel are on site, and that travel to and from the facility by the stack testing team is covered in addition to these costs; therefore, these costs can be considered to be on the low side. Stack testing is an expensive alternative; however when emission factors or mass balance calculations produce results that show regulatory problems, and where there is reason to believe that these results are not representative of actual emissions, stack testing is recommended.

TABLE 2.13 Analytical Techniques

Method	Uses	Sensitivity	Accuracy	Selectivity	Cost
Atomic absorption spectrophotometry	Most metals	ppb	1%	Specific	Low
Gas chromatography	Gases, organics	ppb	Medium/ high	Specific	Medium
Mass spectrophotometry	More than 20,000 gases	ppb	High	Specific	High
Optical microscopy	Particles, crystalline structures	.0.3 mm	Experience of person	Specific	Can be low
Visible spectrophotometry	NO, NO_2, SO_2, organics	ppb	High	Generally specific	Medium

TABLE 2.14 Costs of Testing

Source testing method	Average cost for a single sample	Average cost for triplicate samples
Method 0010, semivolatile organics	$7500	$22,500
TO-01	$3500	$10,000
Method 012, metals	$8500	$25,000
TO-11, aldehydes	$3000	$9,000
TO-14, VOC canisters	$3500	$10,000

2.7 AIR QUALITY MODELING

Estimating the impacts of a yet-unconstructed new source of air pollutant emissions requires air quality dispersion modeling. Air quality models have been in use for several decades. These are computer-based calculation techniques that predict the future impact of constructing and operating new sources of air emissions. The models take into account the source characteristics, meteorological data, and the characteristics of the potential impact area. Modeling can also be used to estimate the impact of existing sources in the absence of monitoring data.

While monitoring data is normally preferable to air quality prediction modeling, it is important to note that there are, even at this date, a limited number of air quality monitoring stations throughout the United States. Unless a project is very near one of these few monitoring locations, the monitored data may not be representative of the existing air quality in the vicinity of the proposed new source. Therefore, dispersion modeling is sometimes used to estimate the existing or background levels of air quality from existing sources in the vicinity. Modeling background concentrations may be important, since one of the tests of acceptability for permitting a new source of air emissions is that it not create a violation of the ambient air quality standards. Knowing the existing baseline concentration in an area requires either representative monitoring data or modeling data of existing sources.

Air quality dispersion models are available from the U.S. EPA and from a number of consultants who have developed proprietary air quality dispersion models. These models vary in their degree of sophistication and they also vary with respect to the type and number of sources for which they can be applied. Likewise, more sophisticated models require more extensive and sophisticated meteorological data for appropriate application.

While it is possible to obtain air quality dispersion models and to apply them, the reader is cautioned that proper application of dispersion models does require experience, and the ability to identify results from models that do not appear to make sense is an important characteristic of an experienced modeler. For example, it is easy to input data into models and to make a mistake regarding the units or dimensions that the models call for. For example, a model that requires the stack height in meters will seriously underpredict impacts if the stack height is input in feet without making the appropriate conversion.

It is important in beginning an air modeling effort to identify the source to be modeled and its characteristics and to come up with a preliminary choice of model for consideration. At this time it is appropriate to develop a modeling protocol that will describe the model to be used and all data to be gathered for input to the model. The modeling protocol is normally submitted to the environmental regulatory agency that will issue the permit for approval before undertaking the modeling. In the event that the agency does not approve of the protocol as submitted, it will usually suggest revisions to the protocol for applicant consideration. In this way a back-and-forth dialogue is generated and appropriate models and parameters are agreed on.

The required components of an air quality modeling protocol include:

- Project description
- Emissions (types and amounts)
- Source description
- Impact area and grid data
- Modeling objectives (i.e., demonstrating compliance with ambient air quality standards or PSD increments or nonattainment area levels of significant impact)
- Model choice (screening model, refined model, single-source model, multisource model, flat terrain model, or complex terrain model)

- Meteorological data choice, either on-site data or data from nearest available weather station

2.7.1 Types of Air Quality Models

Air quality models can be lumped into three general categories: screening models, refined models, and numerical models. Screening models include both simple terrain and complex terrain models but are limited to using hypothetical meteorological data. Screening models typically calculate only a centerline downwind concentration for a source being modeled. The distinction between simple terrain and complex terrain can be understood by examining Fig. 2.6, which shows that terrain elevations below the stack elevation are deemed simple terrain while terrain elevations exceeding the plume elevation, including plume rise, are considered complex terrain situations. The terrain that falls within the plume itself is considered intermediate terrain.

Refined models use hourly meteorological data to calculate the hourly predicted concentration at the receptor grid. These models can also be simple and complex in their treatment of terrain.

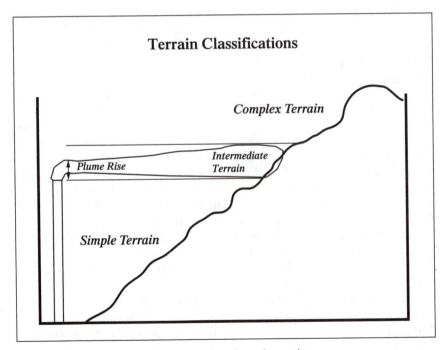

FIGURE 2.6   Distinction between simple terrain and complex terrain.

2.7.2 Meteorological Data

Meteorological data for screening models consist only of the hypothetical worst-case meteorological conditions. For refined modeling, however, meteorological data are required for every hour of a multiyear period. Typically, 5 years of meteorological data are used for refined modeling, although in some instances, with agency approval, 1 year of on-site meteorological data may be sufficient. In fact, 1 year of on-site data may be preferred to 5 years of off-site data if the nearest off-site meteorological data is from a weather station that is not considered representative of the facility being modeled. For example, if the nearest weather data available off site are from a monitoring station 40 to 50 miles away and if there are terrain differences or coastal influences between the monitoring station and the subject facility, the data may be considered unrepresentative. Also, if the facility to be modeled is in a deep river valley and the meteorological data is from a flat terrain site 30, 40, or 50 miles away, the data may not be representative either. It is important, especially under the PSD permitting program, to identify the required meteorological data early on. This is because PSD modeling requires use of 1 year of on-site or 5 years of off-site representative meteorological data. In the event that the off-site meteorological data are not available because of the issues identified above, 1 year of on-site data must be obtained prior to conducting the modeling required for permit application. Since 1 year of meteorological data are needed, and since several months would be necessary to locate and construct a meteorological observing station with its minimum 10-m tower, it is apparent that 15 to 18 months' delay would be required to obtain the data and conduct the modeling necessary to apply for a PSD permit. Note that this could be a fatal issue with respect to a plant expansion or for new construction. Therefore, the sooner an applicant is aware that a proposed action may require a PSD permit, the appropriateness of available meteorological data should be ascertained. In the event that there is nearby data from a similar terrain regime, it is likely that these data would be acceptable to the state or EPA for use in the PSD modeling. In the event that the data are more distant or that there are terrain considerations between the monitoring site and the subject site, it would be important to review the available data as described in the modeling protocol with the appropriate agency to be sure that the data would be acceptable. The last thing an applicant needs is to spend a lot of time, effort, and money conducting PSD modeling only to submit the analysis to the agency and be told the data used was not acceptable.

Documenting the data, the models, and all assumptions in the modeling protocol and getting agency acceptance before the studies are done is the single most important thing that an applicant can do to assure that the results of the modeling will be acceptable to the agency.

Upper air data may also be required. For modeled facilities with short stack heights, i.e., less than 150 to 200 feet, upper air data are normally not required. This is because upper air inversions are not likely to prove limiting to the dispersion from short stacks. For taller stacks, such as those on coal-fired power plants, upper air inversions can prove limiting to dispersion; upper air data are then necessary and refined dispersion modeling must be conducted. Figure 2.7 shows the normal lapse rate in the atmosphere in surface and upper air inversions. In order to determine the height of a potential upper level inversion, it is necessary to have upper air soundings, which are taken by the national weather service twice a day at a very limited number of observing stations (principally major airports). The good news is that the EPA modeling *Guidance* allows one to interpolate between upper air observing stations so as to estimate the upper level inversion height when one is observed to exist. Further, the *Guidance* allows the applicant to interpolate for each hour of the 12 hours between upper air soundings.

2.7.3 Required Modeling Data for the Impact Area

Air quality modeling uses a grid of prediction locations, or receptors. Note that these are not monitoring locations but are rather locations on a map surrounding the proposed facility at which predicted concentrations are calculated. With respect to monitors, however, if one or more exists within the potential impact area, receptors should be situated to correspond with the monitor locations. Doing

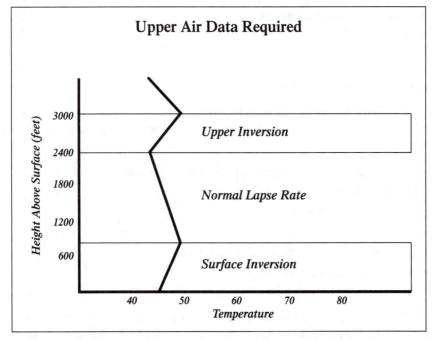

FIGURE 2.7   Normal lapse rate in the atmosphere in surface and upper air inversions.

this will allow the modeler to identify the predicted impact at the monitor and will facilitate calculating the total estimated air quality concentration in the area so that the modeled concentration can be added to the monitored concentration. Doing this, in effect, uses the monitor to calculate the background concentration to which the added impact of the new construction will be added. The cumulative concentration can be compared to the ambient air quality standards in order to demonstrate compliance with the NAAQS.

Additional receptors should be selected to correspond to *sensitive receptors*—facilities such as hospitals, nursing homes, schools, and other places where young people, old people, or people with respiratory illnesses may be present. Sensitive receptors are identified so that a calculation of the projected ambient air quality impact on these especially sensitive elements of the population can be made.

Next, determine the climatalogical statistics for the region. This information is required because the meteorological data used in the modeling must be representative of long-term conditions for the area. Suppose, for example, 1 year of on-site data are used, but that 1 year corresponds to a period of abnormal weather conditions, such as atypical wind patterns caused by a displacement of the jet stream to the north or south of its normal position. Those atypical conditions might render the variability and frequency of occurrence of different wind conditions nonrepresentative of the long-term record. It is the applicant's responsibility in applying for PSD permits, or conducting modeling for other purposes such as risk screen modeling, to demonstrate compliance with acceptable ambient levels of toxic air pollutants in states permitting actions to justify and to defend its choice of meteorological data.

2.7.4 Required Modeling Data for the Source

In order to conduct prediction modeling for a source of air emissions, it is necessary to know a lot about the source. For the release point itself, it is necessary to know:

- Height of the release point above ground, that is, the height of the stack
- Height above the building roof
- Temperature of the release
- Velocity of the release
- Flow rate or volumetric flow rate of the release
- Release point diameter and area
- Mass emissions per unit time

The horizontal length and width of the building may also be necessary because, for short stacks on large buildings, building-effect downwash can occur in moderate to high windspeed conditions. This phenomenon results in emissions being trapped in a cavity downwind of the building and creates high predicted (and actual) concentrations very near to the building. The solution to this problem is to increase the stack to such a height that it is outside of the cavity region created by the building wake. Figure 2.8 shows the building wake circulation region.

2.7.4.1 Specific Models. Appendix E is a list of regulatory models currently recommended by U.S. EPA and generally available either at no cost or low cost from the EPA at its Web page *www.epa.gov.* These models include a wide variety, from simple screenings to complex models. They also include numerical models, which for the most part, are not used in permitting actions. Numerical models solve the physical equations of the atmosphere and require extensive initial state data on the atmosphere on multiple levels and at multiple locations. The initial state data include information on many variables of the atmosphere so that the expansion, contraction, and flow of the atmosphere can be modeled over three-dimensional space. These equations are also used to estimate the dispersion, transport, and reaction of pollutants. Numerical prediction models are used typically by regulatory agencies for purposes of estimating the effectiveness of transportation control strategies and other state-implementation-plan-related measures in improving air quality concentrations in a region.

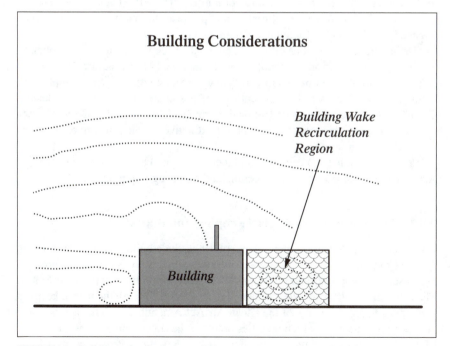

FIGURE 2.8 Building wake circulation region.

2.8 RECORDING AND RECORD KEEPING REQUIREMENTS

Having the correct permits to operate a facility is only the first step in maintaining a clear record of compliance. It is important to maintain records so that you can demonstrate, whenever necessary, that you are operating in accordance with your permits and that you have the proper permits to conduct the business that you are undertaking. The importance of record keeping and reporting cannot be understated. Since environmental regulatory agencies cannot be in all places at all times, and since their money spent on enforcement is limited, although it may not seem this way to the regulated community, agencies rely on record keeping and reporting to track the effectiveness of their regulatory programs.

There are a number of environmental regulations for the air media that require record keeping. These regulations require that certain monitoring activities be undertaken, that records of the monitoring be maintained, and that reports be filed on an annual basis. The Title V facility-wide permitting program is a major driver for record keeping under the 1990 Amendments to the Clean Air Act. Under the Title V program, we have seen the development of enhanced monitoring, only to be replaced by compliance assurance monitoring (CAM), supplemented now by the "credible evidence" policy recently promulgated by EPA. *Credible evidence* is a major new EPA enforcement initiative that promises to totally change how compliance is tracked and enforcement is undertaken.

Certifications are another important element of record keeping. A number of environmental regulations, particularly in the air quality area, require knowledgeable senior staff of companies to certify they are in compliance with applicable regulations. This leads to the concept of the "designated felon," since under the Clean Air Act Amendments of 1990, all criminal violations of the Clean Air Act are now felonies. This makes it much more attractive for the government to prosecute criminal violations.

Table 2.15 is a list of programs (and their acronyms) that, under the Clean Air Act Amendments, lead to record keeping and reporting requirements.

2.8.1 The Necessity of Record Keeping and Reporting

Under the Clean Air Act Amendments of 1990, the administrator of the EPA is required to transmit to Congress a report containing an inventory of the production, use, consumption, and emissions of sources throughout the United States. The EPA generates the air data contained in this report through the record keeping and reporting requirements of the Clean Air Act Amendments. This is a bottom-up reporting scheme in which facilities report to the states, and the states report to EPA. This multitude of reports is intended to allow the state agencies to trace the effectiveness of their regulatory programs and the degree of compliance by

TABLE 2.15   Clean Air Act Programs

Acronym	Stands for...
SIP	State Implementation Plan
NSR	New Source Review
PSD	Prevention of Significant Deterioration
LAER	Lowest Achievable Emission Rate
BACT	Best Available Control Technology
RACT	Reasonably Available Control Technology
NSPS	New Source Performance Standards
NESHAPs	National Emission Standards for Hazardous Air Pollutants
TCM	TCM
ERC	ERC

the regulated community. At the federal level, the reports allow the EPA to track the effectiveness of the state programs and the extent to which delegated programs are achieving their objectives.

State agencies, U.S. EPA, and state and federal courts rely on record keeping and reporting to determine whether a facility is in compliance with its permit terms and conditions and whether progress is being made where violations have been identified and consent agreements with compliance schedules put in place. Reports give states and federal agencies solid evidence and background data for estimating emissions and ambient concentration levels.

2.8.2   Criteria for Records and Reports

Records must be maintained for at least 5 years from the date of each sample, measurement, or report. Environmental permit holders must maintain records of all source testing or monitoring performed at the facility that is required by a Title V operating permit, or by any state permit to construct and operate. Manufacturers have responsibility to establish and maintain records and make reports available at their facility for inspection during normal business hours or to the state regulatory agency or EPA on request. Additionally, facilities have an obligation to allow reasonable access by the state and federal agencies and to assist them in inspecting the facility and the records and reports maintained therein.

Records should contain the following:

- Date, description of source operation, and time of sampling
- Dates analyses were performed by laboratory
- Company name, person name responsible for performing the sampling or analysis
- Analytical technique or method used at the laboratory to analyze the samples
- Results of the laboratory analysis

- Operating conditions of the source or facility being tested and documentation that its operating parameters were within the levels allowed under the applicable permit
- All calibration and maintenance records and all original continuous monitoring instrumentation results, including both digital records and chart records where available.

2.8.3 Benefits of Record Keeping and Reporting

While it may seem that record keeping and reporting are used primarily to show noncompliance with applicable regulations, there are benefits to record keeping and reporting. For example, record keeping may sometimes be used in lieu of monitoring. Take, for example, a surface coating operation that may be subject to a monitoring requirement. Monitoring such operations can be extremely expensive, especially if an agency dictates use of continuous emission monitors. Record keeping, on the other hand, may consist simply of inventory record keeping of the amount of surface coating liquids purchased and applied. Typically, purchasing records are not adequate, and a daily log kept by the operators of the amount of gallons of each surface coating taken from storage to be applied must be maintained. This, however, is much easier and less expensive than installing continuous emission monitors and, as an added benefit, is less likely to show violations of permit conditions. This is because continuous emission monitors track the minute-by-minute emissions of a source. Many permits are (incorrectly) written so that the maximum hourly emission rate is really the average emission rate, thereby inviting a noncompliance situation. This is because, for an average emission rate, the source will obviously sometimes exceed the emission rate and at other times be below the emission rate. On average, for an annual period, the average would be correct. It is important in structuring a permit application, therefore, to identify the maximum hourly emission rate, taking into account the maximum physical capacity of the source and the worst-case liquid in this case being applied. As a note to the application, the annual emission rate is figured by taking into account the hours of operation and actual production level of the equipment. Continuous emission monitors (CEMs), when used, will show the minute-by-minute variation in emission rates. It is not uncommon for a continuous emission monitor to show exceedances of permit limits. Frequently, inspectors will not make the connection that the permit specifies a maximum hourly emission rate, let's say 5 pounds/h. Seeing a continuous emission monitor at a certain parts per million level that translates into, say, 7 lb/h, an inspector may use this as a basis for citing a violation of the permit. This in itself is improper, since the correct way to interpret the CEM data is to average the readings over a complete hour and to see if the 1-hour average exceeds the permit limit. It should be apparent from this analysis that record keeping of the gallons of paint applied in a day is preferable to monitoring, and lessens the possibility that a violation will be recorded. Under record keeping, typically a maximum gallons per day application rate needs to be demonstrated through purchasing or more often now, through a log of the amount applied.

Another benefit of monitoring by record keeping is that it allows a facility to monitor compliance and to self-correct problems before violations can occur. It also allows a facility to strive toward compliance via past records and performance. Another benefit of record keeping is that, should a violation occur, record keeping for past months and years may be very helpful in showing an agency that this is an abnormal event and that, in the past, an excellent record of compliance with permit conditions has been maintained, as shown through the record keeping maintained at the site.

2.8.4 Regulatory Drivers

Record keeping and reporting have been required in numerous regulations for many years. For example:

- National Emission Standard for Hazardous Air Pollutants (NESHAPs)
- New Source Performance Standards (NSPSs)
- Prevention of Significant Deterioration (PSD)
- New Source Review (nonattainment area)

Title V of the 1990 Clean Air Act Amendments specifically addresses permit provisions for monitoring and record keeping. Under Title V, EPA's first draft policy regarding monitoring and record keeping was enhanced monitoring (EM). Enhanced monitoring required that any emission unit at a major facility whose emissions exceeded half the major source threshold was required to be monitored by a continuous emission monitor. The premise of enhanced monitoring was that the source was out of compliance unless monitoring data showed it to be continuously in compliance. This, in effect, is a guilty until proven innocent premise. The regulated community objected strenuously to both the expense of this and the inference that they needed to demonstrate continuous compliance every minute, every hour of the day and year. As a result, EPA promulgated compliance assurance monitoring (CAM). CAM required facilities to undertake monitoring principally when pollution control devices were present. CAM reversed the premise and was supposed to show only on an occasional basis that the facility was in compliance with its permit terms and conditions. It was no longer necessary for the facility to show continuous compliance, and the presumption of noncompliance unless monitoring data showed compliance was reversed. While CAM was an improvement over enhanced monitoring, in practice, it has not proved to be as beneficial as one might expect. Initially, EPA proposed that monitoring data that merely showed that a pollution control device was installed and was operating would be sufficient. EPA suggested that if a scrubber was installed, for example, data showing that the circulating water was circulating and that the pumps were running would be sufficient to show that the control device was present and operating. As the Title V program developed, however, EPA began backpedaling on CAM and is now requiring more record keeping and monitoring than was originally envisioned under this program. In fact, EPA is rejecting Title V permits from states if they do not contain more monitoring than was required in the original permits for major sources before the Title V program.

The credible evidence program, discussed in Sec. 2.9, "Liabilities and Enforcement," is yet another tightening of the regulatory program involving record keeping and reporting.

2.8.5 Detailed Look at Records and Reports

40 CFR 63.122 and 63.123 Subpart G require owners and operators of storage tanks to visually inspect the internal floating roof, primary and secondary seals, manhole and roof hatch seals, gaskets, and membranes each time the storage tank is empty or is degassed each year. This basically requires operators of tanks covered by these subparts to completely empty the tank and to visually inspect the seals each year. Furthermore, records of these inspections must be maintained and be made available to the state agency or EPA. Requirements for such inspections would be made a part of the facility's air permit for applicable tanks.

40 CFR 63.181 Subpart F, Leak Detection and Reporting, requires that all covered equipment have an identification number and be physically tagged. It also requires documentation of any visible, audible, or olfactory evidence of fluid loss.

2.8.6 Types of Systems Used to Maintain Records and Reports

Records and reports are of no value if they cannot be found. Remember that the source has an obligation to assist agencies in reviewing records and reports when inspectors visit the facility. Records may be kept in computer files or on paper. Paper records should be in a permanently bound logbook. The reason for this is to demonstrate that any unfavorable notes or data in records have not been removed. If the logbook is not permanently bound, it is not possible to document that notes made by employees have not been removed from the record book.

Records stored on computers or other electronic media such as microfilm, microfiche, or magnetic tape should be easily accessible on site when the facility is inspected. They should be backed up daily, and they should be generated by automatic data collection systems.

2.8.7 Frequency of Reporting

Various regulatory programs and regulations impose different reporting requirements. For example, Title VI for Class I and II Ozone Depleting Substances requires quarterly reports on the amounts of ozone-depleting substances imported, exported, or produced. Utility and nonutility boilers are required to report quarterly on sulfur oxide emissions or sulfur content of fuel consumed. Tests for motor vehicles are required to be conducted on an annual basis, and reports must be submitted by states for covered fleets on an annual basis. Storage tanks subject to inspection must file their reports within 30 days of the periodic annual inspection.

Under Title V, monitoring reports must be submitted every 6 months unless otherwise stipulated in the operating permit, but a facility may be able to negotiate with the state or EPA regarding how frequently reports must be filed. However, in addition, under Title V a facility is required to promptly report deviations from the permit, even if they are a result of problems or emergencies at the facility. Moreover, states may impose additional reporting requirements, and there is an annual compliance certification that must be filed.

2.8.8   How to Prepare for an Annual Compliance Certification

The annual compliance certification must be signed by a senior person at the facility, one who is also knowledgeable about facility operation. It is common that the person required to sign the certification does not have direct responsibility for the emitting sources, and, therefore, records showing compliance with the permit will help show the person who signs the certification that the source is, in fact, in compliance with its permit limits. If the person who signs the certification cannot verify this, he or she runs the risk of signing a false certification, and could be subject to personal criminal liability under the Clean Air Act. In order to make it easier for a corporate official to sign the certification and to demonstrate compliance in preparing the certification, it is most helpful to implement a system to obtain careful and accurate monitoring data and to have accurate record keeping at all levels of management responsible for the operation of the source. It is also helpful to have continual data collection, whether it be from continual emission monitors, from record keeping, or from regularly scheduled evaluations or tests. It is also important to document any emergencies and shutdowns that may have occurred throughout the reporting period and to be able to document that permit conditions were still maintained or, where variations from permit conditions occurred, that these were reported to the agencies and promptly corrected.

2.9   LIABILITIES AND ENFORCEMENT

Having proper permits for a facility that emits air pollutants, and operating within the requirements of those permits, is one of the principal ways to avoid regulatory liabilities, enforcement actions, and penalties, which can range up to tens or hundreds of thousands of dollars or more and can also include being required to cease operations. Finding that a facility is in violation of a permit, or does not have permits that are required for it, puts the operator into the category of knowing violator. However, this is not the test of whether an operator is a knowing violator. The test, under the Clean Air Act, is that the operator knew or should have known, that it was in violation of the Act. This is a much more stringent test in that the determination of "should have known" is not precise and relies on interpretation of the regulatory agency or the courts. The addition recently of EPA's credible evidence policy further complicates what "should have known" is.

The trend in enforcement recently is toward bigger cases, with civil judicial referrals. This is because, under the Clean Air Act Amendments of 1990, EPA is limited in its penalties to $25,000 a day or $250,000 in total, but EPA can impose multimillion dollar fines by referring cases to the judicial system. Under the Clean Air Act Amendments of 1990, the administrator is free to seek higher penalties as long as the action has the approval of the attorney general. There are a number of programs under the Clean Air Act Amendments of 1990 that further complicate maintaining a good compliance status. These include the EPA field citation or traffic ticket provisions. This allows EPA inspectors to visit a facility at any time and to write field citations for "minor violations." These minor violations can carry with them a $5000 per day assessment. Also, the presumption is that the noncompliance continues until the respondent affirmatively demonstrates compliance.

Another onerous program of the 1990 Amendments is the Citizen Award Program, which is being implemented despite the lack of a final rule from the EPA on how this program should work. This program is basically a bounty hunter program that provides citizens up to $10,000 for reporting violations of the Act.

Citizen suits are another enforcement mechanism available under the Clean Air Act Amendments of 1990. In the past, EPA barred wholly past violations from enforcement; if a facility had a period of noncompliance, but it was corrected prior to an agency discovering the violation, the facility was "off the hook." The facility was not subject to penalty assessment, since by the time the violation had been discovered by the agency, the violation had been corrected. Under the 1990 Amendments, there is no bar for these wholly past violations. The legislature, in the 1990 Amendments, removed the precedent in the Gwaltney Clean Water Act case.

Now, under the 1990 Amendments, wholly past violations still expose companies to enforcement. What this means is that annual reports such as the R form reports under EPCRA, annual emission reports, and compliance certifications under the Title V program, among many other types of reports, may indicate violations of permits. Take, for example, the emissions report, which discloses an annual emission level of 12 tpy for a particular facility. Reviewing the permits for this source indicates that permits limit the source to 10 tpy. The facility has, in effect, reported a violation. In the past, if the facility recognized the problem and did not emit more than the permit limit in the future, it would be barred from enforcement action for the past year in which a violation occurred. Under the 1990 Amendments, however, this past violation is not barred from enforcement. Environmental groups have ever-broader access to annual reports from companies from electronic media such as the Internet. This means that environmental groups have access to monitoring data, as they never had in the past. This makes it possible for them to review monitoring data for facilities, to obtain permit applications through the Freedom of Information Act or other sources, and to compare the permit limits to the annual or semiannual reports. Where annual reported data exceed permit limits, expect a citizen's suit. The problem with this is that it may be the report that is wrong.

In many instances, facilities put less time and effort into filing their reports than they do into filing permit applications. Take, for example, a facility where permits have been in place for a number of years, and a new person is assigned the responsibility for the annual air emissions report. This person, without checking the permit file, estimates annual emissions using current emission factors. The person does an accurate job of estimating emissions, but neglects to check the permit application to determine what the applicable emission limits are for the subject source. The employee reports the annual emissions, taking into account emission factors and utilization levels. It turns out that the reported emission rate exceeds the permit limits. How can this happen? Clearly the permit application was in error or the person comparing the annual emissions report made a mistake. However, there is a third possibility: that the emission factors have changed between the time that the permit application was prepared and the time that the annual emissions report was prepared. This situation happened in New Jersey recently, when emission factors for boilers were changed by the U.S. EPA. EPA based new information on small and midsize boilers to modify the AP-42 emission factors. As a result, when new emission factors were applied to estimate annual emissions for existing boilers, they were found to exceed permit limits. What should a facility do? In some instances it is possible to use the old emission factors, since the boiler has not really changed—it's just that the emission factors have changed. This is somewhat dangerous, however, in that the emission factors changed for a reason. EPA has new data on boiler emissions and took these data into account in revising the emission factors. In addition, after 1 year of this ambiguity, the state revised its emission reporting software program, the program that sources use to file emission reports electronically, to incorporate the new emission factors. Using the new program results in emissions estimates that in many cases exceed permit limits. The only options that are left to a facility are (1) to demonstrate that alternative emission factors are more appropriate than those contained in the state's program or (2) to modify its permits to show new emissions limits. The latter option may be preferable; however, in some instances modifying the permit will result in a significant emissions increase. Even though the subject boiler has not been modified and has not changed its fuel, using new emission factors will show a higher annual emission value than the old emission factors. This is a theoretical emissions increase. For sources located in or near nonattainment areas, this can create problems. It may also create problems for the source in meeting state-of-the-art requirements if these are imposed in the area.

2.9.1 Audit Privilege/Immunity

EPA has changed policy a number of times with respect to audit privilege and immunity. At issue is the question of whether it is advantageous to the environment for facilities to review their compliance with regulations and permits and to self-report problems that are identified. Facilities would like to be free to do this and have immunity from enforcement action should they identify a compliance issue and self-correct it within a reasonable period of time. EPA's position is that self-audit should be encouraged; however, problems that are discovered as a result of an audit should not necessarily be free from enforcement. Currently, EPA has a penalty mitigation policy that is intended to encourage self-auditing and to allow facilities to have the benefit of reduced penalty assessments for problems found, reported, and self-corrected in a timely manner. In practice, however, EPA's penalty mitigation policy is of limited value since the requirements are very stringent for penalty mitigation. The policy requires that the facility identify the problem, not as part of a routine program of reinspection, but as a result of a specific audit.

In addition, the problem must be reported promptly, and it must be corrected within 30 to 60 days of being identified. Many air emission problems cannot be corrected this quickly, since, if a source is in violation of a permit and the permit needs to be modified, a new permit application must be prepared, submitted to the state, and undergo a sometimes lengthy permit review by the state. In other instances, coming into compliance may require installing pollution control equipment—a very time-consuming process. Sources are required to first identify required control equipment and conduct an engineering feasibility study, prepare a conceptual design, obtain quotations for fabrication and installation of the equipment, and then to install and make the equipment operational. This sometimes takes as long as several years, but it would be difficult in any instance to accomplish it in less than 6 to 12 months. This guidance was published on December 22, 1995, and may be found in FR 60, No. 246, pp. 66706–66712.

2.9.2 Credible Evidence

Credible evidence is EPA's latest attempt at enforcement "flexibility." It was originally part of the proposed enhanced monitoring rule, and the final rule was issued on February 24, 1997 (62 FR 8314). Credible evidence represents a fundamental change in enforcement policy. It authorizes the use of any "credible" evidence, other than compliance test methods, to prove noncompliance. EPA's position notwithstanding, the credible evidence regulations significantly increase the stringency of the underlying standards. They also don't take into account emission variability under different operating conditions.

Before credible evidence, EPA was assessing violations on the basis of compliance test methods that were codified in the *Federal Register.* This means that a source could be cited for violations of an emission standard under federal emission requirements only if a reference test showed a violation of the emission standard. Under part of the enhanced monitoring proposal, EPA decided that the administrator could impose violations and undertake enforcement action on the basis of credible evidence that is not necessarily based on a compliance test method. The EPA bases this position on statutory language that gives EPA the ability to bring administrative, civil, or criminal actions "on the basis of information available to the administrator." The statute does not specify on the basis of performance tests alone.

2.9.2.1 What Is Credible Evidence?

Credible evidence is not defined by the regulations or by EPA. Credible evidence includes any evidence that may indicate a problem with the operation of a facility. This could include, for example, purchasing records showing that the wrong grade of fuel or sulfur content of fuel was purchased, or that paint with the wrong volatile content was purchased for surface coating. It could include observations by someone outside the facility who notices smoke or some other air emission and interprets it as a problem. Or it could include data from a test that is not a reference test but just a spot check on the operation of a source.

Who decides if the evidence submitted is "credible"? It is not the EPA administrator or the state inspector; the final judge of whether the evidence is credible is just that, a judge. EPA has decided that the judicial system should be the final arbiter on whether evidence presented is credible. This is an inherent problem in the policy.

EPA takes the position that credible evidence does not increase the stringency of an emissions standard. This does not appear to be the case. Before credible evidence, compliance with an emissions standard could be demonstrated only through a referenced test method that would be undertaken at most once a year, and sometimes only once every several years, at permit renewal. Under credible evidence, however, any other evidence such as fuel purchasing records or smoke seen from the stack can be used to bring enforcement action, and, as noted above, a judge is the arbiter of whether the evidence is in fact credible.

There is another key problem with the credible evidence policy. This involves the inference that if a source at a facility is found to be in violation of a permit through credible evidence, the violation is assumed to have existed every day preceding the obtaining of credible evidence. To what prior date is this noncompliance assumed to extend? It is assumed to extend backward in time to the date of the most recent test showing that the source was in compliance, or in the absence of such test, to the day the source began its first operation. Therefore, evidence of a violation obtained in 1999 of a source that has never been tested but that has been in operation for 15 years could be used to infer noncompliance for every day for the last 15 years. Looking at a potential penalty of $25,000 per day for each day of violation, we can see a potential fine in the hundreds of millions of dollars, and in this case $136 million!

Does EPA seriously expect that such megadollar fines would be collected from the regulated community? Probably not. However, EPA will use credible evidence as a key enforcement tool under the Clean Air Act and will use these large potential fines as a wake-up call to industry to put it on notice that noncompliance as determined under credible evidence can expose companies to extremely high potential fines. The response the EPA hopes for is clearly that industry will operate in compliance and will document compliance on a regular basis to minimize the enforcement exposure under this program.

A final concern under credible evidence is the potential for frivolous lawsuits to be brought by members of the public or by environmental groups under this policy. EPA is not concerned that this will be a problem because EPA intends to bring lawsuits itself only for violations that:

• Create threat and harm to the public health or environment
• Shows significant duration or magnitude
• Present a pattern of noncompliance
• Involve a refusal to provide information
• Involve criminal conduct
• Allow a source to reap a windfall

A flaw with this line of thinking, however, is that such objectives in the application of credible evidence do not necessarily apply to the public or to environmental groups, which may use credible evidence as a tool to bring citizen suits against facilities.

CHAPTER 3
WATER QUALITY

John A. Green
Director of Safety and Health
Tate & Lyle North American Sugars Inc.

3.1 OVERVIEW

Water pollution control laws, including those associated with the Clean Water Act and its related regulations, are broad and encompass many, but not all, industrial operations. The basic mechanism by which a regulatory agency (e.g., the U.S. Environmental Protection Agency (EPA) or an authorized state permitting agency) enforces the prohibitions of an environmental statute is the issuance of permits. Basically a permit is an authorization, license, or equivalent control document issued by the EPA or an "approved state" to implement the requirements of a specific set of regulations. In the context of this chapter, a "permit" would include a National Pollutant Discharge Elimination System (NPDES) "general permit" (40 CFR 122.28).

The various types of permits that are available under the Clean Water Act to control discharges of pollutants into surface waters or discharges of dredged or fill material into wetlands and navigable waters include, but are not limited to, the following:

- Pretreatment permits
- Sewage sludge use and disposal permits
- Stormwater discharge permits
- Dredge and fill permits
- General permits
- Well injection—underground injection control permits

Although each of these permits will be covered in this chapter, the emphasis of our discussions will be on the NPDES and its related stormwater discharge permit program.

3.2 THE NATIONAL POLLUTANT DISCHARGE ELIMINATION SYSTEM

The cornerstone of the nation's water pollution control efforts is the National Pollutant Discharge Elimination System. This program implements the prohibition by the Clean Water Act of discharging pollutants from a point source to the waters of the United States. Under the NPDES program 40 CFR Part 122, all industrial and municipal facilities that discharge wastewater directly into U.S. waters must obtain a permit. Specifically, the Clean Water Act requires NPDES permits for discharges from point sources such as municipal wastewater treatment plants, industries, animal feedlots, aquatic animal production facilities, and mining operations. The mechanisms by which water pollution control programs operate can be found in the program's issuance of permits which specify effluent limitations for each individual industrial and municipal discharger, a compliance schedule, monitoring and reporting requirements, and any other terms and conditions necessary to protect water quality. Wastewater permits are required for all discharges of wastes to waters of the United States (or state waters). Any industrial wastewater discharge to a publicly owned treatment works (POTW or sewer district) and, in some cases, the operation of other wastewater disposal systems, whether or not they have a discharge, such as storage basins, land farming, and recycling systems, also require an appropriate wastewater permit.

The EPA is authorized to issue NPDES permits with conditions the agency "determines are necessary to carry out the provisions" of the Clean Water Act. The various states (and U.S. territories) may assume administration of the system if their programs are approved by the EPA. Currently, over 40 states have EPA-approved NPDES programs. For example, the state of Louisiana was authorized to implement the NPDES program on Aug. 27, 1996. With EPA headquarters and the agency's 10 regional offices, there are thus over four dozen agencies dealing directly and closely with NPDES permits. The remainder of the states and territories review proposed permits, so that in fact the number is even higher.

Originally, NPDES permits concentrated on controlling and/or minimizing the discharge of traditional pollutants (e.g., pH, oxygen-depleting substances, temperature, solids, etc.). However, the current focus of permitting agencies is on removing toxic pollutants from industrial sources and complying with water-quality–based standards and technology-based effluent limitations.

3.2.1 NPDES Permits

Although, as stated earlier, various types of permits are available under the Clean Water Act to control discharges of pollutants into surface waters or discharges of dredged or fill material into wetlands and navigable waters, the focus of this section is confined to NPDES direct wastewater discharge permits (including stormwater associated with industrial activity) with a limited discussion on pretreatment permits. For general industry, the wastewater permits usually required under the NPDES program are those authorizing the direct discharge of wastewaters (including stormwater discharges associated with industrial activity) into the waters of the United States and indirect discharge to a facility which, in turn, treats and discharges the wastewater into the waters of the United States.

3.2.2 When Must an NPDES Permit Be Obtained?

An NPDES permit must be obtained before wastewater is discharged into the waters of the United States. Any material added to water or any physical characteristic imparted to water (heat, pH, color) is a pollutant. A plant must obtain an NPDES permit if it is a point source, that is, a specific and distinct discharge point. A rain ditch, pipe, or container is a point source. A lawn, parking lot, hillside, or field is a nonspecific area source. Process wastewater and stormwater runoff must be permitted. Many times an NPDES permit will be required for noncontact cooling water discharged to the waters of the United States. Although a facility may have more than one discharging point source, a single NPDES permit is usually issued for the entire facility. This permit will contain both general information pertaining to all point sources (usually referred to as outfalls and sequentially numbered as 001, 002, 003, etc.) and specific discharge limitations pertaining to each individual one!

An application for an NPDES permit must be filed before construction begins. If your plant is already operating and has no permit, you are legally required to contact your permitting authority immediately unless you have correspondence in your files that explicitly exempts your operations. The authority must receive applications for renewal of a permit at least 180 days prior to expiration of the old one. If an application for renewal is received on time, the old permit continues in force while the director writes and approves the new one. If it is not filed on a timely basis, the discharge in question could be operating without a permit after midnight the day it expires. This would create the possibility of significant enforcement action.

3.2.3 Scope of the NPDES Program's Permit Requirements

The NPDES program requires permits for the discharge of "pollutants" from any "point source" into "waters of the United States." The terms "pollutant," "point source," and "waters of the United States" are listed below as defined in 40 CFR 122.2.

- *Point source* means any discernible, confined, and discrete conveyance, including but not limited to any pipe, ditch, channel, tunnel, conduit, well, discrete fissure, container, rolling stock, concentrated animal feeding operation, landfill leachate collection system, and vessel or other floating craft from which pollutants are or may be discharged. This term does not include return flows from irrigated agriculture or agricultural stormwater runoff.

- *Pollutant* means dredged spoil, solid waste, incinerator residue, filter backwash, sewage, garbage, sewage sludge, munitions, chemical wastes, biological materials, radioactive materials [except those regulated under the Atomic Energy Act of 1954, as amended (42 U.S.C. 2011 et seq.)], heat, wrecked or discarded equipment, rock, sand, cellar dirt, and industrial, municipal, and agricultural waste discharged into water. It does not mean (*a*) sewage from vessels or (*b*) water, gas, or other material which is injected into a well to facilitate production of oil or gas, or water derived in association with oil and gas production and disposed of in a well, if the well used either to facilitate production or for disposal purposes is approved by authority of the state in which the well is located, and if the state determines that the injection or disposal will not result in the degradation of ground or surface water resources.

- *Waters of the United States* means: (*a*) All waters which are currently used, were used in the past, or may be susceptible to use in interstate or foreign commerce, including all waters which are subject to the ebb and flow of the tide; (*b*) all interstate waters, including interstate "wetlands"; (*c*) all other waters such as intrastate lakes, rivers, streams (including intermittent streams), mudflats, sandflats, "wetlands," sloughs, prairie potholes, wet meadows, playa lakes, or natural ponds the use, degradation, or destruction of which would affect or could affect interstate or foreign commerce including any such waters:

  (1) Which are or could be used by interstate or foreign travelers for recreational or other purposes

  (2) From which fish or shellfish are or could be taken and sold in interstate or foreign commerce

  (3) Which are used or could be used for industrial purposes by industries in interstate commerce

  (*d*) all impoundments of waters otherwise defined as waters of the United States under this definition; (*e*) tributaries of waters identified in paragraphs (*a*) through (*d*) of this definition; (*f*) the territorial sea; and (*g*) "wetlands" adjacent to waters (other than waters that are themselves wetlands) identified in paragraphs (*a*) through (*f*) of this definition. Waters of the United States do not include prior converted cropland. Notwithstanding the determination of an area's status as prior converted cropland by any other federal agency, for the purposes of the Clean

Water Act, the final authority regarding Clean Water Act jurisdiction remains with the EPA.

Note: Waste treatment systems, including treatment ponds or lagoons designed to meet the requirements of CWA [other than cooling ponds as defined in 40 CFR 423.11(m) which also meet the criteria of this definition], are not waters of the United States. This exclusion applies only to manmade bodies of water which neither were originally created in waters of the United States (such as disposal areas in wetlands) nor resulted from the impoundment of waters of the United States.

The following are point sources requiring NPDES permits for discharges:

- 40 CFR 122.1(*b*)(2)(i)—concentrated animal feeding operations as defined in 122.23
- 40 CFR 122.1(*b*)(2)(ii)—concentrated aquatic animal production facilities as defined in 122.24
- 40 CFR 122.1(*b*)(2)(iii)—discharges into aquaculture projects as set forth in 122.25
- 40 CFR 122.1(*b*)(2)(iv)—discharges of stormwater as set forth in 122.26
- 40 CFR 122.1(*b*)(2)(v)—silvicultural point sources as defined in 122.27

In addition to the limitations placed on your discharge stream (flow, concentrations of specific pollutants, mass limits, pH, temperature, etc.), the CWA permits have special boilerplate provisions. These clauses are: (1) bypass, (2) upset, (3) halt or abate, (4) duty to mitigate, (5) notification, and (6) operation and maintenance.

According to the general conditions specified in an NPDES permit, no waste stream may be diverted from any part of a treatment facility. When this happens, it is called *bypass,* and it is prohibited where it causes the discharge limits to be exceeded. Bypass is allowed in emergency conditions in order to avoid severe property damage, personal injury, or loss of life, but the burden of proof is on the permittee. Bypass is allowed if it does not cause discharge limits to be exceeded but only for essential maintenance to ensure efficient operation. You may not routinely bypass for any particular reason.

An *upset* is the unintentional temporary violation of discharge limitations due to factors beyond your control. An upset is an acceptable defense against noncompliant discharges provided the permit is technology-based, and if a notice was filed within 24 hours of the upset event. Permit violations caused by operational error, improper design, inadequate facilities, lack of preventive maintenance, carelessness, or improper operation are not upsets.

An NPDES permit requires the permittee to control its discharge within limits at all times. That is the implication of the discharge limits in its permit. If the discharge cannot be controlled, the operation must be shut down. Occasionally production can be curtailed to expedite getting back into compliance without a complete shutdown.

If preventive measures have been taken and the permittee's discharge is still non-compliant and the environment is damaged in some way, then the permittee has a *duty to mitigate*. This clause in an NPDES permit requires all reasonable steps to be taken to minimize or prevent environmental damage and prompt mitigation once it does occur.

If the permittee performs any physical alterations or adds equipment, the permitting agency must be provided with *notification*. Permit authorities have interpreted this clause to mean alterations or additions which could significantly change the nature of the discharge or which would increase the quantity of pollutants.

Permittees are required to provide *proper operation and maintenance* for treatment systems (e.g., an operator's daily log sheet and maintenance checklists or other administrative evidence that maintenance is being routinely performed).

3.2.4 NPDES Permit Requirement Exclusions

The following discharges do not require NPDES permits:

- 40 CFR 122.3(*a*)—Any discharge of sewage from vessels, effluent from properly functioning marine engines, laundry, shower, and galley sink wastes, or any other discharge incidental to the normal operation of a vessel. This exclusion does not apply to rubbish, trash, garbage, or other such materials discharged overboard nor to other discharges when the vessel is operating in a capacity other than as a means of transportation such as when used as an energy or mining facility, a storage facility, or a seafood processing facility, or when secured to a storage facility or a seafood processing facility, or when secured to the bed of the ocean, contiguous zone, or waters of the United States for the purpose of mineral or oil exploration or development.

- 40 CFR 122.3(*b*)—Discharges of dredged or fill material into waters of the United States which are regulated under section 404 of the CWA.

- 40 CFR 122.3(*c*)—The introduction of sewage, industrial wastes, or other pollutants into publicly owned treatment works by indirect dischargers. Plans or agreements to switch to this method of disposal in the future do not relieve dischargers of the obligation to have and comply with permits until all discharges of pollutants to waters of the United States are eliminated. [See also 122.47(*b*).] This exclusion does not apply to the introduction of pollutants to privately owned treatment works or to other discharges through pipes, sewers, or other conveyances owned by a state, municipality, or other party not leading to treatment works.

- 40 CFR 122.3(*d*)—Any discharge in compliance with the instructions of an on-scene coordinator pursuant to 40 CFR Part 300 (the National Oil and Hazardous Substances Pollution Contingency Plan) or 33 CFR 153.10(*e*) (Pollution by Oil and Hazardous Substances).

- 40 CFR 122.3(*e*)—Any introduction of pollutants from nonpoint-source agricultural and silvicultural activities, including stormwater runoff from orchards, cultivated crops, pastures, range lands, and forest lands, but not discharges from concentrated animal feeding operations as defined in 122.23, discharges from concentrated aquatic animal production facilities as defined in 122.24, discharges to aquaculture projects as defined in 122.25, and discharges from silvicultural point sources as defined in 122.27.

- 40 CFR 122.3(*f*)—Return flows from irrigated agriculture.

- 40 CFR 122.3(*g*)—Discharges into a privately owned treatment works, except as the director may otherwise require under 122.44(*m*).

3.2.5 NPDES Permit Application Forms

Applicants for EPA-issued NPDES permits must submit their applications on EPA's permit application forms or approved state forms when available. Most of the information requested on these application forms is required by federal and/or state regulations. The basic information required in the general form (Form 1) and the additional information required for NPDES applications (Forms 2a through 2d) are listed in 40 CFR 122.21. Applicants for state-issued permits must use state forms, which must require at a minimum the information listed in these sections. Listed below is a partial summary of the various such EPA application forms. (*Note:* Refer to Appendix F for a complete set of these forms.)

- Form 1: General information
- Form 2c: Detailed information on existing sources
- Form 2d: Detailed information on new sources and new discharges
- Form 2e: Information on discharges comprised solely of noncontact cooling water
- Form 2f: Information on stormwater discharges associated with industrial activity

3.2.6 NPDES Technical Regulations

The NPDES permit program has separate additional regulations. These separate regulations are used by permit-issuing authorities to determine what requirements must be placed in permits if they are issued. These separate regulations are located at 40 CFR Parts 125, 129, 133, 136, 40 CFR Subchapter N (Parts 400 through 460), and 40 CFR Part 503.

3.2.7 NPDES Permit Application Process

Normally, an application for an NPDES wastewater permit must be submitted at least 180 days (6 months) before the regulated activity begins. If the applicant is located in a state that has received authority to implement the NPDES program, it must submit its application to the appropriate state director. (*Note:* If a current NPDES permit is in place, the renewal submission must be submitted at least 180 days prior to the expiration of the current permit!) When the office writing the permit receives the application, it will make a preliminary determination of completeness. You will be required to supply missing information before the permitting process continues. When complete, the application will be reviewed to make a preliminary determination of applicability.

A draft permit is forwarded to the applicant for review and comment. The applicant has 30 days to request a meeting to discuss changes. Minor changes may be discussed over the telephone, but typically a face-to-face meeting is required for more substantial changes. A letter covering a marked-up draft permit may also be appropriate in some cases.

Once the applicant and the permit writing office agree on the draft (silence on the part of the one seeking the permit infers tacit agreement), public comment is invited. The permit writing office files a public notice of the proposed permit (usually in the local newspaper) and the public has 30 days to file a petition for a public hearing. If the permitting authority decides that public interest is significant, a hearing will be scheduled, giving the public at least 30 days notice. After public input, the permit writing office recommends approval or denial of the permit to the permit authority. Final decisions of the permit authority may be appealed in court. Requirements have been established for public participation in EPA and state permit issuance and enforcement and related variance proceedings, and in the approval of state NPDES programs. These requirements carry out the purposes of the public participation requirements of 40 CFR Part 25 (Public Participation) and supersede the requirements of that part as they apply to actions covered under Parts 122,123, and 124.

3.2.8 Signatories to Permit Applications and Reports

(*a*) *Applications.* All permit applications must be signed as follows:

(1) *For a corporation:* by a responsible corporate officer. A responsible corporate officer means: (i) A president, secretary, treasurer, or vice-president of the corporation in charge of a principal business function, or any other person who performs similar policy- or decision-making functions for the corporation, or (ii) the manager of one or more manufacturing, production, or operating facilities employing more than 250 persons or having gross annual sales or expenditures exceeding $25 million (in second-quarter 1980 dollars), if authority to sign documents has been assigned or delegated to the manager in accordance with corporate procedures.

[*Note:* EPA does not require specific assignments or delegations of authority to responsible corporate officers identified in 122.22(*a*)(1)(i). The agency will presume that these responsible corporate officers have the requisite authority to sign permit applications unless the corporation has notified the director to the contrary. Corporate procedures governing authority to sign permit applications may provide for assignment or delegation to applicable corporate positions under 122.22(*a*)(1)(ii) rather than to specific individuals.]

(2) *For a partnership or sole proprietorship:* by a general partner or the proprietor, respectively.

(3) *For a municipality, state, federal, or other public agency*: by either a principal executive officer or ranking elected official. For purposes of this section, a principal executive officer of a federal agency includes (i) the chief executive officer of the agency, or (ii) a senior executive officer having responsibility for the overall operations of a principal geographic unit of the agency (e.g., regional administrators of EPA).

(*b*) All reports required by permits and other information requested by the director must be signed by a person described in paragraph (*a*) of this section or by a duly authorized representative of that person. A person is a duly authorized representative only if:

(1) The authorization is made in writing by a person described in paragraph (*a*) of this section.

(2) The authorization specifies either an individual or a position having responsibility for the overall operation of the regulated facility or activity such as the position of plant manager, operator of a well or a well field, superintendent, position of equivalent responsibility, or an individual or position having overall responsibility for environmental matters for the company. (A duly authorized representative may thus be either a named individual or any individual occupying a named position.)

(3) The written authorization is submitted to the director.

(*c*) *Changes to authorization.* If an authorization under paragraph (*b*) of this section is no longer accurate because a different individual or position has responsibility for the overall operation of the facility, a new authorization satisfying the requirements of paragraph (*b*) of this section must be submitted to the director prior to or together with any reports, information, or applications to be signed by an authorized representative.

(*d*) *Certification.* Any person signing a document under paragraph (*a*) or (*b*) of this section must make the following certification: "I certify under penalty of law that this document and all attachments were prepared under my direction or supervision in accordance with a system designed to assure that qualified personnel properly gather and evaluate the information submitted. Based on my inquiry of the person or persons who manage the system, or those persons directly responsible for gathering the information, the information submitted is, to the best of my knowledge and belief, true, accurate, and complete. I am aware that there are significant penalties for submitting false information, including the possibility of fine and imprisonment for knowing violations."

3.2.9 Completeness

The director cannot issue a permit before receiving a complete application for a permit except for NPDES general permits. An application for a permit is complete when the director receives an application form and any supplemental information, which are completed to his or her satisfaction. The completeness of any application for a permit will be judged independently of the status of any other permit application or permit for the same facility or activity. For EPA-administered NPDES programs, an application, which is reviewed under 40 CFR 124.3, is complete when the director receives either a complete application or the information listed in a notice of deficiency.

3.2.10 Continuation of Expiring Permits

When EPA is the permit-issuing authority, the conditions of an expired permit continue in force under 5 USC 558(c) until the effective date of a new permit (see 40 CFR 124.15) if:

(1) The permittee has submitted a timely application under 40 CFR 122.21 which is a complete [under 122.21(e)] application for a new permit.

(2) The regional administrator, through no fault of the permittee, does not issue a new permit with an effective date under 40 CFR 124.15 on or before the expiration date of the previous permit (for example, when issuance is impracticable due to time or resource constraints).

An EPA-issued permit does not continue in force beyond its expiration date under federal law if at that time a state is the permitting authority. States authorized to administer the NPDES program may continue either EPA- or state-issued permits until the effective date of the new permits, if state law allows. Otherwise, the facility or activity is operating without a permit from the time of expiration of the old permit to the effective date of the state-issued new permit.

3.2.11 Conditions Applicable to All Permits

The following conditions apply to all NPDES permits. Additional conditions applicable to NPDES permits can be found in 40 CFR 122.42. All conditions applicable to NPDES permits have to be incorporated into the permits either expressly or by reference. If incorporated by reference, a specific citation to the appropriate federal regulations (or the corresponding approved state regulations) must be given in the permit.

(*a*) *Duty to comply.* The permittee must comply with all conditions of this permit. Any permit noncompliance constitutes a violation of the Clean Water Act and is grounds for enforcement action; for permit termination, revocation and reissuance, or modification; or denial of a permit renewal application.

(1) The permittee must comply with effluent standards or prohibitions established under section 307(*a*) of the Clean Water Act toxic pollutants and with standards for sewage sludge use or disposal established under section 405(*d*) of the CWA within the time provided in the regulations that establish these standards or prohibitions or standards for sewage sludge use or disposal, even if the permit has not yet been modified to incorporate the requirement.

(2) The Clean Water Act provides that any person who violates section 301, 302, 306, 307, 308, 318, or 405 of the act, or any permit condition or limitation implementing any such sections in a permit issued under section 402, or any requirement imposed in a pretreatment program approved under sections 402(*a*)(3) or 402(*b*)(8) of the act is subject to a civil penalty not to exceed $25,000 per day for each violation. The Clean Water Act provides that any person who negligently violates section 301, 302, 306, 307, 308, 318, or 405 of the act, or any condition or limitation implementing any of such sections in a permit issued under section 402 of the act, or any requirement imposed in a pretreatment program approved under section 402(*a*)(3) or 402(*b*)(8) of the act is subject to criminal penalties of $2500 to $25,000 per day of violation, or imprisonment of not more than 1 year, or both. In the case of a second or subsequent conviction for a negligent violation, a person shall be subject to criminal penalties of not more than $50,000 per day of violation, or by imprisonment of not more than 2 years, or both. Any person who knowingly violates such sections or such conditions or limitations is subject to criminal penalties of $5000 to $50,000 per day of violation, or imprisonment for not more than 3 years, or both. In the case of a second or subsequent conviction for a knowing violation, a person shall be subject to criminal penalties of not more than $100,000 per day of violation, or imprisonment of not more than 6 years, or both. Any person who knowingly violates section 301, 302, 306, 307, 308, 318, or 405 of the act, or any permit condition or limitation implementing any of such sections in a permit issued under section 402 of the act, and who knows at that time that he or she thereby places another person in imminent danger of death or serious bodily injury, may, upon conviction, be subject to a fine of not more than $250,000 or imprisonment of not more than 15 years, or both. In the case of a second or subsequent conviction for a knowing endangerment violation, a person shall be subject to a fine of not more than $500,000 or imprisonment of not more than 30 years, or both. An organization, as defined in section 309(*c*)(3)(B)(iii) of the CWA, is, upon conviction of violating the imminent danger provision, subject to

a fine of not more than $1,000,000 and can be fined up to $2,000,000 for second or subsequent convictions.

(3) Any person may be assessed an administrative penalty by the administrator for violating section 301, 302, 306, 307, 308, 318, or 405 of this act, or any permit condition or limitation implementing any of such sections in a permit issued under section 402 of the act. Administrative penalties for Class I violations are not to exceed $10,000 per violation, with the maximum amount of any Class I penalty assessed not to exceed $25,000. Penalties for Class II violations are not to exceed $10,000 per day for each day during which the violation continues, with the maximum amount of any Class II penalty not to exceed $125,000.

(b) *Duty to reapply.* If the permittee wishes to continue an activity regulated by this permit after the expiration date of this permit, the permittee must apply for and obtain a new permit.

(c) *Need to halt or reduce activity not a defense.* It is not a defense for a permittee in an enforcement action that it would have been necessary to halt or reduce the permitted activity in order to maintain compliance with the conditions of this permit.

(d) *Duty to mitigate.* The permittee must take all reasonable steps to minimize or prevent any discharge or sludge use or disposal in violation of this permit which has a reasonable likelihood of adversely affecting human health or the environment.

(e) *Proper operation and maintenance.* The permittee must at all times properly operate and maintain all facilities and systems of treatment and control (and related appurtenances) which are installed or used by the permittee to achieve compliance with the conditions of this permit. Proper operation and maintenance also includes adequate laboratory controls and appropriate quality assurance procedures. This provision requires the operation of backup or auxiliary facilities or similar systems, which are installed by a permittee only when the operation is necessary to achieve compliance with the conditions of the permit.

(f) *Permit actions.* This permit may be modified, revoked and reissued, or terminated for cause. The filing of a request by the permittee for a permit modification, revocation and reissuance, or termination, or a notification of planned changes or anticipated noncompliance does not stay any permit condition.

(g) *Property rights.* This permit does not convey any property rights of any sort, or any exclusive privilege.

(h) *Duty to provide information.* The permittee must furnish to the director, within a reasonable time, any information which the director may request to determine whether cause exists for modifying, revoking and reissuing, or terminating this permit or to determine compliance with this permit. The permittee shall also furnish to the director upon request copies of records required to be kept by this permit.

(i) *Inspection and entry.* The permittee shall allow the director or an authorized representative (including an authorized contractor acting as a representative of the administrator) upon presentation of credentials and other documents as may be required by law, to:

(1) Enter upon the permittee's premises where a regulated facility or activity is located or conducted, or where records must be kept under the conditions of this permit.

(2) Have access to and copy, at reasonable times, any records that must be kept under the conditions of this permit.

(3) Inspect at reasonable times any facilities, equipment (including monitoring and control equipment), practices, or operations regulated or required under this permit.

(4) Sample or monitor at reasonable times, for the purposes of assuring permit compliance or as otherwise authorized by the Clean Water Act, any substances or parameters at any location.

(*j*) *Monitoring and records.*

(1) Samples and measurements taken for the purpose of monitoring shall be representative of the monitored activity.

(2) Except for records of monitoring information required by this permit related to the permittee's sewage sludge use and disposal activities, which must be retained for a period of at least 5 years (or longer as required by 40 CFR Part 503), the permittee must retain records of all monitoring information, including all calibration and maintenance records and all original strip-chart recordings for continuous monitoring instrumentation, copies of all reports required by this permit, and records of all data used to complete the application for this permit, for a period of at least 3 years from the date of the sample, measurement, report, or application. This period may be extended by request of the director at any time.

(3) Records of monitoring information shall include:

(i) The date, exact place, and time of sampling or measurements
(ii) The individual(s) who performed the sampling or measurements
(iii) The date(s) analyses were performed
(iv) The individual(s) who performed the analyses
(v) The analytical techniques or methods used
(vi) The results of such analyses

(4) Monitoring results must be conducted according to test procedures approved under 40 CFR Part 136 or, in the case of sludge use or disposal, approved under 40 CFR Part 136 unless otherwise specified in 40 CFR Part 503, unless other test procedures have been specified in the permit.

(5) The Clean Water Act provides that any person who falsifies, tampers with, or knowingly renders inaccurate any monitoring device or method required to be maintained under this permit shall, upon conviction, be punished by a fine of not

more than $10,000 or by imprisonment for not more than 2 years, or both. If a conviction of a person is for a violation committed after a first conviction of such person under this paragraph, punishment is a fine of not more than $20,000 per day of violation, or imprisonment of not more than 4 years, or both.

(*k*) *Signatory requirement*

(1) All applications, reports, or information submitted to the director must be signed and certified (see 122.22).

(2) The CWA provides that any person who knowingly makes any false statement, representation, or certification in any record or other document submitted or required to be maintained under this permit, including monitoring reports or reports of compliance or noncompliance, shall, upon conviction, be punished by a fine of not more than $10,000 per violation, or imprisonment for not more than 6 months per violation, or both.

(*l*) *Reporting requirements*

(1) *Planned changes.* The permittee is required to give notice to the director as soon as possible of any planned physical alterations or additions to the permitted facility. Notice is required only when:

(i) The alteration or addition to a permitted facility may meet one of the criteria for determining whether a facility is a new source in 122.29(*b*).

(ii) The alteration or addition could significantly change the nature or increase the quantity of pollutants discharged. This notification applies to pollutants which are subject neither to effluent limitations in the permit nor to notification requirements under 122.42(*a*)(1).

(iii) The alteration or addition results in a significant change in the permittee's sludge use or disposal practices, and such alteration, addition, or change may justify the application of permit conditions that are different from or absent in the existing permit, including notification of additional use or disposal sites not reported during the permit application process or not reported pursuant to an approved land application plan.

(2) *Anticipated noncompliance.* The permittee must give advance notice to the director of any planned changes in the permitted facility or activity which may result in noncompliance with permit requirements.

(3) *Transfers.* This permit is not transferable to any person except after notice to the director. The director may require modification or revocation and reissuance of the permit to change the name of the permittee and incorporate such other requirements as may be necessary under the Clean Water Act. (See 122.61; in some cases, modification or revocation and reissuance is mandatory.)

(4) *Monitoring reports.* Monitoring results must be reported at the intervals specified elsewhere in this permit.

(i) Monitoring results must be reported on a Discharge Monitoring Report (DMR) or forms provided or specified by the director for reporting results of monitoring of sludge use or disposal practices.

(ii) If the permittee monitors any pollutant more frequently than required by the permit using test procedures approved under 40 CFR Part 136 or, in the case of sludge use or disposal, approved under 40 CFR Part 136 unless otherwise specified in 40 CFR Part 503, or as specified in the permit, the results of this monitoring shall be included in the calculation and reporting of the data submitted in the DMR or sludge reporting form specified by the director.

(iii) Calculations for all limitations which require averaging of measurements shall utilize an arithmetic mean unless otherwise specified by the director in the permit.

(5) *Compliance schedules.* Reports of compliance or noncompliance with, or any progress reports on, interim and final requirements contained in any compliance schedule of this permit shall be submitted no later than 14 days following each schedule date.

(6) *Twenty-four-hour reporting*

(i) The permittee must report any noncompliance which may endanger health or the environment. Any information shall be provided orally within 24 hours from the time the permittee becomes aware of the circumstances. A written submission must also be provided within 5 days of the time the permittee becomes aware of the circumstances. The written submission shall contain a description of the noncompliance and its cause; the period of noncompliance, including exact dates and times, and if the noncompliance has not been corrected, the anticipated time it is expected to continue; and steps taken or planned to reduce, eliminate, and prevent reoccurrence of the noncompliance.

(ii) The following needs to be included as information which must be reported within 24 hours under this paragraph.

(A) Any unanticipated bypass which exceeds any effluent limitation in the permit [see 40 CFR 122.41(g)].
(B) Any upset which exceeds any effluent limitation in the permit.
(C) Violation of a maximum daily discharge limitation for any of the pollutants listed by the director in the permit to be reported within 24 hours [see 122.44(g)].

(iii) The director may waive the written report on a case-by-case basis for reports under paragraph (l)(6)(ii) of this section if the oral report has been received within 24 hours.

(7) *Other noncompliance.* The permittee must report all instances of noncompliance not reported under paragraphs (l)(4), (5), and (6) of this section, at the time

monitoring reports are submitted. The reports must contain the information listed in paragraph (l)(6) of this section.

(8) *Other information.* Where the permittee becomes aware that it failed to submit any relevant facts in a permit application, or submitted incorrect information in a permit application or in any report to the director, it must promptly submit such facts or information.

(*m*) *Bypass*

(1) *Definitions*

(i) "Bypass" means the intentional diversion of waste streams from any portion of a treatment facility.

(ii) "Severe property damage" means substantial physical damage to property, damage to the treatment facilities which causes them to become inoperable, or substantial and permanent loss of natural resources which can reasonably be expected to occur in the absence of a bypass. Severe property damage does not mean economic loss caused by delays in production.

(2) *Bypass not exceeding limitations.* The permittee may allow any bypass to occur which does not cause effluent limitations to be exceeded, but only if it also is for essential maintenance to assure efficient operation. These bypasses are not subject to the provisions of paragraphs (3) and (4) of this section of the act.

(3) *Notice*

(i) Anticipated bypass. If the permittee knows in advance of the need for a bypass, it must submit prior notice, if possible at least 10 days before the date of the bypass.

(ii) Unanticipated bypass. The permittee must submit notice of an unanticipated bypass as required in paragraph (l)(6) of this section (24-hour notice).

(4) *Prohibition of bypass*

(i) Bypass is prohibited, and the director may take enforcement action against a permittee for bypass, unless:

(A) Bypass was unavoidable to prevent loss of life, personal injury, or severe property damage.

(B) There were no feasible alternatives to the bypass, such as the use of auxiliary treatment facilities, retention of untreated wastes, or maintenance during normal periods of equipment downtime. This condition is not satisfied if adequate backup equipment should have been installed in the exercise of reasonable engineering judgment to prevent a bypass which occurred during normal periods of equipment downtime or preventive maintenance.

(C) The permittee submitted notices as required under paragraph (*m*)(3) of this section.

(ii) The director may approve an anticipated bypass, after considering its adverse effects, if the director determines that it will meet the three conditions listed above in paragraph (i).

(n) Upset

(1) *Definition.* "Upset" means an exceptional incident in which there is unintentional and temporary noncompliance with technology-based permit effluent limitations because of factors beyond the reasonable control of the permittee. An upset does not include noncompliance to the extent caused by operational error, improperly designed treatment facilities, inadequate treatment facilities, lack of preventive maintenance, or careless or improper operation.

(2) *Effect of an upset.* An upset constitutes an affirmative defense to an action brought for noncompliance with such technology-based permit effluent limitations if the requirements of paragraph (*n*)(3) of this section are met. No determination made during administrative review of claims that noncompliance was caused by upset, and before an action for noncompliance, is final administrative action subject to judicial review.

(3) *Conditions necessary for a demonstration of upset.* A permittee who wishes to establish the affirmative defense of upset must demonstrate, through properly signed, contemporaneous operating logs, or other relevant evidence, that:

(i) An upset occurred and that the permittee can identify the cause(s) of the upset.

(ii) The permitted facility was at the time being properly operated.

(iii) The permittee submitted notice of the upset as required in paragraph (1)(6)(ii)(B) of this section (24-hour notice).

(iv) The permittee complied with any remedial measures required under paragraph (*d*) of this section.

(4) *Burden of proof.* In any enforcement proceeding the permittee seeking to establish the occurrence of an upset has the burden of proof.

3.2.12 Establishing Additional Permit Conditions (applicable to state programs, see 40 CFR 123.25)

(*a*) In addition to conditions required in all permits (122.41 and 122.42), the director shall establish conditions, as required on a case-by-case basis, to provide for and assure compliance with all applicable requirements of CWA and regulations. These include conditions under 122.46 (duration of permits), 122.47(*a*) (schedules of compliance), 122.48 (monitoring), and for EPA permits only

122.47(*b*) (alternate schedule of compliance) and 122.49 (considerations under federal law).

(*b*) (1) For a state-issued permit, an applicable requirement is a state statutory or regulatory requirement which takes effect prior to final administrative disposition of a permit. For a permit issued by EPA, an applicable requirement is a statutory or regulatory requirement (including any interim final regulation) which takes effect prior to the issuance of the permit [except as provided in 124.86(*c*) for NPDES permits being processed under subpart E or F of Part 124]. Section 124.14 (reopening of comment period) provides a means for reopening EPA permit proceedings at the discretion of the director where new requirements become effective during the permitting process and are of sufficient magnitude to make additional proceedings desirable. For state- and EPA-administered programs, an applicable requirement is also any requirement which takes effect prior to the modification or revocation and reissuance of a permit, to the extent allowed in 122.62.

(2) New or reissued permits, and to the extent allowed under 122.62 modified or revoked and reissued permits, incorporate each of the applicable requirements referenced in 122.44 and 122.45.

(*c*) *Incorporation.* All permit conditions are incorporated either expressly or by reference. If incorporated by reference, a specific citation to the applicable regulations or requirements must be given in the permit.

3.2.13 Record Keeping

Except for information required by applicants for EPA-issued permits, all new sources, and "sludge-only facilities," in addition to all POTWs and other "treatment works treating domestic sewage," including "sludge-only facilities," who must complete Forms 1 and either 2*b* or 2*c* of the consolidated permit, which shall be retained for a period of at least 5 years from the date the application is signed (or longer as required by 40 CFR Part 503), applicants must keep records of all data used to complete permit applications and any supplemental information submitted under this section for a period of at least 3 years from the date the application is signed.

3.2.14 Duration of Permits (applicable to state programs, see 40 CFR 123.25)

(*a*) NPDES permits are effective for a fixed term not to exceed 5 years.

(*b*) Except as provided in 122.6, the term of a permit may not be extended by modification beyond the maximum duration specified in this section.

(*c*) The director may issue any permit for a duration that is less than the full allowable term under this section.

(*d*) A permit may be issued to expire on or after the statutory deadline set forth in section 301(*b*)(2)(A), (C), and (E) if the permit includes effluent limitations to meet the requirements of section 301(b)(2)(A), (C), (D), (E), and (F), whether or not applicable effluent limitations guidelines have been promulgated or approved.

(e) A determination that a particular discharger falls within a given industrial category for purposes of setting a permit expiration date under paragraph (d) of this section is not conclusive as to the discharger's inclusion in that industrial category for any other purposes and does not prejudice any rights to challenge or change that inclusion at the time that a permit based on that determination is formulated.

3.2.15 Notices, Hearings, and Appeals

The permit writing office will post notice of the pending permit once the draft is written, while the permit is in review. Interested parties have 30 days to submit written comments or to file a petition for a public hearing. The permitting authority decides whether sufficient public interest for a hearing exists. If so, it will schedule one giving a minimum of 30 days notice. After the hearing, all comments from the hearing and written comments received are considered with regard to the permit request. The permitting authority makes a decision to approve or deny the permit. Final decisions can be appealed to chancery court or court of equity or equivalent.

3.2.16 Required Facility-Specific NPDES Permit Application Information

All applicants for NPDES permits are required to provide the following facility-specific information to the director, using the application form provided by the director):

- The activities conducted by the applicant which require it to obtain an NPDES permit.
- Name, mailing address, and location of the facility for which the application is submitted.
- Up to four Standard Industrial Classification (SIC) codes which best reflect the principal products or services provided by the facility.
- The operator's name, address, telephone number, ownership status, and status as federal, state, private, public, or other entity.
- Whether the facility is located on Indian lands.
- A listing of all permits or construction approvals received or applied for under any of the following programs:
  (i) Hazardous Waste Management program under RCRA.
  (ii) UIC program under SDWA.
  (iii) NPDES program under CWA.
  (iv) Prevention of Significant Deterioration (PSD) program under the Clean Air Act.

(v) Nonattainment program under the Clean Air Act.

(vi) National Emission Standards for Hazardous Pollutants (NESHAPS) pre-construction approval under the Clean Air Act.

(vii) Ocean dumping permits under the Marine Protection Research and Sanctuaries Act.

(viii) Dredge or fill permits under section 404 of CWA.

(ix) Other relevant environmental permits, including state permits.

- A topographic map (or other map if a topographic map is unavailable) extending 1 mile beyond the property boundaries of the source, depicting the facility and each of its intake and discharge structures; each of its hazardous waste treatment, storage, or disposal facilities; each well where fluids from the facility are injected underground; and those wells, springs, other surface water bodies, and drinking water wells listed in public records or otherwise known to the applicant in the map area.

- A brief description of the nature of the business.

Additional facility-specific and outfall-specific information is required of the following five classes of applicants for each point source to be covered by the permit:

- Existing manufacturing, commercial mining, and silvicultural dischargers

- Manufacturing, commercial, mining, and silvicultural facilities which discharge only nonprocess wastewater

- New and existing concentrated animal feeding operations and aquatic animal production facilities

- New and existing POTWs

- New manufacturing, commercial, mining, and silvicultural dischargers

3.2.17 Additional Application Requirements for Existing Manufacturing, Commercial, Mining, and Silvicultural Dischargers

Existing manufacturing, commercial mining, and silvicultural dischargers applying for NPDES permits, except for those facilities subject to the requirements of 40 CFR 122.21(*h*), shall provide the following information to the director, using application forms provided by the director.

Outfall location. The latitude and longitude to the nearest 15 seconds and the name of the receiving water.

Line drawing. A line drawing of the water flow through the facility with a water balance, showing operations contributing wastewater to the effluent and treatment units. Similar processes, operations, or production areas may be indicated as a single unit, labeled to correspond to the more detailed identification of this section. The water balance must show approximate average flows at intake and discharge points and between units, including treatment units. If a water balance cannot be determined (for example, for certain mining activities), the applicant may provide instead a pictorial description of the nature and amount of any sources of water and any collection and treatment measures.

Average flows and treatment. A narrative identification of each type of process, operation, or production area which contributes wastewater to the effluent for each outfall, including process wastewater, cooling water, and stormwater runoff; the average flow which each process contributes; and a description of the treatment the wastewater receives, including the ultimate disposal of any solid or fluid wastes other than by discharge. Processes, operations, or production areas may be described in general terms (for example, "dye-making reactor," "distillation tower"). For a privately owned treatment works, this information shall include the identity of each user of the treatment works. The average flow of point sources composed of stormwater may be estimated. The basis for the rainfall event and the method of estimation must be indicated.

Intermittent flows. If any of the discharges to be covered by the permit of this section are intermittent or seasonal, a description of the frequency, duration, and flow rate of each discharge occurrence (except for stormwater runoff, spillage, or leaks).

Maximum production. If an effluent guideline promulgated under section 304 of CWA applies to the applicant and is expressed in terms of production (or other measure of operation), a reasonable measure of the applicant's actual production reported in the units used in the applicable effluent guideline. The reported measure must reflect the actual production of the facility as required by 122.45(b)(2).

Improvements. If the applicant is subject to any present requirements or compliance schedules for construction, upgrading, or operation of waste treatment equipment, an identification of the abatement requirement, a description of the abatement project, and a listing of the required and projected final compliance dates.

Effluent characteristics. Information on the discharge of pollutants specified in this paragraph (except information on stormwater discharges, which is to be provided as specified in 40 CFR 122.26). When "quantitative data" for a pollutant are required, the applicant must collect a sample of effluent and analyze it for the pollutant in accordance with analytical methods approved under 40 CFR Part 136. When no analytical method is approved, the applicant may use any suitable method but must provide a description of the method.

When an applicant has two or more outfalls with substantially identical effluents, the director may allow the applicant to test only one outfall and report that the quantitative data also apply to the substantially identical outfalls.

An applicant must provide quantitative data for certain pollutants known or believed to be present. This does not apply to pollutants present in a discharge solely as the result of their presence in intake water; however, an applicant must report such pollutants as present. Grab samples must be used for pH, temperature, cyanide, total phenols, residual chlorine, oil and grease, fecal coliform, and fecal streptococcus. For all other pollutants, 24-hour composite samples must be used. However, a minimum of one grab sample may be taken for effluents from holding ponds or other impoundments with a retention period greater than 24 hours. In addition, for discharges other than stormwater discharges, the director may waive composite sampling for any outfall for which the applicant demonstrates that the use of an automatic sampler is infeasible and that the minimum of four grab samples will be a representative sample of the effluent being discharged.

For stormwater discharges, all samples must be collected from the discharge resulting from a storm event that is greater than 0.1 inch and at least 72 hours from the previously measurable (greater than 0.1 inch rainfall) storm event. Where feasible, the variance in the duration of the event and the total rainfall of the event should not exceed 50 percent from the average or median rainfall event in that area. For all applicants, a flow-weighted composite shall be taken for either the entire discharge or for the first 3 hours of the discharge. The flow-weighted composite sample for a stormwater discharge may be taken with a continuous sampler or as a combination of a minimum of three sample aliquots taken in each hour of discharge for the entire discharge or for the first 3 hours of the discharge, with each aliquot being separated by a minimum period of 15 minutes (applicants submitting permit applications for stormwater discharges under 40 CFR 122.26(*d*) may collect flow-weighted composite samples using different protocols with respect to the time duration between the collection of sample aliquots, subject to the approval of the director). However, a minimum of one grab sample may be taken for stormwater discharges from holding ponds or other impoundments with a retention period greater than 24 hours. For a flow-weighted composite sample, only one analysis of the composite of aliquots is required.

For stormwater discharge samples taken from discharges associated with industrial activities, quantitative data must be reported for the grab sample taken during the first 30 minutes (or as soon thereafter as practicable) of the discharge for all pollutants specified in 40 CFR 122.26(*c*)(1). For all stormwater permit applicants taking flow-weighted composites, quantitative data must be reported for all pollutants specified in 40 CFR 122.26 except pH, temperature, cyanide, total phenols, residual chlorine, oil and grease, fecal coliform, and fecal streptococcus.

The director may allow or establish appropriate site-specific sampling procedures or requirements, including sampling locations, the season in which the sampling takes place, the minimum duration between the previous measurable storm event and the storm event sampled, the minimum or maximum level of precipitation required for an appropriate storm event, the form of precipitation sampled (snow melt or rainfall), protocols for collecting samples under 40 CFR Part 136, and additional time for submitting data on a case-by-case basis. An applicant is expected to "know or have reason to believe" that a pollutant is present in an effluent based on an evaluation of the expected use, production, or storage of the pollutant, or on any previous analyses for the pollutant. (For example, any pesticide

TABLE 3.1 NPDES Primary Industry Categories

Adhesives and sealants	Ore mining
Aluminum forming	Organic chemicals manufacturing
Auto and other laundries	Paint and ink formulation
Battery manufacturing	Pesticides
Coal mining	Petroleum refining
Coil coating	Pharmaceutical preparations
Copper forming	Photographic equipment and supplies
Electrical and electronic components	Plastics processing
Electroplating	Plastic and synthetic materials manufacturing
Explosives manufacturing	Porcelain enameling
Foundries	Printing and publishing
Gum and wood chemicals	Pulp and paper mills
Inorganic chemicals manufacturing	Rubber processing
Iron and steel manufacturing	Soap and detergent manufacturing
Leather tanning and finishing	Steam electric power plants
Mechanical products manufacturing	Textile mills
Nonferrous metals manufacturing	Timber products processing

manufactured by a facility may be expected to be present in contaminated stormwater runoff from the facility.)

Every applicant must report quantitative data for every outfall for the following pollutants: Biochemical Oxygen Demand (BOD_5), Chemical Oxygen Demand, Total Organic Carbon, Total Suspended Solids, Ammonia (as N), Temperature (both winter and summer), and pH.

Each applicant with processes in one or more primary industry category (see Table 3.1) contributing to a discharge must report quantitative data for the following pollutants in each outfall containing process wastewater:

(A) The organic toxic pollutants in the fractions designated in Table I of Appendix D of 40 CFR Part 122 (NPDES Permit Application Testing Requirements) for the applicant's industrial category or categories unless the applicant qualifies as a small business. Table II of Appendix D lists the organic toxic pollutants in each fraction. The fractions result from the sample preparation required by the analytical procedure which uses gas chromatography and mass spectrometry. A determination that an applicant falls within a particular industrial category for the purposes of selecting fractions for testing is not conclusive as to the applicant's inclusion in that category for any other purposes.

(B) The pollutants listed in Table III of Appendix D of 40 CFR 121 (the toxic metals, cyanide, and total phenols).

Each applicant must indicate whether it knows or has reason to believe that any of the pollutants in Table IV of Appendix D (certain conventional and nonconventional pollutants) is discharged from each outfall. If an applicable effluent limitation guideline either directly limits the pollutant or, by its express terms, indirectly limits the pollutant through limitations on an indicator, the applicant must report quantitative data. For every pollutant discharged which is not so limited in an effluent limitation guideline, the applicant must either report quantitative data or briefly describe the reasons the pollutant is expected to be discharged.

Each applicant must indicate whether it knows or has reason to believe that any of the pollutants listed in Table II or Table III of Appendix D (the toxic pollutants and total phenols) for which quantitative data are not otherwise required under paragraph $(g)(7)(ii)$ of this section is discharged from each outfall. For every pollutant expected to be discharged in concentrations of 10 ppb or greater the applicant must report quantitative data. For acrolein, acrylonitrile, 2,4 dinitrophenol, and 2-methyl-4,6 dinitrophenol, where any of these four pollutants are expected to be discharged in concentrations of 100 ppb or greater, the applicant must report quantitative data. For every pollutant expected to be discharged in concentrations less than 10 ppb, or in the case of acrolein, acrylonitrile, 2,4 dinitrophenol, and 2-methyl-4,6 dinitrophenol, in concentrations less than 100 ppb, the applicant must either submit quantitative data or briefly describe the reasons the pollutant is expected to be discharged. An applicant qualifying as a small business is not required to analyze for pollutants listed in Table II of Appendix D (the organic toxic pollutants).

Each applicant must indicate whether it knows or has reason to believe that any of the pollutants in Table V of Appendix D (certain hazardous substances and asbestos) are discharged from each outfall. For every pollutant expected to be discharged, the applicant must briefly describe the reasons the pollutant is expected to be discharged, and report any quantitative data it has for any pollutant.

Each applicant must report qualitative data, generated using a screening procedure not calibrated with analytical standards, for 2,3,7,8-tetrachlorodibenzo-*p*-dioxin (TCDD) if it:

(A) Uses or manufactures 2,4,5-trichlorophenoxy acetic acid (2,4,5,-T); 2-(2,4,5-trichlorophenoxy) propanoic acid (Silvex, 2,4,5,-TP); 2-(2,4,5-trichlorophenoxy) ethyl, 2,2-dichloropropionate (Erbon); O,O-dimethyl O-(2,4,5-trichlorophenyl) phosphorothioate (Ronnel); 2,4,5-trichlorophenol (TCP); or hexachlorophene (HCP).

(B) Knows or has reason to believe that TCDD is or may be present in an effluent $122.21(g)(8)$.

Small business exemption. An applicant which qualifies as a small business under one of the following criteria is exempt from the requirement to submit quantitative data for the pollutants listed in Table II of Appendix D (the organic toxic pollutants):

(i)   For coal mines, a probable total annual production of less than 100,000 tons per year.

(ii) For all other applicants, gross total annual sales averaging less than $100,000 per year (in second-quarter 1980 dollars).

Use of manufactured toxics. A listing of any toxic pollutant which the applicant currently uses or manufactures as an intermediate or final product or by-product. The director may waive or modify this requirement for any applicant if the applicant demonstrates that it would be unduly burdensome to identify each toxic pollutant and the director has adequate information to issue the permit.

Biological toxicity tests. An identification of any biological toxicity tests which the applicant knows or has reason to believe have been made within the last 3 years on any of the applicant's discharges or on receiving water in relation to a discharge.

Contract analyses. If a contract laboratory or consulting firm performed any of the analyses required by paragraph (g)(7) of this section, the identity of each laboratory or firm and the analyses performed.

Additional information. In addition to the information reported on the application form, applicants must provide to the director, at his or her request, such other information as the director may reasonably require to assess the discharges of the facility and to determine whether to issue an NPDES permit. The additional information may include additional quantitative data and bioassays to assess the relative toxicity of discharges to aquatic life and requirements to determine the cause of the toxicity.

3.2.18 Additional Application Requirements for Manufacturing, Commercial, Mining, and Silvicultural Facilities Which Discharge Only Nonprocess Wastewater

Except for stormwater discharges, all manufacturing, commercial, mining, and silvicultural dischargers applying for NPDES permits which discharge only nonprocess wastewater not regulated by an effluent limitations guideline or new source performance standard shall provide the following information to the director, using application forms provided by the director:

(1) *Outfall location.* Outfall number, latitude and longitude to the nearest 15 seconds, and the name of the receiving water.

(2) *Discharge date* (for new dischargers). Date of expected commencement of discharge.

(3) *Type of waste.* An identification of the general type of waste discharged, or expected to be discharged upon commencement of operations, including sanitary wastes, restaurant or cafeteria wastes, or noncontact cooling water. An identification of cooling water additives (if any) that are used or expected to be used upon commencement of operations, along with their composition if existing composition is available.

(4) *Effluent characteristics.* Quantitative data for the pollutants or parameters listed below, unless testing is waived by the director. The quantitative data may be data collected over the past 365 days, if they remain representative of current operations, and must include maximum daily value, average daily value, and number of measurements taken. The applicant must collect and analyze samples in accordance with 40 CFR Part 136. Grab samples must be used for pH, tem-

perature, oil and grease, total residual chlorine, and fecal coliform. For all other pollutants, 24-hour composite samples must be used. New dischargers must include estimates for the pollutants or parameters listed below instead of actual sampling data, along with the source of each estimate. All levels must be reported or estimated as concentration and as total mass, except for flow, pH, and temperature.

(A) Biochemical Oxygen Demand (BOD_5)

(B) Total Suspended Solids (TSS)

(C) Fecal Coliform (if believed present or if sanitary waste is or will be discharged)

(D) Total Residual Chlorine (if chlorine is used)

(E) Oil and Grease

(F) Chemical Oxygen Demand (COD) (if noncontact cooling water is or will be discharged)

(G) Total Organic Carbon (TOC) (if noncontact cooling water is or will be discharged)

(H) Ammonia (as N)

(I) Discharge flow

(J) pH

(K) Temperature (winter and summer)

The director may waive the testing and reporting requirements for any of the pollutants or flow listed in the above paragraph if the applicant submits a request for such a waiver before or with the application which demonstrates that information adequate to support issuance of a permit can be obtained through less stringent requirements.

If the applicant is a new discharger, it must complete and submit Item IV of Form 2e [see 40 CFR 122.21(*h*)(4)] by providing quantitative data in accordance with that section no later than 2 years after commencement of discharge. However, the applicant need not complete those portions of Item IV requiring tests which have already been performed and reported under the discharge monitoring requirements of the NPDES permit.

The above requirements that an applicant must provide quantitative data or estimates of certain pollutants do not apply to pollutants present in a discharge solely as a result of their presence in intake water. However, an applicant must report such pollutants as present. Net credit may be provided for the presence of pollutants in intake water if the requirements of 122.45(*g*) are met.

(5) *Flow.* A description of the frequency of flow and duration of any seasonal or intermittent discharge (except for stormwater runoff, leaks, or spills).

(6) *Treatment system.* A brief description of any system used or to be used.

(7) *Optional information.* Any additional information the applicant wishes to be considered, such as influent data for the purpose of obtaining "net" credits pursuant to 40 CFR 122.45(*g*).

(8) *Certification.* Signature of certifying official under 40 CFR 122.22.

3.2.19 Additional Application Requirements for New and Existing Concentrated Animal Feeding Operations and Aquatic Animal Production Facilities

New and existing concentrated animal feeding operations (defined in 40 CFR 122.23) and concentrated aquatic animal production facilities (defined in 40 CFR 122.24) must provide the following information to the director, using the application form provided by the director:

(1) *For concentrated animal feeding operations:*

- The type and number of animals in open confinement and housed under roof
- The number of acres used for confinement feeding
- The design basis for the runoff diversion and control system, if one exists, including the number of acres of contributing drainage, the storage capacity, and the design safety factor

(2) *For concentrated aquatic animal production facilities:*

- The maximum daily and average monthly flow from each outfall
- The number of ponds, raceways, and similar structures
- The name of the receiving water and the source of intake water
- For each species of aquatic animals, the total yearly and maximum harvestable weight, the calendar month of maximum feeding, and the total mass of food fed during that month

3.2.20 Additional Application Requirements for New and Existing POTWs

The following types of POTWs must provide the results of valid whole-effluent biological toxicity testing to the director:

- All POTWs with design influent flows equal to or greater than 1 million gallons per day
- All POTWs with approved pretreatment programs or POTWs required to develop a pretreatment program

In addition to the POTWs listed above, the EPA may require other POTWs to submit the results of toxicity tests with their permit applications, based on consideration of the following factors:

- The variability of the pollutants or pollutant parameters in the POTW effluent (based on chemical-specific information, the type of treatment facility, and types of industrial contributors)

- The dilution of the effluent in the receiving water (ratio of effluent flow to receiving stream flow)

- Existing controls on point or nonpoint sources, including total maximum daily load calculations for the water body segment and the relative contribution of the POTW

- Receiving stream characteristics, including possible or known water quality impairment, and whether the POTW discharges to a coastal water, one of the Great Lakes, or a water designated as an outstanding natural resource

- Other considerations (including but not limited to the history of toxic impact and compliance problems at the POTW), which the director determines could cause or contribute to adverse water quality impacts.

On Aug. 4, 1999, the Environmental Protection Agency (EPA) amended permit application requirements and application forms for publicly owned treatment works (POTWs) and other treatment works treating domestic sewage (TWTDS). TWTDS include facilities that generate sewage sludge, provide commercial treatment of sewage sludge, manufacture a product derived from sewage sludge, or provide disposal of sewage sludge.

In addition to promulgating new Federal Forms 2A and 2S, this rule consolidated POTW application requirements, including information regarding toxics monitoring, whole-effluent toxicity (WET) testing, industrial user and hazardous waste contributions, and sewer collection system overflows. The most significant revisions required toxic monitoring by major POTWs (and other pretreatment POTWs) and limited pollutant monitoring by minor POTWs. EPA believed that permitting authorities need this information in order to issue permits that adequately protect the nation's water resources.

Form 2A replaces existing Standard Form A and Short Form A to account for changes in the National Pollutant Discharge Elimination System (NPDES) program since the forms were issued in 1973.

These regulations also clarified the requirements for TWTDS and allow the permitting authorities to obtain the information needed to issue permits that meet the requirements of the 40 CFR Part 503 sewage sludge use or disposal regulations. Form 2S replaces the existing Interim Sewage Sludge Form. Form 2S is similar to the Interim Sewage Sludge Form but requires less information.

EPA revised these regulations to ensure that permitting authorities obtain the information necessary to issue permits that protect the environment in the most efficient manner. The forms make it easier for permit applicants to provide the necessary information with their applications and minimize the need for additional follow-up requests from permitting authorities.

This rule making finalized two sets of application requirements and corresponding permit application forms and provides instructions for each. Section 122.21(j) contains application requirements pertaining to wastewater treatment and discharge into and from publicly owned treatment works (POTWs). The requirements are incorporated into the new Form 2A, which replaces Standard Form A and Short Form A, both of which were developed in 1973. Section 122.21(q) contains application requirements pertaining to generation, treatment, and disposal of sewage sludge at POTWs and other treatment works treating domestic sewage (TWTDS). These requirements are incorporated into the new Form 2S, which replaces the Interim Sewage Sludge Permit Application Form.

EPA promulgated these application regulations and published the new forms for the following five reasons:

- First, to address changes to the NPDES program model since 1973. The NPDES program applicable to POTWs has changed significantly since that time, specifically in the areas of toxics control and water quality–based permitting and pretreatment programs.

- Second, the rule consolidates application requirements from existing regulations into a "modular" permit application form, thereby streamlining and clarifying the process for permit applicants.

- Third, the revisions provide permit writers with the information necessary to develop appropriate NPDES permits consistent with requirements of the Clean Water Act and thus also provide greater certainty for permittees that compliance with their permits constitutes compliance with the CWA.

- Fourth, the rule allows NPDES permitting authorities to waive certain information requirements where information is already available to the permitting authority.

- Fifth, to provide a platform for electronic data transmission.

EPA will use the forms in states where the agency administers the NPDES and/or sewage sludge programs. Authorized states may choose to use these forms because the forms will provide the required application information. Authorized states can also elect to use forms of their own design so long as the information requested includes at least the information required by the final permit application regulations. EPA and state authorities may request additional information from permit applicants whenever necessary to establish appropriate permit limits and conditions. See CWA Sec. 308 and 402(b)(2)(B).

In the December 1995 proposal, EPA asked for comment on whether the forms and instructions should be included with the final rule-making package. EPA received numerous comments that said that the forms and instructions should be published so they could be available for all to review along with the regulation. EPA has changed the forms significantly in response to comments in order to facilitate electronic reporting. The final forms and instructions are included as an appendix to this rule making published in the *Federal Register* of Aug. 4, 1999 (64 FR 42434), but are not printed in the CFR.

(1) *Overview Form 2A.* Prior to this rule, NPDES permitting authorities generally gathered POTW data using Form 1, Standard Form A, and Short Form A. While all these forms are approved federal forms, the NPDES regulations did not require use of the forms by POTWs when applying for a permit. Standard Form A was intended to be used by all POTWs with a design flow equal to or exceeding 1 million gallons per day (mgd) and contains questions about the facility and collection system, discharges to and from the facility (including information on some specific pollutant parameters), planned improvements, and implementation schedules. Short Form A was intended for use by all POTWs with a design flow of less then 1 mgd. It contains only 15 questions of a summary nature, and asks for virtually no information on specific pollutants. Many states used one or both of the federal forms, but a number of states have developed forms that request information not included in the federal forms.

The newly promulgated Form 2A still contains two parts, but the Basic Application Information has been subdivided to reduce the requirements for facilities with a design flow under 0.1 mgd. The "Basic Application Information for All Applicants" part includes information about the collection system and the treatment plant, general information concerning the types of discharges from the treatment plant, identification of outfalls, and effluent monitoring data from the plant for 6 parameters. The requirements are expanded to include effluent monitoring for 14 parameters and several additional questions for POTWs with design flows greater than or equal to 0.1 mgd but less than 1.0 mgd and without pretreatment programs. Larger POTWs and pretreatment POTWs must submit the information requested in the "Supplemental Application Information" part of Form 2A, which requires effluent monitoring data for metals and organic compounds, as well as the parameters required for smaller POTWs. This part also requires results of whole-effluent toxicity tests. information on significant industrial users, and information on combined sewer overflows (CSOs), if applicable.

(2) *Applicability to Privately Owned and Federally Owned Treatment Works.* As in the case of existing Standard Form A and Short Form A, Form 2A and the application requirements at 122 21(j) are required only for POTWs. NPDES permitting authorities have the discretion to use the form on a case-by-case basis for treatment works that are not owned by a state or municipality. As previously discussed, the NPDES program has evolved considerably since EPA promulgated Standard Form A and Short Form A in 1973. The program can clearly be applied to facilities that are similar to POTWs but which do not meet the regulatory definition of POTWs. Although not owned by states or municipalities, such facilities nevertheless may receive predominantly domestic wastewater, provide physical and/or biological treatment, and discharge effluent to waters of the United States. Such facilities include federally owned treatment works (FOTWs) and privately owned treatment works that treat primarily domestic wastewater.

Federal and state permitting authorities use a number of mechanisms for obtaining NPDES permit application information from non-POTW treatment works. These mechanisms include Standard Form A, Short Form A, Form 2C ("Existing Manufacturing, Commercial, Mining, and Silvicultural Operations"), and Form 2E ("Facilities Which Do Not Discharge Process Wastewater"). Form 2A is often the most appropriate application form for non-POTW treatment works.

EPA does not, however, require the Form 2A information from non-POTW treatment works. Despite many functional similarities to POTWs, such facilities do not share the same regulatory requirements. Non-POTW treatment works are not required under the CWA, for example, to develop pretreatment programs. The CWA does not require such facilities to meet secondary treatment requirements, though permits for such facilities often apply secondary treatment–based limits after a best professional judgment evaluation has been performed by the permit writer. NPDES regulations do not require such facilities to report results of whole-effluent toxicity testing with their permit applications. For these facilities, uniformly requiring the same information required in Form 2A might be unnecessary. EPA allows the director to require such facilities to comply with the POTW application requirements (e.g., through Form 2A) on a case-by-case basis. This discretion provides NPDES permit writers with the information necessary to develop permits for facilities that may operate similarly to POTWs but that do not meet the regulatory definition.

(3) *Form 2S.* EPA also finalized a new form, Form 2S, to collect information on sewage sludge from treatment works treating domestic sewage (TWTDS). The term "treatment works treating domestic sewage" is a broad one, intended to reach facilities that generate sewage sludge or effectively change its pollutant characteristics as well as facilities that control its disposal. The term includes all POTWs and other facilities that treat domestic wastewater. It also includes facilities that do not treat domestic wastewater but that treat or dispose of sewage sludge, such as sewage sludge incinerators, composting facilities, commercial sewage sludge handlers that process sludge for distribution, and sites used for sewage sludge disposal. In addition, EPA may designate a facility a TWTDS when the facility's sludge quality or sludge handling, use, or disposal practices have the potential to adversely affect public health and the environment. Individual septic tanks or similar devices are not considered TWTDS.

EPA recognized that the term "biosolids" is now being used by professional organizations and other stakeholders in place of "sewage sludge" to emphasize that it is a resource that can be recycled beneficially. EPA intends to work with these stakeholders to define the term "biosolids" consistent with the definition of "sewage sludge" in the CWA. Until then, EPA will continue to refer to sewage sludge in its regulations.

Form 2S consists of two sections. Part 1 asks for limited background information rather than a complete permit application. Only the information in Part 1 must be submitted by "sludge-only" facilities (i.e., facilities that do not discharge wastewater to surface waters) unless the permit writer determines that the information in Part 2 must also be provided. It is intended to give the permitting authority enough information to decide whether or not to issue a permit to that facility. The information in Part 2 must be submitted by all TWTDS with an NPDES permit and "sludge-only" facilities that have been asked by the permitting authority to submit a complete permit application.

(4) *Reasons for separate Form 2A and Form 2S.* EPA published two separate forms for municipal wastewater discharges and for sewage sludge for several reasons. First, the requirements represented by the two forms differ in their applicability. The NPDES permit application requirements collected in Form 2A apply only to POTWs; the sewage sludge information requirements collected in Form 2S apply to all TWTDS, not just POTWs. Most facilities that generate, treat, or dispose of sewage sludge are POTWs and will be required to submit both application forms. Several thousand TWTDS, however, do not discharge to surface waters and therefore are not required to have NPDES permits. Thus, such TWTDS are subject to sewage sludge requirements (Form 2S) but not to NPDES requirements (Form 2A).

Second, separate application forms are also appropriate because wastewater and sewage sludge may be regulated by different permitting authorities. In 43 states and territories, the NPDES program is administered at the state level through an EPA-approved NPDES program. There are currently only three states that administer an EPA-approved sewage sludge program. Therefore, until more states are authorized to administer the federal sewage sludge program, POTWs in most NPDES states will obtain NPDES permits from the state permitting authority (by submitting Form 2A or a similar state form to the state) and sewage sludge permits from EPA (by submitting Form 2S to the EPA regional office). Separate application forms will facilitate this bifurcated permitting process. In addition, even when a state sludge permitting program is approved, the program will not necessarily be administered by the state's NPDES permitting authority. For example, a POTW in a state with both NPDES and sewage sludge permitting authority could receive its NPDES permit from the water pollution control agency and its sewage sludge permit from a solid waste management agency. Separate Forms 2A and 2S will also facilitate permitting in this situation.

Applicants are allowed to photocopy other forms or reference information that they know was previously submitted to the same permitting authority. Authorized states are free to create their own state forms as long as the forms request the same minimum information.

(5) *Electronic application forms.* Consistent with recent amendments to the Paperwork Reduction Act, the EPA is developing electronic data submission as an alternative format for permit application. The use of electronic media should help to streamline the application process and to reduce the amount of repetition associated with completing application forms that are currently available only in hard copy. As previously noted, the elimination of redundant reporting is one of the

goals of the rule making. EPA's first step in the submission of electronic data is the development of an electronic version of the application form. The agency has developed such an electronic version, which is available by contacting the persons listed on the Internet from the EPA Home Page (www.epa.gov). The application forms will be made available in Word and Windows Wizard formats and will include instructions that guide the applicant through the form. Some authorized states are also considering electronic reporting. EPA believes that providing the forms in an easily manipulated software will also assist states that want to use electronic permit applications.

3.2.21 Additional Application Requirements for New Sources and New Discharges

New manufacturing, commercial, mining, and silvicultural dischargers applying for NPDES permits [except for manufacturing, commercial, mining, and silvicultural facilities which discharge only nonprocess wastewater or new discharges of stormwater associated with industrial activity which are subject to the requirements of §122.26(c)(1), except as provided by §122.26(c)(1)(ii)] shall provide the following information to the director, using the application forms provided by the director:

(1) *Expected outfall location.* The latitude and longitude to the nearest 15 seconds and the name of the receiving water.

(2) *Discharge dates.* The expected date of commencement of discharge.

(3) *Flows, sources of pollution, and treatment technologies:*

- *Expected treatment of wastewater.* Description of the treatment that the wastewater will receive, along with all operations contributing wastewater to the effluent, average flow contributed by each operation, and the ultimate disposal of any solid or liquid wastes not discharged.

- *Line drawing.* A line drawing of the water flow through the facility with a water balance as described in §122.21(g)(2).

- *Intermittent flows.* If any of the expected discharges will be intermittent or seasonal, a description of the frequency, duration, and maximum daily flow rate of each discharge occurrence (except for stormwater runoff, spillage, or leaks).

(4) *Production.* If a new source performance standard promulgated under section 306 of CWA or an effluent limitation guideline applies to the applicant and is expressed in terms of production (or other measure of operation), a reasonable measure of the applicant's expected actual production reported in the units used in the applicable effluent guideline or new source performance standard as required by 40 CFR §122.45(b)(2) for each of the first 3 years. Alternative estimates may also be submitted if production is likely to vary.

(5) *Effluent characteristics.* The requirements in paragraphs (h)(4)(i), (ii), and (iii) of this section that an applicant must provide estimates of certain pollutants

expected to be present do not apply to pollutants present in a discharge solely as a result of their presence in intake water; however, an applicant must report such pollutants as present. Net credits may be provided for the presence of pollutants in intake water if the requirements of 40 CFR 122.45(g) are met. All levels (except for discharge flow, temperature, and pH) must be estimated as concentration and as total mass.

Each applicant must report estimated daily maximum, daily average, and source of information for each outfall for the following pollutants or parameters. The director may waive the reporting requirements for any of these pollutants and parameters if the applicant submits a request for such a waiver before or with the application which demonstrates that information adequate to support issuance of the permit can be obtained through less stringent reporting requirements.

(A) Biochemical Oxygen Demand (BOD)

(B) Chemical Oxygen Demand (COD)

(C) Total Organic Carbon (TOC)

(D) Total Suspended Solids (TSS)

(E) Flow

(F) Ammonia (as N)

(G) Temperature (winter and summer)

(H) pH

Each applicant must report estimated daily maximum, daily average, and source of information for each outfall for the following pollutants, if the applicant knows or has reason to believe they will be present or if they are limited by an effluent limitation guideline or new source performance standard either directly or indirectly through limitations on an indicator pollutant: all pollutants in Table IV of Appendix D of 40 CFR Part 122 (certain conventional and nonconventional pollutants).

Each applicant must report estimated daily maximum, daily average, and source of information for the following pollutants if it knows or has reason to believe that they will be present in the discharges from any outfall:

(A) The pollutants listed in Table III of Appendix D (the toxic metals, in the discharge from any outfall: total cyanide, and total phenols).

(B) The organic toxic pollutants in Table II of Appendix D (except bis (chloromethyl) ether, dichlorofluoro-methane and trichlorofluoro-methane). This requirement is waived for applicants with expected gross sales of less than $100,000 per year for the next 3 years, and for coal mines with expected average production of less than 100,000 tons of coal per year.

The applicant is required to report that 2,3,7,8-tetrachlorodibenzo-P-Dioxin (TCDD) may be discharged if it uses or manufactures one of the following compounds, or if it knows or has reason to believe that TCDD will or may be present in an effluent:

(A) 2,4,5-trichlorophenoxy acetic acid (2,4,5-T) (CAS #93-76-5)

(B) 2-(2,4,5-trichlorophenoxy) propanoic acid (Silvex, 2,4,5-TP) (CAS 93-72-1)

(C) 2-(2,4,5-trichlorophenoxy) ethyl 2,2-dichloropropionate (Erbon) (CAS 136-25-4)

(D) 0,0-dimethyl 0-(2,4,5-trichlorophenyl) phosphorothioate (Ronnel) (CAS 299-84-3)

(E) 2,4,5-trichlorophenol (TCP) (CAS 95-95-4)

(F) Hexachlorophene (HCP) (CAS 70-30-4)

Each applicant must report any pollutants listed in Table V of Appendix D (certain hazardous substances) if it believes they will be present in any outfall (no quantitative estimates are required unless they are already available).

No later than 2 years after the commencement of discharge from the proposed facility, the applicant is required to complete and submit Items V and VI of NPDES application Form 2c [see 40 CFR 122.21(g)]. However, the applicant need not complete those portions of Item V requiring tests which have already been performed and reported under the discharge monitoring requirements of the NPDES permit.

(6) *Engineering report.* Each applicant must report the existence of any technical evaluation concerning wastewater treatment, along with the name and location of similar plants of which it has knowledge.

(7) *Other information.* Any optional information the permittee wishes to have considered.

(8) *Certification.* Signature of certifying official under 40 CFR 122.22.

3.2.22 Special Provisions for Applications from New Sources

(1) The owner or operator of any facility which may be a new source (as defined in 122.2) and which is located in a state without an approved NPDES program must comply with the provisions of this paragraph.

(2) Before beginning any on-site construction as defined in 122.29, the owner or operator of any facility which may be a new source must submit information to the Regional Administrator so that he or she can determine if the facility is a new source. The Regional Administrator may request any additional information needed to determine whether the facility is a new source.

The Regional Administrator makes an initial determination whether the facility is a new source within 30 days of receiving all necessary information under paragraph (k)(2)(i) of this section.

(3) The Regional Administrator then issues a public notice in accordance with 124.10 of the new source determination. If the Regional Administrator has deter-

mined that the facility is a new source, the notice will state that the applicant must comply with the environmental review requirements of 40 CFR 6.600 et seq.

(4) Any interested person may challenge the Regional Administrator's initial new source determination by requesting an evidentiary hearing under subpart E of 40 CFR Part 124 within 30 days of issuance of the public notice of the initial determination. If all parties to the evidentiary hearing on the determination agree, the Regional Administrator may defer the hearing until after a final permit decision is made, and consolidate the hearing on the determination with any hearing on the permit.

3.3 STORMWATER PERMITS

The issues associated with the discharge of contaminated stormwater runoff have long been a concern to both federal and state regulatory agencies. In 1972, the Federal Water Pollution Control Act [also referred to as the Clean Water Act (CWA)] was amended to provide that the discharge of any pollutant to waters of the United States from any point source is unlawful, except if the discharge is in compliance with a National Pollutant Discharge Elimination System (NPDES) permit.

3.3.1 Background of EPA's Stormwater Permit Program

The Clean Water Act (CWA) generally prohibits facilities from discharging pollutants, including stormwater associated with certain industrial activity, through a "point source" to "waters of the United States" without a permit issued under the NPDES program. Section 402(p) of the CWA establishes a framework authorizing EPA to address stormwater discharges. Under Section 402(p), EPA is authorized to require a permit for any stormwater discharge "associated with industrial activity."

EPA originally issued stormwater management regulations in May 1973 which exempted uncontaminated stormwater from permitting unless the source was one specifically designated by a permitting authority. These rules were almost immediately challenged by the Natural Resources Defense Council (NRDC). In March 1976, owing to a court case that ruled in favor of NRDC, the rules were revised to provide coverage of all stormwater discharges except uncontaminated rural runoff. After several more lawsuits by trade associations and environmental groups, EPA revised the regulations in June 1979 and again in May 1980 to require testing for toxic pollutants as defined by the 1977 CWA. July 1982 marked a settlement agreement with industry petitioners who got a narrowed definition of stormwater point source and reduced application requirements by defining two different groups of applicants (Group I and Group II). Meanwhile EPA agreed not to enforce certain discharge limitations.

In November of the same year, EPA proposed a new rule based on the recent settlement agreements which was issued in final form in September 1984 and which defined stormwater point source in the same broad terms of the 1976 and 1980 rules. This rule also established a 6-month deadline for applications. Once again EPA braced for mountains of applications and proposed a rule in March 1985 which again referred to Groups I and II but reduced the data requirements in applications for Group I filers in an attempt to lighten the work load. Application deadlines were again extended for both groups. Then on Aug. 12, 1985, EPA reopened the comment period on the rule proposed in March of that year, and during this period the group application was proposed. EPA also placed industrial pretreaters discharging to POTWs into Group I. The final rule was issued on Aug. 29, and applications were due on Dec. 31, 1987, for Group I and June 30, 1989, for Group II. In February 1987, Congress passed the Water Quality Act of 1987 clarifying its own policy of stormwater management. But in December, before the application deadline, a court remanded the regulations. In a final rule issued in February 1988, EPA deleted the stormwater regulations.

EPA had estimated at that time that there were about 100,000 facilities nationwide discharging stormwater associated with industrial activity (not including oil and gas exploration and production operations) as described under phase I of the stormwater program. The large number of facilities addressed by the regulatory definition of "stormwater discharge associated with industrial activity" has placed a tremendous administrative burden on EPA and states with authorized NPDES programs to issue and administer permits for these discharges.

On Nov. 16, 1990 (55 FR 47990 as amended at 56 FR 12100, Mar. 21, 1991; 56 FR 56554, Nov. 5, 1991; 57 FR 11412, Apr. 2, 1992; 57 FR 60447, Dec. 18, 1992), EPA published final regulations which defined the term "stormwater discharge associated with industrial activity" and addressed point source discharges of stormwater from 11 major categories of industrial activities. EPA's definition of stormwater discharges that are "associated with industrial activity" relied principally on SIC codes.

Stormwater discharge associated with industrial activity means the discharge from any conveyance which is used for collecting and conveying stormwater and which is directly related to manufacturing, processing, or raw materials storage areas at an industrial plant. The term does not include discharges from facilities or activities excluded from the NPDES program under 40 CFR Part 122. For the categories of industries identified in paragraphs (i) through (x) of this paragraph, the term includes, but is not limited to, stormwater discharges from industrial plant yards; immediate access roads and rail lines used or traveled by carriers of raw materials, manufactured products, waste material, or by-products used or created by the facility; material handling sites; refuse sites; sites used for the application or disposal of process wastewaters (as defined at 40 CFR Part 401); sites used for the storage and maintenance of material handling equipment; sites used for residual treatment, storage, or disposal; shipping and receiving areas; manufacturing buildings; storage areas (including tank farms) for raw materials, and intermediate and finished products; and areas where industrial activity has taken place in the past and significant materials remain and are exposed to stormwater. For the categories of industries identified in paragraph $(b)(14)(xi)$ of this paragraph, the term includes only stormwater discharges from all the areas (except access roads and rail lines) where material handling equipment or activities, raw materials, intermediate products, final products, waste materials, by-products, or industrial machinery are exposed to stormwater. Material handling activities include the storage, loading and unloading, transportation, or conveyance of any raw material, intermediate product, finished product, by-product, or waste product. The term excludes areas located on plant lands separate from the plant's industrial activities, such as office buildings and accompanying parking lots as long as the drainage from the excluded areas is not mixed with stormwater drained from the above described areas. Industrial facilities [including industrial facilities that are federally, state, or municipally owned or operated that meet the description of the facilities listed in paragraphs (i) to (xi) of this section] include those facilities designated under the provisions of paragraph $(a)(1)(v)$ of this section. The following categories of facilities are considered to be engaging in "industrial activity" for purposes of the regulations:

- Facilities subject to stormwater effluent limitations guidelines, new source performance standards, or toxic pollutant effluent standards under 40 CFR subchapter N [except facilities with toxic pollutant effluent standards which are exempted under category (xi)]

- Facilities classified as Standard Industrial Classifications 24 (except 2434), 26 (except 265 and 267), 28 (except 283), 29, 311, 32 (except 323), 33, 3441, 373; 40 CFR 122.26(b)(14)(iii)

- Facilities classified as SICs 10 through 14 (mineral industry) including active or inactive mining operations [except for areas of coal-mining operations no longer meeting the definition of a reclamation area under 40 CFR 434.11(1) because the performance bond issued to the facility by the appropriate SMCRA authority has been released, or except for areas of non-coal-mining operations which have been released from applicable state or federal reclamation requirements after Dec. 17, 1990] and oil and gas exploration, production, processing, or treatment operations, or transmission facilities that discharge stormwater contaminated by contact with or that has come into contact with any overburden, raw material, intermediate products, finished products, by-products, or waste products located on the site of such operations (inactive mining operations are mining sites that are not being actively mined, but which have an identifiable owner-operator; inactive mining sites do not include sites where mining claims are being maintained prior to disturbances associated with the extraction, beneficiation, or processing of mined materials, or sites where minimal activities are undertaken for the sole purpose of maintaining a mining claim).

- Hazardous waste treatment, storage, or disposal facilities, including those that are operating under interim status or a permit under subtitle C of RCRA.

- Landfills, land application sites, and open dumps that receive or have received any industrial wastes (waste that is received from any of the facilities described under this subsection) including those that are subject to regulation under subtitle D of RCRA.

- Facilities involved in the recycling of materials, including metal scrapyards, battery reclaimers, salvage yards, and automobile junkyards, including but limited to those classified as SICs 5015 and 5093.

- Steam electric power generating facilities, including coal handling sites.

- Transportation facilities classified as SICs 40, 41, 42 (except 4221-25), 43, 44, 45, and 5171 which have vehicle maintenance shops, equipment cleaning operations, or airport deicing operations. Only those portions of the facility that are either involved in vehicle maintenance (including vehicle rehabilitation, mechanical repairs, painting, fueling, and lubrication), equipment cleaning operations, airport deicing operations, or which are otherwise identified under paragraphs (b)(14) (i)–(vii) or (ix)–(xi) of this section are associated with industrial activity.

- Treatment works treating domestic sewage or any other sewage sludge or wastewater treatment device or system, used in the storage, treatment, recycling, and reclamation of municipal or domestic sewage, including land dedicated to the disposal of sewage sludge that are located within the confines of the facility, with a design flow of 1.0 mgd or more, or required to have an approved pretreatment program under 40 CFR Part 403. Not included are farmlands, domestic gardens, or lands used for sludge management where sludge is beneficially reused and which are not physically located in the confines of the facility, or areas that are in compliance with section 405 of the CWA.

- Construction activity including clearing, grading, and excavation activities except operations that result in the disturbance of less than 5 acres of total land area which are not part of a larger common plan of development or sale. Facilities under Standard Industrial Classifications 20, 21, 22, 23, 2434, 25, 265, 267, 27, 283, 285, 30, 31 (except 311), 323, 34 (except 3441), 35, 36, 37 (except 373), 38, 39, 4221-25 [and which are not otherwise included within categories (ii) to (x)].

These regulations also set forth NPDES permit application requirements for stormwater discharges associated with industrial activity and stormwater discharges from certain municipal separate storm sewer systems.

To provide a reasonable and rational approach to addressing this permitting task, the agency developed a strategy for issuing permits for stormwater discharges associated with industrial activity. In developing this strategy, the agency recognized that the CWA provides flexibility in the manner in which NPDES permits are issued and has used this flexibility to design a workable permitting system. In accordance with these considerations, the permitting strategy (described in more detail in 57 FR 11394) describes a four-tier set of priorities for issuing permits for these discharges:

- Tier I: baseline permitting—one or more general permits will be developed to initially cover the majority of stormwater discharges associated with industrial activity.
- Tier II: watershed permitting—facilities within watersheds shown to be adversely impacted by stormwater discharges associated with industrial activity will be targeted for individual or watershed-specific general permits.
- Tier III: industry-specific permitting—specific industry categories will be targeted for individual or industry-specific general permits.
- Tier IV: facility-specific permitting—a variety of factors will be used to target specific facilities for individual permits.

Industrial activities from all of these categories with the exception of construction activities participated in the group application process. The information contained in the group applications indicated that type and amount of pollutants discharged in stormwater vary from industrial activity to industrial activity because of the variety of potential pollutant sources present in different industrial activities, as well as the variety of pollution prevention measures commonly practiced by each of the regulated industries. To facilitate the process of developing permit conditions for each of the 1200 group applications submitted, EPA classified groups into 29 industrial sectors where the nature of industrial activity, type of materials handled, and material management practices employed were sufficiently similar for the purposes of developing permit conditions.

In September 1992, EPA promulgated its baseline general permits authorizing stormwater discharges associated with industrial activity in states where it is the permitting authority.

The promulgation of the baseline general permit was essentially a response to the group application options. Group applications were submitted in two parts. Part 1 of the application was originally due by Sept. 30, 1991, with Part 2 of the application being due by Oct. 1, 1992. In Part 1 of the application, all participants were identified and information on each facility was included, such as industrial activities, significant materials exposed to stormwater, and material management activities. For Part 1 of the application, groups also had to identify sampling subgroups to submit sampling data for Part 2. Over 1200 groups with over 60,000 member facilities submitted Part 1 applications. Upon review of the Part 1 application, if the EPA determined that the application was an appropriate grouping of facilities with complete information provided on each participant, and a suitable sampling subgroup was proposed, the application was approved.

Part 2 of the application consisted of sampling data from each member of the sampling subgroup identified in Part 1 of the application. In drafting the baseline general permit, EPA reviewed both parts of the applications and formulated the permit language. NPDES-authorized states were provided the data from the group applications. Authorized NPDES states have the authority to propose and finalize either individual permits for each facility included in the application located in the state, or general permits, if the state has general permit authority. If the state felt additional information was needed from the applicants, the state could ask each or any of the applicants for more information on their facility and/or discharge.

On Sept. 29, 1995, the U.S. Environmental Protection Agency issued its multisector general permit authorizing discharges of stormwater from those industrial facilities included in the industrial sectors that participated in the group application process. The 29 industries included in the new multisector general permit are listed in Table 3.2. EPA developed specific requirements for each industrial sector based on information gathered during the group application process. Obligations imposed under the multisector general permit usually are less onerous than those formally required under the baseline general permit. The new multisector general permit has been issued in all states where EPA is the permitting authority under the NPDES program. The permit was also provided to states with authorized NPDES programs and now serves as a model in their stormwater permitting activities. In order to obtain coverage under the multisector general permit, a notice of intent originally had to be submitted to EPA by Dec. 28, 1995.

EPA's baseline general stormwater permit program expired in September 1997. The multisector general permit has now replaced the baseline general permit for all industries. Dischargers must now obtain coverage under an individual NPDES permit or the multisector general permit. To obtain coverage under the multisector general permit, dischargers had to, at a minimum, file a notice of intent more than 48 hours before the expiration of the baseline general permit. Facilities that initially obtained a baseline general permit had to quickly choose to comply with the requirements of either an individual NPDES permit or the multisector general permit.

On July 11, 1997, EPA proposed modifications to the multisector general permit to incorporate those facilities originally not covered by the permit. Table 3.3 sets forth the proposed new facility categories. In addition, EPA proposed a new catch-all type sector (sector AD) to cover certain facilities subject to stormwater permitting but not eligible for the other sectors. EPA has proposed the inclusion of a number of additional facilities within the multisector general permit program. As of the writing of this book, EPA has not yet finalized this proposal. Consequently, its scope is not addressed in the discussions below.

On Sept. 30, 1998 (63 FR 52429), the Regional Administrators of EPA Regions I, II, III, IV, VI, IX, and X provided final notice of modifications to EPA's final NPDES Storm Water Multi-Sector General Permit (MSGP) which was first issued on Sept. 29, 1995 (60 FR 50804), and amended on Feb. 9, 1996

TABLE 3.2 Industrial Sectors Covered by Original Multisector General Permits

---

Timber products
Paper and allied products manufacturing
Chemical and allied products manufacturing
Asphalt paving and roofing materials manufacturers and lubricant manufacturers
Glass, clay, cement, concrete, and gypsum product manufacturing
Primary metals
Metal mining (ore mining and dressing)
Coal mines and coal mining–related facilities
Oil and gas extraction
Mineral mining and dressing
Hazardous waste treatment, storage, or disposal facilities
Landfills and land application sites
Automobile salvage yards
Scrap recycling
Steam electric power generating facilities
Land transportation
Water transportation
Ship and boat building or repairing yards
Air transportation
Treatment works
Food and kindred products
Textile mills, apparel, and other fabric product manufacturing
Furniture and fixtures
Printing and publishing
Rubber, miscellaneous plastic products, and miscellaneous manufacturing industries
Leather tanning and finishing
Fabricated metal products
Transportation equipment, industrial or commercial machinery
Electronic, electrical, photographic, and optical goods

(61 FR 5248), Feb. 20, 1996 (61 FR 6412), and Sept. 24, 1996 (61 FR 50020). EPA has modified the MSGP to authorize stormwater discharges from previously excluded facilities so that they may be covered by the MSGP after expiration of EPA's Baseline Industrial General Permit. EPA also finalized the following limited specific changes to the MSGP as published on Sept. 29, 1995 (60 FR 50804):

- Authorization of mine dewatering discharges from construction sand and gravel, industrial sand, and crushed stone mines in EPA Regions I, II, and X.

- Inclusion in Sector A of the MSGP of the effluent limitation guideline in 40 CFR Part 429, Subpart I for discharges resulting from spray-down of lumber and wood products in storage yards (wet decking).

- Clarification that Sectors X and AA authorize discharges from all facilities in major SIC groups 27 and 34, respectively.

TABLE 3.3 Proposed Placement of Additional Facilities

SIC Code	Facility	Sector
2833–2836	Medicinal chemicals and botanical products; pharmaceutical preparations, in vitro and in vivo diagnostic substances; biological products, except diagnostic substances	C
3922	Petroleum refining	I
3131	Boot and shoe cut stock and findings	V
3142–3144	House slippers, men's dress, street, and work shoes; women's dress, street, and work shoes	V
3149	Footwear, except rubber	V
3151	Leather gloves and mittens	V
3161	Luggage and cases	V
3171	Women's handbags and purses	V
3172	Personal leather goods	V
3199	Leather goods not classified elsewhere	V
3231	Glass products made of purchased glass	E
3261	Vitreous china plumbing fixtures and china and earthenware fittings and bathroom accessories	E
3274	Lime, agricultural/building lime, dolomite, and lime plaster	E
3281	Cut stone and stone products, benches, blackboards, table tops, pedestals, etc.	E
3291	Abrasive products	E
3292	Asbestos products, tiles, building materials, except paper, insulating pipe coverings	E
3296	Mineral wool, insulation	E
3299	Nonmetallic mineral products not classified elsewhere, plaster of Paris and papier-mâché, etc.	E
4221-4225	Warehousing facilities without trucking services	P
LF	Open dumps	L
	Regulated industrial stormwater discharges not covered by any other sector	AD

- Addition of new Sector AD to the MSGP to authorize discharges from Phase I facilities which may not fall into one of the original sectors of the permit, and selected Phase II discharges which are designated for permitting in accordance with 40 CFR 122.26(g)(1)(i).
- Modification of inspection requirements in Sector I for inactive oil and gas extraction facilities which are remotely located and unstaffed.
- Addition of new Addendum I to provide guidance and information to assist applicants with determining permit eligibility concerning protection of historic properties.

• Update of the county and species list of endangered and threatened species found in Addendum H, and provide a listing of additional sources to reference for future updates to the list.

The Regional Administrators also provided a final notice that the agency is not reissuing the NPDES stormwater Baseline Industrial General Permit which was issued on Sept. 9, 1992 (57 FR 41236) or Sept. 25, 1992 (57 FR 44438), depending on the geographic area of applicability, and to terminate this permit (with the limited exceptions discussed in Section I below) upon final modification of the multisector permit. As a result, all industrial facilities previously permitted under the Baseline Industrial General Permit, except as otherwise specified in the notice, are now required to seek stormwater permit coverage under the modified MSGP by Dec. 29, 1998 (within 90 days after the publication of the final notice) or submit an application for an individual NPDES permit.

This action also provided notice for the issuance of the final NPDES for stormwater discharges associated with industrial activity for American Samoa and the Commonwealth of the Northern Mariana Islands (CNMI). The geographic area of coverage of the MSGP being revised includes American Samoa and CNMI on the list of areas for which discharges may be authorized.

3.3.2 Stormwater Permitting Options

There are three potential options an applicant may choose for obtaining permit coverage for stormwater discharges associated with industrial activity:

• The first option is to submit an individual application consisting of EPA Form 1 (EPA Form 3510-1, General Information) and Form 2F (EPA Form 3520-2F).

• The second option is to participate in a group application.

• The third option is to file a notice of intent (NOI) to be covered under a multisector general permit according to the requirements.

In addition to these three permit application options, there are other specific stormwater permit application requirements for small businesses, construction activities, oil and gas operations, and mining operations.

3.3.2.1 Individual Permits. Owners of facilities that discharge stormwater associated with industrial activity that do not participate in a group application or obtain coverage under a multisector general permit must submit an individual application by completing at the very least EPA Form 3510-1 (General Information) and Form 3510-2F. EPA Form 3510-1 requires general information about the facility, including the name and address of the facility, the facility type (i.e., SIC Code), and a map showing specified features. This is the normal NPDES approach to permitting. These applications were initially due on Oct. 1, 1992, for existing facilities. The individual application is similar to an NPDES permit application. A site map and estimate of impervious areas (areas where stormwater will run off instead of soaking into the ground) is required. Also, any materials which may be exposed to stormwater must be identified. Associated outdoors materials management practices and on-site disposal practices must be discussed in the application. The location and description of existing structural and nonstructural controls to reduce pollutants in stormwater runoff must also be described. Structural controls include catchment basins, containment ponds, diversion dikes, etc. Nonstructural controls amount to management practices and policies on housekeeping, where work is performed, training, etc. All stormwater outfalls must be evaluated for unpermitted, nonstormwater discharges, and a certification that this evaluation has been performed must be included with the application. Information about spills or leaks of toxic or hazardous pollutants within the previous 3 years must also be submitted with the application. Finally, quantitative data must be submitted for samples collected on-site during storm events.

Applicants for discharges composed entirely of stormwater must submit Form 1 and Form 2F. Applicants for discharges composed of stormwater and nonstormwater must submit Form 1, Form 2C, and Form 2F (see the following forms section). Applicants for new sources or new discharges (as defined in 40 CFR 122.2) composed of stormwater and nonstormwater submit Form 1, Form 2D, and Form 2F.

3.3.2.2 Group Permits.   The group application procedure is a permitting option available to groups of facilities discharging stormwater associated with industrial activities that have similar operations or waste streams, are in the same effluent guideline subcategory, or have discharges that are sufficiently similar in loading and concentrations to be appropriate for this type of coverage. Dischargers meeting these criteria may collectively submit a two-part group application instead of each filing an individual permit application. Part 1 of the group application must contain the specific information listed directly below while Part 2 of the group application must contain sufficient quantitative data so that both parts can be evaluated as a complete NPDES application for each discharger. At the very least, such a group must consist of at least four participants meeting the above criteria that are listed in nine subdivisions, based on the facility location relative to the nine precipitation zones depicted in Fig. 3.1.

The first part of the group application includes the following information:

• A narrative description that summarizes the industrial activities of the individual participants of the group application and explains why the participants are collectively similar in the context of their processes, operations, materials management practices, and stormwater discharges to be eligible for coverage by a group permit.

• An inventory of significant stored materials exposed to precipitation, and materials management practices used to minimize contact by these materials with precipitation and stormwater runoff.

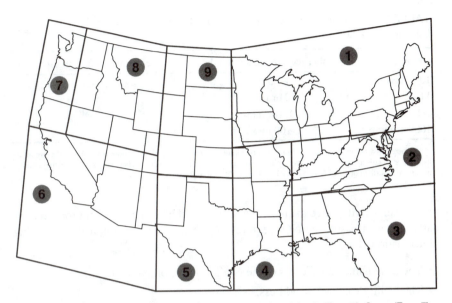

Not Shown: Alaska (Zone 7); Hawaii (Zone 7); Northern Mariana Islands (Zone 7); Guam (Zone 7); American Samoa (Zone 7); Trust Territory of the Pacific Islands (Zone 7); Puerto Rico (Zone 3); Virgin Islands (Zone 3).

FIGURE 3.1 Rainfall zones of the United States.

- An identification of 10 percent of the dischargers participating in the group application from which quantitative data will be submitted in Part 2 along with a description of why these facilities selected to perform sampling and analysis are representative of the group.

Table 3.4 illustrates the various sizes of a group that may be formed, as well as the number of facilities in a designated group required to submit quantitative data. The chart also identifies the minimum number of facilities required to submit quantitative data for Part 2 for each precipitation zone included in the group application.

Part 2 of the group application consists of sampling data from each member of the subgroup identified in Part 1 reported on the applicable data reporting portions of Form 2F. The required certification must also be completed.

3.3.2.3 Multisector General Permits. The third option for obtaining permit coverage for stormwater dischargers associated with industrial activity originally is to submit a Notice of Intent (NOI) to the appropriate permitting agency seeking coverage under a baseline general permit. This general permitting option complements the individual and group application procedures for controlling stormwater discharges associated with industrial activities. By availing themselves of this third permitting option, dischargers of stormwater associated with

TABLE 3.4   Sizes and Number of Facilities in a Group Required to Submit Quantitative Data

Group size	Number required to submit data	Precipitation zone requirements
10 or less	5	None
11–99	10 or more	*
>100	10% or more	*
1000 or more	100 max	*

*There must be two dischargers from each precipitation zone in which 10 or more members of the group are located, or one discharger from each precipitation zone in which nine or fewer members are located. Testing requirements are described in 40 CFR 122.26(c)(1)(i)(E) and 40 CFR 122.21(g)(7).

industrial activity agree to abide by provisions of a "generic" permit for a certain area. In most cases, the NOI requires applicants only to submit the following information:

- The legal name and address of the facility owner or operator.
- The facility's name and address.
- The type of facility (SIC code) or discharges.
- The name of the receiving streams.

The multisector general permit contains some special conditions that are similar to those in the baseline permit. With a few minor exceptions, nonstormwater discharges are not authorized under either permit. Both permits require that releases of certain quantities of hazardous substances and oil be reported to the National Response Center and the permitting authority and that a required stormwater pollution prevention plan be amended to prevent future discharges.

NOI requirements for general permits usually do not require the collection of monitoring data and address only general facility-specific information.

3.3.3   Special Permit Application Requirements

As stated earlier, federal stormwater regulations of 40 CFR 122.26 provide for special regulatory provisions for the following categories.

3.3.3.1   Small Businesses.   An applicant which meets one of the following criteria qualifies as a "small business."

- For coal mines, a probable total annual production of less than 100,000 tons per year.
- For all other applicants, gross total annual sales averaging less than $100,000 per year (in second-quarter 1980 dollars). [*Note:* Sales figures for years after 1980 should be indexed to the second quarter of 1980 by using the gross national

product price deflator (second quarter of 1980 = 100). This index is available in National Income and Product Accounts of the U.S. Department of Commerce, Bureau of Economic Analysis.]

As a qualified "small business," the applicant is exempt from the following requirements for certain organic toxic pollutants listed on Form 2F:

- To submit quantitative data for the pollutants listed in Table I of Appendix D to 40 CFR Part 122 (the organic toxic pollutants).

- To analyze for the organic toxic pollutants in the fractions designated in Table II of Appendix D to 40 CFR Part 122 in each outfall containing process wastewater. (*Note:* Table II of Appendix D to 40 CFR Part 122 lists the organic toxic pollutants in each fraction.)

Small businesses with stormwater discharges associated with industrial activity are subject to other requirements of Form 1 and Form 2F (see forms section), including requirements to sample for specified conventional pollutants and other specified constituents.

3.3.3.2 Construction Activities. Construction activities include clearing, grading, and excavation activities except operations that result in the disturbance of less than 5 acres of total land area which are not part of a larger common plan of development or sale. The application requirements for operators of stormwater discharges associated with industrial activity from construction activities include Form 1 (EPA's General Information Form) and a narrative description of:

- The location (including a map) and the nature of the construction activity

- The total area of the site and the area of the site that is expected to undergo excavation during the life of the permit

- Proposed measures, including best management practices, to control pollutants in stormwater discharges during construction, including a brief description of applicable state and local erosion and sediment control requirements

- Proposed measures to control pollutants in stormwater discharges that will occur after construction operations have been completed, including a brief description of applicable state and local stormwater management controls

- An estimate of the runoff coefficient of the site and the increase in impervious area after the construction addressed in the permit application is completed, the nature of fill material, and existing data describing the soil or the quality of the discharge

- The name of the receiving water body

At this time, a stormwater permit is not required for construction activities that disturb less than 5 acres of land.

3.3.3.3   Mining Operations.   Several specific regulatory provisions apply to stormwater discharges associated with industrial activity from mining operations [40 CFR 122.26(*a*)(2)]. The permitting authority may not require a permit for discharges of stormwater runoff from mining operations that are composed entirely of flows that are from conveyances or systems of conveyances (including, but not limited to, pipes, conduits, ditches, and channels) used for collecting and conveying precipitation runoff, and which are not contaminated by contact with any overburden, raw material, intermediate products, finished product, by-product, or waste products located on the site of such operations.

Owners of any construction activity that disturbs 5 acres or more of total land area must apply for a stormwater discharge permit. After construction, roads for mining operations would not be covered by the stormwater permit regulations unless stormwater runoff from such roads mixes with stormwater that is contaminated by contact with overburden, raw materials, intermediate products, finished products, by-products, or waste products. In cases where roads are constructed out of materials such as overburden or by-products, an NPDES stormwater discharge permit is required.

3.3.3.4   Oil and Gas Operations.   [40 CFR 122.26(*c*)(1)(iii)]: The operator of an existing or new discharge composed entirely of stormwater from an oil or gas exploration, production, processing, or treatment operation, or transmission facility is not required to submit a permit application under most circumstances. A facility is covered only if it: (1) has had a discharge of stormwater resulting in the discharge of a reportable quantity for which notification is or was required under the Comprehensive Environmental Response, Compensation, and Liability Act of 1980 (also known as Superfund) any time since Nov. 16, 1987; or (2) has had a discharge of stormwater resulting in the discharge of a reportable quantity for which notification is or was required at any time under the Clean Water Act since Nov. 16, 1987; or (3) contributes to a violation of a water quality standard.

Under the Clean Water Act, a reportable quantity is the amount of oil that violates applicable water quality standards or causes a film or sheen upon or a discoloration of the surface water or adjoining shorelines, or causes a sludge or emulsion to be deposited beneath the surface of the water or upon adjoining shorelines.

The reportable quantities for other substances are listed in 40 CFR 117.3 and 40 CFR 302.4 in terms of pounds released over any 24-hour period. A listing of the hazardous substances with reportable quantities is given in Appendix G of this book.

3.3.4   Types of Discharges Covered

Coverage under the multisector general permit is available to stormwater discharges from industrial activities represented by the group application process. However, coverage under this permit is not restricted to participants in the group application process. To limit coverage under this general permit to only those who participated in the group application process would not be appropriate for administrative, environmental, and national consistency reasons. The administrative burden for EPA to develop separate general permits for nongroup members would be excessive, unnecessary, and wasteful of tax dollars. EPA would also need to use the same information in the development of such permits. The permits would be essentially the same. The time spent in this process would leave many facilities unregulated for some number of additional months. This would not address the environmental concerns of the Clean Water Act. Likewise, group members are not precluded from seeking coverage under other available stormwater permits such as EPA's "baseline" general permits for Storm Water Discharges Associated with Industrial Activity (57 FR 41175 and 57 FR 44412). Group members must consider, however, that the deadlines for preparing and implementing the pollution prevention plan required under the baseline permit have already expired for existing facilities. Therefore, group members that seek coverage under the baseline general permit must have a pollution prevention plan developed and implemented prior to NOI submittal.

Unlike the baseline general permits, the multisector general permit does not exclude all stormwater discharges subject to effluent limitation guidelines. Four types of stormwater discharges subject to effluent limitation guidelines may be covered under the multisector general permit if they are not already subject to an existing or expired NPDES permit. These discharges include contaminated stormwater runoff from phosphate fertilizer manufacturing facilities, runoff associated with asphalt paving or roofing emulsion production, runoff from material storage piles at cement manufacturing facilities, and coal pile runoff at steam electric generating facilities. The permit does not, however, authorize all stormwater discharges subject to effluent guidelines. Stormwater discharges subject to effluent guidelines under 40 CFR Part 436 or for mine drainage under 40 CFR Part 440 are not covered under the multisector general permit nor are discharges subject to effluent guidelines for acid or alkaline mine drainage under 40 CFR Part 434.

3.3.5 Notification Requirements

General permits for stormwater discharges associated with industrial activity require the submittal of a completed Notification of Intent (NOI) prior to the authorization of such discharges [see 40 CFR 122.28(b)(2)(i), Apr. 2, 1992 (57 FR 11394)]. Consistent with these regulatory requirements, the multisector general permit establishes NOI requirements that operate in addition to the Part 1 and Part 2 group application requirements. To be covered under this permit, facilities, including members of an approved group, must submit an NOI and other required information within 90 days of the effective date of the permit. Conversely, when a facility is able to eliminate a covered stormwater discharge, a Notice of Termination (NOT) must also be filed.

(*Note:* Refer to Appendix F for copies of the appropriate NOI and NOT forms.)

3.3.5.1 Contents of NOIs

• The operator's name, address, telephone number, and status as federal, state, private, public, or other entity.

• Street address of the facility for which the notification is submitted. Where a street address for the site is not available, the location can be described in terms of the latitude and longitude of the facility to the nearest 15 seconds, or the quarter, section, township, and range (to the nearest quarter section) of the approximate center of the site.

• An indication of whether the facility is located on federal Indian reservations.

• Up to four 4-digit SIC codes that best represent the principal products or activities provided by the facility. For hazardous waste treatment, storage, or disposal facilities, land disposal facilities that receive or have received any industrial waste, steam electric power generating facilities, or treatment works treating domestic sewage, a two-character code must be provided.

• The permit number of any NPDES permit for any discharge (including non-stormwater discharges) from the site that is currently authorized by an NPDES permit.

• The name of the receiving water(s), or if the discharge is through a municipal separate storm sewer, the name of the municipal operator of the storm sewer and the receiving water(s) for the discharge through the municipal separate storm sewer.

• The analytical monitoring status of the facility (monitoring or not).

• For a co-permittee, if a stormwater general permit number has been issued, it should be included.

• A certification that the operator of the facility has read and understands the eligibility requirements for the permit and that the operator believes the facility to be in compliance with those requirements.

• Identify type of permit requested (either baseline general, multisector, or construction); longitude and latitude; indication of presence of endangered species; indication of historic preservation agreement; signed certification stating compliance with the National Historic Preservation Act, Endangered Species Act, and the new source performance standard requirements.

• For any facility that begins to discharge stormwater associated with industrial activity after (insert date 270 days after permit finalization), a certification that a stormwater pollution prevention plan has been prepared for the facility in accordance with Part IV of the permit. (A copy of the plan should not be included with the NOI submission.)

In addition, a determination as to whether there are any endangered species identified in an addendum to the permit that would be "in proximity" to the stormwater discharge and a certification that they would not be adversely affected should be included in the NOI form. *Dischargers that cannot make such a certification must apply for an individual permit.*

The NOI must be signed in accordance with the signatory requirements of 40 CFR 122.22. A complete description of these signatory requirements is provided in the instructions accompanying the NOI. Completed NOI forms must be submitted to the Storm Water Notice of Intent (4203), 401 M Street SW, Washington, D.C. 20460.

3.3.5.2 Notice of Intent Submittal Deadlines. Except for the special circumstances discussed below, dischargers who intend to obtain coverage under this permit for a stormwater discharge from an industrial activity that is in existence prior to the date 90 days after permit issuance must submit an NOI on or before the date 90 days after permit issuance, and facilities that begin industrial activities after the date 90 days after permit issuance are required to submit an NOI at least 2 days prior to the commencement of the new industrial activity.

A discharger is not precluded from submitting an NOI at a later date. However, in such instances, EPA may bring appropriate enforcement actions.

The stormwater regulations (40 CFR 122.27) required that facilities that discharge stormwater associated with an industrial activity submit an application for permit coverage on or before Oct. 1, 1992, except industrial activities owned or operated by a medium municipality, which had until May 17, 1993. The multisector general permit does not extend that application deadline. EPA intends that most of the facilities that will seek coverage under the final version of the multisector general permit are members of groups with approved applications; facilities that submitted a Notice of Intent to be covered by EPA's baseline general permit and now wish to switch to coverage under the multisector general permit; or have submitted a complete individual application but have not yet received an individual permit.

EPA may deny coverage under this permit and require submittal of an individual NPDES permit application based on a review of the completeness and/or content of the NOI or other information (e.g., Endangered Species Act compliance, National Historic Preservation Act compliance, water quality information, compliance history, history of spills, etc.). Where EPA requires a discharger authorized under this general permit to apply for an individual NPDES permit (or an alternative general permit), EPA will notify the discharger in writing that a permit application (or different NOI) is required by an established deadline. Coverage under this industry general permit will automatically terminate if the discharger fails to submit the required permit application in a timely manner. Where the discharger does submit a requested permit application, coverage under this general permit will automatically terminate on the effective date of the issuance or denial of the individual NPDES permit or the alternative general permit as it applies to the individual permittee.

3.3.5.3   Notice of Termination (NOT).   Where a discharger is able to eliminate the stormwater discharges associated with industrial activity from a facility, the discharger may submit a Notice of Termination (NOT) form (or photocopy thereof) provided by the director. The NOT form requires the following information:

- Name, mailing address, and location of the facility for which the notification is submitted. Where a street address for the site is not available, the location of the approximate center of the site must be described in terms of the latitude and longitude to the nearest 15 seconds, or the section, township, and range to the nearest quarter.

- The name, address, and telephone number of the operator addressed by the Notice of Termination.

- The NPDES permit number for the stormwater discharge associated with industrial activity identified by the NOT.

- An indication of whether the stormwater discharges associated with industrial activity have been eliminated or the operator of the discharges has changed.

- The following certification: "I certify under penalty of law that all stormwater discharges associated with industrial activity from the identified facility that are authorized by an NPDES general permit have been eliminated or that I am no longer the operator of the industrial activity. I understand that by submitting this Notice of Termination I am no longer authorized to discharge stormwater associated with industrial activity under this general permit, and that discharging pollutants in stormwater associated with industrial activity to waters of the United States is unlawful under the Clean Water Act where the discharge is not authorized by an NPDES permit. I also understand that the submittal of this notice of termination does not release an operator from liability for any violations of this permit or the Clean Water Act."

NOTs are to be sent to the Storm Water Notice of Termination (4203), 401 M Street, SW, Washington, D.C. 20460.

The NOT must be signed in accordance with the signatory requirements of 40 CFR 122.22. A complete description of these signatory requirements is provided in the instructions accompanying the NOT.

3.3.6   Stormwater Pollution Prevention Plans

The multisector general permit uses pollution prevention plans as the primary method of identifying and controlling sources of contamination that could potentially affect the quality of stormwater discharges. Pollution prevention plan requirements vary by industrial sector. Facilities previously covered by the baseline general permit must revise their pollution prevention plans in accordance with the requirements of the multisector general permit. Facilities currently eligible for the multisector general permit must prepare and implement revised plans before filing an NOI. Facilities that were not previously eligible for the multisector general permit must revise their existing plans and implement them within 90 days at the effective date of the modified multisector general permit. All facilities intending to be covered by the multisector general permit for stormwater discharges associated with industrial activity must prepare and implement a stormwater pollution prevention plan. The stormwater permit addresses pollution prevention plan requirements for a number of categories of industries. The following is a discussion of the common plan requirements for all industries.

The pollution prevention approach in the multisector general permit focuses on two major objectives: (1) to identify sources of pollution potentially affecting the quality of stormwater discharges associated with industrial activity from the facility; and (2) to describe and ensure implementation of practices to minimize and control pollutants in stormwater discharges associated with industrial activity from the facility and to ensure compliance with the terms and conditions of this permit.

The stormwater pollution prevention plan requirements in the general permit are intended to facilitate a process whereby the operator of the industrial facility thoroughly evaluates potential pollution sources at the site and selects and implements appropriate measures designed to prevent or control the discharge of pollutants in stormwater runoff. The process involves the following four steps: (1) formation of a team of qualified plant personnel who will be responsible for preparing the plan and assisting the plant manager in its implementation; (2) assessment of potential stormwater pollution sources; (3) selection and implementation of appropriate management practices and controls; and (4) periodic evaluation of the effectiveness of the plan to prevent stormwater contamination and comply with the terms and conditions of this permit. The authorization to include best management practices in the permit to control or abate the discharge of pollutants is derived from 40 CFR 144.45(k).

EPA believes the pollution prevention approach is the most environmentally sound and cost-effective way to control the discharge of pollutants in stormwater runoff from industrial facilities. This position is supported by the results of a comprehensive technical survey EPA completed in 1979. The survey found that two classes of management practices are generally employed at industries to control the nonroutine discharge of pollutants from sources such as stormwater runoff, drainage from raw material storage and waste disposal areas, and discharges from places where spills or leaks have occurred. The first class of management practices includes those that are low in cost, applicable to a broad class of industries and substances, and widely considered essential to a good pollution control program. Some examples of practices in this class are good housekeeping, employee training, and spill response and prevention procedures. The second class includes management practices that provide a second line of defense against the release of pollutants. This class addresses containment, mitigation, and cleanup. Since publication of the 1979 survey, EPA has imposed management practices and controls in NPDES permits on a case-by-case basis. The agency also has continued to review the appropriateness and effectiveness of such practices, as well as the techniques used to prevent and contain oil spills. Experience with these practices and controls has shown that they can be used in permits to reduce pollutants in stormwater discharges in a cost-effective manner. In keeping with both the present and previous administration's objective to attain environmental goals through pollution prevention, pollution prevention has been and continues to be the cornerstone of the NPDES permitting program for stormwater. EPA has developed guidance entitled "Storm Water Management for Industrial Activities: Developing Pollution Prevention Plans and Best Management Practices," September 1992, to assist permittees in developing and implementing pollution prevention measures.

3.3.6.1 Pollution Prevention Team. As a first step in the process of developing and implementing a stormwater pollution prevention plan, permittees are required to identify a qualified individual or team of individuals to be responsible for developing the plan and assisting the facility or plant manager in its implementation. When selecting members of the team, the plant manager should draw on the expertise of all relevant departments within the plant to ensure that all aspects of plant operations are considered when the plan is developed. The plan must clearly describe the responsibilities of each team member as they relate to specific components of the plan. In addition to enhancing the quality of communication between team members and other personnel, clear delineation of responsibilities will ensure that every aspect of the plan is addressed by a specified individual or group of individuals. Pollution prevention teams may consist of one individual where appropriate (e.g., in certain small businesses with limited stormwater pollution potential).

3.3.6.2 Description of Potential Pollution Sources. Each stormwater pollution prevention plan must describe activities, materials, and physical features of the facility that may contribute significant amounts of pollutants to stormwater runoff or, during periods of dry weather, result in pollutant discharges through the separate storm sewers or stormwater drainage systems that drain the facility. This assessment of stormwater pollution risk will support subsequent efforts to identify and set priorities for necessary changes in materials, materials management practices, or site features, as well as aid in the selection of appropriate structural and nonstructural control techniques. Some operators may find that significant amounts of pollutants are running onto the facility property. Such operators should identify and address the contaminated run-on in the stormwater pollution prevention plan. If the run-on cannot be addressed or diverted by the permittee, the permitting authority should be notified. If necessary, the permitting authority may require the operator of the adjacent facility to obtain a permit.

3.3.6.3 Stormwater Pollution Plan Elements. The stormwater pollution prevention plans generally must describe the following elements:

a. Drainage. The plan must contain a map of the site that shows the location of outfalls covered by the permit (or by other NPDES permits), the pattern of stormwater drainage, an indication of the types of discharges contained in the drainage areas of the outfalls, structural features that control pollutants in runoff, surface water bodies (including wetlands), places where significant materials are exposed to rainfall and runoff, and locations of major spills and leaks that occurred in the 3 years prior to the date of the submission of an NOI to be covered under this permit. The map also must show areas where the following activities take place: fueling, vehicle and equipment maintenance and/or cleaning, loading and unloading, material storage (including tanks or other vessels used for liquid or waste storage), material processing, and waste disposal. For areas of the facility that generate stormwater discharges with a reasonable potential to contain significant amounts of pollutants, the map must indicate the probable direction of stormwater flow and the pollutants likely to be in the discharge. Flows with a significant potential to cause soil erosion also must be identified. In order to increase the readability of the map, the inventory of the types of discharges contained in each outfall may be kept as an attachment to the site map.

b. Inventory of exposed materials. Facility operators are required to carefully conduct an inspection of the site and related records to identify significant materials that are or may be exposed to stormwater. The inventory must address materials that within 3 years prior to the date of the submission of an NOI to be covered under this permit have been handled, stored, processed, treated, or disposed of in a manner to allow exposure to stormwater. Findings of the inventory must be documented in detail in the pollution prevention plan. At a minimum, the plan must describe the method and location of on-site storage or disposal; practices used to minimize contact of materials with rainfall and runoff; existing structural and nonstructural controls that reduce pollutants in runoff; and any treatment the runoff receives before it is discharged to surface waters or a separate storm sewer system. The description must be updated whenever there is a significant change in the types or amounts of materials, or material management practices, that may affect the exposure of materials to stormwater.

c. Significant spills and leaks. The plan must include a list of any significant spills and leaks of toxic or hazardous pollutants that occurred in the 3 years prior to the date of the submission of an NOI to be covered under this permit. Significant spills include, but are not limited to, releases of oil or hazardous substances in excess of quantities that are reportable under Section 311 of CWA (see 40 CFR 110.10 and 40 CFR 117.21) or Section 102 of the Comprehensive Environmental Response, Compensation and Liability Act (CERCLA) (see 40 CFR 302.4). Significant spills may also include releases of oil or hazardous substances that are not in excess of reporting requirements and releases of materials that are not classified as oil or a hazardous substance.

The listing should include a description of the causes of each spill or leak, the actions taken to respond to each release, and the actions taken to prevent similar spills or leaks in the future. This effort will aid the facility operator as she or he examines existing spill prevention and response procedures and develops any additional procedures necessary to fulfill the requirements of Part XI of the permit.

d. Nonstormwater discharges. Each pollution prevention plan must include a certification, signed by an authorized individual, that discharges from the site have been tested or evaluated for the presence of nonstormwater discharges. The certification must describe possible significant sources of nonstormwater, the results of any test and/or evaluation conducted to detect such discharges, the test method or evaluation criteria used, the dates on which tests or evaluations were performed, and the on-site drainage points directly observed during the test or evaluation. Acceptable test or evaluation techniques include dye tests, television surveillance, observation of outfalls or other appropriate locations during dry weather, water balance calculations, and analysis of piping and drainage schematics.

Except for flows that originate from fire-fighting activities, sources of non-stormwater that are specifically identified in the permit as being eligible for authorization under the general permit must be identified in the plan. Pollution prevention plans must identify and ensure the implementation of appropriate pollution prevention measures for the nonstormwater discharge.

EPA recognizes that certification may not be feasible where facility personnel do not have an outfall, access hole, or other point of access to the conduit that ultimately receives the discharge. In such cases, the plan must describe why certification was not feasible. Permittees who are not able to certify that discharges have been tested or evaluated must notify the director in accordance with Part XI of the permit.

e. Sampling data. Any existing data on the quality or quantity of stormwater discharges from the facility must be described in the plan, including data collected for Part 2 of the group application process. These data may be useful for locating areas that have contributed pollutants to stormwater. The description should include a discussion of the methods used to collect and analyze the data. Sample collection points should be identified in the plan and shown on the site map.

f. Summary of potential pollutant sources. The description of potential pollution sources culminates in a narrative assessment of the risk potential that sources of pollution pose to stormwater quality. This assessment should clearly point to activities, materials, and physical features of the facility that have a reasonable potential to contribute significant amounts of pollutants to stormwater. Any such activities, materials, or features must be addressed by the measures and controls subsequently described in the plan. In conducting the assessment, the facility operator must consider the following activities: loading and unloading operations; outdoor storage activities; outdoor manufacturing or processing activities; significant dust or particulate generating processes; and on-site waste disposal practices. The assessment must list any significant pollution sources at the site and identify the pollutant parameter or parameters (i.e., biochemical oxygen demand, suspended solids, etc.) associated with each source.

3.3.6.4 Measures and Controls. Following completion of the source identification and assessment phase, the permit requires the permittee to evaluate, select, and describe the pollution prevention measures, best management practices (BMPs), and other controls that will be implemented at the facility. BMPs include processes, procedures, schedules of activities, prohibitions on practices, and other management practices that prevent or reduce the discharge of pollutants in stormwater runoff.

EPA emphasizes the implementation of pollution prevention measures and BMPs that reduce possible pollutant discharges at the source. Source reduction measures include, among others, preventive maintenance, chemical substitution, spill prevention, good housekeeping, training, and proper materials management. Where such practices are not appropriate to a particular source or do not effectively reduce pollutant discharges, EPA supports the use of source control measures and BMPs such as material segregation or covering, water diversion, and dust control. Like source reduction measures, source control measures and BMPs are intended to keep pollutants out of stormwater. The remaining classes of BMPs, which involve recycling or treatment of stormwater, allow the reuse of stormwater or attempt to lower pollutant concentrations prior to discharge.

The pollution prevention plan must discuss the reasons each selected control or practice is appropriate for the facility and how each will address one or more of the potential pollution sources identified in the plan. The plan also must include a schedule specifying the time or times during which each control or practice will be implemented. In addition, the plan should discuss ways in which the controls and practices relate to one another and, when taken as a whole, produce an integrated and consistent approach for preventing or controlling potential stormwater contamination problems. The permit requirements included for the various industry sectors in Part XI of the multisector general permit generally require that the portion of the plan that describes the measures and controls address the following minimum components.

When "minimize/reduce" is used relative to pollution prevention plan measures, EPA means to consider and implement best management practices that will result in an improvement over the baseline conditions as it relates to the levels of pollutants identified in stormwater discharges with due consideration to economic feasibility and effectiveness.

a. Good housekeeping. Good housekeeping involves using practical, cost-effective methods to identify ways to maintain a clean and orderly facility and keep contaminants out of separate storm sewers. It includes establishing protocols to reduce the possibility of mishandling chemicals or equipment and training employees in good housekeeping techniques. These protocols must be described in the plan and communicated to appropriate plant personnel.

b. Preventive maintenance. Permittees must develop a preventive maintenance program that involves regular inspection and maintenance of stormwater management devices and other equipment and systems. The program description should identify the devices, equipment, and systems that will be inspected; provide a schedule for inspections and tests; and address appropriate adjustment, cleaning, repair, or replacement of devices, equipment, and systems. For stormwater management devices such as catch basins and oil/water separators, the preventive maintenance program should provide for periodic removal of debris to ensure that the devices are operating efficiently. For other equipment and systems, the program should reveal and enable the correction of conditions that could cause breakdowns or failures that may result in the release of pollutants.

   c. Spill prevention and response procedures. Based on an assessment of possible spill scenarios, permittees must specify appropriate material handling procedures, storage requirements, containment or diversion equipment, and spill cleanup procedures that will minimize the potential for spills and in the event of a spill enable proper and timely response. Areas and activities that typically pose a high risk for spills include loading and unloading areas, storage areas, process activities, and waste disposal activities. These activities and areas, and their accompanying drainage points, must be described in the plan. For a spill prevention and response program to be effective, employees should clearly understand the proper procedures and requirements and have the equipment necessary to respond to spills.

   d. Inspections. In addition to the comprehensive site evaluation, facilities are required to conduct periodic inspections of designated equipment and areas of the facility. Industry-specific requirements for such inspections, if any, are discussed in Section VIII of the fact sheet. When required, qualified personnel must be identified to conduct inspections at appropriate intervals specified in the plan. A set of tracking or follow-up procedures must be used to ensure that appropriate actions are taken in response to the inspections. Records of inspections must be maintained. These periodic inspections are different from the comprehensive site evaluation, even though the former may be incorporated into the latter. Equipment, area, or other inspections are typically visual and are normally conducted on a regular basis, e.g., daily inspections of loading areas. Requirements for such periodic inspections are specific to each industrial sector in the multisector general permit, whereas the comprehensive site compliance evaluation is required of all industrial sectors. Area inspections help ensure that stormwater pollution prevention measures (e.g., BMPs) are operating and properly maintained on a regular basis. The comprehensive site evaluation is intended to provide an overview of the entire facility's pollution prevention activities.

   e. Employee training. The pollution prevention plan must describe a program for informing personnel at all levels of responsibility of the components and goals of the stormwater pollution prevention plan. The training program should address topics such as good housekeeping, materials management, and spill response procedures. Where appropriate, contractor personnel also must be trained in relevant aspects of stormwater pollution prevention. A schedule for conducting training must be provided in the plan. Several sections in Part XI of the multisector general permit specify a minimum frequency for training of once per year. Others indicate that training is to be conducted at an appropriate interval. EPA recommends that facilities conduct training annually at a minimum. However, more frequent training may be necessary at facilities with high turnover of employees or where employee participation is essential to the stormwater pollution prevention plan.

f. Record keeping and internal reporting procedures. The pollution prevention plan must describe procedures for developing and retaining records on the status and effectiveness of plan implementation. At a minimum, records must address spills, monitoring, and inspection and maintenance activities. The plan also must describe a system that enables timely reporting of stormwater management-related information to appropriate plant personnel.

g. Sediment and erosion control. The pollution prevention plan must identify areas that, owing to topography, activities, soils, cover materials, or other factors, have a high potential for significant soil erosion. The plan must identify measures that will be implemented to limit erosion in these areas.

h. Management of runoff. The plan must contain a narrative evaluation of the appropriateness of traditional stormwater management practices (i.e., practices other than those that control pollutant sources) that divert, infiltrate, reuse, or otherwise manage stormwater runoff so as to reduce the discharge of pollutants. Appropriate measures may include, among others, vegetative swales, collection and reuse of stormwater, inlet controls, snow management, infiltration devices, and wet detention/retention basins.

Based on the results of the evaluation, the plan must identify practices that the permittee determines are reasonable and appropriate for the facility. The plan also should describe the particular pollutant source area or activity to be controlled by each stormwater management practice. Reasonable and appropriate practices must be implemented and maintained according to the provisions prescribed in the plan.

In selecting stormwater management measures, it is important to consider the potential effects of each method on other water resources, such as groundwater. Although stormwater pollution prevention plans primarily focus on stormwater management, facilities must also consider potential groundwater pollution problems and take appropriate steps to avoid adversely impacting groundwater quality. For example, if the water table is unusually high in an area, an infiltration pond may contaminate a groundwater source unless special preventive measures are taken. Under EPA's July 1991 Ground Water Protection Strategy, states are encouraged to develop Comprehensive State Ground Water Protection Programs (CSGWPPs). Efforts to control stormwater should be compatible with state groundwater objectives as reflected in CSGWPPs.

3.3.6.5 *Comprehensive Site Compliance Evaluation.* The permit requires that the stormwater pollution prevention plan describe the scope and content of the comprehensive site evaluations that qualified personnel will conduct to (1) confirm the accuracy of the description of potential pollution sources contained in the plan, (2) determine the effectiveness of the plan, and (3) assess compliance with the terms and conditions of the permit. Note that the comprehensive site evaluations are not the same as periodic or other inspections described for certain industries under the fact sheet. However, in the instances when frequencies of inspections and the comprehensive site compliance evaluation overlap, they may be combined allowing for efficiency, as long as the requirements for both types of inspections are met. The plan must indicate the frequency of comprehensive evaluations (which must be at least once a year), except where comprehensive site evaluations are shown in the plan to be impractical for inactive mining sites, owing to remote location and inaccessibility. The individual or individuals who will conduct the comprehensive site evaluation must be identified in the plan and should be members of the pollution prevention team. Material handling and storage areas and other potential sources of pollution must be visually inspected for evidence of actual or potential pollutant discharges to the drainage system. Inspectors also must observe erosion controls and structural stormwater management devices to ensure that each is operating correctly. Equipment needed to implement the pollution prevention plan, such as that used during spill response activities, must be inspected to confirm that it is in proper working order.

The results of each comprehensive site evaluation must be documented in a report signed by an authorized company official. The report must describe the scope of the comprehensive site evaluation, the personnel making the comprehensive site evaluation, the date(s) of the comprehensive site evaluation, and any major observations relating to implementation of the stormwater pollution prevention plan. Comprehensive site evaluation reports must be retained for at least 3 years after the date of the evaluation. Based on the results of each comprehensive site evaluation, the description in the plan of potential pollution sources and measures and controls must be revised as appropriate within 2 weeks after each comprehensive site evaluation, unless indicated otherwise in Section XI of the permit. Changes in procedural operations must be implemented on the site in a timely manner for nonstructural measures and controls not more than 12 weeks after completion of the comprehensive site evaluation. Procedural changes that require construction of structural measures and controls are allowed up to 3 years for implementation. In both instances, an extension may be requested from the director.

3.3.6.6   Consistency with Other Plans.   Stormwater pollution prevention plans may reference the existence of other plans for Spill Prevention Control and Countermeasure (SPCC) plans developed for the facility under Section 311 of the CWA or Best Management Practices (BMP) programs otherwise required by an NPDES permit for the facility as long as such requirement is incorporated into the stormwater pollution prevention plan.

3.3.7   Monitoring and Reporting Requirements

The permit contains three general types of monitoring requirements: analytical monitoring or chemical monitoring; compliance monitoring for effluent guidelines compliance, and visual examinations of stormwater discharges. Actual monitoring requirements for a given facility under the permit will vary depending upon the industrial activities that occur at a facility and the criteria for determining monitoring used to develop the permit. Through increased analytical or visual monitoring the permittee may be able to better ascertain the effectiveness of its pollution prevention plan.

Analytical monitoring requirements involve laboratory chemical analyses of samples collected by the permittee. The results of the analytical monitoring are quantitative concentration values for different pollutants, which can be easily compared to the results from other sampling events and other facilities, or to national benchmarks. Compliance monitoring requirements are imposed under the multisector general permit to ensure that discharges subject to numerical effluent limitations under the stormwater effluent limitations guidelines are in compliance with those limitations.

Visual examinations of stormwater discharges are the least burdensome type of monitoring requirement under the permit. Almost all of the industrial activities are required to perform visual examinations of their stormwater discharges when they are occurring on a quarterly basis.

3.3.7.1 Analytical Monitoring Requirements.
The multisector general permit requires analytical monitoring for discharges from certain classes of industrial facilities. EPA believes that industries may reduce the level of pollutants in stormwater runoff from their sites through the development and proper implementation of a stormwater pollution prevention plan discussed in the multisector general permit. Analytical monitoring is a means by which to measure the concentration of a pollutant in a stormwater discharge. Analytical results are quantitative and therefore can be used to compare results from discharge to discharge and to quantify the improvement in stormwater quality attributable to the stormwater pollution prevention plan, or to identify a pollutant that is not being successfully controlled by the plan. EPA realizes there are greater cost burdens associated with analytical monitoring in comparison to visual examinations. The multisector general permit only requires analytical monitoring for the industry sectors or subsectors that demonstrated a potential to discharge pollutants at concentrations of concern.

To determine the industry sectors and subsectors that would be subject to analytical monitoring requirements contained in the sections listed in Table 3.2, EPA reviewed the data submitted in the group application process. First, EPA divided the Part 1 and Part 2 application data by the industry sectors listed in Table 3.2. Where a sector was found to contain a wide range of industrial activities or potential pollutant sources, it was further subdivided into the industry subsectors listed in Table 3.3. Next, EPA reviewed the information submitted in Part 1 of the group applications regarding the industrial activities, significant materials exposed to stormwater, and the material management measures employed. This information helped identify potential pollutants that may be present in the stormwater discharges. Then, EPA entered into a database the sampling data submitted in Part 2 of the group applications. Those data were arrayed according to industrial sector and subsector for the purposes of determining when analytical monitoring would be appropriate. Data received by EPA prior to Jan. 1, 1993 (3 months after the application deadline), were entered into EPA's database. Some additional data that were submitted even after Jan. 1, 1993, were also entered into the database to bolster the data set for some sectors or subsectors (e.g., the auto salvage industry). All data submitted even later by group applicants which were not loaded into the database were reviewed by EPA during development of the permit. EPA notes that preliminary copies of the database were distributed to the public upon request in advance of a complete screening of the quality of the data set. These copies of the database contained a variety of errors that were screened and removed prior to EPA statistical analysis and evaluation of the results.

The multisector general permit requires limited discharge monitoring during certain years of permit coverage (typically years 2 and 4) for those sectors which EPA determines have the potential to discharge pollutants at or above concentrations of concern. EPA has proposed to require facilities that transfer to the modified multisector general permit to conduct such monitoring only during year 4 (Oct. 1, 1988, to Sept. 30, 1988). Some facilities will find that the monitoring obligations imposed under the multisector general permit are significantly less onerous than the corresponding requirements of the baseline general permit. In fact, facilities can avoid any obligation to monitor for one or more pollutants by certifying that there are no exposed sources for that particular pollutant or pollutants.

The modified multisector general permit also requires that facilities visually examine grab samples of stormwater for indicators of stormwater pollution on a quarterly basis. Those samples must be collected within the first 30 minutes after the stormwater discharge begins, or as soon as practicable, but under no circumstances longer than 1 hour after the discharge begins. EPA has proposed that facilities transferring to the multisector general permit will be required to begin such examinations during the first full calendar quarter in which they are covered by the multisector general permit.

3.3.7.2 Compliance Monitoring. In addition to the analytical monitoring requirements for certain sectors, the multisector general permit contains monitoring requirements for discharges which are subject to effluent limitations. These discharges must be sampled annually and tested for the parameters which are limited by the permit. Discharges subject to compliance monitoring include coal pile runoff, contaminated runoff from phosphate fertilizer manufacturing facilities, runoff from asphalt paving and roofing emulsion production areas, material storage pile runoff from cement manufacturing facilities, and mine dewatering discharges from crushed stone, construction sand and gravel, and industrial sand mines located in Texas, Louisiana, Oklahoma, New Mexico, and Arizona. All samples are to be grab samples taken within the first 30 minutes of discharge where practicable, but in no case later than the first hour of discharge. Where practicable, the samples shall be taken from the discharges subject to the numeric effluent limitations prior to mixing with other discharges. Monitoring for these discharges is required to determine compliance with numeric effluent limitations.

3.3.7.3 Quarterly Visual Examination of Stormwater Quality. In order to provide a tool for evaluating the effectiveness of the pollution prevention plan, the permit requires the majority of industries covered under the multisector general permit to perform quarterly visual examinations of stormwater discharges. EPA believes these visual examinations will assist with the evaluation of the pollution prevention plan. This section provides a general description of the monitoring and reporting requirements under the multisector general permit. The visual examination provides a simple low-cost means of assessing the quality of stormwater discharge with immediate feedback. Most facilities covered under the multisector general permit are required to conduct a quarterly visual examination of stormwater discharges associated with industrial activity from each outfall, except discharges exempted under the representative discharge provision. The visual examination of stormwater outfalls should include any observations of color, odor, clarity, floating solids, settled solids, suspended solids, foam, oil sheen, or other obvious indicators of stormwater pollution. No analytical tests are required to be performed on these samples.

The examination of the sample must be made in well-lit areas. The visual examination is not required if there is insufficient rainfall or snowmelt to runoff or if hazardous conditions prevent sampling. Whenever practicable, the same individual should carry out the collection and examination of discharges throughout the life of the permit to ensure the greatest degree of consistency possible in recording observations. Grab samples for the examination shall be collected within the first 30 minutes (or as soon thereafter as practical, but not to exceed 1 hour) of when the runoff begins discharging. Reports of the visual examination include the examination date and time, examination personnel, visual quality of the stormwater discharge, and probable sources of any observed stormwater contamination. The visual examination reports must be maintained on-site with the pollution prevention plan.

When conducting a stormwater visual examination, the pollution prevention team or team member should attempt to relate the results of the examination to potential sources of stormwater contamination on the site. For example, if the visual examination reveals an oil sheen, the facility personnel (preferably members of the pollution prevention team) should conduct an inspection of the area of the site draining to the examined discharge to look for obvious sources of spilled oil, leaks, etc. If a source can be located, then this information allows the facility operator to immediately conduct a cleanup of the pollutant source and/or to design a change to the pollution prevention plan to eliminate or minimize the contaminant source from occurring in the future.

To be most effective, the personnel conducting the visual examination should be fully knowledgeable about the stormwater pollution prevention plan, the sources of contaminants on the site, the industrial activities conducted exposed to stormwater, and the day-to-day operations that may cause unexpected pollutant releases.

Other examples include: if the visual examination results in an observation of floating solids, the personnel should carefully examine the solids to see if they are raw materials, waste materials, or other known products stored or used at the site. If an unusual color or odor is sensed, the personnel should attempt to compare the color or odor to the colors or odors of known chemicals and other materials used at the facility. If the examination reveals a large amount of settled solids, the personnel may check for unpaved, unstabilized areas or areas of erosion. If the examination results in a cloudy sample that is very slow to settle out, the personnel should evaluate the site draining to the discharge point for fine particulate material, such as dust, ash, or other pulverized, ground, or powdered chemicals.

If the visual examination results in a clean and clear sample of the stormwater discharge, this may indicate that no visible pollutants are present. This would be an indication of a high-quality result; however, the visual examination will not provide information about dissolved contamination. If the facility is in a sector or subsector required to conduct analytical (chemical) monitoring, the results of the chemical monitoring, if conducted on the same sample, would help to identify the presence of any dissolved pollutants and the ultimate effectiveness of the pollution prevention plan. If the facility is not required to conduct analytical monitoring, it may do so if it chooses to confirm the cleanliness of the sample.

While conducting the visual examinations, personnel should constantly be attempting to relate any contamination that is observed in the samples to the sources of pollutants on site. When contamination is observed, the personnel should be evaluating whether or not additional BMPs should be implemented in the pollution prevention plan to address the observed contaminant and, if BMPs have already been implemented, evaluating whether or not these are working correctly or need maintenance. Permittees may also conduct more frequent visual examinations than the minimum quarterly requirements, if they so choose. By doing so, they may improve their ability to ascertain the effectiveness of their plan. Using this guidance, and employing a strong knowledge of the facility operations, EPA believes that permittees should be able to maximize the efffectivenss of their stormwater pollution prevention efforts through conducting visual examination which gives direct, frequent feedback to the facility operator or pollution prevention team on the quality of the stormwater discharge.

EPA believes that this quick and simple assessment will help the permittee to determine the effectiveness of his or her plan on a regular basis at very little cost. Although the visual examination cannot assess the chemical properties of the stormwater discharged from the site, the examination will provide meaningful results upon which the facility may act quickly. EPA recommends that the visual examination be conducted at different times than the chemical monitoring, but is not requiring this. In addition, more frequent visual examinations can be conducted if the permittee so chooses. In this way, better assessments of the effectiveness of the pollution prevention plan can be achieved. The frequency of this visual examination will also allow for timely adjustments to be made to the plan. If BMPs are performing ineffectively, corrective action must be implemented. A set of tracking or follow-up procedures must be used to ensure that appropriate actions are taken in response to the examinations. The visual examination is intended to be performed by members of the pollution prevention team. This hands-on examination will enhance the staff's understanding of the site's stormwater problems and the effects of the management practices that are included in the plan.

3.3.7.4   SARA Title III, Section 313 Facilities.   Today's permit regulations do not contain any special monitoring requirements for facilities subject to the Toxic Release Inventory (TRI) reporting requirements under Section 313 of the EPCRA. EPA has reviewed data submitted by facilities in the group application and has determined that stormwater monitoring requirements are more appropriately based upon the industrial activity or significant material exposed than upon a facility's status as a TRI reporter under Section 313 of the EPCRA. This determination is based upon a comparison of the data submitted by TRI facilities included in the group application process to data from group application sampling facilities that were not found on the TRI list. The data indicate that there are no consistent differences in the level of water priority chemicals present in samples from TRI facilities when compared to the samples from facilities not subject to TRI reporting requirements.

EPA has included a revised Appendix A that lists 44 additional water priority chemicals that meet the definition of a section 313 water priority chemical or chemical categories requirements as defined by EPA in the permit under Part X, Definitions.

3.3.7.5   Reporting and Retention Requirements.   Permittees are required to submit all analytical monitoring results obtained during the second and fourth year of permit coverage within 3 months of the conclusion of the second and fourth year of coverage of the permit. For each outfall, one Discharge Monitoring Report form must be submitted per storm event sampled. For facilities conducting monitoring beyond the minimum requirements an additional Discharge Monitoring Report form must be filed for each analysis. The permittee must include a measurement or estimate of the total precipitation, volume of runoff, and peak flow rate of runoff for each storm event sampled. Permittees subject to compliance monitoring requirements are required to submit all compliance monitoring results annually on the twenty-eighth day of the month following the anniversary of the publication of this permit. Compliance monitoring results must be submitted on signed Discharge Monitoring Report forms. For each outfall, one Discharge Monitoring Report form must be submitted for each storm event sampled.

Permittees are not required to submit records of the visual examinations of stormwater discharges unless specifically asked to do so by the director. Records of the visual examinations must be maintained at the facility. Records of visual examination of stormwater discharge need not be lengthy. Permittees may prepare typed or handwritten reports using forms or tables which they may develop for their facility. The report need only document the date and time of the examination; the name of the individual making the examination; and any observations of color, odor, clarity, floating solids, suspended solids, foam, oil sheen, and other obvious indicators of stormwater pollution.

The location for submittal of all reports is contained in the permit. Pursuant to the requirements of 40 CFR 122.41(j), the multisector general permit requires permittees to retain all records for a minimum of 3 years from the date of the sampling, examination, or other activity that generated the data.

3.3.7.6 Sample Type.   The discussion below is a general description of the sample type required for monitoring under the multisector general permit. Certain industries have different requirements, however, so permittees should check the industry-specific requirements in Part XI of the multisector general permit to confirm these requirements. Grab samples may be used for all monitoring unless otherwise stated. All such samples shall be collected from the discharge resulting from a storm event that is greater than 0.1 inch in magnitude and that occurs at least 72 hours from the previously measurable (greater than 0.1 inch rainfall) storm event. The required 72-hour storm event interval may be waived by the permittee where the preceding measurable storm event did not result in a measurable discharge from the facility. The 72-hour requirement may also be waived by the permittee where the permittee documents that less than a 72-hour interval is representative for local storm events during the season when sampling is being conducted. The grab sample must be taken during the first 30 minutes of the discharge. If the collection of a grab sample during the first 30 minutes is impracticable, a grab sample can be taken during the first hour of the discharge, and the discharger must submit with the monitoring report a description of why a grab sample during the first 30 minutes was impracticable. A minimum of one grab is required. Where the discharge to be sampled contains both stormwater and nonstormwater, the facility must sample the stormwater component of the discharge at a point upstream of the location where the nonstormwater mixes with the stormwater, if practicable.

3.3.8 Special Requirements for Stormwater Discharges Associated with Industrial Activity from Facilities Subject to EPCRA Section 313 Requirements

The multisector general permit contains special requirements for certain permittees subject to reporting requirements under Section 313 of the EPCRA [also known as Title III of the Superfund Amendments and Reauthorization Act

(SARA)]. EPCRA Section 313 requires operators of certain facilities that manufacture (including import), process, or otherwise use listed toxic chemicals to report annually their releases of those chemicals to any environmental media. Listed toxic chemicals include more than 500 chemicals and chemical classes listed at 40 CFR Part 372 (including chemicals added Nov. 30, 1994).

The criteria for facilities that must report under Section 313 are given at 40 CFR 372.22. A facility is subject to the annual reporting provisions of Section 313 if it meets all three of the following criteria for a calendar year: it is included in SIC codes 20 through 39; it has 10 or more full-time employees; and it manufactures (including imports), processes, or otherwise uses a chemical listed in 40 CFR 372.65 in amounts greater than the "threshold" quantities specified in 40 CFR 372.25.

There are more than 300 individually listed Section 313 chemicals, as well as 20 categories of toxic release inventory (TRI) chemicals for which reporting is required. EPA has the authority to add to and delete from this list. The agency has identified approximately 175 chemicals that it is classifying for the purposes of this general permit as "Section 313 water priority chemicals." For the purposes of this permit, Section 313 water priority chemicals are defined as chemicals or chemical categories that (1) are listed at 40 CFR 372.65 pursuant to EPCRA Section 313; (2) are manufactured, processed, or otherwise used at or above threshold levels at a facility subject to EPCRA Section 313 reporting requirements; and (3) meet at least one of the following criteria: (i) are listed in Appendix D of 40 CFR Part 122 on either Table II (organic priority pollutants), Table III (certain metals, cyanides, and phenols), or Table V (certain toxic pollutants and hazardous substances); (ii) are listed as a hazardous substance pursuant to Section 311(b)(2)(A) of the CWA at 40 CFR 116.4; or (iii) are pollutants for which EPA has published acute or chronic toxicity criteria. A list of the water priority chemicals is provided in Addendum F to the multisector general notice. In the multisector general permit, EPA is not extending the special requirements to facilities that store liquid chemicals in aboveground tanks or handle liquid chemicals in areas exposed to precipitation if such facilities are not subject to EPCRA Section 313 reporting requirements.

3.3.8.1 Summary of Special Requirements. The special requirements in the multisector general permit for facilities subject to reporting requirements under EPCRA Section 313 for a water priority chemical, except those that are handled and stored only in gaseous or nonsoluble liquids or solids (at atmospheric pressure and temperature), state that stormwater pollution prevention plans, in addition to the baseline requirements for plans, must contain special provisions addressing areas where Section 313 water priority chemicals are stored, processed, or otherwise handled. These requirements reflect the best available technology for controlling discharges of water priority chemicals in stormwater. The permit provides that appropriate containment, drainage control, and/or diversionary structures must be provided for such areas. An exemption from the special provisions for

Section 313 facilities will be granted if the facility can certify in the pollution prevention plan that all water priority chemicals handled or used are gaseous or nonsoluble liquids or solids (at atmospheric pressure and temperature). At a minimum, one of the following preventive systems or its equivalent must be used: curbing, culverting, gutters, sewers, or other forms of drainage control to prevent or minimize the potential for stormwater run-on to come into contact with significant sources of pollutants; or roofs, covers, or other forms of appropriate protection to prevent storage piles from exposure to stormwater and wind.

In addition, the permit establishes requirements for priority areas of the facility. Priority areas of the facility include the following: liquid storage areas where stormwater comes into contact with any equipment, tank, container, or other vessel used for Section 313 water priority chemicals; material storage areas for Section 313 water priority chemicals other than liquids; truck and rail car loading and unloading areas for liquid Section 313 water priority chemicals; and areas where Section 313 water priority chemicals are transferred, processed, or otherwise handled.

The permit provides that site runoff from other industrial areas of the facility that may contain Section 313 water priority chemicals or spills of Section 313 water priority chemicals must incorporate the necessary drainage or other control features to prevent the discharge of spilled or improperly disposed material and to ensure the mitigation of pollutants in runoff or leachate. The permit also establishes special requirements for preventive maintenance and good housekeeping, facility security, and employee training.

In the proposed permit, EPA proposed to require facilities subject to EPCRA Section 313 requirements to have a registered professional engineer (PE) certify their pollution prevention plans every 3 years. However, in response to commentors' concerns, EPA revised the permit to eliminate the PE certification requirement. Instead, the permit now requires facilities subject to the special requirements to satisfy the pollution prevention plan signature requirements in Part IV.B.1 of the permit. Instead of certifying the plan every 3 years, facilities subject to EPCRA Section 313 requirements must amend the pollution prevention plan only when significant modifications are made to the facility, such as the addition of material handling areas or chemical storage units.

3.3.8.2 Requirements for Priority Areas.
The permit provides that drainage from priority areas should be restrained by valves or other positive means to prevent the discharge of a spill or other excessive leakage of Section 313 water priority chemicals. Where containment units are employed, such units may be emptied by pumps or ejectors; however, these must be manually activated. Flapper-type drain valves must not be used to drain containment areas, as these will not effectively control spills. Valves used for the drainage of containment areas should, as far as is practical, be of manual, open-and-closed design. If facility drainage does not meet these requirements, the final discharge conveyance of all in-facility storm sewers must be equipped to be equivalent with a diversion sys-

tem that could, in the event of an uncontrolled spill of Section 313 water priority chemicals, return the spilled material or contaminated stormwater to the facility. Records must be kept of the frequency and estimated volume (in gallons) of discharges from containment areas.

Additional special requirements are related to the types of industrial activities that occur within the priority area. These requirements are summarized below.

1. Liquid storage areas. Where stormwater comes into contact with any equipment, tank, container, or other vessel used for Section 313 water priority chemicals, the material and construction of tanks or containers used for the storage of a Section 313 water priority chemical must be compatible with the material stored and conditions of storage, such as pressure and temperature. Liquid storage areas for Section 313 water priority chemicals must be operated to minimize discharges of Section 313 chemicals. Appropriate measures to minimize discharges of Section 313 chemicals may include secondary containment provided for at least the entire contents of the largest single tank plus sufficient freeboard to allow for precipitation, a strong spill contingency and integrity testing plan, and/or other equivalent measures. A strong spill contingency plan would typically contain, at a minimum, a description of response plans, personnel needs, and methods of mechanical containment (such as use of sorbents, booms, collection devices, etc.), steps to be taken for removal of spill chemicals or materials, and procedures to ensure access to and availability of sorbents and other equipment. The testing component of the plan would provide for conducting integrity testing of storage tanks at set intervals such as once every 5 years, and conducting integrity and leak testing of valves and piping at a minimum frequency, such as once per year. In addition, a strong plan would include a written and actual commitment of labor power, equipment, and materials required to comply with the permit and to expeditiously control and remove any quantity of spilled or leaked chemicals that may result in a toxic discharge.

2. Other material storage areas. Material storage areas for Section 313 water priority chemicals other than liquids that are subject to runoff, leaching, or wind must incorporate drainage or other control features to minimize the discharge of Section 313 water priority chemicals by reducing stormwater contact with Section 313 water priority chemicals.

3. Truck and rail car loading and unloading areas. Truck and rail car loading and unloading areas for liquid Section 313 water priority chemicals must be operated to minimize discharges of Section 313 water priority chemicals. Appropriate measures to minimize discharges of Section 313 chemicals may include the placement and maintenance of drip pans (including the proper disposal of materials collected in the drip pans) where spillage may occur (such as hose connections, hose reels, and filler nozzles) when making and breaking hose connections; a strong spill contingency and integrity testing plan; and/or other equivalent measures.

4. Other transfer, process, or handling areas. Processing equipment and materials handling equipment must be operated to minimize discharges of Section 313 water priority chemicals. Materials used in piping and equipment must be compatible with the substances handled. Drainage from process and materials handling areas must minimize stormwater contact with Section 313 water priority chemicals. Additional protection such as covers or guards to prevent exposure to wind, spraying, or releases from pressure relief vents to prevent a discharge of Section 313 water priority chemicals to the drainage system, and overhangs or door skirts to enclose trailer ends at truck loading and unloading docks must be provided as appropriate. Visual inspections or leak tests must be provided for overhead piping conveying Section 313 water priority chemicals without secondary containment.

3.3.8.3 Certification. The multisector general permit allows facilities to provide a certification, signed in accordance with Part VII.G (signatory requirements) of this permit, that all Section 313 water priority chemicals handled and/or stored on-site are only in gaseous or nonsoluble liquid or solid (at atmospheric pressure and temperature) forms in lieu of the additional requirements in Part VI.E.2 of the multisector general permit. By allowing such a certification, EPA hopes to limit the application of the special requirements Part IV.E.2 of the permit to those facilities with 313 water priority chemicals that truly have the potential to contaminate stormwater discharges associated with industrial activity.

3.3.9 Notices and Hearings

EPA has 45 days to review and comment on individual stormwater applications. The public is given 30 days to make written comments after the posting of a notice about the draft permit. EPA has 90 days to review a state's draft general permit. Interested persons may request to receive regular lists of pending permits. All information on the permits not classified as confidential is available for public inspection during normal business hours at the permitting authority. During the public comment period any person or agency may file a petition with the authority for a public hearing. Persons aggrieved by the issuance of a stormwater permit may appeal the action within 30 days after issuance of the permit.

3.4 INDIRECT DISCHARGE (PRETREATMENT) PERMITS

An indirect discharge (or pretreatment) permit is required of an industrial facility that discharges wastewaters into the public sewer that contain pollutants that the receiving sewage plant cannot handle. Typically, the local publicly owned treatment works (POTW) receives a state operating permit from its state and an NPDES permit to discharge effluent into nearby waters of the United States. Since POTWs cannot deal with all industrial wastewaters, the industry must treat the

TABLE 3.5 When Is a Pretreatment Permit Required?

A pretreatment permit must be obtained if the facility's discharge into a public sewer:
- ☐ Is subject to federal pretreatment standards
- ☐ Would cause interference with the proper operation of the POTW
- ☐ Would pass through the POTW, causing it to violate state water quality standards
- ☐ Would contaminate the POTW's sludges so that they could not be properly disposed as nonhazardous solid waste
- ☐ Amounts to 25,000 gallons per day or more, even if none of the above apply
- ☐ Amounts to 5% or more of the POTW's intake, even if none of the above apply

---

wastewater before discharging it into the sewer (hence, pretreatment). Not every industrial facility has to pretreat wastewater but most do. A pretreatment permit is required for the reasons listed in Table 3.5.

The local POTW is usually responsible for developing and implementing a pretreatment permit program. While some POTWs have developed and received approval for administering permit programs, most have not. The state compliance authorities, and in a few cases the regional EPA office, administer the pretreatment permit program in most states.

Wastewater treatment (or pretreatment) facilities must be operated at maximum expected efficiency at all times. Monitoring, record keeping, and other parameters stated in the permit are legal requirements, and all terms and conditions in the permit must be complied with exactly. Process modifications, which may result in an increased volume of wastewater or an increased pollutant load, must be reported to the permit authority in advance for approval.

3.5 SEWAGE SLUDGE USE AND DISPOSAL RESTRICTION

The primary objectives of the sewage sludge permitting regulations are found in 40 CFR Part 503 as the protection of human health and the environment whenever sewage sludge is beneficially applied to land, disposed of in a surface disposal site, or incinerated. These regulations establish standards, which consist of general requirements, pollutant limits, management practices, and operational standards, for the final use or disposal of sewage sludge generated during the treatment of domestic sewage in a treatment works. Standards are included in Part 503 for sewage sludge applied to the land, placed on a surface disposal site, or fired in a sewage sludge incinerator. Also included in Part 503 are pathogen and alternative vector attraction reduction requirements for sewage sludge applied to the land or placed on a surface disposal site.

In addition, the standards in these regulations include the frequency of monitoring and record-keeping requirements when sewage sludge is applied to the land, placed on a surface disposal site, or fired in a sewage sludge incinerator. Also included in this part are reporting requirements for Class I sludge management facilities, publicly owned treatment works (POTWs) with a design flow rate equal to or greater than 1 million gallons per day, and POTWs that serve 10,000 people or more.

The requirements of this program apply to:

- Any person who prepares sewage sludge, applies sewage sludge to the land, or fires sewage sludge in a sewage sludge incinerator and to the owner-operator of a surface disposal site
- Sewage sludge applied to the land, placed on a surface disposal site, or fired in a sewage sludge incinerator
- The exit gas from a sewage sludge incinerator stack
- Land where sewage sludge is applied, to a surface disposal site, and to a sewage sludge incinerator

The requirements in this part may be implemented through a permit:

- Issued to a "treatment works treating domestic sewage," as defined in 40 CFR 122.2, in accordance with 40 CFR Parts 122 and 124 by EPA or by a state that has a state sludge management program approved by EPA in accordance with 40 CFR Part 123 or 40 CFR Part 501
- Issued under subtitle C of the Solid Waste Disposal Act; Part C of the Safe Drinking Water Act; the Marine Protection, Research, and Sanctuaries Act of 1972; or the Clean Air Act. "Treatment works treating domestic sewage" shall submit a permit application in accordance with either 40 CFR 122.21 or an approved state program.

No person shall use or dispose of sewage sludge through any practice for which requirements are established in these regulations except in accordance with such requirements.

3.6 DREDGE AND FILL PERMITS

Another example of a permit program provided by the Clean Water Act (Section 404) and designed to protect the quality of the nation's water is the dredge and fill permit program. This program is jointly administered by EPA and the U.S. Army Corps of Engineers. Applicable regulations can be found in 40 CFR Part 230.

Section 404 of the Clean Water Act requires a permit to be obtained prior to discharging any dredge or fill material into wetlands or navigable waters. The provisions of Section 404 define the discharge of dredged or fill material as the addition of fill material into U.S. waters, and includes the following 11 classes of materials: (1) fill for structures such as sewage treatment plants, intake and outfall pipes associated with power plants, and subaqueous utility lines; (2) property protection and/or reclamation devices such as riprap, groins, seawalls, breakwaters, and retaining walls; (3) site-development fills for recreational, industrial, commercial, residential, and other uses; (4) causeways or road fills; (5) dams and dikes; (6) artificial islands; (7) the building of any structure or impoundment requiring rock, sand, dirt, or other materials for its construction; (8) beach protection; (9) levees; (10) placement of fill that is necessary for the construction of any structure; and (11) artificial reefs.

The regulations found in 40 CFR 230 are divided into eight subparts. Subpart A presents those provisions of general applicability, such as purpose and definitions. Subpart B establishes the four conditions which must be satisfied in order to make a finding that a proposed discharge of dredged or fill material complies with the guidelines. Section 230.11 of Subpart B sets forth factual determinations which are to be considered in determining whether or not a proposed discharge satisfies the Subpart B conditions of compliance. Subpart C describes the physical and chemical components—guidance as to how proposed discharges of dredged or fill material may affect these components. Subparts D through F detail the special characteristics of particular aquatic ecosystems in terms of their values, and the possible loss of these values due to discharges of dredged or fill material. Subpart G prescribes a number of physical, chemical, and biological evaluations and testing procedures to be used in reaching the required factual determinations. Subpart H details the means to prevent or minimize adverse effects. Subpart I concerns advanced identification of disposal areas.

General permit for a category of activities involving the discharge of dredged or fill material complies with these regulations if it meets the applicable restrictions on the discharge in 230.10 and if the permitting authority determines that:

• The activities in such category are similar in nature and similar in their impact upon water quality and the aquatic environment.

• The activities in such category will have only minimal adverse effects when performed separately.

• The activities in such category will have only minimal cumulative adverse effects on water quality and the aquatic environment.

To reach these determinations, the permitting authority must set forth in writing an evaluation of the potential individual and cumulative impacts of the category of activities to be regulated under the general permit. While some of the information necessary for this evaluation can be obtained from potential permittees and others through the proposal of general permits for public review, the evaluation must be completed before any general permit is issued, and the results must be published with the final permit.

This evaluation must be based upon consideration of the prohibitions listed in 230.10(*b*) and the factors listed in 230.10(*c*) and shall include documented information supporting each factual determination in 230.11 of these regulations [consideration of alternatives in 230.10(*a*) are not directly applicable to general permits].

The evaluation must include a precise description of the activities to be permitted under the general permit, explaining why they are sufficiently similar in nature and in environmental impact to warrant regulation under a single general permit based on Subparts C through F of these regulations. Allowable differences between activities which will be regulated under the same general permit shall be specified. Activities otherwise similar in nature may differ in environmental impact because of their location in or near ecologically sensitive areas, areas with unique chemical or physical characteristics, areas containing concentrations of toxic substances, or areas regulated for specific human uses or by specific land or water management plans (e.g., areas regulated under an approved Coastal Zone Management Plan). If there are specific geographic areas within the purview of a proposed general permit (called a draft general permit under a state 404 program), which are more appropriately regulated by individual permit due to the considerations cited in this paragraph, they shall be clearly delineated in the evaluation and excluded from the permit. In addition, the permitting authority may require an individual permit for any proposed activity under a general permit where the nature or location of the activity makes an individual permit more appropriate.

3.7 WELL INJECTION—UNDERGROUND INJECTION CONTROL

Another permitting program under the federal Safe Drinking Water Act is the one that regulates underground injection. When part of a discharger's process wastewater is not being discharged into waters of the United States or contiguous zone because it is disposed into a well, into a POTW, or by land application, thereby reducing the flow or level of pollutants being discharged into waters of the United States, applicable effluent standards and limitations for the discharge in an NPDES permit must be adjusted to reflect the reduced raw waste resulting from such disposal. Effluent limitations and standards in the permit must be calculated by one of the following methods:

- If none of the waste from a particular process is discharged into waters of the United States, and effluent limitations guidelines provide separate allocation for wastes from that process, all allocations for the process shall be eliminated from calculation of permit effluent limitations or standards.

- In all cases other than those described in paragraph $(a)(1)$ of this section, effluent limitations shall be adjusted by multiplying the effluent limitation derived by applying effluent limitation guidelines to the total waste stream by the amount of wastewater flow to be treated and discharged into waters of the United States, and dividing the result by the total wastewater flow. Effluent limitations and standards so calculated may be further adjusted under Part 125, Subpart D [Criteria and Standards for Determining Fundamentally Different Factors under

Sections 301(b)(1)(A), 301(b)(2)(A) and (E) of the act] to make them more or less stringent if discharges to wells, publicly owned treatment works, or by land application change the character or treatability of the pollutants being discharged to receiving waters. This method may be algebraically expressed as

$$P = E * N/T$$

where P is the permit effluent limitation, E is the limitation derived by applying effluent guidelines to the total waste stream, N is the wastewater flow to be treated and discharged to waters of the United States, and T is the total wastewater flow.

The above paragraph does not apply to the extent that promulgated effluent limitations guidelines:

- Control concentrations of pollutants discharged but not mass.
- Specify a different specific technique for adjusting effluent limitations to account for well injection, land application, or disposal into POTWs.
- The methods listed above for adjusting a discharger's NPDES permit do not alter a discharger's obligation to meet any more stringent permit condition requirements established under 40 CFR 122.41, 122.42, 122.43, and 122.44.

CHAPTER 4
INSECTICIDES, FUNGICIDES, AND RODENTICIDES

Brian L. Lubbert
Project Manager
RTP Environmental Associates, Inc.

4.1 HISTORY

Since the beginning of recorded history, human beings have been plagued by pests and have attempted to develop means to control the effects of pests on food storage, the spread of disease by pets, and the general negative impact on society. Through the twentieth century, increases in agricultural production have also increased the need to protect against a wide variety of pests that damage crops and animals. In the United States, the increase in the use of pesticides quickly led to legislation which regulated their use.

4.1.1 Early Twentieth-Century Regulations

In 1910, the United States government, under the Department of Agriculture, promulgated the Insecticide Act of 1910. The act regulated interstate commerce associated with the marketing of pesticides (intrastate commerce was still regulated at the state level).

The Federal Insecticide Act of 1910 was one of the earliest consumer protection laws, although it covered only insecticides and fungicides. It was designed to protect the farmer from substandard or fraudulent products. In 1938, the Pure Food Law was amended to include pesticides on food. One of the most beneficial effects of the regulation was its requirement of adding artificial color to white insecticides such as lead arsenate and sodium fluoride to prevent their use as flour or other cooking ingredients that looked similar to their physical properties. This was the first legislation to protect the consumer from pesticide-contaminated food.

Following the end of World War II, increasingly common advances in technology created a wide variety of pesticides and other methods of pest control and generated a larger market to supply an increasingly demanding consumer. With the increased volume and complexity of pesticides being manufactured and used on the foodstuffs of the United States, the need for stronger controls *on the use* of these pesticides became evident.

4.1.2 The 1947 Federal Insecticide, Fungicide and Rodenticide Act

In 1947, the United States government strengthened and replaced the Insecticide Act of 1910 with the Federal Insecticide, Fungicide and Rodenticide Act (FIFRA). The act regulated the production and products sold in interstate commerce, whether being imported or being offered for export, and if the product was intended for use as an insecticide, fungicide, herbicide, or rodenticide. The Act of 1947 covered two things: labeling and premarket registration. This registration was required on a product-by-product basis. Registration elements included the labeling and data requirements set forth under the regulations. In general, the requirements included:

- Product name
- Name and address of the manufacturer or distributor
- Ingredient statement
- Net content statement
- Any necessary warning and caution statements
- Directions for use

The "necessary warnings and caution statements" under the regulation were considered adequate if they would prevent injury to persons handling, applying, or otherwise exposed to the product. In addition, the registrant needed to demonstrate to the agency that the proposed labeling was adequate. Specific, but limited, testing was also required and the test data needed to include:

- A statement of the composition of the product which described both the physical and chemical properties of all active ingredients

- Toxicological tests that demonstrated the product could be used safely when all warnings and cautions were followed

- Tests to demonstrate that the proposed product would effectively control pests as claimed in the labeling without causing unacceptable damage to the crop treated.

In addition, in consultation with the Food and Drug Administration (FDA), studies were required to determine if any residuals on food were acceptable. These acceptable levels of residuals were covered under the authority of the Federal Food, Drug and Cosmetic Act of 1938 (FFDCA).

4.1.3 FIFRA Amendments

Over the next 40 years amendments to FIFRA strengthened the Act of 1947 and increased its scope on the regulated community. First, the 1958 amendments established residual clearances. Specifically, when the proposed use involved food or feed crops or related commodities, the U.S. Department of Agriculture (USDA) under FIFRA coordinated registrations with the establishment of tolerances by the FDA. Subsequent amendments required by the USDA and FDA both increased labeling and data requirements. For example, the 1964 amendments mandated the use of caution words such as WARNING, DANGER, CAUTION, and KEEP OUT OF REACH OF CHILDREN.

Additionally, in 1959 and 1964, major changes in regulations and policies within the USDA allowed for the removal from market of a number of chemicals and pesticides. The 1964 amendments also included the requirements that manufacturers had to remove all safety claims from their labels. The 1972 amendments included the following changes:

- Require that the user must follow the instruction label.

- Violations can result in heavy fines and imprisonment.

- All pesticides must be classified as restricted use pesticides or general use pesticides (later that became unclassified pesticide).

- Anyone applying or supervising the use of restricted use pesticides must be certified by the state.

- Pesticide manufacturing plants must be registered and inspected by the EPA.

- States may register pesticide products on a limited basis.

- All pesticide products must be registered by the EPA.

- When registering, the manufacturer is required to provide scientific evidence that a product is effectively control tested and labeled, will not injure humans, crops, livestock, wildlife, and the environment, and cannot result in illegal residues in food or feed.

4.1.4　Expanding Environmental Protection Agency Power

In 1970, the U.S. Environmental Protection Agency (EPA), replacing the USDA, began to administer FIFRA and the Pesticide Residual Clearance Provisions of the FFDCA. Prior to EPA involvement, the USDA had begun removing pesticides from the registration list. The removal from the registration list effectively made the pesticides illegal to use and subsequently brought suits to prevent the removal of restricted pesticides from the market. While the EPA initiated and continued the actions of the USDA, the number of cases contested by registrants grew. The cases eventually began to turn on the weight of the potential harm or harm prevention associated with the use of these pesticides. The concept of risk-benefit analysis was first used and later became imperative to the settling of cases. After the decision rendered in 1971 by the United States Appellate Court of the District of Columbia under *Environmental Defense Fund vs. William Ruckerhaus et al.*, risk-benefit analysis became a condition of registration.

Further amendments continued to strengthen FIFRA: the Federal Environmental Pesticide Control Act of 1972 amended FIFRA and gave EPA extended and new powers over pesticides. These new powers included the regulation of the actual use of registered products. In addition, the amendments allowed the EPA to usurp state regulatory activities. The EPA also gained the right to require reregistration of all pesticides by 1976. The EPA also had to develop guidelines that allowed for classification of pesticides for both general and restricted use.

In the *Federal Register* of July 3, 1975, the EPA published regulations for the enforcement of FIFRA. The EPA regulations allowed for provisions for applicants to issue refutable presumptions against registration (RPAR) if certain risk criteria were met or exceeded. However, RPAR became ineffective because the parameters of the regulation were so strict and the time frame for a response was so short that the entire procedure became completely impractical.

In agreement with the *Environmental Defense Fund vs. Ruckerhaus et al.*, the regulations incorporated the concept of weighing benefits versus risks in determining if a product should be allowed to be registered. The results of such a concept ultimately placed a number of agricultural pesticides in jeopardy because there was no evidence of risk or benefits data available. Specifically, prior to the EPA's role in regulating pesticides, the USDA had never previously required that information documenting the benefits of pest control be submitted to the agency. The EPA's request in the mid-1970s ultimately developed a void of data. Previously, the USDA simply assumed that the pest control was adequate if it was claimed as such in the labeling. The subsequent absence of data put the entire pesticide manufacturing industry in a time predicament. Consequently, the National Agricultural Pesticide Impact Assessment Program (NAPIAP) was formally charted by the Secretary of Agriculture on Oct. 19, 1976. In the previous 7 years, suspension hearings were conducted for a number of pesticides including DDT, aldrin, chlordane, mercury, and 245 T. A number of problems developed including a lack of specific biological data and a lack of pesticide evaluation tests that evaluated alternatives and documented yield and/or quality changes.

FIFRA was again amended in 1975, 1978, 1980, and 1981. These amendments generally clarified previous amendments and were designed to improve the registration process in working with the NAPIAP. One of the most significant changes was the setting of standards for active ingredients *individually,* rather than for a specific product. In addition, a conditional registration was allowed, expediting the registration procedure for multiple party registration. In addition, all pre-1975 products were required to be reregistered.

The reregistration procedure has become a trademark of the FIFRA amendments. Once again, in 1988 FIFRA was amended to require EPA to reregister existing pesticides that were originally registered before current scientific and regulatory standards were formally established. Furthermore, the Endangered Species Act required that all federal agencies ensure that any action they carry out or authorize is not likely to jeopardize the continued existence of any species on the endangered species list. FIFRA is therefore required to ensure that the registration of pesticides does not endanger or jeopardize species on the endangered list. This includes the endangered species' habitats.

4.1.5 The Delaney Paradox

No summary of the history of FIFRA would be complete without a description of the Delaney paradox. The Delaney paradox was caused by a clause in the Federal Food Drug and Cosmetic Act of 1958, which stated that no additive will be deemed safe if it is found to induce cancer when ingested by human or animal, and directs the FDA not to approve such food additives. The language had been interpreted to mean a zero risk standard for any cancer-causing food additive, including residuals from pesticides found in processed foods. The paradox is evident when one observes that alternative pesticides used on raw or processed food could actually pose a higher noncancer risk to the public. The EPA had allowed the same pesticides to be used on raw or processed food based on their determination that the risk had actually been negligible. Some connotations of the paradox had also included the fact that the FFDCA prohibited a zero-tolerance policy, while FIFRA allowed for a risk and benefits analysis. The new legislation currently provides for tolerances for pesticide residuals in all types of foods, whether it is raw or processed. The implications are that EPA must determine that the tolerances are safe as defined as a reasonable certainty that no harm will result from the aggregate exposure to the pesticide. The ultimate benefit of the clarification from the legislation in the EPA's eyes will allow the EPA to devote the resources that had previously been consumed by the Delaney-related activities to be used for higher-priority public health and environmental protection issues.

4.2 CURRENT TOPICS

4.2.1 Registration Rates

In 1998, the EPA had registered 27 new pesticides, which included 13 conventional pesticides, 12 biopesticides, and 2 antimicrobial pesticides. Of these 27, 14 are considered to be "safer" pesticides. "Safer" or reduced-risk pesticides have low risk to human health, low toxicity to nontarget organisms, low use rates, low groundwater contamination potential, and low pest-resistance potential, and show compatibility with integrated test management or are biopesticides.

4.2.2 Emergency Conditions for Unregistered Use

Section 18 of the FIFRA authorizes EPA to allow states to permit the use of pesticides for unregistered use for a limited time if EPA determines that emergency conditions exist. Such justifications for exemptions include new tests, the expansion of the range of tests, or the cancellation or removal from the market of a previously registered pesticide product. Most requests for emergency exemptions are made by state-led agricultural agencies, although USDA and U.S. Department of Interior (USDI) also request exemptions.

In 1998, the Office of Pesticide Programs (OPP) authorized 601 requests for the emergency exemption, of which 410 were authorized, 27 were denied, and states withdrew 67. As of the end of the year, 97 requests were still pending. The average emergency exception request took 56 days in 1998. The EPA is actually required to process the request within 50 days of receipt. During the 50-day time period, EPA must perform a multidisciplinary risk assessment of the requested use, relying largely on data that have already been reviewed for the pesticide. A dietary risk assessment, an occupational risk assessment, an ecological and environmental risk assessment, and an assessment of the emergency are conducted before a decision is made. For the past several years, EPA has also evaluated the risk to the most sensitive subpopulation (normally infants and children) in its dietary risk assessments.

The OPP reregistered 169 products, granted 424 cancellations, amended 54 registrations, and suspended 127 products in 1998. EPA under FIFRA must review the health effects on humans, the environment of all pesticides with active ingredients registered prior to 1984, to determine if they meet today's standards.

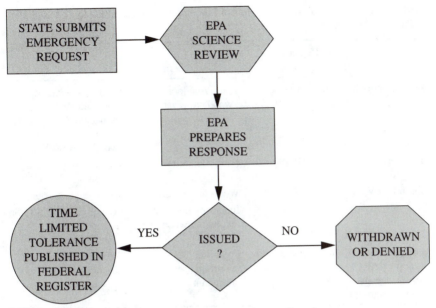

FIGURE 4.1   FIFRA Section 18 Emergency Exemption Process Overview

4.2.3   Reregistration List Expands

In 1998, the OPP completed 13 reregistrations, bringing the total to 184. According to the EPA, reregistration will be complete by 2002. OPP must also reevaluate tolerance settings for the maximum legal amount of pesticide residual permissible on food.

4.2.4   New Tolerances

FQPA requires that EPA reassess all 9700 tolerances over 10 years. These tolerances must use the new "reasonable certainty of no harm" safety standard in the law. The EPA has placed a priority on pesticides that appear to pose the greatest potential risk to the public. These include organelle phosphates, carbonates, probable and possible human carcinogens (see Tables 4.1–4.9), organic chlorine pesticides, and high-hazard inert ingredients. In 1998, over 1400 tolerances were reevaluated against a new standard. Of the previously issued tolerances, 874 were revoked. The FDA now requires the EPA to address the risks associated with infants and children and to publish a specific safety finding before tolerance can be established. The current additional safety factor is tenfold to ensure that tolerances are safe for both children and infants, and requires a collection of better data to prove otherwise. OPP's antimicrobial program has made significant progress in implementing its pesticides program, according to the EPA.

TABLE 4.1   Group A Pesticides—Human Carcinogens Not Classified by OPP (Office of Pesticide Programs)

Active ingredients	Registration date	Use pattern	Regulatory status
Arsenic, inorganic	1972	Indoor/outdoor; wood preservatives; food use	Food uses voluntarily canceled 1993; remaining uses are ant baits and wood preservatives
Benzene	1948	Food use; animal use	Voluntary cancellation 7/31/85
Chromium VI		Wood preservative	See Arsenic, inorganic
Coal tar (coke oven emissions)	1948	Indoor/outdoor use; wood/nonwood use	All uses canceled 5/15/86 except wood use (label restrictions)
Ethylene oxide group I (IARC, 1994); see also under Group B1	1957	Food use; nonfood use; fumigant	Tolerances reassessed by 8/99

Note: Since the evaluation of many of these chemicals is an ongoing process, the information on this list is subject to change.

TABLE 4.2   Group B1 Pesticides—Probable Human Carcinogens with Limited Human Evidence Not Classified by OPP

Active ingredients	Registration date	Use pattern	Regulatory status
Acrylonitrile	1/1/76	Stored tobacco; food processing areas; termiticide	Canceled 10/28/80
Cadmium	1959	Lawn/turf	Last product canceled 12/29/86
Creosote	1948	Wood preservative	Classified as restricted use in 1985
Ethylene oxide (OHEA, 1985); see also under Group A	1957	Food use; nonfood use	Tolerances reassessed by 8/99
Formaldehyde	1948	Food use	Tolerances reassessed by 8/99

Note: Since the evaluation of many of these chemicals is an ongoing process, the information on this list is subject to change.

TABLE 4.3 Group B2 Pesticides—Probable Human Carcinogens with Sufficient Evidence in Animals and Inadequate or No Evidence in Humans Classified by OPP

Active ingredients	Registration date	Use pattern	Regulatory status
Acetochlor	3/11/94	Food use (corn and soybeans)	Tolerances reassessed by 8/99
Aciflurofen, sodium salt	9/20/88	Food use	Tolerances reassessed by 8/99
Amitrole	1960	Outdoor use (right-of-ways)	RED issued 11/22/96
Cacodylic acid	1954	Food use	Tolerances reassessed by 8/99
Captafol	1962	Food use	All products voluntarily canceled 5/15/87; tolerances revocation by 8/99
Captan	1949	Food use	Tolerances reassessed by 8/99
Chlordimeform	1970	Food use (cotton only)	Voluntary cancellation 9/19/88
Chloroaniline	1967	Seed treatment; foliar treatment	All uses canceled FR 3702, 1977
Cyproconazole	12/22/93	Food use	Tolerances reassessed by 8/99
Daminozide (Alar)	1963	Food uses; nonfood uses	Tolerances revocation 3/19/90
1,2-Dichloropropene (Telone)	1960	Food uses	No tolerances established at the time of registration under USDA; currently in special review for groundwater issues
1,1-Dimethyl hydrazine (UDMH)	1960	Food uses; nonfood uses	Tolerances revocation by 3/19/90
Dipropyl isocinchomeronate (MGK 326)	1958	Nonfood use (topical animal use)	Routine registration activities; no pending use
Fenoxycarb	1985	Food uses; nonfood use (nonbearing fruit trees); indoor/outdoor; domestic animal	EUP on pears expired 8/97; new uses (pome fruit, citrus, tree nuts, pasture grasses, and cucurbits established)
Folpet	1948	Food use	Tolerances reassessed by 8/99

TABLE 4.3   Group B2 Pesticides—Probable Human Carcinogens with Sufficient
Evidence in Animals and Inadequate or No Evidence in Humans Classified by OPP
(*Continued*)

Active ingredients	Registration date	Use pattern	Regulatory status
Furmecyclox	Pending registration	Wood preservative	No action from company since 1987
Haloxyfop-methyl	Pending registration	No food uses	
Lactofen	3/18/87	Food use	Tolerances reassessed by 8/99
Mancozeb	1962	Food use	Tolerances reassessed by 8/99
Maneb	1954	Food use	Tolerances reassessed by 8/99
Metam sodium	1955	Food use	Tolerances reassessed by 8/02
Orthophenylphenol	1948	Food use	Tolerances reassessed by 8/99
Oxythioquinox	1963	Food use	Tolerances reassessed by 8/99
Procymidone	Import use only	Food use (wine grapes)	Tolerances reassessed by 8/99
Pronamide	10/11/72	Food use	Tolerances reassessed by 8/99
Propargite	1976	Food use	Tolerances reassessed by 8/99
Propoxur (Baygon)	1976	No food uses; greenhouse; indoor/ outdoor uses	RED to be issued in early 1998
Propylene oxide	1971	Food use	Tolerances reassessed by 8/99
Terrazole	1966	Food use	Tolerances reassessed by 8/99
Thiodicarb	2/10/84	Food use	Tolerances reassessed by 8/99
Triphenyltin hydroxide	2/26/85	Food use	Tolerances reassessed by 8/99

Note: Since the evaluation of many of these chemicals is an ongoing process, the information on this list is subject to change.

TABLE 4.4 Group B2 Pesticides—Probable Human Carcinogens with Sufficient Evidence in Animals and Inadequate or No Evidence in Humans Not Classified by OPP

Active ingredients	Registration date	Use pattern	Regulatory status
Acetaldehyde—this pesticide no longer has registered food uses, and OPP is in the process of revoking tolerances associated with this chemical			
Aldrin	1949	Food use; nonfood use; termiticide	All uses including termiticide uses canceled 1989
Aniline—not registered as active ingredient but as an analog			
Aramite	1951	Use on roses	All products canceled by 1984
Azobenzene—there are no records indicating registered or pending pesticide products			
Bis(chloroethyl) ether—there are no records indicating registered or pending pesticide products			
Carbon tetrachloride	1950	Food use; fumigant	Some uses voluntarily canceled 1985; remaining uses canceled 1988
Chlordane	5/18/78	Food use; termiticide	All uses canceled by 1988
Chloroform	4/23/59	Food use	Canceled 11/10/83
1,2-Dibromo-3-chloropropane (DBCP)	2/26/73	Food use; nonfood use; soil fumigant	Canceled 11/8/88
Dibromoethane, 1,2 (EDB) (=Ethylene dibromide)	10/17/56	Fumigant	All uses canceled 11/10/83 except mangos (expired 9/30/87) and papaya (voluntarily canceled 3/87); remaining uses are vault fumigation and quarantine fumigation of nursery stock
Dichloro diphenyl trichloroethane (DDT)	1952	Food use; nonfood use	Food use canceled 1972; nonfood use canceled 12/19/88
1,2-Dichloroethane	1950	Fumigant	Product canceled 1985
Dichloromethane	1973	Food use; nonfood use; indoor/outdoor	Some uses canceled 1986; tolerances revocation 8/99
Dieldrin	1948	Food use; termiticide	All uses canceled by 1989
Di(2-ethylhexyl)phthalate—according to the Special Review Rainbow Report this chemical has been canceled due to nonsupport of uses			

TABLE 4.4 Group B2 Pesticides—Probable Human Carcinogens with Sufficient Evidence in Animals and Inadequate or No Evidence in Humans Not Classified by OPP (*Continued*)

Active ingredients	Registration date	Use pattern	Regulatory status
Epichlorohydrin—there are no records indicating registered or pending pesticide products			
Ethylene thiourea	Inert ingredient only	In various chemical formulations	This inert no longer in pesticide products by 1987
Heptachlor	1952	Food use; nonfood use; termiticide	All uses canceled 1988 except fire ant use; RED issued in 1992 for fire ant use
Heptachlor epoxide	1952	Food use; nonfood use; termiticide	All uses canceled 1988 except fire ant use; RED issued in 1992 for fire ant use
Hexachlorobenzene	9/28/71	Seed treatment	Voluntarily canceled 7/6/84
Hexachloro cyclohexane, tech.	1960	Food use	Some products canceled 7/21/78; others reformulated to lindane
Lindane	9/2/52	Food use; nonfood use; indoor/ outdoor use	Tolerances reassessed by 8/99
Methylene chloride (see Dichloromethane)			
Mirex	7/30/75	Technical formulation only	Canceled 7/1/87
Pentachlorophenol	1948	No food uses; wood preservative	Wood uses canceled in 1986; nonwood use restricted labeling and currently in reregistration process
Perchlorethylene	1959	Nonfood use; indoor/outdoor	Canceled in 1992 for nonsupport through reregistration
Polychlorinated biphenyls—considered contaminants; elimination of all active and inert uses 1970			
Propiolactone—there are no records indicating registered or pending pesticide products			
Toxaphene	1974	Food use	Canceled 1982
Trichlorethylene	1949	Indoor/outdoor	Products canceled 1985
Trichlorphenol 2,4,6	1960	Disinfectant	Canceled 1987

Note: Since the evaluation of many of these chemicals is an ongoing process, the information on this list is subject to change.

TABLE 4.5   Group C Pesticides—Possible Human Carcinogens Classified by OPP

Active ingredients	Registration date	Use pattern	Regulatory status
Amitraz	3/31/75	Food use	Tolerances reassessed by 8/99
Asulam	10/24/75	Food use	Tolerances reassessed by 8/99
Atrazine	1959	Food use; nonfood use	Tolerances reassessed by 8/99
Benomyl	1969	Food use; indoor/ outdoor use	Tolerances reassessed by 8/99
Bifenthrin	10/2/85	Food use	Tolerances reassessed by 8/02
Bromacil	1963	Food use; nonfood use; indoor/outdoor use	Tolerances reassessed by 8/99
Bromoxynil	4/25/88	Food use	Tolerances reassessed by 8/99
Calcium cyanamide	2/25/72	Food use; nonfood use	Canceled 1984
Carbaryl	1963	Food use; nonfood use	Tolerances reassessed by 8/99
Clofentezine	3/10/89	Food use; nonfood use	Tolerances reassessed by 8/02
Cyanazine	12/6/71	Food use	Tolerances reassessed by 8/99
Cypermethrin	6/12/84	Food use	Tolerances reassessed by 8/99
Dacthal	1958	Food use; nonfood use	Tolerances reassessed by 8/99
Dichlobenil	1982	Food use; outdoor use	Tolerances reassessed by 8/99
Dichlorvos (DDVP)	1948	Food use; nonfood use	Tolerances reassessed by 8/99
Diclofop-methyl	3/31/80	Food use	Tolerances reassessed by 8/99
Dicofol	1957	Food use; outdoor use	Tolerances reassessed by 8/99
Difenoconazole	8/4/94	Food use (wheat use only)	Tolerances reassessed by 8/99
Dimethenamid (SAN 682H)	3/5/98	Food use	Tolerances reassessed by 8/06
Dimethipin (Harvade)	8/13/82	Plant growth regulator	Tolerances reassessed by 8/02
Dimethoate	1962	Food use; nonfood use	Tolerances reassessed by 8/99

TABLE 4.5   Group C Pesticides—Possible Human Carcinogens Classified by OPP
(*Continued*)

Active ingredients	Registration date	Use pattern	Regulatory status
Dinoseb	1976	Food use; nonfood use; indoor/outdoor use	All uses canceled 6/9/89
Ethalfluralin	12/5/83	Food use; nonfood use; indoor/outdoor use	Tolerances reassessed by 8/99
Ethofenprox	Pending registration	Nonfood use	
Fenbuconazole	2/5/92	Food use	Tolerances reassessed by 8/99
Fipronil	5/1/96	Food uses (corn only); domestic animals; indoor bait stations	Corn use registered 11/14/97; other uses pending
Fluometuron	8/7/73	Food use (grapes and cotton only)	Tolerances reassessed by 8/02
Fomesafen	4/10/87	Food use (soybeans only)	Tolerances reassessed by 8/99
Hexaconazole	9/26/96 (import use only)	Food use (import use on bananas)	Tolerances reassessed by 8/99
Hexythiazox (Savey)	4/13/89	Food use; nonfood use	Tolerances reassessed by 8/99
Hydramethylnon (Amdro)	8/20/80	Food use; nonfood use	Tolerances reassessed by 8/02
Hydrogen cyanamide	11/29/93	Plant growth regulator (grapes and kiwi)	New uses under consideration
Imazalil	7/18/83	Food use	Tolerances reassessed by 8/99
Isoxaben	6/16/89	Food use	Tolerances reassessed by 8/06
Linuron	1/2/67	Food use; outdoor use	Tolerances reassessed by 8/02
2-Mercapto benzothiazole	1956	Aquatic indoor	Products canceled due to maintenance fees and DCI; tolerances revocation by 8/02
Methidathion	2/12/76	Food use	Tolerances reassessed by 8/99
Methyl 2-benzimidazole carbamate (MBC)	5/3/90	No food uses; sanitizer; algicide	RED scheduled after 2002
Metolachlor	4/1/77	Food use	Tolerances reassessed by 8/99
Molinate	1965	Food use (rice only)	Tolerances reassessed by 8/99

TABLE 4.5 Group C Pesticides—Possible Human Carcinogens Classified by OPP (*Continued*)

Active ingredients	Registration date	Use pattern	Regulatory status
Nitrofen	12/7/66	Food use	Voluntary cancellation 9/15/83
Norflurazon	1/21/74	Food use	Tolerances reassessed by 8/02
N-Octyl bicycloheptene dicarboximide (MGK-264)	3/15/50	Food use; nonfood use; indoor/outdoor use	Tolerances reassessed by 8/02
Oryzalin	3/15/74	Food use	Tolerances reassessed by 8/99
Oxadiazon	5/24/77	Food use	Tolerances reassessed by 8/99
Oxadixyl	2/27/92	Food use	Tolerances reassessed by 8/99
Oxyfluorfen	5/17/75	Food use	Tolerances reassessed by 8/99
Paradichlorobenzene	1949	Nonfood use; indoor/outdoor use	Regular registration activities
Parathion	1948	Food use	Tolerances reassessed by 8/99
Pendimethalin	7/29/75	Food use	Tolerances reassessed by 8/99
Pentachloro-nitrobenzene	1958	Food use; nonfood use; greenhouse (indoor)	Tolerances reassessed by 8/99
Permethrin	4/11/79	Food use; nonfood use	Tolerances reassessed by 8/99
Phosmet	1966	Food use; nonfood use	Tolerances reassessed by 8/99
Phosphamidon	7/26/60	Food use	Tolerances revocation by 8/99; products canceled 6/15/84
Piperonyl butoxide	8/29/57	Food use; nonfood use; greenhouse (indoor)	Tolerances reassessed by 8/02
Prochloraz	An experimental use permit only issued 11/10/88	Food use	Temporary tolerances expired 11/21/89
Prodiamine	2/7/92	Food use	Tolerances reassessed by 8/99

TABLE 4.5 Group C Pesticides—Possible Human Carcinogens Classified by OPP (*Continued*)

Active ingredients	Registration date	Use pattern	Regulatory status
Propazine	1961	Food use; outdoor use	Tolerances reassessed by 8/99
Propiconazole	8/14/81	Food use	Tolerances reassessed by 8/99
4-Pyridazine carboxylic acid, 2-(4-chlorophenyl)-3-ethyl-2,5-dihydro-5-oxo-, potassium salt (MON 21200)-post FQPA	2/26/97	Food use (wheat only)	Regular registration activities; no new pending uses
Pyrithiobac-sodium	9/29/95	Food use (cotton only)	Tolerances reassessed by 8/99
Simazine	1959	Food use; nonfood use	Tolerances reassessed by 8/99
Tebuconazole	7/15/74	Food use (peanut)	Tolerances reassessed by 8/99
Terbutryn	7/29/74	Nonfood use	Tolerances reassessed by 8/99 (canceled 1/22/91)
2-(Thiocyano-methylthio) benzothiazole (TCMB)	12/11/69	Food use; wood preservative	Tolerances reassessed by 8/99
Triadimefon	8/27/79	Food use	Tolerances reassessed by 8/99
Triadimenol	4/28/89	Food use	Tolerances reassessed by 8/99
Triallate	2/15/62	Food use	Tolerances reassessed by 8/99
Tribenuron methyl	6/30/89	Food use	Tolerances reassessed by 8/02
Tridiphane	2/10/86	Food use	Tolerances reassessed by 8/02
Trifluralin	1/15/73	Food use	Tolerances reassessed by 8/99
Triflusulfuron-methyl	2/4/92	Food use (sugar beets)	Tolerances reassessed by 8/02
Uniconazole	7/3/91	Plant growth regulator; no food uses	Normal registration activity
Vinclozolin	2/11/81	Food use	Tolerances reassessed by 8/99

Note: Since the evaluation of many of these chemicals is an ongoing process, the information on this list is subject to change.

TABLE 4.6 Group C Pesticides—Possible Human Carcinogens Not Classified by OPP

Active ingredients	Registration date	Use pattern	Regulatory status
Acrolein	11/26/75	Aquatic food crops (agricultural irrigation waters)	Restricted use chemical; regular registration activities
Hexachloroethane—there are no records indicating registered or pending pesticide products			
Isophorone	Inert ingredient	Food use	Company has submitted for review information rebutting carcinogen classification
Methylphenol, 3-(*m*-cresol)	3/10/80	Food use; wood preservative	Currently in reregistration process
Tetrachloroethane, 1,1,2-	9/15/78	Nonfood uses	Canceled 7/87
Trichloroethane, 1,1,2-	12/28/49	Nonfood use	Canceled 1992

Note: Since the evaluation of many of these chemicals is an ongoing process, the information on this list is subject to change.

TABLE 4.7 Other Pesticides—Not Applicable under Current Classification Classified by OPP

Active ingredients	Registration date	Use pattern	Regulatory status
Aliette	1988	Food uses	RED issued 1990; all products reregistered 1992
Melamine	3/88	Disinfectant; sanitizers	RED scheduled for 2002

Note: Since the evaluation of many of these chemicals is an ongoing process, the information on this list is subject to change.

4.3 APPLYING FOR PESTICIDES IN THE UNITED STATES

The Federal Insecticide, Fungicide and Rodenticide Act requires the registration of all pesticides. If the proposed use of the pesticide results in residuals in or on raw or agricultural commodities or processed foods and/or feed, a tolerance or an exemption from a tolerance is required under the Federal Food, Drug and Cosmetic Act. The registrant must supply EPA specific test data and analytical methods to establish the tolerance level.

TABLE 4.8 Other Pesticides Not Applicable under Current Classification Not Classified by OPP

Active ingredients	Registration date	Use pattern	Regulatory status
Rhodamine B	Inert ingredient; date unknown due to change in data systems	Seed treatment formulations only	Currently reviewing residue data to ascertain dietary exposure; tolerances reassessment by 8/99
THPS (tetrakis) (hydroxymethyl) phosphonium sulfate (no classification, but accepted NTP negative studies)	10/12/95	Indoor commercial use (antimicrobial uses)	Normal registration activities

Note: Since the evaluation of many of these chemicals is an ongoing process, the information on this list is subject to change.

TABLE 4.9 Classifications under Proposed Revised Guidelines Classified by OPP

Active ingredients	Registration date	Use pattern	Regulatory status
Likely			
Chlorothalonil	1966	Food use; wood preservative	Tolerances reassessed by 8/99
Ethoprop	4/88	Food uses	Tolerances reassessed by 8/99
Iprodione	1980	Food use	Tolerances reassessed by 8/99
Isoxaflutole	Pending registration	Food use (corn)	
Propachlor	9/3/82	Food uses	Tolerances reassessed by 8/99
Known/Likely			
Diuron	5/28/71	Food use	Tolerances reassessed by 8/02
Likely: High Doses/Not Likely: Low Doses			
Alachlor	4/75	Food use	Tolerances reassessed by 8/99
Tribuphos (DEF)	6/29/61	Food use (cotton)	Tolerances reassessed by 8/99
Cannot Be Determined, But Suggestive			
Benoxacor	Inert	Food use	Time-limited tolerances expires 2/98

Note: Since the evaluation of many of these chemicals is an ongoing process, the information on this list is subject to change.

These analytical methods enable the agency to establish tolerances after determining the maximum pesticide residuals of concern which could be present in, or treated on, raw agricultural commodities or processed foods and/or feeds. These analytical methods are used by the FDA for tolerance enforcement. An independent laboratory validation trial is required for the first tolerance petition, including temporary tolerances for residuals of a pesticide, in or on a raw material commodity or processed food or feed.

An independent laboratory validation (ILV) trial is not required if EPA deems an enforcement analytical method superior to the currently accepted enforcement method. Nor is an ILV trial normally required for confirmatory methods. However, the agency always has the discretion to rule that an ILV trial be required for those methods on a case-by-case basis. The independent laboratory should be selected by the same performance standards of any other scientific project. Validation trials must be conducted under FIFRA good laboratory practices. The method must be performed as written with no significant modifications. When an ILV trial is successful, the petitioner must submit the following information:

1. The address and contact person for the independent laboratory
2. The description of the analytical method
3. All recovery and control values for all commodities that were obtained for representative chromatographs for all ILV trials performed
4. Descriptions of the instruments used and operating parameters
5. The description of all problems encountered
6. All steps considered critical
7. Number of person hours required to complete one set of samples
8. The number of calendar days required for one set of samples
9. The contacts between the independent laboratory and the method developers or others familiar with the method

Further information may be obtained from the USEPA Analytical Chemistry Laboratory, Bldg 306, East Beltsville, Md. 20705, telephone 301-504-8225 or USEPA Health Effects Division, 7509C 401 M Street SW, Washington, D.C. 20460, telephone 703-305-5826. Additional contact information may be identified through several divisions of the Office of Pesticide Programs. The registration division is responsible for product registrations, amendments, registrations, tolerances, experimental use permits, and emergency exemptions for all pesticides not assigned to Biopesticides and Pollution Prevention division or the antimicrobial division. General information may be obtained from Linda Arrington at 703-305-5446.

The Biopesticides and Pollution Prevention division is responsible for risk and benefit assessment and risk management functions for microbial pesticides, tolerance reassessment, biochemical pesticides, plant pesticides, and the pesticide environmental stewardship program. The division director's office can be reached at 703-308-8712. The antimicrobial division is responsible for all regulatory activities associates with antimicrobial pesticides including product registrations, amendments, and registrations. This department can be reached at 703-308-6214.

4.4 WHO MUST REGISTER AND WHICH PESTICIDES ARE SUBJECT TO REGISTRATION

Pesticides must be registered if they are:

1. Intended for preventing, destroying, repelling, or mitigating any pests
2. Intended for use as a plant regulator, defoliate, or desiccant

The word *intent* has a broad range of definitions. Under FIFRA, *intent* to use as a pesticide includes the following:

1. *Intent* for use as a pesticide by claims either on the label, in claims, or in other statements
2. *Intent* for use as a pesticide by the composition of the material
3. *Intent* for use as a pesticide by mode of action of the product as distributed or sold
4. *Intent* for use as a pesticide by the substance consisting of one or more active ingredients and having no significant commercially valuable use as distributed or sold other than for pesticidal purposes or the manufacturing of a pesticide
5. *Intent* by actual or constructive knowledge that the substance will be used as a pesticide

For instance, if a manufacturer makes a claim, either on the label or even verbally, regarding the pesticide's utility, then the product is subject to FIFRA and its registration requirements. Also, if the pesticide composition, even in the absence of claims, is made up of ingredients that are well known or have no other nonpesticidal utility, then the project is to be considered a pesticide. An example that the EPA uses is that a company markets a granular 2-4D product that has labeling identifying the presence of 2-4D. The instructions indicate how to apply the material to lawns, a certain dosage rate, a warning regarding overapplication, yet although there is not a claim that broadleaf weeds will be killed, *it still is a pesticide.* Even if pesticide claims are not made for the product and the active ingredient is not currently recognized as a pesticide, the mode of action of the product may be pesticidal in nature. For example, from the example above, even if the product had not been 2-4D, but instead was a currently nonrecognized pesticide active ingredient, the mode of action clearly identifies that it is intended to act as a pesticide in some manner. The only exception to this is if there is a utility that necessitates the same mode of action that the product in use may not be a pesticide.

4.4.1 Exemptions

There are a number of exemptions of products not subject to FIFRA. (However, it is important to be aware that this exemption may not necessarily apply to the Food, Drug and Cosmetics Act.) The FIFRA exemptions include the minimum-risk pesticides listed below.

4.4.1.1 Minimum Risk Pesticides. Certain pesticides are exempt from the rule if they have been considered a minimum-risk pesticide as defined in 40 CFR 152.25. These minimum-risk pesticides include:

Castor oil (U.S.P. or equivalent)

Cedar oil

Cinnamon and cinnamon oil

Citric acid

Citronella and citronella oil

Cloves and clove oil

Corn gluten meal

Corn oil

Cottonseed oil

Dried blood

Eugenol

Garlic and garlic oil

Geraniol

Geranium oil

Lauryl sulfate

Lemongrass oil

Linseed oil

Malic acid

Mint and mint oil

Peppermint and peppermint oil

2-Phenethyl propionate (2-phenylethyl propionate)

Potassium sorbate

Putrescent whole egg solids

Rosemary and rosemary oil

Sesame (includes ground sesame plant) and sesame oil

Sodium chloride (common salt)

Sodium lauryl sulfate

Soybean oil

Thyme and thyme oil

White pepper

Zinc metal strips (consisting solely of zinc metal and impurities)

4.4.1.2 Multipurpose Substances without Pesticidal Claims. Multipurpose substances which have no pesticidal claims include materials that provide a physical barrier against a pest. For example, sphagnum moss when used as a plant growth medium to retard damping off or latex tree dressing that contains claims of preventing the entrance of insects or fungus are exempt from registration.

Other examples include (1) cocoa or pine bark mulch, which claims suppression of weed growth or (2) black plastic or tar paper used to suppress weeds or repel insects. In addition, materials that are naturally pest-resistant are exempt if the claims do not contain pest mitigation effects beyond the material itself. For example, cypress redwood lumber for an outdoor deck that has claims of repellency resistance to wood-boring insects and wood rot fungus is not considered a pesticide.

Materials that have been treated with a pesticide to protect the material itself are not considered pest mitigation beyond the material, provided the treating pesticide is registered. For example, shower curtains treated with a fungicide to retard mildew growth, wood preservatives for lumber, caulks with mildew-side, paints with antimicrobes added for in-can preservation, and leather or other fabrics treated with sanitizer compound are not to be considered pesticides. However, the pesticides used on these materials must be registered and have the appropriate instructions for listing in an appropriate manner.

Paints that claim to have resistant qualities to mildew are acceptable; however, paint that claims to have protection qualities for the prevention of mold spores beneath the paint surfaces is not under the exemption. In addition, all paints used in canneries, breweries, hospitals, or other areas where prevention of bacteria or mold pose a health risk are not subject to the above exemption and are therefore considered pesticides.

4.4.1.3 Treatment of Living Organisms on Human or Animal Bodies. Other products *not to be considered pesticides* are those that are intended to be used on living bodies, either human or animal. Such products include athlete's foot remedies, dandruff medications, lice, soaps and lotions, additives for treatment of fish diseases, dermal disinfectants, and some other antimicrobial (meaning fungicides, disinfectants, or virusicides). However, insecticides are not included in this living body exemption. The most common examples are mosquito repellents. In addition, flee and tick remedies for pets and humans are not exempted. They are considered pesticides and must be registered.

4.4.1.4 Mortuary and Preservation Supplies. One of the unique exemptions is for mortuary supplies intended to prevent mold and bacteria on human cadavers. The rationale is that professional embalmers and morticians are trained in their use, no other person of the living public would be exposed to these products, and thus they do not require the protection afforded by the registration.

In addition, animals and animal organ preservatives such as formaldehyde are not required to undergo registration and are exempt from consideration as pesticides.

4.4.1.5 Products Used in Processing Food, Cosmetics, and Pharmaceuticals. Antimicrobial products used solely in processing in foods or feeds in beverages or pharmaceutical products including grinding, milling, cracking, or other processes that cause a physical change in the commodity are methods that meet the definition of process and therefore are regulated by the FDA and not the EPA. However, drawing, husking, or shelling material does not meet the definition of processed, so that pesticides used during these processes are regulated by FIFRA. Similar to processed food, cosmetics, and pharmaceuticals in general are regulated by the FDA.

4.4.1.6 Pheromones and Leaf Coverings. Pheromones labeled for use only in pheromone traps and ferments sole active ingredient are intraspecies communicators and thus not subject to the regulation. One exemption to this is intraspecies communicators. One limitation to the pheromone exemption is fox urine which is used to repel rabbits and thus is not a pheromone but an allomone and is therefore subject to FIFRA.

Plant or leaf coatings designed to protect against frost or retard water loss or transaction are not pesticides.

4.4.1.7 Cleaning Agent. Deodorizers, bleaches, soaps, and other cleaning agents which do not bear claims to sanitize or disinfect are not considered to be subject to registration; however, any bleach that contains more than 5.25 percent sodium hydrochloride must be registered if the label states that the bacteria will be killed at certain doses. Perhaps paradoxically, an identical bleach does not need to be registered if the label does not contain a bactericide claim, even though it delivers the same benefits at appropriate doses.

4.4.1.8 Attractants. Attractants intended for survey or census purposes and not intended for control or mitigation of the pest are not controlled by this regulation.

4.4.1.9 Plant Nutrients. One of the more obtuse exemptions is for plant nutrients, fertilizers, trace elements, and other products intended solely for providing necessary nutritional needs for plants for vigorous growth. Compounds such as auxins, cytokinins, and gibberellins have no other use other than as plant regulators and are therefore not regulated and are exempted by this regulation.

The EPA should be contacted for exemptions on a case-by-case basis for products that contain auxins, since they may meet criteria for vitamin hormone horticulture products as defined in 40 CFR 152-25D.

4.5 CADDY (COMPUTER-AIDED DOSSIER AND DATA SUPPLY)

CADDY stands for computer-aided dossier and data supply, the OPP newly accepted format for submission of electronically formatted pesticide registration documents. CADDY uses an electronic submission format, which requires all studies to be submitted on CD ROM. The system requires that the files written to CD must be set up in a defined format and structure. In addition to the supplied files, text files are also added to service indices for use by the retrieval software. There is no single software package required to assemble CADDY CD ROM; rather, it is suggested that companies use one of the many varieties of software available on the market today to build their CADDY CD ROM sets. Software that is required includes word processor spreadsheets, database programs, and scanning and graphic packages. Once the CD ROM sets are sent to the EPA, EPA staff uses the retrieval software to evaluate the data on-line. Because the CADDY system is new, registrants may experience, according to the EPA, a learning curve in completing their first CD ROM. The EPA currently does not require submissions on CADDY CD ROM sets; however, registrants may find it advantageous to submit on CD ROM rather than on multiple paper copies. EPA considers the electronic submittal system an investment in the future. In addition, EPA considers the cost savings from storage and mailing of studies to be significant. The data requirements are also compatible with European member states and Canadian pesticide regulatory authorities at this time. A set of web pages on this issue may be found at www.epa.gov/opppmsd1/CADDY/.

4.5.1 How to Submit Pesticide Studies Using CADDY Format

The following is a summary of EPA's guidance on submitting pesticide studies using the CADDY format. Note again that submission of studies in electronic form is an option and not a requirement.

4.5.2 Procedure

Any company planning to submit studies to the EPA using the CADDY format must obtain a range of master record identifiers (MRIDs) for that particular submission. These identifiers may be obtained by submitting to EPA a notification that the company plans to submit studies using the CADDY format. This notification must identify:

• Company name and address
• Type of regulatory action(s) in support of which the studies will be submitted
• Name of the chemical in support of which the studies will be submitted

- A count of the studies which will be submitted
- The approximate date on which the studies will be submitted

This document must be submitted to:

USEPA, Office of Pesticide Programs
Document Processing Desk (CADDY)
Mail Code 7504C
401 M St SW
Washington, D.C. 20460

EPA will reserve a range of MRIDs, and notify the company in writing concerning that range. The range of MRIDs will all have the same first six digits (for example, 789123). The company will use these same first six digits, adding as the seventh and eighth digits a two-digit sequence number to identify each study within the submission. The transmittal document, bibliography, and all administrative material should be labeled with 00 as the seventh and eighth digits of the MRID. If a company plans to submit more than 99 studies, EPA will reserve two ranges of MRIDs for that submission. The company must use each of these MRIDs when submitting both the hard copy and the CADDY CD sets. The MRID must be printed in the upper right corner on the first page of each hard-copy study. Each MRID assigned to a hard-copy version of each study must also correspond to the electronic version of that same study.

Two hard copies of the study and one copy of the CADDY CD set must be submitted. Each study must conform to the requirements specified in PRN 86-5 with respect to format, legibility, confidentiality claims, and good laboratory practices statements. The OPP will check each study for compliance with the requirements of PRN-86-5 and will notify the company regarding acceptance or rejection. If any studies do not comply with PRN 86-5, they must be brought into compliance before the submission is released for further action.

After all studies have been determined to be in compliance with PRN 86-5, EPA will notify the company and will request additional CADDY CD sets incorporating all corrections. The required number of CD ROM sets will vary depending on the number and nature of the studies being submitted. The CADDY CD ROM sets must comply with the CADDY Format Specification version 1.1.

When submitting the additional CADDY CD ROM sets, the company must also submit a signed and dated statement to indicate that the contents of the studies contained on all of the CADDY CD sets are a true and accurate duplicate of the paper copies of the studies. This statement must read as follows: "I certify that the study images contained on the enclosed compact disk are true and accurate duplicates of the previously submitted paper copy of each study contained in this submission. I acknowledge that any knowingly false or misleading statements may be punishable by fine or imprisonment under applicable law." An authorized company representative must sign and date this statement.

The OPP will declare as incomplete any CADDY submission which contains a certification statement that deviates from this wording or does not contain an appropriate signature and date.

The OPP will also declare as incomplete any CADDY submission in which the MRID assigned to the paper copy of a study is not correctly associated with the electronic image of that same study or if the CADDY submission does not meet the technical specifications identified in the CADDY Application Guide.

4.5.3 Supplemental Files

Supplemental electronic files such as databases, spreadsheets, text files, and images which supplement the study reports may also be included with the CADDY CD ROM sets. These files must be contained within the FILES directory as specified in the CADDY format specification document. Guidance regarding acceptable formats of these types of files is not expected to remain static. Hence this guidance will be made available through the following Internet site: www.epa.gov/CADDY/suppfiles, and will be updated periodically to coincide with advances in available technology and software.

Additional information regarding CADDY guidance can be received by contacting:

John Jamula
USEPA, Office of Pesticide Programs
Mail Code 7504C (CADDY)
401 M St SW
Washington, D.C. 20460

or

Joanne Martin
USEPA, Office of Pesticide Programs
Mail Code 7502C
401 M St. SW
Washington, D.C. 20460

In Appendix G you will find CADDY supplemental files guidance for your reference.

CHAPTER 5
HAZARDOUS MATERIALS TRANSPORTATION

John A. Green

Director of Safety & Health
Tate & Lyle North American Sugars Inc.

5.1 OVERVIEW

The requirements of the Hazardous Materials Transportation Act (HMTA) are enforced by the Department of Transportation (DOT), which regulates the shipment of hazardous substances by air, highway, or rail. The HMTA covers not only transporters of hazardous materials but also those who prepare shipments for transport, as well as those responsible for the containers used to transport hazardous materials. Under HMTA, the DOT maintains a classification scheme for hazardous materials and sets standards for proper handling of those materials, including packaging, labeling, and shipping requirements.

In the context of environmental permitting, the focus of the HMTA is on its registration and fee requirements for all persons subject to the act who either transport hazardous materials or offer hazardous materials for shipment. Since these applicability provisions are quite encompassing, the reader should carefully determine whether or not he or she meets the definition of a "hazmat employer" and whether or not he or she employs any "hazmat employees." If either of these two conditions is met, it is very likely that the registration statement requirements would be applicable.

5.2 DEFINITIONS

A "hazardous material" means a substance or material which has been determined by the Secretary of Transportation to be capable of posing an unreasonable risk to health, safety, and property when transported in commerce, and which has been so designated. The term includes hazardous substances, hazardous wastes, marine pollutants, and elevated-temperature materials as defined in this section, materials designated as hazardous under the provisions of 49 CFR 172.101, and materials that meet the defining criteria for hazard classes and divisions contained in 49 CFR Part 173.

A "hazardous substance" for the purposes of this chapter, means a material, including its mixtures and solutions, that

1. Is listed in Appendix A to 49 CFR 172.101
2. Is in a quantity, in one package, which equals or exceeds the reportable quantity (RQ) listed in Appendix A to 49 CFR 172.101
3. When in a mixture or solution:
   a. For radionuclides, conforms to paragraph 6 of Appendix A to 172.101
   b. For other than radionuclides, is in a concentration by weight which equals or exceeds the concentration corresponding to the RQ of the material, as shown in Table 5.1

TABLE 5.1 Concentration by Weight

RQ, lb (kg)	Percent	ppm
5000 (2270)	10	100,000
1000 (454)	2	20,000
10 (45.4)	0.2	2,000
10 (4.54)	0.02	200
1.0 (0.454)	0.002	20

The term does not include petroleum, including crude oil or any fraction thereof which is not otherwise specifically listed or designated as a hazardous substance in Appendix A to 172.101 of this subchapter, and the term does not include natural gas, natural gas liquids, liquefied natural gas, or synthetic gas usable for fuel (or mixtures of natural gas and such synthetic gas).

"Hazardous waste," for the purposes of this chapter, means any material that is subject to the Hazardous Waste Manifest Requirements of the U.S. Environmental Protection Agency specified in 40 CFR Part 262.

A "hazmat employee" means a person who is employed by a hazmat employer and who in the course of employment directly affects hazardous materials transportation safety. This term includes an owner-operator of a motor vehicle which transports hazardous materials in commerce. This term includes an individual, including a self-employed individual, employed by a hazmat employer who, during the course of employment:

1. Loads, unloads, or handles hazardous materials

2. Manufactures, tests, reconditions, repairs, modifies, marks, or otherwise represents containers, drums, or packagings as qualified for use in the transportation of hazardous materials

3. Prepares hazardous materials for transportation

4. Is responsible for safety of transporting hazardous materials

5. Operates a vehicle used to transport hazardous materials

A "hazmat employer" means a person who uses one or more of its employees in connection with:

1. Transporting hazardous materials in commerce.

2. Causing hazardous materials to be transported or shipped in commerce.

3. Representing, marking, certifying, selling, offering, manufacturing, reconditioning, testing, repairing, or modifying containers, drums, or packagings as qualified for use in the transportation of hazardous materials. This term includes an owner-operator of a motor vehicle which transports hazardous materials in commerce.

5.3 REGULATORY HISTORY

The Hazardous Materials Transportation Act was enacted by Public Law 93-633 on Jan. 3, 1975, giving DOT the authority to regulate shipments of hazardous substances by air, highway, or rail.

On Nov. 16, 1990, the HMTA was substantially revised by the Hazardous Materials Transportation Uniform Safety Act of 1990. Among the more substantial changes made to the HMTA were:

• Registration requirements and fees for certain hazardous materials shippers and carriers

• Improvements to placarding requirements

• The requirement of a training program for all public employees involved with hazardous materials

• Motor carrier safety permits

- Fees on railroads
- An increase in allowable civil penalties
- Inspections of vehicles carrying highway route–controlled quantities of radioactive materials

The 1990 amendments also allowed federal regulation to preempt state regulation in the following areas:

- Hazardous materials classification, description, and designation
- Packing, repacking, handling, labeling, marking, and placarding of hazardous materials
- Shipping document preparation, execution, placement, and use; written notification, recording and reporting of unintentional releases of hazardous materials
- Certified package design, manufacture, fabrication, marking, maintenance, reconditioning, repairing, and testing

The law was recodified on July 5, 1994, without substantive amendment by the "Revision of Title 49 USC Act" (Public Law 103-272) and amended on Aug. 26, 1994, by Public Law 103-311, which amended HMTA to exempt foreign offerors from U.S. hazmat registration fees and called for the evaluation of performance-oriented standards for fiber drum packagings in hazmat transport.

5.4 SUMMARY OF PROVISIONS

The Hazardous Materials Transportation Act provides protection against the risks to life and property inherent in the transportation of hazardous material in commerce by improving the regulatory and enforcement authority of the DOT.

The DOT designates material (including an explosive, radioactive material, etiologic agent, flammable or combustible liquid or solid, poison, oxidizing or corrosive material, and compressed gas) or a group or class of material as hazardous when the DOT decides that transporting the material in commerce in a particular amount and form may pose an unreasonable risk to health and safety or property.

The DOT prescribes regulations for the safe transportation of hazardous material in intrastate, interstate, and foreign commerce. These regulations apply to a person:

1. Transporting hazardous material in commerce
2. Causing hazardous material to be transported in commerce
3. Manufacturing, fabricating, marking, maintaining, reconditioning, repairing, or testing a package or container that is represented, marked, certified, or sold by that person as qualified for use in transporting hazardous material in commerce

These regulations shall govern safety aspects of the transportation of hazardous material the DOT considers appropriate.

The act directs DOT to designate material (including an explosive, radioactive material, etiologic agent, flammable or combustible liquid or solid, poison, oxidizing or corrosive material, and compressed gas) or a group or class of material as hazardous when transportation of such material might pose an unreasonable risk to health and safety or property.

5.5 APPLICABILITY

The requirements of the HMTA apply to all persons transporting hazardous material in commerce; persons causing hazardous material to be transported; or persons who manufacture, fabricate, mark, maintain, recondition, repair, or test packagings or containers that are represented, marked, certified, or sold by that person as qualified for use in transporting hazardous material in commerce. As stated earlier, care must be taken in determining a facility's applicability to the registration requirements of the HMTA, especially in reviewing the nature of the waste materials a facility ships off-site for disposal.

In addition, any person under contract with a department, agency, or instrumentality of the United States government that transports or causes to be transported hazardous material, or manufactures, fabricates, marks, maintains, reconditions, repairs, or tests a package or container that the person represents, marks, certifies, or sells as qualified for use in transporting hazardous material must comply with the requirements of the HMTA and its prescribed regulations.

5.6 EXEMPTIONS AND EXCLUSIONS

The DOT may issue an exemption from this chapter or a regulation prescribed under section 5103(b), 5104, 5110, or 5112 of the HMTA to a person transporting or causing to be transported hazardous material in a way that achieves a safety level:

1. At least equal to the safety level required under this chapter
2. Consistent with the public interest and this chapter if a required safety level does not exist

An exemption under the HMTA is effective for not more than 2 years and may be renewed on application to the DOT.

When applying for an exemption or renewal of an exemption under the HMTA, the person must provide a safety analysis prescribed by the DOT that justifies the exemption. The DOT shall publish in the *Federal Register* notice that an application for an exemption has been filed and shall give the public an opportunity to inspect the safety analysis and comment on the application. The HMTA does not require the release of information protected by law from public disclosure.

The DOT shall issue or renew the exemption for which an application was filed or deny such issuance or renewal within 180 days after the first day of the month following the date of the filing of such application, or the DOT shall publish a statement in the *Federal Register* of the reason why the DOT's decision on the exemption is delayed, along with an estimate of the additional time necessary before the decision is made.

Exclusions, in any part, from this chapter and regulations can be prescribed under this chapter for:

1. A public vessel (as defined in section 2101 of title 46)
2. A vessel exempted under section 3702 of title 46 from chapter 37 of title 46
3. A vessel to the extent it is regulated under the Ports and Waterways Safety Act of 1972 (33 U.S.C. 1221 et seq.)

5.7 HANDLING HAZARDOUS MATERIALS

DOT (49 USC 5106) establishes procedures for handling hazardous materials, including the following criteria:

1. A minimum number of personnel
2. Minimum levels of training and qualifications for personnel
3. The kind and frequency of inspections
4. Equipment for detecting, warning of, and controlling risks posed by the hazardous material
5. Specifications for the use of equipment and facilities used in handling and transporting the hazardous material
6. A system of monitoring safety procedures for transporting the hazardous material

The DOT also determines frequency of inspections; the type of equipment required to detect, warn of, and control risks posed by hazardous material; specifications for use of such equipment and facilities handling hazardous material; and safety procedures for transportation of such material.

5.8 REGISTRATION AND FEES

Persons who transport certain types of hazardous material must file a registration statement with DOT. HMTA also allows DOT to establish and impose by regulation and collect an annual fee of at least $250 but not more than $5000 from each person required to file a registration statement under the HMTA. The DOT determines the amount of the fee on at least one of the following bases:

1. Gross revenue from transporting hazardous material
2. The type of hazardous material transported or caused to be transported
3. The amount of hazardous material transported or caused to be transported
4. The number of shipments of hazardous material
5. The number of activities that the person carries out for which filing a registration statement is required under the HMTA
6. The threat to property, individuals, and the environment from an accident or incident involving the hazardous material transported or caused to be transported
7. The percentage of gross revenue derived from transporting hazardous material

Motor carriers that transport or cause to be transported by motor vehicle must carry a safety permit issued by DOT.

5.8.1 Registration and Fee Applicability

The registration and fee requirements of this subpart apply to any person who offers for transportation, or transports, in foreign, interstate, or intrastate commerce:

1. Any highway route–controlled quantity of a Class 7 (radioactive) material, as defined in 173.403 of this chapter.
2. More than 25 kg (55 lb) of a Division 1.1, 1.2, or 1.3 (explosive) material (see 49 CFR 173.50) in a motor vehicle, rail car, or freight container.
3. More than 1 liter (1.06 quarts) per package of a material extremely toxic by inhalation [i.e., "material poisonous by inhalation," as defined in 49 CFR 171.8 of this chapter, that meets the criteria for "hazard zone A," as specified in 49 CFR 173.116(a) or 173.133(a)].
4. A hazardous material in a bulk packaging (see 171.8 of this chapter) having a capacity equal to or greater than 13,248 liters (3500 gallons) for liquids or gases or more than 13.24 cubic meters (468 cubic feet) for solids.
5. A shipment in other than a bulk packaging of 2268 kg (5000 lb) gross weight or more of one class of hazardous materials for which placarding of a vehicle, rail car, or freight container is required for that class, under the provisions of subpart F of 49 CFR part 172 of this chapter. For applicability of this subpart, the term "shipment" means the offering or loading of a hazardous material at one loading facility using one transport vehicle, or the transport of that transport vehicle.

5.8.2   Exceptions to Registration and Fee Applicability

The following are excerpted from the requirements of this subpart:

1. An agency of the federal government
2. A state agency
3. An agency of a political subdivision of a state
4. An employee of any of those agencies in paragraphs (a)(1) through (a)(3) of the HMTA with respect to the employee's official duties
5. A hazmat employee (including, for purposes of this subpart, the owner-operator of a motor vehicle that transports in commerce hazardous materials, if that vehicle at the time of those activities is leased to a registered motor carrier under a 30-day or longer lease as prescribed in 49 CFR part 1057 or an equivalent contractual agreement)
6. A person domiciled outside the United States who offers solely from a location outside the United States hazardous materials for transportation in commerce, provided that the country of which such a person is a domiciliary does not require persons domiciled in the United States who solely offer hazardous materials for transportation to the foreign country from places in the United States to file a registration statement or to pay a registration fee

Upon making a determination that persons domiciled in the United States who offer hazardous materials for transportation to a foreign country solely from places in the United States must file registration statements or pay fees to that foreign country, the U.S. Competent Authority will provide notice of such determination directly to the Competent Authority of that foreign country and by publication in the *Federal Register*. Persons who offer hazardous materials for transportation to the United States from that foreign country must file a registration statement and pay the required fee no later than 60 days following publication of the determination in the *Federal Register*.

5.8.3   General Registration Requirements

Except as provided in 49 CFR 107.616(d), each person subject to this subpart must submit a complete and accurate registration statement on DOT Form F 5800.2 not later than June 30 for each registration year, or in time to comply with paragraph (b) of the HMTA, whichever is later. (*Note:* Refer to Appendix H of this book for a copy of DOT Form F 5800.2.)

You must file a registration statement with the DOT under the HMTA if the person is transporting or causing to be transported in commerce any of the following:

1. A highway route–controlled quantity of radioactive material
2. More than 25 kg of a class A or B explosive in a motor vehicle, rail car, or transport container

3. More than 1 liter in each package of a hazardous material the DOT designates as extremely toxic by inhalation

4. Hazardous material in a bulk packaging, container, or tank, as defined by the DOT, if the packaging, container, or tank has a capacity of at least 3500 gallons or more than 468 cubic feet

5. A shipment of at least 5000 lb (except in a bulk packaging) of a class of hazardous material for which placarding of a vehicle, rail car, or freight container is required under regulations prescribed under this chapter

In addition, the DOT may require any of the following persons to file a registration statement with the DOT under the HMTA:

1. A person transporting or causing to be transported hazardous material in commerce and not required to file a registration statement under paragraph (1) of the HMTA

2. A person manufacturing, fabricating, marking, maintaining, reconditioning, repairing, or testing a package or container the person represents, marks, certifies, or sells for use in transporting in commerce hazardous material the DOT designates

A person required to file a registration statement under the HMTA may transport or cause to be transported, or manufacture, fabricate, mark, maintain, recondition, repair, or test a package or container for use in transporting hazardous material only if the person has a statement on file as required by the HMTA.

The DOT may waive the filing of a registration statement or the payment of a fee required under the HMTA or both for any person not domiciled in the United States who solely offers hazardous materials for transportation to the United States from a place outside the United States if the country of which such person is a domiciliary does not require persons domiciled in the United States who solely offer hazardous materials for transportation to the foreign country from places in the United States to file registration statements or to pay fees for making such an offer.

After Sept. 15, 1992, no person required to file a registration statement may transport or cause to be transported or shipped hazardous materials, unless such person has on file, in accordance with 107.620, a current annual certificate of registration in accordance with the requirements of this subpart.

A registrant whose name or principal place of business has changed during the year of registration must notify RSPA of that change by submitting an amended registration statement not later than 30 days after the change.

Copies of DOT Form F 5800.2 and instructions for its completion may be obtained from the Hazardous Materials Registration Program, DHM-60, U.S. Department of Transportation, Washington, D.C. 20590-0001 or by calling 617-494-2545 or 202-366-4109.

If the registrant is not a resident of the United States, the registrant must attach to the registration statement the name and address of a permanent resident of the United States, designated in accordance with 107.7, to serve as agent for service of process.

5.8.4 Amount of Registration Fee

Each person subject to the requirements for filing a DOT Form F 5800.2 as specified in this subpart must pay an annual fee of $300 (which includes a $50 processing fee).

5.8.5 Registration Fee Payment Procedures

Except as provided in paragraph (d) of the HMTA, each person subject to the requirements of this subpart must mail the registration statement and payment in full to the U.S. Department of Transportation, Hazardous Materials Registration, P.O. Box 740188, Atlanta, GA 30374-0188. A registrant required to file an amended registration statement under 49 CFR 107.608(c) must mail it to the same address.

Payment must be made by certified check, cashier's check, personal check, or money order in U.S. funds and drawn on a U.S. bank, payable to the U.S. Department of Transportation and identified as payment for the "hazmat registration fee" or by a VISA or MasterCard credit card authorization completed and signed on the registration statement.

Payment must correspond to the annual fee indicated in 49 CFR 107.612.

5.8.6 Temporary Registration

A person may obtain a temporary registration number, valid for 45 days from the date of issuance, through an expedited registration process as follows:

1. Contact RSPA by telephone (800-942-6990 or 617-494-2545) and provide name, principal place of business, and credit card payment information
2. Pay a $350 registration and processing fee (including a $50 expedited handling fee); and 49 CFR 107.616(d)(3)
3. Submit a completed registration statement and proof of payment to RSPA before the expiration date of the temporary registration number

5.8.7 Filing Deadlines and Amendments

As stated above, each person subject to these requirements must submit a complete and accurate registration statement on DOT Form F 5800.2 not later than June 30 for each registration year.

The DOT will define by regulation when and under what circumstances a registration statement must be amended and the procedures to follow in amending the statement.

5.8.8   Certificate of Registration Record-Keeping Requirements

Each person subject to the requirements of this subpart, or its agent designated under 107.608(e), must maintain at its principal place of business for a period of 3 years from the date of issuance of each certificate of registration:

1. A copy of the registration statement filed with RSPA
2. The certificate of registration issued to the registrant by RSPA

After Jan. 1, 1993, each motor carrier subject to the requirements of this subpart must carry a copy of its current certificate of registration issued by RSPA or another document bearing the registration number identified as the "U.S. DOT Hazmat Reg. No." on board each truck and truck tractor (not including trailers and semitrailers) used to transport hazardous materials subject to the requirements of this subpart. The certificate of registration or document bearing the registration number must be made available, upon request, to enforcement personnel.

In addition to the requirements of paragraph (a) of the HMTA, after Jan. 1, 1995, each person who transports by vessel a hazardous material subject to the requirements of this subpart must carry on board the vessel a copy of its current certificate of registration or another document bearing the current registration number identified as the "U.S. DOT Hazmat Reg. No."

Each person subject to this subpart must furnish its certificate of registration (or a copy thereof) and all other records and information pertaining to the information contained in the registration statement to an authorized representative or special agent of DOT upon request.

CHAPTER 6
ENVIRONMENTAL IMPACT STATEMENTS

A. Roger Greenway
Principal
RTP Environmental Associates Inc.

Background

Congress passed the National Environmental Policy Act (NEPA), Public Law 91-190 on Jan. 1, 1970. It is codified at 42 U.S.C. 4321. NEPA was amended twice, once July 3, 1975, as PL94-52 and again Aug. 9, 1975, as PL94-83.

The purpose of NEPA is to declare a national policy which will "encourage productive and enjoyable harmony between man and his environment; to promote efforts which will prevent or eliminate damage to the environment and biosphere and to stimulate the health and welfare of man; to enrich the understanding of the ecological systems and natural resources important to the nation; and to establish a Council on Environmental Quality."

NEPA was passed in an era in which much federal legislation was brief. The entire NEPA act covers but three pages. In those three pages, however, are multitudes of requirements, many implied, not explicitly directed in the act. It is interesting that NEPA has probably done more to increase the awareness of environmental issues and to impose environmental decision making in more permit actions than any other piece of legislation. This is because NEPA required environmental impact statements to be prepared under almost all federal actions, including projects receiving federal funding. Also, NEPA served as a model for states and municipalities to require environmental assessments and environmental impact statements before development or construction of major subdivisions and commercial or industrial properties. It is these environmental impact statements that focus communities on the environmental impacts of development in their midst and allow environmental considerations to be taken into account in decision making at all levels of government.

In order to understand NEPA and its companion legislation, the Environmental Quality Improvement Act of 1970, a review of the basic requirements of NEPA, is required.

6.1 NEPA TITLE I—DECLARATION OF NATIONAL ENVIRONMENTAL POLICY

Section 101 of 42 U.S.C. 4331 states the declaration of national environmental policy which is the heart of the NEPA act and subsequent regulations. In this section:

> The Congress, recognizing the profound impact of man's activity on the interrelations of all components of the natural environment, particularly the profound influence of population growth, high density urbanization, industrial expansion, resource exploitation, and new and expanding technological advances and recognizing further the critical importance of restoring and maintaining environmental quality to the overall welfare and development of man, declares that it is the continuing policy of the federal government, in cooperation with state and local governments, and other concerned public and private organizations, to use all practicable means and measures, including financial and technical assistance, in a manner calculated to foster and promote the general welfare, to create and maintain conditions under which man and nature can exist in productive harmony, and fulfill the social, economic and other requirements of present and future generations of Americans.

In order to carry out the responsibilities of the act, Congress deemed that it was essential that the nation:

1. Fulfill the responsibilities of each generation as trustee of the environment for succeeding generations

2. Assure for all Americans safe, healthful, productive, and aesthetically and culturally pleasing surroundings

3. Attain the widest range of beneficial uses of the environment without degradation, risk to health and safety, or other undesirable or unattended consequences

4. Preserve important historic, cultural, and natural aspects of a natural heritage and maintain wherever possible an environment which supports diversity and variety of individual choices

5. Achieve a balance between population and resource use which will permit a high standard of living and a wide sharing of life's amenities

6. Enhance the quality of renewable resources and approach the maximum attainable recycling of depletable resources

Congress recognized that each person should enjoy a healthful environment and that each person has a responsibility to contribute to the preservation and enhancement of the environment. In order to achieve this, Congress directed in NEPA that the policies, regulations, and public laws of the United States be interpreted and administered in accordance with the policy set forth in the act and required all agencies of the federal government to:

1. Utilize a systematic interdisciplinary approach which will ensure the integrated use of the natural and social sciences and the environmental design arts in planning and decision making which may have an impact on man's environment.
2. Identify and develop methods and procedures, in consultation with the Council on Environmental Quality established by Title II of the act, to help ensure that presently unquantified environmental amenities and values may be given appropriate consideration in decision making along with economic and technical considerations.
3. Include in every recommendation or report for proposals for legislation and other major federal actions, significantly affecting the quality of the human environment, a detailed statement by the responsible official on:
   a. The environmental impact of the proposed action.
   b. Any adverse environmental effects which cannot be avoided should the proposal be implemented.
   c. Alternative to the proposed action.
   d. The interrelationship between local short-term uses of the human environment and the maintenance and enhancement of long-term productivity.
   e. Any irreversible and irretrievable commitments of resources which would be involved in the proposed action should it be implemented. Prior to making any detailed statement, the responsible federal official is required to consult with and obtain the comments of any federal agency which has jurisdiction by law or any special expertise with respect to any environmental impact involved.

NEPA requires that copies of all environmental impact assessments and statements, including the comments and views of appropriate federal, state, and local agencies, be made available to the President and the Council on Environmental Equality and as well to the public. This statement must accompany any proposal for major federal action covered by NEPA through the existing agency review processes.

NEPA provides that for major federal actions which are funded under a program of grants to states, the state may prepare the environmental assessment and impact statement, and that such statements prepared by the state shall not be deemed to be legally insufficient solely by reason of having been prepared by a state agency or official if:

1. The state agency or official has statewide jurisdiction and has the responsibility for such action
2. The responsible federal official furnishes guidance and participates in such participation
3. The responsible federal official independently evaluates each statement prior to its approval and adoption
4. After Jan. 1, 1976, the responsible federal official provides early notification to and solicits the views of any other state or any federal land management entity of any action or alternative which may have significant impacts on such state or affected federal land management entity and if there are any disagreements on such impacts because of a written assessment of such impacts and views for incorporation into the detailed statement

6.2 TITLE II—COUNCIL ON ENVIRONMENTAL QUALITY

Title II of NEPA required the creation of the Council of Environmental Quality in the Executive Office of the President. Part of the function of this panel is to assist the President in preparing a report to Congress which is required annually, on the environmental quality of the nation, setting forth:

1. The status and condition of the natural, major natural, manmade, or alternate environmental classes of the nation, including but not limited to the air, the aquatic, including marine, estuarine, and fresh water, and the terrestrial environment, including the forest dry land, wetland range, urban, suburban, and rural environment
2. Current and foreseeable trends in the quality, management, and utilization of such environment and the effects of those trends on the social, economic, and other requirements of the nation
3. The adequacy of available natural resources for fulfilling human and economic requirements of the nation in the light of expected population pressures
4. A review of the programs and activities of the federal government, the state and local governments, and nongovernmental entities and individuals, with particular reference to their effect on the environment and on the conservation, development, and utilization of natural resources
5. A program for remedying the deficiencies of existing programs and activities, together with recommendations for legislation

The Council on Environmental Quality consists of three members who are appointed by the President who are to serve at his pleasure by and with the advice and consent of the Senate. The president designates one of the members of the council to serve as chairman. Each member is required to be a person who, as a result of training, experience, and attainments, is exceptionally well qualified to analyze and interpret environmental trends and information of all kinds; to appraise programs and activities of the federal government in the light of the policies set forth in Title I of NEPA; to be conscious of, and responsive to, the scientific, economic, social, aesthetic, and cultural needs and interest of the nation, and to formulate and recommend national policies to promote the improvement of the quality of the environment.

The Council on Environmental Quality is authorized to employ officers and employees that it deems necessary to carry out its functions under NEPA. The council also employs and sets the compensation of experts and consultants as it deems necessary for carrying out its functions under the act in accordance with Section 3109 of Title V U.S. code.

The major duties of the Council on Environmental Quality are:

1. To assist and advise the President in the preparation of the environmental quality report required by NEPA Section 201

2. To gather timely and authoritative information concerning the information and trends of the quality of the environment, both current and prospective, to analyze and interpret such information for the purpose of determining whether such conditions and trends are interfering, or likely to interfere with achievement of the policy set forth in Title I of NEPA, and to compile and submit to the President the studies related to such conditions and trends

3. To review and appraise the various programs and agencies of the federal government in the light of the policy set forth in Title I of NEPA for the purpose of determining the extent to which their programs and activities are contributing to the achievement of the policy and to make recommendations to the President

4. To develop and recommend to the President national policies to foster and promote the improvement of environmental quality to meet the conservation, social, and economic health and other requirements and the goals of the nation

5. To conduct investigations, studies, surveys, research, and analysis relating to ecological systems and environmental quality

6. To document and define changes in the natural environment including the plant and animal systems and to accumulate necessary data and other information for continuing analysis of these changes or trends in the interpretation of the underlying causes

7. To report at least once a year to the President on the state and condition of the environment

8. To make and furnish such studies and reports thereon including recommendations on matters of policy and legislation as the President may request

Section 205 of NEPA directs the Council on Environmental Quality in carrying out its duties under NEPA to:

1. Consult with a Citizen's Advisory Committee on environmental quality established by Executive Order 11472, dated May 29, 1969, and with such representatives of science, industry, agriculture, labor, conservation organizations, state and local governments, and other groups as it deems advisable

2. To utilize to the fullest extent possible the services, facilities, and information (including statistical information) of public and private agencies and organizations and individuals in order that the duplication of effort and expense may be avoided, thus assuring that the council's activities will not unnecessarily overlap or conflict with similar activities authorized by law and performed by established agencies.

The three members of the Council on Environmental Quality are full-time appointments, and the chairman of the council is compensated at a pay rate provided for level 2 employees of the executive schedule of pay rates. The other members of the council are compensated at level 4 of the executive pay schedule.

6.3 THE ENVIRONMENTAL QUALITY IMPROVEMENT ACT OF 1970

A companion act to NEPA is the Environmental Quality Improvement Act of 1970. The declaration of purpose for this act is that Congress finds that:

1. Humans have caused changes in the environment.

2. Many of these changes may affect the relationship between humans and their environment.

3. Population increases and urban concentrations contribute directly to pollution and the degradation of our environment.

Congress declares in the Environmental Improvement Act of 1970 that there is a national policy for the environment which provides for the enhancement of environmental quality. This policy is evidenced by statutes heretofore enacted (NEPA) relating to the prevention, abatement, and control of environmental pollution, water and land resources, transportation, and economic and regional development. The Environmental Improvement Act has two main thrusts, first to assure that each federal department and agency conducting or supporting public works activities which affect the environment would implement the policies established under existing law, particularly NEPA, CAA, Water Pollution Control Act, and other acts; and second, to authorize an Office of Environmental Quality which notwithstanding the other provision of law would provide the professional and administrative staff for the Council of Environmental Quality established under NEPA

(Public Law 91-190). The Environmental Quality Improvement Act established the Office of Environmental Quality, and the chairman on the Council of Environmental Quality established under NEPA is designated as the director of that office. The compensation of a deputy director was specified in the act to be fixed by the President at a rate not in excess of the annual rate of compensation payable to the deputy director of the Bureau of the Budget. The directors are authorized under the Environment Quality Improvement Act to employ such officers and employees (including experts and consultants) as may be necessary to enable the office to carry out its functions under this title under the Environmental Quality Improvement Act and under NEPA, except that they may employ no more than 10 specialists and other experts without regard to the provisions of Title V U.S. code concerning maximum rates of compensation and pay.

The main functions of the Office of Environmental Quality are to:

1. Provide professional and administrative staff and support for the Council on Environmental Quality established under NEPA

2. Assist federal agencies and departments in appraising the effectiveness of the existing and proposed facilities, programs, policies, and activities of the federal government which affect environmental quality

3. Review the adequacy of existing systems for monitoring and predicting environmental changes in order to achieve effective coverage and efficient use of research facilities and other resources

4. Promote the advancement of scientific knowledge of the effects of actions and technology on the environment and encourage the development of the means to prevent or reduce adverse effects on the health and well-being of humans

5. Assist and coordinate among the federal departments and agencies those programs and activities which affect, protect, and improve environmental quality

6. Assist the federal department and agencies in the development of an interrelationship of environmental quality criteria, and standards established through the federal government

7. Collect, analyze, and interpret data and information on environmental quality, ecological research, and evaluation

6.4 NATIONAL AND ENVIRONMENTAL POLICY ACT PROCEDURES

The environmental impact statement requirements have evolved over the years since 1969 and 1970. Initially, environmental impact statements were strictly prepared by federal agencies. Projects which received federal funding were required to prepare environmental impact assessments (EIAs) which studied the projects, evaluated them, and provided data to the government with respect to the criteria defined in NEPA for identifying baseline conditions, evaluating alternatives, and projecting the impacts of the proposed action. A lead agency would then take the environmental impact assessment and convert it into an environmental impact statement (EIS). The process has evolved over the years, and now many states and even local municipalities require that environmental impact statements be prepared by applicants for major projects. In these instances, the applicant prepares an environmental impact assessment or statement and submits it to an agency. The environmental impact statement is one factor among others used by the agency in determining the acceptability of a proposed project.

The term environmental assessment is also used in state and local regulations and in actions involving permits where environmental considerations need to be addressed. Some jurisdictions distinguish between an environmental review (or report), environmental assessment, and an environmental impact statement depending on the magnitude of the project, with the implication that environmental impact statements are prepared for the larger projects, environmental assessments are prepared for midsize projects and in environmental reviews, or environmental reports are required for smaller actions.

In part because NEPA was such a short piece of legislation, and because implementing agencies had different views on how it should be implemented, EPA issued in 1979 the National Environmental Policy Act Procedures. The National Environmental Policy Act Procedures states as its purpose and policy that EPA, to the extent applicable, must prepare environmental impact statements on those major actions determined to have significant impact on the quality of the human environment. These procedures provide for the identification and analysis of the environmental impacts of the EPA-related activities in the preparation and processing of environmental impact statements. Section 6.102, on Applicability, describes the requirements under which the EPA is required to prepare environmental impact statements for its own programs. These include:

1. Subpart A sets forth requirements for an overview of environment assessment requirements for EPA legislative proposals.

2. Subpart B describes requirements for the content of an EIS prepared pursuant to Subparts E, F, G, H, and I.

3. Subpart C describes requirements for coordination of all environmental laws during the environmental review undertaken pursuant to Subparts E, F, G, H, and I.

4. Subpart D describes the public information requirements, which must be undertaken in conjunction with the environmental review requirements under E, F, G, H, and I.

5. Subpart E describes the environmental review requirements for wastewater treatment construction grants program issued under Title II of the Clean Water Act.

6. Subpart F describes the environmental review requirements for New Source,

National Pollutant Discharge Elimination System (NPDES) permits issued under Section II of the Clean Water Act.

7. Subpart G describes the environmental review requirements for research and development programs undertaken by EPA.

8. Subpart H describes the environmental review requirements for solid waste demonstration projects undertaken by EPA.

9. Subpart I describes the environmental review requirements for the construction of special-purpose facilities and facility renovations by EPA.

Applicability B—Legislative Proposals

The Council on Environmental Quality regulations require EPA to prepare an environmental impact statement for legislative proposals developed by EPA which can significantly affect the quality of the human environment. The regulations require EPA to prepare a draft EIS to be signed by a responsible EPA official concurrently with the development of a legislative proposal.

6.5 CRITERIA FOR INITIATING AN ENVIRONMENTAL IMPACT STATEMENT

An environmental impact statement is required for all of the types of actions noted above. The criteria for actually initiating an EIS include the following:

1. That the federal action may significantly affect the pattern and type of land use (industrial, commercial, agricultural, recreational, residential) or growth and distribution of population.

2. The effects resulting from any structure or facility constructed or operated under the proposed action may conflict with local, regional, or state land use plans or policies.

3. The proposed action may have a significant adverse effect on wetlands including indirect and cumulative effects for any major part of a structure or facility constructed or operated on under the proposed action may be located in wetlands.

4. The proposed action may significantly affect threatened or endangered species or their habitats identified in the Department of Interior's list or a state's list, or a structure or facility constructed or operated under the proposed action may be located in such a habitat.

5. The implementation of the proposed action or plan may directly cause or induce changes that significantly

   - Displace population
   - Alter the character of existing residential areas
   - Adversely affect the floodplain

- Adversely affect a significant amount of important farmlands as defined in Section 6.302 of the act. Section 6.302 states the requirements for an executive summary which is required to be prepared and to describe in sufficient detail (10 to 15 pages) the critical aspects of the EIS so that the reader can become familiar with the proposed action and its net effects. The executive summary is required to focus on:

1. The existing problem
2. A brief description of each alternative evaluated (including the preferred and no action alternatives) along with a listing of environmental impacts, possible mitigation measures relating to each alternative and any areas of controversy (including issues raised by governmental agencies and the public)
3. Any major conclusions

The body of the EIS is required to include statements covering:

1. The purpose and need to summarize clearly and specifically the underlining purpose and need to which EPA is responding
2. Alternatives, including the proposed action which EPA has considered, including the alternative of no action

Subpart C of the National Environmental Policy Act Procedures requires EPA to coordinate with other environmental reviews and other agency requirements. These include landmarks and historical and archaeological sites under which EPA is required to coordinate with the requirements of the Historic Sites Act of 1935 (16 U.S.C. 461 et seq., the National Historic Preservation Act of 1966 as Amended at 16 U.S.C. 470, the Archeological and Historic Preservation Act of 1974 and Executive Order 11593 entitled "Protection and Enhancement of the Cultural Environment." These acts and the executive order require NEPA to focus on review procedures for projects involving:

1. National natural landmarks
2. Historic, architectural, archaeological, and cultural sites
3. Historic and prehistoric and archaeological data

6.6 RELATIONSHIP OF ENVIRONMENTAL IMPACT STATEMENTS TO PERMITTING

The environmental impact statement, or assessment, depending on the state or local regulations, is one of a number of required submissions for site plan approval for minor or major subdivision and for other actions involving construction. Where construction permits are required, the environmental impact statement for a major subdivision or construction project is an essential submission in many parts of the country. The environmental impact statement in most municipalities will be reviewed by the environmental commission for completeness and for adequacy. Where the environmental impact statement or assessment is incomplete or does not adequately review and describe the potential impacts and alternatives considered, it will be required to be redone or improved before an application for construction approval can go forward.

In Appendix J in the back of the book you will find a table of contents of an environmental impact statement. Note that in each of the environmental areas identified, it is required to describe the existing or baseline conditions, for example, air quality, water quality, noise, wetlands, to project the impact of the project on each of these baseline conditions and to describe alternatives, including the no action alternative, and their impacts, for comparison purposes.

Since procedures for projecting impact vary widely from medium to medium and with the regulatory requirements in a given location, it is important to be sure that the level of detail and level of sophistication in these analyses is proportional to the action being evaluated. The amount of scrutiny needed for a subdivision of six houses does not measure up to the level of scrutiny and detail required for a major regional shopping mall or major new interstate highway construction project. These require much more diligence in defining the baseline conditions, and much more care must be taken in estimating the impacts of the action. Additionally, much more thoroughness must be given to the assessments of the alternatives to these actions.

Analytical Techniques

The analysis of the baseline environmental conditions can be based on either:

1. An analysis of available data and reports prepared describing the natural resources of the area, for example, the natural resources inventories prepared by many municipalities
2. By actual observation or monitoring and/or sampling of the environment.

Observation or monitoring is more expensive, but since natural resource inventories are not always up to date, and since they may not adequately describe the environmental conditions at a particular site, on-site monitoring or observations are often required. This is particularly true for air quality and noise impacts, which vary widely from location to location.

Impact estimation techniques are available for many of the environmental media. These include air quality dispersion models to predict the impact of yet unbuilt sources of air emissions on the air environment, water discharge models to predict the impact of as yet nonexistent water discharges, noise models to predict the impact of noise emissions on the environment, and other techniques.

CHAPTER 7
OIL POLLUTION

John A. Green
Director of Safety and Health
Tate & Lyle North American Sugars Inc.

7.1 STATUTORY OVERVIEW, OBJECTIVES, AND PURPOSE

Unlike other federal statutes whose purpose is to control the release of pollutants into the environment (e.g., the Clean Water Act and the Clean Air Act), the Oil Pollution Act of 1990 does not actually rely on any formalized permit programs (e.g., the NPDES program) to provide the mechanisms by which its mandates and prohibitions are enforced. Rather, the requirements of the OPA 90 both expand oil and hazardous substances spill prevention and preparedness activities and improve regulated facility response capabilities. In doing so, the OPA 90 has established an inferred sense of environmental permitting by which a regulated facility (meaning either certain "marine transfer" facilities or a "substantial harm" facility) cannot operate unless it has developed and implemented both a facility response plan, which must be submitted for approval by the EPA, and a SPCC plan, which must be certified by a registered PE and maintained on-site. It is on this context that this chapter will focus. (*Note:* Submission of a SPCC plan to the EPA for approval is not required.)

It is strongly suggested that the reader obtain a complete copy of the oil pollution prevention regulations found in 40 CFR Part 112, including all appendixes, to reference while reviewing the specifics found in this chapter.

Congress passed the Oil Pollution Act of 1990 (OPA 90) (PL 101-380) in response to major oil spills. The OPA 90 was enacted to expand oil and hazardous substances spill prevention and preparedness activities, improve response capabilities, and ensure that owners or operators of certain vessels or facilities that drill, produce, gather, store, process, refine, transfer, distribute, or consume oil pay the costs associated with the cleanup and disposal of discharged oil. The act also established an expanded research and development program, as well as a new oil spill liability trust fund under Section 311 of the Clean Water Act.

As amended by Section 4202 of the OPA 90, Section 311 of the water act requires that regulated facilities have a fully prepared and implemented spill prevention, control, and countermeasure plan which is certified by a registered professional engineer. Facilities must implement the plan, including carrying out the spill prevention and control measures established for the type of facility or operations, such as measures for containing a spill (e.g., berms). In the event that a facility cannot implement containment measures, the facility must develop and incorporate a strong spill contingency plan into their countermeasure plan. In addition, facility owners or operators must conduct employee training on the contents of a SPCC plan. The SPCC plan must be prepared within 6 months of the date a facility commences operations and implemented within 1 year of the date operations begin.

Section 4202(a)(6) of the Oil Pollution Act of 1990, Public Law 101-380, amends section 311(j) of the Federal Water Pollution Control Act (FWPCA), also known as the Clean Water Act (CWA), and under CWA section 311(j)(5) [see 33 U.S.C. 1321(j)(5)] directs the President to issue regulations that require owners or operators of tank vessels, offshore facilities, and certain onshore facilities to prepare and submit to the President plans for, among other things, responding, to the maximum extent practicable, to a worst-case discharge of oil and to a substantial threat of such a discharge. In doing so, it established requirements, and an implementation schedule, for facility response plans (FRPs) and periodic inspections of discharge-removal equipment.

Section 311(j)(1)(C) of the CWA authorizes the President to issue regulations establishing procedures, methods, equipment, and other requirements to prevent discharges of oil from vessels and facilities and to contain such discharges. [See 33 U.S.C. 1321(j)(1)(C).] The President has delegated the authority to regulate non-transportation-related onshore facilities under sections 311(j)(1)(C) and 311(j)(5) of the CWA to the U.S. Environmental Protection Agency (EPA or the agency). [See Executive Order (E.O.) 12777, section 2(b)(1), 56 FR 54757 (Oct. 22, 1991), superseding E.O. 11735, 38 FR 21243.] By this same E.O., the President has delegated similar authority over transportation-related onshore facilities, deepwater ports, and vessels to the U.S. Department of Transportation (DOT), and authority over other offshore facilities, including associated pipelines, to the U.S. Department of the Interior (DOI). A memorandum of understanding (MOU) among EPA, DOI, and DOT effective Feb. 3, 1994, has redelegated the responsibility to regulate certain off-shore facilities located in and along the Great Lakes, rivers, coastal wetlands, and the Gulf Coast barrier islands from DOI to EPA. [See E.O. 12777 2(i) regarding authority to redelegate.] The MOU is included as Appendix B to 40 CFR Part 112. An MOU between the Secretary of Transportation and the EPA administrator, dated Nov. 24, 1971 (36 FR 24080, Dec. 18, 1971), establishes the definitions of non-transportation-related facilities and transportation-related facilities. The definitions from the MOU are currently included in Appendix A to 40 CFR Part 112.

Section 311(j)(5) of the FWPCA requires the preparation and submission of response plans from all onshore facilities that could reasonably be expected to cause either "substantial" or "significant and substantial" harm to the environment by discharging oil into or on the navigable waters, adjoining shorelines, or exclusive economic zone of the United States. Response plans must also be consistent with the National Oil and Hazardous Substances Pollution Contingency Plan (NCP) (40 CFR Part 300) and applicable area contingency plans (ACPs).

As amended by OPA 90, section 311(j)(5) directs the President to issue regulations implementing the new FWPCA requirements for facility response plans. The President delegated this authority, in part, to the Secretary of Transportation (DOT) by E.O. 12777 (3 CFR/cfrc, 1991 Comp.; 56 FR 54757). The Secretary of Transportation, in 49 CFR 1.46(m) (57 FR 8581; Mar. 11, 1992), further delegated, to the Commandant of the Coast Guard, the authority to regulate marine transportation–related (MTR) onshore facilities, and deepwater ports subject to the Deepwater Ports Act of 1974, as amended (33 U.S.C. 1501, et seq.). This rule addresses only MTR facilities that handle, store, or transport oil. Oil spill response plan regulations for vessels are the subject of a separate rulemaking project (CGD 91-034).

A major objective of the OPA 90 amendments to the FWPCA was to create a national planning and response system. OPA 90 requires the President to develop nationwide criteria for determining those facilities which could reasonably be expected to cause substantial harm to the environment. The OPA 90 Conference Report (Report 101-653) states that the criteria should result in a broad requirement for facility owners or operators to prepare and submit response plans. Those facilities identified by the President as having the potential to cause "substantial harm" are the regulated facilities that are required to submit response plans. (*Note:* Refer to section 7.2 for the definition of a "substantial harm facility.")

7.2 DEFINITIONS

"Oil": Section 311(a)(1) of the FWPCA defines "oil" as including, but not limited to, petroleum, fuel oil, sludge, oil refuse, and oil mixed with waste other than dredge spoils [33 U.S.C. 1321(a)(1)]. While the most common oils are the various petroleum oils (e.g., crude oil, gasoline, diesel, etc.), nonpetroleum oils such as animal fats (e.g., tallow, lard, etc.), vegetable oils (e.g., corn oil, sunflower seed oil, palm oil, etc.), and other nonpetroleum oils, such as turpentine, are included within the ambit of this regulation when handled, stored, or transported by a marine transportation–related facility.

The term "discharge" includes, but is not limited to, any spilling, leaking, pumping, pouring, emitting, emptying, or dumping, but excludes (A) discharges in compliance with a permit under section 402 of this act, (B) discharges resulting from circumstances identified and reviewed and made a part of the public record with respect to a permit issued or modified under section 402 of this act, and subject to a condition in such permit, and (C) continuous or anticipated intermittent discharges from a point source, identified in a permit or permit application under section 402 of this act, which are caused by events occurring within the scope of relevant operating or treatment systems.

"Marine transfer" area means that part of a waterfront facility handling oil or hazardous materials in bulk between the vessel, or where the vessel moors, and the first manifold or shutoff valve on the pipeline encountered after the pipeline enters the secondary containment required under 40 CFR 112.7 or 49 CFR 195.264 inland of the terminal manifold or loading arm or, in the absence of secondary containment, to the valve or manifold adjacent to the bulk storage tank, including the entire pier or wharf to which a vessel transferring oil or hazardous materials is moored.

"Substantial harm facility": As required by 40 CFR 112.20(f)(1) and the flowchart in Appendix L to this chapter, a facility is a "substantial harm facility" if either of the following two criteria are met:

1. The facility transfers oil over water to or from vessels and has a total oil storage capacity greater than or equal to 42,000 gallons
2. The facility's total oil storage capacity is greater than or equal to 1 million gallons, and one or more of the following is true:
   a. The facility does not have secondary containment for each aboveground storage area sufficiently large to contain the capacity of the largest aboveground storage tank within each storage area plus sufficient freeboard to allow for precipitation.
   b. The facility is located at a distance (as calculated using the appropriate formula in Appendix L or a comparable formula) such that a discharge from the facility could cause injury to fish and wildlife and sensitive environments.
   c. The facility is located at a distance (as calculated using the appropriate formula in Appendix L or a comparable formula) such that a discharge from the facility would shut down operations at a public drinking water intake.
   d. The facility has had a reportable spill greater than or equal to 10,000 gallons within the last 5 years.

7.3 SPILL PREVENTION, CONTROL, AND COUNTERMEASURE PLANS

As amended by Section 4202 of the OPA 90, Section 311 of the water act requires that regulated facilities have a fully prepared and implemented spill prevention, control, and countermeasure (SPCC) plan which is certified by a registered professional engineer. As amended by the OPA 90, Clean Water Act Section 311 includes minimum requirements for a SPCC plan. The plan must:

- Be consistent with the requirements of the national oil and hazardous substances pollution contingency plan and area contingency plans.
- Identify the qualified individual having full authority to implement removal actions, and require immediate communications between that individual and the appropriate federal official and the persons providing removal personnel and equipment.
- Identify and ensure by contract or other approved means the availability of private personnel and equipment necessary to remove, to the maximum extent practicable, a worst-case discharge (including a discharge resulting from fire or explosion), and to mitigate or prevent a substantial threat of such a discharge.
- Describe the training, equipment testing, periodic unannounced drills, and response actions of persons at the facility to be carried out under the plan to ensure the safety of the facility and to mitigate or prevent a discharge or the substantial threat of a discharge.
- Be updated periodically.

The SPCC plan needs to be a carefully thought out plan, prepared in accordance with good engineering practices, and which has the full approval of management at a level with authority to commit the necessary resources. If the plan calls for additional facilities or procedures, methods, or equipment not yet fully operational, these items should be discussed in separate paragraphs, and the details of installation and operational startup should be explained separately.

A SPCC plan must be developed, certified, and implemented if a facility meets both of the following requirements:

- The underground buried storage capacity of the facility exceeds 42,000 gallons of oil.
- The storage capacity, which is not buried, of the facility exceeds 1320 gallons, and the capacity of a single container is in excess of 660 gallons.

Facilities must implement the plan, including carrying out the spill prevention and control measures established for the type of facility or operations, such as measures for containing a spill (e.g., berms). In the event that a facility cannot implement containment measures, the facility must develop and incorporate a strong spill contingency plan into their countermeasure plan. In addition, facility owners or operators must conduct employee training on the contents of an SPCC plan.

Facilities must prepare an SPCC plan within 6 months of the date they begin operations and implement the plan within 1 year of the date operations begin.

7.3.1 SPCC Plan Elements

The complete SPCC plan must follow the sequence outlined below, and include a discussion of the facility's conformance with the following appropriate guidelines.

7.3.1.1 Recent Spill Events. A facility which has experienced one or more spill events within 12 months prior to the effective date of the regulations should include a written description of each such spill, corrective action taken, and plans for preventing recurrence.

7.3.1.2 Potential Equipment Failure. Where experience indicates a reasonable potential for equipment failure (such as tank overflow, rupture, or leakage), the plan should include a prediction of the direction, rate of flow, and total quantity of oil which could be discharged from the facility as a result of each major type of failure.

7.3.1.3 Containment and/or Diversionary Structures. Appropriate containment and/or diversionary structures or equipment to prevent discharged oil from reaching a navigable watercourse should be provided. One of the following preventive systems or its equivalent should be used as a minimum:

- Onshore facilities
- Dikes, berms, or retaining walls sufficiently impervious to contain spilled oil
- Curbing
- Culverting, gutters, or other drainage systems
- Weirs, booms, or other barriers
- Spill diversion ponds
- Retention ponds
- Sorbent materials
- Offshore facilities
- Curbing, drip pans
- Sumps and collection systems

When it is determined that the installation of structures or equipment listed in 40 CFR 112.7(c) to prevent discharged oil from reaching the navigable waters is not practicable from any onshore or offshore facility, the owner or operator should clearly demonstrate such impracticability and provide the following:

- A strong oil spill contingency plan following the provision of 40 CFR Part 109.

- A written commitment of worker power, equipment, and materials required to expeditiously control and remove any harmful quantity of oil discharged.

7.3.1.4 Conformance with Other Applicable Guidelines. In addition to the minimal prevention standards listed under 40 CFR 112.7(c), sections of the plan should include a complete discussion of conformance with the following applicable guidelines, other effective spill prevention, and containment procedures (or, if more stringent, with state rules, regulations, and guidelines).

7.3.1.4.1 Facility Drainage (Onshore), Excluding Production Facilities

1. Drainage from diked storage areas should be restrained by valves or other positive means to prevent a spill or other excessive leakage of oil into the drainage system or in-plant effluent treatment system, except where plan systems are designed to handle such leakage. Pumps or ejectors may empty diked areas; however, these should be manually activated and the condition of the accumulation should be examined before starting to be sure no oil will be discharged into the water.

2. Flapper-type drain valves should not be used to drain diked areas. Valves used for the drainage of diked areas should, as far as practical, be of manual open-and-closed design. When plant drainage drains directly into watercourses and not into wastewater treatment plants, retained stormwater should be inspected before drainage.

3. Plant drainage systems from undiked areas should, if possible, flow into ponds, lagoons, or catchment basins designed to retain oil or return it to the facility. Catchment basins should not be located in areas subject to periodic flooding.

4. If plant drainage is not engineered as above, the final discharge of all in-plant ditches should be equipped with a diversion system that could, in the event of an uncontrolled spill, return the oil to the plant.

5. Where drainage waters are treated in more than one treatment unit, natural hydraulic flow should be used. If pump transfer is needed, two "lift" pumps should be provided, and at least one of the pumps should be permanently installed when such treatment is continuous. In any event, whatever techniques are used, facility drainage systems should be adequately engineered to prevent oil from reaching navigable waters in the event of equipment failure or human error at the facility.

7.3.1.4.2 Bulk Storage Tanks (Onshore), Excluding Production Facilities

1. No tank should be used for the storage of oil unless its material and construction are compatible with the material stored and conditions of storage such as pressure and temperature.

2. All bulk storage tank installations should be constructed so that a secondary means of containment is provided for the entire contents of the largest single tank plus sufficient freeboard to allow for precipitation. Diked areas should be sufficiently impervious to contain spilled oil. Dikes, containment curbs, and pits are commonly employed for this purpose, but they may not always be appropriate. An alternative system could consist of a complete drainage trench enclosure arranged so that a spill could terminate and be safely confined in an in-plant catchment basin or holding pond.

3. Drainage of rainwater from the diked area into a storm drain or an effluent discharge that empties into an open watercourse, lake, or pond, and bypassing the in-plant treatment system may be acceptable if:

    a. The bypass valve is normally sealed closed.

    b. Inspection of the runoff rainwater ensures compliance with applicable water quality standards and will not cause a harmful discharge as defined in 40 CFR Part 110.

    c. The bypass valve is opened, and resealed following drainage under responsible supervision.

    d. Adequate records are kept of such events.

4. Buried metallic storage tanks represent a potential for undetected spills. A new buried installation should be protected from corrosion by coatings, cathodic protection, or other effective methods compatible with local soil conditions. Such buried tanks should at least be subjected to regular pressure testing.

5. Partially buried metallic tanks for the storage of oil should be avoided, unless the buried section of the shell is adequately coated, since partial burial in damp earth can cause rapid corrosion of metallic surfaces, especially at the earth-air interface.

6. Aboveground tanks should be subject to periodic integrity testing, taking into account tank design (floating roof, etc.) and using such techniques as hydrostatic testing, visual inspection, or a system of nondestructive shell thickness testing. Comparison records should be kept where appropriate, and tank supports and foundations should be included in these inspections. In addition, the outside of the tank should frequently be observed by operating personnel for signs of deterioration, leaks which might cause a spill, or accumulation of oil inside diked areas.

7. To control leakage through defective internal heating coils, the following factors should be considered and applied, as appropriate.

    a. The steam return or exhaust lines from internal heating coils which discharge into an open watercourse should be monitored for contamination, or passed through a settling tank, skimmer, or other separation or retention system.

    b. The feasibility of installing an external heating system should also be considered.

8. New and old tank installations should, as far as practical, be fail-safe engineered or updated into a fail-safe engineered installation to avoid spills. Consideration should be given to providing one or more of the following devices:

    a. High-liquid-level alarms with an audible or visual signal at a constantly manned operation or surveillance station; in smaller plants an audible air vent may suffice.

b. Considering size and complexity of the facility, high-liquid-level pump cutoff devices set to stop flow at a predetermined tank content level.

c. Direct audible or code signal communication between the tank gauger and the pumping station.

d. A fast response system for determining the liquid level of each bulk storage tank such as digital computers, telepulse, or direct vision gauges or their equivalent.

e. Liquid-level-sensing devices should be regularly tested to ensure proper operation.

9. Plant effluents, which are discharged into navigable waters, should have disposal facilities observed frequently enough to detect possible system upsets that could cause an oil spill event.

10. Visible oil leaks, which result in a loss of oil from tank seams, gaskets, rivets, and bolts sufficiently large to cause the accumulation of oil in diked areas should be promptly corrected.

11. Mobile or portable oil storage tanks (onshore) should be positioned or located so as to prevent spilled oil from reaching navigable waters. A secondary means of containment, such as dikes or catchment basins, should be furnished for the largest single compartment or tank. These facilities should be located where they will not be subject to periodic flooding or washout.

7.3.1.4.3 Facility Transfer Operations, Pumping, and In-Plant Process (Onshore), Excluding Production Facilities

1. Buried piping installations should have a protective wrapping and coating and should be cathodically protected if soil conditions warrant. If a section of buried line is exposed for any reason, it should be carefully examined for deterioration. If corrosion damage is found, additional examination and corrective action should be taken as indicated by the magnitude of the damage. An alternative would be the more frequent use of exposed pipe corridors or galleries.

2. When a pipeline is not in service or is in standby service for an extended time, the terminal connection at the transfer point should be capped or blank-flanged, and marked as to origin.

3. Pipe supports should be properly designed to minimize abrasion and corrosion and allow for expansion and contraction.

4. All aboveground valves and pipelines should be subjected to regular examinations by operating personnel, at which time the general condition of items, such as flange joints, expansion joints, valve glands and bodies, catch pans, pipeline supports, locking of valves, and metal surfaces should be assessed. In addition, periodic pressure testing may be warranted for piping in areas where facility drainage is such that a failure might lead to a spill event.

5. Vehicular traffic granted entry into the facility should be warned verbally or by appropriate signs to be sure that the vehicle, because of its size, will not endanger aboveground piping.

7.3.1.4.4 *Facility Tank Car and Tank Truck Loading and Unloading Rack (Onshore)*

1. Tank car and tank truck loading and unloading procedures should meet the minimum requirements and regulation established by the Department of Transportation.

2. Where rack area drainage does not flow into a catchment basin or treatment facility designed to handle spills, a quick drainage system should be used for tank truck loading and unloading areas. The containment system should be designed to hold at least maximum capacity of any single compartment of a tank car or tank truck loaded or unloaded in the plant.

3. An interlocked warning light or physical barrier system, or warning signs, should be provided in loading and unloading areas to prevent vehicular departure before complete disconnect of flexible or fixed transfer lines.

4. Prior to filling and departure of any tank car or tank truck, the lowermost drain and all outlets of such vehicles should be closely examined for leakage, and if necessary, tightened, adjusted, or replaced to prevent liquid leakage while in transit.

7.3.1.4.5 *Oil Production Facilities (Onshore)*

1. Definition. An onshore production facility may include all wells, flowlines, separation equipment, storage facilities, gathering lines, and auxiliary non-transportation-related equipment and facilities in a single geographical oil or gas field operated by a single operator.

2. Oil production facility (onshore) drainage.

   a. At tank batteries and central treating stations where an accidental discharge of oil would have a reasonable possibility of reaching navigable waters, the dikes or equivalent required under 40 CFR 112.7(c)(1) should have drains closed and sealed at all times except when rainwater is being drained. Prior to drainage, the diked area should be inspected. Accumulated oil on the rainwater should be picked up and returned to storage or disposed of in accordance with approved methods.

   b. Field drainage ditches, road ditches, and oil traps, sumps, or skimmers, if such exist, should be inspected at regularly scheduled intervals for accumulation of oil that may have escaped from small leaks. Any such accumulations should be removed.

3. Oil production facility (onshore) bulk storage tanks.

   a. No tank should be used for the storage of oil unless its material and construction are compatible with the material stored and the conditions of storage.

   b. All tank battery and central treating plant installations should be provided with a secondary means of containment for the entire contents of the largest single tank if feasible, or alternate systems such as those outlined in 40 CFR 112.7(c)(1). Drainage from undiked areas should be safely confined in a catchment basin or holding pond.

c. All tanks containing oil should be visually examined by a competent person for condition and need for maintenance on a scheduled periodic basis. Such examination should include the foundation and supports of tanks that are above the surface of the ground.

d. New and old tank battery installations should, as far as practical, be fail-safe engineered or updated into a fail-safe engineered installation to prevent spills. Consideration should be given to one or more of the following:

(1) Adequate tank capacity to assure that a tank will not overfill should a pumper-gauger be delayed in making the regular rounds.

(2) Overflow equalizing lines between tanks so that a full tank can overflow to an adjacent tank.

(3) Adequate vacuum protection to prevent tank collapse during a pipeline run.

(4) High-level sensors to generate and transmit an alarm signal to the computer where facilities are a part of a computer production control system.

4. Facility transfer operations, oil production facility (onshore).

a. All aboveground valves and pipelines should be examined periodically on a scheduled basis for general condition of items such as flange joints, valve glands and bodies, drip pans, pipeline supports, pumping well polish rod stuffing boxes, bleeder, and gauge valves.

b. Salt water (oil field brine) disposal facilities should be examined often, particularly following a sudden change in atmospheric temperature, to detect possible system upsets that could cause an oil discharge.

c. Production facilities should have a program of flowline maintenance to prevent spills from this source. The program should include periodic examinations, corrosion protection, flowline replacement, and adequate records, as appropriate, for the individual facility.

7.3.1.4.6 Oil Drilling and Workover Facilities (Onshore)

1. Mobile drilling or workover equipment should be positioned or located so as to prevent spilled oil from reaching navigable waters.

2. Depending on the location, catchment basins or diversion structures may be necessary to intercept and contain spills of fuel, crude oil, or oily drilling fluids.

3. Before drilling below any casing string or during workover operations, a blowout prevention (BOP) assembly and well control system should be installed that is capable of controlling any well head pressure that is expected to be encountered while that BOP assembly is on the well. Casing and BOP installations should be in accordance with state regulatory agency requirements.

7.3.1.4.7 Oil Drilling, Production, or Workover Facilities (Offshore)

1. Definition: "An oil drilling, production or workover facility (offshore)" may include all drilling or workover equipment, wells, flowlines, gathering lines, platforms, and auxiliary non-transportation-related equipment and facilities in a single geographical oil or gas field operated by a single operator.

2. Oil drainage collection equipment should be used to prevent and control small oil spillage around pumps, glands, valves, flanges, expansion joints, hoses, drain lines, separators, treaters, tanks, and allied equipment. Drains on the facility should be controlled and directed toward a central collection sump or equivalent collection system sufficient to prevent discharges of oil into the navigable waters of the United States. Where drains and sumps are not practicable, oil contained in collection equipment should be removed as often as necessary to prevent overflow.

3. For facilities employing a sump system, sump and drains should be adequately sized and a spare pump or equivalent method should be available to remove liquid from the sump and assure that oil does not escape. A regular scheduled preventive maintenance inspection and testing program should be employed to assure reliable operation of the liquid removal system and pump startup device. Redundant automatic sump pumps and control devices may be required on some installations.

4. In areas where separators and treaters are equipped with dump valves whose predominant mode of failure is in the closed position and where pollution risk is high, the facility should be specially equipped to prevent the escape of oil. This could be accomplished by extending the flare line to a diked area if the separator is near shore, equipping it with a high-liquid-level sensor that will automatically shut in wells producing to the separator, parallel redundant dump valves, or other feasible alternatives to prevent oil discharges.

5. Atmospheric storage or surge tanks should be equipped with high-liquid-level sensing devices or other acceptable alternatives to prevent oil discharges.

6. Pressure tanks should be equipped with high- and low-pressure sensing devices to activate an alarm and/or control the flow or other acceptable alternatives to prevent oil discharges.

7. Tanks should be equipped with suitable corrosion protection.

8. A written procedure for inspecting and testing pollution prevention equipment and systems should be prepared and maintained at the facility. Such procedures should be included as part of the SPCC plan.

9. Testing and inspection of the pollution prevention equipment and systems at the facility should be conducted by the owner or operator on a scheduled periodic basis commensurate with the complexity, conditions, and circumstances of the facility or other appropriate regulations.

10. Surface and subsurface well shut-in valves and devices in use at the facility should be sufficiently described to determine method of activation or control, e.g., pressure differential, change in fluid or flow conditions, combination of pressure and flow, manual or remote control mechanisms. The owner or operator should keep detailed records for each well, while not necessarily part of the plan.

11. Before drilling below any casing string, and during workover operations, a BOP assembly and well control system should be installed that is capable of controlling any well-head pressure that is expected to be encountered while that BOP assembly is on the well. Casing and BOP installations should be in accordance with state regulatory agency requirements.

12. Extraordinary well control measures should be provided should emergency conditions, including fire, loss of control, and other abnormal conditions, occur. The degree of control system redundancy should vary with hazard exposure and probable consequences of failure. It is recommended that surface shut-in systems have redundant or "fail close" valving. Subsurface safety valves may not be needed in producing wells that will not flow but should be installed as required by applicable state regulations.

13. In order that there will be no misunderstanding of joint and separate duties and obligations to perform work in a safe and pollution-free manner, written instructions should be prepared by the owner or operator for contractors and subcontractors to follow whenever contract activities include servicing a well or systems appurtenant to a well or pressure vessel. Such instructions and procedures should be maintained at the offshore production facility. Under certain circumstances and conditions such contractor activities may require the presence at the facility of an authorized representative of the owner or operator who would intervene when necessary to prevent a spill event.

14. All manifolds (headers) should be equipped with check valves on individual flowlines.

15. If the shut-in well pressure is greater than the working pressure of the flowline and manifold valves up to and including the header valves associated with that individual flowline, the flowline should be equipped with a high-pressure sensing device and shut-in valve at the wellhead unless provided with a pressure relief system to prevent overpressuring.

16. All pipelines appurtenant to the facility should be protected from corrosion. Methods used, such as protective coatings or cathodic protection, should be discussed.

17. Submarine pipelines appurtenant to the facility should be adequately protected against environmental stresses and other activities such as fishing operations.

18. Submarine pipelines appurtenant to the facility should be in good operating condition at all times and inspected on a scheduled periodic basis for failures. Such inspections should be documented and maintained at the facility.

7.3.1.4.8 Inspections and Records. Inspections required by this part should be in accordance with written procedures developed for the facility by the owner or operator. These written procedures and a record of the inspections, signed by the appropriate supervisor or inspector, should be made part of the SPCC plan and maintained for a period of 3 years.

7.3.1.4.9 Security (Excluding Oil Production Facilities)

1. All plants handling, processing, and storing oil should be fully fenced, and entrance gates should be locked and/or guarded when the plant is not in production or is unattended.

2. The master flow and drain valves and any other valves that will permit direct outward flow of the tank's content to the surface should be securely locked in the closed position when in nonoperating or nonstandby status.

3. The starter control on all oil pumps should be locked in the "off" position or located at a site accessible only to authorized personnel when the pumps are in a nonoperating or nonstandby status.

4. The loading and unloading connections of oil pipelines should be securely capped or blank-flanged when not in service or standby service for an extended time. This security practice should also apply to pipelines that are emptied of liquid content either by draining or by inert gas pressure.

5. Facility lighting should be commensurate with the type and location of the facility. Consideration should be given to (*a*) Discovery of spills occurring during hours of darkness, both by operating personnel, if present, and by nonoperating personnel (the general public, local police, etc.) and (*b*) prevention of spills occurring through acts of vandalism.

7.3.1.4.10   Personnel, Training, and Spill Prevention Procedures
1. Owners or operators are responsible for properly instructing their personnel in the operation and maintenance of equipment to prevent the discharges of oil and applicable pollution control laws, rules, and regulations.

2. Each applicable facility should have a designated person who is accountable for oil spill prevention and who reports to line management.

3. Owners or operators should schedule and conduct spill prevention briefings for their operating personnel at intervals frequent enough to assure adequate understanding of the SPCC plan for that facility. Such briefings should highlight and describe known spill events or failures, malfunctioning components, and recently developed precautionary measures.

7.3.2   SPCC Plan Review and Approval Requirements

Owners or operators of facilities subject to the OPA 90 and CWA Section 311 requirements must keep a certified copy of the SPCC plan available at the facility for EPA on-site review, if the facility is attended at least 8 hours a day. If the facility is not attended, then the SPCC plan must be kept at the nearest company office.

The owner or operator must review the SPCC plan at least once every 3 years. Reviews must be documented and certified by a professional engineer.

Following the triennial review, the OPA 90 requires the owner or operator to amend the SPCC plan within 6 months to incorporate more effective controls and prevention technology if the technology will significantly reduce the likelihood of a release and if the technology has been field proven at the time of the review. The owner or operator also must amend the SPCC plan whenever there is a change in the facility design, construction, operation, or maintenance that materially affects the facility's potential for discharge into navigable waters, states, or adjoining shorelines. Such amendments must be fully implemented no later than 6 months after the change occurs. In addition, an EPA regional administrator may require amendments to the SPCC plan following a single discharge at a facility in excess of 1000 gallons, or following two discharges within any 12-month period. A registered professional engineer must certify amendments to an SPCC plan.

7.4 FACILITY RESPONSE PLANS

Under the OPA 90, only those facilities that pose substantial harm to the environment and meet the substantial harm criteria under the act are required to prepare a facility response plan in the event of a worst-case discharge. Section 4202(a) of the OPA amends CWA section 311(j) to require regulations for owners or operators of facilities to prepare and submit "a plan for responding, to the maximum extent practicable, to a worst case discharge, and to a substantial threat of such a discharge, of oil or a hazardous substance." This requirement applies to all offshore facilities and any onshore facility that, "because of its location, could reasonably be expected to cause substantial harm to the environment by discharging into or on the navigable waters, adjoining shorelines, or the exclusive economic zone" ("substantial harm facilities"). As stated in the Feb. 17, 1993, proposed rule (58 FR 8824), FRPs addresses only plans for responding to discharges of oil. Unlike the spill prevention and counter control measures found in 40 CFR Part 112, a FRP does not address plans for preventing the discharges of oil.

Under CWA and the Comprehensive Environmental Response, Compensation, and Liability Act (CERCLA), the United States has developed a National Oil and Hazardous Substances Pollution Contingency Plan (NCP) (40 CFR Part 300) and has established area committees to develop area contingency plans (ACPs) as elements of a comprehensive oil and hazardous substance spill response system. As amended by the OPA, CWA section 311(j)(5)(C) sets forth certain minimum requirements for facility response plans. The plans must:

- Be consistent with the requirements of the NCP and ACPs.
- Identify the qualified individual having full authority to implement removal actions, and require immediate communications between that individual and the appropriate federal official and the persons providing removal personnel and equipment.
- Identify and ensure by contract or other approved means the availability of private personnel and equipment necessary to remove, to the maximum extent practicable, a worst-case discharge (including a discharge resulting from fire or explosion), and to mitigate or prevent a substantial threat of such a discharge.
- Describe the training, equipment testing, periodic unannounced drills, and response actions of persons at the facility, to be carried out under the plan to

ensure the safety of the facility and to mitigate or prevent a discharge or the substantial threat of a discharge; and be updated periodically.

Under section 311(j)(5)(D), additional review and approval provisions apply to response plans prepared for offshore facilities and for onshore facilities that, because of their location, "could reasonably be expected to cause significant and substantial harm to the environment by discharging into or on the navigable waters or adjoining shorelines or the exclusive economic zone" ("significant and substantial harm facilities"). Under authority delegated in E.O. 12777, EPA is responsible for the following activities for each of these response plans at non-transportation-related onshore facilities:

- Promptly reviewing the response plan
- Requiring amendments to any plan that does not meet the section 311(j)(5) requirements
- Approving any plan that meets these requirements
- Reviewing each plan periodically thereafter

Section 311(j)(5) also requires that, in a facility response plan, an owner or operator identify and ensure by contract or other means approved by the President the availability of private personnel and equipment sufficient to remove, to the maximum extent practicable, a worst-case discharge and to mitigate or prevent substantial threat of such a discharge.

7.4.1 Criteria for Determining "Substantial Harm"

A facility could, because of its location, reasonably be expected to cause substantial harm to the environment by discharging oil into or on the navigable waters or adjoining shorelines, if it meets any of the following criteria applied in accordance with the flowchart contained in Appendix L to this book.

1. The facility transfers oil over water to or from vessels and has a total oil storage capacity greater than or equal to 42,000 gallons.
2. The facility's total oil storage capacity is greater than or equal to 1 million gallons, and one of the following is true:
   a. The facility does not have secondary containment for each aboveground storage area sufficiently large to contain the capacity of the largest aboveground oil storage tank within each storage area plus sufficient freeboard to allow for precipitation.
   b. The facility is located at a distance (as calculated using the appropriate formula in Appendix C to 40 CFR Part 112 or a comparable formula) such that a discharge from the facility could cause injury to fish and wildlife and sensitive environments. For further description of fish and wildlife and sensi-

tive environments, see Appendixes I, II, and III of the "Guidance for Facility and Vessel Response Plans: Fish and Wildlife and Sensitive Environments" (see Appendix E to 40 CFR Part 112, section 10, for availability) and the applicable Area Contingency Plan prepared pursuant to section 311(j)(4) of the Clean Water Act.

   c. The facility is located at a distance (as calculated using the appropriate formula in Appendix C to 40 CFR Part 112 or a comparable formula) such that a discharge from the facility would shut down a public drinking water intake.

   d. The facility has had a reportable oil spill in an amount greater than or equal to 10,000 gallons within the last 5 years.

To determine whether a facility could, because of its location, reasonably be expected to cause substantial harm to the environment by discharging oil into or on the navigable waters or adjoining shorelines, the regional administrator shall consider the following:

1. Type of transfer operation

2. Oil storage capacity

3. Lack of secondary containment

4. Proximity to fish and wildlife and sensitive environments and other areas determined by the regional administrator to possess ecological value

5. Proximity to drinking water intakes

6. Spill history

7. Other site-specific characteristics and environmental factors that the regional administrator determines to be relevant to protecting the environment from harm by discharges of oil into or on navigable waters or adjoining shorelines

7.4.2 Model Facility-Specific Response Plan

Owners or operators of facilities regulated under this part which pose a threat of substantial harm to the environment by discharging oil into or on navigable waters or adjoining shorelines are required to prepare and submit facility-specific response plans to EPA in accordance with the provisions of Appendix F to 40 CFR 112. This appendix further describes the required elements in 40 CFR 112.20(h).

Response plans must be sent to the appropriate EPA regional office. Figure 7.1 shows each EPA regional office and the address where owners or operators must submit their response plans. Those facilities deemed by the regional administrator (RA) to pose a threat of significant and substantial harm to the environment will have their plans reviewed and approved by EPA. In certain cases, information required in the model response plan is similar to information currently maintained in the facility's SPCC plan as required by 40 CFR 112.3. In these cases,

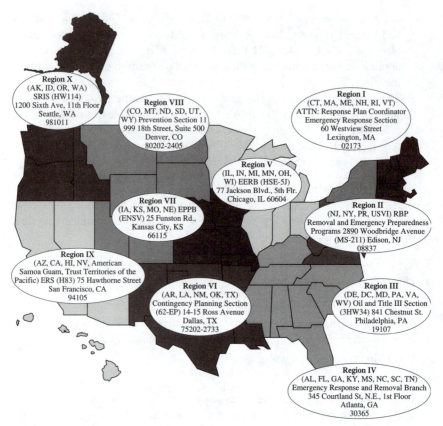

FIGURE 7.1   EPA regional offices for FRP submittal.

owners or operators may reproduce the information and include a photocopy in the response plan.

A complex may develop a single response plan with a set of core elements for all regulating agencies and separate sections for the non-transportation-related and transportation-related components, as described in 40 CFR 112.20(h). Owners or operators of large facilities that handle, store, or transport oil at more than one geographically distinct location (e.g., oil storage areas at opposite ends of a single continuous parcel of property) need to, as appropriate, develop separate sections of the response plan for each storage area.

7.4.2.1   *Emergency Response Action Plan.*

Several sections of the response plan need to be co-located for easy access by response personnel during an actual emergency or oil spill. This collection of sections will be called the Emergency Response Action Plan. The agency intends that the action plan contain only as much information as is necessary to combat the spill and be arranged so response actions are not delayed. The action plan must be arranged in a number of ways. For example, the sections of the Emergency Response Action Plan may be photocopies or condensed versions of the forms included in the associated sections of the response plan. Each Emergency Response Action Plan section may be tabbed for quick reference. The action plan shall be maintained in the front of the same binder that contains the complete response plan or it shall be contained in a separate binder. In the latter case, both binders shall be kept together so that the qualified individual and appropriate spill response personnel can access the entire plan. The Emergency Response Action Plan shall be made up of the following sections:

1. Qualified individual information (Section 7.4.2.2), partial

2. Emergency notification phone list (Section 7.4.2.3.1), partial

3. Spill response notification form (Section 7.4.2.3.1), partial

4. Response equipment list and location (Section 7.4.2.3.2), complete

5. Response equipment testing and deployment (Section 7.4.2.3.3), complete

6. Facility response team (Section 7.4.2.3.4), partial

7. Evacuation plan (Section 7.4.2.3.5), condensed

8. Immediate actions (Section 7.4.2.7.1), complete

9. Facility diagram (Section 7.4.2.9), complete

7.4.2.2 Facility Information. The facility information form is designed to provide an overview of the site and a description of past activities at the facility. Much of the following information may be obtained from the facility's existing SPCC plan.

- Facility name and location: Enter facility name and street address. Enter the address of corporate headquarters only if corporate headquarters is physically located at the facility. Include city, county, state, zip code, and phone number.

- Latitude and longitude: Enter the latitude and longitude of the facility. Include degrees, minutes, and seconds of the main entrance of the facility.

- Wellhead protection area: Indicate if the facility is located in or drains into a wellhead protection area as defined by the Safe Drinking Water Act of 1986 (SDWA). The response plan requirements in the Wellhead Protection Program are outlined by the state or territory in which the facility resides.

- Owner-operator: Write the name of the company or person operating the facility and the name of the person or company that owns the facility, if the two are different. List the address of the owner, if the two are different.

- Qualified individual: Write the name of the qualified individual for the entire facility. If more than one person is listed, each individual indicated in this section needs to have full authority to implement the facility response plan. For each individual, list name, position, home and work addresses (street addresses, not post office boxes), emergency phone number, and specific response training experience.

- Date of oil storage startup: Enter the year in which the storage oil commenced.

- Current operation: Briefly describe the facility's operations and include standard industrial classification (SIC) code.

- Dates and type of substantial expansion: Include information on expansions that have occurred at the facility. Examples of such expansions include but are not limited to throughput expansion, addition of a product line, change of a product line, and installation of additional oil storage capacity. The data provided must include all facility historical information and detail the expansion of the facility. An example of substantial expansion is any material alteration of the facility which causes the owner or operator of the facility to reevaluate and increase the response equipment necessary to adequately respond to a worst-case discharge from the facility.

- Date of last update.

Table 7.1 shows the information contained in the facility information form.

7.4.2.3 Emergency Response Information. The information provided in this section will describe what will be needed in an actual emergency involving the discharge of oil or a combination of hazardous substances and oil discharge. The Emergency Response Information section of the plan must include the following components:

1. The information provided in the Emergency Notification Phone List in Section 7.4.2.3.1 identifies and prioritizes the names and phone numbers of the organizations and personnel that need to be notified immediately in the event of an emergency. This section shall include all the appropriate phone numbers for the facility. These numbers must be verified each time the plan is updated. The contact list must be accessible to all facility employees to ensure that, in case of a discharge, any employee on-site could immediately notify the appropriate parties.

2. The Spill Response Notification Form (Appendix K) creates a checklist of information that shall be provided to the National Response Center (NRC) and other response personnel. All information on this checklist must be known at the time of notification or be in the process of being collected. This notification form is based on a similar form used by the NRC. (*Note:* Do not delay spill notification to collect the information on the list.)

TABLE 7.1   Facility Information Form

Facility name:

Location (street address):

City: State: Zip:

County: Phone number:

Latitude: _____ Degrees _____ Minutes _____ Seconds

Longitude: _____ Degrees _____ Minutes _____ Seconds

Wellhead protection area:

Owner:

Owner location (street address) (if different from facility address):

City: State: Zip:

County:

Phone number:

Operator (if not owner):

Qualified individual(s) (attach additional sheets if more than one):

Name:

Position:

Work address:

Home address:

Emergency phone number:

Date of oil storage startup:

Current operation:

Date(s) and type(s) of substantial expansion(s):

(Attach additional sheets if necessary)

---

3. Section 7.4.2.3.2 provides a description of the facility's list of emergency response equipment and location of the response equipment. When appropriate, the amount of oil that emergency response equipment can handle and any limitations (e.g., launching sites) must be described.

4. Section 7.4.2.3.3 provides information regarding response equipment tests and deployment drills. Response equipment deployment exercises shall be conducted to ensure that response equipment is operational and the personnel who would operate the equipment in a spill response are capable of deploying and operating it. Only a representative sample of each type of response equipment needs to be deployed and operated, as long as the remainder is properly maintained. If appropriate, testing of response equipment may be conducted while it is being deployed. Facilities without facility-owned response equipment must ensure that the oil spill removal organization that is identified in the response plan to provide this response equipment certifies that the deployment exercises have been met. Refer to the National Preparedness for Response Exercise Program (PREP) Guidelines (see Appendix E to 40 CFR Part 112, Section 10, for availability), which satisfy Oil Pollution Act (OPA) response exercise requirements.

5. Section 7.4.2.3.4 lists the facility response personnel, including those employed by the facility and those under contract to the facility for response activities, the amount of time needed for personnel to respond, their responsibility in the case of an emergency, and their level of response training. Three different forms are included in this section. The emergency response personnel list shall be composed of all personnel employed by the facility whose duties involve responding to emergencies, including oil spills, even when they are not physically present at the site. An example of this type of person would be the building engineer-in-charge or plant fire chief. The second form is a list of the emergency response contractors (both primary and secondary) retained by the facility. Any changes in contractor status must be reflected in updates to the response plan. Evidence of contracts with response contractors shall be included in this section so that the availability of resources can be verified. The last form is the facility response team list, which shall be composed of both emergency response personnel (referenced by job title and position) and emergency response contractors, included in one of the two lists described above, that will respond immediately upon discovery of an oil spill or other emergency (i.e., the first people to respond). These are to be persons normally on the facility premises or primary response contractors. Examples of these personnel would be the facility HAZMAT spill team 1, facility fire engine company 1, production supervisor, or transfer supervisor. Company personnel must be able to respond immediately and adequately if contractor support is not available.

6. Section 7.4.2.3.5 lists factors that must, as appropriate, be considered when preparing an evacuation plan. These include:

- Location of stored materials
- Hazards of the material
- Wind direction and speed
- Evacuation routes
- Shelter locations

See Section 7.4.2.3.5 for a more detailed list.

7. Section 7.4.2.3.6 references the responsibilities of the qualified individual for the facility in the event of an emergency.

The information provided in the emergency response section will aid in the assessment of the facility's ability to respond to a worst-case discharge and will identify additional assistance that may be needed. In addition, the facility owner or operator may want to produce a wallet-size card containing a checklist of the immediate response and notification steps to be taken in the event of an oil discharge.

7.4.2.3.1 Notification. Notification must be made as noted in Tables 7.2 and 7.3.

TABLE 7.2 Notification Forms and Information

Date of last update:
Report's name:
Date:
Facility name:
Owner name:
Facility identification number:
Date and time of each NRC notification:

TABLE 7.3 Emergency Notification List

---

1. National response center (NRC)   1-800-424-8802
2. U.S. Coast Guard
3. Regional response center
4. Qualified individual:
     Night phone
5. Company response team
     Night phone
6. Federal on-scene coordinator (OSC) and/or regional response center (RRC):
     Night phone
     Pager
7. Local response team (fire department/cooperatives):
8. Fire marshal:
     Night phone
9. State emergency response commission
     Night phone
10. State police:
11. Local emergency planning committee:
12. Local water supply system:
     Night phone
13. Local television/radio station for evacuation notification:
14. Hospitals:

Date of last update:

7.4.2.3.2   Response Equipment List

TABLE 7.4   Facility Response Equipment List

---

1. Skimmers/pumps—operational status:

Impact:

Number of injuries: Number of deaths:

Type, model, and year

Number:

Capacity: gal/min:

Daily effective rate:

Storage location(s):

Date fuel last changed:

2. Boom—operational status:

Type:

Model:

Year:

Number:

Size (length), ft:

Containment area, ft^2:

Storage Location:

3. Chemicals stored (dispersants listed on EPA's NCP product schedule)

Type	Amount	Date purchased	Treatment capacity	Storage location

Were appropriate procedures used to receive approval for use of dispersants in accordance with the NCP (40 CFR 300.910) and the area contingency plan (ACP), where applicable? (Y/N).

Date authorized:

Name and state of on-scene coordinator (OSC) authorizing use:

4. Date of last update

Dispersant dispensing equipment—operational status:			
Type and year	Capacity	Storage location	Response time, min

TABLE 7.4   Facility Response Equipment List (*Continued*)

Sorbents—operational status:

Type and year purchased:

Amount:

Absorption capacity, gal:

Storage location(s):

Hand tools		
Type and year	Quantity	Storage location

Communication equipment (include operating frequency and channel and/or cellular phone numbers)—operational status		
Type and year	Quantity	Storage location

Fire fighting and personnel protective equipment—operational status:		
Type and year	Quantity	Storage location

Other (e.g., heavy equipment, boats and motors)—operational status:		
Type and year	Quantity	Storage location

7.4.2.3.3   *Response Equipment Testing and Deployment*

TABLE 7.5   Response Equipment Testing and Deployment Drill Log

Date of last update:

Last inspection or response equipment test:

Date:

Inspection frequency:

Last deployment drill date:

Deployment frequency:

Oil spill removal organization certification (if applicable):

7.4.2.3.4 Personnel

TABLE 7.6 Emergency Response Personnel Company Personnel

Name	Phone*	Response time	Responsibility during response action	Response training (type and date)

Date of last update:

Emergency response contractors			
Contractor	Phone	Response time	Contractor responsibility†
1.			
2.			
3.			
4.			
5.			
6.			
7.			
8.			
9.			
10.			
11.			
12.			
13.			

Date of last update:

TABLE 7.6 Emergency Response Personnel Company Personnel (*Continued*)

	Facility response team	
	Response time, min	Phone or pager no. (day/evening)

Team member
1. Qualified individual:

2.

3.

4.

5.

6.

7.

8.

9.

10.

11.

12.

13.

14.

15.

16.

17.

Date of last update:

*Phone number to be used when person is not on-site.
†Include evidence of contracts and agreements with response contractors to ensure the availability of personnel and response equipment.
Note: If the facility uses contracted help in an emergency response situation, the owner or operator must provide the contractors' names and review the contractors' capacities to provide adequate personnel and response equipment.

7.4.2.3.5 Evacuation Plans

7.4.2.3.5.1 FACILITY-WIDE EVACUATION PLAN: Based on the analysis of the facility, as discussed elsewhere in the plan, a facility-wide evacuation plan must be developed. In addition, plans to evacuate parts of the facility that are at a high risk of exposure in the event of a spill or other release must be developed. Evacuation routes must be shown on a diagram of the facility. When developing evacuation plans, consideration must be given to the following factors, as appropriate:

1. Location of stored materials
2. Hazard imposed by spilled material
3. Spill flow direction
4. Prevailing wind direction and speed
5. Water currents, tides, or wave conditions (if applicable)
6. Arrival route of emergency response personnel and response equipment
7. Evacuation routes
8. Alternative routes of evacuation
9. Transportation of injured personnel to nearest emergency medical facility
10. Location of alarm and notification systems
11. The need for a centralized check-in area for evacuation validation (roll call)
12. Selection of a mitigation command center
13. Location of shelter at the facility as an alternative to evacuation

7.4.2.3.5.2 EVACUATION PLAN DEVELOPMENT RESOURCES: One resource that may be helpful to owners or operators in preparing this section of the response plan is the *Handbook of Chemical Hazard Analysis Procedures* by the Federal Emergency Management Agency (FEMA), Department of Transportation (DOT), and EPA. It is available from FEMA, Publication Office, 500 C Street, SW, Washington, D.C. 20472, (202) 646-3484.

7.4.2.3.5.3 EXISTING COMMUNITY EVACUATION PLANS: As specified in 40 CFR 112.20(h)(1)(vi), the facility owner or operator must reference existing community evacuation plans, as appropriate.

7.4.2.3.6 Qualified Individual's Duties.

The duties of the designated qualified individual are specified in 40 CFR 112.20(h)(3)(ix). The qualified individual's duties must be described and be consistent with the minimum requirements in 40 CFR 112.20(h)(3)(ix). In addition, the qualified individual must be identified with the facility information in section 7.4.2.2 of the response plan.

7.4.2.4 Hazard Evaluation.

This section requires the facility owner or operator to examine the facility's operations closely and to predict where discharges could occur. Hazard evaluation is a widely used industry practice that allows facility owners or operators to develop a complete understanding of potential hazards and the response actions necessary to address these hazards. The *Handbook of Chemical Hazard Analysis Procedures,* prepared by the EPA, DOT, and FEMA, and the *Hazardous Materials Emergency Planning Guide* (NRT-1), prepared by the National Response Team, are good references for conducting a hazard analysis. Hazard identification and evaluation will assist facility owners or operators in planning for potential discharges, thereby reducing the severity of discharge impacts that may occur in the future. The evaluation also may help the operator identify and correct potential sources of discharges. In addition, special hazards to workers and emergency response personnel's health and safety shall be evaluated, as well as the facility's oil spill history.

7.4.2.4.1 Hazard Identification. The tank and surface impoundment (SI) forms, or their equivalent, that are part of this section must be completed according to the directions below.

"Surface impoundment" means a facility or part of a facility which is a natural topographic depression, manmade excavation, or diked area formed primarily of earthen materials (although it may be lined with manmade materials), which is designed to hold an accumulation of liquid wastes or wastes containing free liquids, and which is not an injection well or a seepage facility. Similar worksheets, or their equivalent, must be developed for any other type of storage containers.

1. List each tank at the facility with a separate and distinct identifier. Begin aboveground tank identifiers with an "A" and belowground tank identifiers with a "B," or submit multiple sheets with the aboveground tanks and belowground tanks on separate sheets.
2. Use gallons for the maximum capacity of a tank, and use square feet for the area.
3. Using the appropriate identifiers and the following instructions, fill in the appropriate forms:
   a. Tank or SI number—Using the aforementioned identifiers (A or B) or multiple reporting sheets, identify each tank or SI at the facility that stores oil or hazardous materials.
   b. Substance stored—For each tank or SI identified, record the material that is stored therein. If the tank or SI is used to store more than one material, list all of the stored materials.
   c. Quantity stored—For each material stored in each tank or SI, report the average volume of material stored on any given day.
   d. Tank type or surface area and year—For each tank, report the type of tank (e.g., floating top) and the year the tank was originally installed. If the tank has been refabricated, the year that the latest refabrication was completed must be recorded in parentheses next to the year installed. For each SI, record the surface area, the impoundment, and the year it went into service.
   e. Maximum capacity—Record the operational maximum capacity for each tank and SI. If the maximum capacity varies with the season, record the upper and lower limits.

f. Failure and cause—Record the cause and date of any tank or SI failure which has resulted in a loss of tank or SI contents.
4. Using the numbers from the tank and SI forms, label a schematic drawing of the facility. This drawing must be identical to any schematic drawings included in the SPCC plan.
5. Using knowledge of the facility and its operations, describe the following in writing:
   a. The loading and unloading of transportation vehicles that risk the discharge of oil or release of hazardous substances during transport processes. These operations may include loading and unloading of trucks, railroad cars, or vessels. Estimate the volume of material involved in transfer operations if the exact volume cannot be determined.
   b. Day-to-day operations that may present a risk of discharging oil or releasing a hazardous substance. These activities include scheduled venting, piping repair or replacement, valve maintenance, transfer of tank contents from one tank to another, etc. (not including transportation-related activities). Estimate the volume of material involved in these operations if the exact volume cannot be determined.
   c. The secondary containment volume associated with each tank and/or transfer point at the facility. The numbering scheme developed on the tables, or an equivalent system, must be used to identify each containment area. Capacities must be listed for each individual unit (tanks, slumps, drainage traps, and ponds), as well as the facility total.
   d. Normal daily throughput for the facility and any effect on potential discharge volumes that a negative or positive change in that throughput may cause.

7.4.2.4.2 Vulnerability Analysis. The vulnerability analysis must address the potential effects (i.e., to human health, property, or the environment) of an oil spill. Attachment C-III to Appendix C of 40 CFR Part 112 (Appendix L of this book) provides a method that owners or operators shall use to determine appropriate distances from the facility to fish and wildlife and sensitive environments. Owners or operators can use a comparable formula that is considered acceptable by the EPA. If a comparable formula is used, documentation of the reliability and analytical soundness of the formula must be attached to the response plan cover sheet. This analysis must be prepared for each facility and, as appropriate, must discuss the vulnerability of:

1. Water intakes (drinking, cooling, or other)
2. Schools
3. Medical facilities
4. Residential areas
5. Businesses

TABLE 7.7   Hazard Identification Form

Tank No.	Substance stored	Quantity, gal	Tank type and year	Maximum capacity, gal	Failure and cause
Oil and hazardous substance hazard identification tanks*					

Date of last update:

	Hazard identification surface impoundments (SIs)				
	Substance stored	Quantity, gal	Surface area and year	Maximum capacity, gal	Failure and cause
SI No.					

Date of last update:

*Tank = any container that stores oil.
Attach as many sheets as necessary.

6. Wetlands or other sensitive environments
7. Fish and wildlife
8. Lakes and streams
9. Endangered flora and fauna
10. Recreational areas
11. Transportation routes (air, land, and water)
12. Utilities
13. Other areas of economic importance (e.g., beaches, marinas) including terrestrially sensitive environments, aquatic environments, and unique habitats

7.4.2.4.3   Analysis of the Potential for an Oil Spill.   Each owner or operator must analyze the probability of a spill occurring at the facility. This analysis incorporates factors such as oil spill history, horizontal range of a potential spill, vulnerability to natural disaster, and, as appropriate, other factors such as tank age. This analysis will provide information for developing discharge scenarios for a worst-case discharge and small and medium discharges and aid in the development of techniques to reduce the size and frequency of spills. The owner or operator may need to research the age of the tanks and the oil spill history at the facility.

7.4.2.4.4   Facility Reportable Oil Spill History.   Briefly describe the facility's reportable oil spill history for the entire life of the facility to the extent that such information is reasonably identifiable, including:

1. Date of discharge(s)

2. List of discharge causes

3. Material(s) discharged

4. Amount discharged in gallons

5. Amount of discharge that reached navigable waters, if applicable

6. Effectiveness and capacity of secondary containment

7. Cleanup actions taken

8. Steps taken to reduce possibility of recurrence

9. Total oil storage capacity of the tank(s) or impoundment(s) from which the material discharged

10. Enforcement actions

11. Effectiveness of monitoring equipment

12. Description(s) of how each oil spill was detected

The information solicited in this section may be similar to requirements in 40 CFR 112.4(a). Any duplicate information required by 40 CFR 112.4(a) may be photocopied and inserted.

7.4.2.5   Discharge Scenarios.   In this section, the owner or operator is required to provide a description of the facility's worst-case discharge, as well as a small and medium spill, as appropriate. A multilevel planning approach was chosen by EPA because the response actions to a spill (i.e., necessary response equipment, products, and personnel) are dependent on the magnitude of the spill. Planning for lesser discharges is necessary because the nature of the response may be qualitatively different depending on the quantity of the discharge. The facility owner or operator also needs to discuss the potential direction of the spill pathway.

7.4.2.5.1   Small and Medium Discharges

7.4.2.5.1.1   MULTI-LEVEL PLANNING REQUIREMENTS: To address multilevel planning requirements, the owner or operator must consider types of facility-specific spill scenarios that may contribute to a small or medium spill. The scenarios need to account for all the operations that take place at the facility, including but not limited to:

1. Loading and unloading of surface transportation

2. Facility maintenance

3. Facility piping
4. Pumping stations and sumps
5. Oil storage tanks
6. Vehicle refueling
7. Age and condition of facility and components

7.4.2.5.1.2 ADDITIONAL FACTORS: The scenarios must also consider factors that affect the response efforts required by the facility. These include but are not limited to:

1. Size of the spill
2. Proximity to downgradient wells, waterways, and drinking water intakes
3. Proximity to fish and wildlife and sensitive environments
4. Likelihood that the discharge will travel offsite (i.e., topography, drainage)
5. Location of the material spilled (i.e., on a concrete pad or directly on the soil)
6. Material discharged
7. Weather or aquatic conditions (i.e., river flow)
8. Available remediation equipment
9. Probability of a chain reaction of failures
10. Direction of spill pathway

7.4.2.5.2 Worst-Case Discharge

7.4.2.5.2.1 WORST-CASE DISCHARGE VOLUME: In this section, the owner or operator must identify the worst-case discharge volume at the facility. Worksheets for production and nonproduction facility owners or operators to use when calculating worst-case discharge are presented in Appendix D to 40 CFR Part 112. When planning for the worst-case discharge response, all of the aforementioned factors listed in the small- and medium-discharge section of the response plan must be addressed.

7.4.2.5.2.2 "PERMANENTLY MANIFOLDED STORAGE TANKS": For onshore storage facilities and production facilities, permanently manifolded oil storage tanks are defined as tanks that are designed, installed, and/or operated in such a manner that the multiple tanks function as one storage unit (i.e., multiple tank volumes are equalized). In this section of the response plan, owners or operators must provide evidence that oil storage tanks with common piping or piping systems are not operated as one unit. If such evidence is provided and is acceptable to the EPA, the worst-case discharge volume will be based on the combined oil storage capacity of all manifold tanks or the oil storage capacity of the largest single oil storage tank within the secondary containment area, whichever is greater. For permanently manifolded oil storage tanks that function as one storage unit, the worst-case discharge will be based on the combined oil storage capacity of all manifolded tanks or the oil storage capacity of the largest single tank within a secondary containment area, whichever is greater. For purposes of the worst-case discharge calculation, permanently manifolded oil storage tanks that are separated by internal divisions for each tank are considered to be single tanks and individual manifolded tank volumes are not combined.

7.4.2.6 Discharge Detection Systems. In this section, the facility owner or operator needs to provide a detailed description of the procedures and equipment used to detect discharges. A section on spill detection by personnel and a discussion of automated spill detection, if applicable, shall be included for both regular operations and after-hours operations. In addition, the facility owner or operator shall discuss how the reliability of any automated system will be checked and how frequently the system will be inspected.

7.4.2.6.1 Discharge Detection by Personnel. In this section, facility owners or operators must describe the procedures and personnel that will detect any spill or uncontrolled discharge of oil or release of a hazardous substance. A thorough discussion of facility inspections must be included. In addition, a description of initial response actions shall be addressed. This section shall reference section 7.4.2.3.1 of the response plan for emergency response information.

7.4.2.6.2 Automated Discharge Detection. In this section, facility owners or operators must describe any automated spill detection equipment that the facility has in place. This section must include a discussion of overfill alarms, secondary containment sensors, etc. A discussion of the plans to verify an automated alarm and the actions to be taken once verified must also be included.

7.4.2.7 Plan Implementation. In this section, facility owners or operators must explain in detail how to implement the facility's emergency response plan by describing response actions to be carried out under the plan to ensure the safety of the facility and to mitigate or prevent discharges described in section 7.4.2.5 of the response plan. This section needs to include the identification of response resources for small, medium, and worst-case spills; disposal plans; and containment and drainage planning. A list of those personnel who would be involved in the cleanup must be identified. Procedures that the facility will use, where appropriate or necessary, to update their plan after an oil spill event and the time frame to update the plan must be described.

7.4.2.7.1 Response Resources for Small, Medium, and Worst-Case Spills
7.4.2.7.1.1 IDENTIFICATION AND DESCRIPTION OF RESPONSE PLAN IMPLEMENTATION: Once the spill scenarios have been identified in section 7.4.2.5 of the response plan, the facility owner or operator needs to identify and describe implementation of the response actions. The facility owner or operator needs to demonstrate accessibility to the proper response personnel and equipment to effectively respond to all of the identified spill scenarios. The determination and demonstration of adequate response capability are presented in Appendix E to 40 CFR Part 112. In addition, steps to expedite the cleanup of oil spills must be discussed. At a minimum, the following items must be addressed:

1. Emergency plans for spill response
2. Additional response training
3. Additional contracted help
4. Access to additional response equipment and experts
5. Ability to implement the plan, including response training and practice drills

7.4.2.7.2 Disposal Plans

7.4.2.7.2.1 DESCRIPTION OF MATERIALS' DISPOSAL: Facility owners or operators must describe how and where the facility intends to recover, reuse, decontaminate, or dispose of materials after a discharge has taken place. The appropriate permits required to transport or dispose of recovered materials according to local, state, and federal requirements must be addressed. Materials that must be accounted for in the disposal plan, as appropriate, include:

1. Recovered product
2. Contaminated soil
3. Contaminated equipment and materials, including drums, tank parts, valves, and shovels
4. Personnel protective equipment
5. Decontamination solutions
6. Adsorbents
7. Spent chemicals

7.4.2.7.2.2 APPLICABILITY TO OTHER PLANS: These plans must be prepared in accordance with federal [e.g., the Resource Conservation and Recovery Act (RCRA)], state, and local regulations, where applicable. A copy of the disposal plans from the facility's SPCC plan may be inserted with this section, including any diagrams in those plans.

7.4.2.7.3 Containment and Drainage Planning. A proper plan to contain and control a spill through drainage may limit the threat of harm to human health

TABLE 7.8 Oil Spill Response—Immediate Actions

1. Stop the product flow	Act quickly to secure pumps, close valves, etc.
2. Warn personnel	Enforce safety and security measures
3. Shut off ignition sources	Motors, electric circuits, open flames, etc.
4. Initiate containment	Around the tank and/or in the water with oil boom
5. Notify NRC	1-800-424-8802
6. Notify OSC	
7. Notify as appropriate	

TABLE 7.9 Materials' Disposal Chart

Material	Disposal facility	Location	RCRA permit and manifest no.

and the environment. This section must describe how to contain and control a spill through drainage, including:

1. The available volume of containment (use the information presented in section 7.4.2.4.1 of the response plan)
2. The route of drainage from oil storage and transfer areas
3. The construction materials used in drainage troughs
4. The type and number of valves and separators used in the drainage system
5. Sump pump capacities
6. The containment capacity of weirs and booms that might be used and their location (see section 7.4.2.3.2)
7. Other cleanup materials

In addition, facility owners or operators must meet the inspection and monitoring requirements for drainage contained in 40 CFR 112.7(e). A copy of the containment and drainage plans that are required in 40 CFR 112.7(e) may be inserted in this section, including any diagrams in those plans.

(*Note:* The general permit for stormwater drainage may contain additional requirements.)

7.4.2.8 Self-Inspection, Drills and Exercises, and Response Training. The owner or operator must develop programs for facility response training and for drills and exercises according to the requirements of 40 CFR 112.21. Logs must be kept for facility drills and exercises, personnel response training, and spill prevention meetings. Much of the record-keeping information required by this section is also contained in the SPCC plan required by 40 CFR 112.3. These logs may be included in the facility response plan or kept as an annex to the facility response plan.

7.4.2.8.1 Facility Self-Inspection. Pursuant to 40 CFR 112.7(e)(8), each facility must include the written procedures and records of inspections in the SPCC plan. The inspection must include the tanks, secondary containment, and response equipment at the facility. Records of the inspections of tanks and secondary containment required by 40 CFR 112.7(e) must be cross-referenced in the response plan. The inspection of response equipment is a new requirement in this plan. Facility self-inspection requires two steps: (1) a checklist of things to inspect and (2) a method of recording the actual inspection and its findings. The date of each inspection shall be noted. These records are required to be maintained for 5 years.

7.4.2.8.1.1 TANK INSPECTION: The tank inspection checklist presented below has been included as guidance during inspections and monitoring. Similar requirements exist in 40 CFR 112.7(e). Duplicate information from the SPCC plan may be photocopied and inserted in this section. The inspection checklist consists of the items listed in Table 7.10.

7.4.2.8.1.2 RESPONSE EQUIPMENT INSPECTION: Using the emergency response equipment list provided in section 7.4.2.3.2 of the response plan, describe each type of response equipment, checking for the following:

Response equipment checklist:

1. Inventory (item and quantity)

2. Storage location

3. Accessibility (time to access and respond)

4. Operational status and condition

5. Actual use and testing (last test date and frequency of testing)

6. Shelf life (present age, expected replacement date)

Note any discrepancies between this list and the available response equipment.

7.4.2.8.1.3 SECONDARY CONTAINMENT INSPECTION: Inspect the secondary containment (as described in sections 7.4.2.4.1 and 7.4.2.7.2 of the response plan), checking the following (Table 7.12).

7.4.2.8.2 Facility Drills and Exercises

1. CWA section 311(j)(5), as amended by the OPA, requires the response plan to contain a description of facility drills and exercises. According to 40 CFR 112.21(c), the facility owner or operator must develop a program of facility response drills and exercises, including evaluation procedures. Following the PREP guidelines (see Appendix E to 40 CFR Part 112, section 10, for availability) would satisfy a facility's requirements for drills and exercises under this part. Alternately, under 40 CFR 112.21(c), a facility owner or operator may develop a program that is not based on the PREP guidelines. Such a program is subject to approval by the U.S. EPA regional administrator based on the description of the program provided in the response plan.

2. The PREP guidelines specify that the facility conduct internal and external drills and exercises. The internal exercises include qualified individual notification drills, spill management team tabletop exercises, equipment deployment

TABLE 7.10 Tank Inspection Checklist

Action item	(Yes/No)

1. Check tanks for leaks, specifically looking for:

   Drip marks

   Discoloration of tanks

   Puddles containing spilled or leaked material

   Corrosion

   Cracks

   Localized dead vegetation

2. Check foundation for:

   Cracks

   Discoloration

   Puddles containing spilled or leaked material

   Settling

   Gaps between tank and foundation

   Damage caused by vegetation roots

3. Check piping for:

   Droplets of stored material

   Discoloration

   Corrosion

   Bowing of pipe between supports

   Evidence of stored material seepage from valves or seals

   Localized dead vegetation

Tank/surface impoundment inspection log			
Inspector	Tank/SI ID no.	Date	Comments

TABLE 7.11 Response Equipment Inspection Log

[Note: Use section 7.4.2.3.2 of the response plan as a checklist]

Inspector	Date	Comments

TABLE 7.12 Secondary Containment Checklist

Observed conditions	Comments

1. Dike or berm system
   a. Level of precipitation in dike and available capacity
   b. Operational status of drainage valves
   c. Dike or berm permeability
   d. Debris
   e. Erosion
   f. Permeability of the earthen floor of diked area
   g. Location and status of pipes, inlets, drainage beneath tanks, etc.

2. Secondary containment
   a. Cracks
   b. Discoloration
   c. Presence of spilled or leaked material (standing liquid)
   d. Corrosion
   e. Valve conditions

3. Retention and drainage ponds
   a. Erosion
   b. Available capacity
   c. Presence of spilled or leaked material
   d. Debris
   e. Stressed vegetation

[*Note:* During inspection, make note of discrepancies in any of the above-mentioned items, and report them immediately to the proper facility personnel. Similar requirements exist in 40 CFR 112.7(e). Duplicate information from the SPCC plan may be photocopied and inserted in this section.]

exercises, and unannounced exercises. External exercises include area exercises. Credit for an area or facility-specific exercise will be given to the facility for an actual response to a spill in the area if the plan was utilized for response to the spill and the objectives of the exercise were met and were properly evaluated, documented, and self-certified.

3. Section 40 CFR 112.20(h)(8)(ii)requires the facility owner or operator to provide a description of the drill and exercise program to be carried out under the response plan. Qualified individual notification drill and spill management team tabletop drill logs (Table 7.13) shall be provided in sections 7.4.2.8.2.1 and 7.4.2.8.2.2, respectively. These logs may be included in the facility response plan or kept as an annex to the facility response plan. See section 7.4.2.3.3 for equipment deployment drill logs. Table 7.14 provides a sample tabletop exercise log.

7.4.2.8.2.1   QUALIFIED INDIVIDUAL NOTIFICATION DRILL LOGS

TABLE 7.13   Drill Logs—Qualified Individual Notification

Date:

Company:

Qualified individual(s)

Emergency scenario description:

Evaluation:

Changes to be implemented:

Timetable for implementation:

7.4.2.8.2.2   SPILL MANAGEMENT TEAM TABLETOP EXERCISE LOGS

TABLE 7.14   Tabletop Exercise Log

Date:

Company:

Qualified individual(s):

Emergency scenario:

Evaluation:

Changes to be implemented:

Timetable for implementation:

7.4.2.8.3 Response Training. Section 40 CFR112.21(a) requires facility owners or operators to develop programs for facility response training. Facility owners or operators are required by 40 CFR 112.20(h)(8)(iii) to provide a description of the response training program to be carried out under the response plan. A facility's training program can be based on the U.S. Coast Guard's *Training Elements for Oil Spill Response,* to the extent applicable to facility operations, or another response training program acceptable to the RA. The training elements are available from Petty Officer Daniel Caras at (202) 267-6570 or fax 267-4085/4065. Personnel response training logs and discharge prevention meeting logs must be included in sections 7.4.2.8.3.1 and 7.4.2.8.3.2 of the response plan, respectively. These logs (Table 7.15) may be included in the facility response plan or kept as an annex to the facility response plan.

7.4.2.8.3.1 PERSONNEL RESPONSE TRAINING LOGS

TABLE 7.15 Personnel Response Training Log

Name	Response training date and number of hours	Prevention training date and number of hours

7.4.2.8.3.2 DISCHARGE PREVENTION MEETING LOGS: Table 7.16 provides a sample format for Discharge Prevention Meeting Log.

TABLE 7.16 Discharge Prevention Meeting Log

Date:		
Attendees:		
Subject and issue identified	Required action	Implementation date

7.4.2.9 Diagrams.   The facility-specific response must include the following diagrams. Additional diagrams that would aid in the development of response plan sections may also be included.

1. The site plan diagram should, as appropriate, include and identify:
   - The entire facility to scale
   - Above- and belowground bulk oil storage tanks
   - The contents and capacities of bulk oil storage tanks
   - The contents and capacity of drum oil storage areas
   - The contents and capacities of surface impoundments
   - Process buildings
   - Transfer areas
   - Secondary containment systems (location and capacity)
   - Structures where hazardous materials are stored or handled, including materials stored and capacity of storage
   - Location of communication and emergency response equipment
   - Location of electrical equipment which contains oil
   - For complexes only, the interface(s) (i.e., valve or component) between the portion of the facility regulated by EPA and the portion(s) regulated by other agencies. In most cases, this interface is defined as the last valve inside secondary containment before piping leaves the secondary containment area to connect to the transportation-related portion of the facility (i.e., the structure used or intended to be used to transfer oil to or from a vessel or pipeline). In the absence of secondary containment, this interface is the valve manifold adjacent to the tank nearest the transfer structure as described above. The interface may be defined differently at a specific facility if agreed to by the EPA.

2. The site drainage plan diagram should, as appropriate, include:
   - Major sanitary and storm sewers, access holes, and drains
   - Weirs and shutoff valves
   - Surface water receiving streams
   - Fire fighting water sources
   - Other utilities
   - Response personnel ingress and egress
   - Response equipment transportation routes
   - Direction of spill flow from discharge points

3. The site evacuation plan diagram should, as appropriate, include:
   - Site plan diagram with evacuation route(s)
   - Location of evacuation regrouping areas

7.4.2.10 Security.   According to 40 CFR 112.7(e)(9), facilities are required to maintain a certain level of security, as appropriate. In this section, a description of the facility security shall be provided and include, as appropriate:

1. Emergency cutoff locations (automatic or manual valves)
2. Enclosures (e.g., fencing, etc.)
3. Guards and their duties, day and night
4. Lighting
5. Valve and pump locks
6. Pipeline connection caps

The SPCC plan contains similar information. Duplicate information may be photocopied and inserted in this section.

7.4.3 Response Plan Cover Sheet

A three-page form has been developed to be completed and submitted to the EPA by owners or operators who are required to prepare and submit a facility-specific response plan. The cover sheet (Appendix M to this book) must accompany the response plan to provide the agency with basic information concerning the facility. This section will describe the response plan cover sheet and provide instructions for its completion.

7.4.3.1 General Information. Owner-operator of facility: Enter the name of the owner of the facility (if the owner is the operator). Enter the operator of the facility if otherwise. If the owner-operator of the facility is a corporation, enter the name of the facility's principal corporate executive. Enter as much of the name as will fit in each section.

1. Facility name: Enter the proper name of the facility.
2. Facility address: Enter the street address, city, state, and zip code.
3. Facility phone number: Enter the phone number of the facility.
4. Latitude and longitude: Enter the facility latitude and longitude in degrees, minutes, and seconds.
5. Dun and Bradstreet number: Enter the facility's Dun and Bradstreet number if available (this information may be obtained from public library resources).
6. SIC code: Enter the facility's SIC code as determined by the Office of Management and Budget (this information may be obtained from public library resources).
7. Largest oil storage tank capacity: Enter the capacity in *gallons* of the largest aboveground oil storage tank at the facility.
8. Maximum oil storage capacity: Enter the total maximum capacity in *gallons* of all aboveground oil storage tanks at the facility.
9. Number of oil storage tanks: Enter the number of all aboveground oil storage tanks at the facility.

10. Worst-case discharge amount: Using information from the worksheets in Appendix D, enter the amount of the worst-case discharge in *gallons.*

11. Facility distance to navigable waters: Mark the appropriate line for the nearest distance between an opportunity for discharge (i.e., oil storage tank, piping, or flowline) and a navigable water.

7.4.3.2 Applicability of Substantial Harm Criteria. Using the flowchart provided in the attachment to Appendix K in the back of this book, mark the appropriate answer to each question. Explanations of referenced terms can be found in the appendix also. If a comparable formula to the ones described in the attachment is used to calculate the planning distance, documentation of the reliability and analytical soundness of the formula must be attached to the response plan cover sheet.

7.4.3.3 Certification. Complete this block after all other questions have been answered.

7.4.4 Acronyms

The following list contains the acronyms most commonly used in oil pollution prevention.

ACP: Area contingency plan

ASTM: American Society for Testing and Materials

CHRIS: Chemical hazards response information system

CWA: Clean Water Act

DOC: Department of Commerce

DOI: Department of Interior

DOT: Department of Transportation

EPA: Environmental Protection Agency

FEMA: Federal Emergency Management Agency

FR: *Federal Register*

HAZMAT: Hazardous materials

LEPC: Local emergency planning committee

MMS: Minerals Management Service (part of DOI)

NCP: National Oil and Hazardous Substances Pollution Contingency Plan

NOAA: National Oceanic and Atmospheric Administration (part of DOC)

NRC: National Response Center

NRT: National Response Team

OPA: Oil Pollution Act of 1990

OSC: On-scene coordinator

PREP: National Preparedness for Response Exercise Program

RA: Regional administrator

RCRA: Resource Conservation and Recovery Act

RRC: Regional response centers

RRT: Regional response team

RSPA: Research and Special Programs Administration

SARA: Superfund Amendments and Reauthorization Act

SERC: State Emergency Response Commission

SDWA: Safe Drinking Water Act of 1986

SI: Surface impoundment

SIC: Standard industrial classification

SPCC: Spill prevention, control, and countermeasures

USCG: U.S. Coast Guard

7.4.5 FRP Submission Deadlines

The CWA and the OPA 90 require that owners or operators of "substantial harm facilities" submit their response plans to EPA (as delegated by the President in E.O. 12777) by Feb. 18, 1993, or stop handling, storing, or transporting oil. In addition, under CWA section 311(j)(5) and OPA section 4202(b)(4), a facility required to prepare and submit a response plan under the OPA may not handle, store, or transport oil after Aug. 18, 1993, unless: (1) in the case of a facility for which a plan is reviewed by EPA, the plan has been approved by EPA; and (2) the facility is operating in compliance with the plan. The statute provides that a "significant and substantial harm facility" may be allowed to operate without an approved response plan for up to 2 years after the facility submits a plan for review (no later than Feb. 18, 1995, when this provision expired), if the owner or operator certifies that he or she has ensured by contract or other approved means the availability of private personnel and equipment necessary to respond, to the maximum extent practicable, to a worst-case discharge of oil, or a substantial threat of such a discharge. Owners or operators of "substantial harm facilities" are not required to have their plans approved by EPA but were required to operate in compliance with their plans after Aug. 18, 1993.

Owners or operators of existing facilities that were in operation on or before Feb. 18, 1993, who failed to submit a facility response plan to meet the OPA requirements by Feb. 18, 1993, must submit a response plan that meets the requirements of this rule to the EPA by the effective date of the final rule. [See 40 CFR 112.20(a)(1)(ii).] EPA recognizes that such facilities may have prepared and submitted to the RA some form of a response plan after the statutory deadline. Owners or operators may submit revised portions of the plan to bring the plan into compliance with the final rule requirements. Regional administrators will review plans for "significant and substantial harm facilities" for initial approval within a reasonable time. Such plans will be reviewed periodically thereafter on a schedule established by the RA provided that the period between plan reviews does not exceed 5 years. RAs may choose to stagger such plan reviews.

Owners or operators of facilities that commenced operations after Feb. 18, 1993, but before the effective date of the final rule must submit a response plan that meets the requirements of this final rule to the RA by its effective date. EPA recognizes that such facilities may have prepared and submitted some form of a response plan to the EPA prior to the publication of the final rule. Owners or operators may submit revised portions of the plan to bring the plan into compliance with the final rule requirements. [See 40 CFR 112.20(a)(2)(i).] RAs will review plans for "significant and substantial harm facilities" for initial approval within a reasonable time. The plans will then be placed on the region's review cycle as described in the preceding paragraphs.

The agency recognizes that identification of "substantial harm facilities" will continue to occur as new facilities come on-line and existing facilities newly meet the criteria for substantial harm as a result of a change in operations or site characteristics. EPA is requiring in 40 CFR 112.20(a)(2)(ii) and (iii) that: (1) newly constructed facilities (facilities that come into existence after the effective date of the final rule) that meet the applicability criteria must prepare and submit a response plan in accordance with the final rule prior to the start of operations (adjustments to the response plan to reflect changes that occur at the facility during the startup phase of operations must be submitted to the regional administrator after an operational trial period of 60 days); and (2) existing facilities that become subject to the response plan requirements as the result of a planned change in operations (after the effective date of the final rule) must prepare and submit a response plan in accordance with the final rule prior to the implementation of changes at the facility. RAs will review plans submitted for such newly designated "substantial harm facilities" to determine if a facility is a "significant and substantial harm facility." RAs will review for approval plans for "significant and substantial harm facilities" within a reasonable time and then place the plans on the region's review cycle as discussed previously.

An existing facility, however, may become subject to the response plan requirements through one or a combination of unplanned events, such as a reportable spill or the identification of fish and wildlife and sensitive environments adjacent to the site during the ACP refinement process. In the event of such an unplanned change, the owner or operator is required to prepare and submit a response plan to the RA within 6 months of when the change occurs [see 40 CFR 112.20(a)(2)(iv)]. The agency believes that allowing 6 months from when a change caused by an unplanned event occurs to prepare and submit a plan is reasonable.

Under 40 CFR 112.20(g)(2), facility owners or operators are required to review appropriate sections of the NCP and ACP annually and revise their response plans accordingly. In addition, 40 CFR 112.20(d)(1) requires the owner or operator of a facility for which a response plan is required to resubmit relevant portions of the plan within 60 days of each material change in the plan. For "substantial harm facilities," regions will review such changes to determine if the facility should be reclassified as a "significant and substantial harm facility." For "significant and substantial harm facilities," the regions will review such changes for approval as described in 40 CFR 112.20(d)(4).

7.5   ENFORCEMENT AND PENALTIES

Under the OPA, facility owners or operators who fail to comply with section 311(j) requirements are subject to new administrative penalties and more stringent judicial penalties than those imposed previously under the CWA. Section 4301(b) of the OPA amends CWA section 311(b) to authorize a civil judicial penalty of $25,000 per day of violation for failure to comply with regulations under CWA section 311(j). In addition to these civil penalties, OPA section 4301(b) amends CWA section 311(b) to authorize administrative penalties for failure to comply with section 311(j) regulations of up to $10,000 per violation, not to exceed $25,000 for Class I penalties, and up to $10,000 per day per violation, not to exceed $125,000 for Class II penalties. The differences between "Class I" and "Class II" administrative penalties are the amounts of the potential penalties and the hearing procedures used (for instance, Class II procedures will generally ensure the owner or operator a more extensive opportunity to be heard through the proceedings). These revised penalty provisions are applicable to violations occurring after the Aug. 18, 1990, enactment of the OPA. Violations occurring before enactment of the OPA remain subject to penalty provisions originally set forth in CWA section 311.

REFERENCES

CONCAWE, *Methodologies for Hazard Analysis and Risk Assessment in the Petroleum Refining and Storage Industry,* CONCAWE's Risk Assessment Ad-hoc Group, 1982.

U.S. Department of Housing and Urban Development, Siting of HUD-Assisted Projects Near Hazardous Facilities: Acceptable Separation Distances from Explosive and Flammable Hazards, prepared by the Office of Environment and Energy, Environmental Planning Division, Department of Housing and Urban Development, Washington, D.C., 1987.

Handbook of Chemical Hazard Analysis Procedures, U.S. DOT, FEMA, and U.S. EPA.

Technical Guidance for Hazards Analysis: Emergency Planning for Extremely Hazardous Substances, U.S. DOT, FEMA, and U.S. EPA.

The National Response Team, *Hazardous Materials Emergency Planning Guide,* Washington, D.C., 1987.

The National Response Team, Oil Spill Contingency Planning, National Status: A Report to the President, Government Printing Office, Washington, D.C., 1990.

Offshore Inspection and Enforcement Division, Minerals Management Service, Offshore Inspection Program: National Potential Incident of Noncompliance (PINC) List, Reston, Va., 1988.

CHAPTER 8
POLLUTION PREVENTION

Brian L. Lubbert
Project Manager
RTP Environmental Associates, Inc.

In 1984 the Hazardous and Solid Waste Amendments (HSWA) amended the Resource Conservation and Recovery Act (RCRA). The emphasis of HSWA was to establish a national policy that focused on some reduction of hazardous waste. The new waste reduction provisions included requirements for a waste generator to describe efforts to reduce the volume and toxicity of waste, and to certify that a program exists to reduce the volume and toxicity of waste in an economic manner.

It had become clear that industry, and EPA's waste management practices, which focused on "end-of-pipe" waste, were not effective. EPA found that additional measures were needed to address releases of waste into the air, water, and soil, as the amendments to RCRA were not sufficient.

8.1 POLLUTION PREVENTION ACT

The Pollution Prevention Act of 1990's intent was to focus industry, government, and public attention on the need to reduce the amount of pollution before it is generated. The Pollution Prevention Act essentially made national environmental policy, establishing a new and progressive way to reduce pollution. This initiative allowed EPA to focus on pollution prevention as source reduction, which included preventing or reducing wastes prior to emissions into the environment. Often these means of reducing the amount of pollution proved to be cost-effectively changing production, operation, and raw material use in a way that minimized waste from the facility's bottom line. It had been felt that opportunities to reduce pollution at the source were often not realized because of the way existing regulations had been incorporated into compliance. Simply slapping on a control device to control emissions was not necessarily the most effective way to reduce emissions from the source, whereas changing the raw materials could in essence increase production, decrease waste, and minimize pollution. In addition, the Pollution Prevention Act included practices to conserve natural resources by increasing efficiency in the use of energy, water, and other natural resources, and protected the resource itself through conservation.

Practices that have been incorporated in pollution prevention have also been associated with reducing energy demand—practices such as using energy-efficient lighting, increasing recycling both in and out of process, decreasing the demand for raw materials—thereby decreasing pollution at the source of pollution.

The Pollution Prevention Act was codified by the United States Code Title 42, Public Health and Welfare, Chap. 133, "Pollution Prevention." In the codification, Congress documented several findings that led to the utility of the Act. Congress found that the United States annually produces millions of tons of pollution and spends tens of millions of dollars a year trying to control the pollution after the source. Congress found that there were significant opportunities to reduce pollution at the source through changes in production, operation, and raw material use that were cost-effective to the facility and could actually be cost-beneficial. Congress also believed that changes could establish substantial savings in both raw material, pollution control, and liability costs, as well as reduce workers' exposure to both health and safety hazards. However, the main intent was to protect the environment. Congress felt that source reduction was fundamentally more desirable than waste management and control of pollution after it had been generated, despite the fact that EPA had historically paid little attention to reduction from sources. Congress therefore mandated that as the initial step to prevent pollution through reduction at the source, the EPA must establish a source reduction program to collect information, provide financial assistance to states, implement the activities needed to comply with the Act, and disseminate the information to the regulated community, in an effort to share the new wealth of information established by facilities that have tried to minimize pollution at the source. Congress went on to say that it declared that the national policy of the United States was to prevent or reduce pollution at its source whenever feasible: Pollution that cannot be prevented should be recycled in an environmentally safe manner, whenever feasible; pollution that cannot be prevented or recycled should be treated in an environmentally safe manner whenever feasible; and disposal or other release into the environment should be employed only as a last resort and should be conducted in a safe manner.

8.1.1 EPA's Role

EPA was authorized to establish an office that was independent of the agency's single-medium program offices, and that had authority to review and advise single-medium programs on activities to promote a multimedia approach to source reduction. The theory was that simply passing the pollution from one medium type to another was neither environmentally friendly nor cost-effective, and that the reduction at the source itself was the most likely beneficial way to decrease pollution. It was determined that EPA should function in such a way to develop and implement a strategy to promote source reduction. Since 1990, EPA has made matching grants available to states for programs to promote the use of reduction techniques by businesses. Appendix M is a list of State Pollution Prevention Technical Assistance Contacts to assist facilities in implementing a pollution prevention plan.

8.1.2 EPA Goals

The U.S. EPA has been committed to source reduction as the primary management strategy to prevent pollution. The Waste Minimization National Plan established three goals to minimize the most persistent, bioaccumulative, and toxic pollutants:

- Reduce, as a nation, the most toxic constituents by 25 percent by 2000 and 50 percent by 2005
- Avoid transferring across environmental media
- Ensure constituents are reduced at their source whenever possible

EPA has also established five objectives to promote voluntary and regulatory compliance:

- Develop a framework to set national priorities and develop screening tools for identifying priorities and constituents of concern
- Promote multimedia environmental benefits and prevent cross-media transfers
- Demonstrate preference for source reduction by shifting attention to the generators to reduce waste generators at source

• Define and clearly track progress and promote accountability by EPA, states, and industry

• Involve citizens in minimization and implementation decisions

8.2 STATE IMPLEMENTATION

8.2.1 Massachusetts

States have implemented a pollution prevention strategy in a number of different ways, but all emphasize the multimedia aspect of the regulation. Specifically, in Massachusetts, the state agency developed a cross-media inspection program to inhibit the transfer of pollution from one medium to another. The inspection program basically incorporated the pollution reduction requirements directly into enforcement procedures. The initial pilot program received the 1991 Ford Foundation Award for innovation in state and local government. The plan basically trained teams from state agencies to inspect facilities for air, water, and hazardous waste compliance while providing technical assistance with pollution prevention at the same time. The plan was so successful that it has been adopted by other states in varying degrees because of its cost-effectiveness in its initial setting. The Massachusetts approach has been institutionalized in four regional offices.

8.2.2 Wisconsin

The Wisconsin Department of Natural Resources (DNR) has begun to integrate its pollution prevention elements to focus on multimedia permitting and compliance. The DNR hopes that inspections and permit writing that coordinate air, water, and waste programs will institutionalize prevention activities and avoid shifting the pollutants from one medium source to another. The DNR plans to select a number of industrial categories to focus on. To evaluate permits when issued, a review facility plans to determine how pollution prevention techniques can be utilized. The DNR will also rate permits and coordinate inspections to minimize pollution across all three media.

8.2.3 New Jersey

New Jersey has developed a two-tiered approach, which incorporates pollution prevention inspections at a facility and also provides assistance with compliance in air, water, and solid waste regulations. The Pollution Prevention Department basically takes a hands-on approach to provide this assistance to facilities in order to minimize the generation of pollution. Specifically, the New Jersey Department of Environmental Protection (DEP) has attempted to help facilities identify "missed" sources of pollution and to analyze data used in emission statements and hazardous waste generation reports, to document that in fact the numbers are accurate. This multimedia approach to all facilities either demonstrates that they are accurately tracking emissions from one media string to another, or identifies problems with their emission calculations prior to single-medium enforcement. Initially, problems found during a pollution prevention inspection will allow facilities to develop a compliance plan, which will minimize enforcement action.

The second phase of the New Jersey program is to incorporate facilities permitting into one multimedia permit that covers the entire facility. Such an innovative approach will allow DEP to strictly monitor cross-media issues at one time. New Jersey has also begun to develop a multimedia environmental database that will be utilized to facilitate environmental compliance on a facility level instead of a single-medium level; therefore environmental inspectors will be able to have the facility's environmental record at their fingertips during an inspection that covers all types of media.

8.2.4   Other Agencies

Pollution prevention strategies have also been incorporated into other independent agencies, separate from state and local. These include the all-Indian Pueblo Council and the University of New Mexico's Institutionalized Pollution Prevention Program in public governments. The multimedia approach has integrated the ethic of pollution prevention into Indian tribal policy. The coordinator will try to attempt to reduce pollution on federal lands.

8.3   THE ELEMENTS OF A POLLUTION PREVENTION PLAN

Pollution prevention programs (PPPs) vary from state to state but a number of elements are considered a necessity. Specifically, the pollution prevention program begins with management support. A written policy is essential in establishing a program to incorporate current activities and maintain the program. Executive involvement in the program will accommodate increased knowledge of the potential for cost reduction associated with disposal, regulatory compliance, and workers' health and safety.

The PPP should use a coordinator who can develop awareness in a task force team. The coordinator generally acts as a liaison with upper management, and the task force generally comprises representatives from individual departments (maintenance, production, purchasing, legal, etc.). It is advisable to have the task force meet regularly to effectively develop the program and written plan.

In the preliminary PPP assessment, it is important to decide which waste streams are to be reduced or eliminated. The decision usually concerns quantity, characteristics, and disposal costs of waste generated. Employee involvement is invaluable in the assessment, as current waste reduction measures may have already begun on the employees' own initiative.

Assessments are to be conducted to determine both the economic and technical feasibility of the PPP ideas. Complete replacement or significant capital investment may not be practical. However, economic feasibility may prove that an apparently costly capital expenditure may recover its cost in a relatively short period of time.

The written PPP incorporates the results of the above preliminary assessments, options proposed, and feasibility assessments. Project proposals, rejected options, and goals are included in a state-specific format.

The maintenance of the program mandates periodic progress measurements. These measurements are to be conducted on a routine basis. In addition, the review of new or advanced technologies is to be continued. Specifically, previously rejected strategies may be revised to incorporate more inexpensive technologies. The program can be sustained by bringing in new task force members and personnel and publicizing success stories. The job of the pollution prevention team is never complete, nor should management expect that its job be completed. The pollution prevention program should become an integral part of a facility's cost savings measures. The following section details a step-by-step procedure to implement a pollution prevention program.

8.4   CASE STUDY

NOTE: This section has been modified slightly from its original source and is used with permission from the New Jersey Department of Environmental Protection. The original document can be found at *http://www.state.nj.us/dep/opppc/guide/step1.htm.*

8.4.1   Step 1—Understand Pollution Prevention

The first step to implementing a pollution prevention program is understanding what pollution prevention is. Defining pollution prevention is a necessity because pollution prevention is similar in name but different in meaning to many other concepts, including "waste minimization."

Pollution prevention is defined as the reduction or elimination of the need for hazardous substances per unit of product, or the reduction or elimination of the generation of hazardous substances where they are generated within a process. This means minimizing the use and generation of hazardous substances within production processes so they never have the chance to be released into the workplace or environment.

With a few exceptions, the activities a facility undertakes to reduce nonproduct output are pollution prevention. Typically, these activities fall into one of five categories:

- Substituting hazardous substances with nonhazardous or less hazardous ones
- Product redesign
- Production process efficiency improvements
- In-process recycling
- Improved operation or maintenance

Material substitution and product redesign can eliminate a hazardous substance from a process. Process changes, in-process recycling, and improved maintenance can substantially reduce the need for hazardous chemicals, though they seldom result in the complete elimination of a nonproduct output stream.

These changes will reduce or eliminate the risks that hazardous substances pose to employees, consumers, the environment, and human health. When hazardous substances are avoided through prevention, the costs and risks associated with disposal and treatment may never arise. These features make pollution prevention economically and environmentally superior to pollution treatment and disposal.

8.4.1.1 What Is Not Pollution Prevention. Understanding what pollution prevention is *not* can clarify what it is. First, any kind of pollution treatment is not pollution prevention. Second, because pollution prevention operates at the production process level, recycling that takes place outside of a process is not pollution prevention. Third, because pollution prevention reflects improvements in an operation rather than changes in market conditions, if a waste becomes a marketable coproduct through shifting market conditions, its reduction is not pollution prevention. Finally, the Pollution Prevention Rule explicitly states that pollution prevention never increases or transfers risk between workers, consumers, and the environment.

Specific activities that do not qualify as prevention include increased treatment, out-of-process recycling, and disposal. Sometimes these are the only options available, but while they may be appropriate, they can never be pollution prevention. This difference provides a key to distinguishing pollution prevention from other forms of hazardous substance management: A reduction in the amount of a hazardous substance generated is usually considered pollution prevention; dealing with these same materials once they exist, no matter how effectively, is not.

8.4.1.2 In-Process versus Out-of-Process Recycling. It is important that a facility understands the difference between in-process and out-of-process recycling because these two environmental management techniques exist at the boundary between what is pollution prevention and what is not.

In-process recycling *is* pollution prevention. It occurs when a hazardous substance that would otherwise be generated as nonproduct output is returned to a production process using dedicated, fixed, and physically integrated equipment so that nonproduct output and multimedia releases are reduced. Accumulation of material prior to any in-process recycling activity must occur on the same production schedule as the product. These types of recycling systems are more typical for "continuous" processes. Nevertheless, some forms of recycling in "batch" or "campaign" operations are also in-process recycling.

Consider a batch production process that uses cyclohexane as solvent and yields an easily separable product. After the product is separated, the cyclohexane is transferred by hard pipes to a storage tank. After four batches, all the cyclohexane in the tank is transferred via hard piping to a still and is recovered by distillation. The recovered cyclohexane is then piped to another storage tank, from which it is piped back to the original reactor as needed. This activity would meet the definition of in-process recycling.

Certain activities and equipment cannot be part of an in-process recycling system. Containers, such as 55-gallon drums, that are directly handled by workers cannot be used. Pipe connectors and fittings cannot rely solely on friction or other nonmechanical means. All connections must be fixed (i.e., soldered, bolted, or positively connected in another way).

Out-of-process recycling *is not* pollution prevention. It includes both on-site and off-site activities where nonproduct output is transferred, stored, and recovered for use in processes that are not directly connected with fixed equipment that is physically integrated with the recovery system and the process where the nonproduct output was generated. An example of an off-site activity is sending a chlorinated solvent used in degreasing to an outside vendor who reclaims the material. In that case, the facility's need for the solvent remains undiminished. Likewise, regenerating sulfuric acid off site and returning it to the facility is out-of-process recycling, because the facility's sulfuric acid needs remain the same. Such recycling is valuable, but it is not pollution prevention.

On-site out-of-process recycling activities include any on-site recycling or reclamation activities that do not meet the definition of in-process recycling. An example is a central distillation process where different solvents are transferred in drums, stored prior to reclamation, and are used in other processes after reclamation.

Out-of-process recycling is an excellent environmental management technique that has many, but not all, of the benefits of pollution prevention. The state generally recognizes the importance of out-of-process recycling in meeting the environmental and economic goals of industrial facilities. This chapter, however, is designed to help companies find pollution prevention techniques before they settle on out-of-process recycling systems. After the opportunities for pollution prevention have been fully investigated and implemented where feasible, out-of-process recycling is the best environmental option.

8.4.1.3 *Different Types of Output.*

There are generally two types of material that leave a process: product and nonproduct output. *Product* is the desired result of a process, to be directly packaged, if necessary, and sold. Processes may have more than one product. Sometimes, this definition is expanded with two other terms to accurately describe what happens at a facility or in the marketplace.

Intermediate product describes the case of a desired result at the end of a process that requires further work before it can be sold. *Coproduct* describes output from a process that is sold only part of the time and that is nonproduct output during the rest of the time.

Nonproduct output encompasses the rest of what leaves a process. Reduction in nonproduct output *per unit of product* provides a consistent year-to-year measure of progress in pollution prevention. It is a useful measure because it tracks hazardous substances at their source, that is, before out-of-process recycling, storage, and treatment. It also includes fugitive releases.

Reducing the amount of nonproduct output that a production process generates per *unit of product* is one way of measuring progress in pollution prevention. It is a useful measure because it is always determined before out-of-process recycling, storage, and treatment and because it includes fugitive releases. Note that a chemical that is the desired result of a process is still nonproduct output if it leaves the process in any way other than in a product stream, such as in a fugitive release or as a small amount of product lost in a waste stream.

The definition of nonproduct output hinges on what "product" means. *Product* is the desired result of a process, to be directly packaged, if necessary, and sold. Processes may have more than one product. Sometimes the definition of product is not sufficient to describe what happens at a facility or in the marketplace. There are two other terms to cover these situations. First, the term *intermediate product* describes the case of a desired result at the end of a process that requires further work before it can be sold. The second term is *coproduct,* which describes output from a process that is sold only part of the time and which is nonproduct output the rest of the time. A firm can reduce the amount of nonproduct output it generates by finding a market for it and selling it as a coproduct, but, by definition, this is not considered pollution prevention.

8.4.1.4 Environmental Management Hierarchy. The final element of understanding pollution prevention is to see how it fits together with other environmental management techniques. These techniques form a nationally recognized hierarchy for contending with hazardous substances. The *environmental management hierarchy* categorizes environmental management options as follows (in descending order of importance):

Pollution prevention

Out-of-process recycling

Efficient treatment

Safe disposal

Therefore, when a manager considers how to cope with nonproduct output, pollution prevention should be the first option on the list. Out-of-process recycling is next best and should be considered when viable pollution prevention options run out. Once these possibilities are exhausted, safe and efficient treatment or disposal remain as acceptable options.

The goal of this pollution prevention program is to make pollution prevention the environmental protection system that facilities consider first. The Pollution Prevention Rule does this by directing industrial efforts to the very top of the environmental management hierarchy. Businesses can do this too, by emphasizing pollution prevention in their corporate decisions and policies. By doing so, companies can expect improvements throughout their operations, accompanied by good news on the bottom line.

CASE STUDY: Top Shelf's president has a look at pollution prevention.
John Stevens, president of Top Shelf Wallcoverings, had been hearing a lot about the possible economic benefits of something called pollution prevention at the trade association meetings that he regularly attended and decided to find out more about the subject. One of the speakers recommended several different references that he could easily obtain. He found out that this was probably not just a passing fad, but might be a helpful approach for his business. He talked the issue over with his management team at their weekly luncheon meeting, and they decided to explore the approach. While Top Shelf management always prided itself on its quality manufacturing process, the company was using several tons of materials on various government hazardous chemical/pollutant lists, and meeting ever more stringent environmental requirements was getting very expensive. In fact, it had been some time since anyone reviewed carefully how these hazardous chemicals were being used at the plant and how waste containing them was being generated. Maybe, they thought, a modest effort at reviewing their use and process losses of these materials would be worthwhile. So they decided to commit the company to one round of pollution prevention planning.

8.4.2   Step 2—Establish a Pollution Prevention Policy

This step deals with corporate *pollution prevention policies*. The experience of many firms has proved that a written and formally adopted policy is a key to successfully accomplishing prevention goals. While there are no specific requirements for a written policy statement in the Pollution Prevention Rule, the owner or operator of a facility must certify that it is the policy of the facility to achieve the goals established in the pollution prevention plan.

An effective pollution prevention effort needs to have top-level corporate commitment. A simple and effective way of demonstrating commitment is by adopting a pollution prevention policy. Policies will differ from company to company. Some firms may have existing policies such as total quality management that pollution prevention should be coordinated with. Nevertheless, there are a number of features that belong in every firm's policy. They should be gathered together in a pollution prevention policy statement.

8.4.2.1   Contents of a Pollution Prevention Policy Statement.   There is a lot to consider when planning the form and content of a pollution prevention policy.

Usually, the best policies are simple and straightforward, but there are several items that should be included. They are:

- A focused definition of pollution prevention and emphasis on it as the firm's primary environmental management option (see step 1)
- Clear evidence of high level of corporate commitment (see step 3)
- A statement of the objectives of the policy
- A plan to go beyond compliance
- A commitment to progress
- Accountability for progress (see step 3)
- A demonstration of appropriate leadership (see step 3)
- Employee involvement and incentives (see step 3)

Some additional features to consider include:

- Reasons for the policy
- Coordination with energy conservation, water conservation, total quality management efforts, and initiatives to reduce the generation of nonhazardous waste
- A description of how progress will be reported (step 11)

Pollution prevention policies are most effective when they are formally considered and developed to mesh with the firm's overall management style. Notice and review of the policy should follow the same procedures used to disseminate other corporate policies. For example, some businesses use an employee handbook to keep their workers up to date. Others use company newsletters, while still others circulate copies of policy statements at staff meetings. The important point is that everyone at the facility should know that the firm has a strong commitment to pollution prevention.

Policy statements demonstrate to employees and the public that the firm is serious and plans to take action to reduce hazardous substance use, nonproduct output generation, and hazardous substance release. Effective policies clearly identify pollution prevention as the company's preferred approach to environmental management, to be fully explored before recycling, treatment, storage, or disposal are considered.

The policy should explain the company's prevention efforts in terms of continuous improvement in production processes rather than a one-time review of the facility. Your firm might also want to consider including a commitment to cut nonhazardous substances as well as hazardous ones in your policy. Perhaps the policy will relate pollution prevention efforts to other programs such as quality management, water conservation, or energy conservation. Such initiatives have a direct relationship to pollution prevention, often involving the same type of process inspection, organization, and commitment to ongoing progress. Although pollution prevention should be coordinated with other programs, do not lose sight of the primary importance of reducing the use and generation of hazardous substances.

CASE STUDY: Top Shelf writes a pollution prevention policy statement.
John Stevens decided to start things rolling by making pollution prevention company policy. He asked the plant manager, Sarah French, to help him draft a pollution prevention policy that would supplement the existing corporate policy.

Together, they decided that Top Shelf had a good environmental record, but had never formalized it into a program. After several drafts, they settled on the following as a statement of their pollution prevention policy:

"Top Shelf Wallpaper, Incorporated, is committed to a policy of protecting the environment. Our management and employees are dedicated to and responsible for carrying out this policy. Pollution prevention is a way for this company to take our commitment to the environment beyond permit compliance, by adopting techniques within our production processes that reduce the company's need to use and generate hazardous substances. Therefore, we will work together to implement pollution prevention wherever possible. We will systematically and regularly look for pollution prevention in existing processes and through new process design, new maintenance procedures, and product research. These measures will provide a safer environment for both our workers and our community by reducing hazardous substances in the workplace, in the air, in the water, and on the land."

8.4.2.2   Anticipating Obstacles to Your Pollution Prevention Policy.   Pollution prevention often involves fundamental changes in the way parts of your firm operate. Usually, these changes have surprising benefits that include cost reduction and product improvement. Nevertheless, you may encounter internal resistance during your program's starting phases, beginning when you circulate a new pollution prevention policy. Good planning and creative thinking can overcome such resistance.

Typically, skeptics are concerned that:

- *New operating procedures may reduce the rate of production.* It is unusual for a pollution prevention change to significantly lower the production rate. If this does occur, it may be related to start-up and the production rate may increase as familiarity is gained with the new operating procedures. Finally, reduced operating costs achieved through pollution prevention often overcome losses from a slightly slowed rate of production.

- *Changes in the product may change customer acceptance.* It is often possible, with good research and development, to reformulate a product without significantly changing its characteristics. Working with customers during the development process is one approach to gaining final acceptance. Showing customers how the new product is manufactured in a more environmentally safe manner is often a selling point. Sometimes, an environmentally safer product is requested by consumers.

- *There are no process alternatives.* There are many resources for finding alternatives. Publications are available that present ideas that are specific to certain industries.

- *Changes will alter our compliance status.* If there is a question as to how a pollution prevention change will affect an existing permit or how it will impact compliance with another law (such as the federal Clean Air Act), call the local environmental agency for clarification.

You may encounter problems like these as you develop your pollution prevention plan, but they can be overcome with careful planning, analysis, and creative thinking.

8.4.3   Step 3—Leadership and Staffing

This step describes the staffing and leadership of your pollution prevention program. The way a pollution prevention program is structured and staffed has a large impact on its success. There are no mandated staffing elements in the Pollution Prevention Rules.

8.4.3.1   Elements of the People Component.   While an appropriate pollution prevention definition and policy are needed to focus your program, it is people, more than words, that bring about high levels of achievement. There are five elements of the people component:

• Top management leadership

• Senior management oversight, including process managers

• Incentives and involvement of employees, especially operators

• Planning by a multidisciplinary team

• Accountability for the different parts of the program

8.4.3.2   Top Management Leadership.   The level of commitment of the president, chief operating officer, and/or chief executive officer can make or break a pollution prevention program. Ideally, the program will be initiated from the very top or at least have strong support at that level. As a first step, the company policy should be issued by or strongly endorsed by management to demonstrate its commitment to pollution prevention. Ongoing support will be needed throughout to reinforce the initiative of those implementing the program.

8.4.3.3   Senior Management Oversight of the Program.   After your company's management has developed a policy (see step 2), your program should have a clear leader (or leaders) who will spearhead the program. For your program to be effective, it needs to be led by someone who has knowledge of pollution prevention principles and environmental management, coupled with knowledge and responsibility for your facility's production processes.

Pollution prevention planning should be a formally assigned part of the leader's job so that he or she can devote the time necessary to develop an effective program. In order to fulfill this new responsibility, the leader should have sufficient authority to put together a pollution prevention team, to gather needed information, and to make decisions about what pollution prevention options to implement.

8.4.3.4 Employee Involvement. The leader's first task should be to involve employees from all parts of the facility in the pollution prevention program. Since they are directly involved with production, process operators are often especially valuable sources of ideas for reducing nonproduct output (NPO). The method of encouraging employee involvement should conform to the culture and management style of your firm. Some firms may integrate pollution prevention into "total quality management" (TQM) teams, others may use worker-management teams. Incentives, such as awards programs or bonuses, are also good ways to spur employee involvement.

> **CASE STUDY: A team leader looks for volunteers.** The president of Top Shelf posted the newly developed policy around the facility. Sarah, the plant manager, was assigned the task of forming a pollution prevention team and setting up the program. As a follow-up to the posting of the written policy, she presented the policy at a communications meeting. Communications meetings are a common forum at Top Shelf for exchange between managers and production workers. John, the president, expressed his enthusiasm for the program in his introductory remarks at this meeting.
>
> After describing the new program, Sarah explained that she was putting a team together to implement the program. She hoped to assemble a diverse group and asked for interested employees to volunteer to participate. After the meeting, she answered questions and pointed out to several people that their participation could only help their prospects for advancement at the company.

8.4.3.5 Your Pollution Prevention Planning Team. Efforts to encourage employee involvement should coincide with the formation of a pollution prevention planning team. The pollution prevention team is a group of company personnel who will take charge of the pollution prevention activities at your facility. A list of their possible responsibilities appears in "CASE STUDY: The team assembled."

The size of the team you select will depend on the size of your facility. A small facility may find that a "team" of two people is sufficient. All firms should strive to have more than one person on the team in order to get a mix of insights and perspectives. A large facility will benefit from a broad, more diverse group of people and may also find it useful to create separate expert assessment teams to deal with particular processes or sources of nonproduct output.

Your team should have representatives from every facet of your facility's operations. Team members should include people familiar with the company's products and production processes, people familiar with current environmental practices, people with technical expertise in areas related to pollution prevention, people with an understanding of environmental regulations, people involved in your company's finances and marketing, and people with good interpersonal skills. At a smaller facility, one person may represent several of these categories.

Finally, your facility may benefit from using outside consultants or experts from a different facility in your company who can offer new and different viewpoints and ideas. However, pollution prevention planning is most effective when managed in house, since no one knows the facility's processes better than those who work with them every day.

CASE STUDY: The team assembled. Employees at Top Shelf read the posted memos stating the pollution prevention policy and asking for pollution prevention team members. They discussed the plant manager's presentation, and several interested employees volunteered. The president also asked several other managers to join the plant manager on the team. The assembled team included the following:

1. **Plant manager** (Sarah French). As leader of the pollution prevention program, the plant manager heads up the pollution prevention team. She has been with the company for close to 20 years. It was decided that she would lead the team because pollution prevention is closely tied to production processes. She will, however, have to work closely with the environmental manager.
2. **Environmental manager** (Thomas Brown). Since he is responsible for ensuring that the firm is in compliance with environmental regulations, the environmental manager is very familiar with release, permit, discharge, right-to-know, hazardous waste management, and other data. He is the team member who is most familiar with the Pollution Prevention Rules.
3. **Supervisor of maintenance and facilities** (Travis Fox). This member of the team has worked his way up through the ranks of the company over the years. His insight on the facility will be a valuable tool for information on the facility's current processes.
4. **Sales manager** (Emily Cruz). This member is responsible for more than just sales; she is also the facility's "finance whiz." She has a great deal of information on current costs and has raw material purchase and product sales records at her fingertips.
5. **Production workers** (Jerry Davis and Samantha Sweeny). Two production workers joined the team. These members can provide accurate descriptions of current production practices as well as suggest ideas on new approaches to implement the pollution prevention plan. They are the ones who fill out batch sheets on the factory floor. They will be most able to gauge the pollution prevention plan's compatibility with current work practices and supply feedback on front line effects of the changes.

8.4.3.6 Production Management Accountability. It is unlikely that your firm's pollution prevention program will succeed without the means to measure progress and to make your production managers accountable for the pollution prevention effort in their area. They are the ones that will be responsible on a day-to-day basis for implementing pollution prevention initiatives and for identifying additional initiatives on a continuing basis. With this increased responsibility, there should be rewards for pollution prevention accomplishments—most of which will improve the company's profitability.

Pollution prevention works because creative thinkers can find opportunities to protect the environment and save money at almost every facility. People are what provide the driving force that uncovers opportunities at your facility. Assembling a group of creative people with diverse backgrounds and knowledge is half the battle in doing pollution prevention.

8.4.4   Step 4—Identify Your Processes and Sources

In this step, your pollution prevention team will locate where hazardous substances are used or generated throughout the facility. This will lead your team to the *production processes* and *sources* that belong in your plan. Once the relevant processes and sources are found, the team will need to identify and describe them, usually by developing a *process flow diagram,* so they can be easily understood in later steps in the pollution prevention planning process.

Effective pollution prevention, like pollution control, depends on how familiar planners and designers are with the system in which it will operate. Facilities that release hazardous substances usually have pollution control equipment that treats the nonproduct output from the production process. When production processes were installed or treatment equipment was added, a designer decided how hazardous substances would leave the production process and built a system to collect those substances and treat them. An effective pollution control system requires accurate data on the types and amounts of hazardous substances it will be treating.

Pollution prevention planning requires a similar depth of knowledge about a facility's *production processes,* because pollution prevention usually takes place at the process level. To establish that knowledge base, your team must identify the processes that use and generate hazardous substances throughout the facility and the exit points, or *sources,* where nonproduct output leaves.

8.4.4.1   What Is a Production Process?
A *production process* is one or more activities that lead to one or more products (or intermediate products). Processes can create a product directly, create an intermediate product, or produce a result that is necessary for production to continue. Processes may produce coproducts incidentally, but coproducts alone do not define processes. For the purposes of the plan, your team should divide production activities into the simplest activity-product combinations available. Specifically, processes that lead to isolated intermediate products should be thought of as separate from the processes that use the intermediate product. If your team does not divide its operations into simple component processes, it risks hiding opportunities for pollution prevention inside the engineer's black box.

8.4.4.2   Identifying Production Processes.
Usually, common sense will lead your team to the best process identifications for pollution prevention planning. By starting with a list of your products and working backward from that list, your team will be able to trace processes from end to beginning. Doing this will reveal the activity-product combinations that delineate production processes for pollution prevention planning purposes. Include intermediate products in your team's list of products so they can also be used to identify processes. Intermediate products should be easy to find since they are inputs (raw materials) that are made at the facility rather than purchased and brought on site. Identify the activities that lead to intermediate products as separate processes.

8.4.4.3 Process Flow Diagrams. Process flow diagrams (PFDs) are a valuable tool for identifying and describing processes since they display input and output information in a visual format. Obtaining or creating such diagrams now will simplify process identification, and the PFD will prove useful throughout the rest of the planning process.

Simple block diagrams of each process that show the flow from production step to production step will serve your team well. Piping and instrumentation diagrams (P&IDs) or schematic equipment diagrams are useful, but if your team does not already have access to them, there is no need to create them now. The necessary components of a process flow diagram are *raw material inputs, products,* and *nonproduct output* streams connected by blocks that provide an explanation of the *steps* that turn input into product and nonproduct output. At this stage, quantities for these streams are not vital since they will be determined later by *materials accounting* in step 6. Pay special attention to hazardous substance inputs since your team will be tracking them through processes to the point where they are consumed or exit as a component in a nonproduct output or product stream.

Each separate production process has its own identifier, a name or number used as a reference. As your team completes its process identifications, it should assign identifiers and record them, together with a description of the process and a flow diagram. The team will add to this information as planning continues.

If there are processes whose flow diagrams, inputs, and products are similar to one another, they may be good candidates for *grouping,* the next step. Grouping allows similar processes to be collected together and considered a single process for the purposes of the plan, thus streamlining data collection and recording. Processes that have been grouped together have their own separate identifier in the plan. Therefore, if it appears that your team will be grouping production processes later, it may make sense to wait before assigning identifiers to processes that are likely to be grouped.

Processes that do not use or generate a covered hazardous substance may be included in the plan. Identifications made in this step will be used throughout the plan to understand processes, to gather source information, to group sources and processes into manageable sets, and as a basis for learning the more detailed information needed to find pollution prevention opportunities.

8.4.4.4 What Is a Source? In the vocabulary of pollution prevention, *sources* are points or locations in a production process through which hazardous substances exit. Whenever nonproduct output leaves a process, it goes through a source. This view is different from the conventional one of sources as places where a permitted release leaves a facility and enters the environment. Sources are where nonproduct output leaves a production process *prior* to treatment. Pipes or ducts from a process to a treatment system are sources, as are leaks that allow fugitive emissions. One location may host different sources during the steps that make up a process. For instance, a single vent might release one substance during one step of a process and another substance during a later step. Pollution prevention might operate in different ways for each substance, so two sources would be identified even though they occur at the same place.

Pollution prevention may take place at both the source and process level. A spray coating operation is a process that might be ripe for pollution prevention in the form of a switch from an organic solvent to a nonhazardous aqueous solvent. At the same time, planners could consider the individual sources within the coating process for pollution prevention as well. One such source might be a spray booth, the location where coating takes place. If a different spray nozzle arrangement could be devised to minimize overspray at that location, then that would be pollution prevention at the source level. Source identification puts such possibilities on the table.

8.4.4.5 Finding Sources of Nonproduct Output for Pollution Prevention Planning. By creating process flow diagrams, your team has taken a step toward finding *sources,* since flow diagrams show both product and nonproduct output leaving the process. Wherever nonproduct output leaves a process, there is a source, so the team can consider each nonproduct output stream and write down what is known about how it leaves the process. For instance, nonproduct output may be piped to a large combined treatment system, or treated in a wet scrubber dedicated to a particular process. It is likely that some nonproduct output escapes through valves and other fittings as fugitive releases.

Your team can do a qualitative materials accounting check to ensure that it has not overlooked a nonproduct output stream completely. Make a qualitative (rather than a more complicated quantitative) determination of whether the substances that are inputs to a process show up in either product, pass through a source as nonproduct output, or are consumed. When input substances are consumed (i.e., chemically altered) in a process, the team should still be able to find evidence of the consumed input in either the product or as a nonproduct output stream at a source. Doing this provides some assurance that you have found the needed components for a later materials accounting or mass balance.

CASE STUDY: Identifying production processes at Top Shelf. Although it is a small company, Top Shelf had 101 product lines, with different pattern combinations that allowed them to expand to over 2000 different wallpapers. They decided not to spend much time analyzing 13 of these product lines, because they were latex-based and used only a small amount of one hazardous substance. Two of the facility's 14 printing machines are dedicated to making latex-based wallpaper. Reducing hazardous substances was part of the reason the company began using latex.

The remaining 88 product lines all used solvents that were covered hazardous substances. Therefore, all of the production processes that made these product lines had to be addressed in the plan.

To make these products, combinations of five different organic solvents are used to prepare various inks. The inks are pumped into the printing machines and applied to PVC sheets in attractive patterns. The machines vary in age and, therefore, in design. The 10 oldest are original to the facility and were built in 1970. Four new machines were added in 1989. Two of the new machines are the ones dedicated to latex wallpaper production. The other 12 are used as needed to make any wallpaper that requires an organic solvent. Products from any product line can be made on any machine. Short runs are made on newer machines because they are more flexible and can handle several different product lines efficiently, while the older machines need a longer run to be efficient.

Top Shelf's pollution prevention team realized that although all the printing equipment was different, the steps in almost every production process were the same, although the inputs and outputs might change. They wrote a description of a typical process and drew a simple process flow diagram. Their description and diagram identified several sources of nonproduct output, including open mixing drums, ink reservoirs and troughs, pumps, the "coppers" that apply ink to PVC sheets, and doctor blades that wipe excess ink from the coppers after they have been dipped in an ink trough.

Finally, the team realized that the facility did produce one intermediate product, a cotton gauze–backed PVC wallpaper sheet. This sheet was made through a separate process that glued cotton gauze to regular PVC sheets before sending them on to be printed as usual. The glue was 60 percent methyl isobutyl ketone (MiBK), so it had to be identified and described in the plan.

8.4.4.6 *Facility Walk-through.*
The information that the team gathers from process flow diagrams and its own knowledge of the facility may present a clear picture of the facility's overall hazardous substance involvement, but it may not. Information on sources can be especially difficult to collect on paper since fugitive sources of nonproduct output are inherently unrecorded. Often, the best way to truly understand process information is to walk through the facility and follow each process from one end to the other. This also gives people on the pollution prevention team who do not routinely visit the production processes a chance to get a feel for what is involved at each one. If several similar processes are run on the same machinery, then a walk-through that follows one of them from beginning to end may stand in for the others, as they will probably be *grouped* in step 5.

A facility walk-through is most effective when it follows operations from the point where hazardous substances first enter the facility through to where products and nonproduct outputs are generated and then moved off site. This may mean observing operations at several different times to get a complete picture.

Before a walk-through, the team should:

- Develop a list of information that it would like to have.
- Determine the best times to visit all phases of the operation.
- Prepare to talk to individual workers throughout the facility.
- Plan for whatever safety measures may be necessary on the plant floor.

During a walk-through, the team visits as much of the facility as it can, asking questions of the people who work with the production processes every day, taking note of where one step of an operation stops and another begins, and getting a feel for the facility's processes. A walk-through is especially valuable for understanding information that is confusing on paper, correcting flow diagrams, and discovering unknown sources.

Areas where nonproduct output leaves the process in an unusual way, such as leaks or open solvent vessels, should be carefully noted. These sources are of the type that are not planned for and therefore do not show up on process flow diagrams. They should be added to the relevant diagrams when the walk-through is complete.

CASE STUDY: Top Shelf conducts a walk-through.   The team members decided to conduct a walk-through to check the process and source identifications they had made. They had all reviewed the process flow diagrams and many of them assumed that all the sources were identified. However, Jerry and Samantha, who worked with the machinery every day, knew of some sources that were not indicated in the diagram. They wanted to show the rest of the team these additional sources and to look for others as well.

The team members planned their walk-through to take place over 3 days because, in that time period, they could see every activity at the facility, including receiving new raw materials and shipping product. They also prepared questions for the people on the shop floor. Emily, who worked in the sales office, got a refresher on safety procedures, which everyone attended.

The team, led by Jerry and Samantha, found several sources that were not recorded as outputs in the flow diagram, including:

- There was almost always ink remaining in troughs and reservoirs at the end.
- Evaporated solvents were concentrated around ink troughs and mixing drums. The nonproduct output from these sources was vented to an afterburner.
- Appreciable amounts of acetone and methyl ethyl ketone (MEK) were used to clean the printing machines between runs.
- After a run, ink that was not used as an input was usually sent back to storage. Upon investigating the storeroom, the team found several containers of leftover ink that were dried up from sitting too long.

The last two activities, cleaning and storage, did not seem to fit logically as a step within the process, although they were significant sources of nonproduct output. At that time, the team decided to identify cleaning the machinery as a separate production process and to deal with storage later. The team made process flow diagrams for the processes they found and updated the existing diagrams in light of the source data they had found.

At the end of the walk-through, the team discussed how to analyze the afterburner being used to treat some of the hazardous nonproduct output. They decided to analyze its operations separately.

8.4.4.7 *Unusual Activities.*

At the end of the facility walk-through, your team members should understand the mix of products, inputs, outputs, and activities that make up the processes at your facility. At the same time, your team may have observed some unusual activities that do not fit neatly into either the definition of production process or source. For example, it may be necessary to use hazardous substances to periodically clean some machinery. The machinery is part of a larger process, but cleaning it may not seem to be part of that process, even though cleaning is occasionally necessary. Hazardous substances are used to clean periodically, but doing so does not create any product. What is the best way to describe such a situation?

There are two ways to handle this. First, identify the cleaning activity as a process by itself. Instead of a product, the process creates a "desired result":

cleaned machinery. Second, the cleaning operation could be considered a *source* that is part of the process machinery being cleaned. When looked at this way, cleaning the machinery is a periodic step in the overall process.

Each way of identifying such unusual activities will result in a different basis for measuring pollution prevention progress later. As discussed in Sec. 8.4.6.4, pollution prevention progress is recorded through a ratio that measures how efficient processes are in utilizing hazardous substances *per unit of product,* with a goal of diminishing the amount of hazardous substance used or generated as nonproduct output for each unit of product made. If activities like cleaning, storage, material transfer, or maintenance are identified as processes, then progress is measured for each time the activity occurs. Measuring progress this way can hide the advantage of reducing the number of times the activities occur. If, however, they are identified as sources within larger processes, then the use and nonproduct output resulting from activities like cleaning will be measured by the amount required to produce a unit of product for the whole process, thus showing that reducing the need for such activities is a worthwhile pollution prevention technique.

8.4.5   Step 5—Grouping

Grouping makes pollution prevention planning easier by combining several similar processes or sources and treating them as a single aggregate process or source throughout your pollution prevention plan.

In the previous step, activity-product combinations were used to identify and define "production processes." Then, sources within those processes were found by locating the exit points for nonproduct output. While the numbers of processes and sources found may be large, many of them may be very similar. The same or similar raw materials may be used to produce several *similar products* in separate "batch" production processes or in parallel continuous production lines. These processes may also use the same or similar equipment. For example, one mixing vessel can be used to produce several different kinds of fragrances with only minor differences in the mixes of the same raw materials.

At the same time, nonproduct output may escape from *similar equipment* within production processes, such as through many valves of the same design. These situations indicate that there are production processes and/or sources that can be gathered together to make planning easier. It makes sense to treat similar processes and sources as if they were a single process or source.

Combining similar processes or sources together into a composite process or source is called *grouping*. Grouping focuses your attention on whether your similar operations are being run consistently. You may find that techniques that work well in one area are not being followed elsewhere. It can also highlight other pollution prevention opportunities for specific uses of hazardous substances. For example, you may find that a hazardous substance is used only for cleaning in between batches of different products. If similar products could be identified and run in sequence in a "group," you could reduce the amount of times cleaning is required and reduce your use of that substance. Finally, grouping reduces the workload surrounding pollution prevention because it shrinks the number of processes and sources the team must study by identifying "grouped processes" or "grouped sources" that represent their component processes or sources in the pollution prevention plan. Grouping does not eliminate anything from consideration in the plan, but it does organize what must be considered in a more manageable way.

Grouping is not a required step in pollution prevention planning, so your team should use it judiciously. Beware of inappropriate grouping, since badly grouped processes and sources will make later work confusing rather than streamlined.

8.4.5.1 Grouping Processes.   As your team worked through the previous step (Step 4—Identify Your Processes and Sources), production processes were defined around a product, intermediate product, or some other desired result. Products and desired results are the place to start looking for opportunities to group as well, because processes that produce similar results often can be grouped successfully. If those processes also use similar raw materials, then successful grouping is even more likely. Other similarities, like the function of a specific chemical (as a reactant or catalyst) or the use of similar equipment, can confirm the decision to group processes together.

Be aware that inappropriate grouping may cause problems. In grouping, the object is to collect several processes together which are similar enough in terms of their products, material use, and process steps to be treated as a single process. Grouping simplifies process evaluation by minimizing the number of times data need to be collected or recorded and by encouraging the discovery of pollution prevention techniques that will work for all the components of the grouped process. Logically grouped processes allow this; poorly grouped processes create situations where the data collected for a grouped process does not apply to some of its components. Keep in mind that once you have grouped processes together, they will remain grouped throughout the plan.

As an example of grouping processes *inappropriately,* consider a paint manufacturer that produces several colors of both oil-based and latex-based paints. Using color as the only criterion for grouping would be inappropriate. It could lead to yellow oil-based paint and yellow latex-based paint being in the same grouped process, for example. Logically, the latex and oil products should be in separate groups since they are manufactured with different types of chemicals. Color could be a criterion to further group the processes, within the latex and oil groups, to address any concerns with heavy metal content of the pigments, which may vary by color.

Another example of *inappropriate* grouping could involve a chemical manufacturer making organic polymers by adding different functional groups to a base polymer. It would be inappropriate to include a product whose active ingredient was added through an alkylation step with one that is added through sulfonation. The raw materials in these reactions are sufficiently different that these processes should be treated separately.

The Pollution Prevention Rules prohibit grouping production processes together with treatment or control processes. Pollution treatment processes are special because grouping them with production processes can blur the line between treatment and prevention. This is the only restriction on how your team can group. *Let the rule of common sense prevail.*

CASE STUDY: Top Shelf deals with 1056 production processes through grouping. Top Shelf's team had completed three generic process flow diagrams that described wallpaper making, gluing cotton gauze to PVC, and cleaning process machinery. However, the diagrams were more depictions of the basic tasks or steps at the facility than they were schematics of actual processes, which involved may details that were not pictured. To continue its planning, the team knew it would need more detail than that. To fill in detail for the wallpaper-making process, the team had three alternatives: (1) it could consider the way every product in the catalog was made as a separate process, (2) it could assume that the differences between products were inconsequential to their manufacture and consider only one simplified process, or (3) it could look for a level of detail between these extremes. Grouping plays a big part in setting that level of detail.

Sarah considered the options. The first option (considering each product separately) might apply to a facility that made a smaller number of less similar products, but at Top Shelf it would mean analyzing hundreds, if not thousands, of processes, many of which would be very similar. The second option (identify the facility as only one process) had some appeal since it might mean less work, but upon examination, this did not seem to be so. If all the product manufacturing techniques were aggregated into a single process, that process would not only make different products, but would use drastically different hazardous substances at different times to do so. She believed that this would make analysis of the process complicated and might hide opportunities for pollution prevention.

That left her with the task of finding grouped processes for the team to analyze. She enlisted the help of Travis, the maintenance supervisor, in finding these groups, since he was very familiar with the quirks of the facility's machinery and the problems associated with making several of their products. Initially, they thought of ink color, solvent type, design pattern, equipment (printing machines), and brand of dye resin.

Travis felt strongly that the groups they ended up with should differentiate between the two generations of printing machinery at the facility. While the two sets worked very similarly, the older one was generally less efficient, which indicated that there were opportunities for pollution prevention in the older equipment (short of replacing the machines altogether). Sarah agreed that their grouping decision should differentiate between machines.

Sarah's main concern, however, was with the way the facility used hazardous substances themselves. She wanted to do substance-specific analysis in her planning because she suspected that the facility could optimize some of its solvent mixtures to reduce use, nonproduct output, and costs. The facility used five organic solvents to solubilize dye resins. There were close to four dozen separate formulation recipes for different dyes, but as Sarah and Travis examined the different solvent mixtures, patterns emerged.

For instance, all of the mixtures of only MEK and MiBK had a ratio of between 3:1 and 5:1, MEK to MiBK. Other mixtures also had only small variations in the ratio of solvents. By assuming that slight variations in the percentages of solvents in similar mixtures did not make a significant difference in finding pollution prevention options for those mixtures, they were able to break out three solvent combinations that generalized the solvent use for formulating dyes at the entire facility. These three combinations accounted for all but two of the dye formulations. These two formulations, designated D23 and D37, were so dissimilar that they had to be tracked separately.

With the decision to group solvent mixtures as well as equipment, Sarah and Travis had completed their grouping decision, and they had done so in a way that made sense at their facility. Their work yielded 10 grouped processes by dividing different subsets out of the overall task of wallpaper making. First, they'd separated the older generation of equipment from the newer, and then they'd separated different solvent combinations from one another. There were also two nonpapermaking processes that were not grouped: gluing cotton gauze to PVC and cleaning process machinery. These processes, like D23 and D37, would be examined separately.

Sarah sent a memo to the other team members explaining the process definition and grouping decisions and asking for their comments. Thomas looked over the memo and decided that the grouping decision made sense and was consistent with the grouping criteria laid out in the Pollution Prevention Rules.

8.4.5.2 Grouping Sources.   In some instances it may be practical to group sources. Sources, as discussed in step 4, are the locations within processes where nonproduct output exits. The advantage of source grouping is similar to that for process grouping; several sources may be grouped together and treated as a single source. As with processes, grouped sources are treated as a single source.

The Pollution Prevention Rules set up pollution prevention tracking at the process level, so sources need to be related to processes to be tracked. Therefore, when sources are grouped, they must be grouped within the boundaries of a process or grouped process. This ensures that the sources can be tracked, and that their nonproduct output can be consistently counted. There may be equipment at a facility that could be grouped as sources, but because it is necessary to track pollution prevention at the process level to measure progress, such grouping must not be attempted. To keep tracking simple, it is done at the process level. Source-level tracking would allow grouping across process boundaries, but it would complicate tracking and reporting incredibly.

Except for the restriction limiting source grouping to the sources within a process, the criteria for grouping sources are the same as for processes. The emphasis, however, is more on *equipment* similarities than on chemical similarities. To get a feel for the utility of grouping via sources, consider a source grouping example for an oil refinery. At such a facility, there are many processes that produce many similar products. Some of these processes will have sufficient similarities to justify grouping them. Within one of those grouped processes, there could be hundreds of sources, but many of them will stem from very similar equipment, perhaps from a certain kind of valve. Some of the valves may be unusual in some way, but the rest could all be grouped together and treated as a single source within the grouped process for the rest of the plan. As such source grouping is repeated where appropriate, the number of sources for this process becomes more and more manageable.

Some source grouping can be tricky. For instance, similar vents should make up more than one group if their functions are significantly different, even though they may use nearly identical equipment. Analyzer vents may not present the same pollution prevention opportunities as flare vents or combustion vents. Reduced sample size, a typical prevention technique for analyzer vents, is obviously not applicable in the other two cases. Common sense indicates that these sources need to be grouped separately.

A final issue related to grouping is how it affects the work your team has completed. It is a good idea to revisit your process identifications and update them in light of the grouping you've done. If there is a process flow diagram, update it to show that your grouped processes and sources have replaced the processes and sources they are composed of.

In summary, the key to grouping is that processes or sources should be very similar if they are to be treated the same in the pollution prevention plan. If your team keeps this idea in mind, it should have no trouble making sound grouping decisions.

8.4.6 Step 6—Inventory and Record the Firm's Use and Nonproduct Output

In this step, your pollution prevention team will find *throughput* data for the facility and for the processes identified in steps 4 and 5. Facility-level data provide your team with a general understanding of how hazardous substances move through the facility. Process-level data focus more closely on the places where pollution prevention opportunities will be found. Process-level data can be found through *materials accounting* or *mass balancing*. These techniques track substances through each step of a process, and may locate unknown sources of nonproduct output along the way. Your team will also assess the *total costs* of using and generating hazardous substances at the process level.

Pollution prevention opportunities arise through understanding how and where hazardous substances move and function through the facility and its processes. In step 4, your team tracked hazardous substances qualitatively to identify the processes that belong in your plan and to find sources of nonproduct output within those processes. Qualitative information, however, is not adequate for pollution prevention planning; amounts are needed. The quantities of hazardous substances that enter a facility are spread among various processes as inputs. These inputs travel through process steps and leave processes as nonproduct output or as part of a product. Unless pollution prevention intervenes, that nonproduct output is either recycled out-of-process, treated, or allowed to escape as a fugitive emission. Regardless, hazardous substances eventually leave the facility, completing a throughput cycle. These data will do three things: (1) confirm or improve the understanding your team has of facility operations, (2) provide a sound way of prioritizing processes for more detailed analysis, and (3) establish core data on which to base a more detailed analysis. The data are found by accounting for every hazardous substance as it moves through its throughput cycle, starting with the whole facility and working down through processes.

Financial information collected early in the planning process can help focus the program. A comprehensive financial analysis may show that costs that are usually attributed to general facility overhead would be better accounted for as the price of using and generating hazardous substances in a particular process. Pollution prevention can reduce those costs. This financial analysis will complete Part I of your plan.

8.4.6.1   Elements of the Pollution Prevention Plan.   The best pollution prevention plans all contain certain information that has proven effective in identifying cost-effective pollution prevention opportunities. Most state rules on pollution prevention planning require facilities to collect that information, *but any effective plan will contain it.* The state generally does not prescribe how your facility should collect the information nor its format in your plan. The elements of a plan can be broken down into six categories:

1. *Personnel information:* An identification of those responsible for pollution prevention planning at the facility, and their certification of the plan.

2. *Facility-wide data:* What hazardous substances the plan covers and how those substances flow through the facility. These throughput data are also reported on the release and pollution prevention report. See step 10.

3. *Process identification:* What processes at the facility involve hazardous substances (as found in steps 4 and 5), how much of what product they make, and what unit is associated with this amount.

4. *Process-level inventory data:* The use of each hazardous substance, the generation of nonproduct output, the amount recycled, and the amount released for each process.

5. *Hazardous waste information:* The wastes generated at each process and how they are handled.

6. *A comprehensive financial analysis* that explores the real costs of using and generating hazardous substances.

Your team is likely to use available information for many of these categories. Other elements, notably those in categories 4 and 6, are specific to pollution prevention planning and may require special effort, direct measurement, and analysis to obtain. The remainder of this step explains each of the six categories in more detail.

8.4.6.2   Who Is Responsible for the Plan?   Top-level company officials (often the plant manager and the CEO, president, vice president, or owner) should understand and endorse the plan and its goals. Ideally they have followed the pollution prevention program throughout the planning process, perhaps as members of the pollution prevention team. These officials must certify their knowledge and acceptance of the plan and its goals. The name of an employee representative is also recorded in this section.

8.4.6.3   Facility-Wide Data.   The facility-level information on the overall use and generation of nonproduct output for each hazardous substance at the facility shows your team the big picture. It demonstrates where the largest hazardous substance use and generation is, which focuses process and source-level analyses. It also gives your team a gauge for measuring how successful it has been when subsequent process-level analysis is complete.

8.4.6.3.1   Where Will the Team Find Facility-Level Information?   Many of the records the company maintains will provide facility-level information. Therefore, the team need not make direct measurements at this point, although that option is certainly available. Some typical information sources include:

Bills of lading. The logs of material brought on site over the past year, and the product and waste shipped off site.

Blueprints. Plans of the facility include original design specifications, such as storage capacities.

Compliance data. Discharge monitoring reports, VOC inventories, hazardous waste manifests, and hazardous waste generator reports.

Release and inventory reporting records. Current and previous release and throughput inventory reporting forms [Toxic Release Inventory (TRI) form R and state release and pollution prevention reports (if applicable)]. See step 10.

Purchase records. The type and amounts of hazardous substances brought on site as determined by what the company paid for.

Process flow diagrams. Detailed schematic diagrams that show typical hazardous substance flows.

Sales records. The amount of product sent off site as recorded through invoices.

Waste hauling invoices. The amount and type of hazardous waste sent off site as recorded by haulers.

With data like these, the team can start assessing the throughput of each hazardous substance at the facility. Assessing throughput means tracking *inputs* through to where they become outputs. For a whole facility, *materials accounting* is the best way to do this.

Materials accounting means finding a general balance between the inputs and outputs of each separate hazardous substance at the facility, on the premise that all the materials entering a facility must come out in some form or another. By examining existing records, exercising engineering judgment, and gathering new monitoring data as necessary, your team should account for each hazardous substance going in and coming out of the facility during a reporting year.

There are four ways a hazardous substance is counted as facility inputs during a reporting year. Hazardous substances are inputs when they are:

1. *Stored at the facility on the first day of the reporting year.* To account for inventory from year to year, the amount of a hazardous substance stored at a facility when the reporting period begins is considered an input while the amount left in storage at the end of the period is considered an output. Beginning inventory should therefore equal the ending inventory of the previous year.

2. *Brought on site as nonrecycled raw materials.* The amount of new substance that your facility brings on site to use in its operations is an obvious input.

3. *Manufactured as products, coproducts, or nonproduct output.* Creating a hazardous substance on site is conceptually the same as bringing it to the facility from off site.

4. *Recycled outside of processes and used on site as raw materials.* Materials that are recycled, either on site or off site, and used in facility operations are essentially the same as nonrecycled raw materials. In measuring input, the origin of a substance doesn't matter; any material used as a raw material is an input. A goal of pollution prevention is to develop lean, efficient processes that use and produce the minimum amount of hazardous substances necessary. Out-of-process recycling, while reducing the amount of a hazardous substance that a facility purchases as raw material, does not reduce the *demand* for that substance within the facility.

The second half of facility-level materials accounting involves measuring outputs. The Pollution Prevention Rules define four ways a hazardous substance can be counted as facility outputs during the reporting year. Hazardous substances are outputs when they are:

1. *Stored at the facility on the last day of the reporting year.* The difference between the amount stored on site on the first day of the reporting year and the amount stored on the last day accounts for changes in inventory over the reporting period.

2. *Consumed at the facility.* Hazardous substances that are molecularly altered are said to be consumed. When a hazardous material is consumed, it no longer exists at the facility, and must be counted as an output. The materials it becomes may be inputs of another hazardous substance (see *Manufactured as products, coproducts, or nonproduct output,* above).

3. *Shipped off site as a product.* Hazardous substances that are shipped as product leave the facility as an output. If a substance is molecularly altered to become a product, however, it should be counted as consumed, not shipped as a product.

4. *Generated as nonproduct output.* Hazardous substances that are not consumed or part of a product are considered nonproduct output.

8.4.6.3.2 Quantifying Facility-Level Nonproduct Output. Nonproduct output, the last type of output, is a quantity most facilities do not routinely measure. Traditionally, their regulatory compliance has been based on what is released to the environment after treatment, rather than on nonproduct output, which is what leaves processes prior to treatment. Nonproduct output, however, is a quantity that managers should become familiar with because tracking it reveals trends in both environmental management and operating efficiency.

Nonproduct output can be determined in several ways. First, it can be measured directly as it leaves processes. However, it is usually difficult to use this method to find all the nonproduct output generated at an entire facility. Other methods infer nonproduct output from known facility-level data. If no recycling is taking place, nonproduct output can be estimated by relating emissions to the efficiency of the treatment system used. A disadvantage of this method is that it does not account for fugitive emissions, since they are not treated.

Another method uses information already reported to the U.S. Environmental Protection Agency on the federal form R. In the form R, facilities must report the quantities of a hazardous substance that are released as fugitive emissions, treated on and off site, recycled on and off site, and used for energy recovery on and off site. These are the components of nonproduct output, so their sum is the facility-level nonproduct output total. The accuracy of this quantity is dependent on the accuracy and completeness of the components.

A final method is to infer nonproduct output from the materials accounting. If the materials accounting has been accurately completed for everything but nonproduct output, then the difference between the inputs and the known outputs should provide a reasonable nonproduct output figure. It is recommended that your team calculate nonproduct output several different ways to find a consistent answer.

When nonproduct output is known, the total inputs and total outputs for each hazardous substance at the facility should roughly equal one another, completing the materials accounting. In a facility-level accounting, inputs and outputs should be close, although this approach is not as exacting as a mass balance. Your team must choose the level of accuracy that will satisfy your firm's needs. If there is a gross discrepancy between inputs and outputs, then your team has lost track of some of your hazardous substances. Perhaps there is a large source of nonproduct output that was overlooked in step 4 or quantities consumed were counted a second time as being shipped in product. If the reason for the discrepancy cannot be found easily, process-level analysis may locate the problem later.

8.4.6.3.3 Quantifying Facility-Level Use. Your team can also estimate the facility-level use of each hazardous substance from its material accounting data. Facility-wide use includes more than the amount of a hazardous substance purchased as raw material; it is the amount of the substance entering the facility as any of the four inputs discussed above (stored on day one, brought on site, manufactured on site, and recycled) minus the amount of substance left in storage at the end of the reporting period. Note that the materials accounting equation for nonproduct output (all inputs minus the amount stored, shipped in product, and consumed) is very similar to the *use* calculation (all inputs minus the amount stored). In fact, if none of a hazardous substance is consumed or shipped in product, *use* equals nonproduct output. Ideally, many solvents that a facility uses for cleaning will be accounted for in this way, since they are not involved in chemical reactions or product formulation.

This calculation of use highlights a difference between out-of-process recycling and pollution prevention. Out-of-process recycling can reduce the amount of raw materials that the company purchases, but it cannot reduce a process's appetite for those materials. In other words, out-of-process recycling allows a facility to self-generate or regenerate some of the raw material it needs, but it does not reduce the demand for raw material per unit of product. Pollution prevention *can* reduce those needs by making processes more efficient.

After the facility-level accounting has been completed and the results are recorded in the plan, the team should meet and discuss what the data imply about the company's use of hazardous substances, its record keeping, and its priorities for reducing the use and generation of hazardous substances. As your team moves through this step and on to process-level questions, keep in mind that the process-level data should add up to equal the facility-wide totals.

CASE STUDY: Top Shelf collects Part I facility-level information.

Thomas was ready, at this point, to develop use and nonproduct output figures. He worked on the nonproduct output of toluene first, using a materials accounting method. First, he summed up all the facility-level inputs that the facility was reporting for the year (starting inventory, quantity brought on site, quantity manufactured on site, and quantity recovered from on-site out-of-process recycling). The quantities manufactured on site and recycled on site were both zero. Second, he subtracted the sum of the known outputs (ending inventory, quantity shipped in product, quantity consumed) from the summed inputs. The quantity shipped in product and the quantity consumed were both zero. The result of subtracting known outputs from the inputs was the only unknown output, nonproduct output.

Thomas repeated this operation in the same fashion for the other hazardous substances. One exception was MiBK. The process that glued cotton gauze to PVC sheets used MiBK as a solvent. Approximately 15 percent of that MiBK was trapped in the glue and shipped in product. Quantities shipped in product are not nonproduct output. Thomas adjusted the facility-wide totals for MiBK to account for the amount shipped in product and recorded the facility-level nonproduct output for all substances.

Next, he worked on use numbers, once again starting with toluene. Since toluene was used as a solvent that was not incorporated in product, he assumed that none of it was consumed and that only inconsequential amounts were shipped as product. The facility did not recycle any toluene on site. With these variables zeroed out of the use and nonproduct output equations, use should equal nonproduct output. The assumptions held true for all the hazardous substances except MiBK. Although the MiBK shipped in product was not counted as nonproduct output, it did count toward use. Thomas counted the MiBK shipped in product in the facility-level use totals.

Finally, as a check, Thomas compared the facility-level nonproduct output numbers he'd calculated against the total of the quantities he recorded in EPA's form R. The answers to these questions, he realized, were the components of nonproduct output (the quantities treated, recycled, used for energy recovery on or off site, and the quantities released). The total of the components was within 10 percent of the nonproduct output he'd found using materials accounting. This was close, but meant that there were minor accounting problems since the results of the two methods should theoretically equal one another. Right off the bat he thought of two possible sources of error. First, fugitive air emissions were sometimes used as a catchall to account for discrepancies that may not be caused by actual leaks. That leeway made tracking less precise than it would have been if there were no fugitive emissions. Second, there were several large waste streams that were not sampled systematically. Questionable data on the hazardous substances within those waste streams could easily introduce a wobble in the nonproduct output amounts. Thomas believed that the analyses the team would do later at the process and source level would close the 10 percent gap, so he presented the materials accounting data to the team and explained his findings.

8.4.6.4   *Process Identification and Assigning Units of Product.*   In steps 4 and 5, your team divided the operations at the facility into processes and grouped processes. The team described them, perhaps with a flow diagram, and assigned them unique identifiers. That information is the beginning of the required process-level information of Part I of the plan, and should be recorded in the plan at this point, if this hasn't already been done. Two other sets of information must also be recorded about each process, (1) whether and how the process was grouped and (2) product data.

If your team has grouped any of the facility's processes, then your plan must include a description of the grouping decision. The description gives your team a way of linking the grouped process in the plan to the physical processes of which it is composed.

Your team used products to identify processes; products are also the key to meaningful process-level analysis and reporting. Identify your processes' products and record a *unit of product* for each in Part I of your plan. The unit of product is what makes nonproduct output and use comparable from year to year because it separates changes due to pollution prevention from those due to increasing or decreasing production. When the use or nonproduct output of a process is reported on a *per unit of product basis,* an efficiency ratio is established. Your team will be able to reliably measure the effect of pollution prevention using *efficiency ratios* because they eliminate fluctuations in use and nonproduct output that are caused by shifting production levels. Regardless of production levels, pollution prevention will reduce the amount of hazardous substance used or generated per unit of product, since the process is functioning more efficiently.

Choosing a unit of product is a long-term decision. The Rules require that production units remain the same from year to year. Changing them would make year-to-year pollution prevention measurement inaccurate. The unit of product must therefore consistently reflect what a process does. Choose one for every product and intermediate product your covered processes produce. While this seems simple in the abstract, it can be difficult for certain kinds of processes. The simplest type of product to define a unit for is one that is discrete and can be counted. Aluminum cans are an example. Measuring hazardous substance use or nonproduct output per can makes sense.

Sometimes, the nature of a process makes it difficult to define an appropriate unit. For instance, it may not make sense to define a unit of product for an electroplating process as *items plated* if the items vary in size and shape. Instead, the most appropriate unit might be the number of square feet of material plated, which, unfortunately, is more difficult to measure and track. The New Jersey Office of Pollution Prevention is preparing packages to assist industry groups, such as electroplaters, with problems that are specific to their operations.

Activities that do not make a product directly, but that take place during facility operations, are another special case. If your team identified such an activity as a source in another process, then it is part of that process and does not have its own unit of product. If your team identified the activity as a separate process, then a unit of product is needed. For instance, a cleaning activity could have *cleaned coating machines* as its unit of product. That unit of product would establish a meaningful efficiency ratio for measuring the use and nonproduct output of each hazardous substance used or generated through the activity.

Finally, units of product should measure what the process actually produces and should be consistent with units of input. Units of product that are based on money are generally poor, since they introduce fluctuations due to the value of money and explain very little about the process. Likewise, units of product that are inconsistent with units of input make it difficult to relate the amount of product to its components. If a product contains a hazardous substance that was measured in pounds as an input, then measuring the unit of product in gallons clouds materials accounting. Make these units as consistent as possible.

CASE STUDY: Unit of product.   Sarah, the plant manager, assigned units of product to each of the production processes the team had identified. For the 10 grouped processes, this was quite simple; each of them produced wallpaper, which the facility measured and sold by the yard, at a standard width of 62 inches. The production process that glued cotton gauze to PVC sheets was also easy to find a unit of product for. Sarah decided that the intermediate product of that process was cotton-backed PVC sheets, which was measured in yards, just as was finished wallpaper.

Choosing a unit of product for the two cleaning processes was more difficult. Top Shelf did not sell the result of these processes, nor did they create anything that went on to become a product the company sold. Their unit of product, therefore, would have to be unconventional. She thought this over and decided that *cleaned machines* might be the best way to measure the result of these processes. Nevertheless, she had begun thinking about production process efficiency, and realized that cleaned machines would only help measure how efficient the cleaning process was, but would not measure improvements achieved if the team found ways of cleaning the printing and gluing machines less frequently. The solution to this, she reasoned, is to divide up the cleaning processes and include their hazardous substance use and generation in the numbers of the processes they clean. Doing that redefined the cleaning activities as sources within the printing and gluing processes. Reductions at those sources would be measured per yard of wallpaper or cotton-backed PVC. More efficient cleaning would still show up as reduction in nonproduct output and use, and increased efficiency from cleaning less frequently would also show up.

Sarah discussed this idea with the rest of the team, who backed her strategy. They revised their process definitions by deleting cleaning activities as separate processes and adding them as sources in the other processes.

8.4.6.5   *Gathering Process-Level Chemical-Specific Data.*   Process and source data are the mainstay of pollution prevention planning. Obtaining it can lead to

cost-effective pollution prevention investments. *Materials accounting* and *mass balancing* are two methods of gathering these data. The Pollution Prevention Rules require that facilities complete a materials accounting for all production processes. Mass balancing can be used as a second stage to clarify complex processes and to fill in any information gaps left by materials accounting. If there are any *grouped processes* or *grouped sources,* use them in materials accounting and mass balancing since this is the work that grouping is designed to simplify. Like facility-level materials accounting, process-level data gathering tracks individual hazardous substances as they move through processes.

8.4.6.5.1   Stage 1: Materials Accounting at the Process Level.   Materials accounting at the process level parallels materials accounting at the facility level. Begin with existing process-specific records, including: measured rates of flow in and out of a continuous process, batch sheets, product yields, and product specifications. Using a process flow diagram as a guide, find values for the inputs and outputs to each process. A successful materials accounting will establish a general balance between how much of a hazardous substance enters a process and how much leaves the process as output. If necessary, your team may want to account for nonproduct output by finding the material flow through *sources*; however, the detailed analysis could be delayed until later. Your team should choose the level of accuracy that will satisfy your firm's needs.

Your team should seek nonproduct output in the process inventory as components of the facility-wide nonproduct output that has already been measured. In other words, for each hazardous substance, the nonproduct output found in all processes should add up to the total facility-wide nonproduct output. Process-level nonproduct output information is used to target processes for further analysis in the next step. The portion of total nonproduct output contributed by each process will be an indicator of which processes to target for further analysis in the next step.

Sometimes, your team will find that nonproduct output that is indicated by the known inputs and outputs cannot be found leaving a process. It is important to hunt down these unexplained losses because they often are opportunities for pollution prevention. If hazardous substance inputs do not show up in the product stream, are not consumed, and cannot be accounted for as nonproduct output, then your team should look for additional sources that may have been overlooked in step 4.

8.4.6.5.2   Stage 2 (Optional): The Process-Level Mass Balance.   A second stage of the process-level analysis that your team might use is a detailed *mass balance* of the flow of hazardous substances through your processes. Mass balances offer greater accuracy, but also require greater work, than materials accounting. In a basic form, mass balances are defined by the statement:

$$\text{Mass in} = \text{mass out} + \text{accumulation}$$

In other words, anything that goes into a process and does not remain, must exit the process. This statement is similar to the general balance that your team tried to achieve through materials accounting, but a mass balance requires closure of the statement. *Closure* means that inputs must equal outputs (plus accumulation). To achieve this accuracy, samples and measurement replace existing records and estimation. Closure also means that the entire process should ultimately be balanced on a pound-for-pound basis, rather than accounting for each substance alone as in materials accounting.

Mass balances can be time-consuming. They often require direct measurements and sampling, and always require some expertise to determine how the reactions of inputs lead to known outputs. Because of the resources a mass balance requires, some facilities choose to rely on materials accounting which, while not as accurate as a mass balance, may provide adequate information for process evaluation. Later, after your team has targeted some processes as the ones for which pollution prevention is likely, it may be worthwhile to conduct a mass balance for those targeted processes.

If your materials accounting yielded questionable information, however, it is a good idea to do a mass balance now. The careful measurement required in a mass balance should clear up problems in the materials accounting. The process of sampling and measuring flows itself sometimes leads to improvements in process control and efficiency while yielding the data needed for a mass balance. If so, remember to record those improvements as pollution prevention in step 8. At the end of this stage, your team will have collected almost all of the Part I information. Only a few elements remain.

CASE STUDY: Part I for processes.   The team members found the facility-level use and nonproduct output numbers revealing. Most of them were surprised to learn how high the totals were. Nevertheless, they knew that if it they wanted to reduce the amounts of hazardous substances the facility was using and generating, they would have to do it at the process level. They needed reasonable estimates of what happened to hazardous substances at each of the processes they'd identified. The team was skeptical about whether they could produce a representative process-level picture without collecting lots of extremely detailed information. Thomas and Sarah decided to work together to find out.

The data on hand that best explained how hazardous substances were used in production processes were the solvent formulas used in grouping. By using the solvent formulas, Sarah and Thomas felt they could get reasonable use and nonproduct output estimates by back-calculating from the amount of product made at each production process. They invited Emily Cruz, the firm's financial manager, to join their mini think tank. Emily had already started developing a spreadsheet that would take order/production figures and categorize them by the processes that made them. There were two components to achieving this task. The first was easy; each product could be assigned a solvent formula that corresponded to the process. The second component was more difficult. Emily needed to know whether an order was completed on new or old equipment to decide which production process to assign the order to. She was worried about finding this information since any wallpaper could be made on any machine.

Fortunately, when an order came in and was sent to the plant floor, the plant manager assigned a tracking number to it that was used to move the order through the printing process to the warehouse, and from there to the customer. The number included customer identification, a design code, and a code that routed the order to a specific machine. The plant manager coded these orders this way as part of production scheduling, which was always hectic since Top Shelf sometimes worked with a just-in-time inventory system. When an order came off a printing machine and was sent to the warehouse, its tracking number was recorded electronically. All Emily had to do was search a spreadsheet of completed orders to find the numbers that ended in either 11 or 12; they had been completed on new machines.

With Emily's spreadsheet, Sarah and Thomas believed they could relate the solvent formula and production data to find process-level *use* estimates by using the following formula:

Hazardous substance used = (yards of product) $\times$ (pounds of solvent/yard) $\times$ (% hazardous substance in solvent)

However, they did not know what the pounds of solvent per yard of product quantity would be for each production process. They decided to estimate the amount for the next batches of wallpaper made on both the new and the old equipment. These batches gave them the numbers they needed for the two types of equipment: 0.025 lb/yard for the old equipment and 0.02 lb/yard for the new equipment. Plugging these values into their formula gave them the annual use numbers they were looking for.

Next, the team needed to find process-level nonproduct output numbers. Nonproduct output was a new reporting concept for which they had no data at all at the process level. However, they believed that nonproduct output would equal use since there was no recycling and, except for the gluing process, no hazardous substance shipped in product. They adjusted the nonproduct output numbers for MiBK in the gluing process to account for the 15 percent MiBK that was always left in the product.

To check their work, Thomas summed up the process-level estimates for each hazardous substance and compared the sums to the facility-level totals. The sums for three of the solvents were in rough agreement with the facility-level totals, but acetone and MEK fell short by close to 30 percent. Thomas could not figure out why, so he asked other team members for their thoughts. Travis knew off the top of his head that MEK and acetone were used for cleaning. He pointed out that the method they'd used to find process-level data focused only on production and did not account for cleaning, even though they had decided to include cleaning as sources within the production processes. He told the group that the missing MEK and acetone must have been used for cleaning, but that he was surprised that they used as much as 30 percent for this activity. He had always thought the number was closer to 10 percent. The team divided up the solvent quantities into the processes that used them, on the basis of the level of production for each process. They revised their process-level estimates for acetone and MEK and recorded them in the plan.

8.4.6.6 Hazardous Waste Information. This category covers how *hazardous waste* is managed at both the facility and process level. Since nonproduct output often results in hazardous waste, this information is important to your planning. Under the Pollution Prevention Rules, your team must record the amount of hazardous waste produced during the year for each process and for the facility as a whole. It must also record how that hazardous waste is handled, either through *recycling,* or by a treatment, storage, and disposal (TSD) facility. Most of the information required for these categories is already reported by the facility in the manifests for hazardous waste shipments and in annual hazardous waste generator reports. The process-specific data should show up through materials accounting or mass balances.

8.4.6.7   Financial Analysis of Current Processes.   When it has gathered information for the previous five categories, your team may have a new appreciation for the company's involvement with hazardous substances. It is beneficial to find a measure of the real costs of that involvement as well. Your team already knows basic financial data, through the business records that contributed to the process-level hazardous substance inventory.

Purchase prices and disposal fees are part of those records and tell part of the story, but your team may be surprised at how many other costs are attributed to general facility overhead which would be more realistically accounted for as a cost of hazardous substance use and generation at a particular process. Assessing these hidden costs will help the company make better investment decisions. These costs are intended to be included with the costs normally assigned to a process, such as raw material costs, energy costs, and labor.

Finding these costs gives your team a basis for analyzing the cost-effectiveness of pollution prevention options. Knowing these costs is the first step in completing a total cost assessment, which is recommended, though not required, by the Pollution Prevention Rules. Total cost assessment is a managerial decision-making tool that can evaluate the return that pollution prevention or other investments will have on a process. An advantage of total cost assessment is that costs that are seldom counted in other financial analyses are built into this system.

All the costs which are directly linked to hazardous substance management and generation should be considered in a total cost assessment. These include all those required by the Pollution Prevention Rules, plus some others. In some instances, such as hazardous waste disposal, the costs are accounted for, but may be detached from the specific processes that cause them. Allocate those costs to the processes that generate them. Any reasonable formula for assigning nonproduct output costs to specific processes is better than lumping them together in a single overhead account, because overhead costs hide opportunities for savings.

Some types of hazardous substance costs may not be recorded anywhere. These are costs to your facility that are caused by one process, but are accounted for as a cost of a different process. Untangling such accounts will both demonstrate the total costs of nonproduct output at the facility and pinpoint where the most profitable opportunities for pollution prevention investments may be.

Your team's sources of cost data will be found all over the facility, including: purchasing, materials management, financial management, environmental protection, and production. While it may be difficult to disaggregate the costs from each department and associate them with individual processes, the time spent finding nonproduct output costs now will save time and dollars later in the plan when your team considers pollution prevention investment options. The Pollution Prevention Rules require that facilities complete a "comprehensive financial analysis" of the cost of using and generating hazardous substances for each production process.

8.4.6.8   Relation to Ongoing Reporting.   A facility and process-level inventory should be kept up to date and available in the future. The sources used to gather data for Part I analysis should be built into a framework that can be used repeatedly for reporting to the state through plan summaries (see step 10) and release and pollution prevention reports. Once your pollution prevention program is in effect, progress toward achieving reductions in nonproduct output generation and hazardous substance use will be recorded in a companion section to Part I called Part IB. This is the same information that is reported to the state in the pollution prevention plan progress report. Once the plan has been in effect for a year and progress has begun, this information is recorded. Section 8.4.11 (Step 11—Tracking and Reporting Progress) explains the relation of the plan progress report to Part IB of the plan.

8.4.7 Step 7—Targeting

Targeting means prioritizing your processes and sources to determine which ones to examine in Part II of the plan. Many factors will enter into your decision, including: the prospects for reducing your use, nonproduct output generation, and release of hazardous substances; the opportunity for significant cost savings; and the relative ease of dealing with one source or process over another. The Pollution Prevention Rules require that, together with any other considerations that enter into your decision, you target at least 90 percent of your use, 90 percent of your generation, or 90 percent of your release of hazardous substances (Table 8.1).

8.4.8 Step 8—Finding and Analyzing Pollution Prevention Options

In this step, your team will think creatively to devise, analyze, and choose pollution prevention options for processes or sources it targeted. Your team may need

TABLE 8.1 Relative Percent of Hazardous Substance Nonproduct Output

Hazardous process identifiers	Hazardous substance nonproduct output, lb	Percentage of nonproduct output
M1/E1	71,267	32.45
M1/E2	11,830	5.39
M2/E1	46,781	21.30
M2/E2	10,096	4.60
M3/E1	51,167	23.30
M3/E2	9,149	4.17
D23/E1	7,895	3.60
D23/E2	1,419	0.65
D37/E1	5,409	2.46
D37/E2	789	0.36
Gauze glue	3,800	1.73
Total	219,601	100.00

to develop additional detailed information for the targeted processes and sources to find prevention options and to pick the ones that are technically feasible and fiscally sound.

The second half of the plan is about finding and implementing investments in pollution prevention. Most of the data your team will need has been developed in the previous steps. However, there is some information, specifically source-level nonproduct output data, that your team will want for targeted processes, which is not necessary in Part I. Also, if your team chose to put off mass balancing before, doing it now can expand the number of pollution prevention opportunities it is likely to find.

8.4.8.1   *Quantifying Source-Level Nonproduct Output.*   In completing the previous steps, your team identified, grouped, and collected data on sources. Sources, as the points where nonproduct output leaves processes, are excellent places to look for pollution prevention opportunities, but your team will need to know a good deal about them. Your team has already identified them, and the hazardous substances that pass through them should be known from process-level materials accounting. However, the annual quantities of nonproduct output that are generated at each source are probably not known.

The Pollution Prevention Rules require your team to find source-level hazardous substance quantities for the targeted processes. Knowing these amounts will lead to pollution prevention where it can do the most good. Also, knowing the quantities of nonproduct output generated at each source will be necessary if your team decides to conduct a mass balance for its targeted processes.

Mass balancing is not required by the Pollution Prevention Rules, but if you apply a mass balance method to your *targeted* processes and sources, you may find more pollution prevention options than if you do not. This tool gives your team a detailed view of your targeted processes that is not matched by any other kind of analysis. Mass balancing was discussed in step 4. If you intend to do a mass balance for pollution prevention planning, it is recommended that you do it before continuing this step.

Your team does not have to use a mass balance to find its source-level nonproduct output—it can use the simpler materials accounting system. The advantage of a materials accounting approach is that it is simpler; the disadvantage is that it is less accurate. Once source-level nonproduct output has been determined for all the sources in each of the production processes being examined in Part II, the team can begin to brainstorm for available pollution prevention options at those sources and processes.

CASE STUDY: Top Shelf completes its data collection (sources).   The plant manager called a lunchtime meeting of the team to start the ball rolling on Part II of the plan. She turned the meeting over to the environmental manager to explain to everyone where they were in the planning process.

Thomas explained, "At our last meeting, we targeted 10 processes on the basis of the amounts of various solvents they use. We're going to look for pollution prevention options for those 10 processes, so we'll need detailed data on how hazardous substances leave them as nonproduct output. We're required to quantify the amount of nonproduct output generated at each source in these targeted processes. I've talked this over with Sarah, and she has an idea about how we'll get source-level numbers."

Sarah explained her approach, "There are basically two forms of nonproduct output flowing through the sources in our targeted production processes: liquids and air emissions. The liquids are easier to find and easier to measure. Let's quantify this large, easy-to-find nonproduct output first, then we'll move on to the rest. If we have good numbers for the liquid nonproduct output sources, I think Thomas and I will be able to come up with decent estimates of air nonproduct output source data."

The team members agreed with Sarah's approach. They reviewed process flow diagrams for the targeted processes and listed the liquid sources of nonproduct output, which included mixing vats, dye reservoirs, pump liners, piping, and ink trays. Because the steps of each process were essentially the same, the sources were qualitatively similar, but varied in composition and quantity of substances used, and needed to be quantified separately.

The team members had to estimate the hazardous substance nonproduct output from the sources in each targeted process. They chose to do so by calculating amounts based on representative runs of some of the processes. These runs required more detailed analysis than was conducted previously in step 6. Now, the team members would need to do some actual measurements for the sources they identified. Travis, who sometimes bore the brunt of Sarah's production schedule headaches, asked that they minimize the amount of time the team members spent on the plant floor measuring liquid nonproduct output. He suggested that they could probably estimate the amounts they needed if they measured four runs on both the new and old machines: a long one, a short one, one using the formula for the highest-vapor-pressure solvent, and one using the formula for the lowest-vapor-pressure solvent. Then, the unknowns could be inferred from the other measured data. Sarah said she couldn't guarantee that they'd have all the data they needed without checking out some other runs, but she promised they'd take the measurements when there was a lull in orders.

Over several weeks they collected data for the runs Travis had recommended. For each run, the team was careful to measure the liquid residue from both production sources and cleaning sources. When all the representative runs were completed, the team was able to make inferences and calculate liquid nonproduct output amounts for all the targeted processes. Fortunately, the volatility of solvent mixtures did not have a significant effect on the amount of liquid nonproduct output leaving each source.

When the liquid nonproduct output measurement phase was completed, the team members met again to see if, as Sarah hoped, they could infer air source data now that they had other source numbers. They agreed that the numbers for the liquid sources were quite good, so the difference between the liquid sources and the total solvent used in the runs they'd measured should approximately equal the amount of nonproduct output leaving the process from air sources. This assumption was bolstered by their discovery that the volatility of the solvent formulas did not seem to make a significant difference in the amount of liquid they measured in otherwise similar runs. Therefore, they did not have to worry about very different evaporation rates between solvent mixtures.

The team agreed that the largest of the three air sources was drying wallpaper once the ink had been applied. During that step of the process, the object was to drive solvents out of the paper, leaving only ink. While the team agreed that this was the largest of the air sources, it also seemed to be very difficult to estimate its magnitude directly.

At this point, Thomas remembered that the air permit applications for the printing equipment were based on source-specific information. He gathered them together. Applications for the new equipment were up to date and contained data on emission rates in pounds per hour both before and after treatment. He realized that the before treatment data were actually the nonproduct output information he was looking for. In reviewing the calculations, he saw they were based on exposed surface area, and the worst-case high-vapor-pressure solvent. He was able to use the other solvent mixes and time per batch to come up with reasonable estimates of the nonproduct output for each hazardous substance generated in the targeted processes.

When the data were recorded and the team had a chance to look it over, it appeared that nonproduct output was generated in two ways. First, some nonproduct output was generated in constant amounts every time a run was completed, regardless of the run's size. For instance, the pump liners usually had a constant amount of liquid left at the end of each run, except for the processes run on new machines, which consistently had approximately 50 percent less left in the liner. In fact, the team realized that 15 of the 17 sources were a function of the number of batch changeovers. Second, the remaining nonproduct output was generated in a direct relation to the size of the run. Drying wallpaper was one of these second kinds of sources. Only two air sources (PA2 and PA3) were related to the yards of wallpaper produced.

Next, the team used data from the representative runs to calculate annual quantities of nonproduct output generated from each source. They realized they needed to use different methods to calculate nonproduct output for the different sources. For most sources related to the number of batch changeovers, the team needed information on the batches run for each process. Emily was able to use the new production data for each process. The new data were used in the calculations. The two air sources that were directly related to the yards of wallpaper produced were easy. All they had to do was multiply the nonproduct output per yard developed in the representative runs by the total yards of wallpaper produced for the process.

By adding together all the source data for each process, the team developed new process-level data. They compared the new process-level data to the original estimates from step 6. Most of the estimates were close, but the new estimate for acetone in the old equipment was much smaller than the original.

At this point, the team realized there was an important difference in the estimates. In step 6 the team based its estimates on production, while in step 8 it used the number of batches for each process. This highlighted a trend in the production scheduling. The team knew the new equipment was more efficient and usually scheduled shorter batches on one of the four new machines. This made the average batch size for the new equipment much smaller than the old. At the same time, this meant that many of the sources in the new equipment generated more nonproduct output per yard of wallpaper produced than the older equipment. The original estimates, which essentially assumed that batch sizes were the same for all processes, significantly overestimated losses for the old equipment, while the estimates for new equipment underestimated those losses.

The team felt that the new estimates were more accurate and used them in their next steps as they prepared to look at developing pollution prevention options for each source.

After the necessary data were collected, Thomas prioritized the sources by the amount of nonproduct output generated at each targeted production process. He distributed copies of this data to the team members and asked them to use them to prepare for the next meeting by thinking up pollution prevention options for those sources.

8.4.8.2 *Generating Options.*   The exciting part of pollution prevention planning begins here. This is where your team stops collecting data and begins actively looking for ways to reduce your facility's involvement with hazardous substances.

Think about how your team will find pollution prevention options. A good way to get started is to have a member of the pollution prevention team present each targeted process or source to the rest of the group, perhaps together with a schematic or process flow diagram. From this starting point, the team will develop its ideas. Their understanding of how the targeted production process or source functions is vital to developing potential pollution prevention options. Detailed narrative descriptions of the targeted processes or sources provide this understanding. These descriptions include information about any activity that occurs in the process, the overall methods used to achieve the desired result, and the specific techniques used in that method. Once the descriptions are complete, gather your team together to begin identifying pollution prevention options.

Think creatively . . . and fundamentally. Pollution prevention techniques fall along a continuum from fundamental changes of processes and sources to increased efficiency in what already exists. Your team should look for ideas all along this continuum. At one end, there are options that address fundamental questions about your firm, like: What do you sell? Who are your customers? Do you sell a general product to a wide array of users or do you deal with a set of customers, providing them with specific supplies that might be interchanged with something similar or better? Depending on the answer to questions like these, you may be able to eliminate some of your sources (or even processes) altogether by reformulating products or by selling your customers another product your firm makes that does not involve the targeted source or process, but which will serve their needs.

For instance, a paint manufacturer could achieve major pollution prevention progress if it moved away from oil-based paints toward water-based latex paints. That kind of pollution prevention comes from asking fundamental questions about your firm. In the case of the paint manufacturer the question might be "Do we make paint or do we make oil-based paint, and what is the difference to us and our customers?"

If it is impossible to make this type of fundamental change, there are many pollution prevention options that leave processes essentially the same, but alter the hazardous substances they use. For instance, a process that uses an organic solvent might function just as well using a nonhazardous aqueous solvent.

Finally, at the other end of the continuum are options that involve the same chemicals in the same process, but use them more efficiently, thus reducing the use and/or nonproduct output in the process. Equipment modifications, changes in operating parameters, and improved maintenance ("housekeeping") fall into this category.

From these general methods, your team needs to find specific prevention measures for the targeted processes and sources. The team can use any problem-solving system, including answering some targeted questions, conducting a brainstorming session, and looking to outside sources of information. As your team looks for options, start the search using the work that has been done in the previous steps. The nonproduct output, process, and source data that have already been collected is an important and useful base from which to begin looking for available pollution prevention options at the facility.

The answers to a series of questions about your facility may lead to pollution prevention options. Such questions can help the team think fundamentally about pollution prevention and how it relates to a targeted process or source under consideration. Some of these questions include:

- Can we meet our customers' needs with an altered product that generates or uses smaller amounts of hazardous substances?
- Why must we use this particular material?
- Are there simple changes in operations that will prevent pollution?
- Can we substitute less-hazardous or nonhazardous substances for ones we are using now?
- Are there equipment modifications or upgrades we can make to reduce non-product output?

After discussing the issues and recording the ideas raised by these questions, your team can consider questions directed toward more-specific pollution prevention techniques:

- Are our maintenance procedures and schedules optimized?
- Do the equipment operators use the most efficient procedures or would retraining be appropriate?
- How efficient are our housekeeping procedures?
- Are raw materials delivered in optimum quantities at optimum times?
- Do production runs and schedules optimize material usage?

Questions like these will focus your team's thinking on topics that will lead to pollution prevention ideas. They will also lead to new questions that apply more closely to the processes at your facility.

Employees who work with your targeted processes and sources should be encouraged to submit their pollution prevention ideas or to get actively involved in the brainstorming sessions. Develop an easy way for them to make suggestions and offer a bonus for workers who come up with ideas that are used. Or publicize your pollution prevention efforts with an event, such as a facility-wide pollution prevention contest.

Brainstorming is an excellent way of tearing down the obstacles to employee involvement and creative thinking. Bring your pollution prevention team together with the individuals who work with the targeted process or source. The basic principle of brainstorming is that everyone gets an opportunity to suggest "outlandish" ideas and that those ideas are not eliminated before there is time to realize that they may not be so outlandish after all. Each person in the session should come up with as many ideas as possible to share with the rest of the group. Every idea is written down, but ideas are not evaluated at this point. Evaluation is put off until later to ensure that nothing stifles creative thinking during the brainstorming.

Finally, there are many places to get started with *seed ideas*. The EPA and the New Jersey State Technical Assistance Program have descriptions of options that have worked at facilities similar to yours. Industry trade groups are also a good place to turn. These sources will provide your team with assurance that it has not overlooked a simple, proved technique already used by another firm.

CASE STUDY: Top Shelf finds its available options.   Sarah and Thomas prepared for the next team meeting by assembling a folder for each team member. The folders held flow diagrams and tables of process-specific and source-specific nonproduct output data for each hazardous substance at the 10 targeted production processes. They wanted the team to be able to refer to these data during the next meeting, when it would brainstorm for pollution prevention options.

At that meeting, Sarah announced that the team would work through the 10 processes and their sources, and record any option that might qualify as pollution prevention. The team members had prepared for this meeting by noting ideas they had come up with during the preceding weeks. Jerry pointed out that his ideas applied to all the printing machines. He suggested that the team didn't need to go process by process since so many ideas applied universally. Sarah said she wanted to go through the targeted processes one by one because there still could be ideas that applied to only certain solvent formulas, such as raw material substitutions and product reformulation. Nevertheless, she agreed that some ideas were broadly applicable, so they'd be marked on her master copy of prevention options and automatically carried over to each targeted process.

Starting with process M1/E1, the team came up with ideas to reduce nonproduct output at the sources in each targeted process. They also had some ideas that reduced nonproduct output at all the sources in a process, such as production schedule changes and raw material substitutions.

During brainstorming, every idea was noted and the processes and sources it applied to were recorded. The ideas were not discussed during the session, but afterward Thomas said that he did not think ideas 4, 13, 15, or 18 would qualify as pollution prevention because they involved out-of-process recycling. The other ideas were pollution prevention and a revised table became the list of available options in the plan, which shows what options might be feasible at each targeted production process.

8.4.8.3 Analyzing Your Options.   When your team has found all the pollution prevention ideas it can, it should begin to evaluate those ideas. The first step is to screen them to be sure that they represent true pollution prevention techniques. People often have different understandings of what pollution prevention is, so your team may have included in its list of options some concepts that do not fit the definition of pollution prevention your facility is working under. For instance, most kinds of recycling and reuse are not pollution prevention in several states.

Do not discard the ideas that do not fit the definition of pollution prevention. Set them aside; your team may implement them outside of the pollution prevention plan or they may be worked into the plan if viable pollution prevention options are not found for a targeted process or source. All the options that do meet the definition of pollution prevention must be recorded in the plan as *available options*. The team will choose the options it believes the company should invest in from the available options list.

When the team has a list of true prevention approaches, go through the individual alternatives, discuss each one, and eliminate those that are fanciful or plainly unworkable. If your team is unsure about whether to eliminate an option, carry it over to the next step, where a more detailed analysis will reveal the answer.

To decide which alternatives among those remaining will be implemented, a *feasibility analysis* is required for each one. A feasibility analysis for pollution prevention planning consists of two parts: technical analysis and economic analysis. These analyses may be conducted at the same time, although information from the technical analysis may provide cost data for the financial analysis. If the team has found an obviously worthwhile option that it plans to implement, it is not necessary to do a detailed feasibility analysis.

8.4.8.4 Technical Feasibility Analysis. *Is it possible?* That is the first question you need to answer about a pollution prevention option. A more complete form of the question is, "Will our facility be able to use this in our process and will it reduce our use and/or generation of hazardous substances?" This can be easy to determine for ideas that involve changes in procedure, but for a process or equipment change, laboratory research and pilot plant level testing may be needed before you know whether an idea will work.

People from all phases of plant operations should be involved in the technical analysis. They will be the ones who design tests and experiments to show whether an idea will work and what its effect on use and nonproduct output will be. Throughout the technical analysis, financial managers will collect cost data to feed into the next phase, determining financial feasibility.

A first step in answering the question, "Is it possible?" is to know what "it" consists of. An identification of the pollution prevention option that describes how it relates to the processes and sources of nonproduct output it affects will tell your team what the repercussions of the option could be. For instance, one option that appears in the wallpaper case study is installing a closer-fitting roller trough on an older piece of equipment. A description of this option shows that this would require several new pieces of equipment, changes in procedure, and personnel retraining. There might also be new maintenance procedures. Other options could involve different energy needs or space configurations on the shop floor. Your team should learn if your facility can accommodate these kinds of changes. Impractical options should be abandoned.

At the end of your technical analysis your team will have a list of changes that could be made at the facility if money were of no concern. In the next section your team will work out which of them the company can afford to implement. The answer may be all of them or it may be only a portion of them.

8.4.8.5 Financial Feasibility. For those projects that prove to be technically feasible, the next step is to measure their financial feasibility. The essential question here is, *"Will this project be profitable?"* This is where the benefits and costs of an option are translated into concrete financial terms, the language that top management is accustomed to hearing. Then a choice will be made among the many investments competing for limited capital. A comprehensive financial analysis is required by the Pollution Prevention Rules at this point. Such an analysis, which compares to that done for each process in Part I, will highlight the potential for savings through pollution prevention. The results of this comparison must appear in your plan.

Total cost assessment is a financial tool that compares pollution prevention options against the way things are done now and against other prevention possibilities. This tool extends the boundaries of project financial analysis to account for the less tangible, indirect, and longer-term costs and savings typical of pollution prevention investments. This tool also allocates these costs and savings to specific processes and product lines. Total cost assessment uses three types of information for each potential project:

1. *Current operating costs* for a specific process or source, including both direct and indirect, obvious and less obvious costs.

2. *Capital costs* for the alternative technology, including all necessary changes upstream and downstream of the direct process change.

3. *Operating costs and savings* over the life of the proposed project, again including both direct and indirect, obvious and less obvious costs and savings.

Much of the data for item 1 is already in your plan from the financial analysis done at step 6. By combining these costs with those the team finds for items 2 and 3 above, your team will have a basis for calculating several indicators of profitability. These range from a simple payback period to the more complicated, but much preferred, internal rate of return (IRR) and net present value (NPV).

These values are ones which your firm's management may already use to measure against a threshold or "hurdle rate" when it decides whether to make any kind of investment. They are seldom applied to environmental management projects, however, because such projects are thought of as something necessary to remain in business (by staying in compliance with the law) rather than as an opportunity to turn a profit. *Pollution prevention often does turn a profit, so a business needs to think about pollution prevention investments differently from mandatory pollution control investments.* Total cost assessments can show your firm's management which prevention projects are most worthwhile and how they stack up against each other and against other capital investments the facility is considering.

Management always uses its own best judgment in making capital budgeting decisions, guided by experience and intuition about what the long-term effects of a proposed investment will be. This is particularly true for pollution prevention projects for which many costs and savings sometimes are difficult to quantify.

With a completed financial feasibility analysis, your team is ready to choose among a set of pollution prevention options which the firm's technical staff can implement and which the firm's financial managers are satisfied with.

8.4.8.6   Selecting Options to Implement.   The options that made it through the feasibility analysis should all be worthwhile investments. The technical analysis shows that they are possible, and the financial analysis shows that they meet the company's requirements for profitable investments. Ideally, the firm would do all of them, but usually the options need to be given the OK by the company's managers before any action is taken.

Sometimes, management needs to choose between options that cannot be implemented together. In that case, a decision needs to be made about which one to do. A completed feasibility analysis gives managers criteria with which to pick among the options. The magnitude of nonproduct output reductions, the amount of money an option will save, the time it will take to realize a payback, and any other issues stemming from the feasibility analysis can inform management's decision.

If options cannot be done simultaneously, because the company does not have the resources to do all at once, a good way of dealing with them is to wait to decide on them until step 9, where an *implementation schedule* may provide a way around such resource conflicts. Finally, remember that the Pollution Prevention Rules do not require a facility to implement any pollution prevention at all, although most companies will implement the options that turn a profit.

CASE STUDY: Top Shelf chooses investments from its options using feasibility analysis.   When the team had run out of ideas and had confirmed that the ones they had were pollution prevention, it was time to pick among them, according to whether the techniques would fit into the facility's operations and finances. There were 16 techniques that qualified as pollution prevention, although some techniques helped at several sources and were counted as an option at each source. The next meeting started the feasibility analysis.

The first question on the agenda was whether any techniques were obviously impossible. Everyone agreed that each idea had potential for success; some, in fact, seemed obviously worthwhile. For instance, Travis asked whether they could skip detailed analysis of the new cleaning system for coppers. It seemed obvious to him that it would reduce nonproduct output, improve worker safety, and clean more quickly than the current method. He pointed out that the manufacturer's specifications showed the amount of solvent a cleaning run would use, which compared favorably to the source information from the current operations and gave the data necessary to set numerical reduction goals. Finally, he noted that cost comparisons could be made against current hazardous waste shipping charges, since all the waste from the cleaning processes was manifested separately. Thomas reminded the team that the Pollution Prevention Rules required them to have an analysis made up of certain elements in the plan. Travis replied that he was sure he could estimate those numbers in under a day. The team took him up on this and adjourned until the next day.

At the next day's meeting, Travis distributed what he called a "focused feasibility analysis," based on existing information. The team reviewed it and noted that the new cleaning system would reduce use and nonproduct output by about 2000 lb/year, resulting in a savings of raw material purchases and waste disposal costs. The team decided to recommend to John Stevens, the president, that the facility invest in the cleaning system.

The team members couldn't make such easy decisions with the other options; they needed more information. For instance, several options would change equipment at different steps in the production processes. Individually, none of the options appeared to be disruptive, but the team was worried that collectively they might adversely affect production. Sarah decided that Jerry, Samantha, and Travis should work together as an assessment team to analyze these options. They would look at the impact of these techniques on product quality, production speed, turnaround time, worker training, use and nonproduct output reduction, and anything else that could impact production. Travis would call up some equipment manufacturers and metal fabricators to see if their ideas could be put into practice. He'd also get cost estimates and evaluate the savings the facility could realize by making these investments.

Assessment teams seemed to be a good way of doing other feasibility analyses. Every option was assigned to a lead person who was an expert on the most important issue associated with the option. For particularly complex options, others worked with the lead person. Emily took the lead on a color matching computer, which would be a huge capital outlay. Jerry and Samantha had been working on different dye formulations, so they took the lead on raw material substitution. Optimizing production schedules turned out to have so many cross-cutting issues that Sarah took the lead on it, herself. In this way, every option would be assessed by somebody.

They agreed on what information was necessary on each option and decided that it should all be presented in the same format to simplify comparison. They also decided that the same information should be found for their current practices. Each lead person would be responsible for assembling the necessary data. They also decided that when finance numbers were murky, the assessment teams would carry out some form of total cost assessment to clarify things.

After they'd had a chance to collect their data, another meeting was called. It was clear from the beginning that two of the options, the color matching computer and substituting raw materials, could not be assessed quickly or without deeper technical investigation. The team agreed to continue working on these options, but not write them into their plan yet. If their investigations proved the ideas worthwhile, the options could be added through a plan modification or at the 5-year revision. Several team members were very interested in these options and asked that the team formally investigate them, and include a time schedule for the activities needed to come to a decision within the plan.

The other assessments yielded more concrete results. Travis's team felt the team could make decisions on several of the options. For instance, the source-level data collected during representative runs showed that pumps in the new equipment produced 50 percent less nonproduct output than the pumps in the old equipment. Option 1, refitting the old pump liners with new, smaller ones could reduce nonproduct output by thousands of pounds of solvent per year which would save the facility over $50,000 each year. Replacing the pump liners was a one-time cost of $250 liner. The group agreed with the assessment team that this was an obviously worthwhile investment.

The other assessment teams reported their findings to the group, which then decided whether or not to recommend that the company invest in each option. The pollution prevention team chose to implement some options that eliminated others from further consideration. For example, mixing ink directly in the pump reservoir made separate mixing drums unnecessary, so the options treating mixing drums differently dropped out.

Finally, Sarah reported on her investigations into alternative production scheduling. She said that every time they changed products on a machine, a set amount of nonproduct output was always generated. Sarah had estimated that every time equipment was set up for a new product, about 10 to 13 lb of nonproduct output was generated. Anytime they could avoid a changeover, nonproduct output would be reduced by about that much. If they could reduce the number of changeovers by 5 percent, then they could reduce their nonproduct output by thousands of pounds. To reduce changeovers, the facility would have to make longer runs of the same products and possibly dedicate some lines to certain products, as had been done with the latex product lines. The downside of such changes was that they would reduce the facility's capacity for fast turnaround on niche products and would increase their inventory of popular products.

Sarah could tell that this idea made the team nervous. Everyone was used to the facility's just-in-time production system. Many of the facility's systems were designed to serve the needs of just-in-time production. No one wanted to abandon that system, because it allowed them to carry niche products and serve more customers. Emily, however, had sales figures for the last 5 years and pointed out that the facility had consistently sold over 100,000 yards of its three most popular products each year.

For one product, batch size varied from 700 yards to 7000 yards and the average batch size was less than 1400 yards. Sarah estimated that if production planning were improved, the average batch size for these popular products could feasibly be doubled. This would, of course, reduce nonproduct output for the product tremendously.

The team was nervous about this option because it was difficult, if not impossible, to predict market demand. Yet, the numbers were compelling and, theoretically, it wouldn't require revamping the present system. The idea was to expand the production planning window to combine several small orders into one large batch. The members decided to test the idea with a couple of the more popular products. The only real danger was that a popular product would stop selling and wind up in storage for a while, and warehouse space for a few products would not present any significant problems.

The team members summarized their discussions by listing the pollution prevention investments they wanted to implement along with the expected nonproduct output reductions and cost savings. Of the 20 options they had generated, six remained which they planned to implement. They would present their findings to John Stevens, whose final approval would set things in motion.

8.4.9 Step 9—Develop Numerical Goals

Numerical goals can be the driving force that rallies the company around the pollution prevention program. The development of these goals is dependent on which options, among the feasible ones identified in step 8, the company implements. Since the goals are based on a 5-year planning cycle, an *implementation schedule* impacts goal setting. The Pollution Prevention Rules require that facilities have goals for reducing the use and nonproduct output of each hazardous substance, per unit of product, which the facility uses or manufactures above the threshold. Your team will have completed its plan when it has chosen pollution prevention options to implement and set up goals based on those options.

Every option that made it through the complete feasibility analysis in step 8 is an investment opportunity in pollution prevention. Each option is not only physically possible, but fiscally worthwhile. The facility would theoretically benefit by adopting all of the techniques that have made it this far. Nevertheless, resources, time, and capital may keep the facility from adopting such a wholesale approach.

In this step, your team will decide when to make these investments, and, on the basis of that decision, set goals for achievements in pollution prevention.

8.4.9.1 Scheduling Options Implementation. At least two things constrain companies from investing in every pollution prevention option that appears promising: (1) the availability of capital and (2) the timing of pollution prevention implementation as it relates to scheduling other activities at the facility. Fortunately, like many quality management programs, pollution prevention is done in cycles, so over the long run, good ideas that are superseded by others can be implemented eventually. *Implementation schedules* are a way of planning around resource and timing problems. If the facility gets ahead of schedule, or decides to supplement its pollution prevention program, it can modify its plan to include new options between required 5-year plan revisions.

8.4.9.2 First Constraint: Money. While each pollution prevention idea that gets to this stage is economically feasible, your firm may not have the capital to do all of them at once. If the firm is in a position to make such a wholesale investment, then the issues of timing that appear below are what will govern implementation. Unfortunately, some businesses will not be able to commit money to every pollution prevention opportunity at once. Fortunately, many pollution prevention opportunities are inexpensive to implement. In fact, some changes, like changes in procedure, may be virtually cost-free.

The case may arise, however, when your team needs to choose between options that each require enough capital to make them mutually exclusive in the near term. When this happens, the firm may still be able to implement several of the options in one planning cycle by staggering them in your implementation schedule (see below); otherwise, management will have to decide which ones to implement during the current 5-year planning cycle. To help make this decision, your team may need to revisit information on the selected options, such as a total cost assessment, which will show the relative economic benefit of both options, and the technical feasibility analysis, which will show the relative environmental benefit.

8.4.9.3 Second Constraint: Time. Time is another factor that may have an effect on what to do in the near term and what to begin later in the cycle (or in the next cycle). For instance, if an option requires changing a component in a production line, it may mean temporarily shutting down that line. Finding the right moment to do that will affect your team's decision on when to implement certain options. Fortunately, some investments are simple and quick to implement. For instance, a change in the way hazardous materials are handled and stored to reduce spills could be implemented through on-the-job employee training. Other measures, however, may be more complicated, requiring research and development and structural changes on the shop floor. These changes might delay production schedules or pull people away from other projects at the facility. They should be coordinated with planned equipment maintenance or changes in production "campaigns."

Your team needs to factor these constraints into its pollution prevention program decisions. An implementation schedule provides a framework for making those decisions. Simple investments, which require little or no capital and time, will almost naturally be the first ones undertaken at the facility. These can help build confidence in your program because they usually provide quick, tangible, and money-saving results.

A useful procedure is to sort the prevention techniques your team would like to adopt into a hierarchy that accounts for their expense and complexity, factors that may relate to whether they are people-oriented solutions (changes in procedure) or machine-oriented solutions (changes in processes). Your team can use a hierarchy such as this to develop an effective and fiscally responsible implementation schedule for the facility. Estimate the time and capital it will take to install each option and schedule its installation to avoid disrupting other processes at the facility. Record the schedule in the plan.

When the implementation schedule is completed, your team can estimate when the benefits of pollution prevention will appear as reduced use and nonproduct output generation of hazardous substances. Those estimates are the basis for your team's pollution prevention goals for the facility and for processes, which are required by the Pollution Prevention Rules.

CASE STUDY: Top Shelf sets its implementation schedule.   The team at Top Shelf knew its first pollution prevention plan was almost complete. The feasibility analyses of the pollution prevention options were completed, giving the team a list of options that the facility could profitably invest in. Although all of the options looked good, the team knew they would need an implementation plan to see them through to completion.

The team decided the first start date for the implementation schedule would be in 2 months, on July 1. That would give time to secure the approval of top management. It also coincided with the date the plan summary was due to the state and was traditionally a vacation time, when the facility reduced its workload and did yearly maintenance on its equipment, a good time to install pollution prevention equipment. The team members began scheduling.

The first item they looked at was money. Most of the options were inexpensive and could be started with current operating funds. The copper cleaning system, however, was an unbudgeted expense. Emily examined her accounts and decided that the company could allocate the capital for the system by late September without pinching other parts of the budget. According to the supplier, the system could be in place and running 5 months after the order was made.

The other scheduling concerns were time constraints. One option, optimizing the production schedules, was an ongoing challenge for Sarah that she would start right away and continue to work on throughout the planning cycle. The remaining four options involved equipment modifications. Because the same people would be working on implementing these options, the team decided to split them up. Refitting the ink tray and changing procedures to begin reservoir mixing could probably be finished in 3 to 4 months, so the team members decided to start equipment modification with them. On the basis of that schedule, they set the start date for recutting the coppers and refitting the pump liners for November, when the first modifications would be finished.

Recutting the coppers and refitting the pump liners would take months to complete, because the coppers could not be replaced all at once and the pumps had to be sent away for the refit.

8.4.9.4 Pollution Prevention Goals.

Why should facilities have goals? Because goals excite people. In the time since the pollution prevention policy was established and your team was formed, the employees may not have heard much about the pollution prevention program. An official announcement of the options to be implemented and the goals that the facility plans to achieve through those options is an excellent way to rekindle support and excitement for the program. The Pollution Prevention Rules require that plans include both process-level and facility-level goals for each hazardous substance at the facility.

Goals should be easily understood, easily measured, supported by the people they affect, and realistically achievable. The baseline information from Part I (step 6) may have shown that there were inefficiencies in particular processes that could be improved through pollution prevention, but, until Part II was completed, there was no way of estimating what realistic goals for reducing use and nonproduct output at those processes might be. Also, there was no way of combining the separate process-level reduction goals into a hazardous substance reduction goal for the entire facility. Now, your team knows what options it will implement and has scheduled their implementation; it can set reasonable 5-year goals. Over time, goals will facilitate the measurement of progress. If the facility falls short of a goal, that will indicate where more work might be directed.

8.4.9.5 Setting Production Process Use and Nonproduct Output Reduction Goals.

The Part II technical analysis should provide a good estimate of the reduction in pounds of annual use and nonproduct output generation each pollution prevention technique is expected to produce. These expected reductions can translate into goals at the process level. Look at the implementation schedule to see when process and source improvements will manifest themselves as use and nonproduct output reductions. Base the process-level goals on the annual use and nonproduct output levels expected at each production process after 5 years. If every feasible option will be implemented during the 5-year planning period, then the total of the expected nonproduct output and use reductions found through the feasibility analysis in step 8 for each process will be the 5-year goal for that process. If some of the options are not implemented, or will not have an effect until after the 5-year planning cycle is over, then the effect of those options should not be included in the process-level goals. Since the goals are not legally binding, your team can be realistically ambitious. Goals that indicate high expectations will encourage continuous improvement of pollution prevention ideas.

Process-level goals are indexed to the unit of product that your team chooses for each process in step 6. In step 6, the team developed data for the quantities of hazardous substance used and generated as nonproduct output at each process. The goals your team develops in this step are based on reducing the amount of hazardous substance used or generated as nonproduct output for each unit of product produced at each process. This indexing separates changes in use and nonproduct output due to pollution prevention from changes due to fluctuations in production levels. Ultimately, your team will express its production process goals as percent reductions in use and nonproduct output, insulated from changes in production. "CASE STUDY: The team decides on goals" in this step demonstrates how to make such calculations.

8.4.9.6 Setting Facility-Level Use and Nonproduct Output Reduction Goals.
Hazardous substance reductions at the facility level are an important indicator of how well process-level pollution prevention is working. Therefore, facilities must set goals for the whole facility as well as for processes. Facility-level reduction goals are expressed as the amount the facility plans to have reduced its annual use and nonproduct output generation of each hazardous substance after 5 years of pollution prevention. These goals are expressed two ways. First, they are expressed as the difference, in pounds, between the quantity of a hazardous substance used or generated during base year and the quantity used or generated during the last year of the 5-year planning period. Second, they are expressed as the percentage by which the base year use or nonproduct output generation has been reduced by the end of the 5-year planning period for each hazardous substance. Note that these reductions are based on the cumulative effect of pollution prevention implemented at each targeted production process over the 5-year planning period. However, the process-level goals cannot be added directly to calculate the facility goals because they are based on pounds reduced *per unit of product* to account for changes in production at the process level.

However, your team should recognize that there is a relationship between the process-level goals and the goals it chooses for the whole facility. To calculate the facility goals, your team can multiply the quantity of product produced at each process during the base year by the process-level goal (expressed in reductions per unit of product). This will give your team the expected reduction in annual use and nonproduct output generation at each process after 5 years. Gathering these expected quantities for each hazardous substance at all targeted processes will allow your team to assemble facility-level goals for each hazardous substance.

The level of production chosen to find the facility-level goals can impact the accuracy of the goals. Usually, the base year level of production is used, because it is difficult to predict what production will be 5 years out. Since production is likely to change, the plan progress report will use a production index to help account for these changes in production (see step 11). If production levels change drastically, the facility can always revise its goals during the planning period.

CASE STUDY: The team decides on goals.   The team met to wrap things up for this planning cycle by choosing goals for the plan. The team was pleased with the work it had done, and was excited about presenting a complete plan to the president and the rest of the company. The goals they would come up with during this meeting would show what the team expected pollution prevention to accomplish at the facility.

Fortunately, the goals followed directly from the work the team had already done. The team had tables for each targeted process. These tables showed the reductions that each pollution prevention technique was expected to bring about. The tables also categorized the reductions by hazardous substance. The implementation schedule showed that all of these reductions should manifest themselves before the end of the 5-year planning cycle, so they should all be included in the goals. For each hazardous substance, the team added together the 5-year expected reductions in use and nonproduct output at each source within each process. These totals, when expressed on a per unit of product basis, became the team's goals for each process. Thomas pointed out that the goals would actually be reported as percent reductions from the base year use and nonproduct output per unit of product to the goal year use and nonproduct output per unit of product (i.e., a 35 percent reduction in acetone per unit of product, etc.). However, they needed the raw numbers to calculate the percentage reduction and to develop a facility-level goal.

Ultimately, the team arrived at 38 NPO process-level goals, one for each hazardous substance used at each of the 10 targeted production processes. In this case, all of the NPO process-level goals can also be used as use goals because the hazardous substances are "otherwise used" and NPO generation is equivalent to use.

The team based its five facility-level goals (one for each hazardous substance) on successfully achieving the pollution prevention it planned at each process. They used the process-level goals for each hazardous substance (MiBK, MEK, nitropropane, toluene, and acetone) and converted them back from the per unit of product basis since facility-level goals were reported as raw reductions. The team converted the goals by multiplying them by the production at each process during the base year. Finally, they added the results for each hazardous substance together. These totals represented the amount that annual use and nonproduct output for each hazardous substance would be reduced after 5 years if production remained at the base year level. Sarah looked this over and was concerned because she knew that production would fluctuate (she hoped it would go up). If this happened, then the facility would almost surely fail to meet its goals, since use and nonproduct output are linked to production levels. Thomas, however, had been examining the progress report forms he'd have to fill out and realized that they included a production index which would allow the facility to track its goal against the base production levels. He explained this to the other team members and they settled on the facility-level goals (see step 10 for information on how the goals are reported in the plan summary).

Since Top Shelf does not generally consume or produce hazardous substances on site, its nonproduct output reduction goals are usually the same as its use reduction goals. (This is true for most cases involving hazardous substances that are "otherwise used," e.g., as solvents or processing aids.)

There is one exception for Top Shelf. MiBK is incorporated into a glue gauze in one nontargeted process. While there are no process-level goals, this use as a formulation component results in facility-level use of MiBK being greater than the quantity of nonproduct output generated. (This is true for most cases involving hazardous substances that are manufactured or processed.)

8.4.10   Step 10—Summarizing the Plan

A plan summary provides a convenient way of showing the public, management, and regulators what pollution prevention planning the facility has done without revealing all the details of the full plan. Most states have developed plan summary forms that must be filled out by covered facilities, creating a consistent format for reporting summary information.

Your facility's pollution prevention planning is important to many groups, including senior managers, stockholders, the state, and the neighboring community. Nevertheless, they do not need to see the complete plan to understand and appreciate what the facility is doing to protect the environment (and save money) through pollution prevention. A summary of the plan is a valuable tool for briefing people inside and outside the facility. A public summary is also a concrete demonstration of the firm's commitment to protecting the environment through pollution prevention.

Most states will provide covered facilities with plan summary forms to complete. The plan summary consists of information that your team uncovered when it analyzed pollution prevention options (Part II of the plan): the pollution prevention methods selected, the schedule for doing them, and the 5-year reduction goals for use and nonproduct output both at the process and the facility level. To put this information in context, plan summaries include ranges for reporting the amounts of hazardous substances used in the targeted processes, and generic descriptions of all the covered production processes and targeted sources at the facility. This information presents a picture of the business conducted at the facility, but can do so without giving away confidential information. Likewise, the process-level goals in the summary are not reported as the raw numbers your team found in steps 6 and 8, but as percent reductions per unit of product instead. If your team believes that the generic process descriptions and the reporting of process-level goals as percent reductions will still reveal sensitive information, there are provisions in the Pollution Prevention Rules that allow the facility to make this information confidential.

8.4.10.1 Completing a Plan Summary Form.   There are four sections on a state's plan summary. They cover administrative information for the facility, facility-level goals for each hazardous substance used or manufactured above threshold at the facility, process information for each process involved with a covered hazardous substance, and pollution prevention information and goals for each targeted production process.

The person who fills out these forms for the facility will be familiar with the administrative information, which is required on other state reporting forms. There are, however, other elements on the plan summary form that are new. The forms ask for the reduction goals for the hazardous substances used and generated as nonproduct output at both the facility and process levels. Your team established these goals in step 9. Your facility will submit these goals on a separate facility-level information section for each hazardous substance. On that same section, facilities may optionally report numerical data on pollution prevention for the hazardous substance implemented between 1987 and the base year and qualitative descriptions of pollution prevention achievements before 1987.

Generic nomenclature is also used in the process description section to describe every process that involves a covered hazardous substance at the facility. The descriptions will give those who use the plan summary an understanding of what the facility does, without revealing too much about specific operations. It will also put your team's targeting decision in the context of the processes that could have been targeted.

The process-level goals are reported in the plan summary as well, in a section that includes both the goals and a schedule for starting and completing the pollution prevention techniques used to achieve those goals. The schedule uses the generic nomenclature of the EPA's form R to describe the pollution prevention techniques.

Detailed instructions for the plan summary will be included in the reporting package that covered facilities will receive before the summaries are due.

8.4.10.2 Confidentiality on Site and in Summaries.

Preparing a pollution prevention plan sometimes raises confidentiality concerns. The plan should be available to the pollution prevention team and to the managers whose processes are affected by it. It may, however, contain confidential information as part of its inventories or process descriptions. It's possible that such information is not together in one place anywhere else at the facility, so it makes sense to protect any sensitive information it contains. At the same time, the plan must be available to the state inspectors, who are required to treat any information in a plan they review on site as confidential information.

Like the actual plan, a plan summary may contain data that the facility feels should be kept confidential. If your team is creating a summary on its own for senior management or stockholders, then it can control what goes in it, but the summaries that covered facilities prepare for the state must contain specific information. If your team or managers believe that any of the information you would submit in a plan summary is so sensitive that it should remain secret, then the firm will want to file a confidentiality claim with the state to prevent this information from becoming publicly available.

A confidentiality claim allows a facility to submit a preliminary public copy of its plan summary (or progress report) in which potentially confidential information is blacked-out or deleted. The facility also submits a complete summary as well. The blacked out version is what will be made public, while the other is kept as confidential information by the state. Confidentiality claims may not be filed for information pertaining to a hazardous substance's releases into the environment or into a wastewater treatment system. If a confidentiality claim is filed, the company should be able to show that it has taken all reasonable measures to protect the secrecy of the information, that disclosure of the information would be likely to cause the company economic harm, and that it has met the claim substantiation criteria.

CASE STUDY: Thomas summarizes the plan for the team. Once the implementation schedule and goals had been chosen, it fell to Thomas to complete the plan summary forms. He completed the administrative data for the facility easily. Next, he had to fill out a facility-level summary for each covered hazardous substance, one for MiBK, MEK, nitropropane, toluene, and acetone. These sections focused on the facility-level goals the facility had established in step 9.

The next section was a description of each process and grouped process the team had identified by the end of step 5. This meant filling out a section for the gauze gluing operation, which the team had not targeted, as well as a section for each of the targeted processes. Thomas used the narrative description to provide an overall picture of each process, and then carefully described the steps of the processes using the state's generic nomenclature. The nomenclature included terms like formulating, printing, drying, and cleaning which Thomas felt accurately described the steps of the process as the team had defined them through process flow diagrams.

Finally, Thomas filled out goal sheets for each of the targeted production processes, a large subset of the processes he'd described in the previous section. For each process, he reported the facility's goals for use and nonproduct output reductions, and an implementation schedule for starting and completing those options.

8.4.11 Step 11—Tracking and Reporting Progress

Progress reports can be valuable tools for keeping your pollution prevention planning effort on track and keeping company managers, the public, and the state up to date on whether your program is meeting its goals. The state will provide progress report forms that covered facilities must complete each year after submitting their plan summary. The state has combined the pollution prevention progress report with the community right to know release and pollution prevention report.

Is it working? Your firm has made a commitment and allocated resources to pollution prevention. Once the program is under way, the team must answer a question from top executives, the public, and the state: Is the program meeting its goals? The progress report will help answer that question. The state's progress report form is built around the goals your team reports in the plan summary. Learning your facility's progress toward those goals means tracking reductions (or increases) in nonproduct output generation and hazardous substance use. Although it is not part of the report submitted to the state, financial progress should also be tracked so facility managers will know whether the investment potential of pollution prevention is being reached. Such information can guide adjustments to the plan, possibly paving the way for more pollution prevention in this planning cycle, or focusing your team's search for new techniques in the next cycle.

By tracking progress, the team can show how changes due to pollution prevention relate to the goals the firm has set. If the reductions fall short of the goals, then the team will need to find and report the reasons for the lack of progress. Perhaps there has been a delay in the equipment modifications your firm undertook or planned process changes were not properly carried out by personnel. In this way, progress reports will feed back into the pollution prevention program, allowing the team to make adjustments as the plan is carried out.

8.4.11.1 Financial Progress.   A final area where your facility will see progress is in the money saved and spent through pollution prevention, although it is *not* reported on a department form. When assessing the financial feasibility of your pollution prevention options, your team made estimates of the economic impacts of carrying out various options. At this point, you should be able to directly measure how costs have changed for your targeted processes. The cost accounting framework your team set up to complete the requirements of the plan may be very help-

ful in assessing economic progress. Once the firm has begun to realize the financial benefits of pollution prevention in real savings, interest in pollution prevention will increase throughout the company. In addition, knowledge of which pollution prevention measures are most cost-effective will improve your analyses in the future.

8.4.11.2 Confidentiality.

8.4.11.2 Confidentiality. The confidentiality provisions that apply to plan summaries, as described in step 10, also apply to progress reports. A facility manager who feels that the disclosure of the information is likely to cause the company economic harm may submit a confidentiality claim for a progress report. When the department receives a public request to see a progress report for which a confidentiality claim has been filed, it will assess the claim and determine whether it is justified according to the confidentiality provisions of the Polution Prevention Rules.

CASE STUDY: Reporting the first year's progress. Top Shelf's team was responsible for putting the plan into practice. The team members worked hard during the year after submitting the plan summary to the department. Their implementation schedule kept them busy preparing and installing pollution prevention options. Emily completed work on the spreadsheet she'd designed to track the production, use, and nonproduct output generation at each process, which, after the first year was over, allowed Thomas to fill out the progress report forms without very much hassle.

Thomas had a great deal of experience completing the administrative and facility-level throughput data sections of the progress report because they were the same as the state's release and pollution prevention report, which they replaced. The sections that indicated progress toward the facility and process goals were new, but since the company now tracked its efficiency numbers systematically, they were not much trouble either. During the first year, Top Shelf had not made any changes that required a plan update or modification.

8.4.12 Step 12—Update Your Plan

The last step in the planning process is to start again. Pollution prevention should be ongoing, providing continuing environmental and economic benefits to the companies that pursue it.

Pollution prevention teams do not retire; pollution prevention plans are not completed. In the same way that a company manager is always on the lookout for ways to improve business, the pollution prevention team should always be hunting for new opportunities. More often than not pollution prevention opportunities *are* ways to improve business.

By establishing a pollution prevention policy, the firm cleared the way for doing pollution prevention. Now your team's task is to turn its accomplishments into a stable planning framework. There are several reasons for doing pollution prevention this way:

- Initial successes will provide an incentive to do more.
- When long-term projects succeed, resources will become available to start new projects.

• When problems arise for one option, a stable planning structure provides a way to look for alternatives.

Continuing reporting and revision requirements under the Pollution Prevention Rules are another reason for your team to keep its prevention activities current. For these reasons, part of your team's pollution prevention strategy should be one of continuous improvement. Therefore, the final step of pollution prevention planning is to begin again.

Develop a cycle of pollution prevention action and reevaluation. Reevaluation may show that changes in technology or finances have made something feasible that did not appear so in previous planning cycles. Through such checks the company can maintain pollution prevention programs over the long term without exhausting the feasible and financially rewarding options. It is important to find concrete ways of spurring continuing progress, perhaps by offering new employee incentives, or by reviewing past successes and presenting them as the record to beat.

Continuing planning is required by the Pollution Prevention Rules. They require plan revisions every 5 years, yearly updates of certain information, and modifications when significant changes occur that affect the plan. Nevertheless, these requirements should not limit your team from updating and improving its plan more frequently. If your program seems to call for a shorter interval, then follow your program. The Pollution Prevention Rules are designed to encourage planning. More frequent revisions are within the spirit of that design.

If your team does decide to update its plan between 5-year revisions, it can explain how the update would affect the plan summary in a special section of the yearly progress report. That way, progress toward your new goals will be made public through the progress report and your facility will get the credit it deserves.

Beginning again is the way to make your firm's program an ongoing success rather than a brief flurry of pollution prevention techniques. As the planning cycles go by and your team gets more comfortable with pollution prevention, new ideas are almost sure to crop up. (SOURCE: New Jersey Department of Environmental Protection, reproduced with permission.)

8.5 EXAMPLE OF A FEDERAL AGENCY'S POLLUTION PREVENTION STRATEGY

8.5.1 Department of Veterans Affairs (VA) Summary of Pollution Prevention Strategy[1]

The VA's strategy commits it to environmental leadership and preventing pollution by reducing the use of hazardous materials, as well as reducing releases of environmental pollutants to as low as it's reasonably achievable.

---

[1]SOURCE: *www.epa.gov.*

The VA's pollution prevention strategy:

- Requires VA facilities to continue to participate with federal, state, and local officials in emergency planning and community right-to-know activities.
- Requires VA facilities to reduce the use of toxic and hazardous substances and the resulting generation of waste by reviewing facility operations, procedures, and unit processes to determine the potential for source reduction.
- Directs VA facilities to include pollution prevention in the development of facility guidance, policy, and operating procedures.
- Requires the development and maintenance of facility-specific comprehensive inventories of toxic chemicals, extremely hazardous substances, and hazardous chemicals.
- Commits VA facilities to promote pollution prevention awareness through training and education, or outreach/awareness programs.
- Directs VA facilities to implement acquisition and procurement policies and life cycle costing practices that promote pollution prevention, reduce waste, minimize effects on natural resources and encourage economically efficient market demands for items using recovered material.
- Directs VA facilities to purchase environmentally preferable products, when possible.

The pollution prevention strategy also includes specific actions to be implemented by Veterans Health Administration organizations (including Construction Management, Operations and Environmental Management Service), VA regional offices, Acquisition and Material Management, National Cemetery System, and Veterans Health Administration.

Agency commitments to prevention strategy includes two elements not included in Executive Order 12856. These are:

- A plan to review the use of certain pesticides to prevent pollution that could result from these chemicals and establish an annual goal to reduce the use of toxic pesticides in the National Cemetery System.
- A commitment to evaluate the effectiveness of alternative sterilants to ethylene oxide (ETO) and, if appropriate, establish and implement a plan to reduce the use of ETO at VA Medical Centers and other health care facilities.

There are no Department of Veterans Affairs (VA) facilities covered under Executive Order 12856.

CHAPTER 9
CONTAMINATED SITES

John A. Green
Director, Environmental & Quality Assurance
Tate & Lyle North American Sugars Inc.

9.1 EMERGENCY PREPAREDNESS AND COMMUNITY RIGHT-TO-KNOW PERMITTING CONSIDERATIONS

In 1987 and 1988 EPA issued the first of its regulations (40 CFR §§ 355, 370, and 372) that addressed the responsibilities of facility owners/operators to meet specific planning and reporting requirements regarding the inventory, use, and release of a broad spectrum of designated hazardous materials used in their facilities. Since the intent of this rule making was to minimize and/or eliminate the potential for the release of these materials, these regulations did not establish specific permitting provisions as other environmental statutes did. However, their incorporated notification, reporting, and record keeping requirements form a definitive basis for inferred permitting responsibilities related to the Emergency Preparedness and Community Right-to-Know Act (EPCRA). Just as owners/operators of permitted hazardous waste transportation, storage, and disposal (TSD) facilities (refer to in Chap. 11 of this book) must meet stringent requirements, owners and/or operators of EPCRA-covered facilities must comply with many similar requirements that are usually associated with a permitting process. These can be found in the planning, notification, and required reporting provisions of 40 CFR §§ 355, 370, and 372. For example, owners/operators of facilities covered by these regulations essentially are not permitted to operate their facilities without having notified the appropriate agency or agencies of the amounts of these potentially hazardous chemicals in their workplaces. In addition, releases of these chemicals, including those resulting from their use in processing and manufacturing, must be reported to the appropriate agencies.

Unlike the provisions of other environmental statutes, Sec. 313 of EPCRA requires that owners/operators of covered facilities report annually on any and all releases of certain toxic chemicals into the environment. This reporting requirements is not limited solely to the unintentional and/or accidental release of a toxic chemical, such as a spill or tank rupture. Rather, it covers any manner in which the toxic chemical enters the air, water, or ground. These types of releases include other waste management activities that are permitted under separate programs, such as off-site shipments, discharges into water, and discharges to a publicly owned treatment works (POTWs).

This chapter concentrates on the EPCRA notification, reporting, and record keeping requirements that parallel other established permitting provisions.

9.2   REGULATORY OVERVIEW

The Emergency Planning and Community Right-to-Know Act was signed into law on October 17, 1986. Also known as Title III, it is a free-standing title of the Superfund Amendments and Reauthorization Act of 1986 (SARA), which amended the Comprehensive Environmental Response, Compensation, and Liability Act of 1980 (CERCLA). Unlike previous environmental statutes, EPCRA requires companies to publicly disclose specific quantitative information concerning their chemical inventories, processes, chemical consumption, and releases as well as their emergency response plans. EPCRA is intended both to provide the public access to information relating to the hazardous chemicals located in their communities and to use this information as a basis to formulate and implement local emergency response plans.

9.3   EPCRA REGULATORY PROVISIONS

EPCRA establishes these regulatory provisions in the following four sections:

* Emergency Planning (Secs. 301 to 303)
* Emergency Release Notification (Sec. 304)
* Hazardous Chemical Inventory Reporting (Secs. 311 and 312)
* Routine Toxic Chemical Release Reporting—Emissions Inventory (Sec. 313)

9.3.1   Emergency Planning

The requirements of this section apply to any facility at which there is present an amount of any extremely hazardous substance equal to or in excess of its threshold planning quantity, or designated, after public notice and opportunity for comment, by the commission or the governor for the state in which the facility is located. For purposes of this section, an amount of any extremely hazardous substance means the total amount of an extremely hazardous substance present at any one time at a facility at concentrations greater than 1 percent by weight, regardless of location, number of containers, or method of storage.

Section 301 of the act establishes the framework for emergency planning by state and local governments. It calls for the creation of a state emergency response commission in each state and territory, as well as local emergency planning committees. This section calls on these local panels to work with representatives of facilities covered by the law on emergency response plans. The owner or operator of a facility subject to this section must designate a facility representative to participate in the local emergency planning process as a facility emergency response coordinator. The owner or operator has to notify the local emergency planning committee of this selection.

In addition, the owner or operator of a facility subject to this section must inform the local emergency planning committee of any changes occurring at the facility that may be relevant to emergency planning. On request of the local emergency planning committee, the owner or operator of a facility subject to this section also must promptly provide to the committee any information necessary for development or implementation of the local emergency plan.

Subtitle A of the Act establishes the framework for local emergency planning. Under Sec. 302, a facility must notify its state emergency response commission (SERC) and participate, as necessary, with the local emergency planning committee (LEPC) in the local emergency planning process only if it has present an extremely hazardous substance (EHS) in excess of its threshold planning quantity (TPQ). Any facility that produces, uses, or stores any of these listed chemicals in a quantity greater than the designated TPQ must meet all emergency planning requirements. In order to determine if the TPQ of a listed substance is equaled or exceeded, a facility must consider all on-site forms of the substance in which its concentration exceeds 1 percent by weight. Extremely hazardous substances that exist as solids are actually subject to two TPQs, such as 500 lb/10,000 lb. The lower quantity applies only when the substance meets one of the following three conditions:

- It is a powder with a particle size of less than 100 μm
- It meets National Fire Protection Association (NFPA) criteria for a reactivity rating of 2, 3, or 4
- It is handled in solution or molten form

Although obtaining a storage permit is not mandated by EPCRA, Sec. 302 of the Act requires facilities that store or use extremely hazardous substances in excess of the threshold planning quantity to notify the state emergency response commission and participate, as necessary, in local emergency planning activities. EPCRA defines a *facility* as "all buildings, equipment, structures, and other stationary items which are located on a single site or on contiguous or adjacent sites and which are owned or operated by the same person (or by any person which controls, is controlled by, or under common control with, such person)." The term also includes natural containment structures if chemicals are placed in or removed from such structures by human means.

Section 302 directed EPA to publish the list of extremely hazardous substances as an interim final rule within 30 days of the enactment of EPCRA. Section 302(a)(2) required that the list be identical to the list compiled by EPA in 1985 as part of EPA's Chemical Emergency Preparedness Program. Under Sec. 302(a)(4), EPA is authorized to revise the list, but in undertaking any such revision, EPA must take into account the "toxicity, reactivity, volatility, dispersibility, combustibility, or flammability of a substance." The list of extremely hazardous substances can be found in the App. A to 40 CFR Part 355. EPA also used the list in implementing provisions in the 1990 Clean Air Act for the prevention of chemical accidents. From this list, EPA had to use at least 100 chemicals (more if warranted) in determining which facilities will be covered by regulations for chemical accident prevention (40 CFR Part 68).

The term *toxicity* is defined to include "any short- or long-term health effects, which may result from a short-term exposure to the substance."

EPA published a list of 402 extremely hazardous substances on November 17, 1986 (51 FR 41570). On the same day, EPA proposed the deletion of 40 substances from the EHS list on the basis that their original listing was in error. On April 22, 1987, 52 FR 13388, EPA announced that it was deferring the proposed delisting of these substances, pending an evaluation of the long-term effects from short-term exposure to each of them. This deferral was in response to comments from members of the public who argued that the proposed rule was premature. On November 23, 1987, the District Count for the District of Columbia, in *A.L. Laboratories, Inc. v. Environmental Protection Agency,* issued an order requiring EPA to remove several substances from the EHS list, reasoning that Congress did not intend to include in the statutorily designated list substances listed due to "clerical error."

Section 303 requires each local emergency planning committee to work with facility representatives and complete a comprehensive emergency response plan. The plan must be updated annually. The law specifies elements that the local emergency planning committee must include in the emergency plan.

9.3.2 Emergency Notification

Section 304 of EPCRA establishes requirements for immediate reporting of certain releases of EHSs and hazardous substances (HSs) listed under the Comprehensive Environmental Response, Compensation, and Liability Act (CERCLA) to SERCs and LEPCs, similar to the release reporting provisions of CERCLA Sec. 103. Although similar, CERCLA Sec. 103 and EPCRA Sec. 304

differ somewhat in purpose. CERCLA provides generally for federal planning and coordination of entities and for federal contingency plans. CERCLA Sec. 103 requires federal notification for any release of a hazardous substance in an amount equal to or in excess of its reportable quantity (RQ). EPCRA is designed to protect the public in the event of dangerous chemical releases through the establishment of local and state emergency response capability. EPCRA Sec. 304 requires, in addition to any federal notification, notification to state and local authorities for any release of an EHS in an amount equal to or in excess of its RQ. The potential hazards posed by EHSs make state and local notification critical to effective and timely emergency response. EHSs are acutely toxic chemicals, which cause both severe short- and long-term health effects after a single, brief exposure. In many cases, local and state authorities may be the first and only responders to the release of an EHS.

Notifications are required if a release of an EHS or HS is equal to or above the reportable quantity. Section 304(a) of EPCRA provides that chemicals on the EHS list that do not have an RQ assigned to them by regulation will have a reportable quantity of 1 lb. Currently, 204 EHSs have the statutory 1-lb RQ. On August 30, 1989 (54 FR 35988), EPA proposed to modify the statutory RQs for 232 EHSs using a proposed modification of the CERCLA RQ methodology.

Facilities must immediately notify the local emergency planning committee and the state emergency response commission if there is a release of a hazardous substance that exceeds the reportable quantity. Section 304 also requires a written follow-up notice.

This section does not apply to:

- Any release that results in exposure to persons solely within the boundaries of the facility.

- Any release that is a federally permitted release as defined in Sec. 101(10) of CERCLA.

- Any release that is continuous and stable in quantity and rate under the definitions in 40 CFR 302.8(b). Exemption from notification under this subsection does not include exemption from:

    Initial notifications as defined in 40 CFR 302.8(d) and (e).

    Notification of a "statistically significant increase," defined in 40 CFR 302.8(b) as any increase above the upper bound of the reported normal range, which is to be submitted to the community emergency coordinator for the local emergency planning committee for any area likely to be affected by the release and to the state emergency response commission of any state likely to be affected by the release.

    Notification of a "new release" as defined in 40 CFR 302.8(g)(1).

    Notification of a change in the normal range of the release as required under 40 CFR 302.8(g)(2).

- Any release of a pesticide product exempt from CERCLA Sec. 103(a) reporting under Sec. 103(e) of CERCLA.
- Any release not meeting the definition of release under Sec. 101(22) of CERCLA, and therefore exempt from Sec. 103(a) reporting.
- Any radionuclide release that occurs

Naturally in soil from land holdings such as parks, golf courses, or other large tracts of land.

Naturally from land disturbance activities, including farming, construction, and land disturbance incidental to extraction during mining activities, except that which occurs at uranium, phosphate, tin, zircon, hafnium, vanadium, monazite, and rare earth mines. Land disturbance incidental to extraction includes: land clearing; overburden removal and stockpiling; excavating, handling, transporting, and storing ores and other raw (not beneficiated or processed) materials; and replacing in mined-out areas coal ash, earthen materials from farming or construction, or overburden or other raw materials generated from the exempted mining activity.

From the dumping and transportation of coal and coal ash (including fly ash, bottom ash, and boiler slags), including the dumping and land spreading operations that occur during coal ash uses.

From piles of coal and coal ash, including fly ash, bottom ash, and boiler slags.

(*Note:* Releases of CERCLA hazardous substances are subject to the release reporting requirements of CERCLA Sec. 103, codified at 40 CFR, Part 302, in addition to the requirements of this part. Some releases that occur at facilities which do not have to be reported under Sec. 304 may still have to be reported under CERCLA. Release reporting under Sec. 304 is in addition to release notification under CERCLA.)

Any notice of release required under this section must include the following to the extent known at the time of notice and so long as no delay in notice or emergency response results:

- The chemical name or identity of any substance involved in the release.
- An indication of whether the substance is an extremely hazardous substance.
- An estimate of the quantity of any such substance that was released into the environment.
- The time and duration of the release.
- The medium or media into which the release occurred.
- Any known or anticipated acute or chronic health risks associated with the emergency and, where appropriate, advice regarding medical attention necessary for exposed individuals.
- Proper precautions to take as a result of the release, including evacuation (unless such information is readily available to the community emergency coordination pursuant to the emergency plan).

- The names and telephone numbers of the persons to be contacted for further information.

As soon as practicable after a release which requires notice, the owner or operator has to provide a written follow-up emergency notice (or notices, as more information becomes available) setting forth and updating the information required under Par. (b)(2) of Sec. 304, and including additional information with respect to:

- Actions taken to respond to and contain the release
- Any known or anticipated acute or chronic health risks associated with the release
- Where appropriate, advice regarding medical attention necessary for exposed individuals

An owner or operator of a facility from which there is a transportation-related release may meet these notification requirements by providing the information indicated in the above paragraph to the 911 operator, or in the absence of a 911 emergency telephone number, to the operator. For purposes of this paragraph, a transportation-related release means a release during transportation, or storage incident to transportation if the stored substance is moving under active shipping papers and has not reached the ultimate consignee.

9.3.3 Hazardous Chemical Inventory Reporting Requirements

As with the provisions of other environmental statutes that mandate permits, the reporting requirements of EPCRA are designed to provide information to appropriate state, local, and federal officials on the type, amount, location, use, disposal, and release of chemicals at certain facilities. There are three reporting provisions included in the law.

Section 311 applies to facilities that are subject to the Occupational Safety and Health Act of 1970 (OSHA) and regulations implemented under that Act. Facilities covered under OSHA's Hazard Communication standard (29 CFR 1910.1200) must submit material safety data sheets (MSDSs), or a list of the chemicals for which the facility is required to have an MSDS, to local emergency planning committees, state emergency response commissions, and local fire departments. This is a one-time reporting requirement, with the information to have been submitted by October 17, 1987. If lists of chemicals, rather than MSDSs, are submitted, then the owner/operator of the facility must submit the MSDSs to the local planning committee, on the local committee's request. Updates to the lists or the MSDS submissions are due within 3 months of the time the owner/operator is first required to prepare or have available an MSDS for a specific hazardous chemical under OSHA regulations. If there is significant new information on an MSDS that was previously submitted, a revised MSDS must be submitted.

Under Sec. 311, the Environmental Protection Agency can establish threshold quantities for hazardous chemicals. If a facility has an amount of the hazardous chemical that is below the threshold, then no reporting is required.

On July 26, 1990, EPA issued the final thresholds for Secs. 311 and 312 reporting. The threshold quantities triggering reporting requirements were kept at 10,000 lb for hazardous substances. Threshold quantities for extremely hazardous substances were kept at 500 lb or the threshold planning quantity listed for that substance under Sec. 302, whichever is lower.

The following criteria are to be used for calculating these quantities:

- For hazardous chemicals:

  If the reporting is on each component of the mixture that is a hazardous chemical, then the concentration of the hazardous chemical, in weight percent (greater than 1 percent or 0.1 percent if carcinogenic) shall be multiplied by the mass (in pounds) of the mixture to determine the quantity of the hazardous chemical in the mixture.

  If the reporting is on the mixture itself, the total quantity of the mixture shall be reported.

- Aggregation of extremely hazardous substances. To determine whether the reporting threshold for an extremely hazardous substance has been equaled or exceeded, the owner or operator of a facility shall aggregate the following:

  The quantity of the extremely hazardous substance present as a component in all mixtures at the facility and

  All other quantities of the extremely hazardous substance present at the facility

  If the aggregate quantity of an extremely hazardous substance equals or exceeds the reporting threshold, the substance shall be reported.

- If extremely hazardous substances are being reported and are components of a mixture at a facility, the owner or operator of a facility may report either:

  The mixture, as a whole, even if the total quantity of the mixture is below its reporting threshold, or

  The extremely hazardous substance components of the mixture.

If you are required to submit or have available an MSDS under Sec. 311, then you also are required to submit additional information on the chemicals present at the facility under Sec. 312. Starting March 1, 1988, and annually thereafter, the owner or operator of such a facility must submit an inventory form.

The inventory form must contain an estimate of the maximum amount of the hazardous chemicals present at the facility during the preceding year, an estimate of the average daily amount of hazardous chemicals at the facility, and the location of these chemicals at the facility.

Section 312 of the Act calls for two reporting "tiers." Under Tier I, general information on the amount and location of hazardous chemicals at the facility is required. Tier II seeks more detailed information on each chemical. Tier II information does not have to be submitted unless the state commission, the local planning committee, or the fire department with jurisdiction over the facility requests it. Filing is done using software available for free from the EPA Web page (EPA.gov).

9.3.4 Toxic Chemical Emissions Reports

This part sets forth requirements for the submission of information relating to the release of toxic chemicals under Sec. 313 of EPCRA from facilities that meet certain criteria. The information collected under this part is intended to inform the general public and the communities surrounding covered facilities about releases of toxic chemicals, to assist research, to aid in the development of regulations, guidelines, and standards, and for other purposes. This part also sets forth requirements for suppliers to notify persons to whom they distribute mixtures or trade name products containing toxic chemicals that they contain such chemicals.

Facilities subject to this reporting requirement must complete a toxic chemical release form for specified chemicals. The form must be submitted annually by July 1 to EPA and state officials reflecting releases during each preceding calendar year (the first form was due July 1, 1988). These forms ask for a great deal of information. For each listed chemical or chemical category, the facility must provide information on the maximum amount present at the location, the treatment or disposal methods used, and the annual quantity released into the environment.

9.3.4.1 Key Definitions. *Manufacture* means to produce, prepare, import, or compound a toxic chemical. Manufacture also applies to a toxic chemical that is produced coincidentally during the manufacture, processing, use, or disposal of another chemical or mixture of chemicals, including a toxic chemical that is separated from that other chemical or mixture of chemicals as a by-product, and a toxic chemical that remains in that other chemical or mixture of chemicals as an impurity.

Mixture means any combination of two or more chemicals, if the combination is not, in whole or in part, the result of a chemical reaction. However, if the combination was produced by a chemical reaction but could have been produced without a chemical reaction, it is also treated as a mixture. A mixture also includes any combination that consists of a chemical and associated impurities.

Process means the preparation of a toxic chemical, after its manufacture, for distribution in commerce:

• In the same form or physical state as, or in a different form or physical state from, that in which it was received by the person so preparing such substance.

- Alternatively, as part of an article containing the toxic chemical. *Process* also applies to the processing of a toxic chemical contained in a mixture or trade name product.

Otherwise use means to use a toxic chemical, including a toxic chemical contained in a mixture or other trade name product or waste, in any way that is not covered by the terms *manufacture* or *process*. To otherwise use a toxic chemical does not include disposal, stabilization (without subsequent distribution in commerce), or treatment for destruction unless the toxic chemical that was disposed, stabilized, or treated for destruction was either:

- Received from off site for the purposes of further waste management.
- Manufactured as a result of waste management activities on materials received from off site for the purposes of further waste management activities. Relabeling or redistributing of the toxic chemical where no repackaging of the toxic chemical occurs does not constitute otherwise use or processing of the toxic chemical.

Release means any spilling, leaking, pumping, pouring, emitting, emptying, discharging, injecting, escaping, leaching, dumping, or disposing into the environment (including the abandonment or discarding of barrels, containers, and other closed receptacles) of any toxic chemical.

9.3.4.2 *Reporting and Notification*

9.3.4.2.1 Thresholds for Reporting. Except for the alternative thresholds detailed below, the threshold amounts for purposes of reporting under § 372.30 for toxic chemicals are as follows:

- With respect to a toxic chemical manufactured (including imported) or processed at a facility, the threshold is 25,000 lb of the chemical manufactured or processed for the year.
- With respect to a chemical otherwise used at a facility, the threshold is 10,000 lb of the chemical used for the applicable calendar year.
- With respect to activities involving a toxic chemical at a facility, when more than one threshold applies to the activities, the owner or operator of the facility must report if it exceeds any applicable threshold and must report on all activities at the facility involving the chemical, except as provided in § 372.38.

When a facility manufactures, processes, or otherwise uses more than one member of a chemical category listed in § 372.65(c), the owner or operator of the facility must report if it exceeds any applicable threshold for the total volume of all the members of the category involved in the applicable activity. Any such report must cover all activities at the facility involving members of the category.

A facility may process or otherwise use a toxic chemical in a recycle/reuse operation. To determine whether the facility has processed or used more than an applicable threshold of the chemical, the owner or operator of the facility shall count the amount of the chemical added to the recycle/reuse operation during the calendar year. In particular, if the facility starts up such an operation during a calendar year, or in the event that the contents of the whole recycle/reuse operation are replaced in a calendar year, the owner or operator of the facility shall also count the amount of the chemical placed into the system at these times.

A toxic chemical may be listed in § 372.65 with the notation that only persons who manufacture the chemical, or manufacture it by a certain method, are required to report. In that case, only owners or operators of facilities that manufacture that chemical as described in § 372.65 in excess of the threshold applicable to such manufacture in § 372.25 are required to report. In completing the reporting form, the owner or operator is only required to account for the quantity of the chemical so manufactured and releases associated with such manufacturing, but not releases associated with subsequent processing or use of the chemical at that facility. Owners and operators of facilities that solely process or use such a chemical are not required to report for that chemical.

A toxic chemical may be listed in § 372.65 with the notation that it is in a specific form (e.g., fume, dust, solution, or friable) or of a specific color (e.g., yellow or white). In that case, only owners or operators of facilities that manufacture, process, or use that chemical in the form or of the color specified in § 372.65 in excess of the threshold applicable to such activity in § 372.25 are required to report. In completing the reporting form, the owner or operator is required only to account for the quantity of the chemical manufactured, processed, or used in the form or color specified in § 372.65 and for releases associated with the chemical in that form or color. Owners or operators of facilities that solely manufacture, process, or use such a chemical in a form or color other than those specified by § 372.65 are not required to report for that chemical.

Metal compound categories are listed in § 372.65(c). For purposes of determining whether any of the thresholds specified in § 372.25 are met for metal compound category, the owner or operator of a facility must base the threshold determination on the total amount of all members of the metal compound category manufactured, processed, or used at the facility. In completing the release portion of the reporting form for releases of the metal compounds, the owner or operator is required only to account for the weight of the parent metal released. Any contribution to the mass of the release attributable to other portions of each compound in the category is excluded.

9.3.4.2.1.1 ALTERNATIVE THRESHOLD AND CERTIFICATION: With respect to manufacturing, processing, or otherwise using a toxic chemical, the owner or operator of a facility may apply an alternative threshold of 1 million lb/year to that chemical if the owner or operator calculates that the facility would have an annual reportable amount of that toxic chemical not exceeding 500 lb for the combined total quantities released at the facility, disposed within the facility, treated at the facility (as represented by amounts destroyed or converted by treatment processes), covered at the facility as a result of recycle operations, combusted for the purpose of energy recovery at the facility, and amounts transferred from the facility to off-site locations for the purpose of recycle, energy recovery, treatment, and/or disposal. These volumes correspond to the sum of amounts reportable for data elements on EPA form R (EPA form 9350-1; revised December 4, 1993) as Part II Column B or Secs. 8.1 (quantity released), 8.2 (quantity used for energy recovery on-site), 8.3 (quantity used for energy recovery off site), 8.4 (quantity recycled on site), 8.5 (quantity recycled off site), 8.6 (quantity treated on site), and 8.7 (quantity treated off site).

If an owner or operator of a facility determines that the owner or operator may apply the alternative reporting threshold specified in Par. (a) of this section for a specific toxic chemical, the owner or operator is not required to submit a report for that chemical under § 372.30, but must submit a certification statement that contains the information required in § 372.95. The owner or operator of the facility must also keep records as specified in § 372.10(d).

Threshold determination provisions of § 372.25 and exemptions pertaining to threshold determinations in § 372.38 are applicable to the determination of whether the alternative threshold has been met.

Each certification statement under this section for activities involving a toxic chemical that occurred during a calendar year at a facility must be submitted to EPA and to the state in which the facility is located on or before July 1 of the next year.

Each owner or operator who determines that the owner/operator may apply the alternative threshold as specified under § 372.27(a) must retain the following records for a period of 3 years from the date of the submission of the certification statement as required under § 372.27(b):

- A copy of each certification statement submitted by the person under § 372.27(b).
- All supporting materials and documentation used by the person to make the compliance determination that the facility or establishment is eligible to apply the alternative threshold as specified in § 372.27.
- Documentation supporting the certification statement submitted under § 372.27(b) including:

Data supporting the determination of whether the alternative threshold specified under § 372.27(a) applies for each toxic chemical.

Documentation supporting the calculation of annual reportable amount, as defined in § 372.27(a), for each toxic chemical, including documentation supporting the calculations and the calculations of each data element combined for the annual reportable amount.

Receipts or manifests associated with the transfer of each chemical in waste to off-site locations.

9.3.4.2.2 Reporting Requirements and Schedule for Reporting. For each toxic chemical known by the owner or operator to be manufactured (including imported), processed, or otherwise used in excess of an applicable threshold quantity in § 372.25 at its covered facility described in § 372.22 for a calendar year, the owner or operator must submit to EPA and to the state in which the facility is located a completed EPA form R (EPA form 9350-1).

The owner or operator of a covered facility is required to report on a toxic chemical that the owner or operator knows is present as a component of a mixture or trade name product that the owner or operator receives from another person, if that chemical is imported, processed, or otherwise used by the owner or operator in excess of an applicable threshold quantity in § 372.25 at the facility as part of that mixture or trade name product.

The owner or operator knows that a toxic chemical is present as a component of a mixture or trade name product if: (1) the owner or operator knows or has been told the chemical identity or Chemical Abstracts Service registry number of the chemical and the identity or number corresponds to an identity or number in § 372.65 or (2) the owner or operator has been told by the supplier of the mixture or trade name product that the mixture or trade name product contains a toxic chemical subject to Sec. 313 of the Act.

To determine whether a toxic chemical that is a component of a mixture or trade name product has been imported, processed, or otherwise used in excess of an applicable threshold in § 372.25 at the facility, the owner or operator shall consider only the portion of the mixture or trade name product that consists of the toxic chemical and that is imported, processed, or otherwise used at the facility, together with any other amounts of the same toxic chemical that the owner or operator manufactures, imports, processes, or otherwise uses at the facility as follows:

• If the owner or operator knows the specific chemical identity of the toxic chemical and the specific concentration at which it is present in the mixture or trade name product, the owner or operator shall determine the weight of the chemical imported, processed, or otherwise used as part of the mixture or trade name product at the facility and shall combine that with the weight of the toxic chemical manufactured (including imported), processed, or otherwise used at the facility other than as part of the mixture or trade name product. After combining these amounts, if the owner or operator determines that the toxic chemical was manufactured, processed, or otherwise used in excess of an applicable threshold in § 372.25, the owner or operator shall report the specific chemical identity and all releases of the toxic chemical on EPA form R in accordance with the instructions referred to in Subpart E of this part.

• If the owner or operator knows the specific chemical identity of the toxic chemical and does not know the specific concentration at which the chemical is present in the mixture or trade name product, but has been told the upper-bound concentration of the chemical in the mixture or trade name product, the owner or operator shall assume that the toxic chemical is present in the mixture or trade name product at the upper-bound concentration; shall determine whether the chemical has been manufactured, processed, or otherwise used at the facility in excess of an applicable threshold as provided in Par. (b)(3)(i) of this section; and shall report as provided in Par. (b)(3)(i) of this section.

- If the owner or operator knows the specific chemical identity of the toxic chemical, does not know the specific concentration at which the chemical is present in the mixture or trade name product, has not been told the upper-bound concentration of the chemical in the mixture or trade name product, and has not otherwise developed information on the composition of the chemical in the mixture or trade name product, then the owner or operator is not required to factor that chemical in that mixture or trade name product into threshold and release calculations for that chemical.

- If the owner or operator has been told that a mixture or trade name product contains a toxic chemical, does not know the specific chemical identity of the chemical and knows the specific concentration at which it is present in the mixture or trade name product, the owner or operator shall determine the weight of the chemical imported, processed, or otherwise used as part of the mixture or trade name product at the facility. Since the owner or operator does not know the specific identity of the toxic chemical, the owner or operator shall make the threshold determination only for the weight of the toxic chemical in the mixture or trade name product. If the owner or operator determines that the toxic chemical was imported, processed, or otherwise used as part of the mixture or trade name product in excess of an applicable threshold in § 372.25, the owner or operator shall report the generic chemical name of the toxic chemical, or a trade name if the generic chemical name is not known, and all releases of the toxic chemical on EPA form R in accordance with the instructions referred to in Subpart E of this part.

- If the owner or operator has been told that a mixture or trade name product contains a toxic chemical, does not know the specific chemical identity of the chemical, and does not know the specific concentration at which the chemical is present in the mixture or trade name product, but has been told the upper-bound concentration of the chemical in the mixture or trade name product, the owner or operator shall assume that the toxic chemical is present in the mixture or trade name product at the upper-bound concentration, shall determine whether the chemical has been imported, processed, or otherwise used at the facility in excess of an applicable threshold as provided in Par. (b)(3)(iv) of this section, and shall report as provided in Par. (b)(3)(iv) of this section.

- If the owner or operator has been told that a mixture or trade name product contains a toxic chemical, does not know the specific chemical identity of the chemical, does not know the specific concentration at which the chemical is present in the mixture or trade name product, including information they have themselves developed, and has not been told the upper bound concentration of the chemical in the mixture or trade name product, the owner or operator is not required to report with respect to that toxic chemical.

A covered facility may consist of more than one establishment. The owner or operator of such a facility at which a toxic chemical was manufactured (including imported), processed, or otherwise used in excess of an applicable threshold may submit a separate form R for each establishment or for each group of establishments within the facility to report the activities involving the toxic chemical at each establishment or group of establishments, provided that activities involving that toxic chemical at all the establishments within the covered facility are reported. If each establishment or group of establishments files separate reports, then for all other chemicals subject to reporting at that facility it must also submit separate reports. However, an establishment or group of establishments does not have to submit a report for a chemical that is not manufactured (including imported), processed, otherwise used, or released at that establishment or group of establishments.

Each report under this section for activities involving a toxic chemical that occurred during a calendar year at a covered facility must be submitted on or before July 1 of the next year. The first such report for calendar year 1987 activities must be submitted on or before July 1, 1988.

9.3.4.2.3 Reporting Form. The most current version of EPA form R (EPA form 9350-1 and subsequent revisions) and the instructions for completing this form may be obtained by writing to the Section 313 Document Distribution Center, P.O. Box 12505, Cincinnati, OH 45212. EPA also encourages facilities subject to this part to submit the required information to EPA by using magnetic media (computer disk or tape) in lieu of form R. Instructions for submitting and using magnetic media may also be obtained from the address given in this paragraph.

Information elements reportable on EPA form R or equivalent magnetic media format include the following:

1. An indication of whether the report:
   a. Claims chemical identity as trade secret.
   b. Covers the entire facility or part of a facility.
2. Signature of a senior management official certifying the following: "I hereby certify that I have reviewed the attached documents and, to the best of my knowledge and belief, the submitted information is true and complete and that amounts and values in this report are accurate based upon reasonable estimates using data available to the preparer of the report."
3. Facility name and address including the toxic chemical release inventory facility identification number if known.
4. Name and telephone number for both a technical contact and a public contact.

5. The four-digit SIC codes for the facility or establishments in the facility.
6. Latitude and longitude coordinates for the facility.
7. The following facility identifiers:
   a. Dun and Bradstreet identification number.
   b. EPA identification number (RCRA ID number).
   c. National Pollution Discharge Elimination System (NPDES) permit number.
   d. Underground Injection Well Code (UIC) identification number.
8. The names of receiving streams or water bodies to which the chemical is released.
9. Name of the facility's parent company and its Dun and Bradstreet identification number.
10. Name and Chemical Abstracts Service (CAS) number (if applicable) of the chemical reported.
11. If the chemical identity is claimed as a trade secret, a generic name for the chemical.
12. A mixture component identity if the chemical identity is not known.
13. An indication of the activities and uses of the chemical at the facility.
14. An indication of the maximum amount of the chemical on site at any point in time during the reporting year.
15. Information on releases of the chemical to the environment consisting of an estimate of total releases in pounds per year (releases of less than 1000 lb/year may be indicated in ranges) from the facility plus an indication of the basis of estimate for the following:
   a. Fugitive or nonpoint air emissions.
   b. Stack or point air emissions.
   c. Discharges to receiving streams or water bodies including an indication of the percent of releases due to stormwater.
   d. Underground injection on site.
   e. Releases to land on site.
16. Information on transfers of the chemical in wastes to off-site locations as follows:
   a. For transfers to publicly owned treatment works (POTWs):
      (1) The name and address (including county) of each POTW to which the chemical is transferred.
      (2) An estimate of the amount of the chemical transferred in pounds per year (transfers of less than 1000 lb/year may be indicated as a range) and an indication of the basis of the estimate.
   b. For transfers to other off-site locations:
      (1) The name, address (including county), and EPA identification number (RCRA ID number) of each off-site location, including an indication of whether the location is owned or controlled by the reporting facility or its parent company.
      (2) An estimate of the amount of the chemical in waste transferred in

pounds per year (transfers of less than 1000 lb/year may be indicated in ranges) to each off-site location, and an indication of the basis for the estimate and an indication of the type of treatment or disposal used.

17. The following information relative to waste treatment:
   a. An indication of the general type of wastestream containing the reported chemical.
   b. The treatment method applied to the wastestream.
   c. An indication of the concentration of the chemical in the wastestream prior to treatment.
   d. An estimate in percent of the efficiency of the treatment plus an indication of whether the estimate is based on operating data.
   e. An indication (use is optional) of whether treatments listed are part of a treatment sequence.

18. Pollution prevention data (reporting is optional) that includes the type of pollution prevention modification, quantity of the chemical in the wastes prior to treatment and disposal (for both the current and prior reporting year), a production index, and the reason for the pollution prevention action. This optional reporting expires after the 1990 reporting year.

9.3.4.2.4   Reporting Exemptions

9.3.4.2.4.1   DE MINIMIS CONCENTRATIONS OF A TOXIC CHEMICAL IN A MIXTURE: If a toxic chemical is present in a mixture of chemicals at a covered facility and the toxic chemical is in a concentration in the mixture which is below 1 percent of the mixture, or 0.1 percent of the mixture in the case of a toxic chemical which is a carcinogen as defined in 29 CFR 1910.1200(d)(4), a person is not required to consider the quantity of the toxic chemical present in such mixture when determining whether an applicable threshold has been met under § 372.25 or determining the amount of release to be reported under § 372.30. This exemption applies whether the person received the mixture from another person or the person produced the mixture, either by mixing the chemicals involved or by causing a chemical reaction, which resulted in the creation of the toxic chemical in the mixture. However, this exemption applies only to the quantity of the toxic chemical present in the mixture. If the toxic chemical is also manufactured (including imported), processed, or otherwise used at the covered facility other than as part of the mixture or in a mixture at higher concentrations, in excess of an applicable threshold quantity set forth in § 372.25, the person is required to report under § 372.30.

9.3.4.2.4.2   ARTICLES: If a toxic chemical is present in an article at a covered facility, a person is not required to consider the quantity of the toxic chemical present in the article when determining whether an applicable threshold has been met under § 372.25 or determining the amount of release to be reported under § 372.30. This exemption applies whether the person received the article from another person or the person produced the article. However, this exemption applies only to the quantity of the toxic chemical present in the article. If the toxic chemical is manufactured (including imported), processed, or otherwise used at the covered facility other than as part of the article, in excess of an applicable threshold quantity set forth in § 372.25, the person is required to report under § 372.30. Persons potentially subject to this exemption should carefully review the definitions of article and release in § 372.3. If a release of a toxic chemical occurs as a result of the processing or use of an item at the facility, that item does not meet the definition of article.

9.3.4.2.4.3 USES: If a toxic chemical is used at a covered facility for a purpose described in this paragraph (c), a person is not required to consider the quantity of the toxic chemical used for such purpose when determining whether an applicable threshold has been met under § 372.25 or determining the amount of releases to be reported under § 372.30. However, this exemption applies only to the quantity of the toxic chemical used for the purpose described in this paragraph (c). If the toxic chemical is also manufactured (including imported), processed, or otherwise used at the covered facility other than as described in this paragraph (c), in excess of an applicable threshold quantity set forth in § 372.25, the person is required to report under § 372.30:

- Use as a structural component of the facility.

- Use of products for routine janitorial or facility grounds maintenance. Examples include use of janitorial cleaning supplies, fertilizers, and pesticides similar in type or concentration to consumer products.

- Personal use by employees or other persons at the facility of foods, drugs, cosmetics, or other personal items containing toxic chemicals, including supplies of such products within the facility such as in a facility-operated cafeteria, store, or infirmary.

- Use of products containing toxic chemicals for the purpose of maintaining motor vehicles operated by the facility.

- Use of toxic chemicals present in process water and noncontact cooling water as drawn from the environment or from municipal sources, or toxic chemicals present in air used either as compressed air or as part of combustion.

9.3.4.2.4.4 ACTIVITIES IN LABORATORIES: If a toxic chemical is manufactured, processed, or used in a laboratory at a covered facility under the supervision of a technically qualified individual as defined in § 720.3(ee) of Title III, a person is not required to consider the quantity so manufactured, processed, or used when determining whether an applicable threshold has been met under § 372.25 or determining the amount of release to be reported under § 372.30. This exemption does not apply in the following cases:

- Specialty chemical production
- Manufacture, processing, or use of toxic chemicals in pilot-plant-scale operations
- Activities conducted outside the laboratory

9.3.4.2.4.5 CERTAIN OWNERS OF LEASED PROPERTY: The owner of a covered facility is not subject to reporting under § 372.30 if such owner's only interest in the facility is ownership of the real estate on which the facility is operated. This exemption applies to owners of facilities such as industrial parks, all or part of which are leased to persons who operate establishments within Standard Industrial Classification codes 20 through 39, where the owner has no other business interest in the operation of the covered facility. This provision is in 40 CFR 372.38(f).

9.3.4.2.4.6 REPORTING BY CERTAIN OPERATORS OF ESTABLISHMENTS ON LEASED PROPERTY SUCH AS INDUSTRIAL PARKS: If two or more persons who do not have any common corporate or business interest (including common ownership or control) operate separate establishments within a single facility, each such person shall treat the establishments it operates as a facility for purposes of this part. The determinations in § 372.22 and § 372.25 shall be made for those establishments. If any such operator determines that its establishment is a covered facility under § 372.22 and that a toxic chemical has been manufactured (including imported), processed, or otherwise at the establishment in excess of an applicable threshold in § 372.25 for a calendar year, the operator shall submit a report in accordance with § 372.30 for the establishment. For purposes of this paragraph (f), a common corporate or business interest includes ownership, partnership, joint ventures, ownership of a controlling interest in one person by the other, or ownership of a controlling interest in both persons by a third person.

9.3.4.2.4.7 COAL EXTRACTION ACTIVITIES: If a toxic chemical is manufactured, processed, or otherwise used in extraction by facilities in SIC code 12, a person is not required to consider the quantity of the toxic chemical so manufactured, processed, or otherwise used when determining whether an applicable threshold has been met under § 372.25 or 372.27, or determining the amounts to be reported under § 372.30.

9.3.4.2.4.8 METAL MINING OVERBURDEN: If a toxic chemical that is a constituent of overburden is processed or otherwise used by facilities in SIC code 10, a person is not required to consider the quantity of the toxic chemical so processed or otherwise used when determining whether an applicable threshold has been met under § 372.25 or § 372.27, or determining the amounts to be reported under § 372.30.

9.3.4.2.5 Supplier Notification Requirement. Generally speaking, a supplier is a person who owns or operates a facility or establishment that meets all of the following conditions:

- Is in Standard Industrial Classification codes 20 through 39 as set forth in Par. (b) of § 372.22

- Manufactures (including imports) or processes a toxic chemical

- Sells or otherwise distributes a mixture or trade name product containing the toxic chemical, to (1) a facility described in § 372.22 or (2) to a person who in turn may sell or otherwise distributes such mixture or trade name product to a facility described in § 372.22(b)

Such a supplier must notify each person to whom the mixture or trade name product is sold or otherwise distributed from the facility or establishment in accordance with Par. (b) of this section.

The notification required must be in writing and must include:

- A statement that the mixture or trade name product contains a toxic chemical or chemicals subject to the reporting requirements of Sec. 313 of Title III of the Superfund Amendments and Reauthorization Act of 1986 and 40 CFR Part 372.
- The name of each toxic chemical, and the associated CAS registry number of each chemical if applicable, as set forth in § 372.65.
- The percent by weight of each toxic chemical in the mixture or trade name product.

Notification under this section must be provided as follows:

- For a mixture or trade name product containing a toxic chemical listed in § 373.65 with an effective date of January 1, 1987, the person shall provide the written notice described in Par. (b) of this section to each recipient of the mixture or trade name product with at least the first shipment of each mixture or trade name product to each recipient in each calendar year beginning January 1, 1989.
- For a mixture or trade name product containing a toxic chemical listed in § 372.65 with an effective date of January 1, 1989 or later, the person shall provide the written notice described in paragraph (b) of this section to each recipient of the mixture or trade name product with at least the first shipment of the mixture or trade name product to each recipient in each calendar year beginning with the applicable effective date.
- If a person changes a mixture or trade name product for which notification was previously provided under Par. (b) of Section 313 by adding a toxic chemical, removing a toxic chemical, or changing the percent by weight of a toxic chemical in the mixture or trade name product, the person shall provide each recipient of the changed mixture or trade name product a revised notification reflecting the change with the first shipment of the changed mixture or trade name product to the recipient.
- If a person discovers (1) that a mixture or trade name product previously sold or otherwise distributed to another person during the calendar year of the discovery contains one or more toxic chemicals and (2) that any notification provided to such other persons in that calendar year for the mixture or trade name product either did not properly identify any of the toxic chemicals or did not accurately present the percent by weight of any of the toxic chemicals in the mixture or trade name product, the person shall provide a new notification to the recipient within 30 days of the discovery that contains the information described

in Par. (b) of this section and identifies the prior shipments of the mixture or product in that calendar year to which the new notification applies.

- If a MSDS is required to be prepared and distributed for the mixture or trade name product in accordance with 29 CFR 1910.1200, the notification must be attached to or otherwise incorporated into such MSDS. When the notification is attached to the MSDS, the notice must contain clear instructions that the notifications must not be detached from the MSDS and that any copying and redistribution of the MSDS shall include copying and redistribution of the notice attached to copies the MSDS subsequently redistributed.

9.3.4.2.6 Notification Exceptions. Notifications are not required in the following instances:

- If a mixture or trade name product contains no toxic chemical in excess of the applicable de minimis concentration as specified in § 372.38(a)
- If a mixture or trade name product is one of the following:

  An article as defined in § 372.3

  Foods, drugs, cosmetics, alcoholic beverages, tobacco, or tobacco products packaged for distribution to the general public

  Any consumer product, as the term is defined in the Consumer Product Safety Act (15 USC 1251 et seq.), packaged for distribution to the general public

9.3.4.3 Record Keeping. Each person subject to the reporting requirements must retain the following records for a period of 3 years from the date of the submission of a report under § 372.30:

- A copy of each report submitted by the person under § 372.30
- All supporting materials and documentation used by the person to make the compliance determination that the facility or establishments is a covered facility under § 372.22 or § 372.45
- Documentation supporting the report submitted under § 372.30 including:

  Documentation supporting any determination that a claimed allowable exemption under § 372.38 applies

  Data supporting the determination of whether a threshold under § 372.25 applies for each toxic chemical

  Documentation supporting the calculations of the quantity of each toxic chemical released to the environment or transferred to an off-site location

  Documentation supporting the use indications and quantity onsite reporting for each toxic chemical, including dates of manufacturing, processing, or use

  Documentation supporting the basis of estimate used in developing any release or off-site transfer estimates for each toxic chemical

Receipts or manifests associated with the transfer of each toxic chemical in waste to off-site locations

Documentation supporting reported waste treatment methods, estimates of treatment efficiencies, ranges of influent concentration to such treatment, the sequential nature of treatment steps, if applicable, and the actual operating data, if applicable, to support the waste treatment efficiency estimate for each toxic chemical

Suppliers of toxic chemicals who are subject to the notification requirements of this part must retain the following records for a period of 3 years from the date of the submission of a notification under § 372.45:

- All supporting materials and documentation used by the person to determine whether a notice is required under § 372.45
- All supporting materials and documentation used in developing each required notice under § 372.45 and a copy of each notice

Records retained under this section must be maintained at the facility to which the report applies or from which a notification was provided. Such records must be readily available for purposes of inspection by EPA.

9.3.4.4 Compliance and Enforcement. Violators of the requirements of this part shall be liable for a civil penalty in an amount not to exceed $25,000 each day for each violation as provided in Sec. 325(c) of Title III.

This reporting requirements applies to owners and operators of facilities that have 10 or more full-time employees that are in SIC codes 20 through 39 (that is, manufacturing facilities) and that manufactured, processed, imported, or otherwise used a listed toxic chemical in excess of specified threshold quantities.

On May 1, 1997, EPA added seven industry groups to the list of facilities subject to the reporting requirements Sec. 313 of EPCRA and Sec. 6607 of the Pollution Prevention Act of 1990 (62 FR 23834). These industry groups are metal mining [SIC code 10 (except 1011, 1081, and 1094)]; coal mining [SIC code 12 (except 1241)]; electric utilities [SIC codes 4911, 4931, and 4939 (limited to facilities that combust coal and/or oil for the purpose of generating electricity for distribution in commerce)]; commercial hazardous waste treatment [SIC code 4953 (limited to facilities regulated under RCRA Subtitle C)]; chemical and allied products—wholesale (SIC code 5171); petroleum bulk terminals and plans (also known as stations)—wholesale (SIC code 5171); and solvent recovery services [SIC code 7389 (limited to facilities primarily engaged in solvent recovery services on a contract or fee basis)].

This list of toxic chemicals subject to reporting consisted initially of chemicals listed for similar reporting purposes by the states of Maryland and New Jersey. The original list was composed of 328 entries, including 20 categories of chemicals. EPA added 286 chemicals and pesticides on November 30, 1994 (59 FR

61432). State governors and the public also may petition the administrator to add or delete chemicals from the list.

EPA uses the information generated from the Sec. 313 reporting forms to maintain a national toxic chemical inventory that is computer-accessible to the general public.

9.3.4.4.1 Trade Secrets. Trade secret protection procedures are outlined in Sec. 322 of Title III. These provisions apply to trade secret claims made under Sec. 303 emergency planning and Secs. 311, 312, and 313 regarding planning information, community right-to-know requirements, and toxic chemical release reporting. The specific chemical identity of an extremely hazardous substance or toxic chemical can be withheld for specific reasons. Even if the chemical identity is withheld, the generic class or category of the chemical must be provided.

Even if information can be withheld legally from the public, Sec. 323 of the Act requires that it not be withheld from health professionals who require the information for diagnostic purposes, or from local health officials who require the information for assessment activities. In these cases, the person receiving the information must be willing to sign a confidentiality agreement with the facility.

9.3.4.4.2 Enforcement. Enforcement procedures are included in Sec. 325. This section provides for:

- Civil penalties for facility owners or operators who fail to comply with emergency planning requirements
- Civil, administrative, and criminal penalties for owners or operators who fail to comply with the emergency notification requirements following the release of a listed hazardous substance
- Civil and administrative penalties for owners or operators who fail to comply with the reporting requirements in Secs. 311, 312, and 313
- Civil and administrative penalties for trade secret claims that are ruled frivolous
- Criminal penalties for the disclosure of trade secret information

CHAPTER 10
TOXIC SUBSTANCES

John A. Green
Director of Safety & Health
Tate & Lyle North American Sugars Inc.

Robert W. Pender
Project Director
RTP Environmental Associates, Inc.

10.1 TOXIC SUBSTANCES CONTROL ACT (TSCA)

This chapter discusses the permitting implications of TSCA as well as the licensing, training, and notifications required for the following four environmental topics:

- The control of toxic substances, including polychlorinated biphenyls (PCBs)
- Asbestos-containing materials
- Lead-based paint
- Radon

10.1.1 Overview

The Toxic Substances Control Act was enacted by Public Law 94-469 on October 11, 1976, to establish a new system for identifying and evaluating the environmental and health effects of existing chemicals and any new substances entering the U.S. market. TSCA was originally enacted in response to environmental damage from two types of chemicals: polychlorinated biphenyls (PCBs) and chlorofluorocarbons (CFCs).

The four major purposes of the TSCA are:

- To screen new chemicals to see if they pose a risk
- To require testing of chemicals identified as possible risks
- To gather information on existing chemicals
- To control chemicals proved to pose a risk

The law requires any company planning to manufacture or import a new chemical to submit to EPA a premanufacture notice that contains information on the identity, use, anticipated production or import volume, workplace hazards, and disposal characteristics of the substance. EPA can also require manufacturers and processors to report unpublished health and safety studies.

Since its promulgation, TSCA has been amended three times, each resulting in an additional statutory title. Consequently, it now contains four separate titles:

- Title I—The Toxic Substances Control Act places on manufacturers the responsibility to provide data on the environmental and health effects of chemical substances. Title I also gives EPA the authority to regulate the manufacture, use, distribution, and disposal of chemical substances.
- Title II—The Asbestos Hazard Emergency Response Act requires school systems to inspect for and abate asbestos hazards found in school buildings. It was added on October 22, 1986, by Public Law 99-519.
- Title III—The Indoor Radon Abatement Act was added on October 28, 1988, by Public Law 100-551.
- Title IV—The Lead-Based Paint Exposure Reduction Act was added by Public Law 102-550 on October 28, 1992.

10.1.2 TSCA Permitting Requirements

The most commonly utilized mechanism of environmental enforcement is the permit. The focus of TSCA, which regulates the production and importation of industrial chemicals, is a proactive one. Unlike the enforcement mechanisms provided by other environmental statutes (e.g., the Clean Air Act and the Clean Water Act) that control and/or minimize adverse environmental impacts caused by the discharge of chemical substances (i.e., pollutants), the provisions of TSCA do not establish any quantitative permitting requirements.

Instead, TSCA requires that all new chemicals produced in or imported into the United States be evaluated for health and environmental effects. To accomplish this objective TSCA provides a variety of other enforcement mechanisms to assure compliance with its requirements. Under TSCA, EPA has the authority to:

- Require that appropriate records, files, papers (excluding financial data, sales and pricing data, personnel data, and research data, unless such data are described with reasonable specificity in the inspection notice) be properly maintained and available on request to the administrator.

- Require the testing of "new" chemicals that have the potential to present a significant exposure risk to humans and/or the environment.

- Require that a premanufacture review of such "new" chemicals be completed prior to their introduction into the marketplace.

- Require import certification to ensure that all "new" chemical substances imported into the United States comply with the provisions of TSCA.

While other environmental permits [e.g., National Pollution Discharge Elimination System (NPDES) permits, Title V air operating permits] establish specific quantitative limits on industrial discharges, the "permitting" requirements associated with TSCA-related activities are inferred. In other words, if the above required activities are not satisfactorily completed, the owners and/or operators associated with regulated chemical substances are not *permitted* to proceed with their manufacture, use, distribution or disposal. In addition, the TSCA-based regulations found in 40 CFR, Parts 700 through 799, either establish operating criteria that must be met or refer to existing permitting requirements of another environmental statute. For example, owners or operators of facilities that store PCB wastes must register their facility with their EPA administrator and the storage facility must meet specific design criteria. These requirements closely parallel the requirements of a permitted Resource Conservation and Recovery Act (RCRA) hazardous waste storage facility. Also, the removal of asbestos-containing materials (ACM) can be performed and managed only by certified and/or licensed professionals, and the waste from such operations can be disposed of only at a facility having a valid permit to receive such wastes.

It is usually the individual states and/or local municipalities that require the owner or operator of such a site to obtain permits, licenses, or approvals prior to initiating construction, modification, or operation of the source.

The training and licensing requirements for these environmental concerns have come at different time periods. PCBs and asbestos were the first to be regulated, when requirements for licensing, training, storage, and disposal were established, followed most recently with different types of proposed requirements for managing lead-based-paint activities in a facility. There are also training requirements and licensing requirements when radon is to be dealt with.

These statutory requirements, which constitute the focus of these "inferred" permitted TSCA-related activities, are discussed in more detail below.

10.1.3 TSCA Required General Activities

10.1.3.1 Chemical Substances Inventory. EPA has compiled a comprehensive inventory of chemical substances existing in U.S. commerce. If a chemical is not on the inventory, it is considered a "new" chemical. Copies of the inventory can be obtained from EPA. Companies must notify EPA before manufacturing/importing a new—that is, unlisted—chemical. Manufacturers and importers of already-listed substances are required to report current production data every 4 years.

10.1.3.2 Premanufacture Notification. Under Sec. 5 of TSCA, any person who is intending to manufacture or import a chemical substance must first determine if the chemical substance is listed on the TSCA inventory. Companies are required to notify EPA at least 90 days before manufacturing or importing a new chemical substance. The premanufacture notification (PMN) program allows EPA 90 days in which to screen new chemicals. Incomplete or ambiguous PMNs can delay this review period. At the end of the review period, a company can manufacture or import the new substance provided EPA does not take action against it. The PMN must include the following information:

• The identity of the chemical substance, the amount to be manufactured, its categories of use, its potential by-products, employee exposure data, and anticipated disposal methods

• Any known test data on the chemical substance's effects on human health and/or the environment

• A description of any other such data that is known or "reasonably ascertainable" by the persons submitting the information

EPA can designate a use of an existing chemical as a significant new use based on the anticipated extent and type of exposure to humans and the environment. A manufacturer or processor of a chemical for such a significant new use also must report to EPA 90 days before manufacturing the chemical for that use.

Under certain circumstances, chemicals can be exempt from full 90-day PMN review; an example is small quantities used for purposes of research and development. If necessary, EPA can issue an order and seek a court injunction to keep a new chemical off the market.

10.1.3.3 "New" Chemical Testing. As stated above, EPA has the authority to require companies to test "new" substances before they are introduced into the marketplace—that is, to perform premanufacture testing—as well as to test substances that are already on the marketplace. Chemicals are recommended for testing by EPA and an Interagency Testing Committee. This group is made up of representatives of eight federal agencies: EPA, Occupational Safety and Health Administration, Council on Environmental Quality, National Institute for Occupational Safety and Health, National Institute for Environmental Health

Sciences, National Cancer Institute, National Science Foundation, and Department of Commerce.

Prior to issuing a requirement for the testing of such a chemical, EPA must ascertain whether:

- The chemical may present an unreasonable risk to health or the environment, or substantial human or environmental exposure to the substance is expected
- Insufficient data and experience exist to determine or predict the health and environmental effects of the chemical
- Testing of the chemical is necessary to develop such data

Testing might be required to evaluate the characteristics of a chemical, such as persistence or acute toxicity, or to clarify its health and environmental effects, including carcinogenic, mutagenic, behavioral, and synergistic effects. All testing must be conducted in accordance with EPA's Good Laboratory Practice standards.

10.1.4 Polychlorinated Biphenyls (PCBs)

EPA, which regulates polychlorinated biphenyls under Title I of the Toxic Substances Control Act, has determined that PCBs at concentrations greater than 50 parts per million (ppm) present an unreasonable risk to human health and the environment. Also, EPA has determined that an exposure to PCBs at any concentration also might pose a significant risk to humans and the environment.

Substances regulated under provisions of 40 CFR, Part 761, include solvents, oils, waste oils, heat transfer fluids, hydraulic fluids, paints or coatings, sludges, slurries, sediments, dredge spoils, soils, materials containing PCBs as a result of spills, and other chemical substances or combinations thereof containing PCBs. PCBs can be found in liquid, nonliquid, and multiphasic (combinations of liquid and nonliquid) forms. The following criteria should be used to determine PCB concentration to calculate which provisions apply:

- Nonliquid PCB concentrations must be measured on a dry weight basis.
- Liquid PCB concentrations must be measured on a wet weight basis. (Liquid PCBs containing more than 0.5 percent by weight nondissolved material should be analyzed as multiphasic.)
- Multiphasic PCB concentrations must separate the nondissolved materials into nonliquid and liquid PCBs and analyze them separately. The highest concentration level will determine the applicable provisions.

Unless otherwise noted, PCB concentrations are to be determined on a weight-per-weight basis (e.g., milligrams/kilogram) or, for liquids, on a weight-per-volume basis (e.g., milligrams/liter) if the density of the liquid is also reported.

As a result, the agency has promulgated the regulations found in 40 CFR, Part 761, that:

- Prohibit the manufacture, processing, distribution, and use of PCBs, except under specific, limited conditions
- Require notification of PCB waste activity
- Establish a manifest system to track and control PCB waste
- Set minimum standards for the storage and disposal of PCBs and PCB items
- Impose notification and cleanup requirements for PCB spills

[NOTE: EPA intends to address the issues of a ban on the export of PCBs for disposal and the authorized continued use of some non-liquid PCBs in future rule making, according to the PCB rule finalized June 29, 1998 (63 FR 35384).]

Most provisions of 40 CFR, Part 761, apply only if PCBs are present in concentrations above a specified level.

- Provisions that apply to PCBs at concentrations of <50 ppm apply also to contaminated surfaces at PCB concentrations of <10 $\mu g/100$ cm^2.
- Provisions that apply to PCBs at concentrations of 50 to <500 ppm apply also to contaminated surfaces at PCB concentrations of >10 mg/100 cm^2 to <100 mg/100 cm^2.
- Provisions that apply to PCBs at concentrations of ≥500 ppm also apply to contaminated surfaces at PCB concentrations of ≥100 mg/100 cm^2.

(NOTE: References to weights or volumes of PCBs apply to the total weight or total volume of the material that contains the regulated concentrations of PCBs, not the calculated weight or volume of the PCB molecules contained in the material.)

EPA regulations under TSCA require that certain PCB waste generators notify EPA, receive identification numbers, and prepare manifests for off-site shipment of PCB wastes. EPA used the RCRA hazardous waste cradle-to-grave tracking system as its model for the PCB tracking and manifest program.

The PCB wastes subject to TSCA requirements are PCBs and PCB items that no longer serve their intended purpose and must be disposed of, including waste from the cleanup of a spill of material that contained PCBs at a concentration of 50 ppm or greater and laboratory samples no longer used for analytical or enforcement purposes.

The notification requirement is designed to provide EPA with information on the location of PCB waste handlers—generators, storage facilities, transporters, disposal facilities—and enhance compliance and enforcement. Notification also is a prerequisite to issuance of unique identification numbers that waste handlers must include on manifests and other reports that are part of the waste tracking system. PCB generators are prohibited from offering their wastes to storage or disposal facilities or transporters without EPA identification numbers.

Examples of EPA-required approvals for activities associated with PCB wastes are listed below for storage facilities and disposal at chemical waste landfills. The reader should be cognizant that there are additional approval (i.e., permitting) requirements associated with the disposal of PCB wastes at incinerators, high-efficiency boilers, and smelters. Reference should be made to 40 CFR, Parts 761.70 through 761.72, for a listing of these requirements.

10.1.4.1 PCB Storage Facilities. After July 1, 1978, the storage of waste PCBs and PCB items is permitted only in approved storage facilities for which the owners or operators comply with the following storage unit requirements:

* The facilities must meet the following criteria:

  Adequate roof and walls to prevent rainwater from reaching the stored PCBs and PCB items.

  An adequate floor that has continuous curbing with a minimum 6-inch-high curb. The floor and curbing must provide a containment volume equal to at least 2 times the internal volume of the largest PCB article or PCB container or 25 percent of the total internal volume of all PCB articles or PCB containers stored there, whichever is greater. PCB/radioactive wastes are not required to be stored in an area with a minimum 6-inch-high curbing. However, the floor and curbing must still provide a containment volume equal to at least 2 times the internal volume of the largest PCB container or 25 percent of the total internal volume of all PCB containers stored there, whichever is greater.

  No drain valves, floor drains, expansion joints, sewer lines, or other openings that would permit liquids to flow from the curbed area.

  Floors and curbing constructed of portland cement, concrete, or a continuous, smooth, nonporous surface as defined at § 761.3, which prevents or minimizes penetration of PCBs.

  Not located at a site that is below the 100-year flood water elevation.

* The storage of PCBs and PCB items designated for disposal in a storage unit other than one meeting the above design requirements is not permitted unless the unit meets one of the following conditions:

  Is permitted by EPA under Sec. 3004 of RCRA to manage hazardous waste in containers.

  Qualifies for interim status under section 3005 of RCRA to manage hazardous waste in containers, meets the requirements for containment at § 264.175 of this chapter, and spills of PCBs are cleaned up in accordance with Subpart G of the Act.

  Is permitted by a state authorized under Sec. 3006 of RCRA to manage hazardous waste in containers, and spills of PCBs are cleaned up in accordance with Subpart G of Title I.

  Is approved or otherwise regulated pursuant to a State PCB waste management program no less stringent in protection of health or the environment than the applicable TSCA requirements found in Title I.

Is subject to a TSCA Coordinated Approval, which includes provisions for storage of PCBs, issued pursuant to 40 CFR § 761.77.

Has a TSCA PCB waste management approval, which includes provisions for storage of PCB remediation wastes, issued pursuant to 40 CFR § 761.61(c) or § 761.62(c).

10.1.4.2 *Manifest Requirements for Generators.*

Whenever a generator turns its PCB waste over to a transporter for delivery to an off-site storage or disposal facility, a manifest must be prepared. The manifest that tracks PCB wastes is the same manifest that is used to track RCRA wastes.

A generator of PCB waste must prepare a manifest for all PCB waste offered for transport to a disposal or storage facility unless the PCB waste is either:

- In a concentration < 50 ppm, provided that the waste has not been diluted
- Being transported to a facility owned or operated by the generator

Transporters are not permitted to accept PCB waste from a generator unless it is accompanied by a manifest signed by the generator.

10.1.4.3 *Disposal of Waste PCBs and PCB Items at Chemical Waste Landfills.*

Prior to the disposal of any PCBs and PCB items in a chemical waste landfill, the owner or operator of the landfill must receive written approval of the agency regional administrator for the region in which the landfill is located. The approval must be obtained in the following manner:

- *Initial report.* The owner or operator must submit to the regional administrator an initial report that contains:

The location of the landfill

A detailed description of the landfill, including general site plans and design drawings

An engineering report describing the manner in which the landfill complies with the requirements for chemical waste landfills specified in Par. (b) of this section

Sampling and monitoring equipment and facilities available

Expected waste volumes of PCBs

General description of waste materials other than PCBs that are expected to be disposed of in the landfill

Landfill operations plan as required in Par. (b) of this section

Any local, state, or federal permits or approvals

Any schedules or plans for complying with the approval requirements of these regulations.

• *Other information.* In addition to the information listed above, the regional administrator may require the owner or operator to submit any other information that the regional administrator finds to be reasonably necessary to determine whether a chemical waste landfill should be approved.

10.2 ASBESTOS HAZARD ABATEMENT ACT

10.2.1 Overview

Asbestos is a hydrated mineral silicate. It is mined in its ore form and transported to mills where it is processed and manufactured into various types of products. These products have included pipe insulation and boiler insulation, commonly called *thermal system insulation.* Asbestos has been used in fireproofing material, both in spray-on fireproofing and troweled-on fireproofing, as well as in different harder-type products such as floor tiles and transite board. There are records that have shown that asbestos was used as far back as the days of the Romans, who reportedly utilized asbestos for napkins. They would throw them into the fire, the fire would burn off any food particles that remained, and the napkins would be reused.

Asbestos was utilized quite heavily after World War II; it was very popular for its thermal properties. Since that time, studies have shown that there may be some health hazards in asbestos-containing materials (ACMs).

10.2.2 Regulatory History

In 1971 the first regulatory information pertaining to asbestos was promulgated. The Clean Air Act (CAA) and the National Emissions Standards for Hazardous Air Pollutants (NESHAPs), 40 CFR, Part 61, stated that asbestos presented a significant risk to human health as a result of the air emissions from its sources, and therefore is considered a hazardous air pollutant. Utilizing previous studies that were undertaken, the authors of the regulations determined asbestos to be a hazardous air pollutant (HAP), and set up guidelines for exposure.

In 1986 the Toxic Substance Control Act provided information and guidance that pertained to providing additional inspections and taking precautions with asbestos-containing materials. The U.S. Congress determined that the EPA's rule on inspections for local education agencies and notifications of any asbestos-containing materials that are found in school buildings was not good enough. EPA's rule did not include standards for proper identification of asbestos-containing material, nor for appropriate response actions for those materials that are found in a building. They also found that there was not a uniform program for accrediting persons involved with asbestos identification and abatement. This meant that there was no proper training given to those people going out and inspecting buildings and determining the amount, types, and conditions of asbestos-containing materials. Also, there was no formal training for those individuals who performed removal of asbestos-containing materials or who supervised that removal.

Subchapter 2 of TSCA, the Asbestos Hazard Emergency Response Act (AHERA), Subsection 2641, was developed to provide the establishment of federal regulations that would require inspections for asbestos-containing materials and implementation of appropriate response actions with respect to asbestos-containing materials in schools. This subchapter was also to mandate safe and complete periodic reinspections of school buildings to see if there was a change in the condition of those materials in the proceeding years. The Act also called for a study to determine the extent of the danger that may be posed to human health by asbestos located in both public and commercial buildings.

The regulations of 40 CFR, Part 763, Subpart E—Asbestos-Containing Materials in Schools, were promulgated in October 1987. This rule requires local education agencies to identify friable and nonfriable ACM in public and private elementary and secondary schools by visually inspecting school buildings for such materials, sampling such materials if they are not assumed to be ACM, and having samples analyzed by appropriate techniques referred to in this rule. The rule also requires local education agencies to submit management plans to the governor of their state by October 12, 1988; begin to implement the plans by July 9, 1989; and complete implementation of the plans in a timely fashion. In addition, local education agencies are required to use persons who have been accredited to conduct inspections and reinspections, develop management plans, or perform response actions. The rule also includes record-keeping requirements. Local education agencies may contractually delegate their duties under this rule, but they remain responsible for the proper performance of those duties. Local education agencies are encouraged to consult with EPA regional asbestos coordinators, or if applicable, a state's lead agency designated by the state governor, for assistance in complying with this rule. This was the final rule and notice, which provided the guidance for the type of training that would be required, how asbestos surveys would need to be performed, and other activities, such as the development of management plans. No one can perform these activities without undergoing proper training and receiving proper AHERA accreditation. AHERA provides accreditation for those who perform various tasks. AHERA provides accreditation for inspectors, management planners, and project designers. When asbestos is to be disturbed for removal or repair activities, asbestos workers, asbestos supervisors, and asbestos contractors must all receive proper AHERA accreditation.

Those who are accredited under the AHERA program must undergo refresher training every year in order to update their accreditations. This is to be performed every year with a new accreditation given after each training. If the refresher training lapses more than 6 months, those individuals must take the initial required training again, which is anywhere from a 3- to 5-day course, depending on the type of accreditation they are receiving.

10.2.3 Permitting Requirements Associated with Asbestos Activities

Again the actual requirements that form the basis for *permitted* activities are either inferred or they are based on certification and/or licensing requirements as well as transport and disposal criteria. Simply stated, these asbestos-related activities are not permitted unless certain requirements are met. Since these training requirements are specifically required by Title II of TSCA, the emphasis of this chapter will be on them. (NOTE: These training and certification requirements are also found in App. C to 40 CFR, Part 763.)

Since the transportation of ACM waste is covered separately by the Department of Transportation (49 CFR Part 173, Subpart J), and disposal of ACM waste is covered separately by the National Emissions Standards for Hazardous Air Pollutants (40 CFR, Part 61, Subpart M), this chapter will only briefly discuss them.

10.2.3.1 Training. In 1987, EPA developed a model accreditation plan (MAP) for states, which provides guidance for training individuals who perform various types of tasks related to ACMs. This training must be given to those persons who inspect school facilities for ACMs, those who prepare management plans for schools, and those who design and carry out response actions, such as contractors and project designers. Training has been established to give individuals the knowledge and background on the information they needed in order to work with ACMs.

In 1990, the AHERA regulation was updated. The requirements of AHERA were utilized to develop the Asbestos School Hazard Abatement Reauthorization Act of November 1980 (ASHARA). This Act requires training for those individuals who would be performing asbestos-related tasks in public or commercial buildings, not only in schools. There have been no separate training requirements developed for ASHARA. This act requires that those performing various asbestos-related tasks receive AHERA training and refresher training to keep their accreditations up to date.

10.2.3.1.1 Training for Accreditation. Training requirements for purposes of accreditation are specified both in terms of required subjects of instruction and in terms of length of training. Each initial training course has a prescribed curriculum and number of days of training. One day of training equals 8 hours, including breaks and lunch. Course instruction must be provided by EPA or state-approved instructors. EPA or state instructor approval must be based on a review of the instructor's academic credentials and/or field experience in asbestos abatement.

Beyond the initial training requirements, individual states can consider requiring additional days of training for purposes of supplementing hands-on activities or for reviewing relevant state regulations. States also can wish to consider the relative merits of a worker apprenticeship program. Further, they might consider more stringent minimum qualification standards for the approval of training instructors. EPA recommends that the enrollment in any given course be limited to 25 students so that adequate opportunities exist for individual hands-on experience. States have the option to provide initial training directly or approve other entities to offer training. Training/accreditation requirements have been established for the following six disciplines: workers, contractors and/or supervisors, inspectors, management planners, project designers, and project monitors.

Training requirements for each of the six accredited disciplines are outlined below. Persons in each discipline perform a different job function and distinct role. Inspectors identify and assess the condition of asbestos-containing building material (ACBM), or suspect ACBM. Asbestos-containing building material means surfacing ACM, thermal system insulation ACM, or miscellaneous ACM that is found in or on interior structural members or other parts of a school building.

Management planners use data gathered by inspectors to assess the degree of hazard posed by ACBM in schools to determine the scope and timing of appropriate response actions needed for schools. Project designers determine how asbestos abatement work should be conducted. Lastly, workers and contractor/supervisors carry out and oversee abatement work. In addition, a recommended training curriculum is also presented for a sixth discipline, which is not federally accredited, that of project monitor. Each accredited discipline and training curriculum is separate and distinct from the others. A person seeking accreditation in any of the five accredited MAP disciplines cannot attend two or more courses concurrently, but may attend such courses sequentially.

In several instances, initial training courses for a specific discipline (e.g., workers or inspectors) require hands-on training. For asbestos abatement contractor/supervisors and workers, hands-on training should include working with asbestos-substitute materials, fitting and using respirators, use of glovebags, donning protective clothing, and constructing a decontamination unit as well as other abatement work activities.

10.2.3.1.1.1 WORKERS: A person must be accredited as a worker to carry out any of the following activities with respect to friable ACBM in a school or public and commercial building: (1) a response action other than a "small-scale, short-duration" (SSD) activity, (2) a maintenance activity that disturbs friable ACBM other than an SSSD activity, or (3) a response action for a major fiber release episode.

Small-scale, short-duration activities are tasks such as, but not limited to:

• Removal of asbestos-containing insulation on pipes.

• Removal of small quantities of asbestos-containing insulation on beams or above ceilings.

• Replacement of an asbestos-containing gasket on a valve.

• Installation or removal of a small section of drywall.

• Installation of electrical conduits through or proximate to ACMs.

All persons seeking accreditation as asbestos abatement workers must complete at least a 4-day training course as outlined below. The 4-day worker training course must include lectures, demonstrations, at least 14 hours of hands-on training, individual respirator fit testing, course review, and an examination. Hands-on training must permit workers to have actual experience performing tasks associated with asbestos abatement. A person who is otherwise accredited as a contractor/supervisor may perform in the role of a worker without possessing separate accreditation as a worker.

Because of cultural diversity associated with the asbestos workforce, EPA recommends that states adopt specific standards for the approval of foreign language courses for abatement workers. EPA further recommends the use of audiovisual materials to complement lectures, where appropriate.

The training course must adequately address the following topics:

- *Physical characteristics of asbestos.* Identification of asbestos, aerodynamic characteristics, typical uses, and physical appearance, and a summary of abatement control options.

- *Potential health effects related to asbestos exposure.* The nature of asbestos-related diseases; routes of exposure; dose-response relationships and the lack of a safe exposure level; the synergistic effect between cigarette smoking and asbestos exposure; the latency periods for asbestos-related diseases; a discussion of the relationship of asbestos exposure to asbestosis, lung cancer, mesothelioma, and cancers of other organs.

- *Employee personal protective equipment.* Classes and characteristics of respirator types; limitations of respirators; proper selection, inspection; donning, use, maintenance, and storage procedures for respirators; methods for field testing of the facepiece-to-face seal (positive- and negative-pressure fit checks); qualitative and quantitative fit testing procedures; variability between field and laboratory protection factors that alter respiratory fit (e.g., facial hair); the components of a proper respiratory protection program; selection and use of personal protective clothing; use, storage, and handling of nondisposable clothing; and regulations covering personal protective equipment.

- *State-of-the-art work practices.* Proper work practices for asbestos abatement activities, including descriptions of proper construction; maintenance of barriers and decontamination enclosure systems; positioning of warning signs; lockout of electrical and ventilation systems; proper working techniques for minimizing fiber release; use of wet methods; use of negative pressure exhaust ventilation equipment; use of high-efficiency particulate air (HEPA) vacuums; proper cleanup and disposal procedures; work practices for removal, encapsulation, enclosure, and repair of ACM; emergency procedures for sudden releases; potential exposure situations; transport and disposal procedures; and recommended and prohibited work practices.

- *Personal hygiene.* Entry and exit procedures for the work area; use of showers; avoidance of eating, drinking, smoking, and chewing (gum or tobacco) in the work area; and potential exposures, such as family exposure.

- *Additional safety hazards.* Hazards encountered during abatement activities and how to deal with them, including electrical hazards, heat stress, air contaminants other than asbestos, fire and explosion hazards, scaffold and ladder hazards, slips, trips, falls, and confined spaces.

- *Medical monitoring.* OSHA and EPA Worker Protection Rule requirements for physical examinations, including a pulmonary function test, chest x-rays, and a medical history for each employee.

- *Air monitoring.* Procedures to determine airborne concentrations of asbestos fibers, focusing on how personal air sampling is performed and the reasons for it.

- *Relevant federal, state, and local* regulatory requirements, procedures, and standards, with particular attention directed at relevant EPA, OSHA, and state regulations concerning asbestos abatement workers.

- *Establishment* of respiratory protection programs.

- *Course review.* A review of key aspects of the training course.

10.2.3.1.1.2 CONTRACTOR/SUPERVISORS: A person must be accredited as a contractor/supervisor to supervise any of the following activities with respect to friable ACBM in a school or public and commercial building: (1) A response action other than an SSSD activity, (2) a maintenance activity that disturbs friable ACBM other than an SSSD activity, or (3) a response action for a major fiber release episode. All persons seeking accreditation as asbestos abatement contractor/supervisors must complete at least a 5-day training course as outlined below. The training course must include lectures, demonstrations, at least 14 hours of hands-on training, individual respirator fit testing, course review, and a written examination. Hands-on training must permit supervisors to have actual experience performing tasks associated with asbestos abatement.

EPA recommends the use of audiovisual materials to complement lectures, where appropriate. Asbestos abatement supervisors include those persons who provide supervision and direction to workers performing response actions. Supervisors may include those individuals with the position title of foreman, working foreman, or leadman pursuant to collective bargaining agreements. At least one supervisor is required to be at the work site at all times while response actions are being conducted. Asbestos workers must have access to accredited supervisors throughout the duration of the project.

The contractor/supervisor training course must adequately address the following topics:

- *The physical characteristics of asbestos and asbestos-containing materials.* Identification of asbestos, aerodynamic characteristics, typical uses, physical appearance, a review of hazard assessment considerations, and a summary of abatement control options.

- *Potential health effects related to asbestos exposure.* The nature of asbestos-related diseases; routes of exposure; dose-response relationships, and the lack of a safe exposure level; synergism between cigarette smoking and asbestos exposure; and latency period for diseases.

- *Employee personal protective equipment.* Classes and characteristics of respirator types; limitations of respirators; proper selection, inspection, donning, use, maintenance, and storage procedures for respirators; methods for field testing of the facepiece-to-face seal (positive- and negative-pressure fit checks); qualitative and quantitative fit testing procedures; variability between field and laboratory protection factors that alter respiratory fit (e.g., facial hair); the components of a proper respiratory protection program; selection and use of personal protective clothing; use, storage, and handling of nondisposable clothing; and regulations covering personal protective equipment.

- *State-of-the-art work practices.* Proper work practices for asbestos abatement activities, including descriptions of proper construction and maintenance of barriers and decontamination enclosure systems; positioning of warning signs; lockout of electrical and ventilation systems; proper working techniques for minimizing fiber release; use of wet methods; use of negative pressure exhaust ventilation equipment; use of HEPA vacuums; and proper cleanup and disposal procedures. Work practices for removal, encapsulation, enclosure, and repair of ACM; emergency procedures for unplanned releases; potential exposure situations; transport and disposal procedures; and recommended and prohibited work practices. New abatement-related techniques and methodologies may be discussed.

- *Personal hygiene.* Entry and exit procedures for the work area; use of showers; and avoidance of eating, drinking, smoking, and chewing (gum or tobacco) in the work area. Potential exposures, such as family exposure, must also be included.

- *Additional safety hazards.* Hazards encountered during abatement activities and how to deal with them, including electrical hazards, heat stress, air contaminants other than asbestos, fire and explosion hazards, scaffold and ladder hazards, slips, trips, falls, and confined spaces.

- *Medical monitoring.* OSHA and EPA Worker Protection Rule requirements for physical examinations, including a pulmonary function test, chest x-rays, and a medical history for each employee.

- *Air monitoring.* Procedures to determine airborne concentrations of asbestos fibers, including descriptions of aggressive air sampling, sampling equipment and methods, reasons for air monitoring, types of samples, and interpretation of results. EPA recommends that transmission electron microscopy (TEM) be used for analysis of final air clearance samples, and that sample analyses be performed by laboratories accredited by the National Voluntary Laboratory Accreditation Program (NVLAP) of the National Institute of Standards and Technology (NIST).

- *Relevant federal, state, and local* regulatory requirements, procedures, and standards, including:

  Requirements of TSCA Title II

  National Emission Standards for Hazardous Air Pollutants (40 CFR, Part 61), Subparts A (General Provisions) and M (National Emission Standard for Asbestos)

  OSHA standards for permissible exposure to airborne concentrations of asbestos fibers and respiratory protection (29 CFR 1910.134)

  OSHA Asbestos Construction Standard (29 CFR 1926.58)

  EPA Worker Protection Rule (40 CFR, Part 763, Subpart G).

- *Respiratory protection programs* and medical monitoring programs.

- *Insurance and liability issues.* Contractor issues, worker's compensation coverage and exclusions, third-party liabilities and defenses, insurance coverage and exclusions.

- *Record keeping for asbestos abatement projects.* Records required by federal, state, and local regulations; records recommended for legal and insurance purposes.

- *Supervisory techniques for asbestos abatement activities.* Supervisory practices to enforce and reinforce the required work practices and discourage unsafe work practices.

- *Contract specifications.* Discussions of key elements that are included in contract specifications.

- *Course review.* A review of key aspects of the training course.

10.2.3.1.1.3 INSPECTOR: All persons who inspect for ACBM in schools or public and commercial buildings must be accredited. All persons seeking accreditation as an inspector must complete at least a 3-day training course as outlined below. The course must include lectures, demonstrations, 4 hours of hands-on training, individual respirator fit testing, course review, and a written examination.

EPA recommends the use of audiovisual materials to complement lectures, where appropriate. Hands-on training should include conducting a simulated building walk-through inspection and respirator fit testing. The inspector training course must adequately address the following topics:

- *Background information on asbestos.* Identification of asbestos, and examples and discussion of the uses and locations of asbestos in buildings; physical appearance of asbestos.

- *Potential health effects related to asbestos exposure.* The nature of asbestos-related diseases; routes of exposure; dose-response relationships and the lack of a safe exposure level; the synergistic effect between cigarette smoking and asbestos exposure; the latency periods for asbestos-related diseases; a discussion of the relationship of asbestos exposure to asbestosis, lung cancer, mesothelioma, and cancers of other organs.

- *Functions/qualifications and role of inspectors.* Discussions of prior experience and qualifications for inspectors and management planners; discussions of the functions of an accredited inspector as compared to those of an accredited management planner; discussion of inspection process including inventory of ACM and physical assessment.

- *Legal liabilities and defenses.* Responsibilities of the inspector and management planner; a discussion of comprehensive general liability policies, claims made, and occurrence policies, environmental and pollution liability policy clauses; state liability insurance requirements; bonding and the relationship of insurance availability to bond availability.

- *Understanding building systems.* The interrelationship between building systems, including: an overview of common building physical plan layout; heat, ventilation, and air conditioning (HVAC) system types, physical organization, and where asbestos is found on HVAC components; building mechanical systems, their types and organization, and where to look for asbestos on such systems; inspecting electrical systems, including appropriate safety precautions; reading blueprints and as-built drawings.

- *Public/employee/building occupant relations.* Notifying employee organizations about the inspection; signs to warn building occupants; tact in dealing with occupants and the press; scheduling of inspections to minimize disruptions; and education of building occupants about actions being taken.

- *Preinspection planning and review of previous inspection records.* Scheduling the inspection and obtaining access; building record review; identification of probable homogeneous areas from blueprints or as-built drawings; consultation with maintenance or building personnel; review of previous inspection, sampling, and abatement records of a building; the role of the inspector in exclusions for previously performed inspections.

- *Inspecting for friable and nonfriable ACM and assessing the condition of friable ACM.* Procedures to follow in conducting visual inspections for friable and nonfriable ACM; types of building materials that may contain asbestos; touching materials to determine friability; open return air plenums and their importance in HVAC systems; assessing damage, significant damage, potential damage, and potential significant damage; amount of suspected ACM, both in total quantity and as a percentage of the total area; type of damage; accessibility; material's potential for disturbance; known or suspected causes of damage or significant damage; and deterioration as assessment factors.

- *Bulk sampling/documentation of asbestos.* Detailed discussion of the "Simplified Sampling Scheme for Friable Surfacing Materials" (EPA 560/5-85-030a, October 1985); techniques to ensure sampling in a randomly distributed manner for other than friable surfacing materials; sampling of nonfriable materials; techniques for bulk sampling; inspector's sampling and repair equipment; patching or repair of damage from sampling; discussion of polarized light microscopy; choosing an accredited laboratory to analyze bulk samples; quality control and quality assurance procedures. EPA's recommendation that all bulk samples collected from school or public and commercial buildings be analyzed by a properly accredited laboratory.

- *Inspector respiratory protection and personal protective equipment.* Classes and characteristics of respirator types; limitations of respirators; proper selection, inspection, donning, use, maintenance, and storage procedures for respirators; methods for field testing of the facepiece-to-face seal (positive- and negative-pressure fit checks); qualitative and quantitative fit testing procedures; variability between field and laboratory protection factors that alter respiratory fit (e.g., facial hair); the components of a proper respiratory protection program; selection and use of personal protective clothing; use, storage, and handling of nondisposable clothing.

- *Record keeping and writing the inspection report.* Labeling of samples and keying sample identification to sampling location; recommendations on sample labeling; detailing of ACM inventory; photographs of selected sampling areas and examples of ACM condition; information required for inclusion in the management plan required for school buildings under TSCA Title II, Sec. 203(i)(1). EPA recommends that states develop and require the use of standardized forms for recording the results of inspections in schools or public or commercial buildings, and that the use of these forms be incorporated into the curriculum of training conducted for accreditation.

- *Regulatory review.* The following topics should be covered: National Emission Standards for Hazardous Air Pollutants (NESHAPs; 40 CFR, Part 61, Subparts A and M); EPA Worker Protection Rule (40 CFR, Part 763, Subpart G); OSHA Asbestos Construction Standard (29 CFR 1926.58); OSHA respirator requirements (29 CFR 1910.134); the Asbestos-Containing Materials in Schools Rule (40 CFR, Part 763, Subpart E); applicable state and local regulations; differences between federal and state requirements where they apply; and the effects, if any, on public and nonpublic schools or commercial or public buildings.

- *Field trip.* This includes a field exercise, including a walk-through inspection; on-site discussion about information gathering and the determination of sampling locations; on-site practice in physical assessment; classroom discussion of field exercise.

- *Course review.* A review of key aspects of the training course.

10.2.3.1.1.4 MANAGEMENT PLANNER: All persons who prepare management plans for schools must be accredited. All persons seeking accreditation as management planners must complete a 3-day inspector training course as outlined above and a 2-day management planner training course. Possession of current and valid inspector accreditation must be a prerequisite for admission to the management planner training course. The management planner course must include lectures, demonstrations, course review, and a written examination.

EPA recommends the use of audiovisual materials to complement lectures, where appropriate. TSCA Title II does not require accreditation for persons performing the management planner role in public and commercial buildings. Nevertheless, such persons may find this training and accreditation helpful in preparing them to design or administer asbestos operations and maintenance programs for public and commercial buildings.

The management planner training course must adequately address the following topics:

- *Course overview.* The role and responsibilities of the management planner; operations and maintenance programs; setting work priorities; protection of building occupants.
- *Evaluation/interpretation of survey results.* Review of TSCA Title II requirements for inspection and management plans for school buildings as given in Sec. 203(i)(1) of TSCA Title II; interpretation of field data and laboratory results; comparison of field inspector's data sheet with laboratory results and site survey.
- *Hazard assessment.* Amplification of the difference between physical assessment and hazard assessment; the role of the management planner in hazard assessment; explanation of significant damage, damage, potential damage, and potential significant damage; use of a description (or decision tree) code for assessment of ACM; assessment of friable ACM; relationship of accessibility, vibration sources, use of adjoining space, and air plenums and other factors in hazard assessment.
- *Legal implications.* Liability; insurance issues specific to planners; liabilities associated with interim control measures, in-house maintenance, repair, and removal; use of results from previously performed inspections.
- *Evaluation and selection of control options.* Overview of encapsulation, enclosure, interim operations and maintenance, and removal; advantages and disadvantages of each method; response actions described via a decision tree or other appropriate method; work practices for each response action; staging and prioritizing of work in both vacant and occupied buildings; the need for containment barriers and decontamination in response actions.
- *Role of other professionals.* Use of industrial hygienists, engineers, and architects in developing technical specifications for response actions; any requirements that may exist for architect sign-off of plans; team approach to design of high-quality job specifications.
- *Developing an operations and maintenance (O&M) plan.* Purpose of the plan; discussion of applicable EPA guidance documents; what actions should be taken by custodial staff; proper cleaning procedures; steam cleaning and HEPA vacuuming;

reducing disturbance of ACM; scheduling O&M for off-hours; rescheduling or canceling renovation in areas with ACM; boiler room maintenance; disposal of ACM; in-house procedures for ACM—bridging and penetrating encapsulants; pipe fittings; metal sleeves; polyvinyl chloride (PVC), canvas, and wet wraps; muslin with straps, fiber mesh cloth; mineral wool, and insulating cement; discussion of employee protection programs and staff training; case study in developing an O&M plan (development, implementation process, and problems that have been experienced).

- *Regulatory review.* Focusing on the OSHA Asbestos Construction Standard found at 29 CFR 1926.58; the National Emission Standards for Hazardous Air Pollutants (NESHAPs) found at 40 CFR, Part 61, Subparts A (General Provisions) and M (National Emission Standard for Asbestos); EPA Worker Protection Rule found at 40 CFR, Part 763, Subpart G; TSCA Title II; applicable state regulations.

- *Record keeping for the management planner.* Use of field inspector's data sheet along with laboratory results; ongoing record keeping as a means to track asbestos disturbance; procedures for record keeping. EPA recommends that states require the use of standardized forms for purposes of management plans and incorporate the use of such forms into the initial training course for management planners.

- *Assembling and submitting the management plan.* Plan requirements for schools in TSCA Title II Sec. 203(i)(1); the management plan as a planning tool.

- *Financing abatement actions.* Economic analysis and cost estimates; development of cost estimates; present costs of abatement versus future operation and maintenance costs; Asbestos School Hazard Abatement Act grants and loans.

- *Course review.* A review of key aspects of the training course.

10.2.3.1.1.5 PROJECT DESIGNER: A person must be accredited as a project designer to design any of the following activities with respect to friable ACBM in a school or public and commercial building: (1) A response action other than an SSSD maintenance activity, (2) a maintenance activity that disturbs friable ACBM other than an SSSD maintenance activity, or (3) a response action for a major fiber release episode. All persons seeking accreditation as a project designer must complete at least a minimum 3-day training course as outlined below. The project designer course must include lectures, demonstrations, a field trip, course review, and a written examination.

EPA recommends the use of audiovisual materials to complement lectures, where appropriate. The abatement project designer training course must adequately address the following topics:

- *Background information on asbestos.* Identification of asbestos; examples and discussion of the uses and locations of asbestos in buildings; physical appearance of asbestos.

- *Potential health effects related to asbestos exposure.* Nature of asbestos-related diseases; routes of exposure; dose-response relationships and the lack of a safe exposure level; the synergistic effect between cigarette smoking and asbestos exposure; the latency period of asbestos-related diseases; a discussion of the relationship between asbestos exposure and asbestosis, lung cancer, mesothelioma, and cancers of other organs.

- *Overview of abatement construction projects.* Abatement as a portion of a renovation project; OSHA requirements for notification of other contractors on a multiemployer site (29 CFR 1926.58).

- *Safety system design specifications.* Design, construction, and maintenance of containment barriers and decontamination enclosure systems; positioning of warning signs; electrical and ventilation system lockout; proper working techniques for minimizing fiber release; entry and exit procedures for the work area; use of wet methods; proper techniques for initial cleaning; use of negative-pressure exhaust ventilation equipment; use of HEPA vacuums; proper cleanup and disposal of asbestos; work practices as they apply to encapsulation, enclosure, and repair; use of glove bags and a demonstration of glove bag use.

- *Field trip.* A visit to an abatement site or other suitable building site, including on-site discussions of abatement design and building walk-through inspection. Include discussion of rationale for the concept of functional spaces during the walk-through.

- *Employee personal protective equipment.* Classes and characteristics of respirator types; limitations of respirators; proper selection, inspection, donning, use, maintenance, and storage procedures for respirators; methods for field testing of the facepiece-to-face seal (positive- and negative-pressure fit checks); qualitative and quantitative fit testing procedures; variability between field and laboratory protection factors that alter respiratory fit (e.g., facial hair); the components of a proper respiratory protection program; selection and use of personal protective clothing; use, storage, and handling of nondisposable clothing.

- *Additional safety hazards.* Hazards encountered during abatement activities and how to deal with them, including electrical hazards, heat stress, air contaminants other than asbestos, fire, and explosion hazards.

- *Fiber aerodynamics and control.* Aerodynamic characteristics of asbestos fibers; importance of proper containment barriers; settling time for asbestos fibers; wet methods in abatement; aggressive air monitoring following abatement; aggressive air movement and negative-pressure exhaust ventilation as a cleanup method.

- *Designing abatement solutions.* Discussions of removal, enclosure, and encapsulation methods; asbestos waste disposal.

- *Final clearance process.* Discussion of the need for a written sampling rationale for aggressive final air clearance; requirements of a complete visual inspection;

and the relationship of the visual inspection to final air clearance. EPA recommends the use of TEM for analysis of final air clearance samples. These samples should be analyzed by laboratories accredited under the NIST NVLAP.

- *Budgeting/cost estimating.* Development of cost estimates; present costs of abatement versus future operation and maintenance costs; setting priorities for abatement jobs to reduce costs.

- *Writing abatement specifications.* Preparation of and need for a written project design; means and methods specifications versus performance specifications; design of abatement in occupied buildings; modification of guide specifications for a particular building; worker and building occupant health/medical considerations; replacement of ACM with nonasbestos substitutes.

- *Preparing abatement drawings.* Significance and need for drawings, use of as-built drawings as base drawings; use of inspection photographs and on-site reports; methods of preparing abatement drawings; diagramming containment barriers; relationship of drawings to design specifications; particular problems related to abatement drawings.

- *Contract preparation* and administration.

- *Legal/liabilities/defenses.* Insurance considerations; bonding; hold-harmless clauses; use of abatement contractor's liability insurance; claims made versus occurrence policies.

- *Replacement.* Replacement of asbestos with asbestos-free substitutes.

- *Role of other consultants.* Development of technical specification sections by industrial hygienists or engineers; the multidisciplinary team approach to abatement design.

- *Occupied buildings.* Special design procedures required in occupied buildings; education of occupants; extra monitoring recommendations; staging of work to minimize occupant exposure; scheduling of renovation to minimize exposure.

- *Relevant federal, state, and local* regulatory requirements, procedures and standards, including, but not limited to:

  Requirements of TSCA Title II

  National Emission Standards for Hazardous Air Pollutants (40 CFR, Part 61) Subparts A (General Provisions) and M (National Emission Standard for Asbestos)

  OSHA Respirator Standard (29 CFR 1910.134)

  EPA Worker Protection Rule (40 CFR, Part 763, Subpart G)

  OSHA Asbestos Construction Standard (29 CFR 1926.58)

  OSHA Hazard Communication Standard (29 CFR 1926.59)

- *Course review.* A review of key aspects of the training course.

10.2.3.1.1.6 PROJECT MONITOR: EPA recommends that states adopt training and accreditation requirements for persons seeking to perform work as project monitors. Project monitors observe abatement activities performed by contractors and generally serve as a building owner's representative to ensure that abatement work is completed according to specification and in compliance with all relevant statutes and regulations. They may also perform the vital role of air monitoring for purposes of determining final clearance. EPA recommends that a state seeking to accredit individuals as project monitors consider adopting a minimum 5-day training course covering the topics outlined below. The course outlined below consists of lectures and demonstrations, at least 6 hours of hands-on training, course review, and a written examination. The hands-on training component might be satisfied by having the student simulate participation in or performance of any of the relevant job functions or activities.

EPA recommends that the project monitor training course adequately address the following topics:

- *Roles and responsibilities of the project monitor.* Definition and responsibilities of the project monitor, including regulatory/specification compliance monitoring, air monitoring, conducting visual inspections, and final clearance monitoring.

- *Characteristics of asbestos and asbestos-containing materials.* Typical uses of asbestos; physical appearance of asbestos; review of asbestos abatement and control techniques; presentation of the health effects of asbestos exposure, including routes of exposure, dose-response relationships, and latency periods for asbestos-related diseases.

- *Federal asbestos regulations.* Overview of pertinent EPA regulations, including: NESHAPs, 40 CFR, Part 61, Subparts A and M; AHERA, 40 CFR, Part 763, Subpart E; and the EPA Worker Protection Rule, 40 CFR, Part 763, Subpart G. Overview of pertinent OSHA regulations, including: Construction Industry Standard for Asbestos, 29 CFR 1926.58; Respirator Standard, 29 CFR 1910.134; and the Hazard Communication Standard, 29 CFR 1926.59. Applicable state and local asbestos regulations; regulatory interrelationships.

- *Understanding building construction and building systems.* Building construction basics, building physical plan layout; understanding building systems (HVAC, electrical, etc.); layout and organization, where asbestos is likely to be found on building systems; renovations and the effect of asbestos abatement on building systems.

- *Asbestos abatement contracts, specifications, and drawings.* Basic provisions of the contract; relationships between principal parties, establishing chain of command; types of specifications, including means and methods, performance, and proprietary and nonproprietary; reading and interpreting records and abatement drawings; discussion of change orders; common enforcement responsibilities and authority of project monitor.

- *Response actions and abatement practices.* Prework inspections; prework considerations, precleaning of the work area, removal of furniture, fixtures, and equipment; shutdown/modification of building systems; construction and maintenance of containment barriers, proper demarcation of work areas; work area entry/exit, hygiene practices; determining the effectiveness of air filtration equipment; techniques for minimizing fiber release, wet methods, continuous cleaning; abatement methods other than removal; abatement area cleanup procedures; waste transport and disposal procedures; contingency planning for emergency response.

- *Asbestos abatement equipment.* Typical equipment found on an abatement project; air filtration devices, vacuum systems, negative pressure differential monitoring; HEPA filtration units, theory of filtration, design/construction of HEPA filtration units, qualitative and quantitative performance of HEPA filtration units, sizing the ventilation requirements, location of HEPA filtration units, qualitative and quantitative tests of containment barrier integrity; best available technology.

- *Personal protective equipment.* Proper selection of respiratory protection; classes and characteristics of respirator types, limitations of respirators; proper use of other safety equipment, protective clothing selection, use, and proper handling, hard/bump hats, safety shoes; breathing air systems, high pressure versus low pressure, testing for Grade D air, determining proper backup air volumes.

- *Air monitoring strategies.* Sampling equipment, sampling pumps (low versus high volume), flow-regulating devices (critical and limiting orifices), use of fibrous aerosol monitors on abatement projects; sampling media, types of filters, types of cassettes, filter orientation, storage and shipment of filters; calibration techniques, primary calibration standards, secondary calibration standards, temperature/pressure effects, frequency of calibration, record keeping and field work documentation, calculations; air sample analysis, techniques available and limitations of AHERA on their use, transmission electron microscopy (background to sample preparation and analysis, air sample conditions that prohibit analysis, EPA's recommended technique for analysis of final air clearance samples), phase contrast microscopy (background to sample preparation, and AHERA's limits on the use of phase contrast microscopy), what each technique measures; analytical methodologies, AHERA TEM protocol, National Institute of Occupational Safety and Health (NIOSH) 7400, OSHA reference method (nonclearance), EPA recommendation for clearance (TEM); sampling strategies for clearance monitoring, types of air samples (personal breathing zone versus fixed-station area), sampling location and objectives (preabatement, during abatement, and clearance monitoring), number of samples to be collected, minimum and maximum air volumes, clearance monitoring (post–visual inspection) (number of samples required, selection of sampling locations, period of sampling, aggressive sampling, interpretations of sampling results, calculations), quality assurance; special sampling problems, crawl spaces, acceptable samples for laboratory analysis, sampling in occupied buildings (barrier monitoring).

- *Safety and health issues other than asbestos.* Confined-space entry, electrical hazards, fire and explosion concerns, ladders and scaffolding, heat stress, air contaminants other than asbestos, fall hazards, hazardous materials on abatement projects.

- *Conducting visual inspections.* Inspections during abatement, visual inspections using the ASTM E1368 document; conducting inspections for completeness of removal; discussion of "How clean is clean?"

- *Legal responsibilities and liabilities of project monitors.* Specification enforcement capabilities; regulatory enforcement; licensing; powers delegated to project monitors through contract documents.

- *Record keeping and report writing.* Developing project logs/daily logs (what should be included, who sees them); final report preparation; record keeping under federal regulations.

- *Workshops* (6 hours spread over 3 days). Contracts, specifications, and drawings. This workshop could consist of each participant being issued a set of contracts, specifications, and drawings and then being asked to answer questions and make recommendations to a project architect or engineer, or the building owner, based on given conditions and these documents.

- *Air monitoring strategies/asbestos abatement equipment.* This workshop could consist of developing sampling strategies for simulated abatement sites (e.g., occupied buildings, industrial situations). Through demonstrations and exhibition, the project monitor may also be able to gain a better understanding of the function of various pieces of equipment used on abatement projects (air filtration units, water filtration units, negative pressure monitoring devices, sampling pump calibration devices, etc.).

- *Conducting visual inspections.* This workshop could consist, ideally, of an interactive video in which a participant is "taken through" a work area and asked to make notes of what is seen. A series of questions will be asked to stimulate the participant's recall of the area. A series of two or three videos, with different site conditions and different degrees of cleanliness, could be used.

10.2.3.1.2 State Training Requirements. Many states have adopted AHERA regulation and require those individuals performing work in those states to have the proper AHERA training and accreditation. Many of these states have their own licenses or accreditation, which can be received if individuals show proof of their AHERA accreditation. There are some states, though, that provide their own training, which candidates must take in order to receive various types of licenses or accreditation to work in those states. As an example, to work in New York, you must take a course that is approved by the State of New York in order to receive its accreditation. To receive a New York City asbestos investigator license, a training course approved by the New York City Department of Environmental Protection is required. The State of New Jersey does not have any of its own inspection licensing requirements; the AHERA training is required, and AHERA accreditation is necessary to perform inspections in this state. There are licensing requirements for those individuals who remove ACMs, as well as the contractors that employ these individuals.

The State of New Jersey gets somewhat more stringent when a project is performed within a public building. For those individuals who are going to oversee an asbestos removal project in a public building, such as a government building or school, additional training is required. In order to oversee a project, an individual must be an asbestos safety technician (AST) in order to oversee this work, whereas in a commercial building or privately owned building, this type of accreditation is not required in the State of New Jersey.

Different states, depending on the amount of material to be removed, require their own notifications. As an example, in the State of New Jersey, the limit of notification is 3 linear or 3 square feet. Other states, such as Indiana, follow NESHAPs, under which, if a project is going to be 160 square feet or 260 linear feet or more, the state requires proper notification. In such states, the levels and limits of work notifications have been established to follow the NESHAPs regulation, and to be no more stringent.

Compared to many of the air pollutants found in industry and the environment, asbestos is more highly regulated with respect to training, work procedures, notification, and disposal requirements. Much of the training is similar for each state. Some states, though, differ in their approach to who requires training and for the different types of asbestos-related materials. The majority of the licensing requirements deal with the proper inspection of a building as well as training for those who are going to be doing removal work.

10.2.3.2 ACM Waste Transport.   Although there are no regulatory specifications regarding permitting or design of transport vehicles, it is recommended that vehicles used for transport of bagged containerized asbestos waste have an enclosed carrying compartment or utilize a canvas covering sufficient to contain the transported waste, prevent damage to containers, and prevent fiber release. Transport of large quantities of asbestos waste is commonly conducted in a 20-cubic-yard roll-off box, which should also be covered. Vehicles that use compactors to reduce waste volume should not be used because these will cause the waste bags to rupture. Vacuum trucks sometimes used to transport waste slurry must be inspected to ensure that water is not leaking from the truck.

10.2.3.3 ACM Waste Disposal.   Disposal involves the isolation of asbestos waste material in order to prevent fiber release to air or water. Landfilling is recommended as an environmentally sound isolation method because asbestos fibers are virtually immobile in soil. EPA has established asbestos disposal requirements for active and inactive disposal sites under NESHAPs (40 CFR, Part 61, Subpart M) and specifies general requirements for solid waste disposal under RCRA (40 CFR, Part 257). Advance EPA notification of the intended disposal site is required by NESHAPs.

10.2.3.4   Receiving Asbestos Waste.   A landfill approved for receipt of asbestos waste must require notification by the waste hauler that the load contains asbestos. The landfill operator must inspect the loads to verify that asbestos waste is properly contained in leaktight containers and labeled appropriately. The appropriate EPA regional asbestos NESHAPs contact must be notified if the landfill operator believes that the asbestos waste is in a condition that may cause significant fiber release during disposal. In situations when the wastes are not properly containerized, the landfill operator needs to thoroughly soak the asbestos with a water spray prior to unloading, rinse out the truck, and immediately cover the wastes with nonasbestos material prior to compacting the waste in the landfill.

10.3   INDOOR RADON ABATEMENT ACT

Title 15 of the Toxic Substances Control Act (TSCA), the section dealing with commerce and trade, lists indoor radon abatement statutes that have been promulgated and need to be followed during any type of radon activity. The section describes recommended policies when dealing with radon in various types of situations.

Subchapter 3, Title 15 of Chap. 53 of the TSCA has as a national long-term goal that the radon level within buildings should be as low as that of the ambient air outside the buildings; in other words, the air inside a building should be as free of radon as the open areas outside it.

Section 2661 of Subchapter 3 includes information on the recommended policy for dealing with "assisted housing." The purpose of this section is to require the Department of Housing and Urban Development (HUD) to develop an effective departmental policy for dealing with radon contamination, and to follow and utilize any EPA guidelines and standards to ensure that the occupants of this type of housing are not exposed to hazardous levels of radon. This section also requires HUD to assist the EPA in reducing radon in this type of housing. *Assisted housing* is multifamily housing owned by HUD, or public housing and Indian housing assisted under the United States Housing Act of 1937. Housing that is receiving project-based assistance under Sec. 8 of the United States Housing Act, housing that is being assisted under Sec. 236 of the National Housing Act (NHA), and housing that is receiving assistance under Section 221 of the NHA is considered assisted-living housing. Section 2661 also states that it is up to the Secretary of Housing and Urban Development to develop and recommend to Congress a policy for dealing with radon contamination and to specify a program for education, research, and testing on, and mitigation of, radon hazards in housing covered by this section. All of this information was to be developed, coordinated, and acted on by November 7, 1988.

Subchapter 3, concerning indoor radon abatement, stated that no later than June 1, 1989, the EPA was to publish and make available to the public an updated version of a document titled "A Citizen's Guide to Radon." It is the responsibility of the administrator of the EPA to revise and publish the guide as necessary in the future. The guide includes a description of a series of action levels, the health risk associated with different levels of radon exposure, and the health risks to potentially sensitive populations of different levels of radon. The guide also covers the cost and the technological feasibility of reducing radon concentrations within buildings and the relationship between short-term and long-term testing techniques.

Section 2664 of Subchapter 3 states that the administrator of the EPA, by June 1, 1990, was to develop construction standards and techniques for controlling radon levels within new buildings. Some assistance was to be provided by those organizations that have experience in this type of construction and development of new buildings and codes. The standards and techniques for this construction were to apply to different geographic locations, as there are different construction types in the different areas of the country, different weather, different geology, and many different variables that could effect radon levels in new buildings. The entire United States was studied and standards were to be developed by using different locations throughout.

Section 2265 of Subchapter 3 requires the administrator of the EPA, or another federal department that the EPA designates, to develop and implement activities designed to assist the various states with their radon programs. Included in this was the establishment of a clearinghouse of radon-related information such as mitigation studies, public information materials, and surveys of radon levels. Also included was assisting the states with a voluntary proficiency program that rates the effectiveness of radon measurement devices and methods. The administrator was also to rate the effectiveness of private firms and individuals that offer radon-related architecture design, engineering, surveying, or mitigation services. The proficiency program was required to be in operation by October 28, 1989. The assistance was also to include the design and implementation of training seminars for state and local officials and private professional firms that deal with radon, and this training was to address topics such as monitoring, analysis, mitigation, health effects, public information, and public design to assist in mitigation of buildings. To assist the states with their radon programs, some cooperative projects between the EPA and the states' radon programs were required. These types of projects include home evaluation programs, where the EPA evaluates homes in the various states and demonstrates different mitigation methods in these homes. Also required was establishment of a national database, with all the data organized by state, outlining location and amounts of radon in the particular states.

If requested by a particular state, the EPA or other designated agency may provide that state with technical assistance in the development or the implementation of a radon program. This assistance could include the design and implementation of surveys, identification of the location and occurrence of radon within a state, the design and implementation of public information and educational programs, and the design and implementation of state programs to control radon in existing or new structures. The assistance could deal with the assessment of mitigation alternatives in an unusual or unconventional structure, or it could be the design or implementation of methods for radon measurement and mitigation for nonresidential buildings, housing, and child care facilities. The EPA is also to provide appropriate information about the technology and methods of radon assessment and mitigation to professional organizations that represent private firms involved with building design, engineering, and construction. Also, since 1989, the EPA submits to Congress a plan that identifies the assistance they will provide under Subchapter 3. This plan will outline those personnel and the financial resources necessary to do this work. The EPA is also authorized to use a certain amount of funds to establish efficiency rating programs and training seminars.

The EPA is required to conduct research to develop tests and evaluate radon measurement methods and protocols. The EPA also conducted a study to determine the feasibility of establishing a mandatory proficiency-testing program. This program requires that any product offered for sale or device used in connection with a radon service meets minimum performance criteria that have been established. The program addresses procedures for ordering the recall of any products sold that measure radon, and do not meet performance criteria. The program requires that any operator of a device, or persons employing certain techniques used in connection with a radon service, meet a minimum level of proficiency.

It is the responsibility of each state each year to apply for a grant for funds to assist with the development and implementation of programs for the assessment and mitigation of radon. The application for the grant is to include the following information:

- A description of the seriousness and extent of radon exposure in the state
- Identification of the state agency with primary responsibility for the radon programs that will receive the grant money
- A description of the rules and responsibilities of the state agency that administers the radon program
- A list of the activities and programs related to radon that are performed in the state each year
- A proposed budget specifying federal and state funding of each activity
- A 3-year plan that outlines long-range goals and objectives, the tasks that need to be performed to achieve them, and resource requirements to do this work

Those activities that a state may perform to be eligible for some type of grant assistance include surveys of radon levels, development of public information and educational materials about radon, implementation of programs to control radon existing in new structures purchased by the state, purchase of radon measuring equipment and devices, purchase or maintenance of analytical equipment connected with radon measurement and analysis, payment for the development of courses related to radon that are given to state or local employees, and assistance with the payment of the general overhead of the agency in that state that is performing the radon duties. Also there can be grants for the development of a data storage and management system for radon information, for payment for the cost of demonstrating radon mitigation methods and technologies, and for a toll-free radon hotline in the state to provide information and technical assistance.

The EPA, when it comes to providing these grants, may request information, data, and reports developed by the state to determine if there should be continuing eligibility, that is, if the state is keeping up with its radon program. Any state that does receive funds must provide the EPA with all radon-related information about its activities, including the results of radon surveys and mitigation design, demonstration projects, and risk communication studies. Any state receiving funds is to maintain and make available to the public a list of firms and individuals within that state that have received a passing rating under the EPA's proficiency rating program.

Subchapter 3 also states that the EPA was to conduct a study for the purpose of determining the extent of radon contamination within the nation's school buildings. The EPA was to identify and compile a list of the areas within the United States in which there is a high probability that schools have elevated levels of radon.

Under this subchapter, the EPA may make a grant to those entities that are looking to establish and operate regional radon training centers. The purpose of the training center is to develop information and provide training to federal and state officials, special and private firms, and the public regarding the health risks posed by radon and to demonstrate methods for measurement and mitigation. This subchapter also requires that federal buildings conduct a radon survey for the purpose of determining the extent of radon contamination in that federal building. The EPA was to identify and compile a list of areas within the United States where there is a high probability of radon within federal building locations. In putting this list together, the EPA made determinations of radon levels based on geological data, data from high radon levels in homes and in any other structures near federal buildings, and the physical characteristics of federal buildings.

Subchapter 3 gives the EPA the authority to have states establish their own radon programs. Some states have done that and some states have not. As an example, the State of New Jersey requires that anyone who performs any type of radon testing or radon mitigation be properly trained and properly licensed to do this work. The State of New Jersey certifies radon measurement businesses, radon measurement specialists, radon measurement technicians, radon mitigation businesses, radon mitigation specialists, and radon mitigation technicians.

Radon testing is becoming routine in the sale of residential property. Elevated radon levels often lead to the installation of radon mitigation systems.

10.4   LEAD-BASED-PAINT (LBP) EXPOSURE ABATEMENT ACT

10.4.1   The Hazards of LBP and Federal Efforts to Reduce Exposure

The Centers for Disease Control and Prevention (CDC) has estimated that approximately 900,000 children, or about 4.4 percent of children under the age of 6, may have unacceptably high levels of lead in their blood. Lead exposure in young children is of particular concern, because children absorb lead more readily than adults and their nervous systems are particularly vulnerable to the effects of lead. Common sources of lead exposure for children include contaminated dust and paint chips from deteriorating LBP in older homes and from renovation activities that disturb LBP. Children with high levels of lead in their bodies can suffer from learning disabilities, behavioral problems, and mental retardation. The effects of long-term lead exposure or poisoning in children are well documented: higher school failure rates, reductions in lifetime earnings due to permanent loss of intelligence, and increased social pathologies. Fetuses are also at risk, as lead can pass from a pregnant woman's bloodstream to the developing child. There is also some indication that lead exposure contributes to high blood pressure and reproductive and memory problems in adults. Lead has no known use in the body and is difficult to remove from blood and bones in cases where medical intervention is necessary.

Over the past two decades, the federal government has taken a number of steps to address the problems of lead exposure. In 1978, the Consumer Product Safety Commission banned the residential use of paint containing more than 0.06 percent lead by weight on interior and exterior surfaces, toys, and furniture. EPA placed controls on lead in gasoline in 1978 and lowered the maximum levels of lead permitted in public water systems (40 CFR, Parts 141 and 142). The CDC has set and lowered blood lead levels of concern several times, most recently in 1991. The Department of Housing and Urban Development (HUD) began in 1986 to abate lead hazards in public housing that is being renovated or in structures occupied by a child with elevated blood lead levels. These efforts, and those of state and local agencies and the private sector, have reduced the incidence of lead poisoning.

It is estimated that more than half the housing stock in the United States (an estimated 64 million pre-1980 homes) still contain some LBP. Further, the LBP Hazard Reduction and Financing Task Force established by HUD pursuant to Sec. 1015 of Title X (the LBP Hazard Reduction Act of 1992) estimates that between 5 and 15 million housing units contain hazards associated with the presence of LBP.

In response to the health threat, Congress enacted the Residential LBP Hazard Reduction Act of 1992 (hereinafter referred to as Title X of the Housing and Community Development Act of 1992 or as Title X), Pub. L. No. 102-550, 106 Stat. 3897. The purposes of Title X include: (1) to develop a national strategy to build the infrastructure necessary to eliminate LBP hazards in all housing as expeditiously as possible, (2) to reorient the national approach to the presence of LBP in housing to implement a broad program to evaluate and reduce LBP hazards in the nation's housing stock, and (3) to encourage effective action to prevent childhood lead poisoning by establishing a framework for LBP hazard evaluation and reduction and by ending confusion pertaining to reasonable standards of care (Pub. L. 102-550, Title X, Sec. 1003, codified at 42 USC 4851a).

To further these goals, Title X requires that HUD provide public housing authorities and other owners of federally assisted properties with guidelines for evaluating and reducing lead hazards in their properties. Title X also amended TSCA by adding a new Title IV, which directs EPA to promulgate standards to govern: (1) the training and certification of individuals engaged in LBP activities, (2) the accreditation of training programs, and (3) the process by which LBP activities are conducted by certified individuals [TSCA Sec. 402(a), 15 USC 2682(a)]. TSCA Title IV also directs EPA to identify by regulation LBP hazards, lead-contaminated dust, and lead-contaminated soil (TSCA Sec. 403, 15 USC 2683). States and Indian tribes may seek to administer and enforce these requirements (TSCA Sec. 404, 15 USC 2684).

As a result of the enactment of Title X, there is an increasing effort to reduce the hazards posed by LBP in residential housing and other buildings. Although there are a number of methods to reduce LBP exposure, abatements (which under TSCA Title IV involve any set of measures designed to eliminate permanently LBP hazards) are typically conducted in situations where LBP exposure has resulted in elevated blood lead levels in children and in other situations where permanent removal of LBP is desired. Abatement efforts frequently result in the production of LBP waste that may currently be subject to regulatory controls under Subtitle C of the Resource Conservation and Recovery Act (RCRA). The EPA has spent considerable resources working with health specialists, environmental groups, the lead abatement industry, and state and local governments to develop regulatory options for lead abatement activities. The EPA believes that there is an overwhelming consensus that action should be taken as quickly as possible to reduce lead exposure hazards to young children.

The Lead-Based Paint Hazard Reduction and Financing Task Force established by HUD pursuant to Sec. 1015 of Title X (42 USC 4852a), representing the spectrum of interests affected by LBP issues, released final recommendations on evaluating and reducing LBP hazards in private housing on July 11, 1995. Their report is entitled "Putting the Pieces Together: Controlling Lead Hazards in the Nation's Housing." In addition, a letter from the task force to EPA administrator Carol Browner dated April 13, 1994, specifically recommended that the agency "shift regulation of discarded architectural components from the hazardous waste regulatory program to a tailored management program under TSCA §§ 402/404." The task force recommendations enjoy the support of a broad range of the groups and interests affected by LBP activities and regulations. The agency has given substantial weight to the task force recommendations in the development of this proposal. EPA has developed and is proposing a regulatory approach it believes will work both to speed the conduct of lead abatement and deleading activities (by lowering costs) and, at the same time, ensure that LBP debris from all activities is managed and disposed of in a safe, reliable, and effective manner.

10.4.2 TSCA Title IV

On December 18, 1998, EPA published a proposed rule under the authority of Secs. 402 and 404 of TSCA (15 USC 2682 and 2684). Section 402 of TSCA, LBP Activities Training and Certification, directs EPA to promulgate regulations governing the training and certification of individuals engaged in LBP activities, the accreditation of training programs, and standards for conducting LBP activities. Section 404 of TSCA, Authorized State Programs, provides authority for EPA to authorize states to administer and enforce the requirements established by the agency under Sec. 402 of TSCA.

10.4.2.1 LBP Activities. The term *LBP activities* includes, among other activities, abatements in target housing [15 USC 2682(b)(1)]. TSCA Sec. 401(1) defines *abatement* as "any set of measures designed to permanently eliminate LBP hazards" including, among other things, all "clean-up, disposal, and post-abatement clearance testing activities" [15 USC 2681(1)(B)]. Because the term *abatement* includes all cleanup and disposal activities, TSCA Title IV provides the EPA with clear legal authority to promulgate regulations establishing standards for the management and disposal of LBP (including any LBP found on debris) resulting from the abatement of target housing. TSCA Title IV defines *target housing* generally to mean any housing constructed prior to 1978, except for housing for the elderly or those with disabilities (unless any child who is less than 6 years of age resides or is expected to reside in such housing for the elderly or persons with disabilities) or any 0-bedroom dwelling [TSCA Sec. 401(17), 15 USC 2681].

In addition to target housing, the LBP Activities Training and Certification Rule (40 CFR, Part 745) included in the TSCA Sec. 402 requirements a subcategory of public buildings called *child-occupied facilities.* A child-occupied facility is defined as "a building, or portion of a building, constructed prior to 1978, visited regularly by the same child, 6 years of age or under, on at least 2 different days within any week (Sunday through Saturday period), provided that each day's visit lasts at least 3 hours and the combined weekly visits last at least 6 hours, and the combined annual visits last at least 60 hours. Child-occupied facilities may include, but are not limited to, day-care centers, preschools and kindergarten classrooms." Thus, EPA is also covering "child-occupied facilities" in EPA's proposal of December 18, 1998, that are consistent with the LBP Training and Certification Rule.

TSCA Sec. 402 excludes homeowners who conduct LBP activities (including abatement, renovation, and remodeling activities) themselves in target housing that they own, unless the housing is occupied by a person or persons other than the owner or the owners' immediate family while the LBP debris is being generated.

In the case of public buildings constructed before 1978 and commercial buildings, TSCA Sec. 402 defines the term *LBP activities* to include deleading and demolition. *Deleading* is defined to mean "activities conducted by a person who offers to eliminate LBP or LBP hazards or to plan such activities." Management and disposal of LBP debris from public and commercial buildings are among the activities a person conducts to eliminate LBP or LBP hazards, and, therefore, are considered to constitute deleading activities under TSCA Sec. 402(b)(2). Although Sec. 402(b)(2) uses terms such as *identification* and *deleading* instead of the terms used in 402(a), such as *inspection, risk assessment,* and *abatement,* EPA believes that, given the similarity of the population to be protected and the nature of the risk they face, the Sec. 402(b)(2) terms can be understood to include the same types of LBP activities as specified in Sec. 402(b)(1). *Deleading* under Sec. 402(b)(2) is equivalent to *abatement* under Sec. 402(b)(1). Therefore, management and disposal of LBP debris from deleading and demolition are among the LBP activities EPA has the authority to regulate in public buildings and commercial buildings under TSCA Sec. 402.

10.4.2.2 LBP Hazards.   TSCA Sec. 402(c) addresses LBP risks associated with renovation and remodeling activities in target housing, public buildings, and commercial buildings. EPA was directed under Sec. 402(c)(1) to develop guidelines for conducting such activities. These guidelines, "Reducing Lead Hazards When Remodeling Your Home" (EPA 747-R-94-002), were published in April 1994 (updated September 1997) and are available through the National Lead Information Center (telephone: 1-800-424-LEAD). EPA was also directed under Sec. 402(c)(2) to conduct a study of the extent to which renovation and remodeling activities create an "LBP hazard" on a regular or occasional basis. EPA has not completed this study; however, the study does not examine management or disposal of LBP debris. EPA is authorized under Sec. 402(c)(3) of TSCA to apply the standards developed under Sec. 402(a) of TSCA for LBP activities to renovation and remodeling activities that create LBP hazards. EPA has determined for this proposal that improper management and disposal of LBP debris, including debris from renovation and remodeling activities, constitutes an LBP hazard and has included LBP debris from renovation and remodeling activities within the scope of EPA's proposal. The proposed rule determination that improper management and disposal of LBP debris constitutes an LBP hazard is included in the regulatory text of this proposal.

EPA's proposal also includes certain restrictions on the reuse of LBP debris. The proposed restrictions are designed to prevent the transfer of LBP hazards from one structure to another. For example, this proposal would prohibit reuse of LBP debris, which would be identified as an "LBP hazard."

10.4.2.3 LBP Permitting Requirements.   Again, the permitting requirements mandated by Title IV of TSCA are nonspecific and refer more to a "management-by-rule" approach than anything else. This chapter discusses the proposed standards in EPA's proposal of December 18, 1998.

10.4.2.3.1 Certification. Section 402(a)(1) of TSCA directs EPA (the Agency) to promulgate regulations, which ensure that individuals engaged in LBP activities are properly trained, that training programs are accredited, and that contractors engaged in such activities are certified. EPA's action proposes standards for the management and disposal of LBP debris that take into account reliability, effectiveness, and safety. It does not, however, create training requirements for individuals engaged in the management and disposal of LBP debris.

The Agency believes that the activities covered by this proposal, and the requirements governing them, do not warrant any specialized training. These activities and requirements are similar, if not identical, to the types of waste management activities already being conducted by generators, transporters, and disposal facility owner/operators and parties reusing LBP debris. The proposed requirements are designed to be as simple as possible while continuing to meet the TSCA Sec. 402 standard of "taking into account reliability, effectiveness, and safety." The addition of training requirements would add to the burden of conducting LBP debris management and disposal activities without providing a measurable reduction in risk of exposure to LBP hazards.

The primary reason for requiring the certification of individuals is to ensure that the individual has received proper training. However, because the Agency would not require specialized training for the management and disposal of LBP debris, § 745.315 proposes to certify all individuals who comply with the requirements of the rule. Certification would be extended only to individuals and firms engaged in management and disposal of LBP debris. To perform other LBP activities, individuals and firms would need to be certified in accordance with TSCA Secs. 402 and 404 rules (40 CFR, Part 745). This "certification by rule" for management and disposal of LBP debris allows the Agency to efficiently fulfill the TSCA Sec. 402 mandate noted above to "ensure that...contractors engaged in such activities are certified" without sacrificing safety, effectiveness, or reliability. EPA is proposing under Sec. 402 of TSCA to establish a clear regulatory environment covering the management and disposal of LBP debris from abatements, deleading, demolitions, renovations, and remodeling of target housing, public buildings, and commercial buildings. The TSCA standards being proposed today represent a commonsense approach to management and disposal of LBP debris, one that addresses the problems associated with current RCRA regulation of LBP debris.

10.4.2.3.2 Training and Certification Activities. On August 29, 1996, in 61 FR 45778, FRL-5389-9, EPA promulgated a rule under Secs. 402 and 404 of TSCA (hereafter, the LBP training and certification rule) addressing the conduct of certain LBP activities in target housing and child-occupied facilities (40 CFR, Part 745). The LBP training and certification rule requires that individuals and firms conducting specified LBP activities in target housing and child-occupied facilities receive training from accredited training programs and be certified to conduct LBP activities. The rule also contains standards for conducting LBP activities. The LBP training and certification rule did not specifically address the management and disposal of LBP debris. EPA's proposal of December 18, 1998, would create standards under TSCA for the management and disposal of LBP debris and clarifies that other LBP wastes remain subject to RCRA management and disposal requirements.

On August 6, 1999, EPA published a final rule in the *Federal Register* (64 FR 42849) amending the procedural requirements for training and certification of workers involved in lead-based-paint activities in target housing and child-occupied facilities by extending the effective dates for certification of individuals and firms and use of work practice standards that are contained in the final regulations promulgated under Sec. 402 of TSCA. The extension applies only in those states and Indian tribes in which EPA is operating the federal lead-based-paint program. EPA has extended these effective dates in order to provide additional time for individuals to become trained and certified to conduct lead-based-paint activities safely, reliably, and effectively. EPA believed that the extension of the effective dates would result in successful implementation of the federal program and ensure the availability of a well-qualified workforce to perform risk assessments, abatements, and other lead-based-paint activities.

1. *To whom does this action apply?* You may be potentially affected by this action if you operate a training program required to be accredited under 40 CFR 745.225, or if you are a professional, individual, or firm who must be certified to conduct lead-based-paint activities in accordance with 40 CFR 745.226. Potentially affected categories and entities may include, but are not limited to, those listed in Table 10.1. This list is not intended to be exhaustive, but rather provides a guide for readers regarding entities likely to be affected by this action. Other types of entities not listed above could also be affected. The Standard Industrial Classification (SIC) codes have been provided to assist you and others in determining whether or not this action applies to certain entities. To determine whether you or your business is affected by this action, you should carefully examine this action and the applicability provisions in 40 CFR, Part 745, Subpart L.

2. *How can I get additional information, including copies of this final rule and other related documents?* Copies of this final rule and certain other available documents may be obtained electronically from the EPA Internet home page at *http://www.epa.gov/.* On the home page, select "Laws and Regulations" and then look up the entry for this final rule under the "Federal Register" listings at *http://www.epa.gov/fedrgstr/.*

3. *What does this amendment do?* In 1996, EPA published the final TSCA Sec. 402/404 rule for training and certification of workers, accreditation of training programs, and model state programs for lead-based-paint activities in target housing and child-occupied facilities (61 FR 45778, August 29, 1996) (FRL-5389-9). At that time, the implementation of the federal program was delayed until August 29, 1998, to allow states and Indian tribes to apply and receive authorization to run their own EPA-approved lead-based-paint programs based on the model program.

TABLE 10.1 Who Is Affected by Procedural Requirements for Training Workers Involved in Lead-Based-Paint Activities

Type of eligibility	SIC code	Examples of entities
Lead abatement professionals	1799, 8734	Workers, supervisors, inspectors, risk assessors, and project designers engaged in lead-based-paint activities. Firms engaged in lead-based-paint activities.
Training programs	1799, 8331, 8742, 8748	Training programs providing training services in lead-based-paint activities.

The final rule provided for an additional phase-in period to allow the regulated community to come into compliance after the federal program became effective in nonauthorized states and tribes on August 29, 1998. After March 1, 1999, training programs could no longer provide, offer, or claim to provide training or refresher training for lead-based-paint activities defined at § 745.223 without being accredited by EPA according to the requirements of § 745.225. The rule also stated that after August 30, 1999, no individuals or firms could perform, offer, or claim to perform lead-based-paint activities as defined at § 745.223 without certification from EPA under § 745.226 to conduct those activities. A special provision at § 745.226(d) was effective only until August 30, 1999, and allowed individuals to seek certification based on prior training and completion of a refresher course and a certification exam (if applicable). Additionally, after August 30, 1999, all lead-based-paint activities were to be performed according to the work practice standards at § 745.227.

The federal program under Part 745, Subpart L, became effective August 31, 1998, in all nonauthorized states and tribes. The accreditation requirements at § 745.225 became effective March 1, 1999, and all training providers must now be accredited by EPA to offer lead-based-paint activities courses in the federal program.

Although EPA has been reviewing applications for accreditation, there have been several unavoidable delays that have slowed the process of approving a sufficient number of training providers to accommodate the number of individuals seeking certification prior to the August 30, 1999, date. Two important items, the model training courses and the fee schedule, were not made available by EPA to training providers in a sufficiently timely fashion to allow them to prepare their application packages well in advance of the deadlines.

In the preamble to the final rule, EPA indicated that it would make model training courses available in advance for training providers to use in developing their programs (61 FR 45778, at 45783). Under § 745.225(b)(1)(iii), training providers who used EPA model training materials may submit an abbreviated application package for accreditation and thus potentially accelerate the accreditation process. However, EPA was unable to make all model training materials immediately available to training providers. The updating of some of these courses to reflect the course curricula in § 745.225(d) was initially delayed. The development of a new model course for the project designer discipline has not yet been completed. EPA has also changed distributors for the model training course materials. The course materials were not available to the regulated community while a new distributor was being sought and contract arrangements finalized.

EPA was also delayed in promulgating the final fee rule setting out the fee schedule for accreditation of training providers and certification of contractors. The final fee rule was effective June 11, 1999 (64 FR 31092, June 9, 1999) (FRL-6058-6). Prior to its publication, training providers were unsure as to the fee structure and may have delayed preparing accreditation applications while waiting for the fee rule to be finalized.

Because of these delays, some areas of the United States where EPA is running the federal program have insufficient training courses currently available for the number of individuals seeking certification. In some areas this is due to a lack of training provider applicants to provide training. In other areas, this is due to a backlog of training provider applications needing review by EPA. Despite the fact that the August 30, 1999, deadline has passed, EPA has received only a few certification applications because of the difficulty for many individuals to take the courses needed prior to applying for certification. The lack of refresher courses has been a particular problem for those who wish to use the certification based on prior training provisions at § 745.226(d) that require completion of an EPA-accredited refresher course. Additionally, EPA has not made the certification exam available for inspectors, risk assessors, and supervisors who are required by § 745.226(b)(1)(ii) to pass a certification exam after completing the training courses.

EPA believed that it was necessary to extend the effective dates for certification and work practice standards to March 1, 2000, to allow for successful implementation of the federal program. This will allow EPA time to accredit sufficient training providers to accommodate the many individuals who must be certified. In particular, EPA wished to accommodate those individuals who have years of experience conducting lead-based-paint activities and choose to use the certification based on prior training provisions at § 745.226(d), which were also extended until March 1, 2000. These individuals must take refresher courses, which EPA expects to be available in greater numbers with the extended effective date. Once individuals take the appropriate courses, inspectors, risk assessors, and supervisors must also complete the appropriate certification exams. EPA had those exams in place and available for all those who sought to take the exams prior to the March 1, 2000, deadline. This extension applies only in states and tribes in which EPA is operating the federal lead-based-paint program under Part 745, Subpart L. It does not affect states and tribes operating EPA-authorized programs under Part 745, Subpart Q. Additionally, in the federal program, the extension for use of work practice standards does not apply once an individual is certified by EPA, because § 745.226(a)(4) states that individuals who have received EPA certification must conduct lead-based-paint activities in compliance with the appropriate work practice standards in § 745.227.

Without the extension of the effective dates, EPA did not believe that it would be possible to have effective implementation of the federal lead-based-paint program and ensure that individuals are well trained in conducting lead-based-paint activities in target housing and child-occupied facilities. EPA was concerned that under the original deadline, individuals who could not be certified, because of the lack of available training course and certification exams, would not be able to legally perform lead-based-paint activities after August 30, 1999. This would reduce the availability of a well-qualified workforce to conduct lead-based-paint activities. EPA believed that it was more appropriate to extend the effective dates to allow an appropriate amount of time for individuals to complete the necessary prerequisites and receive certification. EPA worked to assist the regulated community in coming into compliance by the March 1, 2000, deadline.

10.4.2.3.3   Summary of Management and Disposal Standards

10.4.2.3.3.1   SCOPE OF PROPOSED STANDARDS: This proposal would apply to persons who generate, store, transport, reuse, transfer for reuse, reclaim, and/or dispose of LBP debris from the following structures and activities: (1) abatement, demolition, renovation, and remodeling in target housing and child-occupied facilities and (2) deleading, demolition, renovation, and remodeling in public buildings and commercial buildings. The definition of LBP debris at § 745.303 of the regulatory text does not include concentrated LBP wastes such as LBP chips, dust, blast media, solvents, sludges, and treatment residues. Such wastes would remain subject to RCRA requirements.

This proposal would not apply to LBP debris generated by persons who conduct abatement or renovation and remodeling activities themselves in target housing in which they reside. Such debris may also be exempt from RCRA Subtitle C requirements under the household hazardous waste exclusion. Under this TSCA proposal, if a homeowner hires an individual or firm to perform abatement, demolition, or renovation activities and LBP debris is created, the individual or firm would be considered to be a generator of LBP debris. In such cases, the individual or firm would be responsible for compliance with the generator requirements in this proposal rather than the homeowner.

One important distinction between this proposal and current RCRA Subtitle C requirements is that this proposal would apply to all LBP debris (as defined at § 745.303), whereas RCRA Subtitle C requirements apply only if LBP debris is a waste and is determined to be "hazardous." The comprehensive coverage of this TSCA proposal would resolve the current problems involved in conducting the TCLP test on heterogeneous LBP debris and in leaving largely unregulated large quantities of "nonhazardous" LBP debris. This proposal would have the effect of subjecting all LBP debris to one commonsense regulatory scheme including management controls that take into account the risks that LBP debris poses to humans, particularly children—even if LBP debris has not been found to be "hazardous" under the TCLP test.

10.4.2.3.3.2   DISPOSAL/RECLAMATION OPTIONS: Section 745.309 of this TSCA proposal would allow disposal of LBP debris in a variety of facilities, specifically:

- Construction and demolition landfills
- Nonmunicipal landfills that accept conditionally exempt small quantity, generated waste
- Hazardous waste disposal facilities, including hazardous waste incinerators and landfills
- In the case of incineration, facilities subject to specified Clean Air Act requirements

Under this TSCA proposal, LBP debris would be able to be reclaimed (either for recovery of lead, or for energy combustion value) only in facilities which meet the Clean Air Act requirements specified at § 745.309(b) of this proposal.

10.4.2.3.3.3   CONTROLS ON TRANSPORTATION, STORAGE, AND REUSE: The Agency has included proposed controls on the transportation, storage, reuse, and transfer for reuse of LBP debris in §§ 745.308 and 745.311. If finalized, this proposed rule would stipulate that, when LBP debris is stored for more than 72 hours, there must be access limitations, and that LBP debris must not be stored for more than 180 days (§ 745.311). There are also proposed limitations on when LBP debris may be transferred for reuse (§ 745.311). In addition, the proposal would require that LBP debris be transported in covered vehicles to prevent any inadvertent release of LBP chips or dust (§ 745.308).

10.4.2.3.3.4   NOTIFICATION AND RECORD KEEPING: In order to promote compliance and provide for effective enforcement of the standards contained in this proposal, the Agency has included a proposed requirement that when LBP debris is transferred from one party to another, the recipient should be notified in writing that the material is LBP debris [§ 745.313(a)]. Both parties to any transfer of LBP debris would also be required to keep a copy of the notification on record for 3 years [§ 745.313(b)].

Bibliography

U.S. Department of Health and Human Services, Centers for Disease Control. February 21, 1997. "Update: Blood Lead Levels—United States, 1991–1994." *Morbidity and Mortality Weekly Report.* Vol. 46, No. 7.

HUD. 1994. Department of Housing and Urban Development, National Housing Survey. Washington, D.C.

Lead-Based Paint Hazard Reduction and Financing Task Force. July 1995. Putting the Pieces Together: Controlling Lead Hazards in the Nation's Housing. HUD-1547-LBP.

Task Force on Lead-Based Paint Hazard Reduction and Financing. April 13, 1994. Letter to Honorable Carol Browner, Administrator, USEPA. Washington, D.C.

CHAPTER 11
HAZARDOUS MATERIALS

Richard Booth
Associate
RTP Environmental Associates, Inc.

John A. Green
Director of Safety and Health
Tate & Lyle North American Sugars, Inc.

Paul E. Neil
Principal
RTP Environmental Associates, Inc.

11.1 OVERVIEW

With the Resource Conservation and Recovery Act (RCRA), uniform waste management practices were elevated to a national level. Enacted in 1976, the law was seen as a reaction to the ever-mounting increase in waste quantities, the environmental and health impacts of unsound disposal practices, and the loss of recoverable materials, from both a material and an energy standpoint.

As encompassing as all of these issues are, Subtitle C of the law, the management of hazardous wastes, produced the most pronounced response and change for industry and the regulatory agencies. This aspect is the primary subject of this chapter. U.S. EPA required almost 6 years after the act had become law to finally develop and implement the first complete set of regulations covering this area. The result was a very complex system of regulations covering the initial generation of hazardous waste to its final disposal, a system, in other words, of managing waste from "cradle to grave." The regulations are divided in the Code of Federal Regulations (CFR) Title 40 into the following parts:

1. Part 260   Hazardous Management System: General

2. Part 261   Identification and Listing of Hazardous Waste

3. Part 262   Standards Applicable to Generators of Hazardous Waste

4. Part 263   Standards Applicable to Transporters of Hazardous Waste

5. Part 264   Standards for Owners and Operators of Hazardous Waste Treatment, Storage, and Disposal Facilities

6. Part 265   Interim Status Standards for Owners and Operators of Hazardous Waste Treatment, Storage, and Disposal Facilities

7. Part 266   Standards for Management of Specific Hazardous Waste and Specific Types of Hazardous Waste Management Facilities

8. Part 268   Land Disposal Restrictions

9. Part 270   EPA Administered Permit Programs: The Hazardous Waste Permit Program

10. Part 271   Requirements for State Hazardous Waste Management Programs

11. Part 272   Approved State Hazardous Waste Management Programs

12. Part 273   Standards for Universal Waste Management

13. Part 279   Standards for the Management of Used Oil

For generators which manifest their waste to permitted transporters and permitted disposal facilities, the following parts generally apply: 260, 261, 262, 266, 268, 273, and 279. Transporters of hazardous waste are subject to Part 263. Generators or facilities which treat, store, or dispose of hazardous waste are subject to Parts 264, 265, 266, 268, 273, and 279. Whichever regulations are more stringent, local, state, or federal, will apply.

11.2   GENERATORS OF SOLID WASTE—WASTE DETERMINATION

This section reviews the permitting requirements and general responsibilities of generators of solid waste. The first decision required to be made is whether the material is a RCRA solid waste. While simple in scope, the regulatory answer has caused many people many surprises. U.S. EPA has divided all known materials into three categories: (1) garbage, refuse, or sludge; (2) solid, liquid, semi-liquid, or contained gaseous material (which is discarded, served its intended purpose, or a manufacturing or mining by-product); or (3) something else. Group 1 materials are a RCRA solid waste. Group 3 materials are not a solid waste, and Group 2 materials are a RCRA solid waste unless specifically exempted in 12 categories.

These exempted solid wastes include (i) domestic sewage, (ii) a CWA point source discharge, (iii) irrigation return flow, (iv) NRC source, special nuclear or by-product material, (v) in situ mining waste, (vi) pulping liquors, (vii) spent sulfuric acid, (viii) secondary materials that are reclaimed and returned to the original process, (ix) spent wood preserving solutions, (x) several listed coke by-product wastes, (xi) dross residue from K061 treatment, (xii) recovered oil. The list is general in nature. The specific exemptions in 40 CFR 261.4(a) should be consulted in making any specific determination regarding a discarded waste. At this point of the decision tree, it doesn't matter what your intentions are regarding the future or immediate use of the solid waste; it has become a RCRA solid waste. However, in 40 CFR 261.2(e), EPA has noted that some materials are not solid waste when recycled when they are used or reused as (1) ingredients in an industrial process to make a produce, (2) effective substitutes for commercial products, or (3) returned to the original process from which they are generated. As always, there are qualifications on these specific exemptions. The next question that has to be answered is whether the RCRA solid waste is hazardous. There are 12 categories of solid waste which EPA has defined as not being hazardous wastes. They include (1) household waste, (2) solid wastes returned to the soil as fertilizers, (3) mining overburden returned to the mine site, (4) combustion waste, (5) waste associated with the exploration, development, or production of crude oil, natural gas, or geothermal energy, (6) certain chromium wastes, (7) certain wastes from the extraction, beneficiation, and processing of ores and minerals, (8) cement kiln dust, (9) arsenical-treated wood or wood products, (10) petroleum-contaminated media and debris, (11) injected groundwater from hydrocarbon recovery operations, (12) used chlorofluoro refrigerants, (13) non-terne-plated used oil filters, and (14) used oil distillation bottoms. There are some states that have deleted or modified these exceptions. As always, the most stringent regulation applies. Both the federal rule in 40 CFR 261.4(b) and its corresponding state and local regulation should be reviewed in making any final determination.

If the solid waste is not excluded, then the next step is to determine if it is a listed waste. EPA has prepared four lists of solid waste which have already been determined to be hazardous waste. The first list, called the "F" wastes, are hazardous waste from nonspecific sources. There are over 40 wastes in this group, each corresponding to a source type. The second list, called the "K" waste, is from specific sources. There are over 150 wastes in this group. The third and fourth lists are under the heading of discarded commercial chemical products, off-specification species, container residues, and spill residues. The "P" wastes are acute hazardous wastes. Over 100 waste codes are listed. The "U" wastes are toxic hazardous wastes. Over 200 waste codes are listed in this group. A specific listed waste can be petitioned to be excluded from the list using the procedures of 40 CFR 260.22, and several petitioners have been successful in delisting their wastes. If the solid waste is not listed, the next step is to determine if the waste exhibits any of the four characteristics of a hazardous waste. The four characteristics are ignitability, corrosivity, reactivity, and toxicity. In this case, the waste is normally tested by an outside laboratory using the established procedures listed in 40 CFR 261. However, a generator can utilize their specific knowledge about the waste to determine if it could possibly exceed any of the limits or exhibit any of the characteristics of a hazardous waste. An assessment, along with documentation of all of the solid waste streams present, should be maintained. A typical example is included at the end of this chapter. This form or similar written documentation is recommended for each waste stream present at a facility.

There are additional regulations which can apply to specific wastes. For example, the wastes in Table 11.1 are regulated under different provisions.

Fluorescent lights are under review by EPA and may be included in the future in the Standards for Universal Waste Management (40 CFR 273). Management of solid waste which is determined to be nonhazardous is normally regulated by state and local regulations, following Subtitle D of RCRA.

TABLE 11.1 Regulated Wastes

Waste	Regulation
Used oil	40 CFR 279
Batteries	40 CFR 273
Pesticides	40 CFR 273
Mercury thermostats	40 CFR 273

11.3 GENERATORS OF HAZARDOUS WASTE—QUANTIFICATION

Once a generator of solid waste has determined that the waste is hazardous, an important step is to estimate its, or in the case of several hazardous waste streams, their total monthly generation rate. EPA has devised three levels of control, dependent upon the monthly generation rate. They are given in Table 11.2.

The classification is dependent upon the waste generated in the month, so that a CESQG could become a SQG for a period and then go back to being a CESQG. In addition, the level of regulatory requirements also changes dependent upon the generator's status.

11.4 REQUIREMENTS FOR CONDITIONALLY EXEMPT SMALL QUANTITY GENERATORS

The regulatory requirements for CESQGs are listed in 40 CFR 261.5. Basically, EPA has exempted these generators from most of the regular requirements as long as the CESQG complies with the following general set of special requirements:

1. Conduct a hazardous waste determination
2. Accumulate no more than 1000 kg at any time
3. Ensure delivery to a permitted storage and treatment or disposal facility, including a facility which beneficially uses or reuses its waste

TABLE 11.2 Control Levels of Generators

Classification of generator	Generator level
1. Conditionally exempt small quantity generator (CESQG)	Less than 100 kg per month*
2. Small quantity generator (SQG)	Less than 1000 kg per month
3. Hazardous waste generator (HWG)	Greater than 1000 kg per month

*And less than 1 kg per month of acute hazardous waste; the acute hazardous wastes are the wastes listed "P," along with several listed "F."

The CESQG is exempt from Parts 262 through 266, 268, 270, and the notification requirements of Section 3010 of RCRA. Thus, a CESQG is not required to have an EPA ID number or use a hazardous manifest. However, for record-keeping aspects and to ensure that a generator has complied with all of the CESQG requirements, many CESQGs have instituted use of a manifest for disposal of hazardous waste. For CESQGs, on the manifest form the initials "CESQG" would be substituted for an EPA ID number. Of the remaining parts of the regulations which apply to CESQGs (Parts 273, universal waste and 279, used oil) the universal waste part has a section which allows the CESQG to manage the waste at their option (40 CFR 275.5). Part 279 has no such exemption, and Section 279.10(b)(3) notes that mixtures of used oil and hazardous waste are subject to regulation as used oil. Refer to the section on used oil.

11.5 REQUIREMENTS FOR SMALL QUANTITY AND HAZARDOUS WASTE GENERATORS

EPA has made the regulations less stringent for SQGs in the area of record keeping by excluding the need for a biennial report. Otherwise, both SQG and HWG follow the same management practices of handling hazardous waste. Part 262, EPA Regulation for Hazardous Waste Generators, is divided into seven subparts:

A. General

B. The manifest

C. Pretransport requirements

D. Record keeping and reporting

E. Exports of hazardous waste

F. Import of hazardous waste

G. Farmers

11.5.1 General

In the first subpart, EPA lays out two special requirements: (1) a generator must determine if a waste is hazardous, as described in 40 CFR 262, and (2) obtain an EPA ID number. The methodology for determining if a solid waste is hazardous is outlined in Chapter 11, Part 11.2 of this book. Also, EPA requires generators to review Parts 264 (Standards for Owners and Operators of Hazardous Waste Treatment Facilities), 265 (Interim Status Standards for Owners and Operators of Hazardous Waste Treatment Facilities), and 268 (Land Disposal Restrictions) for applicability.

EPA Form 8700-12 is used to request an EPA ID number. The forms can be obtained by calling the EPA regional offices or state hazardous waste regulatory offices. The form requests information on the name, location, and contact at the facility, the type of waste activity, description of the waste, and a certification. The form package also lists where the completed form should be sent, dependent upon facility location. Note that a generator must not offer hazardous waste to a transporter or to a treatment, storage, or disposal facility which has not received an EPA ID number.

11.5.2 The Manifest

A hazardous waste generator who is sending hazardous waste off-site is required to use either the EPA Form 8700-22 or a state equivalent manifest form. The manifest lists information regarding the generator and identifies the waste from both an EPA and a Department of Transportation perspective. Also listed on the form is the transporter and designated facility information. The manifest is typically designed as a multipage form, of which individual pages are retained and signed by the generator, transporter, and waste receiving facility. The waste receiving facility receives two copies of the final manifest copies, one which it keeps; the other is sent back to the generator when the waste is accepted by the designated facility. SQGs do not need to use manifests where the waste is retained under a contractual agreement pursuant to several requirements listed in 40 CFR 262.20(e).

11.5.3 Pretransport Requirements

These requirements are divided into two general areas: (1) packaging procedures and (2) accumulation time. The packaging procedures are consistent with Department of Transportation regulations under 49 CFR Parts 172, 173, 178, and 179. In general, the procedures refer to the labeling and marking of waste containers and placarding requirements. Specifically, the containers must be labeled with the words "Hazardous Waste," and the start date of hazardous waste accumulation must be clearly marked and visible on each container. The generator also has to comply with Subparts C and D of 40 CFR 265, with 265.16, and with 40 CFR 268.7(a)(7) (Table 11.3).

The accumulation time requirements are divided into several general aspects: (1) hazardous waste management procedures, (2) time requirements, (3) point of generation procedures, and (4) small quantity generator requirements. The intent of these rules is to require a facility to move their hazardous waste off-site as expeditiously as possible. Generators have found this area of hazardous waste management to present the greatest risk to notice of violations or noncompliance. The general requirements of the hazardous waste management procedures are shown in Table 11.4.

TABLE 11.3   Rule Categories

Rule	General Description
Subpart C	Preparedness and prevention
Subpart D	Contingency plan and emergency procedures
265.16	Personnel training
268.7(a)(7)	Waste analysis and record keeping as part of the land disposal restrictions

TABLE 11.4   Hazardous Waste Management Procedures

Category	Regulation
Containers	Subpart I/40 CFR 265
Tanks	Subpart J/40 CFR 265
Drip pads	Subpart W/40 CFR 265
Containment building	Subpart DD/40 CFR 265

Thus, for a generator that collects their hazardous waste in a 55-gallon drum, which is a normal practice, a generator has to comply with the following general requirements:

1. The container must be in good condition.
2. The container or liner must be compatible with the waste to be stored.
3. The container must always be closed when not in use.
4. The container must not be handled in a manner which causes rupture or leaks.
5. The operator must inspect the container area on a weekly basis (a weekly written inspection form is recommended for compliance with this condition).
6. There are special requirements for ignitable, reactive, or incompatible wastes.

The other areas, tanks, drip pads, and containment buildings have similar requirements. With regard to the time requirements, when accumulation of hazardous waste on-site occurs, EPA requires proper disposal of the waste within 90 days from the start of collection of the hazardous waste unless a generator falls into a special category. Table 11.5 lists the classifications. A generator that exceeds the time limits becomes an operator of a storage facility and is subject to the requirements of 40 CFR Parts 264 and 265. Note that each category has specific conditions which need to be complied with to avoid being classified as a storage facility.

11.5.4   Record Keeping and Reporting

EPA requires that signed hazardous waste manifests (those received from the designated TDS facility), biennial reports, and waste analyses and studies be kept for

TABLE 11.5 Classifications of Waste

Type of generator	Time limitation
Generator	90 days or less
Generator with EPA exemption	Up to 30-day extension
Point of generation	90 days after filling container
Small quantity generator	180 days or less
Small quantity generator with transport distance of 200 miles or more	270 days or less
Small quantity generator with EPA exemption	Up to 30-day extension

3 years. The biennial report is submitted by a generator on Form 8700-13A by Mar. 1 of each even year. This report covers the generator's activities during the previous year. SQGs and CESQGs do not need to submit a biennial report.

However, if a SQG has not received a signed copy of the manifest from the designated facility within 60 days, the state or EPA is required to be notified. A HWG has a shorter time period; at 35 days, the generator must contact the transporter and/or waste receiving facility to ascertain the status of the manifest form. At more than 45 days, a formal exception report must be submitted to EPA or to the state.

In addition, all generators need to comply with the record-keeping requirements of 40 CFR 268.7(a)(7), which basically requires retention of land ban disposal notices for 5 years. In general, generators have stapled copies of the land ban notices to their original manifests to comply with this requirement.

11.5.5 Exports and Imports of Hazardous Waste, and Farmers

These activities and generators have specific requirements in order to comply with their activities. Refer to 262.50 through 262.70 for the specific regulations.

11.6 TREATMENT, STORAGE, AND DISPOSAL FACILITIES (TSDF) PERMITS, STATUTORY OVERVIEW

The management of hazardous wastes is covered by the complex regulations issued under the Resource Conservation and Recovery Act of 1976 (RCRA), as significantly amended by the Hazardous and Solid Waste Amendments of 1984.

Under RCRA a system of "cradle-to-grave" management regulatory requirements has been established to cover the following hazardous waste management activities:

- Identification notification
- Generation

- Transport
- Storage
- Disposal

It is this "cradle-to-grave" concept that mandated EPA to develop and implement a system of permits that establishes standards and criteria for the above activities. Any "facility" that treats, stores, or disposes of hazardous waste is regulated under RCRA Section 3004 and is required to obtain a permit under RCRA Section 3005. A "facility" is defined by EPA regulations as all contiguous land and structures, other appurtenances, and improvements on the land. (*Note:* All regulatory references made in this chapter, such as "part 261" or 264, refer to sections found in Title 40 of the Code of Federal Regulations.)

Section 3004 of RCRA requires the administrator of EPA to develop regulations applicable to owners and operators of hazardous waste treatment, storage, or disposal facilities, as necessary to protect human health and the environment. RCRA requires a permit for the "treatment," "storage," and "disposal" of any "hazardous waste" as identified or listed in 40 CFR Part 261. Section 3005 of RCRA requires EPA to:

- Promulgate regulations requiring each person owning or operating an existing facility or planning to construct a new facility for the treatment, storage, or disposal of hazardous waste identified or listed under this subtitle to have a permit issued pursuant to this section
- To establish requirements for permit applications

In addition to hazardous wastes treatment, storage, and disposal facilities (TSDFs) covered under the regulations of 40 CFR 264 and 265, owners and operators of certain other facilities are required to obtain RCRA permits as well as permits under other programs for certain aspects of the facility operation. RCRA permits are required for:

- Injection wells that dispose of hazardous waste, and associated surface facilities that treat, store, or dispose of hazardous waste (see 270.64). However, the owner and operator with a UIC permit in a state with an approved or promulgated UIC program will be deemed to have a RCRA permit for the injection well itself if they comply with the requirements of 270.60(b) (permit-by-rule for injection wells).
- Treatment, storage, or disposal of hazardous waste at facilities requiring an NPDES permit. However, the owner and operator of a publicly owned treatment works receiving hazardous waste will be deemed to have a RCRA permit for that waste if they comply with the requirements of 270.60(c) (permit-by-rule for POTWs).
- Barges or vessels that dispose of hazardous waste by ocean disposal and onshore hazardous waste treatment or storage facilities associated with an ocean disposal operation. However, the owner and operator will be deemed to have a RCRA permit for ocean disposal from the barge or vessel itself if they comply with the requirements of 270.60(a) (permit-by-rule for ocean disposal barges and vessels).

Generators of hazardous wastes are categorized by either those that treat, store, or dispose of hazardous waste on site or those that have these functions done off-site. Those generators that treat, store, or dispose of hazardous waste on-site must comply with the same standards and permit requirements as treatment, storage, or disposal facilities. However, as explained in Section 11.6.4 generators of hazardous wastes are permitted to store hazardous waste on-site for a period of up to 90 days without having to obtain a permit as a storage facility. The generator must follow EPA-imposed requirements for temporary on-site storage.

Obtaining a permit to construct a facility is not required under RCRA if such a facility is constructed pursuant to an approval issued by the EPA under section 6(e) of the Toxic Substances Control Act for the incineration of polychlorinated biphenyls (PCBs). Any person owning or operating such a facility may, at any time after operation or construction of such facility has begun, file an application for a permit pursuant to RCRA authorizing such facility to incinerate hazardous waste identified or listed.

All submittals for a TSDF must be completed on EPA's hazardous waste permit application, Form 8700-23, consisting of Part A and Part B. While EPA supplies a form for Part A, the detailed information required in Part B of the permit application must be submitted in a narrative form. All Part B applications must contain certain general information that demonstrates compliance with the standards set out in the regulations for interim and permitted facilities. In addition, Part B applications must contain certain specific information depending on the type of waste management units at the facility. If an applicant can demonstrate that providing some of the required information is not feasible, EPA has the authority to make allowances for such less than complete submissions on a case-by-case basis. A professional engineer must certify drawings, specifications, engineering studies, and certain other required technical information. Detailed discussion of the information required in both Parts A and B of a TSDF permit application can be found in Sections 11.6.9 and 11.6.10, respectively.

Submittal of a facility owner or operator's permit application initiates the process for obtaining a RCRA permit. Permit applications are sent to either EPA or state authorities, depending on whether the state has obtained authorization to administer its own hazardous waste permit program. EPA's general permit procedures or comparable procedures also apply to state-administered RCRA programs. Once EPA receives your permit application, the agency:

- Determines if the application is complete
- Prepares a draft permit or issues a notice of intent to deny
- Gives public notice of a draft permit
- Allows for a 45-day public comment period
- Hears appeals

[*Note:* In states where EPA is the permitting authority (i.e., the state has not been authorized to issue permits under 40 CFR Part 271), facility owner-operators must notify and give the public a chance to comment prior to the submission of a Part B permit application.]

Although the discussions in this chapter will concentrate on the permits associated with the TSDFs covered by the regulations found in 40 CFR 264, 265, and 270, the reader should be cognizant of other "special" forms of TSDF permits that are issued by the EPA. These include:

• Permits by rule

• Emergency permits

• Hazardous waste incinerator permits

• Permits for land treatment demonstration using field tests or laboratory analyses

• Interim permits for UIC wells research

• Development and demonstration permits

• Permits for boilers and industrial furnaces burning hazardous waste

Discussion on these "special" permits can be found in Section 11.6.8.

Owners and operators of hazardous waste management units must have permits during the active life (including the closure period) of the unit. Owners or operators of surface impoundments, landfills, land treatment units, and waste pile units that received wastes after July 26, 1982, or that certified closure (according to 265.115) after Jan. 26, 1983, must have postclosure permits unless they demonstrate closure by removal as provided under 270.1(c)(5) and (6). If a postclosure permit is required, the permit must address applicable part 264 Groundwater Monitoring, Unsaturated Zone Monitoring, Corrective Action, and Postclosure Care Requirements. The denial of a permit for the active life of a hazardous waste management facility or unit does not affect the requirement to obtain a postclosure permit.

Each application for a permit required under section 3005 of RCRA must contain any information required under regulations promulgated by the EPA, including information respecting:

• Estimates with respect to the composition, quantities, and concentrations of any hazardous waste identified or listed under the regulations, or combinations of any such hazardous waste and any other solid waste, proposed to be disposed of, treated, transported, or stored, and the time, frequency, or rate of which such waste is proposed to be disposed of, treated, transported, or stored

• The site at which such hazardous waste or the products of treatment of such hazardous waste will be disposed of, treated, transported, or stored.

Permits for such facilities are issued by the EPA (or a state, if applicable), upon a determination of compliance by a facility for which a permit is applied for under this section with the requirements of this section and section 3004. If permit applicants propose modification of their facilities, or if the EPA (or the state) determines that modifications are necessary to conform to the requirements under this section and section 3004, the permit must specify the time allowed to complete the modifications.

Any permit under these regulations is issued for a fixed term, not to exceed 10 years in the case of any land disposal facility, storage facility, or incinerator or other treatment facility. Each permit for a land disposal facility must be reviewed 5 years after date of issuance or reissuance and must be modified as necessary to assure that the facility continues to comply with the currently applicable requirements of section 3004. Nothing precludes the EPA from reviewing and modifying a permit at any time during its term. Reviews of any application for a permit renewal must consider improvements in the state of control and measurement technology as well as changes in applicable regulations. Each permit issued under this section must contain such terms and conditions as the EPA (or the state) determines necessary to protect human health and the environment.

Two categories of TSDFs have been established under RCRA for permit purposes: permitted and interim. Permitted facilities are those that have been issued a permit to treat, store, or dispose of hazardous waste. Interim facilities are those that were in existence on Nov. 19, 1980, complied with notification requirements, and filed a permit application form.

Existing facilities also can become eligible for interim status because of statutory changes, regulatory changes, or actions by the owner-operators that render the facility subject to RCRA permit requirements. Once eligible, such facilities must comply with notification requirements and file a Part A permit application to establish interim status.

Facilities under interim status must comply with a separate set of regulations established by EPA (40 CFR Part 265). Once a facility receives a permit, interim status is terminated and the facility becomes subject to regulation as a permitted facility (40 CFR Part 264).

Facilities without interim status must cease waste management operations until a permit is issued. Facilities ineligible for interim status include:

• New facilities

• Existing facilities that did not qualify for interim status

• Facilities that have been denied permits previously

Recognizing that EPA would require a period of time to issue permits to all facilities, Congress provided, under section 3005(e) of RCRA, that qualifying owners and operators could obtain "interim status" and be treated as having been issued permits until EPA takes final administrative action on their permit applications. The privilege of continuing hazardous waste management operations during interim status carries with it the responsibility of complying with appropriate portions of the section 3004 standards. A more detailed discussion on "interim status" can be found in Section 11.6.12.

EPA has issued numerous regulations to implement RCRA requirements for hazardous waste management facilities. These include the standards of 40 CFR Part 264 (which apply to hazardous waste management units at facilities that have been issued RCRA permits), Part 265 (which apply to hazardous waste management units at interim status facilities), and Part 270 (which provide standards for permit issuance).

The regulations in part 270 cover basic EPA permitting requirements, such as application requirements, standard permit conditions, and monitoring and reporting requirements. These regulations are part of a regulatory scheme implementing RCRA set forth in different parts of the Code of Federal Regulations. Table 11.6 indicates where the regulations implementing RCRA appear in the Code of Federal Regulations.

In the 1984 Hazardous and Solid Waste Amendments (HSWA) to RCRA, Congress expanded EPA's authority to address releases from all solid waste management units (SWMUs) at hazardous waste management facilities. Section 3004(u) of HSWA required that any permit issued under section 3005(c) of RCRA to a treatment, storage, or disposal facility after Nov. 8, 1984, address corrective action for releases of hazardous wastes or hazardous constituents from any SWMU at the facility. Section 3004(v) authorized EPA to require corrective action beyond the facility boundary where appropriate. Section 3008(h) provided EPA with authority to issue administrative orders or bring court action to require corrective action or other measures, as appropriate, when there is or has been a release of hazardous waste or (under EPA's interpretation) of hazardous constituents from a facility authorized to operate under section 3005(e).

Later, in an Aug. 10, 1989, memorandum entitled Coordination of Corrective Action through Permits and Orders [OSWER Directive 9502.1989(04)], EPA

TABLE 11.6   RCRA Regulations

Section of RCRA	Coverage	Final regulation
Subtitle C	Overview and definitions	40 CFR Part 260
3001	Identification and listing of hazardous waste	40 CFR Part 261
3002	Generators of hazardous waste	40 CFR Part 262
3003	Transporters of hazardous waste	40 CFR Part 263
3004	Standards for HWM facilities	40 CFR Parts 264, 265, 266, 267
3005	Permit requirements for HWM facilities	40 CFR Parts 270, 124
3006	Guidelines for state programs	40 CFR Part 271
3010	Preliminary notification of HWM activity	(Public notice) 45, 12746 Feb. 26, 1980

clarified that interpretation by stating that a section 3008(h) order cannot be issued to a facility after final disposition of the permit application.

In practice, the corrective action process is highly site-specific and involves direct oversight by the reviewing agency. Unlike the closure process, which provides two options (closure with waste in place and closure by complete removal and decontamination), the corrective action process provides considerable flexibility to EPA to decide on remedies that reflect the conditions and the complexities of each facility. For example, depending on the site-specific circumstances, remedies may attain media cleanup standards through various combinations of removal, treatment, engineering, and institutional controls.

EPA has codified corrective action requirements at sections 264.101, 264.552, and 264.553, and currently implements these requirements through the permitting process. EPA also implements corrective action by issuing corrective action orders under section 3008(h) of RCRA. In addition, to facilitate the corrective action process, EPA proposed more extensive corrective action regulations on July 27, 1990, under a new Part 264 Subpart S (see 55 FR 30798). The July 27, 1990, Subpart S proposal set forth EPA's interpretation of the statutory requirements at that time. Later, EPA promulgated several sections of that proposal related to temporary units, corrective action management units, and the definition of "facility" (see 58 FR 8658, Feb. 16, 1993).

(*Note:* Since the discussions in this chapter reference numerous sections of the regulations found in Title 40 of the CFR, it is strongly suggested that the reader have available a current copy of these regulations, especially parts 260 through 270.)

11.6.1 Definitions

The following definitions apply to TSDF permits.

Administrator means the administrator of the U.S. Environmental Protection Agency or an authorized representative.

Application means the EPA standard national forms for applying for a permit, including any additions, revisions, or modifications to the forms; or forms approved by EPA for use in approved states, including any approved modifications or revisions. Application also includes the information required by the director under 270.14 through 270.29 (contents of Part B of the RCRA application).

Approved program or approved state means a state which has been approved or authorized by EPA under Part 271.

Aquifer means a geological formation, group of formations, or part of a formation that is capable of yielding a significant amount of water to a well or spring.

Closure means the act of securing a hazardous waste management facility pursuant to the requirements of 40 CFR Part 264.

Component means any constituent part of a unit or any group of constituent parts of a unit which are assembled to perform a specific function (e.g., a pump seal, pump, kiln liner, kiln thermocouple).

Corrective action management unit, or CAMU, means an area within a facility that is designated by the regional administrator under part 264 Subpart S, for the purpose of implementing corrective action requirements under 264.101 and RCRA section 3008(h). A CAMU can only be used for the management of remediation wastes pursuant to implementing such corrective action requirements at the facility.

CWA means the Clean Water Act (formerly referred to as the Federal Water Pollution Control Act or Federal Water Pollution Control Act amendments of 1972) PL 92-500, as amended by PL 92-217 and PL 95-576; 33 U.S.C. 1251 et seq.

Director means the regional administrator or the state director, as the context requires, or an authorized representative. When there is no approved state program, and there is an EPA administered program, director means the regional administrator. When there is an approved state program, director normally means the state director. In some circumstances, however, EPA retains the authority to take certain actions even when there is an approved state program. In such cases, the term director means the regional administrator and not the state director.

Disposal means the discharge, deposit, injection, dumping, spilling, leaking, or placing of any hazardous waste into or on any land or water so that such hazardous waste or any constituent thereof may enter the environment or be emitted into the air or discharged into any waters, including groundwater.

Disposal facility means a facility or part of a facility at which hazardous waste is intentionally placed into or on the land or water, and at which hazardous waste will remain after closure. The term disposal facility does not include a corrective action management unit into which remediation wastes are placed.

Draft permit means a document prepared under 124.6 indicating the director's tentative decision to issue or deny, modify, revoke and reissue, terminate, or reissue a permit. A notice of intent to terminate a permit, and a notice of intent to deny a permit, as discussed in 124.5, are types of draft permits. A denial of a request for modification, revocation and reissuance, or termination, as discussed in 124.5, is not a draft permit. A proposed permit is not a draft permit.

Elementary neutralization unit means a device which:

- Is used for neutralizing wastes only because they exhibit the corrosivity characteristic defined in 261.22, or are listed in subpart D of Part 261 only for this reason

- Meets the definition of tank, tank system, container, transport vehicle, or vessel in 260.10

Emergency permit means a RCRA permit issued in accordance with 270.61.

Environmental Protection Agency (EPA) means the U.S. Environmental Protection Agency.

EPA means the U.S. Environmental Protection Agency.

Existing hazardous waste management (HWM) *facility or existing facility* means a facility which was in operation or for which construction commenced on or before Nov. 19, 1980. A facility has commenced construction if:

- The owner or operator has obtained the federal, state, and local approvals or permits necessary to begin physical construction

- A continuous on-site, physical construction program has begun

- The owner or operator has entered into contractual obligations which cannot be canceled or modified without substantial loss—for physical construction of the facility to be completed within a reasonable time

Facility mailing list means the mailing list for a facility maintained by EPA in accordance with 40 CFR 124.10(c)(1)(ix).

Facility or activity means any HWM facility or any other facility or activity (including land or appurtenances thereto) that is subject to regulation under the RCRA program. Federal, state, and local approvals or permits necessary to begin physical construction means permits and approvals required under federal, state, or local hazardous waste control statutes, regulations, or ordinances.

Final authorization means approval by EPA of a state program which has met the requirements of section 3006(b) of RCRA and the applicable requirements of Part 271, subpart A.

Functionally equivalent component means a component which performs the same function or measurement and which meets or exceeds the performance specifications of another component.

Generator means any person, by site location, whose act or process produces "hazardous waste" identified or listed in 40 CFR Part 261.

Groundwater means water below the land surface in a zone of saturation.

Hazardous waste means a hazardous waste as defined in 40 CFR 261.3.

Hazardous waste management facility (*HWM facility*) means all contiguous land and structures, other appurtenances, and improvements on the land, used for treating, storing, or disposing of hazardous waste. A facility may consist of several treatment, storage, or disposal operational units (for example, one or more landfills, surface impoundments, or combinations of them).

HWM facility means hazardous waste management facility.

Injection well means a well into which fluids are being injected.

In operation means a facility which is treating, storing, or disposing of hazardous waste. Interim authorization means approval by EPA of a state hazardous waste program which has met the requirements of section 3006(g)(2) of RCRA and applicable requirements of Part 271, subpart B.

Major facility means any facility or activity classified as such by the regional administrator or, in the case of approved state programs, the regional administrator in conjunction with the state director.

Manifest means the shipping document originated and signed by the generator which contains the information required by subpart B of 40 CFR Part 262.

National Pollutant Discharge Elimination System means the national program for issuing, modifying, revoking and reissuing, terminating, monitoring and enforcing permits, and imposing and enforcing pretreatment requirements, under sections 307, 402, 318, and 405 of the CWA. The term includes an approved program.

NPDES means national pollutant discharge elimination system.

New HWM facility means a hazardous waste management facility which began operation or for which construction commenced after Nov. 19, 1980.

Off-site means any site which is not on-site.

On-site means on the same or geographically contiguous property which may be divided by public or private right(s)-of-way, provided the entrance and exit between the properties is at a crossroads intersection and access is by crossing as opposed to going along the right(s)-of-way. Noncontiguous properties owned by the same person but connected by a right-of-way which the person controls and to which the public does not have access is also considered on-site property.

Owner or operator means the owner or operator of any facility or activity subject to regulation under RCRA.

Permit means an authorization, license, or equivalent control document issued by EPA or an approved state to implement the requirements of 40 CFR, Part 270 and Parts 271 and 124. Permit includes permit by rule (270.60), and emergency permit (270.61). Permit does not include RCRA interim status (subpart G of 40 CFR Part 270) or any permit which has not yet been the subject of final agency action, such as a draft permit or a proposed permit.

Permit-by-rule means a provision of these regulations stating that a facility or activity is deemed to have a RCRA permit if it meets the requirements of the provision.

Person means an individual, association, partnership, corporation, municipality, state or federal agency, or an agent or employee thereof.

Physical construction means excavation, movement of earth, erection of forms or structures, or similar activity to prepare an HWM facility to accept hazardous waste.

POTW means publicly owned treatment works.

Publicly owned treatment works (POTW) means any device or system used in the treatment (including recycling and reclamation) of municipal sewage or industrial wastes of a liquid nature which is owned by a state or municipality. This definition includes sewers, pipes, or other conveyances only if they convey wastewater to a POTW providing treatment.

RCRA means the Solid Waste Disposal Act as amended by the Resource Conservation and Recovery Act of 1976 (PL 94-580, as amended by PL 95-609 and PL 96-482, 42 U.S.C. 6901 et seq.).

Regional administrator means the regional administrator of the appropriate regional office of the Environmental Protection Agency or the authorized representative of the regional administrator.

Remedial action plan (RAP) means a special form of RCRA permit that a facility owner or operator may obtain instead of a permit issued under 270.3 through 270.66, to authorize the treatment, storage, or disposal of hazardous remediation waste (as defined in 260.10) at a remediation waste management site.

Schedule of compliance means a schedule of remedial measures included in a permit, including an enforceable sequence of interim requirements (for example, actions, operations, or milestone events) leading to compliance with the act and regulations.

SDWA means the Safe Drinking Water Act (PL 95-523, as amended by PL 95-1900; 42 U.S.C. 3001 et seq.).

Site means the land or water area where any facility or activity is physically located or conducted, including adjacent land used in connection with the facility or activity.

State means any of the 50 states, the District of Columbia, Guam, the Commonwealth of Puerto Rico, the Virgin Islands, American Samoa, and the Commonwealth of the Northern Mariana Islands.

State director means the chief administrative officer of any state agency operating an approved program, or the delegated representative of the state director. If responsibility is divided among two or more state agencies, state director means the chief administrative officer of the state agency authorized to perform the particular procedure or function to which reference is made.

State-EPA agreement means an agreement between the regional administrator and the state which coordinates EPA and state activities, responsibilities, and programs.

Storage means the holding of hazardous waste for a temporary period, at the end of which the hazardous waste is treated, disposed, or stored elsewhere.

Transfer facility means any transportation-related facility including loading docks, parking areas, storage areas, and other similar areas where shipments of hazardous waste are held during the normal course of transportation.

Transporter means a person engaged in the off-site transportation of hazardous waste by air, rail, highway, or water.

Treatment means any method, technique, or process, including neutralization, designed to change the physical, chemical, or biological character or composition of any hazardous waste so as to neutralize such wastes or so as to recover energy or material resources from the waste or so as to render such waste nonhazardous or less hazardous; safer to transport, store, or dispose of; or amenable for recovery, amenable for storage, or reduced in volume.

UIC means the underground injection control program under Part C of the Safe Drinking Water Act, including an approved program.

Underground injection means a well injection.

Underground source of drinking water (*USDW*) means an aquifer or portion of an aquifer:

- Which supplies any public water system
- Which contains a sufficient quantity of groundwater to supply a public water system
- Which currently supplies drinking water for human consumption
- Which contains fewer than 10,000 mg per liter total dissolved solids
- Which is not an exempted aquifer

USDW means underground source of drinking water.
Wastewater treatment unit means a device which:

- Is part of a wastewater treatment facility which is subject to regulation under either section 402 or 307(b) of the Clean Water Act
- Receives and treats or stores an influent wastewater which is a hazardous waste as defined in 261.3, or generates and accumulates a wastewater treatment sludge which is a hazardous waste as defined in 261.3, or treats or stores a wastewater treatment sludge which is a hazardous waste as defined in 261.3
- Meets the definition of tank or tank system in 260.10

11.6.2 Scope of the RCRA TSDF Permit Program

RCRA requires a permit for the "treatment," "storage," and "disposal" of any "hazardous waste" as identified or listed in 40 CFR Part 261. The terms "treatment," "storage," "disposal," and "hazardous waste" are defined in Section 11.6.1. Owners and operators of hazardous waste management units must have permits during the active life (including the closure period) of the unit. Owners and operators of surface impoundments, landfills, land treatment units, and waste pile units that received waste after July 26, 1982, or that certified closure (according to 265.115) after Jan. 26, 1983, must have postclosure permits unless they demonstrate closure by removal or decontamination as provided under 270.1(c)(5) and (6), or obtain an enforceable document in lieu of a postclosure permit, as provided under paragraph (c)(7) of Part 270.1. If a postclosure permit is required, the permit must address applicable 40 CFR Part 264 groundwater monitoring, unsaturated zone monitoring, corrective action, and postclosure care requirements. The denial of a permit for the active life of a hazardous waste management facility or unit does not affect the requirement to obtain a postclosure permit under Part 270.

11.6.3 TSDF Permit Application Submittal Time Frames

Owner-operators of TSDFs are required to obtain permits within 90 days after the wastes they handle have been identified and listed by EPA. In addition, TSDFs must have contingency plans for emergencies that might arise from accidental spills or releases.

An owner or operator of a new TSDF must submit both parts of the permit application at least 180 days before construction or operation of the facility is scheduled to begin. Construction or operation of a new TSDF is prohibited until the owner or operator obtains a permit unless the facility is an incinerator constructed to burn polychlorinated biphenyls and approved under TSCA.

Facilities in existence on Nov. 19, 1980, and subject to RCRA regulations at that time, should have already filed Part A of their permit application form in order to obtain interim status.

For facilities in existence on Nov. 19, 1980, but not subject to RCRA regulations until after Nov. 19, 1980, the Part A permit application deadlines are different.

Existing facilities that become subject to RCRA regulations because of regulatory revisions must file Part A within 6 months of the publication date of the revised RCRA regulations.

Existing facilities that qualify for interim status because of statutory changes or actions by the owner-operators must file Part A of the permit application within 30 days of the facility's becoming subject to new RCRA regulations.

In general, Part B of the permit application is due when requested by EPA or an authorized state. The owner or operator must be given at least 6 months to submit Part B once the request has been made. However, facilities that qualified for interim status prior to Nov. 8, 1984, already were required to submit Part B applications. Interim status has been terminated for facilities that failed to meet the applicable deadlines.

Further, existing land disposal facilities that become subject to permit requirements due to statutory or regulatory changes after Nov. 8, 1984, must submit Part B of the permit application 12 months after the date the facility becomes subject to the permit requirements.

11.6.4 Specific TSDF Permit Exclusions

The following are not required to obtain a RCRA TSDF permit:

- Generators who accumulate hazardous waste on-site for less than the time periods provided in 40 CFR 262.34

- Farms who dispose of hazardous waste pesticides from their own use as provided in 40 CFR 262.70

- Facilities operated solely for the treatment, storage, or disposal of hazardous waste excluded from regulations under Part 270 by 40 CFR 261.4 or 261.5 (small generator exemption)

- Owners or operators of totally enclosed treatment facilities as defined in 40 CFR 260.10

- Owners and operators of elementary neutralization units or wastewater treatment units as defined in 40 CFR 260.10

- Transporters storing manifested shipments of hazardous waste in containers meeting the requirements of 40 CFR 262.30 at a transfer facility for a period of 10 days or less

- Persons adding absorbent material to waste in a container (as defined in 260.10) and persons adding waste to absorbent material in a container, provided that these actions occur at the time waste is first placed in the container; and 264.17(b), 264.171, and 264.172 are complied with
- Universal waste handlers and universal waste transporters (as defined in 40 CFR 260.10) managing the wastes listed below. These handlers are subject to regulation under 40 CFR 273: (A) batteries as described in 40 CFR 273.2, (B) pesticides as described in 40 CFR 273.3, and (C) thermostats as described in 40 CFR 273.4.

[*Note:* Generators can accumulate hazardous waste on-site for up to 90 days without a permit or without having interim status provided they place the waste:

- In a container that complies with TSDF interim status standards
- In a tank that complies with TSDF interim status standards, except for the interim standards that apply to testing and analysis of wastes in tanks or to certain closure requirements
- On a drip pad that complies with TSDF interim status standards, provided the generator maintains certain records at the facility
- In a containment building that meets the TSDF interim status requirements for such facilities, provided the generator maintains certain records at the facility

In addition to storing the waste in one of these four types of containers, the generator must

- Mark the date the accumulation began on each tank or container and label or mark it with the words "Hazardous Waste"
- Comply with EPA's rules for emergency preparedness and contingency planning, personnel training, and waste analysis—generators that accumulate hazardous waste for up to 90 days in tanks, containers, or containment buildings must develop and follow a written waste analysis plan that describes the procedures the generator will follow in treating the waste. The waste analysis plan must:

  - Be kept on-site in the generator's records
  - Be based on a detailed chemical and physical analysis of a representative sample of the prohibited waste
  - Be filed with the EPA regional administrator at least 30 days prior to treatment
  - Comply with applicable notification requirements

A generator can also accumulate as much as 55 gallons of hazardous waste or 1 quart of acutely hazardous waste in containers stored near the point of generation. Such an accumulation does not require a TSDF permit or interim status or compliance with the 90-day time limit requirements described above, provided that the containers are:

- In good condition, made of or lined with materials compatible with the wastes, and kept closed
- Marked with the words "Hazardous Waste" or with other words that identify the contents of the containers.]

In addition, a person is not required to obtain an RCRA permit for treatment or containment activities taken during immediate response to any of the following situations:

- A discharge of a hazardous waste
- An imminent and substantial threat of a discharge of hazardous waste
- A discharge of a material which, when discharged, becomes a hazardous waste
- An immediate threat to human health, public safety, property, or the environment from the known or suspected presence of military munitions, other explosive material, or an explosive device, as determined by an explosive or munitions emergency response specialist as defined in 40 CFR 260.10

11.6.5 TSDF Permit Application Requirements

As stated earlier, a facility owner or operator must obtain a TSDF permit under RCRA in order to treat, store, or dispose of hazardous wastes on-site. The facility must operate under the conditions of the permit during its term as well as under any new requirements that are issued or become effective by statute.

EPA's Hazardous Waste Permit Application, Form 8700-23, consists of Part A and Part B. Submittal of a facility owner or operator's permit application initiates the process for obtaining a RCRA permit. Permit applications are sent either to EPA or state authorities, depending on whether the state has obtained authorization to administer its own hazardous waste permit program. EPA's general permit procedures or comparable procedures also apply to state-administered RCRA programs.

Owner-operators of treatment, storage, and disposal facilities that are not ultimate users are subject to all applicable regulations when they store recyclable materials used in a manner constituting disposal. TSDFs that dispose of recyclable materials on land—ultimate users—must comply with all applicable TSD requirements under 40 CFR Parts 264 and 265, permit requirements under 40 CFR Parts 124 and 270, and RCRA notification requirements.

11.6.5.1 TSDF Permit Application Process. Any person who is required to have a TSDF permit (including new applicants and permittees with expiring permits) must complete, sign, and submit an application to the director as described in this section and 270.70 through 270.73. Persons currently authorized with interim status must apply for permits when required by the director. Persons covered by RCRA TSDF permits by rule (270.60) need not apply. Procedures for applications, issuance, and administration of emergency permits are found exclusively in 270.61. Procedures for application, issuance, and administration of research, development, and demonstration permits are found exclusively in 270.65.

11.6.5.2   Who Applies?   When a TSDF or activity is owned by one person but is operated by another person, it is the operator's duty to obtain a permit, except that the owner must also sign the permit application.

11.6.5.3   TSDF Permit Application Completeness.   The director cannot issue a TSDF permit before receiving a complete application for a permit except for permits by rule or emergency permits. An application for a permit is complete when the director receives an application form and any supplemental information which is completed to his or her satisfaction. An application for a TSDF permit is complete notwithstanding the failure of the owner or operator to submit the exposure information described in paragraph (j) of this 40 CFR Part 270.10. The director may deny a TSDF permit for the active life of a hazardous waste management facility or unit before receiving a complete application for a permit.

11.6.5.4   Information Requirements.   All applicants for RCRA TSDF permits must provide information set forth in 270.13 and applicable sections in 270.14 through 270.29 to the director, using the application form provided by the director.

11.6.5.5   Existing   Hazardous   Waste   Facilities   and   Interim   Status Qualifications.   Owners and operators of existing hazardous waste management facilities or of hazardous waste management facilities in existence on the effective date of statutory or regulatory amendments under the act that render the facility subject to the requirement to have a RCRA permit must submit Part A of their permit application no later than:

- 6 months after the date of publication of regulations which first require them to comply with the standards set forth in 40 CFR Part 265 or 266

- 30 days after the date they first become subject to the standards set forth in 40 CFR Part 265 or 266, whichever first occurs

- For generators generating greater than 100 kilograms but less than 1000 kilograms of hazardous waste in a calendar month and treats, stores, or disposes of these wastes on-site, by Mar. 24, 1987

The EPA may, by publication in the *Federal Register,* extend the date by which owners and operators of specified classes of existing hazardous waste management facilities must submit Part A of their permit application if they find that (i) there has been substantial confusion as to whether the owners and operators of such facilities were required to file a permit application and (ii) such confusion is attributed to ambiguities in EPA's Parts 260, 261, 265, or 266 regulations.

The EPA may, by compliance order issued under section 3008 of RCRA, extend the date by which the owner and operator of an existing hazardous waste management facility must submit Part A of their permit application.

The owner or operator of an existing hazardous waste management facility may be required to submit Part B of their permit application. The state director may require submission of Part B (or equivalent completion of the state RCRA application process) if the state in which the facility is located has received interim or final authorization; if not, the regional administrator may require submission of Part B. Any owner or operator is allowed at least 6 months from the date of request to submit Part B of the application. Any owner or operator of an existing hazardous waste management facility may voluntarily submit Part B of the application at any time. Notwithstanding the above, any owner or operator of an existing hazardous waste management facility must submit a Part B permit application in accordance with the dates specified in 270.73. Any owner or operator of a land disposal facility in existence on the effective date of statutory or regulatory amendments under this act that render the facility subject to the requirement to have a RCRA permit must submit a Part B application in accordance with the dates specified in 270.73.

Failure to furnish a requested Part B application on time, or to furnish in full the information required by the Part B application, is grounds for termination of interim status under Part 124.

11.6.5.6 New HWM Facilities.

11.6.5.6 *New HWM Facilities.* Except as noted below, no person can begin physical construction of a new HWM facility without having submitted Parts A and B of the permit application and having received a finally effective RCRA TSDF permit.

An application for a TSDF permit for a new hazardous waste management facility (including both Parts A and B) may be filed any time after promulgation of those standards in Part 264, subpart I applicable to such facility. The application must be filed with the regional administrator if at the time of application the state in which the new hazardous waste management facility is proposed to be located has not received interim or final authorization for permitting such facility, otherwise it must be filed with the state director. Except as provided in the next paragraph, all applications must be submitted at least 180 days before physical construction is expected to commence.

Notwithstanding these requirements, a person may construct a facility for the incineration of polychlorinated biphenyls pursuant to an approval issued by the EPA under section (6)(e) of the Toxic Substances Control Act, and any person owning or operating such a facility may, at any time after construction or operation of such facility has begun, file an application for a RCRA permit to incinerate hazardous waste authorizing such facility to incinerate waste identified or listed under Subtitle C of RCRA.

11.6.5.7 *Updating Permit Applications.* If any owner or operator of a TSDF has filed Part A of a permit application and has not yet filed Part B, the owner or operator must file an amended Part A application:

- With the regional administrator if the facility is located in a state which has not obtained interim authorization or final authorization, within 6 months after the promulgation of revised regulations under Part 261 listing or identifying additional hazardous wastes, if the facility is treating, storing, or disposing of any of those newly listed or identified wastes.

- With the state director, if the facility is located in a state which has obtained interim authorization or final authorization, no later than the effective date of regulatory provisions listing or designating wastes as hazardous in that state in addition to those listed or designated under the previously approved state program, if the facility is treating, storing, or disposing of any of those newly listed or designated wastes.

- As necessary to comply with provisions of 270.72 for changes during interim status or with the analogous provisions of a state program approved for final authorization or interim authorization. Revised Part A applications necessary to comply with the provisions of 270.72 must be filed with the regional administrator if the state in which the facility in question is located does not have interim authorization or final authorization; otherwise it must be filed with the state director (if the state has an analogous provision).

11.6.5.8   Reapplications.   Any HWM facility with an effective TSDF permit must submit a new application at least 180 days before the expiration date of the effective permit, unless permission for a later date has been granted by the director. (The director cannot grant permission for applications to be submitted later than the expiration date of the existing permit.)

11.6.5.9   Record Keeping.   Applicants must keep records of all data used to complete permit applications and any supplemental information submitted under 270.10(d), 270.13, 270.14 through 270.21 for a period of at least 3 years from the date the application is signed.

11.6.5.10   Exposure Information.   After Aug. 8, 1985, any Part B TSDF permit application submitted by an owner or operator of a facility that stores, treats, or disposes of hazardous waste in a surface impoundment or a landfill must be accompanied by information, reasonably ascertainable by the owner or operator, on the potential for the public to be exposed to hazardous wastes or hazardous constituents through releases related to the unit. At a minimum, such information must address:

- Reasonably foreseeable potential releases from both normal operations and accidents at the unit, including releases associated with transportation to or from the unit

- The potential pathways of human exposure to hazardous wastes or constituents resulting from the releases described under paragraph (j)(1)(i) of 40 CFR Part 270.10

- The potential magnitude and nature of the human exposure resulting from such releases

By Aug. 8, 1985, owners and operators of a landfill or a surface impoundment who have already submitted a Part B application needed to submit the exposure information required above.

11.6.5.11 Signatories to TSDF Permit Applications and Reports

11.6.5.11.1 TSDF Permit Applications. All permit applications must be signed as follows:

11.6.5.11.1.1 FOR A CORPORATION: By a responsible corporate officer. For the purpose of this section, a responsible corporate officer means (i) a president, secretary, treasurer, or vice-president of the corporation in charge of a principal business function, or any other person who performs similar policy—or decision-making functions for the corporation, or (ii) the manager of one or more manufacturing, production, or operating facilities employing more than 250 persons or having gross annual sales or expenditures exceeding $25 million (in second-quarter 1980 dollars), if authority to sign documents has been assigned or delegated to the manager in accordance with corporate procedures.

[*Note:* EPA does not require specific assignments or delegations of authority to responsible corporate officers identified in 270.11(a)(1)(i). The agency will presume that these responsible corporate officers have the requisite authority to sign permit applications unless the corporation has notified the director to the contrary. Corporate procedures governing authority to sign permit applications may provide for assignment or delegation to applicable corporate positions under 270.11(a)(1)(ii) rather than to specific individuals.]

11.6.5.11.1.2 FOR A PARTNERSHIP OR SOLE PROPRIETORSHIP: By a general partner or the proprietor, respectively.

11.6.5.11.1.3 FOR A MUNICIPALITY, STATE, FEDERAL, OR OTHER PUBLIC AGENCY: By either a principal executive officer or ranking elected official. For purposes of this section, a principal executive officer of a federal agency includes (i) the chief executive officer of the agency or (ii) a senior executive officer having responsibility for the overall operations of a principal geographic unit of the agency (e.g., regional administrators of EPA).

11.6.5.11.2 Reports. All reports required by permits and other information requested by the director must be signed by a person as described above in this section or by a duly authorized representative of that person. A person is a duly authorized representative only if:

- The authorization is made in writing by a person described above
- The authorization specifies either an individual or a position having responsibility for overall operation of the regulated facility or activity such as the position of plant manager, operator of a well or a well field, superintendent, or position of equivalent responsibility (a duly authorized representative may thus be either a named individual or any individual occupying a named position)

• The written authorization is submitted to the director.

11.6.5.11.3 Changes to Authorization. If an authorization as stated above is no longer accurate because a different individual or position has responsibility for the overall operation of the facility, a new authorization satisfying the requirements of paragraph (b) of 40 CFR Part 270.11 must be submitted to the director prior to or together with any reports, information, or applications to be signed by an authorized representative.

11.6.5.11.4 Certification. Effective June 1, 1999, any person signing a document under paragraph (a) or (b) of 40 CFR Part 270.11 must make the following certification:

> I certify under penalty of law that this document and all attachments were prepared under my direction or supervision according to a system designed to assure that qualified personnel properly gather and evaluate the information submitted. Based on my inquiry of the person or persons who manage the system, or those persons directly responsible for gathering the information, the information submitted is, to the best of my knowledge and belief, true, accurate, and complete. I am aware that there are significant penalties for submitting false information, including the possibility of fine and imprisonment for knowing violations.

For remedial action plans (RAPs) under subpart H of 40 CFR Part 270 if the operator certifies according to the above paragraph of this section, then the owner may choose to make the following certification instead of the above certification:

> Based on my knowledge of the conditions of the property described in the RAP and my inquiry of the person or persons who manage the system referenced in the operator's certification, or those persons directly responsible for gathering the information, the information submitted is, upon information and belief, true, accurate, and complete. I am aware that there are significant penalties for submitting false information, including the possibility of fine and imprisonment for knowing violations.

11.6.6 TSDF Permit Conditions

11.6.6.1 Conditions Applicable to All TSDF Permits. The following conditions apply to all RCRA permits and must be incorporated into the permits either expressly or by reference. If incorporated by reference, a specific citation to these regulations (or the corresponding approved state regulations) must be given in the permit.

• Duty to comply. The permittee must comply with all conditions of the permit, except that the permittee need not comply with the conditions of this permit to the extent and for the duration such noncompliance is authorized in an emergency permit (see 270.61). Any permit noncompliance, except under the terms of an emergency permit, constitutes a violation of the appropriate act and is grounds for enforcement action; for permit termination, revocation and reissuance, or modification; or for denial of a permit renewal application.

- Duty to reapply. If the permittee wishes to continue an activity regulated by this permit after the expiration date of this permit, the permittee must apply for and obtain a new permit.
- Need to halt or reduce activity not a defense. It cannot be a defense for a permittee in an enforcement action that it would have been necessary to halt or reduce the permitted activity in order to maintain compliance with the conditions of this permit.
- In the event of noncompliance with the permit, the permittee must take all reasonable steps to minimize releases to the environment and must carry out such measures as are reasonable to prevent significant adverse impacts on human health or the environment.
- Proper operation and maintenance. The permittee must, at all times, properly operate and maintain all facilities and systems of treatment and control (and related appurtenances) which are installed or used by the permittee to achieve compliance with the conditions of this permit. Proper operation and maintenance includes effective performance, adequate funding, adequate operator staffing and training, and adequate laboratory and process controls, including appropriate quality assurance procedures. This provision requires the operation of backup or auxiliary facilities or similar systems only when necessary to achieve compliance with the conditions of the permit.
- Permit actions. The permit may be modified, revoked and reissued, or terminated for cause. The filing of a request by the permittee for a permit modification, revocation and reissuance, or termination, or a notification of planned changes or anticipated noncompliance does not stay any permit condition.
- Property rights. The permit does not convey any property rights of any sort, or any exclusive privilege.
- Duty to provide information. The permittee must furnish to the director, within a reasonable time, any relevant information which the director may request to determine whether cause exists for modifying, revoking and reissuing, or terminating this permit, or to determine compliance with this permit. The permittee must also furnish to the director, upon request, copies of records required to be kept by this permit.
- Inspection and entry. The permittee must allow the director, or an authorized representative, upon the presentation of credentials and other documents as may be required by law to:
  - Enter at reasonable times upon the permittee's premises where a regulated facility or activity is located or conducted, or where records must be kept under the conditions of this permit.

- Have access to and copy, at reasonable times, any records that must be kept under the conditions of this permit.

- Inspect at reasonable times any facilities, equipment (including monitoring and control equipment), practices, or operations regulated or required under this permit.

- Sample or monitor at reasonable times, for the purposes of assuring permit compliance or as otherwise authorized by RCRA, any substances or parameters at any location.

- Monitoring and records.

  Samples and measurements taken for the purpose of monitoring must be representative of the monitored activity.

  The permittee must retain records of all monitoring information, including all calibration and maintenance records and all original strip-chart recordings for continuous monitoring instrumentation, copies of all reports required by this permit, the certification required by 264.73(b)(9), and records of all data used to complete the application for this permit, for a period of at least 3 years from the date of the sample, measurement, report, certification, or application. This period may be extended by request of the director at any time. The permittee must maintain records from all groundwater monitoring wells and associated groundwater surface elevations, for the active life of the facility, and for disposal facilities for the postclosure care period as well.

- Records for monitoring information must include:

  The date, exact place, and time of sampling or measurements

  The individual(s) who performed the sampling or measurements

  The date(s) analyses were performed

  The individual(s) who performed the analyses

  The analytical techniques or methods used

  The results of such analyses

- Signatory requirements. All applications, reports, or information submitted to the director must be signed and certified (see 270.11).

- Reporting requirements.

- Planned changes. The permittee must give notice to the director as soon as possible of any planned physical alterations or additions to the permitted facility.

- Anticipated noncompliance. The permittee must give advance notice to the director of any planned changes in the permitted facility or activity which may result in noncompliance with permit requirements. For a new facility, the permittee may not treat, store, or dispose of hazardous waste; and for a facility being modified, the permittee may not treat, store, or dispose of hazardous waste in the modified portion of the facility except as provided in 270.42, until:

The permittee has submitted to the director by certified mail or hand delivery a letter signed by the permittee and a registered professional engineer stating that the facility has been constructed or modified in compliance with the permit

The director has inspected the modified or newly constructed facility and finds it is in compliance with the conditions of the permit

Within 15 days of the date of submission of the letter in paragraph (l)(2)(i) of 40 CFR Part 270.30, the permittee has not received notice from the director of his or her intent to inspect, prior inspection is waived, and the permittee may commence treatment, storage, or disposal of hazardous waste

Transfers. The permit is not transferable to any person except after notice to the director. The director may require modification or revocation and reissuance of the permit to change the name of the permittee and incorporate such other requirements as may be necessary under RCRA (see 270.40).

Monitoring reports. Monitoring results must be reported at the intervals specified elsewhere in this permit.

Compliance schedules. Reports of compliance or noncompliance with, or any progress reports on, interim and final requirements contained in any compliance schedule of this permit must be submitted no later than 14 days following each schedule date.

24-hour reporting.

The permittee must report any noncompliance which may endanger health or the environment orally within 24 hours from the time the permittee becomes aware of the circumstances, including:

Information concerning release of any hazardous waste that may cause an endangerment to public drinking water supplies.

Any information of a release or discharge of hazardous waste or of a fire or explosion from the HWM facility, which could threaten the environment or human health outside the facility.

The description of the occurrence and its cause must include:

Name, address, and telephone number of the owner or operator

Name, address, and telephone number of the facility

Date, time, and type of incident

Name and quantity of material(s) involved

The extent of injuries, if any

An assessment of actual or potential hazards to the environment and human health outside the facility, where this is applicable

Estimated quantity and disposition of recovered material that resulted from the incident

- A written submission must also be provided within 5 days of the time the permittee becomes aware of the circumstances. The written submission must contain a description of the noncompliance and its cause; the period of noncompliance including exact dates and times, and if the noncompliance has not been corrected, the anticipated time it is expected to continue; and steps taken or planned to reduce, eliminate, and prevent recurrence of the noncompliance. The director may waive the 5-day written notice requirement in favor of a written report within 15 days.

- Manifest discrepancy report: If a significant discrepancy in a manifest is discovered, the permittee must attempt to reconcile the discrepancy. If not resolved within 15 days, the permittee must submit a letter report, including a copy of the manifest, to the director (see 40 CFR 264.72).

- Unmanifested waste report: This report must be submitted to the director within 15 days of receipt of unmanifested waste (see 40 CFR 264.76).

- Biennial report: A biennial report must be submitted covering facility activities during odd-numbered calendar years (see 40 CFR 264.75).

- Other noncompliance. The permittee must report all instances of noncompliance not reported under paragraphs (l)(4), (5), and (6) of 40 CFR Part 270.30, at the time monitoring reports are submitted. The reports must contain the information listed in the paragraph above that details the 24-hour reporting requirements.

- Other information. Where the permittee becomes aware that it failed to submit any relevant facts in a permit application, or submitted incorrect information in a permit application or in any report to the director, it must promptly submit such facts or information.

- Information repository. The director may require the permittee to establish and maintain an information repository at any time, based on the factors set forth in 40 CFR 124.33(b). The information repository will be governed by the provisions in 40 CFR 124.33(c) through (f).

11.6.6.2 Requirements for Recording and Reporting of Monitoring Results.
All permits must specify:

- Requirements concerning the proper use, maintenance, and installation, when appropriate, of monitoring equipment or methods (including biological monitoring methods when appropriate).

- Required monitoring including type, intervals, and frequency sufficient to yield data which are representative of the monitored activity including, when appropriate, continuous monitoring.

- Applicable reporting requirements based upon the impact of the regulated activity and as specified in Parts 264, 266, and 267. Reporting can be no less frequent than specified in the above regulations.

11.6.6.3   Establishing Permit Conditions

- In addition to conditions required in all permits (270.30), the director can establish conditions, as required on a case-by-case basis, in permits under 270.50 (duration of permits), 270.33(a) (schedules of compliance), 270.31 (monitoring), and for EPA-issued permits only, 270.33(b) (alternate schedules of compliance), and 270.3 (considerations under federal law).

  Each RCRA permit must include permit conditions necessary to achieve compliance with the act and regulations, including each of the applicable requirements specified in Parts 264 and 266 through 268. In satisfying this provision, the EPA may incorporate applicable requirements of Parts 264 and 266 through 268 directly into the permit or establish other permit conditions that are based on these parts.

  Each permit issued under section 3005 of the act must contain terms and conditions as the administrator or state director determines necessary to protect human health and the environment.

- For a state-issued permit, an applicable requirement is a state statutory or regulatory requirement which takes effect prior to final administrative disposition of a permit. For a permit issued by EPA, an applicable requirement is a statutory or regulatory requirement (including any interim final regulation) which takes effect prior to the issuance of the permit [except as provided in 124.86(c) for RCRA permits being processed under subpart E or F of Part 124]. Section 124.14 (reopening of comment period) provides a means for reopening EPA permit proceedings at the discretion of the director where new requirements become effective during the permitting process and are of sufficient magnitude to make additional proceedings desirable. For state- and EPA-administered programs, an applicable requirement is also any requirement which takes effect prior to the modification or revocation and reissuance of a permit, to the extent allowed in 270.41.

- New or reissued permits, and to the extent allowed under 270.41, modified or revoked and reissued permits, must incorporate each of the applicable requirements referenced in this section and in 40 CFR 270.31.

- Incorporation. All permit conditions must be incorporated either expressly or by reference. If incorporated by reference, a specific citation to the applicable regulations or requirements must be given in the permit.

11.6.6.4 Schedules of Compliance

- The permit may, when appropriate, specify a schedule of compliance leading to compliance with the act and regulations.
- Time for compliance. Any schedules of compliance under this section must require compliance as soon as possible.

- Interim dates. Except as provided in paragraph (b)(1)(ii) of 40 CFR Part 270.33, if a permit establishes a schedule of compliance which exceeds 1 year from the date of permit issuance, the schedule must set forth interim requirements and the dates for their achievement.

  The time between interim dates cannot exceed 1 year.

  If the time necessary for completion of any interim requirement is more than 1 year and is not readily divisible into stages for completion, the permit must specify interim dates for the submission of reports of progress toward completion of the interim requirements and indicate a projected completion date.

  Reporting. The permit must be written to require that no later than 14 days following each interim date and the final date of compliance, the permittee must notify the director in writing of its compliance or noncompliance with the interim or final requirements.

  Alternative schedules of compliance. A RCRA permit applicant or permittee may cease conducting regulated activities (by receiving a terminal volume of hazardous waste and, for treatment and storage HWM facilities, closing pursuant to applicable requirements; and, for disposal HWM facilities, closing and conducting postclosure care pursuant to applicable requirements) rather than continue to operate and meet permit requirements as follows:

  If the permittee decides to cease conducting regulated activities at a given time within the term of a permit which has already been issued:

    The permit may be modified to contain a new or additional schedule leading to timely cessation of activities.

    The permittee must cease conducting permitted activities before noncompliance with any interim or final compliance schedule requirement already specified in the permit.

  If the decision to cease conducting regulated activities is made before issuance of a permit whose term will include the termination date, the permit must contain a schedule leading to termination which will ensure timely compliance with applicable requirements.

If the permittee is undecided whether to cease conducting regulated activities, the director may issue or modify a permit to contain two schedules as follows:

- Both schedules must contain an identical interim deadline requiring a final decision on whether to cease conducting regulated activities no later than a date which ensures sufficient time to comply with applicable requirements in a timely manner if the decision is to continue conducting regulated activities.
- One schedule must lead to timely compliance with applicable requirements.
- The second schedule must lead to cessation of regulated activities by a date which will ensure timely compliance with applicable requirements.
- Each permit containing two schedules must include a requirement that after the permittee has made a final decision under paragraph (b)(3)(i) of 40 CFR Part 270.33 it must follow the schedule leading to compliance if the decision is to continue conducting regulated activities, and follow the schedule leading to termination if the decision is to cease conducting regulated activities.

The applicant's or permittee's decision to cease conducting regulated activities must be evidenced by a firm public commitment satisfactory to the director, such as resolution of the board of directors of a corporation.

11.6.7   Special Forms of Permits

11.6.7.1   Permits by Rule.   Notwithstanding any other provision of 40 CFR Part 270 or Part 124, the following must be deemed to have a RCRA permit if the conditions listed are met:

Ocean disposal barges or vessels. The owner or operator of a barge or other vessel which accepts hazardous waste for ocean disposal, if the owner or operator:

- Has a permit for ocean dumping issued under 40 CFR Part 220 (Ocean Dumping, authorized by the Marine Protection, Research, and Sanctuaries Act, as amended, 33 U.S.C. 1420 et seq.)
- Complies with the conditions of that permit
- Complies with the following hazardous waste regulations:

40 CFR 264.11, identification number

40 CFR 264.71, use of manifest system

40 CFR 264.72, manifest discrepancies

40 CFR 264.73(a) and (b)(1), operating record

40 CFR 264.75, biennial report

40 CFR 264.76, unmanifested waste report

- Injection wells. The owner or operator of an injection well disposing of hazardous waste, if the owner or operator:

    Has a permit for underground injection issued under Part 144 or 145

    Complies with the conditions of that permit and the requirements of 144.14 (requirements for wells managing hazardous waste)

    For UIC permits issued after Nov. 8, 1984:

- Complies with 40 CFR 264.101
- Where the UIC well is the only unit at a facility which requires a RCRA permit, complies with 40 CFR 270.14(d)

    Publicly owned treatment works. The owner or operator of a POTW which accepts for treatment hazardous waste, if the owner or operator:

- Has an NPDES permit
- Complies with the conditions of that permit
- Complies with the following regulations

    40 CFR 264.11, identification number

    40 CFR 264.71, use of manifest system

    40 CFR 264.72, manifest discrepancies

    40 CFR 264.73(a) and (b)(1), operating record

    40 CFR 264.75, biennial report

    40 CFR 264.76, unmanifested waste report

    For NPDES permits issued after Nov. 8, 1984, 40 CFR 264.101

If the waste meets all federal, state, and local pretreatment requirements which would be applicable to the waste if it were being discharged into the POTW through a sewer, pipe, or similar conveyance.

11.6.7.2 Emergency Permits.   Notwithstanding any other provision of 40 CFR, Part 270 or Part 124, in the event the director finds an imminent and substantial endangerment to human health or the environment, the director may issue a temporary emergency permit: (1) to a nonpermitted facility to allow treatment, storage, or disposal of hazardous waste or (2) to a permitted facility to allow treatment, storage, or disposal of a hazardous waste not covered by an effective permit.

This emergency permit:

- May be oral or written. If oral, it must be followed in 5 days by a written emergency permit.
- Cannot exceed 90 days in duration.

- Must clearly specify the hazardous wastes to be received, and the manner and location of their treatment, storage, or disposal.

- May be terminated by the director at any time without process if he or she determines that termination is appropriate to protect human health and the environment.

- Must be accompanied by a public notice published under 124.10(b) including:

  Name and address of the office granting the emergency authorization

  Name and location of the permitted HWM facility

  A brief description of the wastes involved

  A brief description of the action authorized and reasons for authorizing it

  Duration of the emergency permit

- Must incorporate, to the extent possible and not inconsistent with the emergency situation, all applicable requirements of 40 CFR, Part 270 and 40 CFR Parts 264 and 266.

11.6.7.3 Hazardous Waste Incinerator Permits

- For the purposes of determining operational readiness following completion of physical construction, the director must establish permit conditions, including but not limited to allowable waste feeds and operating conditions, in the permit to a new hazardous waste incinerator. These permit conditions will be effective for the minimum time required to bring the incinerator to a point of operational readiness to conduct a trial burn, not to exceed 720 hours of operating time for treatment of hazardous waste. The director may extend the duration of this operational period once, for up to 720 additional hours, at the request of the applicant when good cause is shown. The permit may be modified to reflect the extension according to 270.42.

- Applicants must submit a statement, with Part B of the permit application, which suggests the conditions necessary to operate in compliance with the performance standards of 264.343 during this period. This statement should include, at a minimum, restrictions on waste constituents, waste feed rates, and the operating parameters identified in 264.345.

- The director will review this statement and any other relevant information submitted with Part B of the permit application and specify requirements for this period sufficient to meet the performance standards of 264.343 based on his or her engineering judgment.

- For the purposes of determining feasibility of compliance with the performance standards of 264.343 and of determining adequate operating conditions under 264.345, the director must establish conditions in the permit for a new hazardous waste incinerator to be effective during the trial burn.

Applicants must propose a trial burn plan, prepared under paragraph (b)(2) of 40 CFR, Part 270.62 with Part B of the permit application.

The trial burn plan must include the following information:

An analysis of each waste or mixture of wastes to be burned which includes:

Heat value of the waste in the form and composition in which it will be burned.

Viscosity (if applicable) or description of the physical form of the waste.

Organic constituents listed in Part 261, appendix VIII, which are present in the waste to be burned, except that the applicant need not analyze for constituents listed in Part 261, appendix VIII, which would reasonably not be expected to be found in the waste. The constituents excluded from analysis must be identified, and the basis for the exclusion stated. The waste analysis must rely on analytical techniques specified in "Test Methods for Evaluating Solid Waste, Physical/Chemical Methods," EPA Publication SW-846, as incorporated by reference in 260.11 and 270.6, or other equivalent.

An approximate quantification of the hazardous constituents identified in the waste, within the precision produced by the analytical methods specified in "Test Methods for Evaluating Solid Waste, Physical/Chemical Methods," EPA Publication SW-846, as incorporated by reference in 260.11 and 270.6, or their equivalent.

A detailed engineering description of the incinerator for which the permit is sought including:

Manufacturer's name and model number of incinerator (if available)

Type of incinerator

Linear dimensions of the incinerator unit including the cross-sectional area of the combustion chamber

Description of the auxiliary fuel system (type and feed)

Capacity of prime mover

Description of automatic waste feed cutoff system(s)

Stack gas monitoring and pollution control equipment

Nozzle and burner design

Construction materials

Location and description of temperature, pressure, and flow indicating and control devices

A detailed description of sampling and monitoring procedures, including sampling and monitoring locations in the system, the equipment to be used, sampling and monitoring frequency, and planned analytical procedures for sample analysis.

A detailed test schedule for each waste for which the trial burn is planned including date(s), duration, quantity of waste to be burned, and other factors relevant to the director's decision under paragraph (b)(5) of 40 CFR, part 270.62.

A detailed test protocol, including, for each waste identified, the ranges of temperature, waste feed rate, combustion gas velocity, use of auxiliary fuel, and any other relevant parameters that will be varied to affect the destruction and removal efficiency of the incinerator.

A description of, and planned operating conditions for, any emission control equipment which will be used.

Procedures for rapidly stopping waste feed, shutting down the incinerator, and controlling emissions in the event of an equipment malfunction.

Such other information as the director reasonably finds necessary to determine whether to approve the trial burn plan in light of the purposes of this paragraph and the criteria in paragraph (b)(5) of 40 CFR, Part 270.62.

- The director, in reviewing the trial burn plan, must evaluate the sufficiency of the information provided and may require the applicant to supplement this information, if necessary, to achieve the purposes of this paragraph.

- Based on the waste analysis data in the trial burn plan, the director will specify as trial principal organic hazardous constituents (POHCs), those constituents for which destruction and removal efficiencies must be calculated during the trial burn. These trial POHCs will be specified by the director based on an estimate of the difficulty of incineration of the constituents identified in the waste analysis, their concentration or mass in the waste feed, and for wastes listed in part 261, subpart D, the hazardous waste organic constituent or constituents identified in Appendix VII of that part as the basis for listing.

- The director will approve a trial burn plan if he or she finds that:

The trial burn is likely to determine whether the incinerator performance standard required by 264.343 can be met.

The trial burn itself will not present an imminent hazard to human health or the environment.

The trial burn will help the director to determine operating requirements to be specified under 264.345.

The director must send a notice to all persons on the facility mailing list as set forth in 40 CFR 124.10(c)(1)(ix) and to the appropriate units of state and local government as set forth in 40 CFR 124.10(c)(1)(x) announcing the scheduled commencement and completion dates for the trial burn. The applicant may not commence the trial burn until after the director has issued such notice.

This notice must be mailed within a reasonable time period before the scheduled trial burn. An additional notice is not required if the trial burn is delayed due to circumstances beyond the control of the facility or the permitting agency.

This notice must contain:

The name and telephone number of the applicant's contact person

The name and telephone number of the permitting agency's contact office

The location where the approved trial burn plan and any supporting documents can be reviewed and copied

An expected time period for commencement and completion of the trial burn

During each approved trial burn (or as soon after the burn as is practicable), the applicant must make the following determinations:

A quantitative analysis of the trial POHCs in the waste feed to the incinerator

A quantitative analysis of the exhaust gas for the concentration and mass emissions of the trial POHCs, oxygen (O_2) and hydrogen chloride (HCl)

A quantitative analysis of the scrubber water (if any), ash residues, and other residues, for the purpose of estimating the fate of the trial POHCs

A computation of destruction and removal efficiency (DRE), in accordance with the DRE formula specified in 264.343(a)

If the HCl emission rate exceeds 1.8 kilograms of HCl per hour (4 pounds per hour), a computation of HCl removal efficiency in accordance with 264.343(b)

A computation of particulate emissions, in accordance with 264.343(c)

An identification of sources of fugitive emissions and their means of control

A measurement of average, maximum, and minimum temperatures and combustion gas velocity

A continuous measurement of carbon monoxide (CO) in the exhaust gas

Such other information as the director may specify as necessary to ensure that the trial burn will determine compliance with the performance standards in 264.343 and to establish the operating conditions required by 264.345 as necessary to meet that performance standard

The applicant must submit to the director a certification that the trial burn has been carried out in accordance with the approved trial burn plan, and must submit the results of all the determinations required in paragraph (b)(6) of 40 CFR, Part 270.62. This submission must be made within 90 days of completion of the trial burn, or later if approved by the director.

All data collected during any trial burn must be submitted to the director following the completion of the trial burn.

All submissions required by this paragraph must be certified on behalf of the applicant by the signature of a person authorized to sign a permit application or a report under 270.11.

Based on the results of the trial burn, the director will set the operating requirements in the final permit according to 264.345. The permit modification must proceed according to 270.42.

- For the purposes of allowing operation of a new hazardous waste incinerator following completion of the trial burn and prior to final modification of the permit conditions to reflect the trial burn results, the director may establish permit conditions, including but not limited to allowable waste feeds and operating conditions sufficient to meet the requirements of 264.345, in the permit to a new hazardous waste incinerator. These permit conditions will be effective for the minimum time required to complete sample analysis, data computation, and submission of the trial burn results by the applicant, and modification of the facility permit by the director.

- Applicants must submit a statement, with part B of the permit application, which identifies the conditions necessary to operate in compliance with the performance standards of 264.343, during this period. This statement should include, at a minimum, restrictions on waste constituents, waste feed rates, and the operating parameters in 264.345.

- The director will review this statement and any other relevant information submitted with part B of the permit application and specify those requirements for this period most likely to meet the performance standards of 264.343 based on engineering judgment.

- For the purpose of determining feasibility of compliance with the performance standards of 264.343 and of determining adequate operating conditions under 264.345, the applicant for a permit for an existing hazardous waste incinerator must prepare and submit a trial burn plan and perform a trial burn in accordance with 270.19(b) and paragraphs (b)(2) through (b)(5) and (b)(7) through (b)(10) of 40 CFR, Part 270.62 or, instead, submit other information as specified in 270.19(c). The director must announce his or her intention to approve the trial burn plan in accordance with the timing and distribution requirements of paragraph (b)(6) of 40 CFR, Part 270.62. The contents of the notice must include the name and telephone number of a contact person at the facility; the name and telephone number of a contact office at the permitting agency; the location where the trial burn plan and any supporting documents can be reviewed and copied; and a schedule of the activities that are required prior to permit issuance, including the anticipated time schedule for agency approval of the plan and the time period during which the trial burn would be conducted. Applicants submitting information under 270.19(a) are exempt from compliance with 40 CFR 264.343 and 264.345 and therefore are exempt from the requirement to conduct a trial burn. Applicants who submit trial burn plans and receive approval before submission of a permit application must complete the trial burn and submit the results, specified in paragraph (b)(7) of 40 CFR, Part 270.62, with part B of the permit application. If completion of this process conflicts with the date set for submission of the part B application, the applicant must contact the director to establish a later date for submission of the part B application or the trial burn results. Trial burn results must be submitted prior to issuance of the permit. When the applicant submits a trial burn plan with part B of the permit application, the director will specify a time period prior to permit issuance in which the trial burn must be conducted and the results submitted.

11.6.7.4 Permits for Land Treatment Demonstrations Using Field Test or Laboratory Analyses

- For the purpose of allowing an owner or operator to meet the treatment demonstration requirements of 264.272, the director may issue a treatment demonstration permit. The permit must contain only those requirements necessary to meet the standards in 264.272(c). The permit may be issued either as a treatment or disposal permit covering only the field test or laboratory analyses, or as a two-phase facility permit covering the field tests, or laboratory analyses, and design, construction operation, and maintenance of the land treatment unit.

   The director may issue a two-phase facility permit if he or she finds that, based on information submitted in part B of the application, substantial, although incomplete or inconclusive, information already exists upon which to base the issuance of a facility permit.

   If the director finds that not enough information exists upon which to establish permit conditions to attempt to provide for compliance with all of the requirements of subpart M, he or she must issue a treatment demonstration permit covering only the field test or laboratory analyses.

- If the director finds that a phased permit may be issued, he or she will establish, as requirements in the first phase of the facility permit, conditions for conducting the field tests or laboratory analyses. These permit conditions will include design and operating parameters (including the duration of the tests or analyses and, in the case of field tests, the horizontal and vertical dimensions of the treatment zone), monitoring procedures, postdemonstration cleanup activities, and any other conditions which the director finds may be necessary under 264.272(c). The director will include conditions in the second phase of the facility permit to attempt to meet all subpart M requirements pertaining to unit design, construction, operation, and maintenance. The director will establish these conditions in the second phase of the permit based upon the substantial but incomplete or inconclusive information contained in the part B application.

The first phase of the permit will be effective as provided in 124.15(b).

The second phase of the permit will be effective as provided in paragraph (d) of 40 CFR, Part 270.63.

- An owner or operator who has been issued a two-phase permit and has completed the treatment demonstration must submit to the director a certification, signed by a person authorized to sign a permit application or report under 270.11, that the field tests or laboratory analyses have been carried out in accordance with the conditions specified in phase one of the permit for conducting such tests or analyses. The owner or operator must also submit all data collected during the field tests or laboratory analyses within 90 days of completion of those tests or analyses unless the director approves a later date.

- A director who determines that the results of the field tests or laboratory analyses meet the requirements of 264.272 will modify the second phase of the permit to incorporate any requirements necessary for operation of the facility in compliance with Part 264, subpart M, based upon the results of the field tests or laboratory analyses.

This permit modification may proceed under 270.42 or otherwise will proceed as a modification under 270.41(a)(2). If such modifications are necessary, the second phase of the permit will become effective only after those modifications have been made.

If no modifications of the second phase of the permit are necessary, the director will give notice of a final decision to the permit applicant and to each person who submitted written comments on the phased permit or who requested notice of the final decision on the second phase of the permit. The second phase of the permit then will become effective as specified in 124.15(b).

11.6.7.5 Interim Permits for UIC Wells. The director may issue a permit under these regulations to any Class I UIC well (see 144.6) injecting hazardous wastes within a state in which no UIC program has been approved or promulgated. Any such permit must apply and ensure compliance with all applicable requirements of 40 CFR Part 264, subpart R (RCRA standards for wells), and must be for a term not to exceed 2 years. No such permit can be issued after approval or promulgation of a UIC program in the state. All permits under these regulations must contain a condition providing that it will terminate upon final action by the director under a UIC program to issue or deny a UIC permit for the facility.

11.6.7.6 Research, Development, and Demonstration Permits

- The EPA may issue a research, development, and demonstration permit for any hazardous waste treatment facility which proposes to utilize an innovative and experimental hazardous waste treatment technology or process for which permit standards for such experimental activity have not been promulgated under Part 264 or 266. Any such permit must include such terms and conditions as will assure protection of human health and the environment. Such permits:

> Must provide for the construction of such facilities as necessary, and for operation of the facility for not longer than 1 year unless renewed as provided in paragraph (d) of 40 CFR, Part 270.65.

> Must provide for the receipt and treatment by the facility of only those types and quantities of hazardous waste which the EPA deems necessary for purposes of determining the efficacy and performance capabilities of the technology or process and the effects of such technology or process on human health and the environment.

> Must include such requirements as the EPA deems necessary to protect human health and the environment (including, but not limited to, requirements regarding monitoring, operation, financial responsibility, closure, and remedial action), and such requirements as the EPA deems necessary regarding testing and providing of information to the EPA with respect to the operation of the facility.

- For the purpose of expediting review and issuance of permits under this section, the EPA may, consistent with the protection of human health and the environment, modify or waive permit application and permit issuance requirements in Parts 124 and 270 except that there may be no modification or waiver of regulations regarding financial responsibility (including insurance) or of procedures regarding public participation.

- The EPA may order an immediate termination of all operations at the facility at any time it determines that termination is necessary to protect human health and the environment.

- Any permit issued under this section may be renewed not more than three times. Each such renewal must be for a period of not more than 1 year.

11.6.7.7 Permits for Boilers and Industrial Furnaces Burning Hazardous Waste

- General. Owners and operators of new boilers and industrial furnaces (those not operating under the interim status standards of 266.103) are subject to paragraphs (b) through (f) of 40 CFR, Part 270.66. Boilers and industrial furnaces operating under the interim status standards of 266.103 are subject to paragraph (g) of 40 CFR, Part 270.66.

- Permit operating periods for new boilers and industrial furnaces. A permit for a new boiler or industrial furnace must specify appropriate conditions for the following operating periods:

    Pretrial burn period. For the period beginning with initial introduction of hazardous waste and ending with initiation of the trial burn, and only for the minimum time required to bring the boiler or industrial furnace to a point of operational readiness to conduct a trial burn, not to exceed 720 hours of operating time when burning hazardous waste, the director must establish in the pretrial burn period of the permit conditions, including but not limited to, allowable hazardous waste feed rates and operating conditions. The director may extend the duration of this operational period once, for up to 720 additional hours, at the request of the applicant when good cause is shown. The permit may be modified to reflect the extension according to 270.42.

    > Applicants must submit a statement, with part B of the permit application, that suggests the conditions necessary to operate in compliance with the standards of 266.104 through 266.107 during this period. This statement should include, at a minimum, restrictions on the applicable operating requirements identified in 266.102(e).

    > The director will review this statement and any other relevant information submitted with part B of the permit application and specify requirements for this period sufficient to meet the performance standards of 266.104 through 266.107 based on engineering judgment.

    Trial burn period. For the duration of the trial burn, the director must establish conditions in the permit for the purposes of determining feasibility of compliance with the performance standards of 266.104 through 266.107 and determining adequate operating conditions under 266.102(e). Applicants must propose a trial burn plan, prepared under paragraph (c) of 40 CFR, part 270.66, to be submitted with part B of the permit application.

    Posttrial Burn period. For the period immediately following completion of the trial burn, and only for the minimum period sufficient to allow sample analysis, data computation, and submission of the trial burn results by the applicant, and review of the trial burn results and modification of the facility permit by the director to reflect the trial burn results, the director will establish the operating requirements most likely to ensure compliance with the performance standards of 266.104 through 266.107 based on engineering judgment.

    > Applicants must submit a statement, with part B of the application, that identifies the conditions necessary to operate during this period in compliance with the performance standards of 266.104 through 266.107. This

statement should include, at a minimum, restrictions on the operating requirements provided by 266.102(e).

The director will review this statement and any other relevant information submitted with part B of the permit application and specify requirements for this period sufficient to meet the performance standards of 266.104 through 266.107 based on engineering judgment.

Final permit period. For the final period of operation, the director will develop operating requirements in conformance with 266.102(e) that reflect conditions in the trial burn plan and are likely to ensure compliance with the performance standards of 266.104 through 266.107. Based on the trial burn results, the director will make any necessary modifications to the operating requirements to ensure compliance with the performance standards. The permit modification must proceed according to 270.42.

- Requirements for trial burn plans. The trial burn plan must include the following information. The director, in reviewing the trial burn plan, will evaluate the sufficiency of the information provided and may require the applicant to supplement this information, if necessary, to achieve the purposes of this paragraph:

  An analysis of each feed stream, including hazardous waste, other fuels, and industrial furnace feed stocks, as fired, that includes:

    Heating value, levels of antimony, arsenic, barium, beryllium, cadmium, chromium, lead, mercury, silver, thallium, total chlorine and chloride, and ash

    Viscosity or description of the physical form of the feed stream

  An analysis of each hazardous waste, as fired, including:

    An identification of any hazardous organic constituents listed in appendix VIII, part 261, that are present in the feed stream, except that the applicant need not analyze for constituents listed in appendix VIII that would reasonably not be expected to be found in the hazardous waste. The constituents excluded from analysis must be identified and the basis for this exclusion explained. The waste analysis must be conducted in accordance with analytical techniques specified in "Test Methods for Evaluating Solid Waste, Physical/Chemical Methods," EPA Publication SW-846, as incorporated by reference in 260.11 and 270.6, or their equivalent.

    An approximate quantification of the hazardous constituents identified in the hazardous waste, within the precision produced by the analytical methods specified in "Test Methods for Evaluating Solid Waste, Physical/Chemical Methods," EPA Publication SW-846, as incorporated by reference in 260.11 and 270.6, or other equivalent.

A description of blending procedures, if applicable, prior to firing the hazardous waste, including a detailed analysis of the hazardous waste prior to blending, an analysis of the material with which the hazardous waste is blended, and blending ratios.

A detailed engineering description of the boiler or industrial furnace, including:

Manufacturer's name and model number of the boiler or industrial furnace

Type of boiler or industrial furnace

Maximum design capacity in appropriate units

Description of the feed system for the hazardous waste and, as appropriate, other fuels and industrial furnace feedstocks

Capacity of hazardous waste feed system

Description of automatic hazardous waste feed cutoff system(s)

Description of any air pollution control system

Description of stack gas monitoring and any pollution control monitoring systems

- A detailed description of sampling and monitoring procedures including sampling and monitoring locations in the system, the equipment to be used, sampling and monitoring frequency, and planned analytical procedures for sample analysis.

- A detailed test schedule for each hazardous waste for which the trial burn is planned, including date(s), duration, quantity of hazardous waste to be burned, and other factors relevant to the director's decision under paragraph (b)(2) of 40 CFR, Part 270.66.

- A detailed test protocol, including, for each hazardous waste identified, the ranges of hazardous waste feed rate and, as appropriate, the feed rates of other fuels and industrial furnace feedstocks, and any other relevant parameters that may affect the ability of the boiler or industrial furnace to meet the performance standards in 266.104 through 266.107.

A description of, and planned operating conditions for, any emission control equipment that will be used

Procedures for rapidly stopping the hazardous waste feed and controlling emissions in the event of an equipment malfunction

Such other information as the director reasonably finds necessary to determine whether to approve the trial burn plan in light of the purposes of this paragraph and the criteria in paragraph (b)(2) of 40 CFR, Part 270.66

- Trial burn procedures.

  A trial burn must be conducted to demonstrate conformance with the standards of 266.104 through 266.107 under an approved trial burn plan.

  The director will approve a trial burn plan if he or she finds that:

  The trial burn is likely to determine whether the boiler or industrial furnace can meet the performance standards of 266.104 through 266.107.

  The trial burn itself will not present an imminent hazard to human health and the environment.

  The trial burn will help the director to determine operating requirements to be specified under 266.102(e).

  The information sought in the trial burn cannot reasonably be developed through other means.

  The director must send a notice to all persons on the facility mailing list as set forth in 40 CFR 124.10(c)(1)(ix) and to the appropriate units of state and local government as set forth in 40 CFR 124.10(c)(1)(x) announcing the scheduled commencement and completion dates for the trial burn. The applicant may not commence the trial burn until after the director has issued such notice.

  This notice must be mailed within a reasonable time period before the trial burn. An additional notice is not required if the trial burn is delayed due to circumstances beyond the control of the facility or the permitting agency.

  This notice must contain:

  The name and telephone number of applicant's contact person

  The name and telephone number of the permitting agency contact office

  The location where the approved trial burn plan and any supporting documents can be reviewed and copied

  An expected time period for commencement and completion of the trial burn

  The applicant must submit to the director a certification that the trial burn has been carried out in accordance with the approved trial burn plan, and must submit the results of all the determinations required in paragraph (c) of 40 CFR, Part 270.66. This submission must be made within 90 days of completion of the trial burn, or later if approved by the director.

  All data collected during any trial burn must be submitted to the director following completion of the trial burn.

  All submissions required by this paragraph must be certified on behalf of the applicant by the signature of a person authorized to sign a permit application or a report under 270.11.

- Special procedures for DRE trial burns. When a DRE trial burn is required under 266.104(a) the director will specify (based on the hazardous waste analysis data and other information in the trial burn plan) as trial principal organic hazardous constituents (POHCs) those compounds for which destruction and removal efficiencies must be calculated during the trial burn. These trial POHCs will be specified by the director based on information including his or her estimate of the difficulty of destroying the constituents identified in the hazardous waste analysis, their concentrations or mass in the hazardous waste feed, and, for hazardous waste containing or derived from wastes listed in part 261, subpart D, the hazardous waste organic constituent(s) identified in Appendix VII of that part as the basis for listing.

- Determinations based on trial burn. During each approved trial burn (or as soon after the burn as is practicable), the applicant must make the following determinations:

  A quantitative analysis of the levels of antimony, arsenic, barium, beryllium, cadmium, chromium, lead, mercury, thallium, silver, and chlorine and chloride, in the feed streams (hazardous waste, other fuels, and industrial furnace feedstocks)

  When a DRE trial burn is required under 266.104(a)

  A quantitative analysis of the trial POHCs in the hazardous waste feed

  A quantitative analysis of the stack gas for the concentration and mass emissions of the trial POHCs

  A computation of destruction and removal efficiency (DRE), in accordance with the DRE formula specified in 266.104(a)

  When a trial burn for chlorinated dioxins and furans is required under 266.104(e), a quantitative analysis of the stack gas for the concentration and mass emission rate of the 2,3,7,8-chlorinated tetra-octa congeners of chlorinated dibenzo-p-dioxins and furans, and a computation showing conformance with the emission standard

  When a trial burn for particulate matter, metals, or HCl/Cl2 is required under 266.105, 266.106(c) or (d), or 266.107(b)(2) or (c), a quantitative analysis of the stack gas for the concentrations and mass emissions of particulate matter, metals, or hydrogen chloride (HCl) and chlorine (Cl2), and computations showing conformance with the applicable emission performance standards

  When a trial burn for DRE, metals, or HCl/Cl2 is required under 266.104(a), 266.106(c) or (d), or 266.107(b)(2) or (c), a quantitative analysis of the scrubber water (if any), ash residues, other residues, and products for the purpose of estimating the fate of the trial POHCs, metals, and chlorine and chloride

An identification of sources of fugitive emissions and their means of control

A continuous measurement of carbon monoxide (CO), oxygen, and where required, hydrocarbons (HC), in the stack gas

Such other information as the director may specify as necessary to ensure that the trial burn will determine compliance with the performance standards in 266.104 through 266.107 and to establish the operating conditions required by 266.102(e) as necessary to meet those performance standards

- Interim status boilers and industrial furnaces. For the purpose of determining feasibility of compliance with the performance standards of 266.104 through 266.107 and of determining adequate operating conditions under 266.103, applicants owning or operating existing boilers or industrial furnaces operated under the interim status standards of 266.103 must either prepare and submit a trial burn plan and perform a trial burn in accordance with the requirements of this section or submit other information as specified in 270.22(a)(6). The director must announce his or her intention to approve of the trial burn plan in accordance with the timing and distribution requirements of paragraph (d)(3) of 40 CFR, Part 270.66. The contents of the notice must include the name and telephone number of a contact person at the facility; the name and telephone number of a contact office at the permitting agency; the location where the trial burn plan and any supporting documents can be reviewed and copied; and a schedule of the activities that are required prior to permit issuance, including the anticipated time schedule for agency approval of the plan and the time periods during which the trial burn would be conducted. Applicants who submit a trial burn plan and receive approval before submission of the part B permit application must complete the trial burn and submit the results specified in paragraph (f) of 40 CFR, Part 270.66, with the part B permit application. If completion of this process conflicts with the date set for submission of the part B application, the applicant must contact the director to establish a later date for submission of the part B application or the trial burn results. If the applicant submits a trial burn plan with part B of the permit application, the trial burn must be conducted and the results submitted within a time period prior to permit issuance to be specified by the director.

11.6.8   Contents of Part A of the Permit Application

Part A of the RCRA application must include the following information:

- The activities conducted by the applicant which require it to obtain a TSDF permit under RCRA
- Name, mailing address, and location, including latitude and longitude of the facility for which the application is submitted
- Up to four SIC codes which best reflect the principal products or services provided by the facility

- The operator's name, address, telephone number, ownership status, and status as federal, state, private, public, or other entity
- The name, address, and phone number of the owner of the facility
- Whether the facility is located on Indian lands
- An indication of whether the facility is new or existing and whether it is a first or revised application
- For existing facilities, (1) a scale drawing of the facility showing the location of all past, present, and future treatment, storage, and disposal areas; and (2) photographs of the facility clearly delineating all existing structures; existing treatment, storage, and disposal areas; and sites of future treatment, storage, and disposal areas
- A description of the processes to be used for treating, storing, and disposing of hazardous waste, and the design capacity of these items
- A specification of the hazardous wastes listed or designated under 40 CFR part 261 to be treated, stored, or disposed of at the facility, an estimate of the quantity of such wastes to be treated, stored, or disposed of annually, and a general description of the processes to be used for such wastes
- A listing of all permits or construction approvals received or applied for under any of the following programs:
  - Hazardous Waste Management program under RCRA
  - UIC program under the SWDA
  - NPDES program under the CWA
  - Prevention of Significant Deterioration (PSD) program under the Clean Air Act
  - Nonattainment program under the Clean Air Act
  - National Emission Standards for Hazardous Pollutants (NESHAPS) preconstruction approval under the Clean Air Act
  - Ocean dumping permits under the Marine Protection Research and Sanctuaries Act
  - Dredge or fill permits under section 404 of the CWA
  - Other relevant environmental permits, including state permits
  - A topographic map (or other map if a topographic map is unavailable) extending 1 mile beyond the property boundaries of the source
  - Depicting the facility and each of its intake and discharge structures
  - Each of its hazardous waste treatment, storage, or disposal facilities
  - Each well where fluids from the facility are injected underground
  - Those wells, springs, other surface water bodies, and drinking water wells listed in public records or otherwise known to the applicant within $1/4$ mile of the facility property boundary

- A brief description of the nature of the business
- For hazardous debris, a description of the debris category(ies) and contaminant category(ies) to be treated, stored, or disposed of at the facility

11.6.9 Contents of Part B of the Permit Application

11.6.9.1 General Requirements. Part B of the TSDF permit application consists of the general information requirements of this section, and the specific information requirements in 270.14 through 270.29 applicable to the TSDF facility. The part B information requirements presented in 270.14 through 270.29 reflect the standards promulgated in 40 CFR Part 264. These information requirements are necessary in order for EPA to determine compliance with the part 264 standards. If owners and operators of HWM facilities can demonstrate that the information prescribed in part B cannot be provided to the extent required, the director may make allowance for submission of such information on a case-by-case basis. Information required in part B must be submitted to the director and signed in accordance with requirements in 270.11. Certain technical data, such as design drawings and specifications, and engineering studies must be certified by a registered professional engineer. For postclosure permits, only the information specified in 270.28 is required in part B of the permit application.

11.6.9.2 General Information Requirements. The following information is required for all TSDF facilities, except as 264.1 provides otherwise:

- A general description of the facility.
- Chemical and physical analyses of the hazardous waste and hazardous debris to be handled at the facility. At a minimum, these analyses must contain all the information which must be known to treat, store, or dispose of the wastes properly in accordance with Part 264.
- A copy of the waste analysis plan required by 264.13(b) and, if applicable, 264.13(c).
- A description of the security procedures and equipment required by 264.14, or a justification demonstrating the reasons for requesting a waiver of this requirement.
- A copy of the general inspection schedule required by 264.15(b). Include, where applicable, as part of the inspection schedule, specific requirements in 264.174, 245.193(i), 264.195, 264.226, 264.254, 264.273, 264.303, 264.602, 264.1033, 264.1052, 264.1053, 264.1058, 264.1084, 264.1085, 264.1086, and 264.1088.
- A justification of any request for a waiver(s) of the preparedness and prevention requirements of Part 264, subpart C.

- A copy of the contingency plan required by Part 264, subpart D. Note: Include, where applicable, as part of the contingency plan, specific requirements in 264.227, 264.255, and 264.200.

- A description of procedures, structures, or equipment used at the facility to:

  - Prevent hazards in unloading operations (for example, ramps, special forklifts)
  - Prevent runoff from hazardous waste handling areas to other areas of the facility or environment, or to prevent flooding (for example, berms, dikes, trenches)
  - Prevent contamination of water supplies
  - Mitigate effects of equipment failure and power outages
  - Prevent undue exposure of personnel to hazardous waste (for example, protective clothing)
  - Prevent releases to atmosphere

- A description of precautions to prevent accidental ignition or reaction of ignitable, reactive, or incompatible wastes as required to demonstrate compliance with 264.17 including documentation demonstrating compliance with 264.17(c)

- Traffic pattern, estimated volume (number, types of vehicles) and control [for example, show turns across traffic lanes, and stacking lanes (if appropriate); describe access road surfacing and load bearing capacity; show traffic control signals]

- Facility location information

  In order to determine the applicability of the seismic standard [264.18(a)] the owner or operator of a new facility must identify the political jurisdiction (e.g., county, township, or election district) in which the facility is proposed to be located. [*Note:* If the county or election district is not listed in appendix VI of Part 264, no further information is required to demonstrate compliance with 264.18(a).]

  If the facility is proposed to be located in an area listed in appendix VI of Part 264, the owner or operator must demonstrate compliance with the seismic standard. This demonstration may be made using either published geologic data or data obtained from field investigations carried out by the applicant. The information provided must be of such quality to be acceptable to geologists experienced in identifying and evaluating seismic activity. The information submitted must show that either:

  No faults which have had displacement in Holocene time are present, or no lineations which suggest the presence of a fault (which have displacement in Holocene time) within 3000 feet of a facility are present, based on data from:

Published geologic studies

Aerial reconnaissance of the area within a 5-mile radius from the facility

An analysis of aerial photographs covering a 3000-foot radius of the facility

If needed to clarify the above data, a reconnaissance based on walking portions of the area within 3000 feet of the facility

If faults (to include lineations) which have had displacement in Holocene time are present within 3000 feet of a facility, no faults pass within 200 feet of the portions of the facility where treatment, storage, or disposal of hazardous waste will be conducted, based on data from a comprehensive geologic analysis of the site. Unless a site analysis is otherwise conclusive concerning the absence of faults within 200 feet of such portions of the facility, data must be obtained from a subsurface exploration (trenching) of the area within a distance no less than 200 feet from portions of the facility where treatment, storage, or disposal of hazardous waste will be conducted. Such trenching must be performed in a direction that is perpendicular to known faults (which have had displacement in Holocene time) passing within 3000 feet of the portions of the facility where treatment, storage, or disposal of hazardous waste will be conducted. Such investigation must document, with supporting maps and other analyses, the location of faults found. (*Note:* The *Guidance Manual for the Location Standards* provides greater detail on the content of each type of seismic investigation and the appropriate conditions under which each approach or a combination of approaches would be used.)

- Owners and operators of all facilities must provide an identification of whether the facility is located within a 100-year floodplain. This identification must indicate the source of data for such determination and include a copy of the relevant Federal Insurance Administration (FIA) flood map, if used, or the calculations and maps used where an FIA map is not available. Information must also be provided identifying the 100-year flood level and any other special flooding factors (e.g., wave action) which must be considered in designing, constructing, operating, or maintaining the facility to withstand washout from a 100-year flood. [*Note:* Where maps for the National Flood Insurance Program produced by the Federal Insurance Administration (FIA) of the Federal Emergency Management Agency are available, they will normally be determinative of whether a facility is located within or outside of the 100-year floodplain. However, where the FIA map excludes an area (usually areas of the floodplain less than 200 feet in width), these areas must be considered and a determination made as to whether they are in the 100-year floodplain. Where FIA maps are not available for a proposed facility location, the owner or operator must use equivalent mapping techniques to determine whether the facility is within the 100-year floodplain and if so located, what the 100-year flood elevation would be.]

- Owners and operators of facilities located in the 100-year floodplain must provide the following information:

  Engineering analysis to indicate the various hydrodynamic and hydrostatic forces expected to result at the site as a consequence of a 100-year flood

  Structural or other engineering studies showing the design of operational units (e.g., tanks, incinerators) and flood protection devices (e.g., floodwalls, dikes) at the facility and how these will prevent washout

  If applicable, a detailed description of procedures to be followed to remove hazardous waste to safety before the facility is flooded, including:

  Timing of such movement relative to flood levels, including estimated time to move the waste, to show that such movement can be completed before floodwaters reach the facility

  A description of the location(s) to which the waste will be moved and demonstration that those facilities will be eligible to receive hazardous waste in accordance with the regulations under Parts 270, 271, 124, and 264 through 266

  The planned procedures, equipment, and personnel to be used and the means to ensure that such resources will be available in time for use

  The potential for accidental discharges of the waste during movement

- Existing facilities *not* in compliance with 264.18(b) must provide a plan showing how the facility will be brought into compliance and a schedule for compliance.

- An outline of both the introductory and continuing training programs by owners or operators to prepare persons to operate or maintain the HWM facility in a safe manner as required to demonstrate compliance with 264.16. A brief description of how training will be designed to meet actual job tasks in accordance with requirements in 264.16(a)(3).

- A copy of the closure plan and, where applicable, the postclosure plan required by 264.112, 264.118, and 264.197. Include, where applicable, as part of the plans, specific requirements in 264.178, 264.197, 264.228, 264.258, 264.280, 264.310, 264.351, 264.601, and 264.603.

- For hazardous waste disposal units that have been closed, documentation that notices required under 264.119 have been filed.

- The most recent closure cost estimate for the facility prepared in accordance with 264.142 and a copy of the documentation required to demonstrate financial assurance under 264.143. For a new facility, a copy of the required documentation may be submitted 60 days prior to the initial receipt of hazardous wastes, if that is later than the submission of the part B.

- Where applicable, the most recent postclosure cost estimate for the facility pre-pared in accordance with 264.144 plus a copy of the documentation required to demonstrate financial assurance under 264.145. For a new facility, a copy of the required documentation may be submitted 60 days prior to the initial receipt of hazardous wastes, if that is later than the submission of the part B.

- Where applicable, a copy of the insurance policy or other documentation which comprises compliance with the requirements of 264.147. For a new facility, doc-umentation showing the amount of insurance meeting the specification of 264.147(a) and, if applicable, 264.147(b), that the owner or operator plans to have in effect before initial receipt of hazardous waste for treatment, storage, or disposal. A request for a variance in the amount of required coverage, for a new or existing facility, may be submitted as specified in 264.147(c).

- Where appropriate, proof of coverage by a state financial mechanism in com-pliance with 264.149 or 264.150.

- A topographic map showing a distance of 1000 feet around the facility at a scale of 2.5 centimeters (1 inch) equal to not more than 61.0 meters (200 feet). Contours must be shown on the map. The contour interval must be sufficient to clearly show the pattern of surface water flow in the vicinity of and from each operational unit of the facility, for example, contours with an interval of 1.5 meters (5 feet), if relief is greater than 6.1 meters (20 feet), or an interval of 0.6 meter (2 feet), if relief is less than 6.1 meters (20 feet). Owners and operators of HWM facilities located in mountainous areas should use large contour intervals to adequately show topographic profiles of facilities. The map must clearly show the following:

    Map scale and date

    100-year floodplain area

    Surface waters including intermittent streams

    Surrounding land uses (residential, commercial, agricultural, recreational)

    A wind rose (i.e., prevailing wind speed and direction)

    Orientation of the map (north arrow)

    Legal boundaries of the HWM facility site

    Access control (fences, gates)

    Injection and withdrawal wells both on-site and off-site

    Buildings; treatment, storage, or disposal operations; or other structure (recreation areas, runoff control systems, access and internal roads, storm, sanitary, and process sewerage systems, loading and unloading areas, fire control facilities, etc.)

Barriers for drainage or flood control

Location of operational units within the HWM facility site, where hazardous waste is (or will be) treated, stored, or disposed (include equipment cleanup areas). (*Note:* For large HWM facilities EPA will allow the use of other scales on a case-by-case basis.)

- Applicants may be required to submit such information as may be necessary to enable the regional administrator to carry out his or her duties under other federal laws as required in 270.3.

- For land disposal facilities, if a case-by-case extension has been approved under 268.5 or a petition has been approved under 268.6, a copy of the notice of approval for the extension or petition is required.

- A summary of the preapplication meeting, along with a list of attendees and their addresses, and copies of any written comments or materials submitted at the meeting, as required under 124.31(c).

11.6.9.3 Additional Information Requirements. The following additional information regarding protection of groundwater is required from owners or operators of hazardous waste facilities containing a regulated unit except as provided in 264.90(b):

- A summary of the groundwater monitoring data obtained during the interim status period under 265.90 through 265.94, where applicable.
- Identification of the uppermost aquifer and aquifers hydraulically interconnected beneath the facility property, including groundwater flow direction and rate, and the basis for such identification (i.e., the information obtained from hydrogeological investigations of the facility area).
- On the topographic map required under paragraph (b)(19) of 40 CFR, Part 270.14, a delineation of the waste management area, the property boundary, the proposed "point of compliance" as defined under 264.95, the proposed location of groundwater monitoring wells as required under 264.97, and to the extent possible, the information required in paragraph (c)(2) of 40 CFR, Part 270.14.
- A description of any plume of contamination that has entered the groundwater from a regulated unit at the time that the application was submitted that:

Delineates the extent of the plume on the topographic map required under paragraph (b)(19) of 40 CFR, Part 270.14

Identifies the concentration of each appendix IX, of Part 264, constituent throughout the plume or identifies the maximum concentrations of each appendix IX constituent in the plume

- Detailed plans and an engineering report describing the proposed groundwater monitoring program to be implemented to meet the requirements of 264.97.

- If the presence of hazardous constituents has not been detected in the groundwater at the time of permit application, the owner or operator must submit sufficient information, supporting data, and analyses to establish a detection monitoring program which meets the requirements of 264.98. This submission must address the following items specified under 264.98:

- A proposed list of indicator parameters, waste constituents, or reaction products that can provide a reliable indication of the presence of hazardous constituents in the groundwater

- A proposed groundwater monitoring system

- Background values for each proposed monitoring parameter or constituent, or procedures to calculate such values

- A description of proposed sampling, analysis, and statistical comparison procedures to be utilized in evaluating groundwater monitoring data

- If the presence of hazardous constituents has been detected in the groundwater at the point of compliance at the time of the permit application, the owner or operator must submit sufficient information, supporting data, and analyses to establish a compliance monitoring program which meets the requirements of 264.99. Except as provided in 264.98(h)(5), the owner or operator must also submit an engineering feasibility plan for a corrective action program necessary to meet the requirements of 264.100, unless the owner or operator obtains written authorization in advance from the regional administrator to submit a proposed permit schedule for submittal of such a plan. To demonstrate compliance with 264.99, the owner or operator must address the following items:

- A description of the wastes previously handled at the facility

- A characterization of the contaminated groundwater, including concentrations of hazardous constituents

- A list of hazardous constituents for which compliance monitoring will be undertaken in accordance with 264.97 and 264.99

- Proposed concentration limits for each hazardous constituent, based on the criteria set forth in 264.94(a), including a justification for establishing any alternate concentration limits

- Detailed plans and an engineering report describing the proposed groundwater monitoring system, in accordance with the requirements of 264.97

- A description of proposed sampling, analysis, and statistical comparison procedures to be utilized in evaluating groundwater monitoring data.

- If hazardous constituents have been measured in the groundwater which exceed the concentration limits established under 264.94 Table 1, or if groundwater monitoring conducted at the time of permit application under 265.90 through 265.94 at the waste boundary indicates the presence of hazardous constituents from the facility in groundwater over background concentrations, the owner or operator must submit sufficient information, supporting data, and analyses to establish a corrective action program which meets the requirements of 264.100. However, owners or operators are not required to submit information to establish a corrective action program if they demonstrate to the regional administrator that alternate concentration limits will protect human health and the environment after considering the criteria listed in 264.94(b). An owner or operator who is not required to establish a corrective action program for this reason must instead submit sufficient information to establish a compliance monitoring program which meets the requirements of 264.99 and paragraph (c)(6) of 40 CFR, Part 270.14. To demonstrate compliance with 264.100, the owner or operator must address, at a minimum, the following items:

   A characterization of the contaminated groundwater, including concentrations of hazardous constituents

   The concentration limit for each hazardous constituent found in the groundwater as set forth in 264.94

   Detailed plans and an engineering report describing the corrective action to be taken

   A description of how the groundwater monitoring program will demonstrate the adequacy of the corrective action

   The permit may contain a schedule for submittal of the information required in the two bullet points listed immediately above provided the owner or operator obtains written authorization from the regional administrator prior to submittal of the complete permit application.

11.6.9.4 Information Requirements for Solid Waste Management Units. The following information is required for each solid waste management unit at a facility seeking a permit:

   The location of the unit on the topographic map required under paragraph (b)(19) of 40 CFR, Part 270.14

Designation of type of unit

General dimensions and structural description (supply any available drawings)

When the unit was operated

Specification of all wastes that have been managed at the unit, to the extent available

The owner or operator of any facility containing one or more solid waste management units must submit all available information pertaining to any release of hazardous wastes or hazardous constituents from such unit or units.

The owner-operator must conduct and provide the results of sampling and analysis of groundwater, land surface, and subsurface strata, surface water, or air, which may include the installation of wells, where the director ascertains it is necessary to complete a RCRA facility assessment that will determine if a more complete investigation is necessary.

11.6.9.5 Specific Part B Information Requirements for Containers. Except as otherwise provided in 264.170, owners or operators of facilities that store containers of hazardous waste must provide the following additional information:

- A description of the containment system to demonstrate compliance with 264.175. Show at least the following:

    Basic design parameters, dimensions, and materials of construction

    How the design promotes drainage or how containers are kept from contact with standing liquids in the containment system

    Capacity of the containment system relative to the number and volume of containers to be stored

    Provisions for preventing or managing runon

    How accumulated liquids can be analyzed and removed to prevent overflow

    For storage areas that store containers holding wastes that do not contain free liquids, a demonstration of compliance with 264.175(c), including:

    Test procedures and results or other documentation or information to show that the wastes do not contain free liquids

    A description of how the storage area is designed or operated to drain and remove liquids or how containers are kept from contact with standing liquids

- Sketches, drawings, or data demonstrating compliance with 264.176 (location of buffer zone and containers holding ignitable or reactive wastes) and 264.177(c) (location of incompatible wastes), where applicable. Where incompatible wastes are stored or otherwise managed in containers, a description of the procedures used to ensure compliance with 264.177(a) and (b) and 264.17(b) and (c).

• Information on air emission control equipment as required in 270.27.

11.6.9.6 Specific Part B Information Requirements for Tank Systems.
Except as otherwise provided in 264.190, owners and operators of facilities that
use tanks to store or treat hazardous waste must provide the following additional
information:

• A written assessment that is reviewed and certified by an independent, qualified,
  registered professional engineer as to the structural integrity and suitability for
  handling hazardous waste of each tank system, as required under 264.191 and
  264.192

• Dimensions and capacity of each tank

• Description of feed systems, safety cutoff, bypass systems, and pressure controls (e.g., vents)

• A diagram of piping, instrumentation, and process flow for each tank system

• A description of materials and equipment used to provide external corrosion
  protection, as required under 264.192(a)(3)(ii)

• For new tank systems, a detailed description of how the tank system(s) will be
  installed in compliance with 264.192(b), (c), (d), and (e)

• Detailed plans and description of how the secondary containment system for
  each tank system is or will be designed, constructed, and operated to meet the
  requirements of 264.193(a), (b), (c), (d), (e), and (f)

• For tank systems for which a variance from the requirements of 264.193 is
  sought [as provided by 264.193(g)]:

  Detailed plans and engineering and hydrogeological reports, as appropriate,
  describing alternate design and operating practices that will, in conjunction
  with location aspects, prevent the migration of any hazardous waste or hazardous constituents into the groundwater or surface water during the life of
  the facility

  A detailed assessment of the substantial present or potential hazards posed
  to human health or the environment should a release enter the environment

• Description of controls and practices to prevent spills and overflows, as required
  under 264.194(b)

• For tank systems in which ignitable, reactive, or incompatible wastes are to be
  stored or treated, a description of how operating procedures and tank system and
  facility design will achieve compliance with the requirements of 264.198 and
  264.199.

• Information on air emission control equipment as required in 270.27.

11.6.9.7 Specific Part B Information Requirements for Surface Impoundments.   Except as otherwise provided in 264.1, owners and operators of facilities that store, treat, or dispose of hazardous waste in surface impoundments must provide the following additional information:

- A list of the hazardous wastes placed or to be placed in each surface impoundment

- Detailed plans and an engineering report describing how the surface impoundment is designed and is or will be constructed, operated, and maintained to meet the requirements of 264.19, 264.221, 264.222, and 264.223, addressing the following items:

- The liner system (except for an existing portion of a surface impoundment). If an exemption from the requirement for a liner is sought as provided by 264.221(b), submit detailed plans and engineering and hydrogeological reports, as appropriate, describing alternate design and operating practices that will, in conjunction with location aspects, prevent the migration of any hazardous constituents into the groundwater or surface water at any future time.

- The double liner and leak (leachate) detection, collection, and removal system, if the surface impoundment must meet the requirements of 264.221(c). If an exemption from the requirements for double liners and a leak detection, collection, and removal system or alternative design is sought as provided by 264.221(d), (e), or (f), submit appropriate information.

- If the leak detection system is located in a saturated zone, submit detailed plans and an engineering report explaining the leak detection system design and operation, and the location of the saturated zone in relation to the leak detection system.

- The construction quality assurance (CQA) plan if required under 264.19.

- Proposed action leakage rate, with rationale, if required under 264.222, and response action plan, if required under 264.223.

- Prevention of overtopping.

- Structural integrity of dikes.

- A description of how each surface impoundment, including the double liner system, leak detection system, cover system, and appurtenances for control of overtopping, will be inspected in order to meet the requirements of 264.226(a), (b), and (d). This information must be included in the inspection plan submitted under 270.14(b)(5).

- A certification by a qualified engineer which attests to the structural integrity of each dike, as required under 264.226(c). For new units, the owner or operator must submit a statement by a qualified engineer that he or she will provide such a certification upon completion of construction in accordance with the plans and specifications.

- A description of the procedure to be used for removing a surface impoundment from service, as required under 264.227(b) and (c). This information should be included in the contingency plan submitted under 270.14(b)(7).

- A description of how hazardous waste residues and contaminated materials will be removed from the unit at closure, as required under 264.228(a)(1). For any wastes not to be removed from the unit upon closure, the owner or operator must submit detailed plans and an engineering report describing how 264.228(a)(2) and (b) will be complied with. This information should be included in the closure plan and, where applicable, the postclosure plan submitted under 270.14(b)(13).

- If ignitable or reactive wastes are to be placed in a surface impoundment, an explanation of how 264.229 will be complied with.

- If incompatible wastes, or incompatible wastes and materials will be placed in a surface impoundment, an explanation of how 264.230 will be complied with.

- A waste management plan for EPA Hazardous Waste FO20, FO21, FO22, FO23, FO26, and FO27 describing how the surface impoundment is or will be designed, constructed, operated, and maintained to meet the requirements of 264.231. This submission must address the following four items as specified in 264.231:

  The volume, physical, and chemical characteristics of the wastes, including their potential to migrate through soil or to volatilize or escape into the atmosphere

  The attenuative properties of underlying and surrounding soils or other materials

  The mobilizing properties of other materials codisposed with these wastes

  The effectiveness of additional treatment, design, or monitoring techniques

- Information on air emission control equipment as required in 270.27.

11.6.9.8 Specific Part B Information Requirements for Waste Piles. Except as otherwise provided in 264.1, owners and operators of facilities that store or treat hazardous waste in waste piles must provide the following additional information:

- A list of hazardous wastes placed or to be placed in each waste pile [40 CFR 270.18(b)]

- If an exemption is sought to 264.251 and subpart F of part 264 as provided by 264.250(c) or 264.90(2), an explanation of how the standards of 264.250(c) will be complied with or detailed plans and an engineering report describing how the requirements of 264.90(b)(2) will be met

Detailed plans and an engineering report describing how the waste pile is designed and is or will be constructed, operated, and maintained to meet the requirements of 264.19, 264.251, 261.252, and 264.253, addressing the following items:

- The liner system (except for an existing portion of a waste pile), if the waste pile must meet the requirements of 264.251(a). If an exemption from the requirement for a liner is sought as provided by 264.251(b), submit detailed plans, and engineering and hydrogeological reports, as appropriate, describing alternate designs and operating practices that will, in conjunction with location aspects, prevent the migration of any hazardous constituents into the groundwater or surface water at any future time.

- The double liner and leak (leachate) detection, collection, and removal system, if the waste pile must meet the requirements of 264.251(c). If an exemption from the requirements for double liners and a leak detection, collection, and removal system or alternative design is sought as provided by 264.251(d), (e), or (f), submit appropriate information.

- If the leak detection system is located in a saturated zone, submit detailed plans and an engineering report explaining the leak detection system design and operation, and the location of the saturated zone in relation to the leak detection system.

- The construction quality assurance (CQA) plan if required under 264.19.

  Proposed action leakage rate, with rationale, if required under 264.252, and response action plan, if required under 264.253

  Control of runon

  Control of runoff

  Management of collection and holding units associated with runon and runoff control systems

  Control of wind dispersal of particulate matter, where applicable

- A description of how each waste pile, including the double liner system, leachate collection and removal system, leak detection system, cover system, and appurtenances for control of runon and runoff, will be inspected in order to meet the requirements of 264.254(a), (b), and (c). This information must be included in the inspection plan submitted under 270.14(b)(5).

- If treatment is carried out on or in the pile, details of the process and equipment used, and the nature and quality of the residuals.

- If ignitable or reactive wastes are to be placed in a waste pile, an explanation of how the requirements of 264.256 will be complied with.

- If incompatible wastes or incompatible wastes and materials will be placed in a waste pile, an explanation of how 264.257 will be complied with.

- A description of how hazardous waste residues and contaminated materials will be removed from the waste pile at closure, as required under 264.258(a). For any waste not to be removed from the waste pile upon closure, the owner or operator must submit detailed plans and an engineering report describing how 264.310(a) and (b) will be complied with. This information should be included in the closure plan and, where applicable, the postclosure plan submitted under 270.14(b)(13).

- A waste management plan for EPA Hazardous Waste FO20, FO21, FO22, FO23, FO26, and FO27 describing how a waste pile that is not enclosed [as defined in 264.250(c)] is or will be designed, constructed, operated, and maintained to meet the requirements of 264.259. This submission must address the following items as specified in 264.259:

  The volume, physical, and chemical characteristics of the wastes to be disposed in the waste pile, including their potential to migrate through soil or to volatilize or escape into the atmosphere

  The attenuative properties of underlying and surrounding soils or other materials

  The mobilizing properties of other materials codisposed with these wastes

  The effectiveness of additional treatment, design, or monitoring techniques

11.6.9.9 Specific Part B Information Requirements for Incinerators. Except as 264.340 provides otherwise, owners and operators of facilities that incinerate hazardous waste must fulfill the following requirements:

- When seeking an exemption under 264.340(b) or (c) (ignitable, corrosive, or reactive wastes only).

- Documentation that the waste is listed as a hazardous waste in Part 261, subpart D, solely because it is ignitable (Hazard Code I) or corrosive (Hazard Code C) or both.

- Documentation that the waste is listed as a hazardous waste in Part 261, subpart D, solely because it is reactive (Hazard Code R) for characteristics other than those listed in 261.23(a)(4) and (5), and will not be burned when other hazardous wastes are present in the combustion zone.

- Documentation that the waste is a hazardous waste solely because it possesses the characteristic of ignitability, corrosivity, or both, as determined by the tests for characteristics of hazardous waste under Part 261, subpart C.

- Documentation that the waste is a hazardous waste solely because it possesses the reactivity characteristics listed in 261.23(a)(1), (2), (3), (6), (7), or (8), and that it will not be burned when other hazardous wastes are present in the combustion zone.

- Submit a trial burn plan or the results of a trial burn, including all required determinations, in accordance with 270.62.

- In lieu of a trial burn, the applicant may submit the following information:

  An analysis of each waste or mixture of wastes to be burned including:

  Heat value of the waste in the form and composition in which it will be burned.

  Viscosity (if applicable), or description of physical form of the waste.

  An identification of any hazardous organic constituents listed in Part 261, appendix VIII, which are present in the waste to be burned, except that the applicant need not analyze for constituents listed in Part 261, appendix VIII, which would reasonably not be expected to be found in the waste. The constituents excluded from analysis must be identified and the basis for their exclusion stated. The waste analysis must rely on analytical techniques specified in "Test Methods for Evaluating Solid Waste, Physical/Chemical Methods," EPA Publication SW-846, as incorporated by reference in 260.11 and 270.6, or their equivalent.

  An approximate quantification of the hazardous constituents identified in the waste, within the precision produced by the analytical methods specified in "Test Methods for Evaluating Solid Waste, Physical/Chemical Methods," EPA Publication SW-846, as incorporated by reference in 260.11 and 270.6.

  A quantification of those hazardous constituents in the waste which may be designated as POHCs based on data submitted from other trial or operational burns which demonstrate compliance with the performance standards in 264.343.

  A detailed engineering description of the incinerator, including:

  Manufacturer's name and model number of incinerator

  Type of incinerator

  Linear dimension of incinerator unit including cross-sectional area of combustion chamber

  Description of auxiliary fuel system (type and feed)

  Capacity of prime mover

  Description of automatic waste feed cutoff system(s)

  Stack gas monitoring and pollution control monitoring system

  Nozzle and burner design

  Construction materials

  Location and description of temperature, pressure, and flow indicating devices and control devices

  A description and analysis of the waste to be burned compared with the waste for which data from operational or trial burns are provided to support the contention that a trial burn is not needed. The data should include those items listed in paragraph (c)(1) of 40 CFR, Part 270.19. This analysis should specify the POHCs which the applicant has identified in the waste for which a permit is sought, and any differences from the POHCs in the waste for which burn data are provided.

The design and operating conditions of the incinerator unit to be used, compared with that for which comparative burn data are available.

A description of the results submitted from any previously conducted trial burn(s) including:

Sampling and analysis techniques used to calculate performance standards in 264.343

Methods and results of monitoring temperatures, waste feed rates, carbon monoxide, and an appropriate indicator of combustion gas velocity (including a statement concerning the precision and accuracy of this measurement)

The expected incinerator operation information to demonstrate compliance with 264.343 and 264.345 including:

Expected carbon monoxide (CO) level in the stack exhaust gas

Waste feed rate

Combustion zone temperature

Indication of combustion gas velocity

Expected stack gas volume, flow rate, and temperature

Computed residence time for waste in the combustion zone

Expected hydrochloric acid removal efficiency

Expected fugitive emissions and their control procedures

Proposed waste feed cutoff limits based on the identified significant operating parameters

Such supplemental information as the director finds necessary to achieve the purposes of this paragraph.

Waste analysis data, including that submitted in paragraph (c)(1) of 40 CFR, Part 270.19, sufficient to allow the director to specify as permit principal organic hazardous constituents (permit POHCs) those constituents for which destruction and removal efficiencies will be required.

- The director will approve a permit application without a trial burn if he or she finds that:

The wastes are sufficiently similar

The incinerator units are sufficiently similar, and the data from other trial burns are adequate to specify (under 264.345) operating conditions that will ensure that the performance standards in 264.343 will be met by the incinerator.

11.6.9.10   Specific Part B Information Requirements for Land Treatment Facilities.   Except as otherwise provided in 264.1, owners and operators of facilities that use land treatment to dispose of hazardous waste must provide the following additional information:

- A description of plans to conduct a treatment demonstration as required under 264.272. The description must include the following information:

    The wastes for which the demonstration will be made and the potential hazardous constituents in the waste.

    The data sources to be used to make the demonstration (e.g., literature, laboratory data, field data, or operating data).

    Any specific laboratory or field test that will be conducted, including:

    The type of test (e.g., column leaching, degradation)

    Materials and methods, including analytical procedures

    Expected time for completion

    Characteristics of the unit that will be simulated in the demonstration, including treatment zone characteristics, climatic conditions, and operating practices

- A description of a land treatment program, as required under 264.271. This information must be submitted with the plans for the treatment demonstration, and updated following the treatment demonstration. The land treatment program must address the following items:

    The wastes to be land treated.

    Design measures and operating practices necessary to maximize treatment in accordance with 264.273(a) including:

    Waste application method and rate

    Measures to control soil pH

    Enhancement of microbial or chemical reactions

    Control of moisture content

    Provisions for unsaturated zone monitoring, including:

    Sampling equipment, procedures, and frequency

    Procedures for selecting sampling locations

    Analytical procedures

    Chain of custody control

Procedures for establishing background values

Statistical methods for interpreting results

The justification for any hazardous constituents recommended for selection as principal hazardous constituents, in accordance with the criteria for such selection in 264.278(a)

A list of hazardous constituents reasonably expected to be in, or derived from, the wastes to be land treated based on waste analysis performed pursuant to 264.13.

The proposed dimensions of the treatment zone.

- A description of how the unit is or will be designed, constructed, operated, and maintained in order to meet the requirements of 264.273. This submission must address the following items:

  Control of runon.

  Collection and control of runoff.

  Minimization of runoff of hazardous constituents from the treatment zone.

  Management of collection and holding facilities associated with runon and runoff control systems.

  Periodic inspection of the unit. This information should be included in the inspection plan submitted under 270.14(b)(5).

  Control of wind dispersal of particulate matter, if applicable.

- If food-chain crops are to be grown in or on the treatment zone of the land treatment unit, a description of how the demonstration required under 264.276(a) will be conducted including:

  Characteristics of the food-chain crop for which the demonstration will be made

  Characteristics of the waste, treatment zone, and waste application method and rate to be used in the demonstration

  Procedures for crop growth, sample collection, sample analysis, and data evaluation

  Characteristics of the comparison crop including the location and conditions under which it was or will be grown

- If food-chain crops are to be grown, and cadmium is present in the land-treated waste, a description of how the requirements of 264.276(b) will be complied with.

- A description of the vegetative cover to be applied to closed portions of the facility, and a plan for maintaining such cover during the postclosure care period, as required under 264.280(a)(8) and 264.280(c)(2). This information should be included in the closure plan and, where applicable, the postclosure care plan submitted under 270.14(b)(13).

- If ignitable or reactive wastes will be placed in or on the treatment zone, an explanation of how the requirements of 264.281 will be complied with.

- If incompatible wastes, or incompatible wastes and materials, will be placed in or on the same treatment zone, an explanation of how 264.282 will be complied with.

- A waste management plan for EPA Hazardous Waste FO20, FO21, FO22, FO23, FO26, and FO27 describing how a land treatment facility is or will be designed, constructed, operated, and maintained to meet the requirements of 264.283. This submission must address the following items as specified in 264.283:

  The volume, physical, and chemical characteristics of the wastes, including their potential to migrate through soil or to volatilize or escape into the atmosphere

  The attenuative properties of underlying and surrounding soils or other materials

  The mobilizing properties of other materials codisposed with these wastes

  The effectiveness of additional treatment, design, or monitoring techniques

11.6.9.11   Specific Part B Information Requirements for Landfills.   Except as otherwise provided in 264.1, owners and operators of facilities that dispose of hazardous waste in landfills must provide the following additional information:

- A list of the hazardous wastes placed or to be placed in each landfill or landfill cell.
- Detailed plans and an engineering report describing how the landfill is designed and is or will be constructed, operated, and maintained to meet the requirements of 264.19, 264.301, 264.302, and 264.303, addressing the following items:

  The liner system (except for an existing portion of a landfill), if the landfill must meet the requirements of 264.301(a). If an exemption from the requirement for a liner is sought as provided by 264.301(b), submit detailed plans, and engineering and hydrogeological reports, as appropriate, describing alternate designs and operating practices that will, in conjunction with location aspects, prevent the migration of any hazardous constituents into the groundwater or surface water at any future time.

  The double liner and leak (leachate) detection, collection, and removal system, if the landfill must meet the requirements of 264.301(c). If an exemption from the requirements for double liners and a leak detection, collection, and removal system or alternative design is sought as provided by 264.301(d), (e), or (f), submit appropriate information.

If the leak detection system is located in a saturated zone, submit detailed plans and an engineering report explaining the leak detection system design and operation, and the location of the saturated zone in relation to the leak detection system.

The construction quality assurance (CQA) plan if required under 264.19.

Proposed action leakage rate, with rationale, if required under 264.302, and response action plan, if required under 264.303.

Control of runon.

Control of runoff.

Management of collection and holding facilities associated with runon and runoff control systems.

Control of wind dispersal of particulate matter, where applicable.

- A description of how each landfill, including the double liner system, leachate collection and removal system, leak detection system, cover system, and appurtenances for control of runon and runoff, will be inspected in order to meet the requirements of 264.303(a), (b), and (c). This information must be included in the inspection plan submitted under 270.14(b)(5).

- A description of how each landfill, including the liner and cover systems, will be inspected in order to meet the requirements of 264.303(a) and (b). This information should be included in the inspection plan submitted under 270.14(b)(5).

- Detailed plans and an engineering report describing the final cover which will be applied to each landfill or landfill cell at closure in accordance with 264.310(a), and a description of how each landfill will be maintained and monitored after closure in accordance with 264.310(b). This information should be included in the closure and postclosure plans submitted under 270.14(b)(13).

- If ignitable or reactive wastes will be landfilled, an explanation of how the standards of 264.312 will be complied with.

- If incompatible wastes, or incompatible wastes and materials will be landfilled, an explanation of how 264.313 will be complied with.

- If bulk or noncontainerized liquid waste or wastes containing free liquids is to be landfilled prior to May 8, 1985, an explanation of how the requirements of 264.314(a) will be complied with.

- If containers of hazardous waste are to be landfilled, an explanation of how the requirements of 264.315 or 264.316, as applicable, will be complied with.

• A waste management plan for EPA Hazardous Waste FO20, FO21, FO22, FO23, FO26, and FO27 describing how a landfill is or will be designed, constructed, operated, and maintained to meet the requirements of 264.317. This submission must address the following items as specified in 264.317:

The volume, physical, and chemical characteristics of the wastes, including their potential to migrate through soil or to volatilize or escape into the atmosphere

The attenuative properties of underlying and surrounding soils or other materials

The mobilizing properties of other materials codisposed with these wastes

The effectiveness of additional treatment, design, or monitoring techniques

11.6.9.12 Specific Part B Information Requirements for Boilers and Industrial Furnaces Burning Hazardous Waste

• Trial burns:

General. Except as noted below, owners and operators that are subject to the standards to control organic emissions provided by 266.104, standards to control particulate matter provided by 266.105, standards to control metals emissions provided by 266.106, or standards to control hydrogen chloride or chlorine gas emissions provided by 266.107 must conduct a trial burn to demonstrate conformance with those standards and must submit a trial burn plan or the results of a trial burn, including all required determinations, in accordance with 270.66.

A trial burn to demonstrate conformance with a particular emission standard may be waived under provisions of 266.104 through 266.107 and paragraphs (a)(2) through (a)(5) of 40 CFR, part 270.22.

The owner or operator may submit data in lieu of a trial burn, as prescribed in paragraph (a)(6) of 40 CFR, part 270.22.

Waiver of trial burn for DRE:

Boilers operated under special operating requirements. When seeking to be permitted under 266.104(a)(4) and 266.110 that automatically waive the DRE trial burn, the owner or operator of a boiler must submit documentation that the boiler operates under the special operating requirements provided by 266.110.

Boilers and industrial furnaces burning low risk waste. When seeking to be permitted under the provisions for low risk waste provided by 266.104(a)(5) and 266.109(a) that waive the DRE trial burn, the owner or operator must submit:

Documentation that the device is operated in conformance with the requirements of 266.109(a)(1).

Results of analyses of each waste to be burned, documenting the concentrations of nonmetal compounds listed in appendix VIII of part 261, except for those constituents that would reasonably not be expected to be in the waste. The constituents excluded from analysis must be identified and the basis for their exclusion explained. The analysis must rely on analytical techniques specified in "Test Methods for Evaluating Solid Waste, Physical/Chemical Methods" (incorporated by reference, see 260.11).

Documentation of hazardous waste firing rates and calculations of reasonable, worst-case emission rates of each constituent identified in paragraph (a)(2)(ii)(B) of 40 CFR, Part 270.22 using procedures provided by 266.109(a)(2)(ii).

Results of emissions dispersion modeling for emissions identified in the above paragraph using modeling procedures prescribed by 266.106(h). The director will review the emission modeling conducted by the applicant to determine conformance with these procedures. The director will either approve the modeling or determine that alternate or supplementary modeling is appropriate.

Documentation that the maximum annual average ground level concentration of each constituent identified in the above paragraph of this section and quantified in conformance with this section does not exceed the allowable ambient level established in appendixes IV or V of Part 266. The acceptable ambient concentration for emitted constituents for which a specific reference air concentration has not been established in appendix IV or risk-specific dose has not been established in appendix V is 0.1 micrograms per cubic meter, as noted in the footnote to appendix IV.

Waiver of trial burn for metals: When seeking to be permitted under the Tier I (or adjusted Tier I) metals feed rate screening limits provided by 266.106(b) and (e) that control metals emissions without requiring a trial burn, the owner or operator must submit:

Documentation of the feed rate of hazardous waste, other fuels, and industrial furnace feedstocks

Documentation of the concentration of each metal controlled by 266.106(b) or (e) in the hazardous waste, other fuels, and industrial furnace feedstocks, and calculations of the total feed rate of each metal

Documentation of how the applicant will ensure that the Tier I feed rate screening limits provided by 266.106(b) or (e) will not be exceeded during the averaging period provided by that paragraph

Documentation to support the determination of the terrain-adjusted effective stack height, good engineering practice stack height, terrain type, and land use as provided by 266.106(b)(3) through (b)(5)

Documentation of compliance with the provisions of 266.106(b)(6), if applicable, for facilities with multiple stacks

Documentation that the facility does not fail the criteria provided by 266.106(b)(7) for eligibility to comply with the screening limits

Proposed sampling and metals analysis plan for the hazardous waste, other fuels, and industrial furnace feedstocks

Waiver of trial burn for particulate matter. When seeking to be permitted under the low-risk waste provisions of 266.109(b) which waives the particulate standard (and trial burn to demonstrate conformance with the particulate standard), applicants must submit documentation supporting conformance with paragraphs (a)(2)(ii) and (a)(3) of 40 CFR, part 270.22.

Waiver of trial burn for HCl and Cl_2. When seeking to be permitted under the Tier I (or adjusted Tier I) feed rate screening limits for total chloride and chlorine provided by 266.107(b)(1) and (e) that control emissions of hydrogen chloride (HCl) and chlorine gas (Cl_2) without requiring a trial burn, the owner or operator must submit:

Documentation of the feed rate of hazardous waste, other fuels, and industrial furnace feedstocks

Documentation of the levels of total chloride and chlorine in the hazardous waste, other fuels, and industrial furnace feedstocks, and calculations of the total feed rate of total chloride and chlorine

Documentation of how the applicant will ensure that the Tier I (or adjusted Tier I) feed rate screening limits provided by 266.107(b)(1) or (e) will not be exceeded during the averaging period provided by that paragraph

Documentation to support the determination of the terrain-adjusted effective stack height, good engineering practice stack height, terrain type, and land use as provided by 266.107(b)(3)

Documentation of compliance with the provisions of 266.107(b)(4), if applicable, for facilities with multiple stacks

Documentation that the facility does not fail the criteria provided by 266.107(b)(3) for eligibility to comply with the screening limits

Proposed sampling and analysis plan for total chloride and chlorine for the hazardous waste, other fuels, and industrial furnace feedstocks

Data in lieu of trial burn. The owner or operator may seek an exemption from the trial burn requirements to demonstrate conformance with 266.104 through 266.107 and 270.66 by providing the information required by 270.66 from previous compliance testing of the device in conformance with 266.103, or from compliance testing or trial or operational burns of similar boilers or industrial furnaces burning similar hazardous wastes under similar conditions. If data from a similar device are used to support a trial burn waiver, the design and operating information required by 270.66 must be provided for both the similar device and the device to which the data are to be applied, and a comparison of the design and operating information must be provided. The director will approve a permit application without a trial burn if he or she finds that the hazardous wastes are sufficiently similar, the devices are sufficiently similar, the operating conditions are sufficiently similar, and the data from other compliance tests, trial burns, or operational burns are adequate to specify (under 266.102) operating conditions that will ensure conformance with 266.102(c). In addition, the following information must be submitted:

For a waiver from any trial burn:

A description and analysis of the hazardous waste to be burned compared with the hazardous waste for which data from compliance testing, or operational or trial burns are provided to support the contention that a trial burn is not needed

The design and operating conditions of the boiler or industrial furnace to be used, compared with that for which comparative burn data are available

Such supplemental information as the director finds necessary to achieve the purposes of this paragraph

For a waiver of the DRE trial burn, the basis for selection of POHCs used in the other trial or operational burns which demonstrate compliance with the DRE performance standard in 266.104(a). This analysis should specify the constituents in appendix VIII, Part 261, that the applicant has identified in the hazardous waste for which a permit is sought, and any differences from the POHCs in the hazardous waste for which burn data are provided.

- Alternative HC limit for industrial furnaces with organic matter in raw materials. Owners and operators of industrial furnaces requesting an alternative HC limit under 266.104(f) must submit the following information at a minimum:

Documentation that the furnace is designed and operated to minimize HC emissions from fuels and raw materials

Documentation of the proposed baseline flue gas HC (and CO) concentration, including data on HC (and CO) levels during tests when the facility produced normal products under normal operating conditions from normal raw materials while burning normal fuels and when not burning hazardous waste

Test burn protocol to confirm the baseline HC (and CO) level including information on the type and flow rate of all feed streams, point of introduction of all feed streams, total organic carbon content (or other appropriate measure of organic content) of all nonfuel feed streams, and operating conditions that affect combustion of fuel(s) and destruction of hydrocarbon emissions from nonfuel sources

Trial burn plan to:

Demonstrate that flue gas HC (and CO) concentrations when burning hazardous waste do not exceed the baseline HC (and CO) level

Identify the types and concentrations of organic compounds listed in appendix VIII, part 261, that are emitted when burning hazardous waste in conformance with procedures prescribed by the director

Implementation plan to monitor over time changes in the operation of the facility that could reduce the baseline HC level and procedures to periodically confirm the baseline HC level

Such other information as the director finds necessary to achieve the purposes of this paragraph

- Alternative metals implementation approach. When seeking to be permitted under an alternative metals implementation approach under 266.106(f), the owner or operator must submit documentation specifying how the approach ensures compliance with the metals emissions standards of 266.106(c) or (d) and how the approach can be effectively implemented and monitored. Further, the owner or operator must provide such other information that the director finds necessary to achieve the purposes of this paragraph.

- Automatic waste feed cutoff system. Owners and operators must submit information describing the automatic waste feed cutoff system, including any pre-alarm systems that may be used.

- Direct transfer. Owners and operators that use direct transfer operations to feed hazardous waste from transport vehicles (containers, as defined in 266.111) directly to the boiler or industrial furnace must submit information supporting conformance with the standards for direct transfer provided by 266.111.

- Residues. Owners and operators that claim that their residues are excluded from regulation under the provisions of 266.112 must submit information adequate to demonstrate conformance with those provisions.

11.6.9.13 Specific Part B Information Requirements for Miscellaneous Units. Except as otherwise provided in 264.600, owners and operators of facilities that treat, store, or dispose of hazardous waste in miscellaneous units must provide the following additional information:

- A detailed description of the unit being used or proposed for use, including the following:

Physical characteristics, materials of construction, and dimensions of the unit

Detailed plans and engineering reports describing how the unit will be located, designed, constructed, operated, maintained, monitored, inspected, and closed to comply with the requirements of 264.601 and 264.602

For disposal units, a detailed description of the plans to comply with the postclosure requirements of 264.603

- Detailed hydrologic, geologic, and meteorological assessments and land use maps for the region surrounding the site that address and ensure compliance of the unit with each factor in the environmental performance standards of 264.601. If applicants can demonstrate that they do not violate the environmental performance standards of 264.601 and the director agrees with such demonstration, preliminary hydrologic, geologic, and meteorological assessments will suffice.

- Information on the potential pathways of exposure of humans or environmental receptors to hazardous waste or hazardous constituents and on the potential magnitude and nature of such exposures.

- For any treatment unit, a report on a demonstration of the effectiveness of the treatment based on laboratory or field data.

- Any additional information determined by the director to be necessary for evaluation of compliance of the unit with the environmental performance standards of 264.601.

11.6.9.14 Specific Part B Information Requirements for Process Vents. Except as otherwise provided in 264.1, owners and operators of facilities that have process vents to which subpart AA of Part 264 applies must provide the following additional information:

- For facilities that cannot install a closed-vent system and control device to comply with the provisions of 40 CFR 264 subpart AA on the effective date that the facility becomes subject to the provisions of 40 CFR 264 or 265 subpart AA, an implementation schedule as specified in 264.1033(a)(2).

- Documentation of compliance with the process vent standards in 264.1032, including:

  Information and data identifying all affected process vents, annual throughput and operating hours of each affected unit, estimated emission rates for each affected vent and for the overall facility (i.e., the total emissions for all affected vents at the facility), and the approximate location within the facility of each affected unit (e.g., identify the hazardous waste management units on a facility plot plan).

Information and data supporting estimates of vent emissions and emission reduction achieved by add-on control devices based on engineering calculations or source tests. For the purpose of determining compliance, estimates of vent emissions and emission reductions must be made using operating parameter values (e.g., temperatures, flow rates, or concentrations) that represent the conditions that exist when the waste management unit is operating at the highest load or capacity level reasonably expected to occur.

Information and data used to determine whether or not a process vent is subject to the requirements of 264.1032.

- Where an owner or operator applies for permission to use a control device other than a thermal vapor incinerator, catalytic vapor incinerator, flare, boiler, process heater, condenser, or carbon adsorption system to comply with the requirements of 264.1032, and chooses to use test data to determine the organic removal efficiency or the total organic compound concentration achieved by the control device, a performance test plan as specified in 264.1035(b)(3).
- Documentation of compliance with 264.1033, including:

A list of all information references and sources used in preparing the documentation.

Records, including the dates, of each compliance test required by 264.1033(k).

A design analysis, specifications, drawings, schematics, and piping and instrumentation diagrams based on the appropriate sections of "APTI Course 415: Control of Gaseous Emissions" (incorporated by reference as specified in 260.11) or other engineering texts acceptable to the regional administrator that present basic control device design information. The design analysis must address the vent stream characteristics and control device operation parameters as specified above.

A statement signed and dated by the owner or operator certifying that the operating parameters used in the design analysis reasonably represent the conditions that exist when the hazardous waste management unit is or would be operating at the highest load or capacity level reasonably expected to occur.

A statement signed and dated by the owner or operator certifying that the control device is designed to operate at an efficiency of 95 weight percent or greater unless the total organic emission limits of 264.1032(a) for affected process vents at the facility can be attained by a control device involving vapor recovery at an efficiency less than 95 weight percent.

11.6.9.15   Specific Part B Information Requirements for Equipment.   Except as otherwise provided in 264.1, owners and operators of facilities that have equipment to which subpart BB of Part 264 applies must provide the following additional information:

- For each piece of equipment to which subPart BB of Part 264 applies:

  Equipment identification number and hazardous waste management unit identification

  Approximate locations within the facility (e.g., identify the hazardous waste management unit on a facility plot plan)

  Type of equipment (e.g., a pump or pipeline valve)

  Percent by weight total organics in the hazardous waste stream at the equipment

  Hazardous waste state at the equipment (e.g., gas and vapor or liquid)

  Method of compliance with the standard (e.g., "monthly leak detection and repair" or "equipped with dual mechanical seals")

- For facilities that cannot install a closed-vent system and control device to comply with the provisions of 40 CFR 264 subpart BB on the effective date that the facility becomes subject to the provisions of 40 CFR 264 or 265 subpart BB, an implementation schedule as specified in 264.1033(a)(2).

- Where an owner or operator applies for permission to use a control device other than a thermal vapor incinerator, catalytic vapor incinerator, flare, boiler, process heater, condenser, or carbon adsorption system and chooses to use test data to determine the organic removal efficiency or the total organic compound concentration achieved by the control device, a performance test plan as specified in 264.1035(b)(3).

- Documentation that demonstrates compliance with the equipment standards in 264.1052 to 264.1059. This documentation must contain the records required under 264.1064. The regional administrator may request further documentation before deciding if compliance has been demonstrated.

- Documentation to demonstrate compliance with 264.1060 must include the following information:

  A list of all information references and sources used in preparing the documentation.

  Records, including the dates, of each compliance test required by 264.1033(j).

  A design analysis, specifications, drawings, schematics, and piping and instrumentation diagrams based on the appropriate sections of "ATPI Course 415: Control of Gaseous Emissions" (incorporated by reference as specified in 260.11) or other engineering texts acceptable to the regional administrator that present basic control device design information. The design analysis must address the vent stream characteristics and control device operation parameters as specified in 264.1035(b)(4)(iii).

A statement signed and dated by the owner or operator certifying that the operating parameters used in the design analysis reasonably represent the conditions that exist when the hazardous waste management unit is operating at the highest load or capacity level reasonably expected to occur.

- A statement signed and dated by the owner or operator certifying that the control device is designed to operate at an efficiency of 95 weight percent or greater.

11.6.9.16 Specific Part B Information Requirements for Drip Pads.   Except as otherwise provided by 264.1, as amended, owners and operators of hazardous waste treatment, storage, or disposal facilities that collect, store, or treat hazardous waste on drip pads must provide the following additional information:

- A list of hazardous wastes placed or to be placed on each drip pad.
- If an exemption is sought to subpart F of Part 264, as provided by 264.90, detailed plans and an engineering report describing how the requirements of 264.90(b)(2) will be met.
- Detailed plans and an engineering report describing how the drip pad is or will be designed, constructed, operated, and maintained to meet the requirements of 264.573, including the as-built drawings and specifications. This submission must address the following items as specified in 264.571:

    The design characteristics of the drip pad

    The liner system

    The leakage detection system, including the leak detection system and how it is designed to detect the failure of the drip pad or the presence of any releases of hazardous waste or accumulated liquid at the earliest practicable time

    Practices designed to maintain drip pads

    The associated collection system

    Control of runon to the drip pad

    Control of runoff from the drip pad

    The interval at which drippage and other materials will be removed from the associated collection system and a statement demonstrating that the interval will be sufficient to prevent overflow onto the drip pad

    Procedures for cleaning the drip pad at least once every 7 days to ensure the removal of any accumulated residues of waste or other materials, including but not limited to rinsing, washing with detergents or other appropriate solvents, or steam cleaning and provisions for documenting the date, time, and cleaning procedure used each time the pad is cleaned

Operating practices and procedures that will be followed to ensure that tracking of hazardous waste or waste constituents off the drip pad due to activities by personnel or equipment is minimized

Procedures for ensuring that, after removal from the treatment vessel, treated wood from pressure and nonpressure processes is held on the drip pad until drippage has ceased, including record-keeping practices

Provisions for ensuring that collection and holding units associated with the runon and runoff control systems are emptied or otherwise managed as soon as possible after storms to maintain design capacity of the system

If treatment is carried out on the drip pad, details of the process equipment used, and the nature and quality of the residuals

A description of how each drip pad, including appurtenances for control of runon and runoff, will be inspected in order to meet the requirements of 264.573. This information should be included in the inspection plan submitted under 270.14(b)(5)

A certification signed by an independent qualified, registered professional engineer, stating that the drip pad design meets the requirements of paragraphs (a) through (f) of 264.573

A description of how hazardous waste residues and contaminated materials will be removed from the drip pad at closure, as required under 264.575(a). For any waste not to be removed from the drip pad upon closure, the owner or operator must submit detailed plans and an engineering report describing how 264.310(a) and (b) will be complied with. This information should be included in the closure plan and, where applicable, the postclosure plan submitted under 270.14(b)(13)

11.6.9.17 Specific Part B Information Requirements for Air Emission Controls for Tanks, Surface Impoundments, and Containers

- Except as otherwise provided in 40 CFR 264.1, owners and operators of tanks, surface impoundments, or containers that use air emission controls in accordance with the requirements of 40 CFR Part 264, subpart CC must provide the following additional information:

  Documentation for each floating roof cover installed on a tank subject to 40 CFR 264.1084(d)(1) or 40 CFR 264.1084(d)(2) that includes information prepared by the owner or operator or provided by the cover manufacturer or vendor describing the cover design, and certification by the owner or operator that the cover meets the applicable design specifications as listed in 40 CFR 264.1084(e)(1) or 40 CFR 264.1084(f)(1).

Identification of each container area subject to the requirements of 40 CFR Part 264, subpart CC and certification by the owner or operator that the requirements of this subpart are met.

Documentation for each enclosure used to control air pollutant emissions from tanks or containers in accordance with the requirements of 40 CFR 264.1084(d)(5) or 40 CFR 264.1086(e)(1)(ii) that includes records for the most recent set of calculations and measurements performed by the owner or operator to verify that the enclosure meets the criteria of a permanent total enclosure as specified in "Procedure T—Criteria for and Verification of a Permanent or Temporary Total Enclosure" under 40 CFR 52.741, appendix B.

Documentation for each floating membrane cover installed on a surface impoundment in accordance with the requirements of 40 CFR 264.1085(c) that includes information prepared by the owner or operator or provided by the cover manufacturer or vendor describing the cover design, and certification by the owner or operator that the cover meets the specifications listed in 40 CFR 264.1085(c)(1).

Documentation for each closed-vent system and control device installed in accordance with the requirements of 40 CFR 264.1087 that includes design and performance information as specified in 270.24(c) and (d).

An emission monitoring plan for both Method 21 in 40 CFR Part 60, appendix A and control device monitoring methods. This plan must include the following information: monitoring point(s), monitoring methods for control devices, monitoring frequency, procedures for documenting exceedances, and procedures for mitigating noncompliances.

When an owner or operator of a facility subject to 40 CFR Part 265, subpart CC cannot comply with 40 CFR Part 264, subpart CC by the date of permit issuance, the schedule of implementation required under 40 CFR 265.1082.

11.6.9.18 Specific Part B Information Requirements for Postclosure Permits. For postclosure permits, the owner or operator is required to submit only the information specified in 270.14(b)(1), (4), (5), (6), (11), (13), (14), (16), (18), and (19), (c), and (d), unless the regional administrator determines that additional information from 270.14, 270.16, 270.17, 270.18, 270.20, or 270.21 is necessary. The owner or operator is required to submit the same information when an alternative authority is used in lieu of a postclosure permit as provided in 270.1(c)(7).

11.6.10 Changes to TSDF Permits

11.6.10.1 Transfer of TSDF Permits

- A permit may be transferred by the permittee to a new owner or operator only if the permit has been modified or revoked and reissued [under 270.40(b) or

270.41(b)(2)] to identify the new permittee and incorporate such other requirements as may be necessary under the appropriate act.

- Changes in the ownership or operational control of a facility may be made as a Class 1 modification with prior written approval of the director in accordance with 270.42. The new owner or operator must submit a revised permit application no later than 90 days prior to the scheduled change. A written agreement containing a specific date for transfer of permit responsibility between the current and new permittees must also be submitted to the director. When a transfer of ownership or operational control occurs, the old owner or operator must comply with the requirements of 40 CFR Part 264, subpart H (Financial Requirements) until the new owner or operator has demonstrated that he or she is complying with the requirements of that subpart. The new owner or operator must demonstrate compliance with subpart H requirements within 6 months of the date of the change of ownership or operational control of the facility. Upon demonstration to the director by the new owner or operator of compliance with subpart H, the director will notify the old owner or operator that he or she no longer needs to comply with subpart H as of the date of demonstration.

11.6.10.2 Modification or Revocation and Reissuance of Permits. When the director receives any information [for example, inspects the facility, receives information submitted by the permittee as required in the permit (see 270.30), receives a request for revocation and reissuance under 124.5, or conducts a review of the permit file], he or she may determine whether one or more of the causes listed in paragraphs (a) and (b) of 40 CFR, Part 270.41, for modification, or revocation and reissuance or both exist. If cause exists, the director may modify or revoke and reissue the permit accordingly, subject to the limitations of paragraph (c) of 40 CFR, Part 270.41, and may request an updated application if necessary. When a permit is modified, only the conditions subject to modification are reopened. If a permit is revoked and reissued, the entire permit is reopened and subject to revision and the permit is reissued for a new term [see 40 CFR 124.5(c)(2)]. If cause does not exist under this section, the director cannot modify or revoke and reissue the permit, except on request of the permittee. If a permit modification is requested by the permittee, the director can approve or deny the request according to the procedures of 40 CFR 270.42. Otherwise, a draft permit must be prepared and other procedures in Part 124 (or procedures of an authorized state program) followed.

- Causes for modification. The following are causes for modification, but not revocation and reissuance, of permits; the following may be causes for revocation and reissuance, as well as modification, when the permittee requests or agrees.

  Alterations. These are material and substantial alterations or additions to the permitted facility or activity which occurred after permit issuance which justify the application of permit conditions that are different or absent in the existing permit.

Information. The director has received new information. Permits may be modified during their terms for this cause only if the information was not available at the time of permit issuance (other than revised regulations, guidance, or test methods) and would have justified the application of different permit conditions at the time of issuance.

New statutory requirements or regulations. The standards or regulations on which the permit was based have been changed by statute, through promulgation of new or amended standards or regulations, or by judicial decision after the permit was issued.

Compliance schedules. The director determines good cause exists for modification of a compliance schedule, such as an act of God, strike, flood, or materials shortage or other events over which the permittee has little or no control and for which there is no reasonably available remedy.

Notwithstanding any other provision in this section, when a permit for a land disposal facility is reviewed by the director under 270.50(d), the director can modify the permit as necessary to assure that the facility continues to comply with the currently applicable requirements in Parts 124, 260 through 266, and 270.

- Causes for modification or revocation and reissuance. The following are causes to modify or, alternatively, revoke and reissue a permit:

  Cause exists for termination under 270.43, and the director determines that modification or revocation and reissuance is appropriate.

  The director has received notification [as required in the permit, see 270.30(l)(3)] of a proposed transfer of the permit.

- Facility siting. Suitability of the facility location will not be considered at the time of permit modification or revocation and reissuance unless new information or standards indicate that a threat to human health or the environment exists which was unknown at the time of permit issuance.

11.6.10.3   Permit Modification at the Request of the Permittee.   (*Note:* Refer to the attached Appendix I to 40 CFR 270.42 for a listing of Class 1, Class 2, and Class 3 TSDF permit modifications found in Appendix O in the back of this book.)

- Class 1 modifications.

  Except as provided in paragraph (a)(2) of 40 CFR, Part 270.42, the permittee may put into effect Class 1 modifications listed in appendix I of 40 CFR, Part 270.42, under the following conditions:

The permittee must notify the director concerning the modification by certified mail or other means that establish proof of delivery within 7 calendar days after the change is put into effect. This notice must specify the changes being made to permit conditions or supporting documents referenced by the permit and must explain why they are necessary. Along with the notice, the permittee must provide the applicable information required by 270.13 through 270.21, 270.62, and 270.63.

The permittee must send a notice of the modification to all persons on the facility mailing list, maintained by the director in accordance with 40 CFR 124.10(c)(viii), and the appropriate units of state and local government, as specified in 40 CFR 124.10(c)(ix). This notification must be made within 90 calendar days after the change is put into effect. For the Class I modifications that require prior director approval, the notification must be made within 90 calendar days after the director approves the request.

Any person may request the director to review, and the director may for cause reject, any Class 1 modification. The director must inform the permittee by certified mail that a Class 1 modification has been rejected, explaining the reasons for the rejection. If a Class 1 modification has been rejected, the permittee must comply with the original permit conditions.

Class 1 permit modifications identified in appendix I by an asterisk may be made only with the prior written approval of the director.

For a Class 1 permit modification, the permittee may elect to follow the procedures in 270.42(b) for Class 2 modifications instead of the Class 1 procedures. The permittee must inform the director of this decision in the notice required in 270.42(b)(1).

- Class 2 modifications.

For Class 2 modifications, listed in Appendix I of 40 CFR, Part 270.42, the permittee must submit a modification request to the director that:

Describes the exact change to be made to the permit conditions and supporting documents referenced by the permit

Identifies that the modification is a Class 2 modification

Explains why the modification is needed

Provides the applicable information required by 270.13 through 270.21, 270.62, and 270.63

The permittee must send a notice of the modification request to all persons on the facility mailing list maintained by the director and to the appropriate units of state and local government as specified in 40 CFR 124.10(c)(ix) and must publish this notice in a major local newspaper of general circulation.

This notice must be mailed and published within 7 days before or after the date of submission of the modification request, and the permittee must provide to the director evidence of the mailing and publication. The notice must include:

Announcement of a 60-day comment period, in accordance with 270.42(b)(5), and the name and address of an agency contact to whom comments must be sent

Announcement of the date, time, and place for a public meeting held in accordance with 270.42(b)(4)

Name and telephone number of the permittee's contact person

Name and telephone number of an agency contact person

Location where copies of the modification request and any supporting documents can be viewed and copied

The following statement: "The permittee's compliance history during the life of the permit being modified is available from EPA contact person."

The permittee must place a copy of the permit modification request and supporting documents in a location accessible to the public in the vicinity of the permitted facility.

The permittee must hold a public meeting no earlier than 15 days after the publication of the notice required in paragraph (b)(2) of 40 CFR, part 270.42, and no later than 15 days before the close of the 60-day comment period. The meeting must be held to the extent practicable in the vicinity of the permitted facility.

The public must be provided 60 days to comment on the modification request. The comment period will begin on the date the permittee publishes the notice in the local newspaper. Comments should be submitted to the EPA contact identified in the public notice.

No later than 90 days after receipt of the notification request, the director must:

Approve the modification request, with or without changes, and modify the permit accordingly

Deny the request

Determine that the modification request must follow the procedures in 270.42(c) for Class 3 modifications for the following reasons:

There is significant public concern about the proposed modification

The complex nature of the change requires the more extensive procedures of Class 3

Approve the request, with or without changes, as a temporary authorization having a term of up to 180 days

Notify the permittee that he or she will decide on the request within the next 30 days

If the director notifies the permittee of a 30-day extension for a decision, the director must, no later than 120 days after receipt of the modification:

Approve the modification request, with or without changes, and modify the permit accordingly

Deny the request

Determine that the modification request must follow the procedures in 270.42(c) for Class 3 modifications for the following reasons:

There is significant public concern about the proposed modification

The complex nature of the change requires the more extensive procedures of Class 3

Approve the request, with or without changes, as a temporary authorization having a term of up to 180 days

If the director fails to make one of the decisions specified in paragraph (b)(6)(ii) of 40 CFR, Part 270.42 by the 120th day after receipt of the modification request, the permittee is automatically authorized to conduct the activities described in the modification request for up to 180 days, without formal agency action. The authorized activities must be conducted as described in the permit modification request and must be in compliance with all appropriate standards of 40 CFR, Part 265. If the director approves, with or without changes, or denies the modification request during the term of the temporary or automatic authorization provided for in paragraphs (b)(6)(i), (ii), or (iii) of 40 CFR, Part 270.42, such action cancels the temporary or automatic authorization.

In the case of an automatic authorization under paragraph (b)(6)(iii) of 40 CFR, Part 270.42, or a temporary authorization under paragraph (b)(6) (i)(D) or (ii)(D) of 40 CFR, Part 270.42, if the director has not made a final approval or denial of the modification request by the date 50 days prior to the end of the temporary or automatic authorization, the permittee must within 7 days of that time send a notification to persons on the facility mailing list, and make a reasonable effort to notify other persons who submitted written comments on the modification request, that:

The permittee has been authorized temporarily to conduct the activities described in the permit modification request

Unless the director acts to give final approval or denial of the request by the end of the authorization period, the permittee will receive authorization to conduct such activities for the life of the permit

If the owner-operator fails to notify the public by the date specified in paragraph (b)(6)(iv)(A) of 40 CFR, Part 270.42, the effective date of the permanent authorization will be deferred until 50 days after the owner-operator notifies the public.

Except as provided in paragraph (b)(6)(vii) of 40 CFR, Part 270.42, if the director does not finally approve or deny a modification request before the end of the automatic or temporary authorization period or reclassify the modification as a Class 3, the permittee is authorized to conduct the activities described in the permit modification request for the life of the permit unless modified later under 270.41 or 270.42. The activities authorized under this paragraph must be conducted as described in the permit modification request and must be in compliance with all appropriate standards of 40 CFR, Part 265.

In making a decision to approve or deny a modification request, including a decision to issue a temporary authorization or to reclassify a modification as a Class 3, the director must consider all written comments submitted to the EPA during the public comment period and must respond in writing to all significant comments in his or her decision.

With the written consent of the permittee, the director may extend indefinitely or for a specified period the time periods for final approval or denial of a modification request or for reclassifying a modification as a Class 3.

The director may deny or change the terms of a Class 2 permit modification request under paragraphs (b)(6)(i) through (iii) of 40 CFR, Part 270.42, for the following reasons:

The modification request is incomplete

The requested modification does not comply with the appropriate requirements of 40 CFR part 264 or other applicable requirements

The conditions of the modification fail to protect human health and the environment

The permittee may perform any construction associated with a Class 2 permit modification request beginning 60 days after the submission of the request unless the director establishes a later date for commencing construction and informs the permittee in writing before day 60.

- Class 3 modifications.

For Class 3 modifications listed in the attached appendix I of 40 CFR, part 270.42, the permittee must submit a modification request to the director that:

Describes the exact change to be made to the permit conditions and supporting documents referenced by the permit

Identifies that the modification is a Class 3 modification

Explains why the modification is needed

Provides the applicable information required by 40 CFR 270.13 through 270.22, 270.62, 270.63, and 270.66

The permittee must send a notice of the modification request to all persons on the facility mailing list maintained by the director and to the appropriate units of state and local government as specified in 40 CFR 124.10(c)(ix) and must publish this notice in a major local newspaper of general circulation. This notice must be mailed and published within 7 days before or after the date of submission of the modification request, and the permittee must provide to the director evidence of the mailing and publication. The notice must include:

Announcement of a 60-day comment period, and a name and address of an agency contact to whom comments must be sent

Announcement of the date, time, and place for a public meeting on the modification request, in accordance with 270.42(c)(4)

Name and telephone number of the permittee's contact person

Name and telephone number of an agency contact person

Location where copies of the modification request and any supporting documents can be viewed and copied

The following statement: "The permittee's compliance history during the life of the permit being modified is available from the EPA contact person."

The permittee must place a copy of the permit modification request and supporting documents in a location accessible to the public in the vicinity of the permitted facility.

The permittee must hold a public meeting no earlier than 15 days after the publication of the notice required in paragraph (c)(2) of 40 CFR, part 270.42, and no later than 15 days before the close of the 60-day comment period. The meeting must be held to the extent practicable in the vicinity of the permitted facility.

The public must be provided at least 60 days to comment on the modification request. The comment period will begin on the date the permittee publishes the notice in the local newspaper. Comments should be submitted to the EPA contact identified in the notice.

After the conclusion of the 60-day comment period, the director must grant or deny the permit modification request according to the permit modification procedures of 40 CFR, Part 124. In addition, the director must consider and respond to all significant written comments received during the 60-day comment period.

- Other modifications.

In the case of modifications not explicitly listed in the attached appendix I of 40 CFR, Part 270.42, the permittee may submit a Class 3 modification request to the EPA, or he or she may request a determination by the director that the modification should be reviewed and approved as a Class 1 or Class 2 modification. If the permittee requests that the modification be classified as a Class 1 or 2 modification, he or she must provide the EPA with the necessary information to support the requested classification.

The director must make the determination described in paragraph (d)(1) of 40 CFR, Part 270.42, as promptly as practicable. In determining the appropriate class for a specific modification, the director must consider the similarity of the modification to other modifications codified in Appendix I and the following criteria:

Class 1 modifications apply to minor changes that keep the permit current with routine changes to the facility or its operation. These changes do not substantially alter the permit conditions or reduce the capacity of the facility to protect human health or the environment. In the case of Class 1 modifications, the director may require prior approval.

Class 2 modifications apply to changes that are necessary to enable a permittee to respond, in a timely manner, to:

Common variations in the types and quantities of the wastes managed under the facility permit

Technological advancements

Changes necessary to comply with new regulations, where these changes can be implemented without substantially changing design specifications or management practices in the permit

Class 3 modifications substantially alter the facility or its operation.

- Temporary authorizations.

Upon request of the permittee, the director may, without prior public notice and comment, grant the permittee a temporary authorization in accordance with this subsection. Temporary authorizations must have a term of not more than 180 days.

The permittee may request a temporary authorization for:

Any Class 2 modification meeting the criteria in paragraph (e)(3)(ii) of 40 CFR, Part 270.42

Any Class 3 modification that meets the criteria in paragraph (3)(ii) (A) or (B) of 40 CFR, Part 270.42, or that meets the criteria in paragraphs (3)(ii)(C) through (E) of 40 CFR, Part 270.42, and provides improved management or treatment of a hazardous waste already listed in the facility permit

The temporary authorization request must include:

A description of the activities to be conducted under the temporary authorization

An explanation of why the temporary authorization is necessary

Sufficient information to ensure compliance with 40 CFR Part 264 standards

The permittee must send a notice about the temporary authorization request to all persons on the facility mailing list maintained by the director and to appropriate units of state and local governments as specified in 40 CFR 124.10(c)(ix). This notification must be made within 7 days of submission of the authorization request.

The director must approve or deny the temporary authorization as quickly as practical. To issue a temporary authorization, the director must find:

The authorized activities are in compliance with the standards of 40 CFR, Part 264.

The temporary authorization is necessary to achieve one of the following objectives before action is likely to be taken on a modification request:

To facilitate timely implementation of closure or corrective action activities

To allow treatment or storage in tanks or containers, or in containment buildings in accordance with 40 CFR, Part 268

To prevent disruption of ongoing waste management activities

To enable the permittee to respond to sudden changes in the types or quantities of the wastes managed under the facility permit

To facilitate other changes to protect human health and the environment

A temporary authorization may be reissued for one additional term of up to 180 days provided that the permittee has requested a Class 2 or 3 permit modification for the activity covered in the temporary authorization, and:

The reissued temporary authorization constitutes the director's decision on a Class 2 permit modification in accordance with paragraph (b)(6)(i)(D) or (ii)(D) of 40 CFR, Part 270.42

The director determines that the reissued temporary authorization involving a Class 3 permit modification request is warranted to allow the authorized activities to continue while the modification procedures of paragraph (c) of 40 CFR, Part 270.42, are conducted

- Public notice and appeals of permit modification decisions. (1) The director must notify persons on the facility mailing list and appropriate units of state and local government within 10 days of any decision under this section to grant or deny a Class 2 or 3 permit modification request. The director must also notify such persons within 10 days after an automatic authorization for a Class 2 modification goes into effect under 270.42(b)(6)(iii) or (v).

The director's decision to grant or deny a Class 2 or 3 permit modification request under this section may be appealed under the permit appeal procedures of 40 CFR 124.19.

An automatic authorization that goes into effect under 270.42(b)(6)(iii) or (v) may be appealed under the permit appeal procedures of 40 CFR 124.19; however, the permittee may continue to conduct the activities pursuant to the automatic authorization until the appeal has been granted pursuant to 124.19(c), notwithstanding the provisions of 124.15(b).

• Newly regulated wastes and units.

The permittee is authorized to continue to manage wastes listed or identified as hazardous under Part 261, or to continue to manage hazardous waste in units newly regulated as hazardous waste management units, if:

The unit was in existence as a hazardous waste facility with respect to the newly listed or characterized waste or newly regulated waste management unit on the effective date of the final rule listing or identifying the waste, or regulating the unit.

The permittee submits a Class 1 modification request on or before the date on which the waste or unit becomes subject to the new requirements.

The permittee is in compliance with the applicable standards of 40 CFR, Parts 265 and 266.

The permittee also submits a complete Class 2 or 3 modification request within 180 days of the effective date of the rule listing or identifying the waste, or subjecting the unit to RCRA Subtitle C management standards.

In the case of land disposal units, the permittee certifies that each such unit is in compliance with all applicable requirements of Part 265 for groundwater monitoring and financial responsibility on the date 12 months after the effective date of the rule identifying or listing the waste as hazardous, or regulating the unit as a hazardous waste management unit. If the owner or operator fails to certify compliance with all these requirements, he or she will lose authority to operate under this section.

New wastes or units added to a facility's permit under this subsection do not constitute expansions for the purpose of the 25 percent capacity expansion limit for Class 2 modifications.

• Military hazardous waste munitions treatment and disposal. The permittee is authorized to continue to accept waste military munitions notwithstanding any permit conditions barring the permittee from accepting off-site wastes, if:

The facility was in existence as a hazardous waste facility, and the facility was already permitted to handle the waste military munitions, on the date when the waste military munitions became subject to hazardous waste regulatory requirements

On or before the date when the waste military munitions become subject to hazardous waste regulatory requirements, the permittee submits a Class 1 modification request to remove or amend the permit provision restricting the receipt of off-site waste munitions

The permittee submits a complete Class 2 modification request within 180 days of the date when the waste military munitions became subject to hazardous waste regulatory requirements

- Permit modification list. The director must maintain a list of all approved permit modifications and must publish a notice once a year in a statewide newspaper that an updated list is available for review.
- Combustion facility changes to meet part 63 MACT standards. The following procedures apply to hazardous waste combustion facility permit modifications requested under the attached appendix I of 40 CFR, Part 270.42, section l(9).

    Facility owners or operators must comply with the notification of intent to comply (NIC) requirements of 40 CFR 63.1211 before a permit modification can be requested.

    If the director does not approve or deny the request within 90 days of receiving it, the request must be deemed approved. The director may, at his or her discretion, extend this 90-day deadline one time for up to 30 days by notifying the facility owner or operator.

11.6.10.4   Termination of TSDF Permits

- The following are causes for terminating a permit during its term or for denying a permit renewal application:

    Noncompliance by the permittee with any condition of the permit

    The permittee's failure in the application or during the permit issuance process to disclose fully all relevant facts, or the permittee's misrepresentation of any relevant facts at any time

    A determination that the permitted activity endangers human health or the environment and can only be regulated to acceptable levels by permit modification or termination

- The director must follow the applicable procedures in Part 124 or state procedures in terminating any permit under this section.

11.6.11   Expiration and Continuation of TSDF Permits

11.6.11.1   Duration of TSDF Permits

- RCRA permits are effective for a fixed term not to exceed 10 years.
- Except as provided in 270.51, the term of a permit cannot be extended by modification beyond the maximum duration specified in this section.
- The director may issue any permit for a duration that is less than the full allowable term under this section.
- Each permit for a land disposal facility must be reviewed by the director 5 years after the date of permit issuance or reissuance and must be modified as necessary, as provided in 270.41.

11.6.11.2   Continuation of Expiring TSDF Permits

- EPA permits. When EPA is the permit-issuing authority, the conditions of an expired permit continue in force under 5 U.S.C. 558(c) until the effective date of a new permit (see 124.15) if:

  The permittee has submitted a timely application under 270.14 and the applicable sections in 270.15 through 270.29 which is a complete [under 270.10(c)] application for a new permit

  The regional administrator, through no fault of the permittee, does not issue a new permit with an effective date under 124.15 on or before the expiration date of the previous permit (for example, when issuance is impracticable due to time or resource constraints)

- Permits continued under this section remain fully effective and enforceable.

- Enforcement. When the permittee is not in compliance with the conditions of the expiring or expired permit, the regional administrator may choose to do any or all of the following:

  Initiate enforcement action based upon the permit which has been continued

  Issue a notice of intent to deny the new permit under 124.6. If the permit is denied, the owner or operator would then be required to cease the activities authorized by the continued permit or be subject to enforcement action for operating without a permit

  Issue a new permit under part 124 with appropriate conditions

  Take other actions authorized by these regulations

• State continuation. In a state with a hazardous waste program authorized under 40 CFR, Part 271, if a permittee has submitted a timely and complete application under applicable state law and regulations, the terms and conditions of an EPA-issued RCRA permit continue in force beyond the expiration date of the permit, but only until the effective date of the state's issuance or denial of a State RCRA permit.

11.6.12 Interim Status

11.6.12.1 Qualifying for Interim Status

• Any person who owns or operates an "existing HWM facility" or a facility in existence on the effective date of statutory or regulatory amendments under the RCRA act that render the facility subject to the requirement to have an RCRA permit will be considered as having interim status and must be treated as having been issued a permit to the extent he or she has:

Complied with the requirements of section 3010(a) of RCRA pertaining to notification of hazardous waste activity

Complied with the requirements of 270.10 governing submission of part A applications

• Failure to qualify for interim status. If EPA has reason to believe upon examination of a part A application that it fails to meet the requirements of 270.13, it must notify the owner or operator in writing of the apparent deficiency. Such notice must specify the grounds for EPA's belief that the application is deficient. The owner or operator has 30 days from receipt to respond to such a notification and to explain or cure the alleged deficiency in the part A application. If, after such notification and opportunity for response, EPA determines that the application is deficient, it may take appropriate enforcement action.

11.6.12.2 Operation During Interim Status

• During the interim status period the facility must not:

Treat, store, or dispose of hazardous waste not specified in part A of the permit application

Employ processes not specified in part A of the permit application

Exceed the design capacities specified in part A of the permit application

• Interim status standards. During interim status, owners or operators must comply with the interim status standards at 40 CFR, Part 265.

11.6.12.3   Changes during Interim Status

• Except as provided in paragraph (b), the owner or operator of an interim status
  facility may make the following changes at the facility:

  Treatment, storage, or disposal of new hazardous wastes not previously
  identified in part A of the permit application (and, in the case of newly listed
  or identified wastes, addition of the units being used to treat, store, or dispose of the hazardous wastes on the effective date of the listing or identification) if the owner or operator submits a revised part A permit application
  prior to such treatment, storage, or disposal.

  Increases in the design capacity of processes used at the facility if the owner
  or operator submits a revised part A permit application prior to such a
  change (along with a justification explaining the need for the change) and
  the director approves the changes because:

    There is a lack of available treatment, storage, or disposal capacity at
    other hazardous waste management facilities

    The change is necessary to comply with a federal, state, or local require-
    ment

  Changes in the processes for the treatment, storage, or disposal of hazardous
  waste or addition of processes if the owner or operator submits a revised part
  A permit application prior to such change (along with a justification
  explaining the need for the change) and the director approves the change
  because:

    The change is necessary to prevent a threat to human health and the envi-
    ronment because of an emergency situation

    The change is necessary to comply with a federal, state, or local require-
    ment

  Changes in the ownership or operational control of a facility if the new owner
  or operator submits a revised part A permit application no later than 90 days
  prior to the scheduled change. When a transfer of operational control of a facility occurs, the old owner or operator must comply with the requirements of 40
  CFR Part 265, subpart H (Financial Requirements) until the new owner or
  operator has demonstrated to the director that he or she is complying with the
  requirements of that subpart. The new owner or operator must demonstrate
  compliance with subpart H requirements within 6 months of the date of the
  change in ownership or operational control of the facility. Upon demonstration
  to the director by the new owner or operator of compliance with subpart H, the
  director will notify the old owner or operator in writing that he or she no longer
  needs to comply with subpart H as of the date of demonstration. All other
  interim status duties are transferred effective immediately upon the date of the
  change in ownership or operational control of the facility.

Changes made in accordance with an interim status corrective action order issued by EPA under section 3008(h) or other federal authority, by an authorized state under comparable state authority, or by a court in a judicial action brought by EPA or by an authorized state. Changes under this section are limited to the treatment, storage, or disposal of solid waste from releases that originate within the boundary of the facility.

Addition of newly regulated units for the treatment, storage, or disposal of hazardous waste if the owner or operator submits a revised part A permit application on or before the date on which the unit becomes subject to the new requirements.

- Except as specifically allowed under this paragraph, changes listed under paragraph (a) of 40 CFR, Part 270.72, may not be made if they amount to reconstruction of the hazardous waste management facility. Reconstruction occurs when the capital investment in the changes to the facility exceeds 50 percent of the capital cost of a comparable entirely new hazardous waste management facility. If all other requirements are met, the following changes may be made even if they amount to a reconstruction:

Changes made solely for the purposes of complying with the requirements of 40 CFR 265.193 for tanks and ancillary equipment.

If necessary to comply with federal, state, or local requirements, changes to an existing unit, changes solely involving tanks or containers, or addition of replacement surface impoundments that satisfy the standards of section 3004(o).

Changes that are necessary to allow owners or operators to continue handling newly listed or identified hazardous wastes that have been treated, stored, or disposed of at the facility prior to the effective date of the rule establishing the new listing or identification.

Changes during closure of a facility or of a unit within a facility made in accordance with an approved closure plan.

Changes necessary to comply with an interim status corrective action order issued by EPA under section 3008(h) or other federal authority, by an authorized state under comparable state authority, or by a court in a judicial proceeding brought by EPA or an authorized state, provided that such changes are limited to the treatment, storage, or disposal of solid waste from releases that originate within the boundary of the facility.

Changes to treat or store, in tanks, containers, or containment buildings, hazardous wastes subject to land disposal restrictions imposed by 40 CFR, Part 268, or RCRA 3004, provided that such changes are made solely for the purpose of complying with 40 CFR, Part 268, or RCRA section 3004.

Addition of newly regulated units under paragraph (a)(6) of 40 CFR, part 270.72.

Changes necessary to comply with standards under 40 CFR, Part 63, Subpart EEE—National Emission Standards for Hazardous Air Pollutants from Hazardous Waste Combustors.

11.6.12.4   Termination of Interim Status.   Interim status terminates when:

- Final administrative disposition of a permit application, except an application for a remedial action plan (RAP) under subpart H of 40 CFR, Part 270, is made.
- Interim status is terminated as provided in 270.10(e)(5).
- For owners or operators of each land disposal facility which has been granted interim status prior to Nov. 8, 1984, on Nov. 8, 1985, unless:

   The owner or operator submits a part B application for a permit for such facility prior to that date

   The owner or operator certifies that such facility is in compliance with all applicable groundwater monitoring and financial responsibility requirements

- For owners or operators of each land disposal facility which is in existence on the effective date of statutory or regulatory amendments under the act that render the facility subject to the requirement to have a RCRA permit and which is granted interim status, 12 months after the date on which the facility first becomes subject to such permit requirement unless the owner or operator of such facility:

   Submits a part B application for a RCRA permit for such facility before the date 12 months after the date on which the facility first becomes subject to such permit requirement

   Certifies that such facility is in compliance with all applicable groundwater monitoring and financial responsibility requirements

- For owners or operators of any land disposal unit that is granted authority to operate under 270.72(a)(1), (2) or (3), on the date 12 months after the effective date of such requirement, unless the owner or operator certifies that such unit is in compliance with all applicable groundwater monitoring and financial responsibility requirements.

- For owners and operators of each incinerator facility which has achieved interim status prior to Nov. 8, 1984, interim status terminates on Nov. 8, 1989, unless the owner or operator of the facility submits a part B application for a RCRA permit for an incinerator facility by Nov. 8, 1986.

• For owners or operators of any facility (other than a land disposal or an incinerator facility) which has achieved interim status prior to Nov. 8, 1984, interim status terminates on Nov. 8, 1992, unless the owner or operator of the facility submits a part B application for a RCRA permit for the facility by Nov. 8, 1988.

11.7 UNIVERSAL WASTE PERMITS

EPA's universal waste management regulations are set forth in 40 CFR, part 273. Since a major goal of these hazardous waste management regulations is to reduce and streamline the regulatory requirements associated with the collection and management of several widely generated wastes (identified as universal wastes), there are very few actual "permitting" requirements. Most if not all of these requirements can be found with respect to "destination facilities" that receive shipments of universal wastes. These facilities may be required, based on their universal waste management activity, to comply with the TSDF permitting requirements found in Section 11.6 of this chapter.

The definition of destination facility was published in the *Federal Register* on May 15, 1995, to clarify this point. In 273.6 of the final rule, destination facility is defined as "a facility that treats, disposes of, or recycles a particular category of universal waste except those management activities described in paragraphs (a) and (c) of 273.13 and 273.33. A facility at which a particular category of universal waste is only accumulated, is not a destination facility for purposes of managing that category of universal waste." By defining a destination facility based on the universal waste management activity conducted there rather than by whether the facility has a RCRA permit for other waste management activities, the final rule states that a facility which only accumulates a particular category of universal waste is a universal waste handler for that particular category of universal waste. However, if this facility also treats, recycles, and/or disposes of another category of universal waste, that facility is a destination facility for that particular category of universal waste and must comply with both the destination facility requirements for that category of waste and all Subtitle C TSDF permitting requirements.

Since these universal waste management regulations are regarded as an adjunct to the hazardous waste management regulations, a discussion of their requirements follows.

11.7.1 Universal Waste Permits, Regulatory Overview

On Feb. 11, 1993, the Environmental Protection Agency proposed new streamlined hazardous waste management regulations governing the collection and management of certain widely generated wastes (batteries, pesticides, and thermostats) known as universal wastes (58 FR 9346).

These widely generated wastes (batteries, pesticides, and thermostats) are known as universal wastes and are subject to separate regulations (40 CFR, part 273) designed to encourage the recycling of these wastes.

EPA issued separate regulations for universal wastes in May 1995 to ease the regulatory burden on retail stores and others that wish to collect or generate these wastes and to reduce the quantity of these wastes going to municipal solid waste landfills or combustors.

[*Note:* EPA proposed July 27, 1994, to designate mercury-containing lamps either as a universal waste or as conditionally excluded from hazardous waste regulation (59 FR 28288). The rule has not yet been finalized. Several states that have been authorized to develop and implement universal waste management programs have chosen to designate mercury-containing lamps and other nonhazardous solid wastes as universal wastes.]

Generally speaking, a universal waste is a hazardous waste that generally displays the following characteristics:

- It is frequently generated in a wide variety of settings other than the industrial settings usually associated with hazardous wastes.

- It is generated by a vast community, the size of which poses implementation difficulties for both those who are regulated and the regulatory agencies charged with implementing the hazardous waste program.

- It may be present in significant volumes in nonhazardous waste management systems.

Previously enacted RCRA regulations have been a major impediment to national collection and recycling campaigns for these wastes. These universal waste regulations greatly ease the regulatory burden on retail stores and others that wish to collect or generate these wastes and should greatly facilitate programs developed to reduce the quantity of these wastes going to municipal solid waste landfills or combustors. They can also assure that the wastes subject to this system will go to appropriate treatment or recycling facilities pursuant to the full hazardous waste regulatory controls. They also can serve as a prototype system to which EPA may add other similar wastes in the future. A petition process is included through which additional wastes could be added to the universal waste regulations in the future.

11.7.2 Universal Waste Definitions

Battery means a device consisting of one or more electrically connected electrochemical cells, which is designed to receive, store, and deliver electric energy. An electrochemical cell is a system consisting of an anode, a cathode, and an electrolyte, plus such connections (electrical and mechanical) as may be needed to allow the cell to deliver or receive electrical energy. The term battery also includes an intact unbroken battery from which the electrolyte has been removed.

Destination facility means a facility that treats, disposes of, or recycles a particular category of universal waste, except those management activities described in 273.13(a) and (c) and 273.33(a) and (c). A facility at which a particular category of universal waste is only accumulated is not a destination facility for purposes of managing that category of universal waste.

FIFRA means the Federal Insecticide, Fungicide, and Rodenticide Act (7 U.S.C. 136-136y).

Generator means any person, by site, whose act or process produces hazardous waste identified or listed in part 261 or whose act first causes a hazardous waste to become subject to regulation.

Large quantity handler of universal waste means a universal waste handler (as defined in this section) who accumulates 5000 kilograms or more total of universal waste (batteries, pesticides, or thermostats, calculated collectively) at any time. This designation as a large quantity handler of universal waste is retained through the end of the calendar year in which 5000 kilograms or more total of universal waste is accumulated.

On-site means the same or geographically contiguous property which may be divided by public or private right-of-way, provided that the entrance and exit between the properties is at a crossroads intersection, and access is by crossing as opposed to going along the right-of-way. Noncontiguous properties owned by the same person but connected by a right-of-way which he or she controls and to which the public does not have access are also considered on-site property.

Pesticide means any substance or mixture of substances intended for preventing, destroying, repelling, or mitigating any pest, or intended for use as a plant regulator, defoliant, or desiccant, except for:

- New animal drugs under FFDCA section 201(w)
- Animal drugs that have been determined by regulation of the Secretary of Health and Human Services not to be a new animal drug
- Animal feeds under FFDCA section 201(xi)

Small quantity handler of universal waste means a universal waste handler who does not accumulate more than 5000 kilograms total of universal waste (batteries, pesticides, or thermostats, calculated collectively) at any time.

Thermostat means a temperature control device that contains metallic mercury in an ampoule attached to a bimetal-sensing element, and mercury-containing ampoules that have been removed from these temperature control devices in compliance with the requirements of 40 CFR 273.13(c)(2) or 273.33(c)(2).

Universal waste means any of the following hazardous wastes that are subject to the universal waste requirements of 40 CFR Part 273:

- Batteries as described in 40 CFR 273.2
- Pesticides as described in 40 CFR 273.3

- Thermostats as described in 40 CFR 273.4

  Universal waste handler means:

- A generator of universal waste
- The owner or operator of a facility, including all contiguous property, that receives universal waste from other universal waste handlers, accumulates universal waste, and sends universal waste to another universal waste handler, to a destination facility, or to a foreign destination

It does not mean:

- A person who treats [except under the provisions of 40 CFR 273.13(a) or (c), or 273.33(a) or (c)], disposes of, or recycles universal waste
- A person engaged in the off-site transportation of universal waste by air, rail, highway, or water, including a universal waste transfer facility
- Universal waste transfer facility means any transportation-related facility including loading docks, parking areas, storage areas, and other similar areas where shipments of universal waste are held during the normal course of transportation for 10 days or less.

  Universal waste transporter means a person engaged in the off-site transportation of universal waste by air, rail, highway, or water.

11.7.3 Goals of the Universal Waste Management Program

In the proposed part 273 regulations, EPA proposed a set of special requirements for universal hazardous wastes which were designed to accomplish three general goals. Goal 1 was to encourage resource conservation, while ensuring adequate protection of human health and the environment. Goal 2 defined in the proposal was to improve implementation of the current subtitle C hazardous waste regulatory program. With Goal 3, by simplifying the requirements and encouraging collection of these hazardous wastes, EPA hoped to provide incentives for individuals and organizations to collect the unregulated portions of these universal waste streams (e.g., from households or CESQGs) and manage them using the same systems developed for the regulated portion. These wastes would therefore be removed from the municipal waste stream and their input of hazardous constituents to municipal landfills, combustors, and composting projects would be minimized. Each of these goals is discussed in more detail below.

11.7.3.1 Goal 1 (Resource Conservation). The first goal for the universal waste rule is to encourage resource conservation. EPA believes that the universal waste management program serves to stimulate achievement of this goal. While this program applies to both universal wastes destined for recycling and those destined for disposal, as proposed, several features of the rule removed major obstacles faced by persons desiring to recycle these wastes. This program reduced the management requirements for generators, consolidation points (in the final rule referred to as small and large quantity handlers of universal waste), and transporters. Destination facilities must continue to meet all requirements, except manifesting requirements, of the subtitle C regulations. By relaxing the standards for these handlers, collection of universal waste is simplified, thereby encouraging participation in collection programs. EPA believed that the ability to access large quantities of universal waste from central collection centers may encourage the development and use of safe and effective ways to recycle these waste streams. Conversely, limiting the rule to universal waste destined for recycling only may discourage the use and development of recycling technologies, as universal waste handlers may be hesitant to participate in a program that requires knowledge that their universal waste is recycled.

11.7.3.2   Goal 2 (Improved Hazardous Waste Management).   The second goal of the universal waste management program is to improve implementation of the hazardous waste program. EPA believed that the universal waste management program can have significant impacts on waste management practices nationwide. Implementation of the hazardous waste program can be improved by the simplified set of requirements set forth in the rule. The provisions are now written such that handlers of universal wastes more easily understand them. EPA believed that this waste management program is protective of human health and the environment, is clear and easily understood by the diverse community which is targeted in the universal waste management program, and does not require expending unreasonable amounts of time and effort to understand the applicable requirements. The universal waste management program also allows the Part 273 regulations to be applied to all universal wastes, regardless of whether they are destined for recycling or disposal. Thus, compliance and enforcement procedures are easier to implement. Finally, because the final rule does not require that universal waste handlers count those universal wastes managed under Part 273 toward their monthly quantity determination, the rule can greatly simplify the procedures used to determine monthly hazardous waste generation rates for universal waste handlers, thus facilitating the implementation of the regulations.

11.7.3.3   Goal 3 (Separation of Universal Wastes).   The third goal of the universal waste management program is to separate universal waste from the municipal waste stream. Under the full Subtitle C regulations, the management of waste differs based on the waste's generation source. That is, waste generated by consumers in their homes is not regulated under RCRA Subtitle C when discarded, because it is excluded from the definition of hazardous waste under 40 CFR 261.4(b)(1). Conversely, the same waste would be subject to RCRA Subtitle C regulation if generated by commercial establishments, industries, and other nonexempt generators. Wastes covered under the universal waste regulations (batteries, pesticides, and mercury thermostats) are examples of wastes that are generated by both groups. Because the waste itself is the same, and therefore looks the same to waste handlers, universal waste that belongs in a hazardous waste system may be entering municipal solid waste landfills or combustors instead. The EPA believes that the rule is practical enough that, as an infrastructure develops for collecting universal waste, all categories of handlers can manage their universal waste under the Part 273 requirements. Therefore, under this program, management of universal waste is material-specific rather than source-specific. Hence universal waste, regardless of the source of generation, should be easily managed.

11.7.4 Summary of Universal Waste Requirements

This section provides a summary of the final universal waste regulations, 40 CFR, part 273. Each of the universal waste requirements is discussed in more detail in the later sections of this chapter.

11.7.4.1 Wastes Covered under the Universal Waste System. Three types of wastes are covered under the universal waste regulations:

- Hazardous waste batteries
- Hazardous waste pesticides that are either recalled or collected in waste pesticide collection programs
- Hazardous waste thermostats

Other wastes may be added to the universal waste regulations in the future, but at this time only these three wastes are included.

11.7.4.2 Participants in the Universal Waste System. There are four types of participants in the universal waste system:

- Small quantity handlers of universal waste
- Large quantity handlers of universal waste
- Universal waste transporters
- Destination facilities

Each of these participants is described below.

Although there are 10 basic universal waste management requirements, individual participants in the universal waste system are not subject to all 10 requirements. Only those requirements that have been determined to be appropriate for a given type of participant are included in the regulations for that participant. Throughout the universal waste regulations, each of these 10 basic requirements is addressed in regulatory sections using the same section headings. For example, the same requirements are addressed in the off-site shipments section for SQHUWs as are addressed in the off-site shipments sections for LQHUWs, transporters, and destination facilities. In some cases not all issues within a section were determined to be necessary for each type of participant, so some sections do not address every issue addressed in other sections with the same heading.

11.7.4.3 *Universal Waste Management Requirements for Small and Large Quantity Handlers of Universal Waste.* There are two types of handlers of universal waste. The first type of handler is a person who generates, or creates, universal waste. This is a person who uses batteries, pesticides, or thermostats and who eventually decides that they are no longer usable and thus are waste. Contractors or repair people who decide that batteries or thermostats are no longer usable and remove them from service also generate universal waste, and thus are handlers of universal waste.

The second type of handler is a person who receives universal waste from generators or other handlers, consolidates the waste, and then sends it on to other handlers, recyclers, or treatment and disposal facilities.

Universal waste handlers accumulate universal waste but do not treat, recycle, or dispose of the waste. Each separate location (e.g., generating location or collecting location) is considered a separate universal waste handler. Thus, if one company has several locations at which universal waste is generated or collected, each location is a separate handler.

There are two sets of regulations for handlers of universal waste. Subpart B of Part 273 sets forth the requirements that small quantity handlers of universal waste must follow. SQHUWs do not accumulate 5000 kilograms or more total (all universal waste categories combined) of universal waste at their location at any time. Subpart C of Part 273 sets forth the requirements that large quantity handlers of universal waste must follow. LQHUWs accumulate 5000 kilograms or more total (all universal waste categories combined) of universal waste at any time. This designation as a large quantity handler of universal waste is retained through the end of the calendar year in which 5000 kilograms or more total of universal waste is accumulated, at any one time.

EPA realized that some handlers of universal waste who would generally qualify as a small quantity handler may have a one-time, or infrequent, occasion to accumulate 5000 kilograms of universal waste, at any one time, on-site, thus requiring them to comply with the large quantity handler regulations in the rule. EPA did not intend to require these handlers to comply with the more stringent large quantity handler requirements during subsequent years in which they do not accumulate 5000 kilograms or more. EPA clarifies in the definition of large quantity handler of universal waste that this designation is retained by the handler for the remainder of the calendar year in which 5000 kilograms or more of universal waste was accumulated. A handler may reevaluate his or her status as a large quantity handler of universal waste in the following calendar year.

Subparts B (small quantity handlers) and C (large quantity handlers) of 40 CFR, Part 273, each include 11 sections. Because most of the requirements are the same for SQHUWs and LQHUWs, they are described together.

11.7.4.3.1 Applicability.   The first sections (40 CFR 273.10 and 273.30) are called "applicability" and explain whom the subpart B and C requirements apply to.

11.7.4.3.2 Prohibitions.   The second sections, "prohibitions" (40 CFR 273.11 and 273.31), prohibit handlers from disposing of, diluting, or treating universal waste except in certain circumstances.

11.7.4.3.3 Notification.   The third section, "notification," is different for SQHUWs and LQHUWs. 40 CFR 273.12 notes that SQHUWs are not required to notify EPA of their universal waste activities and are not required to obtain an EPA identification number; 40 CFR 273.32 requires LQHUWs to notify EPA and to obtain an EPA identification number.

11.7.4.3.4 Waste Management.   The fourth section, "waste management" (40 CFR 273.13 and 273.33), explains the requirements SQHUWs and LQHUWs must follow when handling universal waste. These provisions require that universal waste be managed in a way that prevents releases to the environment, specify packaging requirements for universal wastes, and set forth procedures that must be followed when handling batteries (e.g., sorting battery types, mixing battery types, disassembling battery packs, removing electrolyte, etc.), and when removing mercury-containing ampoules from thermostats.

11.7.4.3.5 Labeling/Marking.   The next sections, "labeling/marking" (40 CFR 273.14 and 273.34), require handlers to label or mark universal wastes or containers of universal waste to identify the type of universal waste (e.g., used batteries and pesticides).

11.7.4.3.6 Accumulation Time Limit.   The "accumulation time limit" sections (40 CFR 273.15 and 273.35) limit the time that handlers may accumulate universal waste to 1 year (with one exception) and require handlers to be able to demonstrate that wastes are not accumulated for more than 1 year.

11.7.4.3.7 Employee Training.   The seventh section "employee training" (40 CFR 273.16 and 273.36), is somewhat different for SQHUWs and LQHUWs. SQHUWs must distribute basic handling and emergency information to employees handling universal waste. LQHUWs must ensure that employees are familiar with waste handling and emergency procedures as appropriate based on their responsibilities.

11.7.4.3.8 Response to Releases.   The eighth section is entitled "response to releases" (40 CFR 273.17 and 273.37) and requires handlers to immediately contain any releases of universal waste and to handle residues appropriately.

11.7.4.3.9 Off-Site Shipments.   The "off-site shipments" sections (40 CFR 273.18 and 273.38) require handlers to send universal waste only to persons within the universal waste system and specify procedures to be followed when a shipment is rejected by the receiving facility.

11.7.4.3.10 Tracking Universal Waste Shipments. The ninth section, "tracking universal waste shipments" (40 CFR 273.19 and 273.39), is different for SQHUWs and LQHUWs. SQHUWs do not have any requirements. LQHUWs must maintain basic records documenting shipments received at the facility and shipments sent from the facility.

11.7.4.3.11 Exports. The last sections, "exports" (40 CFR 273.20 and 273.40), specify notification procedures that must be followed when handlers ship universal wastes to foreign destinations.

11.7.4.4 Waste Management Requirements for Transporters of Universal Waste. The requirements for transporters of universal waste are found in subpart D of part 273. Transporters are persons who transport universal waste from handlers of universal waste to other handlers, destination facilities, or foreign destinations. A transporter may be an independent shipper contracted to transport the waste or may be a handler who self-transports the waste. A universal waste handler who self-transports waste becomes a transporter for those self-transportation activities and is subject to the requirements of subpart D of the universal waste management program.

The universal waste rule does include some specific requirements for transporters. However, the basic approach to transportation under the universal waste system is that no hazardous waste manifests are required, and transporters must comply with the Department of Transportation (DOT) requirements that would be applicable to the waste if it were being transported as a product. For example, if transporting universal waste batteries, the transporter must comply with the appropriate DOT requirements, which are based on whether the particular battery type is a DOT hazardous material and, if so, which DOT hazardous material requirements apply to the specific battery type.

The universal waste transporter requirements consist of seven sections.

11.7.4.4.1 Applicability. The first, "applicability" (40 CFR 273.50), explains to whom the transporter requirements apply.

11.7.4.4.2 "Prohibitions." The second section (40 CFR 273.51) prohibits transporters from disposing of, diluting, or treating universal waste.

11.7.4.4.3 Waste Management. The third section, "waste management" (40 CFR 273.52), explains that transporters must comply with applicable DOT requirements if the waste they are transporting is a hazardous material under DOT regulations.

11.7.4.4.4 Accumulation Time Limits. The fourth section, entitled "accumulation time limits" (40 CFR 273.53), notes that transporters may store waste for up to 10 days at a transfer facility during the course of transportation. Transfer facilities are transportation-related facilities such as loading docks, parking areas, and storage areas. If a transporter stores waste for more than 10 days at one location, the transporter must comply with the appropriate universal waste handler rules while storing the waste.

11.7.4.4.5 Response to Releases. The fifth transporter section, "response to releases" (40 CFR 273.54), requires transporters to immediately contain any releases of universal waste and to handle residues appropriately.

11.7.4.4.6 Off-Site Shipments. "Off-site shipments" (40 CFR 273.55) prohibits transporters from transporting universal waste to any place other than a universal waste handler, destination facility, or foreign destination.

11.7.4.4.7 Exports. Finally, "exports" (40 CFR 273.56) requires transporters to follow certain requirements for exports of hazardous waste.

11.7.4.5 Waste Management Requirements for Destination Facilities of Universal Wastes. The requirements for destination facilities are found in subpart E of part 273. Destination facility means a facility that treats, disposes of, or recycles a particular category of universal waste, except those management activities described in paragraphs (a) and (c) of 273.13 and 273.33. A facility at which a particular category of universal waste is only accumulated is not a destination facility for purposes of managing that category of universal waste.

The universal waste rules include only two specific universal waste requirements for destination facilities. In general, however, these facilities are subject to the same requirements that are applicable to treatment, storage, and disposal facilities under the full hazardous waste regulations. This includes permitting as well as general facility standards and unit specific requirements. In addition to the full hazardous waste requirements, there are three sections specifying universal waste requirements for destination facilities. For the most part these requirements simply mirror universal waste handler requirements for receipt of universal waste, since destination facilities also receive universal waste.

11.7.4.5.1 Standards for Destination Facilities. First, "standards for destination facilities" (40 CFR 273.60) indicates which of the full hazardous waste regulations destination facilities must follow. These are the same full hazardous waste regulations these facilities would be subject to if they were handling nonuniversal hazardous wastes. Specifically, facilities that treat, dispose of, and recycle universal wastes, except for those activities described in paragraphs (a) and (c) of 273.13 and 273.33, are subject to the permitting or interim status requirements of 40 CFR parts 264 or 265. Facilities that recycle universal waste without accumulating the waste before it is recycled are subject to the recycling requirements of 40 CFR 261.6(c)(2).

Second, "off-site shipments" (40 CFR 273.61) sets forth procedures for rejecting a shipment of universal waste. Finally, "tracking universal waste shipments" (40 CFR 273.62) requires destination facilities to retain the same records for receipt of universal waste shipments that LQHUWs are required to retain. By documenting receipt of universal waste shipments, these records complete documentation of shipments sent from handlers.

11.7.4.6 Import Requirements. Subpart F of the universal waste regulations clarifies the requirements for universal wastes that are imported. In general, once universal waste enters the United States it is subject to the same universal waste requirements it would be if it had been generated in the United States.

11.7.4.7 Petitions to Include Other Wastes Under Part 273. Subpart G of Part 273 includes two sections setting forth the procedures to be used to petition the agency to add additional wastes to the universal waste regulations. Further requirements are specified in 40 CFR 260.20 and 260.23.

11.8 NONHAZARDOUS WASTE MANAGEMENT ACTIVITY PERMITS

Unlike the requirements found in the Resource Conservation and Recovery Act (RCRA) that relate to hazardous waste management activities, there are no similar provisions contained in RCRA that specify actual permitting conditions relating to nonhazardous waste management activities. Nor does RCRA actually define a nonhazardous solid waste. Instead, a two-part test is used to determine whether or not a waste is solid and if so, whether or not it is then hazardous. Wastes meeting the definition of a solid waste, but not the definition of a hazardous waste, are considered nonhazardous.

As discussed further in this chapter, EPA's role is to establish the regulatory basis and provide both guidelines and technical assistance for the development of environmentally sound nonhazardous solid waste management practices. Consequently, the actual responsibility for permitting nonhazardous solid waste management programs belongs primarily to state and local governments with their emphasis being on waste management activities relating to disposal. The discussions in this chapter will initially focus on the federal guidelines and criteria issued by the EPA and then focus on examples of actual approved state permitting requirements. Since there are individual program requirements for each approved state, this chapter's discussion of state permitting criteria will be based on examples of typical state requirements commonly found in approved state programs.

11.8.1 Nonhazardous Waste Regulatory Overview

Waste management activities that involve the accumulation, storage, and disposal of nonhazardous solid waste are regulated under Subtitle D of RCRA (42 USC 6941-6949) which provides the mechanisms for coordinating the roles of federal, state, and local governments. A material is considered nonhazardous solid waste if it meets the definition of solid waste but does not meet the definition of a hazardous waste.

Subtitle D of the Resource Conservation and Recovery Act (RCRA) establishes a framework for federal, state, and local government cooperation in controlling the management of nonhazardous solid waste. Since the responsibility for the permitting of nonhazardous waste management activities (i.e., generation, accumulation and storage, disposal) lies primarily with individual states, EPA's role is to establish the overall regulatory direction and provide minimum nationwide standards for protecting human health and the environment. While the EPA provides federal guidelines and technical assistance to states for developing their own environmentally sound waste management practices, the actual planning and direct implementation of solid waste programs under subtitle D are largely state and local functions. RCRA authorizes states to devise programs to deal with state-specific conditions and needs. EPA retains the authority to enforce the appropriate standards in a given state.

Section 4005(c) of RCRA requires that each state adopt and implement "a permit program or other system of prior approval and conditions" (approved state permit program) adequate to assure that each facility that may receive household wastes and other nonhazardous solid wastes will comply with the revised criteria. Under section 4005(c) the primary responsibility for implementing and enforcing the revised criteria rests with the states. EPA is required to "determine whether each state has developed an adequate program" pursuant to section 4005(c).

11.8.2 Solid Wastes—Definitions, Exclusions, Exemptions, and Special Wastes

11.8.2.1 Solid Wastes Regulated by RCRA.   RCRA's definition of solid waste includes not only solid material but also liquid, semisolid, and contained gaseous material. For the purposes of regulation under RCRA, solid wastes are defined as materials that are:

- Disposed of, burned or incinerated, or accumulated, stored, or treated (but not recycled) before being disposed of or burned.
- Recycled, or accumulated, stored, or treated before recycling. Recycling activities include:

  Using the waste in a manner constituting disposal—that is, placing it on the land or using it to produce products that are placed on the land, excluding those ordinarily placed on the land as a matter of use

  Burning the waste to recover energy or to produce a fuel, excluding commercial products that are fuels

  Reclaiming certain wastes to recover or regenerate materials

  Accumulating the waste speculatively—that is, before recycling

- Inherently wastelike—that is, they contain toxic constituents ordinarily not found in raw materials or products, except in small concentrations, and pose a substantial risk to human health and the environment. These wastes ordinarily are disposed of, burned, or incinerated but not recycled for use or reuse, since they are toxic. This category also includes secondary materials that are hazardous wastes fed to a halogen acid furnace.

Some specific solid wastes, including household wastes, are considered nonhazardous and therefore are not regulated under the RCRA hazardous waste regulations. Nonhazardous solid wastes regulated under Subtitle D (42 USC 6941-6949) include those that are:

- Excluded from regulation as a hazardous waste under 40 CFR 261.4(b) (see Nonhazardous Solid Wastes).
- Delisted by EPA from the hazardous waste list (and do not exhibit a characteristic of hazardous waste)
- Excluded under 40 CFR 261.3 even though they are generated from the treatment, storage, or disposal of a hazardous waste

Some specific solid wastes that are actually considered to be hazardous are treated as nonhazardous under specific circumstances. These wastes include waste samples and hazardous wastes generated in a product or raw material storage tank, raw material transport vehicle or vessel, pipeline, or manufacturing unit before they exit the unit from which they were generated.

11.8.2.2 Nonsolid Waste Materials. Some specific materials are excluded from the EPA's definition of solid waste. The following materials are not considered solid wastes under 40 CFR 261.4(a):

- Domestic sewage or any mixture of domestic sewage and other wastes that passes through a sewer system to a publicly owned treatment works for treatment.
- Industrial wastewater discharges that are subject to regulation as point source discharges under the Clean Water Act.
- Irrigation return flows.
- Certain nuclear waste regulated under the Atomic Energy Act.
- Materials subjected to in situ mining techniques which are not removed from the ground as part of the extraction process.
- Reclaimed pulping liquors, unless accumulated speculatively.
- Spent sulfuric acid used to produce virgin sulfuric acid, unless accumulated speculatively.
- Reclaimed secondary materials that are returned to the original process, provided only tank storage is used, the entire process is closed, reclamation does not involve combustion, secondary materials never are stored for over 12 months

without being used, and the material is not used to produce a fuel or products used in a manner constituting disposal.

- Spent wood-preserving solutions that have been reclaimed and are reused for their original intended purpose; and wastewaters from the wood-preserving process that have been reclaimed and are used to treat wood.

- EPA hazardous wastes K060, K087, K141-K145, K147, and K148 and any wastes from coke by-products.

- Processed scrap metal and shredded circuit boards free of mercury switches, mercury relays, nickel-cadmium batteries, and lithium batteries.

- Nonwastewater splash condenser dross residue from the treatment of K061 waste—emission control dust from the production of steel in rotary kilns, electric furnaces, plasma arc furnaces, slag reactors, rotary hearth furnace and electric furnace combinations, or industrial furnaces—in high-temperature metals recovery units, provided that it is shipped in drums and not land disposed before recovery.

- Recovered oil from petroleum refining, exploration, and production which is to be inserted into the petroleum refining process along with normal process streams prior to crude distillation or catalytic cracking. Oil-bearing hazardous wastes and used oil (defined at 40 CFR 279.1) are not considered recovered oil.

- Condensates derived from the overhead gases from kraft mill steam strippers that are used to comply with the pulping process condensate treatment rule [40 CFR 63.446(e)]. The exemption applies only to combustion at the mill generating the condensates.

- Comparable fuels or comparable syngas fuels (40 CFR 261.38).

In addition, other types of waste are regulated outside of RCRA's solid and hazardous waste regulations, including PCB wastes regulated under the Toxic Substances and Control Act at 40 CFR Part 761 and certain batteries, pesticides, and thermostats regulated as universal wastes at 40 CFR Part 273.

11.8.2.3 Nonhazardous Solid Wastes.   EPA does not consider some specific solid wastes to be hazardous wastes. The following solid wastes are nonhazardous solid wastes under 40 CFR 261.4(b):

- Household wastes.
- Agricultural wastes.
- Mining overburden returned to the mine site.
- Certain wastes generated from combustion of coal or other fossil fuels.
- Wastes from the exploration, development, or production of crude oil, natural gas, or geothermal energy.
- Certain wastes containing chromium.

- Mining wastes from the extraction, beneficiation, and processing of ores and minerals, including coal, phosphate rock, and uranium ore overburden (*Note:* Some mineral processing waste is exempt from RCRA standards because of the Bevill Amendment, which subjects certain ore extraction and beneficiation materials to further study before identification as hazardous waste. However, EPA identified in 1991 some mineral processing wastes that cannot claim Bevill status because they exhibit corrosivity, reactivity, or toxicity characteristics due to metal content.)

- Cement kiln dust wastes.

- Arsenical-treated wood or wood products generated by end users.

- Petroleum-contaminated media and debris that fail the test for the toxicity characteristic—hazardous waste codes D018 through D043 only—and are subject to RCRA corrective action.

- Used chlorofluorocarbon refrigerants from totally enclosed heat transfer equipment that will be reclaimed for further use.

- Used oil filters that are non-terne-plated, have not been mixed with waste characterized by EPA as hazardous, and have been gravity hot-drained so that the used oil can be removed from the filter.

Used oil re-refining distillation bottoms that are used as feedstock to manufacture asphalt products.

11.8.2.3.1 Household Waste Exemption. A regulatory exemption is provided by EPA for household waste that can include wastes that have been collected, transported, stored, treated, disposed of, recovered for fuel, or reused. Examples of such wastes include garbage, trash, and sanitary wastes in septic tanks.

The actual sources of household wastes can vary. This type of waste can come from single and multiple residences, hotels, motels, bunkhouses, ranger stations, crew quarters, campgrounds, picnic grounds, and day-use recreation areas. Municipal resource recovery facilities that accept and burn only household wastes and nonhazardous solid wastes from commercial and industrial establishments also are exempt. It is the responsibility of facility owners to develop and implement management procedures to assure that the facility does not accept any hazardous wastes.

A facility's handling of municipal solid waste (e.g., household waste or nonhazardous wastes from commercial facilities) is not considered or classified as treating, storing, disposing of, or otherwise managing hazardous wastes for the purposes of RCRA regulation.

11.8.2.4 Delisted Hazardous Wastes. Wastes listed as hazardous wastes by EPA can be regulated as nonhazardous only if they are delisted by the EPA, a procedure by which EPA removes the specific hazardous waste from one of the lists of hazardous wastes found in 40 CFR, Part 26. Delisting a waste involves specific protocols by which a facility submits a petition to EPA.

11.8.2.5 Specific Exclusions. EPA considers some specific wastes to be non-hazardous even though they are generated from the treatment, storage, or disposal of hazardous wastes.

The following wastes, unless they exhibit a specific hazardous waste characteristic, are in that category:

- Waste pickle liquor sludge generated by lime stabilization of spent pickle liquor from the iron and steel industry (SIC Codes 331 and 332).
- Wastes from burning certain products of oil-bearing hazardous waste.
- Biological treatment sludge from the treatment of organic waste resulting from carbamate and carbamoyl oxime production.
- Nonwastewater residues, such as slag, resulting from high-temperature metals recovery processing of K061, K062, and F006 waste from rotary kilns, flame reactors, electric furnaces, plasma arc furnaces, slag reactors, rotary hearth furnace and electric furnace combinations, or industrial furnaces. A residue also must meet the generic exclusion levels listed in Table 11.7 for that waste and exhibit no characteristics of hazardous waste. For any such waste sent to a permitted land disposal unit, proper notification must be sent to the EPA regional administrator or authorized state agency.

TABLE 11.7 Generic Exclusion Levels

Constituents	Maximum for any single composite sample TCLP, mg/L
Antimony	0.10
Arsenic	0.050
Barium	7.6
Beryllium	0.010
Cadmium	0.050
Chromium (total)	0.33
Cyanide (total), mg/kg	1.8*
Lead	0.15
Mercury	0.009
Nickel	1.0
Selenium	0.16
Silver	0.30
Thallium	0.020
Zinc	70

*F006 only; not applicable to K061 and K062.

11.8.2.6 Special Wastes. [Reference: 40 CFR 261.4(b)] EPA considers some specific wastes, including fly ash, certain mining wastes, cement kiln dust, and gas production drilling muds and fluids to be "special" wastes since they have not been categorized as hazardous or nonhazardous. As such, EPA is evaluating these wastes to determine whether regulation is warranted under an expanded Subtitle D program or Subtitle C.

11.8.3 Federal Definitions

The definitions set forth in section 1004 of the act apply to this chapter. Special definitions of general concern to this part are provided below, and definitions especially pertinent to particular sections of this part are provided in those sections.

Active life means the period of operation beginning with the initial receipt of solid waste and ending at completion of closure activities in accordance with 258.60 of the regulation.

Active portion means that part of a facility or unit that has received or is receiving wastes and that has not been closed in accordance with 258.60 of this part.

Aquifer means a geological formation, group of formations, or portion of a formation capable of yielding significant quantities of groundwater to wells or springs.

Commercial solid waste means all types of solid waste generated by stores, offices, restaurants, warehouses, and other nonmanufacturing activities, excluding residential and industrial wastes.

Director of an approved state means the chief administrative officer of a state agency responsible for implementing the state permit program that is deemed to be adequate by EPA under regulations published pursuant to sections 2002 and 4005 of RCRA.

Disposal means the discharge, deposit, injection, dumping, spilling, leaking, or placing of any solid waste or hazardous waste into or on any land or water so that such solid waste or hazardous waste or any constituent thereof may enter the environment or be emitted into the air or discharged into any waters, including groundwaters.

Domestic septage is either liquid or solid material removed from a septic tank, cesspool, portable toilet, Type III marine sanitation device, or similar treatment works that receives only domestic sewage. Domestic septage does not include liquid or solid material removed from a septic tank, cesspool, or similar treatment works that receives either commercial wastewater or industrial wastewater and does not include grease removed from a grease trap at a restaurant.

Existing MSWLF unit means any municipal solid waste landfill unit that is receiving solid waste as of the appropriate dates specified in 258.1(e). Waste placement in existing units must be consistent with past operating practices or modified practices to ensure good management.

Facility means all contiguous land and structures, other appurtenances, and improvements on the land used for the disposal of solid waste.

Groundwater means water below the land surface in a zone of saturation.

Household waste means any solid waste (including garbage, trash, and sanitary waste in septic tanks) derived from households (including single and multiple residences, hotels and motels, bunkhouses, ranger stations, crew quarters, campgrounds, picnic grounds, and day-use recreation areas).

Indian lands or Indian country means:

1. All land within the limits of any Indian reservation under the jurisdiction of the United States government, notwithstanding the issuance of any patent, and including rights-of-way running throughout the reservation.

2. All dependent Indian communities within the borders of the United States whether within the original or subsequently acquired territory thereof, and whether within or without the limits of the State.

3. All Indian allotments, the Indian titles to which have not been extinguished, including rights-of-way running through the same.

Indian tribe or tribe means any Indian tribe, band, nation, or community recognized by the Secretary of the Interior and exercising substantial governmental duties and powers on Indian lands.

Industrial solid waste means solid waste generated by manufacturing or industrial processes that is not a hazardous waste regulated under subtitle C of RCRA. Such waste may include, but is not limited to, waste resulting from the following manufacturing processes: electric power generation; fertilizer and agricultural chemicals; food and related products and by-products; inorganic chemicals; iron and steel manufacturing; leather and leather products; nonferrous metals manufacturing and foundries; organic chemicals; plastics and resins manufacturing; pulp and paper industry; rubber and miscellaneous plastic products; stone, glass, clay, and concrete products; textile manufacturing; transportation equipment; and water treatment. This term does not include mining waste or oil and gas waste.

Land application unit means an area where wastes are applied onto or incorporated into the soil surface (excluding manure-spreading operations) for agricultural purposes or for treatment and disposal.

Landfill means an area of land or an excavation in which wastes are placed for permanent disposal, and that is not a land application unit, surface impoundment, injection well, or waste pile.

Lateral expansion means a horizontal expansion of the waste boundaries of an existing MSWLF unit.

Leachate means a liquid that has passed through or emerged from solid waste and contains soluble, suspended, or miscible materials removed from such waste.

Municipal solid waste landfill unit means a discrete area of land or an excavation that receives household waste, and that is not a land application unit, surface impoundment, injection well, or waste pile, as those terms are defined under 257.2. A MSWLF unit also may receive other types of RCRA subtitle D wastes, such as commercial solid waste, nonhazardous sludge, conditionally exempt small quantity generator waste, and industrial solid waste. Such a landfill may be publicly or privately owned. A MSWLF unit may be a new MSWLF unit, an existing MSWLF unit, or a lateral expansion.

New MSWLF unit means any municipal solid waste landfill unit that has not received waste prior to Oct. 9, 1993, or prior to Oct. 9, 1997, if the MSWLF unit meets the conditions of 258.1(f)(1).

Open burning means the combustion of solid waste without:

- Control of combustion air to maintain adequate temperature for efficient combustion.
- Containment of the combustion reaction in an enclosed device to provide sufficient residence time and mixing for complete combustion.
- Control of the emission of the combustion products.

Open dump means a facility for the disposal of solid waste that does not comply with this part.

Operator means the person(s) responsible for the overall operation of a facility or part of a facility.

Owner means the person(s) who owns a facility or part of a facility.

Practice means the act of disposal of solid waste.

Runoff means any rainwater, leachate, or other liquid that drains over land from any part of a facility.

Sanitary landfill means a facility for the disposal of solid waste that complies with this part.

Saturated zone means that part of the earth's crust in which all voids are filled with water.

Sewage sludge means solid, semisolid, or liquid residue generated during the treatment of domestic sewage in a treatment works. Sewage sludge includes, but is not limited to, domestic septage; scum or solids removed in primary, secondary, or advanced wastewater treatment processes; and a material derived from sewage sludge. Sewage sludge does not include ash generated during the firing of sewage sludge in a sewage sludge incinerator or grit and screenings generated during preliminary treatment of domestic sewage in a treatment works.

Sludge means any solid, semisolid, or liquid waste generated from a municipal, commercial, or industrial wastewater treatment plant, water supply treatment plant, or air pollution control facility or any other such waste having similar characteristics and effect.

Solid waste means any garbage, refuse, sludge from a waste treatment plant, water supply treatment plant, or air pollution control facility and other discarded material, including solid, liquid, semisolid, or contained gaseous material resulting from industrial, commercial, mining, and agricultural operations, and from community activities, but does not include solid or dissolved materials in domestic sewage, or solid or dissolved material in irrigation return flows or industrial discharges which are point sources subject to permits under section 402 of the Federal Water Pollution Control Act, as amended (86 Stat. 880), or source, special nuclear, or by-product material as defined by the Atomic Energy Act of 1954, as amended (68 Stat. 923).

State means any of the several states, the District of Columbia, the Commonwealth of Puerto Rico, the Virgin Islands, Guam, American Samoa, and the Commonwealth of the Northern Mariana Islands.

State director means the chief administrative officer of the lead state agency responsible for implementing the state permit program for 40 CFR Part 257, subpart B and 40 CFR Part 258 regulated facilities.

Surface impoundment or impoundment means a facility or part of a facility that is a natural topographic depression, human-made excavation, or diked area formed primarily of earthen materials (although it may be lined with human-made materials), that is designed to hold an accumulation of liquid wastes or wastes containing free liquids, and that is not an injection well. Examples of surface impoundments are holding storage, settling, and aeration pits, ponds, and lagoons.

Uppermost aquifer means the geologic formation nearest the natural ground surface that is an aquifer, as well as lower aquifers that are hydraulically interconnected with this aquifer within the facility's property boundary.

Waste management unit boundary means a vertical surface located at the hydraulically downgradient limit of the unit. This vertical surface extends down into the uppermost aquifer.

Waste pile or pile means any noncontainerized accumulation of solid, nonflowing waste that is used for treatment or storage.

11.8.4 Nonhazardous Solid Waste Management, Federal vs. State Responsibilities

EPA's approach to state permitting recognizes the traditional state role in implementing nonhazardous waste best management practices, including those relating to landfill standards and protecting groundwater. EPA fully intends that states will maintain the lead role in implementing these practices. EPA's goal is for all states to apply for and receive approval of their approved nonhazardous waste management programs. States have the flexibility to tailor standards to meet their state-specific conditions. In requiring that an approved state's program be capable of protecting groundwater at the specified point of compliance, an approved state may adopt its own performance standard; it may use the EPA's specific liner design. In addition, an approved state program would allow for the selection of any design the state determines would be capable of preventing contamination of groundwater beyond the drinking water standards. In short, whenever a state develops a program to deal with local conditions, the federal alternative would have only the status of "guidance" and would not be mandatory.

11.8.4.1 State Nonhazardous Solid Waste Management Plan Guidelines.
[Reference: 40 CFR 256.01] Nonhazardous solid waste management activities are
regulated under Subtitle D of RCRA (42 USC 6941-6949), which covers the management of nonhazardous waste and coordinates the roles of federal, state, and
local governments.

As stated earlier, EPA's role is to establish the regulatory basis and provide
technical assistance for the development of environmentally sound waste management practices. The actual permitting of nonhazardous solid waste programs is
usually the responsibility of state and local governments.

Subtitle D of RCRA (42 USC 6941-6949) requires EPA to issue guidelines for
state development and implementation of solid waste management plans that meet
federal criteria for minimum national performance standards.

11.8.5 Federal Nonhazardous Solid Waste Management Activity Permitting Criteria

The following criteria are often taken into consideration by approved states when
issuing permits for nonhazardous solid waste management activities.

11.8.5.1 Nonhazardous Solid Waste Generation. The permitting of solid
waste generator can be contingent on several factors. Federal regulations apply
only if the federal government is the actual generator. Nonfederal residential,
commercial, and institutional generators are not required to follow the federal
guidelines, although EPA encourages them to do so.

However, generators of nonhazardous solid waste, especially commercial generators, usually must comply with a number of state and local measures for disposing of and recycling nonhazardous solid waste. In order to better understand
these potential state permitting requirements, the federal requirements and recommended procedures are listed below.

11.8.5.1.1 Storage Requirements. [Reference: 40 CFR 243.200] Solid
wastes, including material separated for recycling, must be stored so that they do
not pose a fire, health, or safety hazard or provide food or shelter for disease-carrying animals. Such wastes also must be contained or bundled to avoid spillage.

Food wastes must be secured in covered or closed containers that are nonab-
sorbent, leakproof, durable, easily cleanable (if reused), and designed for safe handling.

Storage requirements for bulky wastes stipulate that all doors from large household appliances be removed and that such items be covered to avoid creating an
"attractive nuisance (e.g., a refrigerator with the door attached poses a danger to
children who could become trapped or injured). Solid waste and water should not
accumulate around the bulky wastes.

11.8.5.1.2 Containers. It is the responsibility of the generator to ensure that
a sufficient number of containers, large enough for food wastes, rubbish, and

ashes, are provided to a residence or other entity for accumulation between trash pickups. Such containers must be maintained in a clean condition so as not to attract insects or other animals that could carry diseases.

Reusable containers that are emptied manually must not exceed 75 pounds (34.05 kilograms) when filled and must be able to be serviced without the collector coming into physical contact with the waste. In addition, the design of all buildings or other facilities must provide for solid waste storage that facilitates easy collection.

11.8.5.1.3 Collection Frequency. [Reference: 40 CFR 243.203] Solid wastes and materials separated for recycling should be collected often enough to avoid attracting animals to the collection site. Solid wastes that include food must be collected weekly, bulky wastes at least once every 3 months.

11.8.5.1.4 Source Separation. [Reference: 40 CFR 246.200-246.202] High-grade paper generated at federal office facilities with more than 100 workers must be separated at the source of generation, separately collected, and sold for recycling. High-grade paper includes letterhead, dry copy papers, various business forms, stationary, typing paper, tablet sheets, and computer printout paper and cards.

Separation, collection, and recycling of used newspaper at the source of generation is required at facilities in which more than 500 families reside (typically military communities).

Federal facilities generating 10 or more tons per month of corrugated containers must separately collect and sell such material for recycling.

11.8.5.1.5 Recommended Procedures. Compliance with the following recommended generator procedures is optional. Their purpose is to provide suggestions for carrying out the generator requirements. However, they are not mandatory for the federal government or any state or local entity.

11.8.5.1.5.1 CONTAINER DESIGN: [Reference: 40 CFR 243.200-2] Reusable containers should be constructed of corrosion-resistant metal or other material that does not absorb water, grease, or oil. Such containers should be leakproof and withstand usage without rusting, cracking, or breaking.

The interior should be smooth without interior projections or rough seams that would interfere with cleaning and emptying. The exterior should not contain cracks, holes, or jagged edges that interfere with safe handling.

Containers should be stored on a firm, level, well-drained surface that can accommodate all containers. The surface should be maintained in a clean, spill-free condition.

Reusable containers that are emptied manually should have a capacity of no more than 35 gallons, unless they are mounted on casters and can be rolled to the collection vehicle and tilted. The containers should be constructed with:

- Rounded edges and tapered sides with large openings at the top of the containers
- Two handles or bails located opposite each other

• Tight-fitting cover handles

In addition, the containers should be constructed so that they do not tip over easily. Reusable containers that are emptied mechanically should be designed to avoid spills or leaks during on-site storage, collection, or transport. Such containers should be:

• Easily cleanable

• Designed for easy access for depositing and removing such wastes by gravity or mechanical means

• Accessible to the collection vehicle

Single-use plastic and paper bags should meet the National Sanitation Foundation Standard 31 for polyethylene refuse bags and NSFS 32 for paper refuse bags.

11.8.5.1.5.2 SAFETY: [Reference: 40 CFR 243.201-2] Collection personnel should receive instruction in proper handling of wastes and containers as described in "Operation Responsible: Safe Refuse Collection" available from the National Audiovisual Center, General Services Administration, Washington, D.C. 20409.

Collection employees should be issued appropriate protective equipment, such as:

• Gloves

• Safety glasses

• Respirators

• Footwear

The equipment should meet standards set in 29 CFR 1910.132-1910.137.

11.8.5.1.5.3 SOURCE SEPARATION PROCEDURES: [Reference: 40 CFR 246.200-246.202] The guidelines recommend the following procedures for separating solid wastes:

• *High-grade paper.* Such paper can be separated, collected, and sold for recycling. Depending on business needs, three systems can be used to separate paper: the desktop system, the two-wastebasket system, and the office centralized container system. Under the desktop system, each employee places paper in a container on his or her desk and all other waste in a desk-side wastebasket. In the two-wastebasket system, recyclable paper is placed in a desk-side basket and all other waste is placed in another wastebasket. In the office centralized container system, all employees take their paper to a centralized location in the office.

• *Residential materials.* Depending on the market, residential generators can separate newspapers, mixed paper, glass, cans, and plastic at the source of generation. Separated waste can be picked up at curbside or taken to a central-ized neighborhood location or collection facility.

* *Corrugated containers.* Commercial establishments generating less than 10 tons of waste corrugated containers per month are encouraged to separately collect and sell such waste for recycling. Typically the containers are separated, flattened, and stored at a central location within the facility. The containers should be baled and transportation arranged by the facility or a private hauler or purchaser.

11.8.5.2 Nonhazardous Solid Waste Transportation. The permitting of non-hazardous solid waste transportation activities can also be contingent on several factors. The federal government must follow certain guidelines when arranging for transportation of its own solid waste. Nonfederal residential, commercial, and institutional generators are not required to follow the federal guidelines, although EPA encourages them to do so.

In order to obtain approved state-issued permits, transporters of nonhazardous solid waste, especially commercial transporters, usually must comply with a num-ber of state and local restrictions on how waste is collected and transported.

In order to better understand potential state permitting issues associated with the collection and transportation of nonhazardous solid wastes, this chapter dis-cusses the following relevant federal guidelines.

11.8.5.2.1 Federal Guidelines for the Collection and Transportation of Nonhazardous Solid Wastes. Federal regulations provide guidelines—but few requirements—for collecting and transporting solid waste generated by residen-tial, commercial, and institutional entities. In general, households, businesses, and educational and health care facilities are covered by the guidelines.

In addition, the federal government has established regulations on the type of transportation that can be used for hauling separated wastes for recycling.

Guidelines under 40 CFR Parts 243 and 246 include "requirements"—regula-tions that are mandatory for certain parties—and "recommended procedures"—suggested procedures for carrying out the requirements.

The requirements are mandatory for the federal government only. Other enti-ties are not required to implement the requirements but are encouraged to do so. The recommended procedures are not mandatory for federal agencies or any other governmental entities.

11.8.5.2.1.1 TRANSPORTATION REQUIREMENTS

COLLECTION EQUIPMENT: [Reference: 40 CFR 243.202-1] Vehicles used for col-lection and transportation of solid waste, including material separated for recy-cling, must meet all applicable federal standards, including, but not limited to:

* Motor Carrier Safety Standards (49 CFR Parts 390-396)
* Noise Emission Standards for Motor Carriers Engaged in Interstate Commerce (40 CFR Part 202)

Vehicles owned by the federal government must be operated in accordance with the Federal Motor Vehicle Standards (49 CFR Parts 500-580).

EQUIPMENT CONDITION: The equipment used for collection and transportation must be constructed, operated, and maintained to minimize health and safety hazards to personnel and the public. In addition, such equipment must be kept clean to avoid attracting disease-carrying animals. Vehicles used for the collection or transportation of solid waste must be enclosed or be covered to prevent spillage while in transit.

EQUIPMENT STANDARDS: The following types of collection equipment must meet the American National Standards Institute's ANSI Z245.1, Safety Standards for Refuse Collection Equipment:

- Front-, rear-, and side-loading compaction equipment
- Hoist-type and tilt-frame equipment
- Satellite vehicles
- Special collection compaction equipment
- Stationary compaction equipment

11.8.5.2.1.2   COLLECTION MANAGEMENT: [Reference: 40 CFR 243.204-1] Vehicle operators must conduct the collection of solid wastes and materials in such a way to ensure the recycling of these materials is performed in accordance with all applicable laws, including traffic. The vehicle operator is responsible for:

- Cleaning up any spills related to the operator's activities
- Protecting private and public property from damage resulting from the operator's activities
- Conducting his or her job in a manner that minimizes disturbances in residential neighborhoods

11.8.5.2.1.3   RECOMMENDED PROCEDURES: Compliance with the following procedures is optional, as they provide suggestions for carrying out the collection and transportation requirements. Although they may be considered in the issuance of a permit by an approved state, they are not mandatory for the federal government or any other entity.

EQUIPMENT DESIGN: [Reference: 40 CFR 243.202-2] Enclosed, metal, leak-resistant compactor vehicles should be used whenever possible. The vehicles should include the following safety features:

- Exterior rear-view mirrors
- Backup lights
- Four-way emergency flashers
- Accessible first aid equipment

- An audible reverse warning light
- An accessible fire extinguisher
- Handholds and platforms, if crew members ride outside the cab for short trips

The design of the vehicle and its engine vehicle should be one that emphasizes the conservation of fuel and the minimization of air pollution. Vehicle size should take into account the following:

- Local weight and height limits for the roads the vehicle will travel on
- Turning radius
- Loading height to ensure overhead clearance in transfer stations, service buildings, incinerators, or other facilities

EQUIPMENT OPERATION: [Reference: 40 CFR 243.202-3] It is the responsibility of the vehicle's owner-operator to see that collection vehicles are maintained and serviced according to manufacturers' recommendations, and cleaned thoroughly at least once a week. In addition, the following should be inspected periodically:

- Brakes
- Windshield wipers
- Taillights
- Backup lights
- Audible reverse warning devices
- Tires
- Hydraulic systems

Any items not in working order should be promptly repaired prior to a vehicle's being operated and/or put back into service.

COLLECTION MANAGEMENT: [Reference: 40 CFR 243.204-2] It is the responsibility of vehicle owner-operators to see that all relevant records documenting operation costs are properly maintained. These records should be used to schedule vehicle maintenance and replacement and track expenses, and to evaluate system operations' impact on the environment.

System operations should minimize fuel consumption by:

- Designing efficient collection routes that minimize driving distances, delays, and trips to the disposal site
- Arranging regular tuneups and checking air pressure and condition of compactor equipment
- Using transfer stations when the distance or travel time from collection routes to disposal site is great and use of such stations is more cost-effective

- Arranging for residential generators to place their solid waste containers at the curb

- Requiring commercial generators that do not dispose of food wastes to increase storage capacity rather than increasing the frequency of trash collection

SOURCE SEPARATION: [Reference: 40 CFR 246.200-7, 246.201-6, and 246.202-5] The following transportation arrangements can be made for source-separated items:

- High-grade paper. Transportation can be provided by the generator or a private hauler or a purchaser. Collection should be on a regular schedule.

- Residential waste. Transportation can be provided by the facility or community that generated the waste, a private hauler, or a purchaser.

- Corrugated containers. Transportation can be provided by the facility, a private hauler, or a purchaser. In facilities where goods are delivered from a central warehouse, delivery trucks can return the containers to the central warehouse, where they are baled and hauled to a user.

11.8.5.3   Nonhazardous Solid Waste Disposal.   Under the authority of sections 1008(a)(3) and 4004(a) of subtitle D of RCRA, EPA first promulgated the Criteria for Classification of Solid Waste Disposal Facilities and Practices (40 CFR part 257) on Sept. 13, 1979. These subtitle D criteria established minimum national performance standards necessary to ensure that "no reasonable probability of adverse effects on health or the environment" will result from solid waste disposal facilities or practices. A facility or practice that met the criteria was classified as a "sanitary landfill."

EPA regulations establish minimum national federal criteria under RCRA for all municipal solid waste landfill (MSWLF) units and under the Clean Water Act for all municipal solid waste landfills that dispose of sewage sludge. These federal criteria for nonhazardous solid waste disposal facilities, found at 40 CFR Parts 257 and 258, are expected to protect human health and the environment.

EPA regulations define a "MSWLF" as an area of land that receives household waste and that is not a land application unit, surface impoundment, injection well, or waste pile. Such a unit can be owned publicly or privately and receive other types of RCRA subtitle D wastes, such as commercial solid waste, nonhazardous sludge, small-quantity generator waste, and industrial solid waste.

The scope of the criteria found in 40 CFR part 257 originally included MSWLFs. However, with the promulgation of the part 258 regulations, the scope of the part 257 regulations now applies only to solid waste disposal facilities other than MSWLFs. These facilities include surface impoundments, land application units, waste piles, landfills other than MSWLFs, and landfills that stopped accepting waste after Part 258's effective date, Oct. 9, 1993.

All owner-operators of new MSWLFs, existing MSWLFs, and lateral expansions—except as specified in the regulations—are covered by the criteria found in the 40 CFR Part 258 regulations.

Any municipal solid waste landfills failing to satisfy any of the criteria of Parts 257 or 258 is considered to be an "open dump" for purposes of state solid waste management planning, which are prohibited under RCRA. State plans developed pursuant to the Guidelines for Development and Implementation of State Solid Waste Management Plans (40 CFR Part 256) had to provide for closing or upgrading all existing open dumps within the state.

Practices that do not comply with these criteria are also deemed to constitute "open dumping" for purposes of the federal prohibition on open dumping in section 4005(a). EPA does not have the authority to enforce the prohibition directly (except in situations involving the disposal or handling of sludge from publicly owned treatment works, where federal enforcement of POTW sludge-handling facilities is authorized under the CWA). However, the "open dumping" prohibition is enforced by approved states and other persons under section 7002 of RCRA.

Under 40 CFR, Parts 257 and 258, solid waste covers any garbage; refuse; sludge from a waste treatment plant, water supply treatment plant, or air pollution control facility; and other discarded material, including solid, liquid, semisolid, or contained gaseous material resulting from industrial, commercial, mining, agriculture, and community activities.

Excluded from the definition are solid or dissolved materials in domestic sewage, solid or dissolved materials in irrigation return flows, industrial discharges that are point sources subject to a Clean Water Act permit, or special nuclear or by-product material defined by the Atomic Energy Act of 1954.

11.8.5.3.1 Part 257 Facilities.   As stated above, in addition to applying to land application units, surface impoundments, waste piles, and landfills (except municipal solid waste landfills), the provisions of 40 CFR Part 257 extend to landfills that stopped accepting waste after part 258's effective date, Oct. 9, 1993. Part 257 also sets standards for nonmunicipal nonhazardous waste disposal units that receive waste from small quantity generators considered conditionally exempt under hazardous waste regulations at 40 CFR 261.5. See separate chapter 610, RCRA Definition of Hazardous Waste for a definition of conditionally exempt small quantity generators.

11.8.5.3.1.1 PART 257 CRITERIA: The provisions of 40 CFR Part 257 list criteria that include general environmental performance standards addressing the following eight major topics. Solid waste disposal facilities or practices that violate any of these criteria pose a reasonable probability of adverse effects on health or the environment:

* *Floodplains.* Facilities or practices in floodplains shall not restrict the flow of the base flood, reduce the temporary water storage capacity of the floodplain, or result in washout of solid waste, so as to pose a hazard to human life, wildlife, or land or water resources (257.3-1).

- *Endangered species.* Facilities or practices shall not cause or contribute to the taking of any endangered or threatened species of plants, fish, or wildlife. The facility or practice shall not result in the destruction or adverse modification of the critical habitat of endangered or threatened species as identified in 50 CFR Part 17 (257.3-2).

- *Surface water.* For purposes of section 4004(a) of the act, a facility shall not cause a discharge of:

  Pollutants into waters of the United States that is in violation of the requirements of the National Pollutant Discharge Elimination System (NPDES) under section 402 of the Clean Water Act, as amended

  Dredged material or fill material to waters of the United States that is in violation of the requirements under section 404 of the Clean Water Act, as amended

  A facility or practice shall not cause non-point-source pollution of waters of the United States that violates applicable legal requirements implementing an areawide or statewide water quality management plan that has been approved by the administrator under section 208 of the Clean Water Act, as amended (257.3-3).

- *Groundwater.* A facility or practice shall not contaminate an underground drinking water source beyond the solid waste boundary or beyond a specified alternative boundary (257.3-4).

- *Land application.* A facility or practice concerning application of solid waste to within 1 meter (3 feet) of the surface of land used for the production of food-chain crops shall not exist or occur, unless in compliance with established annual application requirements (257.35).

- *Disease.* The facility or practice shall not exist or occur unless the on-site population of disease vectors is minimized through the periodic application of cover material or other techniques as appropriate so as to protect public health (257.3-6).

- *Air.* The facility or practice shall not engage in open burning of residential, commercial, institutional, or industrial solid waste. This requirement does not apply to infrequent burning of agricultural wastes in the field, silvicultural wastes for forest management purposes, land-clearing debris, diseased trees, debris from emergency cleanup operations, and ordnance.

  For purposes of section 4004(a) of the act, the facility shall not violate applicable requirements developed under a state implementation plan (SIP) approved or promulgated by the administrator pursuant to section 110 of the Clean Air Act, as amended.

  As used in this section, "open burning" means the combustion of solid waste without control of combustion air to maintain adequate temperature for efficient combustion

Containment of the combustion reaction in an enclosed device to provide sufficient residence time and mixing for complete combustion

Control of the emission of the combustion products (257.3-7)

- *Safety.* Explosive gases. The concentration of explosive gases generated by the facility or practice shall not exceed:

Twenty-five percent of the lower explosive limit for the gases in facility structures (excluding gas control or recovery system components)

The lower explosive limit for the gases at the property boundary

Fires. A facility or practice shall not pose a hazard to the safety of persons or property from fires. This may be accomplished through compliance with 257.3-7 and through the periodic application of cover material or other techniques as appropriate.

Bird hazards to aircraft. A facility or practice disposing of putrescible wastes that may attract birds and which occurs within 10,000 feet (3048 meters) of any airport runway used by turbojet aircraft or within 5000 feet (1524 meters) of any airport runway used by only piston-type aircraft shall not pose a bird hazard to aircraft.

Access. A facility or practice shall not allow uncontrolled public access so as to expose the public to potential health and safety hazards at the disposal site (257.3-8).

11.8.5.3.2 Part 258 Facilities. As stated above, all owner-operators of new MSWLFs, existing MSWLFs, and lateral expansions are covered by part 258 regulations.

The purpose of Part 258 is to establish minimum national criteria for municipal solid waste landfills, including MSWLFs used for sludge disposal and disposal of nonhazardous municipal waste combustion (MWC) ash (whether the ash is codisposed or disposed of in an ash monofill). Part 258 sets forth the following minimum national criteria for MSWLF units:

- Location restrictions
- Operations
- Design
- Groundwater monitoring and corrective action
- Closure and postclosure care

The provisions of Part 258 provide approved states flexibility in implementing these criteria, where states are authorized to run the program. A MSWLF unit that does not meet these Part 258 criteria will be considered to be engaged in the practice of "open dumping" in violation of section 4005 of RCRA. MSWLF units that receive sewage sludge and fail to satisfy these criteria will be deemed to be in violation of sections 309 and 405(e) of the Clean Water Act.

The protection of air quality around landfills is also addressed by the criteria under RCRA.

11.8.5.3.2.1 PART 258 CRITERIA: [Reference: 40 CFR 258.1] The following part 258 standards are criteria that are taken into consideration by an approved state when issuing a permit for a MSWLF. These standards apply to owner-operators of new, existing, and expanding MSWLFs that receive waste on or after Oct. 9, 1993—the effective date of the final rule—with the exception of certain facilities that dispose of less than 20 tons of municipal solid waste daily. These MSWLFs became subject to the regulations on Oct. 9, 1997. Other effective dates are set for MSWLFs meeting certain specific conditions set in 40 CFR 258.1.

APPLICABILITY: [Reference: 40 CFR 258.21(d), 40 CFR 258.23(e), 40 CFR 258.60(b)(3)] Public or privately owned MSWLFs that handle 20 or fewer tons of municipal solid waste per day (based on an annual average) were required to comply with the Part 258 requirements by Oct. 9, 1997. However, these small MSWLFs may be entitled to flexibility in complying with daily cover requirements, methane monitoring, and infiltration layers for final cover.

Approved state regulators have the authority to grant additional flexibility when issuing permits for small MSWLFs after taking into account climatic and hydrogeologic conditions and the protection of health and the environment. The state also must allow for public review and comment before authorizing the additional flexibility.

LOCATION RESTRICTIONS: [Reference 40 CFR Part 258.10-16] These federal criteria create siting restrictions for locating a MSWLF in the proximity of airports, wetlands, floodplains, fault areas, seismic impact zones, and unstable areas. Also addressed are requirements for the closure of existing MSWLF units.

MSWLF OPERATING CRITERIA. PROCEDURES FOR EXCLUDING THE RECEIPT OF HAZARDOUS WASTE: [Reference 40 CFR 258.20] All MSWLF unit owners or operators must implement a program at the facility for detecting and preventing the disposal of regulated quantities of hazardous wastes and polychlorinated biphenyl (PCB) wastes. This program must include random inspections of incoming loads, records of any inspections, and training of facility personnel to recognize regulated hazardous waste and PCB wastes, and notification to states with authorized RCRA subtitle C programs or the EPA regional administrator in an unauthorized state if a regulated hazardous waste or PCB wastes are discovered at the facility.

COVER MATERIAL REQUIREMENTS: [Reference 40 CFR 258.21] Owners or operators of all MSWLF units are required to cover disposed solid waste with at least 6 inches of earthen materials at the end of each operating day. Daily cover is necessary to control disease vectors, fires, odors, blowing litter, and scavenging. The director of an approved state can temporarily waive the daily cover requirement during extreme seasonal climate conditions and may allow alternative materials to be used as daily cover material.

DISEASE VECTOR CONTROL: [Reference 40 CFR 258.22] Owners or operators of all MSWLF units are required to prevent or control on-site disease vector populations using appropriate techniques to protect human health and the environment.

EXPLOSIVE GASES CONTROL: [Reference 40 CFR 258.23] Owners or operators of all MSWLF units are required to ensure that the concentration of methane generated by the MSWLF does not exceed 25 percent of the lower explosive limit (LEL) in on-site structures, such as scale houses, or the LEL itself at the facility property boundary. The owner or operator must implement a routine methane monitoring program, with at least a quarterly monitoring frequency. If the methane concentration limits are exceeded, the owner or operator must notify the state director within 7 days that the problem exists and submit and implement a remediation plan within 60 days.

AIR CRITERIA: [Reference 40 CFR 258.24]40 CFR 258.24(a) requires owners or operators of all MSWLF units to comply with applicable requirements of state implementation plans (SIPs) developed under section 110 of the Clean Air Act (CAA). Open burning is prohibited except in limited circumstances, which include the infrequent burning of agricultural wastes, silvicultural wastes, land-clearing debris, diseased trees, or debris from emergency cleanup operations.

ACCESS REQUIREMENTS: [Reference 40 CFR 258.25] 40 CFR 258.25 requires owners or operators of all MSWLF units to control public access to MSWLF units and to prevent illegal dumping of wastes, public exposure to hazards at MSWLFs, and unauthorized vehicular traffic.

RUNON/RUNOFF CONTROL SYSTEMS: [Reference 40 CFR 258.26] Section 258.26 requires owners or operators of all MSWLF units to design, construct, and maintain runon and runoff control systems to prevent flow onto and control flow from the active portion of the MSWLF unit. Runoff from the active portion of the unit must be handled in accordance with the surface water requirements of today's rule.

SURFACE WATER REQUIREMENTS: [Reference 40 CFR 258.27] All MSWLF units must be operated in compliance with National Pollutant Discharge Elimination System (NPDES) requirements, established pursuant to section 402 of the Clean Water Act. Any discharges of a non-point source of pollution from an MSWLF unit into waters of the United States must be in conformance with any established water quality management plan developed under the Clean Water Act.

LIQUIDS RESTRICTIONS: [Reference 40 CFR 258.28] The disposal of bulk or non-containerized liquid wastes in MSWLF units is prohibited, with two exceptions:

• The waste is household waste (other than septic waste)

• The waste is leachate or gas condensate that is derived from the MSWLF unit, and the MSWLF unit is equipped with a composite liner and leachate collection system

Containers of liquid waste can be placed in MSWLF units only when the containers (1) are small containers similar in size to that typically found in household waste; (2) are designed to hold liquids for use other than storage; or (3) hold household waste. "Liquid waste" is defined in today's rule as any waste material determined to contain free liquids as defined by Method 9095, "Paint Filter Liquids Test."

RECORD-KEEPING AND NOTIFICATION REQUIREMENTS: [Reference 40 CFR 258.29] The owner or operator of each MSWLF unit must retain documents and records required under this part near the facility in an operating record. (An alternative location may be approved by the director of an approved state.) These documents are listed in 258.29(a) of today's rule. Upon completion of each document required in the operating record, the owner or operator must notify the state director of its existence and its addition to the operating record. Furthermore, all information contained in the operating record must be furnished upon request or be made available at all reasonable times for inspection by the state director.

The director of an approved state may set alternative schedules for the record-keeping and notification requirements specified in the rule except the notification requirements in 258.10(b) pertaining to the notification of the FAA by owner-operators planning to site a new or lateral expansion of a MSWLF within a 5-mile radius of an airport, and 258.55(g)(1)(iii) pertaining to the notification of persons who own land or reside on land overlying a plume of groundwater contamination.

DESIGN CRITERIA: [Reference 40 CFR 258.40] Facility design requirements applicable to new MSWLF units and lateral expansions have been established in Subpart D. These requirements do not apply to existing units.

These final design criteria provide owners and operators with two basic design options:

- A site-specific design that meets the performance standard in today's rule and is approved by the director of an approved state
- A composite liner design

The first option, which is available in approved states, allows owners or operators to consider site-specific conditions in developing a design that must be approved by the director of an approved state. This design must meet the performance standard in 258.40, which requires that the design ensure that established maximum contaminant levels (MCLs) listed in Table 11.8 will not be exceeded at the relevant point of compliance.

The second option, the composite liner system, is required only for landfills located in states without EPA-approved programs. The composite liner system is designed to be protective in all locations, including poor locations. It consists of a composite liner, including a flexible membrane liner and a compacted soil component, and a leachate collection and removal system.

GROUNDWATER MONITORING AND CORRECTIVE ACTION: [Reference 40 CFR 258.50-58] The requirements in this part apply to MSWLF units, except as provided in the next paragraph of this section.

TABLE 11.8 Maximum Contaminant Levels

Chemical	(mG/L)
Arsenic	0.05
Barium	1.0
Benzene	0.005
Cadmium	0.01
Carbon tetrachloride	0.005
Chromium (hexavalent)	0.05
1,4-Dichlorobenzene	0.075
1,2-Dichloroethane	0.005
1,1-Dichloroethylene	0.007
2,4-Dichlorophenoxy acetic acid	0.1
Endrin	0.0002
Fluoride	4
Lead	0.05
Lindane	0.004
Mercury	0.002
Methoxychlor	0.1
Nitrate	10
Selenium	0.01
Silver	0.05
Toxaphene	0.005
Trichloroethylene	0.005
1,1,1-Trichloromethane	0.2
2,4,5-Trichlorophenoxy acetic acid	0.01
Vinyl chloride	0.002

Groundwater monitoring requirements under 258.51 through 258.55 of this part may be suspended by the director of an approved state for a MSWLF unit if the owner or operator can demonstrate that there is no potential for migration of hazardous constituents from that MSWLF unit to the uppermost aquifer (as defined in 258.2) during the active life of the unit and the postclosure care period. This demonstration must be certified by a qualified groundwater scientist and approved by the director of an approved state, and must be based upon:

• Site-specific field-collected measurements, sampling, and analysis of physical, chemical, and biological processes affecting contaminant fate and transport

• Contaminant fate and transport predictions that maximize contaminant migration and consider impacts on human health and environment

Owners and operators of MSWLF units, except those meeting the conditions stated above, must comply with the groundwater monitoring requirements of this part according to the following schedule unless an alternative schedule is specified under paragraph (d) of this section:

- Existing MSWLF units and lateral expansions less than 1 mile from a drinking water intake (surface or subsurface) must be in compliance with the groundwater monitoring requirements specified in 258.51-258.55 by Oct. 9, 1994.
- Existing MSWLF units and lateral expansions greater than 1 mile but less than 2 miles from a drinking water intake (surface or subsurface) must be in compliance with the groundwater monitoring requirements specified in 258.51-258.55 by Oct. 9, 1995.
- Existing MSWLF units and lateral expansions greater than 2 miles from a drinking water intake (surface or subsurface) must be in compliance with the groundwater monitoring requirements specified in 258.51-258.55 by Oct. 9, 1996.
- New MSWLF units must be in compliance with the groundwater monitoring requirements specified in 258.51-258.55 before waste can be placed in the unit.

The director of an approved state may specify an alternative schedule for the owners or operators of existing MSWLF units and lateral expansions to comply with the groundwater monitoring requirements listed above. This schedule must ensure that 50 percent of all existing MSWLF units are in compliance by Oct. 9, 1994, and all existing MSWLF units are in compliance by Oct. 9, 1996. In setting the compliance schedule, the director of an approved state must consider potential risks posed by the unit to human health and the environment. The following factors should be considered in determining potential risk:

- Proximity of human and environmental receptors
- Design of the MSWLF unit
- Age of the MSWLF unit
- Size of the MSWLF unit
- Types and quantities of wastes disposed including sewage sludge
- Resource value of the underlying aquifer, including:

  Current and future uses
  Proximity and withdrawal rate of users
  Groundwater quality and quantity

Owners and operators of all MSWLF units that meet the conditions listed below must comply with all applicable groundwater monitoring requirements of this part by Oct. 9, 1997.

Once established at a MSWLF unit, groundwater monitoring shall be conducted throughout the active life and postclosure care period of that MSWLF unit as specified in 258.61.

For the purposes of this subpart, a qualified groundwater scientist is a scientist or engineer who has received a baccalaureate or postgraduate degree in the natural sciences or engineering and has sufficient training and experience in groundwater hydrology and related fields as may be demonstrated by state registration, professional certifications, or completion of accredited university programs that enable that individual to make sound professional judgments regarding groundwater monitoring, contaminant fate and transport, and corrective action.

CLOSURE AND POSTCLOSURE CARE: [Reference 40 CFR 258.60 and 61].

CLOSURE CRITERIA: Owners or operators of all MSWLF units must install a final cover system that is designed to minimize infiltration and erosion. The final cover system must be designed and constructed to:

- Have a permeability less than or equal to the permeability of any bottom liner system or natural subsoils present, or a permeability no greater than 1×10^{-5} centimeters per second, whichever is less
- Minimize infiltration through the closed MSWLF by the use of an infiltration layer that contains a minimum 18 inches of earthen material
- Minimize erosion of the final cover by the use of an erosion layer that contains a minimum 6 inches of earthen material that is capable of sustaining native plant growth

The director of an approved state may approve an alternative final cover design that includes:

- An infiltration layer that achieves an equivalent reduction in infiltration as the infiltration layer specified in paragraphs (a)(1) and (a)(2) of this section.
- An erosion layer that provides equivalent protection from wind and water erosion as the erosion layer specified in paragraph (a)(3) of this section.
- The director of an approved state may establish alternative requirements for the infiltration barrier in paragraph (b)(1) of this section, after public review and comment, for any owners or operators of MSWLFs that dispose of 20 tons of municipal solid waste per day or less, based on an annual average. Any alternative requirements established under this paragraph must:

   Consider the unique characteristics of small communities

   Take into account climatic and hydrogeologic conditions

   Be protective of human health and the environment

The owner or operator must prepare a written closure plan that describes the steps necessary to close all MSWLF units at any point during their active life in accordance with the cover design requirements in 258.60(a) or (b), as applicable. The closure plan, at a minimum, must include the following information:

- A description of the final cover, designed in accordance with 258.60(a) and the methods and procedures to be used to install the cover
- An estimate of the largest area of the MSWLF unit ever requiring a final cover as required under 258.60(a) at any time during the active life

- An estimate of the maximum inventory of wastes ever on-site over the active life of the landfill facility

- A schedule for completing all activities necessary to satisfy the closure criteria in 258.60

The owner or operator must notify the state director that a closure plan has been prepared and placed in the operating record no later than the effective date of this part, or by the initial receipt of waste, whichever is later.

Prior to beginning closure of each MSWLF unit as specified in 258.60(f), an owner or operator must notify the state director that a notice of the intent to close the unit has been placed in the operating record.

The owner or operator must begin closure activities of each MSWLF unit no later than 30 days after the date on which the MSWLF unit receives the known final receipt of wastes or, if the MSWLF unit has remaining capacity and there is a reasonable likelihood that the MSWLF unit will receive additional wastes, no later than 1 year after the most recent receipt of wastes. Extensions beyond the 1-year deadline for beginning closure may be granted by the director of an approved state if the owner or operator demonstrates that the MSWLF unit has the capacity to receive additional wastes and the owner or operator has taken and will continue to take all steps necessary to prevent threats to human health and the environment from the unclosed MSWLF unit.

The owner or operator of all MSWLF units must complete closure activities of each MSWLF unit in accordance with the closure plan within 180 days following the beginning of closure as specified in paragraph (f) of this section. Extensions of the closure period may be granted by the director of an approved state if the owner or operator demonstrates that closure will, of necessity, take longer than 180 days and he or she has taken and will continue to take all steps to prevent threats to human health and the environment from the unclosed MSWLF unit.

Following closure of each MSWLF unit, the owner or operator must notify the state director that a certification, signed by an independent registered professional engineer or approved by director of an approved state, verifying that closure has been completed in accordance with the closure plan, has been placed in the operating record.

Following closure of all MSWLF units, the owner or operator must record a notation on the deed to the landfill facility property, or some other instrument that is normally examined during title search, and notify the state director that the notation has been recorded and a copy has been placed in the operating record.

The notation on the deed must in perpetuity notify any potential purchaser of the property that:

- The land has been used as a landfill facility

- Its use is restricted under 258.61(c)(3)

The owner or operator may request permission from the director of an approved state to remove the notation from the deed if all wastes are removed from the facility.

POSTCLOSURE CARE REQUIREMENTS: [Reference 40 CFR 258.61] Following closure of each MSWLF unit, the owner or operator must conduct postclosure care. Postclosure care must be conducted for 30 years, except as provided under paragraph (b) of this section, and consist of at least the following:

- Maintaining the integrity and effectiveness of any final cover, including making repairs to the cover as necessary to correct the effects of settlement, subsidence, erosion, or other events, and preventing runon and runoff from eroding or otherwise damaging the final cover.

- Maintaining and operating the leachate collection system in accordance with the requirements in 258.40, if applicable. The director of an approved state may allow the owner or operator to stop managing leachate if the owner or operator demonstrates that leachate no longer poses a threat to human health and the environment.

- Monitoring the groundwater in accordance with the requirements of subpart E of this part and maintaining the groundwater monitoring system, if applicable.

- Maintaining and operating the gas monitoring system in accordance with the requirements of 258.23.

The length of the postclosure care period may be:

- Decreased by the director of an approved state if the owner or operator demonstrates that the reduced period is sufficient to protect human health and the environment and this demonstration is approved by the director of an approved state.

- Increased by the director of an approved state if the director of an approved state determines that the lengthened period is necessary to protect human health and the environment.

The owner or operator of all MSWLF units must prepare a written postclosure plan that includes, at a minimum, the following information:

- A description of the monitoring and maintenance activities required in 258.61(a) for each MSWLF unit, and the frequency at which these activities will be performed.

- Name, address, and telephone number of the person or office to contact about the facility during the postclosure period.

- A description of the planned uses of the property during the postclosure period. Postclosure use of the property shall not disturb the integrity of the final cover, liner(s), or any other components of the containment system, or the function of the monitoring systems unless necessary to comply with the requirements in this Part 258. The director of an approved state may approve any other disturbance if the owner or operator demonstrates that disturbance of the final cover, liner, or other component of the containment system, including any removal of waste, will not increase the potential threat to human health or the environment.

The owner or operator must notify the state director that a postclosure plan has been prepared and placed in the operating record no later than the effective date of Part 258, Oct. 9, 1993, or by the initial receipt of waste, whichever is later.

Following completion of the postclosure care period for each MSWLF unit, the owner or operator usually must notify the state director that a certification, signed by an independent registered professional engineer or approved by the director of an approved state, verifying that postclosure care has been completed in accordance with the postclosure plan, has been placed in the operating record.

11.8.6 State Nonhazardous Waste Management Activity Permitting Criteria

As was discussed in the previous sections of this chapter, EPA's role is to establish the regulatory basis and provide both guidelines and technical assistance for the development of environmentally sound nonhazardous solid waste management practices. The actual permitting of nonhazardous solid waste programs is usually the responsibility of state and local governments. Consequently, each of the approved states has developed its own regulations for addressing nonhazardous waste management activities. Since it is beyond the scope of this chapter to review each approved state's program, the discussions contained in this section will be focused on listing typical approved state permitting requirements for transporters and landfills.

11.8.6.1 Transporter Permit Criteria. This section discusses typical permit criteria for the collection, transport, and delivery of regulated waste, originating or terminating at a location within an approved state. Regulated medical waste is not covered in this section and may contain restrictions over and above those mentioned here.

Typically, the following waste management activities are not permitted unless the person or persons performing them have received a valid nonhazardous solid waste transporter's permit from an approved state:

- Collect or remove any regulated waste from its point of origin, generation, or occurrence
- Transport any regulated waste
- Deliver any regulated waste to a treatment, storage, or disposal facility, or otherwise dispose of or relinquish possession of any regulated waste other than as specified in such permit
- Land spread or impound any septage within the scope of those activities as authorized by the approved state
- Land spread sewage sludge within the scope of that activity that is authorized by an approved state.

Usually no person who owns or operates a facility at, or premises on which any regulated waste originates, is generated, or occurs, can deliver or otherwise relinquish possession of such waste except to a person who has a valid permit issued by the approved state.

11.8.6.1.1 Key State Definitions. The following generalizations of key terms used by approved states in their permitting processes are listed so that the reader can better understand issues relating to obtaining a nonhazardous solid waste transporter's permit.

Disposal can mean the abandonment, discharge, deposit, injection, dumping, spilling, leaking, or placing of any nonhazardous solid waste on or into any lands or waters of the state so that such nonhazardous solid waste may enter the environment or be emitted into the air or be discharged into any waters, including groundwaters. Disposal can also mean the thermal destruction of waste or hazardous waste and the burning of such wastes as fuel for the purpose of recovering usable energy.

Storage can mean the holding of nonhazardous solid waste for a temporary period, at the end of which the nonhazardous solid waste is processed, recovered, disposed of, or stored elsewhere.

Storage incidental to transport can mean any on-vehicle storage which occurs en route from the point of initial nonhazardous solid waste pickup to the point of final delivery for purposes such as, but not limited to, overnight on-the-road stops, stops for meals and driver comfort, stops at the transporter's facility for weekends immediately prior to shipment, or on-vehicle storage not to exceed a prescribed number of days at the transporter's facility for the express purpose of consolidating loads (where such loads are not removed from their original packages or containers) for delivery to an authorized treatment, storage, or disposal facility.

Vehicle can mean any device or contrivance which is required by law to be registered with a state, province, or the federal government for conveyance over public roads and which actually contains or carries a regulated waste; for example, in the case of a tractor-trailer combination, the trailer is considered to be the vehicle; and in the case of a roll-off container or other removable containment device, it is the mobile flatbed or the undercarriage that is considered to be the vehicle.

A *solid waste* can be any garbage, refuse, sludge or any solid, liquid, semisolid, or contained gaseous material resulting from industrial, commercial, mining, agricultural, community, or other activities, not excluded below, which is discarded, disposed of, burned, or incinerated, including being burned as a fuel for the purpose of recovering usable energy, or is being accumulated, stored, or physically, chemically, or biologically treated in lieu of or prior to being disposed of, burned, or incinerated, or which has served its original intended use and is sometimes discarded, or is a manufacturing or mining by-product and sometimes is discarded.

The following materials are usually not considered to be a "solid waste" by approved states:

- Domestic sewage and any mixture of domestic sewage and other wastes that pass through a sewer system to a publicly owned treatment works for treatment (domestic sewage means untreated sanitary wastes that pass through a sewage system).

- Industrial wastewater discharges that are point source discharges for which a permit has been issued pursuant to article 17 of the Environmental Conservation Law (*Note:* This exclusion applies only to the actual point source discharge. The exclusion does not apply to industrial wastewaters while they are being collected, stored, or treated before discharge, nor does it apply to sludges that are generated by industrial wastewater treatment).

- Irrigation return flows.

- Radioactive materials which are source, special nuclear, or by-product material. For the purposes of this part: source material means uranium and/or thorium, or ores containing by weight 0.05 percent or more of uranium and/or thorium; special nuclear material means plutonium, uranium 233, uranium enriched in uranium 233 or uranium 235, or any material artificially enriched by any of these; any by-product material means radioactive material yielded in or made radioactive by exposure to radiation incident to the process by producing or utilizing special nuclear materials, trailings, or waste produced by the extrication or concentration of uranium or thorium from any ore processed primarily for its source material content.

- Materials subject to in-site mining techniques which are not removed from the ground as part of the extraction process.

A *regulated waste* is a solid waste which is raw sewage, septage, sludge from a sewage or water supply treatment plant, waste oil, or industrial-commercial waste, including hazardous waste.

An *industrial-commercial waste* is any solid waste that originates at, is generated by, or occurs as a result of any industrial or commercial activity. Industrial-commercial wastes include, but are not limited to:

- Liquids such as:

  Acids, alkalies, caustics, leachate, petroleum (and its derivatives), and process or treatment wastewaters

  Sludges, which are semisolid substances resulting from process or treatment operations or residues from storage or use of liquids

- Solids such as:

  Solidified chemicals, paints, or pigments

  Dredge spoil, foundry sand, and the end or by-products of incineration or other forms of combustion, including bottom ash and fly ash

- Contained gaseous materials

- Hazardous waste as defined in Section 371.1(d) of this title

- Any liquid, sludge, septage, solid, semisolid substance or contained gaseous material in which any of the foregoing is intermixed or absorbed, or onto which any of the foregoing is adhered

(*Note:* The classification of "waste oil" as either a hazardous waste on a nonhazardous special waste varies from state to state. It is imperative for the reader to check his or her approved state's classification of this material. "Waste oil" can be used engine lubricating oil and any other oil, including but not limited to fuel oil, motor oil, gear oil, cutting oil, transmission fluid, hydraulic fluid, dielectric fluid, oil storage tank residue, animal oil, and vegetable oil, which has been contaminated by physical or chemical impurities, through use or accident, and has not subsequently been re-refined.)

11.8.6.1.2   State Transporter Permit Exemptions.   The following types of vehicles usually do not require a nonhazardous solid waste transporter's permit issued by an approved state.

- Rail, water, and air carriers are exempt from the requirements of this part.
- Vehicles transporting the following regulated wastes are usually exempted from the requirement of obtaining a nonhazardous solid waste transporter's permit provided that no other regulated waste is intermixed, contained in, or otherwise included with such waste:

  Vegetable oils and greases from restaurants and fast food operations

  Tallow (animal fat)

  Food processing waste destined for use in other food or animal feed processes (except blood)

  Garbage and trash collection from cafeterias

  Food processing residues which are recognizable as part of the plant or vegetable, including, but not limited to, cabbage leaves, bean snips, onion skins, apple pomace, and grape pomace (except brewery wastes)

  Scraps, including, but not limited to plastic, rubber, paper, cardboard, wood chips, glass, and metal

  Grubbing, construction, and renovation debris, such as roots, stumps, bricks, cement, asphalt, blacktop, stone and like materials, except asbestos

  Agricultural waste, including but not limited to crop residues and animal manure productively employed in agriculture

  Nonhazardous dredge or fill material

Nonhazardous bottom and fly ash from incinerators and resource recovery facilities

Foundry sand containing no phenols (less than 1 part per billion)

Empty drums or containers destined for reconditioning or being returned to the original manufacturer

Empty food containers being collected, transported, or stored for recycling or reuse

Samples shopped to laboratories solely for analysis

Scrap lead-acid automotive batteries destined for recovery

Waste transported by a public utility vehicle where the transportation of such waste is incidental to the primary function of the vehicle whenever the waste is brought to a utility-owned collection facility for storage prior to treatment or disposal

Waste collected, transported, or transferred wholly on-site by the person responsible for the origination, generation, or occurrence of such waste, provided that storage, treatment, and disposal of waste upon those premises are authorized pursuant to this title. (As used in this subparagraph, "on-site" means the same or geographically contiguous property. It may be divided by public or private right-of-way, provided the entrance and exit between the properties is at a crossroads intersection, and access is by crossing, as opposed to going along, the right-of-way. Noncontiguous properties owned by the same person, but connected by a right-of-way which that person controls and to which the public does not have access, are also considered onsite property.)

Pesticides, transported by the farmer who generated them, to a pesticide cleanup day collection site authorized by the permitting approved state

Bottom ash from the burning of fossil fuel, provided that:

The ash has been tested for toxicity by the owner or operator of the generating facility pursuant to a testing protocol approved by the commissioner, and certified to be nontoxic

The ash is destined for use by a municipality or other governmental entity as a traction agent on roadways

11.8.6.1.3 State Transporter Permit Application Process. Applications usually must be completed and submitted on forms prescribed by the department and should indicate the type of waste involved, vehicles that the applicant will use, any transfer or storage facilities the applicant will use (except where such transfer or storage is incidental to transport), and the place or places where, and the manner in which, the applicant will finally treat, store, or dispose of the collected waste. The application should also contain such analyses, plans, reports, fees, insurance certificates, and other data as the department may require.

The applicant typically must demonstrate that the proposed disposal site is one authorized by the approved state.

When regulated wastes are to be disposed of at a site that is not owned by the applicant, the application should be accompanied by written permission from the site owner for such activity.

The department receiving the permit application may require inspection of vehicles as a condition of application approval or review during the permit year.

Applications for permit renewals, in order to be timely submitted for purposes of the State Administrative Procedures Act, must be received by the department at least 30 days in advance of the expiration date of the existing permit.

The department may require a form of surety or financial responsibility from a permittee acceptable in form and amount to the department, to ensure compliance with the terms of the permit issued by the approved state.

Policies of insurance and surety bonds required under this section usually may be replaced by other policies of insurance or surety bonds. Policies should state that the liability of the retiring insurer or surety shall terminate on the effective date of the replacement policy of insurance or surety bond or at the end of the prescribed cancellation period, whichever is sooner.

11.8.6.1.4 State Transporter Permitting Standards. The approved state's decision to issue or deny a permit for the transport of regulated waste usually is based on the following considerations:

- The status of any receiving facilities identified in the permit application. A valid nonhazardous solid waste transporter's permit is not usually issued unless each of the receiving facilities is in one of the following categories:

  A facility authorized by the approved state to accept such waste

  A facility operating under an active approved state issued order on consent

  A facility outside the jurisdiction of the approved state, in such case proof of authorization to operate may be required by the issuing department as a condition of application review

  Facility not requiring any state or federal license, permit, or certificate to operate

A waste transporter permit may be denied, revoked, suspended, or modified if the receiving facility has been determined to have violated any law, rule, or regulation or permit condition related to the operation of its treatment, storage, or disposal facility.

A waste transporter permit may be denied, revoked, suspended, or modified based upon the unsuitability of the applicant under the provisions of the approved state's regulations.

11.8.6.1.5 Vehicle Operation Requirements. The operator of any vehicle used for nonhazardous solid waste management activities usually is required to carry the original permit or a legible photocopy of such permit in the vehicle. The operator should be able to present the permit, together with shipping or transporting documents relative to the waste being transported, to authorized representatives of the issuing department or to any law enforcement officers when requested to do so.

A permittee usually must display the full name of the transporter on both sides of each vehicle and the transporter's permit number in figures at least 3 inches high and of a color which contrasts with the background, in a prominent position on each side and the rear of each vehicle used for nonhazardous solid waste management activities.

The operator of any vehicle used for such activities should remain with such vehicle while it is being filled or discharged.

All wastes must be properly contained during transport so as to prevent leaking, blowing, or any other type of discharge into the environment.

A permittee typically has to submit a report to the permitting department annually, or more frequently if the department deems necessary, on forms prescribed by the department. A permittee should retain for at least 3 years the records on which such reports are based and must be able to make such records available, upon request, to the department during normal business hours.

A permittee and the operator of any vehicle used for these activities has to comply with all applicable state and federal laws and all rules and regulations promulgated thereunder. The permittee is responsible for all requirements for all vehicles, including leased vehicles operated under his or her permit.

A permittee usually must conspicuously mark or placard every vehicle, in a manner consistent with approved state transportation law and any rules and regulations promulgated thereunder and any related federal requirements related to the transportation of the regulated waste and its principal hazard.

Permitted vehicles typically are restricted to the transportation of materials not intended for human or animal consumption or for other use by the general public except when properly cleaned in accordance with all applicable federal and state regulations governing decontamination.

11.8.6.1.6 Transfers of Permits. Permits are usually not transferable. Changes of ownership invalidate the provisions of such permits. Any change of address, name, or location of garaged vehicles must be submitted immediately to the permitting department.

11.8.7 MSWLF Permit Criteria

Usually, a person or persons may not construct or operate a sanitary landfill, or materially alter or extend one, without obtaining a permit from the approving authority before any work, including site preparation, is begun. This requirement typically applies to all the following types of sanitary and municipal landfills:

- Municipal landfills
- Land clearing debris landfills

- Rubble landfills
- Industrial waste landfills

11.8.7.1   MSWLF Permit Exclusions.   Permits typically are not required for the following:

- Land disposal areas for solid waste generated from single-family residences within the property limits of these residences when this disposal is not in violation of state or local laws
- Land disposal areas which are not part of a system of refuse disposal for public use
- Land disposal of agricultural waste on agricultural land
- Land disposal of overburden resulting from mining operation
- Filling operations which consist solely of the importation of clean earthen fill containing rock, concrete, nonrefractory brick, and asphalt created as a result of construction excavation activities, mining, or regrading projects provided that:

    If warranted, a county grading permit is obtained

    The filling, grading, and site stabilization is carried out in accordance with the provisions

11.8.7.2   MSWLF State Permit Applications.   Written requests for permits are usually submitted to the approving state authority using the application forms provided by the approving authority. The request briefly describes the nature of the proposal.

11.8.7.3   Preliminary Report.   Multiple copies of a preliminary report (sometimes referred to as a Phase I report) shall be prepared and submitted along with the request for a permit.

  11.8.7.3.1   Contents of the Preliminary Report.   At a minimum, the preliminary report will include a:

- Completed and signed application form as required by the approved state
- Current U.S.G.S. 7.5-minute quadrangle map with the proposed site outlined
- Current topographic map, which is an accurate depiction of the site at the time of application, at a scale not smaller than 1 inch equals 200 feet, which depicts the property boundaries, on-site buildings and structures, and pertinent surficial features including but not limited to:

    Springs

    Seeps

Streams

Rock outcrops

Sink holes

Surface impoundments

Water wells

Forested areas

The location of any buried or overhead power transmission lines, utility pipelines, or storage tanks on the property

- Map which depicts the surrounding zoning and land use within $1/2$ mile of the site boundaries

- Map showing the distribution of the soils at the site

- Narrative description of the soils at the site

- Map showing the geology at the site based on available data

- Narrative description of the geology at the site based on available data

- Description of the proposed activity including:

    Type of facility

    Area served

    Capacity

    Types of waste accepted

11.8.7.3.2   Preliminary Report Review.   Following receipt of the specified number of copies of the required information, the department receiving the permit application will usually distribute one copy to each of the following (or their equivalent, since titles vary from state to state):

- Chief executive officer or the governing body, or both, of a county or municipality in which the activity is proposed
- Local operating agency responsible for solid waste management
- Local health official
- Secretary, Department of Natural Resources
- Director, Water Resources Administration
- Director, Geologic Survey
- U.S. Geological Survey
- Federal Aviation Administration

- Appropriate Soil Conservation District
- U.S. Army Corps of Engineers
- State Highway Administration

A person receiving a copy of the application and supporting information is usually invited to inspect the proposed site and requested to submit comments to the department originally receiving the permit application within a specified time frame, perhaps 20 days from receipt of the report.

The department shall set a date, time, and place for a joint site inspection meeting with interested agencies and the applicant.

When practicable, perhaps within 60 days of receipt of a complete preliminary report, the department usually will:

- Review the preliminary report for completeness. The department will notify the applicant that the preliminary report is complete.

- Make a determination with respect to the site. If the department determines that the site is not suitable for the intended use, the department will usually deny the application. The applicant is notified in writing by the approving authority and is informed of the basis for the denial, and the appeal process. Otherwise the approving authority advises the applicant in writing of any limitations which the preliminary investigation revealed concerning the use of the site, and of any general recommendations. The applicant is advised to proceed with the preparation of a geologic report.

If the department is unable to complete the review within the established schedule, the department shall notify the applicant in writing, perhaps within 30 days of receipt of the information, and inform the applicant of the anticipated time required to complete the review.

11.8.7.4 Site Geology Report. The applicant shall prepare and submit to the approving authority the prescribed number of copies of a geologic report, sometimes known as a Phase II report, describing the soils, geology, meteorology, and hydrology of the proposed site.

11.8.7.4.1 Site Geology Report Certification The report usually must be developed and signed by a geologist or geotechnical engineer. The geologist usually must possess at least a bachelor's degree from an accredited college or university in the field of geology or a related field of earth science. The geotechnical engineer usually must possess at least a bachelor's degree in civil or environmental engineering from an accredited college or university, with the major course of study in geotechnical engineering or related earth sciences, and should be registered as a professional engineer. The geologist or geotechnical engineer may have to have at least 5 years of experience in performing hydrogeologic investigations.

11.8.7.4.2 Site Geology Report Content. The report usually must include the following information in sufficient detail to permit a comprehensive review of the project:

- An up-to-date site-specific topographic map using a contour interval that is practical for that site. The following items shall be shown on the map:

    Surface waters and natural drainage features

    100-year floodplain

    Property lines

    On-site buildings and structures

    Forested and other vegetated areas

    The location of any buried or overhead power transmission lines, utility pipelines, or storage tanks on the property

- A discussion of the geologic formations directly underlying and in close proximity to the site, the present and projected use of these formations as a source of groundwater and minerals, and the hydrogeologic relationship between the formations.

- A survey of all production wells within $1/2$ mile of the site boundary. Each well shall be located on the topographic map and a table shall be developed which specifies all the available pertinent information such as well depth, screen type, productivity, materials encountered, and water level.

- Groundwater contour maps to show the occurrence and direction of groundwater flow beneath the site superimposed on the current topographic map. Three separate groundwater contour maps as specified below shall be constructed for each distinct water-bearing formation occurring within 50 feet of the anticipated lowest elevation of the refuse cell floor, using monthly groundwater elevation data collected from piezometers on the site over a period of not less than 12 months, or derived using a hydrologic simulation or prediction technique approved by the department. Three groundwater contour maps shall be constructed from a set of:

    Water elevations measured or predicted during the month that represents:

    The most elevated groundwater condition

    The most depressed groundwater condition

    The highest observed or predicted groundwater elevations

- Geologic cross sections in sufficient detail, orientation, and number to clearly identify subsurface conditions at the site.

- A bedrock map, except in the coastal plain outside the fall zone, to show the contours of the bedrock surface beneath the site.

- An isopachous map to show the minimum thickness of soil and other unconsolidated sediments above the elevation of groundwater or bedrock, whichever is the higher.

• A discussion of the chemical quality of groundwater in the aquifers beneath the site. The list of chemical parameters usually must include pH, alkalinity, hardness, chloride, specific conductance, nitrate, chemical oxygen demand, arsenic, barium, cadmium, chromium, zinc, lead, mercury, the volatile priority pollutants, and any other pollutants specified by the permitting department.

• A discussion of the potential for the vertical and horizontal movement of pollutants into the waters of the state.

• The results of a fracture trace or aerial photographic lineament analysis (except in the coastal plain) which identifies the relationship between these features and the local groundwater hydrology.

• Test boring logs, well completion reports, piezometric measurements, chemical and physical soil and sediment analyses, and all accompanying geotechnical analyses. All laboratory and field methodologies and procedures typically must be included.

• A preliminary conceptual design of the proposed municipal landfill based on the geotechnical information gathered in C(1)-(11). The landfill design must typically satisfy the following minimum design standards:

A liner system that is designed, constructed, and installed to facilitate collection of leachate generated by the landfill to prevent migration of pollutants out of the landfill to the adjacent subsurface soil, groundwater, or surface water. The liner may be constructed of natural earthern materials which are excavated from on the site or which are imported from another location. The liner may also be constructed of a synthetic or manufactured membrane material. The liner system usually must be:

Constructed of materials that have sufficient strength and thickness to prevent failure due to pressure gradients, physical contact with the waste or leachate, climatic conditions, the stress of installation, and the stress of daily operation.

A minimum of 1 foot of clay or other natural material having an in-place permeability less than or equal to 1×10^{-7} centimeters per second, or one or more unreinforced synthetic membranes with a combined minimum thickness of 50 mil, or a single reinforced synthetic membrane with a minimum thickness of 30 mil, which has a permeability less than or equal to 1×10^{-10} centimeters per second, placed over a prepared subbase with a minimum thickness of 2 feet and a permeability less than or equal to 1×10^{-5} centimeters per second. The approving authority may authorize the installation of a liner system with specifications different from those listed in this subparagraph only upon a successful demonstration by the applicant that the alternate system is capable of collecting and managing the leachate generated at the site, and that the liner system provides an equivalent level of protection to public health and the environment.

Placed upon a foundation or base capable of providing support to the liner and resistance to pressure gradients above and below the liner to prevent failure of the liner due to settlement, compression, uplift, puncture, cutting, or activities at the landfill.

Installed to cover all surrounding earth likely to be in contact with the waste or leachate.

Installed with a minimum slope of 2 percent to facilitate movement of leachate toward the leachate collection system and prevent leachate ponding on the landfill floor.

- The liner system shall be located entirely above the composite high water table and bedrock. A minimum buffer distance, including the thickness of the prepared subbase, usually is required between the bedrock elevation and the maximum expected groundwater elevation, and the bottom of the liner system, as follows.

- A leachate collection and removal system, located immediately above the liner, that is designed, constructed, maintained, and operated to collect and remove leachate from the landfill. Typically the leachate collection and removal system must be:

Constructed of materials that are chemically resistant to the waste managed in the landfill and the leachate expected to be generated, and of sufficient strength and thickness to prevent collapse under the pressures exerted by overlying wastes, waste cover materials, and any equipment used at the landfill

Designed and operated to function without clogging

Designed and operated to ensure that the leachate depth over the liner does not exceed 30 centimeters (1 foot)

11.8.7.4.3 Site Geology Report Review. Following receipt of the specified number of copies of the site geology report, the permitting department usually distributes one copy to each of the following or their equivalents:

- Secretary, Department of Natural Resources
- Director, Water Resources Administration
- Director, Geological Survey
- U.S. Geological Survey
- County Soil Conservation District

- Chief executive officer or the governing body, or both, of the county or municipality in which the activity is proposed
- Local health official
- Local operating agency responsible for solid waste management

A person receiving a copy of the site geology report usually is requested by the approving authority to submit any comments to the department within a prescribed time period, perhaps 30 days of receipt of the report.

The approving authority typically sets a date, time, and place for a joint plan review meeting with interested agencies and the applicant.

When practicable, perhaps as long as 60 days following the meeting, the approving authority will either deny the permit or determine if:

- Sufficient information is available to proceed to the plans and engineering report
- Revisions to the site geology report are needed

If the department is unable to complete the review within the established time schedule, the department will notify the applicant in writing within a prescribed time period (e.g., 30 days of receipt of the information) and inform the applicant of the anticipated time required to complete the review.

11.8.7.5 Plans and Engineering Report.   Multiple complete sets of plans and engineering reports, sometimes known as a Phase III report, covering the proposed project, prepared, signed, and bearing the seal of a registered professional engineer must typically be submitted to the approving state authority.

11.8.7.5.1 Plans and Engineering Report Content.   These plans and specifications usually must include the following information in sufficient detail to permit a comprehensive review of the project:

- A map which designates the property boundaries, the actual area to be used for filling, and all existing and proposed structures.
- A description of any vehicle weighing facilities, any telephones or other communications equipment, any maintenance and equipment storage facilities, any employee safety and sanitary facilities, and any water supply and sewerage systems. Any proposed on-site water supply and sewerage systems shall be approved by the approving authority.
- The location and type of all existing or proposed on-site roads.
- A description of the types of solid waste which will be accepted, the types of solid waste which may not be accepted, and the area and population which will be served by the facility.
- The anticipated quantities of solid waste which will be accepted and the calculations used to determine the useful life of the facility.
- Proposed methods of collecting and reporting data on the quantities and types of solid waste received and for revising facility life expectancy projections.

- The volume and type of available cover material, the calculated volume of earth needed for daily, intermediate, and final cover, the location of earth stockpiles, and provisions for saving topsoil for use as final cover.
- Proposed means of controlling unauthorized access to the site.
- Proposed operating procedures including:

  Hours and days of operation

  Number and types of equipment to be used

  Number of employees and their duties

  Provisions for fire prevention and control

  Means of preventing public health hazards and nuisances from blowing paper, odors, rodents, vermin, noise, and dust

  Proposed method of daily operation including wet weather operation

- The location and depth of solid waste cells and the sequence of filling.

- Natural or artificial screening to be used.

- Methods of controlling on-site drainage, drainage leaving the site, and drainage onto the site from adjoining areas. Erosion and sediment control provisions shall be approved by appropriate approving agency.

- Proposed methods and engineering specifications for the collection, management, and disposal of leachate generated at the facility. The calculations used to determine the anticipated quantities of leachate that will be generated usually should be included.

- A contingency plan for preventing or abating the pollution of the waters of the state. At a minimum, the contingency plan shall address the following items:

  Emergency provisions of potable water to users whose supply may be affected by the landfill

  A spill containment and prevention plan for any leachate collected or stored at the site

  Emergency telephone numbers and contact persons for fires, medical emergencies, spills of hazardous materials, or other emergency situations

- Proposed methods for controlling landfill gas.

- Proposed methods for covering and stabilizing completed areas.

- A system for routinely monitoring the quality of the waters of the state around and beneath the site, including the location and types of monitoring stations and the methods of construction of monitoring wells. Wells shall be installed by a state licensed well driller in accordance with COMAR 26.04.04.

• A statement that "Burning of solid waste is not allowed, except as permitted by the department of the environment."

• A schedule for implementing construction and implementation of the operation plans and engineering specifications once the refuse disposal permit has been issued.

• A landfill closure and postclosure plan to be followed over a period of not less than 5 years after application of final cover.

• The name, address, and telephone number of the person or agency responsible for the maintenance and operation of the site. Changes to this information shall be submitted to the approving authority once affected.

11.8.7.5.2 Other Requirements for Engineering Plans and Specifications. Plans usually are required to be neatly and legibly drawn upon tracing cloth or other acceptable drafting medium so that copies may be obtained. Except when the originals are presented for preliminary judgment, the originals typically must be kept by the party presenting the plans, and true copies, which may be white, blue, or black-line prints, offset prints, or photostats on a white background, shall be submitted to the approving authority. Plans may not be returned after submission unless revisions are necessary. When revisions are necessary, the revisions should be made on the original tracings and new prints submitted. Pencil, crayon, ink, typewritten or similar revisions, changes, or additions on prints, photostats, or other reproductions are not acceptable. Revisions or additions usually may not be made to prints previously or currently submitted to the approving authority. Plans specifically prepared for each installation are typically required.

Size of plans: Drawings usually must meet certain size requirements, typically 24 inches wide and 36 inches long.

Title of plans: In the lower right-hand corner, or along the bottom border of each separate drawing, the proper title is usually placed containing the names of the county, municipality, company, institution, person, persons for whom it is made, the name of the locality to which it refers, a proper description for a full understanding of the nature of the drawing, and the scale, date, and name of the engineer, architect, or surveyor preparing it. On each plan showing a locality or street layout, an arrow should be placed, indicating the direction of north, and plans should be arranged, whenever practicable, so that north is toward the top or to the left of the sheet.

11.8.7.5.3 Plans and Engineering Report Review. Following receipt of complete plans and specifications, the approving state authority usually distributes a copy to each of the interested agencies and sets a date, time, and place for a joint plan review meeting with the approving authority and the applicant.

When practicable, perhaps within 60 days following this meeting, the approving authority will usually either deny the permit or determine if:

- Sufficient information is available to schedule a public hearing on the request for a permit
- Revisions to the plans and specifications are needed

If the permit is denied, the applicant is to be informed of the basis for the denial and the procedures for appeal of this determination.

If the department is unable to complete the review within the established time schedule, the department typically notifies the applicant in writing, perhaps within 30 days of receipt of the information, and informs the applicant of the anticipated time required to complete the review.

11.8.7.6   Other Requirements for Permits

11.8.7.6.1   Public Hearing.   Before issuing a refuse disposal permit for a municipal landfill, the department typically holds a public hearing on the permit application.

11.8.7.6.2   Deed Amendment.   Before being issued a refuse disposal permit, the applicant must usually submit proof that the deed pertaining to the site in question has been amended to stipulate, upon closeout of this operation, construction or excavation on this site may not begin without first obtaining written authorization from the approving authority.

11.8.7.6.3   Bonding.   Before issuance of the permit, the applicant must usually provide proof of the bonding as required.

11.8.7.6.4   Consistency with the County Solid Waste Management Plan. Before issuance of the permit, the applicant must typically provide:

- A statement from the appropriate local governmental agency concerning the consistency of the proposed facility with the approved county comprehensive solid waste management plan. If the local government fails to provide a response within 60 days of receipt of a request for a statement, a copy of the certified letter to the county requesting a statement usually is deemed to satisfy this requirement.
- Proof that the facility is consistent with the approved county comprehensive solid waste management plan.

11.8.7.6.5   Hydrogeologic Siting Criteria.   As a general requirement, a landfill must be located in a hydrologic setting which is compatible with land disposal of solid waste as determined by the approving authority. For sites which are deemed less suitable for landfill siting due to site-specific features, the approving authority may impose more stringent design or operational constraints, or both, to compensate for the deficiencies. Compatibility criteria include but are not limited to the following:

- Hydraulic conductivity of the in situ soils.
- Thickness, quality, and classification of the in situ soils.
- Depth to the water table and aquifer use within 1000 feet of the landfill boundary.
- Potential for subsurface migration of contaminants in rock fractures or highly permeable zones in bedrock or in in situ soil overburden.

11.8.7.6.6 Additional Monitoring Requirements. If the department determines that contamination of waters of the state has occurred or is liable to occur as a result of operation of the landfill, the approving authority may require the permit holder to periodically collect and analyze groundwater or surface water at the permitted site and to submit the results to the approving authority. The approving authority may furthermore specify the following:

Number and location of the sampling stations:

- Frequency of the analyses
- Sampling and analyses procedures
- Pollutants to be monitored
- Reporting period

11.9 UNDERGROUND STORAGE TANK PERMITTING CONSIDERATIONS

In 1988, EPA issued the first of its regulations (40 CFR 280) under Section 9002 of RCRA that addressed the responsibilities of underground storage tank (UST) owner-operators to meet technical standards and corrective action requirements. While these regulations did not establish specific permitting provisions, their incorporated notification, reporting, and record-keeping requirements form a definitive basis for inferred UST owner-operator permitting responsibilities. Just as owner-operators of permitted hazardous waste TSD facilities (refer to Section 11.3) must meet stringent requirements, owners and/or operators of USTs must comply with many similar requirements that are usually associated with a permitting process. These can be found in the application and notification, design and operating criteria and required response and remedial actions provisions of 40 CFR 280. For example, USTs must meet specific design criteria and owner-operators are not permitted to operate UST units without notifying the appropriate agency or agencies within 30 days of installing a new tank. In addition, all releases from such tanks must be reported to the appropriate agencies along with any corrective action that is planned or undertaken to clean up a leak, and certifications of financial responsibility must be given.

RCRA's underground storage tank regulations (40 CFR 280) cover tanks that store oil, gasoline, and other regulated substances. Tanks that store RCRA hazardous wastes are exempt from these regulations and are covered by storage requirements under 40 CFR 264 and 265. Any waste from these hazardous waste storage tanks is subject to RCRA's other hazardous waste management regulations as well.

This chapter will concentrate on the UST notification, reporting, and record-keeping requirements that parallel other established permitting provisions. As stated earlier, UST owners must notify the appropriate government entities within 30 days of installing a new tank in addition to reporting releases and the corrective actions used to address them, as well as providing documentation on tank construction, operation, and financial assurance. Record-keeping requirements for UST owner-operators require documentation of the operation of corrosion protection equipment and of any repairs made to a UST. Owner-operators also must record steps taken to comply with release detection requirements and the results of a site investigation conducted after a tank is closed permanently.

11.9.1 Regulatory Overview

Under the Hazardous and Solid Waste Amendments of 1984, Congress responded to the increasing threat of groundwater contamination posed by leaking underground storage tanks by adding Subtitle I to the Resource Conservation and Recovery Act. Subtitle I required EPA to develop a comprehensive regulatory program for USTs storing petroleum or hazardous substances. Congress directed the agency to publish regulations that would require owners and operators of new tanks and tanks already in the ground to prevent and detect leaks, clean up leaks, and demonstrate that they are financially capable of cleaning up leaks and compensating third parties for resulting damages.

11.9.2 EPA's Underground Storage Tank Regulatory Program

EPA's UST regulations, 40 CFR 280 and 281, apply to any person who owns or operates a UST or UST system. The term "owner" is defined in the statute generally to mean any person who owns a UST used for the storage, use, or dispensing of substances regulated under Subtitle I of RCRA (which includes both petroleum and hazardous substances) [9001(3), 42 USC 6991(3)]. Owners are responsible for complying with the "technical requirements," "financial responsibility requirements," and "corrective action requirements" specified in the statute and regulations. These requirements constitute an inferred UST permitting process and are intended to ensure that USTs are managed and maintained safely, so that they will not leak or otherwise cause harm to human health and the environment. In addition, should a leak occur, the requirements provide that the owner is responsible for addressing the problem. These same requirements apply to any person who "operates" a UST system. The term "operator" is very broad and means "any person in control of, or having responsibility for, the daily operation of the underground storage tank" [9001(4), 42 USC 6991(4)]. As with owners, there may be more than one operator of a tank at a given time. Each owner and operator has obligations under the statute and regulations. In this respect, it is important to understand that a person may have obligations under Subtitle I either as an owner or as an operator, or both.

11.9.2.1   UST Regulatory Program Components.   The following subsections of this chapter describe briefly each of the major components of EPA's UST regulatory program that are applicable to persons who own or operate USTs and UST systems. Each subsection refers to the various technical standards of 40 CFR 280 which include:

- Subpart B—UST systems: design, construction, installation, and notification (including performance standards for new UST systems, upgrading of existing UST systems, and notification requirements)
- Subpart C—general operating requirements (including spill and overfill control, corrosion protection, reporting, and record keeping)
- Subpart D—release detection
- Subpart E—release reporting, investigation, and confirmation
- Subpart G—out-of-service UST systems (including temporary and permanent closure)

11.9.2.1.1   Leak Prevention.   Before EPA regulations were issued, most tanks were constructed of bare steel and were not equipped with release prevention or detection features. 40 CFR 280.21 requires UST owners and operators to ensure that their tanks are protected against corrosion and equipped with devices that prevent spills and overfills no later than Dec. 22, 1998. Tanks installed before Dec. 22, 1988, must be replaced or upgraded by fitting them with corrosion protection and spill and overfill prevention devices to bring them up to new tank standards. USTs installed after Dec. 22, 1988, must be fiberglass-reinforced plastic, corrosion-protected steel, a composite of these materials, or determined by the implementing agency to be no less protective of human health and the environment, and must be designed, constructed, and installed in accordance with a code of practice developed by a nationally recognized association or independent testing laboratory. Piping installed after Dec. 22, 1988, generally must be protected against corrosion in accordance with a national code of practice. All owners and operators must also ensure that releases due to spilling or overfilling do not occur during product transfer and that all steel systems with corrosion protection are maintained, inspected, and tested in accordance with 280.31.

11.9.2.1.2   Leak Detection.   In addition to meeting the leak prevention requirements, owners and operators of USTs must use a method listed in 280.43

through 280.44 for detecting leaks from portions of both tanks and piping that routinely contain product. Deadlines for compliance with the leak detection requirements have been phased in based on the tank's age: The oldest tanks, which are most likely to leak, had the earliest compliance deadlines. Phase-in of the leak detection requirements was completed in 1993, and all UST systems should now be in compliance with these requirements.

11.9.2.1.3　Release Reporting.　UST owners and operators must, in accordance with 280.50, report to the implementing agency within 24 hours, or another reasonable time period specified by the implementing agency, the discovery of any released regulated UST substances, or any suspected release. Unusual operating conditions or monitoring results indicating a release must also be reported to the implementing agency.

11.9.2.1.4　Closure.　Owners or operators who would like to take tanks out of operation must either temporarily or permanently close them in accordance with 40 CFR part 280 subpart G—Out-of-Service UST Systems and Closure. When UST systems are temporarily closed, owners and operators must continue operation and maintenance of corrosion protection and, unless all USTs have been emptied, release detection. If temporarily closed for 3 months or more, the UST system's vent lines must be left open and functioning, and all other lines, pumps, passageways, and ancillary equipment must be capped and secured. After 12 months, tanks that do not meet either the performance standards for new UST systems or the upgrading requirements (excluding spill and overfill device requirements) must be permanently closed, unless a site assessment is performed by the owner or operator and an extension is obtained from the implementing agency. To close a tank permanently, an owner or operator generally must: Notify the regulatory authority 30 days before closing (or another reasonable time period determined by the implementing agency); determine if the tank has leaked and, if so, take appropriate notification and corrective action; empty and clean the UST; and either remove the UST from the ground or leave it in the ground filled with an inert, solid material.

11.9.2.1.5　Notification, Reporting, and Record Keeping.　UST owners who bring a UST system into use after May 8, 1986, must notify state or local authorities of the existence of the UST and certify compliance with certain technical and other requirements, as specified in 280.22. Owners and operators must also notify the implementing agency at least 30 days (or another reasonable time period determined by the implementing agency) prior to the permanent closure of a UST. In addition, owners and operators must keep records of testing results for the cathodic protection system, if one is used; leak detection performance and upkeep; repairs; and site assessment results at permanent closure (which must be kept for at least 3 years).

11.9.2.1.6　Corrective Action Requirements.　Owners and operators of UST systems containing petroleum or hazardous substances must investigate, confirm, and respond to confirmed releases, as specified in 280.51 through 280.67. These requirements include, where appropriate, performing a release investigation when a release is suspected or to determine if the UST system is the source of an off-site impact (investigation and confirmation steps include conducting tests to determine if a leak exists in the UST or UST system and conducting a site check if tests indicate that a leak does not exist but contamination is present); notifying the appropriate agencies of the release within a specified period of time; taking immediate action to prevent any further release (such as removing product from the UST system); containing and immediately cleaning up spills or overfills; monitoring and preventing the spread of contamination into the soil and/or groundwater; assembling detailed information about the site and the nature of the release; removing free product to the maximum extent practicable; investigating soil and groundwater contamination; and in some cases, outlining and implementing a detailed corrective action plan for remediation.

11.9.2.1.7   Financial Responsibility Requirements.   The financial responsibility regulations (40 CFR 280, subpart H) require that UST owners or operators demonstrate the ability to pay the costs of corrective action and to compensate third parties for injuries or damages resulting from the release of petroleum from USTs. The regulations require all owners or operators of petroleum USTs to maintain an annual aggregate of financial assurance of $1 million or $2 million, depending on the number of USTs owned. Financial assurance options available to owners and operators include purchasing commercial environmental impairment liability insurance; demonstrating self-insurance; obtaining guarantees, surety bonds, or letters of credit; placing the required amount into a trust fund administered by a third party; or relying on coverage provided by a state assurance fund.

11.9.2.2   UST Regulatory Program Applicability.   The requirements of 40 CFR part 280 apply to all owners and operators of a UST system as defined in Section 11.5.2.3 of this chapter except as otherwise provided in the following paragraphs.

11.9.2.2.1   Exclusions.   The following UST systems are excluded from the requirements of 40 CFR 280:

- Any UST system holding hazardous wastes listed or identified under Subtitle C of the Solid Waste Disposal Act, or a mixture of such hazardous waste and other regulated substances.

- Any wastewater treatment tank system that is part of a wastewater treatment facility regulated under section 402 or 307(b) of the Clean Water Act.

- Equipment or machinery that contains regulated substances for operational purposes such as hydraulic lift tanks and electrical equipment tanks.

- Any UST system whose capacity is 110 gallons or less.

- Any UST system that contains a de minimis concentration of regulated substances.

- Any emergency spill or overflow containment UST system that is expeditiously emptied after use.

11.9.2.2.2   Deferrals.   The requirements of the above referenced Subparts B, C, D, E, and G do not apply to any of the following types of UST systems:

- Wastewater treatment tank systems
- Any UST systems containing radioactive material that are regulated under the Atomic Energy Act of 1954 (42 U.S.C. 2011 and following)
- Any UST system that is part of an emergency generator system at nuclear power generation facilities regulated by the Nuclear Regulatory Commission under 10 CFR part 50, appendix A
- Airport hydrant fuel distribution systems
- UST systems with field-constructed tanks

The requirements of the above referenced Subpart D do not apply to any UST system that stores fuel solely for use by emergency power generators.

11.9.2.2.3   Interim Prohibition for Deferred UST Systems.   No person is permitted to install a UST system listed in 280.10(c) for the purpose of storing regulated substances unless the UST system (whether of single- or double-wall construction):

- Will prevent releases due to corrosion or structural failure for the operational life of the UST system
- Is cathodically protected against corrosion, constructed of noncorrodible material, steel clad with a noncorrodible material, or designed in a manner to prevent the release or threatened release of any stored substance
- Is constructed or lined with material that is compatible with the stored substance.

Notwithstanding the above requirements, installation of a UST system without corrosion protection is permitted at a site that is determined by a corrosion expert not to be corrosive enough to cause it to have a release due to corrosion during its operating life. Owners and operators must maintain records that demonstrate compliance with the requirements of this paragraph for the remaining life of the tank. (*Note:* The National Association of Corrosion Engineers Standard RP-02-85, "Control of External Corrosion on Metallic Buried, Partially Buried, or Submerged Liquid Storage Systems," may be used as guidance.)

11.9.2.3   UST Regulatory Program Definitions

Aboveground release means any release to the surface of the land or to surface water. This includes, but is not limited to, releases from the aboveground portion of a UST system and aboveground releases associated with overfills and transfer operations as the regulated substance moves to or from a UST system.

Ancillary equipment means any devices including, but not limited to, such devices as piping, fittings, flanges, valves, and pumps used to distribute, meter, or control the flow of regulated substances to and from a UST.

Belowground release means any release to the subsurface of the land and to groundwater. This includes, but is not limited to, releases from the belowground portions of an underground storage tank system and belowground releases associated with overfills and transfer operations as the regulated substance moves to or from an underground storage tank.

Beneath the surface of the ground means beneath the ground surface or otherwise covered with earthen materials.

Cathodic protection tester means a person who can demonstrate an understanding of the principles and measurements of all common types of cathodic protection systems as applied to buried or submerged metal piping and tank systems. At a minimum, such persons must have education and experience in soil resistivity, stray current, structure-to-soil potential, and component electrical isolation measurements of buried metal piping and tank systems. Cathodic protection is a technique to prevent corrosion of a metal surface by making that surface the cathode of an electrochemical cell. For example, a tank system can be cathodically protected through the application of either galvanic anodes or impressed current.

CERCLA means the Comprehensive Environmental Response, Compensation, and Liability Act of 1980, as amended.

Compatible means the ability of two or more substances to maintain their respective physical and chemical properties upon contact with one another for the design life of the tank system under conditions likely to be encountered in the UST.

Connected piping means all underground piping including valves, elbows, joints, flanges, and flexible connectors attached to a tank system through which regulated substances flow. For the purpose of determining how much piping is connected to any individual UST system, the piping that joins two UST systems should be allocated equally between them.

Consumptive use with respect to heating oil means consumed on the premises.

Corrosion expert means a person who, by reason of thorough knowledge of the physical sciences and the principles of engineering and mathematics acquired by a professional education and related practical experience, is qualified to engage in the practice of corrosion control on buried or submerged metal piping systems and metal tanks. Such a person must be accredited or certified as being qualified by the National Association of Corrosion Engineers or be a registered professional engineer who has certification or licensing that includes education and experience in corrosion control of buried or submerged metal piping systems and metal tanks.

Dielectric material means a material that does not conduct direct electric current. Dielectric coatings are used to electrically isolate UST systems from the surrounding soils. Dielectric bushings are used to electrically isolate portions of the UST system (e.g., tank from piping).

Electrical equipment means underground equipment that contains dielectric fluid that is necessary for the operation of equipment such as transformers and buried electrical cable.

Excavation zone means the volume containing the tank system and backfill material bounded by the ground surface, walls, and floor of the pit and trenches into which the UST system is placed at the time of installation.

Existing tank system means a tank system used to contain an accumulation of regulated substances or for which installation commenced on or before Dec. 22, 1988. Installation is considered to have commenced if:

> The owner or operator has obtained all federal, state, and local approvals or permits necessary to begin physical construction of the site or installation of the tank system

> Either a continuous on-site physical construction or installation program has begun

> The owner or operator has entered into contractual obligations—which cannot be canceled or modified without substantial loss—for physical construction at the site or installation of the tank system to be completed within a reasonable time

Farm tank is a tank located on a tract of land devoted to the production of crops or raising animals, including fish, and associated residences and improvements. A farm tank must be located on the farm property. "Farm" includes fish hatcheries, rangeland, and nurseries with growing operations.

Flow-through process tank is a tank that forms an integral part of a production process through which there is a steady, variable, recurring, or intermittent flow of materials during the operation of the process. Flow-through process tanks do not include tanks used for the storage of materials prior to their introduction into the production process or for the storage of finished products or by-products from the production process.

Free product refers to a regulated substance that is present as a nonaqueous phase liquid (e.g., liquid not dissolved in water).

Gathering lines means any pipeline, equipment, facility, or building used in the transportation of oil or gas during oil or gas production or gathering operations.

Hazardous substance UST system means an underground storage tank system that contains a hazardous substance defined in section 101(14) of the Comprehensive Environmental Response, Compensation and Liability Act of 1980 (but not including any substance regulated as a hazardous waste under subtitle C) or any mixture of such substances and petroleum, and which is not a petroleum UST system. Heating oil means petroleum that is No. 1, No. 2, No. 3—light, No. 4—heavy, No. 5—light, No. 5—heavy, and No. 6 technical grades of fuel oil;

other residual fuel oils (including Navy Special Fuel Oil and Bunker C); and other fuels when used as substitutes for one of these fuel oils. Heating oil is typically used in the operation of heating equipment, boilers, or furnaces.

Hydraulic lift tank means a tank holding hydraulic fluid for a closed-loop mechanical system that uses compressed air or hydraulic fluid to operate lifts, elevators, and other similar devices. Implementing agency means EPA or, in the case of a state with a program approved under section 9004 (or pursuant to a memorandum of agreement with EPA), the designated state or local agency responsible for carrying out an approved UST program.

Liquid trap means sumps, well cellars, and other traps used in association with oil and gas production, gathering, and extraction operations (including gas production plants), for the purpose of collecting oil, water, and other liquids. These liquid traps may temporarily collect liquids for subsequent disposition or reinjection into a production or pipeline stream, or may collect and separate liquids from a gas stream.

Maintenance means the normal operational upkeep to prevent an underground storage tank system from releasing product.

Motor fuel means petroleum or a petroleum-based substance that is motor gasoline, aviation gasoline, No. 1 or No. 2 diesel fuel, or any grade of gasohol, and is typically used in the operation of a motor engine.

New tank system means a tank system that will be used to contain an accumulation of regulated substances and for which installation has commenced after Dec. 22, 1988. (See also Existing Tank System.)

Noncommercial purposes with respect to motor fuel means not for resale.

On the premises where stored with respect to heating oil means UST systems located on the same property where the stored heating oil is used.

Operational life refers to the period beginning when installation of the tank system has commenced until the time the tank system is properly closed under Subpart G.

Operator means any person in control of, or having responsibility for, the daily operation of the UST system.

Overfill release is a release that occurs when a tank is filled beyond its capacity, resulting in a discharge of the regulated substance to the environment.

Owner means:

In the case of a UST system in use on Nov. 8, 1984, or brought into use after that date, any person who owns a UST system used for storage, use, or dispensing of regulated substances

In the case of any UST system in use before Nov. 8, 1984, but no longer in use on that date, any person who owned such UST immediately before the discontinuation of its use.

Person means an individual, trust, firm, joint stock company, federal agency, corporation, state, municipality, commission, political subdivision of a state, or any interstate body. "Person" also includes a consortium, a joint venture, a commercial entity, and the United States government.

Petroleum UST system means an underground storage tank system that contains petroleum or a mixture of petroleum with de minimis quantities of other regulated substances. Such systems include those containing motor fuels, jet fuels, distillate fuel oils, residual fuel oils, lubricants, petroleum solvents, and used oils.

Pipe or piping means a hollow cylinder or tubular conduit that is constructed of nonearthen materials.

Pipeline facilities (including gathering lines) are new and existing pipe rights-of-way and any associated equipment, facilities, or buildings.

Regulated substance means any substance defined in section 101(14) of the Comprehensive Environmental Response, Compensation and Liability Act (CERCLA) of 1980 (but not including any substance regulated as a hazardous waste under subtitle C), and petroleum, including crude oil or any fraction thereof that is liquid at standard conditions of temperature and pressure (60°F and 14.7 pounds per square inch absolute).

The term *regulated substance* includes but is not limited to petroleum and petroleum-based substances comprised of a complex blend of hydrocarbons derived from crude oil through processes of separation, conversion, upgrading, and finishing, such as motor fuels, jet fuels, distillate fuel oils, residual fuel oils, lubricants, petroleum solvents, and used oils.

Release means any spilling, leaking, emitting, discharging, escaping, leaching, or disposing from a UST into groundwater, surface water, or subsurface soils.

Release detection means determining whether a release of a regulated substance has occurred from the UST system into the environment or into the interstitial space between the UST system and its secondary barrier or secondary containment around it.

Repair means to restore a tank or UST system component that has caused a release of product from the UST system.

Residential tank is a tank located on property used primarily for dwelling purposes.

SARA means the Superfund Amendments and Reauthorization Act of 1986.

Septic tank is a watertight covered receptacle designed to receive or process, through liquid separation or biological digestion, the sewage discharged from a building sewer. The effluent from such receptacle is distributed for disposal through the soil, and settled solids and scum from the tank are pumped out periodically and hauled to a treatment facility.

Stormwater or wastewater collection system means piping, pumps, conduits, and any other equipment necessary to collect and transport the flow of surface water runoff resulting from precipitation, or domestic, commercial, or industrial wastewater to and from retention areas or any areas where treatment is designated to occur. The collection of stormwater and wastewater does not include treatment except where incidental to conveyance.

Surface impoundment is a natural topographic depression, manmade excavation, or diked area formed primarily of earthen materials (although it may be lined with manmade materials) that is not an injection well.

Tank is a stationary device designed to contain an accumulation of regulated substances and constructed of nonearthen materials (e.g., concrete, steel, plastic) that provide structural support.

Underground area means an underground room, such as a basement, cellar, shaft, or vault, providing enough space for physical inspection of the exterior of the tank situated on or above the surface of the floor.

Underground release means any belowground release.

Underground storage tank or UST means any one or combination of tanks (including underground pipes connected thereto) that is used to contain an accumulation of regulated substances, and the volume of which (including the volume of underground pipes connected thereto) is 10 percent or more beneath the surface of the ground. This term does not include any:

Farm or residential tank of 1100 gallons or less capacity used for storing motor fuel for noncommercial purposes

Tank used for storing heating oil for consumptive use on the premises where stored

Septic tank

Pipeline facility (including gathering lines) regulated under:

The Natural Gas Pipeline Safety Act of 1968 (49 U.S.C. App. 1671, et seq.)

The Hazardous Liquid Pipeline Safety Act of 1979 (49 U.S.C. App. 2001, et seq.)

Which is an intrastate pipeline facility regulated under state laws comparable to the provisions of the law referred to in paragraph (d)(1) or (d)(2) of this definition

Surface impoundment, pit, pond, or lagoon

Stormwater or wastewater collection system

Flow-through process tank

Liquid trap or associated gathering lines directly related to oil or gas production and gathering operations

Storage tank situated in an underground area (such as a basement, cellar, mineworking, drift, shaft, or tunnel) if the storage tank is situated upon or above the surface of the floor

The term *underground storage tank* or *UST* does not include any pipes connected to any tank which is described in paragraphs (a) through (i) of this definition.

Upgrade means the addition or retrofit of some systems such as cathodic protection, lining, or spill and overfill controls to improve the ability of an underground storage tank system to prevent the release of product.

UST system or tank system means an underground storage tank, connected underground piping, underground ancillary equipment, and containment system, if any.

Wastewater treatment tank means a tank that is designed to receive and treat an influent wastewater through physical, chemical, or biological methods.

11.9.2.4 UST Regulatory Program Notification Requirements. The basis for EPA's inferred UST permitting process can be found in the notification, reporting, and record-keeping requirements of 40 CFR 280. In a summary, owner-operators are not permitted to conduct UST operations unless they:

Notify the appropriate agency within 30 days of installing a new tank

Report suspected releases and certain spills or overfills

Maintain required records

11.9.2.4.1 Notification of UST Installation. Owner-operators of underground storage tanks are required to notify EPA and the appropriate state agency within 30 days after they bring a UST into use. This applies to tanks brought into use after May 8, 1986. Notification is to be made by filing a Notification for Underground Storage Tanks, Form 7530-1 (see attachment). Individual states that have been authorized to operate their own UST programs may require submission of revised versions of the EPA form.

Although a single notice can be utilized for registering multiple tanks owned at one location, separate forms are usually required if the tanks are operated in multiple locations.

Forms have to be sent to the UST program manager in the appropriate EPA regional office as well as to the appropriate state UST regulatory agency. Whenever a tank intended for use as UST is sold, the seller must inform the purchaser of the notification requirements.

11.9.2.4.1.1 INFORMATION REQUIRED: The notification form requires general information about the tank owner, including name and address, as well as information on the location of the tanks. The form includes a certification for compliance with the requirements for cathodic protection of steel tanks and piping, financial responsibility, and release detection.

Detailed information on the tanks is required. This includes status, age, capacity, construction material, internal and external protection, piping, and substance stored. Owner-operators must provide details about tank installation, type of release detection equipment used, and method of corrosion protection.

11.9.2.4.2 Notification of UST Closure. Thirty days before beginning either permanent closure or a "change-in-service" of a UST, the owner-operator must notify the appropriate federal and state agencies. A change-in-service is any continued use of a UST system to store a nonregulated substance.

11.9.2.4.3 Notifications of Releases, Spills, and Overfills

11.9.2.4.3.1 INITIAL NOTIFICATION OF RELEASES: EPA's UST regulations require UST owner-operators to report promptly certain types of situations or events that could signal an existing or impending release from their tanks. Owner-operators must report within 24 hours to EPA or an EPA-approved state or local agency:

The discovery of a release from the tank or the presence of a regulated substance at the site or vapors in soil, basements, sewer and utility lines, or nearby surface water.

Unusual operating conditions, such as malfunctioning of equipment, an unexplained presence of water, or sudden loss of the substance from the tank. Defective equipment that is not causing the tank to leak and is repaired or replaced is excluded from the reporting requirement.

Monitoring results from a release detection method that indicate a release might have occurred, unless:

The monitoring device is found to be defective and is repaired, recalibrated, or replaced immediately and additional monitoring does not confirm the initial result.

In case of inventory control, the second month of data on inventory control does not confirm the initial result.

11.9.2.4.3.2 INITIAL NOTIFICATION OF SPILLS AND OVERFILLS: Owner-operators of USTs must report certain spills or overfills to EPA or an EPA-approved state or local agency within 24 hours of the occurrence (or another reasonable time set by the agency), including spills or overfills of:

Petroleum that results in a release to the environment of more than 25 gallons or that cause a sheen on nearby surface water

Hazardous substances that equal or exceed the reportable quantities under CERCLA

Spills or overfills of less than 25 gallons of petroleum or less than a reportable quantity must be immediately contained and cleaned up. However, if cleanup cannot be accomplished within 24 hours or another reasonable time set by the applicable agency, owner-operators immediately must notify EPA or an EPA-approved state or local agency.

(*Note:* If the quantity of a hazardous substance spilled or released from a UST equals or exceeds the reportable quantities established by CERCLA, a report must be filed immediately to the National Response Center.)

11.9.2.4.3.3 FOLLOW-UP NOTIFICATIONS REPORTS: Unless directed to do otherwise by the implementing agency, tank owner-operators must meet certain follow-up reporting requirements when a tank release has been confirmed. These reporting requirements must be met by the following deadlines:

• Within 20 days of a confirmed release, tank owner-operators must submit a report to EPA or the EPA-approved state or local agency summarizing the initial steps taken to correct the problem.

- Within 45 days, owner-operators must assemble detailed information about the site, including data gleaned during confirmation of the release; the nature and extent of the release; potentially affected population, water bodies, wells, or sewer pipes; climatological conditions; and land use. In addition, information on the results of a site check and "free product" investigation is required. Free product refers to any regulated substance present as a nonaqueous phase liquid investigation.
- Within 45 days, owner-operators must report information on the free product removal activities. The report must include:
- Names of those responsible for the removal
- Estimated quantity, type, and thickness of free product observed or measured in wells, boreholes, and excavations.
- Type of removal used.
- Location of discharge activity.
- Type of treatment applied to the discharge.
- Steps taken to obtain permits for any discharge.
- Disposition of the recovered free product.

As soon as practicable, owner-operators must submit information on the full characterization of soil and groundwater contamination if groundwater wells have been affected, free product needs to be recovered, or contaminated soils might be in contact with groundwater.

Within a schedule set by EPA or an EPA-approved state or local agency, owner-operators must report the results of any corrective action plan that is required for cleaning up the site. This requirement applies only to UST owner-operators required to develop and implement corrective action plans.

11.9.2.4.4 Notifications of Financial Responsibility. Owner-operators of USTs also must submit documentation of financial responsibility:

- Within 30 days after the occurrence of a reportable release, and/or
- If the owner or operator fails to obtain alternate coverage as required by this subpart, within 30 days after the owner or operator receives notice of:

Commencement of a voluntary or involuntary proceeding under Title 11 (Bankruptcy), U.S. Code, naming a provider of financial assurance as a debtor

Suspension or revocation of the authority of a provider of financial assurance to issue a financial assurance mechanism

Failure of a guarantor to meet the requirements of the financial test

Other incapacity of a provider of financial assurance

- If the owner or operator fails to obtain alternate assurance within 150 days of finding that he or she no longer meets the requirements of the financial test based on the year-end financial statements, or within 30 days of notification by the director of the implementing agency that he or she no longer meets the requirements of the financial test, the owner or operator must notify the director of such failure within 10 days.

- If a provider of financial responsibility cancels or fails to renew for reasons other than incapacity of the provider as specified in 280.114, the owner or operator must obtain alternate coverage as specified in this section within 60 days after receipt of the notice of termination. If the owner or operator fails to obtain alternate coverage within 60 days after receipt of the notice of termination, the owner or operator must notify the director of the implementing agency of such failure and submit:

  The name and address of the provider of financial assurance

  The effective date of termination

  The evidence of the financial assistance mechanism subject to the termination maintained in accordance with 280.107(b)

11.9.2.4.4.1 EVIDENCE OF FINANCIAL RESPONSIBILITY: An owner or operator must maintain the following types of evidence for the appropriate use in notifications of financial responsibility:

- An owner or operator using an assurance mechanism specified in 280.95 through 280.100 or 280.102 or 280.104 through 280.107 must maintain a copy of the instrument worded as specified.

- An owner or operator using a financial test or guarantee, or a local government financial test or a local government guarantee supported by the local government financial test must maintain a copy of the chief financial officer's letter based on year-end financial statements for the most recent completed financial reporting year. Such evidence must be on file no later than 120 days after the close of the financial reporting year.

- An owner or operator using a guarantee, surety bond, or letter of credit must maintain a copy of the signed standby trust fund agreement and copies of any amendments to the agreement.

- A local government owner or operator using a local government guarantee under 280.106(d) must maintain a copy of the signed standby trust fund agreement and copies of any amendments to the agreement.

- A local government owner or operator using the local government bond rating test under 280.104 must maintain a copy of its bond rating published within the last 12 months by Moody's or Standard & Poor's.

- A local government owner or operator using the local government guarantee under 280.106, where the guarantor's demonstration of financial responsibility relies on the bond rating test under 280.104, must maintain a copy of the guarantor's bond rating published within the last 12 months by Moody's or Standard & Poor's.

- An owner or operator using an insurance policy or risk retention group coverage must maintain a copy of the signed insurance policy or risk retention group coverage policy, with the endorsement or certificate of insurance and any amendments to the agreements.
- An owner or operator covered by a state fund or other state assurance must maintain on file a copy of any evidence of coverage supplied by or required by the state under 280.101(d).
- An owner or operator using a local government fund under 280.107 must maintain the following documents:

A copy of the state constitutional provision or local government statute, charter, ordinance, or order dedicating the fund.

Year-end financial statements for the most recent completed financial reporting year showing the amount in the fund. If the fund is established under 280.107(a)(3) using incremental funding backed by bonding authority, the financial statements must show the previous year's balance, the amount of funding during the year, and the closing balance in the fund.

If the fund is established under 280.107(a)(3) using incremental funding backed by bonding authority, the owner or operator must also maintain documentation of the required bonding authority, including either the results of a voter referendum [under 280.107(a)(3)(i), or attestation by the state attorney general as specified under 280.107(a)(3)(ii)].

A local government owner or operator using the local government guarantee supported by the local government fund must maintain a copy of the guarantor's year-end financial statements for the most recent completed financial reporting year showing the amount of the fund.

An owner or operator using an assurance mechanism specified in 280.95 through 280.107 must maintain an updated copy of a certification of financial responsibility worded as follows.

11.9.2.4.4.2 CERTIFICATION OF FINANCIAL RESPONSIBILITY: An owner or operator must certify compliance with the financial responsibility requirements of this part as specified in the new tank notification form when notifying the appropriate state or local agency of the installation of a new underground storage tank.

The financial assurance mechanism(s) used to demonstrate financial responsibility under subpart H of 40 CFR part 280 is (are) as follows:

For each mechanism, list the type of mechanism, name of issuer, mechanism number (if applicable), amount of coverage, effective period of coverage and whether the mechanism covers "taking corrective action" and/or "compensating third parties for bodily injury and property damage caused by" either "sudden accidental releases" or "nonsudden accidental releases" or "accidental releases."

Signature of owner or operator.

Name of owner or operator.

Title.

Date.

Signature of witness or notary.

Name of witness or notary.

Date.

The owner or operator must update this certification whenever the financial assurance mechanism(s) used to demonstrate financial responsibility change(s).

11.9.2.5   UST Regulatory Program Record-Keeping Requirements.   [Reference: 40 CFR 280.34(c)] EPA's UST regulations require tank owner-operators to maintain a variety of records on their tanks. The records must be made available to agency inspectors either at the UST site or at an alternative site. Permanent closure records can be mailed to the state or federal agency if neither the UST site nor an alternative site is feasible.

11.9.2.5.1   Operating, Release, and Closure Record-Keeping Requirements. Owner-operators of steel USTs must maintain the following records:

- All written performance claims and maintenance schedules that pertain to release detection equipment must be maintained for 5 years.
- The results of any sampling, testing, or monitoring must be maintained for 1 year, except for tank tightness testing, which must be retained only until the next test is conducted.
- Documentation of calibration, maintenance, and repair of permanent on-site equipment must be maintained for 1 year after the service work is completed.
- Any schedules of required calibration and maintenance provided by the manufacturer must be retained for 5 years from the date of installation.
- Documentation of the operation of the corrosion protection equipment, including the last three inspections conducted under 40 CFR 280.31 and the results of testing from the last two inspections, must be maintained.
- Evidence that any repairs or upgrades to a UST system were done in accordance with regulatory requirements must be retained for the remaining life of the UST system.
- Results of the site assessment required for permanent closure under 40 CFR 280.72 must be kept for at least 3 years after permanent closure.

If no corrosion protection equipment is used, owner-operators must keep, for the life of the tank, a corrosion expert's analysis of the potential for corrosion.

11.9.2.5.2 Financial Responsibility Record-Keeping Requirements. In addition to other record-keeping requirements, UST owner-operators must maintain all evidence that they have the financial means to perform corrective action and assure payment for liability stemming from accidental releases at the UST site or at the owner-operator's worksite.

The following types of evidence of financial responsibility must be maintained:

A copy of the financial instrument documentation or agreement and any amendments

For financial tests or guarantees, a copy of the chief financial officer's letter based on the most recent year-end financial statements

For local government bond rating tests, a copy of the local government's bond rating test published by Moody's or Standard & Poor's within the past year

For local government funds, a copy of the local law or ordinance dictating the fund and the most recent year-end financial statements.

CHAPTER 12
WETLANDS

Amy S. Greene
President
Amy S. Greene Environmental Consultants Inc.

12.1 INTRODUCTION

This chapter reviews the federal wetlands permit program in the United States. Any discharge of dredged or fill material into wetlands or other waters of the United States requires a permit from a district office of the U.S. Army Corps of Engineers (the Corps). The Corps was granted authority over placement of dredged or fill material in wetlands under Sec. 404 of the Clean Water Act. The U.S. EPA also has responsibilities for administering the program. The U.S. Fish and Wildlife Service and the National Marine Fisheries Service have advisory roles. State resource agencies also have authority in that they must issue a 401 Water Quality Certificate to validate any 404 permit issued by the Corps.

For federally funded or undertaken activities, additional wetland policies may apply. On September 14, 1977 President Jimmy Carter issued Executive Order (EO) 11990, "Protection of Wetlands" (3 CFR 121). The order was amended by EO 12608 on September 14, 1987 (52 FR 34617). This order directed all federal agencies to avoid undertaking new construction in wetlands unless there is no practicable alternative to such construction and the proposed action includes all practicable measures to minimize harm to wetlands. This order applies to agency actions including land acquisition, management, and sale, and projects and programs that are financed by or conducted by federal agencies.

There are also numerous state and local wetland regulatory programs. These may include regulations of additional activities in wetlands and may also regulate activities in lands adjacent to wetlands, wetland buffer areas. Some programs use different methodologies for the identification of wetlands. Check with your state and local resource protection agencies for details.

12.2 THE IMPORTANCE OF WETLANDS

A number of values and functions relating to environmental protection have been documented for wetlands. Because of their association with surface waters, wetlands serve to filter out pollutants carried in stormwater runoff before discharge to streams. This protects aquatic life as well as drinking water supplies. Wetlands also moderate flooding by providing storage and slow release of storm flows. In certain locations, wetlands absorb storm energy and protect shorelines from erosion. It is estimated that up to 43 percent of endangered and threatened species have nesting, breeding, or feeding habitats in wetlands. Wetlands are extremely productive habitats. Wetlands serve as home to many fish and wildlife species and contribute nutrients and other resources to aquatic ecosystems (streams, estuaries, and bays). They are critical to the life cycle of many commercially and recreationally valuable species of fish, shellfish, and wildlife.

Wetlands have been lost at a rapid rate. Over half of the wetlands in the conterminous United States were lost between the late 1700s and mid-1970s. About 100 million acres of wetlands remain today.[1] Causes of loss and degradation include ditching, diking, farming, creation of ponds and lakes, deposition of dredged material, filling for development, deposition of wastes, and discharge of contaminants. The loss or degradation of wetlands can lead to increased flooding; species decline, extinction, or deformity; and decline in water quality.

Coastal wetlands make up only 5 percent of the wetland types in the continental United States; inland wetlands, such as freshwater swamps, prairie potholes, bogs, and fens, make up the remaining 95 percent of wetland types.

---

[1]Dahl, T. E., and C. E. Johnson. 1991. *Wetland Status and Trends in the Conterminous United States. Mid-1970s to Mid-1980s.* Washington, DC: U.S. Department of the Interior, Fish and Wildlife Service.

12.3 WETLAND IDENTIFICATION AND DELINEATION METHODOLOGY

The definition of wetlands used in the 404 program was adopted in 1977 and is still used today (42 FR 37, 125–126, 37128–37129; July 19, 1977):

> Areas that are saturated or inundated by surface or groundwater for a frequency and duration sufficient to support, and under normal circumstances do support, a prevalence of vegetation typically adapted for life in saturated soil conditions. Wetlands generally include swamps, marshes, bogs, and similar areas.

Wetlands are vegetated lands that are intermediate between submerged lands and dry lands. They may be permanently flooded, periodically flooded by tidal waters, or only seasonally saturated. Examples of wetland communities include: marshes, treeless areas that may include salt marsh cordgrass, cattails, and other herbaceous vegetation; shrub/scrub communities with vegetation such as button bush and shrub dogwoods; and swamps, which are wooded wetland areas, often dominated by red maples, elms, black gum, and skunk cabbage. Some of these areas, particularly the forested swamp communities, are not readily identifiable to the layperson as wetlands. Wetlands are generally associated with the flood-plains of small and large streams and rivers, but may also occur at the headwaters of small streams, within farm fields, and as isolated pockets within upland communities.

The U.S. Fish and Wildlife Service published a classification system for wetlands that is commonly used in the description of wetland communities for mapping and evaluation purposes.[2] This system divides wetland and open waters into marine, estuarine (associated with estuaries), lacustrine (associated with lakes), riverine (associated with rivers and streams), and palustrine (inland wetlands) systems.

It was not until January 1987 that the Corps published a final manual for the mapping of wetland boundaries, the *Wetlands Delineation Manual*.[3] A subsequent manual was produced from the collaboration of four federal agencies including the U.S. Environmental Protection Agency, the U.S. Fish and Wildlife Service, the U.S. Department of Agriculture (USDA) Soil Conservation Service, and the Corps (*Federal Manual for Identifying and Delineating Jurisdictional Wetlands*).[4] This 1989 manual was subsequently withdrawn and the federal agencies adopted the 1987 Corps manual.

---

[2]Cowardin, L. M., V. Carter, F. C. Golet, and E. T. LaRoe. 1979. *Classification of Wetlands and Deepwater Habitats of the United States*. FWS/OBS-79/31. Washington, DC: U.S. Fish and Wildlife Service.

[3]U.S. Army Corps of Engineers. 1987. *Wetlands Delineation Manual*. Environmental Laboratory, U.S. Army Engineers Waterway Experiment Station Technical Report Y-87-1.

[4]Federal Interagency Committee for Wetland Delineation. 1989. *Federal Manual for Identifying and Delineating Jurisdictional Wetlands*. U.S. Army Corps of Engineers, U.S. Environmental Protection Agency, U.S. Fish and Wildlife Service, and USDA Soil Conservation Service, Washington, DC. Cooperative technical publication. 76 pp., plus appendixes.

The methodology for the identification of wetlands specifies the use of the "three-parameter approach" for the identification of wetlands: vegetation, soils, and hydrology. The following three parameters are diagnostic of wetlands: (1) the land is dominated by hydrophytes, (2) the substrate is undrained hydric soil, and (3) the substrate is saturated with groundwater or flooded with surface water for a significant part of the growing season each year. All three parameters must be present in order for an area to be identified as wetland, unless abnormal circumstances are determined to be present.

A *hydrophyte* is any plant "growing in water, soil, or on a substrate that is at least periodically deficient of oxygen as a result of excessive water content."[5] Since most plant species tolerate a range of growing conditions, individual species are not restricted to either wetland or upland communities. The Fish and Wildlife Service has developed a classification scheme[6] that assigns species to wetland indicator classes according to the following rules in Table 12.1. Pluses or minuses given with these classifications indicate a tendency toward the wetter (+) or drier (−) end of the scale. Hydrophytic vegetation is present if greater than 50 percent of the dominant plant species from all strata are OBL, FACW, and/or FAC (including FACW+, FACW−, FAC+, and FAC− species). When 50 percent or more of the dominant species is FACU and/or UPL, the area is considered to be upland (nonwetland). If any area supports hydrophytes, but hydric soils and wetland hydrology are lacking and normal circumstances exist, then an area is considered to be upland. In order to determine the dominance of each plant species, the cover class (based on percent areal cover) is recorded within a $1/100$-acre circular (11.78-ft radius) plot. Plot size may vary, depending on the size of the community. Relative basal area is determined for each canopy species by using a plotless method.

Hydric soils are soils that are saturated, flooded, or ponded long enough during the growing season to develop anaerobic conditions in a major part of the root zone.[7] Soils are considered hydric when they are (1) somewhat poorly drained and

TABLE 12.1   Plant Affinity for Wetland Conditions

Classification	Occurrence in wetlands, %
Obligate (OBL)	>99
Facultative wet (FACW)	67–99
Facultative (FAC)	34–66
Facultative upland (FACU)	1–33
Upland (UPL)	<1

----

[5]Ibid.

[6]Reed, P. B. 1988. "National List of Plant Species that Occur in Wetlands: National Summary." Biol. Report 88(24). Washington, DC: U.S. Fish and Wildlife Service.

[7]U.S. Department of Agriculture. 1987. *Hydric Soils of the United States,* 2d ed. Washington, DC: Soil Conservation Service.

have a seasonal high water table less than 0.5 ft from the surface or (2) poorly drained or very poorly drained and have a seasonal high water table less than 1.0 or 1.5 ft from the surface. This high water table must be present for 2 weeks or more during the growing season. All organic soils (histosols) or mineral soils with a histic epipedon are hydric soils. The National Committee on Hydric Soils periodically updates the criteria for hydric soils.

In the field, a hand-held auger is used for sampling the soil to examine indicators of hydric soils such as low-chroma colors, mottling, organic accumulation, and high water table. Soils are generally examined to a depth of 20 inches. Hydric conditions for mineral soils with low to moderate organic content are most commonly demonstrated by gleying and mottling. Mineral soils are examined with a Munsell Soil Color Chart.[8] These soils are considered hydric when they are gleyed or when the top of the B horizon (the layer of soil immediately below the topsoil) has chroma of 2 or less if mottling is present, or chroma of 1 or less if no mottling is present. Low chroma numbers are an index of the degree of soil reduction as a result of anaerobic soil conditions. These criteria allow most soils to be classified as either hydric or nonhydric. Hydric soils that have been effectively drained may, however, still show low-chroma colors, but are no longer considered to be hydric because they lack wetland hydrology. Low-chroma colors may also not be used as an indicator of hydric soils in those soils that are sandy, deeply colored as a result of their parent materials, or recently formed (i.e., alluvial). These soils must be evaluated more carefully under the procedures for problem area wetlands outlined by the 1987 Corps *Manual*.

Sandy soils may be considered to be hydric if organic materials have accumulated above or in the surface horizon. Dark vertical streaking in subsurface horizons caused by the downward movement of organic matter also indicates a hydric soil. This may be associated with a spodic (B2h) horizon located at the average depth of the water table.

The Department of Agriculture's Natural Resource Conservation Service (NRCS) has developed state- and county-based lists of hydric soil series. Unlisted soils are considered to be nonhydric. However, some phases of unlisted soils may contain hydric inclusions and thus may be associated with wetlands. These cases must be individually verified in the field. Field soil characteristics should be given precedence over how a site is mapped on a county soil survey. Alluvial soils may not show hydric characteristics because of their recent formation, but may be considered to be hydric for the purposes of wetland delineation.

Wetland hydrology encompasses the hydrologic characteristics of areas that are inundated or have saturated soils for sufficient duration to support hydrophytic vegetation. Hydrologic indicators are generally used to determine the presence or absence of a wetland. Of the three technical criteria, wetland hydrology is generally the least exact and most difficult to establish in the field because of annual, seasonal,

---

[8]Kollmorgen Corporation. 1985. Munsell Soil Color Chart.

and daily fluctuations. An area has wetland hydrology if the soil is saturated to the surface by groundwater or ponded or flooded with surface water for 2 weeks or more during the growing season. Saturation to the surface can occur when the water table is 0.5 to 1.5 ft below the surface, depending on soil permeability.

Indicators of wetland hydrology may be divided into recorded data and field data. Recorded data may be obtained from aerial photographs, soil surveys, historical data, floodplain delineations, or tide/stream gauges. In the field, wetland hydrology may be evidenced by visual observation of saturation, inundation, or depth to standing water. However, it is not necessary to directly demonstrate the hydrology. Other field indicators of wetland hydrology include morphological plant adaptations, oxidized root channels, water marks, surface scouring, water-stained leaves, sediment deposits, drift lines, and moss lines.

12.4 DETERMINING THE EXTENT OF WETLANDS ON A PROPERTY

A qualified wetlands consultant should be retained to determine the extent of wetlands on a property by applying the methodology in the 1987 Corps manual. There is currently no federal certification program for wetland delineators. The Corps established a demonstration program for certification of wetland delineators, published in the *Federal Register* on December 30, 1992 (57 FR 62,312; 1992), which operated between 1993 and 1994 in selected states. Certifications were issued to less than 250 people. As of this writing the Corps has not implemented a nationwide program for certification. The Society of Wetland Scientists has a Professional Wetlands Scientist Certification Program (see *www.sws.org*). Consultation with Corps district and state wetlands program personnel could also yield names of qualified consultants.

12.4.1 Wetland Maps

There are no published maps that show the extent of federally regulated wetlands on any property. There are no regulated wetland maps. Several maps can be used as a preliminary test to see if wetlands occur on a property: (1) U.S. Fish and Wildlife Service National Wetland Inventory maps, (2) U.S. Department of Agriculture Natural Resource Conservation Service county soil surveys, and (3) floodplain maps. These maps are not, however, meant to be used for site-specific planning. Many states have produced wetland maps. However, even these maps have been primarily prepared from analysis of aerial photos and existing published maps and are not accepted as the delineation of wetlands under federal jurisdiction for a property.

12.4.2 Wetland Delineation

To determine the location of the wetland line, a wetland scientist initially consults existing mapping, topographic maps, and aerial photographs. The federal three-parameter approach (1987 Corps manual) is used in the field to identify wetlands. Vegetation communities are identified, and soil samples are examined by using hand augers. A Munsell Soil Color Chart is used to examine soil color and determine drainage class and depth of saturation. Indicators of hydrology, such as mud marks on trees or standing water, are identified. A consultant can provide either (1) a preliminary investigation for the presence of wetlands resulting in an estimate of the extent of wetlands or (2) a detailed wetlands delineation resulting in the field marking of the wetlands line.

12.4.3 Jurisdictional Determination

Under a detailed wetland delineation, sequentially numbered flags are established at intervals along the wetland boundary line. The line is usually surveyed by a licensed surveyor. The basis for the line is documented with vegetation logs, soil boring logs, and photographs, and a detailed report is prepared. The report and wetland map is submitted to the Corps district with a request for a jurisdictional determination. The wetland delineation report includes sampling station data logs, photographs, the NRCS soils map, U.S. Fish and Wildlife Service National Wetlands Inventory (NWI) map, a location map, a site description, and resumés of preparers. The map should be a survey showing the wetland line, sampling station locations, and photograph locations and directions. The inclusion of topographic information on the map is also useful. A field inspection by Corps personnel is usually performed and the line is verified. The Corps then issues a jurisdictional determination, setting the wetland boundary line, which remains valid for 2 to 5 years, depending on the state.

12.5 SECTION 404 WETLAND REGULATORY PROGRAM

The U.S. Army Corps of Engineers regulates placement of dredged or fill material into waters of the United States, including wetlands, under Sec. 404 of the Federal Clean Water Act. The U.S. EPA has the responsibility to provide guidance to this program through the Sec. 404(B)1 guidelines, and has the power to veto permits issued by the Corps. In order for any 404 permit to be valid, the state in which the activity is proposed must issue water quality certification in accordance with Sec. 401 of the Clean Water Act. If the project is located in the coastal zone, a coastal zone consistency determination must be made by the state.

States can assume all or a portion of the Federal 404 permit program under either of two mechanisms. A state may assume administration of the program if the EPA determines that the state program is at least as stringent as the federal 404 program. The EPA retains oversight over the program. Only Michigan and New Jersey have assumed the federal 404 program. The other alternative is for the Corps to issue a state programmatic general permit, which authorizes the state to issue permits for activities covered under certain nationwide permits. Pennsylvania and Maryland are examples of states that have implemented a state programmatic general permit program. This program results in "one-stop shopping" and requires an applicant only to submit a permit application to the state agency, not both the state agency and the Corps of Engineers.

12.5.1 Regulated Activities

The discharge of dredged or fill material into waters of the United States, including wetlands, is regulated under the 404 permit program. Waters of the United States include runoff or overflow from a contained land or water disposal area as a regulated activity. Redeposit of dredged material, other than the incidental fallback of material into essentially the same location from where it was removed, is also regulated. In plain language, if you propose to truck fill material onto a site and dump it on top of a wetland area, you require a 404 permit from the Corps of Engineers. In general, placement of a structure in wetlands is not regulated; however, if any placement of fill is necessary to install the structure, the entire activity is regulated. Placement of pilings in wetlands is considered, in most instances, to constitute a discharge of fill material and requires a 404 permit. Exceptions include installation of pilings for linear projects such as bridges, elevated walkways, and powerline structures, and placement of pilings for piers, wharves, and individual houses; these exceptions are not considered fill and do not require a 404 permit. However, a Sec. 10 permit under the Rivers and Harbors Act of 1899 (33 U.S.C. 401 et seq.) would be required if the pilings are placed in navigable waters.

The rules for the federal 404 permit program are contained in 33 CFR, Parts 323 and 328 (Department of Defense, Department of the Army Corps of Engineers), and in 40 CFR, Part 110 et al. (Environmental Protection Agency). The most recent comprehensive changes to the regulations were published in the *Federal Register* on August 25, 1993 (58 FR 163, 45008–45038). The nationwide permits were most recently modified on December 13, 1996 (61 FR 241, 65874–65922). Nationwide Permit 26 has been proposed for replacement in early 2000. The definition of discharge of dredged or fill material has recently been somewhat modified to reflect a court decision. The Corps of Engineers and EPA had previously regulated "incidental fallback" of fill during any excavation activities under the program (the Tulloch Rule). The most recent court decision required the agencies to relinquish jurisdiction over the incidental fallback of dredged material into the same location from where it was excavated. Activities such as mining in a wetland may therefore not be regulated under the program. However, redeposit of dredged material from mechanized land clearing and removal of dirt and gravel from a stream bed and its subsequent redeposit in the waterway are regulated.

12.5.2 Permits

There are two types of permits that can be issued for wetlands disturbance: general permits and individual (or standard) permits. General permits can be granted for certain minor activities in wetlands, subject to certain conditions. A list of permits that are available on a nationwide basis is given in Table 12.2.

If activities in wetlands cannot be avoided, it is good planning to try to limit the activity in wetlands to those covered under one or more nationwide general permits. The approval process is rapid and predictable. The activities authorized under nationwide general permits are considered to have only minor impacts to

TABLE 12.2   List of Nationwide Permits

1. Aids to navigation
2. Structures in artificial canals
3. Maintenance
4. Fish and wildlife harvesting, enhancement, and attraction devices and activities
5. Scientific measurement devices
6. Survey activities
7. Outfall structures
8. Oil and gas structures
9. Structures in fleeting and anchorage areas
10. Mooring buoys
11. Temporary recreational structures
12. Utility line discharges
13. Bank stabilization
14. Road crossings
15. U.S. Coast Guard–approved bridges
16. Return water from upland contained-disposal areas
17. Hydropower projects
18. Minor discharges
19. Minor dredging
20. Oil spill cleanup
21. Surface coal mining activities
22. Removal of vessels
23. Approved categorical exclusions
24. State-administered Sec. 404 programs
25. Structural discharges
26. Headwaters and isolated waters discharges
27. Wetland and riparian restoration and creation activities
28. Modifications of existing marinas
29. Single-family housing
30. Moist soil management for wildlife
31. Maintenance of existing flood control projects
32. Completed enforcement actions
33. Temporary construction, access, and dewatering
34. Cranberry production activities
35. Maintenance dredging of existing basins
36. Boat ramps
37. Emergency watershed protection and rehabilitation
38. Cleanup of hazardous and toxic waste
39. Reserved
40. Farm buildings

wetlands. Generally, compensation for the filling of wetlands is not required for most nationwide permits, although there is a trend to modify that in the future. Nationwide permits commonly utilized for development activities include NP 26 for filling in headwaters and isolated wetlands and waters, NP 7 for stormwater outfalls, NP 14 for road crossings, and NP 12 for utility crossings. As noted above, Nationwide General Permit 26, commonly used for many types of development, is due to be replaced by more restrictive nationwide general permits in early 2000.

Depending on the nationwide permit, notification of the Corps of Engineers, and possibly other federal agencies, may be required prior to performance of the activity. Even if predischarge notification is not required, it is prudent to get written confirmation from the Corps district that a nationwide general permit applies to the proposed activity. It is much more difficult, after an activity has occurred, to prove that you comply with the nationwide general permit conditions. Consult the Corps district and the regulations for the particular nationwide general permit to identify requirements for application and for verification that the project is authorized under nationwide general permits.

An individual 404 permit is required for all other disturbances in wetlands not authorized under nationwide permits. These are very difficult to obtain. The substantive criteria used to evaluate an application for an individual 404 permit are the EPA Sec. 404(b)(1) guidelines. If the proposed activity is water-dependent, and wetlands disturbance is minimized, a permit may be granted. For non-water-dependent uses, it must be proved that there is no other alternative location or design for the proposed project that would either avoid or minimize wetland disturbance. This is obviously very difficult to satisfy. An alternative site to be considered can be on property owned by the applicant or any property that could be acquired in the region. The definition of this region is related to the proposed use. Therefore, the search for alternative sites for a proposed residential development could therefore encompass many square miles. If an alternative site is available for development that would result in less wetland disturbance than the proposed site, the permit would be denied. Off-site and on-site alternatives must be examined to avoid or minimize wetland impacts. "Mitigation," creation of wetlands from uplands, generally at a ratio of 2:1, would be a condition of an individual permit. Mitigation can also be provided in the form of enhancement of existing degraded wetlands or restoration of previously filled or drained wetlands. The ratio of enhancement of an existing wetland to the wetland proposed for filling would generally be greater than 2:1.

The application form used to apply for an individual 404 permit is Corps of Engineers form 4345, Application for a Department of the Army Permit. You can obtain a current copy of the form from one of the 38 Corps of Engineers district regulatory offices. (Copy included in Appendix XX.) Some districts utilize a joint permit application form to provide for simultaneous state and federal reviews. An environmental questionnaire, supplement to ENG form 4345 (copy included in Appendix XX), should also be included in the application. Site plans must accompany the application, and must be 8.5 × 11 inches in size so that the Corps of Engineers can readily copy and distribute them to review agencies and in the public notice. A location map, plan view, and cross sections are required. A water quality certification in accordance with Sec. 401 of the Clean Water Act is required for all individual permits. A coastal zone consistency determination from the state may also be required if the project is located in the coastal zone. It is useful to obtain a determination from the state historic preservation office that the project will not impact any resources listed or eligible for listing on the list of historic places, prior to submission of the application. It is also useful to request a search for records of endangered or threatened species for the project area from the Fish and Wildlife Service prior to preparation of the application. If any records exist, an analysis of the potential impact of the project on endangered or threatened species or their habitats can be included in the application. It is useful to include a proposed wetland mitigation plan in the application.

When the application is received by the Corps district, an acknowledgment of its receipt and the application number assigned to the file is provided to the applicant. A public notice is issued within 15 days of the application receipt. A 15- to 30-day public comment period follows. The proposal may be reviewed by the Corps and the public; special interest groups; and local, state, and federal agencies. As appropriate, the Corps consults with other federal agencies such as the Fish and Wildlife Service, the EPA, and the National Marine Fisheries Service (NMFS). Additional information may be requested from the applicant. A public hearing may be held. The Corps makes a decision, and the permit is either issued or denied.

12.5.3 Federal Wetland Policy

On August 24, 1993, the Clinton administration issued a federal wetland policy. The policy includes an interim goal of no overall net loss of the nation's remaining wetlands, and the long-term goal of increasing the quality and quantity of the nation's wetland resource base. The policy includes a commitment to provide efficient, fair, flexible, and predictable regulatory programs. Also included are plans for nonregulatory programs, such as advance planning, wetland restoration inventory and research, and creation and expansion of public/private cooperative efforts to reduce the reliance on regulatory programs to protect wetlands. A commitment to base wetland policy on the best scientific information is expressed. Reforms proposed and adopted to implement the policy include, for example, a requirement that the Corps of Engineers, the EPA, the Soil Conservation Service, and the Fish and Wildlife Service use the same procedures to identify wetland areas, specifically the 1987 *Wetland Delineation Manual.* Deadlines for permit decisions under the 404 program were adopted, including a 90-day review period for permit decisions. A review of Nationwide Permit 26 was initiated to improve its implementation and develop regional description of the types of waters and the nature of activities in those waters that will not be subject to authorization under NP 26. The policy also endorsed the use of mitigation banking in the 404 program.

A significant result of the federal wetland policy was the issuance of Regulatory Guidance Letter 93-2 on August 23, 1993, which describes flexibility afforded by the 404(b)(1) guidelines based on the relative severity of the environmental impact of a proposed discharge into wetlands.

12.5.4 Exemptions

Section 404(g) of the Clean Water Act generally exempts discharges of dredged or fill material associated with normal farming, ranching, and forestry activities such as plowing, cultivating, minor drainage, and harvesting for the production of food, fiber, and forest products or upland soil and water conservation practices. This exemption pertains to normal farming and harvesting activities that are part of an established, ongoing farming or forestry operation.

If an activity involving a discharge of dredged or fill material represents a "new use" of the wetland, and the activity would result in a reduction in reach or impairment of flow or circulation of regulated waters, including wetlands, the activity is not exempt. Both conditions must be met in order for the activity to be considered nonexempt.

Activities that bring a wetland into farm production where that wetland has not previously been used for farming are not considered part of an established operation, and therefore are not exempt. In general, any discharge of dredged or fill material associated with an activity that converts a wetland to upland is not exempt, and requires a Sec. 404 permit.

However, introduction of a new cultivation technique such as disking between crop rows for weed control may be a new farming activity, but because the farm operation is ongoing, the activity is exempt from permit requirements under Sec. 404. Planting different crops as part of an established rotation, such as soybeans to rice, is exempt.

The Corps of Engineers has issued Nationwide Permit 34 for cranberry production activities. Cranberries are an agricultural crop that is grown in wetlands. NP 34 allows disturbance of up to 10 acres of waters and wetlands for expansion of existing cranberry growing operations. Other activities that may be exempt from wetland permit requirements include normal farming activities and normal harvesting of forest products in accordance with a forest management plan.

12.5.5 Enforcement

The EPA and the Corps of Engineers share enforcement authority of the Sec. 404 program. There are two broad categories of Sec. 404 violations:

- Failure to comply with the terms or conditions of a Sec. 404 permit
- Discharging dredged or fill material to waters of the United States without first obtaining a permit.

In 1989, EPA and the Corps entered into a memorandum of agreement on enforcement to ensure efficient and effective implementation of this shared authority. Under the agreement, the Corps has the lead on Corps-issued permit violation cases. For unpermitted discharge cases, the EPA and the Corps determine the appropriate lead agency on the basis of criteria in the agreement.

Sections 309 and 404 of the Clean Water Act provide the agencies with several formal enforcement mechanisms. Under Sec. 309(a), the EPA can issue administrative compliance orders requiring a violator to stop any ongoing illegal discharge activity and, where appropriate, to remove the illegal discharge and otherwise restore the site. Section 309(g) authorizes the EPA and the Corps to assess administrative civil penalties of no more than $125,000 per violation.

Sections 309(b) and (d) and 404(s) give the EPA and the Corps the authority to pursue civil judicial enforcement actions seeking restoration and other types of injunctive relief, as well as civil penalties. The agencies also have authority under Sec. 309(c) to bring criminal judicial enforcement actions for knowing or negligent violation of Sec. 404.

It has been the author's experience that if the person placing the illegal fill did so without the knowledge that it was a violation of the Clean Water Act, and, when the violation is discovered, the person is cooperative and moves quickly to restore the wetland or waterway in accordance with an approved restoration plan, no fine, or only a small fine, is levied. It is important to contract with an experienced consultant, and often a qualified attorney, to respond to the notice of violation. The Corps will identify that a violation has occurred but may not indicate the exact extent of the fill. They do not prescribe the restoration activities; it is up to the applicant to propose a plan. It therefore becomes the consultant's task to determine if a violation has actually occurred and to identify the extent of the violation. The first step is to determine the extent of wetlands (or waters) prior to the disturbance. (I use the term "forensic ecology" for this endeavor.) The *Wetland Delineation Manual* provides guidance for determining the location of the wetlands boundary prior to the disturbance. Review of published mapping, such as wetland or soil maps that predate the violation, is useful. Analysis of aerial photographs taken prior to the activity can be very helpful in determining the historic location of wetlands as well as the date of the disturbance or prior disturbances, to determine if the activity was regulated at the time it occurred. Aerial photographs can also provide information on whether there was an established agricultural operation at the property that may qualify the activity for an exemption. Field investigation can include examining the species of vegetation that was cleared to determine if it was hydrophytic. Examination of the soil profile is probably the most valuable investigative technique. Taking a soil core or utilizing a backhoe to dig a trench through fill material can result in finding the original soil profile beneath the fill. This profile can then be analyzed through visual observation to determine if it was a hydric soil. Using these techniques, the location of the wetland line prior to the disturbance can be established. Sometimes this can result in a determination that no disturbance to wetlands has occurred.

If any activity in wetlands has occurred, the consultant, often in conjunction with the attorney, investigates whether the activity is exempt from permit requirements or is authorized under any existing permits, such as a nationwide permit that does not require notification. An analysis can also be made to determine if the activity could be authorized under a nationwide or individual permit. The Corps will accept applications for "after-the-fact" permits. If the activity is not exempt and cannot be permitted, the consultant prepares a restoration plan to reestablish the wetland area, which is submitted to the Corps for authorization.

If you have already received a permit but anticipate that you may not meet one of the permit requirements, it is prudent to notify the Corps and work with them to resolve the issue. Take, for example, a project that received an individual permit with a condition to create tidal wetlands as mitigation. Changes in design of the proposed seawall were made during construction that resulted in making the original mitigation design impossible to construct. The Corps was contacted and an alternative proposal was submitted. The Corps accepted the plan. Even though construction of the plan was not completed by the time frame included in the permit, the Corps did not issue a notice of violation because the developer was cooperative and the developer's consultants kept the Corps staff apprised on the progress of the mitigation plan through reports and telephone contact. The result was successful construction of a tidal wetland and completion of the development project.

If the violation was incurred with the knowledge that it was a violation of Sec. 404, significant fines are levied and wetland restoration and mitigation is required in addition to the fines. Repeated resistance to a cease and desist order resulted in the imposition of jail time in a case in Pennsylvania. Liability for a violation is primarily assigned to the owner of a property but can also be placed on the person or contractor performing the activity or even on the engineer or consultant who advised a client to proceed with the violation.

12.6   CASE STUDIES

12.6.1   Case Study 1   A Home at the Beach: Determination of Authorization under Nationwide Permit 14 for a Driveway Crossing

The following real-life example illustrates the workings of the federal 404 wetland program. Judy and Jim Smith were teachers looking forward to their retirement. They decided to search for a homesite along Delaware Bay. Their search in New Jersey revealed that the shoreline had been marching steadily inland and they chose instead to focus their search on the Delaware side of the bay, which appeared to be more stable. They found the perfect lot in a sleepy shore town. The property had a somewhat low area near the roadway but rose up toward the bay. The realty agent assured them the lot was buildable. Prior to purchasing the lot, on the agent's recommendation, they hired a local environmental consultant to assess any wetland limitations on the property. The wetlands consultant prepared a report that identified wetlands on the property adjacent to the access roadway and suggested that wetland permits might be required from both the Delaware Department of Natural Resources and Environmental Control (DNREC) and the U.S. Corps of Engineers. The consultant determined that the wetlands were mapped on the State of Delaware tidal maps as "marsh," but also concluded that the wetlands were not tidally influenced. He believed that he could secure a determination from DNREC to demap the wetlands on the property. Judy and Jim trusted the findings of the consultant and proceeded to purchase the lot.

Judy and Jim contracted with a builder to erect a prefabricated home on pilings. It was the builder's responsibility to obtain all necessary permits. The builder had the environmental consultant contact DNREC.

On March 19, 1996, the consultant submitted a request for a "wetland determination" to DNREC to obtain a decision as to whether or not the wetland was incorrectly mapped as tidal. On May 2, 1996, DNREC responded that the wetland on the property was correctly mapped as tidal, and that both a Delaware tidal wetland permit and a wetland permit from the Corps would be required. DNREC verbally informed the consultant that the chances of getting a permit to build on the wetlands were "slim to none."

Desperate, Judy and Jim contacted another firm for help. Their goal was to build their retirement home on the beach, not to wage legal battles. An initial phone call to DNREC yielded the same line, "chances slim to none." He could not recall any permits being issued by DNREC for a driveway permit, although permits had been issued for public roadways. However, with further discussion the review officer conceded that construction of a driveway over the wetlands could be possible if adequate justification was provided and wetland impacts were minimized. Construction of a house in tidal wetlands, however, was unlikely to be authorized. On July 24, 1996, DNREC agreed to revisit the site to reevaluate their decision regarding whether it was tidal. Apparently the property had not been mapped as tidal in the 1973 and 1979 mappings but was added in the inventory conducted in 1988 after culverts were installed in the adjacent roadway, which apparently reconnected the on-site wetlands with tidal waters. After much difficulty with coordination of the two agencies, DNREC made a site visit on November 26, 1996. They determined that tidal wetlands were present on the property but were much smaller than shown on the tidal wetland maps. The determination was based on the observation of a culvert connection to tidal waters, located beneath the adjacent roadway; presence of fish in a portion of the wetland; the presence of a tidal marsh plant species, *Spartina patens,* along with *Phragmites australis*; and the presence of saturated soil. Other portions of the site supported *Phragmites australis,* marsh elder, and bayberry, and did not exhibit saturated soils during high tide. Because of the small area of tidal wetlands to be crossed, DNREC indicated that a permit might be issued for a driveway crossing.

Finally, a joint field meeting of the Corps, DNREC, and the consultant was held on December 11, 1996, to make a final determination on the character and extent of the wetlands onsite and identify permitting requirements.

The Corps of Engineers Philadelphia District has jurisdiction, through the 404 permit program, over filling activities in tidal and freshwater wetlands throughout the State of Delaware. Activities in tidal waters are regulated under Sec. 10 of the River and Harbors Act. Authorization under both Acts can be requested within a single application. Generally, the Corps will entertain an individual permit application to cross a wetland to access an upland area for development as long as an alternatives analysis shows that there is no other way to access the upland that would result in less wetland filling. That is, the 404(b)1 guidelines must be satisfied. A condition of an individual permit is to provide mitigation, or construction of replacement wetlands, to compensate for the wetlands filled. Better yet, there is a Nationwide Permit 14 for construction of minor road crossings over wetlands. The crossing cannot exceed 200 ft in length measured across the wetlands. No mitigation is required. The state must validate the permit by issuing 401 water quality certification and a coastal zone consistency determination. Historically, the State of Delaware has issued these approvals on a blanket basis for the nationwide permits.

The location of wetlands under jurisdiction of each agency was determined in the field. One set of boundaries was established by applying the 1987 *Wetland Delineation Manual,* as required by the Corps. A smaller area was mapped as tidal wetlands subject to the jurisdiction of DNREC. The boundaries of the two types of wetland areas were marked in the field with number flagging. A local surveyor was contracted to survey the locations of the two lines.

The Corps line resulted in a wetland driveway crossing of less than 200 ft, so the driveway was eligible for authorization under NP 14. Luckily for the Smiths, a previous revision to the nationwide permits allowed them to apply to tidal wetlands in addition to nontidal wetlands. DNREC was generally satisfied that, if the driveway was kept to a minimum width and did not impede the existing flow of water through the existing wetland, a Type II tidal wetlands permit might be issued.

Two permit applications were prepared. The application to the Corps was for confirmation of authorization under Nationwide Permit 14 for road crossings. The application to DNREC was for a Type II tidal wetland permit, Sec. 401 water quality certification and coastal zone consistency determination under Sec. 307(c)(1) of the Coastal Zone Management Act.

The first order of business was to have a surveyor plot the location of the two wetland lines and overlay the location of the proposed house and driveway. Two separate site plans were prepared, one for each application, showing the applicable wetland line, the proposed location of the driveway and house, and the area of regulated wetlands affected. A cross section of the proposed wetland crossing was also prepared. These maps were completed in February 1997.

The consultant submitted an application for both a jurisdictional determination and predischarge notification under Nationwide Permit 14 to the Philadelphia District Corps of Engineers. The request for jurisdictional determination was to obtain concurrence with the wetland boundary line established in the field. Confirmation of authorization under Nationwide Permit 14 was requested to authorize the driveway crossing of wetlands.

The application for NP 14 demonstrated compliance with the permit conditions, which include:

1. The width of the fill is limited to the minimum necessary for the actual crossing.
2. The fill placed in waters of the United States (which include wetlands) is limited to a filled area of no more than $1/3$ acre. Furthermore, no more than a total of 200 linear feet of the fill for the roadway can occur in special aquatic sites, including wetlands.
3. The crossing is culverted, bridged, or otherwise designed to prevent the restriction of, and to withstand, expected high flows and tidal flows, and to prevent the restriction of low flows and the movement of aquatic organisms.
4. The crossing, including all attendant features, both temporary and permanent, is part of a single and complete project for crossing of a water of the United States.
5. For fills in special aquatic sites, including wetlands, the permittee notifies the district engineer in accordance with the "Notification" general condition. The notification must also include a delineation of affected special aquatic sites, including wetlands.

The amount of fill was limited by proposing to limit the driveway width to 20 ft, which was the minimum required to accommodate the heavy equipment (including a crane) necessary to install the pilings and erect the prefabricated house. The length of the Corps-regulated wetland to be crossed was 84 ft and the total wetland area to be filled was 1835 ft^2 (0.04 acre). The driveway was proposed to be constructed of gravel with a surface layer of oyster shells. Additionally a culvert was proposed to be located at the lowest point of elevation to allow the passage of water within the wetland from one side of the driveway to the other.

In addition to these specific conditions that apply to NP 14, the Corps has general conditions that may apply to all nationwide permits. One of these general conditions (number 9) includes the requirement for 401 water quality certification from the state, mentioned above. General Condition 11 prohibits use of NPs for any activity likely to jeopardize the existence of a threatened or endangered species listed or proposed for listing under the Federal Endangered Species Act. Another general condition (number 12) prohibits activities that may affect historic properties that are listed or are eligible for listing in the National Register of Historic Places. General Condition 13 addresses preconstruction notification. Nationwide Permit 14 is one of the nationwide permits that requires preconstruction notification.

General Condition 13, as well as condition (e) in Nationwide Permit 14, requires that, on receipt of notification for Nationwide Permit 14 utilization, the Corps must provide immediate notification to the U.S. Fish and Wildlife Service, state natural resource or water quality agencies, the EPA, the state historic preservation officer, and if appropriate, the National Marine Fisheries Service. These agencies have 5 calendar days to notify the engineer if they intend to provide substantive, site-specific comments. If so contacted, the Corps waits an additional 10 calendar days before making a decision on the notification. The comments are considered in the Corps decision. The Corps has 30 days from receipt of notification from the applicant to make a determination. The Corps can determine that NP 14 applies, or it can require that an individual permit application be submitted.

To expedite agency review of the Smiths' application, at the suggestion of the Corps the consultant submitted copies of the NP 14 application to the review agencies including the EPA, the Fish and Wildlife Service, the Delaware Historic Preservation Office, and DNREC.

The application submitted to the Corps on March 14, 1997, contained the following:

1. NAP form 1891, Request for Department of the Army Jurisdictional Determination

2. Name, address, and phone number of the applicant and the applicant's agent

3. A copy of the U.S. Geodetic Survey quadrangle map showing the project location

4. A description of the proposed activity to be authorized under NP 14 and a demonstration of how the activity complies with the NP 14 conditions

5. Proof of notification of the EPA, the Fish and Wildlife Service, and the Delaware Historic Preservation Office

6. Plan and profile maps showing the location and limits of disturbance of the regulated activity.

On April 25, 1997, the Corps issued a determination that the project was approved under Nationwide Permit 14 for road crossings. The verification of authorization was valid for 2 years from the date of the determination. The jurisdictional determination was finally issued on November 19, 1997, and is valid for 5 years. The project was able to proceed prior to the issuance of the jurisdictional determination.

The application to the Delaware Department of Natural Resources and Environmental Control for the Type II tidal wetlands permit also included a request for an individual 401 water quality certification and a coastal zone consistency determination. These approvals are required to validate the 404 permit. New nationwide permit regulations were issued on December 13, 1996. The State of Delaware was in the process of issuing blanket 401 water quality certification and Sec. 307(c)(1) coastal zone consistency determination for these nationwide permits but had not yet done so at the time of submission of the Smiths' application. They therefore requested these approvals specifically for the project.

The Type II wetlands permit application included documentation of the basis for the tidal wetland delineation, a project description and statement of need, an alternatives analysis, and an environmental impacts analysis. Impacts to tidal wetlands included a 71-ft crossing and disturbance of 1420 ft² (0.03 acre). The alternatives analysis showed that crossing of the wetland was unavoidable and that wetland impacts were minimized. The application was submitted on March 14, 1997. The wetlands Type II permit was issued on May 30, 1997. By that time, Delaware had already issued a coastal zone consistency determination for Nationwide Permit 14 and had waived the requirement for water quality certification for that permit. Individual approval was therefore not necessary.

12.6.2   Case Study 2   Take Me Out to the Ball Game: Individual 404 Wetland Permits for Construction of a Retail Center

A national developer proposed to construct a major retail center in rapidly growing Deptford Township, Gloucestor County, New Jersey. The developer assembled multiple lots into a 27-acre parcel within the town center zone. The property included public land that supported four existing ball fields that were the home of the Deptford Township Little League. The condition of the ball fields and ancillary facilities was very poor. As a condition of the purchase, the developer agreed to construct eight state-of-the-art baseball fields on a 40-acre property, with room for additional fields. A wetland consultant was contracted to delineate wetlands on the two properties and obtain the necessary state and federal approvals for any wetland encroachments.

Almonesson Creek formed the western border of the retail center property. The creek was tidal in the vicinity of the project. Tidal freshwater marsh wetlands, fringed by freshwater forested wetlands, bordered the creek. Steep slopes rose up from the freshwater wetlands. Four wetland swales traversed the property from east to west, discharging to the wetlands adjacent to Almonesson Creek. The soils on the site were sandy and erodible. The swales were either created or expanded by the discharge of stormwater runoff from adjacent industrial and commercial uses onto the property.

The project plans called for filling in a portion of each of the wetland swales, thereby disturbing 0.35 acre of wetlands.

New Jersey has implemented a wetlands permit program under the New Jersey Freshwater Wetlands Protection Act administered by the N.J. Department of Environmental Protection (DEP). Under this wetlands program, in effect since July 1, 1988, N.J. DEP regulates nearly all activities in wetlands as well as lands immediately adjacent to wetlands (wetland transition areas). N.J. DEP has adopted multiple statewide general permits (SGPs), which are somewhat similar to the U.S. Corps of Engineers' nationwide permits. SGP 7 allows filling of up to 1 acre of wetland swales on a project site. No wetland mitigation is required for this permit and no alternatives analysis must be provided. In March of 1994, N.J. DEP's application to U.S. EPA to assume the federal 404 permit program in New Jersey was approved. Only one other state, Michigan, has assumed the 404 permit program.

This means that for most of New Jersey there is "one-stop shopping." A project requires approval only from the N.J. DEP for activities in wetlands. Major projects (disturbance of more than 5 acres, disturbance of federally endangered species, etc.) are excepted and require federal agency review during the N.J. DEP permit application review process. However, the Corps retained jurisdiction over tidal waters including wetlands and freshwater wetlands within 1000 ft of tidal waters.

Under a memorandum of agreement between the Corps and N.J. DEP, the DEP is the lead agency for determining the location and extent of freshwater wetlands in New Jersey, although the Corps has the ability to make its own determination on a given site. Wetlands were field-delineated on the property, and a request for a letter of interpretation to confirm the accuracy of the wetland boundary lines was submitted to N.J. DEP and subsequently approved. An application for Statewide General Permit 7 was also approved by N.J. DEP for filling of 0.35 acre of wetland swales on the property. An application for a jurisdictional determination and confirmation of authorization under Statewide General Permit 26 was submitted to the Corps of Engineers Philadelphia District. At that time NP 26 could authorize filling of less than 1 acre of wetlands without agency notification. That threshold was reduced to $^1/_3$ acre in 1996. The Corps had previously issued NP 26 authorization for very similar features on another project, so the applicants assumed it would apply here. However, a different review officer was assigned to the application. This reviewer determined that NP 26 did not apply and that an individual permit was required. As of this writing, NP 26 applies to filling of headwater streams and their adjacent wetlands and to filling of isolated waters and wetlands, not connected to a stream system. A headwater stream is a nontidal stream that has less than 5-ft^3/s average annual flow. In New Jersey these are generally streams that have an upstream drainage basin area of between 5 to 7 mi^2. One Corps staff person suggested that a headwater stream is one you can jump over. The Philadelphia District reviewer determined that the wetlands within the swales were not adjacent to headwater streams, and no stream was present within the swales. He therefore classified the swales as "adjacent" to Almonesson Creek, a tidal stream. The proposed filling was therefore ineligible for authorization under NP 26.

Another N.J. DEP approval required for the project was a waterfront development permit. As a result of the standards of the Coastal Rules applicable to this permit, as well as the provisions of the N.J. Freshwater Wetlands Protection Act, the project was designed to minimize impacts to environmentally sensitive areas. Steep slopes adjacent to Almonesson Creek were avoided. The creek and bordering tidal marsh and forested wetlands and a 50-ft buffer (transition area) adjacent to the forested wetlands were preserved. Water quality treatment of stormwater in detention basins was incorporated into the plan.

The consultant arranged a preapplication conference with the Corps and other federal agencies in order to present the project and identify application requirements. (The Corps, the U.S. Fish and Wildlife Service, the EPA, and the National Marine Fisheries Service, and sometimes the N.J. DEP, meet monthly to review permit applications and encourage prospective applicants to attend to introduce projects that are the subject of future applications.) Because of the small area of wetlands proposed to be filled, the nature of the wetlands (essentially stormwater conveyance gullies), and the intrusion of these wetlands into the most suitable area for development on the site, the agencies appeared to be receptive to the proposal.

A permit application was prepared that included the following:

- A description of the project.
- An analysis of project compliance with the 404(b)(1) guidelines, describing how the wetlands filling could not be avoided in light of logistics, costs, and overall project purpose and emphasizing the low environmental impact of the proposed activity.
- Corps of Engineers ENG form 4345.
- Environmental questionnaire.
- Photographs of the wetland areas to be filled.
- A site location map and project plans on $8^1/2 \times 11$ inch paper. The plans clearly showed the area and extent of wetland filling proposed. Both plan view and cross section views were provided for each swale.
- Wetland mitigation plan.

The swales supported scrub shrub and emergent wetland vegetation. The wetland mitigation plan was designed to replace these wetlands with similar vegetation composition. A site was selected on the proposed ball field property within an existing agricultural field. Conversion of an upland forest to wetlands is not appropriate, since another type of valuable community would be sacrificed to construct a wetland. The mitigation site drained to Little Timber Creek, which was within the same drainage basin as Almonesson Creek. The site was immediately adjacent to existing wetlands and was only slightly higher in elevation than these wetlands. Little excavation was therefore required to reach the appropriate depth to water table to create wetlands. Native plants were proposed for installation within the mitigation area.

The Corps issued the individual wetlands permit 60 days from submission of the application. The permit was conditioned on implementation of the wetland mitigation plan simultaneously with construction of the retail facility. Monitoring of the mitigation site for 3 years following construction was required and 85 percent cover of wetland vegetation was required after 3 years. Submission of annual monitoring reports to the Corps was mandated. The site was constructed according to the mitigation plan and accepted by the Corps after the 3-year period.

12.7 SOURCES OF INFORMATION

The headquarters of the U.S. Army Corps of Engineers may be contacted for free information about wetlands and copies of the wetland regulations. They may be contacted at:

U.S. Army Corps of Engineers
Office of the Chief of Engineers
20 Massachusetts Ave., NW
Washington, DC 20314-1000
Telephone: 202-761-0200; fax: 202-761-5096

Numerous Web sites provide volumes of information on wetlands. Some useful ones are:

U.S. Army Corps of Engineers regulatory home page: *http://www.usace.army.mil/lrc/reg/*

U.S. Army Corps of Engineers Sacramento District home page: *http://spk.usace.army.mil/*

Philadelphia District home page: *http://www.nap.usace.army.mil*

Wetland Regulation Center: *http://www.wetlands.com*

Society of Wetlands Scientists: *http://www.wetlands.com*

Wetland Training Institute: *http://www.wetlandtraining.com*

CHAPTER 13
FLOODPLAINS

John Thonet
President
Thonet Associates, Inc.

13.1 FLOODPLAIN REGULATION IN THE UNITED STATES

Floodplain regulation throughout the United States is guided by the National Flood Insurance Program. This federal government program, in cooperation with the private insurance industry, provides flood insurance protection to property owners in flood-prone areas within communities that meet minimum floodplain management standards established by the program. In addition, the program provides incentives to communities to adopt and enforce measures designed to reduce flooding risks beyond that required by the program's minimum standards.

The National Flood Insurance Program illustrates the benefits to be achieved by the cooperative efforts of federal, state, and local government and private industry. The program has resulted in the preparation of detailed flood insurance studies and flood hazard area mapping for communities located throughout the country and has provided the technical basis and guidelines needed for floodplain management and regulation in the United States.

13.2 THE NATIONAL FLOOD INSURANCE PROGRAM

The National Flood Insurance Act of 1968 established the National Flood Insurance Program (NFIP). The Act included several purposes relevant to floodplain regulation, including the following[1]:

1. Encourage state and local governments to make appropriate land use adjustments to constrict the development of land that is exposed to flood damage and minimize damage caused by flood losses.

2. Guide the development of proposed future construction, where practicable, away from locations that are threatened by flood hazards.

3. Encourage lending and credit institutions, as a matter of national policy, to assist in furthering the objectives of the flood insurance program.

4. Assure that any federal assistance provided under the program will be related closely to all flood-related programs and activities of the federal government.

5. Authorize continuing studies of flood hazards in order to provide for a constant reappraisal of the flood insurance program and its effect on land use requirements.

The Flood Disaster Act of 1973 amended the National Flood Insurance Program by[2]:

1. Substantially increasing the available limits of flood insurance coverage.

2. Providing for the expeditious identification of, and the dissemination of information concerning, flood-prone areas.

---

[1] *National Flood Insurance Act, as amended.* United States Code, Title 42. The Public Health and Welfare, Chap. 50—National Flood Insurance, Sec. 4001. Congressional Findings and Declaration of Purpose.
[2] *Flood Disaster Protection Act of 1973.* United States Code, Title 42.

3. Requiring states or local communities, as a condition of future federal financial assistance, to participate in the flood insurance program and to adopt adequate floodplain ordinances with effective enforcement provisions consistent with federal standards to reduce or avoid future flood losses.

4. Requiring that property owners purchase flood insurance if:

Their property was located in an area identified as being flood-prone.

They were receiving assistance in their acquisition or improvement of property through a federal program or a federally supervised, regulated, or insured agency or institution, such as a bank or savings and loan association.

5. Adding flood-related erosion protection.

The National Flood Insurance Reform Act of 1994 further amended the National Flood Insurance Act of 1968 and the Flood Disaster Act of 1973 in ways intended to better accomplish the program's efforts to minimize flood damages throughout the country. In particular, this act[3]:

- Expanded flood insurance purchase requirements.
- Increased requirements for lenders to notify borrowers of the need to purchase flood insurance and established penalties for lenders who fail to require flood insurance or follow notification procedures.
- Established a Community Rating System and incentives, in the form of credits on premium rates for flood insurance coverage, for communities that voluntarily adopt and enforce measures designed to reduce flooding risks which go beyond the program's minimum floodplain management standards.
- Established a National Flood Mitigation Fund and Mitigation Assistance Program for planning and carrying out state- and community-sponsored activities designed to reduce the risk of flood damage to structures covered by flood insurance.
- Added provisions for additional insurance to cover the cost of compliance with land use and control measures for:

Properties that are "repetitive loss structures"

Properties that have flood damage in which the cost of repairs equal or exceed 50 percent of the value of the structure at the time of the flood event

Properties that have sustained flood damage on multiple occasions, if it is cost-effective and in the best interests of the National Flood Insurance Fund to require compliance with the land use and control measures

- Added provisions for regularly assessing the need to update the program's flood maps.

---

[3]*National Flood Insurance Reform Act of 1994.* United States Code, Title V—National Flood Insurance Reform.

13.3 APPLICABILITY OF THE NATIONAL FLOOD INSURANCE PROGRAM TO INDIVIDUAL COMMUNITIES THROUGHOUT THE UNITED STATES

13.3.1 The Emergency Flood Insurance Program

Under the National Flood Insurance Program, no federal financial assistance, including mortgage loans from federally regulated lenders, may be provided within areas identified as having special flood hazards unless the community in which the area is located is participating in the program. At the beginning of the program, however, there were no maps identifying "areas of special flood hazards" and hence the Emergency Flood Insurance Program was instituted to permit community participation even before detailed studies and maps of a community's flood-prone areas could be prepared.

Under the emergency program, preliminary maps showing the approximate limits of areas of special flood hazards were prepared and provided to communities throughout the United States. Upon receipt of these preliminary maps, communities qualified for participation in the emergency program by adopting and enforcing floodplain management regulations in accordance with the National Flood Insurance Program's minimum standards.

13.3.2 The Regular Program

Once a community qualified for the emergency program, a detailed flood insurance study (FIS) was prepared for the community including the preparation of a flood insurance rate map (FIRM). The flood elevations, flood hazard area limits, and floodway limits developed in these detailed studies provided the technical basis for more comprehensive floodplain management regulations that were then adopted and enforced by the community as a condition of the community's eligibility to participate in the regular program.

13.4 TECHNICAL BASIS FOR FLOODPLAIN REGULATIONS

The National Flood Insurance Program resulted in floodplain regulations for communities throughout the United States. The technical basis for these regulations is provided by the community's FIS and its accompanying floodplain maps and flood profiles. These detailed flood studies are prepared by qualified study contractors, which are generally private engineering companies or qualified public agencies, under contract to the Federal Emergency Management Agency (FEMA).

A community's flood insurance study generally includes the following information:

13.4.1 Flood Insurance Study Report

- A description of the study area and its principal flood problems and flood protection measures.
- A description of the hydrologic and hydraulic engineering methods used for determining the floodplain information contained in the report.
- A summary of discharges table (see Table 13.1 for an example) or frequency–discharge–drainage area curves showing the peak discharges for the 10-, 50-, 100-, and 500-year floods for each stream studied.
- Tidal frequency curves, where applicable, in areas of coastal flooding.
- Flood profiles for all riverine flooding sources showing each stream's bottom-elevation profile, the locations of all bridges and other structures affecting flow in the stream, and the 10-, 50-, 100-, and 500-year flood elevations plotted as a function of location along each studied stream.
- Floodway data tables (see Table 13.2 for an example) showing:

  The base flood (100-year flood) elevations at each cross-section location, along each flooding source, with and without consideration of the regulatory floodway.

  The floodway width, cross-sectional area, and mean velocity at each cross-section location along each flooding source.
- A floodway schematic (see Fig. 13.1) depicting various terms important to floodplain management regulations such as *100-year flood plain limit, stream channel, floodway, flood fringe, encroachment,* and *surcharge.*

TABLE 13.1 Example of a Summary of Discharges Table

Flooding source and location	Drainage area, mi²	Peak discharges, ft³/s			
		10-year	50-year	100-year	500-year
Allendale Brook					
At confluence with Ho-Ho-Kus Brook	1.40	430	620	680	1015
At upstream corporate limits of the Borough of Waldwick	1.30	405	585	645	960
At Franklin Turnpike	0.95	320	465	510	760
At upstream corporate limits of the Borough of Allendale	0.21	115	155	165	210

TABLE 13.2 Example of a Floodway Data Table

| Flooding source | | Floodway | | | Base flood water surface elevation, ft NGVD* | | | |
Cross section	Distance†	Width, ft	Section area, ft²	Mean velocity, ft/s	Regulatory	Without floodway	With floodway	Increase
Diamond Brook (continued)								
I	3,865	31	176	5.6	50.5	50.5	50.7	0.2
J	4,620	41	295	3.3	53.1	53.1	53.3	0.2
K	5,463	56	300	3.3	55.9	55.9	56.1	0.2
L	5,985	131	501	2.0	56.5	56.5	56.7	0.2
M	6,670	68	257	3.8	57.3	57.3	57.5	0.2
N	7,330	32	98	10.0	60.2	60.2	60.2	0.0
O	8,135	23	162	6.1	65.7	65.7	65.9	0.2
P	8,610	41	192	5.1	71.9	71.9	71.9	0.0
Q	8,860	48	302	3.2	74.3	74.3	74.3	0.0
R	9,250	40	333	2.9	76.5	76.5	76.6	0.1
S	9,790	56	248	4.0	77.1	77.1	77.2	0.1
T	10,770	37	215	3.5	79.0	79.0	79.2	0.2
U	12,010	230	709	1.1	80.5	80.5	80.6	0.1
V	12,700	55	311	2.4	80.7	80.7	80.9	0.2

*National Geodetic Vertical Datum of 1929.
†Feet above confluence with Passaic River.

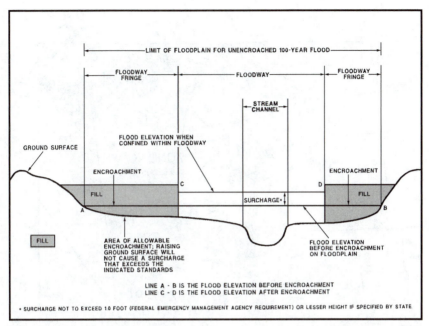

FIGURE 13.1 Floodway schematic.

- Descriptions of the flood insurance rate maps (see Fig. 13.2) that accompany the flood insurance study report and the various flood insurance zone designations shown on those maps. See descriptions below.

13.4.2 Flood Insurance Rate Map (FIRM)

1. A base map showing the community's street layout and the locations of all flooding sources, such as streams and rivers.
2. The locations of special flood hazard areas, which are subject to flooding during a 100-year flood, or the "base flood." Special flood hazard areas are further divided into specific flood insurance rate zones, including the following[4-6]:
   a. Riverine areas of special flood hazard:
      (1) *Zone A*—approximately determined 100-year floodplains for which detailed hydraulic analyses have been performed and for which no flood elevations or depths have been determined.

[4] *Flood Insurance Study Guidelines and Specifications for Study Contractors,* Federal Emergency Management Agency, January 1995, U.S. Government Printing Office 1995-624-336/82337.
[5] Code of Federal Regulation, Title 44—Emergency Management and Assistance, Chap. I—Federal Emergency Management Agency, Part 64—Communities Eligible for the Sale of Insurance, Sec. 64.3—Flood Insurance Maps.
[6] *How to Use a Flood Map to Determine Flood Risk for a Property,* Federal Emergency Management Agency, Map 1995 (FEMA 258).

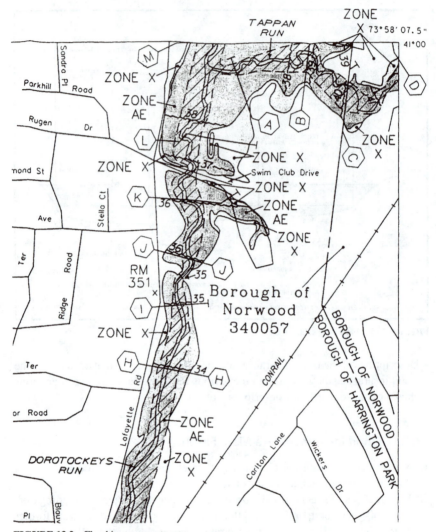

FIGURE 13.2   Flood insurance rate map.

(2) *Zones A1–A30, or Zone AE*—100-year floodplain areas that are determined by detailed study methods and for which 100-year (base flood) elevations have been determined from detailed hydraulic analyses. Zone AE supersedes the numbered A zones in the more recent studies.

(3) *Zone A0*—areas of 100-year shallow flooding and/or unpredictable flow paths (usually sheet flow on sloping terrain) where average depths are between 1 and 3 ft; those depths, derived from detailed hydraulic analyses, are depicted. Zone A0 is also used to depict alluvial fan flood hazards.

(4) *Zone AH* —areas of 100-year shallow flooding and/or unpredictable flow paths (usually ponding areas) where average depths are between 1 and 3 ft. Flood elevations, derived from detailed hydraulic analyses, are depicted.

(5) *Zone A99*—areas within the 100-year floodplain that will be protected by federal flood protection measures such as dikes, dams, and levees, where enough progress on the protection measures has been made to consider it complete for insurance rating purposes.

(6) *Zone AR*—100-year floodplain areas that resulted from decertification of a previously accredited flood protection system that is in the process of being restored to provide protection from the 100-year flood.

b. Coastal areas of special flood hazard:

(1) *Zone V*—approximately determined 100-year coastal (tidal) floodplains (coastal high-hazard area) with additional velocity hazards associated with storm waves for which no detailed hydraulic analyses have been performed and no base flood elevations have been determined.

(2) *Zone V1–V30* or *Zone VE*—100-year coastal (tidal) floodplains (coastal high-hazard area) with additional velocity hazards associated with storm waves for which base flood elevations have been determined from detailed hydraulic analyses. Zone VE supersedes the numbered V zones in the more recent studies.

(3) *Zone V0*—areas of 100-year shallow coastal flooding and/or unpredictable flow profiles, with average depths of between 1 and 3 ft, with additional velocity hazards associated with the storm waves.

c. Areas having only moderate or minimal hazards due to flooding:

(1) *Zone B*—Areas of only moderate hazards due to flooding, including:

(a) Areas between the limits of the 100-year and 500-year flood

(b) Areas subject to 100-year flooding with average depths less than 1 ft

(c) Areas with a contributing drainage area less than 1 mi^2

(d) Areas protected from the 100-year flood by levees

(2) *Zone C*—Areas exhibiting only minimal hazards due to flooding, specifically areas beyond the limits of the 500-year flood.

(3) *Zone X*—A designation that supersedes and combines Zones B and C on more recently prepared FIRMs, thus designating areas of moderate to minimal hazards due to flooding.

d. Other designated areas of flood hazards and flood-related hazards:

(1) *Zone D*—Areas of undetermined, but possible, flood hazards.

(2) *Zones M, N, and P*—Areas of special mudslide hazards, moderate hazards, and undetermined, but possible, mudslide hazards, respectively.

(3) *Zone E*—areas of special flood-related erosion hazards.

3. The locations of valley cross sections used in the hydraulic calculations to determine flood elevations, flood hazard area limits, and floodway limits.

4. Regulatory floodway limits.

5. Base flood elevations at various locations along each flooding source studied, referenced to either the National Geodetic Vertical Datum of 1929 (NGVD) or the North American Vertical Datum of 1988 (NAVD).

13.5 MINIMUM STANDARDS FOR FLOODPLAIN MANAGEMENT

The National Flood Insurance Program establishes minimum standards for floodplain management that must be met by participating communities. The minimum standards vary somewhat depending on the status of the community's FIS and FIRM. However, most communities today have completed and adopted their flood insurance study and flood insurance rate map, thus identifying:

• The locations of the community's riverine and/or coastal areas of special flood hazard as determined by detailed engineering methods.

• Base flood elevations or depths of flooding within areas of special flood hazards.

• The regulatory floodway limits as determined by detailed engineering methods.

• If applicable, the locations of the flood restoration protection areas, mudslide areas, and flood-related erosion-prone areas within the community.

The following summarizes the key flood plain management standards for these communities[7]:

13.5.1 Rules Pertaining to the Use of Base Flood Elevation Data and Floodway Data

1. If a final flood insurance study has been prepared and adopted for the community, the base flood elevation and floodway data contained within that FIS must be used to demonstrate compliance with the community's floodplain management design standards.

2. If base flood elevation and/or floodway data are not available from a community FIS, the community must obtain and review any base flood elevation and floodway data available from federal, state, or other sources. This data must then be used to demonstrate compliance with the community's floodplain management design standards regarding base flood elevations and/or floodways.

3. Base flood elevation data must be provided for all development proposals proposed within designated special flood hazard areas that are greater than 50 lots or 5 acres, whichever is the lesser, regardless of its availability from federal, state, or other sources.

---

[7]Code of Federal Regulations, Title 44—Emergency Management and Assistance, Chap. I—Federal Emergency Management Agency, Part 60—National Flood Insurance Program—Criteria for Land Management and Use (44 CFR 60.1 through 60.8).

13.5.2 General Floodplain Management Requirements

1. Permits are required for all proposed construction or other development, including the placement of manufactured homes, if the proposed construction or other development will be located within a designated special flood hazard area.
2. No permits may be granted for new construction, substantial improvements, or other development (including fill) in the special flood hazard area until a regulatory floodway is designated, unless it can be demonstrated that the proposed development, when combined with all other existing and anticipated development, would not increase the water surface elevation of the base flood more than 1 ft at any point within the community.
3. Permit applications must be reviewed to determine if:
   a. The proposed building sites will be reasonably safe from flooding.
   b. The community's floodplain management standards are met.
   c. All required federal and state approvals have been received.
4. If a proposed building site is located in a flood-prone area, all new construction and substantial improvements must be:
   a. Anchored to prevent flotation, collapse, or lateral movement of the structure.
   b. Constructed with materials and utility equipment resistant to flood damage.
   c. Constructed by methods and practices that minimize flood damage.
   d. Constructed with electrical, heating, ventilation, plumbing, and air conditioning equipment and other service utilities that are designed and/or located to prevent water from entering or accumulating within the components during flooding.
5. If a proposed subdivision or other proposed new development is located in a flood-prone area, it must be reviewed to assure that:
   a. The proposal is consistent with the need to minimize flood damage within the flood-prone area.
   b. Public utilities and facilities, such as sewer, gas, electrical, and water systems, are located and constructed to minimize or eliminate flood damage.
   c. Adequate drainage is provided to reduce exposure to flood hazards.
6. New and replacement water supply systems must be designed to minimize or eliminate infiltration of flood waters into the systems.
7. New and replacement sanitary sewerage systems must be designed to minimize or eliminate infiltration of flood waters into the systems and discharges from the systems into flood waters.
8. On-site waste disposal systems must be located to avoid flooding impacts to the systems or contamination from the systems during flooding.
9. Where base flood elevation data are utilized within the riverine areas of special flood hazard, the community must:
   a. Obtain the elevation of the lowest floor (including basement) of all new or substantially improved structures.
   b. Obtain, if the structure has been floodproofed, the elevation to which the structure was floodproofed.

  c. Maintain a record of all such information with an official designated by the community.
10. Within coastal special flood hazard area zones V1–V30, VE, and V, the community must:
  a. Obtain the elevation of the bottom of the lowest structural member of the lowest floor, excluding pilings and columns, of all new and substantially improved structures, and whether or not such structures contain a basement.
  b. Maintain a record of all such information with the official designated by the community.

13.5.3 Floodplain Management Design Standards Applicable to Riverine Special Flood Hazard Areas

1. Design standard applicable to fill, new construction, substantial improvements, and other development activities in regulatory floodways:
  a. A *regulatory floodway* is the channel of a river or other watercourse and the adjacent land areas that must be reserved in order to discharge the base flood without cumulatively increasing the water surface elevation more than a designated height. It is essentially a right-of-way for flood flows within riverine flood plains. The maximum permitted increase in water surface elevation permitted by the National Flood Insurance Program is 1 ft.
  b. No encroachments, including fill, new construction, substantial improvements, and other development, are permitted within an adopted regulatory floodway that would increase flood levels anywhere in the community during the occurrence of the base flood (100-year) discharge.
2. Design standards applicable to new construction and substantial improvements of residential structures located within riverine flood zones:
  a. Within Zones A1–A30, AH, and AE, require that the lowest floor (including basement) be elevated to or above the base (100-year) flood level.
  b. Within Zone A0, require that the lowest floor (including basement) be elevated above the highest adjacent grade and at least as high as the depth of flooding specified on the community's FIRM.
  c. Within Zones AH and A0, require adequate drainage paths around structures on slopes, to guide floodwaters around and away from proposed structures.
3. Design standards applicable to new construction and substantial improvements of nonresidential structures located within riverine flood zones:
  a. Within Zones A1–A30, AH, and AE, require that either:
    (1) The lowest floor (including basement) be elevated to or above the base flood level.
    (2) Together with attendant utility and sanitary facilities, the structure be designed so that below the base flood level it is floodproofed with walls substantially impermeable to water with structural components able to resist hydrostatic and hydrodynamic loads and effects of buoyancy.

b. Within Zone A0 require that either:

    (1) The lowest floor (including basement) be elevated above the highest adjacent grade, at least as high as the depth of flooding specified on the community's FIRM and at least 2 ft if no depth is specified.

    (2) Together with attendant utility and sanitary facilities be completely floodproofed to that level.

c. Where floodproofing is utilized for a nonresidential structure:

    (1) A registered professional engineer or architect must develop and/or review the structural design, specifications, and plans for construction and certify that the design and methods of construction are in accordance with accepted standards of practice for meeting the above requirements.

    (2) A record of the certification must be maintained with an official designated by the community.

d. Within Zones AH and A0, require adequate drainage paths around structures on slopes, to guide floodwaters around and away from proposed structures.

4. Design standards applicable to fully enclosed flood-prone areas below the lowest floor used solely for parking of vehicles, building access, or storage, other than basements:

    a. New construction or substantial improvements in such fully enclosed areas must be designed to automatically equalize hydrostatic flood forces on exterior walls by allowing for the entry and exit of floodwaters.

    b. Designs for meeting the above requirement must either be certified by a registered professional engineer or architect or meet or exceed the following minimum criteria:

        (1) A minimum of two openings having a total net area of not less than 1 square inch for every square foot of enclosed area subject to flooding must be provided.

        (2) The bottom of all openings may be no higher than 1 ft above grade.

        (3) Openings may be equipped with screens, louvers, valves, or other coverings or devices provided that they permit the automatic entry and exit of floodwaters.

5. Design standards applicable to manufactured homes located within riverine flood zones:

    a. All manufactured homes must be installed by using methods and practices that minimize flood damage, including being elevated and anchored to resist flotation, collapse, or lateral movement. Methods of anchoring may include, but are not limited to, use of over-the-top or frame ties to ground anchors.

    b. Manufactured homes that are placed or substantially improved within Zones A1–A30, AH, and AE must be elevated on a permanent foundation with the lowest floor elevated to or above the base flood elevation. In addition, the manufactured homes must be securely anchored to an adequately anchored foundation system to resist flotation, collapse, and lateral movement. These standards apply to all manufactured homes on sites that are:

(1) Outside a manufactured home park or subdivision.

(2) In a new manufactured home park or subdivision.

(3) In an expansion to an existing manufactured home park or subdivision.

(4) In an existing manufactured home park or subdivision on which a manufactured home has incurred "substantial damage" as the result of a flood.

c. Require that manufactured homes to be placed or substantially improved on sites in an existing manufactured home park or subdivision within Zones A1–A30, AH, and AE, which has not incurred substantial damage in the past as the result of a flood, be elevated so that either:

(1) The lowest floor of the manufactured home is at or above the base flood elevation.

(2) The manufactured home chassis is supported by reinforced piers or other foundation elements of at least equivalent strength that are no less than 36 inches in height above grade and be securely anchored to an adequately anchored foundation system to resist flotation, collapse, and lateral movement.

6. Design standards applicable to recreational vehicles placed within riverine flood zones—recreational vehicles placed on sites within Zones A1–A30, AH, and AE:

a. May not be on the site for more than 180 consecutive days.

b. Must be fully licensed and ready for highway use.

c. Must also meet the same permit requirements as for all other construction and other developments and the same anchoring requirements as for manufactured homes.

7. Design standards applicable to stream channel improvements and relocations:

a. Adjacent communities and the state coordinating office must be notified prior to any alteration or relocation of a watercourse, and copies of such notifications must be submitted to the flood insurance administrator.

b. The flood-carrying capacity within the altered or relocated portion of any watercourse must be maintained.

13.5.4   Floodplain Management Design Standards Applicable to Coastal Areas of Special Flood Hazard

1. Design standards applicable to new construction and substantial improvements within Zones V1–V30, VE, and V:

a. New construction must be located landward of the reach of mean high tide.

b. New construction and substantial improvements must be elevated on pilings or columns so that:

(1) The bottom of the lowest horizontal structural member of the lowest floor, excluding the pilings or columns, is elevated to or above the base flood level.

(2) The pile or column foundation and structure are anchored to resist flotation, collapse, and lateral movement due to the effects of wind and water loads acting simultaneously on all building components. Water loading values used should be those associated with the base flood. Wind loading values used should be those required by state or local building standards. A registered professional engineer or architect must develop or review the structural design, specifications, and plans for the construction, and certify that the design and methods of construction to be used are in accordance with accepted standards of practice for meeting the above requirements.

c. New construction and substantial improvements must have space below the lowest floor either:

(1) Free of obstruction.

(2) Constructed with nonsupporting breakaway walls, open wood latticework, or insect screening designed to collapse under wind and water loads without collapse, displacement, or other structural damage to the elevated portion of the building or supporting foundation system. A breakaway wall must be designed for a safe loading resistance of not less than 10 and no more than 20 pounds per square foot (psf). Use of breakaway walls which exceed a safe loading resistance of 20 psf, either by design or when required by local or state codes, may be permitted only if a registered professional engineer or architect certifies that:

(a) The breakaway wall will collapse from a water load less than that which would occur during the base flood.

(b) The elevated portion of the building and supporting foundation system will not be subject to collapse, displacement, or other structural damage due to the effects of wind and water loads acting simultaneously on all building components, structural and nonstructural. Water loading values used are those associated with the base flood. Wind loading values used are those required by state or local building standards. The enclosed space will be usable solely for parking of vehicles, building access, or storage.

d. The use of fill for structural support of buildings is prohibited.

e. Manmade alteration of sand dunes and mangrove stands is prohibited.

2. Design standards applicable to manufactured homes within coastal areas of special flood hazard—manufactured homes that are placed or substantially improved within Zones V1–V30, V, and VE must meet the following standards:

a. If the manufactured homes are located on any of the following sites, the standards that must be met are the same standards required for all new construction and substantial improvements in Zones V1–V30, V, and VE, described above:

(1) Outside of a manufactured home park or subdivision.

(2) In a new manufactured home park or subdivision.

(3) In an expansion to an existing manufactured home park or subdivision.

(4) In an existing manufactured home park or subdivision on which a manufactured home has incurred "substantial damage" as the result of a flood.

   b. Within an existing manufactured home park or subdivision that has *not* incurred substantial damage to manufactured homes as a result of past flooding, manufactured homes must meet the same standards as for manufactured homes in Zones A1–A30, AH, and AE.

3. Design standards for recreational vehicles placed on sites located within coastal areas of special flood hazard—recreational vehicles placed on sites within Zones V1–V30, V, and VE:

   a. May not be on the site for more than 180 consecutive days.

   b. Must be fully licensed and ready for highway use.

   c. Must also meet the same requirements as for all new construction and substantial improvements in Zones V1–V30, VE, and V.

13.5.5 Floodplain Management Regulations and Design Standards Applicable to Flood Protection Restoration Areas

1. *Eligibility to use the designation Flood Protection Restoration Area.* In order to utilize the designation *Flood Protection Restoration Area,* a community must meet the NFIP's eligibility requirements. A community may be eligible to use this designation if:

   a. Part or all of the community had been protected by a federally funded riverine flood protection system that had been, but is no longer, providing base flood protection on the community's effective FIRM.

   b. The flood protection system has been decertified by a federal agency responsible for flood protection design and construction.

   c. The community is in the process of restoring the flood protection system to once again provide base flood protection.

2. *Compliance standards.* Compliance with the applicable standards for riverine special flood hazard areas and the general floodplain management requirements, outlined above, is required for all new construction and substantial improvements within flood protection restoration areas. In lieu of using the base flood elevations to determine compliance, however, the elevation standard to be used will vary depending on the situation as specified below:

   a. For new construction of structures:

     (1) In "developed areas," as defined by the NFIP, located in Zone AR, AR/A1–A30, AR/AE, AR/AH, AR/A0, or AR/A, and for other areas within Zone AR where the AR flood depth is 5 ft or less, the elevation standard used to measure compliance is the lower of either the AR base flood elevation or the elevation that is 3 ft above the highest adjacent grade.

(2) In Zone AR areas that are *not* designated as developed areas and where the AR flood depth is greater than 5 ft, the elevation standard used to measure compliance is the AR base flood elevation.

(3) In Zones AR/A1–A30, AR/AE, AR/AH, AR/A0, and AR/A, that are *not* designated as developed areas, the elevation standard used to measure compliance is determined by:

    (a) Determining the applicable flood elevation standard for Zone AR as described above.

    (b) Determining the base flood elevation or flood depth for the underlying A1–A30, AE, AH, A0, or A zone.

    (c) Using the higher of the above-determined elevations.

    b. For substantial improvements to existing construction within Zones AR/A1–A30, AR/AE, AR/AH, AR/A0, and AR/A, determine the A1–A30 or AE, AH, A0, or A Zone base flood elevation and use this elevation as the design standard.

3. *Notification requirement.* Permit applicants must be notified of the area's designation as a flood protection restoration area and whether the structure will be elevated or protected to or above the AR flood elevation.

13.5.6 Floodplain Management Design Standards Applicable to Mudslide Areas and Flood-Related Erosion-Prone Areas

1. When mudslide areas and/or flood-related erosion-prone areas have not been identified yet by the federal insurance administrator, but the community has indicated the presence of these hazards, the community must require the following:

    a. Permits for all proposed construction or other development in the community in order to determine if the development is proposed in a mudslide area.

    b. Permits for all proposed construction or other development in the area of flood-related erosion hazard, as it is known to the community.

    c. Review each permit application to determine if the proposed site and improvements will be reasonably safe from the mudslide hazard and/or flood-related erosion and will not cause flood-related erosion hazards or otherwise aggravate existing flood-related erosion hazards.

    d. If the proposed site and improvements are located in an area that may have mudslide hazards, require:

        (1) A site inspection and further review by a qualified person.

        (2) A design that ensures:

            (a) Adequate protection against mudslides.

            (b) No aggravation of existing mudslide hazards by creating on-site or off-site disturbances.

            (c) Drainage, planting, watering, and maintenance that do not endanger slope stability.

  e. If the proposed development is found to be in the path of flood-related erosion or to increase erosion hazard, the development must be relocated or designed with adequate protection measures such that the development will not aggravate the existing erosion hazard.
2. When mudslide areas and/or flood-related erosion-prone areas have been identified by the federal insurance administrator as Zone M and Zone E, respectively, the following requirements apply, in addition to those presented above:
  a. For mudslide hazard areas, a grading ordinance must be developed that:
    (1) Regulates the location of foundation and utility systems of new construction or substantial improvements.
    (2) Provides special requirements for protective measures including such measures as retaining walls, buttress fills, subdrains, diverter terraces, and benchings.
    (3) Requires engineering drawings and specifications to be submitted for corrective measures, accompanied by supporting soils engineering and geology reports.
  b. For flood-related erosion-prone areas, a setback must be required for all new development from ocean, lake, bay, riverfront, or other body of water, to create a safety buffer consisting of a natural vegetative or contour strip.

13.6 VARIANCES AND EXCEPTIONS[8]

13.6.1 Variances

The National Flood Insurance Program does not provide absolute criteria for granting variances from its minimum floodplain management standards. Variances may be granted, however, subject to the following conditions:

1. No variance may be granted within a designated regulatory floodway if any increase in flood levels would occur during the 100-year (base flood) event.
2. A variances may be granted for new construction and substantial improvements to take place on a lot of $1/2$ acre or less if it is contiguous to and surrounded by lots with existing structures constructed below the base flood level, provided that:
  a. A showing of good and sufficient cause is made.
  b. Failure to grant the variance would result in exceptional hardship to the applicant.
    c. Granting the variance will not result in increased flood heights, additional threats to public safety, or extraordinary public expense; will not

---

[8]Code of Federal Regulations, Title 44—Emergency Management and Assistance, Chap. I—Federal Emergency Management Agency, Part 60—National Flood Insurance Program—Criteria for Land Management and Use (44 CFR 60.1 through 60.8).

create nuisances or create fraud on or victimize the public; and will not conflict with existing local laws or ordinances.

d. The variance is the minimum necessary to afford relief, considering the flood hazard.

e. The community does the following:

(1) Notifies the applicant in writing that:

(a) The issuance of the variance will result in increased premium rates for flood insurance up to amounts as high as $25 for $100 of insurance coverage.

(b) Construction below the base flood level increases risks to life and property.

(2) Maintains a record of all variance actions, including justification for their issuance.

(3) Reports all variances issued in its annual or biennial report submitted to the federal insurance administrator.

13.6.2 Exceptions

Certain exceptions from the NFIP's minimum floodplain management standards may be permitted if, because of extraordinary circumstances, local conditions render the application of certain standards the cause for severe hardship and gross inequity for a particular community proposing adoption of the minimum standards.

In these circumstances, the community must write to the federal insurance administrator and explain the nature and extent of and the reasons for the exception request. The request must include sufficient supporting economic, environmental, topographic, hydrologic, and other scientific and technical data, and data relating to the impact on public safety and the environment. The administrator then reviews the community's request for relief and renders a decision.

In particular, standards for floodproofed residential basement below the base flood level in Zones A1–130, AH, A0, and AE, which are not subject to tidal flooding, may be adopted provided that:

1. The areas of special flood hazard in which basements will be permitted are subject to shallow and low-velocity flooding, and there is adequate flood warning time to ensure that all residents are notified of impending floods.

2. The community has adopted floodplain management measures that require that new construction and substantial improvements of residential structures with basements in the above zones must be designed and built so that the basement area, attendant utilities, and sanitary facilities below the floodproofed design level are watertight with walls that are impermeable to the passage of water.

13.7   STATE FLOODPLAIN MANAGEMENT REGULATIONS[9]

The NFIP considers a state to be a community subject to the same regulations as any other community. Accordingly, a state-owned property located in a special flood hazard area must either meet the floodplain management standards of the participating community in which the property is located or establish and enforce floodplain management regulations consistent with the NFIP's minimum standards.

13.8   ADDITIONAL PLANNING STANDARDS FOR FLOOD-PRONE, MUDSLIDE-PRONE, AND FLOOD-RELATED EROSION-PRONE AREAS[9]

The National Flood Insurance Program encourages communities to develop and adopt comprehensive management plans for flood-prone, mudslide-prone, and flood-related erosion-prone areas. The program requires that participating communities consider the following planning standards, though adoption of these additional standards is *not* mandatory.

1. In flood-prone areas:
   a. Development should be permitted only if:
      (1) The development is appropriate given the probability of flood damage and the need to reduce flood losses.
      (2) The development is an acceptable social and economic use of the land in relation to the hazards involved.
      (3) The development does not increase the danger to human life.
   b. Nonessential or improper installation of public utilities and public facilities should be prohibited.
2. In formulating community development goals after the occurrence of a flood disaster, the community should consider:
   a. Preservation of the flood-prone areas for open-space programs.
   b. Relocation of occupants away from flood-prone areas.
   c. Acquisition of frequently flood-damaged structures.
   d. Acquisition of land or land development rights for public purposes consistent with a policy of minimization of future property losses.
3. In formulating community development goals and in adopting floodplain management regulations, communities must consider at least the following factors:

---

[9]Code of Federal Regulations, Title 44—Emergency Management and Assistance, Chap. I—Federal Emergency Management Agency, Part 60—National Flood Insurance Program—Criteria for Land Management and Use (44 CFR 60.1 through 60.8).

a. Human safety.

b. Diversion of development to areas safe from flooding in light of the need to reduce flood damage and to prevent environmentally incompatible floodplain use.

c. Full disclosure to all prospective and interested parties that:

(1) Certain structures are located within flood-prone areas.

(2) Variances have been granted for certain structures located in flood-prone areas.

(3) Premium rates applied to new structures built at elevations below the base flood substantially increase as the elevation decreases.

d. Adverse effects of floodplain development on existing development.

e. Encouragement of floodproofing to reduce flood damage.

f. Flood warning and emergency preparedness plans.

g. Provision for alternative vehicular access and escape routes when normal routes are blocked or destroyed by flooding.

h. Establishment of minimum floodproofing and access requirements for schools, hospitals, nursing homes, orphanages, penal institutions, fire stations, police stations, communications centers, water and sewage pumping stations, and other public or quasi-public facilities already located in the flood-prone area, to enable them to withstand flood damage and to facilitate emergency operations.

i. Improvement of local drainage to control increased runoff that might increase the danger of flooding to other properties.

j. Coordination of plans with neighboring communities' floodplain management programs.

k. All new construction or substantial improvements in areas subject to subsidence must be elevated above the base flood level by an amount equal to expected subsidence for at least a 10-year period.

l. Requiring subdividers in riverine areas to furnish floodway delineations before approving the subdivisions.

m. Prohibition of any alterations or relocation of a watercourse, except as part of an overall drainage basin plan. In the event of an overall drainage basin plan, provide that the flood-carrying capacity within the altered or relocated portion of the watercourse is maintained.

n. Requirement of setbacks for new construction within coastal special flood hazard areas.

o. Requirement of additional elevation above the base flood level for all new construction and substantial improvements within Zones A1–A30, AE, V1–V30, and VE, to protect against such occurrences as wave wash and floating debris, to provide an added margin of safety against floods having a magnitude greater than the base flood, or to compensate for future urban development.

p. Requirement of consistency between state, regional, and local comprehensive plans and floodplain management programs.

q. Requirement of pilings or columns rather than fill, for the elevation of structures within flood-prone areas, in order to maintain the storage capacity of the floodplain and to minimize the potential for negative impacts to sensitive ecological areas.

r. Prohibition, within any floodway or coastal high hazard area, of plants or facilities in which hazardous substances are manufactured.

s. Requirement that a plan for evacuating residents of all manufactured home parks or subdivisions in flood-prone areas be developed, filed with, and approved by appropriate community emergency management authorities.

4. The planning process for communities containing mudslide-prone areas should include:

   a. The existence and extent of the hazard.

   b. The potential effect of inappropriate hillside development.

   c. The means of avoiding the hazard.

   d. The means of adjusting to the hazard.

   e. Coordination of land use, sewer, and drainage regulations and ordinances with fire prevention, floodplain, mudslide, soil, land, and water regulation in neighboring communities.

   f. Planning subdivisions and other developments to avoid exposure to mudslide hazards and the control of public facility and utility extension to discourage inappropriate development.

   g. Public facility location and design requirements with higher site stability and access standards for schools, hospitals, nursing homes, orphanages, correctional and other residential institutions, fire and police stations, communication centers, electric power transformers and substations, water and sewer pumping stations, and any other public or quasi-public institutions located in the mudslide area to enable them to withstand mudslide damage and to facilitate emergency operations.

   h. Provision for emergencies.

5. The planning process for communities containing flood-related erosion-prone areas should include:

   a. The importance of directing future developments to areas not exposed to flood-related erosion.

   b. The possibility of reserving flood-related erosion-prone areas for open-space purposes.

   c. The coordination of all planning for the flood-related erosion-prone areas with planning at the state and regional levels, and with planning at the level of neighboring communities.

   d. Preventive action in E Zones, including setbacks, shore protection works, relocating structures in the path of flood-related erosion, and community acquisition of flood-related erosion-prone properties for public purposes.

   e. Consistency of plans for flood-related erosion-prone areas with comprehensive plans at the state, regional, and local levels.

13.9 STATE COORDINATING AGENCIES[10]

The NFIP encourages states to take a leadership role in encouraging sound floodplain management by agreeing to assign a state coordinating agency to be responsible for coordinating the NFIP's floodplain management efforts in the state. The role of the state coordinating agency is as follows:

1. Enacts whatever legislation is necessary to enable counties and municipalities to regulate development within flood-prone areas.

2. Encourages and assists communities in qualifying for participation in the NFIP.

3. Guides and assists county and municipal agencies in developing, implementing, and maintaining local floodplain management regulations.

4. Provides local governments and the general public with NFIP information and assists communities in disseminating information on minimum elevation requirements for development in flood-prone areas.

5. Assists in the delineation of riverine and coastal flood-prone areas, whenever possible, and provides all relevant technical information to the federal insurance administrator.

6. Recommends priorities for federal floodplain management activities in relation to the needs of county and municipalities within the state.

7. Provides notification to the federal insurance administrator in the event of apparent irreconcilable differences between a community's local floodplain management program and the program's minimum requirements.

8. Establishes minimum state floodplain management regulatory standards consistent with standards established for the NFIP and in conformance with other federal and state environmental and water pollution standards for the prevention of pollution during periods of flooding.

9. Assures coordination and consistency of floodplain management activities with other state, areawide, and local planning and enforcement agencies.

10. Assists in the identification and implementation of flood hazard mitigation recommendations that are consistent with the NFIP's minimum floodplain management standards.

11. Participates in floodplain management training opportunities and other flood hazard preparedness programs whenever practicable.

---

[10]Code of Federal Regulations, Title 44—Emergency Management and Assistance, Chap. I—Federal Emergency Management Agency, Part 60—National Flood Insurance Program—Criteria for Land Management and Use (44 CFR 60.1 through 60.8).

In return for a state's participation in the NFIP as a state coordinating agency, the federal insurance administrator:

- Considers state recommendations prior to implementing NFIP activities affecting state communities.

- Considers state approval or certification of local floodplain management regulations as meeting the NFIP's minimum floodplain management standards.

CHAPTER 14
NUCLEAR MATERIAL LICENSES

Dr. Dean W. Broga
Director, Office of Environmental Health & Safety
Medical College of Virginia

James N. Christman, Esq.
Partner
Hunton & Williams Law Firm

This chapter addresses licenses to use radioactive materials for industrial, medical, or research purposes. These licenses are issued by the U.S. Nuclear Regulatory Commission (NRC) or by the 30 states that have agreed to undertake the licensing role.

The NRC issues licenses for nuclear power reactors, which generate electricity; for commercial nuclear fuel facilities[1]; and for smaller-scale use of "nuclear materials." About 21,600 licenses have been issued for medical, academic, and industrial uses of nuclear material. Some 5900 of these are administered by the NRC, with the other 15,700 administered by the 30 states that participate in the NRC "agreement states" program (discussed below).

The first thing to remember about nuclear materials licenses issued by the NRC is that most of the information you need can be found at the NRC's Web site, particularly *http://www.nrc.gov/NRC/nucmat.html*. If you forget this address, use a Web search engine and type in "nuclear materials"; the correct NRC page will probably turn up as the fourth or fifth hit.

The second thing to remember about nuclear material licenses is that the NRC is publishing a detailed multivolume guidance document called *NUREG-1556*. The volumes of NUREG-1556, which are or will be available at the NRC Web site, are these:

Vol. 1   Portable Gauge Licenses (May 1997)

Vol. 2   Industrial Radiography Licenses (August 1997) (draft)

Vol. 3   Sealed Source and Device Evaluation and Registration (July 1998)

Vol. 4   Fixed Gauge Licenses

Vol. 5   Self-Shielded Irradiator Licenses

Vol. 6   10 CFR Part 36 Irradiator Licenses (March 1998) (draft)

Vol. 7   Academic, Research and Development, and Other Licenses of Limited Scope (May 1998) (draft)

---

[1]The NRC licenses eight major fuel facilities in seven states and two gaseous diffusion uranium enrichment facilities. In addition, a privately owned uranium enrichment facility is to be built in Louisiana.

Vol. 8   Exempt Distribution Licenses (September 1998)

Vol. 9   Medical Use Licenses (August 1998) (draft)

Vol. 11   Licenses of Broad Scope (August 1998) (draft)

The authority for the NRC to license nuclear materials comes from the U.S. Atomic Energy Act. The NRC regulations are found in Title 10 of the Code of Federal Regulations (CFR), primarily in Parts 20–39. In addition, the Food and Drug Administration (FDA) regulates x-ray machines and approves radiopharmaceuticals [see especially the FDA regulations in 21 CFR, Parts 1000–1500 on "Radiological Health" and 21 CFR § 312.140(c)], and the Department of Transportation makes the rules for shipping radioactive materials.

14.1 THE LICENSING PROGRAM

14.1.1 The Law

Federal jurisdiction over devices that use radioactive materials comes from the Atomic Energy Act of 1954, as amended. This federal law requires the Nuclear Regulatory Commission (which used to be called the Atomic Energy Commission) to regulate the possession and use of three types of radioactive materials: (1) source materials, (2) by-product materials, and (3) special nuclear materials. *Source materials* are uranium and thorium and ores containing at least 0.05 percent uranium or thorium. *Special nuclear materials* are plutonium and enriched uranium-233 or uranium-235 used for nuclear fuel. *By-product* materials include the tailings or wastes left over when uranium or thorium is extracted from ore, but for our purposes refer to radioactive materials resulting from fission or activation by neutrons. By-product material licenses are what we will be most concerned with in this chapter.

The Atomic Energy Act is quite general and offers little practical guidance. To get more specific information you must go to the NRC regulations and guidance documents. The most important rules of a general nature for possessing and using nuclear materials are found in the following parts of Title 10 of the Code of Federal Regulations (CFR):

- 10 CFR, Part 2, "Rules of Practice for Domestic Licensing Proceedings and Issuance of Orders"

- 10 CFR, Part 19, "Notices, Instructions and Reports to Workers: Inspection and Investigations"

- 10 CFR, Part 20, "Standards for Protection Against Radiation"

- 10 CFR, Part 21, "Reporting of Defects and Noncompliance"

- 10 CFR, Part 30, "Rules of General Applicability to Domestic Licensing of Byproduct Material"

- 10 CFR, Part 40, "Domestic Licensing of Source Material"
- 10 CFR, Part 71, "Packaging and Transportation of Radioactive Material"
- 10 CFR, Part 150, "Exemptions and Continued Regulatory Authority in Agreement States and in Offshore Waters under Section 274"
- 10 CFR, Part 170, "Fees for Facilities, Materials, Import and Export Licenses and Other Regulatory Services Under the Atomic Energy Act of 1954, as Amended"
- 10 CFR 171, "Annual Fees for Reactor Operating Licenses, and Fuel Cycle Licenses and Materials Licenses, Including Holders of Certificates of Compliance, Registrations, and Quality Assurance Program Approvals and Government Agencies Licensed by NRC"

As indicated above, the basic rules for licensing by-product material are in 10 CFR, Part 30; the basic radiation dose standards are in Part 20.

Certain "general" licenses for particular uses of particular by-product materials are issued in Part 31; these allow you to use the devices named in Part 31 (static elimination devices, for example) so long as you follow the rules, without applying for an individual license.

Four specific types of materials license have their own sections of the rules, as follows:

- Industrial radiography (10 CFR, Part 34)
- Medical use of by-product material (10 CFR, Part 35)
- Irradiators (10 CFR, Part 36)
- Well logging (10 CFR, Part 39)

If you are an NRC licensee or an applicant for an NRC license you need a copy of the regulations. A two-volume bound version of Title 10 of the Code of Federal Regulations (Parts 0–50 and 51–199) can be ordered from the Superintendent of Documents, P.O. Box 371954, Pittsburgh, PA 15250-7954, or by contacting the Government Printing Office (GPO) at *www.gpo.gov.* You can also find the NRC regulations in the "Reference Library" on the NRC Web site. Single copies of the regulations may be requested from the NRC's Regional or Field Offices. A searchable version of 10 CFR can be found on the Web at *http://www.access.gpo.gov/nara/cfr/cfr-table-search.html.* Changes to the regulations, which may be made at any time (though only after public notice), are published in the *Federal Register. Federal Register* notices can be found at *http://www.gpo.ucop.edu/* or in the "Reference Library" on the NRC Web site.

The "law" for licensing purposes consists of the Constitution (rarely pertinent), the statute, and agency regulations. Dipping one level below the regulations you will find a number of NRC "policy" and "guidance" documents, particularly

Regulatory Guides[2] and NUREGs.[3] Traditionally the Regulatory Guides have resembled prescriptive rules, in contrast to the NUREGs, which have been more in the nature of "reports" with detailed discussions of how to comply with the regulations. Now the NRC is replacing Regulatory Guides with NUREGs.[4]

Strictly speaking, guidance documents like NUREGs are not "law"[5]; you do not have to follow them if you can persuade the NRC staff that you have an alternative way of satisfying the regulations. As the NRC itself says, agency guidance documents like NUREGs are "not a substitute for NRC regulations." But as a practical matter the NRC staff may sometimes treat the guidance documents as though they left you, and the staff, no choice.

Nevertheless, as a legal matter, a licensee is not bound to follow NRC guidance *if* it can show that its program meets the regulations in a different manner. For example, NRC guidance documents may say that the radiation safety officer (RSO) should make certain reports in writing. In some circumstances, though, it may be acceptable (with NRC staff approval) to provide oral, rather than written, reports. Similarly, the NRC staff may be flexible about how frequently certain reports must be made or certain procedures followed. It all depends on whether you can show to the staff's satisfaction that your method is adequate to ensure safety and accountability for radioactive materials.

The NUREG-1556 series "Consolidated Guidance about Materials Licenses" provide much useful information. Volumes 1–14 are in final form. Volumes 15 and 16 of NUREG-1556 are still in draft versions; they have been distributed for comment and may be revised to take comments into account. These drafts represent the NRC staff's current position and may be used when preparing requests for a license. Whenever a final NUREG-1556 volume is published, of course, you should use that instead of an outmoded draft version. See www.nrc.gov/nrc/nuregs/index.html for a current list of NUREGs.

---

[2]Regulatory Guides provide guidance to licensees and applicants on acceptable ways of implementing specific parts of the regulations. Many of these date back to the 1970s and 1980s. Only one of the Regulatory Guides for radioactive products, for example, dates from the 1990s.

[3]NUREGs include reports (including those prepared for international agreements), brochures (including directories, manuals, procedural guidances, and newsletters, often intended for internal use at the NRC), conference proceedings, and books. The NRC also sometimes publishes guidance documents called *Temporary Instructions* (TIs). A recent draft TI, for example, attempts to streamline the inspection process for nuclear medicine programs licensed under 10 CFR §§ 35.100, .200, and .300.

[4]For example, Vol. 2 of NUREG-1556, which corresponds with a revision to 10 CFR, Part 34, that was published in May 1997, combines and supersedes earlier guidance, namely draft Regulatory Guide FC 401-4, "Guide for the Preparation of Applications for the Use of Sealed Sources and Devices for Performing Industrial Radiography," and NMSS Policy and Guidance Directive FC 84-15 "Standard Review Plan for Applications for the Use of Sealed Sources and Devices for Performing Industrial Radiography." Likewise Vol. 9 of NUREG-1556, which addresses medical uses, will replace a number of Regulatory Guides. The current draft of Vols. 2 and 9, like most of the other volumes, is available at the NRC's Web site, *http://www.nrc.gov/NRC/NUREGS/SR1556/V2/index.html*.

[5]The law is fairly well established on this point. If an agency document has binding effect or constrains the agency's discretion, then it is effectively a legislative "rule," and a legislative rule has to be subjected to certain procedures, such as publication for public comment. *McLouth Steel Products Corp. v. Thomas,* 838 F.2d 1317, 1320-22 (D.C. Cir. 1988); see also *Montrose Chemical Corp. v. EPA,* 132 F.3d 90 (D.C. Cir. 1998) (informal EPA memorandum did not constitute a regulation).

If you get your license by relying on a draft and later the draft is revised in a way that would give you some advantage, you can ask for an amendment to your license. For example, Vol. 1 of NUREG-1556, which covers licenses for portable gauges, is in effect now. Portable gauge licensees who got their licenses before the guidance was in effect have the option of submitting a complete application using NUREG-1556, Vol. 1, when they file an amendment request. A licensee who wants to take advantage of this option should incorporate the requested change into a complete application, submit it with the amendment fee, and indicate that the complete application is an amendment request to take advantage of the new guidance. When the NRC staff has reviewed the request and resolved any outstanding issues, the staff will amend the license in its entirety without changing the expiration date.

Licensees who want to *renew* their licenses should submit a complete application according to NUREG-1556, Vol. 1. The NRC staff's action will be similar to that for an amendment but include an extension of the license expiration date. By following this procedure, the staff expects all existing portable gauge licenses to be converted to the more performance-based format of NUREG-1556, Vol. 1, within a few years.

14.1.2 Agreement States

Other state or federal licenses may be necessary besides, or in place of, an NRC license. An NRC license, if it is required, does not waive legal obligations to get permits from other agencies. [*Hydro Resources, Inc.* (2929 Coors Road, Suite 101, Albuquerque, NM 87120), CLI-98-16, 48 NRC 119 (1998).[6]]

Thirty states, called *agreement states,* have signed formal agreements with the NRC that give them authority to license and inspect by-product, source, or special nuclear materials used or possessed within their borders. Also, "naturally occurring and accelerator-produced radioactive material" (abbreviated NARM) and x-ray machines are always regulated by the states, not by the NRC. None of the agreements allows the state to regulate nuclear power reactors; the health and safety aspects of reactors are regulated exclusively by the federal government.[7] Figure 14.1 is a map of the agreement states, taken from the NRC's Web site. The NRC has also provided Table 14.1 to tell you which agencies have regulatory authority over which activities. The Office of State Programs at NRC headquarters (1-800-368-5642, extension 415-3340) can help you find the right state agency, often a division of the state health department.

---

[6]References to cases reported as, for example, "48 NRC 119," are to the NRC's "yellow books," in which the NRC publishes decisions of the five NRC commissioners, the administrative judges on the Atomic Safety and Licensing Board Panel, and the directors of the Office of Nuclear Material Safety and Safeguards (for materials licenses) and the Office of Nuclear Reactor Regulation (for nuclear power reactors).

[7]This does not necessarily mean, though, that an injured person will be unable to collect damages in state court, as the *Silkwood* case (discussed below) taught us. Also, the states are allowed to regulate the nonradiological health-and-safety aspects of nuclear facilities, such as the rates to be charged for the electric power.

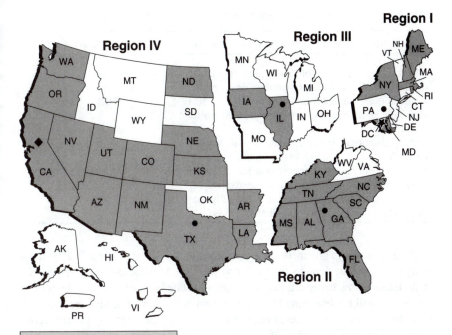

Headquarters
■ Regional Office
◆ Field Office
▨ 30 Agreement States
(approx. 15,800 licensees)
□ 20 Non-Agreement States
(approx. 6,000 licensees)

Note: Alaska and Hawaii are
included in Region IV. Puerto
Rico and Virgin Islands in
Region II

Headquarters
Washington, D.C. 20555-0001
301-415-7000, 1-800-368-5642

Region I
475 Allendale Road
King of Prussia, PA 19406-1415
610-337-5000, 1-800-432-1156

Region II
61 Forsyth Street, SW, Suite 23T85
Atlanta, GA 30303
404-562-4400, 1-800-577-8510

Region III
801 Warrenville Road
Lisle, IL 60532-4351
630-829-9500, 1-800-522-3025

Region IV
611 Ryan Plaza Drive, Suite 400
Arlington, TX 76011-8064
817-860-8100, 1-800-952-9677

Walnut Creek Field Office
1450 Maria Lane
Walnut Creek, CA 94596-5368
510-975-0200, 1-800-882-4672

62-pt1-1029-3501d
072798

FIGURE 14.1 Locations of NRC offices and agreement states.

TABLE 14.1   Who Regulates the Activity?

Applicant and Proposed Location of Work	Regulatory Agency
Federal agency regardless of location [except that Department of Energy and, under most circumstances, its prime contractors are exempt from licensing (10 CFR 30.12)]	NRC
Nonfederal entity in nonagreement state, U.S. territory, or possession	NRC
Nonfederal entity in agreement state at nonfederally controlled site	Agreement state
Nonfederal entity in agreement state at federally controlled site *not* subject to exclusive federal jurisdiction	Agreement state
Nonfederal entity in agreement state at federally controlled site subject to exclusive federal jurisdiction	NRC

If you are in an agreement state, the licensing of all the radionuclides you are likely to use will be done by the state, and you may not need to deal with the NRC at all. Exactly how many different categories of state licenses there will be for different uses will be determined by the state and will differ somewhat from state to state. Since most state licenses require a license fee, the licensing of radioactive materials has become a source of income for state governments; the types of license and fees may have been designed with an eye toward benefiting the state budget.

For the special situation of work at federally controlled sites in agreement states, such as military bases or U.S. Navy ships in port, you need to know whether the property is under exclusive federal jurisdiction, in which case the NRC has regulatory authority despite the property's being in an agreement state. The NRC advises you to ask your local contact at the federal agency that controls the site (for example, the contract officer, base environmental health officer, or district office staff) to help determine the jurisdictional status. An All Agreement States Letter (SP-96-022), available from the NRC, gives guidance on determining jurisdictional status.

If you are in a *non*agreement state, the state (like all states) will regulate accelerator-produced radioisotopes (cobalt-57 and indium-111 for medical use, for example) and x-ray machines, and the NRC will license by-product material. The use of several radioactive materials at a single company, then, may require at least one NRC license and at least one state-issued license. In Virginia, for example (a nonagreement state), medical use of radioactive thallium is licensed by the state, while technetium is licensed by the NRC.

For this reason, in nonagreement states the radiation safety officer has to be knowledgeable about the requirements of both state and federal agencies. If the RSO is not fully cognizant of state and federal rules, it is advisable to get the help of an outside consultant, at least in the initial setting up of the radiation safety program.

X-ray machines used in medicine, as already noted, are regulated by the states. These machines produce x-rays not by using radioactive materials, but rather by accelerating electrons in an x-ray tube so that they strike a heavy metal such as tungsten.[8] They are subject to a Food and Drug Administration inspection after installation, which may be done by a state inspector. Periodic inspections thereafter may have to be done either by a state inspector or by an approved private inspector. X-ray technologists who run the machine may have to be registered with the state, at least if they work somewhere other than a hospital. As always, the laws differ from state to state, so you must refer to the specific laws of your own jurisdiction.

Physicians may find the difference in licensing requirements between x-ray machines and radioactive materials surprising. To use an x-ray machine in most states you need only a medical doctor degree; it is assumed that medical school training equips a doctor to operate x-ray machines (though x-ray technicians who are not MDs are required to take specialized training). To use a fluoroscope unit in their offices, physicians need only to have the machine installed and then send a certificate to the state health department. To use radioactive materials, on the other hand, the NRC will insist that even licensed physicians undergo specialized training. This training requirement may present an unexpected and unwelcome delay for a doctor who wants to start right away to use radioactive materials for diagnosis or treatment.

Even when a nuclear facility is regulated exclusively by the federal NRC, the operator may still be subject to state standards of care, and a person claiming to have been injured by the radiation may be able to recover damages under state law. This was one of the lessons of the Karen Silkwood case. Silkwood, a laboratory analyst at a plant that made nuclear fuel, somehow became contaminated with plutonium. An Oklahoma jury awarded damages for personal injury and property damage and then added $10 million for "punitive" damages, to punish her employer for its lack of care.[9]

The award of damages was upheld, at least in principle, even though the plant operator argued vigorously that it was in "substantial compliance" with federal requirements. One judge, at least, was impressed by the evidence that the plant operator had not followed the best radiation safety practices:

> The Cimarron plant was not designed to satisfy the most simple principles of human factors engineering. The facility was cramped. The plant did not incorporate sufficient safety systems, such as the "state of the art" alarm systems, leak detectors, air monitoring systems, and welded gaskets, into its design.[10]

As we shall see later in this chapter, uncertainty about what practices are "safe enough" can be a problem for NRC licensees.

---

[8]In x-ray machines, electrons are energized to high voltages, acquiring about 70,000 to 115,000 eV of energy in medical diagnostic machines and millions of electron volts in some therapy machines. When the electrons strike the heavy-metal target (usually tungsten) in the x-ray tube, their energy is partly converted into x-ray photons, which produce x-ray images or destroy cancer. J. Shapiro, *Radiation Protection: A Guide for Scientists and Physicians* 12 (2d ed. 1981).

[9]*Silkwood v. Kerr-McGee Corp.,* 769 F.2d 1451 (10th Cir. 1985); 464 U.S. 238 (1984).

[10]*Silkwood,* 769 F.2d at 1468–69.

14.1.3 Exemptions

A number of uses of radioactive material do not require a license because they are exempt by the terms of 10 CFR § 30.19. These include luminous wristwatches, aiming sights, and smoke and chemical agent detectors.

14.1.4 General Licenses

A "general" license for some uses of radioactive materials is granted by 10 CFR § 31.5. A general license is a license granted by the regulation; so long as you follow the rules in the regulation, you do not need to apply for an individualized license. The general license of § 31.5 authorizes the use of gas chromatographs, density gauge devices, and static elimination devices. Another general license, in 10 CFR § 31.7, authorizes safety devices such as exit signs containing tritium or promethium-147 in aircraft. As noted above, another general license is granted by 10 CFR § 31.11 for in vitro testing by physicians.

14.1.5 License Fees

When you apply for an NRC license (or an amendment to a license) you will have to pay an application fee. The fee schedule for material licenses is printed in 10 CFR § 170.31. For some licenses the fee is "full cost," meaning that the fee is based on how much NRC staff effort is needed to review the application. Fixed fees range from $120 (for source material used for shielding) to $7900 (for export of highly enriched uranium or radioactive waste). The NRC will not issue the license until it receives the fee. If you change your mind and decide to withdraw the application, you must do it before NRC staff starts its technical review, or else the NRC will keep the fee.

Most NRC licenses also require an *annual* fee, as prescribed in 10 CFR § 171.16. Some exemptions from annual fees are provided by 10 CFR § 171.11, and "small entities" may be entitled to reduced annual fees under 10 CFR § 171.16(c). Questions about fees can be phoned to the Office of the Controller (OC) at NRC Headquarters in Rockville, Maryland. Dial 301-415-7554, or else call 800-368-5642 and ask for extension 415-7554.

14.2 APPLYING FOR AN NRC LICENSE

14.2.1 Application Form 313

Application for a federal radioactive materials license is made on NRC form 313. The NRC is moving away from a paper-based process and toward an electronic (computerized) process, and eventually you will be able to file form 313 either on paper, the old-fashioned way, or electronically, but for now the electronic filing system is not operational.

The NRC's instructions for filling out the paper application are as follows:

- Use the most recent guidance in preparing an application. (Get it from the Web site and, to be certain nothing has changed, call the NRC Regional or Field Office.)
- Complete NRC form 313 App. B items 1 to 4, 12, and 13 on the form itself.
- Complete NRC form 313 items 5 to 11 on supplementary pages or use App. C.
- For each separate sheet submitted with the application (except for App. C), identify and key it to the item number on the application or the topic to which it refers.
- Submit all documents, including drawings, if you can, printed on $8^{1}/_{2} \times 11$ inch paper. If you have to submit larger documents, fold them to $8^{1}/_{2} \times 11$ inches.
- Identify each drawing with its drawing number, revision number, title, date, scale, and applicant's name. Indicate if drawings have been reduced or enlarged.
- Avoid submitting proprietary information unless it is absolutely necessary.
- Submit an original application and one copy.
- Retain your own copy of the license application for your files.

The NRC says that, as the electronic licensing process develops, it may provide a way to file applications by diskette or CD-ROM and through the Internet. When those mechanisms become available, additional filing instructions will be provided. Until then, the present paper process has to be used.

Your license application will be available to the public in the NRC's Public Document Room. You will therefore want to avoid including proprietary information such as trade secrets. If you do need to include proprietary information to satisfy the application requirements, there is a procedure for keeping it confidential in 10 CFR § 2.790. If you follow this procedure and satisfy the NRC requirements, the proprietary information will be kept secret from the public, though not, of course, from the NRC staff. Do not submit personal information about employees (their home addresses, social security numbers, and so forth) unless the NRC asks for it.

Because the NRC will run the application through an optical character reader to convert it to electronic format, you should also do the following:

- Submit printed or typewritten text, not handwritten, on smooth, crisp paper that will feed easily into a scanner.
- Use typeface designs that are sans serif, such as Arial, Futura, or Univers.
- Use 12-point or larger font size.
- Avoid stylized characters like script or italic.
- Be sure the print is clear and sharp.

- Be sure there is high contrast between the ink and paper. (Black ink on white paper is best.)

14.2.2 Certification

Form 313 must be signed and dated. Unsigned applications are returned to the applicant for signature. Whoever signs the application must be authorized to make binding commitments and to sign official documents on behalf of the applicant. Signing the application acknowledges management's commitment to and responsibility for the radiation protection program. If the application refers to certain commitments, those commitments become part of the licensing conditions and regulatory requirements.

Remember that it is a criminal offense to make a willful false statement on applications or correspondence, under 18 U.S.C. 1001. Also remember that a license can be revoked for "material false statements" in the application.

14.3 COMMON USES OF RADIOACTIVE MATERIALS

There are many workaday uses of radioisotopes for which state or federal licenses are necessary. Isotopes are used in agriculture to limit insect infestations of crops and to irradiate food; in industry to make smoke detectors, measure materials, and inspect welds; in research to test drugs and perform carbon dating; and to detect explosives. Cobalt-60, molybdenum-99, and thallium-201 are commonly used isotopes.

The applications for radioactive materials that require material licenses, generally speaking, fall into these categories:

- Gauge devices, either fixed or portable
- Industrial radiography
- Irradiators
- Research and development
- Medical diagnosis and treatment
- Well logging

Each of these categories is discussed briefly below.

14.3.1 Gauges

14.3.1.1 Portable Gauges. The first volume of the NRC Guidance Document NUREG-1556 addresses licenses for gauges, also referred to as *portable gauges* or *gauging devices.* (Fixed gauges are dealt with in a different volume, Vol. 4.)

Portable gauges are instruments for measuring such things as moisture, density, asphalt thickness, or liquid level. Their designs vary depending on the intended use; they are alike only in that they contain sealed sources of radiation. Gauges are susceptible to theft and damage by, for example, road construction equipment (though Vol. 1 says that sealed sources have not been damaged even when run over by heavy construction equipment). These risks must be addressed in the licensee's safety program.

14.3.1.2   Fixed Gauges.   Sealed sources of radioisotopes of various sizes are used in fixed gauges to measure density, fill level, or flow rate in pipes, tanks, or other vessels. The sources and their housings are manufactured to stringent specifications in an attempt to make them safe. The radioactive sources in fixed gauges are extremely hazardous if mishandled and can cause skin burns, loss of fingers, or other tissue damage, and possibly even death. Contamination, on the other hand, is usually not a problem.

14.3.2   Industrial Radiography

Radiography means examining the structure of materials by nondestructive methods, using ionizing radiation to make radiographic images. Generally gamma-emitting by-product materials are used, most often cobalt-60 or iridium-192. Other isotopes with unique radiological characteristics (californium-252, for example) may also be used; for example, neutron radiography using californium-252 has improved examinations for flaws in military aircraft.

Industrial radiography is used to develop inspection parameters, perform quality engineering evaluations, inspect parts and materials, and do acceptance testing of finished products. For example, large sealed sources are used to nondestructively test welds and castings for cracks and flaws that might make the product or construction unsafe. Radiography is also used for development and postmortem inspection of assemblies or parts that have not functioned as required.

14.3.3   Irradiators

Irradiation of food to destroy contaminants is a controversial but potentially important use of radioactive materials. About 25 percent of all food is wasted as a result of storage problems, and irradiators can help solve this problem. Irradiators, usually containing cesium-137, are also used to irradiate blood products, small animals, and biological samples such as cells, tissues, transplant organs, and viruses. Gamma radiation from cobalt-60 is used for medical sterilization, rubber latex vulcanization, cross-linking of polymers, and wood composite manufacture.

There are two kinds of irradiators: 10 CFR, Part 36, irradiators and self-shielded irradiators that are not subject to 10 CFR, Part 36. Irradiators include teletherapy units, underwater irradiators (which keep the sealed source under water and require the product being irradiated to be placed in a watertight container and lowered into the water), and commercial wet-source-storage irradiators, which store the sealed source in water and raise it into the air to irradiate the product. Sometimes the product is moved into the irradiation room on a conveyor system.

14.3.4 Academic and Research and Development Licenses (ARDLs)

Unsealed radioactive materials are used to study living systems and to improve or test products or materials in laboratories. Users include biotechnology firms, colleges, and testing labs. Small radioactive sources in gas chromatographs and x-ray fluorescence analyzers are included in this category.

14.3.5 Medical Uses

Unsealed radiopharmaceuticals and the radiation from sealed sources are administered to patients for diagnosing and treating disease. Most large hospitals and some major clinics have nuclear medicine departments, and some clinics and private physicians are licensed for limited therapy treatment of outpatients. A few facilities still have cobalt teletherapy machines.

Radionuclides are used in medicine in several ways:

- Diagnostic studies with unsealed radionuclides
- Therapeutic administrations with unsealed radionuclides
- Diagnostic studies with sealed radionuclides
- Manual brachytherapy (placing radioactive "seeds" in the body to irradiate a tumor)
- Therapeutic administrations with sealed sources and devices [that is, teletherapy, remote afterloaders, and gamma stereotactic radiosurgery units].

In medicine, technetium-99m is one of the most commonly used isotopes. It provides the most sensitive test for infection of the bone, for example. Another example is an imaging process using Tc-99m-labeled methylene diphosphonate to determine how well a transplanted kidney "clears" a radio-labeled substance.[11]

Three types of licenses under 10 CFR, Part 35, are of use in the medical community: general, specific, and broad-based. First, as already noted, a general byproduct license created by 10 CFR § 31.11 permits virtually any physician to conduct in vitro testing using radioactive materials. Under this license a doctor is permitted to do clinical or laboratory testing, using iodine-125 or -131, carbon-14, tritium, iron-59, selenium-75, or mock iodine-125 reference or calibration sources.

---

[11]American Nuclear Society, *Nuclear News* 35 (August 1998).

Second, a specific license is issued to many community hospitals. This type of license is restricted to the use of radioactive materials for clinical applications. Under a specific Part 35 license, a hospital can perform research so long as it complies with federal guidelines. For example, research under a National Institutes of Health (NIH) grant for breast biopsies may be allowable even at community hospitals under a specific license.

Third, a broad scope license is often issued to larger facilities, such as hospitals that perform significant medical research. (Broad scope licenses, which are addressed by Vol. 11 of NUREG-1556, are issued not only for medicine but also in other program areas such as R&D and manufacturing.)

Many physicians feel they have sufficient knowledge to use radioactive materials safely, but caution is called for. Even a trained medical doctor may not be aware of all the risks of radiation or not fully conversant with the correct methods of calculating doses. Moreover, the equipment the doctor uses may have been manufactured in a foreign country and have sparse or confusing instructions, and the instrument controls may be labeled only with cryptic symbols. It may be tempting to tinker with the equipment settings in order to produce, for example, a "better picture," but a better picture usually means administering a higher dose of radiation to a patient. X-ray technicians have been prosecuted for "practicing medicine without a license" because they exceeded the recommended dose in order to speed up the imaging process. Incidents like this illustrate why it may be wise to consult an outside expert, either in setting up a radiation program initially or for performing annual audits of the program thereafter.

Doctors or others using radioactive materials must also remember to use the equipment only in ways authorized by the permit. There is sometimes a tendency by skilled professionals to do what makes sense medically but possibly not legally. For example, the NRC issued a notice of violation in 1997 for using a high dose rate afterloader for surface treatment of skin cancer, when the NRC license did not authorize such use.

As with the rest of its material licensing program, for medical licenses the NRC is moving toward a "more risk-oriented, performance-based" regulatory program that "focuses on those procedures that pose the highest risk."[12] It is proposing changes to the regulations in 10 CFR, Part 35, for this purpose. While this approach will provide more flexibility, it has disadvantages as well. With "performance-based" standards there is no precise formula for deciding what is safe enough—no definitive way to balance factors such as risk, cost, chance of mishap, and frequency of testing equipment.

The problem is that there are differences in practices from one research or treatment institution to another, and this variability makes it hard to know what is an "acceptable" practice. An inexperienced RSO at a small facility who is not familiar with industry practice generally may create procedures that are more elaborate

---

[12] 63 FR 43,516 (Aug. 13, 1998); 63 FR 62,763 col. 1 (Nov. 9, 1998).

and more expensive than necessary. The RSO may, for example, create unnecessary record keeping requirements that are inordinately burdensome. Moreover, employees may resent these procedures because they do not see much value in them, and they may therefore be inclined to ignore or evade them.

A more serious problem, though, is inconsistent approaches to radiological safety from one facility to another. With different facilities following different procedures, confusion arises over what is an industry standard of practice. In a performance-based system, the appropriate standard of care may be established, in effect, not by NRC rules but rather by what competent professionals in the industry are doing. If you do not meet the standard set by the industry generally, you may be subject to enforcement penalties or liability. As noted above, liability in the *Karen Silkwood* case was created not by violating NRC regulations but by violating a state standard of care established by a jury, and the jury was no doubt influenced by evidence of what the "best" facilities were doing.

Thus, if a particular university or hospital is hypercautious when it writes its procedures, it may inadvertently set what appears to be an industry "practice standard" that is unnecessarily stringent and expensive, for others as well as for itself. It may decide, for example, to require radiation exposure badges for people who don't really need them. Or it may decide to calibrate an instrument every day, even though the instrument readout has not drifted in years. Or a pharmacy may require that every package of radiopharmaceuticals be swiped both on the outside and again before the inner package is opened, when it would be less costly and just as effective to swipe only the outer package and have personnel *assume* that the inner package is contaminated. Unnecessary precautions may arise out of a hypercautious approach to risk or a distrust of the science of health physics. Many unnecessary requirements have been created because a technician in some laboratory has come up with a "good idea" for radiation protection, without reference to industry practice.

Of course, overstringent or redundant requirements can also result from a prescriptive, government-mandated regulatory system. For example, for a mammography unit, the Food and Drug Administration requires that the beam quality be measured first by the accrediting body, second by a certified medical physicist, and third by an FDA inspector. These three layers of protection can cost millions of dollars per year.

14.3.6 Well Logging

When an oil or gas well has been drilled, cored, and tested and the well has reached its full depth, it is customary to conduct an electrical logging survey of the hole. Various types of radioactivity logs are used to record natural or self-induced radiation, which vary in different types of rock formations. Two types of logs are in common use. The gamma-ray log measures natural radioactivity. The neutron log, which is more pertinent for our purposes, measures the effect of bombarding the rock with artificial radioactivity.[13] On the basis of the cores, tests, and logs, the petroleum geologist can recommend how to equip and complete the well for production, or else decide that it should be plugged and abandoned.

Neutron logging of oil and gas wells is done by lowering an instrument on a cable into the borehole of the well. The instrument uses a radioactive source to measure geophysical properties (conductivity and radiation) of rocks and fluids at intervals. The data gathered in this way are processed by a computer and displayed as curves. The engineer reading the curves can discern such things as density and the presence of aquifers or particular minerals. Well logging can be less expensive than taking and analyzing cores.

Licensing requirements for well logging are spelled out in 10 CFR §§ 39.1–39.103. They cover the use of licensed materials, including sealed sources, radioactive tracers, radioactive markers, and uranium sinker bars (10 CFR § 39.1).

14.4 ELEMENTS OF A RADIATION SAFETY PROGRAM

No matter which kind of NRC license you have, you will need to meet NRC safety requirements. State requirements will be similar or identical, and for transportation the Department of Transportation rules must be met as well.

The following discussion of the elements of an acceptable safety program is taken primarily from Vol. 1 of NUREG-1556, which applies to portable gauges, but the principles discussed apply generally to all types of NRC licenses. Volume 1 will be used as a model for NRC guidance, some of it still being drafted, for other types of material licenses. If you are applying for one of those other types, such as for industrial radiography, be sure to consult the NRC regulations and the appropriate volume of NUREG-1556, and be sure you have the latest version of the guidance.

As noted above with respect to medical licenses, NUREG-1556 is being designed to be more "risk-informed" and "performance-based" and is supposed to give licensees more flexibility, and to require less information from the license applicant, than older guidance. This new approach will be helpful if it focuses resources on the things that create the greatest dangers. It will be somewhat less desirable to the extent it creates uncertainty because safety practices may vary so much from place to place.

14.4.1 ALARA

ALARA, which means "as low as reasonably achievable," is an important feature of NRC licensing. The basic requirement is 10 CFR § 20.1101, which says that "each licensee must develop, document, and implement a radiation protection program commensurate with the scope and extent of licensed activities" and that "the

---

[13]R. Wheeler and M. Whited, *Oil from Prospect to Pipeline* (4th ed. 1981).

licensee shall use, to the extent practicable, procedures and controls based upon sound radiation protection principles to achieve occupational doses and doses to members of the public that are ALARA." A licensee must periodically review its radiation protection program to be sure that ALARA is being implemented.

ALARA is not a dose limit but a goal. It is based on the assumption, which has been used to set radiation standards for many years, that there is no entirely "safe" dose of radiation.[14] Even the smallest dose, it is assumed, has some probability of causing cancer. As the International Commission on Radiological Protection (ICRP), a body of radiation experts that offer recommendations to regulatory agencies around the world, puts it, "[a]s any exposure may involve some degree of risk, the Commission recommends that all unnecessary exposures be avoided, and that all justifiable exposures be kept as low as reasonably achievable, economic and social considerations being taken into account." Both the Department of Energy and the NRC require an ALARA program at nuclear facilities.

ALARA means that you must meet the specific numerical radiation exposure standards in the regulations, particularly 10 CFR, Part 20,[15] and *in addition* you must reduce the radiation dose as much as you reasonably can, taking into account social, technical, and economic constraints. This means looking at ways of reducing exposure, primarily (1) eliminating or reducing the source of radiation, (2) containing the source, (3) minimizing the time spent in radiation fields, (4) increasing the distance from the radiation source, and (5) using shielding. In essence, you make a list of the things you might do to reduce radiation dose and then ask whether each such measure is feasible and worth it in light of the costs and other disadvantages.

NRC regulatory guides 8.10 ("Operating Philosophy for Maintaining Occupational Radiation Exposures ALARA") and 8.18 ("Information Relevant to Ensuring that Occupational Radiation Exposures that Medical Institutions Will Be ALARA") provide the NRC staff position on ALARA. Other NRC publications provide background information, namely NUREG-0267, "Principles and Practices for Keeping Occupational Radiation Exposures at Medical Institutions ALARA"; NUREG-1134, "Radiation Protection Training for Personnel Employed at Medical Facilities"; and NUREG-1516, "Effective Management of Radioactive Material Safety Programs at Medical Facilities."

14.4.2 Record Keeping

NRC licensees are required to keep certain records. For portable gauge devices, for example, licensees have to maintain permanent records on where licensed

---

[14]The model used to guide safety rulemaking is the linear nonthreshold dose-response model, which holds that the health effect of a dose of radiation is linearly related to the dose and that there is no "threshold" dose below which no effects occur. This theory is not universally believed to be true. In fall 1998, however, the National Council on Radiation Protection and Measurement (NCRP) released for public comment its Draft Report 1-6, "Evaluation of the Linear Nonthreshold Dose-Response Model," which again supports the traditional model.

[15]Part 20 contains the basic standards for protection against ionizing radiation. Subpart C gives the occupational dose limits for workers (for example, an annual dose equivalent of 5 rems), while Subpart D provides dose limits for the general public (0.1 rem per year).

material was used or stored while the license was in force. Acceptable records are sketches or written descriptions of storage or use locations specifically listed in the license. Licensees do not have to maintain this information for temporary job sites, if the radioactive sources have never leaked there.

Keeping complete and accurate records is extremely important to the NRC. The NRC commissioners have said:

> We cannot overstate the importance of a licensee's or an applicant's duty to provide the Commission with accurate information.[16]

Under NRC programs, and virtually every other federal regulatory program, employers in industry are sometimes penalized for failing to record or to report information properly. Often the companies who have to pay these fines complain that there was no threat to safety and no harm to the environment, that the violation was only a "paper violation." Often this is true, and it is good that most violations do not harm anyone. But NRC licensees need to understand that, *from the viewpoint of the regulators,* the paperwork is extremely important. The regulators do not visit a licensee's plant or laboratory often, and the only way they have to know what is happening is to read the required reports and records. When the record keeping is not done properly, it strikes, in the regulator's view, at the very heart of the regulatory program. So there is little point in complaining about "trivial" paperwork violations; they are all-important to the regulator, and it is the regulator who, in the first instance at least, imposes fines and other penalties.

Moreover, an NRC license can be revoked for "material false statements," and it is not necessary that the NRC staff have relied on the false information for it to be "material." [*Virginia Electric & Power Co.,* CLI-76-22, 4 NRC 480 (1976); *aff'd,* 571 F.2d 1289 (4th Cir. 1978).] Also, it is a crime to make false statements to agencies of the federal government.[17]

14.4.3 Financial Assurance and Record Keeping for Decommissioning

Sections 30.34 and 30.35 of 10 CFR require licensees to provide evidence of financial assurance for decommissioning. *Decommission* means to remove a facility or site safely from service (see 10 CFR § 30.4), and the NRC wants to be sure that there is a plan for disposing of radioactive materials safely at the end of the license's life and that money will be available to pay for the disposal.

Thus, for example, licensees using portable gauges must maintain, in an identified location, decommissioning records related to structures and equipment where the gauges are used or stored and to leaking sources. The records important

---

[16]*Dr. Randall C. Orem,* CLI-93-14 (June 4, 1993); see also *Piping Specialists, Inc.,* CLI-92-16 (Dec. 1, 1992).
[17]18 USC § 1001.

to decommissioning must be transferred either to a new licensee (if licensed activities are transferred or assigned) or to the NRC regional office (if the license is terminated).

The requirements for financial assurance are specific to the types and quantities of by-product material authorized on the license. Most portable gauge applicants and licensees do not have to comply with the financial assurance requirements, because the thresholds for sealed sources are 3.7×10^6 gigabecquerels (GBq), or 100,000 curies (Ci), of cesium-137 or 3.7×10^3 GBq (100 Ci) of americium-241 or californium-252. A licensee would have to possess hundreds of gauges [which typically contain only about 0.30 GBq (8 mCi) of cesium-137 and 1.5 GBq (40 mCi) of americium-241] before triggering the financial assurance requirements. A portable gauge license does not ordinarily specify the maximum number of gauges that the licensee may have, but it will have a condition requiring the licensee to limit its gauges to quantities not requiring financial assurance for decommissioning. If the licensee does at some point acquire enough gauges to go over the threshold amounts, it will have to provide evidence of financial assurance.

All portable gauge licensees must maintain records of structures and equipment where gauges are used or stored at locations specifically listed in the license. As-built drawings, with modifications of structures and equipment shown as appropriate, fulfill this requirement. If drawings are not available, licensees may substitute records concerning the areas and locations.

If portable gauge licensees have experienced unusual occurrences (for example, leaking sources or other incidents that involved the spread of contamination), they also need to maintain records about any contamination remaining after cleanup and any contamination that may have spread to inaccessible areas [see 10 CFR §§ 30.34(b), 30.35]. Regulatory Guide 3.66, "Standard Format and Content of Financial Assurance Mechanisms Required for Decommissioning Under 10 CFR Parts 30, 40, 70, and 72," gives detailed guidance.

14.4.4   Sealed Sources and Devices

The NRC or agreement states perform safety evaluations of sealed sources of radiation before authorizing a manufacturer to distribute them. This is true, for example, of portable gauges and radiographic exposure devices. The safety evaluation is documented in a sealed source and device (SSD) registration certificate, also called an SSD registration sheet.

When issuing a portable gauge license or radiography license, the NRC usually provides a generic authorization to allow the licensee to possess and use any sealed source/device combination that has been registered by the NRC or an agreement state. Safety evaluation and registration of SSDs are covered by Vol. 3 of NUREG-1556.

You should check with the supplier of a device you plan to use to be sure that it conforms to the SSD designation registered with the NRC or agreement state. Licensees are not allowed to make changes to a sealed source, device, or source/device combination that would change the description or specifications in its registration certificate, unless the NRC first gives permission with a license amendment.

Information on SSD registration certificates is available on the NRC's Sealed Source and Devices Bulletin Board System. The address is *http://www.hsrd.ornl.gov/nrc/ssdrform/htm.* Information about this bulletin board can also be had from the NRC's registration assistant at 301-415-7231.

14.4.5 Radiation Safety Officer

The NRC feels that to ensure safety and compliance with NRC requirements, adequate involvement of company management is important. The NRC wants to see a senior-level manager with responsibility for overseeing licensed activities (NUREG-1556, Vol. 1 § 3). Thus a management representative must sign the form 313 application and acknowledge management's commitment to safety, compliance with regulations, complete and accurate record keeping, knowledge of what is in the application and the resulting license, devoting adequate resources to radiation safety, and selecting a qualified person to serve as radiation safety officer (RSO), responsible for following safe radiological procedures and ensuring that state and federal safety requirements are met [see 10 CFR § 30.33(a)(3)]. The NRC takes these commitments seriously and has on at least one occasion (for a nuclear power plant) held a public hearing and required a company's chief executive officer to testify under oath in defense of the company's "commitment to safety."

RSOs are required to have adequate training and experience. In the past, the NRC has found the following to be evidence of adequate training and experience, for licensees of portable gauges:

- Portable gauge manufacturer's course for users or for RSOs
- Equivalent course that meets criteria listed in App. D to NUREG-1556, Vol. 1

For more demanding uses of radioactive materials (for example, industrial radiography), the training requirements for the RSO are greater, as might be expected. For an industrial radiography license, for example, the RSO must be a qualified radiographer, have a minimum of 2000 hours (1 year of full-time field experience) of hands-on experience as a qualified radiographer, and have formal training in establishing and maintaining a radiation protection program. *Hands-on experience* means experience in all areas considered to be directly involved in radiography, including taking radiographs, surveying device and radiation areas, transporting the radiography equipment to temporary job sites, posting records, and radiation area surveillance (see NUREG-1556, Vol. 2 § 8.8).

The RSO is responsible for the radiation protection program, and is to have independent authority[18] to stop operations the officer considers unsafe and

---

[18] Independent authority has been a big issue with the NRC in the past, particularly in the licensing of nuclear power reactors. The NRC has demanded, for example, that a licensee present evidence to prove that its corporate structure would not easily allow an operations or financial executive to overrule a quality assurance or safety manager.

sufficient time and commitment from management to fulfill the duties of the office. For irradiator licenses, for example, the NRC says that the RSO should be able, on the officer's own authority, to order unsafe practices to stop:

> Consistent with the staff's long-standing guidance, the RSO should have independent authority to stop operations that he or she considers unsafe and to conduct necessary tests or measurements. The RSO should be relatively independent of production responsibilities, to the extent practicable, considering the size of the staff at the facility. The RSO should report directly to the facility manager. He or she must have sufficient time and commitment from management to fulfill certain duties and responsibilities to ensure that licensed materials are used in a safe manner. [NUREG-1556, Vol. 6 § 8.7.01.]

Typical RSO duties for portable gauges, which are described in App. E to NUREG-1556, Vol. 1, include making sure that the following things get done:

* Stopping unsafe licensed activities
* Proper use, maintenance, and storage of gauges, consistent with the license, the SSD sheet, and the manufacturer's recommendations
* Training the users of the gauges
* Making sure that personnel monitoring devices are used and records of exposures kept
* Securing the gauges
* Notifying the authorities if an accident happens
* Investigating and reporting unusual occurrences
* Annual audits
* Transporting of gauges
* Disposing of radioactive materials
* Maintaining records
* Monitoring the license to keep it up to date

Especially for smaller companies, the choice and supervision of the radiation safety officer is of enormous importance, both to achieve safety and to avoid violating NRC or state regulations. The licensing agencies expect the licensee to be aware of the regulatory requirements and their implications, and much of this responsibility falls upon the RSO. Failure of the RSO to perform the duties of the office may result in civil penalties or civil liability.

The NRC expects the RSO to have the experience, training, and judgment to evaluate radiological safety practices. A smaller company, especially, may be completely dependent on the RSO's judgment, for it may not have another employee with enough radiological expertise to evaluate the RSO's decisions. There may be several years between inspections by the NRC or the state, and if the RSO is not implementing proper safety procedures in the meantime, a large number of regulatory violations can be committed on the officer's watch before an outside inspection uncovers the problem. For this reason, it is a good idea to get qualified outside help in setting up a radiation safety program, and also in auditing the program from time to time.

The NRC says that it is important to notify the NRC as soon as possible whenever the RSO changes. For portable gauge licenses, these notifications are handled as "administrative" amendments to the license, which do not require a fee so long as the application contains certain commitments.

14.4.6 Training

NRC training requirements are found in 10 CFR §§ 19.11, 19.12, 30.7, 30.9, 30.10, and 30.33. Training requirements apply to other employees besides the RSO. The NRC requires people using radioactive materials to have adequate training and experience. For portable gauge users, for example, the NRC has found acceptable the completion of the portable gauge manufacturer's course for users or an equivalent course that meets criteria listed in App. D to NUREG-1556, Vol. 1.

An individual using a gauge is usually referred to as an *authorized user.* Authorized users are responsible for ensuring the surveillance, proper use, security, and routine maintenance of the gauge. Moreover, records of their training need to be kept.

For radiographers and radiographer's assistants, training requirements are described in 10 CFR, Part 34.43, and in Sec. 8.9 and App. G of NUREG-1556, Vol. 2. Radiographers must be certified, and they and their assistants must have annual refresher training and be audited twice a year. After June 27, 1999, license applicants no longer have to describe their initial training and examination program in certain topics [see 10 CFR § 34.43(g)]. Their certification will be verified by the NRC during routine inspections.

The selection of responsible personnel and their training is probably the most important thing you can do to avoid violating NRC requirements. Over the years, many NRC incident reports have described overexposure that occurred (often to the assistant's hands) because a radiographer departed for some reason from proper procedures. It is important to impress on employees who use radioactive materials the importance of complying with all workplace practices and safety requirements.

14.4.7 Safety Audits

Licensees are obligated to review their radiation protection programs every year. This annual audit should aim to ensure the following:

- That the licensee is complying with NRC and DOT regulations and the terms and conditions of the license

- That occupational doses and doses to members of the public are as low as reasonably achievable (ALARA)

- That records of audits and other reviews of program content are maintained for 3 years

Appendix F of NUREG-1556 has a suggested audit program for portable gauges. It may need to be tailored to the individual licensee's circumstances.

The NRC wants auditors to observe actual work in progress. It encourages licensees to consider *unannounced* audits of gauge users in the field to make sure they are following all operating and emergency procedures.

If an audit uncovers problems, they must be promptly corrected. The NRC's Information Notice (IN) 96-28, titled "Suggested Guidance Relating to Development and Implementation of Corrective Action," provides guidance on this subject. The NRC will look at the audit results to see if corrective actions are thorough, timely, and sufficient to prevent recurrence. If violations are identified by the licensee itself, and if the licensee takes corrective action, then the NRC may decide not to cite a violation. Information on the NRC's use of its discretion in issuing violations is found in its "General Statement of Policy and Procedures for NRC Enforcement Actions" (NUREG-1600).

Hiring a knowledgeable outside expert to review the company's radiation safety program at the beginning may be enough by itself to avoid major regulatory problems later on. But the company should also consider having an audit done from time to time by a qualified outside expert. Under some licenses issued under 10 CFR, Part 35, for medical facilities, an outside audit has become virtually a requirement of the license.

Holding regular independent audits will be even more beneficial if the NRC will agree to accept them in place of inspections by the regulatory agencies. The NRC is considering "alternative regulatory oversights" for lower-risk procedures.[19] Eventually this may mean that a licensee's strong self-audit program will allow the NRC to exercise less scrutiny over the licensee's operations.

Licensees are required by 10 CFR § 20.2102(a) to maintain records of audits and other reviews of program content and implementation. The NRC says that audit records are acceptable if they contain the following information:

- Date of audit
- Name of persons who conducted the audit
- Persons contacted by the auditors
- Areas audited
- Audit findings
- Corrective actions
- Follow-up

---

[19]63 FR 62,763 col. 1 (Nov. 9, 1998).

You should not submit your audit program to the NRC for review during the licensing phase.

The NRC has made the following guidance documents available: Manual Chap. 87110, App. A, "Industrial/Academic/Research Inspection Field Notes"; NUREG-1600, "General Statement of Policy and Procedures on NRC Enforcement Actions"; and IN96-28, "Suggested Guidance Relating to Development and Implementation of Corrective Action."

14.4.8 Material Receipt and Accountability

Licensees are obligated to maintain records of receipt, transfer, and disposal of gauges and to conduct physical inventories at intervals not to exceed 6 months (unless some other interval is justified by the applicant) to account for all sealed sources.

The NRC says that licensed materials must be tracked from "cradle to grave." Many licensees record daily uses of gauges in a logbook as part of their accountability program. The NRC has provided suggested operating procedures in App. H of NUREG-1556, Vol. 1.

14.4.9 Occupational Dosimetry

The NRC has set certain radiation dose limits, one set for radiation workers and one for the general public. The worker doses are all multiples of 5, namely 15 rems to the eyes per year; 50 rems to skin, arms, and legs; and 5 TEDE (total effective dose equivalent) to the whole body (see Fig. 14.2).

For portable gauges, licensees have two alternative ways to demonstrate that their employees are being monitored for radiation exposure. The licensee may meet the requirement by maintaining, for inspection by the NRC, documentation demonstrating that unmonitored people are not likely to receive a radiation dose more than 10 percent of the allowable limits in any one year. The allowable limits, as provided by 10 CFR § 20.1201, are the ones mentioned above:

- Skin—0.50 sievert (Sv) equal to 50 rems
- Eyes—0.15 Sv (15 rems)
- Elbows to hands—0.50 Sv (50 rems)
- Knees to feet—0.50 Sv (50 rems)

The TEDE for the whole body is to be no more than 0.05 Sv (5 rems). Part 1 of App. I to NUREG-1556, Vol. 1, tells how to prepare a written evaluation to demonstrate that gauge users are not likely to exceed 10 percent of the applicable limits and thus are not required to have personal dosimetry.

Alternatively, the licensee must provide dosimetry processed and evaluated by a processor recommended by the National Voluntary Laboratory Accreditation

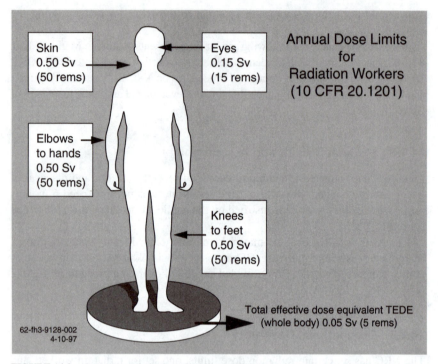

FIGURE 14.2 Annual dose limits for radiation workers.

Program (NVLAP) and exchanged at the frequency recommended by the processor. (These requirements are set out more precisely in 10 CFR §§ 20.1502, 20.1201, 20.1207, and 20.1208.)

Under conditions of routine use, including weekly cleaning and lubricating, the typical portable gauge user does not need a personnel monitoring device (dosimeter), since most of the time, even when a gauge has been run over and damaged, the shielding of the source remains intact. If the user also does not need a dosimeter, the NRC says, if proper emergency procedures are used.

When personnel monitoring is needed, most licensees use either film badges or thermoluminescent dosimeters (TLDs), supplied by an NVLAP-approved processor. Film badges are usually exchanged monthly, because the film fades with time. The exchange frequency for TLDs is usually quarterly.

The licensee should verify that the processor is NVLAP-approved and consult the processor for its recommendation for exchange frequency and proper use. The National Voluntary Laboratory Accreditation Program Directory (National Institute of Standards and Technology Publication 810) is published annually. It is available on the Internet at *http://ts.nist.gov/ts/htdocs/210/214/dosim.htm.*

For radiographers, the licensee must make sure that both radiographers and radiographer's assistants, during all radiographic operations, wear a combination of direct-reading dosimeter (pocket dosimeter or electronic personal dosimeter), an operation alarm ratemeter, and either a film badge or a TLD. The alarm ratemeter is not necessary where other alarming or warning devices are in routine use. The pocket dosimeters must read up to 200 mrems, be recharged at the start of each shift, and be checked every 12 months for correct response. See NUREG-1556, Vol. 2 § 8.16.

14.4.10 Quarterly Maintenance

For radiography licenses, inspection and maintenance must be done every 3 months, according to written procedures. This requirement applies to radiographic exposure devices, source changers, associated equipment, transport and storage containers, and survey instruments. Procedures must also be in place to maintain Type B packaging for radioactive materials to be transported and to make sure that the Type B packages are shipped properly. See NUREG-1556, Vol. 2 § 8.18.

14.4.11 Public Dose

Doses to the general public as well as to the workers are a concern. NRC licensees are required to ensure that licensed portable gauges and radiography devices (for example) will be used, transported, and stored in such a way that members of the public will not receive more than 1 mSv (100 mrems) in 1 year and that the dose from licensed operations in any unrestricted area will not exceed 0.02 mSv (2 mrems) in any one hour. Licensees must also control and maintain constant surveillance over radioactive devices that are not in storage and secure stored gauges from unauthorized removal or use. The regulations are 10 CFR §§ 20.1003, 20.1301, 20.1302, 20.1801, 20.1802, and 20.2107.

Members of the public, in this context, include employees whose assigned duties do not include the use of licensed materials but who nevertheless may work near where the gauges are used or stored. Public dose is controlled, in part, by ensuring that gauges not in use are stored securely (for example, in a locked area) to prevent unauthorized access or use. If gauges are not in storage, then authorized users must maintain constant surveillance to ensure that members of the public (including coworkers) cannot get near the gauges or use them and thus receive unnecessary radiation.

Since a gauge creates a radiation field during storage, it must be stored so that the radiation level in unrestricted areas (for example, an office or the exterior surface of an outside wall) does not exceed the doses listed above. The dose is controlled by decreasing the time spent near the gauge, increasing the distance from the gauge, or using shielding such as brick, concrete, lead, or other solid walls. Gauges should be stored as far away as possible from areas that are occupied by the public.

You can determine the radiation levels adjacent to the storage location either by calculations or by a combination of direct measurements and calculations using some or all of the following: typical known radiation levels provided by the manufacturer, the inverse square law to evaluate the effect of distance on radiation levels, and occupancy factors to account for the actual presence of members of the public and of the gauges. Remember that licensees must demonstrate two things:

- That the radiation dose received by individual members of the public does not exceed 1 mSv (100 mrems) in one calendar year.
- That the radiation dose in unrestricted areas does not exceed 0.02 mSv (2 mrems) in any one hour.

Part 2 of App. I to NUREG-1556, Vol. 1, gives an example. It imagines that a company stores its portable gauges very close to a secretary's desk. For example, the secretary sits 8 ft from one of the gauges, and the gauge manufacturer says that the dose rate is 2 mrems/h at 1 ft from the gauge. The dose to the secretary, who is assumed to sit at the desk 24 hours a day, 365 days a year, is calculated as follows:[20]

$$\text{Hourly dose} = \frac{D_{mfr}\,(d_{mfr})^2}{(d_{sec})^2}$$

FIGURE 14.3 Dose to the secretary calculation.

where D_{mfr} = dose received per hour according to the manufacturer's data, mrems/h
d_{mfr} = distance at which the manufacturer measured the dose, ft
d_{sec} = distance from the gauge to the secretary, ft

In other words, the dose is just the manufacturer's dose multipled by the square of the ratio of the distances. The annual dose is the hourly dose multiplied by the number of hours in the year:

$$\text{Annual dose} = \text{hourly dose} \times 24\,\frac{\text{hours}}{\text{day}} \times 365\,\frac{\text{days}}{\text{year}}$$

FIGURE 14.4 Annual dose calculation.

In this case the hourly dose is 0.031 mrems/h, much less than the allowable 2 mrems, but the annual dose is 272 mrems, more than the allowable 100 mrems/year. In the example in NUREG-1556, the dose from this gauge is then added to the doses from two other gauges, each calculated in the same way, to produce a total.[21]

---

[20]NUREG-1556, Vol. 1, App. I, Part 2, Table I.3.
[21]Table I.6 in NUREG-1556.

Having made these calculations and concluded that the worst-case dose exceeds the NRC standards, the RSO then reviews the assumptions and recognizes that a secretary will occupy the desk far less than 8760 hours per year. If it turns out that the secretary sits at the desk 5 hours a day, works only 3 days a week, and works 52 weeks a year, then the hourly dose of 0.031 mrem/h may be multiplied by a mere 5 hours $\times$ 3 days $\times$ 52 weeks = 780, for a total of only 24.2 mrems/year.

If, after making an initial evaluation, you make changes affecting the storage area (for example, if you change the location of gauges in the storage area, remove shielding, add gauges, change the occupancy of adjacent areas, or move the storage area to a new location), then you must ensure that the gauges are properly secured, perform a new evaluation, and take corrective action as needed.

14.4.12 Operating and Emergency Procedures

According to the rules in 10 CFR §§ 30.34(e), 20.1101, 20.1801, 20.1802, 20.2201–03, and 30.50, licensees of portable gauges must have operating and emergency procedures with the following elements:

- Instructions for using the portable gauge and for performing routine maintenance, according to the manufacturer's recommendations and instructions
- Instructions for maintaining security during storage and transportation
- Instructions to keep the gauge under control and immediate surveillance during use
- Measures to keep radiation exposures ALARA
- Measures to maintain accountability during use
- Measures to control access to a damaged gauge
- Actions to take, and whom to contact, when a gauge has been damaged

Gauges are often damaged by heavy equipment at job sites, and emergency procedures should be available to minimize radiation safety risks when that happens. The area of the crushed gauge must be cordoned off to prevent exposure, and the driver should be kept nearby (though not in the cordoned-off area) to maintain surveillance until emergency response personnel arrive.

Steps must also be taken to prevent loss, theft, or unauthorized use of gauges. The NRC considers security of gauges extremely important, and lack of security is a significant violation for which gauge licensees are fined. Sample procedures for preventing loss or theft are given in App. H to NUREG-1556, Vol. 1.

Some portable gauges are used to make measurements with the unshielded source extended more than 3 ft below ground. Unless precautions are taken, it is possible for the source to be buried under dirt or concrete that collapses around the source during the measurement. Precautionary measures need to be planned in advance in case this happens. To ensure that the hole is free of debris, that debris will not likely reenter the cased hole, and that the source will be able to move freely, the NRC allows licensees to use surface casing from the lowest depth to 12 in. above the surface. If it is not feasible to extend the casing 12 in. above the surface, the licensee may cap the hole and use dummy probes before making measurements with an unshielded source to make sure the hole is free of obstructions. If gauges are used for measurements with the unshielded source extended more than 3 ft beneath the surface, licensees must do the following:

- Use surface casing or alternative procedures to ensure that the source can move freely in the hole
- Provide instructions for procedures for retrieving a stuck source
- Require reporting to the NRC, under 10 CFR § 30.50(b)(2), when a stuck source cannot be retrieved

14.4.13 Radiation Surveys

For radiography licenses, radiation surveys must be performed during use, movement, and storage of the radioactive materials. See NUREG-1556, Vol. 2 § 8.21. Types and frequencies of surveys are listed in Table 8.1 in the NUREG.

14.4.14 Leak Tests

The NRC requires testing to determine whether there is any radioactive leakage from the source in a portable gauge. Testing is acceptable if it is conducted by an organization approved by the NRC or an agreement state or according to procedures approved by the NRC or the state (see 10 CFR § 30.53). The license will require leak tests at intervals approved by the NRC or the agreement state and specified in the SSD registration sheet. The measurement of the leak-test sample is a quantitative analysis requiring that instrumentation used to analyze the sample be capable of detecting 185 Bq (0.005 μCi) of radioactivity.

Manufacturers, consultants, and other organizations may be authorized by the NRC or an agreement state to either perform the entire leak test sequence for other licensees or to provide leak test kits to licensees. If kits are supplied, the licensee is expected to take the leak test sample according to the gauge manufacturer's and kit supplier's instructions and return it to the kit supplier for evaluation and reporting of the results. Licensees may also be authorized to conduct the entire leak test sequence themselves. See draft regulatory guide FC 412-4, "Guide for the Preparation of Applications for the Use of Radioactive Materials in Leak-Testing Services," which is available from the NRC on request.

14.4.15 Instruments

For radiography licenses, radiation surveys are required to measure the radiation fields from the sealed sources. For that reason the licensee must have an adequate number of radiation survey instruments where the radioactive material is kept. The instruments must be calibrated at least every 6 months and also after being serviced. Records must be kept of the calibrations and of equipment problems and maintenance. See 10 CFR §§ 30.33(a)(2), 34.25, 34.31, 34.65; NUREG-1556, Vol. 2 § 8.13.

14.4.16 Controlling Access

Radiography licensees must control access to the areas where radioactive material is being used. Measures such as "danger" signs, locks, and controlling and watching the areas can be used.

Since June 27, 1998, when radiography is done at a temporary job site, at least two qualified people must be there. See NUREG-1556, Vol. 2 § 8.22. Both of them must maintain constant surveillance and be ready to prevent unauthorized entry to the restricted areas.

14.4.17 Maintenance

Licensees must routinely clean and maintain portable gauges according to the manufacturer's recommendations and instructions. For gauges with a source rod, radiation safety procedures for routine cleaning and lubrication of the source rod and shutter mechanism (for example, to remove caked dirt, mud, asphalt, or residues from the source rod and to lubricate the shutter mechanism) must consider ALARA and ensure that the gauge functions as designed and that source integrity is not compromised.

Nonroutine maintenance or repair (beyond routine cleaning and lubrication) that involves detaching the source or source rod from the device, as well as any other activities during which personnel could receive radiation doses exceeding NRC limits, must be performed by the gauge manufacturer or persons specifically authorized by the NRC or the agreement state. Requests for specific authorization to perform nonroutine maintenance or repair must demonstrate that personnel performing the work do the following:

- Have adequate training and experience
- Use equipment and procedures that ensure compliance with regulatory requirements and consider ALARA
- Ensure that the gauge functions as designed and that source integrity is not compromised

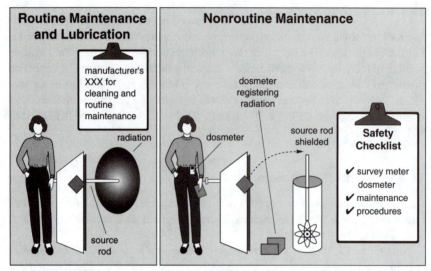

FIGURE 14.5   Routine and nonroutine maintenance of portable gauges.

See 10 CFR §§ 20.1101, 30.34(e). The NRC uses Fig. 14.5 to illustrate routine and nonroutine maintenance of portable gauges.

14.4.18   Transportation

Licensees often transport portable gauges to and from job sites. They are required to comply with Department of Transportation (DOT) regulations, which specify how radioactive materials must be packaged and transported. Applicants for a license must have safety programs for public transport of radioactive material to ensure compliance with DOT regulations. See 10 CFR § 71.5; 49 CFR, Parts 171–178; 10 CFR § 20.1101. "A Review of Department of Transportation Regulations for Transportation of Radioactive Materials" (1983 revision) can be obtained by calling DOT's Office of Hazardous Material Initiatives and Training at 202-366-4900.

During an inspection, the NRC will use the provisions of 10 CFR § 71.5 and a memorandum of understanding with DOT to examine and enforce transportation requirements. Appendix K to NUREG-1556, Vol. 1, lists major DOT regulations and provides a sample shipping paper.

Transportation requirements for hazardous materials are given in a "hazardous materials table," § 172.101 of 49 CFR, which lists eight sorts of "radioactive material," gives the type of packaging required for each, and tells whether a Class 7 "radioactive" hazard warning label must be used. The table also tells us that radioactive materials may be stowed on deck or under deck on a cargo vessel or passenger vessel.

The transportation requirements cover, first of all, shipping papers. A shipment of Class 7 (radioactive) material must include the words "RADIOACTIVE MATERIAL" (unless the words are contained in the proper shipping name) and the name of each radionuclide listed in 49 CFR § 173.453. The shipper must certify that the material is offered for transportation in accordance with DOT regulations by printing, either manually or mechanically, on the shipping paper containing the required shipping description the certification contained in Par. (a)(1) of § 172.204. That is, the shipper must certify that the radioactive materials "are properly classified, described, packaged, marked and labeled, and are in proper condition for transportation according to the applicable regulations of the Department of Transportation"[22] or declare that the contents of the consignment are fully and accurately described and are "classified, packaged, marked, and labeled/placarded, and are in all respects in proper condition for transport according to applicable international and national governmental regulations."[23]

Packaging requirements for radioactive materials are covered by 49 CFR §§ 173.415, .416, .417, .421, .422, .424, .426, .427, .428, and .453. The packages must be properly marked. In particular, they must bear the radioactive white or radioactive yellow label described in 49 CFR § 172.436. Other requirements are given in 49 CFR §§ 172.300, .301, .303, .304, .310, and .324. Labeling requirements are set forth in 49 CFR §§ 172.400, .401, .403, .406, .407, .436, .438, and .440.

Vehicles carrying radioactive materials must be placarded in accordance with 49 CFR §§ 172.500, .502, .504, .506, .516, .519, and .556. Emergency response information must be provided in accordance with 49 CFR §§ 172.600, .602, and .604. Training requirements are contained in 49 CFR §§ 172.702 and .704. Shippers and carriers must have a radiation protection program as spelled out in 49 CFR § 172.800. Finally, requirements for driver training, shipping papers, and general requirements for securing against movement for carriage by public highway are spelled out in 49 CFR §§ 177.816, .817, .834(a), and .842.

14.4.19 Waste Management: Disposal or Transfer of Radioactive Sources

To comply with 10 CFR §§ 20.2001, 30.41, and 30.51, licensed materials must be disposed of by transfer to an authorized recipient. Appropriate records must be maintained.

When disposing of portable gauges or radiography sealed sources (for example), licensees must transfer them to an authorized recipient. Authorized recipients include the original manufacturer of the device, a commercial firm licensed by the NRC or an agreement state to accept radioactive waste, or another licensee authorized to possess the licensed material.

---

[22] 49 CFR § 172.204(a)(1).
[23] 49 CFR § 172.204(a)(2).

Before transferring radioactive material, a licensee must verify that the recipient is authorized to receive it using one of the methods described in 10 CFR § 30.41. In addition, all packages containing radioactive sources must be prepared and shipped in accordance with NRC and DOT regulations. Records of the transfer must be maintained as required by 10 CFR § 30.51.

14.4.20 Obeying the License

Once you have a license for radioactive materials, obviously you have to comply with it, which in turn means you have read it and are familiar with its terms. In particular, you must use only the isotopes, only for the specified purposes, and only in the manner stated in the license. Sources and devices must be used only in accordance with the purposes for which they were designed and in accordance with the manufacturer's recommendations, as specified in an approved sealed source and device (SSD) registration certificate. See 10 CFR § 30.33(a)(1); 10 CFR § 34.1; NUREG-1556, Vol. 2 § 8.7. A license is a legal requirement, and a violation of the license is a violation of the law.

14.5 TRANSFERRING, AMENDING, OR TERMINATING A LICENSE

14.5.1 Transferring a License

Under 10 CFR § 30.34(b) licensees must provide full information and obtain the NRC's written consent before transferring ownership or control of the license ("transferring the license").

Before you transfer ownership or control of a license to another company, you must get the NRC's written consent. Changes in ownership may result from mergers, buyouts, or majority stock transfers. Although NRC says it does not intend to interfere with business decisions, it does want to know, and approve beforehand, any such transfers. See 10 CFR § 30.34(b). It wants to be sure that radioactive materials are possessed only by people with valid NRC licenses, that the materials are properly handled and secured, that the people using the materials are competent and committed to implementing radiological controls, that a clear chain of custody is established to identify who is responsible for final disposal of the gauge, and that public health and safety are protected.

Likewise, if there is a filing of voluntary or involuntary bankruptcy for or against your company, you must notify the appropriate NRC regional administrator in writing and identify the bankruptcy court and the date of filing. See 10 CFR § 30.34(h).

Licensees will be authorized to use sealed sources and devices only if they are registered by the NRC or an agreement state. The NRC or the agreement state performs a safety evaluation of gauges before authorizing a manufacturer to distribute them to specific licensees. The safety evaluation is documented by an SSD registration certificate (SSD registration sheet). When issuing a portable gauge license, the NRC usually gives a generic authorization to allow the licensee to possess and use any sealed source/device combination that has been registered by the NRC or an agreement state.

You should consult with the proposed seller of a device to ensure that it conforms to the sealed source and device designations registered with the NRC or the agreement state, and you should not make any changes to a sealed source, device, or source/device combination that would change the description or specifications from what is indicated in the registration certificate, unless you first get NRC's permission in a license amendment.

Information on SSD registration certificates is available electronically on NRC's Sealed Source and Devices Bulletin Board System (SS&D BBS), which can be accessed free of charge on the Fed World Information Service Network. To get connected, call the Fed World help desk at 703-487-4608. The database is now stored at *http://www.hsrd.ornl.gov/nrc/ssdrform.htm.* For information about the on-line database, contact the registration assistant at 301-415-7231.

14.5.2 Amendments and Renewals to the License

The licensee is required to keep the license current. If any of the information provided in the original application needs to be modified or changed, the licensee must submit an application for a license amendment *before* the change takes place. Also, to continue the license after its expiration date, the licensee must submit an application for a license renewal at least 30 days before the expiration date [see 10 CFR §§ 2.109, 30.36(a)].

To apply for a license amendment, do the following:

- Include the appropriate fee
- Use the most recent guidance in preparing the request
- Submit in duplicate either an NRC form 313 or a letter requesting amendment or renewal
- Provide the license number

For renewals, provide a complete and up-to-date application if many outdated documents are referenced or there have been significant changes in regulatory requirements, NRC guidance, the licensee's organization, or the radiation protection program. Alternatively, describe clearly the exact nature of the changes, additions, and deletions.

Existing portable gauge licensees have the option of submitting a complete application using NUREG-1556, Vol. 1, at the time they file an amendment request. Licensees choosing this option should incorporate the requested change into the complete application, submit it with the appropriate amendment fee, and indicate that the complete application is an amendment request to take advantage of the new guidance.

14.5.3 Terminating a License

Once you decide to end your NRC license, certain requirements must be met under 10 CFR §§ 30.34(b), 30.35(g), 30.36(d) and (j), and 30.51(f). In particular, you are required to notify the NRC in writing within 60 days when the licensed activities have not been conducted for a period of 24 months or a decision is made to cease licensed activities permanently. You must certify the disposition of licensed materials by submitting NRC form 314, "Certificate of Disposition of Materials." You can get this form from the NRC.

Before a license is terminated, you must send the records important to decommissioning, as required by 10 CFR § 30.35(g), to the NRC regional office. If licensed activities are transferred or assigned in accordance with 10 CFR § 30.34(b), records important to decommissioning should be transferred to the new licensee.

14.6 LICENSING PROCEDURES

Ordinarily the *procedure* for getting an NRC material license is straightforward: You submit form 313, the NRC staff reviews it, the NRC staff asks for additional information or improvements to your safety program, you do what the NRC staff asks or else negotiate some compromise, and the license is issued.

14.6.1 Hearings

Occasionally, however, granting or amending a license may become involved in a more elaborate "adjudicatory" proceeding, with a hearing, written evidence, and possibly oral testimony. This happens when any "person whose interest may be affected" files a request for a hearing (10 CFR § 2.1205). The procedures to be followed are laid out in the NRC regulations, 10 CFR §§ 2.1201–.1263.

A request for hearing may be filed by someone other than the license applicant— a neighbor, for example, who fears that the use of radioactive materials will cause personal harm (see 10 CFR § 2.1205). Also, if a hearing is granted, a notice is published in the *Federal Register,* and other interested persons may file petitions to intervene in the proceeding [10 CFR § 2.1205(k)]. It is not uncommon for people who live near a facility to file a petition challenging the issuance of an NRC license or an amendment to a license. See, e.g., *Hydro Resources, Inc.* (2929 Coors Road, Suite 101, Albuquerque, NM 87120), LBP-98-9, 47 NRC 261 (1998). Less common is a challenge from another business, possibly a competitor. [*Quivera Mining Co.* (Ambrosia Lake Facility, Grants, New Mexico), CLI-98-11, 48 NRC 1 (1998).]

A petitioner for a hearing other than the applicant, and a petitioner to intervene in a proceeding, must show "standing," which means a substantial "interest" (meaning a stake) in the license at issue. Almost always the interest claimed is that the radioactive materials will be harmful to the petitioner's health or property. If a hearing is granted, petitioners may raise all sorts of safety and environmental issues, including "environmental justice"—the proposition that a facility creates especial risk or especial harm to ethnic minorities. See *Private Fuel Storage, L.L.C.* (Independent Spent Fuel Storage Installation), CLI-98-13, 48 NRC 26, 35 (1998).

14.6.2 Petitions to Revoke a License

The NRC also provides a procedure for outsiders to request that a license be suspended or revoked after it has been issued. See *Veterans Administration Medical Center* (Philadelphia, Pennsylvania), DD-98-7, 48 NRC 97 (1998).

14.7 ENFORCEMENT

The NRC and the regulatory agencies in agreement states have enforcement programs, which may involve inspections of licensed facilities and, if violations are found, civil penalties or even revocation of the license. For example, the NRC proposed an $8000 penalty against a veterans' hospital for chastising an RSO who had contacted the NRC about possible safety concerns [*Department of Veterans Administration Medical Center* (Philadelphia, Pennsylvania), DD-98-7, 48 NRC 97, 99 (1998).]

Agreement states may be fairly aggressive in their oversight of hospitals or other facilities using radioactive materials, especially since the state may make money by billing licensees for surveys and inspections. Such fees can amount to millions of dollars per year for a large hospital.

The NRC's enforcement policy, NUREG-1600, "General Statement of Policy and Procedures for NRC Enforcement Actions" (rev. 1), is available on the NRC Web site (*http://www.nrc.gov*). The most recent revisions to this policy are summarized at 63 FR 26,630–52 (May 13, 1998). For information on the NRC's radiography inspection procedures, you may refer to Inspection Procedure 87120. The NRC's *Enforcement Manual,* NUREG/BR-0195 (Revision 2), intended for internal use by the NRC, tells how the NRC staff is supposed to implement the NRC enforcement program.

14.7.1 Penalties

The NRC enforcement policy (NUREG-1600) calls for each violation to be categorized into a "severity level." Normally there are four levels of severity, from Level I (the most significant) to IV (least significant). Lesser, "minor" violations below Level IV are possible but are not subject to formal enforcement action, at least by the federal agency. Severity Level I and II violations are considered very serious and usually involve actual or "high potential" impact on the public.

The activity in which the violation occurred is also categorized. For example, the activity may be in Safeguards, Health Physics, Transportation, Fuel Cycle and Materials Operations, Miscellaneous Matters, or Emergency Preparedness. Appendix B to NUREG-1600 gives examples to show what kinds of violations in each activity category are Level I, II, III, or IV violations. For example, in the Health Physics area, a radiation exposure during any year to a worker in excess of 25 rems total effective dose equivalent is a serious, Level I violation.

The enforcement policy also provides a table of base civil penalties ($11,000 for industrial radiographers and other large industrial users; $5500 for academic, medical, and other small material users) and a base civil penalty amount (100 percent of the base penalty for Level I, 80 percent for Level II, and 50 percent for Level III). These amounts may be increased or decreased based on a number of factors discussed in the policy. For example, a licensee normally gets credit if it identified a violation itself instead of waiting for an NRC inspector to find the violation. The lesson here is that it is best to be diligent about finding any violations on your own and, if required, reporting them to the authorities. Penalties, if any, will be lower than if the agency had discovered the violations itself, or learned about them from a whistleblower.

Civil penalties are determined case by case when they are imposed on individuals, rather than companies.

To get an idea of the fines assessed by the NRC recently, you can go to the NRC Office of Enforcement Annual Report for Fiscal Year 1997, available on the NRC Web site. For example, four *proposed* fines are listed for Level I cases, ranging from $5000 to $900,000. (The largest of these were proposed for multiple violations, not single incidents.) For FY 97 there were 3 enforcement items involving academic licenses, 9 involving physicians, 31 for gauge users, 33 against hospitals, one against an irradiator, 9 for radiographers, none for pharmacies, and one for well loggers. Violations included such things as failure to train users and provide TLDs, unauthorized removal of a gauge by a contractor, failure to conduct inventories, loss or improper disposal of radioactive sources, failure to have a written directive for administering iodine-131, leaving a sealed source in an unrestricted area, and unauthorized alterations or repairs to irradiator and safety systems.

14.7.2 Inspections

The NRC has an inspection program that calls for licensed facilities to be inspected periodically, on a schedule that is usually not revealed to the licensee. The frequency of inspections is set by internal agency practices and may be influenced by events. For example, high-dose remote afterloaders (HDRAs) are inspected every year, because such units have a history of causing fatalities. HDRAs are commonly used in oncology, especially to treat cancer of the cervix and some other cancers, because they enable doctors to provide treatment on an outpatient basis that formerly would have had to be done on an inpatient basis. Other facilities may be inspected every 2, 3, 4, or 5 years, with hospitals that provide no therapy perhaps being inspected less frequently.

The somewhat infrequent inspections mean that a facility might inadvertently violate regulations for a long time before being detected by the regulatory agency. The licensee is required to evaluate its own radiation safety program every year, but smaller companies or hospitals may have inexperienced people who have had minimal training, perhaps provided by the manufacturer of the equipment. If repeated violations of the regulations result, the liability can be enormous, and the license can be suspended or revoked.

14.7.3 Allegation Program

Often a safety concern will be made public by a whistleblower, usually a disgruntled employee. The NRC encourages employees to report safety problems through its Allegation Program. NUREG/BR-0240, "Reporting Safety Concerns," written for nuclear workers, tells how workers can report safety concerns to the NRC and how their identities will be protected. This brochure encourages workers to report safety concerns to their employers but also provides a toll-free number (1-800-695-7403) for calling the NRC Safety Hotline.

Retaliation against a worker for reporting a safety problem is not allowed. NUREG/BR-0240 and NRC Management Directive 8.8, "Management of Allegations" (available on the NRC Web site) tell how discrimination cases are handled. The NRC's Office of Investigation investigates allegations of wrongdoing, and discrimination against a whistleblower can result in fines, an order modifying the NRC license, or even referral to the Department of Justice for criminal prosecution. For personal remedies for discrimination the worker must go to the Department of Labor or the state labor department. Discrimination may be found if the worker has been fired, had pay reduced, been given a poor performance appraisal, or reassigned to a lower job by the employer, if it can be established that the employer acted *because* the worker raised safety concerns.

Concerns that are reported to the NRC but that are outside the NRC's jurisdiction will be forwarded to whatever agency is responsible. Examples are concerns about off-site emergency planning, use of NRC-regulated materials in agreement states, control of exempt quantities of licensed material, occupational safety, or disposal of nonnuclear waste. See NUREG/BR-0240, p. 3.

CHAPTER 15
PERMIT APPLICATION PROCEDURES

A. Roger Greenway
Principal
RTP Environmental Associates Inc.

15.1 WHEN DO YOU NEED A PERMIT?

Environmental managers are constantly faced with difficult questions to answer. Just exactly when does something that needs to be done to their facility require an environment permit or a modification of an existing permit? Many environmental managers assume that if emissions or water discharges are not increased by a particular change, either an operation or equipment, that permits are not needed or that revisions to an existing permitted equipment are not required. This is sometime true, however this is *not* always true.

Take, for instance, a woodworking fabrication shop making kitchen cabinets or commercial display cases. Moving a piece of equipment from one production area to another may move its emissions from one dust collection system to a second dust collection system. Removing it from the first dust collection system would not require modifying the permit, however, adding its emissions to a second dust collector would probably require modifying the permit for that piece of equipment. If the owner moved it to a production area and added a new dust collector, this dust collector and a piece of equipment being moved would require permitting assuming it met a threshold for permitting in the state in which it was being operated.

Consider another example, where in a production area a kettle is replaced with a larger kettle. The company takes the position that since the batch sizes are the same, that is the extra size of the kettle does not result in larger batches being made, that the kettle does not require a permit modification. This could be true, depending again on the state in which the facility is located, and assuming that the old kettle is permitted. If the permit for the old kettle limits its production rate to a capacity which is consistent with the old kettle size, and if the new kettle will not be operated at a batch size that exceeds the old batch size, then a permit modification *might* not be required. The rub here is that since the potential emissions from the kettle exceed those of the old kettle due to its larger physical size, a permit modification might still be required. Taking this one step further, assume for a minute that the existing kettle did not have a permit. It was "grandfathered." Now installing a new kettle of a larger size or, in fact, installing a kettle of the same size would require a permit in most states. This would meet the definition of "reconstruction."

Another issue facing facility environmental managers concerns a new pollutant being emitted from a source. Take the example of the kettle discussed above. The kettle now needs to be used to make a new batch product. The batch product will contain raw materials or will generate products which have air emissions which are not listed on an existing permit. These "new" pollutants would require a permit modification. Should this situation occur with a grandfathered kettle, that is one that does not have a permit, and a new product needs to be made in it, this would trigger a loss of grandfathering status for the piece of equipment and a permit would be required. Taking this one step further, situations occur when no change in production occurs, however, due to changes in a supplier of raw materials or due to Material Safety Data Sheets being updated, pollutants are identified by a supplier of raw materials for a process which results in air emissions which are not identified on an existing sources permit or which would cause a grandfathered source to lose its grandfathering status. For example, a supplier of paint to a surface coating operation reports in an updated Materials Safety Data Sheet that a hazardous air pollutant is present in a particular formulation that has been applied in the spray booth for years. Now that the user of the paint has information that a hazardous air pollutant is present in the coating being applied, a permit needs to be obtained if the source has been previously grandfathered, or if previously permitted, the permit needs to be modified to reflect the hazardous air pollutant now identified.

As a practical matter, new pollutants being identified, for permitted pieces of equipment, normally do not require immediate repermitting. Usually, agencies are content to have these pollutants identified at the next time a permit renewal date occurs. It is more difficult in the case of grandfathered sources since permits do not exist so that if a particular source now has a new pollutant emitted, permit applications may be required quickly to assure continued legal operation of the source. This point requires checking the state regulations to see when permits must be obtained, and what events trigger a loss of grandfathering status. Be especially aware that reporting discharges and emissions on R Forms or Emission Statements that are not consistent with permits for the source for which they are being reported, may trigger an enforcement visit from a state agency or a citizen suit from an environmental group, because of the inadequacies or errors in existing permits.

15.1.1 New Equipment

If a facility plans on installing a new source that will emit air pollutants, it needs to carefully review the state air pollution control regulations to identify air permitting requirements. In most states, regulations set deminimus sizes for sources below which permits are not required. For example, many states do not require permits for combustion sources below 1 million btu's. Likewise, many states do not regulate storage tanks below 2000 gallons in capacity. Since deminimus thresholds for permitting vary by state, it is highly recommended that the facility operator consult with a knowledgeable consultant, environmental manager, or with the agency itself to identify if the permits are needed. This is particularly important when hazardous or toxic substances may be a part of the source operation.

15.1.2 Emission Characteristics

In determining whether permit modification or new permits are required, the most important step involves identifying the pollutants to be emitted from the operation. Remember this includes the criteria pollutants, for which there are national ambient air quality standards, as well as hazardous air pollutants. The first thing to do to gain a preliminary estimate of what is emitted from a source is to review the Materials Safety Data Sheets for the raw materials going into the process. For combustion, review pollutant emission factors for fuel oil, natural gas, propane, or other fuel being used. For chemical process operations, review Material Safety Data Sheets for the inputs to the process and review characteristics of the reaction and product produced. Information on how to estimate emissions can be found in Section 15.2 that follows.

15.1.3 Grandfathered Sources

As noted above, many industrial sources in many states have grandfather status. This means that these sources were built and began operation prior to the air pollution control regulations requiring permits for sources of their size and type. Be aware, however, that many such grandfathered sources have actually lost their grandfathering status due to changes in the method of operation or modifications made to the source itself. Consider for example a boiler constructed in 1946 to burn fuel oil which was modified to burn natural gas and fuel oil in the 1980s. No permits were found to exist for this boiler when the facility was inspected during a routine environmental audit. The question was asked as to when the unit was converted to dual fuel capacity and when informed that it was converted in the 1980s, the question was asked who applied for the environmental permits to burn natural gas. Facility managers replied probably the gas company. It turned out the gas company did not apply for the permits for the conversion, assuming no doubt, the facility managers would obtain the permits. The point of this story is that the boiler lost its grandfathered status and permits needed to be applied for immediately. Fortunately the lack of permits was discovered during an environmental audit conducted during a rare amnesty period offered by the state to encourage facilities to identify unpermitted sources that needed permits and permitted sources that needed to have permits modified. It is very likely that most sources that people consider grandfathered, have actually lost their grandfathering status due to the changes in the method of operation, increases in production levels, or new pollutants being identified in connection with the source activity. This is something that requires careful inspection and review any time the permit status of a facility is reviewed.

15.2   HOW TO ESTIMATE EMISSIONS

There are a large number of helpful documents available to assist you in estimating a facility's emissions. One of the most widely used reference documents is the technical series published by the USEPA called *AP-42—Compilation of Air Pollutant Emission Factors*. This reference document is available online by linking from the *epa.gov* homepage. It contains emissions factors for a wide range of sources, combustion types, chemical processes, paper mills, etc. You are always safe as a starting point using AP-42 to estimate facility emission levels. Be aware, however, that AP-42 is sometimes conservative and may not be current for the type of source you operate and may base the emissions guidance in AP-42 on test data from facilities that are not entirely similar to yours. For example, the AP-42 emission factors for pottery firing in kilns report hydrogen fluoride emissions based on a clay-fluoride content that may exceed the fluoride content of clays you are firing in your kilns. If you can show that the input material that you are firing is not similar the input material of the facilities EPA tested, you may be able to use lower emission factors than the AP-42 factors. Be sure, however, the emission factors you use, if you do not use AP-42 factors, are well documented and are defensible. The reason for this is clear. If you base permit applications on emission factors which are not conservative (which do not overstate emissions at least slightly), you run the risk of obtaining a permit and then not being able to show compliance with that permit if you are required to test the source. It is far better to use emission factors that at least slightly overstate your emissions to increase the possibility that you will be successful if a stack test is required. The only exception to this rule is if, using conservative emission factors results in your exceeding a threshold such as a major source threshold for operating permits, you may want to look at the emission factors to see whether more realistic emission factors exist. You may also want to take testing data to identify, based on tests, what your level of emissions really is.

15.2.1 Test Data

Stack tests are usually more accurate than emission factors in identifying your level of emissions. Be aware, however, that stack tests must be conducted in conformance with EPA guidelines and you must be able to operate your facility at maximum capacity. The facility must, of course, also be in existence for you to stack test it. This sounds obvious, however, in many instances you are applying for your permits before the facility is constructed so that it cannot possibly be tested until after permits are obtained and it is built. What to do in these instances if you believe emission factors are not accurate and you prefer to test, is to find a similar facility, preferably one in your company that uses the same equipment to be installed in the subject facility being permitted. If this is not possible, you may be able to use data from other facilities, that use identical or substantially identical equipment and conduct the same operations. Be aware, however, that stack test data must be conducted at full capacity for the equipment, or if it cannot be, permits may not be obtainable for full capacity operation. Normally, when permits are based on stack test data to determine emissions, permits will be obtainable only for the maximum level at which the facility can be tested.

15.3 CONTROL EQUIPMENT

Another key question facility environmental managers must address when looking at permitting new or modified pieces of equipment is whether control equipment is required. Environment managers sometimes assume that any new piece of equipment will require air pollution control equipment. This is not the case. For most pollutants, and in most states, pollution control equipment is required only above a certain size or emission threshold. Be sure to check, therefore, with your state regulators and your state regulations to see if pollution control equipment is required for the source type and size you plan to construct or modify.

If air pollution control equipment or water pollution control equipment is required, be sure to identify the required capabilities for this equipment. For example, for particulates, removal of 99.5% of particles may be required, while for volatile organic substances, removal of 85 to 90% (or more) of the VOCs may be required.

Keep in mind however, the difference between capture of pollutants and destruction, or of pollutants collected. If a spray booth emits 100 lb per hour through spraying a paint film and if a pollution control device can remove 90% of the VOCs sent to it, will emissions to the atmosphere be reduced by 90%? The answer is no. There is another factor to be considered, that is, the factor of collection efficiency. In a spray booth, paint is sprayed in front of a filter bank. Fans draw the spray through the filters and possibly through VOC treatment systems and discharge exhaust gases to the atmosphere. The filter bank, however, does not collect 100% of the fumes from the spray booth. Some of these fumes waft here and there in the work environment and are not captured by the hood. This is called the capture efficiency. Typically, hoods will capture 85 to 90% of the vapors emitted in front of or under them. The remaining 10 to 15% of vapors become fugitive emissions. The 85 to 90% of vapors that are collected are transferred to the pollution control device (if one exists) before they are discharged to the atmosphere, and it is at this stage that the pollution control device may destroy 90 or 95% of the emissions sent to it. The resultant emissions to the atmosphere are the uncontrolled emissions, taking into account the collection efficiency and control efficiency of the control device resulting, in this instance, in an overall collection and destruction efficiency of about 85%.

What about filters and spray booths as VOC control devices? Many people incorrectly assume that filters on a spray booth will reduce VOC emissions to the atmosphere. They do not. The reason they do not is because paint particles or droplets collected on the filters contain VOCs, so they are not emitted to the atmosphere immediately. However, as the paint droplets dry on the filters, the VOCs are released and become air emissions through the stack if the fan is running, or as fugitive emissions back into the room, if they are released after the spray booth fan is turned off. In any instance they become air emissions.

15.4 EMISSION GUARANTEES

Another important issue for the environmental manager in applying for permits for a source that may include a pollution control device is, what are the manufacturer representations regarding emissions from the source or emission controls for the control device. Be aware that manufacturers look at this in at least two different ways. First, manufacturers make statements that the pollution control device will remove 99% of particles. This is considered the manufacturers, "projection." Manufacturers if pressed, will guarantee a level of emissions reduction. For example a manufacturer may guarantee that the pollution control device will remove 95% of emissions of all particles over a certain size range. Note that in this instance, for the projection, no size range was needed. This should alert the reader to the fact that projections are more marketing hype than useful information to base permitting decisions on. The danger for the environmental manager in applying for a permit based on a manufacturer's projection is that if the pollution control device does not meet the projection, you could find yourself in a situation with a permit that cannot be met. Guarantees, on the other hand, represent levels at which manufacturers will stand behind their product and guarantee that it will perform at the level specified. As a result, guaranteed emission levels are always less favorable than projections. Nonetheless, the manufacturer-guarantee levels are ones that should be used in permit applications.

The way that manufacturers write emission performance specifications and control equipment guarantee levels must also be examined carefully. Manufacturers sell equipment for a wide variety of applications and when they guarantee emissions performance, they always make assumptions of the source configuration, emission levels, and all aspects of source utilization. These must be complied with for the emission guarantee to be valid. For example, in guaranteeing a particulate control device will remove 95% of the particles sent to it, the manufacturer will specify what size particle this supplies to. Let's assume for a minute that this is all particles above 25 microns in diameter. If a stack test after the equipment is installed and the facility is operating shows that it is only removing 90% of the particles sent to it, the manufacturer will request a particle size distribution of the material being sent to the control device, to show hopefully, in this case, that the particles being sent to it are smaller than were represented to be in the exhaust at the time the guarantee was sought. Even when the particles are of the size specified when the guarantee was sought, manufacturers will sometimes look for other reasons for not meeting its guaranteed performance level and why others are to blame for this. The equipment was installed improperly, the airflow exceeded the design specification, the distance between the control device and the source was too short, the distance between the control device and the source was too long. You get the idea. The important thing for the owner of a source for which an emission guarantee is sought, is to be sure that the conditions under which the guarantee will be applicable are well thought out and well documented to be the conditions which you will actually need the device to perform under.

15.5 STACK TESTS

When are stack tests required? Stack tests are required for a number of reasons. First, an applicant may decide to do a stack test for a source prior to applying for a permit modification to get a clear understanding of what the emission levels are so that the permits sought can reflect the actual operation of the source.

Another reason stack tests are sometimes undertaken, is when a state agency requires a stack test to demonstrate compliance with a permit that it issued. In most states, stack tests are not required at the time of permit application. Typically, states operate like this. The applicant applies for a permit to construct and operate a source of air pollutants. The state reviews the application and issues a draft permit, the applicant then reviews the draft permit terms and conditions to identify whether they are acceptable to the facility; if they are not, changes are suggested to the state and a negotiation process ensues, resulting, hopefully, in acceptable permit terms for a source. The state then issues a temporary permit for the piece of equipment until such time as it can be inspected and possibly tested. These temporary permits may be 3-month permits which are automatically renewed every 3 months, sometimes for years, until a state inspector can come to the facility to inspect the equipment.

At the time the state is ready to issue a permanent 3- or 5-year permit for the piece of equipment, a stack test is sometimes required. The stack test at this point must show that the permit that has been in existence, on a temporary basis perhaps, for several years is one that the facility can comply with.

When conducting a stack test of a source, there are several key things to consider:

• *Stack test protocol*   this is something that the applicant should develop themselves either in-house or through a consultant. The protocol will describe the details of source operation during a stack test, the testing procedures to be undertaken, the laboratory analyses to be conducted and the method for reporting results. Do not use a stack test protocol prepared by a regulatory agency without a detailed review and after being sure that it is one that you find acceptable. The point is, that even the protocol is negotiable with the agency if, because of particular facts about the source being tested, you feel a different approach to testing is required.

The stack test will require you to operate the source at full-permitted operating levels. This generally means operating the facility at 100% and sometimes at 105% in the case of boiler capacity, so that characteristics at permit-rated capacity can be determined.

The stack test should be scheduled after obtaining bids from several stack testing companies and the state agency should be notified so that they have an opportunity to witness the stack test. The agency may, or may not, actually witness the stack test, but in most cases the state wants to be notified so that it has the opportunity to witness at least a portion of the test. You should also be sure that your facility environmental managers and consultants, if used, are also present during the stack test so that all parties can attest to the operating conditions of the equipment during the test.

What about "bad" results? Let's say a stack test is undertaken and the results come back showing that the source is violating its permits by a particular margin. Does this mean that the permits are invalid or that the source is not meeting its permit limits. It may, or it may not. The source may not have been properly tested, there might have been errors made in laboratory procedures, calibration and/or span gases may have been incorrect, equipment might have drifted during the test, the point being, there can be any number of reasons why a stack test may not be valid. Therefore, before a conclusion is reached that a source is not meeting its permit limits because of a stack test failure, be sure to check all aspects of the stack test to be sure first, that the data are valid. If the data are valid, and the source did not meet its permit limits, then the facility must examine the permits to determine if permit modifications are needed. This also raises questions under EPA's credible evidence policy as to whether the source will be deemed to have been out of compliance with its permit back to the date it first began operation. Therefore, adverse results from a stack test must be considered to be very serious, and must be examined very carefully to be sure that they are valid prior to any further action being taken. Having said this, however, it should be clear that the stack test did show that the source is not meeting its permit, in many instances the facility has a duty to report this, particularly if it is a major source subject to the operating permits regulations, to the state agency and potentially to EPA promptly. If a stack test fails, and the agencies are notified, it is important to schedule an additional test if you feel the test might have been invalid for some reason and to explore ways in which the facility can remain in operation while permit revisions are being undertaken, while the source is modified, or while other available options are explored. This may require the services of an environmental attorney to execute a Consent Agreement with the state agency to allow continued operation while options are being explored and discussed with the agency.

15.6 STATE-OF-THE-ART (SOTA) REQUIREMENTS

One problem in permitting a new source and even more in permitting an existing source that lost grandfathering status or which was modified, is state-of-the-art (SOTA) requirements. Many states require applications for air pollutant permits, incorporate advances in the state of the art of pollution control. The problem with existing sources is that existing sources are, by definition, in existence, and probably do not incorporate the latest-and-greatest in pollution control technology. Therefore, when modifying an existing source, for example putting a new chemical process into a batch process area that is served by a scrubber, it is likely that the scrubber in existence in this area is older and will not meet the removal efficiencies of new scrubbers. In most states, requirements to meet SOTA are based on a threshold. For example, in New Jersey, SOTA requirements apply to sources that emit more than 5 tons per year of emissions.

The problem with SOTA requirements is that they are often "moving targets" and that the requirements change even more quickly than BACT or LAER requirements change, under major New Source Review. Worse, most states do not have clearinghouses which you can consult, via on-line access or through paper inquiries to identify what the latest SOTA requirement is. Some states, notably New Jersey, are now preparing SOTA manuals which they indicate will not be updated more than once per year, which will give applicants guidance as to what SOTA means for their type of source operation.

15.7 PREAPPLICATION MEETINGS

Another question facility environmental managers must address in determining their permit application strategy is whether a preapplication meeting is desirable with an agency. Be advised, preapplication meetings are not always available with agencies. Permit reviewers at environmental agencies do not have the time to meet with every applicant for a permit. Therefore, use your right, if you have one, to request a preapplication meeting sparingly. Reserve it for applications for permits or modifications which are really needed on a fast-track basis. For example, a facility is a batch chemical operation which takes in tolling operations for other chemical companies. It is one facility of a company operating 6 plants across the United States. It is also one of the older facilities and one that the corporate staff have been looking at for potential closure. An opportunity arises for the facility to take on a major new toll product; however, its air permits need to be modified to be able to run the toll processes. If the facility cannot demonstrate its ability to manufacture the product within 6 months, it will lose the opportunity to manufacture the product. In an instance like this, a preapplication meeting with the agency would be warranted, since a case can be made that without the agency's assistance in fast-tracking the permit modifications, the opportunity to manufacture the toll product may be lost and the facility may, in fact, be shut down.

15.8 WHAT IF SOMETHING HAPPENS AFTER FILING THE APPLICATION?

Applicants are sometimes faced with the realization that having filed the permit application for a new source or modification, things have changed. Either that information previously assumed to be correct is discovered to be incorrect, or the operation for which the permit application or modification is submitted has changed and what the facility really needs is a little different. The question facing the applicant in situations like this is, what to do? Should new permit applications be prepared reflecting the changes and submitted, realizing the agency might have spent a month or more reviewing the previous submittal?

The first thing to look at is how long has the existing permit been at the state, and how significant are the changes required. If the permit application has only been at the state agency for 1 or 2 months, and the agency has not completed its

Technical Review of the application, it may be possible, through a telephone call to the agency, to have them stop their review of the application, pending submission of new information. If the facility does not have a history of changing applications in mid-stream, the state *may* be amenable to yielding to the facility's request. There is no guarantee that the agency will be able or willing to do this however, and it is possible that modified applications must be submitted and that these will start the clock again from zero.

If the requested modifications to the permit are minor, for example, an emission level is increased by 20% or emissions are not increased at all but a stack parameter, that is, diameter or height is changed, the agency may be amenable to accepting the modified information and not restarting the clock. The significance of the permit review clock is that in many states, by state law, agencies only have a certain period of time under which to make a decision to make a permit decision. Sometimes this is as short as 30, 60, or 90 days depending on the magnitude of the permit application.

15.9   WHAT IF A PERMIT IS FOUND TO BE NOT ACCEPTABLE?

Through a routine review of facility operations, a permit is found that does not reflect the current source covered by the permit. This could occur for a number of reasons. First, the business has changed and the equipment is being used in ways not anticipated under the permit application. It could be found that the existing permits were erroneously applied for and that the information contained in them is not accurate.

The question facing the environmental manager is, what to do when permits are found that do not reflect the current mode of plant operation? Is the equipment covered by the permit required to be shut down immediately? Can you continue to operate it while you adjust the permits? These are difficult questions that can be answered only by someone experienced in dealing with the state environmental agency. In most instances, if the facility has a good compliance record with the agency, the facility will not be required to be shut down while revised permits are obtained.

The first thing that must be done is to identify what the permit really should look like. This is something that requires significant care. A quick response to this question, may result in modified permits which shortly, in themselves, do not reflect the desired or required operating characteristics of the source. Therefore, think carefully as to what permits really need to look like and what emission characteristics really need to be. If a permit modification does not result in higher emissions, just different emissions, different by virtue of location, different stack parameters, for example, different distribution of emissions during the year, then permit modifications are probably not critical, although they are still needed. It is unlikely that a source would need to be shut down until permit modifications are obtained.

Some state regulatory programs allow 7-day notice changes and other changes for minor modifications which allow a source to continue in operation, while the agency modifies the permit. Technically, however, most permit modifications require the source to be repermitted before it can be operated. Therefore, a source that continues operating a piece of equipment while a permit application for modification is under review, should revert to the mode of operations allowed for in the existing permit until the modification is approved. Doing anything less than this exposes the source to potential enforcement action by the agency.

A working relationship with the state should enable the facility to identify whether it is exposed to significant enforcement risk by operating in a manner consistent with a revised permit or new permit before the permit is approved. To be safe, facilities may want to consider working with an attorney to put into place a consent agreement to allow the source to operate in violation of its existing permits until such time as new permits are approved.

15.10 CONSTRUCTING AT RISK

Constructing at risk means that a company is allowed to construct at risk equipment covered by a permit which has not yet been approved. This is something that each facility must decide for itself if construction at risk is permitted in the State. Having applied for a permit and constructing the equipment at the site will incur a significant expense. How sure are you that the permits are approvable? If the permits are not approved, the equipment cannot be operated. Therefore, constructing at risk carries significant risks for the facility. The reason people construct at risk, however, is that sometimes permit reviews can take months and even years; therefore if a company waited for the permits to be approved to execute contracts to buy equipment, have it delivered, and have it installed before it could be operated it could lose significant market opportunities. Therefore, companies routinely file permits for new equipment, order the equipment and install it at risk, betting that the permits will be approved, and that they can immediately begin operating the equipment. Here again, experience with the agency which is responsible for granting the permit applications will give you a comfort factor as to whether permits really are achievable for the subject equipment.

15.11 DO YOU NEED A CONSULTANT?

Permit applications have gotten increasingly complex. Many states are moving toward electronic permit preparation. At first glance, this approach would seem to make permit applications easier (many people think it's like preparing your taxes using turbo tax). In most cases, electronic preparation does not. The reason is that the amount of information required in electronic permit applications often exceeds that in paper applications, mainly because the electronic permit application software has been developed over a period of years and incorporates the brainstorming of many agency staff in terms of what information it would like to have from applicants. Many of these permit application packages (*see* Chapter 17 for description of New Jersey's AIMS/RADIUS system), are extremely complex. That is not to say that experienced environmental managers at a facility could not master the packages and prepare their own permit applications, but it is not advisable to learn on a fast-track project.

A question of whether an application is prepared by a consultant, in-house, or by a team including in-house and consultant personnel, is one that the facility must decide. There are benefits to each approach.

15.12   *DO YOU NEED AN ATTORNEY?*

The question here is really, *if* you need an attorney. Is an in-house attorney experienced enough with environmental law, regulations, and with the agency you are dealing with, to effectively represent your facility in an environmental permitting issue? Agencies have particular ways of dealing with the regulated community, and an attorney that represents many facilities on permitting issues is probably more experienced with what agencies will accept, what kinds of conditions are allowable in consent agreements, and can more effectively achieve your desired result. Given the potential penalties for noncompliance and the high stakes (including losing the ability to operate the facility) should permits not be approved, it is highly recommended that if permit compliance issues arise, that a competent environmental attorney at least be available, should enforcement action be initiated by the agency.

CHAPTER 16
COORDINATING WITH OTHER REGULATORY PROGRAMS

John A. Green
Director of Safety and Health
Tate & Lyle North American Sugars Inc.

Common or duplicative regulatory requirements associated with environmental permitting can best be coordinated by relying on:

- The overlap between OSHA's Process Safety Management (PSM) Program and EPA's Risk Management Program (RMP)
- The National Response Team's (NRT) Integrated Contingency Plan (ICP) Guidance.

This chapter will briefly discuss the relationship between RMP and PSM and then go into greater detail with ICP, which is probably the most explicit example of how regulatory requirements between different federal agencies can be efficiently coordinated.

16.1 COORDINATION BETWEEN OSHA'S PROCESS SAFETY MANAGEMENT PROGRAM AND EPA'S RISK MANAGEMENT PROGRAM STANDARD

The requirements of both the OSHA Process Safety Management (PSM) Program and EPA's Risk Management Program (RMP) are intended to eliminate or mitigate the consequences of releases of extremely hazardous substances. These standards emphasize the application of management controls to the risks associated with handling or working near such hazardous chemicals.

Since both the PSM and the RMP standards have been developed in fulfillment of obligations under the Clean Air Act Amendments (CAAA) of 1990, Sec. 304(a), there are many common elements between them. While both deal with the prevention of releases of extremely/highly hazardous substances, the OSHA PSM standard is designed to protect workers within a covered facility, while the focus of the EPA RMP standard is to protect the general public outside a covered facility. Although the requirements of both these programs overlap, it is the stated goals of both agencies, the Occupational Safety and Health Administration and the Environmental Protection Agency, that duplicative compliance burdens and requirements should be simplified and minimized. EPA has coordinated with OSHA and the Department of Transportation (DOT) in developing the RMP standard. To the extent possible, covered facilities will not face inconsistent requirements under these agencies' rules.

In this chapter we examine the relationships between these two standards, review in detail the requirements of the OSHA PSM program, and determine how their respective compliance requirements can be mutually dependent on one another.

16.1.1 Common Elements Between OSHA's PSM and EPA's RMP

EPA has developed seven specific elements for its RMP "Program 2" prevention program: safety information (§ 68.48), hazard review (§ 68.50), operating procedures (§ 68.52), training (§ 68.54), maintenance (§ 68.56), compliance audits (§ 68.58), and incident investigation (§ 68.60). Most Program 2 processes are likely to be relatively simple and located at smaller businesses. EPA believes owners or operators of Program 2 processes can successfully prevent accidents without a program as detailed as the OSHA PSM, which was primarily designed for the chemical industry. EPA combined and tailored elements common to OSHA's PSM and EPA's notice of proposed rule making (NPRM) to generate Program 2 requirements and applied them to non-petrochemical industry processes. EPA is also developing model risk management programs (and RMPs) for several industry sectors that will have Program 2 processes. These model guidances will help sources comply by providing standard elements that can be adopted to a specific source. EPA expects that many Program 2 processes will already be in compliance with most of the requirements through compliance with other federal regulations, state laws, industry standards and codes, and good engineering practices.

EPA's "Program 3" prevention program includes the requirements of the OSHA PSM standard, 29 CFR 1910.119(c) through (m) and (o), with minor wording changes to address statutory differences. This makes it clear that one accident prevention program to protect workers, the general public, and the environment will satisfy both OSHA and EPA. For elements that are in both the EPA and OSHA rules, EPA has used OSHA's language verbatim, except for the following changes: the replacement of the terms *highly hazardous substance, employer, standard,* and *facility* with *regulated substance, owner or operator, part or rule,* and *stationary source*; the deletion of specific references to workplace impacts or to *safety and health*; changes to specific schedule dates; and changes to references within the standard. The *safety and health* and *workplace impacts* references occur in OSHA's PSM standard in process safety information [29 CFR 1910.119 (d)(2)(E)], process hazard analysis [29 CFR 1910.119(e)(3)(vii)], and incident investigation [29 CFR 1910.119(m)(1)].

Under the provisions of EPA's RMP, process hazard analyses (PHAs) conducted for OSHA are considered adequate to meet EPA's requirements. They will be updated on the OSHA schedule (i.e., by the fifth anniversary of their initial completion). This approach will eliminate any need for duplicative analyses. Documentation for a PHA developed for OSHA will be sufficient to meet EPA's purposes.

EPA anticipates that sources whose processes are already in compliance with OSHA PSM will not need to take any additional steps or create any new documentation to comply with EPA's Program 3 prevention program. Any PSM modifications necessary to account for protection of public health and the environment along with protection of workers can be made when PSM elements are updated under the OSHA requirements.

16.1.2 Primary Enforcement Authority, EPA versus OSHA

Under the EPA RMP program, changes are included and designed to ensure that OSHA retains its oversight of actions designed to protect workers while EPA retains its oversight of actions to protect public health and the environment.

EPA has modified the OSHA definition of catastrophic release, which serves as the trigger for an incident investigation, to include events "that present imminent and substantial endangerment to public health and the environment." As a result, this rule requires investigation of accidental releases that pose a risk to the public or the environment, whereas the OSHA rule does not. EPA recognizes that catastrophic accidental releases primarily affect the workplace and that this change will have little effect on incident investigation programs already established. However, EPA needs to ensure that deviations that could have had only an off-site impact are also addressed.

In the CAAA of 1990, Congress adopted a number of provisions aimed at reducing the number and severity of chemical accidents. The accident prevention requirements for which EPA is responsible are included primarily in the new Subsection (r) of Sec. 112 of the Clean Air Act (42 U.S.C. 7412), which also establishes a chemical safety and hazard investigation board to investigate chemical accidents. Under Subsection (r) and other CAAA provisions, EPA will conduct research on topics related to chemical accident prevention.

EPA and OSHA are working closely to coordinate interpretation and enforcement of PSM and accident prevention programs. Under a recently signed memorandum of understanding, OSHA and EPA have jointly assumed the responsibilities for conducting investigations of major facility chemical accidents. The fundamental objectives of the EPA/OSHA chemical accident investigation program is to determine and report to the public the facts, conditions, circumstances, and cause or probable root cause of any chemical accident that results in a fatality, serious injury, substantial property damage, or serious off-site impact, including a large-scale evacuation of the general public. The ultimate goal of this joint accident investigation is to determine the root cause in order to reduce the likelihood of recurrence, minimize the consequences associated with accidental releases, and to make chemical production, processing, handling, and storage safer. Since both the PSM and RMP final rules are consistent with the mandate of the CAAA, it is anticipated that joint inspection activities related to the PSM standard will now arise between OSHA, the EPA, and the Chemical Safety and Hazard Investigation Board.

EPA activities include developing new and refining existing criteria with OSHA for selection of accidents for joint investigation or independent investigation by the lead agencies, enhancing investigation techniques of significant chemical accidents, and improving training to EPA, OSHA, and other parties on accident investigation techniques. To assist these operations, EPA and OSHA are now conducting activities to support an external expert panel to review accident investigation reports and make recommendations for further prevention and safety. Although this panel, the Chemical Safety and Hazard Investigation Board, was originally mandated by the 1990 Clean Air Act Amendments, there have never been any serious concerted efforts to fully implement it until recently because of the White House's political opposition and budget cuts.

16.2 THE NATIONAL RESPONSE TEAM'S INTEGRATED CONTINGENCY PLAN

On June 6, 1996 the U.S. Environmental Protection Agency, as the chair of the National Response Team (NRT), announced the availability of the NRT's Integrated Contingency Plan Guidance ("one plan"). This guidance is intended to be used by facilities to prepare emergency response plans. The intent of the NRT is to provide a mechanism for consolidating multiple plans that facilities may have prepared to comply with various regulations into one functional emergency response plan or integrated contingency plan. This chapter contains the suggested ICP outline as well as guidance on how to develop an ICP and demonstrate compliance with various regulatory requirements. The policies set out in this notice are intended solely as guidance.

16.2.1 Regulatory Review

Section 112(r)(10) of the Clean Air Act required the President to conduct a review of federal release prevention, mitigation, and response authorities. The presidential review was delegated to EPA, in coordination with agencies and departments that are members of the National Response Team. The intent of NRT is to provide a mechanism for consolidating multiple plans that facilities may have prepared to comply with various regulations into one functional emergency response plan or integrated contingency plan. A number of statutes and regulations, administered by several federal agencies, include requirements for emergency response planning. A particular facility may be subject to one or more of the following federal regulations:

- EPA's Oil Pollution Prevention Regulation [SPCC (Spill Prevention Control and Countermeasure) and Facility Response Plan Requirements]—40 CFR, Part 112.7(d) and 112.20–.21
- Mineral Management Service's Facility Response Plan Regulation—30 CFR, Part 254
- Research and Special Programs Administration's Pipeline Response Plan Regulation—49 CFR, Part 194
- United States Coast Guard's Facility Response Plan Regulation—33 CFR, Part 154, Subpart F
- EPA's Risk Management Program Regulation—40 CFR, Part 68
- OSHA's Emergency Action Plan Regulation—29 CFR 1910.38(a)
- OSHA's Process Safety Standard—29 CFR 1910.119
- OSHA's HAZWOPER Regulation—29 CFR 1910.120
- EPA's Resource Conservation and Recovery Act Contingency Planning Requirements—40 CFR Part 264, Subpart D, 40 CFR Part 265, Subpart D, and 40 CFR 279.52

In addition, facilities may also be subject to state emergency response planning requirements that this guidance does not specifically address. Facilities are encouraged to coordinate development of their ICP with relevant state and local agencies to ensure compliance with any additional regulatory requirements.

Individual agencies' planning requirements and plan review procedures are not changed by the advent of the ICP format option. This one-plan guidance has been developed to assist facilities in demonstrating compliance with the existing federal emergency response planning requirements referenced above. Although it does not relieve facilities from their current obligations, it has been designed specifically to help meet those obligations. Adherence to this guidance is not required in order to comply with federal regulatory requirements. Facilities are free to continue maintaining multiple plans to demonstrate federal regulatory compliance; however, the NRT believes that an integrated plan prepared in accordance with this guidance is a preferable alternative.

This one-plan guidance was developed through a cooperative effort among numerous NRT agencies, state and local officials, and industry and community representatives. The NRT and the agencies responsible for reviewing and approving federal response plans to which the ICP option applies agree that integrated response plans prepared in the format provided in this guidance will be acceptable and be the federally preferred method of response planning. A distinct advantage of the ICP is that a single functional plan is preferable to multiple plans regardless of the specific format chosen. While they are acceptable, other formats may not allow the same ease of coordination with external plans. In any case, whatever formats a facility chooses, no individual NRT agency will require an integrated response planning format differing from the ICP format described here. The NRT anticipates that future development of all federal regulations addressing emergency response planning will use the ICP Guidance. Also, developers of state and local requirements will be encouraged to be consistent with this document.

The ICP Guidance does not change existing regulatory requirements; rather, it provides a format for organizing and presenting material currently required by the regulations. Individual regulations are often more detailed than the ICP Guidance. To ensure full compliance, facilities should continue to read and comply with all of the federal regulations that apply to them. Furthermore, facilities submitting an ICP (in whatever format) for agency or department review will need to provide a cross-reference to existing regulatory requirements so that plan reviewers can verify compliance with these requirements. The guidance contains a series of matrices designed to assist owners and operators in consolidating various plans and documenting compliance with federal regulatory requirements. The matrices can be used as the basis for developing a cross-reference to various regulatory requirements.

16.2.2 Integrated Contingency Plan Purpose

The purpose of the ICP is to minimize duplication in the preparation and use of emergency response plans at the same facility and to improve economic efficiency for both the regulated and regulating communities. Facility expenditures for the preparation, maintenance, submission, and update of a single plan should be much lower than for multiple plans.

The use of a single emergency response plan per facility eliminates confusion for facility first responders, who often must decide which of their plans is applicable to a particular emergency. The use of a single integrated plan should also improve coordination between facility response personnel and local, state, and federal emergency response personnel.

The adoption of a standard plan format facilitates integration of plans within a facility, in the event that large facilities may need to prepare separate plans for distinct operating units. The ICP concept also allows coordination of facility plans with plans that are maintained by local emergency planning committees (LEPCs), area committees (ACs), cooperatives, and mutual aid organizations. In some cases, there are specific regulatory requirements to ensure that facility plans are consistent with external planning efforts.

16.2.3 Scope

The ICP can be used by any facility subject to federal contingency planning regulations and is also recommended for use by other facilities to improve emergency preparedness through planning. In this context, the term *facility* is meant to have a wide connotation and may include, but is not limited to, any mobile or fixed onshore or offshore building, structure, installation, equipment, pipe, or pipeline.

Facility hazards need to be addressed in a comprehensive and coordinated manner. Accordingly, this guidance is broadly constructed to allow for facilities to address a wide range of risks in a manner tailored to the specific needs of the facility. This includes both physical and chemical hazards associated with events such as chemical releases, oil spills, fires, explosions, and natural disasters.

16.2.4 Organization of the ICP

The ICP format is organized into three main sections:

- An introductory section
- A core plan
- A series of supporting annexes

It is important to note that the elements contained in these sections are not new concepts, but accepted emergency response activities that are currently addressed in various forms in existing contingency planning regulations. The goal of the NRT is not to create new planning requirements, but to provide a mechanism to consolidate existing concepts into a single functional plan structure. This approach would provide a consistent basis for addressing emergency response concerns as it gains widespread use among facilities.

16.2.4.1 ICP Introduction Section. The introduction section of the plan is designed to provide facility response personnel, outside responders, and regulatory officials with basic information about the plan and the entity it covers. It calls for a statement of purpose and scope, a table of contents, information on the current revision date of the plan, general facility information, and the key contacts for plan development and maintenance. This section should present the information in a brief factual manner.

16.2.4.2 ICP Core Plan Section. The structure of the core plan and annexes is based on the structure of the National Interagency Incident Management System (NIIMS) Incident Command System (ICS). NIIMS ICS is a nationally recognized system currently in use by numerous federal, state, and local organizations. It is a type of response management system that has been used successfully in a variety of emergency situations, including releases of oil or hazardous substances. It provides a commonly understood framework that allows for effective interaction among response personnel. Organizing the ICP along the lines of the NIIMS ICS will allow the plan to dovetail with established response management practices, thus facilitating its use during an emergency.

The core plan is intended to contain essential response guidance and procedures. Annexes would contain more detailed supporting information on specific response management functions. The core plan should contain frequent references to the response-critical annexes to direct response personnel to parts of the ICP that contain more detailed information on the appropriate course of action for responders to take during various stages of a response. Facility planners need to find the right balance between the amount of information contained in the core plan versus the response critical annexes (Annexes 1 through 3).

16.2.4.3 ICP Supporting Annexes. Information required to support response actions at facilities with multiple hazards will likely be contained in the annexes. Planners at facilities with fewer hazards may choose to include most, if not all, information in the core plan. Other annexes (i.e., Annexes 4 through 8) are dedicated to providing information that is noncritical at the time of a response (e.g., cross-references to demonstrate regulatory compliance and background planning information). Consistent with the goal of keeping the size of the ICP as manageable as practicable, it is not necessary for a plan holder to provide its field responders with all the compliance documentation (i.e., Annexes 4 through 8) that it submits to regulatory agencies. Similarly, it may not be necessary for a plan holder to submit all annexes to every regulatory agency for review.

Basic headings are consistent across the core plan and annexes to facilitate ease of use during an emergency. These headings provide a comprehensive list of elements to be addressed in the core plan and response annexes and may not be relevant to all facilities. Planners should address those regulatory elements that are applicable to their particular facilities. Planners at facilities with multiple hazards will need to address most, if not all, elements included in this guidance. Planners at facilities with fewer hazards may not need to address certain elements. If planners choose to strictly adopt the ICP outline contained in this guidance but are not required by regulation to address all elements of the outline, they may simply indicate "not applicable" for those items where no information is provided.

16.2.5 Core Plan

The core plan is intended to reflect the essential steps necessary to initiate, conduct, and terminate an emergency response action: recognition, notification, and initial response, including assessment, mobilization, and implementation. This section of the plan is designed to be concise and easy to follow. A rule of thumb is that the core plan should fit in the glovebox of a response vehicle. The core plan need not detail all procedures necessary under these phases of a response, but should provide information that is time-critical in the earliest stages of a response and a framework to guide responders through key steps necessary to mount an effective response. The response action section should be convenient to use and understandable at the appropriate skill level.

The NRT recommends the use of checklists or flowcharts wherever possible to capture these steps in a concise, easy-to-understand manner. The core plan should be constructed to contain references to appropriate sections of the supporting annexes for more detailed guidance on specific procedures. The NRT anticipates that, for a large, complex facility with multiple hazards, the annexes will contain a significant amount of information on specific procedures to follow. For a small facility with a limited number of hazard scenarios, the core plan may contain most, if not all, of the information necessary to carry out the response, thus obviating the need for more detailed annexes. The checklists, depending on their size and complexity, can be in either the core or the support section.

The core plan should reflect a hierarchy of emergency response levels. A system of response levels is commonly used in emergency planning for classifying emergencies according to seriousness and assigning an appropriate standard response or series of response actions to each level. Both complex and simple industrial facilities use a system of response levels for rapidly assessing the seriousness of an emergency and developing an appropriate response. This process allows response personnel to match the emergency and its potential impacts with appropriate resources and personnel. The concept of response levels should be considered in developing checklists or flowcharts designed to serve as the basis for the core plan. Note that for those facilities subject to planning requirements under OPA, response levels in the core plan may not necessarily correspond to discharge planning amounts (e.g., average most probable discharge, maximum most probable discharge, and worst-case discharge).

Facility owners and operators should determine appropriate response levels based on (1) the need to initiate time-urgent response actions to minimize or prevent unacceptable consequences to the health and safety of workers, the public, or the environment and (2) the need to communicate critical information concerning the emergency to off-site authorities. The consideration and development of response levels should, to the extent practicable, be consistent with similar efforts that may have been taken by the LEPC, local area committee, or mutual aid organization. Response levels, which are used in communications with off-site authorities, should be fully coordinated and use consistent terminology.

16.2.6 Annexes

The annexes are designed to provide key supporting information for conducting an emergency response under the core plan as well as to document compliance with regulatory requirements not addressed elsewhere in the ICP. Annexes are not meant to duplicate information that is already contained in the core plan, but to augment core plan information. The annexes should relate to the basic headings of the core plan. To accomplish this, the annexes should contain sections on facility information, notification, and a detailed description of response procedures under the response management system (i.e., command, operations, planning, logistics, and finance). The annexes should also address issues related to postaccident investigation, incident history, written follow-up reports, training and exercises, plan critique and modification process, prevention, and regulatory compliance, as appropriate.

The ICP is based on the NIIMS ICS. If facility owners or operators choose to follow fundamental principles of the NIIMS ICS, then they may adopt NIIMS ICS by reference rather than having to describe the system in detail in the plan. The owner or operator should identify where NIIMS ICS documentation is kept at the facility and how it will be accessed if needed by the facility or requested by the reviewing agency. Regardless of the response management system used, the plan should include:

- An organization chart
- Specific job descriptions
- A description of information flow ensuring liaison with the on-scene coordinator (OSC) and a description of how the selected response management system integrates with a unified command
- If a system other than NIIMS ICS is used, a description of how it differs from NIIMS, or a detailed description of the system used

The NRT anticipates that the use of linkages (i.e., references to other plans) in developing annexes will serve several purposes. Linkages will facilitate integration with other emergency plans within a facility (until such plans can be fully incorporated into the ICP) and with external plans, such as LEPC plans and area contingency plans (ACPs). Linkages will also help ensure that the annexes do not become too cumbersome. The use of references to information contained in external plans does not relieve facilities from regulatory requirements to address certain elements in a facility-specific manner and to have information readily accessible to responders. When determining what information may be linked by reference and what needs to be contained in the ICP, response planners should carefully consider the time-critical nature of the information. If instructions or procedures will be needed immediately during an incident response, they should be presented for ready access in the ICP. The following information would not normally be well-suited for reference to documents external to the ICP: core plan elements, facility and locality information (to allow for quick reference by responders on the layout of the facility and the surrounding environment and mitigating actions for the specific hazards present), notification procedures, details of response management personnel's duties, and procedures for establishing the response management system. Although linkages provide the opportunity to utilize information developed by other organizations, facilities should note that many LEPC plans and ACPs might not currently possess sufficient detail to be of use in facility plans or the ICP. This information may need to be developed by the facility until detailed applicable information from broader plans is available.

In all cases, referenced materials must be readily available to anticipated plan users. Copies of documents that have been incorporated by reference need not be submitted unless it is required by regulation. The appropriate sections of referenced documents that are unique to the facility, those that are not nationally recognized, those that are required by regulation, and those that could not reasonably be expected to be in the possession of the reviewing agency should be provided when the plan is submitted for review and/or approval. Discretion should be used when submitting documents containing proprietary data. It is, however, necessary to identify in the ICP the specific section of the document being incorporated by reference, where the document is kept, and how it will be accessed if needed by the facility or requested by the reviewing agency. In addition, facility owners or operators are reminded to take note of submission requirements of specific regulations when determining what materials to provide an agency for review, as it may not be necessary to submit all parts of an ICP to a particular agency.

As discussed previously, this discussion contains a series of matrices designed to assist owners and operators in the plan consolidation process and in the process of ensuring and documenting compliance with regulatory requirements. The matrix in Sec. 16.2.9 of this chapter displays areas of current regulations that align with the suggested elements contained in the guidance document. When addressing each element of the ICP outline (Sec. 16.2.8), plan drafters can refer to this matrix to identify specific regulatory requirements related to that element. The matrices in Sec. 16.2.9 to this guidance display regulatory requirements as contained in each of the regulations listed in the NRT policy statement above (which are applicable to many facilities) along with an indication of where in the suggested ICP outline these requirements should be addressed. If a facility chooses to follow the ICP outline, these matrices can be included as Annex 8 to a facility's

ICP to provide the necessary cross-reference for plan reviewers to document compliance with various regulatory requirements. To the extent that a plan deviates from the suggested ICP outline, plan drafters will have to alter the matrices to ensure that the location of regulatory requirements within the ICP is clearly identified for plan reviewers.

16.2.7 Integrated Contingency Plan Elements

Presented below is a list of elements to be addressed in the ICP and a brief explanation, displayed in italicized text, of the nature of the information to be contained in that section of the ICP. Section 16.2.8 presents the complete outline of the ICP without the explanatory text. As discussed previously, the elements are organized into three main sections: plan introduction, core plan, and response annexes.

16.2.7.1 Section I—Plan Introduction Elements

16.2.7.1.1 Purpose and Scope of Plan Coverage. This section should provide a brief overview of facility operations and describe in general the physical area and nature of hazards or events to which the plan is applicable. This brief description will help plan users quickly assess the relevancy of the plan to a particular type of emergency in a given location. This section should also include a list of which regulations are being addressed in the ICP.

16.2.7.1.2 Table of Contents. This section should clearly identify the structure of the plan and include a list of annexes. This will facilitate rapid use of the plan during an emergency.

16.2.7.1.3 Current Revision Date. This section should indicate the date that the plan was last revised to provide plan users with information on the currency of the plan. More detailed information on plan update history (i.e., a record of amendments) may be maintained in Annex 6 (Response Critique and Plan Review and Modification Process).

16.2.7.1.4 General Facility Identification Information. This section should contain the following information to provide a brief profile of the facility and its key personnel to facilitate rapid identification of key administrative information:

- Facility name
- Owner/operator/agent (include physical and mailing address and phone number)
- Physical address of the facility (include county/parish/borough, latitude/longitude, and directions)
- Mailing address of the facility (correspondence contact)
- Other identifying information (e.g., ID numbers, SIC code, oil storage start-up date)
- Key contacts for plan development and maintenance

- Phone numbers for key contacts
- Facility phone number
- Facility fax number

16.2.7.2   Section II—Core Plan Elements

16.2.7.2.1   Discovery. This section should address the initial action the persons discovering an incident will take to assess the problem at hand and access the response system. Recognition, basic assessment, source control (as appropriate), and initial notification of proper personnel should be addressed in a manner that can be easily understood by everybody in the facility. The use of checklists and flowcharts is highly recommended.

16.2.7.2.2   Initial Response. This section should provide for activation of the response system following discovery of the incident. It should include an established 24-hour contact point (i.e., that person and alternate who will be called to set the response in motion) and instructions for that person about who to call and what critical information to pass. Plan drafters should also consider the need for bilingual notification. It is important to note that different incident types require that different parties be notified. Appropriate federal, state, and local notification requirements should be reflected in this section of the ICP. Detailed notification lists may be included here or in Annex 2, depending on the variety of notification schemes that a facility may need to implement. For example, the release of an extremely hazardous substance will require more extensive notifications [i.e., to state emergency response commissions (SERCs) and LEPCs] than a discharge of oil. Even though no impacts or awareness are anticipated outside the site, immediate external notifications are required for releases of CERCLA and EPCRA substances. Again, the use of forms, such as flowcharts, checklists, and call-down lists, is recommended.

This section should instruct personnel in the implementation of a response management system for coordinating the response effort. More detailed information on specific components and functions of the response management system (e.g., detailed hazard assessment, resource protection strategies) may be provided in annexes to the ICP. The section should cover:

- Procedures for internal and external notifications (i.e., contact, organization name, and phone number of facility emergency response coordinator, facility response team personnel, federal, state, and local officials)
- Establishment of a response management system
- Procedures for preliminary assessment of the situation, including an identification of incident type, hazards involved, magnitude of the problem, and resources threatened
- Procedures for establishment of objectives and priorities for response to the specific incident, including:

Immediate goals/tactical planning (e.g., protection of workers and public as priorities)

Mitigating actions (e.g., discharge/release control, containment, and recovery, as appropriate)

Identification of resources required for response

- Procedures for implementation of tactical plan
- Procedures for mobilization of resources

This part of the plan should then provide information on problem assessment, establishment of objectives and priorities, implementation of a tactical plan, and mobilization of resources. In establishing objectives and priorities for response, facilities should perform a hazard assessment using resources such as Material Safety Data Sheets (MSDSs) or the Chemical Hazard Response Information System (CHRIS) manual. The *Hazardous Materials Emergency Planning Guide* (NRT-1), developed by the NRT to assist community personnel with emergency response planning, provides guidance on developing hazard analyses. If a facility elects to provide detailed hazard analysis information in a response annex, then a reference to that annex should be provided in this part of the core plan.

Mitigating actions must be tailored to the type of hazard present. For example, containment might be applicable to an oil spill (e.g., use of booming strategies) but would not be relevant to a gas release. The plan holder is encouraged to develop checklists, flowcharts, and brief descriptions of actions to be taken to control different types of incidents. Relevant questions to ask in developing such materials include:

- What type of emergency is occurring?
- What areas/resources have been or will be affected?
- Do we need an exclusion zone?
- Is the source under control?
- What type of response resource is needed?

16.2.7.2.3 Sustained Actions. This section should address the transition of a response from the initial emergency stage to the sustained action stage, where more prolonged mitigation and recovery actions progress under a response management structure. The NRT recognizes that most incidents can be handled by a few individuals without implementing an extensive response management system. This section of the core plan should be brief and rely heavily on references to specific annexes to the ICP.

16.2.7.2.4 Termination and Follow-up Actions. This section should briefly address the development of a mechanism to ensure that the person in charge of mitigating the incident can, in coordination with the federal or state OSC as necessary, terminate the response. In the case of spills, certain regulations may become effective once the emergency is declared over. The section should describe how the orderly demobilization of response resources would occur. In addition, follow-up actions associated with termination of a response (e.g., accident investigation, response critique, plan review, written follow-up reports) should also be outlined in this section. Plan drafters may reference appropriate annexes to the ICP in this section of the core plan.

16.2.7.3   Section III—Annexes

16.2.7.3.1   Annex 1, Facility and Locality Information.   This annex should provide detailed information to responders on the layout of the facility and the surrounding environment. The use of maps and drawings to allow for quick reference is preferable to detailed written descriptions. These should contain information critical to the response such as the location of discharge sources, emergency shutoff valves and response equipment, and nearby environmentally and economically sensitive resources and human populations (e.g., nursing homes, hospitals, and schools). The ACP and LEPC plans may provide specific information on sensitive environments and populations in the area. EPA regional offices, Coast Guard marine safety offices, and LEPCs can provide information on the status of efforts to identify such resources. Plan holders may need to provide additional detail on sensitive areas near the facility. In addition, this annex should contain other facility information that is critical to response; it should complement but not duplicate information contained in Part 4 of the plan introduction section containing administrative information on the facility. The annex should contain, for example:

- Facility maps
- Facility drawings
- Facility description/layout, including identification of facility hazards and vulnerable resources and populations on and off the facility that may be impacted by an incident.

16.2.7.3.2   Annex 2, Notification.   This annex should detail the process of making people aware of an incident (i.e., whom to call, when the call must be made, and what information/data to provide on the incident). The incident commander is responsible for ensuring that notifications are carried out in a timely manner but is not necessarily responsible for making the notifications. ACPs, regional contingency plans (RCPs), and LEPC plans should be consulted and referenced as a source of information on the roles and responsibilities of external parties that are to be contacted. This information is important to help company responders understand how external response officials fit into the picture. Call-down lists must be readily accessible to ensure rapid response. Notification lists provided in the core plan need not be duplicated here but need to be referenced:

- Internal notifications
- Community notifications
- Federal and state agency notifications

16.2.7.3.3   Annex 3, Response Management System.   This annex should contain a general description of the facility's response management system as well as specific information necessary to guide or support the actions of each response management function (i.e., command, operations, planning, logistics, and finance) during a response.

16.2.7.3.3.1   GENERAL: If facility owners or operators choose to follow the fundamental principles of NIIMS ICS (see discussion of annexes above), then they may adopt NIIMS ICS by reference rather than describe the response management system in detail in the plan. In this section of Annex 3, planners should briefly address either (1) basic areas where their response management system is at variance with NIIMS ICS or (2) how the facility's organization fits into the NIIMS ICS structure. This may be accomplished through a simple organizational diagram.

If facility owners or operators choose not to adopt the fundamental principles of NIIMS ICS, this section should describe in detail the structure of the facility response management system.

Regardless of the response management system used, this section of the annex should include the following information:

• Organizational chart

• Specific job description for each position [NOTE: OPA 90 planning requirements for marine transfer facilities (33 CFR 154.1035) require job descriptions for each spill management team member regardless of the response management system employed by the facility]

• A detailed description of information flow

• Description of the formation of a unified command within the response management system

16.2.7.3.3.2   COMMAND:

• *List* facility incident commander and qualified individual (if applicable) by name and/or title and provide information on their authorities and duties. This section of Annex 3 should describe the command aspects of the response management system that will be used (i.e., reference NIIMS ICS or detail the facility's response management system). The locations of predesignated command posts should also be identified.

• *Information* (i.e., internal and external communications). This section of Annex 3 should address how the facility will disseminate information internally (i.e., to facility/response employees) and externally (i.e., to the public). For example, this section might address how the facility would interact with local officials to assist with public evacuation and other needs. Items to consider in developing this section include press release statement forms, plans for coordination with the news media, community relations plan, needs of special populations, and plans for families of employees.

- *Safety.* This section of Annex 3 should include a process for ensuring the safety of responders. Facilities should reference responsibilities of the safety officer, federal/state requirements (e.g., HAZWOPER), and safety provisions of the ACP. Procedures for protecting facility personnel should be addressed (i.e., evacuation signals and routes, sheltering in place).

- *Liaison—staff mobilization.* This section of Annex 3 should address the process by which the internal and external emergency response teams will interact. Given that parallel mobilization may be carried out by various response groups, the process of integration (i.e., unified command) should be addressed. This includes a process for communicating with local emergency management especially where safety of the general public is concerned.

16.2.7.3.3.3 OPERATIONS: This section of Annex 3 should contain a discussion of specific operational procedures to respond to an incident. It is important to note that response operations are driven by the type of incident. That is, a response to an oil spill will differ markedly from a response to a release of a toxic gas to the air. Plan drafters should tailor response procedures to the particular hazards in place at the facility. A facility with limited hazards may have relatively few procedures. A larger, more complex facility with numerous hazards is likely to have a series of procedures designed to address the nuances associated with each type of incident.

- Operational response objectives
- Discharge or release control
- Assessment/monitoring
- Containment
- Recovery
- Decontamination
- Nonresponder medical needs, including information on ambulances and hospitals
- Salvage plans

16.2.7.3.3.4 PLANNING:

- *Hazard assessment,* including facility hazards identification, vulnerability analysis, and prioritization of potential risks. This section of Annex 3 should present a detailed assessment of all potential hazards present at the facility, an analysis of vulnerable receptors (e.g., human populations, both workers and the general public, environmentally sensitive areas, and other facility-specific concerns) and a discussion of which risks deserve primary consideration during an incident.

- *Protection.* This section of Annex 3 should present a discussion of strategies for protecting the vulnerable receptors identified through the hazard analysis.

Primary consideration should be given to minimizing those risks identified as a high priority. Activities to be considered in developing this section include: population protection; protective booming; dispersant use, in-situ burning, bioremediation; water intake protection; wildlife recovery/rehabilitation; natural remediation; vapor suppression; and monitoring, sampling, and modeling. ACPs and LEPC plans may contain much of this information.

- *Coordination* with natural resource trustees. This section should address coordination with government natural resource trustees. In their role as managers of and experts in natural resources, trustees assist the federal OSC in developing or selecting removal actions to protect these resources. In this role, they serve as part of the response organization working for the federal OSC. A key area to address is interaction with facility response personnel in protection of natural resources.

- *Waste management.* This section should address procedures for the disposal of contaminated materials in accordance with federal, state, and local requirements.

16.2.7.3.3.5 LOGISTICS: This section of the Annex 3 should address how the facility will provide for the operational needs of response operations in each of the areas listed above. For example, the discussion of personnel support should address issues such as: volunteer training, management, overnight accommodations, meals, operational/administrative spaces, and emergency procedures. The NRT recognizes that certain logistical considerations may not be applicable to small facilities with limited hazards.

- Medical needs of responders
- Site security
- Communications (internal and external resources)
- Transportation (air, land, water)
- Personnel support (e.g., meals, housing, equipment)
- Equipment maintenance and support

16.2.7.3.3.6 FINANCE/PROCUREMENT/ADMINISTRATION: This section of Annex 3 should address the acquisition of resources (i.e., personnel and equipment) for the response and monitoring of incident-related costs. Lists of available equipment in the local and regional area and how to procure such equipment as necessary should be included. Information on previously established agreements (e.g., contracts) with organizations supplying personnel and equipment (e.g., oil spill removal organizations) also should be included. This section should also address methods to account for resources expended and to process claims resulting from the incident. The section should cover:

- Resource list
- Personnel management
- Response equipment

- Support equipment
- Contracting
- Claims procedures
- Cost documentation

16.2.7.3.4   Annex 4, Incident Documentation.   This annex should describe the company's procedures for conducting a follow-up investigation of the cause of the accident, including coordination with federal, state, and local officials. This annex should also contain an accounting of incidents that have occurred at the facility, including information on cause, amount released, resources impacted, injuries, response actions, etc. This annex should also include information that may be required to prove that the facility met its legal notification requirements with respect to a given incident, such as a signed record of initial notifications and certified copies of written follow-up reports submitted after a response.

16.2.7.3.5   Annex 5, Training and Exercises/Drills.   This annex should contain a description of the training and exercise program conducted at the facility as well as evidence (i.e., logs) that required training and exercises have been conducted on a regular basis. Facilities may follow appropriate training or exercise guidelines (e.g., National Preparedness for Response Exercise Program Guidelines) as allowed under the various regulatory requirements.

16.2.7.3.6   Annex 6, Response Critique and Plan Review and Modification Process.   This annex should describe procedures for modifying the plan on the basis of periodic plan review or lessons learned through an exercise or a response to an actual incident. Procedures to critique an actual or simulated response should be a part of this discussion. A list of plan amendments (i.e., history of updates) should also be contained in this annex. Plan modification should be reviewed as a part of a facility's continuous improvement process.

16.2.7.3.7   Annex 7, Prevention.   Some federal regulations that primarily address prevention of accidents include elements that relate to contingency planning (e.g., EPA's RMP and SPCC regulations and OSHA's Process Safety Standard). This annex is designed to allow facilities to include prevention-based procedures (e.g., maintenance, testing, in-house inspections, release detection, site security, containment, fail-safe engineering) that are required in contingency planning regulations or that have the potential to impact response activities covered in a contingency plan. The modular nature of the suggested plan outline provides planners with necessary flexibility to include prevention requirements in the ICP. This annex may not need to be submitted to regulatory agencies for review.

16.2.7.3.8   Annex 8, Regulatory Compliance and Cross-Reference Matrices. This annex should include information necessary for plan reviewers to determine compliance with specific regulatory requirements. To the extent that plan drafters did not include regulatory required elements in the balance of the ICP, they should be addressed in this annex. This annex should also include signatory pages to convey management approval and certifications required by the regulations, such as certification of adequate response resources and/or statements of regulatory applicability as required by regulations under OPA authority. Finally, this annex should contain cross-references that indicate where specific regulatory requirements are addressed in the ICP for each regulation covered under the plan. As discussed previously, Sec. 16.2.9 contains a series of matrices designed to fulfill this need in those instances where plan drafters adhere to the outline contained in this guidance.

16.2.8 ICP Outline

The following suggested outline of an ICP has been formatted to facilitate its being cross-referenced as a guide to Sec. 18.2.9, "Regulatory Cross-Comparison Matrices."

Section I—Plan Introduction Elements

1. Purpose and Scope of Plan Coverage
2. Table of Contents
3. Current Revision Date
4. General Facility Identification Information
   a. Facility name
   b. Owner/operator/agent (include physical and mailing address and phone number)
   c. Physical address of the facility (include county/parish/borough, latitude/longitude, and directions)
   d. Mailing address of the facility (correspondence contact)
   e. Other identifying information (e.g., ID numbers, SIC code, oil storage start-up date)
   f. Key contacts for plan development and maintenance
   g. Phone number for key contacts
   h. Facility phone number
   i. Facility fax number

Section II—Core Plan Elements

1. Discovery
2. Initial Response
   a. Procedures for internal and external notifications (i.e., contact, organization name, and phone number of facility emergency response coordinator, facility response team personnel, federal, state, and local officials)
   b. Establishment of a response management system
   c. Procedures for preliminary assessment of the situation, including an identification of incident type, hazards involved, magnitude of the problem, and resources threatened

d. Procedures for establishment of objectives and priorities for response to the specific incident, including:

(1) Immediate goals/tactical planning (e.g., protection of workers and public as priorities)

(2) Mitigating actions (e.g., discharge/release control, containment, and recovery, as appropriate)

(3) Identification of resources required for response

e. Procedures for implementation of tactical plan

f. Procedure for mobilization of resources

3. Sustained Actions

4. Termination and Follow-up Actions

Section III—Annexes

Annex 1. Facility and Locality Information

a. Facility maps

b. Facility drawings

c. Facility description/layout, including identification of facility hazards and vulnerable resources and populations on and off the facility which may be impacted by an incident

Annex 2. Notification

a. Internal notifications

b. Community notifications

c. Federal and state agency notifications

Annex 3. Response Management System

a. General

b. Command

(1) List facility incident commander and qualified individual (if applicable) by name and/or title and provide information on their authorities and duties

(2) Information (i.e., internal and external communications)

(3) Safety

(4) Liaison—staff mobilization

c. Operations

(1) Operational response objectives

(2) Discharge or release control

(3) Assessment/monitoring

(4) Containment

(5) Recovery

(6) Decontamination

(7) Nonresponder medical needs including information on ambulances and hospitals

(8) Salvage plans

    d. Planning

        (1) Hazard assessment, including facility hazards identification, vulnerability analysis, and prioritization of potential risks

        (2) Protection

        (3) Coordination with natural resource trustees

        (4) Waste management

    e. Logistics

        (1) Medical needs of responders

        (2) Site security

        (3) Communications (internal and external resources)

        (4) Transportation (air, land, water)

        (5) Personnel support (e.g., meals, housing, and equipment)

        (6) Equipment maintenance and support

    f. Finance/procurement/administration

        (1) Resource list

        (2) Personnel management

        (3) Response equipment

        (4) Support equipment

        (5) Contracting

        (6) Claims procedures

        (7) Cost documentation

Annex 4. Incident Documentation

    a. Postaccident investigation

    b. Incident history

Annex 5. Training and Exercises/Drills

Annex 6. Response Critique and Plan Review and Modification Process

Annex 7. Prevention

Annex 8. Regulatory Compliance and Cross-Reference Matrices

16.2.9 Regulatory Cross-Comparison Matrices

Note: The designations II and III on the right side of the following tables refer to the appropriate sections of the suggested ICP outline in Sec. 18.2.8.

TABLE 16.1   Resource Conservation Recovery Act [40 CFR, Part 264, Subpart D; Part 265, Subpart D; Part 279.52(b)][a]

Regulatory reference	Appropriate ICP section
264.52 Content of contingency plan:	
(a) Emergency response actions[b]	
(b) Amendments to SPCC plan	
(c) Coordination with state and local response parties[c]	II.2.b; III.3.a
(d) Emergency coordinators	II.2.a; III.2
(e) Detailed description of emergency equipment on site	II.2.d.(3); II.2.e; II.2.f; III.3.f.(1); III.3.f.(3); III.3.f.(4)
(f) Evacuation plan if applicable	III.3.b.(3)
264.53 Copies of contingency plan	
264.54 Amendment of contingency plan	III.6
264.55 Emergency coordinator	II.2.a; III.3.b.(1)
264.56 Emergency procedures:	
(a) Notification	II.2.a; III.2; III.3.b.(2)
(b) Emergency identification/characterization	II.2.c; III.3.c.(3)
(c) Health/environmental assessment	II.2.c; III.3.c.(3)
(d) Reporting	II.2.a; III.2; III.3.c.(3)
(e) Containment	III.3.c.(2); III.3.c.(4)
(f) Monitoring	III.3.b.(3); III.3.c.(3)
(g) Treatment, storage, or disposal of wastes	
(h) Cleanup procedures	
(1) Disposal	III.3.d.(4)
(2) Decontamination	III.3.c.(6)
(i) Follow-up procedures	II.4
(j) Follow-up report	III.4.a
265.52 Content of contingency plan:	
(a) Emergency response actions[d]	
(b) Amendments to SPCC plan	
(c) Coordination with state and local response parties[e]	II.2.b; III.3.a
(d) Emergency coordinators	II.2.a; III.2
(e) Detailed description of emergency equipment on site	II.2.f; III.3.f.(1); III.3.f.(3); III.3.f.(4)
(f) Evacuation plan if applicable	III.3.b.(3)
265.53 Copies of contingency plan	
265.54 Amendment of contingency plan	III.6
265.55 Emergency coordinator	II.2.a; III.3.b.(1)
265.56 Emergency procedures:	
(a) Notification	II.2.a; III.2; III.3.b.(2)
(b) Emergency identification/characterization	II.2.c; III.3.c.(3)
(c) Health/environmental assessment	II.2.c; III.3.c.(3)
(d) Reporting	II.2.a; III.2; III.3.c.(3)
(e) Containment	III.3.c.(2); III.3.c.(4)
(f) Monitoring	III.3.b.(3); III.3.c.(3)
(g) Treatment, storage, or disposal of wastes	III.3.d.(4)

TABLE 16.1   Resource Conservation Recovery Act [40 CFR, Part 264, Subpart D; Part 265, Subpart D; Part 279.52(b)][a] (*Continued*)

Regulatory reference	Appropriate ICP section
(h) Cleanup procedures:	
(1) Disposal	III.3.d.(4)
(2) Decontamination	III.3.c.(6)
(i) Follow-up procedures	II.4
(j) Follow-up report	III.4.a
279.52(b)(2) Content of contingency plan:	
(i) Emergency response actions[f]	
(ii) Amendments to SPCC plan	
(iii) Coordination with state and local response parties[g]	II.2.b; III.3.a
(iv) Emergency coordinators	II.2.a; III.2
(v) Detailed description of emergency equipment on site	II.2.d.(3); II.2.e; II.2.f; III.3.f.(1); III.3.f.(3); III.3.f.(4)
(vi) Evacuation plan if applicable	III.3.b.(3)
(3) Copies of contingency plan	
(4) Amendment of contingency plan	III.6
(5) Emergency coordinator	II.2.a; III.3.b.(1)
(6) Emergency procedures:	
(i) Notification	II.2.a; III.2; III.3.b.(2)
(ii) Emergency identification/characterization	II.2.c; III.3.c.(3)
(iii) Health/environmental assessment	II.2.c; III.3.c.(3)
(iv) Reporting	II.2.a; III.2; III.3.c.(3)
(v) Containment	III.3.c.(2); III.3.c.(4)
(vi) Monitoring	III.3.b.(3); III.3.c.(3)
(vii) Treatment, storage, or disposal of wastes	III.3.d.(4)
(viii) Cleanup procedures:	
(A) Disposal	III.3.d.(4)
(B) Decontamination	III.3.c.(6)
(ix) Follow-up report	III.4.a

[a]Facilities should be aware that most states have been authorized by EPA to implement RCRA contingency planning requirements in place of the federal requirements listed. Thus, in many cases, state requirements may not track this matrix. Facilities must coordinate with their respective states to ensure an ICP complies with state RCRA requirements.

[b]Section 264.56 is incorporated by reference at Sec. 264.52(a).

[c]Incorporates by reference Sec. 264.37.

[d]Section 265.56 is incorporated by reference at Sec. 265.52(a).

[e]Incorporates by reference Sec. 265.37.

[f]Section 279.52(b)(6) is incorporated by reference at Sec. 279.52(b)(2)(i).

[g]Incorporates by reference Sec. 279.52(a)(6).

TABLE 16.2   EPA's Oil Pollution Prevention Regulations (40 CFR 112)

Regulatory reference	Appropriate ICP section
112.7(d)(1) Strong spill contingency plan and written commitment of manpower, equipment, and materials*	
112.20(g) General response planning requirements	III.3.d.(3); III.6
112.20(h) Response plan elements	I.2; III.8
(1) Emergency response action plan (Appendix F1.1):	
(i) Identity and telephone number of qualified individual (F1.2.5)	III.3.b.(1)
(ii) Identity of individuals/organizations to contact if there is a discharge (F1.3.1)	III.2
(iii) Description of information to pass to response personnel in event of a reportable spill (F1.3)	II.2.a
(iv) Description of facility's response equipment and its location (F1.3.2)	II.2.d.(3); III.3.e.(3); III.3.e.(6); III.3.f.(1); III.3.f.(3)
(v) Description of response personnel capabilities (F1.3.4)	II.2.b; III.3; III.3.e.(5); III.3.f.(2)
(vi) Plans for evacuation of the facility and a reference to community evacuation plans (F1.3.5)	III.3.b.(3); III.3.e.(5)
(vii) Description of immediate measures to secure the source (F1.7.1)	II.2.d.(2); III.3.c.(2); III.3.c.(4)
(viii) Diagram of the facility (F1.9)	III.1.a–b
(2) Facility information (F1.2, F2.0)	I.4.b–d; III.1
(3) Information about emergency responses	
(i) Identity of private personnel and equipment to remove to the maximum extent practicable a WCD or other discharges (F1.3.2, F1.3.4)	III.3.c.(2); III.3.c.(4)–(5); III.3.e.(5)
(ii) Evidence of contracts or other approved means for ensuring personnel and equipment availability	III.3.e.(5); III.3.f.(5)
(iii) Identity and telephone of individuals/organizations to be contacted in event of a discharge (F1.3.1)	II.2.a; III.2.b–d; III.3.b.(2)
(iv) Description of information to pass to response personnel in event of a reportable spill (F1.3.1)	II.2.a
(v) Description of response personnel capabilities (F1.3.4)	II.2.b; III.3; III.3.e.(5); III.3.f.(2)
(vi) Description of a facility's response equipment, location of the equipment, and equipment testing (F1.3.2, F1.3.3)	II.2.d.(3); III.3.e.(3); III.3.e.(6); III.3.f.(1); III.3.f.(3)
(vii) Plans for evacuation of the facility and a reference to community evacuation plans as appropriate (F1.3.5)	III.3.b.(3); III.3.e.(5)
(viii) Diagram of evacuation routes (F1.9)	III.3.b.(3)
(ix) Duties of the qualified individual (F1.3.6)	II.2.c; II.2.d.(1); I.2.e; III.2.b–c; III.3.c.(3); III.3.d.(1); III.3.f
(4) Hazard evaluation (F1.4)	II.2.c; III.3.d.(1); III.4.b

TABLE 16.2 EPA's Oil Pollution Prevention Regulations (40 CFR 112) (*Continued*)

Regulatory reference	Appropriate ICP section
(5) Response planning levels (F1.5, F1.5.1, F1.5.2)	II.3.d.(1)
(6) Discharge detection systems (F1.6, F1.6.1, F1.6.2)	II.1
(7) Plan implementation (F1.7)	II.2.d–f; II.3; II.4
(i) Response actions to be carried out (F1.7.1.1)	II.2; III.3.d.(2)
(ii) Description of response equipment to be used for each scenario (F1.7.1.1)	III.3.d.(1)
(iii) Plans to dispose of contaminated cleanup materials (F1.7.2)	III.3.c.(5)–(6)
(iv) Measures to provide adequate containment and drainage of spilled oil (F1.7.3)	III.3.c.(2); III.3.c.(4); III.3.d.(2); III.3.d.(4)
(8) Self-inspection, drills/exercises, and response training (F1.8.1–F1.8.3.2)	III.3.e.(6); III.5
(9) Diagrams (F1.9)	III.1.b
(10) Security systems (F1.10)	III.3.e.(2)
(11) Response plan cover sheet (F2.0)	III.5
112.21 Facility response training and drills/exercises (F1.8.2, F1.8.3)—Appendix F, Facility-Specific Response Plan†	I.2
1.0 Model facility-specific response plan	
1.1 Emergency response action plan	
1.2 Facility information	I.3; I.4.a; I.4.b–c; I.4.h; II.2.a; III.1
1.3 Emergency response information:	
1.3.1 Notification	II.2.a; III.2.a–c
1.3.2 Response equipment list	II.2.d.(3); III.3.e.(3); III.3.f.(1); III.3.f.(3)–(4); III.3.e.(6)
1.3.3 Response equipment testing/deployment	
1.3.4 Personnel	II.2.b; III.3; III.3.f.(2)
1.3.5 Evacuation plans	III.3.b.(3); III.3.e.(5)
1.3.6 Qualified individual's duties	II.2
1.4 Hazard evaluation	II.2.c
1.4.1 Hazard identification	III.1.c; III.3.d.(1)
1.4.2 Vulnerability analysis	II.2.c; III.3.d.(1)
1.4.3 Analysis of the potential for an oil spill	III.3.d.(1)
1.4.4 Facility reportable oil spill history	III.4.b
1.5 Discharge scenarios:	
1.5.1 Small and medium discharges	III.3.d.(1)
1.5.2 Worst-case discharge	III.3.d.(1)
1.6 Discharge detection systems:	
1.6.1 Discharge detection by personnel	II.1
1.6.2 Automated discharge detection	II.1
1.7 Plan implementation	II.2
1.7.1 Response resources for small, medium, and worst-case spills	II.2.d.(3); II.2.f; III.3.c.(3); III.3.d.(2); III.3.f.(1); III.3.f.(3)–(4)

TABLE 16.2   EPA's Oil Pollution Prevention Regulations (40 CFR 112) (*Continued*)

Regulatory reference	Appropriate ICP section
1.7.2 Disposal plans	III.3.c.(5)–(6); III.3.d.(4)
1.7.3 Containment and drainage planning	II.2.d; III.3.c.(4); III.3.d.(2)
1.8 Self-inspection, drills/exercises, and response training:	
1.8.1 Facility self-inspection	III.3.e.(6)
1.8.2 Facility drills/exercises	III.5
1.8.3 Response training	III.5
1.9 Diagrams	I.4; III.1.a–c
1.10 Security	III.3.e.(2)
2.0 Response plan cover sheet	I.4.b; I.4.c; I.4.h

*Nonresponse planning parts of this regulation (e.g., prevention provisions) require a specified format. If a facility is required to develop a strong oil spill contingency plan under this section, the requirement can be met through the ICP.

†The appendix further describes the required elements in 120.20(h). It contains regulatory requirements as well as recommendations.

TABLE 16.3   United States Coast Guard Facility Response Plan

Regulatory reference	Appropriate ICP section
154.1026 Qualified individual and alternate qualified individual	II.2.a; III.3.b.(1)
154.1028 Availability of response resources by contract or other approved means	III.3.f or III.8; III.3.f.(5)
154.1029 Worst-case discharge	III.3.d.(1)
154.1030 General response plan contents:	
(a) The plan must be written in English	
(b) Organization of the plan*	I.2
(c) Required contents	
(d) Sections submitted to COTP	
(e) Cross-references	III.8
(f) Consistency with NCP and ACPs	III.3.d.(3)
154.1035 Significant and substantial harm facilities:	
(a) Introduction and plan content	III.1
(1) Facility's name, physical and mailing address, county, telephone, and fax	I.4.a; I.4.c–d; I.4.h–i
(2) Description of a facility's location in a manner that could aid in locating the facility	I.4.c
(3) Name, address, and procedures for contacting the owner/operator on 24-hour basis	I.4.b; II.2.a
(4) Table of contents	I.2
(5) Cross index, if appropriate	III.8
(6) Record of changes to record information on plan updates	I.3; III.6
(b) Emergency response action plan:	
(1) Notification procedures:	
(i) Prioritized list identifying persons, including name, telephone number, and role in plan, to be notified in event of threat or actual discharge	II.2.a; III.2.a–c
(ii) Information to be provided in initial and follow-up notifications to federal, state, and local agencies	III.3.b; III.2.a–c
(2) Facility's spill mitigation procedures†	II.2.d.(2); III.3.c.(2)
(i) Volumes of persistent and nonpersistent oil groups	
(ii) Prioritized procedures/task delegation to mitigate or prevent a potential or actual discharge or emergencies involving certain equipment/scenarios	II.2
(iii) List of equipment and responsibilities of facility personnel to mitigate an average most probable discharge	II.2.e–f; III.3.f.(3); III.3.c.(1)–(5)
(3) Facility response activities†	II.2.c; II.2.e–f; II.3; II.4; III.3.c.(3)
(i) Description of facility personnel's responsibilities to initiate/supervise response until arrival of qualified individual	II.1; II.2
(ii) Qualified individual's responsibilities/authority	II.2

TABLE 16.3 United States Coast Guard Facility Response Plan

Regulatory Reference	Appropriate ICP section
(iii) Facility or corporate organizational structure used to manage response actions	II.2.b; II.3; III.3.a; III.3.b.(2)–(4); III.3.c; III.3.d.(1); III.3.e–f
(iv) Oil spill response organizations/spill management team available by contract or other approved means	II.2.d.(3); III.3.c.(4)–(5); III.3.e.(6); III.3.f.(1)–(2); III.3.f.(5)
(v) For mobile facilities that operate in more than one COTP, the oil spill response organizations/spill management team in the applicable geography-specific appendix	II.2.d.(3)
(4) Fish and wildlife sensitive environments	III.1.c; III.3.d.(1)–(2)
(i) Areas of economic importance and environmental sensitivity as identified in the ACP that are potentially impacted by a WCD	II.2.c
(ii) List areas and provide maps/charts and describe response actions	
(iii) Equipment and personnel necessary to protect identified areas	II.2.e–f; III.3.f.(3); III.3.c.(1)–(5)
(5) Disposal plan	III.3.d.(4)
(c) Training and exercises	III.5
(d) Plan review and update procedures	III.6
(e) Appendices	I.4.c; III.1.b
(1) Facility specific information	III.1
(2) List of contacts	II.2.a; III.2.a–c; III.3.b.(1)
(3) Equipment lists and records	III.3.e.(3); III.3.e.(6); III.3.f.(1); III.3.f.(3)–(5)
(4) Communications plan	III.3.b.(2)
(5) Site-specific safety and health plan	III.3.b.(3); III.3.c.(7); III.3.e.(1)
(6) List of acronyms and definitions	
(7) A geography-specific appendix	
154.1040 Specific requirements for substantial harm facilities	
154.1041 Specific response information to be maintained on mobile MTR facilities	
154.1045 Groups I–IV petroleum oils	
154.1047 Group V petroleum oils	
154.1050 Training	III.5
154.1055 Drills	III.5
154.1057 Inspection and maintenance of response resources	III.3.e.(6)
154.1060 Submission and approval procedures	
154.1065 Plan revision and amendment procedures	III.6
154.1070 Deficiencies	
154.1075 Appeal process	
Appendix C—Guidelines for determining and evaluating required response resources for facility response plans	III.3.f.(3)
Appendix D—Training elements for oil spill response plans	III.5

*Specific plan requirements for sections listed under 154.1030(b) are contained in 154.1035(a)–(g).

†Sections 154.1045 and 154.1047 contain requirements specific to facilities that handle, store, or transport Group I–IV oils and Group V oils, respectively.

TABLE 16.4   DOT/RSPA Facility Response Plan (49 CFR, Part 194)

Regulatory reference	Appropriate ICP section
194.101 Operators required to submit plans	
194.103 Significant and substantial harm: operator's statement	III.8
194.105 Worst-case discharge	III.3.d.(1)
194.107 General response plan requirements:	
(a) Resource planning requirements	III.3.d
(b) Language requirements	
(c) Consistency with NCP and ACPs	III.3.d.(3); III.8
(d) Each response plan must include:	
(1) Core plan contents:	
(i) An information summary as required in 194.113	I.4; III.1
194.113(a) Core plan information summary:	
(1) Name and address of operator	I.4.b; I.4.d
(2) Description of each response zone	I.4.c
194.113(b) Response zone appendix information summary:	
(1) Core plan information summary	I.4; III.1
(2) Name of SAA submission and approval procedures	III.6
194.121 Response plan review and update procedures	III.6
Appendix—SAA recommended guidelines for the preparation of response plans	I.2
Section 1—Information summary	I.4.b–c; II.2.a; II.2.f; III.8
Section 2—Notification procedures	II.2.a; III.2; III.3.b.(2); III.3.e.(3)
Section 3—Spill detection and on-scene spill mitigation procedures	II.1; II.2.e–f; III.3.c.(2)
Section 4—Response activities	II.2.b; III.3.b.(1)
Section 5—List of contacts	II.2.a
Section 6—Training procedures	III.5
Section 7—Drill procedures	III.5
Section 8—Response plan review and update procedures	III.6
Section 9—Response zone appendices	II.2.b; II.3; III.1.a–c; III.3

TABLE 16.5   OSHA Emergency Action Plans [29 CFR 1910.38(a)] and Process Safety (29 CFR 1910.119)

Regulatory reference	Appropriate ICP section
1910.38(a) Emergency action plan:	
(1) Scope and applicability	III.3.c.(1); III.3.d
(2) Elements:	
(i) Emergency escape procedures and emergency escape route assignments	II.2; II.2.c; III.3.b.(3); III.3.c
(ii) Procedures to be followed by employees who remain to operate critical plant operations before they evacuate	III.3.c
(iii) Procedures to account for all employees after emergency evacuation has been completed	II.2.a; III.3.b.(2); III.3.b.(3); III.3.c; III.4
(iv) Rescue and medical duties for those employees who are to perform them	III.3.b.(3); III.3.c; III.3.c.(7); III.3.e.(1)
(v) The preferred means of reporting fires and other emergencies	II.2.a; III.3.b
(vi) Names or regular job titles of persons or departments who can be contacted for further information or explanation of duties under the plan	I.4.f; II.2.a; III.3.b.(2); III.3.b.(4)
(3) Alarm system*	II.2.a; III.3.c.(3); III.3.e.(3)
(4) Evacuation	II.2.d; III.3.b.(3); III.3.c.(3); III.3.d; III.3.d.(1)
(5) Training	III.3.e.(5); III.5
1910.119 Process safety management of highly hazardous chemicals:	
(e)(3)(ii) Investigation of previous incidents	III.4; III.4.b
(e)(3)(iii) Process hazard analysis requirements	III.3.e.(3)
(g)(1)(i) Employee training in process/operating procedures	III.5
(j)(4) Inspection/testing of process equipment	III.3.e.(6)
(j)(5) Equipment repair	III.3.e.(6)
(l) Management of change(s)	III.5
(m) Incident investigation	III.4.a
(n) Emergency planning and response	I.1; II.1; II.2; II.2.d; III.2; III.2.a; III.2.b
(o)(1) Certification of compliance	III.6
1910.165 Employee alarm systems:	
(b) General requirements	III.3.e.(3)
(b)(1) Purpose of alarm system	III.2; III.2.a
(b)(4) Preferred means of reporting emergencies	III.2
(d) Maintenance and testing	III.3.e.(6)
1910.272 Grain handling facilities:	
(d) Development/implementation of emergency action plan	I.1; III.2

*Section 1910.38(a)(3) incorporates 29 CFR 1910.165 by reference.

TABLE 16.6   OSHA HAZWOPER (29 CFR 1910.120)

Regulatory reference	Appropriate ICP section
1910.120(k) Decontamination	III.3.c.(6)
1910.120(l) Emergency response program	I.1
(1) Emergency response plan:	
(i) An emergency response plan shall be developed and implemented by all employers within the scope of this section to handle anticipated emergencies prior to the commencement of hazardous waste operations.	
(ii) Employers who will evacuate their employees from the workplace when an emergency occurs, and who do not permit any of their employees to assist in handling the emergency, are exempt from the requirements of this paragraph if they provide an emergency action plan complying with Sec. 1910.38(a) of this part.	
(2) Elements of an emergency response plan:	
(i) Preemergency planning and coordination with outside parties	I.4.f; II.2.b; II.2.c; III.2.b; III.2.c; III.3.b.(4); III.3.d
(ii) Personnel roles, lines of authority, and communication	I.4.f; II.2.b; III.2.a; III.2.c; III.3.b.(4); III.3.e.(4)
(iii) Emergency recognition and prevention	II.1; III.7
(iv) Safe distances and places of refuge	III.3.b.(3); III.3.d.(2)
(v) Site security and control	III.3.d.(2); III.3.e.(2)
(vi) Evacuation routes and procedures	II.2.d; III.3.b.(3)
(vii) Decontamination procedures	III.3.c.(6)
(viii) Emergency medical treatment and response procedures	II.2.d; III.3.c.(7); III.3.e.(1)
(ix) Emergency alerting and response procedures	II.2; II.2.a; II.2.f; II.4; III.2; III.2.a; III.2.b; III.2.c; III.3.d
(x) Critique of response and follow-up	II.3; III.4; III.4.a; III.6
(xi) PPE (Personal Protective Equipment) and emergency equipment	III.3.e.(6); III.3.f.(3); III.3.d.(2); III.3.e.(6); III.3.f.(3)
(3) Procedures for handling emergency incidents:	
(i) Additional elements of emergency response plans:	
(A) Site topography, layout, and prevailing weather conditions	III.1.c
(B) Procedures for reporting incidents to local, state, and federal government agencies	II.2.a; III.2
(ii) The emergency response plan shall be a separate section of the Site Safety and Health Plan.	
(iii) The emergency response plan shall be compatible with the disaster, fire, and/or emergency response plans of local, state, and federal agencies.	III.3.e

Regulatory reference	Appropriate ICP section
(iv) The emergency response plan shall be rehearsed regularly as part of the overall training program for site operations.	III.5
(v) The site emergency response plan shall be reviewed periodically and, as necessary, be amended to keep it current with new or changing site conditions or information.	
(vi) An employee alarm system shall be installed in accordance with 29 CFR 1910.165 to notify employees of an emergency situation, to stop work activities if necessary, to lower background noise in order to speed communications, and to begin emergency procedures.	
(vii) On the basis of the information available at time of the emergency, the employer shall evaluate the incident and the site response capabilities and proceed with the appropriate steps to implement the site emergency response plan.	II.2.c; II.2.d
1910.120(p)(8) Emergency response program:	I.1
(i) Emergency response plan	
(ii) Elements of an emergency response plan	
(A) Preemergency planning and coordination with outside parties	I.4.f; II.2.b; III.2.b; III.2.c; III.3.b.(4); III.3.d
(B) Personnel roles, lines of authority, and communication	I.4.f; II.2.b; III.2.c; III.2.c; III.3.b.(4); III.3.e.(4)
(C) Emergency recognition and prevention	II.1; III.7
(D) Safe distances and places of refuge	III.3.b.(3); III.3.d.(2)
(E) Site security and control	III.3.d.(2); III.3.e.(2)
(F) Evacuation routes and procedures	II.2.d; III.3.b.(3)
(G) Decontamination procedures	III.3.c.(6)
(H) Emergency medical treatment and response procedures	II.2.d; III.3.c.(7); III.3.e.(1)
(I) Emergency alerting and response procedures	II.2; II.2.a; II.2.f; I.4; III.2; III.2.a; III.2.b; III.2.c; III.3.d
(J) Critique of response and follow-up	II.3; III.4; III.4.a; III.6
(K) PPE and emergency equipment	III.3.e.(6); III.3.f.(3); III.3.d.(2); III.3.e.(6); III.3.f.(3)
(iii) Training	III.5
(iv) Procedures for handling emergency incidents:	
(A) Additional elements of emergency response plans:	
(1) Site topography, layout, and prevailing weather conditions	III.1.c; III.3.d.(1)
(2) Procedures for reporting incidents to local, state, and federal government agencies	II.2.a; III.2

TABLE 16.6 OSHA HAZWOPER (29 CFR 1910.120)

Regulatory reference	Appropriate ICP section
(B) The emergency response plan shall be compatible and integrated with the disaster, fire and/or emergency response plans of local, state, and federal agencies.	III.3.e
(C) The emergency response plan shall be rehearsed regularly as part of the overall training program for site operations.	
(D) The site emergency response plan shall be reviewed periodically and, as necessary, be amended to keep it current with new or changing site conditions or information.	
(E) An employee alarm system shall be installed in accordance with 29 CFR 1910.165.	
(F) On the basis of the information available at the time of the emergency, the employer shall evaluate the incident and the site response capabilities and proceed with the appropriate steps to implement the site emergency response plan.	II.2.d; II.2.e; III.3.d.(1)
1910.120(q) Emergency response to hazardous substance releases:	
(1) Emergency response plan	III.3.1
(2) Elements of an emergency response plan	
(i) Preemergency planning and coordination with outside parties	I.4.f; II.2.b; II.2.c; III.2.b; III.2.c; III.3.b.(4); III.3.d
(ii) Personnel roles, lines of authority, training, and communication	I.4.f; II.2.b; III.2.b; III.2.c; III.3.b.(4); III.3.e.(4)
(iii) Emergency recognition and prevention	II.1; III.7
(iv) Safe distances and places of refuge	III.3.b.(3); III.3.d.(2)
(v) Site security and control	III.3.d.(2); III.3.e.(2)
(vi) Evacuation routes and procedures	II.2.d; III.3.b.(3)
(vii) Decontamination procedures	III.3.c.(6)
(viii) Emergency medical treatment and response procedures	II.2.d; III.3.c.(7); III.3.e.(1)
(ix) Emergency alerting and response procedures	II.2; II.2.a; II.2.f; II.4; III.2; III.2.a; III.2.b; III.2.c; III.3.d
(x) Critique of response and follow-up	II.3; III.4; III.4.a; III.6
(xi) PPE and emergency equipment	III.3.e.(6); III.3.f.(3); III.3.d.(2); III.3.e.(6); III.3.f.(3)
(xii) Emergency response plan coordination and integration	III.3.e; III.8
(3) Procedures for handling emergency response:	
(i) The senior emergency response official responding to an emergency shall become the individual in charge of a site-specific Incident Command System (ICS).	II.2.b; III.3; III.3.a; III.3.b; III.3.b.(1); III.3.b.(2); III.3.e.(3)

Regulatory reference	Appropriate ICP section
(ii) The individual in charge of the ICS shall identify, to the extent possible, all hazardous substances or conditions present and shall address as appropriate site analysis, use of engineering controls, maximum exposure limits, hazardous substance handling procedures, and use of any new technologies.	II.2.c; II.2.d; III.3.c.(3)
(iii) Implementation of appropriate emergency operations and use of PPE	II.2.c; II.2.d; II.2.e; III.3.c; III.3.c.(1); III.3.d.(1); III.3.d.(2)
(iv) Employees engaged in emergency response and exposed to hazardous substances presenting an inhalation hazard or potential inhalation hazard shall wear positive pressure self-contained breathing apparatus while engaged in emergency response	II.2.d
(v) The individual in charge of the ICS shall limit the number of emergency response personnel at the emergency site, in those areas of potential or actual exposure to incident or site hazards, to those who are actively performing emergency operations	III.3.c; III.3.e.(5)
(vi) Backup personnel shall stand by with equipment ready to provide assistance or rescue	II.2.d; III.3.e.(5)
(vii) The individual in charge of the ICS shall designate a safety official, who is knowledgeable in the operations being implemented at the emergency response site.	II.2.d; III.3.b.(3)
(viii) When activities are judged by the safety official to be an IDLH condition and/or to involve an imminent danger condition, the safety official shall have authority to alter, suspend, or terminate those activities	III.3.b.(3)
(ix) After emergency operations have terminated, the individual in charge of the ICS shall implement appropriate decontamination procedures	III.3.c.(6)
(x) When deemed necessary for meeting the tasks at hand, approved self-contained compressed air breathing apparatus may be used with approved cylinders from other approved self-contained compressed air breathing apparatus provided that such cylinders are of the same capacity and pressure rating	
(4) Skilled support personnel	
(5) Specialist employees	
(6) Training	III.5
(7) Trainers	
(8) Refresher training	
(9) Medical surveillance and consultation	
(10) Chemical protective clothing	
(11) Postemergency response operations	

TABLE 16.7 EPA's Risk Management Program (40 CFR Part 68)

Regulatory reference	Appropriate ICP section
68.20-36 Offsite consequence analysis	III.3.d.(1)
68.42 Five-year accident history	III.4.b
68.50 Hazard review	III.3.d.(1)
68.60 Incident investigation	III.4.a
68.67 Process hazards analysis	III.3.d.(1)
68.81 Incident investigation	III.4.a
68.95(a) Elements of an emergency response program:	
(1) Elements of an emergency response plan:	
(i) Procedures for informing the public and emergency response agencies about accidental releases	II.2.a; III.2
(ii) Documentation of proper first-aid and emergency medical treatment necessary to treat accidental human exposures	III.3.c.(7); III.3.e.(1)
(iii) Procedures and measures for emergency response after an accidental release of a regulated substance	II.1; II.2; II.3; II.4; III.3.a–c
(2) Procedures for the use of emergency response equipment and for its inspection, testing, and maintenance	III.3.e.(6)
(3) Training for all employees in relevant procedures	III.5
(4) Procedures to review and update the emergency response plan	III.6
68.95(b) Compliance with other federal contingency plan regulations	
68.95(c) Coordination with the community emergency response plan	

CHAPTER 17
COORDINATING YOUR STATE REQUIREMENTS

Gail H. Allyn
Attorney at Law
Pitney, Hardin, Kipp & Szuch

Donald F. Elias
Principal
RTP Environmental Associates, Inc.

Brian L. Lubbert
Project Manager
RTP Environmental Associates, Inc.

Richard B. Booth
Associate
RTP Environmental Associates, Inc.

BACKGROUND

Many of the permitting requirements inherent in federal environmental legislation and regulations are implemented at the state level. The United States has a rich history of local rule, and states rule is one of the most cherished rights. However, owing to economic realities in the global competitive environment, states normally do not impose environmental permitting requirements which are more restrictive and more expensive to comply with than those required in federal law. The instances in which states do impose restrictions that go beyond federal requirements generally involve regulatory programs which have been developed at state levels before national programs existed.

In order to provide an example of how states interpret and implement federal and environmental permitting requirements, this chapter describes in detail New Jersey's environmental permitting program and regulations behind the permitting programs and also summarizes key aspects of California's environmental permitting requirements.

It is important for the reader to recognize that since state environmental permitting requirements vary from state to state, the forms used and the precise nature of information required vary, even though the underlining requirements generally parallel very closely, if not exactly, the federal requirements. It is important to check with the appropriate state agency and to obtain permit applications and permit guidance that is up to date as of the time the application is being considered.

It is also important to note that in some instances, counties and even municipalities have their own environmental permitting requirements. Therefore, it is important to be knowledgeable about environmental permitting requirements in your state and local area. This will require at a minimum telephone contact with state, county, and municipal officials to determine what environmental permitting requirements may pertain. For example, in many municipalities in New Jersey, application for a site plan approval or for subdivision approval carries with it a requirement to submit an environmental impact statement that follows the National Environmental Policy Act requirements. This, in and of itself, is not an environmental regulation or law; however, it is a component of land use regulations.

17.1 NEW JERSEY REQUIREMENTS

17.1.1 Introduction

New Jersey's environmental programs are administered by the New Jersey Department of Environmental Protection (NJDEP). Most of the programs are derived from similar federal programs, but often they contain more stringent requirements. This chapter describes the following programs and how they affect the regulated community:

- Water pollution control
- Wetlands
- Clean air
- Oil and hazardous substance spills
- Hazardous waste
- Underground storage tanks
- Industrial site recovery
- Right-to-know and pollution prevention
- Toxic Catastrophe Prevention Act

The chapter is structured to provide an overview of the applicable statutes and regulations, to identify the facilities that are regulated, and to give a brief summary of the enforcement authorities. Also provided are reporting numbers and hotlines, agency addresses, and sources for locating additional information.

New Jersey's programs are continually being updated and revised through legislative and regulatory amendments. Therefore, the information contained in this chapter should be used only as a general guide. When confronted with a particular environmental issue, the reader should consult the current version of all applicable statutes and regulations and/or an environmental attorney or professional. Table 17.1 summarizes New Jersey's environmental statutes and regulations.

17.1.2 Water Pollution Control

Overview. EPA has authorized the NJDEP to administer the federal clean water program under New Jersey's Water Pollution Control Act (WPCA) (N.J.S.A. 58:10A). Generally, NJDEP's responsibility entails designating surface water uses, establishing water quality criteria, and administering and enforcing a permit program to control discharges so the criteria can be attained and/or maintained. The state program exceeds federal requirements in several ways. For example, NJDEP has authority to regulate not only discharges to surface water but also discharges to groundwater [N.J.S.A. 58:10A-4(c); N.J.A.C. 7:14A-7.1 et seq.]. Also,

TABLE 17.1 General Overview of New Jersey Environmental Statutes and Regulations

Statute	Key regulations	General purpose
Air pollution control act N.J.S.A. 26:2C	N.J.A.C. 7:27, N.J.A.C. 27A	Regulates the emission of air pollutants into the atmosphere and establishes control and operating requirements
Brownfield and Contaminated Site Remediation Act N.J.S.A. 58:10B		Provides incentives and increased flexibility to encourage industrial site cleanups
Coastal Area Facility Review Act N.J.S.A. 13:19	N.J.A.C. 7:7, N.J.A.C. 7:7E	Regulates construction activities in coastal areas
Discharge Reporting Act N.J.S.A. 13:1K-16	N.J.A.C. 7:1E-5.2	Establishes requirements for certain industrial facilities to report discharges to the local community
Freshwater Wetlands Protection Act N.J.S.A. 13:9B	N.J.A.C. 7:7A	Protects inland freshwater wetlands from unnecessary or undesirable alteration or disturbance
Industrial Site Recovery Act N.J.S.A. 13:1K-6	N.J.A.C. 7:26B, N.J.A.C. 7:26C, N.J.A.C. 7:26E	Requires assessment and remediation of an industrial establishment upon transfer of property or cessation of operations
Pollution Prevention Act N.J.S.A. 13:1D-35	N.J.A.C. 7:1K	Establishes standards governing regeneration; transportation; and treatment, storage, and disposal of hazardous waste
Spill Compensation and Control Act N.J.S.A. 58:10-23.11	N.J.A.C. 7:1E, N.J.A.C. 7:1J	Compels prevention planning for and reporting and cleanup of discharges of hazardous substances
Solid Waste Management Act N.J.S.A. 13:1E	N.J.A.C. 7:26G, N.J.A.C. 7:26H, N.J.A.C. 7:26I	Regulates generation; transportation; and treatment, storage, and disposal of hazardous waste
Toxic Catastrophe Prevention Act N.J.S.A. 13K-19	N.J.A.C. 7:31	Requires facilities that generate, store, handle, or transport extraordinarily hazardous substances to conduct accident risk assessments and devise and implement risk management and reduction programs
Underground Storage of Hazardous Substances Act N.J.S.A. 58:10A-21	N.J.A.C. 7:14B	Regulates USTs which hold petroleum and hazardous substances

TABLE 17.1   General Overview of New Jersey Environmental Statutes and
Regulations (*Continued*)

Statute	Key regulations	General purpose
Water Pollution Control Act N.J.S.A. 58:10A	N.J.A.C. 7:9, N.J.A.C. 7:9B, N.J.A.C. 7:14, N.J.A.C. 7:14A	Regulates discharges into ground and surface waters of the state under a permit program
Wetlands Act of 1970 N.J.S.A. 13:9A	N.J.A.C. 7:7, N.J.A.C. 7:7E	Regulates construction activity in coastal and tidal wetlands
Worker and Community Right-to-Know Act N.J.S.A. 34:5A	N.J.A.C. 7:1G	Requires disclosure of information regarding hazardous chemicals used or stored at industrial facilities

New Jersey has additional guidelines for toxic effluents and its own minimum treatment requirements for all discharges.

Under the WPCA, the regulated community is required to obtain and comply with a New Jersey Pollutant Discharge Elimination System (NJPDES) permit. The NJPDES permit allows for access on request for inspection and sampling by the permitting agency. The owner or operator must maintain records, monitor effluents, and properly operate and maintain all treatment and control systems (see generally N.J.S.A. 58:10A-6; N.J.A.C. 7:14A).

A facility is required to monitor its discharges and to report the results on a discharge monitoring report (DMR) and/or a baseline report (BR) form (see N.J.S.A. 58:10A-6.f; N.J.A.C. 7:14A-6.8). Monitoring requirements, including parameters and frequency, are included in the facility permit and preprinted DMRs, which NJDEP scans electronically (N.J.A.C. 7:14A-6.8). The samples must be representative of the facility activity, collected in accordance with NJDEP protocols, and analyzed by an NJDEP-certified laboratory. Records of monitoring activities must be maintained for 5 years (see N.J.A.C. 7:14A-6.6).

Additionally, within 2 hours of a discharge that could constitute a threat to human health or the environment, the permit holder must notify NJDEP at (609) 292-7172 of the type of discharge, any investigative activities undertaken, and the measures taken to reduce or eliminate the unpermitted discharge (N.J.A.C. 7:14A-6.10). Within 24 hours of the unpermitted discharge, the permit holder must further notify NJDEP of the duration and source of the noncompliance and measures taken to reduce, eliminate, and prevent a recurrence. A written follow-up report must be submitted within 5 days of the occurrence in accordance with the requirements of N.J.A.C. 7:14A-6.10. The report should be submitted to New Jersey Department of Environmental Protection, Assistant Director of Water and Hazardous Waste Enforcement, Division of Enforcement Field Operations, CN-422, 401 East State Street, Trenton, NJ 08625.

17.1.2.1 General NJPDES Permit Conditions. Dischargers of pollutants to surface and groundwater of the state must acquire a NJPDES permit from NJDEP. As provided by N.J.A.C. 7:14A-2.4(b), permits are required for:

- Discharges to surface water and groundwater
- Discharge from an indirect user
- Land application of municipal wastewaters and industrial wastewaters
- Discharge from solid waste management facilities
- Storage of any liquid or solid pollutant
- Discharge of pollutants into wells
- Discharges from concentrated animal feeding operations
- Discharges from concentrated aquatic animal production facilities
- Discharges from aquaculture projects
- Discharges from silvicultural point sources
- Discharges of stormwater to surface waters
- Discharges from site remediation projects
- Treatment, storage, or disposal of hazardous waste not otherwise regulated
- Certain treatment works

Permit conditions are imposed to comply with federal and state standards (N.J.A.C. 7:14A-2.3). Typically a permit will contain applicable effluent limitations, schedules for compliance, and monitoring and reporting requirements (N.J.A.C. 7:14A-6).

An effluent limitation is a restriction on the amount of a specific pollutant allowed to be discharged from a point source (N.J.S.A. 58:10A-3.f). Effluent limitations are based on a two-tiered approach. First, technology-based limitations are applied, the stringency of which depends on the type of pollutant discharged. Second, water quality–based limitations may be required if the technology-based limitations do not sufficiently protect the quality of the receiving water body. Technology-based effluent limitations are contained in N.J.A.C. 7:14A-13.4. Surface water quality standards and the protocol for determining appropriate water quality–based effluent limitations are described in N.J.A.C. 7:14A-13.6.

In certain circumstances, variances from effluent limitations may be obtained. Conditions and procedures for obtaining a variance are provided in N.J.A.C. 7:14A-11.7 through 7:14A-11.9. Also, note that more stringent limitations are required for certain new and modified sources [New Source Performance Standards (NSPS)]. Generally, modifications to an existing source will not be subject to these more stringent limitations unless the modified discharge is constructed at a site where no other source is located, totally replaces existing equipment causing the discharge, or creates a source substantially independent of the existing source.

If a point source requires a permit, NJDEP may be contacted to arrange for a preapplication conference. Following the conference, an application must be filed with NJDEP Division of Water Resources at least 180 days prior to discharge. Permit applications are available from the NJDEP Bureau of Permit Management.

After review of a complete permit application, NJDEP prepares a draft permit. The public is notified of the NJDEP action and is given an opportunity to request a public hearing. NJDEP responds to comments raised during the review process and issues a response summary upon issuance of its final permit decision. The applicant, or certain other interested parties, may contest the decision of the agency within 30 days by filing a permit appeal. Permit fees are imposed for every permit to provide adequate funding to administer and enforce the program. All application records and data must be maintained for 5 years. Application procedures are provided in N.J.A.C. 7:14A-4.

Permits are granted for a fixed term of up to 5 years, although the permit may be reopened and modified based on certain statutory, regulatory, or permit conditions (N.J.S.A. 58:10A-7). The permit may also be transferred upon meeting certain conditions, including notice to NJDEP (N.J.A.C. 7:14A-16.2). To obtain a renewal permit, a renewal application must be filed at least 180 days prior to the permit's expiration date. The expired permit's conditions will continue in full force until a new permit is issued if a timely and complete application has been filed. A permit may be terminated or a renewal application denied, based on the reasons delineated in N.J.A.C. 7:14A-16.6. These reasons include: (1) noncompliance; (2) failure to pay fees; (3) misrepresentation of or failure to fully disclose relevant facts; or (4) a finding that the discharge endangers human health or the environment to the extent that only permit revocation will adequately address the concerns, does not conform to basin or areawide plans, or is inconsistent with any of the applicable regulations or laws.

17.1.2.2   Discharges to Surface Water.   A discharge requires a permit if it results in a discharge of a pollutant from a point source into the surface water of the state. A point source is defined at N.J.S.A 58:10A-3m as:

> any discernible, confined, and discrete conveyance, including but not limited to any pipe, ditch, channel, tunnel, conduit, well, discrete fissure, container, rolling stock, concentrated animal feeding operation, or vessel or other floating craft, from which pollutants are or may be discharged.

The discharge to surface water (DSW) program also includes regulation of stormwater discharges associated with industrial activity from point or nonpoint sources (N.J.A.C. 7:14A-11.1). DSW permit conditions are established pursuant to N.J.A.C. 7:14A-11.2, and generally require monitoring of the mass for each pollutant and the volume of effluent discharged from each outfall. Stormwater discharge requirements are covered in more detail in N.J.A.C. 7:14A-11.5. A discharger may request a variance from effluent limitations by filing a request under N.J.A.C. 7:14A-11.6.

17.1.2.3 Discharges to Groundwater. NJDEP regulates all discharges of pollutants directly to groundwater or onto land from which a pollutant may flow or drain into groundwater under the discharge to groundwater (DGW) permit program (N.J.A.C. 7:14A-7.3). Regulated facilities include surface impoundments, lagoons, and land disposal of dredge material. The discharge to groundwater permit requires development and implementation of a groundwater protection program (GWPP), and sets allowable groundwater concentration limits for hazardous and nonhazardous pollutants based on New Jersey's Groundwater Quality Standards (N.J.A.C. 7:9-7.6). Operational permits generally require installation, operation, and sampling of groundwater monitoring wells to determine if the permittee's operations are contaminating groundwater (N.J.A.C. 7:14A-7.6).

The Underground Injection Control (UIC) program regulates disposal or discharge of waste through underground injection, and storage of fluids or gases in underground geologic formations [N.J.A.C. 7:14A-8.1(b)]. In general, a NJPDES permit is required for a well or structure when (1) it is deeper than its widest surface opening, and (2) emplacement of fluids is its principal function. Five classes of underground injection wells are regulated under N.J.A.C. 7:14A-8.2.

UIC regulations prohibit the movement of fluids from any injection well into underground sources of drinking water. If detected, corrective action may be undertaken. Certain classes of injection wells require individual UIC permits that contain conditions, including reporting requirements, and standards for construction, operation, and monitoring of the well. Underground injection of hazardous and radioactive wastes is prohibited, but NJDEP may permit reinjection of treated groundwater in connection with remedial activities (N.J.A.C. 7:14A-8.7).

17.1.2.4 Discharges to Publicly Owned Treatment Works (POTWs).
Dischargers of nondomestic (i.e., nonsanitary) pollutants to POTWs are required to meet pretreatment standards prior to discharging to the treatment works (N.J.A.C. 7:14A-21). Pretreatment is required to avoid bypass of a pollutant through the POTW (in excess of its permit limits) or interference with the POTW's waste treatment processes. NJDEP has adopted, at N.J.A.C. 7:14A-21.2, the EPA pretreatment standards contained in 40 CFR 403, consisting of (1) categorical standards at 40 CFR 403.6, (2) prohibited discharges (general and specific) at 40 CFR 403.5, and other requirements. Variances from pretreatment standards may be obtained by following the procedures set forth in N.J.A.C. 7:14A-21.5.

The categorical, or industry-wide, standards are set by numeric limitations for pollutants associated with a particular industry (40 CFR 403.6). Prohibited discharge standards apply to all discharges, regardless of industry category (40 CFR 403.5). The general prohibitions apply to the discharge of any pollutant in quantities that cause a POTW to exceed its discharge permit limit for toxic substances or cause the sludge to exceed criteria for its disposal [40 CFR 403.5(a)]. The specific prohibitions consist of numerical and narrative limits relating to (1) pollutants that create a fire or explosion hazard; (2) pollutants that cause corrosive structural damage to the POTW; (3) solid or viscous pollutants that might obstruct the flow in the POTW and cause an interference; (4) any pollutant released at a flow rate or concentration that causes an interference; and (5) heat that inhibits biological activity in the POTW, causing an interference [40 CFR 403.5(b)].

NJDEP has delegated authority to certain POTWs to operate their own pretreatment programs and to issue permits to industrial users [N.J.A.C. 7:14A-19.3(c)]. These POTWs have authority to (1) establish terms and conditions, including effluent limits, upon their industrial users (i.e., local limits); (2) investigate to ensure compliance; and (3) impose remedies, fines, and penalties on violators. Generally, these local standards are more stringent than the federal requirements. Industries discharging to POTWs without approved pretreatment programs are required to obtain NJDEP authorization. A significant indirect user (SIU) must obtain an individual permit. An SIU is a source that discharges significant quantities of effluent to the treatment works or discharges toxic or hazardous substances (N.J.A.C. 7:14A-1.2). Application for and conditions applicable to SIU permits are contained in N.J.A.C. 7:14A-21.

17.1.2.5   Other Agency Powers.   The WPCA authorizes NJDEP to promulgate and enforce regulations to carry out the act. The Department is further empowered to exercise general supervision of the administration and enforcement of the act and attendant regulations and orders, assess discharger compliance, issue certification for federal licenses or permits granted under the CWA, work with other affected agencies to implement the WPCA, and administer federal and state grants to political subdivisions (see N.J.S.A. 58:10A-4 and 5).

The WPCA also grants to NJDEP and local agencies a right of entry for purposes of inspection, sampling, copying, or photographing on all premises on which a point source is located or on which monitoring equipment or records are kept (N.J.S.A. 58:10A-6.g). Each permitted facility or POTW, other than those discharging only stormwater or noncontact cooling water, must be inspected by NJDEP at least once a year. Inspections must be performed within 6 months of permit application, renewal, or issuance. Facilities discharging into a POTW will be inspected by the local agency having jurisdiction at least once per year. A significant noncomplier will be subject to inspection within 60 days of receipt of the DMR that resulted in the source being identified as a significant noncomplier. Inspections may include representative sampling of the permitted discharge, evaluating maintenance records for treatment equipment, evaluating the permittee's sampling techniques, random checking of laboratory test results from the previous 12 months, and inspecting sample storage facilities (see generally N.J.S.A. 58:10A-6).

Finally, the WPCA authorizes NJDEP to assess civil administrative penalties for violations of the act, as well as criminal penalties for negligent or intentional violations of the act (N.J.S.A. 58:10A-10).

17.2 WETLANDS

Overview

The NJDEP has authority to administer the federal wetlands program in certain designated areas. It does so pursuant to the Wetlands Act of 1970 (N.J.S.A. 13:9A) and the Freshwater Wetlands Protection Act (N.J.S.A. 13:9B), which together govern activity in freshwater, tidal, and coastal wetlands. Regulated activities in wetland areas require an individual permit or compliance with the conditions of a general permit. Permit conditions may contain record-keeping, monitoring, and reporting requirements. New Jersey's freshwater wetlands program regulates more wetlands and uses than does the federal program. The wetlands are divided into three resource value classifications: ordinary, intermediate, or exceptional (N.J.S.A.13:9B-7). The classification determines the allowable type and extent of development and required mitigation.

Tidal wetlands are also regulated under the Coastal Area Facility Review Act (CAFRA) (N.J.S.A. 13:19). The Wetlands Act of 1970 requires NJDEP to inventory and map all of the state's tidal wetlands. The maps are filed with the land offices of the counties in which the wetlands are located. NJDEP is authorized to regulate activities in the mapped wetlands by issuing, revising, or repealing orders. The orders form the basis for issuing permits (see generally N.J.S.A. 13:9A-1).

17.2.1 Coastal and Tidal Wetlands Permits

Coastal and tidal wetlands are regulated under the Wetlands Act of 1970 (N.J.S.A. 13:9A), CAFRA (N.J.S.A. 13:19), and the Waterfront Development Law (N.J.S.A. 12:5-3). Under the Wetlands Act, a permit is required for draining, dredging, excavation, or deposition of material, and erection of structures, driving of pilings, or placing of obstructions in any mapped coastal wetland, whether or not it changes the tidal ebb and flow (N.J.S.A.13:9A-4). CAFRA regulates construction of any building or structure in designated coastal areas (N.J.S.A. 13:19-5). The Waterfront Development Law regulates the filling, dredging, or placing of structures, pilings, or other obstructions in any tidal waterway or in certain upland areas adjacent to tidal waterways (N.J.S.A 12:5-3).

Permitting procedures for each of these programs are detailed in the Coastal Permit Program Rules, set forth at N.J.A.C. 7:7. A preapplication conference is an optional, but recommended, first step in the permitting process (N.J.A.C. 7:7). After the conference, the permit application should be completed. All applications require submission of a completed NJDEP Land Use Regulation Program standard application form (LURP-1), the appropriate fee, verifications of notice to local planning boards and environmental commissions, plans and photographs, environmental impact or compliance statement, and various public notices (N.J.A.C. 7:7-4.2). Specific additional submittals may be required for a complete application, depending on the particular project (N.J.A.C. 7:7-4.2). Mitigation will be required for any activity resulting in disturbance to or loss of wetlands. Type (i.e., restoration, creation, enhancement, or contribution), extent, and location of mitigation will be determined on a case-by-case basis.

The application is available to the public for review. Public hearings are held, as necessary, to comply with N.J.A.C. 7:7-4.5. In general, NJDEP acts on CAFRA applications within 60 days of the public hearing, and acts on all wetland and waterfront development applications within 90 days after the application was declared complete for final review. Once issued, a permittee may request modification in accordance with N.J.A.C. 7:7-4.10, or the permit may be suspended or revoked as provided by N.J.A.C. 7:7-4.11.

17.2.2 Freshwater Wetlands Permits

New Jersey regulates inland freshwater wetlands and adjacent transition areas under the Freshwater Wetlands Protection Act N.J.S.A. 13:9B. Permitting procedures are set forth in N.J.A.C. 7:7A. A property owner may request from NJDEP a letter of interpretation (LOI) setting forth NJDEP's determination of whether the property is a wetland and therefore needs a permit. NJDEP's determination can be modified or revoked by EPA. Activities requiring a permit include those which remove, excavate, disturb, or dredge soil, sand, gravel, or aggregate material; drain or disturb the water level or water table; dump, discharge, or fill with any material; drive pilings; place obstructions; and/or destroy plant life. Most normal farming, silvicultural, and ranching activities are exempt from the regulations. A complete listing of exempted activities and areas is provided in N.J.A.C. 7:7A-2.7 and 7:7A-2.8.

Activities may be conducted pursuant to a general or individual permit, depending on the activity. General permit activities are enumerated in N.J.A.C. 7:7A-9.2. Those planning to use a general permit must notify NJDEP, the county clerk, and the clerk of the municipality where the wetland is located, in writing by certified mail at least 30 days prior to the start of work. The notice must contain a description of the project, a description of the location of the activity (including the county, municipality, and lot and block numbers), and a site plan showing existing structures, wetland boundaries, and proposed structures and/or activities. Within 30 days from receipt of notice, NJDEP will advise whether the activity is covered by the general permit. For a project requiring an individual permit, the applicant must submit a permit application in accordance with N.J.A.C.7:7A-11.1 to the following address: New Jersey Department of Environmental Protection, Land Use Regulation Element, Bureau of

Regulation, CN 401, 5 Station Plaza, Trenton, N.J. 08625, Attn: (insert county where wetland is located) Section Chief.

A preapplication conference is encouraged (N.J.A.C. 7:7A-10.1). Permits are effective for a fixed term not to exceed 5 years. All permits contain the conditions enumerated in N.J.A.C. 7:7A-13.1. Mitigation is a required condition in all individual permits. Permits may be transferred, modified, or revoked as detailed in the regulations.

17.2.3 Other Agency Powers

The Wetlands Act of 1970 empowers the commissioner of NJDEP to adopt, modify, or repeal orders regarding the altering or polluting of coastal wetlands, and to assess penalties for violation (N.J.S.A. 13:9A-2, 9). The orders form the basis for the issuance of permits. The act also requires NJDEP to inventory and map state tidal wetlands (N.J.S.A. 13:9A-1). Regulations issued pursuant to the Wetlands Act of 1970 require that inspection authority is a condition of permits (N.J.A.C. 7:7-1.5).

The Freshwater Wetlands Protection Act authorizes NJDEP to promulgate necessary rules and regulations to implement the act and to assess penalties and compel compliance by administrative, civil, or criminal action (N.J.S.A. 13:9B-21). Regulations issued pursuant to the Freshwater Wetlands Protection Act require that inspection authority (including sampling) and reporting and record-keeping requirements are conditions of all individual and general permits [N.J.A.C. 7:7A-13.1(a)].

17.3 NEW JERSEY'S STREAM ENCROACHMENT REGULATIONS

Construction activities and other man-induced land disturbance within New Jersey's floodplains are regulated by New Jersey's flood hazard area control regulations (N.J.A.C. 7:13-1.1 et seq.), more commonly known as New Jersey's stream encroachment regulations. These regulations meet the National Flood Insurance Program's minimum standards and in some ways exceed those minimum standards.

17.3.1 Purpose and Scope

New Jersey's stream encroachment rules recognize that without proper controls, development in watercourses and their floodplains may do the following:

• Adversely affect the flood-carrying capacity of streams and floodplains

• Subject new facilities to flooding

• Reduce natural flood storage that the floodplain provides

- Degrade the water quality of the receiving water body
- Result in increased sedimentation, erosion, or other environmental damage

The stream encroachment rules were enacted to control development in floodplains in order to avoid or mitigate the above detrimental effects of development on the environment and to ensure the safety, health, and general welfare of the people of the state. Specifically, the rules are intended to:

- Minimize potential on- and off-site damage to public or private property caused by development which, at times of flooding, subjects structures to flooding and increases flood heights and/or velocities both upstream and downstream
- Safeguard the public from the dangers and damages caused by materials being swept onto nearby or downstream lands
- Protect and enhance the public's health and welfare by minimizing the degradation of water quality from point and non-point-source pollution
- Protect wildlife and fisheries by preserving and enhancing water quality and the environment associated with the floodplain and the watercourses that create them

17.3.2 Applicability

New Jersey's stream encroachment rules apply to all construction activity or other manmade land disturbance that takes place within the larger of the following areas, unless specifically exempted:

1. Watercourses, which include rivers, perennial and intermittent streams, and any other surface water runoff conveyance channel with defined bed and banks
2. Regulatory floodplains, which include:
   a. Floodplains resulting from New Jersey's "flood hazard area design flood." These are areas delineated and formally adopted by the New Jersey Department of Environmental Protection (NJDEP) that are subject to flooding during floods with peak flow rates that are 25 percent greater than 100-year peak flow rates.
   b. 100-year floodplains along watercourses that have not had flood hazard areas delineated and formally adopted by NJDEP.
3. Areas adjacent to and within 25 feet of the top of channel banks
4. Areas adjacent to and within 50 feet of the top of channel banks along watercourses that:
   a. Contain deposits of acid-producing soils
   b. Are classified as Category One, FW-1 trout-associated, or FW-2 trout-associated
   c. Are a critical part of the habitat supporting a threatened or endangered species of plant or a current population of any species of threatened or

endangered animal on a permanent or temporary basis, for any purpose such as resting, breeding, or feeding during any portion of its life cycle

d. Are located within documented historic habitat for threatened or endangered species of animals, which habitat remains suitable for breeding, resting, or feeding by those species of animal during any portion of its life cycle

Tidal water bodies, such as the Atlantic Ocean, bays, tidal canals, coves, guts, harbors, inlets, sounds, etc., are not regulated by New Jersey's stream encroachment rules, nor are the lower reaches of specifically named watercourses that flow into tidal water bodies and which are subject to the same tidal flooding characteristics as the tidal water body itself.

Tidally influenced areas, subject to both tidal flooding and flooding caused by the tidal wave traveling up a watercourse, are subject to the stream encroachment rule's "floodway" regulations but not the "flood fringe" regulations. Floodways include the watercourse's channel and portions of the floodplain adjoining the channel that are needed to carry and discharge the flood waters. Flood fringe areas include those portions of the regulatory floodplain that lie outside of the floodway area.

Nonregulated Uses in Floodways.   Certain construction activities and manmade land disturbances are not regulated under New Jersey's stream encroachment rules. These nonregulated uses generally include those which:

- Do not further obstruct flood flow or reduce the cross-sectional area of the floodway
- Do not require the erection of structures
- Do not require channel modifications or relocations
- Do not alter the cross-sectional area of a water-control structure such as a bridge, culvert, or dam that is open to flood waters during the regulatory flood
- Do not increase off-site flood damage potential by raising flood elevations off of the property on which the use is proposed by more than 0.2 foot
- Do not adversely affect floodplain areas and areas proximate to channel banks, Category One FW-1 trout-associated waterways, FW-2 trout-associated waterways, and threatened and endangered species habitats
- Do not cause or contribute to a violation of any applicable state water quality standard or otherwise adversely affect water quality
- Are undertaken with the landowner's express written permission

Specific examples of typical nonregulated uses are provided in the stream encroachment rules.

Nonregulated Uses in Flood Fringe Areas.   Nonregulated uses in flood fringe areas generally include uses which:

- Do not further reduce the volume of flood storage available
- Do not require any hydrologic or hydraulic calculations to determine impact
- Do not adversely affect floodplain areas and areas proximate to channel banks, Category One FW-1 trout-associated waterways, FW-2 trout-associated waterways, and threatened and endangered species habitats
- Do not cause or contribute to a violation of any applicable state water quality standard or otherwise adversely affect water quality
- Are undertaken with the landowner's express written permission

Specific examples of typical nonregulated uses are provided in the stream encroachment rules.

Applicability of Engineering and Environmental Standards. New Jersey's stream encroachment rules include both engineering standards and environmental standards. The engineering standards apply only along watercourses that have a total contributory drainage area greater than 50 acres. The environmental standards apply along all watercourses, regardless of drainage area, except along manmade, but not man-altered, watercourses with a total contributory drainage area less than 50 acres.

17.3.3   Minimum Project Standards

The Regulatory Flood. The regulatory flood for "delineated" watercourses is NJDEP's flood hazard area design flood which represents the 100-year flood flow, increased by 25 percent to allow for future development in the drainage basin. In addition, delineated watercourses also generally include a delineated floodway, based on the 100-year flood. For delineated streams, flood profiles, floodway and flood hazard area mapping, and supporting computer input and output, printouts are generally available from NJDEP's Bureau of Floodplain Management, for use by applicants.

In the event a watercourse has not been delineated by the state, the watercourse is considered to be "nondelineated" and the rules require that applicants establish the boundaries of 100-year floodplains and floodways by a standard step backwater analysis. The 100-year flood flows utilized must be based on the assumption that the entire contributory drainage area is fully developed in accordance with the current zoning plan, to the maximum impervious cover allowed, and in accordance with applicable stormwater management regulations.

Prohibited Uses in Flood Fringe Areas and Floodways. Prohibited uses in flood fringe areas include the disposal or storage of pesticides, industrial, hazardous or solid wastes, radioactive materials, petroleum products, or other hazardous materials.

The uses generally prohibited in floodways include the following:

1. The addition of any fill, new structures, or fences which would raise the existing grade of the receiving area and/or create an obstruction to flow, except as specifically provided in the rules
2. The addition of any solid or hazardous waste or pollutant
3. The discharge, processing, storage, or disposal of pesticides, domestic or industrial waters, radioactive materials, petroleum products, or other hazardous materials, except as specifically authorized by law and pursuant to permits, licenses, and grants from all authorities with jurisdiction over such activities
4. The storage of materials or equipment
5. The construction of individual subsurface sewage disposal systems
6. The construction of off-channel detention/retention basins

Specific exceptions to the prohibited uses listed in item 1 above are provided in the stream encroachment rules, as are the conditions governing the granting of those exceptions. These exceptions include land uses, structures, and sanitary landfills existing or constructed prior to Mar. 20, 1995. The reader is referred to N.J.A.C. 7:13-2.2(b) for the specific conditions applicable to these exceptions.

Regulated Uses.   All uses that are not prohibited are regulated and NJDEP's stream encroachment regulations provide specific engineering and/or environmental standards for various types of construction and land disturbance activities. The construction and land disturbance activities for which specific standards are provided include the following:

• Watercourse cleaning
• Excavation
• Disposal of soils
• Stormwater management
• Channel modification
• Underground utilities in floodplains
• Aboveground utilities in floodplains
• Dams
• Structures
• Fill within the floodplain
• Fill within the central Passaic basin
• Bridges and culverts
• Subsurface sewage disposal systems within flood fringe areas

The reader is referred to N.J.A.C. 7:13-2—Project Standards, for the specific standards applicable to each regulated use.

The project standards provided in NJDEP's stream encroachment regulations meet the minimum standards required for participation in the National Flood Insurance Program (NFIP). In many respects, however, NJDEP's standards significantly exceed the national standards. Some of the more significant differences include the following:

1. *Regulatory flood.* New Jersey's regulatory flood assumes flood flows based on ultimate development within the drainage basin rather than on the existing development conditions utilized by the NFIP.

Ultimate development conditions were approximated for New Jersey's "delineated" streams by increasing the 100-year peak flows by 25 percent and utilizing the increased flows to generate the regulatory flood profiles. Ultimate development conditions for "nondelineated" streams assumes maximum impervious cover within the contributing watershed, in accordance with local land development regulations, including consideration of any applicable stormwater management regulations.

2. *Regulatory floodway limits.* A regulatory floodway is a channel of a watercourse and the adjacent land areas that must be reserved in order to discharge the 100-year flood without cumulatively increasing the water surface elevation more than a designated amount. Floodway limits in New Jersey are based on a 0.2-foot permissible rise in the 100-year flood elevation rather than the 1.0-foot rise permitted under the NFIP. As a result New Jersey's floodways are wider than required by the NFIP.

3. *Standards for fill within the flood fringe area.* The NFIP's standards permit the placement of unlimited amounts of fill material within flood fringe areas. New Jersey's stream encroachment rules limit the amount of fill material placed within flood fringe areas to 20 percent of the total volume of flood storage that was available within the flood fringe area of the flooding source on the property under consideration as of Jan. 31, 1980. In addition, within New Jersey's "central Passaic basin," fill must also be limited such that "zero net fill" results within the entire central Passaic basin. In order to accomplish this, projects in this basin must create a volume of flood storage elsewhere within the basin equal in volume to the amount of fill proposed for the project.

4. *Requirements for structures.* Both the NFIP and New Jersey's stream encroachment regulations require that residential structures be elevated so that the lowest floor, including basement, is at or above the regulatory flood elevation. However, New Jersey's regulatory flood is higher than the NFIP's regulatory flood. In addition, New Jersey's rules require that residential developments or subdivisions that create more than one residence must, where feasible, have at least one driveway or access route at or above the regulatory flood elevation. If a dry route is not feasible, then all on-site roads, parking areas, and driveways must be constructed at or above the regulatory flood elevation to the extent possible.

For commercial and industrial structures, New Jersey's regulations require that, like residential structures, these structures must be elevated so that the lowest floor, including basement, is at or above the regulatory flood elevation. NJDEP may exempt a specific structure if it so chooses, upon written application by the owner or builder, providing evidence to show that the raising of the structure would be economically or physically impracticable and that, instead, the structure would be floodproofed. Under the NFIP, floodproofing structures is permitted for all nonresidential structures.

New Jersey's stream encroachment rules also provide that hospitals, clinics, nursing homes, schools, day care centers, hotels, private residences, and similar buildings be elevated so that the lowest floor, including basement, is at or above the regulatory flood elevation. In addition, such developments must have at least one driveway or access route elevated to or above the regulatory flood elevation. These requirements also exceed those of the NFIP.

5. *Stormwater management.*   The NFIP has no specific requirements for stormwater management facilities or discharges within flood-prone areas. New Jersey's stream encroachment regulations, however, have comprehensive rules for stormwater management facilities that discharge into any areas that lie within NJDEP's jurisdiction under those rules.

6. *Environmental standards.*   The NFIP requirements focus on planning and engineering standards aimed at minimizing flood and loss of life. New Jersey's stream encroachment regulations also incorporate significant provisions aimed at preserving and protecting the environment including provisions aimed at:

• Minimizing the degradation of water quality from point and nonpoint sources of pollution

• Protecting wildlife and fisheries by preserving and enhancing water quality and the environment associated with watercourses and their floodplains

17.3.4   General Environmental Standards

New Jersey's stream encroachment rules provide environmental standards that are applicable along all watercourses except along manmade watercourses with contributing drainage areas less than 50 acres. The environmental standards require the following:

1. Applicants must describe all steps taken to minimize pollution, impairment, or destruction of the environment within regulated areas, during construction and operation of the project.

2. Applicants must describe short-term and long-term environmental impacts and the cumulative impacts of each upon the environment.

3. All projects must be designed in accordance with federal, state, and local statutes, regulations, and ordinances.

4. NJDEP will not approve any regulated activity which it determines is likely to significantly and adversely affect the biota of the watercourse or its water quality including, but not limited to, adverse effects on:
   a. Potable water supplies
   b. Flooding
   c. Drainage
   d. Channel stability
   e. Threatened and endangered species of plants and animals
   f. Current or documented historic habitats for threatened or endangered species
   g. Navigation
   h. Energy production
   i. Municipal, industrial, or agricultural water supplies
   j. Fisheries

Specific environmental standards are provided for each of the following:

- Protection of near watercourse vegetation
- Soil erosion and sediment control
- Projects along trout-associated watercourses
- Projects affecting other fish resources
- Projects exposing deposits of acid-producing soils
- Projects affecting freshwater wetlands
- Projects affecting threatened and endangered species

The reader is referred to N.J.A.C. 7:13-3—General Environmental Standards for the specific environmental standards applicable to the above activities.

As a condition of all approvals, permittees are required to take all actions necessary to minimize adverse environmental impacts to the receiving watercourse and other regulated areas. In addition, permittees must restore temporarily disturbed vegetation, habitats, and land and water features to their preconstruction condition and prevent sedimentation and erosion to the greatest extent possible.

17.3.5 Hardship Waivers

A waiver from strict compliance with the stream encroachment rules may be granted by NJDEP if:

1. NJDEP determines that there is no feasible and prudent alternative to the proposed project, including the "no-action" alternative, which would avoid or substantially reduce any anticipated adverse effects and where the waiver is consistent with the reasonable requirements of the public health, safety, and welfare

2. NJDEP determines that the costs of strict compliance are unreasonably high in relationship to the benefits achieved by strict compliance

3. NJDEP and the applicant agree to alternative requirements that, in NJDEP's judgment, provide better protection to the public health, safety, and welfare

A public hearing regarding the hardship waiver is required if requested by NJDEP or at least five members of the public.

In order for NJDEP to grant a hardship waiver, the applicant must demonstrate all of the following:

1. By reason of extraordinary situation or condition of the property, the strict enforcement of New Jersey's stream encroachment rules would result in exceptional and undue hardship upon the applicant.

2. The waiver will not substantially impair the appropriate use or development of adjacent property and will not pose a threat to the environment or public health, safety, and general welfare.

3. The applicant did not create the exceptional or undue hardship being claimed.

17.3.6   Permit Modification Procedures

Modifications to approved plans and permit conditions may be granted only if NJDEP approves the changes in writing. Except for adding watercourses to watercourse cleaning permits, only items already approved on the original permit may be modified.

NJDEP may not approve modifications to approved plans that would affect the hydraulic capacity of the watercourse, including changes affecting bridges and culverts along those watercourses. If such a modification is desired, a new application must be submitted, reviewed, and approved.

17.3.7   Appeal Procedure

A person who considers himself or herself aggrieved by a stream encroachment permit approval or denial may request a hearing to the NJDEP Office of Legal Affairs. Such requests must be made within 10 days of publication of notice of the decision in the *DEPE Bulletin,* or within 10 days of publication of notice of the decision by the permittee, whichever occurs first.

When NJDEP grants a request for a hearing concerning one of its decisions, the request for the hearing is referred to the Office of Administrative Law (OAL) for a fact-finding hearing if required. Then the NJDEP commissioner issues a final decision adopting, rejecting, or modifying the findings of fact and conclusions of law of the administrative law judge, within a specified time frame.

Pending appeal and the NJDEP commissioner's final decision on the appeal, a person may apply to the commissioner for a stay of the issuance of a permit by written request and for good cause shown in that request.

17.4 CLEAN AIR

Overview

The New Jersey Air Pollution Control Act (NJAPCA), N.J.S.A. 26:2C, regulates the emission of air pollutants into the atmosphere. The NJAPCA covers emissions of any air contaminants, which include solid particles, liquid particles, vapors, or gases. NJDEP is responsible for implementing NJAPCA regulations primarily through a state implementation plan (SIP). Pursuant to the federal Clean Air Act, 42 U.S.C. 7401, New Jersey's SIP must include National Ambient Air Quality Standards (NAAQS), which require attainment and maintenance of six criteria pollutants: ozone, carbon monoxide, particulate matter, lead, sulfur dioxide, and nitrogen oxide. The most critical compliance objectives for New Jersey are attainment of ozone and carbon monoxide standards.

NJDEP oversees and controls air contaminant emissions through its permitting program (N.J.A.C. 7:27-8). Any person who intends to construct, install, or alter any equipment or control apparatus must obtain a preconstruction permit [N.J.A.C. 7:27-8.3(a)], and operating certificate [N.J.A.C. 7:27-8.3(b)], which are issued as one document for new facilities [N.J.A.C. 7:27-8.7(a)]. An operating certificate or any renewal thereof is valid for a period of 5 years from the date of issuance, unless sooner revoked by order of NJDEP [N.J.S.A. 26:2C-9.2.(c)(7)].

NJDEP has implemented a separate permitting system that covers "major facilities," defined as any facility with the potential to emit enumerated air contaminants in amounts equal to or greater than threshold levels established at a per ton annual basis (N.J.A.C. 7:27-22.1). This permitting system requires major facilities to obtain facilitywide permits that incorporate all of the applicable air pollution requirements into one document. While facilities still need to maintain existing preconstruction permits and obtain new permits in accordance with N.J.A.C. 7:27-8, the new operating permit replaces existing operating certificates for affected facilities.

17.4.1 Requirements of the Clean Air Act

The NJAPCA and its implementing regulations establish a sophisticated web of air pollution control and operating requirements. In general, these requirements include:

- Record Keeping: Many records not only must be submitted to NJDEP but must be maintained and retained at the facility for a period of 5 years (e.g., N.J.A.C. 7:27-16.22, 19.6, 19.19, and 21.6). In general, records submitted to NJDEP are deemed to be admissions, so the regulated community must be conscious of the need for accuracy and completeness in all submissions.

- Reporting: In addition to the submissions required under permits, companies must give timely notice of air contaminant releases governed by N.J.S.A. 26:2C-19(e) and make timely reports of events that qualify for an affirmative defense, or the defense may be lost [N.J.S.A. 26:2C-19.2(b)].

- Certifications: Specific certifications are required for certain submissions to NJDEP, which often must be signed by a specified high-ranking official (e.g., N.J.A.C. 7:27-1.39).

- Testing: In addition to permit testing requirements, certain facilities have other testing obligations, such as minimum test frequency requirements under the volatile organic compounds (VOC) leak detection and repair program (e.g., N.J.A.C. 7:27-16.18).

- Inspections: NJDEP has the right to enter a facility to inspect equipment and records for compliance, and penalties may be assessed for failure to permit such an inspection as well as for other noncompliance (e.g., N.J.A.C. 7:27-23.7, 25.5, and 1.31).

- Deadlines: Many new regulatory requirements are being phased in over various periods, and different industries may have different deadlines for compliance. Companies must carefully review the new regulations to determine the deadlines applicable to their specific circumstances.

17.5   MAJOR SOURCES OF VOC and NO_x EMISSIONS

NJDEP also adopted rules for the control and prohibition of pollution from oxides of nitrogen (NO_x) and VOCs to reduce the formation of ground-level ozone due to equipment and source operation emissions of these air contaminants.

17.5.1 NO_x Rule

The rule regulating NO_x emissions (the NO_x rule) affects major NO_x facilities such as utilities, large industrial and commercial boilers, stationary gas turbines, stationary internal combustion engines, and other types of large combustion equipment (see N.J.A.C. 7:27-19). The NO_x rule requires any stationary source or group of sources that emits or has the potential to emit at least 25 tons of NO_x per year to implement reasonably available control technology (RACT) to control NO_x emissions [N.J.A.C. 7:27-19.2(a)]. RACT is defined as the lowest emission limitation that a particular source is capable of meeting by applying air pollution control technology that is reasonably available considering technological and economical feasibility. Various maximum emission limits have been set according to classification or source groups (N.J.A.C. 7:27-19.4–19.13).

Any facility regulated pursuant to the NO_x rule, but for which there are no designated emission limitations, must comply with "facility-specific" NO_x emission limits

(N.J.A.C. 7:27-19.13). In such a situation, a facility may seek written approval from NJDEP for a facility-specific NO_x control plan [N.J.A.C. 7:27-19.13(b)]. The plan will evaluate the available control technologies for technological and economic feasibility, and propose a NO_x emission limit for each affected item of equipment or source operation [N.J.A.C. 7:27-19.13(b), (d)]. Accordingly, depending on the type of facility, source operation, costs, and other factors, regulated facilities should have some flexibility to work out viable and cost-effective RACT alternatives with NJDEP.

A facility subject to the NO_x rule may apply to NJDEP for an exemption if the facility's potential to emit NO_x is less than 25 tons per year, and the facility's potential to emit NO_x on any calendar day from May 1 to Sept. 15 is less than 137 pounds per day [N.J.A.C. 7:27-19.2(f)]. NJDEP intends to propose a future amendment to the NO_x rule which exempts sources that burn natural gas as the primary fuel from the emission limits specified for burning liquid fuels, provided that (1) the liquid fuel is used only when natural gas is not available, and (2) the liquid fuel is not used for more than 500 hours per year.

Additionally, facilities which use an approved fuel switching plan to burn cleaner fuel are not restricted to the emission limits otherwise imposed on those types of facilities. Rather, a formula found at N.J.A.C. 7:27-19.20 is used to calculate the emission limits.

17.5.2 VOC Rule

NJDEP also has adopted rules regulating VOC emissions (the VOC rule) (N.J.A.C. 7:27-16.1 et seq.). Like the NO_x rule, the VOC rule applies to major facilities which emit VOCs and regulates categories or source groups of VOC emissions. The VOC rule provides a nearly identical procedure for determining facility-specific RACT requirements.

The VOC rule outlines definitive, categorical VOC regulations that set forth the following VOC and source operations: pipe coating operations, certain graphic arts source operations (including screen presses and ovens, as well as sheet-fed presses and ovens), fugitive emissions from chemical plants not covered by current N.J.A.C. 7:27-16.6, marine tank vessel operations transferring VOCs other than gasoline, and transfers of VOCs other than gasoline to or from delivery vessels. If a facility is subject to any of these general categories, it does not have to comply with the facility-specific requirements.

17.6 PERMITS

17.6.1 Subchapter 8 Program

A preconstruction permit must be obtained prior to construction, installation, or alteration of any equipment or control apparatus [N.J.A.C. 7:27-8.3(a)], and an operating certificate is required for the operation of such equipment [N.J.A.C. 7:27-8.3(b)]. For new facilities, the preconstruction permit and operating certificate are issued as the same document (N.J.A.C. 7:27-8.7). Application forms and information pertinent to applications may be requested from Bureau of New Source Review, Environmental Regulation Program, N.J. Department of Environmental Protection, 401 East State Street, 2nd floor, CN 027, Trenton, N.J. 08625-0027.

17.6.2 Permits for Minor Facilities in New Jersey

17.6.2.1 History. The first regulation that required permits for certain equipment operated in the state of New Jersey was enacted in 1967 under "Laws on Permits," Chapter 106, P.L. 1967, Title 26:26-9.2.

This law originally required permits for control apparatus, certain surface coating equipment, and all equipment handling 50 lb of material per hour. The law is considered to have become effective on June 15, 1967, and was a precursor to Chapter 9 titled "Permits."

Chapter 9, Permits, subsequently became Subchapter 8 (N.J.A.C. 7:27-8) on Jan. 15, 1968. At that time several additional sources required permits including metal cleaning and surface preparation equipment with emissions into the open air. In addition, storage tanks in excess of 10,000 gallons capacity used for storage of specific materials (acids, solvents, diluents, thinners, inks, colorants, lacquers, enamels, varnishes, and liquid resins) required permits. The Jan. 15, 1968, revisions also required permits for most incinerators, pneumatic material handling equipment, and solid fuel burning equipment with a heat input rate of 1,000,000 Btu per hour.

17.6.2.2 Grandfathered Status. The importance of permit required dates is significant in New Jersey, and any other state for that matter. If the equipment was constructed before and not modified after the "grandfathered date" the equipment does not require a permit in New Jersey under N.J.A.C. 7:27-8. In fact, the current codification specifically exempts such sources if they are still operational under N.J.A.C. 7:27-8.2(d).

The current regulations require "significant" sources to have "permits to construct and certificates to operate" for all new or modified equipment. Sources which are not required to have a permit are considered "insignificant sources." This wording is identical to Subchapter 22, which regulates major sources subject to Title V facility operating permits.

Subchapter 8 has been amended an additional five times: March 1973, June 1976, April 1985, March 1991, and May 1998. Permits are required for equipment only if the relevant equipment was installed or modified after the effective date of revision.

The most frequent example cited is a boiler having a heat input of 1,000,000 Btu per hour or greater using fuel oil or natural gas, which would require a permit if installed (or modified) after the March 1973 revisions. However, a boiler (as previously mentioned) which used solid fuel after June 1967 required a permit.

17.6.2.3 Current Regulations

17.6.2.4 Title V and Subchapter 8. Facilities subject to Title V operating permits are not completely exempt from the Subchapter 8 regulations. For example, for facilities subject to Title V requirements which have not been issued a "facility" operating permit, the source remains subject to Subchapter 8 and must obtain and maintain these permits. In addition, certain equipment at an operating permit facility not subject to operating permit requirements (Title V) remains subject to Subchapter 8.

17.6.2.5 What Needs a Permit Today? In 1998, sweeping revisions occurred to the New Jersey requirements. Permits are currently required for 19 different types of equipment in the state of New Jersey. They include:

- Fuel burning equipment with the maximum rated heat input of 1 million Btu per hour or greater.
- Equipment which emits toxic air contaminants, TXS for short, in quantities greater than 0.1 lb per hour. TXS is comprised of two groups in the state of New Jersey. As of June 12, 1998, Group 1 TXS includes benzene (benzol), carbon tetrachloride (tetrachloromethane), chloroform (trichloromethane), dioxane (1,4-diethylene dioxide; 1,4-dioxane), ethylenamine (aziridine), ethylene dibromide (1,-dibromoethane), ethylene dichloride (1,2-dichloroethane), 1,1,2,2-tetrachloroethane (sym tetrachloroethane), tetrachloroethylene (perchloroethylene), 1,1,2-trichloroethane (vinyl trichloride), and trichloroethylene (trichlorethene). As of June 12, 1998, Group 2 TXS includes methylene chloride (dichloromethane), 1,1,1-trichloroethane (methyl chloroform).
- All dry cleaning equipment.
- All surface cleaners with more than 5 percent VOC or HAPs in the cleaning solution, but only if the surface cleaner is an unheated open top surface cleaner with greater than 6 square feet of capacity or 100 gallons or the surface cleaner is heated and has an open top or is conveyorized, or is a stationary spray cleaning or surface stripping operation using $1/2$ gallon or more of cleaning solution in any 1 hour.

The $1/2$ gallon an hour portion of the regulations has recently been changed to be consistent with other cleaning activities in the regulations such as:

- Equipment used in graphic arts including newspapers, screen printing, etc., in which the quantity of ink fountain solution or cleaning material is greater than 1/2 gallon an hour.
- Any tank greater than 100 gallons which is used in etching, tickling, plating, or chromium electroplating or anodizing.
- Transfer operations involving gasoline or other VOCs which under other regulations are required to have a controlled device other than bottom fill or submerge fill.

Storage tanks and reservoirs are broken up into classifications depending on space and on which material they contain:

- The first includes tanks in excess of 10,000 gallons, storage of liquids except water or distills of air require a permit.
- The second includes 2000 gallons or greater if the material stored is a VOC with a partial pressure greater than 0.02 lb, which is 1.0 millimeters of mercury.
- The third category includes tanks or reservoirs with a capacity in excess of 2000 cubic feet which store solid particles.
- Stationary material handling equipment using a pneumatic bucket or belt conveying systems for which emissions occur require permitting.
- Surface coating equipment, including spray painting, roller coating, or electrostatic deposition, in which the quantity of material used is greater than $1/2$ gallon of liquid per hour. Please note that sources may require permits even though they may not contain VOCs in this subcategory. In addition, certain other combustion sources require permits, no matter what their size, for example:
- All burning of noncommercial fuel, crude oil, or processed by-products in any form requires a permit. All incinerators unless the incinerator serves a one- or two-family dwelling or a multioccupied dwelling containing six or fewer family units, if it is owner occupied.
- In addition, certain equipment which is used for treating groundwater, industrial wastewater, or municipal wastewater systems which includes air stripping, aeration, digestion, thickening, flocculating, surface impounding, and dewatering if the material entering the system contains less than 2 percent solids by weight and has concentrations of VOCs in Group 2 TXS an influent of 35 tbw or total Group 1 TXS concentration of 100 tbw or more. Or if there is a discharge of more than 50 lb per hour of sludge.

In the previous item, wastewater with solid content greater than 2 percent was excluded; however, all equipment used in treating waste soils or sludges including municipal solid waste, recycle materials, or equipment handling influent with solids greater than 2 percent by weight or greater requires permitting. This includes soil cleaning, composting, pelletizing, drying, and transfer station operations.

- In addition, New Jersey requires all sanitary landfill and hazardous waste landfills which have equipment venting the landfill, whether it be closed or operating, to obtain air permits.
- Any control apparatus serving equipment for which a permit or certificate is required also needs to be included on the permit application and is therefore covered by this section of the regulations.
- In New Jersey, any equipment in which the combined weight of all raw materials *used* (not emitted) excluding air and water exceeds 50 pounds in any 1 hour except for equipment previously listed requires a permit.

17.6.2.6   Other Requirements.   If an application proposes construction, installation, reconstruction, or modification of equipment to a significant source, the applicant needs to document the state of the art (SOTA) for the source: If the source has the potential to emit a category of air contaminants at a rate greater than those listed below. Note, however, if a source emits an air contaminant that appears in this table and is also a HAP found in Table 17.2, the lower of the two SOTA thresholds applies.

State-of-the-art manuals are available for a variety of sources. These manuals may be retrieved from the NJDEP web site at www.state.nj.us/dep/aqpp/sota.html. The state-of-the-art manuals are available for:

• Storage and transfer operations
• Refineries
• Asphalt plants
• Pharmaceutical and chemical manufacturing
• Surface cleaners
• Degreasers
• Surface coater and spray booths
• Ethylene oxides
• Sterilizers
• Municipal waste water treatment
• Site remediation
• Bakery ovens

TABLE 17.2   Reporting and SOTA Thresholds (Potential to Emit)

Air contaminant	SOTA threshold, tons/year
Total VOC	5.0
TSP	5.0
PM-10	5.0
NO_x	5.0
CO	5.0
SO_2	5.0
Each TXS	Varies, but < 5.0
Each HAP	Varies, but < 5.0
Greenhouse gas	5.0
Ozone depleting substance	5.0
112(r) contaminant	5.0

- Boilers

- Engines

- Turbines

- Glass plants

- Paint, ink, and adhesive manufacturing industries

- Graphic art industries

SOTA manuals were specifically developed to document exactly what the state of the art was at the time the manual was prepared. Previously it had been difficult for facilities to demonstrate state of the art because the "target" seemed to be moving more quickly than the permitting process. NJDEP determined that it was in their best interest and the regulated communities' best interest to develop documents that would be the final word in the state of the art (SOTA). State-of-the-art manuals are updated routinely on a defined schedule to allow the most recent updates in technology to be incorporated in the manuals. In addition to other requirements and SOTA, NJDEP may come back to the facility and request documentation that the facility applicant meets NSPS requirements, PSD requirements, RACT requirements, and all other applicable state or federal air pollution control standard code rules or regulations. It has been the experience of the author that it is not required to demonstrate these requirements upon initial application as it may be apparently obvious that the source is not applicable. However, if there are any gray areas, to save time, the applicant should consider doing a defined review of these requirements and demonstrate to NJDEP that they are not applicable. If these requirements are applicable, the applicant must document how the significant source meets these standards and to what effect they do apply to the significant source. If the facility is subject to PSD, or if the application requires EPA approval to the state implementation plan, or if the emissions are subject to offset requirements, the department will seek public comment prior to the final decision of the application. The department may also seek public comment whenever the commissioner finds a significant degree of interest in the public regarding the application. Such situations are likely to arise if the applicant is receiving a permit because of complaints from their neighborhood about odors and/or uncontrolled emissions.

17.6.2.7 Applications.   In order to expedite processing of permits, NJDEP has developed an electronically based air information management system (AIMS). The system was designed to allow for the quick electronic turnaround of permit application in a manner which will expedite both the permitting process and the permitting review of the applicants. The AIMS system, often called RADIUS since that is the name of the software used, is distributed to facilities in New Jersey and also found on the web site at www.state.nj.us/dep/aqpp/radius.html.

17.6.2.7.1   Paperless Air Permit Applications (What the Future Will Bring). In the near future, all permit applications will be prepared and submitted in electronic format. The future is now in New Jersey. In January 1998 the New Jersey Department of Environmental Protection released the first CDs that would utilize the state's AIMS with a user input system. The remote AIMS data input user system (RADIUS) was developed to allow the electronic preparation and submittal of preconstruction and operating permit applications to the NJDEP.

NJDEP promoted the software with introductory seminars and training classes on its use. NJDEP also offered 6-month extensions to Title V facilities that submitted their operating permits electronically. NJDEP gave an extension to the last two waves of applicants, citing the fact that the time to review electronic applications would be significantly shorter than for those on paper format. Ultimately the entire permitting structure, with the help of the electronic submissions, will be in one electronic database at NJDEP. This AIMS database will eventually contain a complete listing of permits, permit conditions, emissions limitations, compliance and record-keeping status, listing of notices of violations, and other information useful for departmental inspectors.

17.6.2.7.1.1 PROS AND CONS OF SUBMITTING ON ELECTRONIC FORMAT: The AIMS/RADIUS system is actually a two-tiered system. The management system, AIMS, is utilized by the state. However, facilities use the RADIUS portion of the system to submit their entire permit application, which may include self-drafted permit conditions and limitations. The RADIUS system includes a variety of features beneficial to the user. These include:

- An application wizard to guide the user in the selection of the most appropriate application forms
- A complete set of application forms for every type of air permit
- A requirements library which facilitates the generation of user-defined permit conditions and compliance plans
- An electronic administrative completeness check
- Import and export capabilities which allow the electronic transfer of permit application data

The listing of pros and cons of an electronic management system will undoubtedly have a large list of cons. The main disappointment was that the AIMS/RADIUS software is not set up to be a database for the user at a facility. The structure of the submittal system is only to submit an application. The AIMS/RADIUS software, despite what industry was expecting, is not a facility-wide environmental administrative tool. To give AIMS/RADIUS credit, the system is only in its infancy. The first version of the software was put on-line in January 1998. However, since that time several new and updated versions of the software have been released. Except for the nondatabase structure, most of the complaints with AIMS/RADIUS fell into two categories. The first complaint were the delays in permit approvals which occurred during the transition period, when NJDEP was required to switch over to the new computer system (AIMS). However, the delays were more to do with shutting down the old computer database than with bringing AIMS/RADIUS on-line. The second category of complaints dealt with the bugs in the operating system that facilities were required to use to submit their application. The following is a list of fixes and improvements that the NJDEP has made to the AIMS/RADIUS system over the first 18 months of release.

RADIUS 1.0:   Initial release to industry 1/30/98.

RADIUS 1.2:   Released to industry via patch on web page on 7/98. Errors fixed by this patch:

1. Adjustments were made to how the RADIUS program retrieves data from the RADIUS.db file to prevent the system from crashing when printing the compliance plan section of the permit application.

2. Modified the RADIUS export routine so that it does not create file names containing more than nine characters.

3. Corrected the operating permit administrative completeness check.

4. Corrected the emission point minimum airflow report so it shows the decimal place.

5. Corrected the min and max flow on the emission unit/batch process screen to allow entry up to 9,999,999.9.

6. Modified the EU/BP report to display the expanded values.

7. Enabled the batch print button for detail windows.

RADIUS 1.3:   Released to industry via patch on web page. Errors fixed by this patch:

1. Changes were made to the save function for Formula 1 detail windows which will fix the problem of crashing upon saving.

2. RADIUS import/export with missing attachment. The import was modified to first check to see if the attachment exists before attempting to move it to the appropriate directory. This will ensure that all attachments (except the ones that were not there when the export file was created) would be imported.

3. Access error 39 occurring at the very end of the administrative check while the bottom of the screen indicates that RADIUS is counting the errors and warnings has been corrected.

4. System error 39 occurring while the administrative check is checking the PTE during the administration check has been corrected.

5. Error occurring when attempting to export a file. The bottom of the screen stated RADIUS is "exporting Aims_fac_pte." The message stated: "Unable to save export to facpte.txt Canceling export." Error has been corrected.

6. Error 16: Maximum string size exceeded error during export has been corrected.
7. The problem with printing blank pages in the compliance plan has been corrected.
8. The problem with the 'Submit' option from RADIUS has been corrected.
9. The problem with printing a compliance plan with submittal requirements, modified by a DEP user, has been corrected.
10. Problem with standard requirements from library being modified when saving modified library conditions has been fixed.
11. Problem with compliance report flagging requirements as noncompliant when user selects code of 1 (compliant) has been corrected.
12. Problem with application crashing when using the "save as" feature for a document with more than 250 requirements has been fixed.
13. Problem with attachments being cut from their original directory has been fixed.

The next release was RADIUS 1.7.

RADIUS 1.7:   Released to industry via patch on web page on 10/15/98. The new features incorporated in this patch are:

1. Upon import to AIMS, the user will be told what the source document version is and given a chance to cancel the import.
2. In the compliance plan screen, when user hits the print (or print preview in AIMS), a response window will open with three choices: print all requirements, print the currently selected subject item, or print requirements for a selected subject item type.
3. Removal of additional citations display/processing on the compliance plan reports.
4. The disabling of the add/delete buttons for the citation data window on the compliance plan and requirement library requirement definition screen.
5. Ability to export and import permit document sets and permit applications has been added.

RADIUS 1.8:   Released to industry via patch on web page on 2/3/99. The new features incorporated in this patch are:

1. Added a "Renumber NJID" button on all the inventory screens. This button will allow you to change the NJID number to any unused number without deleting any previously entered data.
2. The compliance plan window includes two new columns to number the total requirement line items as well as the number of requirement line items "included" in the compliance plan. Errors fixed by this patch:

1. Corrects the text when a user selects "None" for the monitoring method, averaging period, and frequency in the compliance plan. Previously RADIUS dis-

played this as "...no monitoring method at no frequency at no averaging period." This will be replaced by a single "None."

2. Corrected a bug that caused the compliance plan to print out of order.

3. Corrected a bug that caused extra pages to print when using the operating scenario detail window for raw materials.

RADIUS 2.0:   Released to industry via patch on web page in August 1999. Background: The NJDEP met with an industry work group for three sessions throughout the month of January 1999. The purpose of this work group was to provide an open forum to discuss how the department can best improve RADIUS to meet user expectations. The meetings proved highly successful, and a list of features were chosen by the work group to be incorporated in the RADIUS 2.0 release.

1. *Win32/Win NT/Win 98 compatibility.*   RADIUS has been converted from a 16-bit to a 32-bit application. It is now compatible with Windows NT and 98 operating systems. The conversion to 32-bit provides a more stable RADIUS platform while greatly increasing the speed of the application.

2. *Emission statement to permit application data transfer.*   Users who prepare emission statements are able to transfer any common information to permit applications.

3. *Support for general permits.*   Users are able to apply for general permits via RADIUS.

4. *Row-level copy capability.*   A feature has been added to enable users to copy entire rows at a time, similar to a spreadsheet.

5. *Selective printing ability.*   A printing interface has been added to allow users to select individual or series of pages to print.

6. *Export file renaming.*   Users are able to name the export file prior to saving. This allows the user multiple exports on the same day without changing the name via a separate windows interface.

7. *Double validation of PIN numbers.*   RADIUS now includes a feature which will require the user to key in the PIN number twice before submitting applications. This will reduce keystroke errors when entering PIN numbers which are masked from the screen.

The RADIUS program was designed in such a manner to allow the NJDEP air program to adapt the information contained within RADIUS to the rapidly changing environmental world. Reference tables, as referred to in RADIUS, are the tool that allows NJDEP to do this. For example, selections contained in the drop-down window for control device type (on the control device inventory screen) may be adjusted to allow the addition of a new control technology. RADIUS users simply need to download the new reference tables from our NJDEP's web page and import them into their copy of RADIUS.

17.6.2.7.1.2 HINTS FOR ELECTRONIC SUBMISSIONS: There are many things that can go wrong with electronic submissions. The following is a checklist of things to think about before you submit an electronic submission whether it be in AIMS/RADIUS or the newest SOTA electronic submission software.

ADMINISTRATIVE COMPLETENESS: Is the document you're preparing administratively complete? The electronic submissions system in this example has an administrative check which determines if all of the relevant blanks have been filled in. Two potential problems with such an administrative check are: (1) just because the blanks are filled out does not mean the blanks are filled out correctly (your best bet is to print out a paper copy and check each number by hand), and (2) the administrative check cannot possibly know all of the combinations of potential forms that may need to be completed (therefore, double check which "special" forms need to be completed for your specific application).

LATEST VERSION OF SOFTWARE: Make sure you are using the latest version of the software. All your hard work might be wasted if the department no longer accepts older version output. The state may not upgrade your submittal to the latest version. In fact you may not be able to upgrade your submission to the latest version once you started working on it. Therefore, check the department home page for software upgrades. While you're there, see if there are new reference table patches (for new technologies and regulations) , electronic help files and instructions, and answers to frequently asked questions (FAQs).

FEES AND SIGNATURE: Make sure you've included the initial operating permit application fee. It is easy to forget to cut the check seeing you're only submitting the diskette or e-mailing your submittal. Also you may need to sign something, and if you are required to, know who must sign it. For example, if a responsible official's signature is required, get it. The state's updated database will likely be cross-referenced to see if the names match up.

PAPERWORK: Don't forget to include paperwork if it's required, for example, a facility plot plan drawing showing all of the emission points identified with the emission point ID from the emission point inventory in your application. Other paper documents or attached files may be required including a simplified block diagram for each emission unit and batch process showing the configuration of all equipment, material flows, control device(s) (if any), and emission point(s) for the emission unit. Label in accordance with the ID numbers for emission units, batch process, equipment, control devices, and emission points used in your application. The diagrams may be submitted on paper or may be included on your diskette if the file format is compatible with the state's system (generally Microsoft office products are the standard).

UPDATING DATABASE: Specific information may have changed since your last submission. You will be required to update the department database. In addition, your facility may have a new facility number to go along with the new database. Enter your facility's new program interest number in the field for facility. Be sure to enter the owner, operator (if different from owner), on-site manager, general contact (for your application), and consultant information (if appropriate) under "Contact Information."

EMISSIONS: It's easy to concentrate on hourly emission rates and forget about annual emissions if the two are not listed on the same page. Remember annual maximum emission limits may be less than simply multiplying the maximum hourly emission rates by the number of hours of operation. Be sure to complete the potential to emit for subject item facility and each emission unit or batch process. The correct units for both subject items are tons per year.

COMPLIANCE PLAN: Be sure to address all relevant subject items in the compliance plan including facility, emission point, insignificant source, fugitive emission, equipment, control device, and emission unit or batch process operating scenario summaries. Be aware, for example, that the AIMS/RADIUS system has pull-down menus which list all the possible applicable permit conditions for a selected type of equipment or control device. Do not simply check yes next to all the potential permit conditions. Some of the "automatic" permit conditions may not be applicable to you. In addition, the permit conditions may be modified to minimize record-keeping and monitoring requirements. The state will obviously review all of your changes; however, by carefully addressing the permit conditions you may make the permit reviewer's job easier.

INVENTORIES AND DETAILS: Be sure that all equipment, emission points, and control devices listed in your application are accounted for in the emission unit/batch process inventory. Complete all details screens for equipment, control devices, and emission unit/operating scenarios. You should review the supplemental data forms in the department's guidance on which fields in the details windows must be completed.

EMISSION UNITS (ALSO CALLED PERMITS):   Note that when combining equipment in an emission unit, it is generally best to treat each existing preconstruction permit and operating certificate as a separate emission unit. A facility intends to combine equipment from multiple preconstruction permits, or combine permitted equipment with grandfathered equipment. The operating permit application generally cannot be used to increase emissions, add new air contaminants, or subject equipment to new applicable requirements, without preconstruction review.

In general, the submission of permits in electronic format will expedite the permitting process and allow facilities to gain some sort of control over potential compliance issues prior to submittal of the application. However, the introduction of a large compliance database will ultimately put more information and more readily available information into the hands of field inspectors and enforcement officers. The final outcome will be more strictly enforced regulations, with a smaller potential for things "to be missed" during facility inspection tours.

17.6.3 Operating Permits for "Major" Sources

In addition to the permitting requirements contained in Title 7, Chapter 27, Subchapter 8, the two remaining major air permit regulations in New Jersey are contained in Subchapters 18 and 22. Subchapter 18 contains the regulations concerning the permitting of sources that emit pollutants for which the area is not in attainment. Subchapter 22 contains the state's operating permits program, which implements the federal Title V regulations.

The requirements for facilities to obtain air pollution control operating permits are contained at N.J.A.C. 7:27-22 et seq. (the "operating permit rule"). The operating permit rule is implemented pursuant to the Federal Clean Air Act and rules promulgated by EPA requiring affected facilities to obtain facility-wide permits that incorporate all of the applicable air pollution requirements into one document. While facilities still need to maintain existing preconstruction permits and obtain new permits in accordance with N.J.A.C. 7:27, Subchapter 8, the new operating permit replaces existing operating certificates for affected facilities.

This extensive subchapter represents the state regulations implementing the Federal Clean Air Act Amendments (CAAA) of 1990 Title V regulations. As noted earlier, the Title V regulations were developed by Congress to remove the grandfathered status of all existing major sources. In essence, the rules require any source of an air pollutant that would be major under any section of the Clean Air Act will be required to obtain an operating permit under Title V. No grandfathered status is available to sources under these provisions.

New Jersey's implementation of these Title V regulations parallels the federal requirements.

17.6.3.1 Applicability. The operating permit rule applies to any facility (including equipment and control apparatus that is grandfathered or exempt from preconstruction permits and fugitive emissions) that has the potential to emit the rates of pollutants in Table 17.3.

The operating permit rule also applies to any facility subject to new source performance standards (NSPS), national emission standards for hazardous air pollutants (NESHAPs), prevention of significant deterioration (PSD), Title IV (acid rain requirements), maximizing achievable control technology (MACT) standards, and any requirement EPA promulgates [N.J.A.C. 7:27-22.2(a)(2)].

N.J.A.C. 7:27-22.2 contains the applicability requirements for the Title V program. As mentioned, these regulations require any facility which would be major under any provision of the act to obtain a Title V operating permit. Therefore, facilities emitting more than 10 tons per year of any HAP or 25 tons per year of any combination of HAPs, any source that would be major under the PSD or nonattainment new source review requirements, any Title V (acid rain) facility, and any other source operation which EPA designates as subject to operating permit requirements pursuant to 40 CFR 70.3(a)5. Additionally, a facility not subject to the New Jersey operating permit regulations can elect to voluntarily obtain an operating permit in lieu of obtaining operating certificates for the equipment or control apparatus. This may be advantageous for facilities with numerous air permits for storage and mixing tanks, in that it would consolidate all of their air per-

TABLE 17.3 Pollutant Rate Chart

Pollutant	Facility emission rate
HAPs (singly)	10 tons per year
HAPs (aggregate)	25 tons per year
HAPs	At any lesser rates EPA establishes
Carbon monoxide	100 tons per year
PM-10	100 tons per year
TSP	100 tons per year
Sulfur dioxide	100 tons per year
Lead	10 tons per year
VOCs	25 tons per year
NO_x	25 tons per year
Any other air contaminant	100 tons

mits into a single operating permit with a single renewal date. New facilities are required to submit operating permit applications within 12 months after the new facility commences operation. Operating permits are currently required to be submitted under the RADIUS electronic submission format for the state.

Contents of the permit application are defined in N.J.A.C. 7:27-22.6. A facility that files a timely and complete application in accordance with N.J.A.C. 7:27-22.5 or 22.30, as appropriate, is covered by an application shield. This shield allows the source to operate until the operating permit has been either approved or rejected. Air quality simulation modeling and risk assessments can be required by the department. Additionally, applicants can voluntarily perform air quality simulation modeling and risk assessments in order to increase their comfort level in signing certifications that they are in compliance with all applicable standards and regulations.

Applications must also contain a compliance plan, which describes the current compliance status of the facility with respect to all applicable requirements and sets forth the methods used to determine the facility's compliance status, including the description of any monitoring, record keeping, reporting, or test methods, and any other information necessary to verify compliance with or enforce any proposed permit condition or any applicable requirement. After the agency issues a proposed operating permit, it is then sent to EPA. Under 40 CFR 70.8, EPA has 45 days following receipt of the permit to raise any objections. Compliance with EPA's comments is mandatory, and the applicant and the department must coordinate with EPA to ensure an acceptable resolution.

One of the most common comments received by EPA relates to the requirement for periodic or compliance assurance monitoring. EPA has been particularly insistent in its review of state operating permits that conditions sufficient to determine continuous compliance be included in the permits. Operating permits also contain provisions to allow changes to insignificant source operation without notifying the department or EPA until renewal of the operating permit, or to do simplified permit changes through 7-day notice changes or a minor modification procedure. These are defined in N.J.A.C. 7:27-22.21 through 22.23.

In addition to facility-specific operating permits, there are provisions in N.J.A.C. 7:27-22.14 for general operating permits. These permits would cover specific source categories and contain general operating conditions. N.J.A.C. 7:27-22.15 includes provision for temporary facility operating permits, which authorize operation at more than one location during the term of an operating permit, provided all locations where a facility may be operated are listed in the operating permit.

A permit shield applies to operating permits. N.J.A.C. 7:27-22.17 contains the requirements that must be met to activate a permit shield. It states, "A permit shield provides that compliance with the relevant conditions of the operating permit shall be deemed compliance with the specific applicable requirements that are in effect on the date of issuance of the draft operating permit, and which form the basis for the conditions in the operating permit, provided that the requirements of this section are met." [N.J.A.C. 7:27-17(a)]

The operating permit program also includes a requirement for SOTA air pollution controls. This is contained in N.J.A.C. 7:27-22.35, which states:

(a) Newly constructed, reconstructed, or modified equipment and control apparatus, which constitutes a significant source operation, shall incorporate advances in the art of air pollution control as developed for the kind and amount of air contaminant emitted by the applicant's equipment and control apparatus as provided in this section.

This parallels the Subchapter 8 requirements for state-of-the-art controls.

The timetable for submitting operating permit applications depends on a facility's standard industrial classification (SIC) code. NJDEP encourages all applicants to submit their completed applications no less than 90 days prior to the applicable application deadline set forth at N.J.A.C. 7:27-22.5 [see N.J.A.C. 7:27-22.4(e)]. Application forms for operating permits, modifications to operating permits, and information pertaining to operating permits may be obtained from New Jersey Department of Environmental Protection, Air Quality Regulation Program, CN 027, Trenton, N.J. 08625-0027, Attn: Operating Permits, (609) 633-8248.

Among the more significant rules that have been adopted by NJDEP are the public comment provisions, application shields, permit shields, alternative operating scenarios, and intrafacility emission trading (see generally N.J.A.C. 7:27-22).

17.6.4   Reduction of Air Contaminant Emissions

Subchapter 18 implements the federal laws pertaining to nonattainment new source review (NSR). This refers to the permitting of sources that are either located in or would have a significant impact in areas of the state that are not in attainment of the ambient air quality standards (AAQS) for specific pollutants. The main purpose of the regulations is to allow the state to maintain reasonable further progress toward achieving and maintaining the ambient air quality standards.

Facilities which take steps to reduce the emission of air contaminants can obtain credit for such reductions. There are two state programs which provide mechanisms to obtain credit for air emission reductions. The first is banking of emission credits pursuant to N.J.A.C. 7:27-18, and the second is generating discrete emission reductions (DERs) pursuant to N.J.A.C. 7:27-30.

The applicability requirements are contained in 7:27-18.2. They are as follows:

7:27-18.2 Facilities subject to this subchapter:

(a) This subchapter applies to certain applications, submitted to the department pursuant to N.J.A.C. 7:27-8 or N.J.A.C. 7:27-22 for authorization to construct, reconstruct, or modify control apparatus or equipment at a facility, if the requirements at (b) or (c) below apply and:

  1. The facility has the potential to emit any of the air contaminants listed below in an amount which is equal to or exceeds the threshold levels in Table 17.4.

  2. The emission increase of an air contaminant, proposed in the application, by itself equals or exceeds the threshold level for that air contaminant set forth in (a)1 above.

(b) For a facility which meets the criteria at (a)1 or 2 above, an application is subject to this subchapter if any allowable emissions proposed in the application would result in a significant net emission increase of any air contaminant listed in Table 3 of N.J.A.C. 7:27-18.7, and if the facility for which the construction, reconstruction, or modification is proposed is located at an area which is any of the following:

TABLE 17.4   Air Contaminant and
Threshold Level Table

Air contaminant	Threshold level
Carbon monoxide	100 tons per year
PM_{10}	100 tons per year
TSP	100 tons per year
Sulfur dioxide	100 tons per year
Nitrogen oxides	25 tons per year
VOC	25 tons per year
Lead	10 tons per year

1. Nonattainment for the respective criteria pollutant corresponding to that air contaminant. The respective criteria pollutant for each air contaminant is listed in the definition of the term "respective criteria pollutant" at N.J.A.C. 7:27-18.

2. Attainment for the respective criteria pollutant, and both (b)i and ii below are true:

    i. The proposed significant net emission increase would result in an increase in the ambient concentration of the respective criteria pollutant in an area that is nonattainment for the respective criteria pollutant, as determined by an air quality impact analysis required under N.J.A.C. 7:27-8.5.

    ii. The increase in the ambient concentration of the respective criteria pollutant equals or exceeds the significant air quality impact level specified in Table 1 in N.J.A.C. 7:27-18.4, in the nonattainment area for the respective criteria pollutant.

3. Attainment for the respective criteria pollutant, and the proposed significant net emission increase would result in an increase in the ambient concentration of the respective criteria pollutant that would:

    i. Equal or exceed the significant air quality impact level.

    ii. Result in a violation of an applicable NAAQS or NJAAQS, as determined by an air quality impact analysis required under N.J.A.C. 7:27-8.5.

The threshold levels are particularly low for combustion sources, especially for nitrogen oxides, where the threshold is only 25 tons per year.

17.6.4.1 Requirements. Any source subject to these rules must comply with five specific requirements, as follows:

1. The first of these requires the use of controls which represent the lowest achievable emission rate (LAER) for the nonattainment pollutants.

2. Certify that all existing facilities which are owned or operated by the applicant or any entity controlling, controlled by, or under common control of the applicant are in compliance with all applicable standards and regulations pursuant to Subchapter 18 and the Federal Clean Air Act, or in conformance with an enforceable compliance schedule approved by the department.

3. Secure emission offsets at specified offset ratios dependent upon pollutant and distance. Table 17.5 contains the minimum offset ratios. The minimum offset ratio for lead is 1.00:1.00.

4. Perform an air quality impact analysis which demonstrates that the increase in ambient concentration would be equal to or less than the significant air quality impact levels defined in 7:27-18.4 and that the predicted impacts combined with the existing concentrations would not cause a violation of a New Jersey or national ambient air quality standard. This air quality impact analysis is not required for emissions of VOCs.

TABLE 17.5 N.J.A.C. 7:27-8.5 Minimum Offset Ratio

Air contaminant	Distance, miles	Minimum offset ratio (reductions:increase)
VOC	0–100	1.3:1.0
	100–250	2.6:1.0
	250–500	5.2:1.0
NO_x	0–100	1.3:1.0
	100–250	2.6:1.0
	250–500	5.2:1.0
SO_2	0–0.5	1.0:1.0
	0.5–1.0	1.5:1.0
	1.0–2.0	2.0:1.0
TSP	0–0.5	1.0:1.0
	0.5–1.0	1.5:1.0
	1.0–2.0	2.0:1.0
PM_{10}	0–0.5	1.0:1.0
	0.5–1.0	1.5:1.0
	1.0–2.0	2.0:1.0
CO	0–0.5	1.0:1.0
	0.5–1.0	1.5:1.0
	1.0–2.0	2.0:1.0

5. Prepare and submit an analysis of alternative sites within New Jersey and of alternative sizes, production processes including pollution prevention measures, and environmental control techniques demonstrating that the benefits of the newly constructed, reconstructed, or modified equipment significantly outweigh the environmental and social impacts imposed as a result of the location, construction, reconstruction, or modification and operation of such equipment.

As outlined in these five steps, a substantial burden is placed upon sources that would emit nonattainment pollutants or significantly impact nonattainment areas. These requirements recognize that the existing air quality exceeds current standards, based on the currently permitted sources, and hence any new sources will have to assist in reducing current emissions through offset ratios and meet more stringent control requirements in order to obtain a permit. The fifth condition, requiring an alternatives analysis, has not been handled rigorously within the state, except for controversial sources, particularly municipal waste combustors (MWC). However, it is a requirement of both the federal and state regulation and cannot be ignored.

17.6.4.2 Banking of Offset Credits. Another important section within Subchapter 18 concerns the "banking" of emission reductions. The regulations for this are contained in 7:27-18.8. These allow sources to formally document emission reductions and place them in a precertified state "bank."

The application to bank the emission reductions must be made no later than 12 months after the emission reduction occurs, and no emission reductions due to shutdown of any equipment or source operation are eligible for banking unless the department has been notified at least 60 days prior to the removal of equipment and provides the department the opportunity to inspect the equipment at least 30 days before it is dismantled. In order for an emission reduction to be creditable, an enforceable operating restriction will be placed on the facility upon approval by the department for banking.

Banked emissions are affected by any new state or federal statute, rule, or regulation. The emission reduction credits are reduced to those that would be equal to the allowable emission limits in effect at the time the banked emission reductions are used. Further, 5 years after the date the emission reduction is submitted for banking, 50 percent of the credits revert to the state. Ten years after the date of submittal for banking, the emission reductions may no longer be used as emission offsets.

Credits for reductions in air emissions pursuant to both banking of emission credits and generating DERs may be used by the facilities generating the credits or transferred to other facilities (N.J.A.C. 7:27-18.8, 30.9). The credits may be used to lower actual emission values in order to conform to air permits or other statutory or regulatory limitations on air emissions. The primary difference between banking of emission credits and generating DERs is that DERs can only be used once, whereas banked emission credits can be used for up to 10 years [N.J.A.C. 7:27-18.8(g)]. Also, unlike banked emission reductions, a shutdown or curtailment in operations cannot be used to generate DERs.

Any person may apply to the DEP for the banking of emission reductions to be applied in the future as emission offsets. The applicant must make the application in writing, submitted on a form obtained from the DEP, containing the following information: name and address of person making the application; chemical name of air contaminant; quality of emission reductions with supporting calculations and documentation; reason for the emission reduction; specification of the equipment or source operations related to the emission reductions; and any additional information reasonably necessary to enable the DEP to determine that a creditable emission reduction has been achieved. Such a form may be requested from New Jersey Department of Environmental Protection, Air Quality Permitting Program, Bureau of New Source Review, P.O. Box 027, Trenton, N.J. 08625-0027.

An application to bank emission reductions must be made no later than 12 months after the emission reduction occurs. No emission reductions due to the shutdown of any equipment or source operation shall be eligible for banking, unless the applicant notifies the department at least 60 days prior to removal of the equipment and provides the department with the opportunity to inspect the equipment or source operation at least 30 days before it is dismantled. Any emission reductions submitted to the department for banking shall, upon their approval by the department for banking, be an enforceable operating restriction for the facility (N.J.A.C. 7:27-18.8).

DERs are administered by the Open Market Emissions Trading (OMET) program. The forms and procedures for registering or obtaining DERs can be obtained from the OMET web page at www.omet.com. This web page also provides valuable reference material regarding the average price of DERs, availability, and names and addresses of registrants.

The amount of DERs is calculated as the difference between actual emissions and baseline emissions. The formula for the calculation is found in N.J.A.C. 7:27-30.5. Certain restrictions apply. For example, none of the following emission reductions is a basis for generation of a DER:

• An emission reduction that results from a shutdown or curtailment.

• An emission reduction that results from modifying or discontinuing activity that violates any federal, state, or local law, regulation, permit, or order.

• An emission reduction that is required to comply with a requirement in the Federal Clean Air Act, the New Jersey Air Pollution Control Act, any regulation, permit, or order pursuant thereto; any air quality emission limit or standard in any applicable law, regulation, permit, or order; or any SIP or federal implementation plan (except to the extent that emissions are reduced below the level required to comply).

• An emission reduction which has been used under any other emissions trading program, or which has previously been the basis for generating a DER.

• An emission reduction occurring at a generator source which received approval from the department of an alternative emission limit (including, without limitation, an emission limitation that is part of an averaging plan) to meet a requirement for reasonably available control technology (RACT) under N.J.A.C. 7:27-16 or 19, except to the extent that the emissions are reduced below the level that would have been required had the approval of the alternative emission limit not been issued.

• An emission reduction from a stationary source that is subject to N.J.A.C. 7:27-16 or 19, but for which the department has not yet established an applicable RACT limit either in the rule or in a facility-specific emissions limit submitted to the EPA as a SIP revision.

• An emission reduction which is accompanied by an increase in emissions of any HAP which exceeds the de minimis level designated for that HAP by the EPA pursuant to 42 U.S.C. 7412(g).

• An emission reduction which is accompanied by a violation of a federal or state law, regulation, order, or permit. For example, if the generator source's actual emissions during any portion of the generation period exceed the maximum quantity of emissions which the generator source's permit authorizes for such portion of the generation period, no DERs shall have been generated.

Additionally, no emission reduction generated before May 1, 1992, is a basis for generation of a DER, and no emission reduction is the basis for generation of a DER until the generator source's emissions are reflected in the emissions inventory submitted by the state to the EPA for inclusion in the SIP, or in the annual major point source emission inventory conducted pursuant to N.J.A.C. 7:27-21 (N.J.A.C. 7:27-30.6).

For each batch of DERs generated, the generator needs to submit a notice and certification of DER generation to the registry within 90 days after the last day of the generation period. If the generator submits the notice late, the generator must reduce the quantity of DERs in the notice by 10 percent immediately, and by an additional 10 percent for each additional 30 days that the notice is late. For example, if a generator has generated 100 tons of DERs but submits the notice 40 days late, the generator can include only 80 tons of DERs in the notice (N.J.A.C. 7:27-30.7).

17.6.5   Other Agency Air Powers

NJDEP has a number of enforcement options including civil penalty assessments under N.J.S.A. 26:2C19(b),(d), administrative orders under N.J.S.A. 26:2C-14, civil action for injunction or penalties under N.J.S.A. 26:2C19(a), and criminal action for penalties or imprisonment under N.J.S.A. 26:2C-19(f). Most commonly, NJDEP has issued combined administrative orders and notices of civil administrative penalty assessment (AONOCAPA), rather than bringing civil or criminal enforcement actions in court. Because of staffing limitations and other considerations, it is expected that this approach will continue for routine cases. The regulated community should be made aware, however, of the potential exposure to criminal prosecution in certain types of cases. They also may face citizens' suits for injunctive relief, penalties, and attorneys' fees.

The New Jersey legislature has provided an affirmative defense to liability for certain air pollution control violations that are due to equipment malfunction, startup or shutdown conditions, or maintenance (N.J.S.A. 26:2C-19.1 to -19.5). This recognizes that equipment may malfunction or fail to perform optimally, even when carefully maintained and operated. It also recognizes that violations may occur due to unforeseeable and unavoidable malfunctions from the inherently intricate nature of mechanical equipment. The affirmative defense is not applicable to violations that are part of a recurrent pattern or cause emissions that pose a potential threat to public health, welfare, or the environment [N.J.S.A. 26:2C19.3,19.5(b)]. In order to preserve the affirmative defense, notice must be given to NJDEP by 5:00 P.M. of the second full calendar day following an occurrence (or following discovery of an occurrence through due diligence) [N.J.S.A. 26:2C-19.2(b)].

Counties and other local authorities may adopt ordinances for air pollution control that are more stringent than the NJAPCA and its regulations, subject to approval by NJDEP (N.J.S.A. 26:2C-22; N.J.S.A. 26:3A2-27). The County Environmental Health Act authorizes county agencies to investigate violations of the NJAPCA and its regulations, and to report violations to NJDEP (N.J.S.A. 26:3A2-25). NJDEP may delegate the administration of certain environmental regulations to the counties [N.J.S.A. 26:3A2-28(b)]. County agencies may issue summonses and complaints, and collect penalties for violations (N.J.S.A. 26:3A2-25).

17.7 OIL AND HAZARDOUS SUBSTANCES SPILLS

Overview

The New Jersey Spill Compensation and Control Act (Spill Act) (N.J.S.A. 58:10-23.11) prohibits the discharge of hazardous substances and petroleum except when in compliance with a federal or state permit; provides for the cleanup and removal of discharged substances; and creates a fund to defray the costs of cleanup and removal undertaken by NJDEP and to compensate the victims of a discharge. A "discharge" is defined under the Spill Act at N.J.S.A. 58:10-23.11b.h as:

> any intentional or unintentional action or omission resulting in the releasing, spilling, leaking, pumping, pouring, emitting, emptying or dumping of hazardous substances into the waters or onto the lands of the State, or into waters outside the jurisdiction of the State when damage may result to the lands, waters or natural resources within the jurisdiction of the State.

The Spill Act requires the development of plans to help prevent discharges and to provide procedures to follow if a discharge does occur (N.J.S.A. 58:10-23.11d2 to 23.11d9). In the event of a discharge, the responsible party must notify NJDEP and implement cleanup and removal activities (N.J.S.A. 58:10-23.11d10 and 23.11e). Notification to the NJDEP must be immediate, i.e., within 15 minutes of the discharge, followed by written confirmation. In the absence of responsible party action, NJDEP may perform the cleanup and removal itself or issue a directive to compel the discharger to do so. If NJDEP undertakes the cleanup and removal, its expenditures constitute a debt of the discharger, and NJDEP may file a lien on all of the discharger's property. Any discharger or other person who performs a cleanup and removal has a statutory right of contribution from all other responsible parties (see generally N.J.S.A. 58:10-23.11f).

In addition, the Discharge Reporting Act, which is related to but independent from the Spill Act, requires owners or operators of certain industrial establishments, or real property that once was the site of an industrial establishment, to report the known or suspected occurrence of any historic hazardous discharge to the local municipality and board of health (N.J.S.A. 13:1K-16).

17.7.1 The Spill Act

The Spill Act requires a "major facility" to have in place a discharge, prevention, control, and countermeasure (DPCC) plan, and a discharge, cleanup, and removal

(DCR) plan ready to be implemented in the event of a spill of a hazardous substance (N.J.S.A. 58:10-23.11d2, 23.11d3). All facilities located on one or more contiguous or adjacent properties owned or operated by the same person with a combined storage capacity of (1) 20,000 gallons or more for hazardous substances other than petroleum or (2) 200,000 gallons or more for hazardous substances of all types will be considered a "major facility" (N.J.S.A. 58:10-23.11b.l). For hazardous substances not usually measured by volume, the director of the Division of Taxation will set an appropriate equivalent measure. Appendix A to N.J.A.C. 7:1E lists regulated hazardous substances. Required contents of DPCC and DCR plans are given in N.J.A.C. 7:1E-4.2 and 7:1E-4.3, respectively. The plans must be certified by a professional engineer, renewed every 5 years with NJDEP, and filed with the local emergency prevention or local emergency planning committees. Amendments to the plans must be made within 30 days of the date of any modification of a facility.

A major facility must also comply with NJDEP facility design and maintenance standards. These standards are contained in N.J.A.C. 7:1E-2. The standards address storage, loading and unloading areas, in-facility piping for hazardous substances, process areas, facility drainage and secondary containment, marine transfer facilities, illumination, flood hazard areas, leak detection and monitoring, housekeeping and maintenance, employee training, security, standard operating procedures, and record keeping.

Spill reporting requirements, however, are not limited to major facilities. All persons responsible for any hazardous substance discharges are required to notify the NJDEP hotline at (609) 292-7172, or if that number is inoperable, to notify the New Jersey state police at (609) 882-2000 (N.J.A.C. 7:1E-5.3). There is no de minimis quantity exemption; all discharges of hazardous substances to land or water must be reported within 15 minutes of when the person responsible knew, or should have known, of the discharge. Notification should include (1) date and time of the discharge; (2) type and quantity of the substance discharged; (3) location of the discharge; (4) names and addresses of responsible parties; (5) actions the responsible parties propose to take to contain, clean up, and remove the substances discharged; and (6) any other information the NJDEP requests (see generally N.J.A.C. 7:1E-5). A written confirmation report must also be sent within 30 days of the initial notice to NJDEP Bureau of Discharge Prevention, CN 424, 401 East State Street, Trenton, N.J. 08625-0424.

The confirmation must include the items listed in N.J.A.C. 7:1E-5.8. Owners and operators of major facilities must also notify NJDEP of any malfunction of a leak detection or other discharge monitoring system or of a discharge prevention, safety system, or device (N.J.A.C. 7:1E-5.5). Within 2 hours of initial notification, the owner or operator must then notify NJDEP that the malfunction has been repaired, an alternative discharge detection system has been activated, or the equipment protected by the discharge detection system has been taken out of service (N.J.A.C. 7:1E-5.5).

The Spill Act provides that "any person who has discharged a hazardous substance, or is in any way responsible for any hazardous substance, shall be strictly liable, jointly and severally, without regard to fault, for all cleanup and removal costs no matter by whom incurred." [N.J.S.A. 58:10-23.11g.c(l)]. All cleanup and removal activities under the Spill Act must conform to the technical requirements for site remediation (N.J.A.C. 7:26E). NJDEP is authorized to contain, clean up, and remove discharges, and then to seek recovery from responsible parties for costs incurred. Alternatively, NJDEP may issue directives requiring responsible parties to perform remedial activities (N.J.S.A. 58:10-23.11f). The Spill Act provides a private right of contribution among responsible parties under the Spill Act [N.J.S.A. 58:10-23.11f.a.(2)]. A contribution plaintiff may file a claim with the court against a discharger who has failed to comply with any directive for an amount equal to three times the cost of the cleanup and removal [N.J.S.A. 58:10-23.11f.a.(3)].

17.8 THE DISCHARGE REPORTING ACT

The Discharge Reporting Act (N.J.S.A. 13:1K-15 to 17) requires that known or suspected discharges of hazardous substances at current or former industrial establishments be reported to the municipality and local board of health in which the discharging facility operates or operated. This reporting requirement applies to both current and previous industrial operations. An owner or operator of an industrial establishment or real property that was once the site of an industrial establishment must file a written report within 10 days of discovering or suspecting a discharge. "Industrial establishment" and "hazardous substance" are defined for purposes of the Discharge Reporting Act in N.J.S.A. 13:1K-15. The reporting procedures are integrated with those under the Spill Act at N.J.A.C. 7:1E-5, so that NJDEP provides oral notification to municipal officials regarding any reports of hazardous substance discharge reported to NJDEP. The act itself still requires that owner-operator to file a written report with the municipality within 10 days.

17.8.1 Other Agency Powers

The Spill Act authorizes NJDEP to promulgate necessary rules and regulations to implement the act and to assess penalties and to compel compliance by administrative, civil, or criminal action (N.J.S.A. 58:10-23.11f). Penalties may be levied for failure to allow lawful entry and inspection. NJDEP may, at its discretion, observe, supervise, or participate in any aspect of containment or cleanup and removal activity [N.J.S.A. 58:10-23.11f.a.(1)]. The Spill Act was amended, effective 1992, to provide a private right of contribution among those defined as responsible parties under the Spill Act [N.J.S.A. 58:10-23.11f.a.(2)]. NJDEP has authority to pursue a discharger who has failed to comply with any directive for an amount equal to three times the cost of the cleanup and removal [N.J.S.A. 58:10-23.11f.a.(l)].

The Discharge Reporting Act does not provide NJDEP with any additional enforcement authority. Rather, it serves as a statutory facilitator of the exchange of information among private parties, NJDEP, and local governments regarding discharges of hazardous substances occurring within the state.

17.9 HAZARDOUS WASTE

Overview

Although Subtitle C of the Federal Resource Conservation and Recovery Act (RCRA), as amended by the Hazardous and Solid Waste Amendments of 1984 (HSWA), regulates hazardous waste, EPA has authorized NJDEP to administer and enforce New Jersey's hazardous waste program (40 CFR 272). The state statute governing hazardous waste is the Solid Waste Management Act (SWMA) (N.J.S.A. 13:1E-1 et seq.). The state program has almost fully incorporated RCRA, but it contains several more stringent requirements. These more restrictive requirements affect generators, transporters, and owners and operators of treatment, storage, and disposal facilities (TSDFs).

17.9.1 Generators

Hazardous waste generators are regulated under N.J.A.C. 7:26G-6, and 40 CFR 262 by incorporation. These regulations require the generator to:

- Obtain an EPA generator identification number
- Determine whether its waste is hazardous
- Adhere to land disposal restrictions
- Manifest waste shipments
- Designate a transporter and a TSDF that each possesses an EPA identification (ID) number
- Prepare and implement a contingency plan
- Perform record-keeping and reporting requirements
- Pay all applicable fees

If a generator qualifies as a "small quantity generator," it need not comply with biennial reporting, formal employee training, and contingency planning requirements. A "small quantity generator" is one who generates less than 100 kilograms of hazardous waste per month, up to 1 kg of acutely hazardous waste per month, or up to 100 kg per month of any residue or contaminated soil, waste, or other debris, resulting from a spill of an acutely hazardous waste. N.J.A.C. 7:26G-5.2 and 40 CFR 261.5 detail the special requirements for small quantity generators.

17.9.2　Obtaining an EPA Generator Identification Number

If a generator's waste activity is regulated under RCRA, it must notify EPA and obtain a generator ID number. EPA Form 8700-12, Notification of Regulated Waste Activity, may be obtained from and should be submitted to USEPA Region II, Air and Waste Management Division, Attn: RCRA Notification, 26 Federal Plaza, Room 505, New York, N.Y. 10278. A separate form should be submitted for each site or location where waste is generated.

17.9.3　Determining Hazardous Waste

It is the generator's duty to determine whether its waste is hazardous. N.J.A.C. 7:26G, Subpart 5 and 40 CFR 261, Subpart B, provide the hazardous waste criteria, identification, and listing regulations. Wastes excluded from regulation are listed in 40 CFR 261.4(a). If not specifically excluded, a solid waste is identified as hazardous either by listing or by exhibiting a hazardous characteristic. All listed hazardous wastes appear on one of three lists in 40 CFR 261, Subpart D: (1) hazardous waste from nonspecific sources (40 CFR 261.31); (2) hazardous waste from specific sources (40 CFR 261.32); and (3) discarded commercial chemical products, off-specification species, containers, and spill residues (40 CFR 261.33). Some of these wastes are considered acutely hazardous wastes, designated by a hazard code of H. Mixtures of nonhazardous waste and listed hazardous waste are treated as hazardous waste [40 CFR 261.2(b)(2)]. Additionally, materials derived from listed hazardous wastes are treated as hazardous wastes. 40 CFR 261.3(c)(2)(i).

In addition to listed hazardous waste, a substance may be considered a hazardous waste if it exhibits one or more of the following characteristics: ignitability, corrosivity, reactivity, or toxicity (40 CFR 261, Subpart C, see specifically 261.21, 261.22, 261.23, and 261.4). The generator may determine if a waste exhibits one or more of these characteristics by either sampling and analysis or knowledge of the waste's properties. Tests to determine whether a waste exhibits a characteristic are contained in 40 CFR 261, Subpart C; see specifically appendix II. A mixture of a nonhazardous waste and characteristic hazardous waste is treated as hazardous only if the mixture exhibits the characteristic [40 CFR 261.3(a)(2)(iii-iv)].

A generator may petition the agency to have its waste delisted [N.J.A.C. 7:26G-4.1(c)(5); 40 CFR 260.20 and 260.22]. Delisting petitions are variances that are waste- and site-specific. The petitioner must demonstrate that the listed waste generated at a particular facility meets the criteria set forth at 40 CFR 261, Subpart D. Delisting petitions must be filed with NJDEP and go through a notice and comment rule-making process (N.J.A.C. 7:26G-4.2; 40 CFR 260.20 and 260.22). A delisting is effective upon publication of final notice granting the petition.

17.9.4 Land Disposal Restrictions

With the exception of the provisions provided in N.J.A.C. 7:26G-11.1, the federal land disposal restrictions (LDRs), contained in 40 CFR, Part 268, apply to RCRA hazardous wastes that are land disposed. EPA may set one of three types of treatment standards, all based on the best demonstrated available technology (BDAT) identified for the waste: (1) a concentration level to be achieved prior to disposal; (2) a specified technology to be used; or (3) a "no land disposal" designation when the waste is no longer generated, is totally recycled, currently is not being land disposed, or no residuals are produced from treatment. The standard must be met prior to land disposal of the waste. Treatability variances, equivalent treatment method petitions, no migration petitions, or delisting may be initiated to avert the strict LDR requirements. Information regarding exceptions is also presented in 40 CFR, Part 268.

17.9.5 On-Site Accumulation

A generator may accumulate hazardous waste on-site without a permit for 90 days or less. Accumulation of hazardous waste is regulated by 40 CFR 262.34(a). Waste must be accumulated in approved containers, clearly marked as hazardous waste, and labeled with the date upon which accumulation began. Extension of the 90-day deadline may be granted at the discretion of the director of the Division of Hazardous Waste Management. A generator may also accumulate waste for more than 90 days when such accumulation is less than 55 gallons of hazardous waste or less than 1 quart of acutely hazardous waste. 40 CFR 262.34(b). Once the waste meets or exceeds either of these volumes, the 90-day regulations begin and the accumulation regulations apply.

17.9.6 Waste Shipments Manifest

The generator must prepare the waste for transportation. This includes placing waste in approved containers and labeling them in accordance with the Department of Transportation requirements. Packaging and labeling requirements are set forth at N.J.A.C. 7:26G-6.1(c)(8) and 40 CFR 262, Subpart C.

Generally, each shipment of waste must be accompanied by a Uniform Hazardous Waste Manifest form [N.J.A.C. 7:26G-4.1(c)(4)(ii), 7:26G-6.1(b) and (c)(3), and appendix to 7:26G-6.; 40 CFR 262.20]. Manifest forms must be obtained from NJDEP for wastes to be shipped to a site within New Jersey

[N.J.A.C. 7:26G-6.1(c)(4); 40 CFR 262.21(a)]. For domestic shipments to sites outside New Jersey in states which supply manifests, the manifest form shall be supplied by the destination state [40 CFR 262.21(a)]. When the waste is shipped from New Jersey and the destination state does not supply the manifest, the NJDEP shall supply the manifest [N.J.A.C. 7:26G-6.1(c)(5); 40 CFR 262.21(b)]. The manifest must include the following:

- Generator name, mailing address, site address, phone number, and EPA ID number
- Transporter name, phone number, New Jersey registration number, and EPA ID number
- TSDF name, address, phone number, and EPA ID number
- Name, type, and quantity of hazardous waste being shipped
- Any special handling instructions
- Proper waste code(s) describing the shipment

[N.J.A.C. 7:26G-6.1(c)(3) and 7:26G Appendix; 40 CFR 262.20(a) and 262 Appendix]. When a waste can be described by more than one code, the waste code hierarchy set forth at N.J.A.C. 7:26G-6.2 controls the coding of the manifest. If listed wastes are mixed, all applicable waste codes must appear on the manifest. The generator also must certify on the manifest that it is implementing an on-site hazardous waste reduction program and has selected the safest available treatment, storage, and disposal methods for its waste. The generator and transporter must sign the manifest by hand prior to release. One copy of the manifest is retained by the generator, and copies are forwarded to the states of origin and destination. Three copies accompany the waste shipment to the TSDF, where the TSDF owner or operator signs the manifest. One copy is then retained by the TSDF and the remaining copies are returned to the transporter and generator.

If a generator has not received its return copy of the manifest with the handwritten signature of the owner or operator of the designated facility within 35 days from the time the waste was accepted by the initial transporter, the generator must contact the transporter and/or the owner or operator of the facility to determine the status of the waste and contact NJDEP at (609) 292-7081 to notify them of the situation [N.J.A.C. 7:26G-6.1(c)(11); 40 CFR 262.42(a)(1)]. Exception reporting is required when the generator has not received the return copy of the manifest within 45 days of the date the waste was accepted by the initial transporter [N.J.A.C. 7:26G-6.1(c)(12); 40 CFR 262.42(a)(2)].

17.9.7 Hazardous Waste Remediation

The newly promulgated Hazardous Remediation Waste Management Requirements (HWIR-Media) RCRA regulations, effective as of June 1, 1999, apply to hazardous remediation wastes treated, stored, or disposed of during cleanup actions, and generally allow greater flexibility in conducting cleanups subject to RCRA than under the old rules [63 Fed. Reg. 65,874 (Nov. 30, 1998)]. Part of RCRA's corrective action program, the HWIR-Media regulations are effective in New Jersey.

The new regulations provide five major changes. First, they create a new remedial action plan (RAP) to allow treatment, storage, and disposal of hazardous remediation waste without a RCRA TSD permit. Second, they allow an owner to obtain a RAP without being subject to facility-wide RCRA corrective action. Third, they create a new kind of unit called a "staging pile" to allow for the storage of hazardous remediation waste up to $2^{1}/2$ years without a RCRA TSD permit. Fourth, they exclude dredged materials from RCRA Subtitle C if they are managed under the Clean Water Act or Ocean Dumping Act. Fifth, they make it easier for states to receive authorization from EPA to update their RCRA programs.

The HWIR creates a new Subpart H of RCRA for remedial action plans, or RAPs. The RAP is a special form of RCRA permit designed to replace the traditional RCRA TSD permit for remediation work. In promulgating the RAP, EPA's goal was to streamline the permit process and thus remove disincentives to remediating hazardous waste cleanups. One of the main benefits of the RAP, as opposed to the traditional TSD permit, is that a RAP does not subject the owner to the RCRA corrective action program. In addition, RAPs should provide for flexibility in responding to the unusual circumstances often encountered at remediation sites and for satisfying public participation and information requirements. It should be noted that RAPs provide only a mechanism for complying with RCRA's requirements for management of remediation wastes; they do not take the place of federal or state programs establishing cleanup goals or selecting remedies. RAPs apply only to remediation waste and cannot be used to manage process waste.

The new HWIR establishes separate facility standards under 40 CFR Part 264 for "remediation waste management sites." The new regulations provide somewhat greater flexibility by implementing general performance standards for remediation site inspections, worker training, preparedness and prevention plans, contingency plans, and emergency procedures.

The new regulations create a new type of waste management unit called a "staging pile," defined as an accumulation of solid, nonflowing remediation waste used only during remedial operations for temporary storage. Staging piles allow storage of waste during remediation for up to 2 years, plus one 180-day extension, without triggering the traditional comprehensive standards for RCRA TSD facilities. Thus, staging piles should allow material to be accumulated in sufficient quantities to efficiently ship it off-site or treat it on-site, without subjecting the owner to detailed RCRA requirements. Staging piles are not considered land disposal units, and therefore, unlike traditional RCRA waste piles, it is not necessary to meet RCRA LDRs before placing waste in a staging pile. It must be emphasized, however, that staging piles are for storage only and not for treatment, storage, or disposal of remediation waste.

17.9.8   Record-Keeping and Reporting Requirements

By Mar. 1 of each even-numbered year, each generator must submit to NJDEP a biennial report of all manifest activities, hazardous waste activities, and waste minimization efforts undertaken (40 CFR 262.41). The report is filed on NJDEP forms [N.J.A.C. 7:26G-6.1(c)(10); 40 CFR 262.41(a)]. Record-keeping requirements include maintaining copies of all manifests and each biennial report and exception report for at least 3 years [N.J.A.C. 7:26G-6.1(c)(a); 40 CFR 262.40]. Additional record keeping is required for exported waste shipments.

17.9.9   Applicable Fees

Fees imposed on generators include those for annual reporting, manifest processing, conducting inspections and compliance reviews, and classifying and delisting wastes. Fees are paid in accordance with N.J.S.A. 13:1E-1 et seq. at the rate provided in the schedule set forth in N.J.A.C. 7:26G, Subchapter 3. Fees for activities related to hazardous waste are to be paid by certified check or money order made payable to Treasurer, State of New Jersey, and are to be submitted to the following address: New Jersey Department of Environmental Protection, Attn: Bureau of Revenue, CN 417, Trenton, N.J. 08625-0417.

17.9.10   Transporters

Prior to operation, a hazardous waste transporter must obtain an approved registration statement from the NJDEP. A hazardous waste transporter license application may be obtained by calling NJDEP at (609) 292-7081 (see generally N.J.A.C. 7:26-7.2). The application for the registration statement must include proof of compliance with the minimum financial responsibility requirements covering public liabilities, property damage, and environmental restoration set out in 49 CFR Part 387, disclosure of any conviction for any criminal offense involving transportation of hazardous waste during the 10-year period prior to the application, vehicle identification numbers and license plates, a copy of the lease for any leased vehicles, and lease or ownership information regarding any waste transfer facility that the applicant intends to operate. Also a disclosure statement as set forth in N.J.A.C. 7:26G-7.4, or an affidavit explaining why an exemption is requested, shall also be included in the application (N.J.A.C. 7:26G-7.2).

In addition, a transporter must not transport hazardous wastes without having first received an EPA identification number from the EPA. Applications for the identification number can be obtained from EPA using EPA Form 8700-12 (N.J.A.C. 7:26G-7.1, 40 CFR 263.11). Nor may a transporter accept hazardous waste from a generator unless it is accompanied by a manifest signed by the generator pursuant to 40 CFR 262.20. Before transporting the waste, the transporter must sign and date the manifest, acknowledging acceptance. The transporter must return a signed copy to the generator before leaving the generator's property.

All vehicles that transport hazardous waste shall comply with the registration requirements as set forth in N.J.A.C. 7:26G-7.2(b). These include displaying a current New Jersey hazardous waste decal and the NJDEP registration number. When the vehicle is sold or otherwise no longer in the possession of the original applicant, the NJDEP must be notified and the decals destroyed. The NJDEP has the right to enter and inspect all vehicles registered to transport hazardous waste while in operation on the highways of the state or at their places of business [N.J.A.C. 7:26G-7.3(b)].

A transporter who delivers a hazardous waste to another transporter or to the designated facility must obtain the date of delivery and the handwritten signature of that transporter or the owner or operator of the facility on the manifest, retain one copy, and give the remaining copies of the manifest to the accepting transporter or facility. The transporter must keep a copy of the manifest for a period of 3 years from the date the waste was accepted by the transporter [40 CFR 263.22(a)]. In the event of a discharge during transportation, the transporter must take appropriate actions to protect human health or the environment. The transporter must notify the National Response Center and the NJDEP at (609) 292-7172 [N.J.A.C. 7:26G-7.1(c)(2)].

17.9.11   Treatment, Storage, and Disposal Facilities

In general, under 40 CFR 264.10-264.19 a TSDF owner or operator must:

- Acquire an operating permit from NJDEP before construction, installation, or modification of a hazardous waste facility
- Obtain an EPA ID number
- Conduct personnel training
- Accept only properly packaged and labeled waste shipments accompanied by a manifest
- Analyze waste
- Process the manifest
- Adhere to permit conditions and regulatory standards
- Maintain daily operating records
- Conduct periodic inspections

TSDFs must comply with general and facility-specific requirements found in N.J.A.C. 7:26G-8 through 11. Facilities exempt from these regulations are found in N.J.A.C. 7:26-8.1. The regulations address facility design, operation, construction, maintenance, closure, and postclosure. Additional requirements apply to facilities located in a 100-year floodplain. The federal RCRA regulations impose substantial financial responsibility requirements.

17.9.12 Waste Analysis

The TSDF owner or operator must obtain a detailed chemical and physical analysis of any waste received (see generally 40 CFR 264.13). The owner or operator is required to draft, implement, and keep a waste analysis plan that describes the methods and frequency of sampling and analysis. Furthermore, the TSDF owner or operator must develop a written contingency plan that addresses procedures for unplanned releases of hazardous waste into the environment (40 CFR 264.50 et seq.). The federal RCRA regulations contain corrective action requirements to remedy past and present releases of hazardous waste.

17.9.13 Processing the Manifest

In processing the manifest, if a significant discrepancy exists between the waste quantity received and that reported on the manifest, the TSDF must attempt to reconcile the difference with the generator or transporter. If the discrepancy is not resolved within 15 days, the TSDF must notify the NJDEP in writing, and provide a copy of the manifest, a letter report, and details of attempts to reconcile the discrepancy. A discrepancy is considered significant when there exists a greater than 10 percent weight variation in bulk wastes, a variation in piece count of batch wastes, or a discrepancy in waste type (see generally 40 CFR 264.72).

17.9.14 Periodic Inspections

Owners and operators must periodically inspect their facilities for malfunctions, deterioration, operator errors, and discharges that may result in harm to human health or the environment. Inspections must be conducted frequently enough to identify problems before they become significant. If any problem is detected, the owner or operator must remedy it (see 40 CFR 264.15).

17.9.15 Maintenance of Operating Records

The state and federal RCRA regulations require owners and operators of TSDFs to maintain operating records, manifests, unmanifested waste reports, biennial reports, incident reports, and facility closure records as set forth in 40 CFR 264 and 265. All monitoring information, permit reports, and data included in the permit application must be retained for at least 3 years. Owners and operators of TSDFs must maintain daily operating records, including descriptions of incidents that trigger the contingency plan, records and results of inspections, and refusals by authorities (local emergency personnel) to enter into emergency response and procedure arrangements. Owners and operators must submit biennial reports including a compilation of the daily records, a summary of all hazardous waste activities, and when applicable, hazardous waste landfill records to the EPA regional administrator (40 CFR 264.75).

17.9.16   Adherence to Permit Conditions and Regulatory Standards

Any noncompliance that may result in imminent harm to human health or the environment must be reported to NJDEP within 24 hours after the owner or operator became aware of the situation. A follow-up report must be provided within 5 days of the notification.

17.9.17   Permits

A permit is required for the treatment, storage, and disposal of any hazardous waste as identified in 40 CFR 261 (N.J.A.C. 7:26G-12.1). Owners and operators of hazardous waste management units must have permits during the active life of the unit. Facilities which do not need permits are listed in 40 CFR 270.1(c)(2).

The TSDF permit application consists of two parts. Part A of the application includes general information such as activities of the applicant, mailing address, location of the facility, SIC codes which describe the principal products, and ownership information. Part B of the application requires more detailed information, including a description of the facility, chemical and physical analyses of the hazardous waste, a copy of the waste analysis plan, a copy of the general inspection schedule, a copy of the contingency plan, precaution information, traffic patterns, and other facility information (40 CFR 270.14). Additional information may be required by NJDEP depending on the type of TSDF facility. For instance, containers, tank systems, waste piles, incinerators, land treatment facilities, industrial furnaces, landfills, impoundments, and drip pads have specific part B requirements (see 40 CFR 270.15-29).

The permit is effective for a fixed term not to exceed 10 years. Except as provided in 40 CFR 270.51, the term of the permit shall not be extended by modification beyond the maximum duration. Upon the expiration of the term of the permit, a new application must be filed (40 CFR 270.50). For land disposal facilities, the permit must be reviewed by the EPA director 5 years after the date of permit issuance or reissuance.

The permit may be modified "for cause." Situations that may trigger a modification include material and substantial alterations made to the facility or operations, change in applicable standards or regulations, and adjustment to the level of financial responsibility (40 CFR 270.42). A permit may be transferred by the permittee to a new owner or operator only if the permit has been modified or revoked and reissued to identify the new permittee [40 CFR 270.40 N.J.A.C. 7:26G-12.1(c)(8)]. The change of ownership or operational control shall not occur until the NJDEP issues approval to the new owner or operator in accordance with the requirements of N.J.S.A. 13:1E-133.

17.9.18 Other Agency Powers

NJDEP is authorized to enter and inspect vehicles transporting or licensed to transport hazardous waste, either while in operation on the state highways or at the owner's or operator's place of business (N.J.A.C. 7:26G-7.3).

NJDEP or the local board of health may inspect the premises of any hazardous waste generator, transporter, or TSDF at any time in order to determine compliance with the SWMA [N.J.S.A. 13:1E-48.20(a)].The inspection may include copying of records pertaining to the waste, inspection, and monitoring operations, and obtaining samples of the waste. RCRA requires that every TSDF be inspected at least once every 2 years. Annual inspections of government (federal, state, or local) owned or operated TSDFs are required. The act specifically provides for weekly inspections of major hazardous waste facilities by NJDEP and for assessment of costs incurred for the inspection. The local board of health or county health department may also conduct a weekly inspection of major facilities (N.J.S.A. 13:1E-42.1).

The SWMA also authorizes NJDEP to assess civil administrative penalties, seek judicial relief in the form of court-ordered civil penalties, or petition the state attorney general to bring a criminal action against the violator [see N.J.S.A. 13:1E-48.20(b)].

17.10 UNDERGROUND STORAGE TANKS

Overview

Underground storage tanks (USTs) that hold petroleum or hazardous substances are regulated by the federal government under RCRA, 42 U.S.C. 6991 et seq.; the state of New Jersey under the New Jersey Underground Storage of Hazardous Substances Act (NJUST Act), N.J.S.A. 58:10A-21 et seq.; and by municipalities under municipal ordinances, which must be expressly approved by NJDEP. In New Jersey, the UST program is implemented by NJDEP pursuant to authority delegated by the federal government in 1991.

The NJUST act is fashioned after RCRA and regulates all tanks containing hazardous substances. NJUST, in general, requires the registration of USTs and establishment of technical standards and permitting for new, existing, and closed tanks. A regulated UST includes both the tank and associated piping. Hazardous substances include motor fuel; petroleum products that are liquids at standard temperature and pressure; RCRA hazardous wastes; "hazardous substances" as defined under the CWA, CERCLA, and the Spill Act; and CWA-designated toxic pollutants (N.J.S.A. 58:10A-22e). Tanks excluded from regulation under N.J.A.C. 7:14B-1.4 include the following:

- Farm or residential tanks with capacities of 1100 gallons or less, used to store motor fuel for noncommercial use
- Tanks with capacities of 2000 gallons or less used to store heating oil for on-site use in a nonresidential building
- Tanks used to store heating oil for on-site use in a residential building
- Septic tanks
- Pipeline facilities regulated under other statutes
- Surface impoundments, pits, ponds, lagoons otherwise regulated
- Stormwater or wastewater collection systems otherwise regulated
- Flow-through process tanks
- Liquid traps or gathering lines related to oil or gas production and gathering operations
- Tanks on or above an underground floor (e.g., basements, cellars, mines, tunnels)
- Wastewater treatment tanks

17.10.1 Tank Registration

In New Jersey, owners or operators of underground storage tanks (USTs) must register their UST facility with NJDEP (N.J.A.C. 7:14B-2.1). A facility is defined as one or more underground storage tank systems on a contiguous piece of property. Information to be provided includes the age, size, contents, and type of construction of all tanks, as well as financial assurance information (N.J.A.C. 7:14B-2.2). Registration forms can be requested from Industrial Site Evaluation Element, Division of Responsible Party Site Remediation, 401 E. State Street, CN-028, Trenton, N.J. 08625, Attn.: UST Registration/Certification.

The registration certificate is effective for a maximum period of 3 years. The New Jersey underground storage tank facility certification questionnaire must be submitted prior to expiration of a registration certificate (N.J.A.C. 7:14B-2.2). The form certifies that the facility is complying with regulatory monitoring, record-keeping, and financial assurance requirements. It also updates any information regarding the registration status or any corrections or changes regarding the tank construction or operation (N.J.A.C. 7:14B-2.2). The certification form is submitted with a registration fee. Fees are assessed according to the schedule in N.J.A.C. 7:14B-3.5.

Tank registration is not transferable. Any change in ownership should be reported to NJDEP, so that a new registration certificate can be issued to the new owner (N.J.A.C. 7:14B-2.3). In addition, certain modifications to the facility may require amendment of a registration certificate (N.J.A.C. 7:14B-2.4).

17.10.2   Performance Standards and Engineering Requirements

NJDEP established performance standards and engineering requirements for new and existing USTs, including system piping. These standards address construction (N.J.A.C. 7:14B-4), operation (N.J.A.C. 7:14B-5), and monitoring (N.J.A.C. 7:14-6). Included in the technical requirements are secondary containment, leak detection, overflow protection, corrosion protection, and spill prevention (see generally N.J.A.C. 7:14B-6.1 through 6.4). All new tanks installed after Sept. 4, 1990, were required to meet the construction standards (N.J.A.C. 7:14B-4.1). Existing tanks had to be upgraded to these standards by Dec. 22, 1998 (N.J.A.C. 7:14B-4.2).

17.10.3   Records Maintenance

Owners and operators of USTs must maintain certain records, including the following: documentation of operation of corrosion protection equipment; documentation of repairs; compliance with release detection requirements; results of remedial investigations; installation checklists; and other records (N.J.A.C. 7:14B-5.6 and 6.7). Records must be available for inspection by NJDEP on request. After a facility is no longer operational, the owner or operator may make a written request to NJDEP for permission to discard the records. Records must be preserved until receipt of NJDEP's written permission to discard them [N.J.A.C. 7:14B-5.6(c) and 6.7(g)].

17.10.4   Release Reporting and Corrective Action

If a release occurs, the owner or operator has a duty to investigate and report the release to NJDEP. The owner or operator must conduct an investigation within 7 days from the date of a suspected release. NJDEP regulations at N.J.A.C. 7:14B-7.1 list several circumstances that constitute a suspected release. These circumstances include discrepancy in inventory, evidence of contamination of adjacent media, unexplained presence of water in the tank, erratic dispensing, sudden loss of product, and precision test results indicating a release. During the investigation, the owner or operator should, as appropriate, perform an inventory analysis, visually inspect the tank, calibrate dispenser meters, check the monitoring system, or perform groundwater analyses (see generally N.J.A.C. 7:14B-7.2).

Any person, upon confirmation that a release occurred, must report the release immediately to NJDEP emergency action hotline at (609) 292-7172 and the local health department. The notification must include the type and quantity of the release, the location of the release, and actions undertaken. Additional notification may be required, depending on the location and type of release (N.J.A.C. 7:14B-7.3). Upon confirmation of a release, corrective action must be implemented through the facility's release response plans. The owner or operator must determine the source of the discharge; discontinue the use of the leaking UST; mitigate fire, safety, or health hazards; conduct a visual inspection to detect above- or belowground discharges; remove hazardous substances from the leaking UST; and repair, replace, or close the UST (see N.J.A.C. 7:14B-8.1). The owner or operator must also implement the discharge mitigation requirements set forth at N.J.A.C. 7:14B-8.2, including:

- All free-floating or sinking product must be removed from above or below the water table in a manner that minimizes the spread of contamination. If this activity results in a discharge regulated by the WPCA, the owner or operator must obtain an NJPDES permit prior to implementing the removal activities.

- Soils contributing to a violation of standards adopted by the NJDEP, resulting in vapor hazards or posing a threat to human health, must be removed or decontaminated.

- For USTs containing gasoline, monitoring wells must be installed to define the groundwater contamination resulting from the release.

- The proximity of the discharge to water wells, surface water bodies, underground structures, and the surrounding population must be determined.

- Any groundwater or soil remediation activities required by the department must be completed.

If the release is to a secondary containment area, the owner or operator must submit a written follow-up report to NJDEP within 30 days describing the corrective action taken. If the release is to the environment, a written follow-up report by a qualified groundwater consultant must be submitted to NJDEP at Bureau of Underground Storage Tanks, 401 East State Street, CN-028, Trenton, N.J. 08625. The report must also be submitted to the local health department within 120 days of the initial release notification.

The follow-up report must include the results of mitigation and cleanup effort; the nature, quantity, and migration route of the hazardous substance; results of the investigation into the surrounding land uses; results of any monitoring and sampling; description of corrective actions taken and future actions planned; signed certification as to the accuracy of the report; and any other information requested by NJDEP. Records generated during corrective action must be maintained indefinitely. Reports must conform to the requirements of N.J.A.C. 7:14B-8.3.

17.10.5 UST Closures

UST closure must be completed according to the requirements set forth at N.J.A.C. 7:14B-9. Tanks that have been empty for more than 12 months require permanent closure, unless an exemption is obtained [N.J.A.C. 7:14B-9.1(c)]. Generally, the owner or operator is required to remove the UST and the monitoring system. Abandonment may be permissible if the UST is under a permanent structure, removal will cause damage to another structure, the UST is inaccessible, or the UST is being converted to a UST for nonregulated substances. When converting a UST, the owner or operator must perform a site assessment and ensure that the UST is properly cleaned.

Closure requirements for USTs containing nonhazardous wastes and petroleum products are set forth at N.J.A.C. 7:14B-9.2. NJDEP must be informed at least 30 days prior to closure of petroleum product USTs. Closure requirements for USTs containing hazardous wastes are set forth at N.J.A.C. 7:14B-9.3. NJDEP closure notification forms must be used. A site closure plan must also be developed. The closure plan consists of an implementation schedule, a site assessment, a tank decommissioning plan, and a plan for removal and disposal of all residual tank contents. Closure must be performed by an individual certified under N.J.A.C. 7:14B-13.

The owner or operator must, within 120 days of initiation of closure activities, submit to NJDEP a site investigation report [N.J.A.C. 7:14B-9.5(a)]. Records generated during tank closure must be maintained indefinitely [see N.J.A.C. 7:14B-9.5(c)].

17.10.6   Permits

An NJDEP permit is required for repairing, installing, expanding, or substantially modifying a UST system. Permitting requirements are set forth in N.J.A.C. 7:14-10. Additionally, for replacing, installing, or substantially modifying a facility, construction permits under the New Jersey Uniform Construction Code must also be obtained. Permit application conditions depend on the type of system and proposed activity. Permit applications must be submitted to NJDEP, Bureau of Field Operations, P.O. Box 435, 401 East State Street, Trenton, N.J. 08625.

17.10.7   Other Agency Powers

The NJUST gives NJDEP the authority to enter any place of business to inspect, photograph, or take samples in UST areas (N.J.S.A. 58:10A-30). It also provides NJDEP with authority to assess civil administrative penalties for violations (N.J.S.A. 58:10A-24.6).

17.11   INDUSTRIAL SITE RECOVERY

Overview

New Jersey's Industrial Site Recovery Act (ISRA) (formerly known as the Environmental Cleanup Responsibility Act, or ECRA) (N.J.S.A. 13:1K-6 et seq.), requires that owners and operators of certain "industrial establishments," prior to cessation of operations or transfer of ownership or operations, notify NJDEP and conduct an environmental assessment of the industrial establishment. Should contamination be identified, the transferor must then take steps to remediate. ISRA can be viewed as a "buyer protection plan" which seeks to ensure that subject sites are investigated and, if necessary, cleaned up by the transferor of the real property or business at a time when a source of funds (i.e., closing proceeds) is most readily available. By requiring compliance on the part of a business ceasing operations, a second goal of ISRA is to prevent contaminated sites from being abandoned and imposing upon third parties the ultimate burden of cleanup.

The Brownfields and Contaminated Site Remediation Act is intended to promote cleanup and development of industrial properties. It introduced more flexibility into the ISRA program, as well as incentives to persons willing to undertake cleanup and redevelopment of vacant or abandoned properties.

17.11.1 Applicability

"Industrial establishments" covered by ISRA are those places of business that involve the generation, manufacture, refining, transportation, treatment, storage, handling, or disposal of hazardous substances or hazardous wastes on-site, and which have an SIC number within major groups 22-39 inclusive, 46-49 inclusive, 51, or 76 [N.J.S.A. 13:1K-8 (definition of "industrial establishment")]. Where more than one type of operation is conducted by a business, ISRA follows the guidance of the Office of Management and Budget's *Standard Industrial Classifications Manual* and looks to the primary activity of the business for purposes of determining applicability. Certain SIC numbers within the referenced major groups are specifically exempted in the regulations, which were originally promulgated under ECRA, N.J.A.C. 7:26B-1 et seq., and which, except where ISRA requires otherwise, are still being enforced by NJDEP. NJDEP is currently in the process of formulating amendments to these regulations to conform to the changes resulting from ISRA.

Also exempt from ISRA's purview are (1) industrial establishments (or portions thereof) that are subject to "operational closure and postclosure maintenance requirements" under the SWMA, the New Jersey Major Hazardous Waste Facilities Siting Act, and the federal Solid Waste Disposal Act, and (2) businesses involved in the production or distribution of agricultural commodities (N.J.S.A. 13:1K-8). Furthermore, certain businesses that would otherwise fit within the definition of an industrial establishment may qualify for a de minimis quantity exemption under ISRA, provided that the owner or operator of the business can certify that the total quantity of hazardous substances and wastes generated, manufactured, refined, transported, treated, stored, handled, or disposed of at any time during the owner's or operator's period of ownership or operation falls below certain specified volumes (N.J.A.C. 7:26B-2.3).

Transactions that fall within the scope of "closing operations" "change in ownership" or "transferring ownership or operations" and trigger the requirements of ISRA are described at N.J.A.C. 7:26B-1.4 to include, but are not limited to, the following:

- Sale or conveyance of real property that is the site of an industrial establishment
- Termination or assignment of a lease, unless the operations of the industrial establishment are not disrupted
- Change in ownership of the industrial establishment (the business) itself, including certain transfers of indirect ownership interests, partnership interests, and shares of stock, that results in a change in the person holding the controlling interest of the owner or operator
- Dissolution of an entity owning (including certain indirect owners) or operating an industrial establishment
- Sale or transfer of more than 50 percent of the assets of the industrial establishment measured over a 5-year period
- A 90 percent reduction in product output, employees, or area of operations of an industrial establishment measured over a 5-year period
- Initiation of Chapter 7 bankruptcy proceedings or the filing of a plan of reorganization under Chapter 11 of the Bankruptcy Code

Both ISRA and its implementing regulations also provide specific exemptions from the covered list of transactions, including, but not limited to, certain corporate reorganizations "not substantially affecting the ownership of the industrial establishment," certain intrafamily transfers, and grants of mortgages, security interests, or other liens on real or personal property (see N.J.S.A. 13:1K-8; N.J.A.C. 7:26B-2.1).

17.11.2 ISRA Compliance Process

An owner or operator of an industrial establishment planning to close operations or transfer ownership or operations must notify the department in writing, no more than 5 days subsequent to closing operations or to its public release of its decision to close operations, whichever occurs first, or within 5 days after the execution of an agreement to transfer ownership or operations, as applicable (N.J.A.C. 7:26B-3.2). The notice to the department must contain information such as the date of the closing of operations; the date of execution of the agreement to transfer ownership or operations and the names, addresses, and telephone numbers of the parties to the transfer (N.J.A.C. 7:26B-3.3). The notice shall be transmitted to the department in the manner and form required by the department and sent to New Jersey Department of Environmental Protection, Division of Responsible Party Site Remediation, Industrial Site Evaluation Element, 401 East State Street, CN 028, Trenton, N.J. 08625-0028, Attn: Initial Notice Section, (609) 633-0708.

Subsequent to the submittal of the notice, the owner or operator of an industrial establishment shall remediate the industrial establishment with oversight by NJDEP pursuant to N.J.A.C. 7:26C. For purposes of ISRA, "remediate" and "remediation" are broadly defined to include all investigatory activities necessary to determine whether there are any known or suspected areas of contamination at a site, in addition to actual cleanup activities at such a site (N.J.S.A. 13:1K-8). The soil, groundwater, and surface remediation restricted use and nonrestricted use criteria shall be selected by the owner or operator and reviewed and approved by the NJDEP (N.J.S.A. 13:1K-9). The technical requirements for site remediation are specified in N.J.A.C. 76E.

Thus, at a minimum, a preliminary assessment (the initial search for and evaluation of current and historic ownership, operational, and environmental information for a particular site) must be performed at each industrial establishment at which the requirements of ISRA are triggered (N.J.A.C. 7:26B-6.1; 7:26E-3.1). In the case of an expedited review application, a preliminary assessment report need not be submitted to NJDEP (N.J.A.C. 7:26B-5.1). However, it is NJDEP's view that an applicant for expedited review nonetheless must take steps, consistent with those of a preliminary assessment, to evaluate the subject facility in order for the complying party to be able to make the certifications necessary for expedited review. If the preliminary assessment reveals no reason to suspect that a discharge of hazardous substances or wastes has occurred at the industrial establishment, then the complying party may submit a "negative declaration" affidavit certifying that there has been no discharge of hazardous substances or wastes or that any such discharges have been remediated in compliance with applicable regulations (N.J.A.C. 7:26B-6.7). If the negative declaration is acceptable to NJDEP, it will be approved by the issuance of a "no further action" (NFA) letter evidencing the satisfaction of the requirements of ISRA for the particular transaction at issue, and the transaction may proceed to closing. The NFA letter will include a covenant not to sue from NJDEP.

If, however, the preliminary assessment reveals the existence or suspected existence of a discharge, then the complying party must further investigate the industrial establishment, perform a site investigation (sampling and/or other investigative activities to determine whether a discharge of a hazardous substance or waste has occurred) (N.J.A.C. 7:26E-3.3); remedial investigation (sampling/investigatory activities to determine the nature and extent of contamination at a site) (N.J.A.C. 7:26E-4.1); and remedial action (the actual cleanup activities) (N.J.A.C. 7:26E-5.1), each as deemed necessary by the results of the prior phase of investigation. Except as described in Section 17.11.3, an ISRA-triggering transaction may close only upon (1) completion of all necessary remedial phases and issuance of a "NFA letter" or (2) approval by NJDEP of a remedial action work plan (if applicable) (N.J.S.A. 13:1K-9b, c).

The above-described compliance process can be lengthy and costly, even for relatively "clean" sites. The New Jersey Legislature has attempted to relieve the regulated community of some of the delay and expense involved in complying with ISRA through various methods. One is the privatization of remedial activities by specifically authorizing complying parties to do much of the remedial investigation required pursuant to ISRA without prior NJDEP approval of each phase of the investigation (N.J.S.A. 13:1K-9.f). Others include the following expedited and limited review mechanisms intended to streamline the compliance process:

- Expedited review: Subject to certain requirements, an owner or operator may apply for expedited review by the department where the industrial establishment has previously undergone an EPA- or NJDEP-approved remediation pursuant to ECRA/ISRA or an "equivalent" remedial program, and the owner or operator can certify that there has been no subsequent discharge at the site that has been neither remediated nor approved in compliance with applicable requirements (N.J.S.A. 13:1K-11.2; N.J.A.C. 7:26B-5.1).

- Limited site review: Subject to certain requirements, limited review is appropriate for those industrial establishments that have previously undergone an EPA- or NJDEP-approved remediation pursuant to ECRA/ISRA or an "equivalent" remedial program, and whose owner or operator can certify that, except for an identified discharge(s) which has yet to be remediated (or which has been remediated but has yet to be approved), there have been no other subsequent discharges at the site that have been neither remediated nor approved in compliance with applicable requirements (N.J.S.A. 13:1K-11.3; N.J.A.C. 7:26B-5.5).

- Area of concern waivers: Subject to the specific requirements of ISRA, area of concern waivers allow specific areas of an industrial establishment to be exempted from further review if such areas have been previously remediated and approved, and provided there have been no subsequent discharges in or from such areas (N.J.S.A. 13:1K-11.4; N.J.A.C. 7:26B-5.2).

- Discharges of minimal environmental concern: Subject to certain requirements (including completion of a preliminary assessment, site investigation, and remedial investigation), a complying party may apply to NJDEP to close on a triggering transaction prior to the issuance of an NFA letter or remedial action work plan approval letter if the results of the investigation reveal the existence of no more than two areas of concern at a site, neither of which is expected to require greater than 6 months to remediate (N.J.S.A. 13:1K-11.7; N.J.A.C. 7:26B-5.6).

In addition to these streamlined review mechanisms, ISRA also includes a mechanism to exempt from compliance certain industrial establishments at which the only area of concern is one or more regulated underground storage tanks, or a discharge from such tanks. Again, ISRA outlines various requirements that must be met in order for the owner or operator of a subject site to be able to avail itself of this provision (N.J.S.A. 13:1K-11.6; N.J.A.C. 7:26B-5.3).

17.11.3 Remediation Agreements

In situations where an owner or operator of an industrial establishment desires to close an ISRA-triggering transaction prior to issuance of a remedial action work plan approval letter or an NFA letter, the owner or operator may enter into a remediation agreement (formerly known as an administrative consent order, or ACO) with NJDEP (N.J.A.C. 7:26B-4.1). Pursuant to the remediation agreement, NJDEP authorizes the transaction to close before the owner or operator fully complies with ISRA, and in turn, the owner or operator certifies that (1) it will complete all necessary remedial actions at the site in a manner consistent with NJDEP regulations and all applicable laws, and (2) it is subject to the provisions of ISRA, including liability for penalties for violations (N.J.S.A. 13:1K-9.e). Because postclosing access to the site will be required by both the transferor and NJDEP to complete the required remedial activities, NJDEP also requires that the application for a remediation agreement be signed by the prospective transferee of the property or business. The transferee thereby specifically acknowledges its obligation to provide the necessary access.

17.11.4 Remediation Funding

Along with execution of a remediation agreement, the owner or operator must also demonstrate a remediation funding source in an amount sufficient to cover the entire estimated cost of remediation (based on an estimate prepared by the owner or operator and approved by NJDEP) [N.J.S.A. 13:1K-9.e.(3); N.J.A.C. 7:26B-6.4]. Under ISRA, the remediation funding source may be in the form of a remediation trust fund, an environmental insurance policy, a line of credit, or if the complying party qualifies, a loan from the Hazardous Discharge Site Remediation Fund (HDSRF) (N.J.S.A. 58:10B-3; N.J.A.C. 7:26C-7.2). Funds from a remediation trust fund, line of credit, and HDSRF loan need not serve merely as security for a complying party's cleanup obligations but may be drawn upon to pay for the remediation activities at the site. The complying party may also satisfy the remediation funding source requirement by self-guaranteeing the remedial work, provided certain financial tests can be met (N.J.A.C. 7:26C-7.7). Whatever source of remediation funding is used, it must be maintained for the duration of the ISRA compliance process. Funding must be increased whenever the estimated cost of remediation increases and may, subject to NJDEP approval, be decreased if the estimated cost of remediation decreases (N.J.A.C. 7:26C-7.9). For all sources of remediation funding except the HDSRF loan and self-guarantee, ISRA also imposes on the complying party an annual surcharge of 1 percent of the funding amount (N.J.A.C. 7:26C-7.8). The funds generated by the surcharge are deposited in the HDSRF.

A remediation funding source is also required if a remedial action work plan must be implemented at the site, even if a complying party is not subject to a remediation agreement. In such a case, the remediation funding source is required within 14 days of NJDEP's approval of the remedial action work plan.

17.12 THE BROWNFIELDS AND CONTAMINATED SITE REMEDIATION ACT

The Brownfields and Contaminated Site Remediation Act, enacted in January 1998, is designed to spur the acquisition and development of vacant or underutilized property that is known or suspected to be contaminated. The definition of "brownfields" used in the act is the same definition used by the federal government and expands upon the state's initial view that "brownfields" would include just vacant property (N.J.S.A. 58:10B-26).

17.12.1 Liability Protection

The Brownfields Act provides liability protection to parties that knowingly acquire a brownfield site, provided the acquiring party then remediates the site to standards imposed by the NJDEP and receives approval of the remediation by NJDEP. Once the remediation is approved, the state must provide a covenant not to sue the remediating party for any damages resulting from the contamination that was remediated (N.J.S.A. 58:10B-13.1). The covenant not to sue, however, does not afford protection to the party that was actually responsible for the discharge. The act also provides statutory protection against suits by third parties based on damages resulting from the prior contamination that was remediated. The only liability that the act does not and cannot afford protection against is liability to the federal government. However, it appears unlikely that the federal government will seek to recover damages from a remediated site that has received state approval.

17.12.2 Selection of Remedies

The Brownfields Act also contains other incentives to developers of brownfield sites aside from the liability protection discussed above. For example, the act permits the acquiring party to choose the remedial action for the site. In the past, if a nonpermanent remedy was proposed, such as leaving contaminated soil in place and capping the area, the remediating party was required to perform a detailed cost assessment of the nonpermanent remedy and the lowest-cost permanent remedy. The nonpermanent remedy could only be used in those situations where the cost of the permanent remedy was more than twice the cost of the nonpermanent remedy. Now, the remedy only has to be demonstrated by the remediating party to be protective of public health, safety, and the environment without regard to costs (N.J.S.A. 58:10B-12). The ability to use institutional and engineering controls can vastly reduce the costs associated with remediations, thereby making brownfield sites more attractive to investors and developers. In addition, despite strong lobbying by various groups to include local involvement in choosing remedial alternatives for sites in their neighborhoods, the Brownfields Act contains no requirement for public participation in the selection of the remedy. It is left entirely to the remediating party.

17.12.3 Financial Incentives

Aside from the exemptions from liability available, the Brownfields Act also provides financial incentives for development of brownfield sites, expanding the tax incentives in a 1996 state law that authorized municipalities to designate certain properties as environmental opportunity zones (N.J.S.A. 54:4-3.153). The 1996 law authorized 10-year property tax abatements for properties within the designated zones which were remediated and returned to commercial or industrial use. The Brownfields Act extends the tax abatement to 15 years if the developer uses an unrestricted use or limited restricted use remedy. In addition, the act permits municipalities to provide the tax abatement to sites that are remediated and developed for residential use.

17.12.4   Redevelopment Agreements

Another financial incentive is available for developers that enter into a redevelopment agreement with the state—a completely new feature (N.J.S.A. 58:10B-27). The decision whether or not to enter into a redevelopment agreement is solely within the discretion of the commissioner of Commerce and Economic Development and the state treasurer and both must agree to enter into any such agreement. While the act provides guidelines as to when such an agreement would be considered, the state can only enter into such an agreement if a finding is made that the state tax revenues to be realized from the redevelopment project will be in excess of the amount necessary to reimburse the developer. If a redevelopment agreement is entered into, the developer may be reimbursed up to a total of 75 percent of the total cost of its remediation (N.J.S.A. 58:10B-28).

17.12.5   Task Force

The Brownfields Act does more than just provide liability protection and financial incentives to developers of brownfield sites, whether in or out of an environmental opportunity zone. The act also creates the brownfields redevelopment task force (N.J.S.A. 58:10B-23). The task force is charged with preparing and updating an inventory of brownfield sites within the state. The inventory, to the extent practicable, shall include an assessment of the contaminants known or suspected at the site, the extent of any remediation performed at the site, the site's proximity to transportation networks, and the availability of infrastructure to support the redevelopment of the site. The task force is also charged with coordinating the state's policy on brownfields redevelopment and actively marketing the sites on the inventory to prospective developers.

The Brownfields Act also requires NJDEP to undertake an investigation to determine the extent of contamination of every aquifer in the state (N.J.S.A. 58:10B-21). Thereafter this information is to be made available to the public via NJDEP's geographic information system. Similarly, NJDEP is required to investigate and map those areas of the state at which large areas of historic fill exist, which information is to be periodically updated (N.J.S.A. 58:10B-22). Again, the information obtained is to be made available to the public via NJDEP's geographic information system. Since nonpermanent remedies which utilize institutional and engineering controls are presumed to be appropriate for areas of historic fill, the mapping of these areas within the state can drastically reduce the time and costs of remediating sites with large areas of fill.

17.12.6 Other Agency Powers

ISRA authorizes NJDEP to adopt rules and regulations to implement the statute, including criteria and minimum standards necessary for the submission, evaluation, and approval of preliminary assessments, site investigations, remedial investigations, and remedial action work plans, and for the implementation thereof (N.J.S.A. 13:1K-10; see generally N.J.A.C. 7:26B; N.J.A.C. 7:26E). ISRA also gives NJDEP a right of reasonable access to a subject property to inspect the premises, review records, and take soil, groundwater, or other samples deemed necessary to verify the submissions made by the owner or operator and/or the owner's or operator's compliance with ISRA (N.J.S.A. 13:lK-10; N.J.A.C. 7:26B-1.9).

Finally, an owner's or operator's failure to comply with ISRA may result in the imposition of one or more of the following penalties: (1) the sale or transfer of the subject property may be voided by either NJDEP or the transferee; (2) the transferee may collect damages from the transferor, who is strictly liable, without regard to fault, for all cleanup and removal costs and for all direct and indirect damages resulting from the failure to implement any necessary cleanup plan; and (3) civil penalties ranging from $250 to $25,000 per day may be sought by NJDEP through a summary proceeding under the Penalty Enforcement Law (N.J.S.A. 13:1K-13; N.J.A.C. 7:26B-1.11).

17.13   RIGHT-TO-KNOW AND POLLUTION PREVENTION

Overview

The term "right-to-know" refers to federal and state legislation requiring disclosure of information regarding hazardous chemicals used or stored at industrial facilities. Right-to-Know requirements are generally divided into two categories. The first category, community information, requires submission of hazardous chemical information to state and local emergency response agencies so that they can properly prepare for emergencies such as fires, explosions, or discharges and spills. The second category, which covers worker or employee safety, requires employers to inform and train employees regarding hazardous substances that they may be exposed to in their workplace. The NJDEP regulates the community information aspects of New Jersey's Worker and Community Right-to-Know Act (N.J.S.A. 34:5A-1 et seq.), while the Department of Health regulates the worker safety aspects of the program.

In addition, the collection and reporting of such hazardous chemical information has become a tool for broader pollution minimization goals. The New Jersey Pollution Prevention Act (N.J.S.A. 13:1D-35, et seq.) seeks to reduce significantly the use of hazardous substances in manufacturing and the resulting generation of hazardous wastes.

17.13.1   New Jersey Workers and Community Right-to-Know Act

The New Jersey Right-to-Know Act (N.J.S.A. 34:5A) requires disclosure of information about hazardous substances to workers and the public. The act applies to employers in certain designated SIC codes. It is administered in part pursuant to regulations adopted by NJDEP at N.J.A.C. 7:1G. NJDEP's community information program consists of two informational surveys: the community right-to-know survey and the release and pollution prevention report.

17.13.2   The Community Right-to-Know Survey

The Community Right-to-Know Survey form requires disclosure of inventory information including quantity, hazard type, and location of hazardous substances at the facility. The hazardous substances that must be reported on the survey form are listed on the New Jersey Environmental Hazardous Substance list (N.J.A.C. 7:1G-2.1). The reporting threshold for hazardous substances on the survey is 500 pounds or the TPQ as under EPCRA 311 and 312, whichever is less. The survey form is also designed to meet the requirements of the federal EPCRA 312 annual inventory form. Therefore, all OSHA hazardous substances at the site at any one time of 10,000 pounds or more, and all EPA extremely hazardous substances under EPCRA 312 that were present at the site in volumes of 500 pounds or the TPQ, whichever is less, should also be reported. As a result, only one form needs to be submitted to satisfy both federal EPCRA 312 and the New Jersey Right-to-Know Environmental Survey (see generally N.J.A.C. 7:1G-3.1).

The community right-to-know survey must be submitted to NJDEP, the local police and fire departments, local emergency planning committee, and the county health department. It must be submitted and updated every year by Mar. 1, reporting on the chemical inventories for the previous calendar year (see N.J.A.C. 7:1G-5.1).

17.13.3   Release and Pollution Prevention Report

The release and pollution prevention report requires employers that are subject to both NJDEP's community information program and federal EPCRA 313 release reporting requirements to submit information on the facility's use and final disposal of chemicals on the New Jersey Environmental Hazardous Substance list (N.J.A.C. 7:1G-2.1) as well as the facility's efforts toward pollution prevention. This report supplements, but does not replace, the federal 313 report. The release information required on the report includes (1) the quantities brought to the site; (2) the quantities consumed on-site, produced on-site, and shipped off-site; (3) information on waste minimization; and (4) information on the final disposal sites of waste material (see N.J.A.C. 7:1G-4.1).

Reporting on the facility's pollution prevention and waste minimization efforts is also required. The reporting thresholds are the same as the reporting thresholds under the federal EPCRA 313 reporting. Reports are due by July 1 of the year following the reporting year [N.J.A.C. 7:1G-5.1(c)]. For New Jersey facilities subject to the federal EPCRA 313, at least two forms must be submitted: the New Jersey release and pollution prevention report and the federal 313 toxic release inventory forms (one for each chemical over the thresholds).

17.13.4   Pollution Prevention Act

The New Jersey Pollution Prevention Act (N.J.S.A. 13:1D-35) focuses on the minimization of hazardous chemical use and hazardous waste generation. It applies to manufacturing facilities in certain SIC codes, which are also subject to the right-to-know requirements. It is implemented through regulations adopted by NJDEP at N.J.A.C. 7:1K.

The main requirement of the act is the preparation and submission of a pollution prevention plan and a pollution prevention summary, which must be updated annually and revised at least every 5 years (N.J.A.C. 7:1K-3.1, 3.7, 3.8). Information to be provided includes facility-level and process-level inventory data about the use, generation, and release of hazardous substances at the facility, and information on the total costs of using and generation hazardous substances. The plan shall also include a technical and financial analysis of pollution prevention options for reducing the use and generation of hazardous substances (N.J.A.C. 7:1K-4.3, 4.5).

A facility is required to set its own minimization goals, but achievement of those goals is not mandated by the act. NJDEP may condition the issuance or renewal of air, water, or solid waste permits, however, on implementation of pollution prevention strategies. NJDEP may also issue facility-wide permits to replace single-media permits, which incorporate pollution prevention elements (N.J.A.C. 7:1K-7). Annual progress reports are required to be submitted by July 1 of each year (N.J.A.C. 7-1K-6.1). The release and pollution prevention report described above in connection with the right-to-know program satisfies the pollution prevention annual reporting requirement. The regulations also provide a mechanism for protection of confidential business information (N.J.A.C. 7:1K-8 through 11).

17.13.5   Other Agency Powers

The New Jersey Right-to-Know Act provides NJDEP with the right to enter a covered facility during normal operating hours to determine compliance with the act [see N.J.S.A. 34:5A-29(b)]. It should also be noted that the New Jersey

Department of Health has a similar right of entry under the worker safety provisions of the act.

NJDEP is authorized to enter and inspect facilities in connection with the pollution prevention program (N.J.S.A. 13:1D-46). NJDEP may also assess penalties for violations of the Pollution Prevention Act program (N.J.S.A. 13:1D-49).

17.14 TOXIC CATASTROPHE PREVENTION ACT

Overview

The Toxic Catastrophe Prevention Act (TCPA) (N.J.S.A. 13:1K), was enacted to protect the public from catastrophic accidents and chemical releases of hazardous substances to the environment by anticipating the circumstances that could result in such releases and requiring precautionary and preemptive actions to prevent such releases [N.J.A.C. 7:31-1.3(a)]. The federal counterpart to the state program is the accidental release prevention (ARP) program at 40 CFR 68, which was promulgated pursuant to sweeping changes to the Clean Air Act in 1990, known as the 1990 amendments. The New Jersey requirements were amended in 1998 to be more consistent with the federal program. The NJDEP funds the TCPA through collection of annual fees from affected facilities. The calculation of the fee is provided in N.J.A.C. 7:31-1.11.

17.14.1 Risk Management Program

The TCPA regulations require a risk management program from all facilities which handle, use, manufacture, store, or have the capability of generating an extraordinarily hazardous substance (EHS) at certain quantities. The names and threshold amounts of EHSs are found at N.J.A.C. 7:31 Subchapter 6.

For new facilities, an owner or operator must execute a consent agreement containing an approved risk management program with the NJDEP before the facility can commence operation (N.J.A.C. 7:31-1.9). The program should contain requirements for safety review of design of equipment, requirements for standard operating procedures, requirements for preventive maintenance programs, requirements for operator training and accident investigation procedures, requirements for risk assessment for specific pieces of equipment or operating alternatives, requirements for emergency response planning, and internal or external audit procedures.

A risk management program should not be confused with a risk management plan (RMP). An RMP is a report required by both EPA and NJDEP. It evaluates the effects of a worst-case accident at the facility and includes an off-site consequence analysis, the 5-year accident history, prevention program information, and emergency response program information. In addition to the federal RMP, the TPCA regulations require an RMP which is more comprehensive than the federal

RMP (N.J.A.C. 7:31-7.2). Every affected facility must generate an RMP by deadlines provided at N.J.A.C. 7:31-7.1. The earliest deadline for submittal of an RMP under the state or federal program for existing facilities is June 21, 1999.

The owner or operator is required to maintain records of calculations, data, and assumptions used in the worst-case analysis, and must also maintain a history of all accidental releases at the facility that resulted in deaths, injuries, or significant property damage. Affected facilities must also develop and implement a management system. The management system must oversee implementation of the risk management program, establish a documentation plan for storage and updating of documentation pursuant to the TCPA regulations, and provide a means of recording the daily quantity of each extraordinarily hazardous substance contained at the facility (N.J.A.C. 7:31-1.1).

17.14.2 Other Agency Powers

Pursuant to TCPA, the NJDEP has the right to enter and inspect or audit any stationary source to determine compliance. The NJDEP may test or sample materials, copy any document, and interview any employee (N.J.A.C. 7:31-8.2).

17.15 CALIFORNIA

17.15.1 Introduction

In terms of environmental regulations and programs, California is one of the most proactive states in the country. Each year the state adopts, amends, and repeals hundreds of individual laws or parts of laws affecting the environmental process and its attendant impacts on business development and growth. In the past, prospective developers seeking environmental approvals and/or permits for new or modified projects have faced a regulatory bureaucracy and process that seemed to have no end. In the early 1990s up to the present, the state legislature at the urging of businesses, citizens, and in some cases even agencies charged with enforcing this legislation have instituted regulatory reforms which have made the environmental process more friendly, more usable, more efficient, and more beneficial to all of the citizens of the state.

Presently in California, most of the numerous environmental agencies are covered under the umbrella of the California Environmental Protection Agency (CAL EPA). Created in 1991, CAL EPA represents an extensive reorganization of state government which essentially unified the state's environmental authority under a single cabinet-level agency. CAL EPA consists of the following state agencies:

1. California Air Resources Board (ARB)
2. Department of Pesticide Regulation
3. Department of Toxic Substances Control (DTSC)

4. California Integrated Waste Management Board (IWMB)

5. Office of Environmental Health Hazard Assessment (OEHHA)

6. State Water Resources Control Board (SWRCB)

7. California Regional Water Quality Control Boards (RWQCB)

In addition, it should be noted, that there are still a number of state agencies with environmental oversight and project approval authority that are not within the umbrella of CAL EPA. Examples of these agencies are:

1. California Energy Commission

2. California Coastal Commission

3. California Coastal Conservancy

4. California Department of Forestry

5. California Department of Conservation

6. California Department of Fish and Game

7. California Department of Food and Agriculture

8. State Lands Commission

Owing to the size of the state and the many differences in topography, climate, population, and other factors, the design of most of the state's environmental programs is based upon a central agency which may or may not have direct control or oversight of a number of local or regional agencies. Examples are as follows:

1. State Air Resources Board—35 independent local air agencies

2. State Water Resources Control Board—9 regional water quality control boards

3. State Department of Toxic Substances Control—4 regional areas

However, one should keep in mind that California is above all a land use planning state. There are currently 58 counties and about 460 incorporated cities, each with its own land use regulations. These local governments have planning and building departments whereby some activities or development are allowed by right while other activities or development require special or conditional use permits.

Project owners or developers need to accomplish the following preproject tasks in order to proceed through the environmental process in an effective and efficient manner:

1. Identify and make contact with the following agencies with jurisdiction over the proposed project site:
   a. Local city or county planning agency
   b. Local air pollution control district (air quality management district)
   c. Regional water quality control board office
   d. Regional office of the department of toxic substances control

2. Once contact is made, these agencies (above) will be able to help the developer identify the potential need to contact other applicable agencies. Typically, the local planning agency will be able to give the owner a preliminary indication as to what agency will assume the lead agency role in the planning and environmental process versus which agency will be considered a responsible agency (as discussed in 17.15.2—CEQA of this book).

3. The owner is encouraged to work with the various agencies on a concurrent schedule. This includes the preparation, submittal, and processing of the various agency applications and data requirements. The developer should note that although concurrent processing is encouraged, final approvals from the various agencies will not occur until such time as the "lead" agency certifies the outcome of the California Environmental Quality Act (CEQA) process, if CEQA applies. In those cases in which CEQA does not apply, the developer should coordinate with the various agencies remaining, i.e., those with project jurisdiction, to complete the necessary environmental processes in a timely and cost-effective manner.

In the following sections, a general discussion will be presented on a number of California environmental agencies in order to acquaint the reader with both the agencies and their programs. Detailed discussion of these agencies, their programs, and regulations is beyond the scope of this book. In the Reference section at the end of this chapter the authors have provided a current listing of many of the identified agencies and their Internet site addresses. Contacting these sites will in most cases make available the following types of information:

1. Main agency addresses and staff directories: regional office locations, maps, and staff directory

2. Agency program overview

3. Agency regulations

4. Agency regulations, workshop, and hearing schedules

5. Agency newsletter and/or periodic reports

6. Agency publications

7. Application forms and permitting guidance documents

17.15.2 CEQA—California Environmental Quality Act

The California Environmental Quality Act, as enacted by the California legislature in 1970 and as amended to date, is considered to be California's most important environmental law. The purpose of CEQA is fourfold.

1. Inform the public and responsible agencies about potential environmental effects of proposed projects

2. Identify ways to avoid or mitigate potentially significant environmental effects

3. Prevent significant damage to the environment by requiring mitigation for projects

4. Disclosure to the public of the justification for overriding consideration for approval of any project with unavoidable, significant environmental effects

The CEQA applies to any or all activities undertaken by the following:

1. Government agencies

2. Activities financed in whole or in part by governmental agencies

3. Private activities which require approval from one or more government agencies

There are essentially three basic tasks to be accomplished within the CEQA process. These are:

1. Identify whether a project is subject to CEQA. This is most often done by a government agency, but it can also be accomplished by a private entity working with a governmental agency.

   Under CEQA, agencies involved in the process are referred to as "lead" or "responsible." The lead agency has the primary authority to implement the CEQA process and ensure that the proper project and environmental documents are prepared, reviewed, and certified. Responsible agencies act concurrently with the lead agency in processing their own specific permits or approvals, but responsible agencies may not issue a project's permits or approvals until the lead agency has approved and certified the necessary environmental documents.

2. If the project is subject to CEQA, then the governmental agency, in cooperation with the project developer, prepares an initial study to determine the potential effects on the environment. The lead agency, or an environmental evaluation committee, decides at a public hearing the nature of these potential environmental effects.

3. Dependent upon the outcome of the initial study, the process may trigger one of three possible options, as follows:
   a. Negative declaration (ND), which indicates that there is no substantial evidence that the project may have a significant effect or impact
   b. Mitigated negative declaration (MND), which indicates that the project can be modified in such a manner as to reduce or avoid the significant impact to a level of less than significant.
   c. Environmental impact report (EIR) if the initial study indicates that the project may have a significant effect and such effect or effects cannot be reduced to an insignificant level.

For NDs and MNDs, the lead agency prepares a draft permit for their respective commission or board, for their approval. Once the board has approved the draft permit, the lead agency can issue the permit. If an EIR is required, a very different process, although similar to the federal NEPA process, is used.

The EIR is a detailed report which analyzes and summarizes the proposed project's environmental effects, potential measures to mitigate the identified significant environmental effects, alternatives to the project, and any associated environmental effects, as well as any growth-inducing potential associated with the project.

Once an EIR is prepared and publicly reviewed, the lead agency then submits the EIR and the draft permit to the respective committee or board for their approval. Normally the board of supervisors (at a county level) or the city council (at a city level) will also need to approve the EIR and draft permit. Subsequent to certification of the EIR, the lead and responsible agencies may finalize the processing of their respective permits or authorizations.

Figure 17.1 presents a generalized flow of the CEQA process.

It should be noted that the CEQA process is subject to AB884 time frames, as discussed in Section 17.15.3. This act clearly defines the time frames for both the agencies and owners as they make their way through the environmental process.

17.15.3 AB884—Assembly Bill 884, Development Project: Environmental Quality

Prior to discussing the environmental and permitting processes on a specific basis, it is necessary to understand the environmental processing time frames currently required as a matter of law within California.

In the early to mid-1970s, as a result of the passage of the California Environmental Quality Act (1970), project developers became increasingly dissatisfied with the environmental permit issuance process. They believed, and rightly so, that the process was somewhat arbitrary and that it could go on for long periods of time with no resolution of issues or outcome. As a result, the legislature enacted provisions within the government code and the public resource code to rectify this situation and bring some semblance of stability and order to the process of environmental review and permit issuance.

The legislation is commonly referred to as AB884 (Assembly Bill 884, McCarthy). This important legislation requires the following:

1. Each lead and responsible agency is to compile a list of information and criteria needed to be included in a development or permit application, and that such information be disseminated to each applicant or to anyone who requests it.

2. The information compiled must also indicate the criteria which the agency will apply in order to determine the completeness of any application submitted to it for a development project.

In addition, AB884 set specific time limits on agencies for review and decision making on development projects. These are as follows:

CEQA PROCESS FLOW CHART

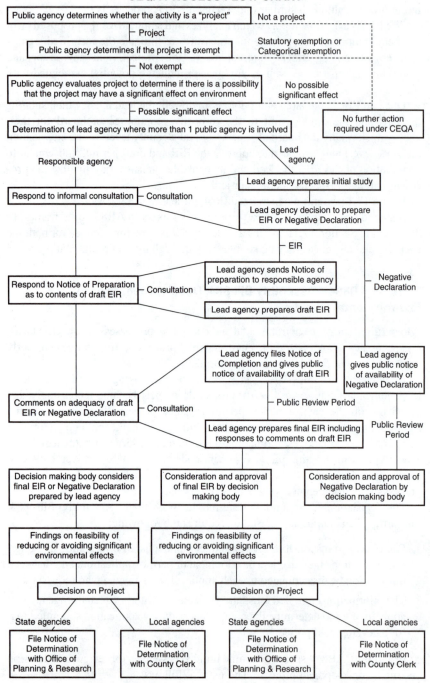

FIGURE 17.1 CEQA process flow chart.

1. Not later than 30 calendar days after any public agency has received an application for a development project, such agency will determine in writing whether such application is complete and will immediately transmit such determination to the applicant for the development project. In the event that the application is determined not to be complete, the agency's determination must specify those parts of the application which are incomplete and indicate the manner in which they can be made complete.

2. After a public agency accepts an application as complete, the agency cannot subsequently request of an applicant any new or additional information which, with respect to a state agency, was not specified in the list prepared pursuant to Section 65940 or, with respect to a local agency, was not required as part of the application. Such agency may, in the course of processing the application, request the applicant to clarify, amplify, correct, or otherwise supplement the information required for the application.

3. Any public agency which is the lead agency for a development project must approve or disapprove an application for a project within 1 year from the date on which an application requesting approval of such project has been received and accepted as complete by the agency.

4. In the event that a combined environmental impact report and environmental impact statement is being prepared on a development project pursuant to Section 21083.6 of the Public Resources Code, a lead agency may waive the time limits established in Section 65950. In any event, the lead agency shall approve or disapprove the project within 60 days after the combined environmental impact report and environmental impact statement has been completed and adopted.

5. A public agency which is a responsible agency for a development project must approve or disapprove such project within whichever of the following periods of time is longer:

   a. Within 180 days from the date on which the lead agency has approved or disapproved such project.

   b. Within 180 days of the date on which completed applications for such projects have been received and accepted as complete by each such responsible agency.

There are several caveats to the time frames delineated above, but these caveats are oriented in favor of the project owner-developer. They are as follows:

- All time limits specified in this article are maximum time limits for approving or disapproving development projects. All public agencies shall, if possible, approve or disapprove development projects in shorter periods of time.

- The time limits established by this article shall not apply in the event that federal statutes or regulations require time schedules which exceed such time limits.

- The time limits established by this article shall not apply to applications to appropriate water where such applications have been protested pursuant to Chapter 4 (commencing with Section 1330) of Part 2 of Division 2 of the Water Code.

- In the event that a lead agency or responsible agency fails to act to approve or to disapprove a project within the time limits required by this article, such failure to act shall be deemed approval of the development project.

- The time limits established by Sections 65950 and 65952 may be extended for a period not to exceed 90 days upon consent of the public agency and the applicant.

Proper implementation of the AB884 process has resulted in developers having a reasonable degree of certainty that the environmental review and permit process is fixed and that an outcome of decision will be reached within a period of time which allows for thorough agency review and balances the project development and implementation costs, and development schedule so as not to put a project in jeopardy.

Figure 17.2 presents a basic flowchart of the AB884 process, and Fig. 17.3 shows the general time lines and constraints for environmental review currently imposed on agencies in California.

17.15.4   Proposition 65—Safe Drinking Water and Toxic Enforcement Act of 1986

The Safe Drinking Water and Toxic Enforcement Act of 1986, better known as Proposition 65, was enacted by the voters of California in November 1986. The act, in and of itself, does not place any permitting requirements on businesses, but rather affected businesses are required to notify both its employees and the affected public of exposures to toxic substances as listed and defined within the act.

As of Apr. 1, 1996, the act contained a listing of over 550 chemicals and/or substances which were classified as either carcinogens or reproductive toxicants or both. The list of substances contains a wide range of chemicals including, but not limited to, dyes, solvents, pesticides, drugs, food additives, and by-products of certain processes. These substances may occur naturally or they may be synthetic or manmade.

Any company or business with 10 or more employees that operates within the state or sells products within the state must comply with the requirements of Proposition 65. Businesses are:

1. Prohibited from knowingly discharging listed chemicals into sources of drinking water.

2. Required to provide a "clear and reasonable" warning before knowingly and intentionally exposing anyone to a listed chemical. Such warning can be given by a variety of means such as labeling a consumer product, posting signs at the workplace, or publishing notices in a newspaper.

Pursuant to Proposition 65, a warning must be given unless an affected business can demonstrate that the exposure proves "no significant risk." No significant risk is defined as follows:

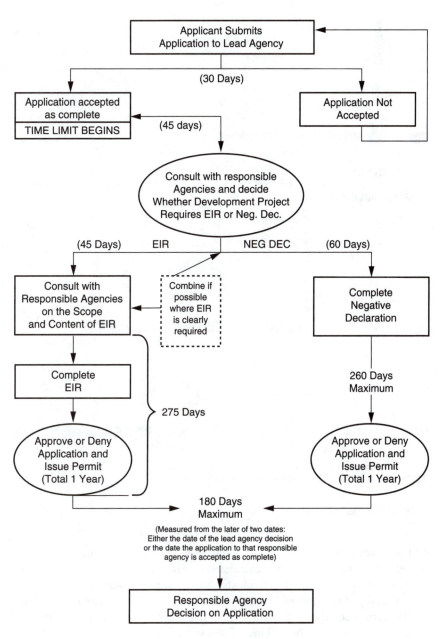

FIGURE 17.2 Basic flowchart for simple development projects.

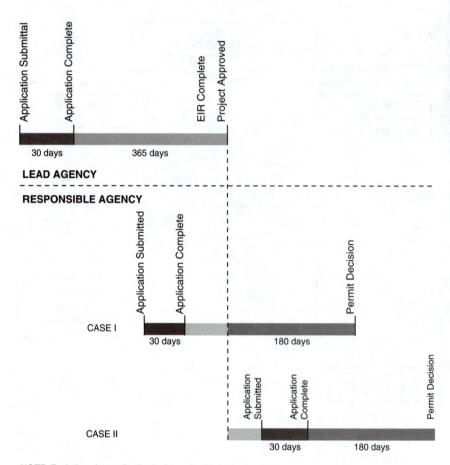

FIGURE 17.3   AB884 permit review time line.

1. For a chemical that is listed as a carcinogen, the "no significant risk" level is defined as the level which is calculated to result in not more than one excess case of cancer in 100,000 individuals exposed over a 70-year lifetime. In other words, if you are exposed to the chemical in question at this level every day for 70 years, theoretically it will increase your chances of getting cancer by no more than 1 case in 100,000 individuals so exposed.

2. For chemicals listed as reproductive toxicants, the no significant risk level is defined as the level of exposure which, even if multiplied by 1000, will not produce birth defects or other reproductive harm. That is, the level of exposure is below the "no observable effect level" (NOEL), divided by 1000. (The "no

observable effect level" is the highest dose level which has not been associated with an observable reproductive harm in humans or test animals.)

When a warning is given by a business, it means one of two things:

1. The business has evaluated the exposure and has concluded that it exceeds the no significant risk level.
2. The business has chosen to provide a warning simply based on its knowledge about the presence of a listed chemical, without attempting to evaluate the exposure. In these cases, exposure could be below the Proposition 65 level of concern, or could even be zero.

The general required content and form of warnings is specified within the act. Warnings are not required to be filed with the state; therefore, each affected business should keep detailed records of its Proposition 65 compliance program and efforts. Proposition 65 has accomplished the following:

1. It has provided an effective mechanism for reducing certain exposures that may not have been adequately controlled under existing federal or state laws.
2. Certain chemicals on the list are no longer used as constituents of some commonly used products.
3. Proposition 65 has resulted in the extensive dissemination of information about listed chemicals and their health effects.

The California state agency charged with the implementation and enforcement of Proposition 65 is the Office of Environmental Health Hazard Assessment (OEHHA).

17.15.5   City or County Permitting Requirements

Each city or county has prepared and adopted a general plan which guides the current and future land development of that jurisdiction. There are five different approvals or permits which may become necessary in developing a project or activity. The five are described and listed in the following subsections.

17.15.5.1   Local General Plan Amendments.   Development in California is required to be consistent with the general plan as implemented by the zoning and subdivision ordinances. If a proposed project is not consistent with the general plan, then a developer can submit an application to amend the general plan. Amendments are adopted by resolution by the local governing body (city council or county board of supervisors).

17.15.5.2   Specific Plan.   A specific plan is basically a detailed plan for the future development of a specific area or activity. If a project or development is located in the specific plan area or is covered by the specific plan activity, the planning departments must also be consistent with the specific plan. Otherwise, an amendment is necessary for the specific plan. The amendment process is identical to the general plan amendment process.

17.15.5.3 Zoning Ordinance Amendments. Each parcel of land in California has a zoning designation which details the potential uses of that land by right or conditionally. If a project or development is not included in the list of potential uses by right or conditionally, then a zoning ordinance amendment becomes necessary. The zoning ordinance amendment follows the same process as a general plan amendment.

17.15.5.4 Special or Conditional Use Permit. If the project or development is not permitted by right in the zone designation but is allowed subject to special conditions, then a special or conditional use permit is required for the project or development. This permit is used where a certain land use is acceptable within the zone yet not at every location within the zone unless conditions are attached. These requirements are specifically tailored to avoid environmental and land use issues associated with the particular proposed project or development. Also, the conditional use permit is an administrative action, unlike the general plan amendment, the special plan amendment, or the zoning ordinance amendment. An administrative zoning body, such as the planning commission, has the authority to approve these permits.

17.15.5.5 Subdivision Map Approval. In California, no one is allowed to subdivide land without local government approval. If a project or development requires a subdivision of land, then the Subdivision Map Act regulates this activity, and the process follows an administrative process similar to that for the conditional use permit process.

17.15.6 Air Resources Board

There are two basic entities responsible for the protection and maintenance of California's air resources. They are the State Air Resources Board (ARB) and the local air pollution control districts (APCDs) and/or air quality management districts (AQMDs).

Presently, there are 23 APCDs and 12 AQMDs, all of which will be referred to as APCDs. These 35 APCDs have jurisdiction over the air programs in the 58 California counties. Several of the APCDs cover multiple county areas. In addition to the APCDs, the state is broken up into regions known as air basins. These basins are delineated and identified based upon similarities in air quality and transport status, i.e., pollutant transport to and from the air basin.

The ARB has primary jurisdiction over nonstationary sources as well as program oversight authority as related to the local APCDs. The local APCDs have primary responsibility for regulation, control, permitting, and enforcement of air regulations on stationary sources (including fugitive sources, indirect sources, and nontraditional sources). The permits issued by the APCDs are the authority to construct and the permit to operate. The authority to construct is required for any project or development which has the potential to emit air pollutants from a stationary source. Each APCD has exceptions to this requirement. However, for those projects that require a permit, the permit must be obtained prior to construction. For major sources, many APCDs have been delegated authority from the U.S. EPA to issue prevention of significant deterioration permits. The APCDs also issue permits to operate. Once a project is constructed, the facility or equipment is inspected and once compliance is verified, the APCD issues a permit to operate. Also, for major sources, most APCDs have been delegated authority from the U.S. EPA to issue Title V permits to operate.

As stated above, stationary source permitting programs are handled exclusively by the various APCDs. These programs are subject to the CEQA and AB884 programs as discussed earlier. In most instances, an APCD will act as a responsible agency under CEQA, but there are occasions when the local districts take on the role as lead agency as it relates to both the CEQA and AB884 processes.

Prior to 1992, the one thing that could be said about the local APCDs is that they were "consistently inconsistent." Since 1992, there has been a concerted effort on the part of both the ARB and the APCDs to make the various local regulations as consistent as possible, on a statewide basis. But there still remain significant differences in basic rules between the urban and rural districts, and it should not be implied that these differences are detrimental to overall air quality, as they are not.

Each of California air district regulations contains the following general types of rules and regulations:

1. Basic nuisance provisions
2. Visible emissions limitations
3. Basic prohibitory rules for particulate matter, sulfur oxides, nitrogen oxides, carbon monoxides, volatile organic compounds, and other specific pollutants
4. Permit processing rules
5. Fee provisions
6. Source or process specific regulations
7. New source review provisions for both major and nonmajor sources for criteria pollutants
8. New source review for air toxics sources
9. Existing source air toxic regulations
10. Hearing board and variance procedures
11. Air pollution episode procedures
12. NSPS delegations

13. NESHAPs delegations
14. Indirect and fugitive source regulations
15. Title IV and V permit programs
16. Other regulations specifically targeted at one or more special problems within that APCD

Permit application, processing, and issuance programs at the local APCDs have been in place since the mid-1970s and are well developed and in most cases very efficient. Project owner-developers are encouraged to meet with applicable APCD staff well in advance of the project initiation date. Preproject meetings can be used to acquire permit application forms, list criteria for development applications, discuss the needs of the APCD staff in processing the potential project application, and determine if any permitting pitfalls exist.

In addition to the programs above, the local APCDs have the responsibility to implement and administer several state-mandated programs which are codified in state law but not within the APCD regulations. One such example is the AB2588 Air Toxics Hot Spots: Information and Assessment Act (AB2588). The AB2588 program is essentially an air toxics identification, quantification, and reporting program which applies to a wide variety of sources based upon emissions of criteria pollutants. Affected sources must identify all processes and emissions points at each facility, quantify the emissions of all identified air toxics from such emissions points (as delineated in the regulation), and report such emissions to the local APCDs on a periodic basis. APCDs then score each facility based on a number of factors as either high, medium, or low priority. All high-priority and some medium point sources are then required to conduct health risk assessments. The outcome of the risk assessments is used by the APCDs in cooperation with the facilities to institute risk reduction and risk management procedures to reduce exposures to the surrounding population to toxic emissions. Project developers should use the preproject meeting as a forum to ascertain what other programs or regulations, not codified in the APCD rules, will be applicable to the project.

17.15.7 State Water Quality Control Board

The State Water Resources Control Board (SWRCB) was created by the California legislature in 1967. The SWRCB mission is to ensure the highest reasonable quality of waters of the state, while allocating those waters to achieve the optimum balance of beneficial uses. The joint authority to oversee water quality as well as water use enables the SWRCB to provide comprehensive protection for all of California's waters.

California is divided into nine regions, each designated as a regional water quality control board. Each RWQCB is charged with the responsibility to develop and enforce water quality regulations and programs which will protect the beneficial use of the state's waters within each specified region. Each RWQCB develops a "basin plan" specific to its hydrologic area. These plans incorporate programs to regulate waste discharge requirements, provide for enforcement procedures, and monitor existing water quality.

The RWQCB's efforts are directed at both surface and groundwater impacts and uses, as well as commercial and industrial discharges, commercial and industrial treatment processes, underground storage tank programs, nonpoint sources of discharges, and watershed management. Prospective project developers should contact the applicable RWQCB with jurisdiction over their project site to ascertain the level of RWQCB involvement in project review and permitting.

The types of permits issued by the RWQCB are as follows:

1. National pollutant discharge elimination system permit—which regulates any surface water discharges.
2. Waste discharge requirements permit—regulates any activity or facility which discharges waste that has the potential to affect groundwater.
3. Underground storage of hazardous substances permit is issued to owners of underground storage tanks. These permits are normally issued by local agencies with RWQCB oversight.

17.15.8 Department of Toxic Substances Control

The Department of Toxic Substances Control (DTSC) is another agency within the California Environmental Protection Agency (CAL EPA). Similar to the ARB, SWRCB, and IWMB, this agency has the primary responsibility to protect public health and the environment from harmful exposure to hazardous wastes and substances, without unnecessarily impacting sustainable growth and development.

In addition to regulating and permitting under the federal hazardous waste (RCRA) regulations, DTSC also regulates non-RCRA hazardous waste. California has adopted more stringent criteria for identifying hazardous wastes. These wastes which are not RCRA waste yet fall into the California criteria become the non-RCRA hazardous wastes and must be managed, in most cases, like RCRA hazardous waste. The criteria for identifying non-RCRA hazardous waste is extensive; the reader is directed to the California Code of Regulations, Title 22, Division 4.5 for additional information. However, a waste profile form presented in Appendix P of this book is based on the California criteria, which can be compared to the RCRA waste characterization form.

In addition to the federal permitting requirements for hazardous waste, the following lists some of the permits, approvals, and reports that are mandatory in California:

1. Hazardous waste generators must submit quarterly hazardous waste tax returns to the state board of equalization.

2. Hazardous waste transporters must register with the Department of Toxic Substances Control and comply with specific procedures.

3. All hazardous waste generators are required to use the California hazardous waste manifest, including small quantity generators.

4. Generators of California extremely hazardous wastes must obtain an extremely hazardous substances disposal permit prior to disposal of the waste.

5. Generators producing more than 12,000 kg per year of hazardous waste are required to prepare source reduction plans every 4 years.

17.15.9   Integrated Waste Management Board

The California Integrated Waste Management Board (IWMB) began operation in 1990 following the passage of the Integrated Waste Management Act by the California legislature in 1989. The act not only established the board but set statewide goals for division of wastes, initiation of enhanced landfill safety requirements, programs for wastes such as tires and used oil, and recycling programs. Most, if not all, of the IWMB's programs are implemented and carried out by the city and county governments, commonly referred to as LEAs (local enforcement agencies). The IWMB is responsible for ensuring that the state has adequate landfill capacity to dispose of all wastes not diverted per state requirements. Additionally the IWMB, in cooperation with local agencies, implements and oversees the following programs:

• Local assistance and planning
• Public education programs
• Market development effort
• Reduce and reuse programs
• State agency waste reduction programs
• Waste tire and used oil programs
• Household hazardous waste programs

The Integrated Waste Management Board issues a solid waste facilities permit which covers landfills, transfer and processing stations, composting facilities, waste to energy facilities, recycling centers, and used tire facilities.

17.15.10   Department of Fish and Game

Project owner-developers who contemplate or require an alteration to an existing streambed or lake as part of the project development should contact the California Department of Fish and Game to acquire the necessary application forms and guidance materials for appropriate alteration permits. Regulatory requirements for such permits can be found in the California Public Resources Code Section 1600 et seq. The need for a permit is triggered by any of the following:

A project will:

- Direct, obstruct, or change the natural flow or the bed channel, or back of any river, stream, or lake designated by the department in which there is at any time an existing fish or wildlife resource or from which these resources derive benefit
- Use material from the streambeds designated by the department
- Result in the disposal or deposition of debris, waste, or other material containing crumbled, flaked, or ground pavement where it can pass into any river, stream, or lake designated by the department

Permit issuance may be impacted by the CEQA process. Permit issuance involves site studies by the department in consultation with any and all local agencies involved in the project. The process also incorporates an arbitration phase if mutual agreement between the department and the agency cannot be reached on the project design and mitigation. Permit fees are required at the time of application filing. Additionally, if the department in its review of the application determines that a threatened or endangered species may be present, the following items may be required:

- A Section 10A or Section 7 consultation with U.S. Fish and Wildlife Services for a "taking" of a federally listed threatened or endangered species
- A 2081 permit issued by the California Department of Fish and Game under the California Endangered Species Act

REFERENCES (ON CALIFORNIA)

Proposition 65 in Plain Language, Cal EPA, OEHHA, Feb. 20, 1998.

Kirvan, B. J., and J. J. Park, "Warning Wars, A Statute Intended to Protect Drinking Water Provides Windfalls to Private Parties," *Los Angeles Daily Journal,* Apr. 26, 1996.

Safe Drinking Water and Toxic Enforcement Act of 1986, Title 22, Division 2, Part 2, Subdivision 1, Chapter 3, 12000 et seq., West Group, 1997.

Guidelines for the Preparation of Air Quality Impact Analyses, Ventura County APCD, Oct. 1989.

California Environmental Quality Act as Amended in 1993, Latham and Watkins, Attorneys at Law, San Diego, Calif., CEQA Seminar, 1993.

AB 884, Government Code, Chapter 4.5, Divisions 1 of Title 7, 65920 et seq., Sept. 1977.

AB 884 List and Criteria Correspondence, CARB, May 23, 1978.

California Environmental Quality Act, Public Resources Code, Division 13, Chapter 4.5, 2100 et seq.

Proposed Amendments to the Air Designation for State and National Ambient Air Quality Standards, CARB, TSD, Nov. 1996.

Christie, Glennys, Editor, *California Public Sector,* 4th Edition, Public Sector Publications, Sacramento, Calif., 1997.

The California Planners 1996 Book of Lists, 1996 Edition, State of California, Office of Planning and Research, Feb. 1996.

California Environmental Laws, 1996 Edition, West Publishing Company, 1996.

AB 2588 Regulation, California Administrative Code Title 17, Subchapter 7.6, Article 93300 et seq., July 1995.

Proposed Amendments to the Emissions Inventory Criteria and Guidelines Report Adopted Pursuant to the Air Toxics "Hot Spots" Information and Assessment Act of 1987, Cal EPA, CARB, TSD, Feb. 1997.

Cal EPA Directory, California Environmental Protection Agency, Sacramento, Calif., Nov. 1996.

Waste Management Reference

CIWMB 1996 Annual Report, California Integrated Waste Management Board, Apr. 20, 1990.

Water References

SWRCB, *Mission Statement Apr. 20, 1998,* State Water Resources Control Board, Cal EPA.

SWRCB, *Protecting California's Water,* State Water Resources Control Board, Cal EPA, Apr. 20, 1998.

Toxic Substances Control References

Department of Toxic Substances Control, *Mission Statement,* DTSC, Apr. 20, 1998.

Department of Toxic Substances Control, *Fact Sheet,* EPA ID Numbers, Document 107, DTSC, Cal EPA, 94-75392, May 1994.

Current California Agency Internet Site Addresses

Air Resources Board, www.arb.ca.gov*.

Integrated Waste Management Board, www.ciwmb.ca.gov.

State Water Resources Control Board, www.swrcb.ca.gov*.

Department of Pesticide Regulation, www.ccdpr.ca.gov.

Office of Environmental Health Hazard Assessment, www.calepa.cabwnet.gov/oehha/.

California Energy Commission, www.energy.ca.gov.

CEQA Information Network, http://ceres.ca.gov/planning*.

Department of Toxic Substance Control, www.calepa.ca.gov/dtsc*.

Streambed Section Reference

Streambed/Lake Alteration Notification, Guidance and Application Package, Department of Fish and Game, Form FG 2023 (November 1987), May 1994.

CHAPTER 18
THE ENVIRONMENT AND THE FUTURE

A. Roger Greenway
Principal
RTP Environmental Associates Inc.

18.1 PERMITTING FOR THE FUTURE

EPA has recognized for some time that environmental permitting in the United States has become a complex and difficult process; many companies that have applied for environmental permits have complained of problems with EPA and state permitting agencies. In 1994, EPA established the Permit Improvement Team (PIT) in an effort to improve the environmental permitting processes. EPA undertook extensive outreach to obtain input from all public stakeholders on issues of concern in specific media permitting programs. This information helped EPA in setting specific areas to identify areas of improvement and to develop recommendations on how improvement could be realized. The Permit Improvement Team developed a concept paper on environmental permitting that included specific activities in six areas aimed at assisting in the transition from the current permitting system to that envisioned as the future approach.

An immediate problem that was identified is that the existing environmental legislation under which the EPA permitting program exists may limit the agency's ability to change the programs. This is because specific requirements in each piece of environmental legislation for environmental permits at the federal and state level prescribe the nature and procedures for obtaining such permits. For EPA to come up with a more logical and cohesive permitting approach would require legislative changes to most, if not all, current environmental legislation.

Another objective of the Permit Improvement Team was to see if the agency's objectives could be met, while at the same time improving the efficiency of the EPA. This is in line with the perceived desire of the public for the government to achieve more with less funding.

A key outcome of PIT's work was the realization that what would be most effective would be "public performance-based permitting." The essence of this is to shift the focus of environmental permitting toward the measurement and assurance of performance, while providing flexibility as to how a permittee would meet performance standards. Under this concept, the focus of the system would be not to simply identify performance requirements, but to identify performance within the public arena. To the extent possible and appropriate, the public should be involved in the setting of performance standards and the measurement and judgment of a permittee's performance. A conclusion of PIT was that too much time and resources are being currently spent in the current system in reviewing the technical means by which a permittee will comply with permit conditions. The team concluded that, while 25 years ago detailed technical reviews were warranted, sufficient progress has been made in verifying technology and increasing corporate and environmental responsibility that it is now appropriate to reevaluate this approach.

The ultimate measure of the performance of EPA's environmental permitting system is the condition of the air, land, and water. Current permitting systems focus primarily on gathering information about a company's compliance, but at the same time generate very little information on the actual effects of the permittee activities on human health and the environment. Environmental permitting systems are also seen to lack the flexibility to restructure and rearrange their priorities in response to environmental performance data, since they are often set up to issue individual permits solely on the basis of the potential impacts of each facility. In reality, what is needed is an approach to writing permits that reflects the fact that the environment, whether it be land, water, or air, is impacted by the sum total of all environmental discharges and emissions.

One approach to shifting the focus is to increase ambient monitoring, while decreasing other compliance information requirements for a facility. This would allow permitting agencies to prioritize permitting information requirements on the basis of real environmental impacts. PIT envisioned taking this approach one step further, which would be to prioritize which facilities are required to get permits, which can receive general permits, and which may require no permits at all. These decisions would be based on certain conditions or levels of emissions and the condition of the environment in the surrounding area.

The goal in reforming the existing permitting system is to make permitting systems less proscriptive and more performance-based. One method of encouraging flexibility is to allow permitting agencies to reduce the number of times permits need to be formally modified. Under the current system, permit modifications are required frequently, for example, every 3 or 5 years, or when facility modifications occur. One of the purposes of making permitting performance-based rather than technology-based is to encourage and allow facilities to pursue innovative technical approaches at preventing pollution at the source. In addition to pollution prevention technologies, the permitting system should encourage use of more cost-effective innovative technologies of any type, where practical and consistent with legal requirements. EPA envisions that the flexibility provided the facilities would be coupled with a public reporting and compliance system that would provide visible and understandable information on whether or not facilities are meeting their permit conditions.

The EPA Permit Improvement Team developed a number of recommendations in the six main areas:

1. Administrative streamlining recommendations
   a. Create a predictable, user-friendly federal permit system
   b. Encourage and implement flexible permitting projects
   c. Tier permitting programs in proportion to environmental significance
   d. Establish computer systems to permit and track results
   e. Provide regional permit process assistance

2. Alternatives to individual permit task force recommendations
   a. Consider the appropriateness of using alternative permit approaches
   b. Consider the performance of state, tribal, and local permit programs that may regulate the same or similar activities
   c. Develop and maintain a clearinghouse of permit alternatives being developed and used in federal, state, tribal, and local programs throughout the country
   d. Specific recommendations for alternatives to individual permits and the air, water, and waste permitting programs

3. Enhance public participation task force recommendations
   a. Develop easy reference guidance for public participation activities
   b. Utilize the environmental justice public participation checklist as guidance to the extent appropriate and feasible
   c. Develop an inventory of mechanisms that promote access to environmental information
   d. Explore, and possibly conduct pilot studies for, the development and use of comprehensive, multimedia community involvement plans
   e. Develop a series of case studies on the effectiveness of public participation activities

4. Performance measures task force recommendations
   a. Develop generic performance measures covering process, results, and customer service:
      (1) *Process.* Time limits and number of pending permits
      (2) *Results.* Environmental indicators and level of compliance

(3) Customer service. Customer satisfaction

  b. Develop generic tracking measures covering processes and results:

    (1) Processes. Overall time required for permit issuance permit application completeness, cost of permitting program, and number of pending renewals (air/water) and interim status Resource Conservation Recovery Act (RCRA) permits

    (2) *Results*. Pollution prevention/innovative technology

5. Pollution prevention incentives task force recommendations

  a. Link performance-based permitting with facility-based permitting, consolidation of permitting requirements, and cross-media permitting

  b. Create industry-sector inventories of regulatory thresholds for permitting

  c. Explore offering alternative emissions tracking in exchange for using pollution prevention practices

  d. Share pollution prevention data with permit applicants and affected communities, and give basic pollution prevention training to permit writers

  e. Develop an enforcement policy to accommodate the possibility that innovative pollution prevention technologies may not perform as expected or might take longer to achieve compliance

  f. In all general permits and permits-by-rule, include language that explains a preference for using pollution prevention approaches and the potential economic benefits of pollution prevention

6. Training task force recommendations

  a. Provide information to the regulated community and others

  b. Provide information on every significant new rule

  c. Define and provide training course on skills and knowledge needed to issue permits

  d. Store and provide critical knowledge

In July of 1996, the Permit Improvements Team published its final draft of a concept paper on environmental permitting and task force recommendations. The purpose of the concept paper was to review EPA's efforts over the past 25 years of searching for the best ways to protect the environment. The paper notes that one of the most successful methods has been EPA's programs requiring industrial and municipal facilities to obtain permits to control pollutant emissions into the air, land, and water. Programs such as the New Source Review program for air emissions, the National Pollutant Discharge Elimination System (NPDES) for water discharges, and the RCRA for hazardous waste management have achieved success for each of their relevant areas in reducing the impacts of industrial and municipal facilities on human health and the environment. At this point, the EPA feels the most significant challenge it faces is to answer the public's demand for more environmental protection at less cost. An additional concern is that traditional permitting methods may be losing their ability to produce further improvements in environmental indicators. This is because, once major government and municipal sources are controlled, there is little more that permitting programs can achieve. The agency must, therefore, look to more innovative ways to provide environmental improvement.

The Permit Improvement Team, in its concept paper, provides a list of recommendations to serve as a guide to the EPA as it strives to improve the issuance, implementation, and compliance in enforcement of environmental permits. PIT recognizes that there is much ongoing work in this area and that this will be an evolutionary process.

Besides PIT, there are a number of other groups within the agency that have been formed to address current environmental challenges through innovative approaches. These include the Common Sense Initiative, Project XL, the Environmental Leadership Program, the Incentives for Environmental Auditing and Self-Policing, the Small Business Policy, and the Incentives for Small Committees Policy. The PIT program, instead of duplicating the efforts of these other programs, intends to build on them and to provide a synergistic environment where each of these programs may achieve improvements in its own area.

The PIT concept paper provides guidance to the EPA for developing a revised approach to environmental permitting: public performance-based permitting. This, in effect, is a major paradigm shift for EPA. It incorporates two concepts in one: (1) the establishment of a defined level of performance to be achieved by the permittee and (2) providing the public with the necessary information to monitor the permitting process and compliance of permitted facilities. Once the final draft of the concept paper has been approved, after the incorporation of any additional comments, the intent is that it be used by EPA in developing revised permitting programs.

EPA envisions that the recommendations of PIT, in addition to being useful to the EPA in revising federal permitting programs, will also be useful to state, tribal, and local governments where these principles may also be helpful.

It is important to note that EPA's relationship with state, tribal, and local environmental agencies will change if PIT's recommendations are adopted. In the past EPA has dictated that state programs must be no less restrictive than the federal permitting programs, but were free to be more restrictive. EPA maintained veto power over all state, tribal, and local permitting actions. Under PIT, however, instead of issuing most permits itself and having veto power over individual permits at the state, tribal, and local level, EPA would establish programs under which state, tribal, and local authorities would perform most of the permitting. Recently, EPA and the states signed an agreement, the National Environmental Performance Partnership System, whose aim is to make EPA's oversight of states less uniform and prescriptive and more based on performance, so that states with more effective programs and proved environmental results might receive less stringent oversight by EPA. A similar approach is being developed for Indian tribes.

18.2   PERMIT IMPROVEMENT INITIATIVES

EPA has recently completed a review of the permitting programs of more than 100 environmental agencies that have permit reform projects under way. This review, entitled "The Inventory of USEPA/State Permit Improvement Initiative," is available via the Internet at *www.epa.gov* by using the search function and entering "permit improvement team" to locate the inventory. The most recent update of this inventory can be found as App. Q.

18.3   PUBLIC PERFORMANCE-BASED PERMITTING

The purpose of environmental permitting is to establish the level of performance needed by facilities or individuals to protect human health and the environment. EPA has, in some cases, set performance standards that are highly prescriptive (including detailed technology or management requirements) and eliminate or severely restrict alternative approaches to achieving compliance. In other cases, EPA based its standard on a technology that can be viewed by the regulated community as the technology of choice. In many air-related instances, however, EPA sets acceptable ambient levels, for example the National Ambient Air Quality Standards, and relies on states to implement state implementation plans and specific permitting requirements to achieve these standards.

It is the conclusion of PIT that, in the past, too much time and resources have been spent on reviewing the technical means by which a permittee will comply with permit conditions. Conducting detailed technical reviews of off-the-shelf technologies has resulted in several negative consequences:

1. Permitting agencies are overloaded with routine detailed paperwork to review. This takes time away from other activities such as verifying the equivalency performance for innovative technologies, and it also causes permit actions to take an unacceptable amount of time and prevents some logical and beneficial order in their priorities.

2. The regulated community, in addition to sometimes being burdened by unwarranted paperwork, a slow permitting process, and unnecessary economic hardships, in some cases has not been given the flexibility or any incentive to seek the kind of technological innovations that could prevent pollution at the source and/or provide better environmental results at lower cost.

3. The permitting process is largely focused on technical issues, sometimes beyond the grasp and interest of the general public. The permitting agency and permittee can spend much time grappling with these issues while the public is usually excluded until such time as the issues have been resolved with the writing of a draft permit.

In order to remedy this situation, PIT proposed a permitting approach called *public performance-based permitting,* or P3. The essence of this approach is to shift the focus of environmental permitting toward a measurement of performance, while providing flexibility as to how a permittee will meet performance standards. The P3 principle includes three different types of performance. The existing permitting programs each contains elements of this principle. The objectives of the permitting programs will be to more fully implement each type of performance:

1. *Environmental results:* How are permit activities actually affecting the environment? To improve knowledge and understanding of this factor, PIT recommended that permitting agencies increase ambient environmental monitoring as a permit condition in selected permits. Additionally, ambient monitoring results will be made available to the public in an understandable format.

2. Facility compliance: How well are permitted facilities complying with their permits over time? To increase the compliance rate, PIT recommended that permitting agencies should:

   • Establish reporting requirements based on a facility's level of compliance.
   • Create incentives for pollution prevention and technological innovation.
   • Provide compliance assistance to facilities that are making good-faith efforts, but are finding it difficult to comply. Furthermore, compliance data should be made available to the public in understandable terms.

3. Agency performance: How good a job are EPA and other environmental permitting agencies doing? To ensure that they continue to protect the environment in the most cost-effective and efficient ways possible, PIT recommends that EPA provide methods that measure the performance of permitting systems and continue to improve these systems on the basis of the performance data received. Information on the performance of all permitting agencies should be publicly reported in understandable terms.

A key theme through each of these three recommendations is monitoring of performance and reporting to the public in understandable terms. Traditionally, permitting agencies have limited public participation to common comment periods and hearings at latter stages of the permitting process. The PIT concept paper sets forth a more open process that provides the public opportunities for early and more meaningful participation. This model builds on recent initiatives and public participation including EPA's RCRA Expanded Public Participation Rule and the Chemical Manufacturing Association's (CMA's) Responsible Care Program.

In addition to increasing public awareness regarding facility operations, these programs can serve as powerful incentives for facilities to reduce their toxic emissions so as to avoid arousing public concern. P3 would extend these concepts to the public reporting of the ambient monitoring results, facility compliance data, and information on how well EPA and permitting agencies are performing.

Furthering effective permitting process for individual permits requires that the public be involved early and intimately enough that their needs and concerns may be incorporated into permits and other aspects of the facility and/or agency policy. An example of how effective this can be is the CMA Responsible Care Program. Under this program, chemical plants are encouraged to establish community advisory panels to which the facility and members of its surrounding community can establish a continuing dialog. Such forms allow the public and the facility new opportunities to educate each other on their respective needs and concerns, and to jointly resolve differences on environmental issues.

PIT envisions public performance-based permitting to change the relationship among agencies, permittees, and the general public. The permitting process is currently often burdened with mistrust and environmental adversarial relationships among all three of these parties. If these relationships can be rebuilt on the basis of trust, partnering accountability, and cooperation, the most serious obstacles to an effective and efficient permitting program system will have been removed.

18.3.1 Environmental Results

The ultimate measure of the performance of EPA's environmental permitting system, as understood by PIT, is the condition of the air, land, and water. While current permitting systems focus primarily on gathering information about permittee's compliance, and comparatively little information on the actual effects of the permit activities on human health and the environment, the P3 process would correct this situation. The P3 process is consistent with other EPA initiatives, including its effort to reduce paperwork requirements by 25 percent.

Since EPA began issuing environmental permits in the early 1970s, there have been instances where either general permits or permits-by-rule have been used. For example, the RCRA permits-by-rule are routinely given to publicly owned treatment works, and there have been a number of general permits developed for wastewater discharges. These permits are issued to replace individual permits and require less administrative oversight, as well as less effort on the part of permit applicants.

The Permit Improvement Team, in its efforts to develop alternatives to individual permits, has identified two criteria to be used to determine when general permits, permits-by-rule, hybrid permits or conditional/de minimis permits might be used. These criteria are:

• Issue permits only where there is a real or potential adverse environmental impact and a regulatory agency needs to be involved.

• Issue individual permits only where there is potential for significant environmental impact or high degree of variability in regulatory requirements for individual facilities.

PIT has also developed a hierarchy of permitting standards. The key in these standards is to make the permitting system less prescriptive and more performance-based, and to continue to achieve environmental objectives with less prescriptive requirements for the applicant. The standards would also: (1) aid the environment by encouraging pollution prevention, (2) help permittees by giving them the opportunity to develop more cost-effective approaches to pollution control and prevention, and (3) help permitting agencies by allowing them to shift resources from extensive engineering and paperwork reviews to focus on establishing ambient monitoring systems, environmental problem analysis, standard setting, and compliance assurance and enforcement.

Permitting based on performance standards rather than on technology or management requirements is not a completely new concept. EPA's National Pollutant Discharge Elimination System uses a similar approach. Performance-based permitting is now the recommended approach, wherever feasible and appropriate, for all EPA permitting programs, and state, tribal, and local governments will be provided the flexibility and guidance to implement similar programs.

EPA has developed a hierarchy of preferred approaches to be used in setting permitting standards:

- Base performance standards on ambient environmental goals
- Base performance standards on technological achievability
- Set technology or management-specific standards

A key assumed benefit of the shift in permitting at the federal (and hopefully state and tribal levels) is to increase a facility's operational flexibility. A key complaint of existing approaches is that operating flexibility is hampered when, before modifications to a facility can be implemented, permits have to be modified or new permits obtained, often requiring months and in some cases years for final approval. The PIT-recommended approach would:

- Alleviate the problem by making agency reviews of permits more performance-based. This would allow reducing review steps to those needed to reasonably demonstrate that the permittee will meet performance standards.
- Permitting agencies would be given leeway to reduce reporting and compliance monitoring requirements that are deemed to be unnecessary or duplicative.
- Permitting agencies would also be allowed to reduce the number of times permits need to be formally modified.
- Permitted facilities would be encouraged to establish mechanisms for conducting regular dialog with the public such as community advisory committees.
- Permitting agencies would be encouraged to use a permitting process to encourage municipal or industrial facilities to practice pollution prevention.
- Permitting agencies would be encouraged to support innovative technologies and pollution prevention.

- Agencies would be encouraged to make distinctions between paperwork violations of little or no consequence to the environment, and permit violations with the potential to damage the environment or human health. Enforcement activities would take into account facility compliance records to help EPA state and tribal permitting agencies better target inspections, enforcement actions, and penalties on the basis of severity of violations.

18.4 GOALS FOR ADMINISTRATIVE STREAMLINING

EPA has also established an administrative streamlining task force whose goal is to improve the permit process by analyzing successful permit programs across the country and recommend permitting process changes designed to apply these successes more broadly. The principal recommendations of this task force include:

- Create a predictable, user-friendly federal permit process:

    Create a single point of contact for all media permits

    Encourage and implement flexible permitting projects

- Tier permitting programs in proportion to environmental significance

- Establish computer systems to integrate facility databases with geographic information systems and permit application systems

- Develop regional permit process assistance

As part of the recommendation above, EPA envisions an electronic database/clearinghouse that would provide access to, and maintain, relevant information necessary for the permit writer in all media, including pollution prevention, toxic use reduction, pollution allocation/total maximum daily load (TMDL) models, and site-specific protocols.

18.4.1 Program-Specific Recommendations

The Permit Improvement Team has developed a series of recommendations for each environmental program area. The principle recommendations are:

1. NPDES—Stormwater:
    a. The development of general permit language that emphasizes pollution prevention (P2) and best management practices (BMP) in nonconstruction, industrial, and multisector general permits.
    b. The establishment of appropriate monitoring requirements, based on industry type, water quality, or capability to implement BMP.

2. NPDES—Process wastewater:
   a. Maintain individual permits in areas where water quality is limited or where TMDLs are necessary or where specific conditions to be addressed in the permit are not amendable to a general permit.
   b. Permit duration should be increased from 5 to 10 years or to the life of the facility.
   c. The use of general permits in nonwater-quality-limited areas and non-TMDL areas should be expanded through policy directives and through development of general permit boilerplates and establishment of a national clearinghouse for general permits.
   d. A permit-by-rule (PBR) should be established for de minimis discharges that establish threshold conditions, below which no reporting would be required. Overall monitoring requirements should be decreased, but include ambient, as well as end-of-pipe monitoring.

3. Toxic Substances Control Act (TSCA):
   a. As policy matter, state the primacy for the PCB disposal program (one-stop shopping).
   b. Consolidate hazardous waste requirements to avoid program duplication.
   c. Utilize EPA grant money for state actions such as PCB and hazardous waste control. The task force also recommended incorporation of PCB combustion requirements into the air permits program.

4. Safe Drinking Water Act/Underground Injection Control Program:
   a. Continue use of authorization by rule for shallow injection wells (Class V wells), provided they comply with certain minimal requirements.
   b. Allow for injection of fluids related to oil and gas production (Class II wells) where appropriate; continue use of area permits.
   c. Require continuing individual permitting for Class I wells (deep wells for industrial, municipal, and hazardous waste injection.

5. RCRA Permit Program: The Permit Improvement Team identified needed improvements to the RCRA permitting program but noted that there are regulatory or statutory barriers to some of the approaches they would like to see implemented. This means that regulations would need to be amended and perhaps legislation enacted to allow some of the recommended changes to the RCRA program. The recommended changes include:
   a. RCRA base program. Maintain individual permits for facilities requiring operating and postclosure land disposal permits.
   b. General permits
      (1) The Office of Solid Waste should establish a general permit boilerplate and promote the use of general permits for noncommercial storage or treatment facilities, including laboratories.
      (2) Extend the generator storage time frames from 90 to 270 days for laboratories as part of regulatory reinvention. For hazardous waste

combustion facilities, EPA should incorporate RCRA requirements into the air permits program where both apply—for example, for combustion emission control requirements.

(3) For state permitting of corrective action programs where feasible.

(4) Prioritize the issuance of corrective action permits and orders by focusing on state programs that are not authorized and do not have cleanup program requirements substantially similar to those of the EPA.

6. Air—New Source Review (NSR) Permit Program: The task force agrees with the Office of Air Quality Planning and Standards (OAQPS) in its NSR reform efforts. Specific recommendations of PIT include:

   a. Implementing plant-wide applicability limit (PAL) policy

   b. Allow states more flexibility to match the level of permitting effort to environmental significance.

   c. Include special provisions to encourage use of innovative technologies.

   d. Promote pollution prevention activities.

   e. Develop more extensive programs for minor sources through the use of:

   (1) Modified definition of potential-to-emit to make it less restrictive.

   (2) Develop and promote the use of general permits by preparing boilerplate language for applicable sources in establishing a national clearinghouse of general permits.

   (3) Develop guidance on de minimis levels for selected minor sources under which no permit would be required.

7. Air—Title V Permit Program: PIT supports EPA's White Paper and supplemental Part 70 proposals described in Chap. 1. It recommends:

   a. Evaluating techniques to take inherent operating limitations into account in determining potential to emit.

   b. Investigating methods to simplify the renewal process to allow for automatic renewal upon a recertification that no facility changes have occurred or new requirements have come into effect since the initial permit issuance:

   (1) Develop and promote the use of general permits for sources with low actual emissions by preparing boilerplate language for applicable sources and establishing a national clearinghouse of general permits.

   (2) Allow a self-implementation alternative for facilities with actual emissions of less than 50 percent of applicable emissions standards (e.g., Title V applicability threshold).

   (3) Implement flexible permits through the use of plantwide applicability limits.

   (4) Allow states more flexibility in deciding the most effective monitoring methods and controls.

   (5) Allow states and tribal permitting authorities more flexibility in establishing additional de minimis levels for selected minor sources, under which no permit would be required. This would require that only sources with true health and environmental risks require permits. For example, in Massachusetts, sources with emissions below 1 tpy do not require an air permit.

Clearly, this last approach of de minimis permit exemptions must be implemented on a state-by-state, region-by-region basis. Allowing all sources of under 1 tpy to be exempt from permitting requirements would be a disaster in highly developed areas where the cumulative impact of a large number of sub-1-tpy sources could be a major contributor to air quality problems.

18.4.2 Measures of Environmental Agency Performance

An interesting component of the Permit Improvement Team recommendations is the focus on improving the performance of environmental permitting agencies. This is a recognition of the fact that, while applicants have been required for years to provide all information considered relevant to permitting decisions by the agencies and to provide these often on very strict timelines, agencies have not been held accountable to similar performance requirements. In some instances, states have adopted walk-through permitting procedures, where an applicant can walk in with a simple permit application and walk it through an agency office to facilitate a quick response. In other instances, states have adopted 30-, 60-, or 90-day permit review windows under which they would commit to provide a determination on a permit application or the permit would be deemed granted. These types of programs have been growing across the United States and the Permit Improvement Team is recommending that agencies be held accountable, and that their performance be documented. PIT has developed generic performance measures for this purpose. The first performance measure involves process performance:

1. *Timeliness:* PIT recommends that each EPA regional office establish processing time goals for each type of permit they issue, expressed as a percentage of applications processed within a specified time frame. Regional offices are encouraged to determine the appropriateness of dividing the permit universe according to the degree of potential environmental impact—for example, minor, significant minor, or major. The total processing time for each permit would be tracked. For Type I permits, the time required from receipt of an application to agency determination that the application is complete would be tracked as follows:

    Percent of time the determination is made within 30 days.

    Percent of time the determination is made between 30 and 60 days.

    Percent of time the determination is made between 60 and 90 days.

In a similar way, the time required from receipt of a complete application to issuance of the proposed permit would be tracked. The time frame here would track as follows:

Percent proposals/decisions made within 60 days.

Percent proposals/decisions made between 60 and 90 days.

Percent proposals/decisions made between 90 and 180 days.

The purpose of these measures, and other similar measures, is to have the regional office focus on each step of the permit review and approval process. The time required to process the permit is influenced by the performance of both the regulatory agency and the permittee, as well as by the level of public comment. To achieve the most rapid processing of a permit possible, the agency and permittee need to work together.

2. *Number of pending permits:* Each regional office that has issuing permit authority is encouraged to establish a goal for the maximum number of permits for new discharges, emissions, or releases pending at any time. The purpose of this measure is to provide a measure for the number of permits for new discharges that have not been processed within a defined time period. This measure of backlog above the goal would trigger an evaluation to determine its cause and how to improve the agency's performance.

3. *Results:*
   a. *Environmental indicators.* The success of environmental permitting programs needs to be evaluated on the basis of the environmental conditions that exist in a particular area. Although permanent discharges are not the only sources of pollutants, they are regulated to limit their impacts so environmental goals can be achieved. Therefore, PIT recommends that all permitting agencies develop specific environmental indicators to be used to evaluate the overall success of their permitting programs. EPA is in the process of developing environmental indicators for the nation. Once these indicators are determined, each regional office will be encouraged to work with the respective state and tribal governments to establish specific indicators for those jurisdictions. It is not clear, as of this writing, what these environmental indicators will be. Clearly, ambient air quality standards have existed for several decades and provide indications of environmental performance. Presumably, an environmental indicator could be the number of times an ambient air quality standard is exceeded, or for areas where no exceedence does occur, the margin between the ambient air quality standard and indicators of ambient air quality as reported by the Ambient Air Quality Monitoring Network in an area.
   b. *Level of compliance.* The compliance status of all permitted facilities is an important performance measure for permitting programs. In order for environmental protection to occur, facilities must be in compliance with their permits. Just issuing the permit does not ensure protection, therefore it is necessary to determine the level of compliance with those permits to help identify what greater clarity of permit conditions is needed and where to provide technical assistance.

CHAPTER 19
ENVIRONMENTAL JUSTICE

Nicole E. Ambrose
Environmental Scientist
RTP Environmental Associates Inc.

19.1 INTRODUCTION TO ENVIRONMENTAL JUSTICE

No person in the United States shall, on the ground of race, color, or national origin, be excluded from participation in, be denied the benefits of, or be subjected to discrimination under any program or activity receiving Federal financial assistance. (*Title VI*)

While the nation promises all Americans clean air and a safe, healthful environment free of toxic waste, segments of the population find themselves forced to live in areas where they encounter disproportionately high and adverse human health and environmental effects from pollution and toxins. Citizens are living in communities saturated with abandoned toxic waste sites, freeways, smokestacks, waste landfills, and wastewater pipes from polluting industries. Are these communities experiencing a significantly reduced quality of life relative to surrounding or comparable levels, and should they be afforded special protection from additional adverse impacts? Some groups believe this is true under the "environmental justice" definition, which applies Title VI of the 1964 Civil Rights Act (forbidding discrimination against minorities with respect to air and water permitting or other environmental projects), so that every person has the right to a safe, healthy, productive, and sustainable environment (*environment* includes the ecological, physical, social, political, aesthetic, and economic environments). This right should then be freely exercised, whereby individual and group identities, needs, and dignities are preserved, fulfilled, and respected in a way that provides for self-actualization and personal and community empowerment. However, states and industry question the fairness of applying this definition to denying a permit for a company that would create 15 percent of air emissions in an area where 85 percent was already being emitted by existing facilities. The concern is that a claim of disparate impact can derail a permit for years, perhaps permanently, and impact community growth, small business, and industry development.

19.1.1 Evolution of the Environmental Justice Movement

Historically, there has been an awareness of a disproportionate burden borne by some minorities and low-income communities and an inequitable distribution of the benefits associated with environmental regulations. These issues have been a topic of discussion and study for more than 25 years. For example, in 1971, the annual report of the President's Council on Environmental Quality presented information on disparities in environmental hazards across socioeconomic groups. These and subsequent studies published during the next two decades evaluated the social justice aspects and showed that economically disadvantaged minority communities were more likely to experience above-average levels of air pollution.[1] Unfortunately, most of these communities lacked political clout, economic means, or awareness of rights and opportunities to participate in environmental decision making to amend these injustices. Before long, many community-action groups emerged nationwide in the 1980s to protest locally unwanted land uses.

Then in 1982, a small, low-income, predominately African-American community characterized the beginning of the environmental justice movement through community activism. A landfill site was created in Warren County, North Carolina, that was going to be used for the disposal of PCB-contaminated soil,

---

[1] Ken Sexton, Alfred Marcus, K. Easter, and Timothy Burkhardt, eds., *Better Environmental Decisions: Strategies for Governments, Businesses, and Communities* (Washington D.C.: Island Press, 1999).

which had been removed from 14 colonies throughout the state. Many civil- and states-rights activists collaborated to stage numerous demonstrations, creating national attention. These demonstrations resulted in the creation of a rally point for those eager to focus on the prejudiced usage of community lands. The protestors not only saw this issue as a threat to the residents' health and all-around well-being, but as a violation of their civil rights. The protest sparked the beginning of a study by the General Accounting Office of eight southern states to determine if a correlation between the location of hazardous waste landfills and the racial and economic status of the surrounding communities existed. The investigation revealed that there was an obvious bias in the placement of landfills: three out of every four landfills were located near predominantly minority communities.[2]

These and other findings prompted many civil and social justice groups across the country to emerge and protest against environmental hazards located in poor communities. As the number of grass roots organizations grew and their activities expanded, they began to come into contact with each other, resulting in the formation of regional assistance networks to help local communities and groups with issues related to environmental justice.[3] Then in 1985, the first national African-American environmental organization, the Center for Environment, Commerce, and Energy, was established and the National Council of Churches' Eco-Justice Working Group began to address environmental issues. Two years later, the United Church of Christ's Commission for Racial Justice released a nationwide study on the sociodemographic characteristics of populations living near waste sites called "Toxic Waste and Race in the United States." This landmark report coined the term "environmental racism" and found that in communities with one or more commercial hazardous waste facilities, the proportion of racial minorities was significantly greater than in communities without such facilities.[4]

Subsequently, over the last decade, attention to the degree to which the distribution of environmental hazards is due to discrimination of particular sectors of our society has been steadily growing. Studies have been conducted by a variety of organizations (National Law etc.), which conclude that certain communities are indeed more vulnerable than others to the health threats from environmental pollution. These studies maintain that the implementation of key environmental laws have not historically provided environmental protection to all citizens. This concern slowly grew into a national movement to assure environmental justice for all segments of our society. In 1990, a group of social scientists, community activists, and civil rights leaders formed the Michigan Coalition, devoted to exploiting environmental justice issues. This group, among other accomplishments, was primarily responsible for issuing the formation of the internal Environmental Equity

---

[2]"History of Environmental Justice," *http://www-personal.umich.edu/-jrajzer/nre/history.html.*

[3]Environmental Justice Resource Center. *People of Color Environmental Groups: 1994–1995 Directory.* Prepared by Robert D. Bullard, Clark Atlanta University, Atlanta, Georgia.

[4]Susan A. Perlin, Ken Sexton, and David W.S. Wong, "An Examination of Race and Poverty for Populations Living Near Industrial Sources of Air Pollution," *Journal of Exposure Analysis and Environmental Epidemoiology,* September 1,1999, pp. 29–48.

Workgroup, whose objective was to review the evidence that racial minority and low-income communities bear a disproportionate environmental risk burden and to make recommendations for agency action on environmental equity issues.[5] The Workgroup examined and integrated environmental justice concerns and activities into EPA's existing environmental programs and released "Environmental Equity: Reducing Risk for All Communities,"[6] which sparked a flurry of press articles, media coverage, and public attention. Soon after, environmental justice was at the forefront of the nation's policy agenda, and policy responses began to be formulated and implemented at federal, state, and local levels. Currently, environmental justice remains a conspicuously political and increasingly litigious topic, raising the question of whether science can make a difference in such a politicized, value-laden debate.[7]

19.2 PRESIDENTIAL ACTIONS—EXECUTIVE ORDER 12898

President Clinton and Vice President Gore made environmental justice a high priority for the new administration. On Earth Day 1993, President Clinton issued a statement on environmental justice, calling for an interagency review of federal, state, and local regulations and enforcement activities that affect low-income and minority communities.[8] On February 11, 1994, President Clinton issued Executive Order 12898, "Federal Actions to Address Environmental Justice in Minority Populations and Low-Income Populations." The presidential memorandum accompanying the order has the primary purpose of ensuring that "each Federal agency shall make achieving environmental justice part of its mission by identifying and addressing, as appropriate, disproportionately high and adverse human health or environmental effects of its programs, policies, and activities on minority populations and low-income populations in the United States and its territories and possessions, the District of Columbia, the Commonwealth of Puerto Rico, and the Commonwealth of the Mariana Islands."[9] The U.S. Environmental Protection Agency's Office of Environmental Justice defines environmental justice as:

> The fair treatment and meaningful involvement of all people regardless of race, color, national origin, or income with respect to the development, implementation, and enforcement of environmental laws, regulations, and policies. Fair treatment means that no group of people, including racial, ethnic, or socioeconomic group

---

[5]"Better Environmental Decisions," p. 427 (footnote 1).

[6]U.S. EPA, 1992.

[7]Ken Sexton and John L. Adgate, "Looking at Environmental Justice from an Environmental Health Perspective," March 1, 1999, pp. 3–8.

[8]"Better Environmental Decisions," p. 429. (footnote 1).

[9]Executive Order 12898: "Federal Actions to Address Environmental Justice in Minority Populations and Low-Income Populations," *http://www.worldnetdaily.com/bluesky_govdocs/eos/eo12898.html*, p. 1.

should bear a disproportionate share of the negative environmental consequences resulting from industrial, municipal, and commercial operations or the execution of federal, state, local, land tribal programs and policies. The goal of this "fair treatment" is not to shift risk among populations, but to identify potential disproportionately high and adverse effects and identify alternatives that may mitigate these impacts.

All agency strategies have to consider enforcement of statutes in areas with minority populations and low-income populations, greater public participation, improvement of research and identification of differential patterns of subsistence use of natural resources. The executive order also requires that agencies conduct activities that substantially affect human health or the environment in a nondiscriminatory manner. In addition, better data collection and research is required, that is, whenever practicable and appropriate, future human health research must look at diverse segments of population and must identify multiple and cumulative exposures.

19.3 MINORITY AND LOW-INCOME POPULATIONS

A fundamental step in environmental justice is to identify and understand where minority and/or low-income communities are located and how the lives and livelihoods of the residents may be impacted. Minority communities and low-income communities are likely to be dependent on their surrounding environment, more susceptible to pollution and environmental degradation, and often less permanent than other communities bearing disproportionately high and adverse effects.

Minority and low-income populations are identified in a similar manner, using census data and various locational and distributional tools. A minority population may be present if the minority population percentage of the affected area (area of proposed project which will or may have an effect on) is "meaningfully greater" than the minority population percentage in the general population or other "appropriate unit of geographic analysis."[10] Low-income populations, on the other hand, are defined uniquely by each state and region. However, many communities are identified with the use of annual statistical poverty thresholds from the Bureau of the Census Current Population Reports, Series P-60 on Income and Poverty.[10]

One of the most important elements in evaluating whether a minority or low-income population is present is selecting a comparison population (or group of individuals living in geographic proximity to one another) to distinguish if there

---

[10]United States Environmental Protection Agency, Office of Federal Activities, "Final Guidance for Incorporating Environmental Justice Concerns in EPA's NEPA Compliance Analyses," http://es.epa.gov-/oeca/ofa/ejepa.html, pp. 7, 8.

are "meaningfully greater" percentages or if either type of group experiences common conditions of environmental exposure. The selection of the appropriate unit of geographic analysis may be a governing body's jurisdiction, a neighborhood census tract, or other similar unit. This is done to prevent artificial dilution or inflation of the affected minority population. Furthermore, a demographic comparison to the next larger geographic area or political jurisdiction should be presented to place population characteristics in context and allow the analyst to judge whether alternatives adequately distinguish among populations. Furthermore, minority communities and destitute areas make up a very small percentage of the total population and are often concentrated in smaller areas within the larger geographically and/or economically defined population surrounding the site in question. Unfortunately, census data can be broken down only to certain levels (i.e., census tracts/blocks), and possible distortion of population breakdowns and pockets of minority or low-income communities may be missed in a traditional census tract–based analysis. Therefore, the results of the analyses unfortunately may not completely reflect the impacts that are possibly inflicted on these smaller communities.

An alternative that assists in clarifying some of these problems is the use of maps, aerial photographs, geographical information systems (GIS), and sociodemographic data. These tools can be useful by providing a visual understanding of the communities, the proximity of the proposed project to these populations, and how potential impacts may be distributed. In addition, these aids can also reveal key natural resources, clusters of facilities or sites that may contribute to cumulative impacts, and depict boundaries of an identified community. Furthermore, they can incorporate the spatial relationship that exists between the community, natural resources, and known pollutant sources.

19.4 THE INTERAGENCY WORKING GROUP AND THE FORMATION OF THE NATIONAL ENVIRONMENTAL JUSTICE ADVISORY COUNCIL

Executive Order 12898 requires the creation of an Interagency Working Group (IWG), that comprises the heads of 17 departments/executive agencies and several White House offices. Administrator Carol Browner has taken a leadership role in helping federal agencies implement the order and chairs the Interagency Working Group, whose responsibilities include:

1. Providing guidance to federal agencies on criteria for identifying disproportionately high and adverse health and environmental effects on minority populations and low-income populations.

2. Assisting in coordinating research by, and stimulating cooperation among, the EPA, the Department of Health and Human Services, the Department of Housing and Urban Development, and other agencies conducting research or

other activities in accordance with Human Health and Environmental Research and Analysis.

3. Examining existing data and studies on environmental justice.

4. Holding public meetings.

5. Coordinating data collection.

6. Developing interagency model projects.

To address these issues, on April 11, 1994, EPA formed the National Environmental Justice Advisory Council (NEJAC), which comprises 25 representatives from academia, business and industry, state, tribal, and local governments, experts in various fields (medicine, technology, socioeconomic), environmental organizations, community groups, and nongovernmental organizations. This council was composed of players from all sides. The federal advisory committee provides independent advice, consultation, and recommendations to the EPA administrator on matters related to environmental justice. The NEJAC is a parent council that has seven subcommittees:[11]

- *Air and Water Subcommittee.* Supplies advice in areas related to air and water.

- *Enforcement Subcommittee.* Provides advice in areas related to environmental justice, including promoting cooperative and supportive relationships aimed at ensuring environmental justice in enforcement and compliance activities at the federal, state, local, and tribal levels; evaluating the development, implementation, and enforcement of national environmental justice policies and programs by EPA; identifying administrative legislative options to improve enforcement and compliance policies, programs, and activities; reviewing research and data management systems to support and strengthen programs; and promoting environmental justice awareness through education and the development of effective outreach strategies involving stakeholders.

- *Indigenous Peoples Subcommittee.* Provides a forum for representatives of indigenous communities, including grass roots organizations from within those communities to bring their environmental justice concerns, recommendations, and advice to the NEJAC to ensure that the concerns are addressed by EPA in a manner that fulfills the trust responsibility, respects tribal sovereignty and the government-to-government relationship, upholds treaties, and promotes tribal self-determination.

- *Health and Research Subcommittee.* Promotes cooperative and supportive relationships aimed at ensuring environmental justice in health and research activities. The subcommittee evaluates the development of measures, criteria, and participatory methods to improve environmental regulations and policy

---

[11]National Environmental Justice Advisory Council (NEJAC), main page Web site: *http://es.epa.goc/oeca/oej/nejac/mainpage.html.*

responses that protect human health, and advocates community education and training as a means to involve communities in solving environmental problems within their communities.

- *International Subcommittee.* Promotes cooperative and supportive relationships aimed at ensuring environmental justice in international activities and increases awareness of issues related to international environmental justice.

- *Public Participation and Accountability Subcommittee.* Promotes the cooperative relationships and development of methods for institutionalizing public participation as it relates to ensuring environmental justice in public participation activities, and establishes definitions of public participation and accountability that include information, education, and community economic development.

- *Waste and Facility Siting Subcommittee.* Promotes cooperative and supportive partnerships aimed at ensuring environmental justice in activities related to waste and facility siting. The subcommittee not only establishes prototypes for public involvement, but also identifies administrative and legislative options to improve policies related to waste and facility siting policies, programs, and activities.

19.5   THE ENVIRONMENTAL JUSTICE STRATEGY

The goals of the strategy are fundamental in its operation and mission. Its principles and initiatives are intended to ensure that every segment of the population, regardless of race, culture, or income lives in a clean, healthy, and sustainable community. The purpose of the strategy is to ensure that EPA programs, policies, and activities integrate environmental justice consistently with the executive order. In fact, environmental justice is just one of the seven guiding principles established in the agency's strategic plan called the *New Generation of Environmental Protection.*[12] For example, in EPA's community base of environmental protection, the agency works with affected communities in fashioning strategies to promote a healthy environment and sustainable economy. The agency has also developed an approach focused on establishing commonsense principles and procedures for conducting the agency's business. The commonsense initiative is a sweeping effort to work with industry on a sector-by-sector basis to address public health and environmental issues. It brings together communities, environmentalists, industries, states, tribes, and others to develop cleaner, cheaper and smarter solutions.[12]

Each EPA office and region was requested to develop a strategy or action plan to address environmental justice concerns in various mission areas such as:

1. Public participation, accountability, partnerships, outreach, and communication stakeholders

---

[12]United States Environmental Protection Agency, "Environmental Justice Strategy, Text of Report," *http://www.epa.gov/docs/oejpubs/strategy/strategy.txt.html*, p. 3.

2. Health and environmental research

3. Data collection analysis and stakeholder access to public information

4. American Indian and indigenous environmental protection

5. Enforcement compliance assurance and regulatory reviews

19.5.1 Public Participation Accountability, Partnerships, Outreach, and Communications with Stakeholders

A comprehensive approach to identifying and addressing environmental justice concerns requires early involvement of affected communities and other stakeholders. Additionally, approaches to effectively address environmental justice issues require the combination of partnership, outreach, and communication with affected communities; leveraging of resources; and coordination among the federal, tribal, state, and local governments, environmental organizations, academic institutions, nonprofit organizations, and the business and industry sectors.[13] The most challenging aspect of this mission is to incorporate the expertise of all these members continually throughout the process.

The main objective in this mission topic is to ensure that there is active public participation and to provide input early in the environmental decision making. EPA is also pledging that public notices relating to human health or the environment are concise, understandable to the community involved, and readily accessible to the public. In addition, EPA is proposing to publish public minutes for meetings in languages other than English, in local and minority-oriented newspapers, and through electronic media including radio and television. Furthermore, EPA will administer appropriate grant programs and examine and promote technical assistance for minority or low-income communities.[13] The mission topic also incorporates an ongoing orientation and training program for EPA's personnel in each region on environmental justice issues including those related to public participation, tribal relations, health research, and data gathering.[13] These training programs will include input from the stakeholder and grass roots organizations.

19.5.2 Health and Environmental Research

Human health and environmental research is a cornerstone of informed decision making to ensure a healthy environment. The main environmental justice objective for EPA's health and environmental research is to improve the scientific basis for decisions by conducting research and related activities to identify and prioritize environmental health risks, as well as pollution prevention opportunities

---

[13]U.S. EPA, pp. 3,4 (footnote 12).

for risk reduction.[14] Therefore, EPA is currently exploring, through pilot projects, resources and strategies necessary to help train community residents to be effective collaborators in their research process. This training includes aspects pertaining to the decision-making processes, research design, questionnaire construction, data collection, and data analysis.[14] In addition, the mission area is working with the scientific community to improve health and risk assessments and incorporate environmental justice, including socioeconomic issues, into the policies and guidance.

By working with the affected stakeholders, EPA will conduct research in the areas where it can make the greatest contribution to environmental justice in a manner to ensure the agency's environmental justice policies are based on sound science. EPA will continue to develop human exposure data and will address exposure in at least three main areas: methods development, model development, and monitoring data.[14]

19.5.3 Data Collection, Analysis, and Access

A sound information resource management foundation is vital to the agency and its ability to provide objective, reliable, and understandable information on programs and stakeholders. By strategically managing and integrating information, the agency will better understand environmental justice issues and make better decisions. EPA will work with the affected communities, state, tribal, and local governments, and others to have the best information available to identify and address disproportionately high and adverse human health and/or environmental effects on minority populations and low-income populations. EPA will examine and expand, as appropriate, its databases to identify major facilities or sites, including federal and nonfederal facilities or sites, that pose a substantial environmental, human health, or economic effect on the surrounding populations. EPA will also work to fill data gaps, including those related to pollution prevention and affected communities and those identified by affected communities with interactive need assessments. EPA will fill these gaps by increasing the accuracy of its locational data for major facilities or sites, potential toxic releases and environmental monitoring points, and affected communities. Furthermore, EPA will also promote the use of geographical information systems (GIS) to enhance identification of disproportionately affected communities.

19.5.4 American Indian, Alaskan Native, and Indigenous Community Programs

Environmental protection for American Indian, Alaskan native, and indigenous communities is a critical part of the agency's mission. This is illustrated by the

---

[14]U.S. EPA, p. 5 (footnote 12).

agency's Indian policy and the establishment of the Tribal Operations Committee, the American Indian Advisory Council, and the American Indian Environmental Office.[15] EPA will continue to work with other federal agencies and federally recognized tribes to effectively protect and improve tribal health and environmental conditions. These activities will include providing outreach, education, training, and technical, financial, and legal assistance to develop, implement, and maintain comprehensive tribal environmental programs that will undertake the remediation of environmental hazards and the development and implementation of tribal environmental codes and tribal-EPA agreements to address tribal needs, programs, delegations, and direct federal implementation.[15] EPA will implement its programs for both American Indians and indigenous communities, recognizing the government-to-government relationship, the Federal Trust responsibility, tribal sovereignty, treaty-protected rights, other tenets of federal Indian law, and in particular historical and cultural needs of tribes and indigenous populations.[15] Furthermore, the human health, environmental research, and other activities involving tribal and indigenous communities will take into account the cultural meaning and use of natural resources. To ensure this consistency, the Office of environmental justice, environmental justice coordinators, the American Indian Environmental Office, the Office of Enforcement and Compliance Assurance, and the Indian coordinators will all work closely together to coordinate these activities.

19.5.5 Enforcement, Compliance Assurance, and Regulatory Review

Strong and effective enforcement of environmental and civil rights laws is fundamental to virtually every mission of EPA. The agency recognizes that conditions affecting covered populations, whether in rural or urban areas, can result in multiple exposures, high-level exposures from a single source, and chronic noncompliance.[16] The pollution comes from diverse sources, including both private and federal facilities. The executive order emphasizes that existing laws, including the National Environmental Policy Act (NEPA) and Title 6 of the Civil Rights Act of 1964, provide opportunities for federal agencies to address environmental hazards in minority communities and low-income communities.[16]

EPA will incorporate environmental justice concerns into its program for ensuring compliance with federal and environmental requirements at both private and federal facilities. The agency will review and advise, as needed, significant policy and guidance documents to address environmental justice issues. Furthermore, EPA will work to ensure that inspection and enforcement actions are sufficient to address those effects.

---

[15]U.S. EPA, p. 7 (footnote 12).
[16]U.S. EPA, p. 8 (footnote 12).

19.6 BUSINESS TODAY

The world is constantly evolving through advances in technologies, emerging concerns, and doing business responsibly. This evolution changes the "rules" for doing business successfully in this country. Facilities are currently designing and operating more efficiently and in an environmentally safer manner because of the demand for a cleaner environment from government regulations, environmental organizations, activists, community groups, and the public as a whole.

Today, a company must have a viable product, build or maintain its facilities in suitable locations for operational considerations, comply with regulatory requirements, involve the a community through public participation about its operations and safety programs, and consider the possible impacts to everyone in the community. Pressure from environmental activists, organizations, and community groups is helping to ensure that facilities are designed to operate more efficiently and in an environmentally safer manner.

Unfortunately, sometimes even if industry complies with all of the "rules," it can still be tied up in legal matters for years or be denied permits. As an example, the Japanese corporation Shintech proposed to build a chemical manufacturing facility in Convent, Louisiana (St. James Parish), along the Louisiana Industrial Corridor. The facility would produce polyvinyl chloride (PVC), vinyl chloride monomer (VCM), ethylene dichloride (EDC), and chlor-alkai. This production would generate 1.5 tons of pollutants each year, including more than 100,000 pounds of vinyl chloride, a Class A human carcinogen.[17] Shintech Inc. attempted to play by the "new rules"; it developed a $700 million state-of-the-art facility, met all state regulatory requirements, and communicated somewhat with the public.

However, the facility chose its location in one of the most polluted areas in the United States for toxic releases into the air, land, and water, an area that has been called "cancer alley." Because of this stigma, a coalition of more than 20 groups decided to give the facility a fight. They were determined not to allow another chemical plant to be built along this corridor, even if the project would have produced economic benefits such as generation of jobs, higher incomes, and tax revenues. However, the cost of pollution to the community amounts to approximately $4500 per child. Nonetheless, Shintech's air, water, and coastal zone permits were granted, with the backing of nearly all the relevant state and local officials, and over 20 groups and most of the local population (which is 80 percent African-American and 40 percent impoverished). When the permits were in review by EPA, the federal agency cited 50 problems with the air permit, and in a precedent-setting decision, rejected the state-granted permit and challenged state officials to deal with "environmental justice" questions.[18] Unfortunately, this controversial

---

[17]"Shintech Action Network," main page, *http://www.leanweb.org/shintech*.

[18]"Shintech Tool Kit," main page: *http://www.leanweb.org/shintools*.

issue may never be solved since the company is currently pursuing relocating the facility instead of waiting for the possible approval by the EPA, stating that the project is not environmentally unjust.

Cases like the Shintech project leave many believing that environmental justice is going to deter economic justice by hindering, and sometimes preventing, the influx of significant capital expenditures from private companies into economically depressed areas.[19] Nevertheless, the possible delay due to investigation and litigation is forcing industry to become "smarter" in choosing site locations.

19.7 FEDERALLY ASSISTED PROGRAMS AND THE REGULATIONS

As previously stated, Title VI prohibits discrimination based on race, color, or national origin under programs or activities receiving federal financial assistance. EPA has adopted the Title VI implementing regulations that prohibit unjustified discriminatory effects that may occur under federally assisted programs or activities.[20] It has been found that discrimination can result from policies and practices that appear on the surface to be beneficial to the public, but have disproportionate effects on segments of the affected community. Seemingly neutral policies or practices that result in discriminatory effects violate EPA's Title VI regulations unless they are justified and there are no less discriminatory alternatives.[20]

States and agencies are often awarded grants from EPA for continuing their environmental programs in compliance with EPA statutes. Attention is drawn to the various conditions that must be met to receive funding from EPA. One condition is that the recipient must comply with the Title VI regulations that prohibit unjustified discriminatory effects to segments of our society. EPA's Title VI regulations define a *recipient* as "any state or its political subdivision, any instrumentality of a state or its political subdivision, or any public or private agency, institution, organization, or other entity, or any person to which federal financial assistance is extended directly or through another recipient...."[20] Title VI creates four recipients in the nondiscrimination obligation that are contractual in nature in exchange for accepting federal funding. Acceptance of EPA funding creates an obligation on the recipient to comply with the regulations for as long as EPA funding is extended.[20]

Under amendments made to Title VI by the Civil Rights Restoration Act of 1978, a *program* or *activity* means all of the operations of a department, agency,

---

[19]Jeffery S. Heaton, "Environmental Justice: Deterrent to Economic Justice," *Environmental Manager*, January 1999, p. 12.

[20]United States Environmental Protection Agency, "Interim Guidance for Investigating Title VI Administrative Complaints Challenging Permits," *http://www.es.epa.gov/oeca/oej/titlevi/html*, p. 2.

special-purpose district, or other instrumentality of a state or of the local government, any part of which is extended federal financial assistance.[21] Hence, unless expressly exempted from Title VI by federal statute, all programs or activities of a department or agency that receives EPA funds is subject to Title VI, including those programs and activities that are not EPA-funded. An example is the issuance of permits by EPA recipients under solid waste programs administered pursuant to Subtitle D of the Resource Conservation Recovery Act (which has not been grant-funded by EPA) or the actions they take under programs that do not derive their authority from EPA statutes, but are part of a program or activity covered by EPA's Title VI regulations if the recipient receives any funding from EPA.[21] In the event EPA finds discrimination in the recipient's permitting program and the recipient is not able to come into compliance voluntarily, EPA is required by its Title VI regulations to initiate procedures to deny, annul, suspend, or terminate EPA funding.[22] EPA also may use any other means authorized by law to obtain compliance, including referring the matter to the Department of Justice for litigation. In appropriate cases, the Department of Justice may file suit seeking injunctive relief; moreover, individuals may file a private right of action in court to force the nondiscrimination requirements in Title VI or EPA's implementing regulations without exhausting the administrative remedies.[22]

19.7.1 Overview of Framework for Processing Complaints

Under EPA's Title VI regulations, the Office of Civil Rights (OCR) will be processing the complaints. The subject of these complaints will be discriminatory intent and/or effect due to permitting. The following list of steps is used in processing a complaint:[23,24]

1. *Acceptance of the complaint.* On receiving a Title VI complaint, OCR will determine whether the complaint states a valid claim. If it finds the complaint valid, the complaint will be accepted for processing within 20 calendar days with acknowledgment of its receipt, and the complainant and the EPA recipient will be so notified. If the OCR does not accept the complaint, it will be rejected or, if appropriate, referred to another federal agency.

2. *Investigation and disparate impact assessment.* Once a complaint is accepted for processing, the Office of Civil Rights will conduct a factual investigation to determine whether the permits at issue will create a disparate impact or

---

[21]U.S. EPA, p. 2 (footnote 20).

[22]U.S. EPA, p. 3 (footnote 20).

[23]U.S. EPA, pp. 3, 4 (footnote 20) and 40 CFR 7.120(d)(1).

[24]United States Environmental Protection Agency, 40 CFR 7.120(d)(1).

add to an existing disparate impact on a racial, ethnic, or low-income population. If, on the basis of its investigation, OCR concludes that there is no disparate impact, the complaint will be dismissed. If the OCR makes an initial finding of a disparate impact, it will notify the recipient and the complainant and seek a response from the recipient within a specified time period. Under appropriate circumstances, OCR may seek comment from the recipient, permittee, and/or complainants on preliminary data analyses before making an initial finding concerning disparate impacts.

3. *Rebuttal mitigation.* The notice of initial finding of a disparate impact will provide the recipient the opportunity to rebut OCR's finding, propose a plan for mitigating the disparate impact, or to justify this impact. If the recipient successfully rebuts the OCR's finding, or if the recipient elects to submit a plan for mitigating the disparate impact, and on the basis of this review, EPA agrees that the disparate impact will be mitigated sufficiently pursuant to the plan, the parties will be so notified. Assuming that assurances are provided about implementation of such a mitigation plan, no further action on the complaint will be required.

4. *Justification.* If the recipient can neither rebut the initial finding of disparate impact nor develop an acceptable mitigation plan, then the recipient may seek to demonstrate that it has a substantial, legitimate interest that justifies the decision to proceed with the permit notwithstanding the disparate impact. Even where a substantial, legitimate justification is preferred, OCR will need to consider whether it can be shown that there is an alternative that would satisfy the stated interest while eliminating or mitigating the disparate impact.

5. *Preliminary finding of noncompliance.* If the recipient fails to review OCR's initial finding of a disparate impact and can neither mitigate nor justify the disparate impact issue, OCR will, within 180 calendar days of the start of the complaint investigation, send the recipient written notice of preliminary finding of noncompliance, with a copy to the grant award official and the Assistant Attorney General for Civil Rights. The Office of Civil Rights notice may include recommendations for the recipient to achieve voluntary compliance and, where appropriate, the recipient's right to engage in voluntary compliance negotiations.

6. *Formal determination of noncompliance.* If, within 50 calendar days of receipt of the notice of the preliminary finding, the recipient does not agree to OCR's recommendations or fails to submit a written response demonstrating that the OCR's initial findings are incorrect or that voluntary compliance can be achieved through other steps, OCR will issue a formal written determination of noncompliance, with a copy to the award official and the Assistant Attorney General for Civil Rights.

7. *Voluntary compliance.* The recipient will have 10 calendar days from the receipt of the formal determination of noncompliance from which to come into voluntary compliance. If the recipient fails to meet this deadline, OCR will start procedures to deny, annul, suspend, or terminate EPA's assistance in accordance with 40 CFR, Part 7, Sec. 130(b) and consider other appropriate action, including referring the matter to the Department of Justice for litigation.

8. *Informal resolution.* EPA Title VI regulations call for OCR to pursue an informal resolution of administrative complaints wherever practicable. Therefore, OCR will discuss, at any point during the process outlined above, offers by recipients to reach an informal resolution, and will, to the extent appropriate, endeavor to facilitate the informal resolution process and involvement of affected stakeholders. Originally, in the interest of conserving EPA investigative resources for truly intractable matters, it will make sense to encourage dialogue at the beginning of the investigation of complaints accepted for processing. Accordingly, in notifying a recipient of the acceptance of the complaint for investigation, OCR will encourage the recipient to engage in an effort to negotiate an informal settlement.

19.7.2 Rejecting or Accepting Complaints for Investigation

It is the general policy of OCR to investigate all administrative complaints that have apparent merit and are complete or properly pleaded.[25] A complete and properly pleaded complaint:[25]

1. Is in writing, signed, and provides an avenue for contacting the signatory (i.e., phone number, address, etc.)
2. Describes the alleged discriminatory acts that violate EPA Title VI regulations (an act of intentional discrimination or one that has the effect of discriminating on the basis of race, color, or national origin)
3. Is filed within 180 calendar days of the alleged discriminatory acts
4. Identifies the EPA recipient that committed the alleged discriminatory acts

EPA Title VI regulations contemplate that OCR will make its determination to accept, reject, or refer (to the appropriate federal agency) a complaint within 20 calendar days of acknowledgement of recipient.[26] Wherever possible, within the 20-day period, OCR will establish whether the person or entity that took the alleged discriminatory act is in fact an EPA recipient as defined by 40 CFR 7.25. If the complaint does not specifically mention that the alleged discriminatory actor is an EPA financial assistance recipient, OCR may presume so for the purpose of deciding whether to set the complaint for further processing.[25]

---

[25]U.S. EPA, p. 4 (footnote 20).

[40]"Interim Guidance for Investigating Title VI Administrative Complaints Challenging Permits," p. 4.

[26]United States Environmental Protection Agency, 40 CFR 7.120 (d)(1).

19.7.3 Time Line for Complaints

Under EPA's Title VI regulations, a complaint must be filed within 180 calendar days of the alleged discriminatory act.[27] EPA interprets this regulation to mean that complaints alleging discriminatory effects resulting from issuance of a permit must be filed with EPA within 180 calendar days of issuance of a final permit. However, OCR may wave the 180-day time limit for good cause.[28]

The waiver for good cause is used as a strategy by EPA. If a complainant exhausts the administrative remedies available under the appeals process, sometimes an early resolution occurs. Under such circumstances, and after considering other factors relevant to the particular case, OCR may also waive the time limit if the complaint is filed within a reasonable time period (within 2 months) after the conclusion of the administrative appeal process.[29] In addition, OCR usually provides the recipient with the information contained in the complaint for consideration in the permit issuance process. OCR may also notify the complainant that the objection is premature, but that OCR is keeping the reproach on file and in active status pending issuance of the final permit by the recipient.[29]

19.7.4 Permit Modifications

EPA believes that permit modifications that reduce adverse impacts and improve the environmental operation of the facility should be encouraged. Modifications that do not cause adverse impacts to human health, the environment, or are considered neutral (e.g., a facility name change) are almost always dismissed. On the other hand, complaints associated with modifications that result in an increase of pollution would not be ignored. These complaints will be researched as to the nature of modified operation and its associated impacts.[29] The complaint must allege and establish adverse impacts specially associated with the modification under review.

19.7.5 Permit Renewals

Generally, permit renewals should be treated and analyzed as if they were new facility permits, since permit renewal is, by definition, an occasion to review the overall operations of a permitted facility and make any necessary changes. Generally, permit renewals are not issued without public notice and an opportunity for the public

---

[27] United States Environmental Protection Agency, 40 Code of Federal Regulations 40 CFR 7.120 (d)(2).

[28] United States Environmental Protection Agency, 40 CFR 7.120 (d)(2).

[29] U.S. EPA, p. 5 (footnote 20).

to challenge the propriety of granting a renewal under the relevant environmental laws and regulations.[30]

19.7.6 Impacts and the Disparate Impact Analysis

Each allegation should incorporate all the circumstances surrounding the site and have the complaint based on facts, not opinion or hearsay. Several reliable techniques and tools are used to evaluate the cases to determine if a disproportional impact is occurring. Therefore, five basic steps are used to determine if a disparate impact exists.[31]

1. *Identifying the affected population.* The first step is to identify the population affected by the permit that triggered the complaint. The affected population is one that suffers adverse impacts of the permitted activity. The impacts investigated must result from the permit or permits at issue.

The adverse impacts from permitted facilities are rarely distributed in a predictable and uniform manner. However, the proximity to a facility will often be a reasonable indicator of where impacts are concentrated. Accordingly, where more precise information is not available, OCR will generally use proximity to a facility to identify adversely affected populations. The proximity analysis should reflect the environmental medium and impact of concern in the case.

2. *Determining the demographics of the affected population.* The second step is to determine the racial and/or ethnic composition of the affected population for the permitted facility at issue in the complaint. To do this, OCR uses demographic mapping technology, such as geographic information systems (GIS). In conducting a typical analysis to determine the affected population, OCR generates data estimating the race and/or ethnicity and density of populations within a certain proximity from a facility or within the distribution pattern for a release/impact based on scientific models. OCR then identifies and characterizes the affected population for the facility at issue. If the affected population for the permit at issue is of the alleged racial or ethnic group or groups named in the complaint, then the demographic analysis is repeated for each facility in the chosen universes of facilities discussed below.

3. *Determining the universes of facilities and the total affected populations.* The third step is to identify which other permitted facilities, if any, are to be included in the analysis and to determine the racial or ethnic composition of the populations affected by those permits. There may be more than one appropriate universe of facilities. OCR will determine the appropriate universe of facilities on the basis of the allocations and facts of a particular case. However, facilities not

---

[30]U.S. EPA, p. 5 (footnote 20).

[31]U.S. EPA, pp. 5–7 (footnote 20).

under the recipient's jurisdiction should not be included in the universe of facilities examined.

During OCR's investigation, if it finds that the universe of facilities selected by the complainant is not supported by the facts, OCR will explain what it has found and provide the complainant the opportunity to support the use of its proposed universe. If the complainant cannot adequately support the proposed universe, then OCR should investigate a universe of facilities on the basis of the facts available and OCR's reasonable interpretation of the theory of the case presented. Once the appropriate universes of facilities is determined, the affected population for each facility in the universe should be added together to form the total affected population.

Originally, OCR would entertain cases only in which the permitted facility at issue was one of several facilities, which together present a cumulative burden, or reflected a burden or a pattern of disparate impact. EPA recognizes the potential for disparate outcomes in this area because most permits control pollution rather than prevent it altogether. Consequently, permits that satisfy the basic public health and environmental protections contemplated under EPA's programs nonetheless bear the potential for discriminatory effects where residual pollution and other recognizable impacts are distributed disproportionately to communities with particular racial or ethnic characteristics. From its experience to date, the EPA believes that this is most likely to be true either (1) where an individual permit contributes to or compounds a preexisting burden being shouldered by a neighboring community, such that the community's cumulative burden is disproportionate when compared to other communities, or (2) where an individual permit is part of a broader pattern in which it has become more likely that certain types of operations, with their accompanying burdens, will be permitted in a community with particular racial or ethnic characteristics.

4. *Conducting a disparate impact analysis.* The next step is to conduct a disparate impact analysis that, at a minimum, includes comparing the racial, ethnic, or income characteristics within the affected population. It will also likely include comparing the racial characteristics of the affected population to the nonaffected population. Such an approach can show whether persons protected under Title VI are being impacted at a disparate rate. EPA generally expects the rates of impact for the affected populations and comparison population to be relatively comparable under properly implemented programs. Since there is no one formula or analysis to be applied, OCR may identify, on a case-by-case basis, other comparisons to determine disparate impact.

5. *Determining the significance of disparity.* The final phase of the analysis is to use arithmetic or statistical analysis to determine whether the disparity is significant under Title VI. OCR will use trained statisticians to evaluate disparity calculations done by investigators. After calculations are reviewed by experts, OCR may make a *prima facie* disparate impact finding subject to the recipient's opportunity to rebut.

19.7.7   Justification

If a preliminary finding of noncompliance has not been successfully rebutted and the disparate impact cannot be successfully mitigated, the recipient will have the opportunity to "justify" the decision to issue the permit notwithstanding the disparate impact, on the basis of the substantial, legitimate interest of the recipient. A legitimate justification is not just demonstrating that the permit complies with the applicable environmental regulations, but that there is some value to the recipient in the permitted activity. Because interests of a state or local environmental agency are influenced and informed by broader interests of the government of which it is a part, OCR entertains justifications based on broader governmental interests (i.e., interests that are not limited by the jurisdiction of the recipient agency).[32] The types of factors that sustain consideration in accessing sufficiency can include, but are not limited to, the seriousness of the disparate impact, whether the permits at issue is a renewal (with demonstrated benefits), permits are for a new facility, and whether any of the articulated benefits associated with a permit can be expected to benefit the particular community that is subject to the Title VI complaint. Most importantly, if a less discriminatory alternative exists, the facility must mandate it to avoid noncompliance. Furthermore, mitigation measures should be considered such as less discriminatory alternatives, including additional permit conditions that would lessen or eliminate the demonstrated adverse disparate impacts.[32]

19.8   CONCLUSION

Through public outcries and implementation of policies over the past two decades, environmental justice has evolved into an important facet in our society. The debate about environmental justice is now about values and fairness to all segments of society, regardless of age, culture, ethnicity, health conditions, gender, race, or socioeconomic status. Everyone seems to agree that environmental justice is a goal to strive toward; however, value judgments vary slightly depending on which side a person is sitting on. This is one of the major difficulties in environmental justice, avoiding the conflicts and difficult decisions. The subject is extremely difficult to define accurately, and the issues raised in this debate force society to confront unpleasant choices about whether to address this problem or that one, or whether to protect one group over another. However, attaining environmental justice depends on defining a set of definitions and explicit decisions about how to put principles into practice. The decisions that are being made today will set procedural fairness as to how our society will define policies and actual unfairness, and how to rectify these injustices and inequities.

---

[32]U.S. EPA, p. 5 (footnote 20).

APPENDIX A

LISTING OF STATE AGENCIES RESPONSIBLE FOR ENVIRONMENTAL PERMITTING

		Alabama		
		Alabama Department of Environmental Management 1400 Coliseum Blvd. Montgomery, AL 36110-2059 General Information: (334) 271-7700 Internet: http://www.adem.state.al.us		
Air:	Department of Air Quality	Ron Gore	Chief	(334) 271-7823
Water:	Water Division	Charles Horn	Chief	(334) 271-7823
Waste:				

		Alaska		
		Alaska Department of Environmental Conservation 410 Willoughby Avenue, Suite 105 Juneau, Alaska 99801-1795 General Information: (907) 465-5010 Internet: http://www.state.ak.us/dec/home.htm		
Air:	Air and Water Quality Division	Tom Chapple	Director	(907) 269-7687
Water:	Air and Water Quality Division	Tom Chapple	Director	(907) 269-7687
Waste:	Spill Prevention & Response Division	Larry Dietrick	Director	(907) 465-5250

		Arizona		
		Arizona Department of Environmental Quality 3033 N. Central Avenue Phoenix, AZ 85012 General Information: (602) 207-2300 Internet: http://www.adeq.state.az.us		
Air:	Air Quality Division	Nancy Wrona	Director	(602) 207-2308
Water:	Water Quality Division	Karen L. Smith	Director	(602) 207-2306
Waste:	Waste Division	Jean Calhoun	Director	(602) 207-2382

		Arkansas		
		Arkansas Department of Pollution Control and Ecology 8001 National Drive P.O. Box 8913 Little Rock, AR 72209-8913 General Information: (501) 682-0744 Internet: http://www.adeq.state.ar.us/		
Air:	Air Division	Keith Michaels	Chief	(501) 682-0730
Water:	Water Division	C.C Bennett	Chief	(501) 682-0654
Waste:	Hazardous Waste Division Solid Waste Division	Mike Bates Dennis Burks	Chief Chief	(501) 682-0831 (501) 682-0600

California			
California Air Resources Board 555 Capitol Mall, Suite 525 Sacramento, CA 95814 General Information: (916) 445-3846 Internet: http://www.calepa.ca.gov			
Air:	California Air Resources Board 2020 L Street, P.O. Box 2815 Sacramento, CA 95812	Alan C. Lloyd	Chairman	(916) 322-5840
Water:	Water Resource Control Board 901 P Street, P.O. Box 100 Sacramento, CA 95812	Stan Martinson Water Quality Division	Chief	(916) 657-0756
Waste:	California Integrated Waste Mgmt. Board Sacramento, CA 95826	Julie Nauman Permitting & Enforcement Division	Deputy Director	(916) 255-2431

Colorado			
Colorado Department of Public Health and Environment 4300 Cherry Creek Drive, South Denver, CO 80246-1530 General Information: (303) 692-2000 Internet: http://www.cdphe.state.co.us/cdphehom.asp			
Air:	Air Pollution Control Division	Margie Perkins	Director	(303) 692-3100
Water:	Water Quality Control Division	David Holm	Director	(303) 692-3500
Waste:	Hazardous Materials & Waste Management Division	Howard Roitman	Director	(303) 692-3300

Connecticut			
Connecticut Department of Environmental Protection 79 Elm Street Hartford, CT 06106 General Information: (860) 424-3000 Internet: http://www.dep.state.ct.us/			
Air:	Air Management	Carmine DiBattista	Bureau Chief	(860) 424-3026
Water:				
Waste:	Waste Management	Richard Barlow	Bureau Chief	(860) 424-3021

Delaware			
Department of Natural Resources and Environmental Control 89 Kings Highway Dover, DE 19901 General Information: (302) 739-4506 Internet: http://www.dnrec.state.de.us/			
Air:	Air & Waste Management Division	(Vacant)	Director	(302) 739-4764
Water:	Water Resources Division	Sergio Huerta	Director	(302) 739-4860
Waste:	Air & Waste Management Division	(Vacant)	Director	(302) 739-4764

District of Columbia

Environmental Health Administration
51 N Street NE
Washington, DC 20002
General Information: (202) 535-2250
Internet: http://www.dchealth.com

Air:		Maurice Knuckles, Ph.D.	Assistant Deputy Director	(202) 442-8983
Water:				
Waste:				

Florida

Florida Department of Environmental Protection
3900 Commonwealth Boulevard
Tallahassee, FL 32399-3000
General Information: (850) 488-1073
Internet: http://www.dep.state.fl.us

Air:	Air Resources Management Division	Clair Fancy Air Regulations	Bureau Chief	(850) 488-1344
Water:	Water Facilities Division	Richard Drew Water Facilities Regulations	Bureau Chief	(850) 487-0563
Waste:	Waste Management Division	Bill Hinkley Solid & Hazardous Waste	Bureau Chief	(850) 488-0300

Georgia

Georgia Department of Natural Resources
205 Butler Street, SE   Suite 1252
Atlanta, GA 30334
General Information: (404) 656-3500
Internet: http://www.georgianet.org/dnr/environ/

Air:	Air Protection	Ronald C. Methier	Branch Chief	(404) 363-7000
Water:	Water Resource Management	Nolton Johnson	Branch Chief	(404) 656-6328
Waste:	Hazardous Waste	Jennifer Kaduck	Branch Chief	(404) 656-7802

Guam

Guam Environmental Protection Agency
P.O. Box 22439
GMF, GU 96921

Air:		Dave Longa	Depty Administrator	(671) 475-1658
Water:				
Waste:				

Hawaii

Hawaii Department of Health, Environmental Management Division
919 Ala Moana Boulevard
Honolulu, HI 96814
General Information: (808) 586-4400
Internet: http://www.state.hi.us/health

Air:	Clean Air	Willy Nagamine	Branch Chief	(808) 586-4200
Water:	Clean Water	Denis Lau	Branch Chief	(808) 586-4309
Waste:	Solid & Hazardous Waste	Steven Chang	Branch Chief	(808) 586-4225
	Wastewater	Dennis Tulang	Branch Chief	(808) 586-4294

Idaho

Division of Environmental Quality
450 W. State Street
P.O Box 83720
Boise, ID 83720-0036
General Information: (208) 334-5500
Internet: http://www.state.id.us/deq/

Air:	Air & Hazardous Waste	Orville Green	Assistant Administrator	(208) 373-0440
Water:	Water Quality & Remediation	Larry Koenig	Assistant Administrator	(208) 373-0407
Waste:	Air & Hazardous Waste	Orville Green	Assistant Administrator	(208) 373-0440

Illinois

Illinois Environmental Protection Agency
P.O Box 83720
Springfield, IL 62794-9276
General Information: (217) 782-7860
Internet: http://www.epa.state.il.us/

Air:	Air Pollution Control Division	Dennis Lawler	Manager	(217) 782-7326
Water:	Water Pollution Control Division	(Vacant)	Manager	(217) 782-1654
Waste:	Remedial Management Division	Gary King	Manager	(217) 782-9407

Indiana

Indiana Department of Environmental Management
Indiana Government Center. North
100 N. Senate Avenue
P.O. Box 6015
Indianapolis, IN 46206-6015
General Information: (317) 232-8603
Internet: http://www.ai.org/idem

Air:	Air Management	Janet McCabe	Assistant Commissioner	(317) 233-6861
Water:	Water Management	Matt Rueff	Assistant Commissioner	(317) 232-8476
Waste:	Solid & Hazardous Waste Management	Bruce Palin	Assistant Commissioner	(317) 232-3210

Iowa

Iowa Department of Natural Resources
Wallace Building
Des Moines, IA 50319-0034
General Information: (515) 281-5145
Internet: http://www.state.ia.us/.government/dnr/

Air:	Air Quality	Pete Hamlin	Bureau Chief	(515) 281-8852
Water:	Water Quality	(Vacant)	Bureau Chief	(515) 281-8869
Waste:	Waste Management	Sharon Timmins	Bureau Chief	(515) 281-4076

Kansas

Kansas Department of Health & Environment
400 SW Eighth Avenue, Suite 200
Topeka, KS 66603-3930
General Information: (785) 296-1500
Internet: http://www.kdhe.state.ks.us

Air:	Air & Radiation	Jan Sides	Director	(785) 296-1593
Water:	Water	Karl Mueldener	Director	(785) 296-5500
Waste:	Waste Management	William Bider	Director	(785) 296-1600

Kentucky

Kentucky Department for Environmental Protection
Fort Boone Plaza
14 Reilly Road
Frankfort, KY 40601
General Information: (502) 564-2150
Internet: http://www.ky.us/agencies/eqc/eqc.html

Air:	Air Quality Division	John Hornback	Director	(502) 564-3383
Water:	Water Division	Jack A. Wilson	Director	(502) 564-3410
Waste:	Waste Management Division	Robert Daniell	Director	(502) 564-6716

Louisiana

Department of Environmental Quality
P.O. Box 82231
Baton Rouge, LA 70884-2231
General Information: (225) 763-5423
Internet: http://www.state.la.us/welcome.htm

Air:	Permits Division	Gustave VonBodungen	Assistant Secretary	(225) 765-0219
Water:	Permits Division	Michael Vince	Administrator	(225) 765-0110
Waste:	Remediation Services	Keith Casanova	Administrator	(225) 765-0489

Maine

Department of Environmental Protection
17 State House Station
Augusta, ME 04333-0017
Internet: http://www.state.me.us/dep/mdephome.htm

Air:	Licensing Division	Bryce Sproul	Director	(207) 287-2437
Water:	Water Resource Regulation Division	Michael Barden	Director	(207) 287-7700
Waste:	Oil & Hazardous Waste Facilities Division	Scott Whittier	Director	(207) 287-7674

Maryland

Maryland Department of the Environment
2500 Broening Highway
Baltimore, MD 21224
General Information: (410) 631-3000
Internet: http://www.mde.state.md.us/

Air:	Air Quality Permits Program	Karen Irons	Administrator	(410) 631-3230
Water:	Waste Management Administration	J.L. Hearn	Director	(410) 631-3567
Waste:	Hazardous Waste Program	Harold Dye	Administrator	(410) 631-3343
	Wastewater Permits Program	J.James Dieter	Manager	(410) 631-3619

Massachusetts

Department of Environmental Protection
One Winter Street, 2nd Floor
Boston, MA 02108
General Information: (617) 292-5915
Internet: http://www.state.ma.us/dep

Air:	Operations & Programs	Edward Kunce	Deputy Commissioner	(617) 292-5915
Water:				
Waste:				

Michigan

Department of Environmental Quality
P.O. Box 30473
Lansing, MI 48909-7973
General Information: (800) 662-9278
Internet: http://www.deq.state.mi.us

Air:	Air Quality Division	Dennis Drake	Chief	(517) 373-7023
Water:	Surface Water Quality Division	David Hamilton	Chief	(517) 373-1949
Waste:	Waste Management Division	Jim Sygo	Chief	(517) 373-2730

Minnesota

Minnesota Pollution Control Agency
500 Lafayette Road
St. Paul, MN 55155-4001
General Information: (651) 296-6157
Internet: http://www.dnr.state.mn.us

Air:	Enforcement Division	William Bernhjelm	Director	(651) 296-4828
Water:	Water Division	Kent Lokkesmoe	Director	(651) 296-4810
Waste:				

Mississippi

Mississippi Department of Environmental Quality
Pollution Control Office
P.O. Box 10385
Jefferson City, MO 65102
General Information: (601) 359-3513
Internet: http://www.deq.state.ms.us

Air:	Air Quality	Dwight Wylie	Branch Chief	(601) 961-5587
Water:	Surface Water	Barry Royals	Branch Chief	(601) 961-5102
Waste:	Hazardous Waste	Jerry Banks	Branch Chief	(601) 961-5221

Missouri

Missouri Department of Natural Resources
P.O. Box 176
Jefferson City, MO 65102
General Information: (573) 751-3443
Internet: http://www.dnr.state.mo.us/homednr.htm

Air:	Environmental Quality Division	John A. Young	Director	(573) 751-0763
Water:				
Waste:				

Montana

Department of Environmental Quality
P.O. Box 200901
Helena, MT 59620-0901
Internet: www.deq.state.mt.us

Air:	Permitting & Compliance Division	Don Vidrine	(406) 444-2467
	Air & Waste Management Bureau	Don Vidrine	(406) 444-2467
Water:	Water Protection Bureau	Bonnie Lovelace	(406) 444-4969
Waste:	Permitting Compliance Division	Don Vidrine	(406) 444-2467
	Air & Waste Management Bureau	Don Vidrine	(406) 444-2467

Nebraska			
	Nebraska Department of Environmental Quality		
	1200 N Street, Suite 400		
	P.O. Box 98922		
	Lincoln, NE 68509-8922		
	General Information: (402) 471-2186		
	Internet: www.deq.state.ne.us/airq.nsf/pages/aqs		

Air:	Programs	Jay Ringerberg	Deputy Director	(402) 471-0001
Water:	Water Quality Division	Patrick Rice	Assistant Director	(402) 471-3098
Waste:				

Nevada			
	Division of Environmental Protection		
	123 W. Nye Lane, Room 230		
	Carson City, NV 89706-0818		
	General Information: (775) 687-4360		
	Internet: www.state.nv.us/ndep/index.htm		

Air:	Environmental Protection Division	Allen Biaggi	Administrator	(775) 687-4670
Water:	Water Planning	Naomi Smith-Duerr	Administrator	(775) 687-3600
Waste:				

New Hampshire			
	Department of Environmental Services		
	Six Hazen Drive		
	Concord, NH 03301		
	General Information: (603) 271-3503		
	Internet: www.des.state.nh.us/descover.htm		

Air:	Air Resource Division	Kenneth A. Colburn	Director	(603) 271-3503
Water:	Water Division	Harry T. Stewart, P.E.	Director	(603) 271-3503
Waste:	Waste Management Division	Philip J. O'Brien, Ph.D.	Director	(603) 271-3503

New Jersey			
	New Jersey State Department of Environmental Protection		
	401 East State Street		
	P.O. Box 423		
	Trenton, NJ 08625-0423		
	General Information: (609) 777-3373		
	Internet: http://www.state.nj.us/dep/		

Air:	Air Quality Permitting	William O'Sullivan	Administrator	(609) 984-1484
Water:	Water Quality Division	Dennis Hart	Director	(609) 292-4543
Waste:	Solid & Hazardous Waste Division	John Castner	Director	(609) 984-6880

New Mexico

New Mexico Environmental Department
1190 St. Francis Drive
P.O. Box 26110
Santa Fe, NM 87502
General Information: (505) 827-2855
Internet: www.nmenv.state.nm.us/

Air:	Air Quality	Cecilia Williams	Bureau Chief	(505) 827-0042
Water:	Surface Water Quality	Jim Davis	Bureau Chief	(505) 827-0187
Waste:	Hazardous & Radioactive Material	James Bearzi	Bureau Chief	(505) 827-1557

New York

New York State Department of Environmental Conservation
50 Wolf Road
Albany, NY 12233
General Information: (518) 457-5400
Internet: http://www.dec.state.ny.us

Air:	Air Resources Division	Robert Warland	Director	(518) 457-7230
Water:	Water Quality & Remediation	Erin Crotty	Deputy Commissioner	(518) 457-6557
Waste:	Solid & Hazardous Materials Division	Steve Hammond	Director	(518) 457-6934

North Carolina

Division of Environmental Management
1601 Mail Service Center
Raleigh, NC 27699-1601
General Information: (919) 733-4984
Internet: http://www.ehnr.state.nc.us/EHNR/

Air:	Air Quality Division	Alan Klimek	Director	(919) 715-6232
Water:	Water Quality Division	Tommy Stevens	Director	(919) 733-7015
Waste:	Waste Management Division	Bill Meyer	Director	(919) 733-4996

North Dakota

North Dakota Department of Health
1200 Missouri Avenue
P.O. Box 5520
Bismarck, ND 58506-5520
General Information: (701) 328-2372
Internet: www.health.state.nd.us/ndhd/default.asp

Air:				
Water:	Water Quality Division	Dennis Fewless	Director	(701) 328-5210
Waste:	Waste Management Division	Neil Knatterud	Director	(701) 328-5166

Ohio

Ohio Environmental Protection Agency
Fountain Square
Columbus, OH 43224-1387
General Information: (614) 265-6565
Internet: http://www.dnrstate.oh.us

Air:

Water:	Water Division	Jim Morris	Chief	(614) 265-6717

Waste:

Oklahoma

Oklahoma Department of Environmental Quality
707 N. Robinson
P.O. Box 1677
Oklahoma City, OK 73101-1677
Internet: www.deq.state.ok.us/

Air:	Air Quality Division	Scott Thomas	Director	(405) 702-4100
Water:	Water Quality Division	Water Quality Division	Director	(405) 702-8100
Waste:	Minimization/Hazardous Waste	Dianne Wilkins	Director	(405) 702-1000

Oregon

Oregon Department of Environmental Quality
811 SW Sixth Avenue
Portland, OR 97204-1390
General Information: (503) 229-5696
Internet: www.deq.state.or.us/

Air:	Air Quality Division	Greg Green	Administrator	(503) 229-5397
Water:	Water Quality Division	Mike Llewelyn	Administrator	(503) 229-5324
Waste:	Waste Management & Cleanup Division	May Wahl	Administrator	(503) 229-5072

Pennsylvania

Pennsylvania Department of Environmental Protection
P.O. Box 2063
Harrisburg, PA 17105-2063
General Information: (717) 783-2300
Internet: www.dep.state.pa.us/

Air:	Air Quality Bureau	James M. Salvaggio	Director	(717) 787-9702
Water:	Water Quality Protection Bureau	Glenn Maurer	Director	(717) 787-2666
Waste:				
	Land Recycling & Waste Mmgt Bureau	James Snyder	Director	(717) 787-9870

Puerto Rico

Puerto Rico Environmental Quality Board
P.O. Box 9066600
San Juan, PR   9066600

Air:	Natural & Environmental Resources Dep.	Daniel Pagan	Secretary	(787) 723-3090
Water:				
Waste:				

Rhode Island

Department of Environmental Management
235 Promenade Street, Suite 425
Providence, RI   02908
General Information: (401) 222-6800
Internet: http://www.state.ri.us/dem

Air:	Air Resources	Stephen Majkut	Chief	(401) 222-2808
Water:	Permitting	Russell J. Chateauneuf	Administrator	(401) 222-2306
Waste:	Waste Management	Leo Hellested	Chief	(401) 222-2797

South Carolina

South Carolina Department of Health and Environmental Control
2600 Bull Street
Columbia, SC   29201
General Information: (803) 898-3432
Internet: www.state.sc.us/dhec/

Air:	Air Compliance Mgmt Division	Richard D. Sharpe	Director	(803) 898-4108
Water:	Water Facilities Permitting Division	Jeffrey P. deBessonet, P.E.	Director	(803) 898-4157
Waste:	Solid Waste Management Division	William W. Culler	Director	(803) 896-4201

South Dakota

Dept. of Env. and Natural Resources, Division of Environmental Services
Joe Foss Bldg
523 E. Capitol Avenue
Pierre, SD   57501-3181
General Information: (605) 773-3151
Internet: www.state.sd.us/state/executive/denr/denr/htm

Air:	Air Quality Program	Jeanne Goodman	Administrator	(605) 773-7171
Water:	Surface Water Quality Program	Tim Tollefsrud	Administrator	(605) 773-3351
Waste:	Waste Management Program	Vonni Kallemeyn	Administrator	(605) 773-3153

	Tennessee			

Department of Environment and Conservation

Life & Casualty Tower
401 Church Street, 21st Floor
Nashville, TN 37243-0435
General Information: (615) 532-0109
Internet: http://www.state.tn.us/environment

Air:	Air Pollution Control	Tracy Carter	Director	(615) 532-0554
Water:	Water Pollution Control	Paul Davis	Director	(615) 532-0625
Waste:				(615) 532-0109

	Texas			

Texas Natural Resource Conservation Commission
12100 Park 35 Circle
P.O. Box 13087
Austin, TX 78711-3087
General Information: (512) 239-5500
Internet: http://www.tnrcc.state.tx.us

Air:	Air Permits	Walter Bradley	Director	(512) 239-5440
Water:	Water Permits & Resource Management	Mike Cowan	Director	(512) 239-4050
Waste:	Waste Permits	Dale Burnett	Director	(512) 239-6787

	Utah			

Department of Environmental Quality
168 North 1950 West
Salt Lake City, UT 84116
General Information: (801)536-4400
Internet: http://www.eq.state.ut.us

Air:	Air Quality Division	Regg Olsen	Manager	(801) 536-4165
		Permitting Branch		
Water:	Water Quality Division	Fred Pehrson	Manager	(801) 538-6076
		Permits, Compliance & Monitoring		
Waste:	Solid and Hazardous Waste Division	Scot Anderson	Manager	(801) 538-6784
		Hazardous Waste Branch		

	Vermont			

Vermont Department of Environmental Conservation
State Complex
103 South Main Street
Waterbury, VT 05671-0401
General Information: (802) 241-3800
Internet: www.anr.state.vt.us/dec/dec/htm

Air:	Air Pollution Control Division	Richard Valentinetti	Director	(802) 241-3800
Water:	Water Quality Division	(Vacant)	Director	(802) 241-3800
Waste:	Waste Management Division	Skip Flanders	Director	(802) 241-3800

Virginia				
Virginia Department of Environmental Quality				
629 East Main Street				
P.O. Box 10009				
Richmond, VA 23240				
General Information: (804) 698-4000				
Internet: www.deq.state.va.us/				
Air:	Air Program	John Daniel	Coordination Director	(804) 698-4311
Water:	Water Program	Larry Lawson	Coordination Director	(804) 698-4108
Waste:	Waste Program	Hassan Vakili	Coordination Director	(804) 698-4155

Washington				
Washington State Department of Ecology				
P.O. Box 47600				
Olympia, WA 98504-7600				
General Information: (360) 407-6000				
Internet: http://www.wa.gov/ecology				
Air:	Air Program	Mary Burg	Manager	(360) 407-6880
Water:	Water Quality Program	Megan White	Manager	(360) 407-6405
Waste:	Solid Waste & Financial Assistance	Cullen Stephenson	Manager	(360) 407-6103

West Virginia				
Bureau of Environment				
10 McJunkin Road				
Nitro, WV 25143-2506				
General Information: (304) 759-0515				
Internet: www.dep.state.wv.us/				
Air:	Air Quality Office	Skip Kropp	Chief	(304) 558-4022
Water:	Water Resource Office	Barbara Taylor	Chief	(304) 558-2107
Waste:	Waste Management Office	B. F. Smith	Chief	(304) 558-5929

Wisconsin				
Wisconsin Department of Natural Resources				
P.O. Box 7921				
Madison, WI 53707				
General Information: (608) 266-2121				
Internet: http://www.dnr.state.wi.us/				
Air:	Air Management Bureau	Lloyd Eagan	Director	(608) 266-0603
Water:	Watershed Management Bureau	Al Shea	Director	(608) 267-2759
Waste:	Waste Management Bureau	Suzanne Bangert	Director	(608) 267-6854

Wyoming
Department of Environmental Quality
Herschler Building, 4th Floor
122 West 25th Street
Cheyenne, WY   82002
General Information: (307) 777-7937
Internet: www.deq.state.wy.us/

Air:	Air Quality Division	Dan Olson	Administrator	(307) 777-7391
Water:	Water Quality Division	Gary Beach	Administrator	(307) 777-7781
Waste:	Solid Hazardous Waste Division	Dave Finley	Administrator	(307) 777-7752

APPENDIX B
WATER POLLUTION CONTROL GLOSSARY

Aeration   A process which promotes biological degradation of organic matter. The process may be passive (as when waste is exposed to air) or active (as when a mixing or bubbling device introduces the air).

Backfill   Earth used to fill a trench or an excavation.

Baffles   Finlike devices installed vertically on the inside walls of liquid waste transport vehicles that are used to reduce the movement of the waste inside the tank.

Berm   An earthen mound used to direct the flow of runoff around or through a structure.

Best Management Practice (BMP)   Schedules of activities, prohibitions of practices, maintenance procedures, and other management practices to prevent or reduce the pollution of waters of the United States. BMPs also include treatment requirements, operating procedures, and practices to control facility site runoff, spillage or leaks, sludge or waste disposal, or drainage from raw material storage.

Biodegradable   The ability to break down or decompose under natural conditions and processes.

Boom   1. A floating device used to contain oil on a body of water. 2. A piece of equipment used to apply pesticides from ground equipment such as a tractor or truck.

Buffer Strip or Zone   Strips of grass or other erosion-resistant vegetation between a waterway and an area of more intensive land use.

By-product   Material, other than the principal product, that is generated as a consequence of an industrial process.

Calibration   A check of the precision and accuracy of measuring equipment.

CERCLA   Comprehensive Environmental Response, Compensation, and Liability Act.

Chock   A block or wedge used to keep rolling vehicles in place.

Clay Lens   A naturally occurring, localized area of clay that acts as an impermeable layer to runoff infiltration.

Concrete Aprons   A pad of nonerosive material designed to prevent scour holes developing at the outlet ends of culverts, outlet pipes, grade stabilization structures, and other water control devices.

Conduit   Any channel or pipe for transporting the flow of water.

Conveyance   Any natural or manmade channel or pipe in which concentrated water flows.

Corrosion   The dissolving and wearing away of metal caused by a chemical reaction such as between water and the pipes that the water contacts, chemicals touching a metal surface, or contact between two metals.

Culvert   A covered channel or a large-diameter pipe that directs water flow below the ground level.

CWA: Clean Water Act   (formerly referred to as the Federal Water Pollution Control Act or Federal Water Pollution Control Act Amendments of 1972).

Denuded   Land stripped of vegetation such as grass, or land that has had vegetation worn down due to impacts from the elements or humans.

Dike An embankment to confine or control water, often built along the banks of a river to prevent overflow of lowlands; a levee.

Director The regional administrator or an authorized representative.

Discharge A release or flow of stormwater or other substance from a conveyance or storage container.

Drip Guard A device used to prevent drips of fuel or corrosive or reactive chemicals from contacting other materials or areas.

Emission Pollution discharged into the atmosphere from smokestacks, other vents, and surface areas of commercial or industrial facilities and from motor vehicle, locomotive, or aircraft exhausts.

Erosion The wearing away of land surface by wind or water. Erosion occurs naturally from weather or runoff but can be intensified by land-clearing practices related to farming, residential or industrial development, road building, or timber cutting.

Excavation The process of removing earth, stone, or other materials.

Fertilizer Materials such as nitrogen and phosphorus that provide nutrients for plants. Commercially sold fertilizers may contain other chemicals or may be in the form of processed sewage sludge.

Filter Fabric Textile of relatively small mesh or pore size that is used to (*a*) allow water to pass through while keeping sediment out (permeable), or (*b*) prevent both runoff and sediment from passing through (impermeable).

Filter Strip Usually long, relatively narrow area of undisturbed or planted vegetation used to retard or collect sediment for the protection of watercourses, reservoirs, or adjacent properties.

Flange A rim extending from the end of a pipe; can be used as a connection to another pipe.

Flow Channel Liner A covering or coating used on the inside surface of a flow channel to prevent the infiltration of water to the ground.

Flowmeter A gauge that shows the speed of water moving through a conveyance.

General Permit A permit issued under the NPDES program to cover a certain class or category of stormwater discharges. These permits allow for a reduction in the administrative burden associated with permitting stormwater discharges associated with industrial activities. For example, EPA is planning to issue two general permits: NPDES General Permits for Storm Water Discharges from Construction Activities that are classified as "Associated with Industrial Activity" and NPDES General Permits for Storm Water Discharges from Industrial Activities that are classified as "Associated with Industrial Activities." EPA is also encouraging delegated states which have an approved general permits program to issue general permits.

Grading The cutting and/or filling of the land surface to a desired slope or elevation.

Hazardous Substance 1. Any material that poses a threat to human health and/or the environment. Hazardous substances can be toxic, corrosive, ignitable, explosive, or chemically reactive. 2. Any substance named required by EPA to be reported if a designated quantity of the substance is spilled in the waters of the United States or if otherwise emitted into the environment.

Hazardous Waste By-products of human activities that can pose a substantial or potential hazard to human health or the environment when improperly managed. Possesses at least one of four characteristics (ignitability, corrosivity, reactivity, or toxicity), or appears on special EPA lists.

Holding Pond A pond or reservoir, usually made of earth, built to store polluted runoff for a limited time.

Illicit Connection Any discharge to a municipal separate storm sewer that is not composed entirely of stormwater except discharges authorized by an NPDES permit (other than the NPDES permit for discharges from the municipal separate storm sewer) and discharges resulting from fire-fighting activities.

Infiltration   1. The penetration of water through the ground surface into subsurface soil or the penetration of water from the soil into sewer or other pipes through defective joints, connections, or accesshole walls. 2. A land application technique where large volumes of wastewater are applied to land, allowed to penetrate the surface and percolate through the underlying soil.

Inlet   An entrance into a ditch, storm sewer, or other waterway.

Intermediates   A chemical compound formed during the making of a product.

Irrigation   Human application of water to agricultural or recreational land for watering purposes.

Jute   A plant fiber used to make rope, mulch, netting, or matting.

Lagoon   A shallow pond where sunlight, bacterial action, and oxygen work to purify wastewater.

Land Application Units   An area where wastes are applied onto or incorporated into the soil surface (excluding manure spreading operations) for treatment or disposal.

Land Treatment Units   An area of land where materials are temporarily located to receive treatment. Examples include sludge lagoons and stabilization ponds.

Landfills   An area of land or an excavation in which wastes are placed for permanent disposal, and which is not a land application unit, surface impoundment, injection well, or waste pile.

Large and Medium Municipal Separate Storm Sewer System   All municipal separate storm sewers that are either: (i) located in an incorporated place (city) with a population of 100,000 or more as determined by the latest Decennial Census by the Bureau of Census (these cities are listed in Appendices F and G of 40 CFR Part 122); or (ii) located in the counties with unincorporated urbanized populations of 100,000 or more, except municipal separate storm sewers that are located in the incorporated places, townships, or towns within such counties (these counties are listed in Appendices H and I of 40 CFR Part 122); or (iii) owned or operated by a municipality other than those described in paragraph (i) or (ii) and that are designated by the director as part of the large or medium municipal separate storm sewer system.

Leaching   The process by which soluble constituents are dissolved in a solvent such as water and carried down through the soil.

Level Spreader   A device used to spread out stormwater runoff uniformly over the ground surface as sheetflow (i.e., not through channels). The purpose of level spreaders is to prevent concentrated, erosive flows from occurring and to enhance infiltration.

Liming   Treating soil with lime to neutralize acidity levels.

Liner   1. A relatively impermeable barrier designed to prevent leachate from leaking from a landfill. Liner materials include plastic and dense clay. 2. An insert or sleeve for sewer pipes to prevent leakage or infiltration.

Liquid Level Detector   A device that provides continuous measures of liquid levels in liquid storage areas or containers to prevent overflows.

Material Storage Areas   Onsite locations where raw materials, products, final products, by-products, or waste materials are stored.

Mulch   A natural or artificial layer of plant residue or other materials covering the land surface which conserves moisture, holds soil in place, aids in establishing plant cover, and minimizes temperature fluctuations.

Noncontact Cooling Water   Water used to cool machinery or other materials without directly contacting process chemicals or materials.

Notice of Intent (NOI)   An application to notify the permitting authority of a facility's intention to be covered by a general permit; exempts a facility from having to submit an individual or group application.

NPDES   EPA's program to control the discharge of pollutants to waters of the United States. See the definition of "National Pollutant Discharge Elimination System" in 40 CFR 122.2 for further guidance.

NPDES Permit   An authorization, license, or equivalent control document issued by EPA or an approved state agency to implement the requirements of the NPDES program.

Oil and Grease Traps   Devices which collect oil and grease, removing them from water flows.

Oil Sheen   A thin, glistening layer of oil on water.

Oil/Water Separator   A device installed, usually at the entrance to a drain, which removes oil and grease from water flows entering the drain.

Organic Pollutants   Substances containing carbon which may cause pollution problems in receiving streams.

Organic Solvents   Liquid organic compounds capable of dissolving solids, gases, or liquids.

Outfall   The point, location, or structure where wastewater or drainage discharges from a sewer pipe, ditch, or other conveyance to a receiving body of water.

Permeability   The quality of a soil that enables water or air to move through it. Usually expressed in inches per hour or inches per day.

Permit   An authorization, license, or equivalent control document issued by EPA or an approved state agency to implement the requirements of an environmental regulation; e.g., a permit to operate a wastewater treatment plant or to operate a facility that may generate harmful emissions.

Permit Issuing Authority (or Permitting Authority)   The state agency or EPA regional office which issues environemntal permits to regulated facilities.

Plunge Pool   A basin used to slow flowing water, usually constructed to a design depth and shape. The pool may be protected from erosion by various lining materials.

Pneumatic Transfer   A system of hoses which uses the force of air or other gas to push material through; used to transfer solid or liquid materials from tank to tank.

Point Source   Any discernible, confined, and discrete conveyance, including but not limited to any pipe, ditch, channel, tunnel, conduit, well, discrete fissure, container, rolling stock, concentrated animal feeding operation, or vessel or other floating craft, from which pollutants are or may be discharged. This term does not include return flows from irrigated agriculture or agricultural stormwater runoff.

Pollutant   Any dredged spoil, solid waste, incinerator residue, filter backwash, sewage, garbage, sewage sludge, munitions, chemical wastes, biological materials, radioactive materials (except those regulated under the Atomic Energy Act of 1954, as amended [42 (U.S.C. 2011 et seq.)], heat, wrecked or discharged equipment, rock, sand, cellar dirt, and industrial, municipal, and agricultural waste discharged into water. It does not mean: (i) sewage from vessels; or (ii) water, gas, or other material which is injected into a well to facilitate production of oil or gas, or water derived in association with oil and gas production and disposed of in a well, if the well used either to facilitate production or for disposal purposes is approved by the authority of the state in which the well is located, and if the state determines that the injection or disposal will not result in the degradation of ground or surface water resources [Section 502(6) of the CWA].

Radioactive materials covered by the Atomic Energy Act are those encompassed in its definition of source, by-product, or special nuclear materials. Examples of materials not covered include radium and accelerator-produced isotopes. [See *Train v. Colorado Public Interest Research Group, Inc.*, 426 U.S. 1 (1976).]

Porous Pavement   A human-made surface that will allow water to penetrate through and percolate into soil (as in porous asphalt pavement or concrete). Porous asphalt pavement is comprised of irregular-shaped crush rock precoated with asphalt binder. Water seeps through into lower layers of gravel for temporary storage, then filters naturally into the soil.

Precipitation   Any form of rain or snow.

Preventive Maintenance Program   A schedule of inspections and testing at regular intervals intended to prevent equipment failures and deterioration.

Process Wastewater   Water that comes into direct contact with or results from the production or use of any raw material, intermediate product, finished product, by-product, waste product, or wastewater.

PVC (Polyvinyl Chloride)   A plastic used in pipes because of its strength; does not dissolve in most organic solvents.

Raw Material   Any product or material that is converted into another material by processing or manufacturing.

RCRA   Resource Conservation and Recovery Act.

Recycle   The process of minimizing the generation of waste by recovering usable products that might otherwise become waste. Examples are the recycling of aluminum cans, wastepaper, and bottles.

Reportable Quantity (RQ)   The quantity of a hazardous substance or oil that triggers reporting requirements under CERCLA or the Clean Water Act. If a substance is released in amounts exceeding its RQ, the release must be reported to the National Response Center, the State Emergency Response Commission, and community emergency coordinators for areas likely to be affected (see Appendix I for a list of RQs).

Residual   Amount of pollutant remaining in the environment after a natural or technological process has taken place, e.g., the sludge remaining after initial wastewater treatment, or particulates remaining in air after the air passes through a scrubbing or other pollutant removal process.

Retention   The holding of runoff in a basin without release except by means of evaporation, infiltration, or emergency bypass.

Retrofit   The modification of stormwater management systems in developed areas through the construction of wet ponds, infiltration systems, wetland plantings, stream bank stabilization, and other BMP techniques for improving water quality. A retrofit can consist of the construction of a new BMP in the developed area, the enhancement of an older stormwater management structure, or a combination of improvement and new construction.

Rill Erosion   The formation of numerous, closely spread streamlets due to uneven removal of surface soils by stormwater or other water.

Riparian Habitat   Areas adjacent to rivers and streams that have a high density, diversity, and productivity of plant and animal species relative to nearby uplands.

Runoff   That part of precipitation, snow melt, or irrigation water that runs off the land into streams or other surface water. It can carry pollutants from the air and land into the receiving waters.

Runon   Stormwater surface flow or other surface flow which enters property other than that where it originated.

Sanitary Sewer   A system of underground pipes that carries sanitary waste or process wastewater to a treatment plant.

Sanitary Waste   Domestic sewage.

SARA   Superfund Amendments and Reauthorization Act.

Scour   The clearing and digging action of flowing water, especially the downward erosion caused by stream water in sweeping away mud and silt from the stream bed and outside bank of a curved channel.

Sealed Gate   A device used to control the flow of liquid materials through a valve.

Secondary Containment   Structures, usually dikes or berms, surrounding tanks or other storage containers and designed to catch spilled material from the storage containers.

Section 313 Water Priority Chemical A chemical or chemical categories which: (1) are listed at 40 CFR 372.65 pursuant to Section 313 of the Emergency Planning and Community Right-to-Know Act (EPCRA) [also known as Title III of the Superfund Amendments and Reauthorization Act (SARA) of 1986]; (2) are present at or above threshold levels at a facility subject to EPCRA Section 313 reporting requirements; and (3) meet at least one of the following criteria: (i) are listed in Appendix D of 40 CFR Part 122 on either Table II (organic priority pollutants), Table III (certain metals, cyanides, and phenols), or Table V (certain toxic pollutants and hazardous substances); (ii) are listed as a hazardous substance pursuant to Section 311(b)(2)(A) of the CWA at 40 CFR 116.4; or (iii) are pollutants for which EPA has published acute or chronic water quality criteria. See Addendum B of this permit. (List is included as Appendix I.)

Sediment Trap A device for removing sediment from water flows; usually installed at outfall points.

Sedimentation The process of depositing soil particles, clays, sands, or other sediments that were picked up by flowing water.

Sediments Soil, sand, and minerals washed from land into water, usually after rain. They pile up in reservoirs, rivers, and harbors, destroying fish-nesting areas and holes of water animals and cloud the water so that needed sunlight might not reach aquatic plants. Careless farming, mining, and building activities will expose sediment materials, allowing them to be washed off the land after rainfalls.

Sheet Erosion Erosion of thin layers of surface materials by continuous sheets of running water.

Sheetflow Runoff which flows over the ground surface as a thin, even layer, not concentrated in a channel.

Shelf Life The time for which chemicals and other materials can be stored before becoming unusable due to age or deterioration.

Significant Materials Include, but are not limited to: raw materials; fuels; materials such as solvents, detergents, and plastic pellets; finished materials such as metallic products; raw materials used in food processing or production; hazardous substances designated under section 101(14) of the Comprehensive Environmental Response, Compensation, and Liability Act (CERCLA); any chemical the facility is required to report pursuant to section 313 of Title III of the Superfund Amendments and Reauthorization Act (SARA); fertilizers; pesticides; and waste products such as ashes, slag, and sludge that have a potential to be released with stormwater discharges [122.26(b)(12)].

Significant Spills Includes, but is not limited to releases of oil or hazardous substances in excess of reportable quantities under Section 311 of the CWA (see 40 CFR 110.10 and CFR 117.21) or Section 102 of CERCLA (see 40 CFR 302.4).

Slag Non-metal-containing waste leftover from the smelting and refining of metals.

Slide Gate A device used to control the flow of water through stormwater conveyances.

Sloughing The movement of unstabilized soil layers down a slope due to excess water in the soils.

Sludge A semisolid residue from any of a number of air or water treatment processes. Sludge can be a hazardous waste.

Soil The unconsolidated mineral and organic material on the immediate surface of the earth that serves as a natural medium for the growth of plants.

Solids Dewatering A process for removing excess water from solids to lessen the overall weight of the wastes.

Source Control A practice or structural measure to prevent pollutants from entering stormwater runoff or other environmental media.

Spent Solvent   A liquid solution that has been used and is no longer capable of dissolving solids, gases, or liquids.

Spill Guard   A device used to prevent spills of liquid materials from storage containers.

Spill Prevention Control and Countermeasures Plan (SPCC)   Plan consisting of structures, such as curbing, and action plans to prevent and respond to spills of hazardous substances as defined in the Clean Water Act.

Stopcock Valve   A small valve for stopping or controlling the flow of water or other liquid through a pipe.

Storm Drain   A slotted opening leading to an underground pipe or an open ditch for carrying surface runoff.

Stormwater   Runoff from a storm event, snow melt runoff, and surface runoff and drainage.

Stormwater Discharge Associated with Industrial Activity   The discharge from any conveyance which is used for collecting and conveying stormwater and which is directly related to manufacturing, processing, or raw materials storage areas at an industrial plant. The term does not include discharges from facilities or activities excluded from the NPDES program under 40 CFR Part 122. For the categories of industries identified in subparagraphs (i) through (x) of this subsection, the term includes, but is not limited to, stormwater discharges from industrial plant yards; immediate access roads and rail lines used or traveled by carriers of raw materials, manufactured products, waste material, or by-products used or created by the facility; material handling sites; refuse sites; sites used for the application or disposal of process waste waters (as defined at 40 CFR 401); sites used for the storage and maintenance of material handling equipment; sites used for residual treatment, storage, or disposal; shipping and receiving areas; manufacturing buildings; storage areas (including tank farms) for raw materials, and intermediate and finished products; and areas where industrial activity has taken place in the past and significant materials remain and are exposed to stormwater. For the categories of industries identified in subparagraph (xi), the term includes only stormwater discharges from all the areas (except access roads and rail lines) that are listed in the previous sentence where material handling equipment or activities, raw materials, intermediate products, final products, waste material, by-products, or industrial machinery are exposed to stormwater. For the purposes of this paragraph, material handling activities include the storage, loading and unloading, transportation, or conveyance of any raw material, intermediate product, finished product, by-product, or waste product. The term excludes areas located on plant lands separate from the plant's industrial activities, such as office buildings and accompanying parking lots as long as the drainage from the excluded areas is not mixed with stormwater drained from the above described areas. Industrial facilities (including industrial facilities that are federally, state, or municipally owned or operated that meet the description of the facilities listed in this paragraph (i)–(xi) include those facilities designated under the provision of 122.26(a)(1)(v). The following categories of facilities are considered to be engaging in "industrial activity" for purposes of this subsection: (i) facilities subject to stormwater effluent limitations guidelines, new source performance standards, or toxic pollutant effluent standards under 40 CFR Subchapter N [except facilities with toxic pollutant effluent standards which are excepted under category (xi) of this paragraph]; (ii) facilities classified as Standard Industrial Classifications 24 (except 2434), 26 (except 265 and 267), 28 (except 283 and 285), 29, 311, 32 (except 323), 33, 3441, 372; (iii) facilities classified as Standard Industrial Classifications 10 through 14 (mineral industry) including active or inactive mining operations (except for areas of coal mining operations no longer meeting the definition of a reclamation area under 40 CFR 434.11(1) because the performance bond issued to the facility by the appropriate SMCRA authority has been released, or except for areas of non-coal-mining operations which have been released from applicable state or federal reclamation requirements after Dec. 17, 1990, and oil and gas exploration, production, processing, or treatment operations, or transmission facilities that discharge stormwater contaminated by contact with or that has come into contact with any overburden, raw material, intermediate products, finished products, by-products, or waste products located on the site of such operations; (inactive mining operations are mining sites that are not being actively mined, but which have an identifiable owner-operator; inactive mining sites do not include sites where mining claims are being maintained prior to disturbances associated with the extraction, beneficiation, or processing of mined materials, nor sites where minimal activities are undertaken for the sole purpose of maintaining a mining claim); (iv) hazardous waste treatment, storage, or disposal facilities, including those that are operating under interim status or a permit under Subtitle C of RCRA; (v) landfills, land application sites, and open dumps that receive or have received any industrial wastes (waste that is received from any of the facilities described under this subsection) including those that are subject to regulation under Subtitle D of RCRA; (vi) facilities involved in the recycling of materials, including metal scrapyards, battery reclaimers, salvage yards, and automobile junkyards, including but limited to those classified as Standard Industrial Classification 5015 and 5093; (vii) steam electric power generating facilities, including coal handling sites; (viii) transportation facilities classified as Standard Industrial Classifications 40, 41, 42 (except 4221-25), 43, 44, 45, and 5171 which have vehicle maintenance shops, equipment cleaning operations, or airport deicing operations. Only those portions of the facility that are either involved in vehicle maintenance (including vehicle rehabilitation, mechanical repairs, painting, fueling, and lubrication), equipment cleaning operations, airport deicing operations, or which are otherwise identified under paragraphs (i)–(vii) or (ix)–(xi) of this subsection are associated with industrial activity; (ix) treatment works treating domestic sewage or any other sewage sludge or wastewater treatment device or system, used in the storage treatment, recycling, and reclamation of municipal or domestic sewage, including land dedicated to the disposal of sewage sludge that is located within the confines of the facility, with a design flow of 1.0 mgd or more, or required to have an approved pretreatment program under 40 CFR 403. Not included are farmlands, domestic gardens, or lands used for sludge management where sludge is beneficially reused and which are not physically located in the confines of the facility, or areas that are in compliance with Section 405 of the CWA; (x) construction activity including clearing, grading, and excavation activities except: operations that result in the disturbance of less than 5 acres of total land area which are not part of a larger common plan of development or sale; (xi) facilities under Standard Industrial Classification 20, 21, 22, 23, 2434, 25, 265, 267, 27, 283, 285, 30, 31 (except 311), 323, 34 (except 3441), 35, 36, 37 (except 373), 38, 39, 4221-25 [and which are not otherwise included within categories (ii)–(x)].

Note: The Transportation Act of 1991 provides an exemption from stormwater permitting requirements for certain facilities owned or operated by municipalities with a population of less than 100,000. Such municipalities must submit stormwater discharge permit applications for only airports, power plants, and uncontrolled sanitary landfills that they own or operate, unless a permit is otherwise required by the permitting authority.

Subsoil The bed or stratum of earth lying below the surface soil.

Sump A pit or tank that catches liquid runoff for drainage or disposal.

Surface Impoundment Treatment, storage, or disposal of liquid wastes in ponds.

Surface Water All water naturally open to the atmosphere (rivers, lakes, reservoirs, streams, wetlands impoundments, seas, estuaries, etc.); also refers to springs, wells, or other collectors which are directly influenced by surface water.

Swale An elongated depression in the land surface that is at least seasonally wet, is usually heavily vegetated, and is normally without flowing water. Swales direct stormwater flows into primary drainage channels and allow some of the stormwater to infiltrate into the ground surface.

Tarp A sheet of waterproof canvas or other material used to cover and protect materials, equipment, or vehicles.

Topography   The physical features of a surface area including relative elevations and the position of natural and human-made features.

Toxic Pollutants   Any pollutant listed as toxic under Section 501(a)(1) or, in the case of "sludge use or disposal practices," any pollutant identified in regulations implementing Section 405(d) of the CWA. Please refer to 40 CFR Part 122 Appendix D.

Treatment   The act of applying a procedure or chemicals to a substance to remove undesirable pollutants.

Tributary   A river or stream that flows into a larger river or stream.

Underground Storage Tanks (USTs)   Storage tanks with at least 10 percent or more of their storage capacity underground (the complete regulatory definition is at 40 CFR Part 280.12).

Waste   Unwanted materials left over from a manufacturing or other process.

Waste Pile   Any noncontainerized accumulation of solid, nonflowing waste that is used for treatment or storage.

Water Table   The depth or level below which the ground is saturated with water.

Waters of the United States   (a) All waters, which are currently used, were used in the past, or may be susceptible to use in interstate or foreign commerce, including all waters which are subject to the ebb and flow of the tide; (b) All interstate waters, including interstate "wetlands"; (c) All other waters such as intrastate lakes, rivers, streams (including intermittent streams), mudflats, sandflats, "wetlands," sloughs, prairie potholes, wet meadows, playa lakes, or natural ponds, the use, degradation, or destruction of which would affect or could affect interstate or foreign commerce including any such waters: (1) which are or could be used by interstate or foreign travelers for recreational or other purposes; (2) from which fish or shellfish are or could be taken and sold in interstate or foreign commerce; or (3) which are used or could be used for industrial purposes by industries in interstate commerce; (d) all impoundment of waters otherwise defined as waters of the United States under this definition; (e) tributaries of waters identified in paragraphs (a) through (d) of this definition; (f) the territorial sea; and (g) "wetlands" adjacent to waters (other than waters that are themselves wetlands) identified in paragraphs (a) through (f) of this definition.

Waste treatment systems, including treatment ponds or lagoons designed to meet the requirements of CWA [other than cooling ponds as defined in 40 CFR 423.11(m) which also meet the criteria of this definition] are not waters of the United States. This exclusion applies only to manmade bodies of water which neither were originally created in waters of the United States (such as disposal area in wetlands) nor resulted from the impoundment of waters of the United States.

Waterway   A channel for the passage or flow of water.

Wetlands   An area that is regularly saturated by surface or groundwater and subsequently is characterized by a prevalence of vegetation that is adapted for life in saturated soil conditions. Examples include swamps, bogs, fens, marshes, and estuaries.

Wet Well   A chamber used to collect water or other liquid and to which a pump is attached.

Wind Break   Any device designed to block wind flow and intended for protection against any ill effects of wind.

APPENDIX C

United States	Office of Solid Waste	EPA 550-B-98-017
Environmental Protection	and Emergency Response	November 1998
Agency	(5104)	www.epa.gov/ceppo

Title III
List of Lists

Consolidated List of Chemicals Subject to the Emergency Planning and Community Right-To-Know Act (EPCRA) and Section 112(r) of the Clean Air Act, as Amended

Title III of the Superfund Amendments
and Reauthorization Act of 1986,
and Title III of the Clean Air Act
Amendments of 1990

- EPCRA Section 302 Extremely Hazardous Substances
- CERCLA Hazardous Substances
- EPCRA Section 313 Toxic Chemicals
- CAA 112(r) Regulated Chemicals For Accidental Release Prevention

TABLE OF CONTENTS

TITLE III LIST OF LISTS
Consolidated List of Chemicals Subject to the Emergency Planning and
Community Right-to-Know Act (EPCRA) and Section 112(r) of the Clean Air Act, as Amended

This consolidated chemical list includes chemicals subject to reporting requirements under Title III of the Superfund Amendments and Reauthorization Act of 1986 (SARA)[1], also known as the Emergency Planning and Community Right-to-Know Act (EPCRA), and chemicals listed under section 112(r) of Title III of the Clean Air Act (CAA) of 1990, as amended. This consolidated list has been prepared to help firms handling chemicals determine whether they need to submit reports under sections 302, 304, or 313 of SARA Title III (EPCRA) and, for a specific chemical, what reports may need to be submitted. It also will also help firms determine whether they will be subject to accident prevention regulations under CAA section 112(r). Separate lists are also provided of Resource Conservation and Recovery Act (RCRA) waste streams and unlisted hazardous wastes, and of radionuclides reportable under the Comprehensive Environmental Response, Compensation, and Liability Act of 1980 (CERCLA). These lists should be used as a reference tool, not as a definitive source of compliance information. Compliance information for EPCRA is published in the Code of Federal Regulations (CFR), 40 CFR Parts 302, 355, and 372. Compliance information for CAA section 112(r) is published in 40 CFR Part 68.

The chemicals on the consolidated list are ordered by Chemical Abstracts Service (CAS) registry number. Categories of chemicals, which generally do not have CAS registry numbers, but which are cited under CERCLA and EPCRA section 313, are placed at the end of the list. For reference purposes, the chemicals (with their CAS numbers) are ordered alphabetically following the CAS-order list.

The list includes chemicals referenced under five federal statutory provisions, discussed below. More than one chemical name may be listed for one CAS number, because the same chemical may appear on different lists under different names. For example, for CAS number 8001-35-2, the names toxaphene (from the section 313 list), camphechlor (from the section 302 list), and camphene, octachloro- (from the CERCLA list) all appear on this consolidated list. The chemical names on this consolidated list generally are those names used in the regulatory programs developed under SARA Title III (EPCRA), CERCLA, and CAA section 112(r), but each chemical may have other synonyms that do not appear on this list.

(1) EPCRA Section 302 Extremely Hazardous Substances (EHSs)

The presence of EHSs in quantities in excess of the Threshold Planning Quantity (TPQ), requires certain emergency planning activities to be conducted. The extremely hazardous substances and their TPQs are listed in 40 CFR Part 355, Appendices A and B.

TPQ. The consolidated list presents the TPQ (in pounds) for section 302 chemicals in the

---

[1] **This consolidated list does not include all chemicals subject to the reporting requirements in sections 311 and 312 of SARA Title III (EPCRA). These hazardous chemicals, for which material safety data sheets (MSDS) must be developed under Occupational Safety and Health Act Hazard Communication Standards, are identified by broad criteria, rather than by enumeration. There are over 500,000 products that satisfy the criteria. See 40 CFR Part 370 for more information.**

column following the chemical name. For chemicals that are solids, there may be two TPQs given (e.g., 500/10,000). In these cases, the lower quantity applies for solids in powder form with particle size less than 100 microns, or if the substance is in solution or in molten form. Otherwise, the 10,000 pound TPQ applies.

EHS RQ. Releases of reportable quantities (RQ) of EHSs are subject to state and local reporting under section 304 of SARA Title III (EPCRA). EPA has promulgated a rule (61 FR 20473, May 7, 1996) that adjusted RQs for EHSs without CERCLA RQs to levels equal to their TPQs. The EHS RQ column lists these adjusted RQs for EHSs not listed under CERCLA and the CERCLA RQs for those EHSs that are CERCLA hazardous substances (see the next section for a discussion of CERCLA RQs).

(2) CERCLA Hazardous Substances

Releases of CERCLA hazardous substances, in quantities equal to or greater than their reportable quantity (RQ), are subject to reporting to the National Response Center under CERCLA. Such releases are also subject to state and local reporting under section 304 of SARA Title III (EPCRA). CERCLA hazardous substances, and their reportable quantities, are listed in 40 CFR Part 302, Table 302.4. Radionuclides listed under CERCLA are provided in a separate list, with RQs in Curies.

RQ. The CERCLA RQ column in the consolidated list shows the RQs (in pounds) for chemicals that are CERCLA hazardous substances. Carbamate wastes under RCRA that have been added to the CERCLA list with statutory one-pound RQs are indicated by an asterisk ("*") following the RQ.

Metals. For metals listed under CERCLA (antimony, arsenic, beryllium, cadmium, chromium, copper, lead, nickel, selenium, silver, thallium, and zinc), no reporting of releases of the solid form is required if the mean diameter of the pieces of the solid metal released is greater than 100 micrometers (0.004 inches). The RQs shown on the consolidated list apply to smaller particles.

Note that the consolidated list does not include all CERCLA regulatory synonyms. See 40 CFR Part 302, Table 302.4 for a complete list.

(3) CAA Section 112(r) List of Substances for Accidental Release Prevention

Under the accident prevention provisions of section 112(r) of the CAA, EPA developed a list of 77 toxic substances and 63 flammable substances. Threshold quantities (TQs) were established for these substances. The list and TQs identify facilities subject to accident prevention regulations. The list of substances and TQs and the requirements for risk management programs for accidental release prevention are found in 40 CFR Part 68. This consolidated list includes both the common name for each listed chemical under section 112(r) and the chemical name, if different from the common name, as separate listings.

The CAA section 112(r) list includes several substances in solution that are covered only in concentrations above a specified level. These substances include: ammonia (concentration 20% or greater) (CAS number 7664-41-7); hydrochloric acid (37% or greater) (7647-01-0); hydrogen fluoride/hydrofluoric acid (50% or greater) (7664-39-3); and nitric acid (80% or greater) (7697-37-2). Hydrogen chloride (anhydrous) and ammonia (anhydrous) are listed, in addition to the solutions of these substances, with different TQs. Only the anhydrous form of sulfur dioxide (7446-09-5) is covered.

These substances are presented on the consolidated list with the concentration limit or specified form (e.g., anhydrous), as they are listed under CAA section 112(r).

TQ. The CAA section 112(r) TQ column in the consolidated list shows the TQs (in pounds) for chemicals listed for accidental release prevention.

(4) EPCRA Section 313 Toxic Chemicals

Emissions, transfers, and waste management data for chemicals listed under section 313 must be reported annually as part of the community right-to-know provisions of SARA Title III (EPCRA) (40 CFR Part 372).

Section 313. The notation "313" in the column for section 313 indicates that the chemical is subject to reporting under section 313 and section 6607 of the Pollution Prevention Act under the name listed. In cases where a chemical is listed under section 313 with a second name in parentheses or brackets, the second name is included on this consolidated list with an "X" in the section 313 column. An "X" in this column also may indicate that the same chemical with the same CAS number appears on another list with a different chemical name. For chemical categories reportable under section 313, category codes for section 313 reporting are listed in this column.

Diisocyanates and PACs. In the November 30, 1994, expansion of the section 313 list, 20 specific chemicals were added as members of the diisocyanate category, and 19 specific chemicals were added as members of the polycyclic aromatic compounds (PAC) category. These chemicals are included in the CAS order listing on this consolidated list. The symbol "#" following the "313" notation in the section 313 column identifies diisocyanates, and the symbol "+" identifies PACs, as noted in footnotes. Chemicals belonging to these categories are
reportable under section 313 by category, rather than by individual chemical name.

Ammonium Salts. The listing for ammonia under section 313 includes anhydrous ammonia and aqueous ammonia from water dissociable salts and other sources. Ten percent of total aqueous ammonia is reportable under this listing.

(5) Chemical Categories

The CERCLA and EPCRA section 313 lists include a number of chemical categories as well as specific chemicals. Categories appear on this consolidated list at the end of the CAS number listing. Specific chemicals listed as members of the diisocyanate and PAC categories under EPCRA section 313 (see section (4) above) are included in the list of specific chemicals by CAS number, not in the category listing. The chemicals on the consolidated list have not been systematically evaluated to determine whether they fall into any listed categories.

Some chemicals not specifically listed under CERCLA may be subject to CERCLA reporting as part of a category. For example, strychnine, sulfate (CAS number 60-41-3), listed under EPCRA section 302, is not individually listed on the CERCLA list, but is subject to CERCLA reporting under the listing for strychnine and salts (CAS number 57-24-9), with an RQ of 10 pounds. Similarly, nicotine sulfate (CAS number 65-30-5) is subject to CERCLA reporting under the listing for nicotine and salts (CAS number 54-11-5, RQ 100 pounds), and warfarin sodium (CAS number 129-06-6) is subject to CERCLA reporting under the listing for warfarin and salts, concentration >0.3% (CAS number 81-81-2, RQ 100 pounds). Note that some CERCLA listings, although they include CAS numbers, are for general categories and are not restricted to the specific CAS number (e.g., warfarin and salts). The CERCLA list also includes a number of generic categories that have not been assigned RQs; chemicals falling into these categories are considered CERCLA hazardous substances, but are not required to be reported under CERCLA unless otherwise listed under CERCLA with an RQ.

A number of chemical categories are subject to EPCRA section 313 reporting. Be aware that certain chemicals listed under EPCRA section 302, CERCLA, or CAA section 112(r) may belong to section 313 categories. For example, mercuric acetate (CAS number 1600-27-7), listed under section 302, is not specifically listed under section 313, but could be reported under section 313 as "Mercury Compounds" (no CAS number).

(6) **RCRA Hazardous Wastes**

The consolidated list includes specific chemicals from the RCRA P and U lists only (40 CFR 261.33). This listing is provided as an indicator that companies may already have data on a specific chemical that may be useful for SARA Title III reporting. It is not intended to be a comprehensive list of RCRA P and U chemicals. RCRA hazardous wastes consisting of waste streams on the F and K lists, and wastes exhibiting the characteristics of ignitability, corrosivity, reactivity, and toxicity, are provided in a separate list. This list also includes carbamate wastes added to the CERCLA list with one-pound statutory RQs (indicated by an asterisk ("*") following the RQ). The descriptions of the F and K waste streams have been abbreviated; see 40 CFR Part 302, Table 302.4, or 40 CFR Part 261 for complete descriptions.

RCRA Code. The letter-and-digit code in the RCRA Code column is the chemical's RCRA hazardous waste code.

TITLE III LIST OF LISTS
CONSOLIDATED LIST OF CHEMICALS SUBJECT TO THE EMERGENCY PLANNING AND
COMMUNITY RIGHT-TO-KNOW ACT (EPCRA) AND SECTION 112(r) OF THE CLEAN AIR ACT, AS AMENDED

CAS Number	Chemical Name	Sec. 302 (EHS) TPQ	Section 304 EHS RQ	CERCLA RQ	CAA 112(r) TQ	Sec 313	RCRA CODE
50-00-0	Formaldehyde	500	100	100	15,000	313	U122
50-00-0	Formaldehyde (solution)	500	100	100	15,000	X	U122
50-07-7	Mitomycin C	500/10,000	10	10			U010
50-14-6	Ergocalciferol	1,000/10,000	1,000				
50-18-0	Cyclophosphamide			10			U058
50-29-3	DDT			1			U061
50-32-8	Benzo[a]pyrene			1		313+	U022
50-55-5	Reserpine			5,000			U200
51-03-6	Piperonyl butoxide					313	
51-21-8	Fluorouracil	500/10,000	500			313	
51-21-8	5-Fluorouracil	500/10,000	500			X	
51-28-5	2,4-Dinitrophenol			10		313	P048
51-43-4	Epinephrine			1,000			P042
51-75-2	2-Chloro-N-(2-chloroethyl)-N-methylethanamine	10	10			X	
51-75-2	Mechlorethamine	10	10			X	
51-75-2	Nitrogen mustard	10	10			313	
51-79-6	Carbamic acid, ethyl ester			100		X	U238
51-79-6	Ethyl carbamate			100		X	U238
51-79-6	Urethane			100		313	U238
51-83-2	Carbachol chloride	500/10,000	500				
52-68-6	Phosphonic acid, (2,2,2-trichloro-1-hydroxyethyl)-,di			100		X	
52-68-6	Trichlorfon			100		313	
52-85-7	Famphur			1,000		313	P097
53-70-3	Dibenz[a,h]anthracene			1		313+	U063
53-96-3	2-Acetylaminofluorene			1		313	U005
54-11-5	Nicotine	100	100	100			P075
54-11-5	Nicotine and salts			100			P075
54-11-5	Pyridine, 3-(1-methyl-2-pyrrolidinyl)-,(S)-	100	100	100			P075
54-62-6	Aminopterin	500/10,000	500				
55-18-5	N-Nitrosodiethylamine			1		313	U174
55-21-0	Benzamide					313	
55-38-9	O,O-Dimethyl O-(3-methyl-4-(methylthio) phenyl) es					X	
55-38-9	Fenthion					313	
55-63-0	Nitroglycerin			10		313	P081
55-91-4	Diisopropylfluorophosphate	100	100	100			P043
55-91-4	Isofluorphate	100	100	100			P043
56-04-2	Methylthiouracil			10			U164
56-23-5	Carbon tetrachloride			10		313	U211
56-25-7	Cantharidin	100/10,000	100				
56-35-9	Bis(tributyltin) oxide					313	
56-38-2	Parathion	100	10	10		313	P089
56-38-2	Phosphorothioic acid, O,O-diethyl-O-(4-nitrophenyl) ester	100	10	10		X	P089
56-49-5	3-Methylcholanthrene			10			U157
56-53-1	Diethylstilbestrol			1			U089
56-55-3	Benz[a]anthracene			10		313+	U018
56-72-4	Coumaphos	100/10,000	10	10			

+ Member of PAC category.

CAS Number	Chemical Name	Sec. 302 (EHS) TPQ	Section 304 EHS RQ	CERCLA RQ	CAA 112(r) TQ	Sec 313	RCRA CODE
57-12-5	Cyanides (soluble salts and complexes)			10			P030
57-14-7	1,1-Dimethyl hydrazine	1,000	10	10	15,000	313	U098
57-14-7	Dimethylhydrazine	1,000	10	10	15,000	X	U098
57-14-7	Hydrazine, 1,1-dimethyl-	1,000	10	10	15,000	X	U098
57-24-9	Strychnine	100/10,000	10	10			P108
57-24-9	Strychnine, and salts			10			P108
57-33-0	Pentobarbital sodium					313	
57-41-0	Phenytoin					313	
57-47-6	Physostigmine	100/10,000	1*	1*			P204
57-57-8	beta-Propiolactone	500	10	10		313	
57-64-7	Physostigmine, salicylate (1:1)	100/10,000	1*	1*			P188
57-74-9	Chlordane	1,000	1	1		313	U036
57-74-9	4,7-Methanoindan,	1,000	1	1		X	U036
57-97-6	7,12-Dimethylbenz[a]anthracene			1		313+	U094
58-36-6	Phenoxarsine, 10,10'-oxydi-	500/10,000	500				
58-89-9	Cyclohexane,	1,000/10,000	1	1		X	U129
58-89-9	Hexachlorocyclohexane (gamma isomer)	1,000/10,000	1	1		X	U129
58-89-9	Lindane	1,000/10,000	1	1		313	U129
58-90-2	2,3,4,6-Tetrachlorophenol			10			
59-50-7	p-Chloro-m-cresol			5,000			U039
59-88-1	Phenylhydrazine hydrochloride	1,000/10,000	1,000				
59-89-2	N-Nitrosomorpholine			1		313	
60-00-4	Ethylenediamine-tetraacetic acid (EDTA)			5,000			
60-09-3	4-Aminoazobenzene					313	
60-11-7	4-Dimethylaminoazobenzene			10		313	U093
60-11-7	Dimethylaminoazobenzene			10		X	U093
60-29-7	Ethane, 1,1'-oxybis-			100	10,000		U117
60-29-7	Ethyl ether			100	10,000		U117
60-34-4	Hydrazine, methyl-	500	10	10	15,000	X	P068
60-34-4	Methyl hydrazine	500	10	10	15,000	313	P068
60-35-5	Acetamide			100		313	
60-41-3	Strychnine, sulfate	100/10,000	10	10			
60-51-5	Dimethoate	500/10,000	10	10		313	P044
60-57-1	Dieldrin			1			P037
61-82-5	Amitrole			10		313	U011
62-38-4	Phenylmercuric acetate	500/10,000	100	100			P092
62-38-4	Phenylmercury acetate	500/10,000	100	100			P092
62-44-2	Phenacetin			100			U187
62-50-0	Ethyl methanesulfonate			1			U119
62-53-3	Aniline	1,000	5,000	5,000		313	U012
62-55-5	Thioacetamide			10		313	U218
62-56-6	Thiourea			10		313	U219
62-73-7	Dichlorvos	1,000	10	10		313	
62-73-7	Phosphoric acid, 2-dichloroethenyl dimethyl ester	1,000	10	10		X	
62-74-8	Fluoroacetic acid, sodium salt	10/10,000	10	10		X	P058
62-74-8	Sodium fluoroacetate	10/10,000	10	10		313	P058
62-75-9	Methanamine, N-methyl-N-nitroso-	1,000	10	10		X	P082
62-75-9	N-Nitrosodimethylamine	1,000	10	10		313	P082
62-75-9	Nitrosodimethylamine	1,000	10	10		X	P082
63-25-2	Carbaryl			100		313	U279

+ Member of PAC category.
* RCRA carbamate waste; statutory one-pound RQ applies until RQs are adjusted.

CAS Number	Chemical Name	Sec. 302 (EHS) TPQ	Section 304 EHS RQ	CERCLA RQ	CAA 112(r) TQ	Sec 313	RCRA CODE
63-25-2	1-Naphthalenol, methylcarbamate			100		X	U279
64-00-6	Phenol, 3-(1-methylethyl)-, methylcarbamate	500/10,000	1*	1*			P202
64-18-6	Formic acid			5,000		313	U123
64-19-7	Acetic acid			5,000			
64-67-5	Diethyl sulfate			10		313	
64-75-5	Tetracycline hydrochloride					313	
64-86-8	Colchicine	10/10,000	10				
65-30-5	Nicotine sulfate	100/10,000	100	100			
65-85-0	Benzoic acid			5,000			
66-75-1	Uracil mustard			10			U237
66-81-9	Cycloheximide	100/10,000	100				
67-56-1	Methanol			5,000		313	U154
67-63-0	Isopropyl alcohol (mfg-strong acid process)					313	
67-64-1	Acetone			5,000			U002
67-66-3	Chloroform	10,000	10	10	20,000	313	U044
67-66-3	Methane, trichloro-	10,000	10	10	20,000	X	U044
67-72-1	Hexachloroethane			100		313	U131
68-12-2	Dimethylformamide			100		X	
68-12-2	N,N-Dimethylformamide			100		313	
68-76-8	2,5-Cyclohexadiene-1,4-dione, 2,3,5-tris(1-aziridinyl)-					X	
68-76-8	Triaziquone					313	
70-25-7	Guanidine, N-methyl-N'-nitro-N-nitroso-			10			U163
70-30-4	Hexachlorophene			100		313	U132
70-69-9	Propiophenone, 4'-amino	100/10,000	100				
71-36-3	n-Butyl alcohol			5,000		313	U031
71-43-2	Benzene			10		313	U019
71-55-6	Methyl chloroform			1,000		X	U226
71-55-6	1,1,1-Trichloroethane			1,000		313	U226
71-63-6	Digitoxin	100/10,000	100				
72-20-8	Endrin	500/10,000	1	1			P051
72-43-5	Benzene, 1,1'-(2,2,2-trichloroethylidene)bis [4-methoxy-			1		X	U247
72-43-5	Methoxychlor			1		313	U247
72-54-8	DDD			1			U060
72-55-9	DDE			1			
72-57-1	Trypan blue			10		313	U236
74-82-8	Methane				10,000		
74-83-9	Bromomethane	1,000	1,000	1,000		313	U029
74-83-9	Methyl bromide	1,000	1,000	1,000		X	U029
74-84-0	Ethane				10,000		
74-85-1	Ethene				10,000	X	
74-85-1	Ethylene				10,000	313	
74-86-2	Acetylene				10,000		
74-86-2	Ethyne				10,000		
74-87-3	Chloromethane			100	10,000	313	U045
74-87-3	Methane, chloro-			100	10,000	X	U045
74-87-3	Methyl chloride			100	10,000	X	U045
74-88-4	Methyl iodide			100		313	U138
74-89-5	Methanamine			100	10,000		
74-89-5	Monomethylamine			100	10,000		
74-90-8	Hydrocyanic acid	100	10	10	2,500	X	P063

* RCRA carbamate waste; statutory one-pound RQ applies until RQs are adjusted.

Page 4

CAS Number	Chemical Name	Sec. 302 (EHS) TPQ	Section 304 EHS RQ	CERCLA RQ	CAA 112(r) TQ	Sec 313	RCRA CODE
74-90-8	Hydrogen cyanide	100	10	10	2,500	313	P063
74-93-1	Methanethiol	500	100	100	10,000	X	U153
74-93-1	Methyl mercaptan	500	100	100	10,000	313	U153
74-93-1	Thiomethanol	500	100	100	10,000	X	U153
74-95-3	Methylene bromide			1,000		313	U068
74-98-6	Propane				10,000		
74-99-7	1-Propyne				10,000		
74-99-7	Propyne				10,000		
75-00-3	Chloroethane			100	10,000	313	
75-00-3	Ethane, chloro-			100	10,000	X	
75-00-3	Ethyl chloride			100	10,000	X	
75-01-4	Ethene, chloro-			1	10,000	X	U043
75-01-4	Vinyl chloride			1	10,000	313	U043
75-02-5	Ethene, fluoro-				10,000		
75-02-5	Vinyl fluoride				10,000		
75-04-7	Ethanamine			100	10,000		
75-04-7	Monoethylamine			100	10,000		
75-05-8	Acetonitrile			5,000		313	U003
75-07-0	Acetaldehyde			1,000	10,000	313	U001
75-08-1	Ethanethiol				10,000		
75-08-1	Ethyl mercaptan				10,000		
75-09-2	Dichloromethane			1,000		313	U080
75-09-2	Methylene chloride			1,000		X	U080
75-15-0	Carbon disulfide	10,000	100	100	20,000	313	P022
75-19-4	Cyclopropane				10,000		
75-20-7	Calcium carbide			10			
75-21-8	Ethylene oxide	1,000	10	10	10,000	313	U115
75-21-8	Oxirane	1,000	10	10	10,000	X	U115
75-25-2	Bromoform			100		313	U225
75-25-2	Tribromomethane			100		X	U225
75-27-4	Dichlorobromomethane			5,000		313	
75-28-5	Isobutane				10,000		
75-28-5	Propane, 2-methyl				10,000		
75-29-6	Isopropyl chloride				10,000		
75-29-6	Propane, 2-chloro-				10,000		
75-31-0	Isopropylamine				10,000		
75-31-0	2-Propanamine				10,000		
75-34-3	1,1-Dichloroethane			1,000		X	U076
75-34-3	Ethylidene Dichloride			1,000		313	U076
75-35-4	1,1-Dichloroethylene			100	10,000	X	U078
75-35-4	Ethene, 1,1-dichloro-			100	10,000	X	U078
75-35-4	Vinylidene chloride			100	10,000	313	U078
75-36-5	Acetyl chloride			5,000			U006
75-37-6	Difluoroethane				10,000		
75-37-6	Ethane, 1,1-difluoro-				10,000		
75-38-7	Ethene, 1,1-difluoro-				10,000		
75-38-7	Vinylidene fluoride				10,000		
75-43-4	Dichlorofluoromethane					313	
75-43-4	HCFC-21					X	
75-44-5	Carbonic dichloride	10	10	10	500	X	P095

CAS Number	Chemical Name	Sec. 302 (EHS) TPQ	Section 304 EHS RQ	CERCLA RQ	CAA 112(r) TQ	Sec 313	RCRA CODE
75-44-5	Phosgene	10	10	10	500	313	P095
75-45-6	Chlorodifluoromethane					313	
75-45-6	HCFC-22					X	
75-50-3	Methanamine, N,N-dimethyl-			100	10,000		
75-50-3	Trimethylamine			100	10,000		
75-55-8	Aziridine, 2-methyl	10,000	1	1	10,000	X	P067
75-55-8	Propyleneimine	10,000	1	1	10,000	313	P067
75-56-9	Oxirane, methyl-	10,000	100	100	10,000	X	
75-56-9	Propylene oxide	10,000	100	100	10,000	313	
75-60-5	Cacodylic acid			1			U136
75-63-8	Bromotrifluoromethane					313	
75-63-8	Halon 1301					X	
75-64-9	tert-Butylamine			1,000			
75-65-0	tert-Butyl alcohol					313	
75-68-3	1-Chloro-1,1-difluoroethane					313	
75-68-3	HCFC-142b					X	
75-69-4	CFC-11			5,000		X	U121
75-69-4	Trichlorofluoromethane			5,000		313	U121
75-69-4	Trichloromonofluoromethane			5,000		X	U121
75-71-8	CFC-12			5,000		X	U075
75-71-8	Dichlorodifluoromethane			5,000		313	U075
75-72-9	CFC-13					X	
75-72-9	Chlorotrifluoromethane					313	
75-74-1	Plumbane, tetramethyl-	100	100		10,000		
75-74-1	Tetramethyllead	100	100		10,000		
75-76-3	Silane, tetramethyl-				10,000		
75-76-3	Tetramethylsilane				10,000		
75-77-4	Silane, chlorotrimethyl-	1,000	1,000		10,000		
75-77-4	Trimethylchlorosilane	1,000	1,000		10,000		
75-78-5	Dimethyldichlorosilane	500	500		5,000		
75-78-5	Silane, dichlorodimethyl-	500	500		5,000		
75-79-6	Methyltrichlorosilane	500	500		5,000		
75-79-6	Silane, trichloromethyl-	500	500		5,000		
75-86-5	Acetone cyanohydrin	1,000	10	10		X	P069
75-86-5	2-Methyllactonitrile	1,000	10	10		313	P069
75-87-6	Acetaldehyde, trichloro-			5,000			U034
75-88-7	2-Chloro-1,1,1-trifluoroethane					313	
75-88-7	HCFC-133a					X	
75-99-0	2,2-Dichloropropionic acid			5,000			
76-01-7	Pentachloroethane			10		313	U184
76-02-8	Trichloroacetyl chloride	500	500			313	
76-06-2	Chloropicrin					313	
76-13-1	Ethane, 1,1,2-trichloro-1,2,2,-trifluoro-					X	
76-13-1	Freon 113					313	
76-14-2	CFC-114					X	
76-14-2	Dichlorotetrafluoroethane					313	
76-15-3	CFC-115					X	
76-15-3	Monochloropentafluoroethane					313	
76-44-8	Heptachlor			1		313	P059
76-44-8	1,4,5,6,7,8,8-Heptachloro-3a,4,7,7a-tetrahydro-4,7-me			1		X	P059

CAS Number	Chemical Name	Sec. 302 (EHS) TPQ	Section 304 EHS RQ	CERCLA RQ	CAA 112(r) TQ	Sec 313	RCRA CODE
76-87-9	Triphenyltin hydroxide					313	
77-47-4	Hexachlorocyclopentadiene	100	10	10		313	U130
77-73-6	Dicyclopentadiene					313	
77-78-1	Dimethyl sulfate	500	100	100		313	U103
77-81-6	Tabun	10	10				
78-00-2	Tetraethyl lead	100	10	10			P110
78-34-2	Dioxathion	500	500				
78-48-8	DEF					X	
78-48-8	S,S,S-Tributyltrithiophosphate					313	
78-53-5	Amiton	500	500				
78-59-1	Isophorone			5,000			
78-71-7	Oxetane, 3,3-bis(chloromethyl)-	500	500				
78-78-4	Butane, 2-methyl-				10,000		
78-78-4	Isopentane				10,000		
78-79-5	1,3-Butadiene, 2-methyl-			100	10,000		
78-79-5	Isoprene			100	10,000		
78-81-9	iso-Butylamine			1,000			
78-82-0	Isobutyronitrile	1,000	1,000		20,000		
78-82-0	Propanenitrile, 2-methyl-	1,000	1,000		20,000		
78-83-1	Isobutyl alcohol			5,000			U140
78-84-2	Isobutyraldehyde					313	
78-87-5	1,2-Dichloropropane			1,000		313	U083
78-87-5	Propane 1,2-dichloro-			1,000		X	U083
78-88-6	2,3-Dichloropropene			100		313	
78-92-2	sec-Butyl alcohol					313	
78-93-3	Methyl ethyl ketone			5,000		313	U159
78-93-3	Methyl ethyl ketone (MEK)			5,000		X	U159
78-94-4	Methyl vinyl ketone	10	10				
78-97-7	Lactonitrile	1,000	1,000				
78-99-9	1,1-Dichloropropane			1,000			
79-00-5	1,1,2-Trichloroethane			100		313	U227
79-01-6	Trichloroethylene			100		313	U228
79-06-1	Acrylamide	1,000/10,000	5,000	5,000		313	U007
79-09-4	Propionic acid			5,000			
79-10-7	Acrylic acid			5,000		313	U008
79-11-8	Chloroacetic acid	100/10,000	100	100		313	
79-19-6	Thiosemicarbazide	100/10,000	100	100		313	P116
79-21-0	Ethaneperoxoic acid	500	500		10,000	X	
79-21-0	Peracetic acid	500	500		10,000	313	
79-22-1	Carbonochloridic acid, methylester	500	1,000	1,000	5,000	X	U156
79-22-1	Methyl chlorocarbonate	500	1,000	1,000	5,000	313	U156
79-22-1	Methyl chloroformate	500	1,000	1,000	5,000	X	U156
79-31-2	iso-Butyric acid			5,000			
79-34-5	1,1,2,2-Tetrachloroethane			100		313	U209
79-38-9	Ethene, chlorotrifluoro-				10,000		
79-38-9	Trifluorochloroethylene				10,000		
79-44-7	Dimethylcarbamyl chloride			1		313	U097
79-46-9	2-Nitropropane			10		313	U171
80-05-7	4,4'-Isopropylidenediphenol					313	
80-15-9	Cumene hydroperoxide			10		313	U096

CAS Number	Chemical Name	Sec. 302 (EHS) TPQ	Section 304 EHS RQ	CERCLA RQ	CAA 112(r) TQ	Sec 313	RCRA CODE
80-15-9	Hydroperoxide, 1-methyl-1-phenylethyl-			10		X	U096
80-62-6	Methyl methacrylate			1,000		313	U162
80-63-7	Methyl 2-chloroacrylate	500	500				
81-07-2	Saccharin (manufacturing)			100		313	U202
81-07-2	Saccharin and salts			100			U202
81-81-2	Warfarin	500/10,000	100	100		X	P001
81-81-2	Warfarin, & salts, conc.>0.3%			100		X	P001
81-88-9	C.I. Food Red 15					313	
82-28-0	1-Amino-2-methylanthraquinone					313	
82-66-6	Diphacinone	10/10,000	10				
82-68-8	PCNB			100		X	U185
82-68-8	Pentachloronitrobenzene			100		X	U185
82-68-8	Quintozene			100		313	U185
83-32-9	Acenaphthene			100			
84-66-2	Diethyl phthalate			1,000			U088
84-74-2	n-Butyl phthalate			10		X	U069
84-74-2	Dibutyl phthalate			10		313	U069
85-00-7	Diquat			1,000			
85-01-8	Phenanthrene			5,000		313	
85-44-9	Phthalic anhydride			5,000		313	U190
85-68-7	Butyl benzyl phthalate			100			
86-30-6	N-Nitrosodiphenylamine			100		313	
86-50-0	Azinphos-methyl	10/10,000	1	1			
86-50-0	Guthion	10/10,000	1	1			
86-73-7	Fluorene			5,000			
86-88-4	ANTU	500/10,000	100	100			P072
86-88-4	Thiourea, 1-naphthalenyl-	500/10,000	100	100			P072
87-62-7	2,6-Xylidine					313	
87-65-0	2,6-Dichlorophenol			100			U082
87-68-3	Hexachloro-1,3-butadiene			1		313	U128
87-68-3	Hexachlorobutadiene			1		X	U128
87-86-5	PCP			10		X	
87-86-5	Pentachlorophenol			10		313	
88-05-1	Aniline, 2,4,6-trimethyl-	500	500				
88-06-2	2,4,6-Trichlorophenol			10		313	
88-72-2	o-Nitrotoluene			1,000			
88-75-5	2-Nitrophenol			100		313	
88-85-7	Dinitrobutyl phenol	100/10,000	1,000	1,000		313	P020
88-85-7	Dinoseb	100/10,000	1,000	1,000		X	P020
88-89-1	Picric acid					313	
90-04-0	o-Anisidine			100		313	
90-43-7	2-Phenylphenol					313	
90-94-8	Michler's ketone					313	
91-08-7	Benzene, 1,3-diisocyanato-2-methyl-	100	100	100	10,000	X	
91-08-7	Toluene-2,6-diisocyanate	100	100	100	10,000	313	
91-20-3	Naphthalene			100		313	U165
91-22-5	Quinoline			5,000		313	
91-58-7	2-Chloronaphthalene			5,000			U047
91-59-8	beta-Naphthylamine			10		313	U168
91-66-7	N,N-Diethylaniline			1,000			

Page 8

CAS Number	Chemical Name	Sec. 302 (EHS) TPQ	Section 304 EHS RQ	CERCLA RQ	CAA 112(r) TQ	Sec 313	RCRA CODE
91-80-5	Methapyrilene			5,000			U155
91-93-0	3,3'-Dimethoxybenzidine-4,4'-diisocyanate					313#	
91-94-1	3,3'-Dichlorobenzidine			1		313	U073
91-97-4	3,3'-Dimethyl-4,4'-diphenylene diisocyanate					313#	
92-52-4	Biphenyl			100		313	
92-67-1	4-Aminobiphenyl			1		313	
92-87-5	Benzidine			1		313	U021
92-93-3	4-Nitrobiphenyl			10		313	
93-65-2	Mecoprop					313	
93-72-1	Silvex (2,4,5-TP)			100			
93-76-5	2,4,5-T acid			1,000			
93-79-8	2,4,5-T esters			1,000			
94-11-1	2,4-D Esters			100		X	
94-11-1	2,4-D isopropyl ester			100		313	
94-36-0	Benzoyl peroxide					313	
94-58-6	Dihydrosafrole			10		313	U090
94-59-7	Safrole			100		313	U203
94-74-6	(4-Chloro-2-methylphenoxy) acetic acid					X	
94-74-6	MCPA					X	
94-74-6	Methoxone					313	
94-75-7	Acetic acid, (2,4-dichlorophenoxy)-			100		X	U240
94-75-7	2,4-D			100		313	U240
94-75-7	2,4-D Acid			100		X	U240
94-75-7	2,4-D, salts and esters			100			U240
94-79-1	2,4-D Esters			100			
94-80-4	2,4-D butyl ester			100		313	
94-80-4	2,4-D Esters			100		X	
94-82-6	2,4-DB					313	
95-47-6	Benzene, o-dimethyl-			1,000		X	U239
95-47-6	o-Xylene			1,000		313	U239
95-48-7	o-Cresol	1,000/10,000	100	100		313	U052
95-50-1	o-Dichlorobenzene			100		X	U070
95-50-1	1,2-Dichlorobenzene			100		313	U070
95-53-4	o-Toluidine			100		313	U328
95-54-5	1,2-Phenylenediamine					313	
95-57-8	2-Chlorophenol			100			U048
95-63-6	1,2,4-Trimethylbenzene					313	
95-69-2	p-Chloro-o-toluidine					313	
95-80-7	2,4-Diaminotoluene			10		313	
95-94-3	1,2,4,5-Tetrachlorobenzene			5,000			U207
95-95-4	2,4,5-Trichlorophenol			10		313	
96-09-3	Styrene oxide			100		313	
96-12-8	DBCP			1		X	U066
96-12-8	1,2-Dibromo-3-chloropropane			1		313	U066
96-18-4	1,2,3-Trichloropropane					313	
96-33-3	Methyl acrylate					313	
96-45-7	Ethylene thiourea			10		313	U116
97-23-4	Dichlorophene					313	
97-23-4	2,2'-Methylenebis(4-chlorophenol					X	
97-56-3	C.I. Solvent Yellow 3					313	

Member of diisocyanate category.

CAS Number	Chemical Name	Sec. 302 (EHS) TPQ	Section 304 EHS RQ	CERCLA RQ	CAA 112(r) TQ	Sec 313	RCRA CODE
97-63-2	Ethyl methacrylate			1,000			U118
98-01-1	Furfural			5,000			U125
98-05-5	Benzenearsonic acid	10/10,000	10				
98-07-7	Benzoic trichloride	100	10	10		313	U023
98-07-7	Benzotrichloride	100	10	10		X	U023
98-09-9	Benzenesulfonyl chloride			100			U020
98-13-5	Trichlorophenylsilane	500	500				
98-16-8	Benzenamine, 3-(trifluoromethyl)-	500	500				
98-82-8	Cumene			5,000		313	U055
98-86-2	Acetophenone			5,000		313	U004
98-87-3	Benzal chloride	500	5,000	5,000		313	U017
98-88-4	Benzoyl chloride			1,000		313	
98-95-3	Nitrobenzene	10,000	1,000	1,000		313	U169
99-08-1	m-Nitrotoluene			1,000			
99-30-9	Dichloran					313	
99-30-9	2,6-Dichloro-4-nitroaniline					X	
99-35-4	1,3,5-Trinitrobenzene			10			U234
99-55-8	5-Nitro-o-toluidine			100		313	U181
99-59-2	5-Nitro-o-anisidine					313	
99-65-0	m-Dinitrobenzene			100		313	
99-98-9	Dimethyl-p-phenylenediamine	10/10,000	10				
99-99-0	p-Nitrotoluene			1,000			
100-01-6	p-Nitroaniline			5,000		313	P077
100-02-7	p-Nitrophenol			100		X	U170
100-02-7	4-Nitrophenol			100		313	U170
100-14-1	Benzene, 1-(chloromethyl)-4-nitro-	500/10,000	500				
100-25-4	p-Dinitrobenzene			100		313	
100-41-4	Ethylbenzene			1,000		313	
100-42-5	Styrene			1,000		313	
100-44-7	Benzyl chloride	500	100	100		313	P028
100-47-0	Benzonitrile			5,000			
100-75-4	N-Nitrosopiperidine			10		313	U179
101-05-3	Anilazine					313	
101-05-3	4,6-Dichloro-N-(2-chlorophenyl)-1,3,5-triazin-2-amine					X	
101-14-4	MBOCA			10		X	U158
101-14-4	4,4'-Methylenebis(2-chloroaniline)			10		313	U158
101-27-9	Barban			1*			U280
101-55-3	4-Bromophenyl phenyl ether			100			U030
101-61-1	4,4'-Methylenebis(N,N-dimethyl)benzenamine					313	
101-68-8	MDI			5,000		X	
101-68-8	Methylenebis(phenylisocyanate)			5,000		313#	
101-77-9	4,4'-Methylenedianiline			10		313	
101-80-4	4,4'-Diaminodiphenyl ether					313	
101-90-6	Diglycidyl resorcinol ether					313	
102-36-3	Isocyanic acid, 3,4-dichlorophenyl ester	500/10,000	500				
103-85-5	Phenylthiourea	100/10,000	100	100			P093
104-12-1	p-Chlorophenyl isocyanate					313	
104-49-4	1,4-Phenylene diisocyanate					313#	
104-94-9	p-Anisidine					313	
105-46-4	sec-Butyl acetate			5,000			

Member of diisocyanate category.
* RCRA carbamate waste; statutory one-pound RQ applies until RQs are adjusted.

CAS Number	Chemical Name	Sec. 302 (EHS) TPQ	Section 304 EHS RQ	CERCLA RQ	CAA 112(r) TQ	Sec 313	RCRA CODE
105-60-2	Caprolactam			5,000			
105-67-9	2,4-Dimethylphenol			100		313	U101
106-42-3	Benzene, p-dimethyl-			100		X	U239
106-42-3	p-Xylene			100		313	U239
106-44-5	p-Cresol			100		313	U052
106-46-7	1,4-Dichlorobenzene			100		313	U072
106-47-8	p-Chloroaniline			1,000		313	P024
106-49-0	p-Toluidine			100			U353
106-50-3	p-Phenylenediamine			5,000		313	
106-51-4	p-Benzoquinone			10		X	U197
106-51-4	Quinone			10		313	U197
106-88-7	1,2-Butylene oxide			100		313	
106-89-8	Epichlorohydrin	1,000	100	100	20,000	313	U041
106-89-8	Oxirane, (chloromethyl)-	1,000	100	100	20,000	X	U041
106-93-4	1,2-Dibromoethane			1		313	U067
106-93-4	Ethylene dibromide			1		X	U067
106-96-7	Propargyl bromide	10	10				
106-97-8	Butane				10,000		
106-98-9	1-Butene				10,000		
106-99-0	1,3-Butadiene			10	10,000	313	
107-00-6	1-Butyne				10,000		
107-00-6	Ethyl acetylene				10,000		
107-01-7	2-Butene				10,000		
107-02-8	Acrolein	500	1	1	5,000	313	P003
107-02-8	2-Propenal	500	1	1	5,000	X	P003
107-05-1	Allyl chloride			1,000		313	
107-06-2	1,2-Dichloroethane			100		313	U077
107-06-2	Ethylene dichloride			100		X	U077
107-07-3	Chloroethanol	500	500				
107-10-8	n-Propylamine			5,000			U194
107-11-9	Allylamine	500	500		10,000	313	
107-11-9	2-Propen-1-amine	500	500		10,000	X	
107-12-0	Ethyl cyanide	500	10	10	10,000		P101
107-12-0	Propanenitrile	500	10	10	10,000		P101
107-12-0	Propionitrile	500	10	10	10,000		P101
107-13-1	Acrylonitrile	10,000	100	100	20,000	313	U009
107-13-1	2-Propenenitrile	10,000	100	100	20,000	X	U009
107-15-3	1,2-Ethanediamine	10,000	5,000	5,000	20,000		
107-15-3	Ethylenediamine	10,000	5,000	5,000	20,000		
107-16-4	Formaldehyde cyanohydrin	1,000	1,000				
107-18-6	Allyl alcohol	1,000	100	100	15,000	313	P005
107-18-6	2-Propen-1-ol	1,000	100	100	15,000	X	P005
107-19-7	Propargyl alcohol			1,000		313	P102
107-20-0	Chloroacetaldehyde			1,000			P023
107-21-1	Ethylene glycol			5,000		313	
107-25-5	Ethene, methoxy-				10,000		
107-25-5	Vinyl methyl ether				10,000		
107-30-2	Chloromethyl methyl ether	100	10	10	5,000	313	U046
107-30-2	Methane, chloromethoxy-	100	10	10	5,000	X	U046
107-31-3	Formic acid, methyl ester				10,000		

Page 11

CAS Number	Chemical Name	Sec. 302 (EHS) TPQ	Section 304 EHS RQ	CERCLA RQ	CAA 112(r) TQ	Sec 313	RCRA CODE
107-31-3	Methyl formate				10,000		
107-44-8	Sarin	10	10				P111
107-49-3	TEPP	100	10	10			P111
107-49-3	Tetraethyl pyrophosphate						
107-92-6	Butyric acid			5,000			
108-05-4	Acetic acid ethenyl ester	1,000	5,000	5,000	15,000	X	
108-05-4	Vinyl acetate	1,000	5,000	5,000	15,000	313	
108-05-4	Vinyl acetate monomer	1,000	5,000	5,000	15,000	X	
108-10-1	Methyl isobutyl ketone			5,000		313	U161
108-23-6	Carbonochloridic acid, 1-methylethyl ester	1,000	1,000		15,000		
108-23-6	Isopropyl chloroformate	1,000	1,000		15,000		
108-24-7	Acetic anhydride			5,000		313	U147
108-31-6	Maleic anhydride			5,000		X	U239
108-38-3	Benzene, m-dimethyl-			1,000		313	U239
108-38-3	m-Xylene			100		313	U052
108-39-4	m-Cresol					313	
108-45-2	1,3-Phenylenediamine			5,000			U201
108-46-3	Resorcinol			1,000		313	U027
108-60-1	Bis(2-chloro-1-methylethyl)ether			1,000		X	U027
108-60-1	Dichloroisopropyl ether			1,000		313	U220
108-88-3	Toluene			100		313	U037
108-90-7	Chlorobenzene	10,000	10,000		15,000		
108-91-8	Cyclohexanamine	10,000	10,000		15,000		
108-91-8	Cyclohexylamine					313	
108-93-0	Cyclohexanol			5,000			U057
108-94-1	Cyclohexanone			5,000		313	U188
108-95-2	Phenol	500/10,000	1,000	1,000			P014
108-98-5	Benzenethiol	500	100	100			P014
108-98-5	Thiophenol			5,000		313	U191
109-06-8	2-Methylpyridine			5,000		X	U191
109-06-8	2-Picoline	500	500		15,000		
109-61-5	Carbonochloridic acid, propylester	500	500		15,000		
109-61-5	Propyl chloroformate				10,000		
109-66-0	Pentane				10,000		
109-67-1	1-Pentene			1,000			
109-73-9	Butylamine	500/10,000	1,000	1,000		313	U149
109-77-3	Malononitrile					313	
109-86-4	2-Methoxyethanol			100			
109-89-7	Diethylamine				10,000		
109-92-2	Ethene, ethoxy-				10,000		
109-92-2	Vinyl ethyl ether				10,000		
109-95-5	Ethyl nitrite				10,000		
109-95-5	Nitrous acid, ethyl ester			1,000			U213
109-99-9	Furan, tetrahydro-			100	5,000		U124
110-00-9	Furan	500	100	100	5,000		
110-16-7	Maleic acid			5,000			
110-17-8	Fumaric acid			5,000			
110-19-0	iso-Butyl acetate			5,000		X	
110-54-3	Hexane			5,000		313	
110-54-3	n-Hexane						

CAS Number	Chemical Name	Sec. 302 (EHS) TPQ	Section 304 EHS RQ	CERCLA RQ	CAA 112(r) TQ	Sec 313	RCRA CODE
110-57-6	trans-1,4-Dichloro-2-butene	500	500				
110-57-6	trans-1,4-Dichlorobutene	500	500			313	
110-75-8	2-Chloroethyl vinyl ether					X	
110-80-5	Ethanol, 2-ethoxy-			1,000			U042
110-80-5	2-Ethoxyethanol			1,000		X	U359
110-82-7	Cyclohexane			1,000		313	U359
110-86-1	Pyridine			1,000		313	U056
110-89-4	Piperidine			1,000		313	U196
111-42-2	Diethanolamine	1,000	1,000		15,000		
111-44-4	Bis(2-chloroethyl) ether			100		313	
111-44-4	Dichloroethyl ether	10,000	10	10		313	U025
111-54-6	Ethylenebisdithiocarbamic acid, salts & esters	10,000	10	10		X	U025
111-69-3	Adiponitrile			5,000		X	U114
111-91-1	Bis(2-chloroethoxy) methane	1,000	1,000				
114-26-1	Phenol, 2-(1-methylethoxy)-, methylcarbamate			1,000		313	U024
114-26-1	Propoxur			100		X	U411
115-02-6	Azaserine			100		313	U411
115-07-1	Propene			1			U015
115-07-1	1-Propene				10,000	X	
115-07-1	Propylene				10,000	X	
115-10-6	Methane, oxybis-				10,000	313	
115-10-6	Methyl ether				10,000		
115-11-7	2-Methylpropene				10,000		
115-11-7	1-Propene, 2-methyl-				10,000		
115-21-9	Trichloroethylsilane				10,000		
115-26-4	Dimefox	500	500				
115-28-6	Chlorendic acid	500	500				
115-29-7	Endosulfan					313	
115-32-2	Benzenemethanol,	10/10,000	1	1			P050
115-32-2	Dicofol			10		X	
115-90-2	Fensulfothion			10		313	
116-06-3	Aldicarb	500	500				
116-14-3	Ethene, tetrafluoro-	100/10,000	1	1		313	P070
116-14-3	Tetrafluoroethylene				10,000		
117-79-3	2-Aminoanthraquinone				10,000		
117-80-6	Dichlone					313	
117-81-7	Bis(2-ethylhexyl)phthalate						
117-81-7	DEHP			100		X	U028
117-81-7	Di(2-ethylhexyl) phthalate			100		X	U028
117-84-0	n-Dioctylphthalate			100		313	U028
117-84-0	Di-n-octyl phthalate			5,000			U107
118-74-1	Hexachlorobenzene			5,000			U107
118-79-6	2,4,6-Tribromophenol			10		313	U127
119-38-0	Isopropylmethylpyrazolyl dimethylcarbamate	500	1*	1*			U408
119-90-4	3,3'-Dimethoxybenzidine			100			P192
119-93-7	3,3'-Dimethylbenzidine			100		313	U091
119-93-7	o-Tolidine			10		313	U095
120-12-7	Anthracene			10		X	U095
120-36-5	2,4-DP			5,000		313	
120-58-1	Isosafrole					313	
				100		313	U141

* RCRA carbamate waste; statutory one-pound RQ applies until RQs are adjusted.

CAS Number	Chemical Name	Sec. 302 (EHS) TPQ	Section 304 EHS RQ	CERCLA RQ	CAA 112(r) TQ	Sec 313	RCRA CODE
120-71-8	p-Cresidine					313	
120-80-9	Catechol			100		313	
120-82-1	1,2,4-Trichlorobenzene			100		313	
120-83-2	2,4-Dichlorophenol			100		313	U081
121-14-2	2,4-Dinitrotoluene			10		313	U105
121-21-1	Pyrethrins			1			
121-29-9	Pyrethrins			1			
121-44-8	Triethylamine			5,000		313	U404
121-69-7	N,N-Dimethylaniline			100		313	
121-75-5	Malathion			100		313	
122-09-8	Benzeneethanamine, alpha,alpha-dimethyl-			5,000			P046
122-34-9	Simazine					313	
122-39-4	Diphenylamine					313	
122-42-9	Propham			1*			U373
122-66-7	1,2-Diphenylhydrazine			10		313	U109
122-66-7	Hydrazine, 1,2-diphenyl-			10		X	U109
122-66-7	Hydrazobenzene			10		X	U109
123-31-9	Hydroquinone	500/10,000	100	100		313	
123-33-1	Maleic hydrazide			5,000			U148
123-38-6	Propionaldehyde			1,000		313	
123-61-5	1,3-Phenylene diisocyanate					313#	
123-62-6	Propionic anhydride			5,000			
123-63-7	Paraldehyde			1,000		313	U182
123-72-8	Butyraldehyde					313	
123-73-9	2-Butenal, (e)-	1,000	100	100	20,000		U053
123-73-9	Crotonaldehyde, (E)-	1,000	100	100	20,000		U053
123-86-4	Butyl acetate			5,000			
123-91-1	1,4-Dioxane			100		313	U108
123-92-2	iso-Amyl acetate			5,000			
124-04-9	Adipic acid			5,000			
124-40-3	Dimethylamine			1,000	10,000	313	U092
124-40-3	Methanamine, N-methyl-			1,000	10,000	X	U092
124-41-4	Sodium methylate			1,000			
124-48-1	Chlorodibromomethane			100			
124-65-2	Sodium cacodylate	100/10,000	100				
124-73-2	Dibromotetrafluoroethane					313	
124-73-2	Halon 2402					X	
124-87-8	Picrotoxin	500/10,000	500				
126-72-7	Tris(2,3-dibromopropyl) phosphate			10		313	U235
126-98-7	Methacrylonitrile	500	1,000	1,000	10,000	313	U152
126-98-7	2-Propenenitrile, 2-methyl-	500	1,000	1,000	10,000	X	U152
126-99-8	Chloroprene			100		313	
127-18-4	Perchloroethylene			100		X	U210
127-18-4	Tetrachloroethylene			100		313	U210
127-82-0	Zinc phenolsulfonate			5,000			
128-03-0	Potassium dimethyldithiocarbamate					313	
128-04-1	Sodium dimethyldithiocarbamate					313	
128-66-5	C.I. Vat Yellow 4					313	
129-00-0	Pyrene	1,000/10,000	5,000	5,000			
129-06-6	Warfarin sodium	100/10,000	100	100			

Member of diisocyanate category.
* RCRA carbamate waste; statutory one-pound RQ applies until RQs are adjusted.

CAS Number	Chemical Name	Sec. 302 (EHS) TPQ	Section 304 EHS RQ	CERCLA RQ	CAA 112(r) TQ	Sec 313	RCRA CODE
130-15-4	1,4-Naphthoquinone			5,000			U166
131-11-3	Dimethyl phthalate			5,000		313	U102
131-52-2	Sodium pentachlorophenate					313	
131-74-8	Ammonium picrate			10			P009
131-89-5	2-Cyclohexyl-4,6-dinitrophenol			100			P034
132-27-4	Sodium o-phenylphenoxide					313	
132-64-9	Dibenzofuran			100		313	
133-06-2	Captan			10		313	
133-06-2	1H-Isoindole-1,3(2H)-dione,			10		X	
133-07-3	Folpet					313	
133-90-4	Benzoic acid, 3-amino-2,5-dichloro-			100		X	
133-90-4	Chloramben			100		313	
134-29-2	o-Anisidine hydrochloride					313	
134-32-7	alpha-Naphthylamine			100		313	U167
135-20-6	Benzeneamine, N-hydroxy-N-nitroso, ammonium salt					X	
135-20-6	Cupferron					313	
136-45-8	Dipropyl isocinchomeronate					313	
137-26-8	Thiram			10		313	U244
137-30-4	Ziram			1*			P205
137-41-7	Potassium N-methyldithiocarbamate					313	
137-42-8	Metham sodium					313	
137-42-8	Sodium methyldithiocarbamate					X	
138-93-2	Disodium cyanodithioimidocarbonate					313	
139-13-9	Nitrilotriacetic acid					313	
139-25-3	3,3'-Dimethyldiphenylmethane-4,4'-diisocyanate					313#	
139-65-1	4,4'-Thiodianiline					313	
140-29-4	Benzyl cyanide	500	500				
140-76-1	Pyridine, 2-methyl-5-vinyl-	500	500				
140-88-5	Ethyl acrylate			1,000		313	U113
141-32-2	Butyl acrylate					313	
141-66-2	Dicrotophos	100	100				
141-78-6	Ethyl acetate			5,000			U112
142-28-9	1,3-Dichloropropane			5,000			
142-59-6	Nabam					313	
142-71-2	Cupric acetate			100			
142-84-7	Dipropylamine			5,000			U110
143-33-9	Sodium cyanide (Na(CN))	100	10	10			P106
143-50-0	Kepone			1			U142
144-49-0	Fluoroacetic acid	10/10,000	10				
145-73-3	Endothall			1,000			P088
148-79-8	Thiabendazole					313	
148-79-8	2-(4-Thiazolyl)-1H-benzimidazole					X	
148-82-3	Melphalan			1			U150
149-30-4	MBT					X	
149-30-4	2-Mercaptobenzothiazole					313	
149-74-6	Dichloromethylphenylsilane	1,000	1,000				
150-50-5	Merphos					313	
150-68-5	Monuron					313	
151-38-2	Methoxyethylmercuric acetate	500/10,000	500				
151-50-8	Potassium cyanide	100	10	10			P098

Member of diisocyanate category.
* RCRA carbamate waste; statutory one-pound RQ applies until RQs are adjusted.

Page 15

CAS Number	Chemical Name	Sec. 302 (EHS) TPQ	Section 304 EHS RQ	CERCLA RQ	CAA 112(r) TQ	Sec 313	RCRA CODE
151-56-4	Aziridine	500	1	1	10,000	X	P054
151-56-4	Ethyleneimine	500	1	1	10,000	313	P054
152-16-9	Diphosphoramide, octamethyl-	100	100	100			P085
156-10-5	p-Nitrosodiphenylamine					313	
156-60-5	1,2-Dichloroethylene			1,000			U079
156-62-7	Calcium cyanamide			1,000		313	
189-55-9	Benzo(rst)pentaphene			10		313+	U064
189-55-9	Dibenz[a,i]pyrene			10		X	U064
189-64-0	Dibenzo(a,h)pyrene					313+	
191-24-2	Benzo[ghi]perylene			5,000			
191-30-0	Dibenzo(a,l)pyrene					313+	
192-65-4	Dibenzo(a,e)pyrene					313+	
193-39-5	Indeno(1,2,3-cd)pyrene			100		313+	U137
194-59-2	7H-Dibenzo(c,g)carbazole					313+	
205-82-3	Benzo(j)fluoranthene					313+	
205-99-2	Benzo[b]fluoranthene			1		313+	
206-44-0	Fluoranthene			100			U120
207-08-9	Benzo(k)fluoranthene			5,000		313+	
208-96-8	Acenaphthylene			5,000			
218-01-9	Benzo(a)phenanthrene			100		313+	U050
218-01-9	Chrysene			100		X	U050
224-42-0	Dibenz(a,j)acridine					313+	
225-51-4	Benz[c]acridine			100			U016
226-36-8	Dibenz(a,h)acridine					313+	
297-78-9	Isobenzan	100/10,000	100				P040
297-97-2	O,O-Diethyl O-pyrazinyl phosphorothioate	500	100	100			P040
297-97-2	Thionazin	500	100	100			P071
298-00-0	Methyl parathion	100/10,000	100	100		313	P071
298-00-0	Parathion-methyl	100/10,000	100	100		X	P094
298-02-2	Phorate	10	10	10			P039
298-04-4	Disulfoton	500	1	1			
300-62-9	Amphetamine	1,000	1,000			313	
300-76-5	Naled			10			U144
301-04-2	Lead acetate			10		X	
301-12-2	S-(2-(Ethylsulfinyl)ethyl) O,O-dimethyl ester phosphor					313	
301-12-2	Oxydemeton methyl	1,000	1	1	15,000	313	U133
302-01-2	Hydrazine			10			U143
303-34-4	Lasiocarpine			10			U035
305-03-3	Chlorambucil					313	
306-83-2	2,2-Dichloro-1,1,1-trifluoroethane					X	
306-83-2	HCFC-123	500/10,000	1	1		313	P004
309-00-2	Aldrin	500/10,000	1	1		X	P004
309-00-2	1,4:5,8-Dimethanonaphthalene,			100			P041
311-45-5	Diethyl-p-nitrophenyl phosphate					313	
314-40-9	Bromacil					X	
314-40-9	5-Bromo-6-methyl-3-(1-methylpropyl)-2,4-(1H,3H)-py	500/10,000	1,000	1,000			P128
315-18-4	Mexacarbate	1/10,000	1				
316-42-7	Emetine, dihydrochloride			10		X	
319-84-6	alpha-BHC			10		313	
319-84-6	alpha-Hexachlorocyclohexane						

+ Member of PAC category.

CAS Number	Chemical Name	Sec. 302 (EHS) TPQ	Section 304 EHS RQ	CERCLA RQ	CAA 112(r) TQ	Sec 313	RCRA CODE
319-85-7	beta-BHC			1			
319-86-8	delta-BHC			1			
327-98-0	Trichloronate	500	500				
329-71-5	2,5-Dinitrophenol			10			
330-54-1	Diuron			100		313	
330-55-2	Linuron					313	
333-41-5	Diazinon			1		313	
334-88-3	Diazomethane			100		313	
353-42-4	Boron trifluoride compound with methyl ether (1:1)	1,000	1,000		15,000		
353-42-4	Boron, trifluoro[oxybis[methane]]-, (T-4)-	1,000	1,000		15,000		
353-50-4	Carbonic difluoride			1,000			U033
353-59-3	Bromochlorodifluoromethane					313	
353-59-3	Halon 1211					X	
354-11-0	HCFC-121a					X	
354-11-0	1,1,1,2-Tetrachloro-2-fluoroethane					313	
354-14-3	HCFC-121					X	
354-14-3	1,1,2,2-Tetrachloro-1-fluoroethane					313	
354-23-4	1,2-Dichloro-1,1,2-trifluoroethane					313	
354-23-4	HCFC-123a					X	
354-25-6	1-Chloro-1,1,2,2-tetrafluoroethane					313	
354-25-6	HCFC-124a					X	
357-57-3	Brucine			100		313	P018
359-06-8	Fluoroacetyl chloride	10	10				
371-62-0	Ethylene fluorohydrin	10	10				
379-79-3	Ergotamine tartrate	500/10,000	500				
422-44-6	1,2-Dichloro-1,1,2,3,3-pentafluoropropane					313	
422-44-6	HCFC-225bb					X	
422-48-0	2,3-Dichloro-1,1,1,2,3-pentafluoropropane					313	
422-48-0	HCFC-225ba					X	
422-56-0	3,3-Dichloro-1,1,1,2,2-pentafluoropropane					313	
422-56-0	HCFC-225ca					X	
431-86-7	1,2-Dichloro-1,1,3,3,3-pentafluoropropane					313	
431-86-7	HCFC-225da					X	
460-19-5	Cyanogen			100	10,000		P031
460-19-5	Ethanedinitrile			100	10,000		P031
460-35-5	3-Chloro-1,1,1-trifluoropropane					313	
460-35-5	HCFC-253fb					X	
463-49-0	1,2-Propadiene				10,000		
463-49-0	Propadiene				10,000		
463-58-1	Carbon oxide sulfide (COS)			100	10,000	X	
463-58-1	Carbonyl sulfide			100	10,000	313	
463-82-1	2,2-Dimethylpropane				10,000		
463-82-1	Propane, 2,2-dimethyl-				10,000		
465-73-6	Isodrin	100/10,000	1	1		313	P060
470-90-6	Chlorfenvinfos	500	500				
492-80-8	Auramine			100		X	U014
492-80-8	C.I. Solvent Yellow 34			100		313	U014
494-03-1	Chlornaphazine			100			U026
496-72-0	Diaminotoluene			10			U221
502-39-6	Methylmercuric dicyanamide	500/10,000	500				

CAS Number	Chemical Name	Sec. 302 (EHS) TPQ	EHS RQ	CERCLA RQ	CAA 112(r) TQ	Sec 313	RCRA CODE
			Section 304				
504-24-5	4-Aminopyridine	500/10,000	1,000	1,000			P008
504-24-5	Pyridine, 4-amino-	500/10,000	1,000	1,000			P008
504-60-9	1,3-Pentadiene			100	10,000		U186
505-60-2	Ethane, 1,1'-thiobis[2-chloro-	500	500			X	
505-60-2	Mustard gas	500	500			313	
506-61-6	Potassium silver cyanide	500	1	1			P099
506-64-9	Silver cyanide			1			P104
506-68-3	Cyanogen bromide	500/10,000	1,000	1,000			U246
506-77-4	Cyanogen chloride			10	10,000		P033
506-77-4	Cyanogen chloride ((CN)Cl)			10	10,000		P033
506-78-5	Cyanogen iodide	1,000/10,000	1,000				
506-87-6	Ammonium carbonate			5,000			
506-96-7	Acetyl bromide			5,000			
507-55-1	1,3-Dichloro-1,1,2,2,3-pentafluoropropane					313	
507-55-1	HCFC-225cb					X	
509-14-8	Methane, tetranitro-	500	10	10	10,000		P112
509-14-8	Tetranitromethane	500	10	10	10,000		P112
510-15-6	Benzeneacetic acid,			10		X	U038
510-15-6	Chlorobenzilate			10		313	U038
513-49-5	sec-Butylamine			1,000			
514-73-8	Dithiazanine iodide	500/10,000	500				
528-29-0	o-Dinitrobenzene			100		313	
532-27-4	2-Chloroacetophenone			100		313	
533-74-4	Dazomet					313	
533-74-4	Tetrahydro-3,5-dimethyl-2H-1,3,5-thiadiazine-2-thione					X	
534-07-6	Bis(chloromethyl) ketone	10/10,000	10				
534-52-1	4,6-Dinitro-o-cresol	10/10,000	10	10		313	P047
534-52-1	Dinitrocresol	10/10,000	10	10		X	P047
534-52-1	4,6-Dinitro-o-cresol and salts			10			P047
535-89-7	Crimidine	100/10,000	100				
538-07-8	Ethylbis(2-chloroethyl)amine	500	500				
540-59-0	1,2-Dichloroethylene					313	
540-73-8	Hydrazine, 1,2-dimethyl-			1			U099
540-84-1	2,2,4-Trimethylpentane			1,000			
540-88-5	tert-Butyl acetate			5,000			
541-09-3	Uranyl acetate			100			
541-25-3	Lewisite	10	10				
541-41-3	Ethyl chloroformate					313	
541-53-7	Dithiobiuret	100/10,000	100	100		X	P049
541-53-7	2,4-Dithiobiuret	100/10,000	100	100		313	P049
541-73-1	1,3-Dichlorobenzene			100		313	U071
542-62-1	Barium cyanide			10			P013
542-75-6	1,3-Dichloropropene			100		X	U084
542-75-6	1,3-Dichloropropylene			100		313	U084
542-76-7	3-Chloropropionitrile	1,000	1,000	1,000		313	P027
542-76-7	Propionitrile, 3-chloro-	1,000	1,000	1,000		X	P027
542-88-1	Bis(chloromethyl) ether	100	10	10	1,000	313	P016
542-88-1	Chloromethyl ether	100	10	10	1,000	X	P016
542-88-1	Dichloromethyl ether	100	10	10	1,000	X	P016
542-88-1	Methane, oxybis[chloro-	100	10	10	1,000	X	P016

CAS Number	Chemical Name	Sec. 302 (EHS) TPQ	Section 304 EHS RQ	CERCLA RQ	CAA 112(r) TQ	Sec 313	RCRA CODE
542-90-5	Ethylthiocyanate	10,000	10,000				
543-90-8	Cadmium acetate			10			
544-18-3	Cobaltous formate			1,000			
544-92-3	Copper cyanide			10			P029
554-13-2	Lithium carbonate					313	
554-84-7	m-Nitrophenol			100			
555-77-1	Tris(2-chloroethyl)amine	100	100				
556-61-6	Isothiocyanatomethane	500	500			X	
556-61-6	Methyl isothiocyanate	500	500			313	
556-64-9	Methyl thiocyanate	10,000	10,000		20,000		
556-64-9	Thiocyanic acid, methyl ester	10,000	10,000		20,000		
557-19-7	Nickel cyanide			10			P074
557-21-1	Zinc cyanide			10			P121
557-34-6	Zinc acetate			1,000			
557-41-5	Zinc formate			1,000			
557-98-2	2-Chloropropylene				10,000		
557-98-2	1-Propene, 2-chloro-				10,000		
558-25-8	Methanesulfonyl fluoride	1,000	1,000				
563-12-2	Ethion	1,000	10	10			
563-41-7	Semicarbazide hydrochloride	1,000/10,000	1,000				
563-45-1	3-Methyl-1-butene				10,000		
563-46-2	2-Methyl-1-butene				10,000		
563-47-3	3-Chloro-2-methyl-1-propene				10,000		
563-68-8	Thallium(I) acetate			100		313	
569-64-2	C.I. Basic Green 4			100			U214
573-56-8	2,6-Dinitrophenol			10		313	
584-84-9	Benzene, 2,4-diisocyanato-1-methyl-	500	100	100	10,000	X	
584-84-9	Toluene-2,4-diisocyanate	500	100	100	10,000	313	
590-18-1	2-Butene-cis				10,000		
590-21-6	1-Chloropropylene				10,000		
590-21-6	1-Propene, 1-chloro-				10,000		
591-08-2	1-Acetyl-2-thiourea			1,000			P002
592-01-8	Calcium cyanide			10			P021
592-04-1	Mercuric cyanide			1			
592-85-8	Mercuric thiocyanate			10			
592-87-0	Lead thiocyanate			10			
593-60-2	Vinyl bromide			10			
594-42-3	Methanesulfenyl chloride, trichloro-			100		313	
594-42-3	Perchloromethyl mercaptan	500	100	100	10,000	X	
594-42-3	Trichloromethanesulfenyl chloride	500	100	100	10,000	313	
597-64-8	Tetraethyltin	500	100	100	10,000	X	
598-31-2	Bromoacetone	100	100				
598-73-2	Bromotrifluoroethylene			1,000			P017
598-73-2	Ethene, bromotrifluoro-				10,000		
606-20-2	2,6-Dinitrotoluene				10,000		
608-93-5	Pentachlorobenzene			100		313	U106
609-19-8	3,4,5-Trichlorophenol			10			U183
610-39-9	3,4-Dinitrotoluene			10			
612-82-8	3,3'-Dimethylbenzidine dihydrochloride					313	
612-82-8	o-Tolidine dihydrochloride					X	

CAS Number	Chemical Name	Sec. 302 (EHS) TPQ	Section 304 EHS RQ	CERCLA RQ	CAA 112(r) TQ	Sec 313	RCRA CODE
612-83-9	3,3'-Dichlorobenzidine dihydrochloride					313	
614-78-8	Thiourea, (2-methylphenyl)-	500/10,000	500				
615-05-4	2,4-Diaminoanisole					313	
615-28-1	1,2-Phenylenediamine dihydrochloride					313	
615-53-2	N-Nitroso-N-methylurethane			1			U178
621-64-7	Di-n-propylnitrosamine			10		X	U111
621-64-7	N-Nitrosodi-n-propylamine			10		313	U111
624-18-0	1,4-Phenylenediamine dihydrochloride					313	
624-64-6	2-Butene, (E)				10,000		
624-64-6	2-Butene-trans				10,000		
624-83-9	Methane, isocyanato-	500	10	10	10,000	X	P064
624-83-9	Methyl isocyanate	500	10	10	10,000	313	P064
625-16-1	tert-Amyl acetate			5,000			
626-38-0	sec-Amyl acetate			5,000			
627-11-2	Chloroethyl chloroformate	1,000	1,000				
627-20-3	2-Pentene, (Z)-				10,000		
628-63-7	Amyl acetate			5,000			
628-86-4	Mercury fulminate			10			P065
630-10-4	Selenourea			1,000			P103
630-20-6	Ethane, 1,1,1,2-tetrachloro-			100		X	U208
630-20-6	1,1,1,2-Tetrachloroethane			100		313	U208
630-60-4	Ouabain	100/10,000	100				
631-61-8	Ammonium acetate			5,000			
636-21-5	o-Toluidine hydrochloride			100		313	U222
639-58-7	Triphenyltin chloride	500/10,000	500			313	
640-19-7	Fluoroacetamide	100/10,000	100	100			P057
644-64-4	Dimetilan	500/10,000	1*	1*			P191
646-04-8	2-Pentene, (E)-				10,000		
675-14-9	Cyanuric fluoride	100	100				
676-97-1	Methyl phosphonic dichloride	100	100				
680-31-9	Hexamethylphosphoramide			1		313	
684-93-5	N-Nitroso-N-methylurea			1		313	U177
689-97-4	1-Buten-3-yne				10,000		
689-97-4	Vinyl acetylene				10,000		
692-42-2	Diethylarsine			1			P038
696-28-6	Dichlorophenylarsine	500	1	1			P036
696-28-6	Phenyl dichloroarsine	500	1	1			P036
709-98-8	N-(3,4-Dichlorophenyl)propanamide					X	
709-98-8	Propanil					313	
732-11-6	Phosmet	10/10,000	10				
757-58-4	Hexaethyl tetraphosphate			100			P062
759-73-9	N-Nitroso-N-ethylurea			1		313	U176
759-94-4	EPTC					X	
759-94-4	Ethyl dipropylthiocarbamate					313	
760-93-0	Methacrylic anhydride	500	500				
764-41-0	2-Butene, 1,4-dichloro-			1		X	U074
764-41-0	1,4-Dichloro-2-butene			1		313	U074
765-34-4	Glycidylaldehyde			10			U126
786-19-6	Carbophenothion	500	500				
812-04-4	1,1-Dichloro-1,2,2-trifluoroethane					313	

* RCRA carbamate waste; statutory one-pound RQ applies until RQs are adjusted.

CAS Number	Chemical Name	Sec. 302 (EHS) TPQ	Section 304 EHS RQ	CERCLA RQ	CAA 112(r) TQ	Sec 313	RCRA CODE
812-04-4	HCFC-123b					X	
814-49-3	Diethyl chlorophosphate	500	500				
814-68-6	Acrylyl chloride	100	100		5,000		
814-68-6	2-Propenoyl chloride	100	100		5,000		
815-82-7	Cupric tartrate			100			
822-06-0	Hexamethylene-1,6-diisocyanate			100		313#	
823-40-5	Diaminotoluene			10			
824-11-3	Trimethylolpropane phosphite	100/10,000	100				U221
834-12-8	Ametryn					313	
834-12-8	N-Ethyl-N'-(1-methylethyl)-6-(methylthio)-1,3,5,-triaz					X	
842-07-9	C.I. Solvent Yellow 14					313	
872-50-4	N-Methyl-2-pyrrolidone					313	
900-95-8	Stannane, acetoxytriphenyl-	500/10,000	500				
919-86-8	Demeton-S-methyl	500	500				
920-46-7	Methacryloyl chloride	100	100				
924-16-3	N-Nitrosodi-n-butylamine			10		313	U172
924-42-5	N-Methylolacrylamide					313	
930-55-2	N-Nitrosopyrrolidine			1			U180
933-75-5	2,3,6-Trichlorophenol			10			
933-78-8	2,3,5-Trichlorophenol			10			
944-22-9	Fonofos	500	500				
947-02-4	Phosfolan	100/10,000	100				
950-10-7	Mephosfolan	500	500				
950-37-8	Methidathion	500/10,000	500				
957-51-7	Diphenamid					313	
959-98-8	alpha - Endosulfan			1			
961-11-5	Phosphoric acid, 2-chloro-1-(2,3,5-trichlorophenyl) eth					X	
961-11-5	Tetrachlorvinphos					313	
989-38-8	C.I. Basic Red 1					313	
991-42-4	Norbormide	100/10,000	100				
998-30-1	Triethoxysilane	500	500				
999-81-5	Chlormequat chloride	100/10,000	100				
1024-57-3	Heptachlor epoxide			1			
1031-07-8	Endosulfan sulfate			1			
1031-47-6	Triamiphos	500/10,000	500				
1066-30-4	Chromic acetate			1,000			
1066-33-7	Ammonium bicarbonate			5,000			
1066-45-1	Trimethyltin chloride	500/10,000	500				
1072-35-1	Lead stearate			10			
1111-78-0	Ammonium carbamate			5,000			
1114-71-2	Butylethylcarbamothioic acid S-propyl ester					X	
1114-71-2	Pebulate					313	
1116-54-7	N-Nitrosodiethanolamine			1			U173
1120-71-4	Propane sultone			10		313	U193
1120-71-4	1,3-Propane sultone			10		X	U193
1122-60-7	Nitrocyclohexane	500	500				
1124-33-0	Pyridine, 4-nitro-, 1-oxide	500/10,000	500				
1129-41-5	Metolcarb	100/10,000	1*	1*			P190
1134-23-2	Cycloate					313	
1163-19-5	Decabromodiphenyl oxide					313	

Member of diisocyanate category.
* RCRA carbamate waste; statutory one-pound RQ applies until RQs are adjusted.

Page 21

CAS Number	Chemical Name	Sec. 302 (EHS) TPQ	Section 304 EHS RQ	CERCLA RQ	CAA 112(r) TQ	Sec 313	RCRA CODE
1185-57-5	Ferric ammonium citrate			1,000			
1194-65-6	Dichlobenil			100			
1300-71-6	Xylenol			1,000			
1303-28-2	Arsenic pentoxide	100/10,000	1	1			P011
1303-32-8	Arsenic disulfide			1			
1303-33-9	Arsenic trisulfide			1			
1306-19-0	Cadmium oxide	100/10,000	100				
1309-64-4	Antimony trioxide			1,000			
1310-58-3	Potassium hydroxide			1,000			
1310-73-2	Sodium hydroxide			1,000			
1313-27-5	Molybdenum trioxide					313	
1314-20-1	Thorium dioxide					313	
1314-32-5	Thallic oxide			100			P113
1314-62-1	Vanadium pentoxide	100/10,000	1,000	1,000			P120
1314-80-3	Sulfur phosphide			100			U189
1314-84-7	Zinc phosphide	500	100	100			P122
1314-84-7	Zinc phosphide (conc. > 10%)	500	100	100			P122
1314-84-7	Zinc phosphide (conc. <= 10%)	500	100	100			U249
1314-87-0	Lead sulfide			10			
1319-72-8	2,4,5-T amines			5,000			
1319-77-3	Cresol (mixed isomers)			100		313	U052
1320-18-9	2,4-D Esters			100		X	
1320-18-9	2,4-D propylene glycol butyl ether ester			100		313	
1321-12-6	Nitrotoluene			1,000			
1327-52-2	Arsenic acid			1			
1327-53-3	Arsenic trioxide	100/10,000	1	1			P012
1327-53-3	Arsenous oxide	100/10,000	1	1			P012
1330-20-7	Xylene (mixed isomers)			100		313	U239
1332-07-6	Zinc borate			1,000			
1332-21-4	Asbestos (friable)			1		313	
1333-74-0	Hydrogen				10,000		
1333-83-1	Sodium bifluoride			100			
1335-32-6	Lead subacetate			10			U146
1335-87-1	Hexachloronaphthalene					313	
1336-21-6	Ammonium hydroxide			1,000			
1336-36-3	PCBs			1		X	
1336-36-3	Polychlorinated biphenyls			1		313	
1338-23-4	Methyl ethyl ketone peroxide			10			U160
1338-24-5	Naphthenic acid			100			
1341-49-7	Ammonium bifluoride			100			
1344-28-1	Aluminum oxide (fibrous forms)					313	
1397-94-0	Antimycin A	1,000/10,000	1,000				
1420-07-1	Dinoterb	500/10,000	500				
1464-53-5	2,2'-Bioxirane	500	10	10		X	U085
1464-53-5	Diepoxybutane	500	10	10		313	U085
1558-25-4	Trichloro(chloromethyl)silane	100	100				
1563-38-8	Carbofuran phenol			1*			U367
1563-66-2	Carbofuran	10/10,000	10	10		313	P127
1582-09-8	Benezeneamine, 2,6-dinitro-N,N-dipropyl-4-(trifluoro			10		X	
1582-09-8	Trifluralin			10		313	

* RCRA carbamate waste; statutory one-pound RQ applies until RQs are adjusted.

CAS Number	Chemical Name	Sec. 302 (EHS) TPQ	Section 304 EHS RQ	CERCLA RQ	CAA 112(r) TQ	Sec 313	RCRA CODE
1600-27-7	Mercuric acetate	500/10,000	500				
1615-80-1	Hydrazine, 1,2-diethyl-			10			U086
1622-32-8	Ethanesulfonyl chloride, 2-chloro-	500	500				
1634-04-4	Methyl tert-butyl ether			1,000		313	
1646-88-4	Aldicarb sulfone			1*			P203
1649-08-7	1,2-Dichloro-1,1-difluoroethane					313	
1649-08-7	HCFC-132b					X	
1689-84-5	Bromoxynil					313	
1689-84-5	3,5-Dibromo-4-hydroxybenzonitrile					X	
1689-99-2	Bromoxynil octanoate					313	
1689-99-2	Octanoic acid, 2,6-dibromo-4-cyanophenyl ester					X	
1717-00-6	1,1-Dichloro-1-fluoroethane					313	
1717-00-6	HCFC-141b					X	
1746-01-6	2,3,7,8-Tetrachlorodibenzo-p-dioxin (TCDD)			1			
1752-30-3	Acetone thiosemicarbazide	1,000/10,000	1,000				
1762-95-4	Ammonium thiocyanate			5,000			
1836-75-5	Benzene, 2,4-dichloro-1-(4-nitrophenoxy)-					X	
1836-75-5	Nitrofen					313	
1861-40-1	Benfluralin					313	
1861-40-1	N-Butyl-N-ethyl-2,6-dinitro-4-(trifluoromethyl) benzen					X	
1863-63-4	Ammonium benzoate			5,000			
1888-71-7	Hexachloropropene			1,000			U243
1897-45-6	1,3-Benzenedicarbonitrile, 2,4,5,6-tetrachloro-					X	
1897-45-6	Chlorothalonil					313	
1910-42-5	Paraquat dichloride	10/10,000	10			313	
1912-24-9	Atrazine					313	
1912-24-9	6-Chloro-N-ethyl-N'-(1-methylethyl)-1,3,5-triazine-2,4					X	
1918-00-9	Dicamba			1,000		313	
1918-00-9	3,6-Dichloro-2-methoxybenzoic acid			1,000		X	
1918-02-1	Picloram					313	
1918-16-7	2-Chloro-N-(1-methylethyl)-N-phenylacetamide					X	
1918-16-7	Propachlor					313	
1928-38-7	2,4-D Esters			100			
1928-43-4	2,4-D 2-ethylhexyl ester					313	
1928-47-8	2,4,5-T esters			1,000			
1928-61-6	2,4-D Esters			100			
1929-73-3	2,4-D butoxyethyl ester			100		313	
1929-73-3	2,4-D Esters			100		X	
1929-82-4	2-Chloro-6-(trichloromethyl)pyridine					X	
1929-82-4	Nitrapyrin					313	
1937-37-7	C.I. Direct Black 38					313	
1982-47-4	Chloroxuron	500/10,000	500				
1982-69-0	3,6-Dichloro-2-methoxybenzoic acid, sodium salt					X	
1982-69-0	Sodium dicamba					313	
1983-10-4	Tributyltin fluoride					313	
2001-95-8	Valinomycin	1,000/10,000	1,000				
2008-46-0	2,4,5-T amines			5,000			
2032-65-7	Mercaptodimethur	500/10,000	10	10		X	P199
2032-65-7	Methiocarb	500/10,000	10	10		313	P199
2074-50-2	Paraquat methosulfate	10/10,000	10				

* RCRA carbamate waste; statutory one-pound RQ applies until RQs are adjusted.

CAS Number	Chemical Name	Sec. 302 (EHS) TPQ	Section 304 EHS RQ	CERCLA RQ	CAA 112(r) TQ	Sec 313	RCRA CODE
2097-19-0	Phenylsilatrane	100/10,000	100				
2104-64-5	EPN	100/10,000	100				
2155-70-6	Tributyltin methacrylate					313	
2164-07-0	Dipotassium endothall					313	
2164-07-0	7-Oxabicyclo(2.2.1)heptane-2,3-dicarboxylic acid, dip					X	
2164-17-2	Fluometuron					313	
2164-17-2	Urea, N,N-dimethyl-N'-[3-(trifluoromethyl)phenyl]-					X	
2212-67-1	1H-Azepine-1 carbothioic acid, hexahydro-S-ethyl ester					X	
2212-67-1	Molinate					313	
2223-93-0	Cadmium stearate	1,000/10,000	1,000				
2231-57-4	Thiocarbazide	1,000/10,000	1,000				
2234-13-1	Octachloronaphthalene					313	
2238-07-5	Diglycidyl ether	1,000	1,000				
2275-18-5	Prothoate	100/10,000	100				
2300-66-5	Dimethylamine dicamba					313	
2303-16-4	Carbamothioic acid,			100		X	U062
2303-16-4	Diallate			100		313	U062
2303-17-5	Triallate			1*		313	U389
2312-35-8	Propargite			10		313	
2439-01-2	Chinomethionat					313	
2439-01-2	6-Methyl-1,3-dithiolo[4,5-b]quinoxalin-2-one					X	
2439-10-3	Dodecylguanidine monoacetate					X	
2439-10-3	Dodine					313	
2497-07-6	Oxydisulfoton	500	500				
2524-03-0	Dimethyl chlorothiophosphate	500	500			313	
2524-03-0	Dimethyl phosphorochloridothioate	500	500			X	
2540-82-1	Formothion	100	100				
2545-59-7	2,4,5-T esters			1,000			
2556-36-7	1,4-Cyclohexane diisocyanate					313#	
2570-26-5	Pentadecylamine	100/10,000	100				
2587-90-8	Phosphorothioic acid,	500	500				
2602-46-2	C.I. Direct Blue 6					313	
2631-37-0	Promecarb	500/10,000	1*	1*			P201
2636-26-2	Cyanophos	1,000	1,000				
2642-71-9	Azinphos-ethyl	100/10,000	100				
2655-15-4	2,3,5-Trimethylphenyl methylcarbamate					313	
2665-30-7	Phosphonothioic acid, methyl-, O-(4-nitrophenyl) O-	500	500				
2699-79-8	Sulfuryl fluoride					313	
2699-79-8	Vikane					X	
2702-72-9	2,4-D sodium salt					313	
2703-13-1	Phosphonothioic acid, methyl-, O-ethyl	500	500				
2757-18-8	Thallous malonate	100/10,000	100				
2763-96-4	5-(Aminomethyl)-3-isoxazolol	500/10,000	1,000	1,000			P007
2763-96-4	Muscimol	500/10,000	1,000	1,000			P007
2764-72-9	Diquat			1,000			
2778-04-3	Endothion	500/10,000	500				
2832-40-8	C.I. Disperse Yellow 3					313	
2837-89-0	2-Chloro-1,1,1,2-tetrafluoroethane					313	
2837-89-0	HCFC-124					X	
2921-88-2	Chlorpyrifos			1			

Member of diisocyanate category.
* RCRA carbamate waste; statutory one-pound RQ applies until RQs are adjusted.

CAS Number	Chemical Name	Sec. 302 (EHS) TPQ	Section 304 EHS RQ	Section 304 CERCLA RQ	CAA 112(r) TQ	Sec 313	RCRA CODE
2944-67-4	Ferric ammonium oxalate						
2971-38-2	2,4-D chlorocrotyl ester			1,000			
2971-38-2	2,4-D Esters			100			
3012-65-5	Ammonium citrate, dibasic			100		X	
3037-72-7	Silane, (4-aminobutyl)diethoxymethyl-			5,000			
3118-97-6	C.I. Solvent Orange 7	1,000	1,000				
3164-29-2	Ammonium tartrate					313	
3165-93-3	4-Chloro-o-toluidine, hydrochloride			5,000			U049
3173-72-6	1,5-Naphthalene diisocyanate			100		313#	
3251-23-8	Cupric nitrate						
3254-63-5	Phosphoric acid, dimethyl 4-(methylthio) phenyl ester			100			
3288-58-2	O,O-Diethyl S-methyl dithiophosphate	500	500				U087
3383-96-8	Temephos			5,000			
3486-35-9	Zinc carbonate					313	
3547-04-4	DDE			1,000			
3569-57-1	Sulfoxide, 3-chloropropyl octyl			5,000			
3615-21-2	Benzimidazole, 4,5-dichloro-2-(trifluoromethyl)-	500	500				
3653-48-3	(4-Chloro-2-methylphenoxy) acetate sodium salt	500/10,000	500				
3653-48-3	Methoxone sodium salt					X	
3689-24-5	Sulfotep					313	
3689-24-5	Tetraethyldithiopyrophosphate	500	100	100			P109
3691-35-8	Chlorophacinone	500	100	100			P109
3697-24-3	5-Methylchrysene	100/10,000	100				
3734-97-2	Amiton oxalate					313+	
3735-23-7	Methyl phenkapton	100/10,000	100				
3761-53-3	C.I. Food Red 5	500	500				
3813-14-7	2,4,5-T amines					313	
3878-19-1	Fuberidazole			5,000			
4044-65-9	Bitoscanate	100/10,000	100				
4080-31-3	1-(3-Chloroallyl)-3,5,7-triaza-1-azoniaadamantane chl	500/10,000	500				
4098-71-9	Isophorone diisocyanate					313	
4104-14-7	Phosacetim	100	100			313#	
4109-96-0	Dichlorosilane						
4109-96-0	Silane, dichloro-				10,000		
4128-73-8	4,4'-Diisocyanatodiphenyl ether				10,000		
4170-30-3	2-Butenal					313#	
4170-30-3	Crotonaldehyde	1,000	100	100	20,000	X	U053
4301-50-2	Fluenetil	1,000	100	100	20,000	313	U053
4418-66-0	Phenol, 2,2'-thiobis[4-chloro-6-methyl-	100/10,000	100	100			
4549-40-0	N-Nitrosomethylvinylamine	100/10,000	100	100			
4680-78-8	C.I. Acid Green 3			10		313	P084
4835-11-4	Hexamethylenediamine, N,N'-dibutyl-					313	
5124-30-1	1,1'-Methylene bis(4-isocyanatocyclohexane)	500	500			313	
5234-68-4	Carboxin					313#	
5234-68-4	5,6-Dihydro-2-methyl-N-phenyl-1,4-oxathiin-3-carbox					313	
5344-82-1	Thiourea, (2-chlorophenyl)-					X	
5385-75-1	Dibenzo(a,e)fluoranthene	100/10,000	100	100			P026
5522-43-0	1-Nitropyrene					313+	
5598-13-0	Chlorpyrifos methyl					313+	
5598-13-0	O,O-Dimethyl-O-(3,5,6-trichloro-2-pyridyl)phosphoro					313	X

+ Member of PAC category.
# Member of diisocyanate category.

CAS Number	Chemical Name	Sec. 302 (EHS) TPO	Section 304 EHS RQ	CERCLA RQ	CAA 112(r) TQ	Sec 313	RCRA CODE
5836-29-3	Coumatetralyl	500/10,000	500				
5893-66-3	Cupric oxalate			100			
5902-51-2	5-Chloro-3-(1,1-dimethylethyl)-6-methyl-2,4(1H,3H)-					X	
5902-51-2	Terbacil					313	
5952-26-1	Ethanol, 2,2'-oxybis-, dicarbamate			1*			U395
5972-73-6	Ammonium oxalate			5,000			
6009-70-7	Ammonium oxalate			5,000			
6369-96-6	2,4,5-T amines			5,000			
6369-97-7	2,4,5-T amines			5,000			
6459-94-5	C.I. Acid Red 114					313	
6533-73-9	Thallium(I) carbonate	100/10,000	100	100			U215
6533-73-9	Thallous carbonate	100/10,000	100	100			U215
6923-22-4	Monocrotophos	10/10,000	10				
7005-72-3	4-Chlorophenyl phenyl ether			5,000			
7287-19-6	N,N'-Bis(1-methylethyl)-6-methylthio-1,3,5-triazine-2,					X	
7287-19-6	Prometryn					313	
7421-93-4	Endrin aldehyde			1			
7428-48-0	Lead stearate			10			
7429-90-5	Aluminum (fume or dust)					313	
7439-92-1	Lead			10		313	
7439-96-5	Manganese					313	
7439-97-6	Mercury			1		313	U151
7440-02-0	Nickel			100		313	
7440-22-4	Silver			1,000		313	
7440-23-5	Sodium			10			
7440-28-0	Thallium			1,000		313	
7440-36-0	Antimony			5,000		313	
7440-38-2	Arsenic			1		313	
7440-39-3	Barium					313	P015
7440-41-7	Beryllium			10		313	
7440-43-9	Cadmium			10		313	
7440-47-3	Chromium			5,000		313	
7440-48-4	Cobalt					313	
7440-50-8	Copper			5,000		313	
7440-62-2	Vanadium (fume or dust)					313	
7440-66-6	Zinc (fume or dust)			1,000		313	
7440-66-6	Zinc			1,000			
7446-08-4	Selenium dioxide			10			
7446-09-5	Sulfur dioxide	500	500				
7446-09-5	Sulfur dioxide (anhydrous)	500	500		5,000		
7446-11-9	Sulfur trioxide	100	100		10,000		
7446-14-2	Lead sulfate			10			P115
7446-18-6	Thallium(I) sulfate	100/10,000	100	100			P115
7446-18-6	Thallous sulfate	100/10,000	100	100			U145
7446-27-7	Lead phosphate			10			
7447-39-4	Cupric chloride			10			
7487-94-7	Mercuric chloride	500/10,000	500				
7488-56-4	Selenium sulfide			10			U205
7550-45-0	Titanium chloride (TiCl4) (T-4)-	100	1,000	1,000	2,500	X	
7550-45-0	Titanium tetrachloride	100	1,000	1,000	2,500	313	

* RCRA carbamate waste; statutory one-pound RQ applies until RQs are adjusted.

CAS Number	Chemical Name	Sec. 302 (EHS) TPQ	Section 304 EHS RQ	CERCLA RQ	CAA 112(r) TQ	Sec 313	RCRA CODE
7558-79-4	Sodium phosphate, dibasic						
7580-67-8	Lithium hydride			5,000			
7601-54-9	Sodium phosphate, tribasic	100	100				
7631-89-2	Sodium arsenate			5,000			
7631-90-5	Sodium bisulfite	1,000/10,000	1	1			
7632-00-0	Sodium nitrite			5,000			
7637-07-2	Borane, trifluoro-			100		313	
7637-07-2	Boron trifluoride	500	500		5,000	X	
7645-25-2	Lead arsenate	500	500		5,000	313	
7646-85-7	Zinc chloride			1			
7647-01-0	Hydrochloric acid			1,000			
7647-01-0	Hydrochloric acid (conc 37% or greater)			5,000			
7647-01-0	Hydrochloric acid (aerosol forms only)			5,000	15,000		
7647-01-0	Hydrogen chloride (anhydrous)			5,000		313	
7647-01-0	Hydrogen chloride (gas only)	500	5,000	5,000	5,000	X	
7647-18-9	Antimony pentachloride	500	5,000	5,000	5,000	X	
7664-38-2	Phosphoric acid			1,000			
7664-39-3	Hydrofluoric acid			5,000		313	
7664-39-3	Hydrofluoric acid (conc. 50% or greater)	100	100	100		X	U134
7664-39-3	Hydrogen fluoride	100	100	100	1,000	X	U134
7664-39-3	Hydrogen fluoride (anhydrous)	100	100	100		313	U134
7664-41-7	Ammonia	100	100	100	1,000	X	U134
7664-41-7	Ammonia (anhydrous)	500	100	100		313	
7664-41-7	Ammonia (conc 20% or greater)	500	100	100	10,000	X	
7664-93-9	Sulfuric acid	500	100	100	20,000	X	
7664-93-9	Sulfuric acid (aerosol forms only)	1,000	1,000	1,000			
7681-49-4	Sodium fluoride	1,000	1,000	1,000		313	
7681-52-9	Sodium hypochlorite			1,000			
7696-12-0	2,2-Dimethyl-3-(2-methyl-1-propenyl)cyclopropaneca			100			
7696-12-0	Tetramethrin					X	
7697-37-2	Nitric acid					313	
7697-37-2	Nitric acid (conc 80% or greater)	1,000	1,000	1,000		313	
7699-45-8	Zinc bromide	1,000	1,000	1,000	15,000	X	
7705-08-0	Ferric chloride			1,000			
7718-54-9	Nickel chloride			1,000			
7719-12-2	Phosphorous trichloride			100			
7719-12-2	Phosphorus trichloride	1,000	1,000	1,000	15,000		
7720-78-7	Ferrous sulfate	1,000	1,000	1,000	15,000		
7722-64-7	Potassium permanganate			1,000			
7722-84-1	Hydrogen peroxide (Conc.> 52%)			100			
7723-14-0	Phosphorus (yellow or white)	1,000	1,000				
7723-14-0	Phosphorus	100	1	1		313	
7726-95-6	Bromine	100	1	1			
7733-02-0	Zinc sulfate	500	500		10,000	313	
7738-94-5	Chromic acid			1,000			
7758-01-2	Potassium bromate			10			
7758-29-4	Sodium phosphate, tribasic					313	
7758-94-3	Ferrous chloride			5,000			
7758-95-4	Lead chloride			100			
7758-98-7	Cupric sulfate			10			
				10			

CAS Number	Chemical Name	Sec. 302 (EHS) TPQ	Section 304 EHS RQ	CERCLA RQ	CAA 112(r) TQ	Sec 313	RCRA CODE
7761-88-8	Silver nitrate			1			
7773-06-0	Ammonium sulfamate			5,000			
7775-11-3	Sodium chromate			10			
7778-39-4	Arsenic acid			1			P010
7778-44-1	Calcium arsenate	500/10,000	1	1			
7778-50-9	Potassium bichromate			10			
7778-54-3	Calcium hypochlorite			10			
7779-86-4	Zinc hydrosulfite			1,000			
7779-88-6	Zinc nitrate			1,000			
7782-41-4	Fluorine	500	10	10	1,000	313	P056
7782-49-2	Selenium			100		313	
7782-50-5	Chlorine	100	10	10	2,500	313	
7782-63-0	Ferrous sulfate			1,000			
7782-82-3	Sodium selenite			100			
7782-86-7	Mercurous nitrate			10			
7783-00-8	Selenious acid	1,000/10,000	10	10			U204
7783-06-4	Hydrogen sulfide	500	100	100	10,000	313	U135
7783-07-5	Hydrogen selenide	10	10		500		
7783-35-9	Mercuric sulfate			10			
7783-46-2	Lead fluoride			10			
7783-49-5	Zinc fluoride			1,000			
7783-50-8	Ferric fluoride			100			
7783-56-4	Antimony trifluoride			1,000			
7783-60-0	Sulfur fluoride (SF4), (T-4)-	100	100		2,500		
7783-60-0	Sulfur tetrafluoride	100	100		2,500		
7783-70-2	Antimony pentafluoride	500	500				
7783-80-4	Tellurium hexafluoride	100	100				
7784-34-1	Arsenous trichloride	500	1	1	15,000		
7784-40-9	Lead arsenate			1			
7784-41-0	Potassium arsenate			1			
7784-42-1	Arsine	100	100		1,000		
7784-46-5	Sodium arsenite	500/10,000	1	1			
7785-84-4	Sodium phosphate, tribasic			5,000			
7786-34-7	Mevinphos	500	10	10		313	
7786-81-4	Nickel sulfate			100			
7787-47-5	Beryllium chloride			1			
7787-49-7	Beryllium fluoride			1			
7787-55-5	Beryllium nitrate			1			
7788-98-9	Ammonium chromate			10			
7789-00-6	Potassium chromate			10			
7789-06-2	Strontium chromate			10			
7789-09-5	Ammonium bichromate			10			
7789-42-6	Cadmium bromide			10			
7789-43-7	Cobaltous bromide			1,000			
7789-61-9	Antimony tribromide			1,000			
7790-94-5	Chlorosulfonic acid			1,000			
7791-12-0	Thallium chloride TlCl	100/10,000	100	100			U216
7791-12-0	Thallous chloride	100/10,000	100	100			U216
7791-21-1	Chlorine monoxide				10,000		
7791-21-1	Chlorine oxide				10,000		

CAS Number	Chemical Name	Sec. 302 (EHS) TPQ	Section 304 EHS RQ	CERCLA RQ	CAA 112(r) TQ	Sec 313	RCRA CODE
7791-23-3	Selenium oxychloride	500	500				
7803-51-2	Phosphine	500	100	100	5,000	313	P096
7803-55-6	Ammonium vanadate			1,000			P119
7803-62-5	Silane				10,000		
8001-35-2	Camphechlor	500/10,000	1	1		X	P123
8001-35-2	Camphene, octachloro-	500/10,000	1	1		X	P123
8001-35-2	Toxaphene	500/10,000	1	1		313	P123
8001-58-9	Creosote			1		313	U051
8003-19-8	Dichloropropane - Dichloropropene (mixture)			100			
8003-34-7	Pyrethrins			1			
8014-95-7	Oleum (fuming sulfuric acid)			1,000	10,000		
8014-95-7	Sulfuric acid (fuming)			1,000	10,000		
8014-95-7	Sulfuric acid, mixture with sulfur trioxide			1,000	10,000		
8065-48-3	Demeton	500	500				
9006-42-2	Metiram					313	
9016-87-9	Polymeric diphenylmethane diisocyanate					313#	
10022-70-5	Sodium hypochlorite			100			
10025-73-7	Chromic chloride	1/10,000	1				
10025-78-2	Silane, trichloro-				10,000		
10025-78-2	Trichlorosilane				10,000		
10025-87-3	Phosphorus oxychloride	500	1,000	1,000	5,000		
10025-87-3	Phosphoryl chloride	500	1,000	1,000	5,000		
10025-91-9	Antimony trichloride			1,000			
10026-11-6	Zirconium tetrachloride				5,000		
10026-13-8	Phosphorus pentachloride	500	500				
10028-15-6	Ozone	100	100			313	
10028-22-5	Ferric sulfate			1,000			
10031-59-1	Thallium sulfate	100/10,000	100	100			
10034-93-2	Hydrazine sulfate					313	
10039-32-4	Sodium phosphate, dibasic			5,000			
10043-01-3	Aluminum sulfate			5,000			
10045-89-3	Ferrous ammonium sulfate			1,000			
10045-94-0	Mercuric nitrate			10			
10049-04-4	Chlorine dioxide				1,000	313	
10049-04-4	Chlorine oxide (ClO2)				1,000	X	
10049-05-5	Chromous chloride			1,000			
10061-02-6	trans-1,3-Dichloropropene					313	
10099-74-8	Lead nitrate			10			
10101-53-8	Chromic sulfate			1,000			
10101-63-0	Lead iodide			10			
10101-89-0	Sodium phosphate, tribasic			5,000			
10102-06-4	Uranyl nitrate			100			
10102-18-8	Sodium selenite	100/10,000	100	100			
10102-20-2	Sodium tellurite	500/10,000	500				
10102-43-9	Nitric oxide	100	10	10	10,000		P076
10102-43-9	Nitrogen oxide (NO)	100	10	10	10,000		P076
10102-44-0	Nitrogen dioxide	100	10	10			P078
10102-45-1	Thallium(I) nitrate			100			U217
10102-48-4	Lead arsenate			1			
10108-64-2	Cadmium chloride			10			

Member of diisocyanate category.

Page 29

CAS Number	Chemical Name	Sec. 302 (EHS) TPQ	Section 304 EHS RQ	CERCLA RQ	CAA 112(r) TQ	Sec 313	RCRA CODE
10124-50-2	Potassium arsenite	500/10,000	1	1			
10124-56-8	Sodium phosphate, tribasic			5,000			
10140-65-5	Sodium phosphate, dibasic			5,000			
10140-87-1	Ethanol, 1,2-dichloro-, acetate	1,000	1,000				
10192-30-0	Ammonium bisulfite			5,000			
10196-04-0	Ammonium sulfite			5,000			
10210-68-1	Cobalt carbonyl	10/10,000	10				
10222-01-2	2,2-Dibromo-3-nitrilopropionamide					313	
10265-92-6	Methamidophos	100/10,000	100				
10294-34-5	Borane, trichloro-	500	500		5,000	X	
10294-34-5	Boron trichloride	500	500		5,000	313	
10311-84-9	Dialifor	100/10,000	100				
10347-54-3	1,4-Bis(methylisocyanate)cyclohexane					313#	
10361-89-4	Sodium phosphate, tribasic			5,000			
10380-29-7	Cupric sulfate, ammoniated			100			
10415-75-5	Mercurous nitrate			10			
10421-48-4	Ferric nitrate			1,000			
10453-86-8	5-(Phenylmethyl)-3-furanyl)methyl					X	
10453-86-8	Resmethrin					313	
10476-95-6	Methacrolein diacetate	1,000	1,000				
10544-72-6	Nitrogen dioxide			10			
10588-01-9	Sodium bichromate			10			
10605-21-7	Carbendazim			1*			U372
11096-82-5	Aroclor 1260			1			
11097-69-1	Aroclor 1254			1			
11104-28-2	Aroclor 1221			10			
11115-74-5	Chromic acid			1			
11141-16-5	Aroclor 1232			1			
12002-03-8	Cupric acetoarsenite	500/10,000	1	1			
12002-03-8	Paris green	500/10,000	1	1			P114
12039-52-0	Selenious acid, dithallium(1+) salt			1,000			
12054-48-7	Nickel hydroxide			10			
12108-13-3	Manganese, tricarbonyl methylcyclopentadienyl	100	100			X	
12122-67-7	Carbamodithioic acid, 1,2-ethanediylbis-, zinc complex					313	
12122-67-7	Zineb			100			
12125-01-8	Ammonium fluoride			5,000			
12125-02-9	Ammonium chloride			100			
12135-76-1	Ammonium sulfide					X	
12427-38-2	Carbamodithioic acid, 1,2-ethanediylbis-, manganese c					313	
12427-38-2	Maneb			1			
12672-29-6	Aroclor 1248			1			
12674-11-2	Aroclor 1016			1,000			
12771-08-3	Sulfur monochloride	100	100				
13071-79-9	Terbufos	100	100				
13171-21-6	Phosphamidon	1,000	1,000			313	
13194-48-4	Ethoprop	1,000	1,000			X	
13194-48-4	Ethoprophos	1,000	1,000			X	
13194-48-4	Phosphorodithioic acid O-ethyl S,S-dipropyl ester					313	
13356-08-6	Fenbutatin oxide					X	
13356-08-6	Hexakis(2-methyl-2-phenylpropyl)distannoxane						

\# Member of diisocyanate category.
* RCRA carbamate waste; statutory one-pound RQ applies until RQs are adjusted.

CAS Number	Chemical Name	Sec. 302 (EHS) TPQ	Section 304 EHS RQ	CERCLA RQ	CAA 112(r) TQ	Sec 313	RCRA CODE
13410-01-0	Sodium selenate	100/10,000	100				
13450-90-3	Gallium trichloride	500/10,000	500				
13463-39-3	Nickel carbonyl	1	10	10	1,000		P073
13463-40-6	Iron carbonyl (Fe(CO)5), (TB-5-11)-	100	100		2,500	X	
13463-40-6	Iron, pentacarbonyl-	100	100		2,500	313	
13474-88-9	1,1-Dichloro-1,2,2,3,3-pentafluoropropane					313	
13474-88-9	HCFC-225cc					X	
13560-99-1	2,4,5-T salts			1,000			
13597-99-4	Beryllium nitrate			1			
13684-56-5	Desmedipham					313	
13746-89-9	Zirconium nitrate			5,000			
13765-19-0	Calcium chromate			10			U032
13814-96-5	Lead fluoborate			10			
13826-83-0	Ammonium fluoborate			5,000			
13952-84-6	sec-Butylamine			1,000			
14017-41-5	Cobaltous sulfamate			1,000			
14167-18-1	Salcomine	500/10,000	500				
14216-75-2	Nickel nitrate			100			
14258-49-2	Ammonium oxalate			5,000			
14307-35-8	Lithium chromate			10			
14307-43-8	Ammonium tartrate			5,000			
14484-64-1	Ferbam					313	
14484-64-1	Tris(dimethylcarbamodithioato-S,S')iron					X	
14639-97-5	Zinc ammonium chloride			1,000			
14639-98-6	Zinc ammonium chloride			1,000			
14644-61-2	Zirconium sulfate			5,000			
15271-41-7	Bicyclo[2.2.1]heptane-2-carbonitrile,	500/10,000	500				
15339-36-3	Manganese, bis(dimethylcarbamodithioato-S,S')-			1*			P196
15646-96-5	2,4,4-Trimethylhexamethylene diisocyanate					313#	
15699-18-0	Nickel ammonium sulfate			100			
15739-80-7	Lead sulfate			10			
15950-66-0	2,3,4-Trichlorophenol			10			
15972-60-8	Alachlor					313	
16071-86-6	C.I. Direct Brown 95					313	
16543-55-8	N-Nitrosonornicotine					313	
16721-80-5	Sodium hydrosulfide			5,000			
16752-77-5	Ethanimidothioic acid, N-[(methylamino)carbonyl]	500/10,000	100	100			P066
16752-77-5	Methomyl	500/10,000	100	100			P066
16871-71-9	Zinc silicofluoride			5,000			
16919-19-0	Ammonium silicofluoride			1,000			
16923-95-8	Zirconium potassium fluoride			1,000			
16938-22-0	2,2,4-Trimethylhexamethylene diisocyanate					313#	
17702-41-9	Decaborane(14)	500/10,000	500				
17702-57-7	Formparanate	100/10,000	1*	1*			P197
17804-35-2	Benomyl			1*		313	U271
18883-66-4	Streptozotocin			1			U206
19044-88-3	4-(Dipropylamino)-3,5-dinitrobenzenesulfonamide					X	
19044-88-3	Oryzalin					313	
19287-45-7	Diborane	100	100		2,500		
19287-45-7	Diborane(6)	100	100		2,500		

\# Member of diisocyanate category.
* RCRA carbamate waste; statutory one-pound RQ applies until RQs are adjusted.

CAS Number	Chemical Name	Sec. 302 (EHS) TPQ	Section 304 EHS RQ	CERCLA RQ	CAA 112(r) TQ	Sec 313	RCRA CODE
19624-22-7	Pentaborane	500	500				
19666-30-9	3-(2,4-Dichloro-5-(1-methylethoxy)phenyl)-5-(1,1-dim					X	
19666-30-9	Oxydiazon					313	
20325-40-0	o-Dianisidine dihydrochloride					X	
20325-40-0	3,3'-Dimethoxybenzidine dihydrochloride					313	
20354-26-1	2-(3,4-Dichlorophenyl)-4-methyl-1,2,4-oxadiazolidine-					X	
20354-26-1	Methazole					313	
20816-12-0	Osmium oxide OsO4 (T-4)-			1,000		X	P087
20816-12-0	Osmium tetroxide			1,000		313	P087
20830-75-5	Digoxin	10/10,000	10				
20830-81-3	Daunomycin			10			U059
20859-73-8	Aluminum phosphide	500	100	100		313	P006
21087-64-9	Metribuzin					313	
21548-32-3	Fosthietan	500	500				
21609-90-5	Leptophos	500/10,000	500				
21725-46-2	Cyanazine					313	
21908-53-2	Mercuric oxide	500/10,000	500				
21923-23-9	Chlorthiophos	500	500				
22224-92-6	Fenamiphos	10/10,000	10				
22781-23-3	Bendiocarb			1*		313	U278
22781-23-3	2,2-Dimethyl-1,3-benzodioxol-4-ol methylcarbamate			1*		X	U278
22961-82-6	Bendiocarb phenol			1*			U364
23135-22-0	Oxamyl	100/10,000	1*	1*			P194
23422-53-9	Formetanate hydrochloride	500/10,000	1*	1*			P198
23505-41-1	Pirimifos-ethyl	1,000	1,000				
23564-05-8	Thiophanate-methyl			1*		313	U409
23564-06-9	(1,2-Phenylenebis(iminocarbonothioyl)) biscarbamic acid					X	
23564-06-9	Thiophanate ethyl					313	
23950-58-5	Benzamide, 3,5-dichloro-N-(1,1-dimethyl-2-propynyl)			5,000		X	U192
23950-58-5	Pronamide			5,000		313	U192
24017-47-8	Triazofos	500	500				
24934-91-6	Chlormephos	500	500				
25154-54-5	Dinitrobenzene (mixed isomers)			100			
25154-55-6	Nitrophenol (mixed isomers)			100			
25155-30-0	Sodium dodecylbenzenesulfonate			1,000			
25167-67-3	Butene				10,000		
25167-82-2	Trichlorophenol			10			
25168-15-4	2,4,5-T esters			1,000			
25168-26-7	2,4-D Esters			100			
25311-71-1	2-((Ethoxy((1-methylethyl)amino)phosphinothioyl)ox					X	
25311-71-1	Isofenphos					313	
25321-14-6	Dinitrotoluene (mixed isomers)			10		313	
25321-22-6	Dichlorobenzene			100		X	
25321-22-6	Dichlorobenzene (mixed isomers)			100		313	
25376-45-8	Diaminotoluene (mixed isomers)			10		313	U221
25376-45-8	Toluenediamine			10		X	U221
25550-58-7	Dinitrophenol			10			
26002-80-2	2,2-Dimethyl-3-(2-methyl-1-propenyl)cyclopropanecar					X	
26002-80-2	Phenothrin					313	
26264-06-2	Calcium dodecylbenzenesulfonate			1,000			

* RCRA carbamate waste; statutory one-pound RQ applies until RQs are adjusted.

CAS Number	Chemical Name	Sec. 302 (EHS) TPQ	Section 304 EHS RQ	CERCLA RQ	CAA Sec 112(r) TQ	Sec 313	RCRA CODE
26419-73-8	Carbamic acid, methyl-,	100/10,000	1*	1*			P185
26471-62-5	Benzene, 1,3-diisocyanatomethyl-			100	10,000	X	U223
26471-62-5	Toluenediisocyanate (mixed isomers)			100	10,000	313	U223
26471-62-5	Toluene diisocyanate (unspecified isomer)			100	10,000	X	U223
26628-22-8	Sodium azide (Na(N3))	500	1,000	1,000		313	P105
26638-19-7	Dichloropropane			1,000			
26644-46-2	N,N'-(1,4-Piperazinediylbis(2,2,2-trichloroethylidene))					X	
26644-46-2	Triforine					313	
26952-23-8	Dichloropropene			100			
27137-85-5	Trichloro(dichlorophenyl)silane	500	500				
27176-87-0	Dodecylbenzenesulfonic acid			1,000			
27314-13-2	4-Chloro-5-(methylamino)-2-[3-(trifluoromethyl)pheny					X	
27314-13-2	Norflurazon					313	
27323-41-7	Triethanolamine dodecylbenzene sulfonate			1,000			
27774-13-6	Vanadyl sulfate			1,000			
28057-48-9	d-trans-Allethrin					313	
28057-48-9	d-trans-Chrysanthemic acid of d-allethrone					X	
28249-77-6	Carbamic acid, diethylthio-, S-(p-chlorobenzyl)					X	
28249-77-6	Thiobencarb					313	
28300-74-5	Antimony potassium tartrate			100			
28347-13-9	Xylylene dichloride	100/10,000	100				
28407-37-6	C.I. Direct Blue 218					313	
28772-56-7	Bromadiolone	100/10,000	100				
29232-93-7	O-(2-(Diethylamino)-6-methyl-4-pyrimidinyl)-O,O-dim					X	
29232-93-7	Pirimiphos methyl					313	
30525-89-4	Paraformaldehyde			1,000			
30558-43-1	Ethanimidothioic acid, 2-(dimethylamino)-N-hydroxy-			1*			U394
30560-19-1	Acephate					313	
30560-19-1	Acetylphosphoramidothioic acid O,S-dimethyl ester					X	
30674-80-7	Methacryloyloxyethyl isocyanate	100	100				
31218-83-4	3-((Ethylamino)methoxyphosphinothioyl)oxy)-2-buten					X	
31218-83-4	Propetamphos					313	
32534-95-5	2,4,5-TP esters			100			
33089-61-1	Amitraz					313	
33213-65-9	beta - Endosulfan			1			
34014-18-1	N-(5-(1,1-Dimethylethyl)-1,3,4-thiadiazol-2-yl)-N,N'-					X	
34014-18-1	Tebuthiuron					313	
34077-87-7	Dichlorotrifluoroethane					313	
35367-38-5	Diflubenzuron					313	
35400-43-2	O-Ethyl O-(4-(methylthio)phenyl)phosphorodithioic ac					X	
35400-43-2	Sulprofos					313	
35554-44-0	1-(2-(2,4-Dichlorophenyl)-2-(2-propenyloxy)ethyl)-1H					X	
35554-44-0	Imazalil					313	
35691-65-7	1-Bromo-1-(bromomethyl)-1,3-propanedicarbonitrile					313	
36478-76-9	Uranyl nitrate			100			
37211-05-5	Nickel chloride			100			
38661-72-2	1,3-Bis(methylisocyanate)cyclohexane					313#	
38727-55-8	Diethatyl ethyl					313	
39156-41-7	2,4-Diaminoanisole sulfate					313	
39196-18-4	Thiofanox	100/10,000	100	100			P045

Member of diisocyanate category.
* RCRA carbamate waste; statutory one-pound RQ applies until RQs are adjusted.

CAS Number	Chemical Name	Sec. 302 (EHS) TPQ	Section 304 EHS RQ	CERCLA RQ	CAA 112(r) TQ	Sec 313	RCRA CODE
39300-45-3	Dinocap					313	
39515-41-8	Fenpropathrin					313	
39515-41-8	2,2,3,3-Tetramethylcyclopropane carboxylic acid					X	
40487-42-1	N-(1-Ethylpropyl)-3,4-dimethyl-2,6-dinitrobenzenamine					X	
40487-42-1	Pendimethalin					313	
41198-08-7	O-(4-Bromo-2-chlorophenyl)-O-ethyl-S-propylphospho					X	
41198-08-7	Profenofos					313	
41766-75-0	3,3'-Dimethylbenzidine dihydrofluoride					313	
41766-75-0	o-Tolidine dihydrofluoride					X	
42504-46-1	Isopropanolamine dodecylbenzene sulfonate			1,000			
42874-03-3	Oxyfluorfen					313	
43121-43-3	1-(4-Chlorophenoxy)-3,3-dimethyl-1-(1H-1,2,4-triazol-					X	
43121-43-3	Triadimefon					313	
50471-44-8	3-(3,5-Dichlorophenyl)-5-ethenyl-5-methyl-2,4-oxazol					X	
50471-44-8	Vinclozolin					313	
50782-69-9	Phosphonothioic acid, methyl-, S-(2-(bis(1-methylethyl	100	100				
51235-04-2	Hexazinone					313	
51338-27-3	2-(4-(2,4-Dichlorophenoxy)phenoxy)propanoic acid, m					X	
51338-27-3	Diclofop methyl					313	
51630-58-1	4-Chloro-alpha-(1-methylethyl)benzeneacetic acid					X	
51630-58-1	Fenvalerate					313	
52628-25-8	Zinc ammonium chloride			1,000			
52645-53-1	3-(2,2-Dichloroethenyl)-2,2-dimethylcyclopropane car					X	
52645-53-1	Permethrin					313	
52652-59-2	Lead stearate			10			
52740-16-6	Calcium arsenite			1*			
52888-80-9	Carbamothioic acid, dipropyl-, S-(phenylmethyl) ester			1*			U387
53404-19-6	Bromacil, lithium salt					313	
53404-19-6	2,4-(1H,3H)-Pyrimidinedione,					X	
53404-37-8	2,4-D 2-ethyl-4-methylpentyl ester					313	
53404-60-7	Dazomet, sodium salt					313	
53404-60-7	Tetrahydro-3,5-dimethyl-2H-1,3,5-thiadiazine-2-thione,					X	
53467-11-1	2,4-D Esters			100			
53469-21-9	Aroclor 1242			1			
53558-25-1	Pyriminil	100/10,000	100				
55285-14-8	Carbosulfan			1*			P189
55290-64-7	2,3,-Dihydro-5,6-dimethyl-1,4-dithiin 1,1,4,4-tetraoxide					X	
55290-64-7	Dimethipin					313	
55406-53-6	3-Iodo-2-propynyl butylcarbamate					313	
55488-87-4	Ferric ammonium oxalate			1,000			
56189-09-4	Lead stearate			10			
57213-69-1	Triclopyr triethylammonium salt					313	
58270-08-9	Zinc, dichloro(4,4-dimethyl-5((((methylamino)carbony	100/10,000	100				
59669-26-0	Thiodicarb			1*		313	U410
60168-88-9	.alpha.-(2-Chlorophenyl)-.alpha.-4-chlorophenyl)-5-pyr					X	
60168-88-9	Fenarimol					313	
60207-90-1	1-(2-(2,4-Dichlorophenyl)-4-propyl-1,3-dioxolan-2-yl)-methyl					X	
60207-90-1	Propiconazole					313	
61792-07-2	2,4,5-T esters			1,000			
62207-76-5	Cobalt, ((2,2'-(1,2-ethanediylbis(nitrilomethylidene))	100/10,000	100				

* RCRA carbamate waste; statutory one-pound RQ applies until RQs are adjusted.

CAS Number	Chemical Name	Sec. 302 (EHS) TPQ	Section 304 EHS RQ	CERCLA RQ	CAA 112(r) TQ	Sec 313	RCRA CODE
62476-59-9	Aciffuorfen, sodium salt						
62476-59-9	5-(2-Chloro-4-(trifluoromethyl)phenoxy)-2-nitrobenzo					313	
63938-10-3	Chlorotetrafluoroethane					X	
64902-72-3	2-Chloro-N-(((4-methoxy-6-methyl-1,3,5-triazin-2-yl)					313	
64902-72-3	Chlorsulfuron					X	
64969-34-2	3,3'-Dichlorobenzidine sulfate					313	
66441-23-4	2-(4-((6-Chloro-2-benzoxazolylen)oxy)phenoxy)propa					313	
66441-23-4	Fenoxaprop ethyl					X	
67485-29-4	Hydramethylnon					313	
67485-29-4	Tetrahydro-5,5-dimethyl-2(1H)-pyrimidinone(3-(4-(tri|					313	
68085-85-8	3-(2-Chloro-3,3,3-trifluoro-1-propenyl)-2,2-Dimethylc					X	
68085-85-8	Cyhalothrin					X	
68359-37-5	Cyfluthrin					313	
68359-37-5	3-(2,2-Dichloroethenyl)-2,2-dimethylcyclopropanecarb					313	
69409-94-5	N-(2-Chloro-4-(trifluoromethyl)phenyl)-DL-valine(+)-					X	
69409-94-5	Fluvalinate					X	
69806-50-4	Fluazifop butyl					313	
69806-50-4	2-(4-(5-(Trifluoromethyl)-2-pyridinyl]oxy]-phenoxy)pr					313	
71751-41-2	Abamectin					X	
71751-41-2	Avermectin B1					313	
72178-02-0	5-(2-Chloro-4-(trifluoromethyl)phenoxy)-N-methylsulf]					X	
72178-02-0	Fomesafen					X	
72490-01-8	Fenoxycarb					313	
72490-01-8	(2-(4-Phenoxy-phenoxy)-ethyl)carbamic acid ethyl ester					313	
74051-80-2	2-(1-(Ethoxyimino)					X	
74051-80-2	Sethoxydim					X	
75790-84-0	4-Methyldiphenylmethane-3,4-diisocyanate					313	
75790-87-3	2,4'-Diisocyanatodiphenyl sulfide					313#	
76578-14-8	2-(4-((6-Chloro-2-quinoxalinyl)oxy]phenoxy) propanoi					313#	
76578-14-8	Quizalofop-ethyl					X	
77501-63-4	5-(2-Chloro-4-(trifluoromethyl)phenoxy)-2-nitro-2-eth					313	
77501-63-4	Lactofen					X	
82657-04-3	Bifenthrin					313	
88671-89-0	.alpha.-Butyl-.alpha.-(4-chlorophenyl)-1H-1,2,4-triazo					313	
88671-89-0	Myclobutanil					X	
90454-18-5	Dichloro-1,1,2-trifluoroethane					313	
90982-32-4	Chlorimuron ethyl					313	
90982-32-4	Ethyl-2-((((4-chloro-6-methoxyprimidin-2-yl)-carbony					313	
101200-48-0	2-(4-Methoxy-6-methyl-1,3,5-triazin-2-yl)-methylamin					X	
101200-48-0	Tribenuron methyl					X	
111512-56-2	1,1-Dichloro-1,2,3,3,3-pentafluoropropane					313	
111512-56-2	HCFC-225eb					313	
111984-09-9	o-Dianisidine hydrochloride					X	
111984-09-9	3,3'-Dimethoxybenzidine hydrochloride					X	
127564-92-5	Dichloropentafluoropropane					313	
128903-21-9	2,2-Dichloro-1,1,1,3,3-pentafluoropropane					313	
128903-21-9	HCFC-225aa					313	
134190-37-7	Diethyldiisocyanatobenzene					X	
136013-79-1	1,3-Dichloro-1,1,2,3,3-pentafluoropropane					313#	
136013-79-1	HCFC-225ea					313	
	Organorhodium Complex (PMN-82-147)	10/10,000	10	**		X	

Member of diisocyanate category.

** This chemical was identified from a Premanufacture Review Notice (PMN) submitted to EPA. The submitter has claimed certain information on the submission to be confidential, including specific chemical identity.

CAS Number	Chemical Name	Sec. 302 (EHS) TPQ	Section 304 EHS RQ	CERCLA RQ	CAA 112(r) TQ	Sec 313	RCRA CODE
	Antimony Compounds			***		N010	
	Arsenic Compounds			***		N020	
	Barium Compounds					N040	
	--Except Barium Sulfate (under 313)			***		N050	
	Beryllium Compounds			***		N078	
	Cadmium Compounds			***			
	Chlordane (Technical Mixture and Metabolites)			***			
	Chlorinated Benzenes			***			
	Chlorinated Ethanes			***			
	Chlorinated Naphthalene			***		N084	
	Chlorinated Phenols			***			
	Chloroalkyl Ethers			***		N084	
	Chlorophenols			***		N090	
	Chromium Compounds			***		N096	
	Cobalt Compounds						
	Coke Oven Emissions			***		N100	
	Copper Compounds						
	--Except copper phthalocyanine compounds (313)##						
	--Except C.I. Pigment Blue 15 (under 313)						
	--Except C.I. Pigment Green 7 (under 313)						
	--Except C.I. Pigment Green 36 (under 313)			***		N106	
	Cyanide Compounds			***			
	DDT and Metabolites			***			
	Dichlorobenzidine					N120	
	Diisocyanates (includes only 20 chemicals)			***			
	Diphenylhydrazine			***			
	Endosulfan and Metabolites			***			
	Endrin and Metabolites					N171	
	Ethylenebisdithiocarbamic acid, salts and esters			***			
	Fine mineral fibers			***		N230	
	Glycol Ethers			***			
	Haloethers			***			
	Halomethanes			***			
	Heptachlor and Metabolites			***			
	Hexachlorocyclohexane (all isomers) CAS 608-73-1			***		N420	
	Lead Compounds			***		N450	
	Manganese Compounds			***		N458	
	Mercury Compounds			***		N495	
	Nickel Compounds					N503	
	Nicotine and salts					N511	
	Nitrate compounds (water dissociable)			***			
	Nitrophenols			***			
	Nitrosamines			***			
	Phthalate Esters					N575	
	Polybrominated Biphenyls (PBBs)					N583	
	Polychlorinated alkanes (C10 to C13)					N590	
	Polycyclic aromatic compounds (includes 19 chems)			***			
	Polycyclic organic matter						

All copper pthalocyanine compounds substituted with only hydrogen and/or bromine or chlorine.
*** Indicates that no RQ is assigned to this generic or broad class, although the class is a CERCLA hazardous substance.
See 50 Federal Register 13456 (April 4, 1985).
Values in Section 313 column represent Category Codes for reporting under Section 313.

CAS Number	Chemical Name	Sec. 302 (EHS) TPQ	Section 304 EHS RQ	CERCLA RQ	CAA 112(r) TQ	Sec 313	RCRA CODE
	Polynuclear Aromatic Hydrocarbons			***			
	Selenium Compounds			***		N725	
	Silver Compounds			***		N740	
	Strychnine and salts					N746	
	Thallium Compounds			***		N760	
	Warfarin and salts					N874	
	Zinc Compounds			***		N982	

*** Indicates that no RQ is assigned to this generic or broad class, although the class is a CERCLA hazardous substance.
See 50 Federal Register 13456 (April 4, 1985).
Values in Section 313 column represent Category Codes for reporting under Section 313.

APPENDIX A
ALPHABETICAL LISTING OF CHEMICAL NAME AND CAS NUMBER

CAS Number	Chemical Name
71751-41-2	Abamectin
83-32-9	Acenaphthene
208-96-8	Acenaphthylene
30560-19-1	Acephate
75-07-0	Acetaldehyde
75-87-6	Acetaldehyde, trichloro-
60-35-5	Acetamide
64-19-7	Acetic acid
94-75-7	Acetic acid, (2,4-dichlorophenoxy)-
108-05-4	Acetic acid ethenyl ester
108-24-7	Acetic anhydride
67-64-1	Acetone
75-86-5	Acetone cyanohydrin
1752-30-3	Acetone thiosemicarbazide
75-05-8	Acetonitrile
98-86-2	Acetophenone
53-96-3	2-Acetylaminofluorene
506-96-7	Acetyl bromide
75-36-5	Acetyl chloride
74-86-2	Acetylene
30560-19-1	Acetylphosphoramidothioic acid O,S-dimethyl
591-08-2	1-Acetyl-2-thiourea
62476-59-9	Acifluorfen, sodium salt
107-02-8	Acrolein
79-06-1	Acrylamide
79-10-7	Acrylic acid
107-13-1	Acrylonitrile
814-68-6	Acrylyl chloride
124-04-9	Adipic acid
111-69-3	Adiponitrile
15972-60-8	Alachlor
116-06-3	Aldicarb
1646-88-4	Aldicarb sulfone
309-00-2	Aldrin
28057-48-9	d-trans-Allethrin
107-18-6	Allyl alcohol
107-11-9	Allylamine
107-05-1	Allyl chloride
7429-90-5	Aluminum (fume or dust)
1344-28-1	Aluminum oxide (fibrous forms)
20859-73-8	Aluminum phosphide
10043-01-3	Aluminum sulfate
834-12-8	Ametryn
117-79-3	2-Aminoanthraquinone
60-09-3	4-Aminoazobenzene
92-67-1	4-Aminobiphenyl
82-28-0	1-Amino-2-methylanthraquinone
2763-96-4	5-(Aminomethyl)-3-isoxazolol
54-62-6	Aminopterin
504-24-5	4-Aminopyridine
78-53-5	Amiton
3734-97-2	Amiton oxalate
33089-61-1	Amitraz
61-82-5	Amitrole
7664-41-7	Ammonia
7664-41-7	Ammonia (anhydrous)
7664-41-7	Ammonia (conc 20% or greater)
631-61-8	Ammonium acetate

CAS Number	Chemical Name
1863-63-4	Ammonium benzoate
1066-33-7	Ammonium bicarbonate
7789-09-5	Ammonium bichromate
1341-49-7	Ammonium bifluoride
10192-30-0	Ammonium bisulfite
1111-78-0	Ammonium carbamate
506-87-6	Ammonium carbonate
12125-02-9	Ammonium chloride
7788-98-9	Ammonium chromate
3012-65-5	Ammonium citrate, dibasic
13826-83-0	Ammonium fluoborate
12125-01-8	Ammonium fluoride
1336-21-6	Ammonium hydroxide
5972-73-6	Ammonium oxalate
6009-70-7	Ammonium oxalate
14258-49-2	Ammonium oxalate
131-74-8	Ammonium picrate
16919-19-0	Ammonium silicofluoride
7773-06-0	Ammonium sulfamate
12135-76-1	Ammonium sulfide
10196-04-0	Ammonium sulfite
3164-29-2	Ammonium tartrate
14307-43-8	Ammonium tartrate
1762-95-4	Ammonium thiocyanate
7803-55-6	Ammonium vanadate
300-62-9	Amphetamine
628-63-7	Amyl acetate
123-92-2	iso-Amyl acetate
626-38-0	sec-Amyl acetate
625-16-1	tert-Amyl acetate
101-05-3	Anilazine
62-53-3	Aniline
88-05-1	Aniline, 2,4,6-trimethyl-
90-04-0	o-Anisidine
104-94-9	p-Anisidine
134-29-2	o-Anisidine hydrochloride
120-12-7	Anthracene
7440-36-0	Antimony
	Antimony Compounds
7647-18-9	Antimony pentachloride
7783-70-2	Antimony pentafluoride
28300-74-5	Antimony potassium tartrate
7789-61-9	Antimony tribromide
10025-91-9	Antimony trichloride
7783-56-4	Antimony trifluoride
1309-64-4	Antimony trioxide
1397-94-0	Antimycin A
86-88-4	ANTU
12674-11-2	Aroclor 1016
11104-28-2	Aroclor 1221
11141-16-5	Aroclor 1232
53469-21-9	Aroclor 1242
12672-29-6	Aroclor 1248
11097-69-1	Aroclor 1254
11096-82-5	Aroclor 1260
7440-38-2	Arsenic
1327-52-2	Arsenic acid
7778-39-4	Arsenic acid

APPENDIX C

ALPHABETICAL LISTING OF CHEMICAL NAME AND CAS NUMBER

CAS Number	Chemical Name
	Arsenic Compounds
1303-32-8	Arsenic disulfide
1303-28-2	Arsenic pentoxide
1327-53-3	Arsenic trioxide
1303-33-9	Arsenic trisulfide
1327-53-3	Arsenous oxide
7784-34-1	Arsenous trichloride
7784-42-1	Arsine
1332-21-4	Asbestos (friable)
1912-24-9	Atrazine
492-80-8	Auramine
71751-41-2	Avermectin B1
115-02-6	Azaserine
2212-67-1	1H-Azepine-1 carbothioic acid, hexahydro-S-ethyl
2642-71-9	Azinphos-ethyl
86-50-0	Azinphos-methyl
151-56-4	Aziridine
75-55-8	Aziridine, 2-methyl
101-27-9	Barban
7440-39-3	Barium
	Barium Compounds
	--Except Barium Sulfate (under 313)
542-62-1	Barium cyanide
22781-23-3	Bendiocarb
22961-82-6	Bendiocarb phenol
1582-09-8	Benezeneamine,
1861-40-1	Benfluralin
17804-35-2	Benomyl
225-51-4	Benz[c]acridine
98-87-3	Benzal chloride
55-21-0	Benzamide
23950-58-5	Benzamide,
56-55-3	Benz[a]anthracene
98-16-8	Benzenamine, 3-(trifluoromethyl)-
71-43-2	Benzene
510-15-6	Benzeneacetic acid,
135-20-6	Benzeneamine, N-hydroxy-N-nitroso, ammonium
98-05-5	Benzenearsonic acid
100-14-1	Benzene, 1-(chloromethyl)-4-nitro-
1897-45-6	1,3-Benzenedicarbonitrile, 2,4,5,6-tetrachloro-
1836-75-5	Benzene, 2,4-dichloro-1-(4-nitrophenoxy)-
584-84-9	Benzene, 2,4-diisocyanato-1-methyl-
91-08-7	Benzene, 1,3-diisocyanato-2-methyl-
26471-62-5	Benzene, 1,3-diisocyanatomethyl-
108-38-3	Benzene, m-dimethyl-
95-47-6	Benzene, o-dimethyl-
106-42-3	Benzene, p-dimethyl-
122-09-8	Benzeneethanamine, alpha,alpha-dimethyl-
115-32-2	Benzenemethanol,
98-09-9	Benzenesulfonyl chloride
108-98-5	Benzenethiol
72-43-5	Benzene, 1,1'-(2,2,2-trichloroethylidene)bis
92-87-5	Benzidine
3615-21-2	Benzimidazole, 4,5-dichloro-2-(trifluoromethyl)-
205-99-2	Benzo[b]fluoranthene
205-82-3	Benzo(j)fluoranthene
207-08-9	Benzo(k)fluoranthene
65-85-0	Benzoic acid

CAS Number	Chemical Name
133-90-4	Benzoic acid, 3-amino-2,5-dichloro-
98-07-7	Benzoic trichloride
100-47-0	Benzonitrile
189-55-9	Benzo(rst)pentaphene
191-24-2	Benzo[ghi]perylene
218-01-9	Benzo(a)phenanthrene
50-32-8	Benzo[a]pyrene
106-51-4	p-Benzoquinone
98-07-7	Benzotrichloride
98-88-4	Benzoyl chloride
94-36-0	Benzoyl peroxide
100-44-7	Benzyl chloride
140-29-4	Benzyl cyanide
7440-41-7	Beryllium
7787-47-5	Beryllium chloride
	Beryllium Compounds
7787-49-7	Beryllium fluoride
7787-55-5	Beryllium nitrate
13597-99-4	Beryllium nitrate
319-84-6	alpha-BHC
319-85-7	beta-BHC
319-86-8	delta-BHC
15271-41-7	Bicyclo[2.2.1]heptane-2-carbonitrile,
82657-04-3	Bifenthrin
1464-53-5	2,2'-Bioxirane
92-52-4	Biphenyl
111-91-1	Bis(2-chloroethoxy) methane
111-44-4	Bis(2-chloroethyl) ether
542-88-1	Bis(chloromethyl) ether
108-60-1	Bis(2-chloro-1-methylethyl)ether
534-07-6	Bis(chloromethyl) ketone
117-81-7	Bis(2-ethylhexyl)phthalate
7287-19-6	N,N'-Bis(1-methylethyl)-6-methylthio-1,3,5-triazine
10347-54-3	1,4-Bis(methylisocyanate)cyclohexane
38661-72-2	1,3-Bis(methylisocyanate)cyclohexane
56-35-9	Bis(tributyltin) oxide
4044-65-9	Bitoscanate
10294-34-5	Borane, trichloro-
7637-07-2	Borane, trifluoro-
10294-34-5	Boron trichloride
7637-07-2	Boron trifluoride
353-42-4	Boron trifluoride compound with methyl ether (1:1)
353-42-4	Boron, trifluoro[oxybis[methane]]-, (T-4)-
314-40-9	Bromacil
53404-19-6	Bromacil, lithium salt
28772-56-7	Bromadiolone
7726-95-6	Bromine
598-31-2	Bromoacetone
35691-65-7	1-Bromo-1-(bromomethyl)-1,3-propanedicarbonitril
353-59-3	Bromochlorodifluoromethane
41198-08-7	O-(4-Bromo-2-chlorophenyl)-O-ethyl-S-propylpho
75-25-2	Bromoform
74-83-9	Bromomethane
314-40-9	5-Bromo-6-methyl-3-(1-methylpropyl)-2,4-(1H,3H)
101-55-3	4-Bromophenyl phenyl ether
598-73-2	Bromotrifluoroethylene
75-63-8	Bromotrifluoromethane
1689-84-5	Bromoxynil

Page A-3

ALPHABETICAL LISTING OF CHEMICAL NAME AND CAS NUMBER

CAS Number	Chemical Name
1689-99-2	Bromoxynil octanoate
357-57-3	Brucine
106-99-0	1,3-Butadiene
78-79-5	1,3-Butadiene, 2-methyl-
106-97-8	Butane
78-78-4	Butane, 2-methyl-
4170-30-3	2-Butenal
123-73-9	2-Butenal, (e)-
25167-67-3	Butene
590-18-1	2-Butene-cis
624-64-6	2-Butene, (E)
624-64-6	2-Butene-trans
106-98-9	1-Butene
107-01-7	2-Butene
764-41-0	2-Butene, 1,4-dichloro-
689-97-4	1-Buten-3-yne
1929-73-3	2,4-D butoxyethyl ester
123-86-4	Butyl acetate
110-19-0	iso-Butyl acetate
105-46-4	sec-Butyl acetate
540-88-5	tert-Butyl acetate
141-32-2	Butyl acrylate
71-36-3	n-Butyl alcohol
78-92-2	sec-Butyl alcohol
75-65-0	tert-Butyl alcohol
109-73-9	Butylamine
78-81-9	iso-Butylamine
513-49-5	sec-Butylamine
13952-84-6	sec-Butylamine
75-64-9	tert-Butylamine
85-68-7	Butyl benzyl phthalate
88671-89-0	.alpha.-Butyl-.alpha.-(4-chlorophenyl)-1H-1,2,4-tri
106-88-7	1,2-Butylene oxide
1114-71-2	Butylethylcarbamothioic acid S-propyl ester
1861-40-1	N-Butyl-N-ethyl-2,6-dinitro-4-(trifluoromethyl)
84-74-2	n-Butyl phthalate
107-00-6	1-Butyne
123-72-8	Butyraldehyde
107-92-6	Butyric acid
79-31-2	iso-Butyric acid
75-60-5	Cacodylic acid
7440-43-9	Cadmium
543-90-8	Cadmium acetate
7789-42-6	Cadmium bromide
10108-64-2	Cadmium chloride
	Cadmium Compounds
1306-19-0	Cadmium oxide
2223-93-0	Cadmium stearate
7778-44-1	Calcium arsenate
52740-16-6	Calcium arsenite
75-20-7	Calcium carbide
13765-19-0	Calcium chromate
156-62-7	Calcium cyanamide
592-01-8	Calcium cyanide
26264-06-2	Calcium dodecylbenzenesulfonate
7778-54-3	Calcium hypochlorite
8001-35-2	Camphechlor
8001-35-2	Camphene, octachloro-

CAS Number	Chemical Name
56-25-7	Cantharidin
105-60-2	Caprolactam
133-06-2	Captan
51-83-2	Carbachol chloride
28249-77-6	Carbamic acid, diethylthio-, S-(p-chlorobenzyl)
51-79-6	Carbamic acid, ethyl ester
26419-73-8	Carbamic acid, methyl-,
12427-38-2	Carbamodithioic acid, 1,2-ethanediylbis-,
12122-67-7	Carbamodithioic acid, 1,2-ethanediylbis-, zinc
2303-16-4	Carbamothioic acid,
52888-80-9	Carbamothioic acid, dipropyl-, S-(phenylmethyl)
63-25-2	Carbaryl
10605-21-7	Carbendazim
1563-66-2	Carbofuran
1563-38-8	Carbofuran phenol
75-15-0	Carbon disulfide
353-50-4	Carbonic difluoride
75-44-5	Carbonic dichloride
79-22-1	Carbonochloridic acid, methylester
108-23-6	Carbonochloridic acid, 1-methylethyl ester
109-61-5	Carbonochloridic acid, propylester
463-58-1	Carbon oxide sulfide (COS)
56-23-5	Carbon tetrachloride
463-58-1	Carbonyl sulfide
786-19-6	Carbophenothion
55285-14-8	Carbosulfan
5234-68-4	Carboxin
120-80-9	Catechol
75-69-4	CFC-11
75-71-8	CFC-12
76-14-2	CFC-114
76-15-3	CFC-115
75-72-9	CFC-13
2439-01-2	Chinomethionat
133-90-4	Chloramben
305-03-3	Chlorambucil
57-74-9	Chlordane
	Chlordane (Technical Mixture and Metabolites)
115-28-6	Chlorendic acid
470-90-6	Chlorfenvinfos
90982-32-4	Chlorimuron ethyl
	Chlorinated Benzenes
	Chlorinated Ethanes
	Chlorinated Naphthalene
	Chlorinated Phenols
7782-50-5	Chlorine
10049-04-4	Chlorine dioxide
7791-21-1	Chlorine monoxide
7791-21-1	Chlorine oxide
10049-04-4	Chlorine oxide (ClO2)
24934-91-6	Chlormephos
999-81-5	Chlormequat chloride
494-03-1	Chlornaphazine
107-20-0	Chloroacetaldehyde
79-11-8	Chloroacetic acid
532-27-4	2-Chloroacetophenone
	Chloroalkyl Ethers
4080-31-3	1-(3-Chloroallyl)-3,5,7-triaza-1-azoniaadamantane

Page A-4

ALPHABETICAL LISTING OF CHEMICAL NAME AND CAS NUMBER

CAS Number	Chemical Name
106-47-8	p-Chloroaniline
108-90-7	Chlorobenzene
510-15-6	Chlorobenzilate
66441-23-4	2-(4-((6-Chloro-2-benzoxazolylen)oxy)phenoxy)p
51-75-2	2-Chloro-N-(2-chloroethyl)-N-methylethanamine
59-50-7	p-Chloro-m-cresol
2971-38-2	2,4-D chlorocrotyl ester
124-48-1	Chlorodibromomethane
75-68-3	1-Chloro-1,1-difluoroethane
75-45-6	Chlorodifluoromethane
5902-51-2	5-Chloro-3-(1,1-dimethylethyl)-6-methyl-2,4(1H,3H
75-00-3	Chloroethane
107-07-3	Chloroethanol
627-11-2	Chloroethyl chloroformate
1912-24-9	6-Chloro-N-ethyl-N'-(1-methylethyl)-1,3,5-triazine-
110-75-8	2-Chloroethyl vinyl ether
67-66-3	Chloroform
74-87-3	Chloromethane
64902-72-3	2-Chloro-N-(((4-methoxy-6-methyl-1,3,5-triazin-2-
27314-13-2	4-Chloro-5-(methylamino)-2-[3-(trifluoromethyl)ph
542-88-1	Chloromethyl ether
51630-58-1	4-Chloro-alpha-(1-methylethyl)benzeneacetic acid
1918-16-7	2-Chloro-N-(1-methylethyl)-N-phenylacetamide
107-30-2	Chloromethyl methyl ether
3653-48-3	(4-Chloro-2-methylphenoxy) acetate sodium salt
94-74-6	(4-Chloro-2-methylphenoxy) acetic acid
563-47-3	3-Chloro-2-methyl-1-propene
91-58-7	2-Chloronaphthalene
3691-35-8	Chlorophacinone
95-57-8	2-Chlorophenol
	Chlorophenols
43121-43-3	1-(4-Chlorophenoxy)-3,3-dimethyl-1-(1H-1,2,4-tria
60168-88-9	.alpha.-(2-Chlorophenyl)-.alpha.-4-chlorophenyl)-
104-12-1	p-Chlorophenyl isocyanate
7005-72-3	4-Chlorophenyl phenyl ether
76-06-2	Chloropicrin
126-99-8	Chloroprene
542-76-7	3-Chloropropionitrile
557-98-2	2-Chloropropylene
590-21-6	1-Chloropropylene
76578-14-8	2-(4-((6-Chloro-2-quinoxalinyl)oxy]phenoxy)
7790-94-5	Chlorosulfonic acid
63938-10-3	Chlorotetrafluoroethane
354-25-6	1-Chloro-1,1,2,2-tetrafluoroethane
2837-89-0	2-Chloro-1,1,1,2-tetrafluoroethane
1897-45-6	Chlorothalonil
95-69-2	p-Chloro-o-toluidine
3165-93-3	4-Chloro-o-toluidine, hydrochloride
1929-82-4	2-Chloro-6-(trichloromethyl)pyridine
75-88-7	2-Chloro-1,1,1-trifluoroethane
75-72-9	Chlorotrifluoromethane
62476-59-9	5-(2-Chloro-4-(trifluoromethyl)phenoxy)-2-nitrobe
72178-02-0	5-(2-Chloro-4-(trifluoromethyl)phenoxy)-N-methyl
77501-63-4	5-(2-Chloro-4-(trifluoromethyl)phenoxy)-2-nitro-2-
69409-94-5	N-(2-Chloro-4-(trifluoromethyl)phenyl)-DL-valine(+
460-35-5	3-Chloro-1,1,1-trifluoropropane
68085-85-8	3-(2-Chloro-3,3,3-trifluoro-1-propenyl)-2,2-Dimeth
1982-47-4	Chloroxuron

CAS Number	Chemical Name
2921-88-2	Chlorpyrifos
5598-13-0	Chlorpyrifos methyl
64902-72-3	Chlorsulfuron
21923-23-9	Chlorthiophos
1066-30-4	Chromic acetate
7738-94-5	Chromic acid
11115-74-5	Chromic acid
10025-73-7	Chromic chloride
10101-53-8	Chromic sulfate
7440-47-3	Chromium
	Chromium Compounds
10049-05-5	Chromous chloride
28057-48-9	d-trans-Chrysanthemic acid of d-allethrone
218-01-9	Chrysene
4680-78-8	C.I. Acid Green 3
6459-94-5	C.I. Acid Red 114
569-64-2	C.I. Basic Green 4
989-38-8	C.I. Basic Red 1
1937-37-7	C.I. Direct Black 38
28407-37-6	C.I. Direct Blue 218
2602-46-2	C.I. Direct Blue 6
16071-86-6	C.I. Direct Brown 95
2832-40-8	C.I. Disperse Yellow 3
3761-53-3	C.I. Food Red 5
81-88-9	C.I. Food Red 15
3118-97-6	C.I. Solvent Orange 7
97-56-3	C.I. Solvent Yellow 3
842-07-9	C.I. Solvent Yellow 14
492-80-8	C.I. Solvent Yellow 34
128-66-5	C.I. Vat Yellow 4
7440-48-4	Cobalt
10210-68-1	Cobalt carbonyl
	Cobalt Compounds
62207-76-5	Cobalt,((2,2'-(1,2-ethanediylbis(nitrilomethylidyne))
7789-43-7	Cobaltous bromide
544-18-3	Cobaltous formate
14017-41-5	Cobaltous sulfamate
	Coke Oven Emissions
64-86-8	Colchicine
7440-50-8	Copper
	Copper Compounds
	--Except copper phthalocyanine compounds
	--Except C.I. Pigment Blue 15 (under 313)
	--Except C.I. Pigment Green 7 (under 313)
	--Except C.I. Pigment Green 36 (under 313)
544-92-3	Copper cyanide
56-72-4	Coumaphos
5836-29-3	Coumatetralyl
8001-58-9	Creosote
120-71-8	p-Cresidine
108-39-4	m-Cresol
95-48-7	o-Cresol
106-44-5	p-Cresol
1319-77-3	Cresol (mixed isomers)
535-89-7	Crimidine
4170-30-3	Crotonaldehyde
123-73-9	Crotonaldehyde, (E)-
98-82-8	Cumene

Page A-5

ALPHABETICAL LISTING OF CHEMICAL NAME AND CAS NUMBER

CAS Number	Chemical Name
80-15-9	Cumene hydroperoxide
135-20-6	Cupferron
142-71-2	Cupric acetate
12002-03-8	Cupric acetoarsenite
7447-39-4	Cupric chloride
3251-23-8	Cupric nitrate
5893-66-3	Cupric oxalate
7758-98-7	Cupric sulfate
10380-29-7	Cupric sulfate, ammoniated
815-82-7	Cupric tartrate
21725-46-2	Cyanazine
	Cyanide Compounds
57-12-5	Cyanides (soluble salts and complexes)
460-19-5	Cyanogen
506-68-3	Cyanogen bromide
506-77-4	Cyanogen chloride
506-77-4	Cyanogen chloride ((CN)Cl)
506-78-5	Cyanogen iodide
2636-26-2	Cyanophos
675-14-9	Cyanuric fluoride
1134-23-2	Cycloate
68-76-8	2,5-Cyclohexadiene-1,4-dione,
108-91-8	Cyclohexanamine
110-82-7	Cyclohexane
2556-36-7	1,4-Cyclohexane diisocyanate
58-89-9	Cyclohexane,
108-93-0	Cyclohexanol
108-94-1	Cyclohexanone
66-81-9	Cycloheximide
108-91-8	Cyclohexylamine
131-89-5	2-Cyclohexyl-4,6-dinitrophenol
50-18-0	Cyclophosphamide
75-19-4	Cyclopropane
68359-37-5	Cyfluthrin
68085-85-8	Cyhalothrin
94-75-7	2,4-D
94-75-7	2,4-D Acid
94-80-4	2,4-D butyl ester
94-11-1	2,4-D Esters
94-79-1	2,4-D Esters
94-80-4	2,4-D Esters
1320-18-9	2,4-D Esters
1928-38-7	2,4-D Esters
1928-61-6	2,4-D Esters
1929-73-3	2,4-D Esters
2971-38-2	2,4-D Esters
25168-26-7	2,4-D Esters
53467-11-1	2,4-D Esters
94-11-1	2,4-D isopropyl ester
1320-18-9	2,4-D propylene glycol butyl ether ester
94-75-7	2,4-D, salts and esters
20830-81-3	Daunomycin
533-74-4	Dazomet
53404-60-7	Dazomet, sodium salt
94-82-6	2,4-DB
96-12-8	DBCP
72-54-8	DDD
72-55-9	DDE

CAS Number	Chemical Name
3547-04-4	DDE
50-29-3	DDT
	DDT and Metabolites
17702-41-9	Decaborane(14)
1163-19-5	Decabromodiphenyl oxide
78-48-8	DEF
117-81-7	DEHP
8065-48-3	Demeton
919-86-8	Demeton-S-methyl
13684-56-5	Desmedipham
1928-43-4	2,4-D 2-ethylhexyl ester
53404-37-8	2,4-D 2-ethyl-4-methylpentyl ester
10311-84-9	Dialifor
2303-16-4	Diallate
615-05-4	2,4-Diaminoanisole
39156-41-7	2,4-Diaminoanisole sulfate
101-80-4	4,4'-Diaminodiphenyl ether
496-72-0	Diaminotoluene
823-40-5	Diaminotoluene
95-80-7	2,4-Diaminotoluene
25376-45-8	Diaminotoluene (mixed isomers)
20325-40-0	o-Dianisidine dihydrochloride
111984-09-9	o-Dianisidine hydrochloride
333-41-5	Diazinon
334-88-3	Diazomethane
226-36-8	Dibenz(a,h)acridine
224-42-0	Dibenz(a,j)acridine
53-70-3	Dibenz[a,h]anthracene
194-59-2	7H-Dibenzo(c,g)carbazole
5385-75-1	Dibenzo(a,e)fluoranthene
132-64-9	Dibenzofuran
192-65-4	Dibenzo(a,e)pyrene
189-64-0	Dibenzo(a,h)pyrene
191-30-0	Dibenzo(a,l)pyrene
189-55-9	Dibenz[a,i]pyrene
19287-45-7	Diborane
19287-45-7	Diborane(6)
96-12-8	1,2-Dibromo-3-chloropropane
106-93-4	1,2-Dibromoethane
1689-84-5	3,5-Dibromo-4-hydroxybenzonitrile
10222-01-2	2,2-Dibromo-3-nitrilopropionamide
124-73-2	Dibromotetrafluoroethane
84-74-2	Dibutyl phthalate
1918-00-9	Dicamba
1194-65-6	Dichlobenil
117-80-6	Dichlone
99-30-9	Dichloran
95-50-1	o-Dichlorobenzene
25321-22-6	Dichlorobenzene
95-50-1	1,2-Dichlorobenzene
541-73-1	1,3-Dichlorobenzene
106-46-7	1,4-Dichlorobenzene
25321-22-6	Dichlorobenzene (mixed isomers)
91-94-1	3,3'-Dichlorobenzidine
	Dichlorobenzidine
612-83-9	3,3'-Dichlorobenzidine dihydrochloride
64969-34-2	3,3'-Dichlorobenzidine sulfate
75-27-4	Dichlorobromomethane

C.48

ALPHABETICAL LISTING OF CHEMICAL NAME AND CAS NUMBER

CAS Number	Chemical Name
110-57-6	trans-1,4-Dichloro-2-butene
110-57-6	trans-1,4-Dichlorobutene
764-41-0	1,4-Dichloro-2-butene
101-05-3	4,6-Dichloro-N-(2-chlorophenyl)-1,3,5-triazin-2-am
1649-08-7	1,2-Dichloro-1,1-difluoroethane
75-71-8	Dichlorodifluoromethane
75-34-3	1,1-Dichloroethane
107-06-2	1,2-Dichloroethane
52645-53-1	3-(2,2-Dichloroethenyl)-2,2-dimethylcyclopropane
68359-37-5	3-(2,2-Dichloroethenyl)-2,2-dimethylcyclopropane
75-35-4	1,1-Dichloroethylene
156-60-5	1,2-Dichloroethylene
540-59-0	1,2-Dichloroethylene
111-44-4	Dichloroethyl ether
1717-00-6	1,1-Dichloro-1-fluoroethane
75-43-4	Dichlorofluoromethane
108-60-1	Dichloroisopropyl ether
75-09-2	Dichloromethane
1918-00-9	3,6-Dichloro-2-methoxybenzoic acid
1982-69-0	3,6-Dichloro-2-methoxybenzoic acid, sodium salt
542-88-1	Dichloromethyl ether
19666-30-9	3-(2,4-Dichloro-5-(1-methylethoxy)phenyl)-5-(1,1-
149-74-6	Dichloromethylphenylsilane
99-30-9	2,6-Dichloro-4-nitroaniline
127564-92-5	Dichloropentafluoropropane
128903-21-9	2,2-Dichloro-1,1,1,3,3-pentafluoropropane
422-48-0	2,3-Dichloro-1,1,1,2,3-pentafluoropropane
422-44-6	1,2-Dichloro-1,1,2,3,3-pentafluoropropane
422-56-0	3,3-Dichloro-1,1,1,2,2-pentafluoropropane
507-55-1	1,3-Dichloro-1,1,2,2,3-pentafluoropropane
13474-88-9	1,1-Dichloro-1,2,2,3,3-pentafluoropropane
431-86-7	1,2-Dichloro-1,1,3,3,3-pentafluoropropane
136013-79-1	1,3-Dichloro-1,1,2,3,3-pentafluoropropane
111512-56-2	1,1-Dichloro-1,2,3,3,3-pentafluoropropane
97-23-4	Dichlorophene
87-65-0	2,6-Dichlorophenol
120-83-2	2,4-Dichlorophenol
51338-27-3	2-(4-(2,4-Dichlorophenoxy)phenoxy)propanoic
696-28-6	Dichlorophenylarsine
50471-44-8	3-(3,5-Dichlorophenyl)-5-ethenyl-5-methyl-2,4-ox
20354-26-1	2-(3,4-Dichlorophenyl)-4-methyl-1,2,4-oxadiazolidi
709-98-8	N-(3,4-Dichlorophenyl)propanamide
35554-44-0	1-(2-(2,4-Dichlorophenyl)-2-(2-propenyloxy)ethyl)
60207-90-1	1-(2-(2,4-Dichlorophenyl)-4-propyl-1,3-dioxolan-2
26638-19-7	Dichloropropane
8003-19-8	Dichloropropane - Dichloropropene (mixture)
78-99-9	1,1-Dichloropropane
78-87-5	1,2-Dichloropropane
142-28-9	1,3-Dichloropropane
26952-23-8	Dichloropropene
542-75-6	1,3-Dichloropropene
10061-02-6	trans-1,3-Dichloropropene
78-88-6	2,3-Dichloropropene
75-99-0	2,2-Dichloropropionic acid
542-75-6	1,3-Dichloropropylene
4109-96-0	Dichlorosilane
76-14-2	Dichlorotetrafluoroethane
34077-87-7	Dichlorotrifluoroethane

CAS Number	Chemical Name
90454-18-5	Dichloro-1,1,2-trifluoroethane
812-04-4	1,1-Dichloro-1,2,2-trifluoroethane
354-23-4	1,2-Dichloro-1,1,2-trifluoroethane
306-83-2	2,2-Dichloro-1,1,1-trifluoroethane
62-73-7	Dichlorvos
51338-27-3	Diclofop methyl
115-32-2	Dicofol
141-66-2	Dicrotophos
77-73-6	Dicyclopentadiene
60-57-1	Dieldrin
1464-53-5	Diepoxybutane
111-42-2	Diethanolamine
38727-55-8	Diethatyl ethyl
109-89-7	Diethylamine
29232-93-7	O-(2-(Diethylamino)-6-methyl-4-pyrimidinyl)-O,O-di
91-66-7	N,N-Diethylaniline
692-42-2	Diethylarsine
814-49-3	Diethyl chlorophosphate
134190-37-7	Diethyldiisocyanatobenzene
117-81-7	Di(2-ethylhexyl) phthalate
3288-58-2	O,O-Diethyl S-methyl dithiophosphate
311-45-5	Diethyl-p-nitrophenyl phosphate
84-66-2	Diethyl phthalate
297-97-2	O,O-Diethyl O-pyrazinyl phosphorothioate
56-53-1	Diethylstilbestrol
64-67-5	Diethyl sulfate
35367-38-5	Diflubenzuron
75-37-6	Difluoroethane
71-63-6	Digitoxin
2238-07-5	Diglycidyl ether
101-90-6	Diglycidyl resorcinol ether
20830-75-5	Digoxin
55290-64-7	2,3,-Dihydro-5,6-dimethyl-1,4-dithiin
5234-68-4	5,6-Dihydro-2-methyl-N-phenyl-1,4-oxathiin-3-car
94-58-6	Dihydrosafrole
	Diisocyanates (includes only 20 chemicals)
4128-73-8	4,4'-Diisocyanatodiphenyl ether
75790-87-3	2,4'-Diisocyanatodiphenyl sulfide
55-91-4	Diisopropylfluorophosphate
115-26-4	Dimefox
309-00-2	1,4:5,8-Dimethanonaphthalene,
55290-64-7	Dimethipin
60-51-5	Dimethoate
119-90-4	3,3'-Dimethoxybenzidine
20325-40-0	3,3'-Dimethoxybenzidine dihydrochloride
91-93-0	3,3'-Dimethoxybenzidine-4,4'-diisocyanate
111984-09-9	3,3'-Dimethoxybenzidine hydrochloride
124-40-3	Dimethylamine
2300-66-5	Dimethylamine dicamba
60-11-7	4-Dimethylaminoazobenzene
60-11-7	Dimethylaminoazobenzene
121-69-7	N,N-Dimethylaniline
57-97-6	7,12-Dimethylbenz[a]anthracene
119-93-7	3,3'-Dimethylbenzidine
612-82-8	3,3'-Dimethylbenzidine dihydrochloride
41766-75-0	3,3'-Dimethylbenzidine dihydrofluoride
22781-23-3	2,2-Dimethyl-1,3-benzodioxol-4-ol
79-44-7	Dimethylcarbamyl chloride

ALPHABETICAL LISTING OF CHEMICAL NAME AND CAS NUMBER

CAS Number	Chemical Name
2524-03-0	Dimethyl chlorothiophosphate
75-78-5	Dimethyldichlorosilane
91-97-4	3,3'-Dimethyl-4,4'-diphenylene diisocyanate
139-25-3	3,3'-Dimethyldiphenylmethane-4,4'-diisocyanate
34014-18-1	N-(5-(1,1-Dimethylethyl)-1,3,4-thiadiazol-2-yl)-N,N'
68-12-2	Dimethylformamide
68-12-2	N,N-Dimethylformamide
57-14-7	1,1-Dimethyl hydrazine
57-14-7	Dimethylhydrazine
55-38-9	O,O-Dimethyl O-(3-methyl-4-(methylthio) phenyl)
7696-12-0	2,2-Dimethyl-3-(2-methyl-1-propenyl)cyclopropan
26002-80-2	2,2-Dimethyl-3-(2-methyl-1-propenyl)cyclopropan
105-67-9	2,4-Dimethylphenol
99-98-9	Dimethyl-p-phenylenediamine
2524-03-0	Dimethyl phosphorochloridothioate
131-11-3	Dimethyl phthalate
463-82-1	2,2-Dimethylpropane
77-78-1	Dimethyl sulfate
5598-13-0	O,O-Dimethyl-O-(3,5,6-trichloro-2-pyridyl)phospho
644-64-4	Dimetilan
25154-54-5	Dinitrobenzene (mixed isomers)
99-65-0	m-Dinitrobenzene
528-29-0	o-Dinitrobenzene
100-25-4	p-Dinitrobenzene
88-85-7	Dinitrobutyl phenol
534-52-1	Dinitrocresol
534-52-1	4,6-Dinitro-o-cresol
534-52-1	4,6-Dinitro-o-cresol and salts
25550-58-7	Dinitrophenol
51-28-5	2,4-Dinitrophenol
329-71-5	2,5-Dinitrophenol
573-56-8	2,6-Dinitrophenol
25321-14-6	Dinitrotoluene (mixed isomers)
121-14-2	2,4-Dinitrotoluene
606-20-2	2,6-Dinitrotoluene
610-39-9	3,4-Dinitrotoluene
39300-45-3	Dinocap
88-85-7	Dinoseb
1420-07-1	Dinoterb
117-84-0	n-Dioctylphthalate
117-84-0	Di-n-octyl phthalate
123-91-1	1,4-Dioxane
78-34-2	Dioxathion
82-66-6	Diphacinone
957-51-7	Diphenamid
122-39-4	Diphenylamine
122-66-7	1,2-Diphenylhydrazine
	Diphenylhydrazine
152-16-9	Diphosphoramide, octamethyl-
2164-07-0	Dipotassium endothall
142-84-7	Dipropylamine
19044-88-3	4-(Dipropylamino)-3,5-dinitrobenzenesulfonamide
136-45-8	Dipropyl isocinchomeronate
621-64-7	Di-n-propylnitrosamine
85-00-7	Diquat
2764-72-9	Diquat
138-93-2	Disodium cyanodithioimidocarbonate
298-04-4	Disulfoton

CAS Number	Chemical Name
514-73-8	Dithiazanine iodide
541-53-7	Dithiobiuret
541-53-7	2,4-Dithiobiuret
330-54-1	Diuron
27176-87-0	Dodecylbenzenesulfonic acid
2439-10-3	Dodecylguanidine monoacetate
2439-10-3	Dodine
120-36-5	2,4-DP
2702-72-9	2,4-D sodium salt
316-42-7	Emetine, dihydrochloride
115-29-7	Endosulfan
959-98-8	alpha - Endosulfan
33213-65-9	beta - Endosulfan
	Endosulfan and Metabolites
1031-07-8	Endosulfan sulfate
145-73-3	Endothall
2778-04-3	Endothion
72-20-8	Endrin
7421-93-4	Endrin aldehyde
	Endrin and Metabolites
106-89-8	Epichlorohydrin
51-43-4	Epinephrine
2104-64-5	EPN
759-94-4	EPTC
50-14-6	Ergocalciferol
379-79-3	Ergotamine tartrate
75-04-7	Ethanamine
74-84-0	Ethane
75-00-3	Ethane, chloro-
107-15-3	1,2-Ethanediamine
75-37-6	Ethane, 1,1-difluoro-
460-19-5	Ethanedinitrile
60-29-7	Ethane, 1,1'-oxybis-
79-21-0	Ethaneperoxoic acid
1622-32-8	Ethanesulfonyl chloride, 2-chloro-
630-20-6	Ethane, 1,1,1,2-tetrachloro-
505-60-2	Ethane, 1,1'-thiobis[2-chloro-
75-08-1	Ethanethiol
76-13-1	Ethane, 1,1,2-trichloro-1,2,2,-trifluoro-
30558-43-1	Ethanimidothioic acid,
16752-77-5	Ethanimidothioic acid, N-[[methylamino)carbonyl]
10140-87-1	Ethanol, 1,2-dichloro-, acetate
110-80-5	Ethanol, 2-ethoxy-
5952-26-1	Ethanol, 2,2'-oxybis-, dicarbamate
74-85-1	Ethene
598-73-2	Ethene, bromotrifluoro-
75-01-4	Ethene, chloro-
79-38-9	Ethene, chlorotrifluoro-
75-35-4	Ethene, 1,1-dichloro-
75-38-7	Ethene, 1,1-difluoro-
109-92-2	Ethene, ethoxy-
75-02-5	Ethene, fluoro-
107-25-5	Ethene, methoxy-
116-14-3	Ethene, tetrafluoro-
563-12-2	Ethion
13194-48-4	Ethoprop
13194-48-4	Ethoprophos
110-80-5	2-Ethoxyethanol

ALPHABETICAL LISTING OF CHEMICAL NAME AND CAS NUMBER

CAS Number	Chemical Name
74051-80-2	2-(1-(Ethoxyimino)
25311-71-1	2-((Ethoxyl((1-methylethyl)amino]phosphinothioyl]
141-78-6	Ethyl acetate
107-00-6	Ethyl acetylene
140-88-5	Ethyl acrylate
31218-83-4	3-((Ethylamino)methoxyphosphinothioyl)oxy)-2-bu
100-41-4	Ethylbenzene
538-07-8	Ethylbis(2-chloroethyl)amine
51-79-6	Ethyl carbamate
75-00-3	Ethyl chloride
541-41-3	Ethyl chloroformate
90982-32-4	Ethyl-2-((((4-chloro-6-methoxyprimidin-2-yl)-carbo
107-12-0	Ethyl cyanide
759-94-4	Ethyl dipropylthiocarbamate
74-85-1	Ethylene
	Ethylenebisdithiocarbamic acid, salts and esters
111-54-6	Ethylenebisdithiocarbamic acid, salts & esters
107-15-3	Ethylenediamine
60-00-4	Ethylenediamine-tetraacetic acid (EDTA)
106-93-4	Ethylene dibromide
107-06-2	Ethylene dichloride
371-62-0	Ethylene fluorohydrin
107-21-1	Ethylene glycol
151-56-4	Ethyleneimine
75-21-8	Ethylene oxide
96-45-7	Ethylene thiourea
60-29-7	Ethyl ether
75-34-3	Ethylidene Dichloride
75-08-1	Ethyl mercaptan
97-63-2	Ethyl methacrylate
62-50-0	Ethyl methanesulfonate
834-12-8	N-Ethyl-N'-(1-methylethyl)-6-(methylthio)-1,3,5,-tri
35400-43-2	O-Ethyl O-(4-(methylthio)phenyl)phosphorodithioic
109-95-5	Ethyl nitrite
40487-42-1	N-(1-Ethylpropyl)-3,4-dimethyl-2,6-dinitrobenzena
301-12-2	S-(2-(Ethylsulfinyl)ethyl) O,O-dimethyl ester
542-90-5	Ethylthiocyanate
74-86-2	Ethyne
52-85-7	Famphur
22224-92-6	Fenamiphos
60168-88-9	Fenarimol
13356-08-6	Fenbutatin oxide
66441-23-4	Fenoxaprop ethyl
72490-01-8	Fenoxycarb
39515-41-8	Fenpropathrin
115-90-2	Fensulfothion
55-38-9	Fenthion
51630-58-1	Fenvalerate
14484-64-1	Ferbam
1185-57-5	Ferric ammonium citrate
2944-67-4	Ferric ammonium oxalate
55488-87-4	Ferric ammonium oxalate
7705-08-0	Ferric chloride
7783-50-8	Ferric fluoride
10421-48-4	Ferric nitrate
10028-22-5	Ferric sulfate
10045-89-3	Ferrous ammonium sulfate
7758-94-3	Ferrous chloride

CAS Number	Chemical Name
7720-78-7	Ferrous sulfate
7782-63-0	Ferrous sulfate
	Fine mineral fibers
69806-50-4	Fluazifop butyl
4301-50-2	Fluenetil
2164-17-2	Fluometuron
206-44-0	Fluoranthene
86-73-7	Fluorene
7782-41-4	Fluorine
640-19-7	Fluoroacetamide
144-49-0	Fluoroacetic acid
62-74-8	Fluoroacetic acid, sodium salt
359-06-8	Fluoroacetyl chloride
51-21-8	Fluorouracil
51-21-8	5-Fluorouracil
69409-94-5	Fluvalinate
133-07-3	Folpet
72178-02-0	Fomesafen
944-22-9	Fonofos
50-00-0	Formaldehyde
107-16-4	Formaldehyde cyanohydrin
50-00-0	Formaldehyde (solution)
23422-53-9	Formetanate hydrochloride
64-18-6	Formic acid
107-31-3	Formic acid, methyl ester
2540-82-1	Formothion
17702-57-7	Formparanate
21548-32-3	Fosthietan
76-13-1	Freon 113
3878-19-1	Fuberidazole
110-17-8	Fumaric acid
110-00-9	Furan
109-99-9	Furan, tetrahydro-
98-01-1	Furfural
13450-90-3	Gallium trichloride
765-34-4	Glycidylaldehyde
	Glycol Ethers
70-25-7	Guanidine, N-methyl-N'-nitro-N-nitroso-
86-50-0	Guthion
	Haloethers
	Halomethanes
353-59-3	Halon 1211
75-63-8	Halon 1301
124-73-2	Halon 2402
354-14-3	HCFC-121
354-11-0	HCFC-121a
306-83-2	HCFC-123
354-23-4	HCFC-123a
812-04-4	HCFC-123b
2837-89-0	HCFC-124
354-25-6	HCFC-124a
1649-08-7	HCFC-132b
75-88-7	HCFC-133a
1717-00-6	HCFC-141b
75-68-3	HCFC-142b
75-43-4	HCFC-21
75-45-6	HCFC-22
128903-21-9	HCFC-225aa

Page A-9

ALPHABETICAL LISTING OF CHEMICAL NAME AND CAS NUMBER

CAS Number	Chemical Name
422-48-0	HCFC-225ba
422-44-6	HCFC-225bb
422-56-0	HCFC-225ca
507-55-1	HCFC-225cb
13474-88-9	HCFC-225cc
431-86-7	HCFC-225da
136013-79-1	HCFC-225ea
111512-56-2	HCFC-225eb
460-35-5	HCFC-253fb
76-44-8	Heptachlor
	Heptachlor and Metabolites
1024-57-3	Heptachlor epoxide
76-44-8	1,4,5,6,7,8,8-Heptachloro-3a,4,7,7a-tetrahydro-4,
118-74-1	Hexachlorobenzene
87-68-3	Hexachloro-1,3-butadiene
87-68-3	Hexachlorobutadiene
	Hexachlorocyclohexane (all isomers) CAS
319-84-6	alpha-Hexachlorocyclohexane
58-89-9	Hexachlorocyclohexane (gamma isomer)
77-47-4	Hexachlorocyclopentadiene
67-72-1	Hexachloroethane
1335-87-1	Hexachloronaphthalene
70-30-4	Hexachlorophene
1888-71-7	Hexachloropropene
757-58-4	Hexaethyl tetraphosphate
13356-08-6	Hexakis(2-methyl-2-phenylpropyl)distannoxane
4835-11-4	Hexamethylenediamine, N,N'-dibutyl-
822-06-0	Hexamethylene-1,6-diisocyanate
680-31-9	Hexamethylphosphoramide
110-54-3	Hexane
110-54-3	n-Hexane
51235-04-2	Hexazinone
67485-29-4	Hydramethylnon
302-01-2	Hydrazine
1615-80-1	Hydrazine, 1,2-diethyl-
57-14-7	Hydrazine, 1,1-dimethyl-
540-73-8	Hydrazine, 1,2-dimethyl-
122-66-7	Hydrazine, 1,2-diphenyl-
60-34-4	Hydrazine, methyl-
10034-93-2	Hydrazine sulfate
122-66-7	Hydrazobenzene
7647-01-0	Hydrochloric acid (conc 37% or greater)
7647-01-0	Hydrochloric acid
7647-01-0	Hydrochloric acid (aerosol forms only)
74-90-8	Hydrocyanic acid
7664-39-3	Hydrofluoric acid
7664-39-3	Hydrofluoric acid (conc. 50% or greater)
1333-74-0	Hydrogen
7647-01-0	Hydrogen chloride (anhydrous)
7647-01-0	Hydrogen chloride (gas only)
74-90-8	Hydrogen cyanide
7664-39-3	Hydrogen fluoride
7664-39-3	Hydrogen fluoride (anhydrous)
7722-84-1	Hydrogen peroxide (Conc.> 52%)
7783-07-5	Hydrogen selenide
7783-06-4	Hydrogen sulfide
80-15-9	Hydroperoxide, 1-methyl-1-phenylethyl-
123-31-9	Hydroquinone

CAS Number	Chemical Name
35554-44-0	Imazalil
193-39-5	Indeno(1,2,3-cd)pyrene
55406-53-6	3-Iodo-2-propynyl butylcarbamate
13463-40-6	Iron carbonyl (Fe(CO)5), (TB-5-11)-
13463-40-6	Iron, pentacarbonyl-
297-78-9	Isobenzan
75-28-5	Isobutane
78-83-1	Isobutyl alcohol
78-84-2	Isobutyraldehyde
78-82-0	Isobutyronitrile
102-36-3	Isocyanic acid, 3,4-dichlorophenyl ester
465-73-6	Isodrin
25311-71-1	Isofenphos
55-91-4	Isofluorphate
133-06-2	1H-Isoindole-1,3(2H)-dione,
78-78-4	Isopentane
78-59-1	Isophorone
4098-71-9	Isophorone diisocyanate
78-79-5	Isoprene
42504-46-1	Isopropanolamine dodecylbenzene sulfonate
67-63-0	Isopropyl alcohol (mfg-strong acid process)
75-31-0	Isopropylamine
75-29-6	Isopropyl chloride
108-23-6	Isopropyl chloroformate
80-05-7	4,4'-Isopropylidenediphenol
119-38-0	Isopropylmethylpyrazolyl dimethylcarbamate
120-58-1	Isosafrole
556-61-6	Isothiocyanatomethane
143-50-0	Kepone
77501-63-4	Lactofen
78-97-7	Lactonitrile
303-34-4	Lasiocarpine
7439-92-1	Lead
301-04-2	Lead acetate
7645-25-2	Lead arsenate
7784-40-9	Lead arsenate
10102-48-4	Lead arsenate
7758-95-4	Lead chloride
	Lead Compounds
13814-96-5	Lead fluoborate
7783-46-2	Lead fluoride
10101-63-0	Lead iodide
10099-74-8	Lead nitrate
7446-27-7	Lead phosphate
1072-35-1	Lead stearate
7428-48-0	Lead stearate
52652-59-2	Lead stearate
56189-09-4	Lead stearate
1335-32-6	Lead subacetate
7446-14-2	Lead sulfate
15739-80-7	Lead sulfate
1314-87-0	Lead sulfide
592-87-0	Lead thiocyanate
21609-90-5	Leptophos
541-25-3	Lewisite
58-89-9	Lindane
330-55-2	Linuron
554-13-2	Lithium carbonate

ALPHABETICAL LISTING OF CHEMICAL NAME AND CAS NUMBER

CAS Number	Chemical Name
14307-35-8	Lithium chromate
7580-67-8	Lithium hydride
121-75-5	Malathion
110-16-7	Maleic acid
108-31-6	Maleic anhydride
123-33-1	Maleic hydrazide
109-77-3	Malononitrile
12427-38-2	Maneb
7439-96-5	Manganese
15339-36-3	Manganese, bis(dimethylcarbamodithioato-S,S')- Manganese Compounds
12108-13-3	Manganese, tricarbonyl methylcyclopentadienyl
101-14-4	MBOCA
149-30-4	MBT
94-74-6	MCPA
101-68-8	MDI
51-75-2	Mechlorethamine
93-65-2	Mecoprop
148-82-3	Melphalan
950-10-7	Mephosfolan
149-30-4	2-Mercaptobenzothiazole
2032-65-7	Mercaptodimethur
1600-27-7	Mercuric acetate
7487-94-7	Mercuric chloride
592-04-1	Mercuric cyanide
10045-94-0	Mercuric nitrate
21908-53-2	Mercuric oxide
7783-35-9	Mercuric sulfate
592-85-8	Mercuric thiocyanate
7782-86-7	Mercurous nitrate
10415-75-5	Mercurous nitrate
7439-97-6	Mercury
	Mercury Compounds
628-86-4	Mercury fulminate
150-50-5	Merphos
10476-95-6	Methacrolein diacetate
760-93-0	Methacrylic anhydride
126-98-7	Methacrylonitrile
920-46-7	Methacryloyl chloride
30674-80-7	Methacryloyloxyethyl isocyanate
10265-92-6	Methamidophos
137-42-8	Metham sodium
74-89-5	Methanamine
75-50-3	Methanamine, N,N-dimethyl-
124-40-3	Methanamine, N-methyl-
62-75-9	Methanamine, N-methyl-N-nitroso-
74-82-8	Methane
74-87-3	Methane, chloro-
107-30-2	Methane, chloromethoxy-
624-83-9	Methane, isocyanato-
115-10-6	Methane, oxybis-
542-88-1	Methane, oxybis[chloro-
594-42-3	Methanesulfenyl chloride, trichloro-
558-25-8	Methanesulfonyl fluoride
509-14-8	Methane, tetranitro-
74-93-1	Methanethiol
67-66-3	Methane, trichloro-
57-74-9	4,7-Methanoindan,

CAS Number	Chemical Name
67-56-1	Methanol
91-80-5	Methapyrilene
20354-26-1	Methazole
950-37-8	Methidathion
2032-65-7	Methiocarb
16752-77-5	Methomyl
94-74-6	Methoxone
3653-48-3	Methoxone sodium salt
72-43-5	Methoxychlor
109-86-4	2-Methoxyethanol
151-38-2	Methoxyethylmercuric acetate
101200-48-0	2-(4-Methoxy-6-methyl-1,3,5-triazin-2-yl)-methyla
96-33-3	Methyl acrylate
74-83-9	Methyl bromide
563-46-2	2-Methyl-1-butene
563-45-1	3-Methyl-1-butene
74-87-3	Methyl chloride
80-63-7	Methyl 2-chloroacrylate
79-22-1	Methyl chlorocarbonate
71-55-6	Methyl chloroform
79-22-1	Methyl chloroformate
56-49-5	3-Methylcholanthrene
3697-24-3	5-Methylchrysene
75790-84-0	4-Methyldiphenylmethane-3,4-diisocyanate
2439-01-2	6-Methyl-1,3-dithiolo[4,5-b]quinoxalin-2-one
101-14-4	4,4'-Methylenebis(2-chloroaniline)
97-23-4	2,2'-Methylenebis(4-chlorophenol)
101-61-1	4,4'-Methylenebis(N,N-dimethyl)benzenamine
5124-30-1	1,1'-Methylene bis(4-isocyanatocyclohexane)
101-68-8	Methylenebis(phenylisocyanate)
74-95-3	Methylene bromide
75-09-2	Methylene chloride
101-77-9	4,4'-Methylenedianiline
115-10-6	Methyl ether
78-93-3	Methyl ethyl ketone
78-93-3	Methyl ethyl ketone (MEK)
1338-23-4	Methyl ethyl ketone peroxide
107-31-3	Methyl formate
60-34-4	Methyl hydrazine
74-88-4	Methyl iodide
108-10-1	Methyl isobutyl ketone
624-83-9	Methyl isocyanate
556-61-6	Methyl isothiocyanate
75-86-5	2-Methyllactonitrile
74-93-1	Methyl mercaptan
502-39-6	Methylmercuric dicyanamide
80-62-6	Methyl methacrylate
924-42-5	N-Methylolacrylamide
298-00-0	Methyl parathion
3735-23-7	Methyl phenkapton
676-97-1	Methyl phosphonic dichloride
115-11-7	2-Methylpropene
109-06-8	2-Methylpyridine
872-50-4	N-Methyl-2-pyrrolidone
1634-04-4	Methyl tert-butyl ether
556-64-9	Methyl thiocyanate
56-04-2	Methylthiouracil
75-79-6	Methyltrichlorosilane

ALPHABETICAL LISTING OF CHEMICAL NAME AND CAS NUMBER

CAS Number	Chemical Name
78-94-4	Methyl vinyl ketone
9006-42-2	Metiram
1129-41-5	Metolcarb
21087-64-9	Metribuzin
7786-34-7	Mevinphos
315-18-4	Mexacarbate
90-94-8	Michler's ketone
50-07-7	Mitomycin C
2212-67-1	Molinate
1313-27-5	Molybdenum trioxide
76-15-3	Monochloropentafluoroethane
6923-22-4	Monocrotophos
75-04-7	Monoethylamine
74-89-5	Monomethylamine
150-68-5	Monuron
2763-96-4	Muscimol
505-60-2	Mustard gas
88671-89-0	Myclobutanil
142-59-6	Nabam
300-76-5	Naled
91-20-3	Naphthalene
3173-72-6	1,5-Naphthalene diisocyanate
63-25-2	1-Naphthalenol, methylcarbamate
1338-24-5	Naphthenic acid
130-15-4	1,4-Naphthoquinone
134-32-7	alpha-Naphthylamine
91-59-8	beta-Naphthylamine
7440-02-0	Nickel
15699-18-0	Nickel ammonium sulfate
13463-39-3	Nickel carbonyl
7718-54-9	Nickel chloride
37211-05-5	Nickel chloride
	Nickel Compounds
557-19-7	Nickel cyanide
12054-48-7	Nickel hydroxide
14216-75-2	Nickel nitrate
7786-81-4	Nickel sulfate
54-11-5	Nicotine
54-11-5	Nicotine and salts
	Nicotine and salts
65-30-5	Nicotine sulfate
1929-82-4	Nitrapyrin
	Nitrate compounds (water dissociable)
7697-37-2	Nitric acid (conc 80% or greater)
7697-37-2	Nitric acid
10102-43-9	Nitric oxide
139-13-9	Nitrilotriacetic acid
100-01-6	p-Nitroaniline
99-59-2	5-Nitro-o-anisidine
98-95-3	Nitrobenzene
92-93-3	4-Nitrobiphenyl
1122-60-7	Nitrocyclohexane
1836-75-5	Nitrofen
10102-44-0	Nitrogen dioxide
10544-72-6	Nitrogen dioxide
51-75-2	Nitrogen mustard
10102-43-9	Nitrogen oxide (NO)
55-63-0	Nitroglycerin

CAS Number	Chemical Name
25154-55-6	Nitrophenol (mixed isomers)
554-84-7	m-Nitrophenol
100-02-7	p-Nitrophenol
88-75-5	2-Nitrophenol
100-02-7	4-Nitrophenol
	Nitrophenols
79-46-9	2-Nitropropane
5522-43-0	1-Nitropyrene
	Nitrosamines
924-16-3	N-Nitrosodi-n-butylamine
1116-54-7	N-Nitrosodiethanolamine
55-18-5	N-Nitrosodiethylamine
62-75-9	N-Nitrosodimethylamine
62-75-9	Nitrosodimethylamine
86-30-6	N-Nitrosodiphenylamine
156-10-5	p-Nitrosodiphenylamine
621-64-7	N-Nitrosodi-n-propylamine
759-73-9	N-Nitroso-N-ethylurea
684-93-5	N-Nitroso-N-methylurea
615-53-2	N-Nitroso-N-methylurethane
4549-40-0	N-Nitrosomethylvinylamine
59-89-2	N-Nitrosomorpholine
16543-55-8	N-Nitrosonornicotine
100-75-4	N-Nitrosopiperidine
930-55-2	N-Nitrosopyrrolidine
1321-12-6	Nitrotoluene
99-08-1	m-Nitrotoluene
88-72-2	o-Nitrotoluene
99-99-0	p-Nitrotoluene
99-55-8	5-Nitro-o-toluidine
109-95-5	Nitrous acid, ethyl ester
991-42-4	Norbormide
27314-13-2	Norflurazon
2234-13-1	Octachloronaphthalene
1689-99-2	Octanoic acid, 2,6-dibromo-4-cyanophenyl ester
8014-95-7	Oleum (fuming sulfuric acid)
888888-88-8	Organorhodium Complex (PMN-82-147)
19044-88-3	Oryzalin
20816-12-0	Osmium oxide OsO4 (T-4)-
20816-12-0	Osmium tetroxide
630-60-4	Ouabain
2164-07-0	7-Oxabicyclo(2.2.1)heptane-2,3-dicarboxylic acid,
23135-22-0	Oxamyl
78-71-7	Oxetane, 3,3-bis(chloromethyl)-
75-21-8	Oxirane
106-89-8	Oxirane, (chloromethyl)-
75-56-9	Oxirane, methyl-
301-12-2	Oxydemeton methyl
19666-30-9	Oxydiazon
2497-07-6	Oxydisulfoton
42874-03-3	Oxyfluorfen
10028-15-6	Ozone
30525-89-4	Paraformaldehyde
123-63-7	Paraldehyde
1910-42-5	Paraquat dichloride
2074-50-2	Paraquat methosulfate
56-38-2	Parathion
298-00-0	Parathion-methyl

ALPHABETICAL LISTING OF CHEMICAL NAME AND CAS NUMBER

CAS Number	Chemical Name
12002-03-8	Paris green
1336-36-3	PCBs
82-68-8	PCNB
87-86-5	PCP
1114-71-2	Pebulate
40487-42-1	Pendimethalin
19624-22-7	Pentaborane
608-93-5	Pentachlorobenzene
76-01-7	Pentachloroethane
82-68-8	Pentachloronitrobenzene
87-86-5	Pentachlorophenol
2570-26-5	Pentadecylamine
504-60-9	1,3-Pentadiene
109-66-0	Pentane
109-67-1	1-Pentene
646-04-8	2-Pentene, (E)-
627-20-3	2-Pentene, (Z)-
57-33-0	Pentobarbital sodium
79-21-0	Peracetic acid
127-18-4	Perchloroethylene
594-42-3	Perchloromethyl mercaptan
52645-53-1	Permethrin
62-44-2	Phenacetin
85-01-8	Phenanthrene
108-95-2	Phenol
114-26-1	Phenol, 2-(1-methylethoxy)-, methylcarbamate
64-00-6	Phenol, 3-(1-methylethyl)-, methylcarbamate
4418-66-0	Phenol, 2,2'-thiobis[4-chloro-6-methyl-
26002-80-2	Phenothrin
58-36-6	Phenoxarsine, 10,10'-oxydi-
72490-01-8	(2-(4-Phenoxy-phenoxy)-ethyl)carbamic acid
696-28-6	Phenyl dichloroarsine
23564-06-9	(1,2-Phenylenebis(iminocarbonothioyl))
95-54-5	1,2-Phenylenediamine
106-50-3	p-Phenylenediamine
108-45-2	1,3-Phenylenediamine
615-28-1	1,2-Phenylenediamine dihydrochloride
624-18-0	1,4-Phenylenediamine dihydrochloride
104-49-4	1,4-Phenylene diisocyanate
123-61-5	1,3-Phenylene diisocyanate
59-88-1	Phenylhydrazine hydrochloride
62-38-4	Phenylmercuric acetate
62-38-4	Phenylmercury acetate
10453-86-8	5-(Phenylmethyl)-3-furanyl)methyl
90-43-7	2-Phenylphenol
2097-19-0	Phenylsilatrane
103-85-5	Phenylthiourea
57-41-0	Phenytoin
298-02-2	Phorate
4104-14-7	Phosacetim
947-02-4	Phosfolan
75-44-5	Phosgene
732-11-6	Phosmet
13171-21-6	Phosphamidon
7803-51-2	Phosphine
52-68-6	Phosphonic
2703-13-1	Phosphonothioic acid, methyl-, O-ethyl
50782-69-9	Phosphonothioic acid, methyl-,

CAS Number	Chemical Name
2665-30-7	Phosphonothioic acid, methyl-, O-(4-nitrophenyl)
7664-38-2	Phosphoric acid
961-11-5	Phosphoric acid,2-chloro-1-(2,3,5-trichlorophenyl)
62-73-7	Phosphoric acid,2-dichloroethenyl dimethyl ester
3254-63-5	Phosphoric acid,dimethyl 4-(methylthio) phenyl
13194-48-4	Phosphorodithioic acid O-ethyl S,S-dipropyl ester
56-38-2	Phosphorothioic
2587-90-8	Phosphorothioic
7719-12-2	Phosphorous trichloride
7723-14-0	Phosphorus
7723-14-0	Phosphorus (yellow or white)
10025-87-3	Phosphorus oxychloride
10026-13-8	Phosphorus pentachloride
7719-12-2	Phosphorus trichloride
10025-87-3	Phosphoryl chloride
	Phthalate Esters
85-44-9	Phthalic anhydride
57-47-6	Physostigmine
57-64-7	Physostigmine, salicylate (1:1)
1918-02-1	Picloram
109-06-8	2-Picoline
88-89-1	Picric acid
124-87-8	Picrotoxin
26644-46-2	N,N'-(1,4-Piperazinediylbis(2,2,2-trichloroethylider
110-89-4	Piperidine
51-03-6	Piperonyl butoxide
23505-41-1	Pirimifos-ethyl
29232-93-7	Pirimiphos methyl
75-74-1	Plumbane, tetramethyl-
	Polybrominated Biphenyls (PBBs)
	Polychlorinated alkanes (C10 to C13)
1336-36-3	Polychlorinated biphenyls
	Polycyclic aromatic compounds (includes only 19
	Polycyclic organic matter
9016-87-9	Polymeric diphenylmethane diisocyanate
	Polynuclear Aromatic Hydrocarbons
7784-41-0	Potassium arsenate
10124-50-2	Potassium arsenite
7778-50-9	Potassium bichromate
7758-01-2	Potassium bromate
7789-00-6	Potassium chromate
151-50-8	Potassium cyanide
128-03-0	Potassium dimethyldithiocarbamate
1310-58-3	Potassium hydroxide
137-41-7	Potassium N-methyldithiocarbamate
7722-64-7	Potassium permanganate
506-61-6	Potassium silver cyanide
41198-08-7	Profenofos
2631-37-0	Promecarb
7287-19-6	Prometryn
23950-58-5	Pronamide
1918-16-7	Propachlor
463-49-0	Propadiene
463-49-0	1,2-Propadiene
75-31-0	2-Propanamine
74-98-6	Propane
75-29-6	Propane, 2-chloro-
78-87-5	Propane 1,2-dichloro-

Page A-13

ALPHABETICAL LISTING OF CHEMICAL NAME AND CAS NUMBER

CAS Number	Chemical Name
463-82-1	Propane, 2,2-dimethyl-
75-28-5	Propane, 2-methyl
107-12-0	Propanenitrile
78-82-0	Propanenitrile, 2-methyl-
1120-71-4	Propane sultone
1120-71-4	1,3-Propane sultone
709-98-8	Propanil
2312-35-8	Propargite
107-19-7	Propargyl alcohol
106-96-7	Propargyl bromide
107-02-8	2-Propenal
107-11-9	2-Propen-1-amine
115-07-1	Propene
115-07-1	1-Propene
590-21-6	1-Propene, 1-chloro-
557-98-2	1-Propene, 2-chloro-
115-11-7	1-Propene, 2-methyl-
107-13-1	2-Propenenitrile
126-98-7	2-Propenenitrile, 2-methyl-
107-18-6	2-Propen-1-ol
814-68-6	2-Propenoyl chloride
31218-83-4	Propetamphos
122-42-9	Propham
60207-90-1	Propiconazole
57-57-8	beta-Propiolactone
123-38-6	Propionaldehyde
79-09-4	Propionic acid
123-62-6	Propionic anhydride
107-12-0	Propionitrile
542-76-7	Propionitrile, 3-chloro-
70-69-9	Propiophenone, 4'-amino
114-26-1	Propoxur
107-10-8	n-Propylamine
109-61-5	Propyl chloroformate
115-07-1	Propylene
75-55-8	Propyleneimine
75-56-9	Propylene oxide
74-99-7	Propyne
74-99-7	1-Propyne
2275-18-5	Prothoate
129-00-0	Pyrene
121-21-1	Pyrethrins
121-29-9	Pyrethrins
8003-34-7	Pyrethrins
110-86-1	Pyridine
504-24-5	Pyridine, 4-amino-
54-11-5	Pyridine, 3-(1-methyl-2-pyrrolidinyl)-,(S)-
140-76-1	Pyridine, 2-methyl-5-vinyl-
1124-33-0	Pyridine, 4-nitro-, 1-oxide
53404-19-6	2,4-(1H,3H)-Pyrimidinedione,
53558-25-1	Pyriminil
91-22-5	Quinoline
106-51-4	Quinone
82-68-8	Quintozene
76578-14-8	Quizalofop-ethyl
50-55-5	Reserpine
10453-86-8	Resmethrin
108-46-3	Resorcinol

CAS Number	Chemical Name
81-07-2	Saccharin (manufacturing)
81-07-2	Saccharin and salts
94-59-7	Safrole
14167-18-1	Salcomine
107-44-8	Sarin
7783-00-8	Selenious acid
12039-52-0	Selenious acid, dithallium(1+) salt
7782-49-2	Selenium
	Selenium Compounds
7446-08-4	Selenium dioxide
7791-23-3	Selenium oxychloride
7488-56-4	Selenium sulfide
630-10-4	Selenourea
563-41-7	Semicarbazide hydrochloride
74051-80-2	Sethoxydim
7803-62-5	Silane
3037-72-7	Silane, (4-aminobutyl)diethoxymethyl-
75-77-4	Silane, chlorotrimethyl-
4109-96-0	Silane, dichloro-
75-78-5	Silane, dichlorodimethyl-
75-76-3	Silane, tetramethyl-
10025-78-2	Silane, trichloro-
75-79-6	Silane, trichloromethyl-
7440-22-4	Silver
	Silver Compounds
506-64-9	Silver cyanide
7761-88-8	Silver nitrate
93-72-1	Silvex (2,4,5-TP)
122-34-9	Simazine
7440-23-5	Sodium
7631-89-2	Sodium arsenate
7784-46-5	Sodium arsenite
26628-22-8	Sodium azide (Na(N3))
10588-01-9	Sodium bichromate
1333-83-1	Sodium bifluoride
7631-90-5	Sodium bisulfite
124-65-2	Sodium cacodylate
7775-11-3	Sodium chromate
143-33-9	Sodium cyanide (Na(CN))
1982-69-0	Sodium dicamba
128-04-1	Sodium dimethyldithiocarbamate
25155-30-0	Sodium dodecylbenzenesulfonate
7681-49-4	Sodium fluoride
62-74-8	Sodium fluoroacetate
16721-80-5	Sodium hydrosulfide
1310-73-2	Sodium hydroxide
7681-52-9	Sodium hypochlorite
10022-70-5	Sodium hypochlorite
124-41-4	Sodium methylate
137-42-8	Sodium methyldithiocarbamate
7632-00-0	Sodium nitrite
131-52-2	Sodium pentachlorophenate
132-27-4	Sodium o-phenylphenoxide
7558-79-4	Sodium phosphate, dibasic
10039-32-4	Sodium phosphate, dibasic
10140-65-5	Sodium phosphate, dibasic
7601-54-9	Sodium phosphate, tribasic
7758-29-4	Sodium phosphate, tribasic

ALPHABETICAL LISTING OF CHEMICAL NAME AND CAS NUMBER

CAS Number	Chemical Name
7785-84-4	Sodium phosphate, tribasic
10101-89-0	Sodium phosphate, tribasic
10124-56-8	Sodium phosphate, tribasic
10361-89-4	Sodium phosphate, tribasic
13410-01-0	Sodium selenate
7782-82-3	Sodium selenite
10102-18-8	Sodium selenite
10102-20-2	Sodium tellurite
900-95-8	Stannane, acetoxytriphenyl-
18883-66-4	Streptozotocin
7789-06-2	Strontium chromate
57-24-9	Strychnine
	Strychnine and salts
57-24-9	Strychnine, and salts
60-41-3	Strychnine, sulfate
100-42-5	Styrene
96-09-3	Styrene oxide
3689-24-5	Sulfotep
3569-57-1	Sulfoxide, 3-chloropropyl octyl
7446-09-5	Sulfur dioxide
7446-09-5	Sulfur dioxide (anhydrous)
7783-60-0	Sulfur fluoride (SF4), (T-4)-
7664-93-9	Sulfuric acid
7664-93-9	Sulfuric acid (aerosol forms only)
8014-95-7	Sulfuric acid (fuming)
8014-95-7	Sulfuric acid, mixture with sulfur trioxide
12771-08-3	Sulfur monochloride
1314-80-3	Sulfur phosphide
7783-60-0	Sulfur tetrafluoride
7446-11-9	Sulfur trioxide
2699-79-8	Sulfuryl fluoride
35400-43-2	Sulprofos
93-76-5	2,4,5-T acid
1319-72-8	2,4,5-T amines
2008-46-0	2,4,5-T amines
3813-14-7	2,4,5-T amines
6369-96-6	2,4,5-T amines
6369-97-7	2,4,5-T amines
93-79-8	2,4,5-T esters
1928-47-8	2,4,5-T esters
2545-59-7	2,4,5-T esters
25168-15-4	2,4,5-T esters
61792-07-2	2,4,5-T esters
13560-99-1	2,4,5-T salts
77-81-6	Tabun
34014-18-1	Tebuthiuron
7783-80-4	Tellurium hexafluoride
3383-96-8	Temephos
107-49-3	TEPP
5902-51-2	Terbacil
13071-79-9	Terbufos
95-94-3	1,2,4,5-Tetrachlorobenzene
1746-01-6	2,3,7,8-Tetrachlorodibenzo-p-dioxin (TCDD)
79-34-5	1,1,2,2-Tetrachloroethane
630-20-6	1,1,1,2-Tetrachloroethane
127-18-4	Tetrachloroethylene
354-14-3	1,1,2,2-Tetrachloro-1-fluoroethane
354-11-0	1,1,1,2-Tetrachloro-2-fluoroethane

CAS Number	Chemical Name
58-90-2	2,3,4,6-Tetrachlorophenol
961-11-5	Tetrachlorvinphos
64-75-5	Tetracycline hydrochloride
3689-24-5	Tetraethyldithiopyrophosphate
78-00-2	Tetraethyl lead
107-49-3	Tetraethyl pyrophosphate
597-64-8	Tetraethyltin
116-14-3	Tetrafluoroethylene
67485-29-4	Tetrahydro-5,5-dimethyl-2(1H)-pyrimidinone(3-(4-(
533-74-4	Tetrahydro-3,5-dimethyl-2H-1,3,5-thiadiazine-2-thi
53404-60-7	Tetrahydro-3,5-dimethyl-2H-1,3,5-thiadiazine-2-thi
7696-12-0	Tetramethrin
39515-41-8	2,2,3,3-Tetramethylcyclopropane carboxylic acid
75-74-1	Tetramethyllead
75-76-3	Tetramethylsilane
509-14-8	Tetranitromethane
1314-32-5	Thallic oxide
7440-28-0	Thallium
563-68-8	Thallium(I) acetate
6533-73-9	Thallium(I) carbonate
7791-12-0	Thallium chloride TlCl
	Thallium Compounds
10102-45-1	Thallium(I) nitrate
7446-18-6	Thallium(I) sulfate
10031-59-1	Thallium sulfate
6533-73-9	Thallous carbonate
7791-12-0	Thallous chloride
2757-18-8	Thallous malonate
7446-18-6	Thallous sulfate
148-79-8	Thiabendazole
148-79-8	2-(4-Thiazolyl)-1H-benzimidazole
62-55-5	Thioacetamide
28249-77-6	Thiobencarb
2231-57-4	Thiocarbazide
556-64-9	Thiocyanic acid, methyl ester
139-65-1	4,4'-Thiodianiline
59669-26-0	Thiodicarb
39196-18-4	Thiofanox
74-93-1	Thiomethanol
297-97-2	Thionazin
23564-06-9	Thiophanate ethyl
23564-05-8	Thiophanate-methyl
108-98-5	Thiophenol
79-19-6	Thiosemicarbazide
62-56-6	Thiourea
5344-82-1	Thiourea, (2-chlorophenyl)-
614-78-8	Thiourea, (2-methylphenyl)-
86-88-4	Thiourea, 1-naphthalenyl-
137-26-8	Thiram
1314-20-1	Thorium dioxide
7550-45-0	Titanium chloride (TiCl4) (T-4)-
7550-45-0	Titanium tetrachloride
119-93-7	o-Tolidine
612-82-8	o-Tolidine dihydrochloride
41766-75-0	o-Tolidine dihydrofluoride
108-88-3	Toluene
25376-45-8	Toluenediamine
584-84-9	Toluene-2,4-diisocyanate

ALPHABETICAL LISTING OF CHEMICAL NAME AND CAS NUMBER

CAS Number	Chemical Name
91-08-7	Toluene-2,6-diisocyanate
26471-62-5	Toluenediisocyanate (mixed isomers)
26471-62-5	Toluene diisocyanate (unspecified isomer)
95-53-4	o-Toluidine
106-49-0	p-Toluidine
636-21-5	o-Toluidine hydrochloride
8001-35-2	Toxaphene
32534-95-5	2,4,5-TP esters
43121-43-3	Triadimefon
2303-17-5	Triallate
1031-47-6	Triamiphos
68-76-8	Triaziquone
24017-47-8	Triazofos
101200-48-0	Tribenuron methyl
75-25-2	Tribromomethane
118-79-6	2,4,6-Tribromophenol
1983-10-4	Tributyltin fluoride
2155-70-6	Tributyltin methacrylate
78-48-8	S,S,S-Tributyltrithiophosphate
52-68-6	Trichlorfon
76-02-8	Trichloroacetyl chloride
120-82-1	1,2,4-Trichlorobenzene
1558-25-4	Trichloro(chloromethyl)silane
27137-85-5	Trichloro(dichlorophenyl)silane
71-55-6	1,1,1-Trichloroethane
79-00-5	1,1,2-Trichloroethane
79-01-6	Trichloroethylene
115-21-9	Trichloroethylsilane
75-69-4	Trichlorofluoromethane
594-42-3	Trichloromethanesulfenyl chloride
75-69-4	Trichloromonofluoromethane
327-98-0	Trichloronate
25167-82-2	Trichlorophenol
15950-66-0	2,3,4-Trichlorophenol
933-78-8	2,3,5-Trichlorophenol
933-75-5	2,3,6-Trichlorophenol
95-95-4	2,4,5-Trichlorophenol
88-06-2	2,4,6-Trichlorophenol
609-19-8	3,4,5-Trichlorophenol
98-13-5	Trichlorophenylsilane
96-18-4	1,2,3-Trichloropropane
10025-78-2	Trichlorosilane
57213-69-1	Triclopyr triethylammonium salt
27323-41-7	Triethanolamine dodecylbenzene sulfonate
998-30-1	Triethoxysilane
121-44-8	Triethylamine
79-38-9	Trifluorochloroethylene
69806-50-4	2-(4-(5-(Trifluoromethyl)-2-pyridinyl]oxy]-phenoxy
1582-09-8	Trifluralin
26644-46-2	Triforine
75-50-3	Trimethylamine
95-63-6	1,2,4-Trimethylbenzene
75-77-4	Trimethylchlorosilane
15646-96-5	2,4,4-Trimethylhexamethylene diisocyanate
16938-22-0	2,2,4-Trimethylhexamethylene diisocyanate
824-11-3	Trimethylolpropane phosphite
540-84-1	2,2,4-Trimethylpentane
2655-15-4	2,3,5-Trimethylphenyl methylcarbamate

CAS Number	Chemical Name
1066-45-1	Trimethyltin chloride
99-35-4	1,3,5-Trinitrobenzene
639-58-7	Triphenyltin chloride
76-87-9	Triphenyltin hydroxide
555-77-1	Tris(2-chloroethyl)amine
126-72-7	Tris(2,3-dibromopropyl) phosphate
14484-64-1	Tris(dimethylcarbamodithioato-S,S')iron
72-57-1	Trypan blue
66-75-1	Uracil mustard
541-09-3	Uranyl acetate
10102-06-4	Uranyl nitrate
36478-76-9	Uranyl nitrate
2164-17-2	Urea, N,N-dimethyl-N'-[3-(trifluoromethyl)phenyl]-
51-79-6	Urethane
2001-95-8	Valinomycin
7440-62-2	Vanadium (fume or dust)
1314-62-1	Vanadium pentoxide
27774-13-6	Vanadyl sulfate
2699-79-8	Vikane
50471-44-8	Vinclozolin
108-05-4	Vinyl acetate
108-05-4	Vinyl acetate monomer
689-97-4	Vinyl acetylene
593-60-2	Vinyl bromide
75-01-4	Vinyl chloride
109-92-2	Vinyl ethyl ether
75-02-5	Vinyl fluoride
75-35-4	Vinylidene chloride
75-38-7	Vinylidene fluoride
107-25-5	Vinyl methyl ether
81-81-2	Warfarin
	Warfarin and salts
81-81-2	Warfarin, & salts, conc.>0.3%
129-06-6	Warfarin sodium
108-38-3	m-Xylene
95-47-6	o-Xylene
106-42-3	p-Xylene
1330-20-7	Xylene (mixed isomers)
1300-71-6	Xylenol
87-62-7	2,6-Xylidine
28347-13-9	Xylylene dichloride
7440-66-6	Zinc
7440-66-6	Zinc (fume or dust)
557-34-6	Zinc acetate
14639-97-5	Zinc ammonium chloride
14639-98-6	Zinc ammonium chloride
52628-25-8	Zinc ammonium chloride
1332-07-6	Zinc borate
7699-45-8	Zinc bromide
3486-35-9	Zinc carbonate
7646-85-7	Zinc chloride
	Zinc Compounds
557-21-1	Zinc cyanide
58270-08-9	Zinc,dichloro(4,4-dimethyl-5((((methylamino)carbo
7783-49-5	Zinc fluoride
557-41-5	Zinc formate
7779-86-4	Zinc hydrosulfite
7779-88-6	Zinc nitrate

ALPHABETICAL LISTING OF CHEMICAL NAME AND CAS NUMBER

CAS Number	Chemical Name
127-82-2	Zinc phenolsulfonate
1314-84-7	Zinc phosphide (conc. > 10%)
1314-84-7	Zinc phosphide (conc. <= 10%)
1314-84-7	Zinc phosphide
16871-71-9	Zinc silicofluoride
7733-02-0	Zinc sulfate
12122-67-7	Zineb
137-30-4	Ziram
13746-89-9	Zirconium nitrate
16923-95-8	Zirconium potassium fluoride
14644-61-2	Zirconium sulfate
10026-11-6	Zirconium tetrachloride

APPENDIX B
RADIONUCLIDES LISTED UNDER CERCLA
FOR REFERENCE ONLY, NOT FOR REGULATORY COMPLIANCE
SEE 40 CFR PART 302, TABLE 302.4, APPENDIX B, FOR MORE INFORMATION

Radionuclide Name	Atomic Number	RQ (curies)	Radionuclide Name	Atomic Number	RQ (curies)
Radionuclides (unlisted)		1	Barium-128	56	10
Actinium-224	89	100	Barium-131	56	10
Actinium-225	89	1	Barium-131m	56	1000
Actinium-226	89	10	Barium-133	56	10
Actinium-227	89	0.001	Barium-133m	56	100
Actinium-228	89	10	Barium-135m	56	1000
Aluminum-026	13	10	Barium-139	56	1000
Americium-237	95	1000	Barium-140	56	10
Americium-238	95	100	Barium-141	56	1000
Americium-239	95	100	Barium-142	56	1000
Americium-240	95	10	Berkelium-245	97	100
Americium-241	95	0.01	Berkelium-246	97	10
Americium-242	95	100	Berkelium-247	97	0.01
Americium-242m	95	0.01	Berkelium-249	97	1
Americium-243	95	0.01	Berkelium-250	97	100
Americium-244	95	10	Beryllium-007	4	100
Americium-244m	95	1000	Beryllium-010	4	1
Americium-245	95	1000	Bismuth-200	83	100
Americium-246	95	1000	Bismuth-201	83	100
Americium-246m	95	1000	Bismuth-202	83	1000
Antimony-115	51	1000	Bismuth-203	83	10
Antimony-116	51	1000	Bismuth-205	83	10
Antimony-116m	51	100	Bismuth-206	83	10
Antimony-117	51	1000	Bismuth-207	83	10
Antimony-118m	51	10	Bismuth-210	83	10
Antimony-119	51	1000	Bismuth-210m	83	0.1
Antimony-120 (16 min)	51	1000	Bismuth-212	83	100
Antimony-120 (5.76 day)	51	10	Bismuth-213	83	100
Antimony-122	51	10	Bismuth-214	83	100
Antimony-124	51	10	Bromine-074	35	100
Antimony-124m	51	1000	Bromine-074m	35	100
Antimony-125	51	10	Bromine-075	35	100
Antimony-126	51	10	Bromine-076	35	10
Antimony-126m	51	1000	Bromine-077	35	100
Antimony-127	51	10	Bromine-080	35	1000
Antimony-128 (10.4 min)	51	1000	Bromine-080m	35	1000
Antimony-128 (9.01 hours)	51	10	Bromine-082	35	10
Antimony-129	51	100	Bromine-083	35	1000
Antimony-130	51	100	Bromine-084	35	100
Antimony-131	51	1000	Cadmium-104	48	1000
Argon-039	18	1000	Cadmium-107	48	1000
Argon-041	18	10	Cadmium-109	48	1
Arsenic-069	33	1000	Cadmium-113	48	0.1
Arsenic-070	33	100	Cadmium-113m	48	0.1
Arsenic-071	33	100	Cadmium-115	48	100
Arsenic-072	33	10	Cadmium-115m	48	10
Arsenic-073	33	100	Cadmium-117	48	100
Arsenic-074	33	10	Cadmium-117m	48	10
Arsenic-076	33	100	Calcium-041	20	10
Arsenic-077	33	1000	Calcium-045	20	10
Arsenic-078	33	100	Calcium-047	20	10
Astatine-207	85	100	Californium-244	98	1000
Astatine-211	85	100	Californium-246	98	10
Barium-126	56	1000	Californium-248	98	0.1

RADIONUCLIDES LISTED UNDER CERCLA
FOR REFERENCE ONLY, NOT FOR REGULATORY COMPLIANCE
SEE 40 CFR PART 302, TABLE 302.4, APPENDIX B, FOR MORE INFORMATION

Radionuclide Name	Atomic Number	RQ (curies)	Radionuclide Name	Atomic Number	RQ (curies)
Californium-249	98	0.01	Curium-245	96	0.01
Californium-250	98	0.01	Curium-246	96	0.01
Californium-251	98	0.01	Curium-247	96	0.01
Californium-252	98	0.1	Curium-248	96	0.001
Californium-253	98	10	Curium-249	96	1000
Californium-254	98	0.1	Dysprosium-155	66	100
Carbon-011	6	1000	Dysprosium-157	66	100
Carbon-014	6	10	Dysprosium-159	66	100
Cerium-134	58	10	Dysprosium-165	66	1000
Cerium-135	58	10	Dysprosium-166	66	10
Cerium-137	58	1000	Einsteinium-250	99	10
Cerium-137m	58	100	Einsteinium-251	99	1000
Cerium-139	58	100	Einsteinium-253	99	10
Cerium-141	58	10	Einsteinium-254	99	0.1
Cerium-143	58	100	Einsteinium-254m	99	1
Cerium-144	58	1	Erbium-161	68	100
Cesium-125	55	1000	Erbium-165	68	1000
Cesium-127	55	100	Erbium-169	68	100
Cesium-129	55	100	Erbium-171	68	100
Cesium-130	55	1000	Erbium-172	68	10
Cesium-131	55	1000	Europium-145	63	10
Cesium-132	55	10	Europium-146	63	10
Cesium-134	55	1	Europium-147	63	10
Cesium-134m	55	1000	Europium-148	63	10
Cesium-135	55	10	Europium-149	63	100
Cesium-135m	55	100	Europium-150 (12.6 hours)	63	1000
Cesium-136	55	10	Europium-150 (34.2 yr)	63	10
Cesium-137	55	1	Europium-152	63	10
Cesium-138	55	100	Europium-152m	63	100
Chlorine-036	17	10	Europium-154	63	10
Chlorine-038	17	100	Europium-155	63	10
Chlorine-039	17	100	Europium-156	63	10
Chromium-048	24	100	Europium-157	63	10
Chromium-049	24	1000	Europium-158	63	1000
Chromium-051	24	1000	Fermium-252	100	10
Cobalt-055	27	10	Fermium-253	100	10
Cobalt-056	27	10	Fermium-254	100	100
Cobalt-057	27	100	Fermium-255	100	100
Cobalt-058	27	10	Fermium-257	100	1
Cobalt-058m	27	1000	Fluorine-018	9	1000
Cobalt-060	27	10	Francium-222	87	100
Cobalt-060m	27	1000	Francium-223	87	100
Cobalt-061	27	1000	Gadolinium-145	64	100
Cobalt-062m	27	1000	Gadolinium-146	64	10
Copper-060	29	100	Gadolinium-147	64	10
Copper-061	29	100	Gadolinium-148	64	0.001
Copper-064	29	1000	Gadolinium-149	64	100
Copper-067	29	100	Gadolinium-151	64	100
Curium-238	96	1000	Gadolinium-152	64	0.001
Curium-240	96	1	Gadolinium-153	64	10
Curium-241	96	10	Gadolinium-159	64	1000
Curium-242	96	1	Gallium-065	31	1000
Curium-243	96	0.01	Gallium-066	31	10
Curium-244	96	0.01	Gallium-067	31	100

RADIONUCLIDES LISTED UNDER CERCLA
FOR REFERENCE ONLY, NOT FOR REGULATORY COMPLIANCE
SEE 40 CFR PART 302, TABLE 302.4, APPENDIX B, FOR MORE INFORMATION

Radionuclide Name	Atomic Number	RQ (curies)	Radionuclide Name	Atomic Number	RQ (curies)
Gallium-068	31	1000	Indium-115m	49	100
Gallium-070	31	1000	Indium-116m	49	100
Gallium-072	31	10	Indium-117	49	1000
Gallium-073	31	100	Indium-117m	49	100
Germanium-066	32	100	Indium-119m	49	1000
Germanium-067	32	1000	Iodine-120	53	10
Germanium-068	32	10	Iodine-120m	53	100
Germanium-069	32	10	Iodine-121	53	100
Germanium-071	32	1000	Iodine-123	53	10
Germanium-075	32	1000	Iodine-124	53	0.1
Germanium-077	32	10	Iodine-125	53	0.01
Germanium-078	32	1000	Iodine-126	53	0.01
Gold-193	79	100	Iodine-128	53	1000
Gold-194	79	10	Iodine-129	53	0.001
Gold-195	79	100	Iodine-130	53	1
Gold-198	79	100	Iodine-131	53	0.01
Gold-198m	79	10	Iodine-132	53	10
Gold-199	79	100	Iodine-132m	53	10
Gold-200	79	1000	Iodine-133	53	0.1
Gold-200m	79	10	Iodine-134	53	100
Gold-201	79	1000	Iodine-135	53	10
Hafnium-170	72	100	Iridium-182	77	1000
Hafnium-172	72	1	Iridium-184	77	100
Hafnium-173	72	100	Iridium-185	77	100
Hafnium-175	72	100	Iridium-186	77	10
Hafnium-177m	72	1000	Iridium-187	77	100
Hafnium-178m	72	0.1	Iridium-188	77	10
Hafnium-179m	72	100	Iridium-189	77	100
Hafnium-180m	72	100	Iridium-190	77	10
Hafnium-181	72	10	Iridium-190m	77	1000
Hafnium-182	72	0.1	Iridium-192	77	10
Hafnium-182m	72	100	Iridium-192m	77	100
Hafnium-183	72	100	Iridium-194	77	100
Hafnium-184	72	100	Iridium-194m	77	10
Holmium-155	67	1000	Iridium-195	77	1000
Holmium-157	67	1000	Iridium-195m	77	100
Holmium-159	67	1000	Iron-052	26	100
Holmium-161	67	1000	Iron-055	26	100
Holmium-162	67	1000	Iron-059	26	10
Holmium-162m	67	1000	Iron-060	26	0.1
Holmium-164	67	1000	Krypton-074	36	10
Holmium-164m	67	1000	Krypton-076	36	10
Holmium-166	67	100	Krypton-077	36	10
Holmium-166m	67	1	Krypton-079	36	100
Holmium-167	67	100	Krypton-081	36	1000
Hydrogen-003	1	100	Krypton-083m	36	1000
Indium-109	49	100	Krypton-085	36	1000
Indium-110 (4.9 hours)	49	10	Krypton-085m	36	100
Indium-110 (69.1 min)	49	100	Krypton-087	36	10
Indium-111	49	100	Krypton-088	36	10
Indium-112	49	1000	Lanthanum-131	57	1000
Indium-113m	49	1000	Lanthanum-132	57	100
Indium-114m	49	10	Lanthanum-135	57	1000
Indium-115	49	0.1	Lanthanum-137	57	10

**RADIONUCLIDES LISTED UNDER CERCLA
FOR REFERENCE ONLY, NOT FOR REGULATORY COMPLIANCE
SEE 40 CFR PART 302, TABLE 302.4, APPENDIX B, FOR MORE INFORMATION**

Radionuclide Name	Atomic Number	RQ (curies)	Radionuclide Name	Atomic Number	RQ (curies)
Lanthanum-138	57	1	Molybdenum-099	42	100
Lanthanum-140	57	10	Molybdenum-101	42	1000
Lanthanum-141	57	1000	Neodymium-136	60	1000
Lanthanum-142	57	100	Neodymium-138	60	1000
Lanthanum-143	57	1000	Neodymium-139	60	1000
Lead-195m	82	1000	Neodymium-139m	60	100
Lead-198	82	100	Neodymium-141	60	1000
Lead-199	82	100	Neodymium-147	60	10
Lead-200	82	100	Neodymium-149	60	100
Lead-201	82	100	Neodymium-151	60	1000
Lead-202	82	1	Neptunium-232	93	1000
Lead-202m	82	10	Neptunium-233	93	1000
Lead-203	82	100	Neptunium-234	93	10
Lead-205	82	100	Neptunium-235	93	1000
Lead-209	82	1000	Neptunium-236 (1.2E 5 yr)	93	0.1
Lead-210	82	0.01	Neptunium-236 (22.5 hours)	93	100
Lead-211	82	100	Neptunium-237	93	0.01
Lead-212	82	10	Neptunium-238	93	10
Lead-214	82	100	Neptunium-239	93	100
Lutetium-169	71	10	Neptunium-240	93	100
Lutetium-170	71	10	Nickel-056	28	10
Lutetium-171	71	10	Nickel-057	28	10
Lutetium-172	71	10	Nickel-059	28	100
Lutetium-173	71	100	Nickel-063	28	100
Lutetium-174	71	10	Nickel-065	28	100
Lutetium-174m	71	10	Nickel-066	28	10
Lutetium-176	71	1	Niobium-088	41	100
Lutetium-176m	71	1000	Niobium-089 (122 minutes)	41	100
Lutetium-177	71	100	Niobium-089 (66 minutes)	41	100
Lutetium-177m	71	10	Niobium-090	41	10
Lutetium-178	71	1000	Niobium-093m	41	100
Lutetium-178m	71	1000	Niobium-094	41	10
Lutetium-179	71	1000	Niobium-095	41	10
Magnesium-028	12	10	Niobium-095m	41	100
Manganese-051	25	1000	Niobium-096	41	10
Manganese-052	25	10	Niobium-097	41	100
Manganese-052m	25	1000	Niobium-098	41	1000
Manganese-053	25	1000	Osmium-180	76	1000
Manganese-054	25	10	Osmium-181	76	100
Manganese-056	25	100	Osmium-182	76	100
Mendelevium-257	101	100	Osmium-185	76	10
Mendelevium-258	101	1	Osmium-189m	76	1000
Mercury-193	80	100	Osmium-191	76	100
Mercury-193m	80	10	Osmium-191m	76	1000
Mercury-194	80	0.1	Osmium-193	76	100
Mercury-195	80	100	Osmium-194	76	1
Mercury-195m	80	100	Palladium-100	46	100
Mercury-197	80	1000	Palladium-101	46	100
Mercury-197m	80	1000	Palladium-103	46	100
Mercury-199m	80	1000	Palladium-107	46	100
Mercury-203	80	10	Palladium-109	46	1000
Molybdenum-090	42	100	Phosphorus-032	15	0.1
Molybdenum-093	42	100	Phosphorus-033	15	1
Molybdenum-093m	42	10	Platinum-186	78	100

RADIONUCLIDES LISTED UNDER CERCLA
FOR REFERENCE ONLY, NOT FOR REGULATORY COMPLIANCE
SEE 40 CFR PART 302, TABLE 302.4, APPENDIX B, FOR MORE INFORMATION

Radionuclide Name	Atomic Number	RQ (curies)	Radionuclide Name	Atomic Number	RQ (curies)
Platinum-188	78	100	Protactinium-230	91	10
Platinum-189	78	100	Protactinium-231	91	0.01
Platinum-191	78	100	Protactinium-232	91	10
Platinum-193	78	1000	Protactinium-233	91	100
Platinum-193m	78	100	Protactinium-234	91	10
Platinum-195m	78	100	Radium-223	88	1
Platinum-197	78	1000	Radium-224	88	10
Platinum-197m	78	1000	Radium-225	88	1
Platinum-199	78	1000	Radium-226	88	0.1
Platinum-200	78	100	Radium-227	88	1000
Plutonium-234	94	1000	Radium-228	88	0.1
Plutonium-235	94	1000	Radon-220	86	0.1
Plutonium-236	94	0.1	Radon-222	86	0.1
Plutonium-237	94	1000	Rhenium-177	75	1000
Plutonium-238	94	0.01	Rhenium-178	75	1000
Plutonium-239	94	0.01	Rhenium-181	75	100
Plutonium-240	94	0.01	Rhenium-182 (12.7 hours)	75	10
Plutonium-241	94	1	Rhenium-182 (64.0 hours)	75	10
Plutonium-242	94	0.01	Rhenium-184	75	10
Plutonium-243	94	1000	Rhenium-184m	75	10
Plutonium-244	94	0.01	Rhenium-186	75	100
Plutonium-245	94	100	Rhenium-186m	75	10
Polonium-203	84	100	Rhenium-187	75	1000
Polonium-205	84	100	Rhenium-188	75	1000
Polonium-207	84	10	Rhenium-188m	75	1000
Polonium-210	84	0.01	Rhenium-189	75	1000
Potassium-040	19	1	Rhodium-099	45	10
Potassium-042	19	100	Rhodium-099m	45	100
Potassium-043	19	10	Rhodium-100	45	10
Potassium-044	19	100	Rhodium-101	45	10
Potassium-045	19	1000	Rhodium-101m	45	100
Praseodymium-136	59	1000	Rhodium-102	45	10
Praseodymium-137	59	1000	Rhodium-102m	45	10
Praseodymium-138m	59	100	Rhodium-103m	45	1000
Praseodymium-139	59	1000	Rhodium-105	45	100
Praseodymium-142	59	100	Rhodium-106m	45	10
Praseodymium-142m	59	1000	Rhodium-107	45	1000
Praseodymium-143	59	10	Rubidium-079	37	1000
Praseodymium-144	59	1000	Rubidium-081	37	100
Praseodymium-145	59	1000	Rubidium-081m	37	1000
Praseodymium-147	59	1000	Rubidium-082m	37	10
Promethium-141	61	1000	Rubidium-083	37	10
Promethium-143	61	100	Rubidium-084	37	10
Promethium-144	61	10	Rubidium-086	37	10
Promethium-145	61	100	Rubidium-087	37	10
Promethium-146	61	10	Rubidium-088	37	1000
Promethium-147	61	10	Rubidium-089	37	1000
Promethium-148	61	10	Ruthenium-094	44	1000
Promethium-148m	61	10	Ruthenium-097	44	100
Promethium-149	61	100	Ruthenium-103	44	10
Promethium-150	61	100	Ruthenium-105	44	100
Promethium-151	61	100	Ruthenium-106	44	1
Protactinium-227	91	100	Samarium-141	62	1000
Protactinium-228	91	10	Samarium-141m	62	1000

RADIONUCLIDES LISTED UNDER CERCLA
FOR REFERENCE ONLY, NOT FOR REGULATORY COMPLIANCE
SEE 40 CFR PART 302, TABLE 302.4, APPENDIX B, FOR MORE INFORMATION

Radionuclide Name	Atomic Number	RQ (curies)	Radionuclide Name	Atomic Number	RQ (curies)
Samarium-142	62	1000	Tantalum-176	73	10
Samarium-145	62	100	Tantalum-177	73	1000
Samarium-146	62	0.01	Tantalum-178	73	1000
Samarium-147	62	0.01	Tantalum-179	73	1000
Samarium-151	62	10	Tantalum-180	73	100
Samarium-153	62	100	Tantalum-180m	73	1000
Samarium-155	62	1000	Tantalum-182	73	10
Samarium-156	62	100	Tantalum-182m	73	1000
Scandium-043	21	1000	Tantalum-183	73	100
Scandium-044	21	100	Tantalum-184	73	10
Scandium-044m	21	10	Tantalum-185	73	1000
Scandium-046	21	10	Tantalum-186	73	1000
Scandium-047	21	100	Technetium-093	43	100
Scandium-048	21	10	Technetium-093m	43	1000
Scandium-049	21	1000	Technetium-094	43	10
Selenium-070	34	1000	Technetium-094m	43	100
Selenium-073	34	10	Technetium-096	43	10
Selenium-073m	34	100	Technetium-096m	43	1000
Selenium-075	34	10	Technetium-097	43	100
Selenium-079	34	10	Technetium-097m	43	100
Selenium-081	34	1000	Technetium-098	43	10
Selenium-081m	34	1000	Technetium-099	43	10
Selenium-083	34	1000	Technetium-099m	43	100
Silicon-031	14	1000	Technetium-101	43	1000
Silicon-032	14	1	Technetium-104	43	1000
Silver-102	47	100	Tellurium-116	52	1000
Silver-103	47	1000	Tellurium-121	52	10
Silver-104	47	1000	Tellurium-121m	52	10
Silver-104m	47	1000	Tellurium-123	52	10
Silver-105	47	10	Tellurium-123m	52	10
Silver-106	47	1000	Tellurium-125m	52	10
Silver-106m	47	10	Tellurium-127	52	1000
Silver-108m	47	10	Tellurium-127m	52	10
Silver-110m	47	10	Tellurium-129	52	1000
Silver-111	47	10	Tellurium-129m	52	10
Silver-112	47	100	Tellurium-131	52	1000
Silver-115	47	1000	Tellurium-131m	52	10
Sodium-022	11	10	Tellurium-132	52	10
Sodium-024	11	10	Tellurium-133	52	1000
Strontium-080	38	100	Tellurium-133m	52	1000
Strontium-081	38	1000	Tellurium-134	52	1000
Strontium-083	38	100	Terbium-147	65	100
Strontium-085	38	10	Terbium-149	65	100
Strontium-085m	38	1000	Terbium-150	65	100
Strontium-087m	38	100	Terbium-151	65	10
Strontium-089	38	10	Terbium-153	65	100
Strontium-090	38	0.1	Terbium-154	65	10
Strontium-091	38	10	Terbium-155	65	100
Strontium-092	38	100	Terbium-156	65	10
Sulfur-035	16	1	Terbium-156m (24.4 hours)	65	1000
Tantalum-172	73	100	Terbium-156m (5.0 hours)	65	1000
Tantalum-173	73	100	Terbium-157	65	100
Tantalum-174	73	100	Terbium-158	65	10
Tantalum-175	73	100	Terbium-160	65	10

RADIONUCLIDES LISTED UNDER CERCLA
FOR REFERENCE ONLY, NOT FOR REGULATORY COMPLIANCE
SEE 40 CFR PART 302, TABLE 302.4, APPENDIX B, FOR MORE INFORMATION

Radionuclide Name	Atomic Number	RQ (curies)	Radionuclide Name	Atomic Number	RQ (curies)
Terbium-161	65	100	Uranium-233	92	0.1
Thallium-194	81	1000	Uranium-234	92	0.1
Thallium-194m	81	100	Uranium-235	92	0.1
Thallium-195	81	100	Uranium-236	92	0.1
Thallium-197	81	100	Uranium-237	92	100
Thallium-198	81	10	Uranium-238	92	0.1
Thallium-198m	81	100	Uranium-239	92	1000
Thallium-199	81	100	Uranium-240	92	1000
Thallium-200	81	10	Vanadium-047	23	1000
Thallium-201	81	1000	Vanadium-048	23	10
Thallium-202	81	10	Vanadium-049	23	1000
Thallium-204	81	10	Xenon-120	54	100
Thorium-226	90	100	Xenon-121	54	10
Thorium-227	90	1	Xenon-122	54	100
Thorium-228	90	0.01	Xenon-123	54	10
Thorium-229	90	0.001	Xenon-125	54	100
Thorium-230	90	0.01	Xenon-127	54	100
Thorium-231	90	100	Xenon-129m	54	1000
Thorium-232	90	0.001	Xenon-131m	54	1000
Thorium-234	90	100	Xenon-133	54	1000
Thulium-162	69	1000	Xenon-133m	54	1000
Thulium-166	69	10	Xenon-135	54	100
Thulium-167	69	100	Xenon-135m	54	10
Thulium-170	69	10	Xenon-138	54	10
Thulium-171	69	100	Ytterbium-162	70	1000
Thulium-172	69	100	Ytterbium-166	70	10
Thulium-173	69	100	Ytterbium-167	70	1000
Thulium-175	69	1000	Ytterbium-169	70	10
Tin-110	50	100	Ytterbium-175	70	100
Tin-111	50	1000	Ytterbium-177	70	1000
Tin-113	50	10	Ytterbium-178	70	1000
Tin-117m	50	100	Yttrium-086	39	10
Tin-119m	50	10	Yttrium-086m	39	1000
Tin-121	50	1000	Yttrium-087	39	10
Tin-121m	50	10	Yttrium-088	39	10
Tin-123	50	10	Yttrium-090	39	10
Tin-123m	50	1000	Yttrium-090m	39	100
Tin-125	50	10	Yttrium-091	39	10
Tin-126	50	1	Yttrium-091m	39	1000
Tin-127	50	100	Yttrium-092	39	100
Tin-128	50	1000	Yttrium-093	39	100
Titanium-044	22	1	Yttrium-094	39	1000
Titanium-045	22	1000	Yttrium-095	39	1000
Tungsten-176	74	1000	Zinc-062	30	100
Tungsten-177	74	100	Zinc-063	30	1000
Tungsten-178	74	100	Zinc-065	30	10
Tungsten-179	74	1000	Zinc-069	30	1000
Tungsten-181	74	100	Zinc-069m	30	100
Tungsten-185	74	10	Zinc-071m	30	100
Tungsten-187	74	100	Zinc-072	30	100
Tungsten-188	74	10	Zirconium-086	40	100
Uranium-230	92	1	Zirconium-088	40	10
Uranium-231	92	1000	Zirconium-089	40	100
Uranium-232	92	0.01	Zirconium-093	40	1

**RADIONUCLIDES LISTED UNDER CERCLA
FOR REFERENCE ONLY, NOT FOR REGULATORY COMPLIANCE
SEE 40 CFR PART 302, TABLE 302.4, APPENDIX B, FOR MORE INFORMATION**

Radionuclide Name	Atomic Number	RQ (curies)
Zirconium-095	40	10
Zirconium-097	40	10

NOTES: m - Signifies a nuclear isomer which is a radionuclide in a higher energy metastable state relative to the parent isotope. Final RQs for all radionuclides apply to chemical compounds containing the radionuclides and elemental forms regardless of the diameter of pieces of solid material.

An adjusted RQ of one curie applies to all radionuclides not otherwise listed. Whenever the RQs in the SARA Title III Consolidated List and this list are in conflict, the lowest RQ applies.

Notification requirements for releases of mixtures or solutions of radionuclides can be found in 40 CFR section 302.6(b).

APPENDIX C
RCRA WASTE STREAMS AND UNLISTED HAZARDOUS WASTES
THE DESCRIPTIONS OF THE WASTE STREAMS HAVE BEEN TRUNCATED.
THIS LIST SHOULD BE USED FOR REFERENCE ONLY
COMPLIANCE INFORMATION CAN BE FOUND IN 40 CFR PART 302 AND TABLE 302.4

RCRA Code	Description	RQ (lbs)
F001	The following spent halogenated solvents used in degreasing:	10
	(a) Tetrachloroethylene (CAS No. 127-18-4, RCRA Waste No. U210)	100
	(b) Trichloroethylene (CAS No. 79-01-6, RCRA Waste No. U228)	100
	(c) Methylene chloride (CAS No. 75-09-2, RCRA Waste No. U080)	1,000
	(d) 1,1,1-Trichloroethane (CAS No. 71-55-6, RCRA Waste No. U226)	1,000
	(e) Carbon tetrachloride (CAS No. 56-23-5, RCRA Waste No. U211)	10
	(f) Chlorinated hydrocarbons	5,000
F002	The following spent halogenated solvents:	10
	(a) Tetrachloroethylene (CAS No. 127-18-4, RCRA Waste No. U210)	100
	(b) Methylene chloride (CAS No. 75-09-2, RCRA Waste No. U080)	1,000
	(c) Trichloroethylene (CAS No. 79-01-6, RCRA Waste No. U228)	100
	(d) 1,1,1-Trichloroethane (CAS No. 71-55-6, RCRA Waste No. U226)	1,000
	(e) Chlorobenzene (CAS No. 108-90-7, RCRA Waste No. U037)	100
	(f) 1,1,2-Trichloro-1,2,2-trifluoroethane (CAS No. 76-13-1)	5,000
	(g) o-Dichlorobenzene (CAS No. 95-50-1, RCRA Waste No. U070)	100
	(h) Trichlorofluoromethane (CAS No. 75-69-4, RCRA Waste No. U121)	5,000
	(i) 1,1,2-Trichloroethane (CAS No. 79-00-5, RCRA Waste No. U227)	100
F003	The following spent non-halogenated solvents and still bottoms from recovery:	100
	(a) Xylene (CAS No. 1330-20-7, RCRA Waste No. U239)	1,000
	(b) Acetone (CAS No. 67-64-1, RCRA Waste No. U002)	5,000
	(c) Ethyl acetate (CAS No. 141-78-6, RCRA Waste No. U112)	5,000
	(d) Ethylbenzene (CAS No. 100-41-4)	1,000
	(e) Ethyl ether (CAS No. 60-29-7, RCRA Waste No. U117)	100
	(f) Methyl isobutyl ketone (CAS No. 108-10-1, RCRA Waste No. U161)	5,000
	(g) n-Butyl alcohol (CAS No. 71-36-3, RCRA Waste No. U031)	5,000
	(h) Cyclohexanone (CAS No. 108-94-1, RCRA Waste No. U057)	5,000
	(i) Methanol (CAS No. 67-56-1, RCRA Waste No. U154)	5,000
F004	The following spent non-halogenated solvents and still bottoms from recovery:	100
	(a) Cresols/cresylic acid (CAS No. 1319-77-3, RCRA Waste No. U052)	1,000
	(b) Nitrobenzene (CAS No. 98-95-3, RCRA Waste No. U169)	1,000
F005	The following spent non-halogenated solvents and still bottoms from recovery:	100
	(a) Toluene (CAS No. 108-88-3, RCRA Waste No. U220)	1,000
	(b) Methyl ethyl ketone (CAS No. 78-93-3, RCRA Waste No. U159)	5,000
	(c) Carbon disulfide (CAS No. 75-15-0, RCRA Waste No. P022)	100
	(d) Isobutanol (CAS No. 78-83-1, RCRA Waste No. U140)	5,000
	(e) Pyridine (CAS No. 110-86-1, RCRA Waste No. U196)	1,000
F006	Wastewater treatment sludges from electroplating operations (w/some exceptions)	10
F007	Spent cyanide plating bath solns. from electroplating	10
F008	Plating bath residues from electroplating where cyanides are used	10
F009	Spent stripping/cleaning bath solns. from electroplating where cyanides are used	10
F010	Quenching bath residues from metal heat treating where cyanides are used	10
F011	Spent cyanide soln. from salt bath pot cleaning from metal heat treating	10
F012	Quenching wastewater sludges from metal heat treating where cyanides are used	10

**RCRA WASTE STREAMS AND UNLISTED HAZARDOUS WASTES
THE DESCRIPTIONS OF THE WASTE STREAMS HAVE BEEN TRUNCATED.
THIS LIST SHOULD BE USED FOR REFERENCE ONLY
COMPLIANCE INFORMATION CAN BE FOUND IN 40 CFR PART 302 AND TABLE 302.4**

RCRA Code	Description	RQ (lbs)
F019	Wastewater treatment sludges from chemical conversion aluminum coating	10
F020	Wastes from prod. or use of tri/tetrachlorophenol or derivative intermediates	1
F021	Wastes from prod. or use of pentachlorophenol or intermediates for derivatives	1
F022	Wastes from use of tetra/penta/hexachlorobenzenes under alkaline conditions	1
F023	Wastes from mat. prod. on equip. previously used for tri\tetrachlorophenol	1
F024	Wastes from production of chlorinated aliphatic hydrocarbons (C1-C5)	1
F025	Lights ends, filters from prod. of chlorinated aliphatic hydrocarbons (C1-C5)	1
F026	Waste from equipment previously used to prod. tetra/penta/hexachlorobenzenes	1
F027	Discarded formulations containing tri/tetra/pentachlorophenols or derivatives	1
F028	Residues from incineration of soil contaminated w/ F020,F021,F022,F023,F026,F027	1
F032	Wastewaters, process residuals from wood preserving using chlorophenolic solns.	1
F034	Wastewaters, process residuals from wood preserving using creosote formulations	1
F035	Wastewaters, process residuals from wood preserving using arsenic or chromium	1
F037	Petroleum refinery primary oil/water/solids separation sludge	1
F038	Petroleum refinery secondary (emulsified) oil/water/solids separation sludge	1
K001	Wastewater treatment sludge from creosote/pentachlorophenol wood preserving	1
K002	Wastewater treatment sludge from prod. of chrome yellow and orange pigments	10
K003	Wastewater treatment sludge from prod. of molybdate orange pigments	10
K004	Wastewater treatment sludge from prod. of zinc yellow pigments	10
K005	Wastewater treatment sludge from prod. of chrome green pigments	10
K006	Wastewater treatment sludge from prod. of chrome oxide green pigments	10
K007	Wastewater treatment sludge from prod. of iron blue pigments	10
K008	Oven residue from prod. of chrome oxide green pigments	10
K009	Dist. bottoms from prod. of acetaldehyde from ethylene	10
K010	Dist. side cuts from prod. of acetaldehyde from ethylene	10
K011	Bottom stream from wastewater stripper in acrylonitrile prod.	10
K013	Bottom stream from acetonitrile column in acrylonitrile prod.	10
K014	Bottoms from acetonitrile purification column in acrylonitrile prod.	5,000
K015	Still bottoms from the dist. of benzyl chloride	10
K016	Heavy ends or dist. residues from prod. of carbon tetrachloride	1
K017	Heavy ends from the purification column in epichlorohydrin prod.	10
K018	Heavy ends from the fractionation column in ethyl chloride prod.	1
K019	Heavy ends from the dist. of ethylene dichloride during its prod.	1
K020	Heavy ends from the dist. of vinyl chloride during prod. of the monomer	1
K021	Aqueous spent antimony catalyst waste from fluoromethanes prod.	10
K022	Dist. bottom tars from prod. of phenol/acetone from cumene	1
K023	Dist. light ends from prod. of phthalic anhydride from naphthalene	5,000
K024	Dist. bottoms from prod. of phthalic anhydride from naphthalene	5,000
K025	Dist. bottoms from prod. of nitrobenzene by nitration of benzene	10
K026	Stripping still tails from the prod. of methyl ethyl pyridines	1,000
K027	Centrifuge/dist. residues from toluene diisocyanate prod.	10
K028	Spent catalyst from hydrochlorinator reactor in prod. of 1,1,1-trichloroethane	1
K029	Waste from product steam stripper in prod. of 1,1,1-trichloroethane	1

RCRA WASTE STREAMS AND UNLISTED HAZARDOUS WASTES
THE DESCRIPTIONS OF THE WASTE STREAMS HAVE BEEN TRUNCATED.
THIS LIST SHOULD BE USED FOR REFERENCE ONLY
COMPLIANCE INFORMATION CAN BE FOUND IN 40 CFR PART 302 AND TABLE 302.4

RCRA Code	Description	RQ (lbs)
030	Column bottoms/heavy ends from prod. of trichloroethylene and perchloroethylene	1
031	By-product salts generated in the prod. of MSMA and cacodylic acid	1
032	Wastewater treatment sludge from the prod. of chlordane	10
033	Wastewaster/scrubwater from chlorination of cyclopentadiene in chlordane prod.	10
034	Filter solids from filtration of hexachlorocyclopentadiene in chlordane prod.	10
035	Wastewater treatment sludges from the prod. of creosote	1
036	Still bottoms from toluene reclamation distillation in disulfoton prod.	1
037	Wastewater treatment sludges from the prod. of disulfoton	1
038	Wastewater from the washing and stripping of phorate production	10
039	Filter cake from filtration of diethylphosphorodithioic adid in phorate prod.	10
040	Wastewater treatment sludge from the prod. of phorate	10
041	Wastewater treatment sludge from the prod. of toxaphene	1
042	Heavy ends/residues from dist. of tetrachlorobenzene in 2,4,5-T prod.	10
043	2,6-Dichlorophenol waste from the prod. of 2,4-D	10
044	Wastewater treatment sludge from manuf. and processing of explosives	10
045	Spent carbon from treatment of wastewater containing explosives	10
046	Wastewater sludge from manuf.,formulating,loading of lead-based initiating compd	10
047	Pink/red water from TNT operations	10
048	Dissolved air flotation (DAF) float from the petroleum refining industry	10
049	Slop oil emulsion solids from the petroleum refining industry	10
050	Heat exchanger bundle cleaning sludge from petroleum refining industry	10
051	API separator sludge from the petroleum refining industry	10
052	Tank bottoms (leaded) from the petroleum refining industry	10
060	Ammonia still lime sludge from coking operations	1
061	Emission control dust/sludge from primary prod. of steel in electric furnaces	10
062	Spent pickle liquor generated by steel finishing (SIC codes 331 and 332)	10
064	Acid plant blowdown slurry/sludge from blowdown slurry from primary copper prod.	10
065	Surface impoundment solids at primary lead smelting facilities	10
066	Sludge from treatment of wastewater/acid plant blowdown from primary zinc prod.	10
069	Emission control dust/sludge from secondary lead smelting	10
071	Brine purification muds from mercury cell process in chlorine production	1
073	Chlorinated hydrocarbon waste from diaphragm cell process in chlorine production	10
083	Distillation bottoms from aniline extraction	100
084	Wastewater sludges from prod. of veterinary pharm. from arsenic compds.	1
085	Distillation or fractionation column bottoms in prod. of chlorobenzenes	10
086	Wastes/sludges from prod. of inks from chromium and lead-containing substances	10
087	Decanter tank tar sludge from coking operations	100
088	Spent potliners from primary aluminum reduction	10
090	Emission control dust/sludge from ferrochromiumsilicon prod.	10
091	Emission control dust/sludge from ferrochromium prod.	10
093	Dist. light ends from prod. of phthalic anhydride by ortho-xylene	5,000
094	Dist. bottoms in prod. of phthalic anhydride by ortho-xylene	5,000
095	Distillation bottoms in prod. of 1,1,1-trichloroethane	100

RCRA WASTE STREAMS AND UNLISTED HAZARDOUS WASTES
THE DESCRIPTIONS OF THE WASTE STREAMS HAVE BEEN TRUNCATED.
THIS LIST SHOULD BE USED FOR REFERENCE ONLY
COMPLIANCE INFORMATION CAN BE FOUND IN 40 CFR PART 302 AND TABLE 302.4

RCRA Code	Description	RQ (lbs)
K096	Heavy ends from dist. column in prod. of 1,1,1-trichloroethane	100
K097	Vacuum stripper discharge from the chlordane chlorinator in prod. of chlordane	1
K098	Untreated process wastewater from the prod. of toxaphene	1
K099	Untreated wastewater from the prod. of 2,4-D	10
K100	Waste leaching soln from emission control dust/sludge in secondary lead smelting	10
K101	Dist. tar residue from aniline in prod. of veterinary pharm. from arsenic compd.	1
K102	Residue from activated carbon in prod. of veterinary pharm. from arsenic compds.	1
K103	Process residues from aniline extraction from the prod. of aniline	100
K104	Combined wastewater streams generated from prod. of nitrobenzene/aniline	10
K105	Aqueous stream from washing in prod. of chlorobenzenes	10
K106	Wastewater treatment sludge from mercury cell process in chlorine prod.	1
K107	Column bottoms from separation in prod. of UDMH from carboxylic acid hydrazides	10
K108	Condensed column overheads and vent gas from prod. of UDMH from -COOH hydrazides	10
K109	Spent filter catridges from purif. of UDMH prod. from carboxylic acid hydrazides	10
K110	Condensed column overheads from separation in UDMH prod. from -COOH hydrazides	10
K111	Product washwaters from prod. of dinitrotoluene via nitration of toluene	10
K112	Reaction by-product water from drying in toluenediamine prod from dinitrotoluene	10
K113	Condensed liquid light ends from purification of toluenediamine during its prod.	10
K114	Vicinals from purification of toluenediamine during its prod from dinitrotoluene	10
K115	Heavy ends from toluenediamine purification during prod. from dinitrotoluene	10
K116	Organic condensate from solvent recovery system in prod. of toluene diisocyanate	10
K117	Wastewater from vent gas scrubber in ethylene bromide prod by ethene bromination	1
K118	Spent absorbent solids in purification of ethylene dibromide in its prod.	1
K123	Process waterwater from the prod. of ethylenebisdithiocarbamic acid and salts	10
K124	Reactor vent scubber water from prod of ethylenebisdithiocarbamic acid and salts	10
K125	Filtration/other solids from prod. of ethylenebisdithiocarbamic acid and salts	10
K126	Dust/sweepings from the prod. of ethylenebisdithiocarbamic acid and salts	10
K131	Wastewater and spent sulfuric acid from the prod. of methyl bromide	100
K132	Spent absorbent and wastewater solids from the prod. of methyl bromide	1,000
K136	Still bottoms from ethylene dibromide purif. in prod. by ethene bromination	1
K140	Floor sweepings, etc., from the production of 2,4,6-tribromophenol	100
K141	Process residues from coal tar recovery in coking	1
K142	Tar storage tank residues from coke prod. from coal or recovery of coke by-prods	1
K143	Process residues from recovery of light oil in coking	1
K144	Wastewater residues from light oil refining in coking	1
K145	Residues from naphthalene collection and recovery from coke by-products	1
K147	Tar storage tank residues from coal tar refining in coking	1
K148	Residues from coal tar distillation. including still bottoms, in coking	1
K149	Distillation bottoms from the prod. of chlorinated toluenes/benzoyl chlorides	10
K150	Organic residuals from Cl gas and HCl recovery from chlorinated toluene prod.	10
K151	Wastewater treatment sludge from production of chlorotoluenes/benzoyl chlorides	10
K156	Organic waste from production of carbamates and carbamoyl oximes	1*
K157	Wastewaters from production of carbamates and carbamoyl oximes (not sludges)	1*

* RCRA carbamate waste; statutory one-pound RQ applies until RQs are adjusted.

RCRA WASTE STREAMS AND UNLISTED HAZARDOUS WASTES
THE DESCRIPTIONS OF THE WASTE STREAMS HAVE BEEN TRUNCATED.
THIS LIST SHOULD BE USED FOR REFERENCE ONLY
COMPLIANCE INFORMATION CAN BE FOUND IN 40 CFR PART 302 AND TABLE 302.4

RCRA Code	Description	RQ (lbs)
K158	Bag house dusts & filter/separation solids from prod of carbamates, carb oximes	1*
K159	Organics from treatment of thiocarbamate waste	1*
K161	Purif. solids/bag house dust/sweepings from prod of dithiocarbamate acids/salts	1*
K169	Crude oil storage tank sediment from refining operations	10
K170	Clarified slurry oil tank sediment of in-line filter/separation solids	1
K171	Spent hydrotreating catalyst	1
K172	Spent hydrorefining catalyst	1
D001	Unlisted hazardous wastes characteristic of ignitability	100
D002	Unlisted hazardous wastes characteristic of corrosivity	100
D003	Unlisted hazardous wastes characteristic of reactivity	100
	Unlisted hazardous wastes characteristic of toxicity:	
D004	Arsenic	1
D005	Barium	1,000
D006	Cadmium	10
D007	Chromium	10
D008	Lead	10
D009	Mercury	1
D010	Selenium	10
D011	Silver	1
D012	Endrin	1
D013	Lindane	1
D014	Methoxychlor	1
D015	Toxaphene	1
D016	2,4-D	100
D017	2,4,5-TP	100
D018	Benzene	10
D019	Carbon tetrachloride	10
D020	Chlordane	1
D021	Chlorobenzene	100
D022	Chloroform	10
D023	o-Cresol	100
D024	m-Cresol	100
D025	p-Cresol	100
D026	Cresol	100
D027	1,4-Dichlorobenzene	100
D028	1,2-Dichloroethane	100
D029	1,1-Dichloroethylene	100
D030	2,4-Dinitrotoluene	10
D031	Heptachlor (and epoxide)	1
D032	Hexachlorobenzene	10
D033	Hexachlorobutadiene	1
D034	Hexachloroethane	100
D035	Methyl ethyl ketone	5,000

* RCRA carbamate waste; statutory one-pound RQ applies until RQs are adjusted.

RCRA WASTE STREAMS AND UNLISTED HAZARDOUS WASTES
THE DESCRIPTIONS OF THE WASTE STREAMS HAVE BEEN TRUNCATED.
THIS LIST SHOULD BE USED FOR REFERENCE ONLY
COMPLIANCE INFORMATION CAN BE FOUND IN 40 CFR PART 302 AND TABLE 302.4

RCRA Code	Description	RQ (lbs)
D036	Nitrobenzene	1,000
D037	Pentachlorophenol	10
D038	Pyridine	1,000
D039	Tetrachloroethylene	100
D040	Trichloroethylene	100
D041	2,4,5-Trichlorophenol	10
D042	2,4,6-Trichlorophenol	10
D043	Vinyl chloride	1

Information Sources

For copies of this or other Title III or CAA 112(r) documents, contact:

U.S. Environmental Protection Agency
National Center for Environmental Publications and Information (NCEPI)
P.O. Box 42419
Cincinnati, OH 45242
1-800/490-9198
FAX: (513) 489-8695

http://www.epa.gov/ncepihom/orderpub.html

Please order using the full publication title and publication number on the title page. The publication number for this document is 550-B-98-017. There is a limit of five titles in a two-week period.

A dBASE version of this consolidated list, with a print program, is available to be downloaded from the Internet at:

http://www.epa.gov/swercepp/tools.html

The dBASE files are provided for users who wish to manipulate the lists or incorporate them into other databases.

A .PDF version of this document, which can be downloaded and printed, is available under General Publications at:

http://www.epa.gov/swercepp/pubs.html

Questions concerning changes to the list or other aspects of Title III of SARA and section 112(r) of the Clean Air Act may be addressed to:

Emergency Planning and Community Right-to-Know Information Hotline
U.S. Environmental Protection Agency (5104)
401 M Street, SW
Washington, DC 20460

1-800-424-9346 or 703-412-9810 (TDD: 800-553-7672)
9:00 am to 6:00 pm, Eastern Time, Monday - Friday.

APPENDIX D

MACT STANDARDS

EPA - Clean Air Act - Title III
2 Year Final MACT Standards

MACT STANDARD Source Categories Affected	CFR Sub Parts	Final Fed Register Date & Citation	Initial Compliance Date	Project Lead	Implementation Lead	Compliance Lead
Dry Cleaning						
• Commercial drycleaning dry-to-dry • Commercial drycleaning transfer machines • Commercial drycleaning transfer machines • Industrial drycleaning dry-to-dry • Industrial drycleaning transfer machines	M	09/22/93 (58FR49354)	12/20/93	George Smith (919) 541-1549 smith.georgef@epa.gov	None	Joyce Chandler (202) 564-7073
Hazardous Organic NESHAP	F, G, H, I	04/22/94 (59FR19402)	10/24/94	Jan Meyer (919) 541-5254 meyer.jan@epa.gov		Jeff Kenknight (202) 564-7033

EPA - Clean Air Act - Title III
4 Year Final MACT Standards

MACT STANDARD Source Categories Affected	CFR Sub Parts	Final Fed Register Date & Citation	Initial Compliance Date	Project Lead	Implementation Lead	Compliance Lead
Aerospace Industry	GG	09/01/95 (60FR45948)	09/01/98	Jim Szykman (919) 541-2452 szykman.jim@epa.gov	Ingrid Ward (919) 541-0300 ward.ingrid@epa.gov	Anthony Raia (202) 564-6045 raia.anthony@epa.gov
Asbestos (delisted)		11/30/95 (60FR61550)	11/30/95	Susan Zapata (919) 541-5167 zapata.susan@epa.gov	none	Tom Ripp (202) 564-7003
Chromium Electroplating • Chromic Acid Anodizing • Chromic Acid Anodizing* • Decorative Chromium Electroplating • Decorative Chromium Electroplating* • Hard Chromium Electroplating • Hard Chromium Electroplating*	N	01/25/95 (60FR49848)	01/25/96 deco 01/25/97 others	Lalit Banker (919) 541-5420 banker.lalit@epa.gov	none	Scott Throwe (202) 564-7013 throwe.scott@epa.gov
Coke Ovens	L	10/27/93 (58FR57898)	11/15/93	Amanda Aldridge (919) 541-5268 aldridge.amanda@epa.gov	none	Maria Malave (202) 564-7027 malave.maria@epa.gov
Commercial Sterilizers • Commercial Sterilization Facilities • Commerical Sterilization Facilities*	O	12/06/94 (59FR62585)	12/06/97	David Markwordt (919) 541-0837 markwordt.david@epa.gov	Gil Wood (919) 541-5272 wood.gil@epa.gov	Karin Leff (202) 564-7068 leff.karin@epa.gov
Degreasing Organic Cleaners (Halogenated Solvent Cleaning) • Halogenated Solvent Cleaners • Halogenated Solvent Cleaners*	T	12/02/94 (59FR61801)	12/02/97	Paul Almodovar (919) 541-0283 almodovar.paul@epa.gov	Ingrid Ward (919) 541-0300 ward.ingrid@epa.gov	Tracy Back (202) 564-7076 back.tracy@epa.gov

D.2

EPA - Clean Air Act - Title III
4 Year Final MACT Standards

MACT STANDARD Source Categories Affected	CFR Sub Parts	Final Fed Register Date & Citation	Initial Compliance Date	Project Lead	Implementation Lead	Compliance Lead
Industrial Cooling Towers	Q	09/08/94 (59FR46339)	03/08/96	Phil Mulrine (919) 541-5289 mulrine.phil@epa.gov	none	Mimi Guernica (202) 564-2415 guernica.mimi@epa.gov
Magnetic Tape	EE	12/15/94 (59FR64580)	12/15/96	Gail Lacy (919) 541-5261 lacy.gail@epa.gov	none	Seth Heminway (202) 564-7017 heminway.seth@epa.gov
Marine Vessels	Y	09/19/95 (60FR48388)	09/19/99	David Markwordt (919) 541-0837 markwordt.david@epa.gov	none	Virginia Lathrop (202) 564-7057 lathrop.virginia@epa.gov
Off-Site Waste Treatment	DD	07/01/96 (61FR34139)	07/01/99	Elaine Manning (919) 541-5499 manning.elaine@epa.gov	Larry Brockman (919) 541-5398 brockman.larry@epa.gov	Walt Derieux (202) 564-7067 derieux.walt@epa.gov
Petro Refineries	CC	08/18/95 (60FR4344)	08/18/98	Jim Durham (919) 541-5672 durham.jim@epa.gov	Larry Brockman (919) 541-5398 brockman.larry@epa.gov	Tom Ripp (202) 564-7003 ripp.tom@epa.gov
Printing/Publishing	KK	05/30/96 (61FR27132)	05/30/99	Dave Salman (919) 541-0859 salman.dave@eoa.gov/td>	none	Ginger Gotliffe (202) 564-7072 gotliffe.ginger@epa.gov
Polymers & Resins I	U	09/05/96 (61FR469006)	03/05/97	Bob Rosensteel (919) 541-5608 rosensteel.bob@epa.gov	Sheila Milliken (919) 541-5398 milliken.sheila@epa.gov	Sally Sasnett (202) 564-7074 sasnett.sally@epa.gov/td>

Polymers & Resins I

- Butyl Rubber
- Epichlorohydrin Elastomers
- Ethylene Propylene Rubber
- Hypalon (TM) Production
- Neoprene Production
- Nitrile Butadiene Rubber
- Polybutadiene Rubber
- Polysulfide Rubber
- Styrene-Butadiene Rubber & Latex

EPA - Clean Air Act - Title III
4 Year Final MACT Standards

MACT STANDARD Source Categories Affected	CFR Sub Parts	Final Fed Register Date & Citation	Initial Compliance Date	Project Lead	Implementation Lead	Compliance Lead
Polymers & Resins II • Epoxy Resins Production • Non-Nylon Polyamides Production	W	03/08/95 (60FR12670)	03/03/98	Randy McDonald (919) 541-5402 mcdonald.randy@epa.gov	none	Sally Sasnett (202) 564-7074 sasnett.sally@epa.gov
Polymers & Resins IV =Acrylonitrile-Butadiene-Styrene =Methyl Methacrylate-Acrylonitrile+ Methyl Methacrylate-Butadiene++ =Polystyrene Styrene Acrylonitrile Polyethylene Terephthalate	JJJ	09/12/96 (61FR48208)	03/12/97	Bob Rosensteel (919) 541-5608 rosensteel.bob@epa.gov	Sheila Milliken (919) 541-2625 milliken.sheila@epa.gov	Sally Sasnett (202) 564-7074 sasnett.sally@epa.gov
Secondary Lead Smelters	X	06/23/95 (60FR32587)	06/23/97	Kevin Cavender (919) 541-2364 cavender.kevin@epa.gov	none	Jane Engert (202) 564-5021 engert.jane@epa.gov
Shipbuilding MACT	II	12/15/95 (60FR64330)	12/16/97	Mohamed Serageldin (919) 541-2379 serageldin.mohamed@epa.gov	none	Suzanne Childress (202) 564-7018 childress.suzanne@epa.gov
Gasoline Distribution	R	12/14/94 (59FR64303)	12/15/97	Steve Shedd (919) 541-5397 shedd.steve@epa.gov	none	Julie Tankersley (202) 564-7002 tankersley.julie@epa.gov
Wood Furniture	JJ	12/07/95 (60FR62930)	11/21/97	Paul Almodovar (919) 541-0283 almodovar.paul@epa.gov	Gil Wood (919) 541-5272 wood.gil@epa.gov	Robert Marshall - (202) 564-7021

D.4

EPA - Clean Air Act - Title III
7 Year Final MACT Standards

MACT STANDARD Source Categories Affected	CFR Sub Parts	Final Fed Register Date & Citation	Initial Compliance Date	Project Lead	Implementation Lead	Compliance Lead
Chromium Chemical Manufacturing		Delisted 06/04/96 (61FR28197)		Iliam Rosario (919) 541-5308 rosario.iliam@epa.gov		
EAF: Stainless>br>& Non-Stainless Steel		Delisted 06/04/96 (61FR28197)		Phil Mulrine (919) 541-5289 mulrine.phil@epa.gov		
Ferroalloys Production	XXX	05/20/99 (64FR27450)		Conrad Chin (919) 541-1512 conrad.chin@epa.gov	none	Maria Malave (202) 546-7027 malave.maria@epa.gov
Flexible Polyurethane Foam Production	III	10/07/98 (63FR53980)		David Svendsgaard (919) 541-2380 svendsgaard.david@epa.gov	Ingrid Ward (919) 541-0300 ward.ingrid@epa.gov	Maria Malave (202) 546-7027 malave.maria@epa.gov
Generic MACT± • Acetal Resins • Hydrogen Fluoride • Polycarbonates Production • Acrylic/Modacrylic Fibers	YY	6/29/99 (64FR34853)		David Markwordt (919) 541-0837 markwordt.david@epa.gov	none	Belinda Breidenbach (202) 564-7022 breidenbach.belinda@epa.gov
Mineral Wool Production	DDD	06/01/99 (64FR29489)		Mary Johnson (919) 541-5025 johnson.mary@epa.gov		Scott Throwe (202) 564-7013 throwe.scott@epa.gov
Nylon 6 Production		delisted 02/12/98 (63FR7155)		Mark Morris (919) 541-5416 morris.mark@epa.gov		
Oil & Natural Gas Production	HH	06/17/99 (64FR32609)		Greg Nizich (919) 541-3078 nizich.greg@epa.gov		Dan Chadwick (202) 564-7054 chadwick.dan@epa.gov

D.5

EPA - Clean Air Act - Title III
7 Year Final MACT Standards

MACT STANDARD Source Categories Affected	CFR Sub Parts	Final Fed Register Date & Citation	Initial Compliance Date	Project Lead	Implementation Lead	Compliance Lead
Pesticide Active Ingredient Production • 4-Chlroo-2-Methyl Acid Production • 2,4 Salts & Esters Production • 4,6-dinitro-o-cresol Production • Butadiene Furfural Cotrimer • Captafol Production • Captan Production • Chloroneb Production • Chlorothalonil Production • Dacthal (tm) production • Sodium Pentachlorophenate Production • Tordon (tm) Acid Production	MMM	06/23/99 (64FR3549)		Lalit Banker (919) 541-5420 banker.lalit@epa.gov	Ingrid Ward (919) 541-0300 ward.ingrid@epa.gov	Steve Howie (202) 564-4146 howie.steve@epa.gov
Pharmaceuticals Production	GGG	09/21/98 (63FR50280)		Randy McDonald (919) 541-5402 mcdonald.randy@epa.gov	Shelia Milliken (919) 541-2625 milliken.shelia@epa.gov	Joanne Berman (202) 564-7064 berman.joanne@epa.gov
Phosphoric Acid/ Phosphate Fertilizers	AA BB	06/10/99 (64FR31358)		Ken Durkee (919) 541-5425 durkee.ken@epa.gov		Steve Howie (202) 564-4146 howie.steve@epa.gov
Polyether Polyols Production	PPP	06/01/99 (64FR29419)		Penny Lassiter (919) 541-5396 lassiter.penny@epa.gov	Ingrid Ward (919) 541-0300 ward.ingrid@epa.gov	Joanne Berman (202) 564-7064 berman.joanne@epa.gov
Portland Cement Manufacturing	LLL	06/14/99 (64FR31898)		Joe Wood (919) 541-5446 wood.joe@epa.gov	Julie McClintock (919) 541-5339 mcclintock.julie@epa.gov	Scott Throwe (202) 564-7013 throwe.scott@epa.gov
Primary Aluminum	LL	10/07/97 (62FR52407)		Steve Fruh (919) 541-2837 fruh.steve@epa.gov		Deborah Thomas (202) 564-5041 thomas.deborah@epa.gov

MACT STANDARD Source Categories Affected	CFR Sub Parts	Final Fed Register Date & Citation	Initial Compliance Date	Project Lead	Implementation Lead	Compliance Lead
Primary Lead Smelting	TTT	06/04/99 (64FR30194)		Kevin Cavender (919) 541-2364 cavender.kevin@epa.gov		Deborah Thomas (202) 564-5041 thomas.deborah@epa.gov
Pulp & Paper (non-combust) MACT I	S	04/15/98 (63FR18504)		Steve Shedd (919) 541-5397 shedd.steve@epa.gov	Gill Wood (919) 541-5515 painter.david@epa.gov	Seth Herminway (202) 563-7017 herminway.seth@epa.gov
Pulp & Paper (non-chem) MACT III	S	04/15/98 (63FR18504)		Elaine Manning (919) 541-5499 manning.elaine@epa.gov	Gil Wood (919) 541-5272 wood.gil@epa.gov	Seth Herminway (202) 563-7017 herminway.seth@epa.gov
Steel Pickling-HCL Process	CCC	06/22/99 (64FR33202)		Jim Maysilles (919) 541-3265 maysillis.jim@epa.gov	Gil Wood (919) 541-5272 wood.gil@epa.gov	Maria Malave (202) 564-7027 malave.maria@epa.gov
Tetrahydrobenzaldehyde Manufacture Formerly known as Butadiene Dimers Production	F	05/12/98 (63FR26078)		John Schaefer (919) 541-0296 schaefer.john@epa.gov		Marcia Mia (202) 564-7042 mia.marcia@epa.gov
Wood Treatment MACT		Delisted 06/04/96 (61FR28197)		Gene Crumpler (919) 541-0881 crumpler.gene@epa.gov		
Wool Fiberglass Manufacturing	NNN	06/14/99 (64FR31695)		Bill Neuffer (919) 541-5435 neuffer.bill@epa.gov		Scott Throwe (202) 565-7013 throwe.scott@epa.gov

MACT STANDARD Source Categories Affected	CFR Sub Parts	Final Fed Register Date & Citation	Initial Compliance Date	Project Lead	Implementation Lead	Compliance Lead
Cyanuric Chloride Production		Delisted 02/12/98 (63FR7155)		Keith Barnett (919) 541-5605 barnett.keith@epa.gov		
Lead Acid Battery Manufacturing		Delisted 05/17/96		Walt Stevenson (919) 541-5264 stevenson.walt@epa.gov		
Natural Gas Transmission and Storage	HHH	06/17/99 (64FR32609)		Greg Nizich (919) 541-3078 nizich.greg@epa.gov	Larry Brockman (919) 541-5398 brockman.larry@epa.gov	Dan Chadwick (202) 564-7054 chadwick.dan@epa.gov

APPENDIX E
AIR QUALITY MODELS

Instructions for accessing the listed models are available at the EPA Website www.epa.gov/ttn/scram/t22.htm.

REGULATORY MODELS

ISC3 Models

ISCST3 -- The ISCST3 Short Term Model (376KB,ZIP)
        Dated: 6/24/99

TEST-ST -- Original ISCST Test Case (20KB,ZIP)
        Dated: 6/24/99

AREATEST -- Area Source Test Case (8KB,ZIP)
        Dated: 6/24/99

DEPTEST -- Deposition Test Case (186KB,ZIP)
        Dated: 6/24/99

GASDTEST -- Gas Dry Deposition Test Case (12KB,ZIP)
        Dated: 6/24/99

TEST-PM -- PM Test Case (800KB,ZIP)
        Dated: 6/24/99

ISCSTSRC -- ISCST Source Code (221KB,ZIP)
        Dated: 6/24/99

ISC3MET -- BINTOASC and METLIST Programs (73KB,ZIP)

ISCLT3 -- ISCLT3 Long Term Model (220KB,ZIP)

TEST-LT -- Original ISCLT Test Case (20KB,ZIP)

AREALT -- Area Source Test Case (8KB,ZIP)

DEPLT -- Deposition Test Case (7KB,ZIP)

ISCLTSRC -- ISCLT3 Source Code (135KB,ZIP)

Urban Airshed Model

UAM1 -- UAM Source Code (96KB,ZIP)

UAM2 -- Preprocessor source Code (162KB,ZIP)

UAM3 -- UAM Conversion Programs (181KB,ZIP)

UAM4 -- UAM Readme, Emissions Prep Sys Source Code (202KB,ZIP)

UAM5 -- UAM Diagnostic Wind Model (DWM) Source Code (66KB,ZIP)

UAM6 -- UAM ROM-UAM Interface Source Code (87KB,ZIP)

EPS2TABB -- UAM Emissions Processing Sys 2.0 (Tables-B) (16KB,ZIP)

CHEMPARM -- UAM Chemical Data Base File (5KB,TXT)

DISPLAY1 -- UAM Display Output File #1 (29KB,ZIP)

DISPLAY2 -- UAM Display Output file #2 (61KB,ZIP)

UAMUNIX -- UAM IBM to UNIX Conversion (9KB,ZIP)

BIOMASS -- UAM County Biomas Conversion (33KB,ZIP)

USERTIPS -- UAM User Tips (26KB,TXT)

UAMQAS-- UAM Quality Assurance System (26KB,ZIP)

UAMPES -- UAM Performance Evaluation system (14KB,ZIP)

UAMPPS -- UAM Postprocessing System (22KB,ZIP)

EPS2FORT -- UAM Emission Preprocessor system (EPS) (31KB,ZIP)

EPS2DATA -- UAM EPS Data Files (72KB,ZIP)

EPS2TEST -- UAM EPS Test Files (Updated 10/2/95) (20KB,ZIP)

EPS2SAS -- UAM EPS Emissions Display System (15KB,ZIP)

UAMTRAIN -- UAM Training, Questions and Answers (10KB,TXT)

EPS2MODS -- Changes to EPS2; See Amendment 4, MCB#7 (29KB,TXT)

UAMCHEM -- UAM (93287) Changes (32KB,ZIP)

ROMUPDTE -- ROM Updates (3KB,ZIP)

BEIS -- BEIS Source Code (30KB,ZIP)

BEIS2 -- BEIS2 Source Code (52KB,ZIP)Updated-10/6/97

CNTLEM -- Expanded Documentation o n use of CNTLEM (16KB,ZIP)

DISPLAY -- Average Concentration Output Program (10KB,ZIP)

ATL-ELPA -- UAM Surrogate Data:Atlanta-ElPaso (36KB,ZIP)

HOUS-NY -- UAM Surrogate Data:Houston-New York (42K,ZIP)

PHIL-STL -- UAM Surrogate Data:Philadelphis-St. Louis (41KB,ZIP)

DOMAINS -- Regulatory Contacts/Coordinators for SIP Applications (4KB,ZIP)

BEIS2TC -- Test case for October 1997 Release of BEIS2 (1.3MB,ZIP)

BEIS2ADR -- Internet addresses for BEIS2 data (.4KB,TXT)

BEISNOTE -- Release note on the compilation of BEIS2. (TXT)

Other Regulatory Models

BLP -- BLP Model (213KB,ZIP) Updated:11/99

CALINE3 -- Caline3 Model (53KB,ZIP)

CDM2 -- CDM2 Model (76KB,ZIP)

EKMA1 -- EKMA Source code and Test Cases (150KB,ZIP)

EKMA2 -- EKMA Model, PC Executable (103KB,ZIP)

EKMA3 -- EKMA Input Generator (214KB,ZIP)

EKMATRAJ -- Trajectory Model for EKMA (Spreadsheet) (18KB,ZIP)

RAM -- RAM Model (144KB,ZIP)

CTDMP1 -- CTDMPLUS Executables 1 and Readme File (170KB,ZIP)

CTDMP2 -- CTDMPLUS Executables 2 (221KB,ZIP)

CTDMP2E -- CTDMPLUS Extended Memory Executables (174KB,ZIP)

CTDMP3 -- CTDMPLUS Fortran Source Code (108KB,ZIP)

CTDMP4 -- METPRO Fortran Source Code (26KB,ZIP)

CTDMP5 -- FITCON, HCRIT Fortran Source Code (24KB,ZIP)

CTDMP6 -- READ62 Fortran Source Code (5KB,ZIP)

CTDMP7 -- CHRIT Fortran Source Code (7KB,ZIP)

CTDMP8 -- Pascal Source Code (64KB,ZIP)

CTDMP9 -- Test Case I/O Files (38KB,ZIP)

CTDMP10 -- Readme and BAT Files (7KB,ZIP)

CAL3QHC -- CAL3QHC Model PC Executable; CAL3QHCR Model included (323KB,ZIP)

OCD5 - OCD version 5 Executable (7.77MB,ZIP)

OCD5FOR - OCD version 5 source code(175KB,EXE)

SCREENING MODELS

COMPLEX1 -- Complex1 Model PC version Executable (129KB,ZIP)

CTSCRN1 -- CTSCRN Model, Readme and Test Case (332KB,ZIP)

CTSCRN1E -- CTSCRN Model Extended Memory Executable (254KB,ZIP)

CTSCRN2 -- CTSCRN Model Source Code (126KB,ZIP)

LONGZ -- LONGZ Model, IBM 3090 Mainframe Version (56KB,ZIP)

RTDM32 -- RTDM32 Model, PC Version, Executable (160KB,ZIP)

SHORTZ -- SHORTZ Model, IBM 3090 Mainframe Version (79KB,ZIP)

VALLEY -- VALLEY Model, IBM 3090 Mainframe Version (23KB,ZIP)

VISCREEN -- VISCREEN Model PC Version, Executable (121KB,ZIP)

TSCREEN1 -- TSCREEN Model, File 1 (568KB,ZIP)

TSCREEN2 -- TSCREEN Model, File 2 (193KB,ZIP)

TSCREEN3 -- TSCREEN Model, File 3 (342KB,ZIP)

TSCREEN4 -- TSCREEN Model, File 4 (182KB,ZIP)

TSCRCODE -- TSCCREEN Source Code (418KB,ZIP)

SCREEN3 -- SCREEN 3 Model, Executable, Source Code and Readme (176KB,ZIP)

OTHER MODELS

ALTMODEL -- Alternative Air Quality Models(232KB,PDF)
Please read the "Explanatory Note" first.

DEGSRC -- DEGADIS Model Source Code (265KB,ZIP)

DEGEXE -- DEGADIS Mode Executable (691KB,ZIP)

CMB7 -- CMB7 Model , PC Version (311KB,ZIP)

RUNPLUVU -- PLUVUE2 Model Executable and Readme File

(606KB,ZIP)

RNPLUVU2 -- PLUVUE2 Source code(164KB,ZIP)

SDM -- SDM Model Executable, Readme and Test Cases #1 and #2
(328KB,ZIP)

MESOCODE -- MESOPUFF Source code (124KB,ZIP)

MESOPUF1 -- MESOPUFF Executable, Test Inputs and Readme
(342KB,ZIP)

MESOPUF2 -- Additional Program Executables for MESOPUFF
(673KB,ZIP)

MESOOUT -- MESOPUFF Test Case Outputs (560KB,ZIP)

OZIPRMOD -- OZIPR - A one-dimensional photochemical box model.
Executable, test inputs, source code (907KB,ZIP)
Please read the readme first. Supporting report can
can be found under Supporting Reports on the Model
Support/Guidance web page- see file: OZIPRPT.

RPM_EXE -- RPM-IV Executables and Readme (808KB,ZIP)

RPM_FOR -- RPM-IV Source Code (199KB,ZIP)

RPM_DATA -- RPM-IV Sample Inputs and Outputs (35KB,ZIP)

RPM_TIPS -- RPM-IV Usertips (1KB,TXT)

RELATED PROGRAMS

CALMPRO -- Meteorological CALMPRO Program (12KB,ZIP, IBM Mainframe)

CHAVG -- Meteorological CHAVG Program (32KB,ZIP, IBM Mainframe Version)

WINDROSE -- WINDROSE Program, PC Version and Test Files (285KB,ZIP)

STAR -- Stability Rose Program and Test Case ; PC Version (216KB,ZIP)

CONCOR -- Converts LL to UTM and UTM to LL (31KB,ZIP)

WRPLOT --WindRose Plotting Program (1087KB,ZIP)

BPIP -- BPIP Executable, Source Code and Readme (91KB,ZIP)

MPRM

MPRM -- MPRM Readme (Revised 12/15/99)

MPRM1 -- MPRM data extraction and quality assurance executable
(249KB,ZIP) (Revised 12/15/99)

MPRM2 -- MPRM program executable (232KB,ZIP) (Revised 12/15/99)

MPRM3 -- MPRM test cases (341KB,ZIP) (Revised 12/15/99)

MPRM4 -- MPRM source code (204KB,ZIP) (Revised 12/15/99)

PCRAMMET (Updated 8/14/98)

PCRAMMET -- PCRAMMET Preprocessor Program executable (296KB,ZIP)
Readme file included. Updated 12/2/99

PCREX1 -- PCRAMMET example using simulated Year 2000 HUSWO data
New 12/2/99 (254KB,ZIP)

PCREX2 -- PCRAMMET SAMSON Year 1998 example (403KB,ZIP)
Updated 12/2/99

PCREX3 -- PCRAMMET SCRAM Year 1998 example (175KB,ZIP)
Updated 12/2/99

PCREX4 -- PCRAMMET CD144/TD3240 variable format year 1998
example (323KB,ZIP) Updated 12/2/99

PCREX5 -- PCRAMMET example using CD144/TD3240 fixed format
Year 1998 example (323KB,ZIP) New 12/2/99

PCRMCODE -- PCRAMMET source code (52KB,ZIP)
Updated 12/2/99

Mixing Height Program (12/6/98)

MIXHTS -- Mixing Height program executable (130KB,ZIP)
Readme file included.

MIXEX1 -- Mixing Height example (827KB,ZIP)

MIXHCODE -- Mixing Height source code (19K,ZIP)

UAM-V

The United States Environmental Protection Agency(US EPA), with the permission of Systems Applications International, Inc. is providing public access to version 1.24 of the UAM-V modeling system for governmental agencies for bona fide governmental purposes. Government users within the U.S. may download the UAM-V modeling system in accordance with the terms provided. Files with the *.exe extension are self extracting ZIP files. After downloading to your PC, file expansion is accomplished by typing the filename (ex. UAMV124.EXE). When appropriate, a README file will provide instructional information.

8/05/99

The UAM-V modeling system (versions 1.24 and 1.30) is now available directly from SAI for general use and at no cost to interested parties. A license agreement is no longer required, although the copyright still applies. To register as a user and obtain your copy of the UAM-V modeling system software, contact Sharon Douglas at sgd@saintl.com or SDouglas@icfconsulting.com. Government user's within the U.S. may still access version 1.24 (the OTAG version) of the modeling system from this webpage. Such users are encouraged to register with SAI to receive information on updates to the code, training workshops, and available databases. The updated version of the UAM-V modeling system (version 1.30) includes process-analysis capabilities, an enhanced chemical mechanism (enhanced treatment of hydrocarbon and toxic species), updated deposition and nested-grid algorithms, a flexible coordinate system (including Lambert conformal), and user-selection of a "standard " or "fast" solver.

> README -- Overview of UAM-V Modeling System (10KB,TXT)
>   Updated 8/05/99
>
> UAMV124 -- UIAM-V Modeling System (376KB,EXE)
>
> UAMVMAN -- UAM-V User's Guide (WP5) (200KB,EXE)
>
> TCREADME -- OTAG Test-Case Data Information (8KB,TXT)
>
> UVPCODE -- UAM-V Post Processing System Source Code (630KB,EXE), revised 5/8/97.
>
> UPSUG -- UAM-V Post Processing System User's Guide (WP5) (52KB,EXE)
>
> UPSUGFIG -- Graphic Images for the UAM-V Post Processor System User's Guide (WP5) (1680KB,EXE)
>
> RAMSUAMV -- Preprocessor to convert RAMS meteorology output to UAMV input format (73KB,EXE)
>
> METEOROL -- Meteorology programs (including SAIMM) for UAM-V (270KB,EXE)

AIRQUAL -- Air quality preprocessor programs (Source Code) (301KB,EXE)

EMISUTIL -- Emission input preprocessor/utility programs for UAM-V (15KB,EXE)

OTHPREPS -- Preprocessors for ozone column, landuse and photoylsis rates (386KB,EXE)

PREPUG -- Documentation for UAM-V (related) preprocessor programs (WP5) (1MB,EXE)

INSTPROC -- UAM-V installation procedures for multiple platforms (WP5) (6KB,ZIP)

BEIS2 -- BEIS2 Source Code (52KB,ZIP)Updated-10/6/97

CB-IVEVOL -- Evolution of CB-IV, differences and principal applications (WP6) (8KB,ZIP)

BEIS2TC -- Test case for October 1997 Release of BEIS2 (1.3MB,ZIP)

OTAG1HR -- UAM-V Model Performance in OTAG Simulations for 1-hour
          Statistics (865KB,ZIP)

OTAG8HR -- UAM-V Model Performance in OTAG Simulations for 8-hour
          Statistics (612KB,ZIP)

BEIS2ADR -- Internet addresses for BEIS2 data (.4KB,TXT)

BEISNOTE -- Release note on the compilation of BEIS2. (TXT)

UAMVASC -- Convert UAM-V input and output binary files to ascii. (26KB,EXE)

UAMVBINR -- Convert UAM-V input and output ascii files generated by UAMVASC
           back to binary. (26KB,EXE)

NON-EPA MODELS

This area contains non-EPA models. These models are not supported or maintained by the U.S. EPA. All available information relating to the model/programs are contained within each ZIPPED file. Questions and requests relating to updates should be directed to the originator.

SLAB - SLAB Dense Gas Dispersion Model(111KB,ZIP)

ADAM - ADAM Dense Gas Dispersion Model (277KB,ZIP)

AFTOX - AFTOX Toxic Chemical Dispersion Model (173KB,ZIP)

AFTOXDOC - AFTOX Model User's Guide, ASCII (26KB,ZIP)

OBOD1315 - Open Burning/Open Detonation Model (744KB,ZIP)
Updated (2/3/00)

APPENDIX F
EPA WATER-PERMITTING FORMS

- EPA 3320-1: Discharge Monitoring Report

- EPA 3510-1: General Information

- EPA 3510-2C: Existing Manufacturing, Commercial, Mining and Silvicultural Operations

- EPA 3510-2D: New Sources and New Dischargers Application for Permit to Discharge Process Wastewater

- EPA 3510-2E: Facilities Which Do Not Discharge Process Wastewater

- EPA 3510-2F: Application for Permit to Discharge Stormwater Discharges Associated with Industrial Activity

- EPA 3510-6: Notice of Intent (NOI) for Storm Water Discharges

- EPA 3510-7: Notice of Termination (NOT) of Coverage Under the NPDES General Permit for Storm Water Discharges Associated with Industrial Activity

- EPA 3510-9: Notice of Intent (NOI) for Storm Water Discharges Associated with Construction Activity

- EPA 3560-3: NPDES Compliance Inspection Report

- EPA 7520-6: Underground Injection Control Permit Application

- EPA 7530-1: Notification for Underground Storage Tanks

- EPA 7550-6: NPDES Application for Permit to Discharge — Short Form A

PERMITTEE NAME/ADDRESS (Include
Facility Name/Location if different)
NAME _____
ADDRESS _____

FACILITY _____
LOCATION _____

NATIONAL POLLUTANT DISCHARGE ELIMINATION SYSTEM (NPDES)
DISCHARGE MONITORING REPORT (DMR)

(17-19)

Form Approved.
OMB No. 2040-0004
Approval expires 5-31-98

PERMIT NUMBER	DISCHARGE NUMBER
(2-16)	

MONITORING PERIOD

	YEAR	MO	DAY		YEAR	MO	DAY
FROM	(20-21)	(22-23)	(24-25)	TO	(26-27)	(28-29)	(30-31)

NOTE: Read instruction before completing this form.

PARAMETER (32-37)		QUANTITY OR LOADING (54-61)			QUALITY OR CONCENTRATION			NO. EX (62-63)	FREQUENCY OF ANALYSIS (64-68)	SAMPLE TYPE (69-70)
		(3 Card Only) (46-53) AVERAGE	MAXIMUM	UNITS	(4 Card Only) (38-45) MINIMUM	AVERAGE	MAXIMUM	UNITS		
	SAMPLE MEASUREMENT									
	PERMIT REQUIREMENT									
	SAMPLE MEASUREMENT									
	PERMIT REQUIREMENT									
	SAMPLE MEASUREMENT									
	PERMIT REQUIREMENT									
	SAMPLE MEASUREMENT									
	PERMIT REQUIREMENT									
	SAMPLE MEASUREMENT									
	PERMIT REQUIREMENT									
	SAMPLE MEASUREMENT									
	PERMIT REQUIREMENT									
	SAMPLE MEASUREMENT									
	PERMIT REQUIREMENT									

QUALITY OR CONCENTRATION (46-53)   (54-61)

NAME/TITLE PRINCIPAL EXECUTIVE OFFICER	I CERTIFY UNDER PENALTY OF LAW THAT I HAVE PERSONALLY EXAMINED AND AM FAMILIAR WITH THE INFORMATION SUBMITTED HEREIN, AND BASED ON MY INQUIRY OF THOSE INDIVIDUALS IMMEDIATELY RESPONSIBLE FOR OBTAINING THE INFORMATION, I BELIEVE THE SUBMITTED INFORMATION IS TRUE, ACCURATE AND COMPLETE. I AM AWARE THAT THERE ARE SIGNIFICANT PENALTIES FOR SUBMITTING FALSE INFORMATION, INCLUDING THE POSSIBILITY OF FINE AND IMPRISONMENT. SEE 18 U.S.C. § 1001 AND 33 U.S.C. § 1319. (Penalties under these statutes may include fines up to $10,000 and or maximum imprisonment of between 6 months and 5 years.)	SIGNATURE OF PRINCIPAL EXECUTIVE OFFICER OR AUTHORIZED AGENT	TELEPHONE	DATE
TYPED OR PRINTED			AREA CODE NUMBER	YEAR MO DAY

COMMENT AND EXPLANATION OF ANY VIOLATIONS (Reference all attachments here)

EPA Form 3320-1 (Rev. 9-88) Previous editions may be used.
STF ENV961F.1

(REPLACES EPA FORM T-40 WHICH MAY NOT BE USED.)

PAGE ____ OF ____

F.2

Please print or type in the unshaded areas only (fill-in areas are spaced for elite type, i.e., 12 characters/inch).

Form Approved. OMB No. 2040-0086. Approval expires 5-31-92.

FORM	U.S. ENVIRONMENTAL PROTECTION AGENCY	1. EPA I.D NUMBER
1 **EPA** GENERAL	**GENERAL INFORMATION** Consolidated Permits Program (Read the "General Instructions" before starting)	S T/A C F D 1 2 13 14 15

LABEL ITEMS		GENERAL INSTRUCTIONS
I. EPA I.D. NUMBER		If a preprinted label has been provided, affix it in the designated space. Review the information carefully; if any of it is incorrect, cross through it and enter the correct data in the appropriate fill-in area below. Also, if any of the preprinted data is absent *(the area to the left of the label space lists the information that should appear)*, please provide it in the proper fill-in area(s) below. If the label is complete and correct, you need not complete Items I, III, V, and VI *(except VI-B which must be completed regardless)*. Complete all items if no label has been provided. Refer to the instructions for detailed item descriptions and for the legal authorizations under which this data is collected.
III. FACILITY NAME		
V. FACILITY MAILING ADDRESS	**PLEASE PLACE LABEL IN THIS SPACE**	
VI. FACILITY LOCATION		

II. POLLUTANT CHARACTERISTICS

INSTRUCTIONS: Complete A through J to determine whether you need to submit any permit application forms to the EPA. If you answer "yes" to any questions, you must submit this form and the supplemental form listed in the parenthesis following the question. Mark "X" in the box in the third column if the supplemental form is attached. If you answer "no" to each question, you need not submit any of these forms. You may answer "no" if your activity is excluded from permit requirements; see Section C of the instructions. See also, Section D of the instructions for definitions of **bold-faced terms**.

SPECIFIC QUESTIONS	MARK "X"			SPECIFIC QUESTIONS	MARK "X"		
	YES	NO	FORM ATTACHED		YES	NO	FORM ATTACHED
A. Is this facility a **publicly owned treatment works** which results in a **discharge** to waters of the U.S.? (FORM 2A)				B. Does or will this facility *(either existing or proposed)* include a **concentrated animal feeding operation** or **aquatic animal production facility** which results in a **discharge** to waters of the U.S.? (FORM 2B)			
	16	17	18		19	20	21
C. Is this a facility which currently results in **discharges** to **waters of the U.S.** other than those described in A or B above? (FORM 2C)				D. Is this a proposed facility *(other than those described in A or B above)* which will result in a **discharge** to **waters of the U.S.**? (FORM 2D)			
	22	23	24		25	26	27
E. Does or will this facility treat, store, or dispose of **hazardous wastes**? (FORM 3)				F. Do you or will you **inject** at this facility industrial or municipal effluent below the lowermost stratum containing, within one quarter mile of the well bore, **underground sources of drinking water**? (FORM 4)			
	28	29	30		31	32	33
G. Do you or will you **inject** at this facility any produced water or other fluids which are brought to the surface in connection with conventional oil or natural gas production, **inject** fluids used for enhanced recovery of oil or natural gas, or **inject** fluids for storage of liquid hydrocarbons? (FORM 4)				H. Do you or will you **inject** at this facility fluids for special processes such as mining of sulfur by the Frasch process, solution mining of minerals, in situ combustion of fossil fuel, or recovery of geothermal energy? (FORM 4)			
	34	35	36		37	38	39
I. Is this facility a proposed **stationary source** which is one of the 28 industrial categories listed in the instructions and which will potentially emit 100 tons per year of any air pollutant regulated under the Clean Air Act and may affect or be located in an **attainment area**? (FORM 5)				J. Is this facility a proposed **stationary source** which is NOT one of the 28 industrial categories listed in the instructions and which will potentially emit 250 tons per year of any air pollutant regulated under the Clean Air Act and may affect or be located in an **attainment area**? (FORM 5)			
	40	41	42		43	44	45

III. NAME OF FACILITY

C 1	SKIP	
15	16 - 29 30	69

IV. FACILITY CONTACT

	A. NAME & TITLE *(last, first, & title)*	B. PHONE *(area code & no.)*
C 2		
15 16	45	46 - 48 49 - 51 52 - 55

V. FACILITY MAILING ADDRESS

	A. STREET OR P.O. BOX
C 3	
15 16	45

	B. CITY OR TOWN	C. STATE	D. ZIP CODE
C 4			
15 16	40	41 42	47 - 51

VI. FACILITY LOCATION

	A. STREET, ROUTE NO. OR OTHER SPECIFIC IDENTIFIER
C 5	
15 16	45

	B. COUNTY NAME
46	70

	C. CITY OR TOWN	D. STATE	E. ZIP CODE	F. COUNTY CODE *(if known)*
C 6				
15 16	40	41 42	47 - 51	52 - 54

EPA Form 3510-1 (8-90)   CONTINUE ON REVERSE

STF ENV594F.1

CONTINUED FROM THE FRONT

VII. SIC CODES *(4 digit in order of priority)*

A. FIRST

C
7
15 16 - 19

(specify)

B. SECOND

C
7
15 16 - 19

(specify)

C. THIRD

C
7
15 16 - 19

(specify)

D. FOURTH

C
7
15 16 - 19

(specify)

VIII. OPERATOR INFORMATION

A. NAME

C
8
15 16

B. Is the name listed in item VIII-A also the owner?

☐ YES ☐ NO

55 66

C. STATUS OF OPERATOR *(Enter the appropriate letter into the answer box; if "Other," specify.)*

F = FEDERAL   M = PUBLIC *(other than federal or state)*
S = STATE   O = OTHER *(specify)*
P = PRIVATE

(specify)

58

D. PHONE *(area code & no.)*

C
A
15 16 - 18 19 - 21 22 - 25

E. STREET OR P.O. BOX

26 - 55

F. CITY OR TOWN

C
B
15 16 - 40

G. STATE 41 42

H. ZIP CODE 47 - 51

IX. INDIAN LAND

Is the facility located on Indian lands?

☐ YES ☐ NO

52

X. EXISTING ENVIRONMENTAL PERMITS

A. NPDES *(Discharges to Surface Water)*

C T
9 N
15 16 17 18 - 30

D. PSD *(Air Emissions from Proposed Sources)*

C T
9 P
15 16 17 18 - 30

B. UIC *(Underground Injection of Fluids)*

C T
9 U
15 16 17 18 - 30

E. OTHER *(specify)*

C T
9
15 16 17 18 - 30

(specify)

C. RCRA *(Hazardous Wastes)*

C T
9 R
15 16 17 18 - 30

E. OTHER *(specify)*

C T
9
15 16 17 18 - 30

(specify)

XI. MAP

Attach to this application a topographic map of the area extending to at least one mile beyond property boundaries. The map must show the outline of the facility, the location of each of its existing and proposed intake and discharge structures, each of its hazardous waste treatment, storage, or disposal facilities, and each well where it injects fluids underground. Include all springs, rivers and other surface water bodies in the map area. See instructions for precise requirements.

XII. NATURE OF BUSINESS *(provide a brief description)*

XIII. CERTIFICATION *(see instructions)*

I certify under penalty of law that I have personally examined and am familiar with the information submitted in this application and all attachments and that, based on my inquiry of those persons immediately responsible for obtaining the information contained in the application, I believe that the information is true, accurate and complete. I am aware that there are significant penalties for submitting false information, including the possibility of fine and imprisonment.

A. NAME & OFFICIAL TITLE *(type or print)*

B. SIGNATURE

C. DATE SIGNED

COMMENTS FOR OFFICIAL USE ONLY

C
C
15 16 - 55

EPA Form 3510-1 (8-90)

EPA I.D. NUMBER *(copy from Item 1 of Form 1)*	Form Approved OMB No. 2040-0086 Approval expires 5-31-92

Please print or type in the unshaded areas only.

FORM 2C EPA NPDES

U.S. ENVIRONMENTAL PROTECTION AGENCY

APPLICATION FOR PERMIT TO DISCHARGE WASTEWATER EXISTING MANUFACTURING, COMMERCIAL, MINING AND SILVICULTURAL OPERATIONS
Consolidated Permits Program

I. OUTFALL LOCATION

For each outfall, list the latitude and longitude of its location to the nearest 15 seconds and the name of the receiving water.

A. OUTFALL NUMBER *(list)*	B. LATITUDE			C. LONGITUDE			D. RECEIVING WATER *(name)*
	1. DEG.	2. MIN.	3. SEC.	1. DEG.	2. MIN.	3. SEC.	

II. FLOWS, SOURCES OF POLLUTION, AND TREATMENT TECHNOLOGIES

A. Attach a line drawing showing the water flow through the facility. Indicate sources of intake water, operations contributing wastewater to the effluent, and treatment units labeled to correspond to the more detailed descriptions in Item B. Construct a water balance on the line drawing by showing average flows between intakes, operations, treatment units, and outfalls. If a water balance cannot be determined *(e.g., for certain mining activities)*, provide a pictorial description of the nature and amount of any sources of water and any collection or treatment measures.

B. For each outfall, provide a description of: (1) All operations contributing wastewater to the effluent, including process wastewater, sanitary wastewater, cooling water, and storm water runoff; (2) The average flow contributed by each operation; and (3) The treatment received by the wastewater. Continue on additional sheets if necessary.

1. OUT-FALLING *(list)*	2. OPERATION(S) CONTRIBUTING FLOW		3. TREATMENT	
	a. OPERATION *(list)*	b. AVERAGE FLOW *(include units)*	a. DESCRIPTION	b. LIST CODES FROM TABLE 2C-1

OFFICIAL USE ONLY *(effluent guidelines sub-categories)*

EPA Form 3510-2C (8-90)   PAGE 1 OF 4   CONTINUE ON REVERSE
STF ENV441F.1

CONTINUED FROM THE FRONT

C. Except for storm runoff, leaks, or spills, are any of the discharges described in Items II-A or B intermittent or seasonal?
☐ YES *(complete the following table)*   ☐ NO *(go to Section III)*

1. OUTFALL NUMBER *(list)*	2. OPERATION(s) CONTRIBUTING FLOW *(list)*	3. FREQUENCY		4. FLOW					
		a. DAYS PER WEEK *(specify average)*	b. MONTHS PER YEAR *(specify average)*	a. FLOW RATE *(in mgd)*		b. TOTAL VOLUME *(specify with units)*		c. DUR-ATION *(in days)*	
				1. LONG TERM AVERAGE	2. MAXIMUM DAILY	1. LONG TERM AVERAGE	2. MAXIMUM DAILY		

III. PRODUCTION

A. Does an effluent guideline limitation promulgated by EPA under Section 304 of the Clean Water Act apply to your facility?
☐ YES *(complete Item III-B)*   ☐ NO *(go to Section IV)*

B. Are the limitations in the applicable effluent guideline expressed in terms of production *(or other measure of operation)?*
☐ YES *(complete Item III-C)*   ☐ NO *(go to Section IV)*

C. If you answered "yes" to Item III-B, list the quantity which represents an actual measurement of your level of production, expressed in the terms and units used in the applicable effluent guideline, and indicate the affected outfalls.

1. AVERAGE DAILY PRODUCTION			2. AFFECTED OUTFALLS *(list outfall numbers)*
a. QUANTITY PER DAY	b. UNITS OF MEASURE	c. OPERATION, PRODUCT, MATERIAL, ETC. *(specify)*	

IV. IMPROVEMENTS

A. Are you now required by any Federal, State or local authority to meet any implementation schedule for the construction, upgrading or operation of wastewater treatment equipment or practices or any other environmental programs which may affect the discharges described in this application? This includes, but is not limited to, permit conditions, administrative or enforcement orders, enforcement compliance schedule letters, stipulations, court orders, and grant or loan conditions.   ☐ YES *(complete the following table)*   ☐ NO *(go to Item IV-B)*

1. IDENTIFICATION OF CONDITION, AGREEMENT, ETC.	2. AFFECTED OUTFALLS		3. BRIEF DESCRIPTION OF PROJECT	4. FINAL COMPLIANCE DATE	
	a. NO.	b. SOURCE OF DISCHARGE		a. REQUIRED	b. PROJECTED

B. OPTIONAL: You may attach additional sheets describing any additional water pollution control programs *(or other environmental projects which may affect your discharges)* you now have underway or which you plan. Indicate whether each program is now underway or planned, and indicate your actual or planned schedules for construction.   ☐ MARK "X" IF DESCRIPTION OF ADDITIONAL CONTROL PROGRAMS IS ATTACHED

EPA I.D. NUMBER *(copy from Item 1 of Form 1)*

CONTINUED FROM PAGE 2

V. INTAKE AND EFFLUENT CHARACTERISTICS

A, B, & C: See instructions before proceeding — Complete one set of tables for each outfall — Annotate the outfall number in the space provided.
NOTE: Tables V-A, V-B, and V-C are included on separate sheets numbered V-1 through V-9.

D. Use the space below to list any of the pollutants listed in Table 2c-3 of the instructions, which you know or have reason to believe is discharged or may be discharged from any outfall. For every pollutant you list, briefly describe the reasons you believe it to be present and report any analytical data in your possession.

1. POLLUTANT	2. SOURCE	1.POLLUTANT	2. SOURCE

VI. POTENTIAL DISCHARGES NOT COVERED BY ANALYSIS

Is any pollutant listed in Item V-C a substance or a component of a substance which you currently use or manufacture as an intermediate or final product or byproduct?

☐ YES *(list all such pollutants below)*   ☐ NO *(go to Item VI-B)*

EPA Form 3510-2C (8-90)   PAGE 3 OF 4   CONTINUE ON REVERSE

STF ENV441F.3

CONTINUED FROM THE FRONT

VII. BIOLOGICAL TOXICITY TESTING DATA

Do you have any knowledge or reason to believe that any biological test for acute or chronic toxicity has been made on any of your discharges or on a receiving water in relation to your discharge within the last 3 years?

☐ YES *(identify the test(s) and describe their purposes below)* ☐ NO *(go to Section VIII)*

VIII. CONTRACT ANALYSIS INFORMATION

Were any of the analyses reported in Item V performed by a contract laboratory or consulting firm?

☐ YES *(list the name, address, and telephone number of, and pollutants analyzed by, each such laboratory or firm below)* ☐ NO *(go to Section IX)*

A. NAME	B. ADDRESS	C. TELEPHONE (area code & no.)	D. POLLUTANTS ANALYZED (list)

IX. CERTIFICATION

I certify under penalty of law that this document and all attachments were prepared under my direction or supervision in accordance with a system designed to assure that qualified personnel properly gather and evaluate the information submitted. Based on my inquiry of the person or persons who manage the system or those persons directly responsible for gathering the information, the information submitted is, to the best of my knowledge and belief, true, accurate, and complete. I am aware that there are significant penalties for submitting false information, including the possibility of fine and imprisonment for knowing violations.

A. NAME & OFFICIAL TITLE *(type or print)*	B. PHONE NO. *(area code & no.)*
C. SIGNATURE	D. DATE SIGNED

STF ENV441F.4

PLEASE PRINT OR TYPE IN THE UNSHADED AREAS ONLY. You may report some or all of this information on separate sheets (use the same format) instead of completing these pages. SEE INSTRUCTIONS.

EPA I.D. NUMBER (copy from Item 1 of Form 1)

OUTFALL NO.

V. INTAKE AND EFFLUENT CHARACTERISTICS (continued from page 3 of Form 2-C)

PART A - You must provide the results of at least one analysis for every pollutant in this table. Complete one table for each outfall. See instructions for additional details.

1. POLLUTANT	2. EFFLUENT								3. UNITS		4. INTAKE (optional)		
	a. MAXIMUM DAILY VALUE		b. MAXIMUM 30 DAY VALUE (if available)		c. LONG TERM AVRG. VALUE (if available)		d. NO. OF ANALYSES	a. CONCEN-TRATION	b. MASS	a. LONG TERM AVERAGE VALUE		b. NO. OF ANALYSES
	(1) CONCENTRATION	(2) MASS	(1) CONCENTRATION	(2) MASS	(1) CONCENTRATION	(2) MASS				(1) CONCENTRATION	(2) MASS	
a. Biochemical Oxygen Demand (BOD)												
b. Chemical Oxygen Demand (COD)												
c. Total Organic Carbon (TOC)												
d. Total Suspended Solids (TSS)												
e. Ammonia (as N)	VALUE		VALUE		VALUE					VALUE		
f. Flow	VALUE		VALUE		VALUE					VALUE		
g. Temperature (winter)	VALUE		VALUE		VALUE				°C	VALUE		
h. Temperature (summer)	VALUE		VALUE		VALUE				°C	VALUE		
i. pH	MINIMUM	MAXIMUM	MINIMUM	MAXIMUM				STANDARD UNITS				

PART B - Mark "X" in column 2-a for each pollutant you know or have reason to believe is present. Mark "X" in column 2-b for each pollutant you believe to be absent. If you mark column 2a for any pollutant which is limited either directly, or indirectly but expressly, in an effluent limitations guideline, you must provide the results of at least one analysis for that pollutant. For other pollutants for which you mark column 2a, you must provide quantitative data or an explanation of their presence in your discharge. Complete one table for each outfall. See the instructions for additional details and requirements.

1. POLLU-TANT AND CAS NO. (if available)	2. MARK "X"			3. EFFLUENT							4. UNITS		5. INTAKE (optional)		
	a. BE-LIEVED PRE-SENT	b. BE-LIEVED AB-SENT		a. MAXIMUM DAILY VALUE		b. MAXIMUM 30 DAY VALUE (if available)		c. LONG TERM AVRG. VALUE (if available)		d. NO. OF ANAL-YSES	a. CONCEN-TRATION	b. MASS	a. LONG TERM AVERAGE VALUE		b. NO. OF ANAL-YSES
				(1) CONCENTRATION	(2) MASS	(1) CONCENTRATION	(2) MASS	(1) CONCENTRATION	(2) MASS				(1) CONCENTRATION	(2) MASS	
a. Bromide (24959-67-9)															
b. Chlorine, Total Residual															
c. Color															
d. Fecal Coliform															
e. Fluoride (16984-48-8)															
f. Nitrate— Nitrite (as N)															

EPA Form 3510-2C (8-90)

PAGE V-1

CONTINUE ON REVERSE

STF BNV/441F.5

F.9

ITEM V-B CONTINUED FROM FRONT

1. POLLUTANT AND CAS NO. (if available)	2. MARK 'X'		3. EFFLUENT							4. UNITS		5. INTAKE (optional)		
	a. BE-LIEVED PRE-SENT	b. BE-LIEVED AB-SENT	a. MAXIMUM DAILY VALUE		b. MAXIMUM 30 DAY VALUE (if available)		c. LONG TERM AVRG. VALUE (if available)		d. NO. OF ANAL-YSES	a. CONCEN-TRATION	b. MASS	a. LONG TERM AVERAGE VALUE		b. NO. OF ANAL-YSES
			(1) CONCENTRATION	(2) MASS	(1) CONCENTRATION	(2) MASS	(1) CONCENTRATION	(2) MASS				(1) CONCENTRATION	(2) MASS	
g. Nitrogen, Total Organic (as N)														
h. Oil and Grease														
i. Phosphorus (as P), Total (7723-14-0)														
j. Radioactivity														
(1) Alpha, Total														
(2) Beta, Total														
(3) Radium, Total														
(4) Radium 226, Total														
k. Sulfate (as SO₄) (14808-79-8)														
l. Sulfide (as S)														
m. Sulfite (as SO₃) (14265-45-3)														
n. Surfactants														
o. Aluminum, Total (7429-90-5)														
p. Barium, Total (7440-39-3)														
q. Boron, Total (7440-42-8)														
r. Cobalt, Total (7440-48-4)														
s. Iron, Total (7439-89-6)														
t. Magnesium, Total (7439-95-4)														
u. Molybdenum, Total (7439-98-7)														
v. Manganese, Total (7439-96-5)														
w. Tin, Total (7440-31-5)														
x. Titanium, Total (7440-32-6)														

EPA Form 3510-2C (8-90)

PAGE V-2

CONTINUE ON PAGE V-3
STF ENV441F8

F.10

CONTINUED FROM PAGE 3 OF FORM 2-C

PART C - If you are a primary industry and this outfall contains process wastewater, refer to Table 2c-2 in the instructions to determine which of the GC/MS fractions you must test for. Mark "X" in column 2-a for all such GC/MS fractions that apply to your industry and for ALL toxic metals, cyanides, and total phenols. If you are not required to mark column 2-a (secondary industries, nonprocess wastewater outfalls, and nonrequired GC/MS fractions), mark "X" in column 2-b for each pollutant you know or have reason to believe is present. Mark "X" in column 2-c for each pollutant you believe is absent. If you mark column 2a for any pollutant, you must provide the results of at least one analysis for that pollutant. If you mark column 2b for any pollutant, you must provide the results of at least one analysis for that pollutant if you know or have reason to believe it will be discharged in concentrations of 10 ppb or greater. If you mark column 2b for acrolein, acrylonitrile, 2,4 dinitrophenol, or 2-methyl-4, 6 dinitrophenol, you must provide the results of at least one analysis for each of these pollutants which you know or have reason to believe that you discharge in concentrations of 100 ppb or greater. Otherwise, for pollutants for which you mark column 2b, you must either submit at least one analysis or briefly describe the reasons the pollutant is expected to be discharged. Note that there are 7 pages to this part, please review each carefully. Complete one table (all 7 pages) for each outfall. See instructions for additional details and requirements.

1. POLLUTANT AND CAS NUMBER (if available)	2. MARK 'X'			3. EFFLUENT							4. UNITS			5. INTAKE (optional)			6. NO. OF ANAL-YSES
	a. TEST-ING RE-QUIR-ED	b. BE-LIEVED PRE-SENT	c. BE-LIEVED AB-SENT	a. MAXIMUM DAILY VALUE		b. MAXIMUM 30 DAY VALUE (if available)		c. LONG TERM AVRG. VALUE (if available)		d. NO. OF ANAL-YSES	a. CONCEN-TRATION	b. MASS		a. LONG TERM AVERAGE VALUE		
				(1) CONCENTRATION	(2) MASS	(1) CONCENTRATION	(2) MASS	(1) CONCENTRATION	(2) MASS					(1) CONCEN-TRATION	(2) MASS	

METALS, CYANIDE, AND TOTAL PHENOLS

1M. Antimony, Total (7440-36-0)																
2M. Arsenic, Total (7440-38-2)																
3M. Beryllium, Total, 7440-41-7)																
4M. Cadmium, Total (7440-43-9)																
5M. Chromium, Total (7440-47-3)																
6M. Copper, Total (7440-50-8)																
7M. Lead, Total (7439-92-1)																
8M. Mercury, Total (7439-97-6)																
9M. Nickel, Total (7440-02-0)																
10M. Selenium, Total (7782-49-2)																
11M. Silver, Total (7440-22-4)																
12M. Thallium, Total (7440-28-0)																
13M. Zinc, Total (7440-66-6)																
14M. Cyanide, Total (57-12-5)																
15M. Phenols, Total																

DIOXIN

| 2,3,7,8 Tetra-chlorodibenzo-P-Dioxin (1764-01-6) | | | | DESCRIBE RESULTS | | | | | | | | | | | | |

CONTINUED FROM THE FRONT

1. POLLUTANT AND CAS NUMBER (if available)	2. MARK "X"			3. EFFLUENT									4. UNITS			5. INTAKE (optional)		
	a. TEST-ING RE-QUIR-ED	b. BE-LIEVED PRE-SENT	c. BE-LIEVED AB-SENT	a. MAXIMUM DAILY VALUE		b. MAXIMUM 30 DAY VALUE (if available)		c. LONG TERM AVRG. VALUE (if available)		d. NO. OF ANAL-YSES	a. CONCEN-TRATION	b. MASS		a. LONG TERM AVERAGE VALUE		b. NO. OF ANAL-YSES
				(1) CONCENTRATION	(2) MASS	(1) CONCENTRATION	(2) MASS	(1) CONCENTRATION	(2) MASS						(1) CONCEN-TRATION	(2) MASS	

GC/IMS FRACTION — VOLATILE COMPOUNDS

1V. Acrolein (107-02-8)																
2V. Acrylonitrile (107-13-1)																
3V. Benzene (71-43-2)																
4V. Bis (Chloromethyl) Ether (542-88-1)																
5V. Bromoform (75-25-2)																
6V. Carbon Tetrachloride (56-23-5)																
7V. Chlorobenzene (108-90-7)																
8V. Chlorodibromomethane (124-48-1)																
9V. Chloroethane (75-00-3)																
10V. 2-Chloroethylvinyl Ether (110-75-8)																
11V. Chloroform (67-66-3)																
12V. Dichlorobromomethane (75-27-4)																
13V. Dichlorodifluoromethane (75-71-8)																
14V. 1,1-Dichloroethane (75-34-3)																
15V. 1,2-Dichloroethane (107-06-2)																
16V. 1,1-Dichloroethylene (75-35-4)																
17V. 1,2-Dichloropropane (78-87-5)																
18V. 1,3-Dichloropropylene (542-75-6)																
19V. Ethylbenzene (100-41-4)																
20V. Methyl Bromide (74-83-9)																
21V. Methyl Chloride (74-87-3)																

EPA Form 3510-2C (8-90)

PAGE V-4

CONTINUE ON PAGE V-5
STF ENV441F.8

F.12

EPA I.D. NUMBER (copy from Item 1 of Form 1)   OUTFALL NUMBER

1. POLLUTANT AND CAS NUMBER (if available)	2. MARK 'X'			3. EFFLUENT						4. UNITS		5. INTAKE (optional)		
	a. TEST-ING RE-QUIR-ED	b. BE-LIEVED PRE-SENT	c. BE-LIEVED AB-SENT	a. MAXIMUM DAILY VALUE		b. MAXIMUM 30 DAY VALUE (if available)		c. LONG TERM AVRG. VALUE (if available)			a. CONCEN-TRATION	b. MASS	a. LONG TERM AVERAGE VALUE		b. NO. OF ANAL-YSES
				(1) CONCENTRATION	(2) MASS	(1) CONCENTRATION	(2) MASS	(1) CONCENTRATION	(2) MASS	d. NO. OF ANAL-YSES			(1) CONCEN-TRATION	(2) MASS	

GC/MS FRACTION — VOLATILE COMPOUNDS (continued)

22V. Methylene Chloride (75-09-2)															
23V. 1,1,2,2-Tetra-chloroethane (79-34-5)															
24V. Tetrachloro-ethylene (127-18-4)															
25V. Toluene (108-88-3)															
26V. 1,2-Trans-Dichloroethylene (156-60-5)															
27V. 1,1,1-Tri-chloroethane (71-55-6)															
28V. 1,1,2-Tri-chloroethane (79-00-5)															
29V. Trichloro-ethylene (79-01-6)															
30V. Trichloro-fluoromethane (75-69-4)															
31V. Vinyl Chloride (75-01-4)															

GC/MS FRACTION — ACID COMPOUNDS

1A. 2-Chloro-phenol (95-57-8)															
2A. 2,4-Dichloro-phenol (120-83-2)															
3A. 2,4-Dimethyl-phenol (105-67-9)															
4A. 4,6-Dinitro-O-Cresol (534-52-1)															
5A. 2,4-Dinitro-phenol (51-28-5)															
6A. 2-Nitrophenol (88-75-5)															
7A. 4-Nitrophenol (100-02-7)															
8A. P-Chloro-M-Cresol (59-50-7)															
9A. Pentachloro-phenol (87-86-5)															
10A. Phenol (108-95-2)															
11A. 2,4,6-Tri-chlorophenol (88-06-2)															

EPA Form 3510-2C (8-90)

CONTINUE ON REVERSE
STF ENV44\F.9

F.13

CONTINUED FROM THE FRONT

1. POLLUTANT AND CAS NUMBER (if available)	2. MARK 'X'			3. EFFLUENT					4. UNITS			5. INTAKE (optional)		
	a. TEST-ING RE-QUIR-ED	b. BE-LIEVED PRE-SENT	c. BE-LIEVED AB-SENT	a. MAXIMUM DAILY VALUE		b. MAXIMUM 30 DAY VALUE (if available)		c. LONG TERM AVRG. VALUE (if available)		d. NO. OF ANAL-YSES	a. CONCEN-TRATION	b. MASS	a. LONG TERM AVERAGE VALUE		b. NO. OF ANAL-YSES
				(1) CONCENTRATION	(2) MASS	(1) CONCENTRATION	(2) MASS	(1) CONCENTRATION	(2) MASS				(1) CONCEN-TRATION	(2) MASS	

GC/MS FRACTION — BASE/NEUTRAL COMPOUNDS

1B. Acenaphthene (83-32-9)															
2B. Acenaphtylene (208-96-8)															
3B. Anthracene (120-12-7)															
4B. Benzidine (92-87-5)															
5B. Benzo (a) Anthracene (56-55-3)															
6B. Benzo (a) Pyrene (50-32-8)															
7B. 3, 4-Benzo-fluoranthene (205-99-2)															
8B. Benzo (ghi) Perylene (191-24-2)															
9B. Benzo (k) Fluoranthene (207-08-9)															
10B. Bis (2-Chloro-ethoxy) Methane (111-91-1)															
11B. Bis (2-Chloro-ethyl) Ether (111-44-4)															
12B. Bis (2-Chloroiso-propyl) Ether (102-60-1)															
13B. Bis (2-Ethyl-hexyl) Phthalate (117-81-7)															
14B. 4-Bromo-phenyl Phenyl Ether (101-55-3)															
15B. Butyl Benzyl Phthalate (85-68-7)															
16B. 2-Chloro-naphthalene (91-58-7)															
17B. 4-Chloro-phenyl Phenyl Ether (7005-72-3)															
18B. Chrysene (218-01-9)															
19B. Dibenzo (a, h) Anthracene (53-70-3)															
20B. 1,2-Dichloro-benzene (95-50-1)															
21B. 1,3-Dichloro-benzene (541-73-1)															

EPA Form 3510-2C (8-90)

PAGE V-6

CONTINUE ON PAGE V-7

STT ENV441F-10

F.14

EPA I.D. NUMBER (copy from Item 1 of Form 1) | OUTFALL NUMBER

1. POLLUTANT AND CAS NUMBER (if available)	2. MARK 'X'			3. EFFLUENT							4. UNITS		5. INTAKE (optional)		
	a. TEST-ING RE-QUIR-ED	b. BE-LIEVED PRE-SENT	c. BE-LIEVED AB-SENT	a. MAXIMUM DAILY VALUE		b. MAXIMUM 30 DAY VALUE (if available)		c. LONG TERM AVRG. VALUE (if available)		d. NO. OF ANAL-YSES	a. CONCEN-TRATION	b. MASS	a. LONG TERM AVERAGE VALUE		b. NO. OF ANAL-YSES
				(1) CONCENTRATION	(2) MASS	(1) CONCENTRATION	(2) MASS	(1) CONCENTRATION	(2) MASS				(1) CONCEN-TRATION	(2) MASS	
GC/MS FRACTION — BASE/NEUTRAL COMPOUNDS (continued)															
22B. 1,4-Dichloro-benzene (106-46-7)															
23B. 3,3'-Dichloro-benzidine (91-94-1)															
24B. Diethyl Phthalate (84-66-2)															
25B. Dimethyl Phthalate (131-11-3)															
26B. Di-N-Butyl Phthalate (84-74-2)															
27B. 2,4-Dinitro-toluene (121-14-2)															
28B. 2,6-Dinitro-toluene (606-20-2)															
29B. Di-N-Octyl Phthalate (117-84-0)															
30B. 1,2-Diphenyl-hydrazine (as Azo-benzene) (122-66-7)															
31B. Fluoranthene (206-44-0)															
32B. Fluorene (86-73-7)															
33B. Hexachloro-benzene (118-74-1)															
34B. Hexachloro-butadiene (87-68-3)															
35B. Hexachloro-cyclopentadiene (77-47-4)															
36B. Hexachloro-ethane (67-72-1)															
37B. Indeno (1,2,3-cd) Pyrene (193-39-5)															
38B. Isophorone (78-59-1)															
39B. Naphthalene (91-20-3)															
40B. Nitrobenzene (98-95-3)															
41B. N-Nitroso-dimethylamine (62-75-9)															
42B. N-Nitrosodi-N-Propylamine (621-64-7)															

EPA Form 3510-2C (8-90)

CONTINUE ON REVERSE
STF ENV441F.11

1. POLLUTANT AND CAS NUMBER (if available)	2. MARK "X"			3. EFFLUENT						4. UNITS			5. INTAKE (optional)		
	a. TEST-ING RE-QUIR-ED	b. BE-LIEVED PRE-SENT	c. BE-LIEVED AB-SENT	a. MAXIMUM DAILY VALUE		b. MAXIMUM 30 DAY VALUE (if available)		c. LONG TERM AVRG. VALUE (if available)		d. NO. OF ANAL-YSES	a. CONCEN-TRATION	b. MASS	a. LONG TERM AVERAGE VALUE		b. NO. OF ANAL-YSES
				(1) CONCENTRATION	(2) MASS	(1) CONCENTRATION	(2) MASS	(1) CONCENTRATION	(2) MASS				(1) CONCEN-TRATION	(2) MASS	
GC/MS FRACTION — BASE/NEUTRAL COMPOUNDS (continued)															
43B. N-Nitro-sodiphenylamine (86-30-6)															
44B. Phenanthrene (85-01-8)															
45B. Pyrene (129-00-0)															
46B. 1,2,4-Tri-chlorobenzene (120-82-1)															
GC/MS FRACTION — PESTICIDES															
1P. Aldrin (309-00-2)															
2P. α -BHC (319-84-6)															
3P. β -BHC (319-85-7)															
4P. γ -BHC (58-89-9)															
5P. δ -BHC (319-86-8)															
6P. Chlordane (57-74-9)															
7P. 4,4'-DDT (50-29-3)															
8P. 4,4'-DDE (72-55-9)															
9P. 4,4'-DDD (72-54-8)															
10P. Dieldrin (60-57-1)															
11P. α -Endosulfan (115-29-7)															
12P. β -Endosulfan (115-29-7)															
13P. Endosulfan Sulfate (1031-07-8)															
14P. Endrin (72-20-8)															
15P. Endrin Aldehyde (7421-93-4)															
16P. Heptachlor (76-44-8)															

EPA Form 3510-2C (8-90)

CONTINUE ON PAGE V-9

STF ENV44F.12

EPA I.D. NUMBER (copy from Item 1 of Form 1) | OUTFALL NUMBER

CONTINUED FROM PAGE V-8

1. POLLUTANT AND CAS NUMBER (if available)	2. MARK "X"			3. EFFLUENT							4. UNITS		5. INTAKE (optional)		
	a. TEST-ING RE-QUIR-ED	b. BE-LIEVED PRE-SENT	c. BE-LIEVED AB-SENT	a. MAXIMUM DAILY VALUE		b. MAXIMUM 30 DAY VALUE (if available)		c. LONG TERM AVRG. VALUE (if available)		d. NO. OF ANAL-YSES	a. CONCEN-TRATION	b. MASS	a. LONG TERM AVERAGE VALUE		b. NO. OF ANAL-YSES
				(1) CONCENTRATION	(2) MASS	(1) CONCENTRATION	(2) MASS	(1) CONCENTRATION	(2) MASS				(1) CONCEN-TRATION	(2) MASS	
GC/MS FRACTION — PESTICIDES (continued)															
17P. Heptachlor Epoxide (1024-57-3)															
18P. PCB-1242 (53469-21-9)															
19P. PCB-1254 (11097-69-1)															
20P. PCB-1221 (11104-28-2)															
21P. PCB-1232 (11141-16-5)															
22P. PCB-1248 (12672-29-6)															
23P. PCB-1260 (11096-82-5)															
24P. PCB-1016 (12674-11-2)															
25P. Toxaphene (8001-35-2)															

PAGE V-9

EPA Form 3510-2C (8-90)

STF ENV441F-13

Form Approved OMB No 2040 0086 Approval Expires 5/31/92

EPA ID Number *(copy from Item 1 of Form 1)*

Please type or print in the unshaded areas only

Form
2D
NPDES

EPA

New Sources and New Dischargers
Application for Permit to Discharge Process Wastewater

I. Outfall Location

For each outfall, list the latitude and longitude, and the name of the receiving water

Outfall Number *(list)*	Latitude			Longitude			Receiving Water *(name)*
	Deg	Min	Sec	Deg	Min	Sec	

II. Discharge Date *(When do you expect to begin discharging?)*

III. Flows, Sources of Pollution, and Treatment Technologies

A.   For each outfall, provide a description of (1) All operations contributing wastewater to the effluent, including process wastewater, sanitary wastewater, cooling water, and stormwater runoff; (2) The average flow contributed by each operation; and (3) The treatment received by the wastewater. Continue on additional sheets if necessary.

Outfall Number	1. Operations Contributing Flow *(list)*	2. Average Flow *(include units)*	3. Treatment *(Description or List Codes from Table 2D-1)*

EPA Form 3510-2D (Rev. 8-90)

STF ENV443F.1

B. Attach a line drawing showing the water flow through the facility. Indicate sources of intake water, operations contributing wastewater to the effluent, and treatment units labeled to correspond to the more detailed descriptions in Item III-A. Construct a water balance on the line drawing by showing average flows between intakes, operations, treatment units, and outfalls. If a water balance cannot be determined (e.g., for certain mining activities), provide a pictorial description of the nature and amount of any sources of water and any collection or treatment measures.

C. Except for storm runoff, leaks, or spills, will any of the discharges described in item III-A be intermittent or seasonal?

☐ Yes *(complete the following table)*   ☐ No *(go to item IV)*

Outfall Number	1. Frequency		2. Flow		
	a Days Per Week *(specify average)*	b Months Per Year *(specify average)*	a Maximum Daily Flow Rate *(in mgd)*	b Maximum Total Volume *(specify with units)*	c Duration *(in days)*

IV. Production

If there is an applicable production-based effluent guideline or NSPS, for each outfall list the estimated level of production projection of actual production level, not design), expressed in the terms and units used in the applicable effluent guideline or NSPS, for each of the first 3 years of operation. If production is likely to vary, you may also submit alternative estimates (attach a separate sheet).

Year	a Quantity Per Day	b Units of Measure	c Operation, Product Material, etc *(specify)*

F.20

CONTINUED FROM THE FRONT	EPA ID Number *(copy from Item 1 of Form 1)*	Outfall Number

V. Effluent Characteristics

A, and B: These items require you to report estimated amounts *(both concentration and mass)* of the pollutants to be discharged from each of your outfalls. Each part of this item addresses a different set of pollutants and should be completed in accordance with the specific instructions for that part. Data for each outfall should be on a separate page. Attach additional sheets of paper if necessary.

General Instructions *(See table 2D-2 for Pollutants)*
Each part of this item requests you to provide an estimated daily maximum and average for certain pollutants and the source of information. Data for all pollutants in Group A, for all outfalls, must be submitted unless waived by the permitting authority. For all outfalls, data for pollutants in Group B should be reported only for pollutants which you believe will be present or are limited directly by an effluent limitations guideline or NSPS or indirectly through limitations on an indicator pollutant.

1 Pollutant	2 Maximum Daily Value *(include units)*	3 Average Daily Value *(include units)*	4 Source *(see instructions)*

CONTINUED FROM THE FRONT	EPA ID Number *(copy from Item 1 of Form 1)*	

C. Use the space below to list any of the pollutants listed in Table 2D-3 of the instructions which you know or have reason to believe will be discharged from any outfall. For every pollutant you list, briefly describe the reasons you believe it will be present.

1 Pollutant	2 Reason for Discharge

VI. Engineering Report on Wastewater Treatment

A. If there is any technical evaluation concerning your wastewater treatment, including engineering reports or pilot plant studies, check the appropriate box below.
☐ Report Available   ☐ No Report

B. Provide the name and location of any existing plant(s) which, to the best of your knowledge, resembles this production facility with respect to production processes, wastewater constituents, or wastewater treatments.

Name	Location

EPA ID Number *(copy from Item one of Form 1)*

VII. Other Information *(Optional)*

Use the space below to expand upon any of the above questions or to bring to the attention of the reviewer any other information you feel should be considered in establishing permit limitations for the proposed facility. Attach additional sheets if necessary.

VIII. Certification

I certify under penalty of law that this document and all attachments were prepared under my direction or supervision in accordance with a system designed to assure that qualified personnel properly gather and evaluate the information submitted. Based on my inquiry of the person or persons who manage the system, or those persons directly responsible for gathering the information, the information submitted is, to the best of my knowledge and belief, true, accurate, and complete. I am aware that there are significant penalties for submitting false information, including the possibility of fine and imprisonment for knowing violations.

A. Name and Official Title *(type or print)*	B. Phone No.
C. Signature	D. Date Signed

	EPA ID Number *(copy from Item 1 of Form 1)*	Form Approved. OMB No. 2040-0086
Please type or print in the unshaded areas only		Approval expires 5-31-92.

Form

2E
NPDES **EPA** Facilities Which Do Not Discharge Process Wastewater

I. Receiving Waters

For this outfall, list the latitude and longitude, and name of the receiving water(s).

Outfall Number *(list)*	Latitude			Longitude			Receiving Water *(name)*
	Deg	Min	Sec	Deg	Min	Sec	

II. Discharge Date *(If a new discharger, the date you expect to begin discharging)*

III. Type of Waste

A. Check the box(es) indicating the general type(s) of wastes discharged.

☐ Sanitary Wastes ☐ Restaurant or Cafeteria Wastes ☐ Noncontact Cooling Water ☐ Other Nonprocess Wastewater *(Identify)*

B. If any cooling water additives are used, list them here. Briefly describe their composition if this information is available.

IV. Effluent Characteristics

A. **Existing Sources** — Provide measurements for the parameters listed in the left-hand column below, unless waived by the permitting authority *(see instructions)*.

B. **New Dischargers** — Provide estimates for the parameters listed in the left-hand column below, unless waived by the permitting authority. Instead of the number of measurements taken, provide the source of estimated values *(see instructions)*.

Pollutant or Parameter	(1) Maximum Daily Value *(Include units)*		(2) Average Daily Value *(last year)* *(include units)*		(3) *(or)* Number of Measurements Taken *(last year)*	(4) Source of Estimate *(if new discharger)*
	Mass	Concentration	Mass	Concentration		
Biochemical Oxygen Demand (BOD)						
Total Suspended Solids (TSS)						
Fecal Coliform *(if believed present or if sanitary waste is discharged)*						
Total Residual Chlorine *(if chlorine is used)*						
Oil and Grease						
*Chemical oxygen demand (COD)						
*Total organic carbon (TOC)						
Ammonia *(as N)*						
Discharge Flow	Value					
pH *(give range)*	Value					
Temperature *(Winter)*		°C		°C		
Temperature *(Summer)*		°C		°C		

*If noncontact cooling water is discharged

EPA Form 3510-2E (8-90)

V. Except for leaks or spills, will the discharge described in this form be intermittent or seasonal? If yes, briefly describe the frequency of flow and duration. ☐ Yes ☐ No

VI. Treatment System *(Describe briefly any treatment system(s) used or to be used)*

VII. Other Information *(Optional)*

Use the space below to expand upon any of the above questions or to bring to the attention of the reviewer any other information you feel should be considered in establishing permit limitations. Attach additional sheets, if necessary.

VIII. Certification

I certify under penalty of law that this document and all attachments were prepared under my direction or supervision in accordance with a system designed to assure that qualified personnel properly gather and evaluate the information submitted. Based on my inquiry of the person or persons who manage the system, or those persons directly responsible for gathering the information, the information submitted is to the best of my knowledge and belief, true, accurate, and complete. I am aware that there are significant penalties for submitting false information, including the possibility of fine and imprisonment for knowing violations.

A. Name & Official Title	B. Phone No. *(area code & no.)*
C. Signature	D. Date Signed

EPA ID Number *(copy from Item I of Form 1)*	Form Approved. OMB No. 2040-0086

Please print or type in the unshaded areas only

Approval expires 5-31-92

Form **2F** NPDES	**EPA**	United States Environmental Protection Agency Washington, DC 20460 **Application for Permit To Discharge Stormwater Discharges Associated with Industrial Activity**

Paperwork Reduction Act Notice

Public reporting burden for this application is estimated to average 28.6 hours per application, including time for reviewing instructions, searching existing data sources, gathering and maintaining the data needed, and completing and reviewing the collection of information. Send comments regarding the burden estimate, any other aspect of this collection of information, or suggestions for improving this form, including suggestions which may increase or reduce this burden to: Chief, Information Policy Branch, PM-223, U.S. Environmental Protection Agency, 401 M St., SW, Washington, DC 20460, or Director, Office of Information and Regulatory Affairs, Office of Management and Budget, Washington, DC 20503.

I. Outfall Location

For each outfall, list the latitude and longitude of its location to the nearest 15 seconds and the name of the receiving water.

A. Outfall Number *(list)*	B. Latitude	C. Longitude	D. Receiving Water *(name)*

II. Improvements

A. Are you now required by any Federal, State, or local authority to meet any implementation schedule for the construction, upgrading or operation of wastewater treatment equipment or practices or any other environmental programs which may affect the discharges described in this application? This includes, but is not limited to, permit conditions, administrative or enforcement orders, enforcement compliance schedule letters, stipulations, court orders, and grant or loan conditions.

1. Identification of Conditions, Agreements, Etc.	2. Affected Outfalls		3. Brief Description of Project	4. Final Compliance Date	
	number	source of discharge		a. req.	b. proj.

B. You may attach additional sheets describing any additional water pollution (or other environmental projects which may affect your discharges) you now have under way or which you plan. Indicate whether each program is now under way or planned, and indicate your actual or planned schedules for construction.

III. Site Drainage Map

Attach a site map showing topography (or indicating the outline of drainage areas served by the outfall(s) covered in the application if a topographic map is unavailable) depicting the facility including: each of its intake and discharge structures; the drainage area of each storm water outfall; paved areas and buildings within the drainage area of each storm water outfall, each known past or present areas used for outdoor storage or disposal of significant materials, each existing structural control measure to reduce pollutants in storm water runoff, materials loading and access areas, areas where pesticides, herbicides, soil conditioners and fertilizers are applied; each of its hazardous waste treatment, storage or disposal units (including each area not required to have a RCRA permit which is used for accumulating hazardous waste under 40 CFR 262.34); each well where fluids from the facility are injected underground; springs, and other surface water bodies which receive storm water discharges from the facility.

STF ENV406F.1

F.26

Continued from the Front

IV. Narrative Description of Pollutant Sources

A. For each outfall, provide an estimate of the area (include units) of impervious surfaces (including paved areas and building roofs) drained to the outfall, and an estimate of the total surface area drained by the outfall.

Outfall Number	Area of Impervious Surface (provide units)	Total Area Drained (provide units)	Outfall Number	Area of Impervious Surface (provide units)	Total Area Drained (provide units)

B. Provide a narrative description of significant materials that are currently or in the past three years have been treated, stored or disposed in a manner to allow exposure to storm water; method of treatment, storage, or disposal; past and present materials management practices employed to minimize contact by these materials with storm water runoff; materials loading and access areas; and the location, manner, and frequency in which pesticides, herbicides, soil conditioners, and fertilizers are applied.

C. For each outfall, provide the location and a description of existing structural and nonstructural control measures to reduce pollutants in storm water runoff; and a description of the treatment the storm water receives, including the schedule and type of maintenance for control and treatment measures and the ultimate disposal of any solid or fluid wastes other than by discharge.

Outfall Number	Treatment	List Codes from Table 2F-1

V. Nonstormwater Discharges

A. I certify under penalty of law that the outfall(s) covered by this application have been tested or evaluated for the presence of nonstormwater discharges, and that all nonstormwater discharges from these outfall(s) are identified in either an accompanying Form 2C or Form 2E application for the outfall.

Name and Official Title (type or print)	Signature	Date Signed

B. Provide a description of the method used, the date of any testing, and the onsite drainage points that were directly observed during a test.

VI. Significant Leaks or Spills

Provide existing information regarding the history of significant leaks or spills of toxic or hazardous pollutants at the facility in the last three years, including the approximate date and location of the spill or leak, and the type and amount of material released.

STF ENV408F.2

EPA ID Number *(copy from Item I of Form 1)*

Continued from Page 2

VII. Discharge Information

A, B, C, & D: See instructions before proceeding. Complete one set of tables for each outfall. Annotate the outfall number in the space provided.
Tables VII-A, VII-B, and VII-C are included on separate sheets numbered VII-1 and VII-2.

E: Potential discharges not covered by analysis - is any pollutant listed in table 2F-2, 2F-3 or 2F-4, a substance or a component of a substance which you currently use or manufacture as an intermediate or final product or byproduct?

☐ Yes *(list all such pollutants below)*   ☐ No *(go to Section IX)*

VIII. Biological Toxicity Testing Data

Do you have any knowledge or reason to believe that any biological test for acute or chronic toxicity has been made on any of your discharges or on a receiving water in relation to your discharge within the last 3 years?

☐ Yes *(list all such pollutants below)*   ☐ No *(go to Section IX)*

IX. Contract Analysis Information

Were any of the analysis reported in item VII performed by a contract laboratory or consulting firm?

☐ Yes *(list the name, address, and telephone number of, and pollutants analyzed by, each such laboratory or firm below)*   ☐ No *(go to Section X)*

A. Name	B. Address	C. Area Code & Phone No.	D. Pollutants Analyzed

X. Certification

I certify under penalty of law that this document and all attachments were prepared under my direction or supervision in accordance with a system designed to assure that qualified personnel properly gather and evaluate the information submitted. Based on my inquiry of the person or persons who manage the system or those persons directly responsible for gathering the information, the information submitted is, to the best of my knowledge and belief, true, accurate, and complete. I am aware that there are significant penalties for submitting false information, including the possibility of fine and imprisonment for knowing violations.

A. Name & Official Title *(type or print)*	B. Area Code and Phone No.
C. Signature	D. Date Signed

EPA Form 3510-2F (Rev. 1-92)   Page 3 of 3

STF ENV408F.3

EPA ID Number *(copy from Item I of Form 1)*	Form Approved. OMB No. 2040-0086
	Approval expires 5-31-92

VII. Discharge Information *(Continued from page 3 of Form 2F)*

Part A - You must provide the results of at least one analysis for every pollutant in this table. Complete one table for each outfall. See instructions for additional details.

Pollutant and CAS Number *(if available)*	Maximum Values *(include units)*		Average Values *(include units)*		Number of Storm Events Sampled	Sources of Pollutants
	Grab Sample Taken During First 20 Minutes	Flow-weighted Composite	Grab Sample Taken During First 20 Minutes	Flow-weighted Composite		
Oil and Grease		N/A				
Biological Oxygen Demand (BOD5)						
Chemical Oxygen Demand (COD)						
Total Suspended Solids (TSS)						
Total Nitrogen						
Total Phosphorus						
pH	Minimum	Maximum	Minimum	Maximum		

Part B - List each pollutant that is limited in an effluent guideline which the facility is subject to or any pollutant listed in the facility's NPDES permit for its process wastewater (if the facility is operating under an existing NPDES permit). Complete one table for each outfall. See the instructions for additional details and requirements.

Pollutant and CAS Number *(if available)*	Maximum Values *(include units)*		Average Values *(include units)*		Number of Storm Events Sampled	Sources of Pollutants
	Grab Sample Taken During First 20 Minutes	Flow-weighted Composite	Grab Sample Taken During First 20 Minutes	Flow-weighted Composite		

STF ENV408F.4

Continued from the Front

Part C - List each pollutant shown in Tables 2F-2, 2F-3, and 2F-4 that you know or have reason to believe is present. See the instructions for additional details and requirements. Complete one table for each outfall.

Pollutant and CAS Number (if available)	Maximum Values (include units)		Average Values (include units)		Number of Storm Events Sampled	Sources of Pollutants
	Grab Sample Taken During First 20 Minutes	Flow-weighted Composite	Grab Sample Taken During First 20 Minutes	Flow-weighted Composite		

Part D - Provide data for the storm event(s) which resulted in the maximum values for the flow weighted composite sample.

1. Date of Storm Event	2. Duration of Storm Event (in minutes)	3. Total rainfall during storm event (in inches)	4 Number of hours between beginning of storm measured and end of previous measurable rain event	5. Maximum flow rate during rain event (gallons/minute or specify units)	6. Total flow from rain event (gallons or specify units)

9. Provide a description of the method of flow measurement or estimate.

EPA Form 3510-2F (Rev. 1-92) Page VII-2

STF ENV408F.5

THIS FORM REPLACES PREVIOUS FORM 3510-6 (8-92)   Form Approved. **OMB No. 2040-0086**
See Reverse for Instructions   Approval expires: 8-31-98

NPDES   United States Environmental Protection Agency
FORM   Washington, DC 20460

EPA **Notice of Intent (NOI) for Storm Water Discharges Associated with Industrial Activity Under a NPDES General Permit**

Submission of this Notice of Intent constitutes notice that the party identified in Section II of this form intends to be authorized by a NPDES permit issued for storm water discharges associated with industrial activity in the State identified in Section III of this form. Becoming a permittee obligates such discharger to comply with the terms and conditions of the permit. ALL NECESSARY INFORMATION MUST BE PROVIDED ON THIS FORM.

I. Permit Selection: You must indicate the NPDES Storm Water general permit under which you are applying for coverage. Check one of these.

Baseline Industrial ☐	Baseline Construction ☐	Multi-Sector (Group Permit) ☐

II. Facility Operator Information

Name: _____   Phone: _____

Address: _____   Statue of Owner/Operator: ☐

City: _____   State: _____   ZIP Code: _____

III. Facility/Site Location

Name: _____   Is the facility located on Indian Lands? (Y or N) ☐

Address: _____

City: _____   State: _____   ZIP Code: _____

Latitude: _____ Longitude: _____ Quarter: _____ Section: _____ Township: _____ Range: _____

IV. Site Activity Information

MS4 Operator Name: _____

Receiving Water Body: _____

If you are filing as a co-permittee, enter storm water general permit number: _____

SIC or Designated Activity Code: Primary: _____   2nd: _____

Is the facility required to submit monitoring data? (1, 2, 3, or 4) ☐

If You Have Another Existing NPDES Permit, Enter Permit Number: _____

Multi-Sector Permit Applicants Only:
Based on the instructions provided in Addendum H of the Multi-Sector permit, are species identified in Addendum H in proximity to the storm water discharges to be covered under this permit, or the areas of BMP construction to control those storm water discharges? (Y or N) ☐

Will construction (land disturbing activities) be conducted for storm water controls? (Y or N) ☐

Is applicant subject to and in compliance with a written historic preservation agreement? (Y or N) ☐

V. Additional Information Required for Construction Activities Only

Project Start Date: _____   Completion Date: _____   Estimated Area to be Disturbed (in Acres): _____   Is the Storm Water Pollution Prevention Plan in compliance with State and/or Local sediment and erosion plans? (Y or N) ☐

VI. Certification: The certification statement in Box 1 applies to all applicants.
The certification statement in Box 2 applies only to facilities applying for the Multi-Sector storm water general permit.

BOX 1
ALL APPLICANTS:
I certify under penalty of law that this document and all attachments were prepared under my direction or supervision in accordance with a system designed to assure that qualified personnel properly gather and evaluate the information submitted. Based on my inquiry of the person or persons who manage the system, or those persons directly responsible for gathering the information, the information submitted is, to the best of my knowledge and belief, true, accurate, and complete. I am aware that there are significant penalties for submitting false information, including the possibility of fine and imprisonment for knowing violations.

BOX 2
MULTI-SECTOR STORM WATER GENERAL PERMIT APPLICANTS ONLY:
I certify under penalty of law that I have read and understand the Part I.B. eligibility requirements for coverage under the Multi-Sector storm water general permit, including those requirements relating to the protection of species identified in Addendum H.

To the best of my knowledge, the discharges covered under this permit, and construction of BMPs to control storm water run-off, are not likely to and will not likely adversely affect any species identified in Addendum H of the Multi-Sector storm water general permit or are otherwise eligible for coverage due to previous authorization under the Endangered Species Act.

To the best of my knowledge, I further certify that such discharges, and construction of BMPs to control storm water run-off, do not have an effect on properties listed or eligible for listing on the National Register of Historic Places under the National Historic Preservation Act, or are otherwise eligible for coverage due to a previous agreement under the National Historic Preservation Act.

I understand that continued coverage under the Multi-Sector general permit is contingent upon maintaining eligibility as provided for in Part I.B.

Print Name: _____   Date: _____

Signature: _____

EPA Form 3510-6 (8-98)

STF ENV533F

THIS FORM REPLACES PREVIOUS FORM 3510-7 (8-92) Form Approved. OMB No. 2040-0086
Please See Instructions Before Completing This Form Approval expires: 8-31-98

NPDES FORM

EPA

United States Environmental Protection Agency
Washington, DC 20460
Notice of Termination (NOT) of Coverage Under a NPDES General Permit for Storm Water Discharges Associated with Industrial Activity

Submission of this Notice of Termination constitutes notice that the party identified in Section II of this form is no longer authorized to discharge storm water associated with industrial activity under the NPDES program. ALL NECESSARY INFORMATION MUST BE PROVIDED ON THIS FORM.

I. Permit Information

NPDES Storm Water General Permit Number: _____

Check Here if You are No Longer the Operator of the Facility: ☐

Check Here if the Storm Water Discharge is Being Terminated: ☐

II. Facility Operator Information

Name: _____ Phone: _____

Address: _____

City: _____ State: _____ ZIP Code: _____

III. Facility/Site Location Information

Name: _____

Address: _____

City: _____ State: _____ ZIP Code: _____

Latitude: _____ Longitude: _____ Quarter: _____ Section: _____ Township: _____ Range: _____

IV. Certification: I certify under penalty of law that all storm water discharges associated with industrial activity from the identified facility that are authorized by a NPDES general permit have been eliminated or that I am no longer the operator of the facility or construction site. I understand that by submitting this Notice of Termination, I am no longer authorized to discharge storm water associated with industrial activity under this general permit, and that discharging pollutants in storm water associated with industrial activity to waters of the United States is unlawful under the Clean Water Act where the discharge is not authorized by a NPDES permit. I also understand that the submittal of this Notice of Termination does not release an operator from liability for any violations of this permit or the Clean Water Act.

Print Name: _____ Date: _____

Signature: _____

Instructions for Completing Notice of Termination (NOT) Form

Who May File a Notice of Termination (NOT) Form

Permittees who are presently covered under an EPA-issued National Pollutant Discharge Elimination System (NPDES) General Permit (including the 1995 Multi-Sector Permit) for Storm Water Discharges Associated with Industrial Activity may submit a Notice of Termination (NOT) form when their facilities no longer have any storm water discharges associated with industrial activity as defined in the storm water regulations at 40 CFR 122.26(b)(14), or when they are no longer the operator of the facilities.

For construction activities, elimination of all storm water discharges associated with industrial activity occurs when disturbed soils at the construction site have been finally stabilized and temporary erosion and sediment control measures have been removed or will be removed at an appropriate time, or that all storm water discharges associated with industrial activity from the construction site that are authorized by a NPDES general permit have otherwise been eliminated. Final stabilization means that all soil-disturbing activities at the site have been completed, and that a uniform perennial vegetative cover with a density of 70% of the cover for unpaved areas and areas not covered by permanent structures has been established, or equivalent permanent stabilization measures (such as the use of riprap, gabions, or geotextiles) have been employed.

Where to File NOT Form

Send this form to the the following address:

Storm Water Notice of Termination (4203)
401 M Street, S.W.
Washington, D.C. 20460

Completing the Form

Type or print, using upper-case letters, in the appropriate areas only. Please place each character between the marks. Abbreviate if necessary to stay within the number of characters allowed for each item. Use only one space for breaks between words, but not for punctuation marks unless they are needed to clarify your response. If you have any questions about this form, telephone or write the Notice of Intent Processing Center at (703) 931-3230.

EPA Form 3510-7 (8-98)

STF ENV498F

F.32 APPENDIX F

NPDES
FORM

United States Environmental Protection Agency
Washington, DC 20460
**Notice of Intent (NOI) for Storm Water Discharges Associated with
CONSTRUCTION ACTIVITY Under a NPDES General Permit**

Submission of this Notice of Intent constitutes notice that the party identified in Section I of this form intends to be authorized by a NPDES permit issued for storm water discharges associated with construction activity in the State/Indian Country Land identified in Section II of this form. Submission of this Notice of Intent also constitutes notice that the party identified in Section I of this form meets the eligibility requirements in Part I.B. of the general permit (including those related to protection of endangered species determined through the procedures in Addendum A of the general permit), understands that continued authorization to discharge is contingent on maintaining permit eligibility, and that implementation of the Storm Water Pollution Prevention Plan required under Part IV of the general permit will begin at the time the permittee commences work on the construction project identified in Secion II below. IN ORDER TO OBTAIN AUTHORIZATION, ALL INFORMATION REQUESTED MUST BE INCLUDED ON THIS FORM. SEE INSTRUCTIONS ON BACK OF FORM.

I. Owner/Operator (Applicant) Information

Name: [ ]   Phone: [ ]

Address: [ ]   Status of Owner/Operator: [P]

City: [ ]   State: [ ]   Zip Code: [ ]

II. Project/Site Information

Is the facility located on Indian Country Lands? Yes ☐ No ☐

Project Name: [ ]

Project Address/Location: [ ]

City: [ ]   State: [ ]   Zip Code: [ ]

Latitude: [ ]   Longitude: [ ]   County: [ ]

Has the Storm Water Pollution Prevention Plan (SWPPP) been prepared?   Yes ☐   No ☐

Optional: Address of location of SWPPP for viewing   ☐ Address in Section I above   ☐ Address in Section II above   ☐ Other address (if known) below:

SWPPP Address: [ ]   Phone: [ ]

City: [ ]   State: [ ]   Zip Code: [ ]

Name of Receiving Water: [ ]

[ ]
Month Day Year
Estimated Construction Start Date

[ ]
Month Day Year
Estimated Completion Date

Based on instruction provided in Addendum A of the permit, are there any listed endangered or threatened species, or designated critical habitat in the project area?

Yes ☐   No ☐

Estimate of area to be disturbed (to nearest acre): [ ]

Estimate of Likelihood of Discharge (choose only one):

1. ☐ Unlikely   3. ☐ Once per week   5. ☐ Continual

2. ☐ Once per month   4. ☐ Once per day

I have satisfied permit eligibility with regard to protection of endangered species through the indicated section of Part I.B.3.e.(2) of the permit (check one or more boxes):

(a) ☐   (b) ☐   (c) ☐   (d) ☐

III. Certification

I certify under penalty of law that this document and all attachments were prepared under my direction or supervision in accordance with a system designed to assure that qualified personnel properly gather and evaluate the information submitted. Based on my inquiry of the person or persons who manage this system, or those persons directly responsible for gathering the information, the information submitted is, to the best of my knowledge and belief, true, accurate, and complete. I am aware that there are significant penalties for submitting false information, including the possibility of fine and imprisonment for knowing violations.

Print Name: [ ]   Date: [ ]

Signature: _____

EPA Form 3510-9 replaced 3510-6 (8-98)

EPA

Instructions – EPA Form 3510-9
**Notice of Intent (NOI) for Storm Water Discharges Associated with
Construction Activity to be Covered Under a NPDES Permit**

Form Approved.   OMB No. 2040-0188

Who Must File a Notice of Intent Form

Under the provisions of the Clean Water Act, as amended, (33 U.S.C. 1251 et.seq.; the Act), except as provided by Part I.B.3 the permit, Federal law prohibits discharges of pollutants in storm water from construction activities without a National Pollutant Discharge Elimination System Permit. Operator(s) of construction sites where 5 or more acres are disturbed, smaller sites that are part of a larger common plan of development or sale where there is a cumulative disturbance of at least 5 acres, or any site designated by the Director, must submit an NOI to obtain coverage under an NPDES Storm Water Construction General Permit. If you have questions about whether you need a permit under the NPDES Storm Water program, or if you need information as to whether a particular program is administered by EPA or a State agency, write to or telephone the Notice of Intent Processing Center at (703) 931-3230.

Where to File NOI Form

NOIs must be sent to the following address:

Storm Water Notice of Intent (4203)
USEPA
401 M. Street, SW
Washington, D.C. 20460

Do not send Storm Water Pollution Prevention Plans (SWPPPs) to the above address. For overnight/express delivery of NOIs, please include the room number 2104 Northeast Mall and phone number (202) 260-9541 in the address.

When to File

This form must be filed at least 48 hours before construction begins.

Completing the Form

OBTAIN AND READ A COPY OF THE APPROPRIATE EPA STORM WATER CONSTRUCTION GENERAL PERMIT FOR YOUR AREA. To complete this form, type or print, using uppercase letters, in the appropriate areas only. Please place each character between the marks (abbreviate if necessary to stay within the number of characters allowed for each item). Use one space for breaks between words, but not for punctuation marks unless they are needed to clarify your response. If you have any questions on this form, call the Notice of Intent Processing Center at (703) 931-3230.

Section I. Facility Owner/Operator (Applicant) Information

Provide the legal name, mailing address, and telephone number of the person, firm, public organization, or any other entity that meet either of the following two criteria: (1) they have operational control over construction plans and specifications, including the ability to make modifications to those plans and specifications; or (2) they have the day-to-day operational control of those activities at the project necessary to ensure compliance with SWPPP requirements or other permit conditions. Each person that meets either of these criteria must file this form. Do not use a colloquial name. Correspondence for the permit will be sent to this address.

Enter the appropriate letter to indicate the legal status of the owner/operator of the project: F = Federal; S = State; M = Public (other than federal or state); P = Private.

Section II. Project/Site Information

Enter the official or legal name and complete street address, including city, county, state, zip code, and phone number of the project or site. If it lacks a street address, indicate with a general statement the location of the site (e.g., Intersection of State Highways 61 and 34). Complete site information must be provided for permit coverage.

The applicant must also provide the latitude and longitude of the facility in degrees, minutes, and seconds to the nearest 15 seconds. The latitude and longitude of your facility can be located on USGS quadrangle maps. Quadrangle maps can be obtained by calling 1-800 USA MAPS. Longitude and latitude may also be obtained at the Census Bureau Internet site: http://www.census.gov/cgi-bin/gazetteer.

Latitude and longitude for a facility in decimal form must be converted to degrees, minutes and seconds for proper entry on the NOI form. To convert decimal latitude or longitude to degrees, minutes, and seconds, follow the steps in the following example.

Convert decimal latitude 45.1234567 to degrees, minutes, and seconds.

1) The numbers to the left of the decimal point are degrees.
2) To obtain minutes, multiply the first four numbers to the right of the decimal point by 0.006. 1234 x .006 = 7.404.
3) The numbers to the left of the decimal point in the result obtained in step 2 are the minutes: 7'.
4) To obtain seconds, multiply the remaining three numbers to the right of the decimal from the result in step 2 by 0.06: 404 x 0.06 = 24.24. Since the numbers to the right of the decimal point are not used, the result is 24".
5) The conversion for 45.1234 = 45° 7' 24".

Indicate whether the project is on Indian Country Lands.

Indicate if the Storm Water Pollution Prevention Plan (SWPPP) has been developed. Refer to Part IV of the general permit for information on SWPPPs. To be eligible for coverage, a SWPPP must have been prepared.

Optional: Provide the address and phone number where the SWPPP can be viewed if different from addresses previously given. Check appropriate box.

Enter the name of the closest water body which receives the project's construction storm water discharge.

Enter the estimated construction start and completion dates using four digits for the year (i.e. 05/27/1998).

Enter the estimated area to be disturbed including but not limited to: grubbing, excavation, grading, and utilities and infrastructure installation. Indicate to the nearest acre; if less than 1 acre, enter "1." Note: 1 acre = 43,560 sq. ft.

Indicate your best estimate of the likelihood of storm water discharges from the project. EPA recognizes that actual discharges may differ from this estimate due to unforeseen or chance circumstances.

Indicate if there are any listed endangered or threatened species, or designated critical habitat in the project area.

Indicate which Part of the permit that the applicant is eligible with regard to protection of endangered or threatened species, or designated critical habitat.

Section III. Certification

Federal Statutes provide for severe penalties for submitting false information on this application form. Federal regulations require this application to be signed as follows:

For a corporation: by a responsible corporate officer, which means: (i) president, secretary, treasurer, or vice president of the corporation in charge of a principal business function, or any other person who performs similar policy or decision making functions, or (ii) the manager of one or more manufacturing, production, or operating facilities employing more than 250 persons or having gross annual sales or expenditures exceeding $25 million (in second-quarter 1980 dollars), if authority to sign documents has been assigned or delegated to the manager in accordance with corporate procedures.

For a partnership or sole proprietorship: by a general partner of the proprietor, or

For a municipality, state, federal, or other public facility: by either a principal executive or ranking elected official. An unsigned or undated NOI form will not be granted permit coverage.

Paperwork Reduction Act Notice

EPA

United States Environmental Protection Agency
Washington, D.C. 20460

NPDES Compliance Inspection Report

Form Approved.
OMB No. 2040-0057
Approval expires 4-30-88

Section A: National Data System Coding

Transaction Code	NPDES	yr/mo/day	Inspection Type	Inspector	Fac Type
1 □ 2 |5| 3 □	11 12 □ 17	18 □	19 □	20 □

Remarks

21 └──┘ 66

Reserved	Facility Evaluation Rating	BI	QA	- - - - - - - - - - - Reserved - - - - - - - - - - -
67 └──┘ 69	70 □	71 □	72 □	73 □ 74 75 └────────┘ 80

Section B: Facility Data

Name and Location of Facility Inspected	Entry Time □ AM □ PM	Permit Effective Date
	Exit Time/Date	Permit Expiration Date

Name(s) of On-Site Representative(s)	Title(s)	Phone No(s)

Name, Address of Responsible Official	Title
	Phone No. Contacted □ Yes □ No

Section C: Areas Evaluated During Inspection
(S = Satisfactory, M = Marginal, U = Unsatisfactory, N = Not Evaluated)

Permit	Flow Measurement	Pretreatment	Operations & Maintenance
Records/Reports	Laboratory	Compliance Schedules	Sludge Disposal
Facility Site Review	Effluent/Receiving Waters	Self-Monitoring Program	Other:

Section D: Summary of Findings/Comments (Attach additional sheets if necessary)

Name(s) and Signature(s) of Inspector(s)	Agency/Office/Telephone	Date

Signature of Reviewer	Agency/Office	Date

Regulatory Office Use Only

Action Taken	Date	Compliance Status
		□ Noncompliance □ Compliance

EPA Form 3560-3 (Rev. 3-85) Previous editions are obsolete.
STF ENV645F

Form Approved. OMB No. 2040-0042. Expires 9-30-86

Form	UNITED STATES ENVIRONMENTAL PROTECTION AGENCY	I. EPA ID NUMBER		
4 EPA UIC	**EPA** UNDERGROUND INJECTION CONTROL PERMIT APPLICATION (Collected under the authority of the Safe Drinking Water Act, Sections 1421, 1422, 40 CFR 144)	U	T/A	C

READ ATTACHED INSTRUCTIONS BEFORE STARTING

FOR OFFICIAL USE ONLY

Application approved mo day year	Date Received mo day year	Permit/Well Number	Comments

II. FACILITY NAME AND ADDRESS

Facility Name

Street Address

City	State	ZIP Code

III. OWNER/OPERATOR AND ADDRESS

Owner/Operator Name

Street Address

City	State	ZIP Code

IV. OWNERSHIP STATUS *(Mark 'x')*

☐ A. Federal   ☐ B. State   ☐ C. Private

☐ D. Public   ☐ E. Other *(Explain)*

V. SIC CODES

VI. WELL STATUS *(Mark 'x')*

☐ A. Operating	Date Started mo day year	☐ B. Modification/Conversion	☐ C. Proposed

VII. TYPE OF PERMIT REQUESTED *(Mark 'x' and specify if required)*

☐ A. Individual ☐ B. Area	Number of Exist- ing wells	Number of Pro- posed wells	Name(s) of field(s) or project(s)

VIII. CLASS AND TYPE OF WELL *(see reverse)*

A. Class(es) *(enter code(s))*	B. Type(s) *(enter code(s))*	C. If class is "other" or type is code 'x,' explain	D. Number of wells per type (if area permit)

IX. LOCATION OF WELL(S) OR APPROXIMATE CENTER OF FIELD OR PROJECT

X. INDIAN LANDS *(Mark 'x')*

C	A. Latitude			B. Longitude			Township and Range							☐ Yes ☐ No
	Deg	Min	Sec	Deg	Min	Sec	Twsp	Range	Sec	¼ Sec	Feet from	Line	Feet from	Line

XI. ATTACHMENTS

(Complete the following questions on a separate sheet(s) and number accordingly; see instructions)

FOR CLASSES I, II, III (and other classes) complete and submit on separate sheet(s) Attachments A—U (pp 2-6) as appropriate. Attach maps where required. List attachments by letter which are applicable and are included with your application:

XII. CERTIFICATION

I certify under the penalty of law that I have personally examined and am familiar with the information submitted in this document and all attachments and that, based on my inquiry of those individuals immediately responsible for obtaining the information, I believe that the information is true, accurate, and complete. I am aware that there are significant penalties for submitting false information, including the possibility of fine and imprisonment. (Ref. 40 CFR 144.32)

A. Name and Title *(Type or Print)*	B. Phone No. *(Area Code and No.)*
C. Signature	D. Date Signed

EPA Form 7520-6 (2-84)

Page 1 of 5

STF ENV452F

	United States	Form Approved.
EPA	**Environmental Protection Agency** Washington, DC 20460	OMB No. 2050-0068. Approval expires 3/31/98

Notification for Underground Storage Tanks

State Agency Name and Address:

STATE USE ONLY

ID NUMBER:

DATE RECEIVED:

TYPE OF NOTIFICATION

☐ A. NEW FACILITY ☐ B. AMENDED ☐ C. CLOSURE

_____ No. of tanks at facility _____ No. of continuation sheets attached

A. Date Entered Into Computer _____

B. Data Entry Clerk Initials _____

C. Owner Was Contacted to Clarify Responses. Comments:

INSTRUCTIONS

Please **type or print in ink** all items except "signature" in section V. This form must be completed for each location containing underground storage tanks. If more than five (5) tanks are owned at this location, photocopy the following sheets, and staple continuation sheets to the form.

GENERAL INFORMATION

Notification is required by Federal law for all underground tanks that have been used to store regulated substances since January 1, 1974, that are in the ground as of May 8, 1986, or that are brought into use after May 8, 1986. The Information requested is required by Section 9002 of the Resource Conservation and Recovery Act, (RCRA), as amended.

The primary purpose of this notification program is to locate and evaluate underground tanks that store or have stored petroleum or hazardous substances. It is expected that the information you provide will be based on reasonably available records, or in the absence of such records, your knowledge, belief, or recollection.

Who Must Notify? Section 9002 of RCRA, as amended, requires that, unless exempted, owners of underground tanks that store regulated substances must notify designated State or local agencies of the existence of their tanks. Owner means—

a) in the case of an underground storage tank in use on November 8, 1984, or brought into use after that date, any person who owns an underground storage tank used for storage, use, or dispensing of regulated substances, and

b) in the case of any underground storage tank in use before November 8, 1984, but no longer in use on that date, any person who owned such tank immediately before discontinuation of its use.

c) if the State so requires, any facility that has undergone any changes to facility information or tank system status (only amended tank information needs to be included).

What Tanks Are Included? Underground storage tank is defined as any one or combination of tanks that (1) is used to contain an accumulation of "regulated substances," and (2) whose volume (including connected underground piping) is 10% or more beneath the ground. Some examples are underground tanks storing: 1. Gasoline, used oil or diesel fuel, and 2. Industrial solvents, pesticides, herbicides or fumigants.

What Tanks Are Excluded? Tanks removed from the ground are not subject to notification. Other tanks excluded from notification are:

1. farm or residential tanks of 1,100 gallons or less capacity used for storing motor fuel for noncommercial purposes;
2. tanks used for storing heating oil for consumptive use on the premises where stored;

3. septic tanks;
4. pipeline facilities (including gathering lines) regulated under the Natural Gas Pipeline Safety Act of 1968, or the Hazardous Liquid Pipeline Safety Act of 1979, or which is an intrastate pipeline facility regulated under State laws;
5. surface impoundments, pits, ponds, or lagoons;
6. storm water or waste water collection systems;
7. flow-through process tanks;
8. liquid traps or associated gathering lines directly related to oil or gas production and gathering operations;
9. storage tanks situated in an underground area (such as a basement, cellar, mineworking, drift, shaft, or tunnel) if the storage tank is situated upon or above the surface floor.

What Substances Are Covered? The notification requirements apply to underground storage tanks that contain regulated substances. This includes any substance defined as hazardous in Section 101 (14) of the Comprehensive Environmental Response, Compensation and Liability Act of 1980 (CERCLA), with the exception of those substances regulated as hazardous waste under Subtitle C of RCRA. It also includes petroleum, e.g., crude oil or any fraction thereof which is liquid at standard conditions of temperature and pressure (60 degrees Fahrenheit and 14.7 pounds per square inch absolute).

Where To Notify? Send completed forms to:

When To Notify? 1. Owners of underground storage tanks in use or that have been taken out of operation after January 1, 1974, but still in the ground, must notify by May 8, 1986. 2. Owners who bring underground storage tanks into use after May 8, 1986, must notify within 30 days of bringing the tanks into use. 3. If the State requires notification of any amendments to facility, send information to State agency immediately.

Penalties: Any owner who knowingly fails to notify or submits false information shall be subject to a civil penalty not to exceed $10,000 for each tank for which notification is not given or for which false information is submitted.

I. OWNERSHIP OF TANK(S)

Owner Name (Corporation, Individual, Public Agency, or Other Entity)

Street Address

City		State	Zip Code

County

Phone Number (include Area Code)

II. LOCATION OF TANK(S)

If required by State, give the geographic location of tanks by degrees, minutes and seconds. Examples Lat. 42, 36, 12 N Long. 85, 24, 17W

Latitude _____ Longitude _____

Facility Name of Company Site Identifier, as applicable

(If same as Section I, mark box here) ☐

Street Address

City		State	Zip Code

County	Municipality

EPA Form 7530-1 (Rev. 8-94) Electronic and Paper versions acceptable.
Previous editions may be used while supplies last.

Page 1 of 5

STF ENV563F.1

EPA	United States **Environmental Protection Agency** Washington, DC 20460	Form Approved. OMB No. 2050-0068. Approval expires 3/31/98

Notification for Underground Storage Tanks

III. TYPE OF OWNER

☐ Federal Government ☐ Commercial

☐ State Government ☐ Private

☐ Local Government

IV. INDIAN LANDS

Tanks are located on land within an Indian Reservation or on other trust lands. ☐

Tanks are owned by native American nation, tribe, or individual. ☐

Tribe or Nation:

V. TYPE OF FACILITY

☐ Gas Station   ☐ Railroad   ☐ Trucking/Transport

☐ Petroleum Distributor   ☐ Federal - Non-Military   ☐ Utilities

☐ Air Taxi (Airline)   ☐ Federal - Military   ☐ Residential

☐ Aircraft Owner   ☐ Industrial   ☐ Farm

☐ Auto Dealership   ☐ Contractor   ☐ Other (Explain) _____

VI. CONTACT PERSON IN CHARGE OF TANKS

Name:	Job Title:	Address:	Phone Number (Include Area Code):

VII. FINANCIAL RESPONSIBILITY

☐ I have met the financial responsibility requirements in accordance with 40 CFR Subpart H

Check All that Apply

☐ Self Insurance   ☐ Guarantee   ☐ State Funds

☐ Commercial Insurance   ☐ Surety Bond   ☐ Trust Fund

☐ Risk Retention Group   ☐ Letter of Credit   ☐ Other Method Allowed - Specify

_____

VIII. CERTIFICATION (Read and sign after completing all sections)

I certify under penalty of law that I have personally examined and am familiar with the information submitted in this and all attached documents, and that based on my inquiry of those individuals immediately responsible for obtaining the information, I believe that the submitted information is true, accurate, and complete.

Name and official title of owner or owner's authorized representative (Print)	Signature	Date Signed

Paperwork Reduction Act Notice

EPA estimates public reporting burden for this form to average 30 minutes per response including time for reviewing instructions, gathering and maintaining the data needed and completing and reviewing the form. Send comments regarding this burden estimate to Chief, Information Policy Branch (2136), U.S. Environmental Protection Agency, 401 M Street, Washington D.C. 20460, marked "Attention Desk Officer for EPA." This form amends the previous notification form as printed in 40 CFR Part 280, Appendix I. Previous editions of this notification form may be used while supplies last.

EPA Form 7530-1 (Rev. 8-94) Electronic and Paper versions acceptable.
Previous editions may be used while supplies last.

Page 2 of 5

STF ENV563F.2

EPA	United States **Environmental Protection Agency** Washington, DC 20460	Form Approved. OMB No. 2050-0068. Approval expires 3/31/98

Notification for Underground Storage Tanks

IX. DESCRIPTION OF UNDERGROUND STORAGE TANKS (Complete for each tank at this location.)

Tank Identification Number	Tank No._____	Tank No._____	Tank No._____	Tank No._____	Tank No._____
1. Status of Tank (Mark only one)					
Currently In Use	☐	☐	☐	☐	☐
Temporarily Out of Use	☐	☐	☐	☐	☐
Permanently Out of Use	☐	☐	☐	☐	☐
Amendment of Information	☐	☐	☐	☐	☐
2. Date of Installation (mo./year)					
3. Estimated Total Capacity (gallons)					
4. Material of Construction (Mark all that apply)					
Asphalt Coated or Bare Steel	☐	☐	☐	☐	☐
Cathodically Protected Steel	☐	☐	☐	☐	☐
Epoxy Coated Steel	☐	☐	☐	☐	☐
Composite (Steel with Fiberglass)	☐	☐	☐	☐	☐
Fiberglass Reinforced Plastic	☐	☐	☐	☐	☐
Lined Interior	☐	☐	☐	☐	☐
Double Walled	☐	☐	☐	☐	☐
Polyethylene Tank Jacket	☐	☐	☐	☐	☐
Concrete	☐	☐	☐	☐	☐
Excavation Liner	☐	☐	☐	☐	☐
Unknown	☐	☐	☐	☐	☐
Other, Please specify	_____	_____	_____	_____	_____
Has tank been repaired?	☐	☐	☐	☐	☐
5. Piping (Material) (Mark all that apply) Bare Steel	☐	☐	☐	☐	☐
Galvanized Steel	☐	☐	☐	☐	☐
Fiberglass Reinforced Plastic	☐	☐	☐	☐	☐
Copper	☐	☐	☐	☐	☐
Cathodically Protected	☐	☐	☐	☐	☐
Double Walled	☐	☐	☐	☐	☐
Secondary Containment	☐	☐	☐	☐	☐
Unknown	☐	☐	☐	☐	☐
Other, Please specify	_____	_____	_____	_____	_____
6. Piping (Type) (Mark all that apply)					
Suction: no valve at tank	☐	☐	☐	☐	☐
Suction: valve at tank	☐	☐	☐	☐	☐
Pressure	☐	☐	☐	☐	☐
Gravity Feed	☐	☐	☐	☐	☐
Has piping been repaired?	☐	☐	☐	☐	☐

EPA	**United States** **Environmental Protection Agency** Washington, DC 20460	Form Approved. OMB No. 2050-0068. Approval expires 3/31/98
	Notification for Underground Storage Tanks	

Tank Identification Number	Tank No. _____	Tank No. _____	Tank No. _____	Tank No. _____	Tank No. _____
7. Substance Currently or Last Stored in Greatest Quantity by					
Volume Gasoline	☐	☐	☐	☐	☐
Diesel	☐	☐	☐	☐	☐
Gasohol	☐	☐	☐	☐	☐
Kerosene	☐	☐	☐	☐	☐
Heating Oil	☐	☐	☐	☐	☐
Used Oil	☐	☐	☐	☐	☐
Other	☐	☐	☐	☐	☐
Please Specify	_____	_____	_____	_____	_____
Hazardous Substance	☐	☐	☐	☐	☐
CERCLA name and/or					
CAS number	_____	_____	_____	_____	_____
Mixture of Substances	☐	☐	☐	☐	☐
Please Specify	_____	_____	_____	_____	_____

X. TANKS OUT OF USE, OR CHANGE IN SERVICE					

1. Closing of Tank
   A. Estimated date last used
      (mo./day/year)					

B. Estimated date tank closed (mo./day/year)					
C. Tank was removed from ground D. Tank was closed in ground E. Tank filled with inert material Describe	☐ ☐	☐ ☐	☐ ☐	☐ ☐	☐ ☐
F. Change in service	☐	☐	☐	☐	☐
2. Site Assessment Completed	☐	☐	☐	☐	☐
Evidence of a leak detected	☐	☐	☐	☐	☐

EPA Form 7530-1 (Rev. 8-94) Electronic and Paper versions acceptable.
Previous editions may be used while supplies last.

Page 4 of 5

STF ENV563F 4

EPA	United States **Environmental Protection Agency** Washington, DC 20460	Form Approved. OMB No. 2050-0068. Approval expires 3/31/98

Notification for Underground Storage Tanks

XI. CERTIFICATION OF COMPLIANCE (COMPLETE FOR ALL NEW AND UPGRADED TANKS AT THIS LOCATION)

Tank Identification Number	Tank No. ____	Tank No. ____	Tank No. ____	Tank No. ____	Tank No. ____

1. Installation

A. Installer certified by tank and piping manufacturers	☐	☐	☐	☐	☐
B. Installer certified or licensed by the implementing agency	☐	☐	☐	☐	☐
C. Installation inspected by a registered engineer	☐	☐	☐	☐	☐
D. Installation inspected and approved by implementing agency	☐	☐	☐	☐	☐
E. Manufacturer's installation checklists have been completed	☐	☐	☐	☐	☐
F. Another method allowed by State agency. Please Specify ____	☐	☐	☐	☐	☐

2. Release Detection (Mark all that apply)

	TANK	PIPING	TANK	PIPING	TANK	PIPING	TANK	PIPING	TANK	PIPING
A. Manual tank gauging	☐		☐		☐		☐		☐	
B. Tank tightness testing	☐		☐		☐		☐		☐	
C. Inventory Controls	☐		☐		☐		☐		☐	
D. Automatic tank gauging	☐		☐		☐		☐		☐	
E. Vapor monitoring	☐	☐	☐	☐	☐	☐	☐	☐	☐	☐
F. Groundwater monitoring	☐	☐	☐	☐	☐	☐	☐	☐	☐	☐
G. Verify monitoring/secondary containment	☐	☐	☐	☐	☐	☐	☐	☐	☐	☐
H. Automatic line leak detectors	☐	☐	☐	☐	☐	☐	☐	☐	☐	☐
I. Line tightness testing	☐	☐	☐	☐	☐	☐	☐	☐	☐	☐
J. Other method allowed by implementing agency. Please specify ____	☐	☐	☐	☐	☐	☐	☐	☐	☐	☐

3. Spill and Overfill Protection

A. Overfill device installed	☐	☐	☐	☐	☐
B. Spill device installed	☐	☐	☐	☐	☐

OATH: I certify the information concerning installation that is provided is section XI is true to the best of my belief and knowledge.

Installer: _____   _____   _____
                        Name   Signature   Date

_____   _____
          Position   Company

NATIONAL POLLUTANT DISCHARGE ELIMINATION SYSTEM
APPLICATION FOR PERMIT TO DISCHARGE - SHORT FORM A

Form Approved
OMB No. 158-R0096

APPLICATION NUMBER

FOR
AGENCY
USE

DATE RECEIVED

| YEAR | MO. | DAY |

To be filed only by municipal wastewater dischargers

Do not attempt to complete this form before reading the accompanying instructions

Please print or type

1. Name of organization responsible for facility _____

2. Address, location, and telephone number of facility producing discharge:
    A. Name _____
    B. Mailing address:
        1. Street address _____
        2. City _____ 3. County _____
        4. State _____ 5. ZIP _____
    C. Location:
        1. Street _____
        2. City _____ 3. County _____
        4. State _____
    D. Telephone No. _____
                Area Code

If all your waste is discharged into a publicly owned waste treatment facility and to the best of your knowledge you are not required to obtain a discharge permit, proceed to item 3. Otherwise proceed directly to item 4.

3. If you meet the condition stated above, check here ☐ and supply the information asked for below. After completing these items, please complete the date, title, and signature blocks below and return this form to the proper reviewing office without completing the remainder of the form.
    A. Name of organization responsible for receiving waste _____
    B. Facility receiving waste:
        1. Name _____
        2. Street address _____
        3. City _____ 4. County _____
        5. State _____ 6. ZIP _____

4. Type of treatment:
    A. ☐ None   B. ☐ Primary   C. ☐ Intermediate   D. ☐ Secondary   E. ☐ Advanced

5. Design flow (average daily) of facility _____ mgd.

6. Percent BOD removal (actual):
    A. ☐ 0-29.9   B. ☐ 30-64.9   C. ☐ 65-84.9   D. ☐ 85-94.9   E. ☐ 95 or more

7. Population served:
    A. ☐ 1-199   B. ☐ 200-499   C. ☐ 500-999   D. ☐ 1,000-4,999
    E. ☐ 5,000-9,999   F. ☐ 10,000 or more

8. Number of separate discharge points:
    A. ☐ 1   B. ☐ 2   C. ☐ 3   D. ☐ 4   E. ☐ 5   F. ☐ 6 or more

9. Description of waste water discharged to surface waters only (check as applicable).

Discharge per operating day	Flow, MGD (million gallons per operating day)							Volume treated before discharging (percent)				
	0-0.0099 (1)	0.01-0.049 (2)	0.05-0.099 (3)	0.1-0.49 (4)	0.5-0.99 (5)	1.0-4.9 (6)	5 or more (7)	None (8)	0.1-34.9 (9)	35-64.9 (10)	65-94.9 (11)	95-100 (12)
A. Average												
B. Maximum												

10. If any waste water, treated or untreated, is discharged to places other than surface waters, check below as applicable.

Waste water is discharged to	Flow, MGD (million gallons per operating day)						
	0-0.0099 (1)	0.01-0.049 (2)	0.05-0.099 (3)	0.1-0.49 (4)	0.5-0.99 (5)	1.0-4.9 (6)	5 or more (7)
A. Deep well							
B. Evaporation lagoon							
C. Subsurface percolation system							
D. Other, specify:							

11. Is any sludge ultimately returned to a waterway?
  A. ☐ yes   B. ☐ no

12. a. Do you receive industrial waste?
  1. ☐ yes   2. ☐ no
  b. If yes, enter approximate number of industrial dischargers into system _____

13. Type of collection sewer system:
  A. ☐ Separate sanitary
  B. ☐ Combined sanitary and storm
  C. ☐ Both separate and combined sewer systems

14. Name of receiving water or waters _____

15. Does your discharge contain or is it possible for your discharge to contain one or more of the following substances: ammonia, cyanide, aluminum, beryllium, cadmium, chromium, copper, lead, mercury, nickel, selenium, zinc, phenols.
  A. ☐ yes   B. ☐ no

I certify that I am familiar with the information contained in the application and that to the best of my knowledge and belief such information is true, complete, and accurate.

_____
Printed Name of Person Signing

_____
Title

_____
Date Application Signed

_____
Signature of Applicant

APPENDIX G
REPORTABLE QUANTITIES OF HAZARDOUS SUBSTANCES

TABLE 117.3—REPORTABLE QUANTITIES OF HAZARDOUS SUBSTANCES DESIGNATED PURSUANT TO SECTION 311 OF THE CLEAN WATER ACT

Material	Category	RQ in pounds (kilograms)
Acetaldehyde	C	1,000 (454)
Acetic acid	D	5,000 (2,270)
Acetic anhydride	D	5,000 (2,270)
Acetone cyanohydrin	A	10 (4.54)
Acetyl bromide	D	5,000 (2,270)
Acetyl chloride	D	5,000 (2,270)
Acrolein	X	1 (0.454)
Acrylonitrile	B	100 (45.4)
Adipic acid	D	5,000 (2,270)
Aldrin	X	1 (0.454)
Allyl alcohol	B	100 (45.4)
Allyl chloride	C	1,000 (454)
Aluminum sulfate	D	5,000 (2,270)
Ammonia	B	100 (45.4)
Ammonium acetate	D	5,000 (2,270)
Ammonium benzoate	D	5,000 (2,270)
Ammonium bicarbonate	D	5,000 (2,270)
Ammonium bichromate	A	10 (4.54)
Ammonium bifluoride	B	100 (45.4)
Ammonium bisulfite	D	5,000 (2,270)
Ammonium carbamate	D	5,000 (2,270)
Ammonium carbonate	D	5,000 (2,270)
Ammonium chloride	D	5,000 (2,270)
Ammonium chromate	A	10 (4.54)
Ammonium citrate dibasic	D	5,000 (2,270)
Ammonium fluoborate	D	5,000 (2,270)
Ammonium fluoride	B	100 (45.4)
Ammonium hydroxide	C	1,000 (454)
Ammonium oxalate	D	5,000 (2,270)
Ammonium silicofluoride	C	1,000 (454)
Ammonium sulfamate	D	5,000 (2,270)
Ammonium sulfide	B	100 (45.4)
Ammonium sulfite	D	5,000 (2,270)
Ammonium tartrate	D	5,000 (2,270)
Ammonium thiocyanate	D	5,000 (2,270)
Amyl acetate	D	5,000 (2,270)
Aniline	D	5,000 (2,270)
Antimony pentachloride	C	1,000 (454)
Antimony potassium tartrate	B	100 (45.4)
Antimony tribromide	C	1,000 (454)
Antimony trichloride	C	1,000 (454)
Antimony trifluoride	C	1,000 (454)

TABLE 117.3—REPORTABLE QUANTITIES OF HAZARDOUS SUBSTANCES DESIGNATED PURSUANT TO
SECTION 311 OF THE CLEAN WATER ACT—Continued

Material	Category	RQ in pounds (kilograms)
Antimony trioxide	C	1,000 (454)
Arsenic disulfide	X	1 (0.454)
Arsenic pentoxide	X	1 (0.454)
Arsenic trichloride	X	1 (0.454)
Arsenic trioxide	X	1 (0.454)
Arsenic trisulfide	X	1 (0.454)
Barium cyanide	A	10 (4.54)
Benzene	A	10 (4.54)
Benzoic acid	D	5,000 (2,270)
Benzonitrile	D	5,000 (2,270)
Benzoyl chloride	C	1,000 (454)
Benzyl chloride	B	100 (45.4)
Beryllium chloride	X	1 (0.454)
Beryllium fluoride	X	1 (0.454)
Beryllium nitrate	X	1 (0.454)
Butyl acetate	D	5,000 (2,270)
Butylamine	C	1,000 (454)
n-Butyl phthalate	A	10 (4.54)
Butyric acid	D	5,000 (2,270)
Cadmium acetate	A	10 (4.54)
Cadmium bromide	A	10 (4.54)
Cadmium chloride	A	10 (4.54)
Calcium arsenate	X	1 (0.454)
Calcium arsenite	X	1 (0.454)
Calcium carbide	A	10 (4.54)
Calcium chromate	A	10 (4.54)
Calcium cyanide	A	10 (4.54)
Calcium dodecylbenzenesulfonate	C	1,000 (454)
Calcium hypochlorite	A	10 (4.54)
Captan	A	10 (4.54)
Carbaryl	B	100 (45.4)
Carbofuran	A	10 (4.54)
Carbon disulfide	B	100 (45.4)
Carbon tetrachloride	A	10 (4.54)
Chlordane	X	1 (0.454)
Chlorine	A	10 (4.54)
Chlorobenzene	B	100 (45.4)
Chloroform	A	10 (4.54)
Chlorosulfonic acid	C	1,000 (454)
Chlorpyrifos	X	1 (0.454)
Chromic acetate	C	1,000 (454)
Chromic acid	A	10 (4.54)
Chromic sulfate	C	1,000 (454)
Chromous chloride	C	1,000 (454)
Cobaltous bromide	C	1,000 (454)
Cobaltous formate	C	1,000 (454)
Cobaltous sulfamate	C	1,000 (454)
Coumaphos	A	10 (4.54)
Cresol	B	100 (45.4)
Crotonaldehyde	B	100 (45.4)
Cupric acetate	B	100 (45.4)
Cupric acetoarsenite	X	1 (0.454)
Cupric chloride	A	10 (4.54)
Cupric nitrate	B	100 (45.4)
Cupric oxalate	B	100 (45.4)
Cupric sulfate	A	10 (4.54)
Cupric sulfate, ammoniated	B	100 (45.4)
Cupric tartrate	B	100 (45.4)
Cyanogen chloride	A	10 (4.54)
Cyclohexane	C	1,000 (454)
2,4-D Acid	B	100 (45.4)
2,4-D Esters	B	100 (45.4)
DDT	X	1 (0.454)
Diazinon	X	1 (0.454)
Dicamba	C	1,000 (454)
Dichlobenil	B	100 (45.4)
Dichlone	X	1 (0.454)
Dichlorobenzene	B	100 (45.4)
Dichloropropane	C	1,000 (454)
Dichloropropene	B	100 (45.4)
Dichloropropene-Dichloropropane (mixture)	B	100 (45.4)

TABLE 117.3—REPORTABLE QUANTITIES OF HAZARDOUS SUBSTANCES DESIGNATED PURSUANT TO
SECTION 311 OF THE CLEAN WATER ACT—Continued

Material	Category	RQ in pounds (kilograms)
2,2-Dichloropropionic acid	D	5,000 (2,270)
Dichlorvos	A	10 (4.54)
Dicofol	A	10 (4.54)
Dieldrin	X	1 (0.454)
Diethylamine	B	100 (45.4)
Dimethylamine	C	1,000 (454)
Dinitrobenzene (mixed)	B	100 (45.4)
Dinitrophenol	A	10 (45.4)
Dinitrotoluene	A	10 (4.54)
Diquat	C	1,000 (454)
Disulfoton	X	1 (0.454)
Diuron	B	100 (45.4)
Dodecylbenzenesulfonic acid	C	1,000 (454)
Endosulfan	X	1 (0.454)
Endrin	X	1 (0.454)
Epichlorohydrin	B	100 (45.4)
Ethion	A	10 (4.54)
Ethylbenzene	C	1,000 (454)
Ethylenediamine	D	5,000 (2,270)
Ethylenediamine-tetraacetic acid (EDTA)	D	5,000 (2,270)
Ethylene dibromide	X	1 (0.454)
Ethylene dichloride	B	100 (45.4)
Ferric ammonium citrate	C	1,000 (454)
Ferric ammonium oxalate	C	1,000 (454)
Ferric chloride	C	1,000 (454)
Ferric fluoride	B	100 (45.4)
Ferric nitrate	C	1,000 (454)
Ferric sulfate	C	1,000 (454)
Ferrous ammonium sulfate	C	1,000 (454)
Ferrous chloride	B	100 (45.4)
Ferrous sulfate	C	1,000 (454)
Formaldehyde	B	100 (45.4)
Formic acid	D	5,000 (2,270)
Fumaric acid	D	5,000 (2,270)
Furfural	D	5,000 (2,270)
Guthion	X	1 (0.454)
Heptachlor	X	1 (0.454)
Hexachlorocyclopentadiene	A	10 (4.54)
Hydrochloric acid	D	5,000 (2,270)
Hydrofluoric acid	B	100 (45.4)
Hydrogen cyanide	A	10 (4.54)
Hydrogen sulfide	B	100 (45.4)
Isoprene	B	100 (45.4)
Isopropanolamine dodecylbenzenesulfonate	C	1,000 (454)
Kepone	X	1 (0.454)
Lead acetate	A	10 (4.54)
Lead arsenate	X	1 (0.454)
Lead chloride	A	10 (4.54)
Lead fluoborate	A	10 (4.54)
Lead fluoride	A	10 (4.54)
Lead iodide	A	10 (4.54)
Lead nitrate	A	10 (4.54)
Lead stearate	A	10 (4.54)
Lead sulfate	A	10 (4.54)
Lead sulfide	A	10 (4.54)
Lead thiocyanate	A	10 (4.54)
Lindane	X	1 (0.454)
Lithium chromate	A	10 (4.54)
Malathion	B	100 (45.4)
Maleic acid	D	5,000 (2,270)
Maleic anhydride	D	5,000 (2,270)
Mercaptodimethur	A	10 (4.54)
Mercuric cyanide	X	1 (0.454)
Mercuric nitrate	A	10 (4.54)
Mercuric sulfate	A	10 (4.54)
Mercuric thiocyanate	A	10 (4.54)
Mercurous nitrate	A	10 (4.54)
Methoxychlor	X	1 (0.454)
Methyl mercaptan	B	100 (45.4)
Methyl methacrylate	C	1,000 (454)
Methyl parathion	B	100 (45.4)

TABLE 117.3—REPORTABLE QUANTITIES OF HAZARDOUS SUBSTANCES DESIGNATED PURSUANT TO
SECTION 311 OF THE CLEAN WATER ACT—Continued

Material	Category	RQ in pounds (kilograms)
Mevinphos	A	10 (4.54)
Mexacarbate	C	1,000 (454)
Monoethylamine	B	100 (45.4)
Monomethylamine	B	100 (45.4)
Naled	A	10 (4.54)
Naphthalene	B	100 (45.4)
Naphthenic acid	B	100 (45.4)
Nickel ammonium sulfate	B	100 (45.4)
Nickel chloride	B	100 (45.4)
Nickel hydroxide	A	10 (4.54)
Nickel nitrate	B	100 (45.4)
Nickel sulfate	B	100 (45.4)
Nitric acid	C	1,000 (454)
Nitrobenzene	C	1,000 (454)
Nitrogen dioxide	A	10 (4.54)
Nitrophenol (mixed)	B	100 (45.4)
Nitrotoluene	C	1,000 (454)
Paraformaldehyde	C	1,000 (454)
Parathion	A	10 (4.54)
Pentachlorophenol	A	10 (4.54)
Phenol	C	1,000 (454)
Phosgene	A	10 (4.54)
Phosphoric acid	D	5,000 (2,270)
Phosphorus	X	1 (0.454)
Phosphorus oxychloride	C	1,000 (454)
Phosphorus pentasulfide	B	100 (45.4)
Phosphorus trichloride	C	1,000 (454)
Polychlorinated biphenyls	X	1 (0.454)
Potassium arsenate	X	1 (0.454)
Potassium arsenite	X	1 (0.454)
Potassium bichromate	A	10 (4.54)
Potassium chromate	A	10 (4.54)
Potassium cyanide	A	10 (4.54)
Potassium hydroxide	C	1,000 (454)
Potassium permanganate	B	100 (45.4)
Propargite	A	10 (4.54)
Propionic acid	D	5,000 (2,270)
Propionic anhydride	D	5,000 (2,270)
Propylene oxide	B	100 (45.4)
Pyrethrins	X	1 (0.454)
Quinoline	D	5,000 (2,270)
Resorcinol	D	5,000 (2,270)
Selenium oxide	A	10 (4.54)
Silver nitrate	X	1 (0.454)
Sodium	A	10 (4.54)
Sodium arsenate	X	1 (0.454)
Sodium arsenite	X	1 (0.454)
Sodium bichromate	A	10 (4.54)
Sodium bifluoride	B	100 (45.4)
Sodium bisulfite	D	5,000 (2,270)
Sodium chromate	A	10 (4.54)
Sodium cyanide	A	10 (4.54)
Sodium dodecylbenzenesulfonate	C	1,000 (454)
Sodium fluoride	C	1,000 (454)
Sodium hydrosulfide	D	5,000 (2,270)
Sodium hydroxide	C	1,000 (454)
Sodium hypochlorite	B	100 (45.4)
Sodium methylate	C	1,000 (454)
Sodium nitrite	B	100 (45.4)
Sodium phosphate, dibasic	D	5,000 (2,270)
Sodium phosphate, tribasic	D	5,000 (2,270)
Sodium selenite	B	100 (45.4)
Strontium chromate	A	10 (4.54)
Strychnine	A	10 (4.54)
Styrene	C	1,000 (454)
Sulfuric acid	C	1,000 (454)
Sulfur monochloride	C	1,000 (454)
2,4,5-T acid	C	1,000 (454)
2,4,5-T amines	D	5,000 (2,270)
2,4,5-T esters	C	1,000 (454)
2,4,5-T salts	C	1,000 (454)

TABLE 117.3—REPORTABLE QUANTITIES OF HAZARDOUS SUBSTANCES DESIGNATED PURSUANT TO SECTION 311 OF THE CLEAN WATER ACT—Continued

Material	Category	RQ in pounds (kilograms)
TDE	X	1 (0.454)
2,4,5-TP acid	B	100 (45.4)
2,4,5-TP acid esters	B	100 (45.4)
Tetraethyl lead	A	10 (4.54)
Tetraethyl pyrophosphate	A	10 (4.54)
Thallium sulfate	B	100 (45.4)
Toluene	C	1,000 (454)
Toxaphene	X	1 (0.454)
Trichlorfon	B	100 (45.4)
Trichloroethylene	B	100 (45.4)
Trichlorophenol	A	10 (4.54)
Triethanolamine dodecylbenzenesulfonate	C	1,000 (454)
Triethylamine	D	5,000 (2,270)
Trimethylamine	B	100 (45.4)
Uranyl acetate	B	100 (45.4)
Uranyl nitrate	B	100 (45.4)
Vanadium pentoxide	C	1,000 (454)
Vanadyl sulfate	C	1,000 (454)
Vinyl acetate	D	5,000 (2,270)
Vinylidene chloride	B	100 (45.4)
Xylene (mixed)	B	100 (45.4)
Xylenol	C	1,000 (454)
Zinc acetate	C	1,000 (454)
Zinc ammonium chloride	C	1,000 (454)
Zinc borate	C	1,000 (454)
Zinc bromide	C	1,000 (454)
Zinc carbonate	C	1,000 (454)
Zinc chloride	C	1,000 (454)
Zinc cyanide	A	10 (4.54)
Zinc fluoride	C	1,000 (454)
Zinc formate	C	1,000 (454)
Zinc hydrosulfite	C	1,000 (454)
Zinc nitrate	C	1,000 (454)
Zinc phenolsulfonate	D	5,000 (2,270)
Zinc phosphide	B	100 (45.4)
Zinc silicofluoride	D	5,000 (2,270)
Zinc sulfate	C	1,000 (454)
Zirconium nitrate	D	5,000 (2,270)
Zirconium potassium fluoride	C	1,000 (454)
Zirconium sulfate	D	5,000 (2,270)
Zirconium tetrachloride	D	5,000 (2,270)

[50 FR 13513, Apr. 4, 1985, as amended at 51 FR 34547, Sept. 29, 1986; 54 FR 33482, Aug. 14, 1989; 58 FR 35327, June 30, 1993; 60 FR 30937, June 12, 1995]

APPENDIX H
CADDY SUPPLEMENTAL FILES GUIDANCE

These pages are intended to provide the latest guidance for the supplemental files. These are extra files that may be included in the \FILES directory on a CADDY volume in a CD-ROM set, per the CADDY Format Specification. They are organized by data requirement type. Corresponding to each data requirement is a listing of file types data evaluators find useful in conducting their evaluations. These pages will be updated from time to time.

SERIES 860—RESIDUE CHEMISTRY TEST GUIDELINES

Data should be provided in the FILES directory of the index volume of the CADDY CD-ROM set. The files should be in a subdirectory called RESIDUE (e.g., /FILES/RESIDUE).

New GLN		Old GLN	
860.1000	Background	170- 1	
860.1100	Chemical identity	171- 2	summarize Phys/Chem properties as specified under Product chemistry GLN
860.1200	Directions for use	171- 3	word processing format
860.1300	Nature of the residue-- plants, livestock	171- 4a, b	Report and data in word processing format; data tables in spreadsheet format also; structures in MPG and/or ISIS-Draw
860.1340	Residue analytical method	171- 4c, d	Report and data in word processing format; data tables in spreadsheet format also; structures in MPG and/or ISIS-Draw
860.1360	Multiresidue method	171- 4m	Report and data in word processing format; data tables in spreadsheet format also; structures in MPG and/or ISIS-Draw
860.1380	Storage stability data	171- 4e	Report and data in word processing format; data tables in spreadsheet format also; structures in MPG and/or ISIS-Draw
860.1400	Water, fish, and irrigated crops	171- f,g,h	Report and data in word processing format; data tables in spreadsheet

			format also; structures in MPG and/or ISIS-Draw
860.1460	Food handling	171- 4i	Report and data in word processing format; data tables in spreadsheet format also; structures in MPG and/or ISIS-Draw
860.1480	Meat/ milk/ poultry/ eggs	171- 4j	Report and data in word processing format; data tables in spreadsheet format also; structures in MPG and/or ISIS-Draw
860.1500	Crop field trials	171- 4k	Report and data in word processing format; data tables in spreadsheet format also; structures in MPG and/or ISIS-Draw
860.1520	Processed food/ feed	171- 4l	Report and data in word processing format; data tables in spreadsheet format also; structures in MPG and/or ISIS-Draw
860.1550	Proposed tolerances	171- 6	word processing format
860.1560	Reasonable grounds in support of the petition	171- 7	word processing format
860.1650	Submittal of analytical reference standards	171- 13	
860.1850	Confined accumulation in rotational crops	165- 1	Report and data in word processing format; data tables in spreadsheet format also; structures in MPG and/or ISIS-Draw
860.1900	Field accumulation in rotational crops	165- 2	Report and data in word processing format; data tables in spreadsheet format also; structures in MPG and/or ISIS-Draw

Word Processing format means Word Perfect 6.1 format for IBM PC (not Mac), plus the word processing format used by the registrant or petitioner, e.g., Microsoft Word. Structures should be submitted in Molecular Presentation Graphics (MPG) and/or ISIS-Draw format.

Spreadsheet format means Lotus Ver 5 for IBM PC (not Mac) format with any formulations for calculations included, plus the format used by the petitioner/ registrant, e.g., Excel.

If tables were created in a different format, such as Microsoft Access, include that file as well.

SERIES 870—HEALTH EFFECTS TEST GUIDELINES

Data should be provided in the FILES directory of the index volume of the CADDY CD-ROM set. The files should be in a subdirectory called HEALTHFX (e.g., /FILES/HEALTHFX). Electronic data that is desired includes report and data in word processing format; data tables in spreadsheet format also; structures in MPG and/or ISIS-Draw. Chemical structures included in the report should be submitted in Molecular Presentation Graphics (MPG) and/or ISIS-Draw format. Word Processing format means Word Perfect 6.1 format for IBM PC (not Mac), plus the word processing format used by the registrant or petitioner, e.g., Microsoft Word. Note: Files and report should be in the standardized review summary format specified by HED for TOX reviews. Spreadsheet format means Lotus Ver 5 for IBM PC (not Mac) format with any formulations for calculations included, plus the format used by the petitioner/registrant, e.g., Excel. If tables were created in a different format, such as Microsoft Access, include that file as well.

New GLN	Name	Old GLN
	Group A-- Acute Toxicity Test Guidelines.	
870.1000	Acute toxicity testing- background	none
870.1100	Acute oral toxicity	81- 1
870.1200	Acute dermal toxicity	81- 2
870.1300	Acute inhalation toxicity	81- 3
870.1350	Acute inhalation toxicity with histopathology	none
	Group B-- Specific Organ/ Tissue Toxicity Test Guidelines.	
870.2400	Acute eye irritation	81- 4
870.2500	Acute dermal irritation	81- 5
870.2600	Skin sensitization	81- 6
	Group C-- Subchronic Toxicity Test Guidelines.	
870.3100		82- 1
870.3150	Subchronic nonrodent oral toxicity-- 90- day	82- 1
870.3200	Repeated dose dermal toxicity-- 21/ 28 days	82-- 2
870.3250	Subchronic dermal toxicity-- 90 days	82- 3
870.3465	Subchronic inhalation toxicity	82- 4
870.3500	Preliminary developmental toxicity screen	none

870.3600	Inhalation developmental toxicity study	none
870.3700	Prenatal developmental toxicity study	83- 3
870.3800	Reproduction and fertility effects	83- 4
	Group D-- Chronic Toxicity Test Guidelines.	
870.4100	Chronic toxicity	83- 1
870.4200	Carcinogenicity	83- 2
870.4300	Combined chronic toxicity/ carcinogenicity	83- 5
	Group E-- Genetic Toxicity Test Guidelines.	
870.5100	Escherichia coli WP2 and WP2 uvrA reverse mutation assays	84- 2
870.5140	Gene mutation in Aspergillus nidulans	84- 2
870.5195	Mouse biochemical specific locus test	84- 2
870.5200	Mouse visible specific locus test	84- 2
870.5250	Gene mutation in Neurospora crassa	84- 2
870.5265	The Salmonella typhimurium reverse mutation assay	84- 2
870.5275	Sex- linked recessive lethal test in Drosophila melanogaster	84- 2
870.5300	Detection of gene mutations in somatic cells in culture	84- 2
870.5375	In vitro mammalian cytogenetics	84- 2
870.5380	In vivo mammalian cytogenetics tests: spermatogonial chromosomal aberrations	84- 2
870.5385	In vivo mammalian cytogenetics tests: Bone marrow chromosomal analysis	84- 2
870.5395	In vivo mammalian cytogenetics tests: Erythrocyte micronucleus assay	84- 2
870.5450	Rodent dominant lethal assay	84- 2
870.5460	Rodent heritable translocation assays	84- 2
870.5500	Bacterial DNA damage or repair tests	84- 2
870.5550	Unscheduled DNA synthesis in mammalian cells in culture	84- 2
870.5575	Mitotic gene conversion in Saccharomyces cerevisiae	84- 2
870.5900	In vitro sister chromatid exchange assay	84- 2
870.5915	In vivo sister chromatid exchange assay	84- 2
	Group F-- Neurotoxicity Test Guidelines.	
870.6100	Delayed neurotoxicity of organophosphorus substances following acute and 28- day exposure	81- 7, 82- 5, 82- 6

870.6200	Neurotoxicity screening battery	81- 8, 82- 7, 83- 1
870.6300	Developmental neurotoxicity study	83- 6
870.6500	Schedule- controlled operant behavior	85- 5
870.6850	Peripheral nerve function	85- 6
870.6855	Neurophysiology: Sensory evoked potentials	none

SERIES 875—OCCUPATIONAL AND RESIDENTIAL EXPOSURE TEST GUIDELINES

Data should be provided in the FILES directory of the index volume of the CADDY CD-ROM set. The files should be in a subdirectory called EXPOSURE (e.g., /FILES/EXPOSURE).

New GLN	Name	Old GLN	Electronic information desired
	Group A-- Applicator Exposure Monitoring Test Guidelines.		
875.1000	Background for application exposure monitoring test guidelines	230	
875.1100	Dermal exposure-- outdoor	231	Report and data in word processing format; data tables in spreadsheet format also; structures in MPG and/or ISIS-Draw
875.1200	Dermal exposure-- indoor	233	Report and data in word processing format; data tables in spreadsheet format also; structures in MPG and/or ISIS-Draw
875.1300	Inhalation exposure-- outdoor	232	Report and data in word processing format; data tables in spreadsheet format also; structures in MPG and/or ISIS-Draw
875.1400	Inhalation exposure-- indoor	234	Report and data in word processing format; data tables in spreadsheet format also; structures in MPG and/or ISIS-Draw
875.1500	Biological monitoring	235	Report and data in word processing format; data tables in spreadsheet format also; structures in MPG and/or ISIS-Draw

875.1600	Application exposure monitoring data reporting	236	Report and data in word processing format; data tables in spreadsheet format also; structures in MPG and/or ISIS-Draw
Group B-- Postapplication Exposure Monitoring Test Guidelines.			
875.2000	Background for postapplication exposure monitoring test guidelines	130, 131	
875.2100	Foliar dislodgeable residue dissipation	132- 1	Report and data in word processing format; data tables in spreadsheet format also; structures in MPG and/or ISIS-Draw
875.2200	Soil residue dissipation	132- 1	Report and data in word processing format; data tables in spreadsheet format also; structures in MPG and/or ISIS-Draw
875.2400	Dermal exposure	133- 3	Report and data in word processing format; data tables in spreadsheet format also; structures in MPG and/or ISIS-Draw
875.2500	Inhalation exposure	133- 4	Report and data in word processing format; data tables in spreadsheet format also; structures in MPG and/or ISIS-Draw
875.2600	Biological monitoring	235	Report and data in word processing format; data tables in spreadsheet format also; structures in MPG and/or ISIS-Draw
875.2800	Descriptions of human activity	133- 1	
875.2900	Data reporting and calculations	134	

Word Processing format means Word Perfect 6.1 format for IBM PC (not Mac), plus the word processing format used by the registrant or petitioner, e.g., Microsoft Word. Structures should be submitted in Molecular Presentation Graphics (MPG) and/or ISIS-Draw format.

Spreadsheet format means Lotus Ver 5 for IBM PC (not Mac) format with any formulations for calculations included, plus the format used by the petitioner/registrant, e.g., Excel.

If tables were created in a different format, such as Microsoft Access, include that file as well.

AQUATIC TOXICITY DATA

General Guidance

- Data files must be in ASCII
- Data should be provided in the FILES directory of the index volume of the CADDY CD-ROM set. The files should be in a subdirectory called AQUATOX. (e.g., /FILES/AQUATOX).
- Data files should have a file extension of ".DAT" (e.g., "FISHGRO1.DAT").
- Missing measurements should be indicated by a dot (".") as in the example below.
- Individual numbers or values on a line are separated by one or more spaces.
- Lines or rows of data are separated by a hard return.
- Character data such as variable names or treatment group names should consist of strings of length not more than 8 characters.
- Ratio endpoints should be represented as decimal numbers rather than as percentages.
- Concerning the treatment column, a separate file is needed to explain or identify what each letter represents.

Specific Guidance

72-3 Oyster Shell Deposition Study. Table should present growth per individual per replicate in a rectangular format containing three columns. The first column is the solvent control/control/treatment level, denoted by a letter such as A, B, C, etc. The second column is the replicate number (e.g., 1 and 2). The third column is the 96-hour shell growth (mm) for each individual oyster.

The following is an example dataset from a hypothetical 96-hour oyster shell growth study.

treatment	replicate	shell growth for individual oysters
A	1	2.04
A	1	2.10
A	1	1.96
A	1	2.06
A	1	etc.
A	2	2.00
A	2	1.89
A	2	1.20
A	2	etc.
B	1	2.10
B	1	1.78
B	1	2.40
B	1	etc.
B	2	2.01
B	2	2.30
B	2	1.89
B	2	etc.

C	1	2.03
C	1	2.03
C	1	1.98
etc.		

72-4 Daphnia Life Cycle Study. For each individual producing daphnid, indicate length and dry weight.

Treatment	Replicate	Length	Weight
A	1	#	#
A	1	#	#
A	1	#	#
A	1	#	#
A	1	etc.	etc.
A	2	#	#
A	2	#	#
A	2	#	#
A	2	#	#
A	2	etc.	etc.
B	1	#	#
B	1	#	#
B	1	#	#
B	1	#	#
B	1	etc.	etc.
B	2	#	#
B	2	#	#
B	2	#	#
B	2	#	#
B	2	etc.	etc.
etc.			

represents a measured value (length or weight) for an individual

72-4 Fish Early Life Stage Study, and 72-5 Fish Full Life Cycle Study. The following provides guidance for submitting data files from the 72-4 Fish Early Life Stage Study, and 72-5 fish full life cycle study in electronic format. The Agency could save time and effort in reviewing data if the individual length and weight of F1 generation fish were provided in electronically readable format.

F1 generation fish lenghts and weights should be provided in a rectangle database. The first column indicates the treatment, usually delineated with an alphabetical designation, i.e., A, B, C, etc. The content of the remaining columns depends on the design of the study, and how often length measurements are made. If there are only "replicates" per treatment, the second column is a number to indicate replicate number, usually one and two unless there are more than two replicates per treatment level. However, if a nested design is used, with, say two aquaria per treatment, and two replicates per aquaria, the second column would designate the aquaria, and the third column would be the replicate. The next column(s) will contain the length measurement for individual fish. If length was only measured at the termination of the study, only one column of length values will be provided. However, if length was measured for individual fish more than once during the study, each such set of measurements would be reported in a column; the first column of which representing the earliest measurement. The final column is the weight of individual fish measured at the end of the study.

A dataset from a hypothetical study with two replicates per treatment, and in which the length was measured only at test termination along with weight would look something like this:

Treatment	Replicate	length	weight
A	1	27	350
A	1	22	329
A	1	29	344
A	1	31	352
A	2	26	346
A	2	32	354
etc.			

A dataset from a hypothetical study with two replicates, and in which the length was measured at day 30 and again at test termination along with weight would look something like this:

Treatment	Replicate	length at 30 days	length at test termination	weight
A	1	22	29	334
A	1	23	28	345
A	1	20	30	350
A	1	19	27	323
A	2	19	25	341
A	2	21	31	339
etc.				

If the study design included aquaria and replicates within treatments, the dataset would look something like this:

Treatment	Aquaria	Replicate	Length at test termination	Weight
A	I	1	#	#
A	I	1	#	#
A	I	1	#	#
etc.				
A	I	2	#	#
A	I	2	#	#
A	I	2	#	#
etc.				
A	II	1	#	#
A	II	1	#	#
A	II	1	#	#
etc.				
A	II	2	#	#
A	II	2	#	#
A	II	2	#	#
etc.				

72-4A & 72-4B TERRESTRIAL ANIMAL TOXICITY STUDIES

Avian reproduction data should be submitted in ASCII files. Data should be provided in the FILES directory of the index volume of the CADDY CD-ROM set. The files should be in a subdirectory called ANIMAL (e.g., /FILES/ANIMAL). The data should be in a rectangular array with one record (i.e., row) containing data for each pen of birds. The records should contain values for 16 variables, each separated by one or more blank space. Missing values must be represented by a period (.). The order of variables, indicating the content of columns in the rectangular array, should be as follows:

    1.3 Treatment Group

    1.3 Eggs Laid

    1.3 Eggs Cracked

    1.3 Eggs Set

    1.3 Viable Embryos

    1.3 Live 3-Week Embryos

    1.3 Normal Hatchlings

    1.3 14-Day-Old Survivors

    1.3 Egg Shell Thickness (mm)

    1.3 Hatchling Weight (g)

1.3 14-Day-Old Survivor Weight (g)

1.3 Food Consumption (g/bird/day)

1.3 Initial Male Weight (g)

1.3 Final Male Weight (g)

1.3 Initial Female Weight (g)

1.3 Final Female Weight (g)

The first variable is a letter or number representing the treatment group. The control should be designated as "A" or "1." Letters or numbers for other treatment groups should be in ascending order according to the dose level. Variables 2-8 should be expressed as totals per pen, and variables 9-16 should be expressed as pen means. Note that the preferred way to express mean food consumption is as grams consumed per bird per day.

Data files should contain only data without any title or column headings. This will allow data to be read directly by the computer program without editing. If desired, the registrant may provide a separate file that includes a title identifying the study and column headings identifying the variables.

SURFACE WATER, GROUNDWATER STUDIES, AND MONITORING DATA

166.1 Small-Scale Prospective Ground Water Monitoring Studies

166.2 Small-Scale Retrospective Ground Water Monitoring Studies

166.3 Large-Scale Retrospective Ground Water Monitoring Studies

General Guidance

- Data files must be ASCII.
- Data should be provided in the FILES directory of the index volume of the CADDY CD-ROM set. The files should be in a subdirectory called WATER (e.g., /FILES/WATER).
- Data files should have a file extension of ".DAT" (e.g., "SITE002.DAT").
- Missing measurements should be indicated by a dot (".") as in the example below.
- Individual numbers or values on a line are separated by one or more spaces.
- Lines or rows of data are separated by a hard return.
- Character data such as variable names or treatment group names should consist of strings of length not more than 8 characters.

- Ratio endpoints such as emergence (the fraction of seeds that emerge) should be represented as decimal numbers rather than as percentages.

Specific Guidance

For each of these study reports, provide general data in ASCII format to include:

- Graphs
- Raw data tables with associated descriptive text
- Input variables and results of statistical analysis

Provide all tables of raw data in ASCII code. Tables should include:

- Information on site location (FIPS code; latitude and longitude)
- Time of sample
- Detection limits
- Analytical methods
- Time of collection (date; hour)
- Sampling method(s)
- pH/redox potential (measured in-situ)
- Water temperature
- Suspended sediments concentrations

It would be helpful if all Surface and Ground Water Modeling submissions would be accompanied with the following in ASCII:

- All model input parameters with source (reference(s)) identified
- Met files
- Model output

TERRESTRIAL AND AQUATIC PLANT STUDIES

General Guidance

- Data files must be ASCII.
- Data should be provided in the FILES directory of the index volume of the CADDY CD-ROM set. The files should be in a subdirectory called PHYTO-TOX (e.g., /FILES/PHYTOTOX).
- Data files must have a file extension of ".DAT" (e.g., "ONION.DAT").

- Missing measurements should be indicated by a dot (".") as in the example below.
- Individual numbers or values on a line are separated by one or more spaces.
- Lines or rows of data are separated by a hard return.
- Character data such as variable names or treatment group names should consist of strings of length not more than 8 characters.
- Ratio endpoints such as emergence (the fraction of seeds that emerge) should be represented as decimal numbers rather than as percentages.

Specific Guidance

Following is how a file might look for a study with two treatment levels and four measurement endpoints:

- Line 1: Title line

BLECKOFOS	ONION	EMERGENCE		
10	23			
%EMERG	%VISTOX	HEIGHT	WEIGHT	
C	2.5	3.0	1.00	0.00
C	1.9	2.8	1.00	0.00
1	1.8	2.2	0.79	0.00
1	1.9	2.5	0.79	0.00
2	.	.	0.00	.
2	1.0	1.9	0.22	1.00

- Line 2: Dose values applied to treated groups
- Line 3: Names of measurement endpoints (<= 8 characters per name)
- Lines 4 ff: data for individual plots

Data columns as follows:

- Column 1: treatment group
- Columns 2-5: values of measurement endpoints. The endpoints are identified on Line 3. (The 3rd name on Line 3 is the name of the 3rd endpoint from left to right, and so on.)

The Supplemental Files should be placed in the \FILES directory of the index volume of the CADDY CD-ROM set. Although the CADDY Format Specification states that the \FILES directory may be used on any volume of a CD-ROM set, placing the directory consistently on the index volume allows for easier location of the files by EPA scientists, especially when numerous volumes are included in a CD-ROM set. For additional ease in locating the files, please use the following directory structure to group files within the \FILES directory:

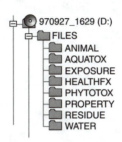

Each subdirectory beneath \FILES corresponds to a particular discipline or grouping of data as laid out in the separate guidance pages linked to above. Other files which may be asked for or included which are not within the scope of these Supplemental Files guidelines may be included in the base \FILES directory. In this circumstance, include an ASCII file called README.TXT to explain what these additional files are.

U.S. DEPARTMENT OF TRANSPORTATION HAZARDOUS MATERIALS REGISTRATION STATEMENT

Exempted from Paperwork Reduction Act by 49 App. U.S.C. 1805(c)(13)

U. S. DEPARTMENT OF TRANSPORTATION
HAZARDOUS MATERIALS REGISTRATION STATEMENT
REGISTRATION YEAR 19____ -19____
(Please Type or Print all Responses)

Initial Registration___   Renewal of Registration___   Amendment to Registration___   Expedited Follow-up___

Current Registration #_____

1. **Registrant** _____

   (Company Name)

   (Place pre-printed label here if provided and if name and address are correct. Otherwise, provide correct information.)

2. **Mailing Address of Principal Place of Business**

   Street or P.O. Box_____ City_____

   County _____ State_____ Zip Code _____ - _____ Country _____

3. **Carrier's US DOT ID Number, ICC Number, or Reporting Railroad Alphabetic Code (if applicable)**

   US DOT ID # _____   ICC # _____   Railroad Alphabetic Code _____

4. **Mode(s) Used to Transport Hazardous Materials:**   Highway____   Rail____   Water____   Air____

5. **Industrial Classification:** Check the primary industry in which the registrant operates. **Mark Only One.**

   Transportation
   - ☐ Carrier
   - ☐ Warehousing
   - ☐ Freight Forwarding, Agent Services
   - ☐ Transportation Repair & Service Facilities

   Manufacturing
   - ☐ Petroleum Refining & Related Industries
   - ☐ Apparel & Other Textile Products
   - ☐ Paper, Wood, & Allied Products
   - ☐ Printing & Publishing
   - ☐ Chemicals & Allied Products
   - ☐ Food & Kindred Products
   - ☐ Rubber & Miscellaneous Plastic Products
   - ☐ Electric & Electronic Equipment

   - ☐ Explosives
   - ☐ Other Manufacturing Industries

   Other
   - ☐ Agriculture & Support Services
   - ☐ Mining - Metal & Non-Metal
   - ☐ Oil & Gas Extraction
   - ☐ Gasoline, Fuel Oil, Propane Sales & Delivery
   - ☐ Construction - All Types
   - ☐ Wholesale or Retail Trade
   - ☐ Non-Transportation Repair Facilities
   - ☐ Hazardous Waste Services - Transportation, Disposal, Treatment
   - ☐ Other_____

6. **Annual Registration Fee.** The combined registration and processing fee is $300.00. (Complete only when submitting initial or renewal registration.)

   Total Amount Enclosed:_____

   Make check or money order in U.S. funds, drawn on a U.S. bank, and payable to "U.S. Department of Transportation," and identified as payment for the "Hazmat Registration Fee."

   ### Method of Payment

   Check ___   Money Order ___   Credit Card: VISA ___   MasterCard ___

   | Card Number: _____ | Expiration Date: _____ |
   | | MO   YR |

   Name as it appears on the card _____

   Authorized Signature _____

   Cardholder acknowledges ordering goods or services in the amount of the total shown hereon and agrees to perform the obligations set forth in the Cardholder's agreement with the issuer. Credit card statement will list this payment as "US DOT Hazmat Regis."

   NOTE: If completing an Expedited Registration, **do not** resubmit credit card information here.

Form DOT F 5800.2   (Revised 04/95)   THIS FORM MAY BE REPRODUCED
STF HAZ811F.1

7. PRIOR-YEAR SURVEY INFORMATION: Hazardous Materials Activities, and States in Which Activity was Conducted. Indicate those activities conducted by the registrant during the previous calendar year (e.g., 1994 for the 1995-96 Registration Year). Mark "A" through "E," as appropriate, to indicate the category or categories and the activity or activities (shipper, carrier, or other) in which the registrant acted. Check all categories and activities that apply. "Other" may be checked to indicate offeror activities not covered under the heading of shipper or carrier, such as freight forwarder or agent. Carriers should circle all states in which they operated as a hazardous materials carrier. Shippers and others engaged in offering hazardous materials should circle only those states from which they offered hazardous materials. They do not need to indicate to or through which states shipments were sent. A list of the states and their abbreviations appears in the accompanying materials. Circle "48 Contiguous States," if appropriate, to indicate that the activity was conducted in all of the 48 contiguous states. If the registrant did not engage in activities covered by "A" through "E" during the previous year, but plans to do so in the current registration year, mark only "F."

A. ____   Offered or transported in commerce any highway route-controlled quantity of a Class 7 (radioactive) material.

     1. Shipper____   2. Carrier ____   3. Other (Freight Forwarder, Agent, etc.)____

     AL AR AZ CA CO CT DE FL GA ID IL IN IA KS KY LA MA MD ME MI MN
     MO MS MT NC ND NE NH NJ NM NV OH OK OR PA RI SC SD TN TX UT
     VT VA WA WV WI WY   48 Contiguous States   AK AS DC GU HI MP PR VI

B. ____   Offered or transported in commerce more than 25 kilograms (55 pounds) of a Division 1.1, 1.2, or 1.3 (explosive) material in a motor vehicle, rail car, or freight container.

     1. Shipper____   2. Carrier ____   3. Other (Freight Forwarder, Agent, etc.)____

     AL AR AZ CA CO CT DE FL GA ID IL IN IA KS KY LA MA MD ME MI MN
     MO MS MT NC ND NE NH NJ NM NV NY OH OK OR PA RI SC SD TN TX UT
     VT VA WA WV WI WY   48 Contiguous States   AK AS DC GU HI MP PR VI

C. ____   Offered or transported in commerce more than 1 liter (1.06 quarts) per package of a material extremely toxic by inhalation (Division 2.3, Hazard Zone A, or Division 6.1, Packing Group 1, Hazard Zone A).

     1. Shipper____   2. Carrier ____   3. Other (Freight Forwarder, Agent, etc.)____

     AL AR AZ CA CO CT DE FL GA ID IL IN IA KS KY LA MA MD ME MI MN
     MO MS MT NC ND NE NH NJ NM NV NY OH OK OR PA RI SC SD TN TX UT
     VT VA WA WV WI WY   48 Contiguous States   AK AS DC GU HI MP PR VI

D. ____   Offered or transported in commerce a hazardous material or hazardous waste in a bulk packaging (see 49 CFR 171.8) having a capacity equal to or greater than 13,248 liters (3,500 gallons) for liquids or gases or more than 13.24 cubic meters (468 cubic feet) for solids.

     1. Shipper____   2. Carrier ____   3. Other (Freight Forwarder, Agent, etc.)____

     AL AR AZ CA CO CT DE FL GA ID IL IN IA KS KY LA MA MD ME MI MN
     MO MS MT NC ND NE NH NJ NM NV OH OK OR PA RI SC SD TN TX UT
     VT VA WA WV WI WY   48 Contiguous States   AK AS DC GU HI MP PR VI

E. ____   Offered or transported in commerce a shipment, in other than a bulk packaging, of 2,268 kilograms (5,000 pounds) gross weight or more of one class of hazardous materials or hazardous waste for which placarding of a vehicle, rail car, or freight container is required.

     1. Shipper____   2. Carrier ____   3. Other (Freight Forwarder, Agent, etc.)____

     AL AR AZ CA CO CT DE FL GA ID IL IN IA KS KY LA MA MD ME MI MN
     MO MS MT NC ND NE NH NJ NM NV NY OH OK OR PA RI SC SD TN TX UT
     VT VA WA WV WI WY   48 Contiguous States   AK AS DC GU HI MP PR VI

F. ____ Did not engage in any of the activities listed in A through E during the previous calendar year.

8. Certification of Information. I certify that, to the best of my knowledge, the above information is true, accurate, and complete.

Certifier's Name _____ Date_____
              (Print)

Title _____ Phone _____

Certifier's Signature _____

FALSE STATEMENTS MAY VIOLATE 18 U.S.C. 1001.

MAIL COMPLETED FORM TO:	U.S. Department of Transportation Hazardous Materials Registration P.O. Box 740188 Atlanta, GA 30374-0188

Please retain a copy of this form for your records.

STF HAZ811F.2

APPENDIX J
TABLE OF CONTENTS OF AN ENVIRONMENTAL IMPACT STATEMENT

TABLE OF CONTENTS

TABLE OF CONTENTS (CONTINUED)

TABLE OF CONTENTS (CONTINUED)

TABLE OF CONTENTS (CONTINUED)

TABLE OF CONTENTS (CONTINUED)

APPENDIX A	SEPTEMBER 28, 1996, AGREEMENT
APPENDIX B	DEPARTMENT OF THE INTERIOR APPROPRIATIONS ACT, PL 105-83 CALIFORNIA STATE LEGISLATURE ASSEMBLY BILL 1986
APPENDIX C	FEBRUARY 27, 1998, PRE-PERMIT APPLICATION AGREEMENT IN PRINCIPLE
APPENDIX D	SCOPING REPORT
APPENDIX E	3-YEAR (INTERIM) AND 47-YEAR (DEFAULT) AQUATIC STRATEGY AND MITIGATION FOR TIMBER HARVEST AND ROADS
APPENDIX F	HEADWATERS FOREST ACQUISITION FINANCIAL PLAN, MAY 5, 1998
APPENDIX G	WATERSHED ANALYSIS FRAMEWORK
APPENDIX H	MITIGATION MEASURES FOR CUMULATIVELY AFFECTED WATERSHEDS
APPENDIX I	EQUIVALENT BUFFER AREA INDEX (EBAI) METHODOLOGY FOR FINE SEDIMENT
APPENDIX J	WETLANDS AND RIPARIAN LANDS AND LARGE WOODY DEBRIS EBAI
APPENDIX K	AQUATIC PROPERLY FUNCTIONING CONDITIONS MATRIX
APPENDIX L	CALIFORNIA WILDLIFE HABITAT RELATIONSHIP DATABASE CROSSWALKS
APPENDIX M	WILDLIFE HABITAT ANALYSES AND DEFINITIONS
APPENDIX N	BACKGROUND INFORMATION ON MARBLED MURRELET FOR THE PALCO HABITAT CONSERVATION PLAN (N1 AND N2)
APPENDIX O	LIST OF INTERESTED INDIAN TRIBES
APPENDIX P	HABITAT CONSERVATION PLAN OPERATING CONSERVATION PROGRAM
APPENDIX Q	SUSTAINED YIELD PLAN
APPENDIX R	MITIGATION MONITORING AND REPORTING PROGRAM
APPENDIX S	IMPLEMENTATION AGREEMENT WITH REGARD TO THE PACIFIC LUMBER COMPANY HABITAT CONSERVATION PLAN
APPENDIX T	RESPONSES TO COMMENTS
APPENDIX U	SUBSTANTIVE COMMENTS ON CD-ROM

APPENDIX K
SPILL RESPONSE NOTIFICATION FORM

Spill Response Notification Form

1. Reporter's Last Name:	
2. Reporter's First Name & Middle Initial	
3. Position:	

4. Phone Numbers:
 	
    Day #
 	
    Night #	

5. Company	
6. Organization type	
7. Street Address	
8. City, State & Zip Code	
9. Were Materials Discharged?	(Y/N)
Confidential?	(Y/N)
10. Meeting Federal Obligations to Report?	(Y/N)
11. Date Called	(Y/N)
12. Calling for Responsible Party	(Y/N)
13.Time Called	(AM/PM)
14.Incident Description	
15. Date of Incident	
16. Time of Incident	(AM/PM)
17. Incident Address/Location:	
18. Neatest City & State	
19. Distance from City	
20. Direction from City	
21. Section, Township & Borough	

22. Container Type	
23. Tank Oil Storage Capacity	
24. Facility Oil Storage Capacity	
25. Facility Longitude (degrees & minutes)	
26. Facility Latitude (degrees & minutes)	

CHRIS Code	Discharged Quantity	Units of Measure	Material Discharged into Water	Quantity	Units of Measure

1. Response Action Taken to Control or Mitigate the Incident.	

2. Were there Evacuations? (Y/N)

3. Number Evacuated:

4. Was there any Damage? (Y/N)

5. Damage in Dollars (approximate):

6. Medium Affected:

7. Description:

8. More Information about Medium:

9. Any information about the incident not recorded elsewhere in the report:

APPENDIX L
SUBSTANTIAL HARM CRITERIA

L.1 INTRODUCTION

The flowchart provided in Fig. L.1 shows the decision tree with the criteria to identify whether a facility "could reasonably be expected to cause substantial harm to the environment by discharging into or on the navigable waters or adjoining shorelines." In addition, the Regional Administrator has the discretion to identify facilities that must prepare and submit facility-specific response plans to EPA.

L.2 DEFINITIONS

Great Lakes means Lakes Superior, Michigan, Huron, Erie, and Ontario, their connecting and tributary waters, the Saint Lawrence River as far as Saint Regis, and adjacent port areas.

Higher Volume Port Areas include

1. Boston, MA
2. New York, NY
3. Delaware Bay and River to Philadelphia, PA
4. St. Croix, VI
5. Pascagoula, MS
6. Mississippi River from Southwest Pass, LA to Baton Rouge, LA
7. Louisiana Offshore Oil Port (LOOP), LA
8. Lake Charles, LA
9. Sabine-Neches River, TX
10. Galveston Bay and Houston Ship Channel, TX
11. Corpus Christi, TX
12. Los Angeles/Long Beach Harbor, CA
13. San Francisco Bay, San Pablo Bay, Carquinez Strait, and Suisun Bay to Antioch, CA
14. Straits of Juan de Fuca from Port Angeles, WA to and including Puget Sound, WA
15. Prince William Sound, AK
16. Others as specified by the Regional Administrator for any EPA Region

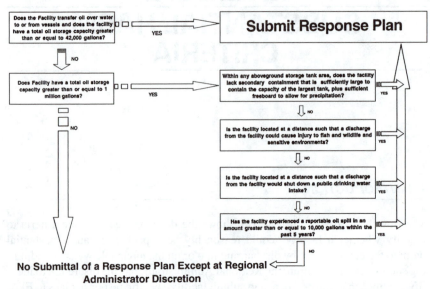

FIGURE L.1   Flowchart of criteria for substantial harm.

Inland Area means the area shoreward of the boundary lines defined in 46 CFR Part 7, except in the Gulf of Mexico. In the Gulf of Mexico, it means the area shoreward of the lines of demarcation (COLREG lines as defined in 33 CFR 80.740-80.850). The inland area does not include the Great Lakes.

Rivers and Canals means bodies of water confined within the inland area, including the Intracoastal Waterways and other waterways artificially created for navigating that have project depths of 12 feet or less.

L.3 DESCRIPTION OF SCREENING CRITERIA FOR THE SUBSTANTIAL HARM FLOWCHART

A facility that has the potential to cause substantial harm to the environment in the event of a discharge must prepare and submit a facility-specific response plan to EPA in accordance with Appendix F to this part. A description of the screening criteria for the substantial harm flowchart is provided below:

Non-Transportation-Related Facilities with a Total Oil Storage Capacity Greater Than or Equal to 42,000 Gallons Where Operations Include Over-Water Transfers of Oil

A non-transportation-related facility with a total oil storage capacity greater than 42,000 gallons that transfers oil over water to or from vessels must submit a response plan to EPA. Daily oil transfer operations at these types of facilities occur between barges and vessels and onshore bulk storage tanks over open water. These facilities are located adjacent to navigable water.

Lack of Adequate Secondary Containment at Facilities with a Total Oil Storage Capacity Greater Than or Equal to 1 Million Gallons

Any facility with a total oil storage capacity greater than or equal to 1 million gallons without secondary containment sufficiently large to contain the capacity of the largest aboveground oil storage tank within each area plus sufficient freeboard to allow for precipitation must submit a response plan to EPA. Secondary containment structures that meet the standard of good engineering practice for the purposes of this part include berms, dikes, retaining walls, curbing, culverts, gutters, or other drainage systems.

Proximity to Fish and Wildlife and Sensitive Environments at Facilities with a Total Oil Storage Capacity Greater Than or Equal to 1 Million Gallons

A facility with a total oil storage capacity greater than or equal to 1 million gallons must submit its response plan if it is located at a distance such that a discharge from the facility could cause injury (as defined in 40 CFR 112.2) to fish and wildlife and sensitive environments. For further description of fish and wildlife and sensitive environments, see Appendices I, II, and III to DOC/NOAA's "Guidance for Facility and Vessel Response Plans: Fish and Wildlife and Sensitive Environments" (see Appendix E to 40 CFR Part 112, section 10, for availability) and the applicable Area Contingency Plan. Facility owners or operators must determine the distance at which an oil spill could cause injury to fish and wildlife and sensitive environments using the appropriate formula presented in Attachment C-111 to 40 CFR Part 112 or a comparable formula.

Proximity to Public Drinking Water Intakes at Facilities with a Total Storage Oil Capacity Greater Than or Equal to 1 Million Gallons

A facility with a total storage capacity greater than or equal to 1 million gallons must submit its response plan if it is located at a distance such that a discharge ftorn the facility would shut down a public drinking water intake, which is analogous to a public water system as described in 40 CFR 143.2(c). The distance at which an oil spill from an SPCC-regulated facility would shut down a public drinking water intake shall be calculated using the appropriate formula presented in Attachment C-III to this 40 CFR Part 112 or a comparable formula.

Facilities That Have Experienced Reportable Oil Spills in an Amount Greater Than or Equal to 10,000 Gallons within the Past 5 Years and That Have a Total Oil Storage Capacity Greater Than or Equal to 1 Million Gallons

A facility's oil spill history within the past 5 years must be considered in the evaluation for substantial harm. Any facility with a total oil storage capacity greater than or equal to 1 million gallons that has experienced a reportable oil spill in an amount greater than or equal to 10,000 gallons within the past 5 years must submit a response plan to EPA.

L.4 CERTIFICATION FOR FACILITIES THAT DO NOT POSE SUBSTANTIAL HARM

If the facility does not meet the substantial harm criteria listed in Fig. L.1, the owner or operator needs to complete and maintain at the facility the certification form. In the event an alternative formula that is comparable to the one in this appendix is used to evaluate the substantial harm criteria, the owner or operator must attach documentation to the certification form that demonstrates the reliability and analytical soundness of the comparable formula and must notify the Regional Administrator in writing that an alternative formula was used.

REFERENCES

Chow, V.T. *Open Chanel Hydraulics*, McGraw Hill, 1959.
USCG IFR (58 FR 7353, February 5, 1993). This document is available through EPA's rulemaking docket.

APPENDIX M

RESPONSE PLAN COVER SHEET AND SUBSTANTIAL HARM CERTIFICATION

Response Plan Cover Sheet

I. General Information

Owner/Operator of Facility:

Facility Name:

Facility Address (street address or route):

City, State, and U.S. Zip Code:

Facility Phone No.:

Latitude (Degrees: North), degrees, minutes, seconds

Dun & Bradstreet Number:

Largest Aboveground Oil Storage Tank Capacity (Gallons):

Number of Aboveground Oil Storage Tanks:

Longitude (Degrees: West): degrees, minutes, seconds

Standard Industrial Classification (SIC) Code:
Maximum Oil Storage Capacity (Gallons):
Worst Case Oil Discharge Amount (Gallons):
Facility Distance to Navigable Water. Mark the appropriate line.
0-1/4 mile   1/4 -1/2 mile   1/2 -1 mile
>1 mile

Response Plan Cover Sheet

II. Applicability of Substantial Harm Criteria

1. Does the facility transfer oil over-water, and does the facility have a total oil storage capacity greater than or equal to 42,000 gallons?

Yes

No

2. Does the facility have a total oil storage capacity greater than or equal to 1 million gallons and, within any storage area, does the facility lack secondary containment that is sufficiently large to contain the capacity of the largest aboveground oil storage tank plus sufficient freeboard to allow for precipitation?

Yes

No

3. Does the facility have a total oil storage capacity greater than or equal to 1 million gallons and is the facility located at a distance (as calculated using the appropriate formula in Appendix C of 40 CFR Part 112 or a comparable formula) such that a discharge from the facility could cause injury to fish and wildlife and sensitive environments?

Yes

No

---

4. Does the facility have a total oil storage capacity greater than or equal to 1 million gallons and is the facility located at a distance (as calculated using the appropriate formula in Appendix C of 40 CFR Part 112 or a comparable formula) such that a discharge from the facility would shut down a public drinking water intake?

Yes

No

---

5. Does the facility have a total oil storage capacity greater than or equal to 1 million gallons and has the facility experienced a reportable oil spill in an amount greater than or equal to 10,000 gallons within the last 5 years?

Yes

No

Response Plan Cover Sheet

III. Certification

I certify under penalty of law that I have personally examined and am familiar with the information submitted in this document, and that based on my inquiry of those individuals responsible for obtaining information, I believe that the submitted information is true, accurate, and complete.

Signature

Name (Please type or print):

Title:

Date:

1. These numbers may be obtained from public library resources.

2. Explanations of the above-referenced terms can be found in Appendix C of 40 CFR Part 112. If a comparable formula to the ones contained in Attachment C-III to Appendix C of 40 CFR Part 112 is used to establish the appropriate distance to fish and wildlife and sensitive environments or public drinking water intakes, documentation of the reliability and analytical soundness of the formula must be attached to this form.

3. For further description of fish and wildlife and sensitive environments, see Appendices I, II, and III to DOC/NOAA's "Guidance for Facility and Vessel Response Plans: Fish and Wildlife and Sensitive Environments" (see Appendix E to this part, section 10, for availability) and the applicable ACP.

REFERENCES:

- CONCAWE, *Methodologies for Hazard Analysis and Risk Assessment in the Petroleum Refining and Storage Industry*, prepared by CONCAWE's Risk Assessment Ad-hoc Group, 1982.
- U.S. Department of Housing and Urban Development, *Siting of HUD-Assisted Projects Near Hazardous Facilities: Acceptable Separation Distances from Explosive and Flammable Hazards,* prepared by the Office of Environment and Energy, Environmental Planning Division, Department of Housing and Urban Development, Washington, D.C., 1987.
- U.S. DOT, FEMA and U.S. EPA, *Handbook of Chemical Hazard Analysis Procedures*, U.S. Government Printing Office, Washington, D.C.
- U.S. DOT, FEMA and U.S. EPA, *Technical Guidance for Hazards Analysis: Emergency Planning for Extremely Hazardous Substances*, U.S. Government Printing Office, Washington, D.C.
- The National Response Team, *Hazardous Materials Emergency Planning Guide*, Washington, D.C., 1987.
- The National Response Team, *Oil Spill Contingency Planning, National Status: A Report to the President*, U.S. Government Printing Office, Washington, D.C., 1990.
- Offshore Inspection and Enforcement Division, Minerals Management Service, Offshore Inspection Program: National Potential Incident of Noncompliance (PINC) List. Reston, Va, 1988.

APPENDIX N
STATE POLLUTION PREVENTION TECHNICAL ASSISTANCE CONTACTS BY REGION

Region 1

US EPA Region 1
Mark Mahoney
JFK Federal Bldg (SPP)
Boston, MA 02203
Ph: 617/ 565-1155
Fx: 617/565-4939

Connecticut DEP
Mary Sherwin
79 Elm St
Hartford, CT 06106
Ph: 860/424-3297
Fx: 860/566-4924

Maine DEP
Ann Pistell
State House Station 17
Augusta, ME 04333
Ph: 207/287-2811
Fx: 207/287-2814

MA DEP - OTA
Rick Reibstein
100 Cambridge St. Rm 2109
Boston, MA 02202
Ph: 617/792-3260
Fx: 617/727-3827

MA STEP Program
Paul Richards
100 Cambridge St. Rm 2000
Boston, MA 02202
Ph: 617/727-9800
Fx: 617/727-2754

Toxics Use Reduction Institute
Janet Clark
One University Avenue
Lowell, MA 01854
Ph: 508/934-3275
Fx: 508/934-3050

New Hampshire DES
Sara Johnson
6 Hazen Drive
Concord, NH 03301
Ph: 603/271-6460

Rhode Island DEM
Richard Enander
235 Promenade St.
Providence, RI 02908
Ph: 401/277-3434
Fx: 401/277-2591

Narragansett Bay Commission
James McCaughey
235 Promenade St.
Providence, RI 02908
Ph: 401/277-6680
Fx: 401/277-2584

Vermont ANR
Gary Gulka
103 South Main St
Waterbury, VT 05671
Ph: 802/241-3626

NEWMOA
Terri Goldberg
129 Portland St, Suite 602
Boston, MA 02114
Ph: 617/367-8558
Fax: 617-367-0449

Region 2

US EPA Region 2
Ed Linky
290 Broadway (2-OPM-PPI)
New York, NY 10007
Ph: 212/ 637-3742
Fx: 212/637-3771

New Jersey DEP
50 Trenton Ave
Frenchtown, NJ 08825
Ph: 609/777-0518
Fx: 609/292-1816

NJ TAP
Laura Battista
138 Warren St
Newark, NJ 07102
Ph: 973/596-5864
Fx: 973/596-6367

New York State DEC
P2 Unit
50 Wolf Rd
Albany, NY 12233
Ph: 518/457-2553
Fx: 518/457-2570

Puerto Rico Environment
Carlos Gonzales
Ph: 809/765-7517 x381
Fx: 809/765-6853

Region 3

US EPA Region 3
Jeff Burke
841 Chestnut Bldg
Philadelphia, PA 19107
Ph: 215/566-2761
Fx: 215/566-2782

Delaware DNR
Andrea Kreiner
PO Box 1401
89 Kings Highway
Dover, DE 19903
Ph: 302/739-3822
Fx: 302/739-6242

MD Dept of Environment
Laura Armstrong
2500 Broening Hwy
Baltimore, MD 21224
Ph: 410/631-4119
Fx: 410/631-4477

PA Dept of Environment
Meredith Hill
PO Box 2063
Harrisburg, PA 17105
Ph: 717/783-0540
Fx: 717/787-8926

PA Technical Assistance
Jack Gido
110 Barbara Bldg II
University Park, PA 16802
Ph: 814/865-0427
Fx: 814/865-5909

West Virginia DEP-OWR
Leroy Gilbert
Rt 3 Box 384
Danese, WV 25831
Ph: 304/484-6269
Fx: 304/558-2780

Virginia DEQ
Sharon K. Baxter
PO Box 10009
Richmond, VA 23240
Ph: 804/698-4344
Fx: 804/698-4277

Region 4

US EPA Region 4
Bernie Hayes
61 Forsyth St SW
Atlanta, GA 30303
Ph: 404/ 562-9430
Fx: 404/562-9066

Alabama DEM - P2 Unit
Gary Ellis
1751 Cong. Dickerson Dr
Montegomery, AL 36130
Ph: 334/213-4303
Fx: 334/213-4399

Florida DEP - P2 Program
Julie Abcarian
2600 Blair Stone Road
Tallahassee, FL 32399
Ph: 850/488-0300
Fx: 850/921-8061

Georgia DNR- P2AD
Jancie Hatcher
7 MLK Jr. Dr. Suite 450
Atlanta, GA 30334
Ph: 404/651-5120
Fx: 404/651-5130

Kentucky DEP
Vicki Pettus
14 Reilly Road
Frankfort, KY 40601
Ph: 502/564-6716

Kentucky P2 Center
Donald Douglass
Acad. Bldg, Room 420
Louisville, KY 40292
Ph: 502/852-0965
Fx: 852-0964

Mississippi DEQ
Thomas E. Whitten
PO Box 10385
Jackson, MS 39209
Ph: 601/961-5171

North Carolina DEHNR
Gary Hunt
PO Box 29569
Raleigh, NC 27626
Ph: 919/715-6500
Fx: 919/715-6794

Tennessee DEC
Angie Pitcock
401 Church St
Nashville, TN 37243
Ph: 615/532-0760

South Carolina DHEC
Robert Burgess
2600 Bull St
Columbia, SC 29208
Ph: 803/734-4761
Fx: 803/734-9934

Region 5

US EPA Region 5
Phil Kaplan
77 West Jackson Blvd
Chicago, IL 60604
Ph: 312/353-4669
Fx: 312/353-4788

Indiana DEM
Cheri Storms
150 West Market St. Suite 703
Indianapolis, IN 46206
Ph: 317/233-1041
Fx: 317/233-5627

Minnesota OEA
Stacy Stinson
520 Lafayette Road North
St. Paul, MN 55155
Ph: 612/215-0296
Fx: 612/297-8709

MN Technology Inc.
Kevin O'Donnell
111 3rd Ave. South
Minneapolis, MN 55401
Ph: 612/672-3446
Fx: 612/497-8475

National Farmstead Program
Liz Nevers
B142 Steenbock Library
Madison WI 53706
Ph: 608/265-2774
Fx: 608/265-2775

Illinois EPA
Kevin Greene
1021 N Grant Ave. East
Springfiled, IL 62794
Ph: 217/782-8700
Fx: 217/782-9142

IN Clean Manufacturing Tech & Safe Materials Institute
Alice Smith
2655 Yeager Rd. Suite103
West Lafayette, IN 47906
Ph: 765/463-4749
Fx: 765/463-3795

Minnesota (MN TAP)
Cindy McComas
1313 5th St SW Suite 207
Minneapolis, MN 55414
Ph: 612/627-4556
Fx: 612/627-4769

Ohio EPA
Michael Kelley
PO Box 1049
Columbus, OH 43216
Ph: 614/644-3469
Fx: 614/728-2807

Illinois Waste Mgmt
Gary Miller
One East Hazelwood Dr.
Champaign, IL 61820
Ph: 217/333-8942
Fx: 217/333-8944

Michigan DEQ
Marcia Horan
PO Box 30457
Lansing, MI 48909
Ph: 517/373-9122
Fx: 517/335-4729

MN Pollution Control Agency
Al Innes
520 Lafayette Road North
St. Paul, MN 55155
Ph: 612/296-7330
Fx: 612/297-8676

University of Wisconsin
Tom Blewett
620 Langdon St, Rm 533
Madison, WI 53703
Ph: 608/262-0936
Fx: 608/262-6250

Wisconsin DNR
Lynn Persson
PO Box 7921
Madison, WI 53707
Ph: 608/267-3763
Fx: 608/267-0496

Region 6

US EPA Region 6
Eli Martinez
1455 Ross Ave Suite 1200
Dallas, TX 75202
Ph: 214/665-2119
Fx: 214/665-7446

Arkansas IDC
Alford Drinkwater
One Capitol Mall
Little Rock, AR 72201
Ph: 501/682-7325
Fx: 501/682-2703

Louisiana DEQ
Gary Johnson
PO Box 82263
Baton Rouge, LA 70884
Ph: 504/765-0739
Fx: 504/765-0742

Louisiana TAP
University of New Orleans
New Orleans, LA
Ph: 504/286-6305
Fax: 504/286-5586

Texas NRCC
Ken Zarker
PO Box 13087 - MC112
Austin, TX 78711
Ph: 512/239-3144
Fx: 512/239-3165

New Mexico ED
Patricia Gallagher
PO Box 26110
Sante Fe, NM 87502
Ph: 505/827-0677
Fx: 505/827-2836

**TX Manufacturing
Assistance Center**
Conrad Soltero
Univ. of TX-El Paso
500 W University, Burges
El Paso, TX 75202
Ph: 915/747-5930
Fx: 915/747-5437

**Gulf Coast Hazardous
Substance Research**
Margaret Aycock
PO Box 10671
Beaumont, TX 77710
Ph: 409/880-8897
Fx: 409/880-1837

Oklahoma DEQ
Dianne Wilkins
1000 NE Tenth St
Oklahoma City, OK 73117
Ph: 405/271-1400
Fx: 405/271-1317

**Lower Colorado River
Authority**
Charles Urdy
PO Box 220
Austin, TX 78767
Ph: 512/473-3200
Fx: 512/473-4066

Region 7

US EPA Region 7
Marc Matthews
726 Minnesota Ave
(ARTD/TSPP)
Kansas City, KS 66101
Ph: 913/551-7517
Fx: 913/551-7065

Iowa DED
Linda King
Small Business Liaison
Ph: 515/242-4761
Fx: 515-242-6338

Missouri DNR - TAP
Becky Shannon
PO Box 176
Jefferson City, MO 65102
Ph: 573/526-6627
Fx: 573/526-5808

Iowa DNR
Brian Tormey
502 E. 9th St
Des Moines, IA 50319
Ph: 515/281-8927
Fx: 515/281-8895

Kansas DHE
Janet Neff
Bldg. 283, Forbes Field
Topeka, KS 66620
Ph: 785/296-0669
Fx: 785/296-3266

**Nebraska Business
Development Center**
Rick Yoder
1135 M St, Suite 200
Lincoln, NE 68508
Ph: 402/472-1183
Fx: 402/472-3363

**Iowa Waste Reduction
Center**
John L. Konefes
1005 Technology Parkway
Cedar Fall, IA 50614
Ph: 319/273-2079
Fax: 319/273-2926

KSU - P2 Institute
Sherry Davis
133 Ward Hall
Manhatten, KS 66506
Ph: 785/532-6501
Fx: 785/532-6952

Nebraska DEQ, P2 Office
Ben Hammerschmidt
PO Box 98922
Lincoln, NE 68509
Ph: 402/471-6988
Fax: 402/471-2909

Region 8

US EPA Region 8
Linda Walters
999 18th St, Suite 500
Denver, CO 80202
Ph: 303/312-6385
Fx: 303/312-6741
**North Dakota Dept of
Health**
Jeffrey L. Burgess
Ph: 701/328-5150
Fx: 701/328-5200

Colorado DHE
Parry Burnap
4300 Cherry Creek Dr
Denver, CO 80222
Ph: 303/692-2975
Fx: 303/782-4969
South Dakota DENR
Dennis Clarke
Joe Foss Bldg.,
523 East Capitol
Pierre, SD 57501-3181
Ph: 605/773-4254
Fx: 605/773-4068
Wyoming DEQ
Patricia Jordan
122 West 25th
Cheyenne, WY 82992
Ph: 307/777-6105
Fx: 307/777-5973

Montana P2 Program
Michael P. Vogel
109 Taylor Hall
Bozeman, MT 59717
Ph: 406/994-3451
Fx: 406/994-5417
Utah DEQ
Sonja Wallace
PO Box 144810
Salt Lake City, UT 84114
Ph: 801/536-4477
Fx: 801/536-4401

Region 9

US EPA Region 9
Bill Wilson
75 Hawthorn St (WST11)
San Francisco, CA 94105
Ph: 415/744-2192
Fx: 415/744-1796

**California Energy
Commission**
David Jones
1519 9th St
Sacramento, CA 95814
Ph: 916/654-4554

Hawaii Dept of Health
Marlyn Aguilar
919 Ala Moana Blvd,
Rm # 212
Honolulu, HI 96814
Ph: 808/586-4373
Fx: 808/586-7509

Arizonia DEQ
Jacquelione Maye
3033 North Central Ave
Phoenix, AZ 85012
Ph: 602/207-4607
Fax: 602/207-2302
**CA Toxic Substance
Control**
Kathy Barwick
PO Box 806
Sacramento, CA 95812
Ph: 916/322-1815
Fx: 916/327-4494

California EPA
Terri Cronin
8800 Cal Center Dr
Sacramento, CA 95826

UCLA P2 Center
Billy Romain
Ph: 310/825-2654
Fx: 310/206-3906

**Nevada Small Business
Development Center**
Kevin Dick
6100 Neil Rd. Suite 200
Reno, NV 89511
Ph: 702/689-6677
Fx: 702/689-6689

Region 10

US EPA Region 10
Carolyn Gangmark
1200 Sixth Ave (01-085)
Seattle, WA 98101
Ph: 206/553-4072
Fx: 206/5538338

Alaska DEC
Marianne See
555 Cordova St
Anchorage, AK 99501
Ph: 907/269-7586
Fx: 907/269-7600

Idaho DEQ
John Bernardo
1410 N Hilton
Boise, ID 83706
Ph: 208/373-0502
Fx: 208/373-0169

Oregon DEQ
Marianne Fitzgerald
811 SW Sixth St
Portland, OR 97204
Ph: 503/229-6457
Fax: 503/229-5850

Washington DEC
Thomas Eaton
PO Box 47600
Olympia, WA 98504
Ph: 360/407-6086
Fx: 360/407-6715

Washington State University
Carol Reisenberg
501 Johnson Tower
Pullman, WA 99164
Ph: 509/335-1576
Fx: 509/335-0949

Pacific Northwest P2 Resource Center
Madeline Sten
1326 Fifth Ave., Suite 650
Seattle, WA 98101
Ph: 206/223-1151
Fx: 206/223-1165

APPENDIX O
CLASSIFICATION OF PERMIT MODIFICATIONS

Modifications	Class
A. General Permit Provisions	
1. Administrative and informational changes..............................	1
2. Correction of typographical errors....	1
3. Equipment replacement or upgrading with functionally equivalent components (e.g., pipes, valves, pumps, conveyors, controls).............................	1
4. Changes in the frequency of or procedures for monitoring, reporting, sampling, or maintenance activities by the permittee	
a. To provide for more frequent monitoring, reporting, sampling, or maintenance...........................	1
b. Other changes......................	2
5. Schedule of compliance:	
a. Changes in interim compliance dates, with prior approval of the Director.............................	11
b. Extension of final compliance date.................................	3
6. Changes in expiration date of permit to allow earlier permit termination, with prior approval of the Director....	11
7. Changes in ownership or operational control of a facility, provided the procedures of §270.40(b) are followed..............................	11
B. General Facility Standards	
b. [Removed]	

c. To incorporate changes associated with underlying hazardous constituents in ignitable or corrosive wastes......	11
d. Other changes......................	2
2. Changes to analytical quality assurance/control plan:	
a. To conform with agency guidance or regulations..........................	1
b. Other changes......................	2
3. Changes in procedures for maintaining the operating record..................	
4. Changes in frequency or content of inspection schedules..................	2
5. Changes in the training plan:	
a. That affect the type or decrease the amount of training given to employees...........................	2
b. Other changes......................	1
6. Contingency plan:	
a. Changes in emergency procedures (i.e., spill or release response procedures).........................	2
b. Replacement with functionally equivalent equipment, upgrade, or relocate emergency equipment listed..............................	1
c. Removal of equipment from emergency equipment list.......................	2
d. Changes in name, address, or phone number of coordinators or other persons or agencies identified in the plan.................................	1

7. Construction quality assurance plan:	
a. Changes that the CQA officer certifies in the operating record will provide equivalent or better certainty that the unit components meet the design specifications..................	
b. Other Changes........................	
Note: When a permit modification (such as introduction of a new unit) requires a change in facility plans or other general facility standards, that change shall be reviewed under the same procedures as the permit modification.	
C. Ground-Water Protection 1. Changes to wells: a. Changes in the number, location, depth, or design of upgradient or downgradient wells of permitted ground-water monitoring system........	2
b. Replacement of an existing well that has been damaged or rendered inoperable, without change to location, design, or depth of the well..........	1
2. Changes in ground-water sampling or analysis procedures or monitoring schedule, with prior approval of the Director...............................	11
3. Changes in statistical procedure for determining whether a statistically significant change in ground-water quality between upgradient and downgradient wells has occurred, with prior approval of the Director.........	11
4. Changes in point of compliance........	12
5. Changes in indicator parameters, hazardous constituents, or concentration limits (including ACLs):	
a. As specified in the groundwater protection standard...................	3

b. As specified in the detection monitoring program...................	2
6. Changes to a detection monitoring program as required by §264.98(j), unless otherwise specified in this appendix..............................	2
7. Compliance monitoring program:	
a. Addition of compliance monitoring program as required by §§264.98(h)(4) and 264.99.........................	3
b. Changes to a compliance monitoring program as required by §264.99(k), unless otherwise specified in this appendix.............................	2
8. Corrective action program: a. Addition of a corrective action program as required by §§264.99(i)(2) and 264.100.........................	3
b. Changes to a corrective action program as required by §264.100(h), unless otherwise specified in this Appendix............................	2
D. Closure	
1. Changes to the closure plan:	
a. Changes in estimate of maximum extent of operations or maximum inventory of waste on-site at any time during the active life of the facility, with prior approval of the Director............................	11
b. Changes in the closure schedule for	

any unit, changes in the final closure schedule for the facility, or extension of the closure period, with prior approval of the Director........	11
c. Changes in the expected year of final closure, where other permit conditions are not changed, with prior approval of the Director.............	11
d. Changes in procedures for decontamination of facility equipment or structures, with prior approval of the Director.........................	11
e. Changes in approved closure plan resulting from unexpected events occurring during partial or final closure, unless otherwise specified in this appendix........................	2
f. Extension of the closure period to allow a landfill, surface impoundment or land treatment unit to receive nonhazardous wastes after final receipt of hazardous wastes under §264.113(d) and (e)...................	2
2. Creation of a new landfill unit as part of closure........................	3
3. Addition of the following new units to be used temporarily for closure activities: a. Surface impoundments................	3
b. Incinerators........................	3
c. Waste piles that do not comply with §264.250(c)...........................	3
d. Waste piles that comply with §264.250(c)...........................	2
e. Tanks or containers (other than specified below)......................	2
f. Tanks used for neutralization, dewatering, phase separation, or component separation, with prior approval of the Director.............	11

g. Staging piles........................	2
E. Post-Closure	
1. Changes in name, address, or phone number of contact in post-closure plan...................................	1
2. Extension of post-closure care period................................	2
3. Reduction in the post-closure care period................................	3
4. Changes to the expected year of final closure, where other permit conditions are not changed........................	1
5. Changes in post-closure plan necessitated by events occurring during the active life of the facility, including partial and final closure....	2
F. Containers	
1. Modification or addition of container units:	
a. Resulting in greater than 25% increase in the facility's container storage capacity, except as provided in F(1)(c) and F(4)(a) below.	3
b. Resulting in up to 25% increase in the facility's container storage capacity, except as provided in F(1)(c) and F(4)(a) below.............	2
c. Or treatment processes necessary to treat wastes that are restricted from land disposal to meet some or all of the applicable treatment standards or to treat wastes to satisfy (in whole or in part) the standard of "use of practically available technology that	11

yields the greatest environmental benefit" contained in §268.8(a)(2)(ii), with prior approval of the Director. This modification may also involve addition of new waste codes or narrative descriptions of wastes. It is not applicable to dioxin-containing wastes (F020, 021, 022, 023, 026, 027, and 028)..........	
2: a. Modification of a container unit without increasing the capacity of the unit.................................	2
b. Addition of a roof to a container unit without alteration of the containment system...................	1
3. Storage of different wastes in containers, except as provided in (F)(4) below: a. That require additional or different management practices from those autho-rized in the permit...................	3
b. That do not require additional or different management practices from those authorized in the permit........ Note: See §270.42(g) for modification procedures to be used for the management of newly listed or identified wastes.	2
4. Storage of treatment of different wastes in containers:	
a. That require addition of units or change in treatment process or management standards, provided that the wastes are restricted from land disposal and are to be treated to meet some or all of the applicable treatment standards, or that are to be treated to satisfy (in whole or in part) the standard of "use of practi-cally available technology that yields the greatest environmental benefit" contained in §268.8(a)(2)(ii). This modification is not applicable to dioxin-containing wastes (F020, 021, 022, 023, 026, 027, and 028)..........	1
b. That do not require the addition of	

units or a change in the treatment process or management standards, and provided that the units have previously received wastes of the same type (e.g., incinerator scrubber water). This modification is not applicable to dioxin-containing wastes (F020, 021, 022, 023, 026, 027, and 028)...............................	11
G. Tanks	
1: a. Modification or addition of tank units resulting in greater than 25% increase in the facility's tank capacity, except as provided in G(1)(c), G(1)(d), and G(1)(e) below...	3
b. Modification or addition of tank units resulting in up to 25% increase in the facility's tank capacity, except as provided in G(1)(d) and G(1)(e) below........................	2
c. Addition of a new tank that will operate for more than 90 days using any of the following physical or chemical treatment technologies: neutralization, dewatering, phase separation, or component separation...	2
d. After prior approval of the Director, addition of a new tank that will operate for up to 90 days using any of the following physical or chemical treatment technologies: neutralization, dewatering, phase separation, or component separa-tion...............................	11
e. Modification or addition of tank units or treatment processes necessary to treat wastes that are restricted from land disposal to meet some or all of the applicable treatment standards or to treat wastes to satisfy (in whole or in part) the standard of "use of practically available technology that yields the greatest environmental benefit" contained in §268.8(a)(2)(ii), with prior approval of the Director. This modification may also involve addition of new waste codes. It is not applicable to dioxin-containing wastes (F020, 021, 022, 023, 026, 027, and 028).........	11

2. Modification of a tank unit or secondary containment system without increasing the capacity of the unit....	2
3. Replacement of a tank with a tank that meets the same design standards and has a capacity within +/- 10% of the replaced tank provided................ --The capacity difference is no more than 1500 gallons, --The facility's permitted tank capacity is not increased, and --The replacement tank meets the same conditions in the permit.	1
4. Modification of a tank management practice.............................	2
5. Management of different wastes in tanks: a. That require additional or different management practices, tank design, different fire protection specifications, or significantly different tank treatment process from that authorized in the permit, except as provided in (G)(5)(c) below........	3
b. That do not require additional or different management practices, tank design, different fire protection specifications, or significantly different tank treatment process than authorized in the permit, except as provided in (G)(5)(d)................	2
c. That require addition of units or change in treatment processes or management standards, provided that the wastes are restricted from land disposal and are to be treated to meet some or all of the applicable treatment standards or that are to be treated to satisfy (in whole or in part) the standard of "use of practically available technology that yields the greatest environmental benefit" contained in	

§268.8(a)(2)(ii). The modification is not applicable to dioxin-containing wastes (F020, 021, 022, 023, 026, 027, and 028)............................	11
d. That do not require the addition of units or a change in the treatment process or management standards, and provided that the units have previously received wastes of the same type (e.g., incinerator scrubber water). This modification is not applicable to dioxin-containing wastes (F020, 021, 022, 023, 026, 027, and 028)............................... Note: See §270.42(g) for modification procedures to be used for the management of newly lilsted or identified wastes.	1
H. Surface Impoundments 1. Modification or addition of surface impoundment units that result in increasing the facility's surface impoundment storage or treatment capacity.............................	3
2. Replacement of a surface impoundment unit..................................	3
3. Modification of a surface impoundment unit without increasing the facility's surface impoundment storage or treatment capacity and without modifying the unit's liner, leak detection system, or leachate collection system......................	2
4. Modification of a surface impoundment management practice....................	2

5. Treatment, storage, or disposal of different wastes in surface impoundments: a. That require additional or different management practices or different design of the liner or leak detection system than authorized in the permit............................	3
b. That do not require additional or different management practices or different design of the liner or leak detection system than authorized in the permit..........................	2
c. That are wastes restricted from land disposal that meet the applicable treatment standards or that are treated to satisfy the standard of "use of practically available technology that yields the greatest environmental benefit" contained in §269.8(a)(2)(ii), and provided that the unit meets the minimum téchnological requirements stated in §268.5(h)(2). This modification is not applicable to dioxin-containing wastes (F020, 021, 022, 023, 026, 027, and 028).................................	1
d. That are residues from wastewater treatment or incineration, provided that disposal occurs in a unit that meets the minimum technological requirements stated in §268.5(h)(2), and provided further that the surface impoundment has previously received wastes of the same type (for example, incinerator scrubber water). This modification is not applicable to dioxin-containing wastes (F020, 021, 022, 023, 026, 027, and 028)..........	1
6. Modifications of unconstructed units to comply with §§264.221(c), 264.222, 264.223, and 264.226(d)................	*1
7. Changes in response action plan: a. Increase in action leakage rate.....	3

b. Change in a specific response reducing its frequency or effectiveness.........................	3
c. Other changes....................... Note: See §270.42(g) for modification procedures to be used for the management of newly listed or identified wastes	2
I. Enclosed Waste Piles. For all waste piles except those complying with §264.250(c), modifications are treated the same as for a landfill. The following modifications are applicable only to waste piles complying with §264.250(c).	
1. Modification or addition of waste pile units: a. Resulting in greater than 25% increase in the facility's waste pile storage or treatment capacity.........	3
b. Resulting in up to 25% increase in the facility's waste pile storage or treatment capacity....................	2
2. Modification of waste pile unit without increasing the capacity of the unit...................................	2
3. Replacement of a waste pile unit with another waste pile unit of the same design and capacity and meeting all waste pile conditions in the permit....	1

4. Modification of a waste pile management practice....................	2
5. Storage or treatment of different wastes in waste piles: a. That require additional or different management practices or different design of the unit....................	3
b. That do not require additional or different management practices or different design of the unit.........	2
6.Conversion of an enclosed waste pile to a containment building unit............ Note: See §270.42(g) for modification procedures to be used for the management of newly listed or identified wastes.	2
J. Landfills and Unenclosed Waste Piles 1. Modification or addition of landfill units that result in increasing the facility's disposal capacity...........	3
2. Replacement of a landfill.............	3
3. Addition or modification of a liner, leachate collection system, leachate detection system, run-off control, or final cover system....................	3
4. Modification of a landfill unit without changing a liner, leachate collection system, leachate detection system, run-off control, or final cover system...............................	2
5. Modification of a landfill management practice..............................	2

6. Landfill different wastes: a. That require additional or different management practices, different design of the liner, leachate collection system, or leachate detection system..............................	3
b. That do not require additional or different management practices, different design of the liner, leachate collection system, or leachate detection system...............................	2
c. That are wastes restricted from land disposal that meet the applicable treatment standards or that are treated to satisfy the standard of "use of practically available technology that yields the greatest environmental benefit" contained in §268.8(a)(2)(ii), and provided that the landfill unit meets the minimum technological requirements stated in §268.5(h)(2). This modification is not applicable to dioxin-containing wastes (F020, 021, 022, 023, 026, 027, and 028)...............................	1
d. That are residues from wastewater treatment or incineration, provided that disposal occurs in a landfill unit that meets the minimum technological requirements stated in §268.5(h)(2), and provided further that the landfill has previously received wastes of the same type (for example, incinerator ash). This modification is not applicable to dioxin-containing wastes (F020, 021, 022, 023, 026, 027, and 028)..........	1
7. Modifications of unconstructed units to comply with §§264.251(c), 264.252, 264.253, 264.254(c), 264.301(c), 264.302, 264.303(c), and 264.304.......	*1
8. Changes in response action plan: a. Increase in action leakage rate.....	3
b. Change in a specific response reducing its frequency or effectiveness........................	3

c. Other changes...................... Note: See §270.42(g) for modification procedures to be used for the management of newly listed or identified wastes.	2
K. Land Treatment 1. Lateral expansion of or other modification of a land treatment unit to increase areal extent...............	3
2. Modification of run-on control system.................................	2
3. Modify run-off control system.........	3
4. Other modifications of land treatment unit component specifications or standards required in permit..........	2
5. Management of different wastes in land treatment units: a. That require a change in permit operating conditions or unit design specifications.......................	3
b. That do not require a change in permit operating conditions or unit design specifications................. Note: See §270.42(g) for modification procedures to be used for the management of newly listed or identified wastes	2
6. Modification of a land treatment unit management practice to: a. Increase rate or change method of waste application....................	3
b. Decrease rate of waste application..........................	1
7. Modification of a land treatment unit management practice to change measures of pH or moisture content, or to enhance microbial or chemical reactions............................	2
8. Modification of a land treatment unit management practice to grow food chain crops, to add to or replace existing permitted crops with different food chain crops, or to modify operating plans for distribution of animal feeds resulting from such crops.............	3

9. Modification of operating practice due to detection of releases from the land treatment unit pursuant to §264.278(g)(2).........................	3
10. Changes in the unsaturated zone monitoring system, resulting in a change to the location, depth, number of sampling points, or replace unsaturated zone monitoring devices or components of devices with devices or components that have specifications different from permit requirements.....	3
11. Changes in the unsaturated zone monitoring system that do not result in a change to the location, depth, number of sampling points, or that replace unsaturated zone monitoring devices or components of devices with devices or components having specifications different from permit requirements.....	2
12. Changes in background values for hazardous constituents in soil and soil-pore liquid......................	2
13. Changes in sampling, analysis, or statistical procedure.................	2
14. Changes in land treatment demonstration program prior to or during the demonstration..............	2
15. Changes in any condition specified in the permit for a land treatment unit to reflect results of the land treatment demonstration, provided performance standards are met, and the Director's prior approval has been received.......	11
16. Changes to allow a second land treatment demonstration to be conducted when the results of the first demonstration have not shown the conditions under which the wastes can be treated completely, provided the conditions for the second demonstration are substantially the same as the conditions for the first demonstration and have received the prior approval of the Director.........................	11

17. Changes to allow a second land treatment demonstration to be conducted when the results of the first demonstration have not shown the conditions under which the wastes can be treated completely, where the conditions for the second demonstration are not substantially the same as the conditions for the first demonstration............................	3
18. Changes in vegetative cover requirements for closure...............	2
L. Incinerators, Boilers, and Industrial Furnaces:	11
1. Changes to increase by more than 25% any of the following limits authorized in the permit: A thermal feed rate limit, a feedstream feed rate limit, a chlorine/chloride feed rate limit, a metal feed rate limit, or an ash feed rate limit. The Director will require a new trial burn to substantiate compliance with the regulatory performance standards unless this demonstration can be made through other means.............................	3
2. Changes to increase by up to 25% any of the following limits authorized in the permit: A thermal feed rate limit, a feedstream feed rate limit, a chlorine/chloride feed rate limit, a metal feed rate limit, or an ash feed rate limit. The Director will require a new trial burn to substantiate compliance with the regulatory performance standards unless this demonstration can be made through other means.............................	2

3. Modification of an incinerator, boiler, or industrial furnace unit by changing the internal size or geometry of the primary or secondary combustion units, by adding a primary or secondary combustion unit, by substantially changing the design of any component used to remove HCl/Cl2, metals, or particulate from the combustion gases, or by changing other features of the incinerator, boiler, or industrial furnace that could affect its capability to meet the regulatory performance standards. The Director will require a new trial burn to substantiate compliance with the regulatory performance standards unless this demonstration can be made through other means...........................	3
4. Modification of an incinerator, boiler, or industrial furnace unit in a manner that would not likely affect the capability of the unit to meet the regulatory performance standards but which would change the operating conditions or monitoring requirements specified in the permit. The Director may require a new trial burn to demonstrate compliance with the regulatory performance standards.......	2
5. Operating requirements: a. Modification of the limits specified in the permit for minimum or maximum combustion gas temperature, minimum combustion gas residence time, oxygen concentration in the secondary combustion chamber, flue gas carbon monoxide and hydrocarbon concentration, maximum temperature at the inlet to the particulate matter emission control system, or operating parameters for the air pollution control system. The Director will require a new trial burn to substantiate compliance with the regulatory performance standards unless this demonstration can be made through other means........................	3

b. Modification of any stack gas emission limits specified in the permit, or modification of any conditions in the permit concerning emergency shutdown or automatic waste feed cutoff procedures or controls....	3
c. Modification of any other operating condition or any inspection or recordkeeping requirement specified in the permit............................	2
6. Burning different wastes: a. If the waste contains a POHC that is more difficult to burn than authorized by the permit or if burning of the waste requires compliance with different regulatory performance standards than specified in the permit. The Director will require a new trial burn to substantiate compliance with the regulatory performance standards unless this demonstration can be made through other means.............................	3
b. If the waste does not contain a POHC that is more difficult to burn than authorized by the permit and if burning of the waste does not require compliance with different regulatory performance standards than specified in the permit............................ Note: See §270.42(g) for modification procedures to be used for the management of newly listed or identified wastes	2
7. Shakedown and trial burn: a. Modification of the trial burn plan or any of the permit conditions applicable during the shakedown period for determining operational readiness after construction, the trial burn period, or the period immediately following the trial burn..............	2

b. Authorization of up to an additional 720 hours of waste burning during the shakedown period for determining operational readiness after construction, with the prior approval of the Director........................	11
c. Changes in the operating requirements set in the permit for conducting a trial burn, provided the change is minor and has received the prior approval of the Director........	11
d. Changes in the ranges of the operating requirements set in the permit to reflect the results of the trial burn, provided the change is minor and has received the prior approval of the Director..............	11
8. Substitution of an alternative type of nonhazardous waste fuel that is not specified in the permit................	1
9. Technology Changes Needed to meet Standards under 40 CFR part 63 (Subpart EEE--National Emission Standards for Hazardous Air Pollutants From Hazardous Waste Combustors), provided the procedures of §270.42(i) are followed	
M. Containment Buildings. 1. Modification or addition of containment building units: a. Resulting in greater than 25% increase in the facility's containment building storage or treatment capacity..............................	3
b. Resulting in up to 25% increase in the facility's containment building storage or treatment capacity.........	2
2. Modification of a containment building unit or secondary containment system without increasing the capacity of the unit...................................	2
3. Replacement of a containment building with a containment building that meets the same design standards provided: a. The unit capacity is not increased............................	

	1
b. The replacement containment building meets the same conditions in the permit................................	1
4. Modification of a containment building management practice....................	2
5. Storage or treatment of different wastes in containment buildings: a. That require additional or different management practices.................	3
b. That do not require additional or different management practices........	2
N. Corrective Action: 1. Approval of a corrective action management unit pursuant to §264.552...	3
2. Approval of a temporary unit or time extension for a temporary unit pursuant to §264.553............................	2
3. Approval of a staging pile or staging pile operating term extension pursuant to §264.554............................	2

----

1 Class 1 Modifications requiring prior Agency approval.

[§270.42 Appendix I table revised at 57 FR 37263, Aug. 18, 1992; amended at 58 FR 8685, Feb. 16, 1993; amended at 58 FR 29886, May 24, 1993; 63 FR 33829, June 19, 1998; 63 FR 65941, Nov. 30, 1998, effective June 1, 1999]

APPENDIX P
WASTE PROFILE FORM

Waste Profile Form for

_____

1. **Source and Nature of Waste**

   Generator Name and Address: _____

   Person completing form: _____ Date: _____
   Description of waste/physical appearance and form: _____
   _____

   Process producing waste: _____
   _____
   _____

   Know ingredients/composition (e.g. from Material Safety Data Sheet for raw materials, give % if known):
   _____
   _____

   Volume generated per unit time: _____ per: _____
   Procedure or basis to assure sample is representative: _____
   _____
   _____

   Date(s) sample was taken: _____ Name of sampler: _____

2. **Federal RCRA List of Hazardous Waste**

   Is the waste a listed RCRA hazardous waste based on its source or composition or is it in a mixture or derived
   from an RCRA listed waste? _____

   Description of RCRA listed waste [40 CFR §261.31, .32, OR .33] _____

   RCRA listed waste number: _____

3. **Testing Facilities**

   Name/address of laboratory(s) conducting testing and areas to be performed: _____
   _____

   Laboratory contact(s): _____ Tel: _____
   State certification in field of testing verified: _____
   specific handling procedures (describe check with lab as to requirements for certain methods):
   _____

4. Characterization of Sample for Physical Hazardous Properties

4.1   Ignitability: Flashpoint is   °F (40 CFR 261.21)

        Basis for value: ☐ Testing: Laboratory: _____

                                          Reference: _____

                                          Data: _____

                           ☐ Knowledge: Source of Information: _____

4.2   Corosivity: pH is: (40 CFR 261.22)

        Basis for value: ☐ Testing: Laboratory: _____

                                            Reference: _____

                                          Date: _____

                           ☐ Knowledge: Source of Information: _____

4.3   Reactivity: Sample contains reactive material as defined in 40 CFR 261.23:

        Basis of determination: ☐ Knowledge   ☐ Other: _____

        Sample contains cyanides or sulfides at levels which may produce toxic gases?   Yes   No

        Basis for determination: ☐ Testing: Laboratory:

                                            Reference: _____

                                          Data: _____

                           ☐ Knowledge: Source of Information: _____

5. Characterization of Sample for Toxicity Characteristics

5.1   Toxicity Characteristic Leaching Procedure (TCLP) [RCRA characteristic]:

        Results: Report any value which exceeds regulatory level in 40 CFR 261.24

Constituent	Regulatory Level	Concentration Determined
_____	_____	_____
_____	_____	_____
_____	_____	_____

        Basis for not testing: ☐ Knowledge   ☐ Other: _____

6. Conclusions

☐ Hazardous Waste, Reason(s):   _____

☐ Nonhazardous, Reason(s):   _____

APPENDIX Q
INVENTORY OF USEPA/STATE PERMIT IMPROVEMENT INITIATIVES

Introduction

I. Multimedia

    A. EPA HQ
    B. Region 1
    C. Region 2
    D. Region 3
    E. Region 5
    F. Region 6
    G. Region 9
    H. Region 10

II. Pollution Prevention (P2)

    A. EPA HQ
    B. Region 1
    C. Region 2
    D. Region 3
    E. Region 5
    F. Region 9
    G. Region 10

III. Air

    A. EPA HQ
    B. Region 1
    C. Region 2
    D. Region 3
    E. Region 5
    F. Region 6
    G. Region 10

IV. Hazardous Waste

    A. EPA HQ
    B. Region 1
    C. Region 2
    D. Region 5
    E. Region 6
    F. Region 9
    G. Region 10

V. Water

    A. EPA HQ
    B. Region 1
    C. Region 2
    D. Region 3
    E. Region 5
    F. Region 6
    G. Region 9

VI. PCBs

    A. EPA HQ
    B. Region 5
    C. Region 9   INVENTORY OF USEPA/STATE PERMIT
                            IMPROVEMENT INITIATIVES

Introduction

The Permits Improvement Team (PIT) was created in July 1994
to examine all of EPA's permitting programs (air, water, and
hazardous waste) and identify how they can be improved. The
Team consists of EPA, state, tribal and local government
officials.

It was recognized early on that a number of permit
improvement initiatives were already completed or underway
at EPA headquarters, the EPA regional offices and state,
tribal and local governments. Rather than "reinvent the
wheel", the PIT began to inventory these initiatives. This
inventory was used to help the PIT identify specific
recommendations on how to improve EPA's permitting programs.

As the PIT's efforts proceeded, it also became evident that the inventory could serve another purpose. That being, to provide a source of information on what was already underway across the country to those involved in permitting at the federal, state, tribal and local level.

This inventory was assembled by the PIT from information provided by EPA and other governmental officials. The process began by having each team member identify ongoing efforts in their jurisdiction. In addition, a general request for permit improvement initiatives was made of each EPA headquarters permitting and regional office. The same request was also made of the state associations for air, water and waste officials. The inventory was assembled and additions continued to be added as the PIT became aware of them.

The inventory is organized first by category of permit improvement initiative (multi-media, pollution prevention, air, water, hazardous waste, PCB's) and then by government agency conducting the initiative (EPA headquarters, EPA region, state, tribal or local government). State, tribal and local government initiatives are presented by EPA region. Table 1 provides a listing of states and their associated region. The inventory is arranged by title of each initiative included. State, tribal and local government initiatives begin with the governmental organization for ease of reference.

This document is envisioned to be continuously updated as new information becomes available to the PIT. The inventory is being placed on the Internet so that it can be accessible to all interested parties. In addition, it has been provided to the state associations electronically, so they can place it on any of their bulletin boards.

If an EPA, state, tribal or local governmental official wants to update and existing entry or provide a new item they should send that information to Lance Miller, Executive Director PIT, USEPA, 2890 Woodbridge Ave., Mail Stop 100, Edison, NJ 08837.

TABLE 1

STATES AND ASSOCIATED REGION

STATE	REGION	STATE	REGION
Alabama	4	Nevada	9
Alaska	10	New Hampshire	1
Arizona	9	New Jersey	2
Arkansas	6	New Mexico	6
California	9	New York	2
Colorado	8	North Carolina	4
Connecticut	1	North Dakota	8
Delaware	3	Ohio	5
District of Columbia	3	Oklahoma	6
Florida	4	Oregon	10
Georgia	4	Pennsylvania	3
Hawaii	9	Rhode Island	1
Idaho	10	South Carolina	4
Illinois	5	South Dakota	8
Indiana	5	Tennessee	4
Iowa	7	Texas	6
Kansas	7	Utah	8
Kentucky	4	Vermont	1
Louisiana	6	Virginia	3
Maine	1	Washington	10
Maryland	3	West Virginia	3
Massachusetts	1	Wisconsin	5
Michigan	5	Wyoming	8
Minnesota	5	American Samoa	9
Mississippi	4	Guam	9
Missouri	7	Puerto Rico	2
Montana	8	Virgin Islands	2
Nebraska	7		

INVENTORY OF USEPA/STATE PERMIT IMPROVEMENT INITIATIVES

I. Multimedia

A. EPA HQ

Common Sense Initiative (CSI):

Work group of industry executives, environmental leaders,
government officials and labor and environmental justice

representatives to focus on 6 pilot industries to find ways
that tougher goals and greater flexibility can result in
"cleaner, cheaper, smarter" performance in 6 areas:
regulation, pollution prevention, reporting, compliance,
permitting, and environmental technology.

Status: Ongoing
Contact: Vivian Daub, 202-260-6790

Customer Service Plans:

Pursuant to Executive Order 12862, "Setting Customer Service
Standards", EPA must establish and implement customer
service standards to carry out the principles of the
National Performance Review. The customer service plan
describes customer service standards, future plans for
customer surveys, and the agency's benchmark for customer
services. Identification of training for managers and
employees to carry out the customer service plans should
also be included. Agency plan and 5 individual pilot plans
for handling inquires, water grants management, conducting
inspections, increasing public access to community right-to-
know information and environmental permitting will be
prepared.

Status: Draft plans due to White House 8/25/95, final
          plans to be published 9/8/95.
Contacts: Shelley Metzenbaum, 202-260-4719, Karl Hausker,
          202-260-4335

Community Environmental Protection Approach:

EPA initiated the Community Environmental Protection
Approach to comprehensively protect entire ecosystems, an
effort which is being led jointly by the Office of Water
(OW), Office of Policy, Planning, and Evaluation (OPPE), and
Office of Administration and Resource Management (OARM). In
simplest terms, community environmental protection is place-
based environmental management driven by the local needs of
communities and ecosystems to achieve our national
environmental protection and human health goals. EPA's
experience with the Great Lakes, Chesapeake Bay, and
National Estuary programs have proven the merits of a place-

based approach. Several elements are necessary to achieve
the community-based approach to environmental problem-
solving that lies at the heart of ecosystem management:
development of a framework for setting priorities and
environmental objectives and for implementing actions with
the stakeholders: development of national tools that can be
used in place-based environmental management; and
collaboration with a range of partners, both public and
private. EPA's role will vary from one ecosystem to another
-- EPA may work as a convener, facilitator or provider of
tools, depending on ecosystem needs.

Status: Ongoing
Contacts: Maurice LeFranc, OPPE 202-260-4908, Ed Hanley,
        OARM 202-260-6980, Menchu Martinez, OW 202-260-
        9818

Environmental Appeals Board Practice Manual:

The Agency's Environmental Appeals Board is currently
completing a manual explaining its operations. The Manual
discusses rules applicable to appeals from Resource
Conservation and Recovery Act (RCRA), Underground Injection
Control (UIC), Prevention of Significant Deterioration (PSD)
and National Pollutant Discharge Elimination System (NPDES)
permits, and gives the reader practical information on how
the appellate process works. It includes sample forms and
pleadings.

Status: Completed, 11/94
Contact: David Heckler, 202-501-7060

Environmental Justice:

Environmental Justice Steering Committee established to
provide Agency vision and overall program direction.
Environmental Justice Policy Working Group formed to develop
policy and coordinate multi-media issues. Each program and
region has developed Environmental Justice Implementation
Plans. National Environmental Justice Advisory Group formed
with representatives of interested parties to provide advice
to the Agency.

Status: Ongoing
Contact: Clarice Gaylord, Office of Environmental Justice,
202-260-0850

Environmental Leadership Program (ELP):

In the June 21, 1994 Federal Register, EPA published a
notice soliciting proposals for ELP pilot projects and
outlining the eligibility criteria for facilities to
participate. These pilot projects will explore ways that
EPA and states might encourage facilities to develop
innovative auditing and compliance programs and reduce the
risk of non-compliance through pollution prevention
practices. These voluntary pilot projects will encourage
industry to take greater responsibility for self-monitoring,
which can lead to improved compliance, pollution prevention,
and environmental protection. EPA will recognize facilities
for outstanding environmental management practices, and give
them an opportunity to examine and address barriers to self-
monitoring and compliance efforts. In addition, the
projects will help EPA design a full-scale leadership
program, and determine if implementing such a program can
help improve environmental compliance.

Status: Ongoing
Contact: Tai-ming Chang, OECA, 202-564-5081

Environmental Management System (EMS) Standards:

EMS provide structured ways to incorporate environmental
principles and goals into a management decision framework.
They can be applied by any organization interested in
improving its overall environmental performance, both in
terms of improved compliance and in moving beyond compliance
in areas such as pollution prevention, life cycle
assessment, and sustainable development. The International
Standards Organization (ISO) is working to develop
international EMS standards, while the National Sanitation
Foundation (NSF) is developing a national EMS standard
primarily for use by firms doing business in the US.

Status: ISO standards to be completed in early 1996. NSF
standards to be completed in early 1995

Contact: Jim Horne, OW, 202-260-5802

Environmental Technology Initiative (ETI):

ETI, which was created by President Clinton in his 1993 State of the Union address, is coordinated by EPA's Innovative Technology Council. ETI is designed to accelerate environmental protection, strengthen America's industrial base and increase exports of U.S. technologies and expertise. ETI awards grants to states, tribes and federal agencies to: adapt EPA's policy, regulatory and compliance framework to promote innovation; strengthen the capacity of technology developers and users to succeed in environmental innovation; strategically invest EPA funds in the development and commercialization of promising new environmental monitoring control, remediation and pollution prevention technologies; and accelerate the diffusion of innovative technologies at home and abroad. Grants are also awarded to assist in private sector commercialization of successful technology projects and to the non-profit sector for innovative technology projects. ETI funding was set at $36 million for FY1994 and $80 million for FY 1995, with increases projected for future years.

Status: Ongoing
Contact: David Berg, OPPE, 202-260-2182 or Connie Sasala,
OPPE, 202-260-9514

Office of Enforcement and Compliance Assurance (OECA):

OECA is developing guidance in several areas that are relative to permitting: multi-media plain English guidance to regulations for the dry cleaning industry; Resource Conservation and Recovery Act (RCRA) Waste Analysis Plan guidance for boiler and incinerator furnaces; applicability and compliance guidance under the New Source Review provisions; and multi-media inspection enforcement program guidance and interim final status report.

Other OECA activities include: completing work with Dun & Bradstreet to establish reliable corporate data linkages through Integrated Data for Enforcement Analysis (IDEA) System; developing demographic and ecosystem targeting methods to support Environmental Justice activities and ecosystem-based efforts such as addressing posted stream segments/contaminated sediments; developing and implementing a formal system for strengthening the participation of state, local, and tribal authorities in the development of OECA planning, priority setting, and policy development; pilot training in environmental justice communities designated to provide information on the basic statutory requirements; developing an ecosystem workplan; and re-engineering reporting using existing data and measures of significant noncompliance and compliance status/rates to incorporate multi-media, sector and environmental justice perspectives.

Status: To be completed in FY 1995
Contact: Refer to September 20, 1994 memorandum from Steven Herman for specific office assignments. Call 202-260-4134

Pesticide Regulation Integration:

The Office of Water (OW) and Office of Pollution Prevention and Toxic Substances (OPPTS) working on a Memorandum of Agreement integrating NPDES and pesticide guidelines to provide permit writers and industry with information on how to incorporate pollution prevention into permits.

Status: Ongoing
Contacts: Donna Reed, OW, 202-260-9532; Karen Angulo, OPPTS, 703-305-5011

Sector Notebook Project:

The Office of Compliance is developing information on 20 industry categories that will facilitate the transition from a single-media approach to a sector orientation and develop comparisons between industries for targeting purposes. Information to be compiled in each Notebook will include: general industry profile and regulatory requirements; environmental compliance profile; existing government/sector interaction; and pollution prevention activities being implemented by each sector.

Status: Final Notebooks to be completed by 9/30/94.

Contacts: Mike Barrette, 703-308-8676 or Greg Waldrip, 703-308-8694

Senior Environmental Enforcement & Compliance Forum (SEECF):

EPA and the Department of Justice created SEECF to work with states and tribes to coordinate environmental enforcement policy between the branches of government.

Status: Ongoing
Contact: Susan Susanke, OECA, 202-260-1008

Small Business Regulatory Reform:

The EPA Small Business Ombudsman, in cooperation with other key EPA offices, the Small Business Administration, Office of Management and Budget's Office of Information and Regulatory Affairs, and five other federal agencies, have concluded a series of working group meetings. Participation in the project also included five small business sectors. As a result of these meetings, these agencies have jointly issued a Findings and Recommendations Report covering a number of environmental regulatory issues that significantly affect the small business community. Subjects covered in this project included recycling and waste disposal, chemicals and metals, trucking and transportation, restaurants, and food processing. The report has received widespread distribution throughout government and EPA. Follow-up activities for EPA implementation of the report's findings and recommendations, as well as implementation actions by other federal agencies that participated, are underway.

Status: Report complete, implementation ongoing
Contact: Karen Brown, OASBO/OSDBU, 703-305-5938

State/EPA Capacity Steering Committee:

Joint policy statement on state/EPA relations developed that specifies the underlying basis for the relationship. This relationship is based on states being the primary environmental managers and EPA's primary roles being standard-setting, program review, research, collection/analysis/ sharing information, and technical assistance. Governing principles are: clear goals and expectations; clear roles and responsibilities; open and honest communication; shared responsibility and accountability for success; mutual respect, trust, and continuous improvement; and mutual commitment to pollution prevention.

Status: Report of the Task Force to Enhance State Capacity issued July 1993; Draft State Capacity Implementation Plan issued 4/21/93; Joint Policy Statement on State/EPA Relations issued 7/14/94, National Environmental Performance Partnership System signed 5/17/95

Contact: Chuck Kent, State/EPA Capacity Implementation Team, OROSLR, 202-260-2462

Sustainable Industries:

Initiated in 1991 by OPPE, the fundamental goal of the Sustainable Industries program is an industry-specific approach to policymaking that focuses on promoting "cleaner, cheaper, smarter" solutions to environmental problems. The Sustainable Industries projects use the "backward mapping" approach, evaluating corporate decision-making factors (both "drivers" and barriers) that affect the environmental and economic performance of firms in the industries studied. Knowledge of these factors was then used to develop policy recommendations that seem most likely to achieve long-term eco-efficiency in the industries studied. Information was drawn from multi-stakeholder groups that included representatives from the industries, EPA, environmental organizations, and other key constituencies.

Status: First phase completed in July 1994, working with the metal finishing, photoimaging, and thermoset plastics industries. Phase 2 of the metal finishing project is being pursued within the CSI; the thermoset and photoimaging projects are continuing within the Sustainable Industries program.

Contact: Bob Benson, OPPE, 202-260-8668

B. Region 1

Region 1 - Environmental Justice:

1) Developing a matrix to better target inspections. The matrix uses criteria that combine regional and national initiatives and geographic targeting. 2) Developing a composite drawing on GIS of storage installations that will display the relative vulnerability of principal water sheds, feeder tributaries, recreational shorelines and natural habitats to potential pollution. Using EPA Region 1 EJ data project classifications will allow for additional analysis to obtain information on the total number of facilities falling into each of the EJ classifications.

Status: 1) Complete; 2) Scheduled for completion 6/94
Contact: Environmental Services Division; 617-860-4316

Connecticut - Environmental Permitting Re-engineering and Restructuring Plan:

CT DEP in consultation with the Environmental Permitting Task Force and their consultant prepared a plan to streamline their permitting processes. The plan includes the DEP's permitting mission and provides specific recommendations to: simplify and standardize workflows; eliminate redundant and nonsubstantive activities; eliminate processing bottlenecks; prioritize applications in a consistent manner; simplify application forms and provide applicants with checklists for completing applications; revise current fee structures, including reviewing fee amounts, fee collection, and other potential revenue sources for the department; streamline and eliminate unnecessary hearings; streamline the ways businesses report spills; share routine inspection reports with subject companies; use private contract services in the review process, and allow for the use of private consultants to certify whether permit applications are complete. Other recommendations focus on communication and outreach; coordination and management; information technology; and human resources and training. Report includes survey of 43 other state permit streamlining efforts.

Status: Final Report issued 3/1/93
Contact: Robert Kaliszewski, Permit Ombudsman CT DEP, 203-
424-3003

Massachusetts - Environmental Protection Integrated Computer
System:

EPICS takes information supplied by 12 separate MA
Department of Environmental Protection (DEP) divisions,
including air emissions, hazardous waste and water supply,
and combines it into a single database. This gives DEP
employees instant access to all of the agency's information
and allows them to search for data on a facility by entering
its name and location. This and a two year cross-training
program have allowed for inspectors to do multi-media
inspections.

Status: Online
Contact: Patricia Deese-Stanton, Assistant Commissioner, MA
DEP 617-292-5853.

Massachusetts Permit Streamlining Legal Advisory Committee:

The Legal Advisory Committee was established by the
Secretaries of Environmental Affairs and Economic Affairs
and included representatives from business and environmental
constituencies. The Committee prepared a report, highlights
of which are:
    * recommendations for outreach (permitting manuals,
preapplication conferences, permitting assistance office,
with regional presence, easy access to agency policies)
    * recommendations for coordination of permitting
(single point of contact, coordination among different
agencies issuing media- specific permits for single project,
permit ombudsmen for projects)
    * recommendations for increased use of general permit
mechanism in appropriate situations with emphasis on
enforcement of permit conditions
    * recommendations for uniform timeframes for steps in
permit process and decisions
    * recommendations for master uniform reporting form to
be submitted on single annual date for all permits with
annual reporting requirement

* recommendation for additional study of issues involved in consolidated or integrated permit approach.

Status: Final Report April 1995
Contact: Marilyn L. Sticklor, Co-Chair, Goulston & Storrs,
        617-482-1776

C. Region 2

New York - Regulatory Reform:

The NY Department of Environmental Conservation (DEC) established an internal regulatory review group to identify regulatory reforms for the agency. The first report includes regulatory reforms in nine areas: eliminating unnecessary reviews/accelerating the permit process; encouraging voluntary site cleanups; expanding programs to help business comply with regulatory requirements; accelerating the permit hearing process; reducing DEC staff review of certain projects by increasing reliance on Professional Engineer certifications provided by applicants; reducing duplicative regulation among federal, state and local governments; improving information systems to reduce workloads on both department staff and the regulated community; eliminating unnecessary reporting requirements; developing resource management plans as a way to establish environmental goals within whole natural systems, ecosystems or areas; and creating a framework for expedited permitting.

Status: Report issued July 1994; second progress report
        due spring of 1995.
Contact: Lou Concra, NY DEC, 518-457-7424

D. Region 3

Delaware - Natural Resources and Environmental Control Permitting Reform Project:

Permitting reform initiated in January 1994 to; maintain or improve environmental protection, achieve clarity in the permitting process, simplify permitting processes, promote efficiency in permitting, and achieve consistency. Flow charts prepared for each permitting process. 135 recommendations developed for specific permits and general action.

Status: Final Report dated August 12, 1994
Contact: Phillip Cherry, 302-739-6400

Maryland - Multi-media Permits:

As an added service to business and industry, the MD Department of the Environment (MDE) offers the opportunity to acquire multi-media permit coverage. This service is not only extended to new business and industry locating in Maryland, but to existing industry where multiple permits are required.

Status: Ongoing
Contact: Dana Bauer, MDE Water Management Division, 410-631-3512

Maryland - Environmental Permits Service Center: The Center is designed to provide the public and regulated community with a consolidated environmental permits process that is more efficient, comprehensive, and proactive in meeting their management objectives. The Center will improve services to businesses, local governments, community organizations and interested citizens by: providing timely permit information and reliable permit guidance; enhancing staff skills through cross-training; improving coordination on air, water, and waste regulatory issues; incorporating pollution prevention into everyday regulatory requirements; including community involvement up front in the regulatory process; and instilling timeliness and accountability for meeting permit milestones.

Status: Center opened 12/1/94
Contact: Mitch McCalmon, MDE, 410-631-3772

Pennsylvania - Process Improvement Team (PIT):

The PIT was formed to study permitting and compliance processes of PA's Department of Natural Resources (DNR) and make recommendations on ways to increase environmental protection while improving the efficiency of our processes, reducing the current backlog of permits, and seeking more effective ways of achieving compliance. The PIT focused on four main program areas: residual waste, National Pollutant Discharge Elimination System (NPDES), sewage sludge and wetlands. The PIT identified ten major issues: administrative changes; automated production and electronic storage for permit documents; data management; public notification and comment; standardizing/streamlining minor sewage permits; watershed toxics analysis and its relationship to the permitting process; increased pollution prevention focus; fee structure; training and communications; and watershed teams. Recommendations were developed for each area.

Status: Final report released 2/24/94. Implementation
     ongoing.
Contact: Stuart Gansell, DNR, 717-787-8184

E. Region 5

Michigan - Auto Project:

Target certain chemicals and develop voluntary effort to review controls.

Status: Ongoing
Contact: Marcia Horan, 517-373-9122

F. Region 6

Region 6 - Multimedia Permits Team:

Interdivisional permits quality action team formed to develop a permit process to ensure high-quality EPA/state coordination in all permit programs. The Team interviewed internal and external customers of the permitting process and developed recommendations in a number of areas. The Team also developed a permit access system, which provides integrated access to EPA's data systems which contain information on permit status.

Status: Recommendations made to Region 6 management;
        Permit Access System has been piloted on a small
        scale
Contact: Gus Chavarria, 214-665-7165

Oklahoma - Uniform Permit Process:

Tiered system developed for four categories: new sources;
existing sources (facility in operation - new permits);
existing sources (facility in operation -permit renewals and
modifications); and plans and certifications. The tiers
have different levels of public notice, public input and
decision-makers.

Status: Online
Contact: Steve Thompson, OKDEQ, 405-271-7339

Texas - Permit Processing Review Initiative:

The TX Natural Resource Conservation Commission (TNRCC)
identified and evaluated issues and/or concerns of agency
stakeholders relating to the fundamental permitting
processes across the agency. Report summarizes 77
recommendations to be implemented and 41 that have been
implemented.

Status: Report dated 8/1/94
Contact: Chris Macomb, Director Watershed Management Div.,
        TNRCC

   G. Region 9

California - Office of Environmental Technology:

California's Environmental Protection Agency (Cal/EPA) is
proposing the formation of an Office of Environmental
Technology which will manage a process for the evaluation
and certification of environmental technology. The concept

has been piloted in the Department of Toxic Substances Control for several hazardous waste technologies.

Status: Authorizing legislation will be sought.
Contact: Michael Kahoe, Deputy Secretary, Cal/EPA, 916-322-5844

California - Permit and Regulatory Reform:

Efforts began in 1992 to reform permitting processes by examining the requirements for a permit, how permits are processed, recordkeeping and paperwork requirements and interaction among permitting, data reporting, enforcement and planning functions. Recommendations developed in 1992 and implemented in 1993-1994 cover: regulatory assistance; one-stop permit centers; establishment of permit teams to coordinate permitting on major projects; regulatory reform task forces for specific industries; programmatic reviews; reducing the number of activities needing individual permits; consolidating monitoring data report requirements; developing a uniform permit tracking system; and legislative reform. A February 1995 report summarizes the status of these reforms.

Status: Varies
Contact: Patrick Dorais, Cal/EPA, 916-322-2858

California - Permit Relief Communities:

Environmental regulation in California is complicated by the array of implementing agencies at the state and local levels. California is proposing the formation of "Permit Relief Communities," each of which would develop a single compliance assurance plan representing all state and local environmental requirements for its community. Companies would be required to comply with the community compliance assurance plan. The initial stage will be development of pilot "Permit Relief Communities" to test the concept. Authorizing legislation is required.

Status: Cal/EPA is working with its regulatory agencies to

organize a work plan; consultation with interest
groups will also be initiated soon.
Contact: Michael Kahoe, Deputy Secretary, Cal/EPA, 916-322-
5844

California - Reduce Individual Permits Through More
Registration Processes:

Promote the use of general permits in each of the media
programs; in addition, share with EPA toxicology information
that California develops as part of its pesticide
registration process, so that EPA's registration of
pesticides is also accelerated.

Status: General permitting has been implemented in the
Hazardous waste and solid waste programs and will
be more fully expanded within the water program in
FY95. Legislation is needed to authorize the

State Board to issue general permits for land
application of treated effluents.
Contact: Paul Blais, Special Assistant to the Secretary,
Cal/EPA, 916-324-7584

California - Unified Environmental Statute:

A comprehensive organic environmental statute that would
address air and water quality, solid and hazardous wastes,
pesticides and scientific risk assessment will be proposed
by Cal/EPA.

Status: A commission has been formed that will put forth a
proposal early in 1995.
Contact: Micheal Kahoe, Deputy Secretary, Cal/EPA, 916-322-
5844

H. Region 10

Alaska - Assistance Project:

Alaska has established an assistance project to consolidate
services to the business community in permits, compliance

and pollution prevention. This project will serve as an initial "one-stop shopping" opportunity for the regulated community, and includes providing information on pollution prevention as a cost-effective way to achieve compliance and reduce the need for permits. This project will also encourage inclusion of P2 in permits and examine multi-media coordination.

Status: Ongoing
Contact: David Wigglesworth, 907-563-6529

Oregon - Cross-Media Risk Assessment Model:

OR's Department of Environmental Quality (DEQ) has developed, with assistance from EPA, a cross-media risk assessment model to be used during permit application review and development. The model evaluates the long-term fate and transport of toxic chemicals discharged to the environment and evaluates alternative approaches during permit review.

Status: Implementation has been initiated.
Contact: Marianne Fitzgerald, DEQ, 503-229-5946

Oregon - Permits Handbook:

DEQ has issued a new permits handbook which provides guidance on obtaining environmental permits and includes a policy statement on pollution prevention. Readers are encouraged to contact regional offices for assistance.

Status: Handbook is available at all DEQ offices.
Contact: Carolyn Young, DEQ, 503-229-6271

Washington - Governor's Task Force on Regulatory Reform:

External task force preparing reports on various activities across state government. Draft reports prepared on alternative approaches, with a focus on voluntary compliance, and on rules review, focusing on a procedure for reviewing priority rules and establishing a streamlined rules repeal procedure.

Status: Final report issued 1/95

Contact: Phil Miller, 360-407-6985

Washington - Joint Aquatic Resources Permit Application:

Piloting use of a new single application form for six different types of federal state and local permits related to projects in shoreline areas.

Status: Ongoing
Contact: Bonnie Shorin, 360-407-7297

Washington - Land use Planning and Permitting Reform:

Integrate and Streamline state land use planning and permitting by: integrating the procedures for comprehensive land use planning and development regulations with procedures for environmental impact review and the procedures for shoreline planning under three different statutes; "front-load" the analysis and mitigation of environmental impacts of land use development during preparation and adoption of comprehensive plans and development regulations by local government; streamline project permit review procedures and create more certainty at the permitting stage by relying on those "front-end" plan and regulatory decisions and eliminating duplicative reconsideration of decisions.

Status: Recommendation developed as part of Task Force on
        Regulatory Reform
Contact: Phil Miller, 360-407-6985

Washington State - Single Permit Feasibility Study:

Determining the feasibility of combining air, water quality, and hazardous waste permits for a single facility into one permit. Piloting with one or more volunteer facilities.

Status: Ongoing
Contact: Hugh O'Neill, 360-407-6118

Washington - State and Local Permit Coordination:

Proposed legislation to create a procedure for coordinating

all required state and local permits through designation of
a lead permit manger and active management of an agreed-upon
permit processing timeline. Coordination procedure would be
made available at an applicant's option.

Status: Recommendation developed as part of Task Force on
        Regulatory Reform
Contact: Keith Phillips, 360-407-6907

II. Pollution Prevention (P2)

    See also;
        Section III.A; Air Permitting Improvements, and
               Flexible Permitting
        Section IV. A; Permitting Subcommittee
        Section IV. G; Alaska Permitting

    A. EPA HQ

Pollution Prevention Information Clearinghouse (PPIC):

The PPIC is dedicated to reducing or eliminating pollutants
through technology transfer, education, and public
awareness. It is operated by EPA's OPPTS. The
Clearinghouse is a free, non-regulatory service that
consists of a telephone reference and referral service, a
distribution center for selected EPA documents, and a
special collection available for interlibrary loan.

Status: On-Line
Contact: Labat-Anderson, Incorporated, under contract for
        EPA, 202-260-1023

Pollution Prevention Integration Initiative (P2IN):

The Office of Prevention, Pesticides and Toxic Substances
(OPPTS) has organized and is expanding its electronic,
Agency-wide network of pollution prevention contacts. The
goal of this network is to: (1) exchange information that
can serve as a "road map" in identifying what prevention
activities are occurring and where and who the experts or
points of contact are; (2) identify how prevention is being
institutionalized into EPA's mainstream activities; (3) put

the Agency in a position to facilitate proactive inter-
office and cross-program communication and information
sharing so as to further pollution prevention opportunities
and integration; (4) generate projects and initiatives --
e.g., the Common Sense Initiative and Environmental
Technology Initiative.

Status: Ongoing
Contacts: John Shoaff, OPPTS 202-260-1831, or Julie Lynch,
OPPTS, 202-260-4000

Pollution Prevention (P2): State Efforts to Integrate P2
into Their Activities:

USEPA report summarizes efforts of 41 states concerning P2.
One of the program elements discussed is permitting, both
multimedia permitting and the integration of P2 into
individual media permits. Twenty-five states have a section
describing their permitting efforts. Other topics covered
are legislation, inspections, enforcement, data integration
and organization.

Status: Report dated 9/93, EPA/742/B-93-002 - "Ongoing
Efforts by State Regulatory Agencies to Integrate
Pollution Prevention into Their Activities"
Contact: Lena Ferris-Hann, Pollution Prevention Div.,
OPPTs, 202-260-2237

Source Reduction Review Project (SRRP):

In the SRRP, the Office of Pollution Prevention and Toxics
(OPPT) has taken the lead in working with other program
offices to incorporate pollution prevention into regulations
the Agency is developing, specifically Clean Water Act
effluent guidelines, Resource Conservation and Recovery Act
(RCRA) hazardous waste listings, and Clean Air Act Maximum
Achievable Control Technology (MACT) standards. During the
three years of this project, changes have occurred in the

way information is collected for rulemaking, analysis is
done and standards are set.

Status: Ongoing for individual rulemakings. A white paper
      on SRRP opportunities, barriers and options for
      overcoming these barriers is being prepared.
Contact: Kathy Davey, Pollution Prevention Div., OPPT, 202-
      260-4164

   B. Region 1.

Connecticut - Publishing & Printing Pollution Prevention
Process:

Develop outreach material to promote pollution prevention in
the Publishing and Printing Industry.

Status: Ongoing
Contact: Mary Sherwin, 203-566-5217

   C. Region 2.

New Jersey - Solvents Emissions at Automobile Plants:

Incorporating pollution prevention planning requirements
targeted at solvent emissions into permits for automobile
manufacturing plants. The permits require the facility to
conduct a waste minimization assessment for solvents, and
contain specific suggestions for actions that should be
taken in the waste minimization assessment. As a condition
of the permit, the facility must also document annual
emissions of solvents and show trends in solvent use per
unit product.

Status: Ongoing
Contact: Louis Mikolajczyk, Chief Bureau of NSR, 609-292-
      9258

New York - Waste Reduction for Electronics Industry:

Waste reduction guidance, re: opportunities, assessments, methods included, four workshops for technical assistance and outreach.

Status: Completed
Contact: John Iannotti, 518-457-7267

D. Region 3.

Maryland - Pollution Prevention:

With support from EPA's Office of Pollution Prevention, Maryland has undertaken several projects aimed at integrating pollution prevention with the regulatory programs. Included are such things as training for Departmental staff, exploring opportunities with banking institutions for financial incentives through loan programs with industry, and expanding on existing efforts to employ pollution prevention credits during negotiated settlement processes.

Status: Ongoing
Contact: Dana Bauer, MDE Water Management Division, 410-631-3512

E. Region 5

Illinois - "Top of the Pile" Review for Voluntary Pollution Prevention Proposals:

Illinois EPA conducts expedited or "top of the pile" review for permit applications, variance petitions, site specific standard or rule petitions that address voluntary pollution or recycling proposals. The proposal can be a new process employing pollution prevention/recycling technology or can be a modification to an existing process. The environmental benefits of the proposal must be quantified relative to the status quo; the proposal must be a "significant departure from previous practice at he facility" and the facility must justify why it needs special processing by Illinois EPA. A statutory modification was passed to enable the agency to conduct these expedited reviews.

Status: Ongoing
Contact: Tom Wallin, Office of Pollution Prevention, 217-782-8700

Ohio - Automotive Pollution Prevention Project:

Reduce the generation and release of persistent toxic chemicals by the automotive industry in the Great Lakes Basin.

Status: Ongoing
Contact: Michael Kelly, 614-644-2980

F. Region 9

Arizona - Environmental Leadership Program:

Pollution prevention with emphasis on streamlining permitting, inspections and self-audits.

Status: Formative
Contact: Beverly Westgard, 602-207-2203

G. Region 10

Alaska - Facility Planning:

Alaska has begun a process to determine the future of pollution prevention facility planning. As a first step, Alaska is reviewing P2 facility planning efforts in Alaska and the nation. Upon completion of this review, Alaska will convene a partnership of public and private entities to determine what will work best for Alaska, and to make formal recommendations to the Commissioner and the legislature.

Status: Ongoing
Contact: David Croxton, 907-563-6529

Alaska - Permit Fees:

Alaska is examining incentives to foster P2 through
reductions in permit fees for facilities that implement P2
planning.

Status: Ongoing
Contact: David Wigglesworth, 907-563-6529

Oregon - Regulatory Guidance:

The Handbook of Regulatory Guidance for Oregon Construction
Contractors was developed to compile the myriad of
environmental regulations the construction industry is
guided by. The document sets the stage for a companion
document entitled "Environmental Handbook for Oregon
Construction Contractors: Best Pollution Prevention
Practices", that highlights best pollution prevention
practices for the construction industry.

Status: Completed
Contact: Sandy Gerkowitz, 503-229-5918

Washington - "Snapshots" Program:

Modelled after "Shopsweeps" (see Section III.G. below) for
the printing industry, but multi-media (air, hazardous waste
and water) with a focus on pollution prevention.
Educational materials on compliance and P2 were developed in
conjunction with the industry and 1200 facilities are
currently being visited by state and local government
personnel.

Status: Ongoing
Contact: Darin Rice, 360-407-6743
III. Air

   A. EPA HQ

Air Permitting Improvements:

The Office of Air Quality Planning and Standards (OAQPS) is

working with a contractor (RTI) to survey what pollution
prevention initiatives are being utilized in the states and
what next steps the Air program should focus on.

Status: Ongoing - draft report due to be completed 8/95,
        final report by 9/95
Contacts: Leo Stander, OAQPS, 919-541-5589; Melissa Malkin,
        RTI, 919-541-6154

Compliance Assurance Monitoring Rule:

Office of Air and Radiation will be proposing a flexible
approach to Clean Air Act monitoring requirements. Rules
would be written to address differences among regulated
facilities instead of a blanket rule for all permittees.
Sources would have to comply with the monitoring
requirements only after regulatory guidance is issued.

Status: Proposal due by 12/95, final rule due 6/96
Contact: Peter Westlin, OAR, 919-541-1058

Flexible Permitting:

Projects with Intel facility in Region 10 and Merck facility
with state of Virginia. Office of Air Quality Planning and
Standards (OAQPS) has work assignment with RTI to look at
examples of innovative air permits in the Regions/states,
with emphasis on pollution prevention. (See Section II.G.
below for more detail on Intel permit.)

Status: Intel permit complete; Merck permit ongoing
Contact: Intel permit: Dave Dellarco, Region 10, 206-553-
        5973; Merck permit: Mike Trutna, OAQPS, 919-541-
        5345 or Jesse Baskir, RTI, 919-541-5882

New Source Review (NSR) Reform:

Subcommittee of industry, environmental organizations, and
state and local agencies developed draft recommendations to
reduce the complexity and perceived impediments to speedy
review of the current systems, while maintaining the
environmental goals and benefits. Recommendations have been
drafted in the following areas: NSR permitting issues

associated with near Class 1 areas, procedures and coordination, determination of adverse impacts, and mitigation of source impacts; Best Available Control Technology/Lowest Achievable Emission Rate (BACT/LAER), BACT/LAER clearinghouse and presumptive BACT, BACT/LAER criteria and innovative control technologies and pollution prevention; impact of existing sources in Class 1 areas; and NSR applicability. Regulations being proposed to incorporate the recommendations.

Status: Recommendations from subgroup completed July 1994; Regulatory proposal under development.
Contact: David Solomon, 919-541-5375

B. Region 1

Region 1 - Air Permitting Improvements:

Working on a number of state initiatives: With CT to develop general permits establishing technical standards to "keep small sources small"; with MA to put technical standards into state Implementation Plans (SIPs) - so that as long as a source fits within a category described, a small source stays small; with MA on a rulemaking regarding operating permits -- to issue small-scale operating permits with restrictions to keep companies out of the full-blown permitting process -- and considering the use of this process for toxics as well as criteria pollutants; and with MA on pollution prevention guidance, to implement the concept of comparing actual to future predicted actual emissions when evaluating pollution prevention changes (vs. the traditional rule in the Prevention of Significant Deterioration (PSD) and New Source Review (NSR) programs of comparing actual to potential emissions, which may discourage pollution prevention).

Status: Ongoing
Contact: Lynne Hamjian, 617-565-4181

C. Region 2

New Jersey - Air Permitting Program:

Seeking to use facility-wide pollution permit as the air operating permit for the first 5-year term of the facility-wide permit. Other initiatives include: technical manuals that provide a comprehensive list of requirements that are applicable to a specific type of equipment; permit application checklists to provide a comprehensive list of items that must be included in an application; holding public hearings when an application is received; categorizing permit applications into levels with different procedures based on complexity; and permit workshops to provide guidance, training and information to applicants.

Status: Complete
Contact: William O'Sullivan, NJ DEP, 609-984-1484

New Jersey - Asphalt Manufacturers Good Operating Practice Requirements in Permits:

New Jersey has been incorporating good operating practice requirements targeted at air pollution control into both New Source Review and other New Jersey air permits for asphalt manufacturers. The permittees must report to the state quarterly on the good operating practices they have undertaken. Examples of effective practices are listed in the permit (e.g., adjusting burner systems to optimize fuel atomization and fuel/air mixing). The facility is not limited to these measures, but they must show that they have taken some measures. The program was developed in conjunction with the industry which helped identify the good operating practices.

Status: Program has been successfully in place 5 years
Contact: Louis Mikolajczyk, Chief, Bureau of NSR, 609-292-9258

New Jersey - Reasonable Available Control Technology (RACT) for VOCs and NOx:

Two RACT rules in New Jersey contain pollution prevention provisions:

a. Sources at major facilities that are required to show compliance with the requirement to collect greater than

90% of VOC emissions are explicitly allowed to use pollution prevention measures in their demonstrations of compliance.

Use of compliant surface coatings with low VOC content is considered to meet RACT and no control devices are then required. Use of low VOC volume percentage solutions meet RACT in graphic arts operations except for screen printing operations, and no control devices are required.

b. The NOx RACT standard for non-utility boilers with heat input of between 20-50 million BTUs per hour requires facilities to annually adjust the combustion process of the boiler. This is a pollution prevention measure to reduce emissions that also saves fuel.

Status: Ongoing
Contact: Louis Mikolajczyk, Chief, Bureau of NSR, 609-292-
9258

D. Region 3

See "Flexible Permitting" under Section II.A. above.

E. Region 5

Region 5 - Data Management:

The implementation of the Title V operating permits programs will require the transfer of enormous amounts of information between the states and Region, including draft, proposed and final permits, applications, amendments and technical support documents. The ability to effectively transfer and manage this information in an electronic format is critical to the success of the state programs and Federal oversight (which must occur on strict timelines). Region 5's Air and Radiation Division (ARD) has formed a data management workgroup to concentrate on the issue of permit information transfer and management. The workgroup is reviewing software options for internal permit review and management purposes and is also working with each state to develop complimentary systems which will allow for electronic transfer of information among offices. In addition, the workgroup is developing an electronic database fact sheet to be used as a cover sheet by the states which will highlight various criteria and assist the Regional permit reviewer in identifying permits of concern; this database could also be used as a management tool for sorting and compiling statistics. The workgroup is developing options for the near term as some of our states are close to Title V approval, and long term solutions which involve applying for various resource assistance grants to fully computerize and integrate the entire process.

Status: Ongoing
Contact: Genevieve Nearmyer, 312-353-4761

Region 5 - Differential Oversight:

Beyond trying to ensure individual permits meet all Clean Air Act requirements, the goal of the Air and Radiation Division (ARD) is to develop competent, self-sustaining state permit programs. To further this goal, ARD developed a system which utilizes the Section 105 grant process and differential oversight to encourage states to improve and/or maintain high quality permit programs. Specifically, a portion of the ARD workload model which allocates Section 105 grant resources among the states is performance-based, and one of the areas of review is NSR permitting. Criteria were developed which evaluate the state permitting programs in five categories, including notification, applicability, permit drafting issue resolution and improvements. Each year, states are evaluated and the portion of available NSR grant monies awarded to each state is dependent on its score. In addition to the resource benefit, states which achieved excellent ratings in all categories receive less real time individual permit review in the next year by the ARD permit review staff.

For future years, since Title V permit fees will cover the costs of state permitting, the resource portion of the differential oversight incentive will no longer be applicable. However, ARD will continue to evaluate state permitting programs, including Title V, using this performance-based approach. The reward for quality programs will continue to be a reduction in real time Federal oversight of individual draft permits.

Status: Ongoing
Contact: Cheryl Newton, 312-353-6730

Region 5 - Permit Review Priorities:

In recognition of the enormous numbers of draft permits which are sent to the Region for review under the state New Source Review construction permit programs, the Air and Radiation Division (ARD) developed a priority scheme which classifies permits into three categories. The categories represent three levels of review priority and cover issues such as geographic area (attainment versus nonattainment, proximity to Class I areas or Native American lands, etc.), synthetic minors, and industry type (e.g., power plants and waste management facilities). This procedure helps individual permit reviewers prioritize their workload to concentrate on issues of concern and best assess overall state permitting. Currently, the ARD is developing a similar priority scheme to help evaluate and prioritize the influx of state operating permit drafts expected in the near future.

Status: Ongoing
Contact: Ronald Van Mersbergen, 312-886-6056

Minnesota - Improvements to Air Emission Permitting Process:

The following streamlining initiatives are being worked on by the MN Pollution Control Agency (MPCA) and Office of the Attorney General:

1) Registration permits/control equipment rule. New rules effective on December 26, 1994 are designed to allow small sources to obtain a very streamlined registration permit instead of the more detailed and time-consuming individual source permits. It is expected that this rule will cover half of the sources that need air emission

permits; these sources account for less than ten percent of Minnesota's emissions. The rule is designed to impose needed emission requirements on these sources without engaging in a long permit issuance process.

The new rule also imposes standards of performance for use of control equipment that would allow sources to take credit for emission reductions caused by control equipment in determining what type of permit is required. These enforceable emission reductions allow sources to obtain less complicated permits by establishing limits on the source through the control equipment rule.

2) Emphasis on general permits. Minnesota is emphasizing general permits to handle most of the sources not handled by the registration permit rule. For example, general permits are being developed for asphalt plants and sand and gravel facilities. These permits spell out the typical requirements for sources in these categories. Individual sources that need special conditions not present in the general permit would need an individual permit, but the general permit would cover almost all sources in a particular category. Minnesota is increasing the usefulness of the general permit by including typical alternative operating scenarios that a source category might want to use. That way, certain common changes at a source do not require a new permit; the general permit includes the new limitations that apply to the new activity as an alternative scenario.

3) 3M flexible permit. For sources too complex or large to include in the registration permit or general permit categories, Minnesota is gaining experience at permitting a source in a way that allows flexibility in operations to be authorized in the permit rather than through numerous amendments over time as the facility changes. One permit of this type has been issued to 3M. In exchange for voluntarily lowering its overall emission limit for the facility, 3M received authorization for a great deal of flexibility in changing its operations as long as it remains below the lower emission limit. Once such flexibility is allowed, the hardest issue is how the changes and resulting emissions will be monitored and recorded. The permit deals extensively with the monitoring and recordkeeping required for the various changes 3M is authorized to make.

4) Streamlining the permit template. Minnesota is trying to reduce the size and complexity of its air emission permits by creating a more streamlined permit template. The intent is to present emission limits, reporting requirements and compliance schedules in table format, and to reduce the paraphrasing of rule text in the permit by simply referring the permittee to the rule (i.e. the permit does not restate the performance test procedures but refers the permittee to the proper rule which sets out those procedures).

Status: Complete/Ongoing
Contact: Ann Seha, Assistant Attorney General, 612-297-8755

F. Region 6

Texas - Air Preconstruction Permit Process:

The TX Natural Resource Conservation Commission (TNRCC) has adopted new rules that provide for a new, more flexible type of permit for new facilities and modifications to comply with air emission standards. Under these rules, the applicant must submit information that demonstrates that the following criteria are met: protection of public health and welfare; measurements of emissions; Best Available Control Technology (BACT); Federal New Source Performance Standards (NSPS); National Emission Standards for Hazardous Air Pollutants (NESHAPS) and Maximum Achievable Control Technology (MACT); performance demonstration; air dispersion modeling or ambient monitoring; and proposed control technology and compliance demonstration.

Status: Rules adopted November 25, 1994
Contact: Mary Ruth Holder, TNRCC, 512-239-1966

Texas - Flexible Air Permit Based on Emissions Cap:

Flexible air permits based on emissions cap are used as an alternative to current preconstruction requirements. One permit is issued for a given plant or site, but it can contain multiple emissions caps or multiple individual emissions limits.

Status: Program began this year and at least one permit

has been written.
Contact: Victoria Shu, TNRCC, NSR, 512-239-1230

G. Region 10

Region 10 and Oregon - Air Permitting Improvements:

In 1993, Intel Corporation, EPA and the Oregon Department of
Environmental Quality (DEQ) joined in a partnership to
evaluate opportunities to incorporate flexibility and
pollution prevention into permits issued under Title V of
the Clean Air Act as amended in 1990. The project created a
Title V permit that will demonstrate the ability of P2 to
perform equally as well as traditional end-of-pipe controls.
The permit contains the following requirements:
  * Emission limits and performance standards
  * Plant Site Emission Limits
  * Reasonably Achievable control Technology (RACT)
     standards
  * Aggregate Hazardous Air Pollutant emission limits
  * Pollution Prevention condition
  * Pre-approved changes
  * Monitoring requirements
  * Reporting requirements
  * General conditions

Status: Final permit issued
Contact: David Dellarco, Region 10, 206-553-4978

Oregon - Title V Permits:

Oregon DEQ instituted a pilot program for a select group of
affected sources in an effort to detect and resolve problems
in the Title V application and permitting process.

Status: Ongoing
Contact: Shelly McIntyre, Oregon Department of Justice,
     503-229-5725
IV. Hazardous Waste

A. EPA HQ

Innovative Approaches to Resource Conservation and Recovery

Act (RCRA) Permitting:

Evaluating past and present suggestions for innovative RCRA permitting. Developing, in accordance with the RCRA Implementation Study and National Performance Review, recommendations for improving the RCRA permitting process. Current initiatives include: updating and compiling permit writer "part B" checklists; examining ways to simplify modification and renewal processes; and submitting an Environmental Technology Initiative proposal for the development of permitting software and innovative inspector tools.

Status: Options Paper 5/24/94, ETI proposal 9/22/94
Contact: Ken Amaditz, OSW, 703-308-7056

RCRA Expanded Public Participation Rule:

Proposed rule (June 94) to provide for earlier and more meaningful public participation.

Status: Final Rule expected late summer 1995
Contact: Tricia Buzzell, OSW, 703-308-8632

RCRA Post Closure Rule:

Proposal to allow permitting agency to substitute enforcement or other authority in lieu of a post-closure permit.

Status: Anticipated final rule early 1996
Contact: Barbara Foster, OSW, 703-308-7057

RCRA Siting Workgroup:

Develop options and recommendations for RCRA policy on facility siting issues, examining both technical and environmental justice concerns.

Status: Ongoing - recommendations expected to be developed
        for consideration by AA in Fall of 1995.
Contact: Vernon Myers, OSW, 703-308-8660

Authorization of Indian Tribes Hazardous Waste Programs
under RCRA Subtitle C:

Rule would give Tribes the authority to implement Subtitle C
in whole or in part. This is different than current
requirements for states, which must adopt the entire
program.

Status: To be proposed in early 1995.
Contact: Felicia Wright, OSW, 703-308-8634

RCRA Waste Minimization National Plan Steering Committee:

Four subcommittees have been established, one of which is
the permitting subcommittee as described below.

Status: Developing a broad strategy for minimizing all
          hazardous waste.
Contact: Donna Perla, 703-308-8402

Permitting Subcommittee:

Examine approaches used to incorporate waste
minimization/pollution prevention into RCRA permits; develop
national guidance and training to promote waste minimization
in RCRA permits; identify and resolve crosscutting and
multi-media pollution prevention permit, inspection and
enforcement issues.

Status: Ongoing
Contact: Jim Lounsbury, 703-308-8463

   B.   Region 1

Region 1 - RCRA Permitting Improvements:

Working with CT on a permit streamlining project to develop
a guidance manual for implementing contingency plans for
treatment, storage and disposal facilities. The guidance
will try to differentiate between highly hazardous
substances vs. more "routine" hazardous wastes - then have

permit requirements to better reflect the real risks.

Status: Ongoing
Contact: John Podgurski, 617-573-9680

C. Region 2

Region 2 - RCRA Prioritization System - Inclusion of
Environmental Justice (EJ) Factors:

Incorporate EJ factors into environmental benefits ranking.
The inclusion of EJ factors will ensure that the Agency
properly evaluates human health and environmental issues for
minority and low-income populations in proximity to RCRA
regulated facilities. The environmental benefits ranking
along with the NCAPS ranking determines the overall facility
priority in RCRA permitting and corrective action program
activities.

Status: Ongoing
Contact: Andrew Bellina, EPA Region 2, 212-637-4110

Region 2 - RCRA Public Involvement:

Begin public involvement process at the planning stage
(e.g., at time of permit application) in order to give the
public enough time for comments. For Environmental Justice
facilities, submit a Citizen's Participation Plan in RCRA
permit applications and corrective action tasks. This plan
would include a list of organizations and concerned citizens
in the community and describe a proactive approach to be
taken by the facility to inform the community of the
proposed actions and how community members can voice their
opinions.

Status: Ongoing
Contact: Andrew Bellina, EPA Region 2, 212-637-4110

Region 2 - Training Waste Program Personnel in Environmental

Justice:

Environmental justice training plan developed for all waste program staff and management. The training includes Regional and state RCRA permitting and corrective action personnel and covers revised public notification outreach procedures which will incorporate environmental justice concerns.

Status: Training began 12/94; continuing
Contact: Wilfredo Palomino, EPA Region 2, 212-637-4179
New York - Hazardous Waste Management for Printers:

Survey printing industry on product processes, waste management methods, regulatory compliance, six workshops, and technical assistance.
Status: Completed
Contact: John Iannotti, 518-457-7267

D. Region 5

Minnesota - Improvements to Hazardous Waste Permitting Program:

The following permit improvement initiatives have been implemented by the MN Pollution Control Agency (MPCA):
   1. Conduct annual facility roundtable for all facility owners and operators to discuss issues of importance to hazardous waste facilities. Facility/state workgroups are dealing with term of permits, incident reporting and agency/county issues.
   2. Publish a semiannual newsletter updating all treatment, storage and disposal facilities (TSDFs) on new informational items such as proposed new rules affecting facilities, etc.
   3. Use reminder letters sent out one year in advance of the permit expiration date.
   4. Revise the public notice to make it more user-friendly by replacing legal jargon with more understandable language.
   5. Make permit conditions more facility-specific.
   6. Include compliance schedules in permits to expedite permit issuance.

7. Re-draft modified permits to include the modified text with the most current revision date on each page. A summary is also included at the beginning of the permit listing each modification and the effective date. This eliminates confusion regarding "current" permit.

8. Provide facilities with a "tip" sheet listing useful information for preparing permit applications, and meet with permittees prior to reissuance to review necessary changes to the permit.

9. Update MPCA permit review checklists as needed.

10. Schedule and coordinate permit drafting and public noticing to avoid a "crunch" with clerical staff tasks.

11. Use computer aided permit boilerplates and public notice documents to save time in permit drafting.

12. Use combined completeness and technical review to reduce the number of iterations required on permit applications.

13. Provide permittees with a "pre-review" of current permit application so as to improve the quality of permit reissuance application submittals and expedite application review.

14. Allow permittee to review draft permit prior to placing public notice so as to resolve any issues they may have well before the permit document is finalized.

Status: Complete/Ongoing
Contact: Bruce Brott, Hazardous Waste Division, 612-297-
        8380

E. Region 6

Region 6 - Customer Service Questionnaire:

A questionnaire is being developed that will be sent to citizens who have participated in the public outreach/comment aspects of the RCRA permit process. The questionnaire is being designed to provide customer feedback on the public's involvement efforts over time.

Status: Piloted in early 1995, results under analysis
Contact: Arnold Ondarza, 214-665-6790

Region 6 - RCRA "Paperless Permit" Initiative:

A LAN-based computerized permit system developed to reduce processing time. After implementation, the average time lag between state signature and EPA signature of jointly-issued permits dropped from 69 days to 8 days. The system also improved the consistency of RCRA permits and has been exported to states as they have become authorized to run the program.

Status: Completed 1991
Contact: Arnold Ondarza, 214-665-6790

Texas - Hazardous Waste Facility Permit Streamlining:

The TX Natural Resource Conservation Commission (TNRCC) has proposed regulations to expand public awareness through published notice of intended applications; requiring more balanced and representative makeup of local review committees; requiring professional facilitators to coordinate the activities of the review committees; and directing prospective applicants to defray the reasonable expenses of committees and facilitators.

Status: Rule proposal 9/30/94.
Contact: 512-239-6087

F. Region 9

Region 9 - Environmental Justice Support:

The Region is providing support to their RCRA staff in addressing environmental justice issues they encounter in permitting decisions. This is done through presentations to permits staff to familiarize them with issues and options, as well as to provide proactive support in identifying potential areas of EJ concern. Grant guidance for states also includes environmental justice work that may be applied in their RCRA permitting programs.

Status: Ongoing
Contact: Karen Scheuermann, Hazardous Waste Management

Division, 415-744-2057

Region 9 - RCRA California Multi-Year Permit Strategy:

A strategy is being developed for addressing remaining
permit and closure activities in priority. The goal is to
create a multi-year plan which can be used as the grant
workplan. This will reduce the level of review required
each year when the grant is negotiated.

Status: Ongoing. Expected completion date: July 1, 1995
Contact: John McCarroll, Hazardous Waste Management
        Division, 415-744-2057

Region 9 - RCRA Permit Grant Streamlining:

The Region is developing a pool of activities from which the
state of California can choose projects (as opposed to
approving a site-specific, activity-specific grant
workplan). The Region is also developing a set of
substitution criteria for the state to follow. This will
provide flexibility for the state to manage its program and
will reduce or eliminate the need for grant workplan
amendments.

Status: Ongoing (To be included in the FY96 grant)
Contact: Paula Bisson, Hazardous Waste Management Division,
        415- 744-2064

California - Recycling Requirements for Hazardous Wastes:

Cal/EPA has revised requirements for addressing hazardous
wastes that are being recycled. Recyclable material will
not be classified as a waste if it is: (1) used or reused as
an ingredient in an industrial process to make a product,
(2) used or reused as a safe and effective substitute for
commercial products, (3) returned to the original process
from which the material is returned as a substitute for raw
material feedstock, and the process uses raw materials as
principal feedstocks.

Status: Effective 1/1/93
Contact: Department of Toxic Substance Control, 916-323-

6042

California - Tiered Permitting of Hazardous Wastestreams:

Tiered permitting system developed for non-RCRA hazardous waste, including: conditionally exempt small quantity treatment; conditionally exempt; conditionally authorized; and permit-by-rule. Fifteen different wastestreams are addressed by the tiered system based on type of treatment employed and quantity and concentration of the waste.

Status: Effective 1/93
Contact: Department of Toxic Substance Control, 916-323-5871

G. Region 10

Alaska - Permitting:

Alaska regularly incorporates pollution prevention planning conditions in its Resource Conservation and Recovery Act permits and is piloting P2 language in wastewater and solid waste permits. As appropriate, local governments seeking new landfills are being required to demonstrate that they have considered all waste management options and are following a prescribed hierarchy of waste management alternatives.

Status: Ongoing
Contact: David Croxton, 907-563-6529

Washington State - Refinery Hazardous Waste Incineration:

Examine and develop options for managing combustible wastes from refineries other than incineration in hazardous waste incinerators.

Status: Formative
Contact: Kim Anderson, 206-407-6931 or Stan Springer, 206-407-6723

Washington State - "Shopsweeps" Program:

"Shopsweeps" was a voluntary program providing technical assistance on the proper management of hazardous wastes to automotive repair shops, conducted in cooperation with the industry. Over 1700 shops were visited in a six month period and of the total compliance recommendations made during the visits, 61% had been complied with by the shops, based on follow-up visits to 5% of the shops.
Status: Completed in 1993
Contact: Darin Rice, 360-407-6743

Washington State - Underground Storage Tank Permit Streamlining:

Reduce permit instructions from 30 pages to one and combine the permit and fee invoice into one form. Beginning to issue permits and fees as part of a master business licensing program.

Status: Ongoing
Contact: Ron Moyer, 360-407-7217

V. Water

   A. EPA HQ

General National Pollutant Discharge Elimination System (NPDES) Permits Clearinghouse:

The Permits Division in the Office of Water maintains a clearinghouse of NPDES general permits issued by EPA and states. Permits for stormwater discharges are not included. The clearinghouse contains information on the permit conditions of each general permit. The purpose of the clearinghouse is to transfer information on existing general permits to permit writers interested in writing new general permits.

Status: Complete and updated periodically
Contact: Brian Bell, 202-260-6057

NPDES Permit Streamlining Activities:

Encourage use of general permits. Thirty-nine states have approved general permits programs. Nineteen states have issued general permits since 1/1/91 for non-stormwater discharges, and all 39 states have issued general permits for stormwater discharges associated with industrial activities. There are about 250 general permits that cover about 82,000 facilities. States or EPA have issued general permits for: stormwater from industrial activities; non-contact cooling water; concentrated animal feeding operations; groundwater cleanup dewatering; underground storage tanks; and hydrostatic testing of oil and gas lines. A national database of general permits and their conditions has been established and is updated quarterly.

Status: Complete and ongoing
Contact: Jim Pendergast, OW, 202-260-9537

NPDES Pretreatment Streamlining:

Straw proposal developed to streamline the procedures associated with developing and maintaining approved pretreatment programs. The proposed streamlining allows incorporating only significant elements of the approved program, and requires a formal modification only where the pretreatment program is made less restrictive or where the Approval Authority so requests.

Status: Straw proposal issued May 1994, results of
        stakeholder meetings being studied
Contact: Elaine Brenner, OW, 202-260-4933

NPDES Rulemaking:

Revising the basic NPDES permit regulations to expand the list of allowable minor permit modifications (those that do not require public notice) and reduce the administrative requirements for general permits.

Status: Temporarily on hold
Contact: Tom Charlton, OW, 202-260-6960

NPDES Watershed Strategy:

The Permits Division in the Office of Water developed a
strategy for integrating NPDES permit issuance into an
overall watershed framework. This strategy provides
examples for permitting programs in EPA Regions on how they
assist states in focusing on true environmental problems in
critical watersheds. The strategy has 6 elements:
state-wide coordination, NPDES permit issuance, monitoring
and assessment, programmatic measures including
environmental indicators, public participation, and
enforcement.

Status: Completed March 1994
Contact: Deborah Nagle, 202-260-2656

NPDES Watershed Successes:

The Permits Division in the Office of Water issued a report
on the status of EPA Regions in assisting states in
integrating NPDES activities into a watershed context. The
report includes many examples of successes; these provide
states/Regions with examples of what might be applicable to
their jurisdictions.

Status: Completed October 1994
Contact: Deborah Nagle, 202-260-2656

National Water Program Agenda for the Future:

The USEPA Office of Water has issued an agenda for the
future (1995). The agenda includes: polling the states and
tribes to identify specific items in day-to-day
implementation that do not make sense; providing
implementation choices when developing programs, guidance
and regulations; and identifying reporting that will be
simplified or eliminated to achieve a 25% reduction in the
reporting burden. This agenda lays out the overall
operating strategy for the Office of Water.

Status: Memorandum issued December 30, 1994 by Robert
         Perciasepe
Contact: Mark Luttner, 202-260-5700

Watershed Policy Committee:

Developing an action plan that includes: enhanced federal agency coordination; integration of EPA initiatives -- NPDES watershed strategy, Comprehensive State Groundwater Protection Programs, state nonpoint source programs -- into state comprehensive programs; expanded use of the tools needed to carry out watershed management such as new methods, models, and monitoring techniques; improved internal EPA coordination; and an aggressive outreach effort to watershed stakeholders.

Status: Announced in 10/7/94 memo from AA Perciasepe; ongoing.
Contact: Janet Pawlukiewicz, 202-260-2194

B. Region 1

Region 1 - Water Program Improvements:

Quality Action Team (QAT) recommendations implemented to improve the permitting process. A focus was to establish procedures to better address the backlog of permit appeals.

Status: Ongoing
Contact: Dianne Chabot-O'Malley, 617-565-3430

Connecticut - General Permit for Wastewater Discharges Generated by Publishing and Printing Activities:

Develop a simplified permitting process for the Publishing and Printing Industry.

Status: Ongoing
Contact: CT DEP, 203-566-7167

C. Region 2

New Jersey - Pollution Discharge Elimination System
Improvements:

Report prepared describing evaluation of program,
identification of problems and proposed actions to improve
the program.

Status: Report dated 11/2/93; implementation ongoing.
Contact: John Laurita, NJ DEP, 609-292-4543

D. Region 3

Maryland - General Permits:

At MD Department of the Environment (MDE), a number of
general permits are being written, including: seafood
processors, well pump tests, hydrostatic pipe and tank
tests, marina facilities, surface coal mines, vehicle car
washes, sand and gravel operations, stormwater, heat pumps,
animal feed lots, cooling water, and swimming pool backwash
water.

Status: Ongoing
Contact: Dana Bauer, Water Management Division, 410-631-
3512

Maryland - NPDES Data Base Enhancements:

In FY 94 and 95 Region III EPA provided grant funds to
Virginia to develop a new software program specifically
tailored to NPDES permit writing and tracking. A copy of
this prototype has been provided to Maryland and other
Region III states in order to facilitate their efforts to
improve upon their own databases. In FY 95 Maryland
received a $10,000 grant to focus on a better computer
application for permit writing.

Status: Ongoing
Contact: Dana Bauer, Water Management Division, 410-631-
3512

Maryland - Permit Backlog Elimination Plans:

Over the last year the MDE Water Management Division has aggressively pursued the reduction of permit backlogs for both municipal and industrial facilities. This is being achieved through a combination of workload redistribution, application analysis and sorting and development of additional general permits. A key element of the analysis and sorting process and workload distribution includes the determination of which permits can be quickly processed using desktop computations and which will require in-stream surveys or more complicated models. Priorities are assigned accordingly and permits which can be issued more timely than others are moved through the work stream more quickly. This has yielded significant results with MDE having eliminated all of the municipal permit backlog in FY95. It is anticipated that similar results will occur with the industrial permit backlog, which is projected to be eliminated by July 1995.

Status: Ongoing
Contact: Dana Bauer, Water Management Division, 410-631-3512

E. Region 5

Region 5 - Dredge and Fill Permitting:

The Region 5 Water Program has: 1) Assisted in the development of general permits for states and counties so that certain sizes and classes of wetlands can be regulated at the local level. 2) Drafted position paper on how to handle wetland permit issues on tribal lands. The system is geared to educate tribes on their responsibilities and provide a framework for the Army Corps of Engineers and EPA to improve delivery of wetlands program to tribes.

Status: 1) Complete; 2) Under review
Contact: Doug Ehorn, Chief Wetlands and Watersheds Section, 312-886-0243.

Region 5 - NPDES Permitting:

In April 1992, a state/EPA quality action team was established to investigate and recommend opportunities for improvement in the NPDES permit process. Recommendation included: 1) Reducing the number of in-line permit reviews done by EPA. Minimum reduction of 50% with focus on high-priority discharges. 2) Periodic reviews of state permit programs by EPA to assess the overall effectiveness of the state program. 3) Joint establishment of priorities during program planning. 4) Establishment of a standing workgroup of EPA and state permits and standards program managers to reach resolution of issues and share information. 5) Preparation, by states, of their decision-making procedures and submission to EPA for review. Any issues identified will be addressed with the particular state or the workgroup if of Regional significance. 6) Identification in annual state program plans of permitting priorities, specific permits (issues or areas) that will be subject to in-line review and permit issuance commitments.

Status: Implementation of recommendations began in FY94.
Contact: Tim Henry, Chief Permits Section, 312-886-6107

Region 5 - Underground Injection Control (UIC) Permitting:

The following actions have been implemented to improve UIC permitting: 1) Have permit writers prepare their own permits. 2) Have all correspondence related to permit issuance go through the permit writer. 3) Revise permit approval process to remove non-value added steps. 4) Issue Class I permits for 8-10 year lifespans to stagger the workload. 5) For Class V sites that would not be able to obtain a permit don't do the call-in. 6) Sign MOU's with other units that provide assistance. 7) Establish a tracking system to cover major workload items. 8) Handle requests for assistance from other units as formal requests, and include in the tracking system.

Status: Implemented
Contact: Rebecca Harvey, Chief Permit Unit, UIC Section,
          312-886-6594

Indiana - NPDES Expert System:

The Indiana Department of Environmental Management has started a project to develop a menu-driven expert system to help permit writers draft permits. This project was started in an effort to provide training to new permit writers in the state. The expert system takes permit writers through the process of writing a permit, cross references all appropriate state regulations and internal procedures, and results in a draft permit.

Status: Ongoing
Contact: Rod Thomson 317-233-8399

Wisconsin - Tiered Fee System for Stormwater Permits:

General stormwater permits are divided into three tiers by SIC code (heavy industry, light industry, and everything else). Tier I permits come with a $200 annual fee and a requirement to develop a stormwater management plan and to do sampling and chemical analysis. Tier II permits have an $100 annual fee and require a management plan but no sampling or chemical analysis. Tier III carries no fee and not planning or testing requirements. If a Tier I or II facility takes measures (usually pollution prevention practices like covering a dumpster) to get to the point where runoff is not contaminated, then it can be classified as Tier III.

Status: Ongoing
Contact: Roger Larson, 608-266-2666

   F. Region 6

Region 6 - NPDES Computerized Permit System:

Computerized permit generation system developed which reduces permit development time and paper consumption.

Status: Completed 1989
Contact: Jack Ferguson, 214-665-7170

   G. Region 9

California - External Program Review Report of the State Water Resources Control Board and the Regional Water Quality Control Boards:

Governor Wilson, in July 1993, requested an external review of mandates and programs of the State Water Board and the nine Regional Water Quality Control Boards. The goal was "... to identify how best the Boards can meet their mandates to protect the water resources of the state, while removing unnecessary red tape that hinders the economic resurgence in California." The review was conducted by members from the regulated community, environmental groups, and other stakeholders.

Status: Final Report issued 6/17/94
Contacts: James Strock, Secretary Cal/EPA, 916-445-3846 or
       John Caffrey, Chairman, State Water Resources
       Control Board

VI. PCBs

   A. EPA HQ

PCB Disposal Rule:

Proposal to reduce disposal costs by billions of dollars, through modifications to the disposal rule.

Status: Proposed 12/6/94, analyzing comments
Contact: Tony Baney, OPPTS, 202-260-3933

   B. Region 5

Region 5 - PCB Permitting:

Region 5 Pesticides and Toxic Substances Branch (PTSB) initiated a process in which Region 5 will issue a permit to a state which will allow the disposal of PCBs in facilities approved by that state. This was possible because the state's requirements for municipal landfills are equivalent to the requirements for PCB landfills under TSCA. PTSB has also modified the draft and final permit review process. A thorough peer review of draft and final permits is conducted, assuring all regulatory and technical requirements are included.

Status: Ongoing
Contact: Tony Martig, 312-353-2291

C. Region 9

Region 9 - PCB Permit Writers Guide:

A guide is being prepared that will define all of the
information, documentation and supporting materials needed
for the review and processing of permit applications to
engage in research and development, conduct storage or
dispose of PCBs. The guide will serve to expedite the
review of applications, identification of deficiencies, and
final decision making.

Status: Ongoing
Contact: Yosh Tokiwa, Air and Toxics Division, Region 9,
          415-744-1109

INDEX

ABOUT THE EDITOR

A. Roger Greenway is Principal of RTP Environmental Associates, Inc., lectures frequently at national and regional conferences on air quality and other environmental topics, and has more than 25 years' experience in environmental permitting, impact assessments, and consulting. A former principal of Dames & Moore, he is a Certified Consulting Meteorologist (CCM), a Qualified Environmental Professional (QEP), and an Accredited Asbestos Inspector, Manager, and Abatement Project Designer. He is author of more than 50 technical publications on environmental subjects.